OMICS
Applications in
Crop Science

OMICS
Applications in
Crop Science

Edited by
Debmalya Barh

CRC Press
Taylor & Francis Group
Boca Raton London New York

CRC Press is an imprint of the
Taylor & Francis Group, an **informa** business

CRC Press
Taylor & Francis Group
6000 Broken Sound Parkway NW, Suite 300
Boca Raton, FL 33487-2742

First issued in paperback 2017

© 2014 by Taylor & Francis Group, LLC
CRC Press is an imprint of Taylor & Francis Group, an Informa business

No claim to original U.S. Government works

Version Date: 20131017

ISBN 13: 978-1-4665-8525-6 (hbk)
ISBN 13: 978-1-138-07476-7 (pbk)

Visit the Taylor & Francis Web site at
http://www.taylorandfrancis.com

and the CRC Press Web site at
http://www.crcpress.com

Dedicated to my beloved grandmother

Ms. Kulabala Samanta

Contents

Foreword ix

Preface...xi

Editor..xiii

Contributors...xv

1 Omics-Based Approaches for Rice Improvement 1
   Somnath Roy, PhD; Amrita Banerjee, PhD; Somnath Bhattacharya, PhD;
   Arunava Pattanayak, PhD; and Kailash C. Bansal, PhD

2 Practical Omics Approaches for Drought Tolerance in Rice 47
   Prashant Vikram, PhD; B.P. Mallikarjuna Swamy, PhD; and Arvind Kumar, PhD

3 Omics Approaches in Maize Improvement 73
   Pawan K. Agrawal, PhD; Navinder Saini, PhD; B. Kalyana Babu, PhD;
   and Jagdish C. Bhatt, PhD

4 Omics Approaches in Pulses.. 101
   Abhishek Bohra, PhD; Uday Chand Jha, PhD; Balwant Singh, PhD;
   Khela Ram Soren, PhD; Indra Prakash Singh, PhD; Sushil Kumar Chaturvedi, PhD;
   Nagaswamy Nadarajan, PhD; and Debmalya Barh, PhD

5 Genetic Engineering in Potato Improvement.. 139
   Elena Rakosy-Tican, PhD

6 Omics Applications in *Brassica* Species 163
   Xiaonan Li, PhD; Nirala Ramchiary, PhD; Vignesh Dhandapani, PhD;
   Su Ryun Choi, PhD; and Yong Pyo Lim, PhD

7 Integrating Omics Approaches in Sugarcane Improvement....................... 191
   Rachayya M. Devarumath, PhD; Sachin B. Kalwade, PhD; Pranali A. Kulkarni, PhD;
   Suman S. Sheelavanthmath, PhD; and Penna Suprasanna, PhD

8 Recent Advances in Temperate Fruit Crops: An Omics Perspective 251
   Md. Abdur Rahim, PhD and Livio Trainotti, PhD

9 Omics Approaches in Tropical Fruit Crops 285
   Kundapura V. Ravishankar, PhD; Kanupriya, PhD; Ajitha-kumar Rekha, PhD;
   Anuradha Upadhyay, PhD; Chinmaiyan Vasugi, PhD; N. Vijayakumari, PhD;
   Pooja Kishnani, PhD; and Makki R. Dinesh, PhD

10 *Catharanthus roseus*: The Metabolome That Represents a Unique Reservoir
   of Medicinally Important Alkaloids under Precise Genomic Regulation............ 325
   Ashutosh K. Shukla, PhD and Suman P.S. Khanuja, PhD

11 Omics of Secondary Metabolic Pathways in *Withania somnifera* Dunal (ashwagandha) 385
*Neelam S. Sangwan, PhD; Laxmi N. Misra, PhD; Sandhya Tripathi, PhD;
and Amit K. Kushwaha, PhD*

12 Genetic Engineering in Ornamental Plants 409
*Rajesh Kumar Dubey, PhD; Simrat Singh, PhD; Gurupkar Singh Sidhu, PhD;
and Manisha Dubey, PhD*

13 Omics Advances in Tea (*Camellia sinensis*) 439
Mainaak Mukhopadhyay, PhD; Bipasa Sarkar, PhD; and Tapan Kumar Mondal, PhD

14 Omics Approaches to Improving Fiber Qualities in Cotton 467
Tianzhen Zhang, PhD; Xiangdong Chen, PhD; and Wangzhen Guo, PhD

15 Forestry and Engineered Forest Trees 491
Mohammed Ellatifi, PhD

16 Application of Omics Technologies in Forage Crop Improvement 523
Suresh Kumar, PhD and Vishnu Bhat, PhD

17 Bioenergy Crops Enter the Omics Era 549
*Atul Grover, PhD; Patade Vikas Yadav, PhD; Maya Kumari, PhD;
Sanjay Mohan Gupta, PhD; Mohommad Arif, PhD; and Zakwan Ahmed, PhD*

18 Molecular Farming in the Decades of Omics 563
Dinesh K. Yadav, PhD; Neelam Yadav, PhD; and S.M. Paul Khurana, PhD

19 Natural Pesticidome Replacing Conventional Pesticides 603
Daiane Hansen, PhD

**20 Intellectual Property Rights in Plant Biotechnology: Relevance, Present
Status, and Future Prospects** 621
*Dinesh Yadav, PhD; Gautam Anand, PhD; Sangeeta Yadav, PhD;
Amit K. Dubey, PhD; Naveen C. Bisht, PhD; and Bijaya K. Sarangi, PhD*

Index 671

Foreword

This valuable book comes out at a very convenient and even crucial time. Plants, both extant and extinct, are the ultimate source of most of our food, fuel, fibers, fabrics, and even our "farmaceuticals." The world population is still increasing, available arable land is decreasing, the climate is becoming more erratic and extreme, and water is an increasingly scarce resource over much of the planet, while in other areas, flooding predominates. The situation is severe, perhaps dire, but the attitude of Chicken Little, "The sky is falling, the sky is falling!" is not the attitude being adopted by scientists around the world. Instead, agronomists, biochemists, biotechnologists, geneticists, horticulturalists, molecular biologists, plant biologists, and others are seeking to adopt the most recent, promising techniques developed from the various fields of *omics*, such as genomics (the complete DNA sequence of an organism), transcriptomics (the array of mRNAs currently being used), proteomics (the array of proteins present), metabolomics (the vast array of metabolites present in a cell, organ, or individual), and phenomics (the phenotypic expression of all of the above). Here, in this book, are the fruits (and seeds, leaves, and roots) of these collaborative efforts to alleviate global suffering.

The particular plants studied include those used for food, such as rice in Chapters 1 and 2, maize in Chapter 3, legumes in Chapter 4, potatoes in Chapter 5, *Brassica* species in Chapter 6, sugarcane in Chapter 7, temperate fruit in Chapter 8, tropical fruit in Chapter 9, tea (the most important crop for an Englishman such as myself) in Chapter 13, medicinal plants in Chapters 10 and 11, fiber (cotton) in Chapter 14, natural resources (forest trees) in Chapter 15, animal feed (forage crops) in Chapter 16, bioenergy crops in Chapter 17, molecular "pharming" in Chapter 18, biopesticides in Chapter 19, beauty (ornamentals) in Chapter 12, and patent rights in Chapter 20. The authors come from five continents: Africa (Morocco), Asia (China, India, the Philippines, Korea), Australia, Europe (Italy, Romania), and South America (Brazil) and represent countries comprising over 3 billion people (>40% of the world's population).

The specific problems they confront and the "omics" approaches employed include genomics, proteomics, metabolomics, and phenomics to identify quantitative trait loci (QTL) in rice (Chapter 1); genomics and phenomics to identify specific QTL for drought resistance in rice (Chapter 2); genomics, proteomics, and metabolomics in maize (Chapter 3); a gamut of approaches to understand pulse crops (Chapter 4); genetic engineering to improve the potato (Chapter 5); functional genomics, transcriptomics, and metabolomics in *Brassica* species (Chapter 6) genomics, transcriptomics, proteomics, and metabolomics in sugarcane (Chapter 7);, transcriptomics, proteomics, and metabolomics approaches for temperature fruit crops, with the complete genome sequences of 11 of these crops already furnished (Chapter 8); transcriptome and proteome analysis in several tropical fruit crops, with the full genome already mapped for four (Chapter 9); genomics, transcriptomics, proteomics, and metabolomics combined with the important medicinal plants *Catharanthus roseus* and *Withania somnifera* and a new "omics", alkaloidomics, which has been specially developed (Chapters 10 and 11); genetic engineering used with ornamental plants to enhance resistance, shelf life, and other desirable traits (Chapter 12); the eagerly anticipated release of the complete genome of tea (*Camellia sinensis* [L.] O. Kuntze), which is expected to bring improvements in this recalcitrant species (Chapter 13); and genomics, transcriptomics, proteomics, and metabolomics in cotton (Chapter 14). The situation in forest trees is somewhat different because they are diverse in nature, normal genetics is very problematic in woody plants and too little is understood about "omics"; the future lies with genetic engineering (Chapter 15). Genomics and metabolomics are currently being used for forage crops, which, after digestion, provide animal protein to a starving world (Chapter 16). Genomics and transcriptomics are aiding in the analysis of the enzymes involved in the production of the four most important molecular biofuel plants (Chapter 17). Molecular farming of plant-derived pharmaceuticals benefits from the use of plant-based expression systems to generate over 100 recombinant proteins (Chapter 18). The use of plant-based pesticides is discussed (Chapter 19) and finally, the role of intellectual property rights, especially in biotechnology, is discussed (Chapter 20).

I highly recommend this book for keeping up to date on the basics of plant biology and the latest happenings at the cutting edge of the field, especially regarding the important crop plants.

Eric Davies, PhD
Professor Emeritus
Department of Plant Biology
North Carolina State University
Raleigh, North Carolina

Preface

The term *omics* is probably derived from the Sanskrit word *Om*, which depicts completeness. During the last two decades the term *omics* has been suffixed with several biological topics to provide complete information on the subject; for example genomics, proteomics, and so on. Although the cutting-edge technologies in this *omics* era that are applied to address various biomedical issues are well documented in several books and much literature, a limited number of books with limited information are available for important agricultural crops.

This book, *Omics Applications in Crop Science*, is an effort to fill the gap and to provide students and researchers in molecular plant biology and crop science with a comprehensive resource covering applications of various *omics* technologies such as genomics, transcriptomics, proteomics, metabolomics to important agronomic, horticultural, medicinal, plantation, fibre, forage, and bioenergy crops. Seventy-five experts from nine countries have combined and shared their practical experience with the latest advancements in the field in developing this unique book. Since the list of major crops is long, this book includes applications of omics technologies in the 16 most important crops. Further, important intellectual properties or patents specific to either technologies or crops are included as appropriate.

The book consists of 20 chapters, and in the first three chapters, omics approaches in the most important cereal crops, rice and maize, are covered. Chapter 1, by Dr. Somnath Roy and colleagues, describes omics-based approaches for rice improvement and Chapter 2, by Dr. Prashant Vikram et al., illustrates various practical omics approaches for drought tolerance in rice. Chapter 3, by Dr. Pawan Kumar Agrawal's group, explains how the omics-based strategies can be used for maize improvement. Chapter 4 is dedicated to pulses. Dr. Abhishek Bohra et al. have given a comprehensive account of omics and omics approaches used in improving chickpea, pigeonpea, cowpea, and lentil in this chapter. Under tuberous crops, the potato is selected for this book, and in Chapter 5, Dr. Elena Rakosy-Tican discusses genetic engineering and strategies in improving this important crop. The Brassicaceae family comprises some important vegetable crops. In Chapter 6, Dr. Xiaonan Li et al. demonstrate the omics and applications of various omics-based approaches in several *Brassica* species. Chapter 7, by Dr. Rachayya M. Devarumath and colleagues, describes how omics strategies can be useful in improving the most important industrial crop, sugarcane. Chapters 8 and 9 deal with fruit crop omics. While Chapter 8, by Dr. Md. Abdur Rahim and Dr. Livio Trainotti, deals with temperate fruits; Chapter 9, by Dr. Kundapura V. Ravishankar et al., provides an in-depth account of tropical fruit omics. Chapters 10 and 11 include omics of medicinal plants. In Chapter 10, Dr. Ashutosh K. Shukla and Dr. Suman P.S. Khanuja discuss the metabolomics of *Catharanthus roseus* and how the medicinally important alkaloids of the plant are produced under precise genomic regulation. Similarly, the omics of another important medicinal plant, *Withania somnifera*, has been documented by Dr. Neelam S. Sangwan and colleagues in Chapter 11. Chapter 12 is on floriculture: Dr. Rajesh Kumar Dubey's group discuss various genetic engineering approaches in improving ornamental and flowering plants. In Chapter 13, Dr. Mainaak Mukhopadhyay and colleagues highlight the omics advances in tea, one of the most important beverage and plantation crops. In Chapter 14, Dr. Tianzhen Zhang's team focus on various omics strategies in improving the fiber qualities of cotton. Chapters 15 and 16 provide omics-related information on very important but neglected plants: forest trees and forage crops. Thanks to Dr. Mohammed Ellatifi for giving a very comprehensive account of forest tree omics and engineered forest trees for human benefit in Chapter 15. Similarly, credit must be given to Dr. Suresh Kumar and Dr. Vishnu Bhat for providing valuable information on the application of omics technologies in forage crop improvement in Chapter 16. In Chapter 17, Dr. Atul Grover and his team document a budding area in plant biotechnology with the title "Bioenergy Crops Enter the Omics Era." Chapter 18, by Dr. Dinesh K. Yadav and colleagues, gives a detailed account on how omics technologies are applicable in molecular farming, along with associated issues such as commercial aspects of molecular farming, clinical trials of plant-produced pharmaceuticals, and regulatory issues. A new term, *natural pesticidome*, has been coined and described by Dr. Daiane Hansen in Chapter 19, describing how they can be used to replace chemical pesticides for a green environment. The book is concluded with an important topic, intellectual property rights (IPR), in Chapter 20. This chapter presents basic as well as various practical IPR issues in plant biotechnology. Dr. Dinesh Yadav and colleagues also provide a long list of patented technologies and plant-specific patents in this chapter.

I believe that this book and its rich content and coverage will be worthwhile to students and researchers in the field of cutting edge plant science. Your suggestions and comments to improve the next edition would be much appreciated.

Debmalya Barh
Editor

Editor

Debmalya Barh (MSc, MTech, MPhil, PhD, PGDM) is the founder and president of the Institute of Integrative Omics and Applied Biotechnology (IIOAB), India, a virtual global platform for multidisciplinary research and advocacy. He is also a consultant biotechnologist with an international reputation. As a postgraduate in horticulture, he started his early career as a horticulturist and later transformed himself as an active researcher in multidisciplinary cutting-edge "omics" fields including plant-related omics. He has worked with nearly 400 researchers from 30–35 countries and has more than 100 high impact publications to his credit. He is a well-known editor for several research reference books in the foremost domains of "omics" published by reputed international publishers. *OMICS: Applications in Biomedical, Agricultural, and Environmental Sciences* is one of these books related to plant omics, published by Taylor & Francis in 2013. He also serves as an editorial and review board member for a number of highly professional international research journals.

Contributors

Pawan K. Agrawal, PhD
Vivekananda Parvatiya Krishi
    Anusandhan Sansthan (VPKAS)
Almora, India

Zakwan Ahmed, PhD
Biotechnology Division
Defence Institute of Bio-Energy Research
Haldwani, India

Gautam Anand, PhD
Department of Biotechnology
Deen Dayal Upadhyaya Gorakhpur
    University
Gorakhpur, India

Mohommad Arif, PhD
Biotechnology Division
Defence Institute of Bio-Energy Research
Haldwani, India

B. Kalyana Babu, PhD
Vivekananda Parvatiya Krishi
    Anusandhan Sansthan
Almora, India

Amrita Banerjee, PhD
Division of Crop Improvement, ICAR
    Research Complex for NEH Region
Shillong, India

Kailash C. Bansal, PhD
National Bureau of Plant Genetic Resources
New Delhi, India

Debmalya Barh, PhD
Institute of Integrative Omics and Applied
    Biotechnology
Purba Medinipur, India

Vishnu Bhat, PhD
Department of Botany
University of Delhi, North Campus
New Delhi, India

Jagdish C. Bhatt, PhD
Vivekananda Parvatiya Krishi
    Anusandhan Sansthan
Almora, India

Somnath Bhattacharya, PhD
Department of Genetics
Mohanpur, India

Naveen C. Bisht, PhD
National Institute of Plant Genome
    Research
Jawaharlal Nehru University
New Delhi, India

Abhishek Bohra, PhD
Indian Institute of Pulses Research
Kanpur, India

Sushil Kumar Chaturvedi, PhD
Indian Institute of Pulses Research
Kanpur, India

Xiangdong Chen, PhD
Cotton Research Institute
Nanjing Agricultural University
Nanjing, China

and

Henan Institute of Science and
    Technology
Xinxiang, China

Su Ryun Choi, PhD
Plant Genomics Institute
Chungnam National University
Daejeon, South Korea

Rachayya M. Devarumath, PhD
Molecular Biology and Genetic
    Engineering Division
Vasantdada Sugar Institute
Pune, India

Vignesh Dhandapani, PhD
Plant Genomics Institute
Chungnam National University
Daejeon, South Korea

Makki R. Dinesh, PhD
Indian Institute of Horticultural Research
Bangalore, India

Amit K. Dubey, PhD
Department of Biotechnology
Deen Dayal Upadhyay Gorakhpur
    University
Gorakhpur, India

Manisha Dubey, PhD
Department of Plant Pathology
Punjab Agricultural University
Ludhiana, India

Rajesh Kumar Dubey, PhD
Department of Floriculture and
    Landscaping
Punjab Agricultural university
Ludhiana, India

Mohammed Ellatifi, PhD
Sylva-World for Development and
    the Protection of Forests and the
    Environment
Casablanca, Morocco

Atul Grover, PhD
Biotechnology Division
Defence Institute of Bio-Energy Research
Haldwani, India

Wangzhen Guo, PhD
Cotton Research Institute
Nanjing Agricultural University
Nanjing, China

Sanjay Mohan Gupta, PhD
Biotechnology Division
Defence Institute of Bio-Energy Research
Haldwani, India

Daiane Hansen, PhD
Federal University of Goiás
Goiânia, Brazil

and

Biological Institute
São Paulo, Brazil

Uday Chand Jha, PhD
Indian Institute of Pulses Research
Kanpur, India

Sachin B. Kalwade, PhD
Molecular Biology and Genetic
    Engineering Division
Vasantdada Sugar Institute
Pune, India

Kanupriya, PhD
Indian Institute of Horticultural
    Research
Bangalore, India

Suman P.S. Khanuja, PhD
Suman Khanuja Innovation Enterprises
Lucknow, India

S.M. Paul Khurana, PhD
Amity Institute of Biotechnology
Gurgaon, India

Pooja Kishnani, PhD
National Reserch Centre for Citrus
Nagpur, India

Pranali A. Kulkarni, PhD
Molecular Biology and Genetic
    Engineering Division
Vasantdada Sugar Institute
Pune, India

Arvind Kumar, PhD
International Rice Research Institute
Los Baños, the Philippines

Suresh Kumar, PhD
Division of Crop Improvement
Indian Grassland and Fodder Research
   Institute
Jhansi, India

and

Division of Biochemistry,
Indian Agricultural Research Institute
New Delhi, India

Maya Kumari, PhD
Biotechnology Division
Defence Institute of Bio-Energy
   Research
Haldwani, India

Amit K. Kushwaha, PhD
Metabolic and Structural Biology
   Department
Central Institute of Medicinal and
   Aromatic Plants, Council of Scientific
   and Industrial Research
Lucknow, India

Xiaonan Li, PhD
Plant Genomics Institute
Chungnam National University
Daejeon, South Korea

Yong Pyo Lim, PhD
Plant Genomics Institute
Chungnam National University
Daejeon, South Korea

Laxmi N. Misra, PhD
Metabolic and Structural Biology
   Department
Central Institute of Medicinal and
   Aromatic Plants, Council of Scientific
   and Industrial Research
Lucknow, India

Tapan Kumar Mondal, PhD
National Research Center on DNA
   Fingerprinting
New Delhi, India

Mainaak Mukhopadhyay, PhD
Biotechnology Laboratory
Faculty of Horticulture
Uttar Banga Krishi Viswavidyalaya
Cooch Behar, India

Nagaswamy Nadarajan, PhD
Indian Institute of Pulses Research
Kanpur, India

Arunava Pattanayak, PhD
Centre for Biotechnology, ICAR Research
   Complex for NEH Region
Shillong, India

Md. Abdur Rahim, PhD
Department of Biology
University of Padova
Padua, Italy

Elena Rakosy-Tican, PhD
Molecular Biology and Biotechnology
   Department
Babes-Bolyai University
Cluj-Napoca, Romania

Nirala Ramchiary, PhD
School of Life Sciences
Jawaharlal Nehru University
New Delhi, India

Kundapura V. Ravishankar, PhD
Indian Institute of Horticultural
   Research
Bangalore, India

Ajitha-kumar Rekha, PhD
Indian Institute of Horticultural
   Research
Bangalore, India

Somnath Roy, PhD
National Bureau of Plant Genetic
   Resources
Shillong, India

Navinder Saini, PhD
Vivekananda Parvatiya Krishi
    Anusandhan Sansthan
Almora, India

Neelam S. Sangwan, PhD
Metabolic and Structural Biology
    Department
Central Institute of Medicinal and
    Aromatic Plants, Council of Scientific
    and Industrial Research
Lucknow, India

Bijaya K. Sarangi, PhD
Environmental Biotechnology Division
National Environmental Engineering
    Research Institute, Council of Scientific
    and Industrial Research
Nehru Marg, India

Bipasa Sarkar, PhD
Department of Chemistry
North Bengal University
Darjeeling, India

Suman S. Sheelavanthmath, PhD
Department of Biotechnology
Sinhgad Science College
Pune, India

Ashutosh K. Shukla, PhD
CSIR-Central Institute of Medicinal and
    Aromatic Plants
Lucknow, India

Gurupkar Singh Sidhu, PhD
School of Agricultural Biotechnology
Punjab Agricultural University
Ludhiana, India

Balwant Singh, PhD
National Research Centre on Plant
    Biotechnology
New Delhi, India

Indra Prakash Singh, PhD
Indian Institute of Pulses Research
Kanpur, India

Simrat Singh, PhD
Department of Floriculture and
    Landscaping
Punjab Agricultural University
Ludhiana, India

Khela Ram Soren, PhD
Indian Institute of Pulses Research
Kanpur, India

Penna Suprasanna, PhD
Plant Biology, Nuclear Agricultural and
    Biotechnology Division
Bhabha Atomic Research Centre
Mumbai, India

B.P. Mallikarjuna Swamy, PhD
International Rice Research Institute
Los Baños, the Philippines

Livio Trainotti, PhD
Department of Biology
University of Padova
Padua, Italy

Sandhya Tripathi, PhD
Metabolic and Structural Biology
    Department
Central Institute of Medicinal and
    Aromatic Plants, Council of Scientific
    and Industrial Research
Lucknow, India

Anuradha Upadhyay, PhD
National Research Centre for Grapes
Pune, India

Chinmaiyan Vasugi, PhD
Indian Institute of Horticultural
    Research
Bangalore, India

Vijayakumari N., PhD
National Reserch Centre for Citrus
Nagpur, India

Prashant Vikram, PhD
Genetic Resource Program
International Maize and Wheat
    Improvement Center (CIMMYT)
Texcoco, México

**Bidhan Chandra Krishi
Viswavidyalaya, PhD**
Department of Genetics
Mohanpur, India

Dinesh Yadav, PhD
Department of Biotechnology
Deen Dayal Upadhyaya Gorakhpur
    University
Gorakhpur, India

and

Australian Centre for Plant Functional
    Genomics
University of Adelaide
Glen Osmond, Australia

Dinesh K. Yadav, PhD
Amity Institute of Biotechnology
Manesar, India

Neelam Yadav, PhD
Amity Institute of Biotechnology
Manesar, India

Patade Vikas Yadav, PhD
Biotechnology Division
Defence Institute of Bio-Energy
    Research
Haldwani, India

Sangeeta Yadav, PhD
Department of Biotechnology
Deen Dayal Upadhyaya Gorakhpur
    University
Gorakhpur, India

Tianzhen Zhang, PhD
Cotton Research Institute
Nanjing Agricultural University
Nanjing, China

1

Omics-Based Approaches for Rice Improvement

Somnath Roy, PhD; Amrita Banerjee, PhD; Somnath Bhattacharya, PhD;
Arunava Pattanayak, PhD; and Kailash C. Bansal, PhD

CONTENTS

1.1 Introduction 1
1.2 Rice Genomics 3
1.3 Marker- and Genomics-Assisted Breeding 4
    1.3.1 Expression Genetics and eQTLs 6
    1.3.2 Advanced Backcross QTL Analysis 7
    1.3.3 Allele Mining or EcoTILLING 8
    1.3.4 Association Mapping 8
1.4 Rice Functional Genomics: From Sequence to Gene Function 9
    1.4.1 Identification of Genes through Functional Genomics Approaches 13
        1.4.1.1 Genes for Agromorphological Traits 13
        1.4.1.2 Genes for Abiotic Stress Tolerance 15
        1.4.1.3 Genes for Biotic Stress Tolerance 18
1.5 Rice Proteomics 20
    1.5.1 Functional Analysis of Rice Proteome 21
        1.5.1.1 Rice Proteome in Different Organs and Developmental Stages 22
        1.5.1.2 Rice Proteome under Biotic and Abiotic Stresses 23
        1.5.1.3 Hormone Response Proteome 25
1.6 Rice Metabolomics 26
1.7 Rice Phenomics 27
1.8 Conclusion 28
References 30

1.1 Introduction

Rice (*Oryza sativa* L.) is the primary staple for half of the world's population and 75% of the population living in poverty. Current global production of milled rice is 465.34 Mt and total consumption of milled rice is 466.12 Mt (Figure 1.1a). Between 1966 and 2000, the Green Revolution increased food production in densely populated developing countries by 125% (Khush 2001). In recent years, a significant increment in rice yield has been achieved through two major genetic interventions: (i) the use of semidwarf genes that improved harvest index, lodging resistance, and nitrogen responsiveness, and (ii) the exploitation of heterosis through the production of hybrids. The improvement in rice yield slowed down considerably during the 1990s compared to the 1980s, and in the past 10 years it has shown little improvement (Figure 1.1b). This slowing pace of yield increase needs to be improved in the context of the high rate of world population growth, climate change, and the reduction in

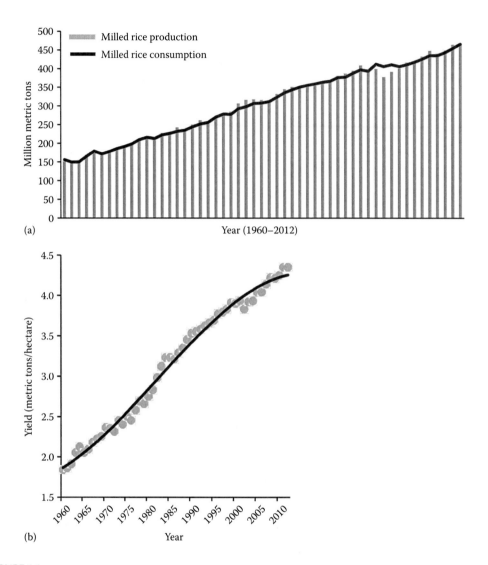

FIGURE 1.1
Global rice statistics: (a) production and consumption of milled rice and (b) rice yield trend from 1960 to 2012. (From USDA PSD, www.fas.usda.gov/psdonline/psdquery.aspx, accessed 13 December 2012.)

cultivable agricultural lands due to urbanization and soil degradation. It has been projected that global food production needs to be increased by 70% by 2050 to meet the demand caused by increasing global population and thereby consumption (Varshney et al. 2011).

Rice has been grown in sustainable high-output agroecosystems for thousands of years. The genus *Oryza* occupies a distinct phylogenetic position in a separate subfamily, the Ehrhartoideae (Kellogg 2001). The principal morphological characteristics of the genus include rudimentary sterile lemmas, bisexual spikelets, and narrow, linear, herbaceous leaves with scabrous margins. The genus *Oryza* includes only about 23 species and is remarkable for the diverse ecological adaptations of its species (Vaughan et al. 2003). Archaeological evidence indicated that rice was domesticated about 10,000 years ago (Khush 1997), similarly to wheat (*Triticum aestivum*) and maize (*Zea mays* ssp. *mays*). The autogamous nature of rice has restricted gene flow and as a result geographically or

ecologically distinct groups of rice show greater genetic differentiation than the alloga-
mous species.

Asian cultivated rice (*Oryza sativa*) evolved from the wild species *O. rufipogon* Griff., an
AA genome wild relative of rice that includes both annual and perennial ecotypes (Oka
1988; Londo et al. 2006), through massive gene losses from its ancestor (Sasaki and Itoh
2010). However, evidence suggests that two primary subspecies of cultivated rice, *O. sativa
indica* and *O. sativa japonica*, are the products of two separate domestication processes from
the ancestral species (Second 1982; Doi et al. 2002; Cheng et al. 2003). The two subspecies
have several morphological and physiological differences. *Indica* varieties tend to be found
throughout the lowlands of tropical Asia, whereas *japonica* types are distributed in upland
hills of southern China, Northeast Asia, Southeast Asia, and Indonesia, further differentiated into *tropical japonica* and *temperate japonica* depending on the growing regions (Khush
1997; Takahashi et al. 1997; Oka 1988; Garris et al. 2005). Although *indica* and *japonica* represent the deepest genetic differentiation within *O. sativa*, five major subpopulations—*indica,
tropical japonica, temperate japonica, aus,* and *aromatic*—are widely recognized (Kovach et al.
2007). The two subspecies of *O. sativa* show significant diversity in single nucleotide polymorphisms (SNPs), intergenic sequences, and individual gene duplications, suggesting
that the genomes have undergone dynamic genome evolution (Garris et al. 2005; Yu et al.
2005). With completion of whole genome sequencing of *indica* and *japonica* rice and subsequent development of new tools for analyzing rice genome, proteome, metabolome, and
phenome, rice research has gained new pace, which has led to several significant scientific
achievements in understanding the crop as a whole.

1.2 Rice Genomics

Genomics is the study of genomes. Rice is an excellent model plant for genomics, second
only to *Arabidopsis* (Izawa and Shimamoto 1996; Goff 1999). The rice genome sequence
provides a platform for organizing information about diverse cereals and understanding
the shared and the independent dimensions of cereal evolution. Characterization of plant
genomes and the genes contained within them will aid geneticists and molecular biologists in their quest to understand cereal biology, and will help plant breeders in their goal
of developing better products.

Rice has been considered as a model for genome research in the cereals because of its
small genome size and global importance as a food crop. It serves as a reference genome
for molecular biological studies in the cereals, for the following reasons: (i) availability
of complete genome sequences of both *indica* (cv 93-11) and *japonica* (cv Nipponbare)
subspecies (Goff et al. 2002; Yu et al. 2002), which enabled investigations of transcriptome activity in a range of tissues and developmental stages (Li et al. 2006; Nobuta et al.
2007); (ii) large genetic resources for comparative genomics (Gramene Maps Database;
www.gramene.org/cmap); (iii) conservation of gene content and gene order among the
cereals (Schmidt 2000); (iv) well-established transformation techniques (Hiei et al. 1994;
Komari et al. 1998); (v) strong genome databases facilitating access and depositing of
information (www.gramene.org); (vi) the high-resolution mapping of epigenetic modifications for chromosomes (Li et al. 2008); and (vii) the genome-wide identification of
genetic variation in gene expression between rice subspecies and their hybrids (Zhang
et al. 2008; Wei et al. 2009).

Whole-genome sequencing of rice, along with the extensive genetic and physical mapping efforts, provides a foundation for genome sequencing of diverse cereals, identifying orthologous genes, and ultimately generating new insights into cereal evolutionary history. In addition to the rice genome sequence, knowledge of the full set of genes will facilitate comprehensive study of the gene complement in rice and related species to see which pathways are shared and which are unique, and how these pathways may have been modified to sustain yield under challenging environments. The *Oryza* Map Alignment Project (OMAP; www.omap.org) was started with the goal of developing an experimentally tractable and cloned model system to understand the evolution, physiology, and biochemistry of the genus *Oryza*. This project constructed physical maps from the bacterial artificial chromosome (BAC) libraries representing 14 with *Oryza* species including diploids and tetraploids. The data have been compared with domesticated *O. sativa* subspecies. The OMAP-generated data were integrated with other genomic databases such as Gramene (www.gramene.org) for rapid access by the research communities. Overall, OMAP provided genome-wide comparisons among *Oryza* species and aided in understanding the collective *Oryza* genome.

The current map-based rice genome sequence assembly (373.2 Mb) covers ~95% of the *japonica* genome (Rice Annotation Project Database [RAP-DB]; http://rapdb.dna.affrc.go.jp), which has been annotated using *ab initio* gene prediction, comparative genomics, and several other computational methods (Sasaki et al. 2002; IRGSP 2005; Yuan et al. 2005; Tanaka et al. 2008). Rice genome annotation can be obtained from the Rice Genome Annotation Project database (RGAP 7; http://rice.plantbiology.msu.edu). The total number of loci as per current updates from RGAP 7 is 55,986, which yielded 66,338 transcripts or gene models. About half of these gene models have been empirically validated by characterizing RNA transcripts that include expressed sequence tags (ESTs), full-length cDNA sequences, gene-expression arrays, massive parallel signature sequencing, and serial analysis of gene expression. These analyses also identified and validated numerous genes (Kikuchi et al. 2003; Juretic et al. 2004). The rate of gene-function discovery in rice is slower than that of the dicot plant *Arabidopsis thaliana* because the process of gene fine-mapping requires crossing genetically divergent rice varieties or subspecies, genotyping of a large number of F_2 mutants, and large greenhouse or field space for phenotyping (Lukowitz et al. 2000). The high level of redundancy (due to the high degree of duplicated genomic segments in rice genome) further complicates rice mutant analysis (Shiu et al. 2004; Tian et al. 2004).

1.3 Marker- and Genomics-Assisted Breeding

The rice DNA markers developed during the late 1980s and 1990s as an output of rice genomics research have greatly facilitated marker-assisted selection (MAS) for improving precision and efficiency of conventional plant breeding in developing new rice varieties (for review, see Collard and Mackill 2007; Jena and Mackill 2008). In the 1990s, restriction fragment length polymorphism (RFLP) and random amplified polymorphic DNA (RAPD) markers were predominantly used for rice breeding (Mohan et al. 1994; Huang et al. 1997). Later these markers were transformed to polymerase chain reaction (PCR)-based markers called sequence tagged site (STS) markers to improve the specificity and reliability (Inoue et al. 1994; Lang et al. 1999). The microsatellites or simple sequence repeat (SSR) markers, being highly reproducible, codominant in inheritance, highly polymorphic, and transferable between populations, became the most preferred markers in rice (McCouch et al. 1997). The updated rice marker information can be accessed from the Gramene database (www.gramene.org/markers). The integration of information on the saturated rice molecular map and genome sequence data with modern biotechnological tools has resulted in the isolation of a large number of genes underlying agriculturally important traits of rice. New genomics tools such as functional markers and bioinformatics platforms could potentially increase the efficiency of crop improvement. Now plant breeders can use improved selection techniques or can fix superior allele combinations in desired genetic backgrounds. Functional markers have been developed (Andersen and Lübberstedt 2003) extensively for plant species in which ESTs or gene sequence data are available (Gupta and Rustgi 2004). It has been demonstrated that functional markers such as SSRs, SNPs, and conserved ortholog sequences (COSs) can be developed by screening the unigene consensus sequences (Rudd et al. 2005). The functional markers are superior to neutral molecular markers owing to complete linkage with trait locus alleles and functional motifs, and allow reliable application of markers in populations without prior mapping. The functional markers can be used to fix alleles in several genetic backgrounds without additional calibration. Functional markers have been used in rice for improvement of many important traits such as SSR markers linked with the *waxy* gene used for selecting amylase content (Zhou et al. 2003) and development of rice blast resistance through MAS for *Pi-1*, *Pi-2*, and *Pi-33* (Chen et al. 2008). Roy et al. (2012) used functional markers for the *BADH2* gene to screen aromatic and nonaromatic rice germplasm for the presence of 8 bp deletion and three SNPs in the exon 7 of chromosome 8 (Bradbury et al. 2005).

Rice molecular markers have been used extensively for assessing the genetic diversity in germplasm collections (see McCouch et al. 2012). However, only a small fraction of the natural genetic diversity available in the world's germplasm repositories has been explored to date. This scenario is expected to change with the advent of high-throughput genotyping and next-generation sequencing technologies (Box 1.1). Recently, systematic

---

BOX 1.1　NEXT-GENERATION SEQUENCING (NGS) TECHNOLOGY

The automated Sanger method of sequencing is considered a first-generation technology, and newer methods are referred to as next-generation sequencing (NGS). These newer technologies constitute various strategies which combine template preparation, sequencing and imaging, and genome alignment and assembly methods (Metzker 2010). NGS technology facilitates identification and tracking of genetic variation more efficiently and precisely. With these technologies thousands of candidate genes can be tracked within large gene bank collections in the same pooled sequencing reaction and it is now possible to resequence candidate genes, entire transcriptomes or entire plant genomes more economically (Varshney et al. 2009). Huang et al. (2009b) developed the first high-throughput genotyping method that uses SNPs detected by whole-genome resequencing using NGS. The NGS method was faster and more precise in recombination breakpoint determination compared to the genetic map-based approach (Huang et al. 2009b). The continuous advancement of NGS technology will facilitate an understanding of complex phenomena, such as heterosis and epigenetics, which have important implications for crop genetics and breeding (Varshney et al. 2009).

characterization of rice genetic diversity has been started (Huang et al. 2011; Zhao et al. 2011). In addition, the DNA markers have made it possible to survey genes in the breeders' germplasm for selecting parental lines to be used in breeding programs (Wang et al. 2007), and also to identify genomic regions under selection (i.e., allelic shifts) of breeding populations.

The marker-assisted backcrossing approaches have frequently been applied in rice improvement by using markers to select for target loci, minimize linkage drag, and recover the recurrent parent genome (Hospital and Charcosset 1997; Hospital 2001). Gene or QTL pyramiding (the process of combining genes/QTLs in progeny usually arising from different parents) has been successful mainly due to the availability of linked molecular markers for rice (Hittalmani et al. 2000; Ashikari and Matsuoka 2006). Gene pyramiding in rice has been particularly useful for combining resistance genes/QTLs against rice blast and bacterial blight diseases into suitable genetic backgrounds. Thus, MAS has been used extensively for improving rice for agronomic performance and against biotic as well as abiotic stresses (summarized in Collard et al. 2008).

The spectacular developments in the field of rice genomics research in the twenty-first century, such as high-throughput equipment for whole-genome sequencing, gene expression and genome characterization, and establishment of bioinformatics platforms and genomics databases, have transformed rice breeding into genomics-assisted breeding. This is a holistic approach that enables breeders to predict phenotype from genotype using different genomic tools and strategies. It greatly accelerates the development of improved cultivars with enhanced resistance or tolerance to biotic and/or abiotic stresses and higher agronomic performance by improving the precision and efficiency of predicting phenotypes from genotypes (Varshney et al. 2006). Genomics-assisted breeding has been demonstrated successfully in many cereals including rice. New approaches have recently been adopted for the functional characterization of allelic variation and the identification of sequence motifs affecting phenotypic variation (Figure 1.2). These approaches include advanced backcross–QTL analysis, expression genetics, allele mining, EcoTILLING, and association mapping.

1.3.1 Expression Genetics and eQTLs

With the rapid progress in gene expression analysis, it is now possible to map the inheritance of the gene expression pattern by analyzing the expression levels of genes or gene clusters within a segregating population. Such analyses determine map positions of the loci regulating the transcript abundance of the genes, which are referred to as expression QTL or eQTL (reviewed in Kliebenstein 2009). Based on the location of the transcript in respect to eQTL influencing the expression of that transcript, eQTLs can be classified as *cis*- or *trans*-acting. The eQTLs thus identified make it possible to identify factors influencing the level of gene expression (Varshney et al. 2005). Whole-genome eQTL analysis would contribute to multifactorial dissection of the expression profile of a given mRNA, cDNA, protein, or metabolite into its underlying genetic components, as well as localization of them on the genetic map (Jansen and Nap 2001). Microarray-based analyses of expression levels of the genes in individuals from a segregating population, and the expression polymorphisms in the mapping population, can also provide high-throughput markers for map construction that can be used for mapping eQTL (West et al. 2007; Potokina et al. 2008). A limited number of eQTL analyses have been reported in rice. Wang et al. (2010) reported microarray-based eQTL analyses in rice shoots at 72 h after germination using recombinant inbred lines (RILs). From this study they obtained a total of 26,051 eQTLs,

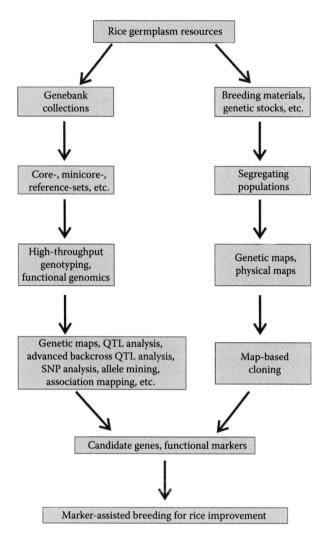

FIGURE 1.2
Rice genomics workflow.

including both *cis*- and *trans*-eQTLs. They also identified 171 eQTL-hot spots in the rice genome, each of which controls transcript variations of many expression traits (e-traits). The correlations between trait QTLs and eQTLs were also detected from this study, by comparing the locations of the QTLs using confidence intervals to identify the number and location of genes affecting trait-related gene expression. During eQTL analysis, the technical and environmental factors can result in the detection of false *hot spots* or hubs of *trans*-acting eQTLs that affect the expression of many more genes than expected by chance (de Koning and Haley 2005).

1.3.2 Advanced Backcross QTL Analysis

Advanced backcross QTL analysis (AB-QTL) is proposed as a method of combining QTL analysis with varietal improvement (Tanksley and Nelson 1996). In this approach, an unadapted donor line (such as wild species or landraces) is backcrossed to a superior cultivar for the discovery and transfer of valuable QTL alleles from wild species. During backcrossing, negative selection is exercised to reduce the frequency of deleterious donor alleles. The segregating BC_2F_2 or BC_2F_3 population is then evaluated for traits of interest and genotyped with polymorphic molecular markers. The data thus obtained are subjected to QTL analysis, resulting in the identification of QTLs while transferring them into adaptive genetic backgrounds (Varshney et al. 2005). The AB-QTL analysis could be effective in detecting additive, dominant, partially dominant, or overdominant QTLs (Tanksley and Nelson 1996). Thus, AB-QTL analysis can expand the opportunities to exploit wild germplasm for quantitative trait improvement, if employed successfully. The problem associated with AB-QTL analysis is that the wild species chromosome segments often mask the magnitude of some favorable effects of certain introgressed alleles, as observed during QTL analysis for grain quality in an advanced backcross population derived from *Oryza sativa* and *O. rufipogon* (Septiningsih et al. 2003). Maintaining an adequate population size in selected backcross populations is difficult in AB-QTL analysis, because of negative selection imposed on the backcross progenies for unfavorable alleles. The AB-QTL approach has been evaluated in rice to identify trait-improving QTLs from wild species such as *O. rufipogon* (Xiao et al. 1998; Moncada et al. 2001; Septiningsih et al. 2003).

1.3.3 Allele Mining or EcoTILLING

Allele mining is a promising approach to dissect naturally occurring allelic variation at candidate genes controlling key agronomic traits. The novel and superior alleles of agronomically superior genes thus identified from a crop genepool will be useful in developing improved cultivars. The process involved in identifying the alleles of a fully characterized gene in a germplasm collection is called allele mining. The genome sequence information of rice has made it possible to devise rapid and inexpensive PCR strategies to isolate useful alleles of well-characterized rice genes from the rice genepool. The PCR amplicons of the target gene from the set of germplasm then sequenced and the allelic variations are detected by sequence analyses. The superior alleles are then identified by comparing sequence information with trait phenotype data. Latha et al. (2004) applied the allele mining approach to identify superior alleles for stress tolerance in rice.

Another strategy of mining based on TILLING, called EcoTILLING, was developed for detecting multiple types of polymorphisms in germplasm collections (Comai et al. 2004). This is a high-throughput technique for the detection of DNA polymorphisms based on heteroduplex mismatch cleavage by an endonuclease CELI and gel electrophoresis (Till et al. 2003). EcoTILLING allows natural alleles at a locus to be characterized across many germplasms, enabling both SNP discovery and haplotyping in a cost-effective manner.

1.3.4 Association Mapping

Our understanding of genetic architecture of complex traits in rice is mainly based on traditional QTL linkage mapping, where biparental populations such as doubled haploids (DHs), F_2, or RILs have been widely used to construct molecular marker maps and to identify genes or QTLs for various traits. The mapping populations used in these studies are the products of just one or a few cycles of meiotic recombination, limiting the resolution of genetic maps, and are impossible to scale to investigate the genomic potential and wide phenotypic variation of rice genetic resources available in public germplasm repositories (Zhao et al. 2011). The narrow genetic base of modern rice cultivars is a serious obstacle to sustaining and improving productivity as these may become vulnerable to abiotic and biotic stresses in a changing climate. The rice germplasm resources conserved *ex situ* in different gene banks worldwide are the important reservoirs of natural genetic variation, originated from a number of historical genetic events as responses to environmental stresses and selection during domestication (Ross-Ibarra et al. 2007). The utilization of these rice germplasm resources in association mapping can provide greater resolution for identifying genes responsible for variation in a quantitative trait (Buckler and Thornsberry 2002; Flint-Garcia et al. 2003). Association mapping detects correlations between genotypes and phenotypes in a sample of individuals based on linkage disequilibrium (LD), providing higher mapping resolution, broader allele coverage, and a cost-effective gene tagging approach. Linkage disequilibrium refers to a historically reduced (nonequilibrium) level of the recombination of specific alleles at different loci controlling particular genetic variations in a population (Abdurakhmonov and Abdukarimov 2008). Different aspects of LD and its impact on association mapping have been reviewed elsewhere (Gupta et al. 2005; Mather et al. 2007). Varshney et al. (2005) and Schulze and McMahon (2002) gave an overview of LD-based association mapping. Here, we will highlight two recent studies on genome-wide association study (GWAS) conducted in rice. Genome-wide association studies of 14 agronomic traits in 517 rice landraces were conducted by Huang and colleagues (2011). In this study, Huang and colleagues identified ~3.6 million SNPs by sequencing the rice landraces and constructed a high-density haplotype map of rice genome. They conducted GWAS for 14 agronomic traits and identified loci explaining ~36% of the phenotypic variance. They demonstrated that GWAS, which is facilitated by second-generation genome sequencing, can be a powerful tool for explaining genomics of complex traits in rice. In another study, Zhao and colleagues (2011) conducted GWAS based on genotyping 44,100 SNP variants across 413 diverse rice accessions. Using crosspopulation-based mapping strategies they identified dozens of common variants for 34 rice traits. They also developed a 44K SNP array that is commercially available and can be used in other GWAS for diverse traits in rice. These genome-wide association mapping studies have made a platform to conduct GWAS in rice germplasm resources using modern phenomics facilities, which will accelerate varietal development and crop improvement.

1.4 Rice Functional Genomics: From Sequence to Gene Function

The term *functional genomics* describes the development and application of global or genome-wide experimental approaches to assess the functions of genes by using the information and reagents provided by structural genomics (Hieter and Boguski 1997). Several approaches have been used to explore the probable function of the genes, as well as to monitor their expression in relation to various other genes of rice (Figure 1.3). Recent advances in rice functional genomics analytical tools and their applications have been discussed in Jung et al. (2008) and are summarized in Box 1.2.

The current rice functional genomics tools suffer from major limitations such as functional redundancy within rice gene families, lack of homozygous progenies for analysis, difficulty managing the most-used rice cultivars in artificial growing conditions, restricted access to genomics resources, and limitations in comparing gene-expression

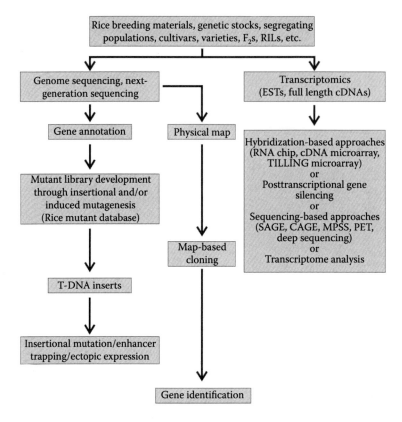

FIGURE 1.3
Flowchart for the exploitation of rice genetic resources for crop improvement by using different genetic and genomic strategies.

profiles. The functional redundancy within gene families can be overcome by generating double or triple mutants or by silencing multiple genes. Collaboration among global research institutes needs to be initiated to expand the number of homozygous progenies on a genomic scale that can also be utilized in generating double or triple mutants. The homozygous lines in genomic analysis will be useful in overcoming the problem of somaclonal variation, which frequently masks true phenotypes. International cooperation is also required in sharing the genetic resources, which will expedite the genomic research. A publicly available and affordable platform like the NSF Rice Multi-Platform Microarray Search tool (www.ricearray.org/element/search_multiple.shtml) needs to be developed to facilitate access to all gene-expression profiles and enhance convenience of usage and consistency among data sets.

The International Rice Functional Genomics Consortium (IRFGC; http://irfgc.irri.org) has been formed to determine the function of all the rice genes; it provides a platform for sharing materials, integrating databases, seeking partnerships, implementing cooperative initiatives, and accelerating delivery of research results to benefit rice production. Several national programs have also been developed to make use of rice genome information (Harris 2002; Tyagi and Khurana 2003; Xue et al. 2003). The availability of common resources will allow broader access and sharing of rice genetic information that is vital to functional genomics research and improvement of rice crops.

BOX 1.2   FUNCTION GENOMICS TOOLS

1. **Map-based cloning:** Genes governing a trait are isolated on the basis of their map position, and thereby isolation from large YAC/BAC libraries. Mapping narrows down the genetic interval containing the mutation or gene underlying the genetic variation. The process also called positional cloning. Different genetic markers (RFLPs, AFLPs, SSRs, etc.) are used to map the gene of interest (GOI).

2. **Gene-indexed mutants:** This approach involves development of lines in which genes are randomly tagged by DNA insertion elements. The process involves insertion of a known segment of DNA into the GOI, which results in either gene knockout or gene overexpression. DNA inserts that create loss-of-function mutants include Transfer-DNA (T-DNA) of *Agrobacterium tumifaciens*, heterologous transposons (Ds and dSpm) and the Tos17 retrotransposons (Jung et al. 2008). On the contrary, the DNA inserts that lead to gain-of-function phenotype contain a strong enhancer element near one end that boosts the gene expression. This activation tagging approach can address the problem of gene redundancy associated with traditional screens for loss-of-function mutations. The FL-cDNA overexpression (FOX) gene-hunting system is another gain-of-function approach developed recently. Targeting-induced local lesions in genomes (TILLING) is a useful supplementary tool to inactivate genes for which insertions are not available, or to obtain partial loss-of-function mutations for evaluation of allelic series. TILLING populations of rice are generated by chemical mutagenesis and point mutants are screened by using single nucleotide polymorphism (SNP) detection assays (Till et al. 2007).

3. **Deletion mutants:** Deletion mutants are generated using induced mutagenesis (Leung et al. 2001; Wang et al. 2004; Wu et al. 2005) and oligonucleotide tiling/microarrays are used to identify genomic deletions. This method has advantages over map-based cloning, which is labor-intensive and time consuming, especially for mutants with phenotypes that need intricate analyses, such as submergence tolerance, metabolic, and cell signaling mutants. Oligonucleotide tiling arrays are a subtype of microarray chips and involve design of short genome-wide probes which are used to hybridize with DNA from the mutant lines, allowing detection of the probe or probes corresponding to a genomic deletion.

4. **EST-profile:** Profiling of gene expression can give vital information about a particular gene. Genome-wide expression profile is generated from ESTs, DNA sequences read from either end of cDNA molecules' profiles. An EST profile can provide information about the expressed part of the genome; however, it has problems related to redundancy. Large-scale EST data sets have generated for rice (Yamamoto and Sasaki 1997; Babu et al. 2002; Sahi et al. 2003).

5. **Gene traps:** Gene entrapment strategies rely on the use of DNA inserts containing reporter gene constructs, whose expression is dependent on *cis*-acting regulatory sequences at the site of insertion. The reporter gene enables identification of gene(s) which might not display an obvious mutant phenotype.

Three basic types of gene traps are constructed using reporter genes such as those encoding β-glucuronidase (GUS) and green fluorescent protein (GFP); these are enhancer trap, promoter trap, and gene trap (Chin et al. 1999; Wu et al. 2003).

6. **Microarray:** Microarray, also termed reverse northern-dot blots technology allows measurements of the expression levels of thousands of genes in a single experiment (Lipshutz et al. 1999). It also allows identification of transcriptionally active regions (TARs) in the genome, and it can also be used for genome-wide polymorphism surveys and the identification of mutants (Ramsay 1998). Several microarray platforms have been developed for two rice subspecies (see Jung et al. 2008). These array platforms can be used to study the gene expression profile at different plant growth stages or under stress conditions (Cooper et al. 2003; Zhu et al. 2003).

7. **Tiling arrays:** Genome tiling arrays are a subtype of microarrays. In contrast to traditional microarrays, oligonucleotide probes are prepared to cover the entire genome or, in other words, to "tile" a continuous path along each chromosome (Johnson et al. 2005). Because of the high resolution and sensitivity, tiling arrays can be used to empirically validate putative gene models and to identify novel transcription units (Li et al. 2007).

8. **Serial analysis of gene expression (SAGE):** SAGE allows a rapid and detailed analysis of thousands of transcripts in single attempt. It provides quantitative information on the abundance of known transcripts and has the capacity to identify novel expressed genes (Velculescu et al. 1995; Saha et al. 2002).

9. **Massive parallel signature sequencing (MPSS):** MPSS is an improvement over SAGE and can identify a large number (>10,000,000) of distinct signatures of cDNA libraries in a single analysis. This method is used to validate predicted gene models and to identify novel genes. MPSS was used to develop a comprehensive expression atlas of rice sequences (rice-MPSS; Nakano et al. 2006). This approach has not been exploited widely to study rice transcriptomics because of the high cost that limits the number of biological replicates that can be performed. As a result, the quantitative significance of the resulting gene-expression profiles cannot be validated statistically.

10. **Gene silencing:** RNA interference (RNAi) technology allows posttranscriptional silencing of target gene or genes and provides enhanced capability to study gene families. RNAi has been used effectively to knockdown multiple genes simultaneously with inverted-repeat constructs that target unique or conserved regions of multiple genes (Miki et al. 2005). Micro-RNAs (miRNAs) which are derived from irregular stem-loop structures often reduce target gene expression by interacting with target mRNAs having sequence complementarity. Artificial miRNAs (amiRNAs) have been developed by site-directed mutagenesis of endogenous miRNA precursors in *Arabidopsis* (Schwab et al. 2006) and this approach entails a limitation on the number of genes that can be silenced simultaneously.

11. **Transient assay systems:** The rice protoplast transient assay system is a useful tool for functional validation of highly prioritized candidate genes. Protoplast transient assays are likely to be more biologically relevant than

data obtained from heterologous cell systems such as bacteria and yeast (Jung et al. 2008). Transient assay systems can be used to monitor subcellular localization of target proteins (Chen et al. 2006), detect protein–protein interactions (Chen et al. 2006), and measure promoter activity (Bart et al. 2006).

12. **Transcriptome analysis:** RNA sequencing based on next-generation DNA sequencing technology provides a cost-effective way for mapping and quantifying the transcriptome (all transcripts). Transcriptome analysis enables us to analyze complex RNA mixtures and the extent of alternative splicing, discover rare transcripts, detect untranslated regions (UTRs), and identify gene fusions. The combined approach of whole-genome transcriptome, predicted pathway, and insertional mutant analysis can be applied to identify and characterize candidate genes (for details, see Jung et al. 2008).

13. **Phylogenomics**: Phylogenomics study analyzes genomic data in a phylogenetic context (Jiao and Deng 2007). This approach is particularly useful in categorizing genes into functional groups based on sequence homology when studying large gene families for which limited phenotypic data are available. Phylogenomics does not rely on sequence information from numerous phenotypically characterized populations, and thus provides a rapid and logical basis for further detailed functional studies (Dardick and Ronald 2006). The Rice Kinase Database (RKD; http://rkd.ucdavis.edu) was created based on a phylogenomics approach and enables analyses of diverse sets of genomic information in phylogenetic context (Dardick et al. 2007).

1.4.1 Identification of Genes through Functional Genomics Approaches

The advances in rice functional genomics can be seen from the increasing number of genes being cloned and fine-mapped. This has greatly facilitated the analysis of important agronomic traits of rice, as well as its response to abiotic and biotic stresses. When the function of a gene or, in other words, when a gene (or genes) underlying a trait is identified, such molecular biological knowledge can be used to generate new rice cultivars with greater yield and improved tolerance to stresses.

1.4.1.1 Genes for Agromorphological Traits

1.4.1.1.1 Plant Height

Introduction of semidwarf cultivars made the Green Revolution possible by making rice plants more responsive to nitrogen fertilizer and by improving the harvest index (the ratio of grain to grain plus straw) and biomass production (Khush 1999). The rice *semi dwarf1* (*sd1*) gene involved in gibberellin (GA) synthesis (Sasaki et al. 2002) has been characterized by map-based cloning. It was found that manipulating GA metabolism in rice can produce high-yielding semidwarf cultivars (Sakamoto et al. 2003). Other genes such as *Gid1*, *Gid2*, and *Slr1* involved in the GA-signaling pathway have also been identified through map-based cloning (Ueguchi-Tanaka et al. 2005; Ikeda et al. 2001; Sasaki et al. 2003, respectively).

The rice *OsBRI1*, an ortholog of *Arabidopsis* semi-dwarf gene *BRASSINOSTEROID INSENSITIVE1* (*BRI1*) (Chono et al. 2003), is reported in d61 rice mutant (Yamamuro et al. 2000). The loss of function of *OsBRI1* in transgenic rice prevents internode elongation and bending of the lamina joint, and shows increased grain yield under high planting density due to erect leaves that capture more sunlight (Sakamoto et al. 2006; Morinaka et al. 2006).

1.4.1.1.2 Tillering

Rice tillers are the lateral branches on the short basal internodes of the main culm or on primary or higher-order tillers, and each tiller has the capacity to generate panicles. Thus, increasing the number of tillers can contribute to yield enhancement. The *MONOCULM1* (*MOC1*) gene regulating tillering or lateral branching in rice was identified by map-based cloning with a loss-of-function mutant *moc1* (Li et al. 2003). Plants with *moc1* produce only a single culm.

In rice, the *OsTB1/FINE CULM1* (*FC1*), which is orthologous to the *Zea mays TEOSINTE BRANCHED1* (*TB1*) gene contributing to the domestication of maize, is found to negatively regulate tillering (Takeda et al. 2003). A few other rice genes for several dwarf mutants with increased tiller number have been characterized, such as the *DWARF3* (*D3*) gene identified by map-based cloning of the *d3* locus (Ishikawa et al. 2005), *high tillering draf1* (*htd1*) (Zou et al. 2006), *DWARF10* (*D10*) identified from *d10* mutant (Arite et al. 2007), *DWARF27* from *d27* mutant (Lin et al. 2009), and *DWARF88* from *d88* mutant (Gao et al. 2009). These genes were found to be the key regulators of tillering in rice and can be manipulated for further improvement of rice yield.

1.4.1.1.3 Plant Architecture

Several genes have been identified to control the architecture of rice plants, which is an important agronomic trait that affects grain yield. The *OsCRY1* gene controlling leaf elongation in rice seedlings was identified by Zhang et al. (2006). The prostrate growth habit in wild rice is controlled by the *PROG1* gene, which was identified by Tan et al. (2008). The *OsIAA1* gene was reported to be responsible for reduced plant height and loose plant architecture (Song et al. 2009). Rice leaf rolling *shallot-like1* (*sll1*) mutant plants with defective development of sclerenchymatous cells on the abaxial side of leaves due to KANADI transcription factor were characterized by Zhang et al. (2009a).

1.4.1.1.4 Floral Organ Development

Rice inflorescence morphology is determined mainly by primary and secondary inflorescence branches. Flower development in rice is found to be regulated by *Leafy hull sterile1* (*Lhs1*), a homeotic mutation in rice MADS box gene identified by T-DNA technology (Jeon et al. 2000). Gao et al. (2010) characterized one *SEPALLATA* (*SEP*)-like gene, *OsMADS34*, involved in the development of inflorescence and spikelets in rice. Another MADS box gene, *OsMADS6*, is found to specify rice floral organ identities and determination of the floral meristem (Li et al. 2010a). Anther development in rice is found to be regulated by the *Undeveloped tapetum1* (*Udt1*) gene, which was identified by T-DNA technology (Jung et al. 2005). The *Udt1* gene regulates early tapetum development in rice.

1.4.1.1.5 Grain Yield Traits

Grains per panicle, grain weight, and grain filling are primary components of grain yield in rice. In recent years, many attempts have been made to characterize QTLs for these traits. The QTL *Grain number1* (*Gn1*), which increases grain number, has been identified on the short arm of chromosome 1 by positional cloning (Ashikari et al. 2005). This gene has two alleles, *Gn1a* and *Gn1b*, encoding a cytokinin oxidase/dehydrogenase (CKX), *OsCKX2* (Ashikari et al. 2005), which preferentially and irreversibly degrades cytokinins (promotes cell division) in inflorescence meristems and thus controls the number of flowers.

Downregulation of *OsCKX2* in transgenic rice having antisense *OsCKX2* cDNA construct led to an increased number of grains (Ashikari et al. 2005). Grain number per panicle is mostly determined by panicle architecture (Sakamoto and Matsuoka 2008), therefore, genes controlling primary and secondary panicle branching and panicle length also regulate grain number per panicle. A major rice grain yield QTL, *DEP1*, was isolated by positional cloning and found to reduce inflorescence internode length and increase grains per panicle (Huang et al. 2009a).

Grain length, width, and thickness determine grain weight and size. In addition to grain weight, grain size is also an important breeding objective as it determines the quality of rice grains. Several QTLs for grain weight have been reported. *GS3*, a major QTL that negatively regulates grain shape, was identified in the pericentric region of chromosome 3 (Fan et al. 2006). It was reported that the candidate gene for the *GS3* encodes a transmembrane protein containing an intercellular phosphatidylethanolamine-binding protein (PEBP)-like domain, an extracellular von Willebrand factor type C (VWFC) module, and a tumor necrosis factor receptor (NGFR) family cysteine-rich domain. A non-sense mutation in the PEBP-like domain resulted in a truncated form of the protein in all large-grain rice cultivars. Song et al. (2007) cloned and characterized the grain-weight-related gene, *GW2*. The candidate gene for *GW2* encodes a RING-type ubiquitin E3 ligase and a loss-of-function *GW2* allele (NIL-*GW2*) resulted in increased grain width and weight. *GW2* negatively regulates grain width by controlling cell division in the rice hulls. The *grain number, plant height and heading date 7* (*Ghd7*) QTL was isolated and characterized by Xue et al. (2008), which could be used to optimize rice cultivars for the regions that they are grown in and for changing climatic conditions.

Grain filling contributes greatly to grain weight and is regulated by QTLs (Takai et al. 2005). The genes and underlying molecular mechanisms controlling grain filling remain unexploited. Wang et al. (2008a) isolated and functionally characterized the rice *GRAIN INCOMPLETE FILLING1* (*GIF1*) gene that encodes cell wall invertase required for carbon partitioning during the early grain filling stage. This gene is reported to be a potential domestication gene and can be used for further crop improvement. Higher night temperatures during the grain filling stage as a consequence of global warming reduced rice grain yield (Peng et al. 2004) and quality by producing chalky grains. Gene-expression analysis by microarray and semiquantitative reverse transcriptase polymerase chain reaction (RT-PCR) during the grain filling stage at higher night temperature revealed the downregulation of several genes for starch and storage protein synthesis, and the upregulation of genes for starch-consuming α-amylases and for heat-shock proteins (Yamakawa et al. 2007). Such studies in addition to QTL analyses will allow breeders to improve grain filling in rice, which is highly influenced by dynamic and complex metabolic processes.

1.4.1.2 Genes for Abiotic Stress Tolerance

Abiotic stresses are the major constraints on rice productivity. Abiotic stresses are those imposed by the physical environment, such as excess or shortage of water, or inadequate or excess minerals in the soils. Under various scenarios of climate change, abiotic stresses are likely to be intensified. With the advent of physiological and molecular understanding of the effects of environmental stresses, omics-based studies are gaining more attention for the identification of genes involved and their expression patterns during the course of stress sensing and response. Systematic omics-based studies will lead to unraveling of coregulated genes (Provart and McCourt 2004). When plants sense stresses, a signal transduction cascade is invoked. Secondary messengers relay the signal through the plant system, ultimately activating stress-responsive genes generating the initial stress response. Stress-induced gene products may be involved either in stress tolerance or in signal transduction. Under different types of stress, tolerance genes enable plants to survive by synthesizing biomolecules and inducing a variety of cellular responses depending on the nature of the stress. It was reported that multiple signaling pathways can be activated during exposure to stress, as under stresses such as drought, cold, and salinity, accumulation of compatible solutes and antioxidants was noted (Hasegawa et al. 2000). Rabbani et al. (2003) studied rice gene expression by transcriptome analyses under stresses of cold, drought, high salinity, and hormone (abscisic acid, ABA) application, and reported upregulation of 73 genes under these stresses. This study also demonstrated the overlap in gene expression under abiotic stresses. Under stress conditions, expressions of many genes are induced, and their products have important roles in stress responses and tolerance (for a comprehensive review, see Todaka et al. 2012). While exploring the crosstalk among different abiotic stress responses, a similar type of plant response against drought and salt stress was registered, with both disrupting the ion and osmotic homeostasis. Thus signaling pathways for these two stresses are expected to be similar. Both stresses resulted in an increase in ABA expression, which is known to be involved in the tolerance to osmotic stress and regulation of the plant–water balance.

1.4.1.2.1 Salt-Stress-Related Genes

High salinity is a major problem in coastal areas, as well as in areas facing overuse of underground water and poor irrigation management. A large number of QTLs have been identified for salt tolerance in rice using different types of DNA markers. The physiological and molecular understanding of salt stress response of rice has been reviewed in Sahi et al. (2006). The *SKC1* QTL on chromosome 1 controls shoot sodium (Na) content. It was subsequently cloned and found to encode a sodium transporter as *OsHKT8* that helps to control K⁺ homeostasis under salt stress (Ren et al. 2005). In another study, Thomson et al. (2010) characterized the *Saltol* QTL on rice chromosome 1, controlling Na⁺/K⁺ homeostasis at the seedling stage, and proposed that *Saltol* QTL may harbor the gene *SKC1* identified by Ren et al. (2005).

Transcription factors (TFs) are powerful tools for genetic engineering, as overexpression of a TF can lead to the upregulation of the whole array of genes under its control (see Todaka et al. 2012 for a list of TFs and their role in stress response). Dubouzet et al. (2003) reported expression of a rice TF *OsDREB2A* (an *Arabidopsis* DREB1-type gene that binds to a *cis*-acting element DRE/CRT) under salt stress and dehydration. *OsZIP23*, a member of the *bZIP* (Basic region/Leu Zipper) TF family, can be useful in genetic improvement against salt stress (Xiang et al. 2008). A stress-responsive *NAC* gene, *SNAC2*, was isolated and characterized by Hu et al. (2008). Expression analysis of the *SNAC2* gene suggested that it might be involved in salt stress tolerance in rice.

1.4.1.2.2 Submergence-Related Genes

Submergence of rice by flash flooding is a common phenomenon in the monsoon season in South and Southeast Asia. It is a major constraint to rice production, as rice cultivars vary in their capacity to tolerate complete submergence. QTL analyses have revealed that a large portion of the variation in plant response to submergence can be explained by the *Sub1* locus on chromosome 9. The *Sub1* locus was characterized by map-based cloning as a large QTL containing a cluster of three genes (*Sub1A*, *Sub1B*, and *Sub1C*) encoding ethylene response factor (ERF)-like TFs (Xu et al. 2006). Only *Sub1A* carries allelic variation across rice varieties, while *Sub1B* and *Sub1C* are invariably present in the *Sub1* region. *Sub1A* had two alleles: *Sub1A-1*, a tolerance-specific allele, and *Sub1A-2*, an intolerance-specific allele. Overexpression of *Sub1A-1* confers tolerance in submergence-intolerant cultivars. These genes control highly conserved hormonal, physiological, and developmental processes that determine the elongation of the rice plant when submerged (Fukao et al. 2006). For the characterization of *Sub1A-1* in conditioning submergence tolerance in rice cultivars, breeding programs have been undertaken to incorporate this gene into popular cultivars through marker-assisted backcrossing.

1.4.1.2.3 Cold-Stress-Related Genes

Rice is a cold-sensitive plant. Low temperatures drastically reduce its production. During early growth stages of the plant, occurrence of cold stress inhibits seedling establishment and leads to nonuniform crop maturation. Spikelet fertility decreases when rice plants perceive cold stress at the booting stage due to the failure of microspore development. Cold stress at the booting stage is a serious problem both at high altitudes and in uplands at low altitudes. Cold tolerance in rice cultivars is a complex trait and its genetic mechanism is not well understood. Many studies have been performed to identify QTLs underlying cold tolerance in the seedling as well as the booting stage (Suh et al. 2010, 2012). A QTL designated as *qCTS12*, located on the short arm of chromosome 12, contributed over 40% of the phenotypic variance (Andaya and Mackill 2003). This QTL was fine mapped using SSR markers, and the most likely genes underlying the QTL are *OsGSTZ1* and *OsGSTZ2* (Andaya and Tai 2006). Saito et al. (2004) reported physical mapping of a booting stage cold tolerance QTL *Ctb1* and also identified the candidate genes. They proposed that *Ctb1* is likely to be associated with anther length, which is one of the major factors in cold tolerance at the booting stage. Another QTL *qCTB7* on chromosome 7 was analyzed by Zhou et al. (2010) and 12 putative candidate genes were identified. A rice gene encoding a calcium-dependent protein kinase, *OsCDPK7*, was found to be induced by cold stress. The extent of tolerance to cold stress was associated with the level of *OsCDPK7* expression (Saijo et al. 2000). Rice TF *MYBS3* was reported to play a commercial role in cold adaptation in rice (Su et al. 2010). Transcription profiling of transgenic rice overexpressing or underexpressing *MYBS3* led to the identification of many genes in the *MYBS3*-mediated cold signaling pathway. Ma et al. (2009) showed that overexpression of *OsMYB3R-2* enhanced cold tolerance in transgenic rice plants by altering cell cycle and regulating expression of stress genes. Overexpression of proteins/enzymes associated with stress responses has been a common practice in improving stress tolerance of crop plants. But constitutive overexpression of these proteins often leads to impaired growth. It was suggested that use of stress-inducible promoters for the expression of these TFs minimizes their negative effects on plant growth (Nakashima et al. 2007), and thus can be utilized in rice improvement.

1.4.1.2.4 Phosphorus-Uptake-Related Genes

Phosphorus (P) is an essential macronutrient for plant growth and development, as it is a constituent of ATP, nucleic acids, phospholipids, and other key molecules of plants. P is one of the limiting factors for plant growth due to its rapid immobilization by soil organic and inorganic components (for review, see Hinsinger et al. 2011). More than half of global rainfed lowland rice is produced on poor and problem soils that have naturally low P-fixing rates. Rice has evolved numerous strategies to optimize soil phosphate (Pi) acquisition from soil solution and its distribution to different organs and subcellular components under Pi-deficient conditions (Vance et al. 2003). Zhou et al. (2008) reported *OsPHR2* as a key regulator for Pi starvation signaling in rice. Recently, identification and functional characterization of a nucleus-localized R2R3-type MYB TF, *OsMYB2P-1*, has been reported in rice (Dai et al. 2012). Expression of *OsMYB2P-1* was induced in rice seedlings by Pi starvation. Overexpression of this gene enhanced tolerance to Pi starvation. Dai et al. (2012) demonstrated that *OsMYB2P-1* is a novel TF associated with Pi starvation signaling in rice. To date, phosphorus uptake 1 (*Pup1*) is the only QTL available to breeders for marker-assisted breeding to improve rice yield under Pi-deficient conditions (Chin et al. 2010, 2011). To gain insight into the functional mechanism of *Pup1*, Gamuyao et al. (2012) identified a *Pup1*-specific protein kinase gene, and named it *PSTOL1* (phosphorus-starvation tolerance1). This gene was identified in the traditional *aus*-type rice cultivar Kasalath and was absent in P-starvation-intolerant modern varieties. Overexpression of *PSTOL1* led to significant increase in grain yield under Pi-deficient conditions in intolerant modern rice varieties.

1.4.1.3 Genes for Biotic Stress Tolerance

Rice is often subjected to a variety of biotic stresses from fungi, bacteria, viruses, insects, and nematodes throughout the life cycle. Under favorable conditions, these pathogens and insects cause severe loss in grain yield. An understanding of the molecular mechanisms in host resistance to biotic stresses is essential to devise control strategies. In the case of rice diseases, significant achievements have been made in elucidating the molecular basis of rice–pathogen interactions. A number of resistance (*R*) genes have been cloned and characterized in the recent past. In response to pathogen attack, plants have evolved defense mechanisms against the pathogens. Different hypotheses have been proposed to explain the host–pathogen interactions (Vanderplank 1982). Rice disease resistance is primarily classified into two main categories: qualitative resistance (or complete resistance) that is conferred by a single *R* gene and is pathogen race-specific, and quantitative resistance (or partial resistance) that is controlled by multiple genes or QTLs.

The qualitative resistance has been successfully deployed in the case of rice blast (caused by the fungus *Magnaporthe oryzae*) and bacterial leaf blight (BLB, caused by a biotrophic pathogen *Xanthomonus oryzae* pv. *oryzae*) diseases, but this resistance can be rapidly overcome due to strong selection pressure against it and the rapid evolution of the pathogens. It is a long-term competitive evolutionary process between the pathogenicity and host resistance.

To date, around 100 rice blast resistance genes have been identified (for review, see Sharma et al. 2012), spanning all rice chromosomes (Figure 1.4). Various genomics tools have been used for the identification and mapping of the *R* genes. The molecular map-based techniques are mostly used for the identification of *R* genes. Availability of high-density molecular maps of rice helps in utilizing available marker information for the identification and localization of the *R* genes. About 30 blast resistance genes have been identified in exploring this strategy. On the other hand, the blast resistance gene *Pid3* has been identified by genome-wide *in silico* comparison of paired NBS–LRR genes and their pseudogene/alleles between the genome sequence of *indica* rice cv 93-11 and *japonica* cv Nipponbare available in the public domain (Shang et al. 2009). The third approach, QTL mapping, including single marker analysis (SMA), standard interval mapping (SIM), and composite interval mapping (CIM), was successfully used to identify 350 QTLs for blast resistance and 23 resistant loci within those QTL regions (Ballini et al. 2008). To date, only 19 blast resistance genes have been cloned out of 100 *R* genes identified and mapped from *indica* and *japonica* rice (Wang et al. 1999; Bryan et al. 2000; Sharma et al. 2005, 2010; Kumar et al. 2007; Rai et al. 2011; Gupta et al. 2012).

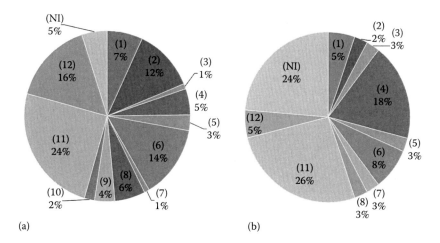

FIGURE 1.4
Chromosome-wise distribution of (a) rice blast (From Sharma, T. R., Rai, A. K., Gupta, S. K., Vijayan, J., Devanna, B. N., and Ray, S., *Agricultural Research*, 1, 37–52, 2012.) and (b) bacterial leaf blight resistance genes. Chromosome number is indicated in parentheses. *Note:* NI, no information on chromosomal location.

Numerous major genes including 25 dominant and 12 recessive genes have been identified for resistance to various strains of *X. Oryzae* pv. *oryzae*, and have been named in a series from *Xa1* to *Xa36* (reviewed in Xia et al. 2012). More than 25 bacterial blight resistance (*R*) genes were mapped on rice chromosomes (Figure 1.4), with some of the genes being allelic or tightly linked with each other (Kinoshita 1994, 1995; Lin et al. 1996; Xiang et al. 2006). The chromosomal locations for the rest of the bacterial blight resistance genes are still unknown. A total of six *R* genes against BLB have been cloned and characterized. Among these, four were dominant, *Xa1, Xa21, Xa3/Xa26, Xa27* (Song et al. 1995; Yoshimura et al. 1998; Sun et al. 2004; Gu et al. 2005), and the other two were recessive, *xa5* (Iyer and McCouch 2004) and *xa13* (Chu et al. 2004). The characterized *R* genes confer either expressive resistance (*xa13* and *Xa27*)—gene expression plays a direct role in resisting BLB pathogen—or interactive resistance (*xa5, Xa1, Xa21, Xa3/Xa26*): interaction between host receptor and pathogen elicitor restricts pathogen attack.

Quantitative resistance is broad-spectrum resistance and can be pathogen race-non-specific (Kou and Wang 2010). For rice sheath blight (caused by soil-borne necrotrophic pathogen *Rhizoctonia solani*), false smut (cause by *Ustilaginoidea virens*), and bacterial streak (caused by *Xanthomonus oryzae* pv. *oryzicola*) no *R* genes have been reported. As a result, only quantitative resistance breeding can be practiced against these diseases. In recent years, researchers have identified resistance QTLs against sheath blight (Pinson et al. 2005), false smut (Li et al. 2008), and bacterial streak (Chen et al. 2006). Among the viral diseases of rice, major *R* genes were reported for *Rice yellow mottle virus* (Alber et al. 2006) and *Rice stripe virus* (Wang et al. 2010). The molecular bases of rice quantitative disease resistance have recently been reviewed in Kou and Wang (2012). A candidate gene hypothesis-based strategy of validation and functional analysis of QTL has been established to characterize minor resistance QTLs (Hu et al. 2008). Information generated from the characterization of resistance genes/QTLs will elucidate molecular basis of resistance against pathogens and will ultimately contribute to rice improvement through resistance breeding.

Rice is attacked by a large number of insects. Among the insect pests, plant hoppers, stem borers, and gall midges are most important. There are six types of plant hopper, viz. brown planthopper (BPH), white backed planthopper (WBPH), green leafhopper (GLH),

TABLE 1.1

BPH-Resistant *R* Genes Isolated from Wild Species

Gene	Wild Source	Reference
Bph10	*Oryza australiensis*	Ishii et al. (1994)
Bph12(t)	*O. latifolia*	Yang et al. (2002)
Bph13(t)	*O. eichingeri*	Liu et al. (2001)
	O. officinalis	Renganayaki et al. (2002)
Bph14, Bph15	*O. officinalis*	Huang et al. (2001)
Bph18(t)	*O. australiensis*	Jena et al. (2006)
Bph20(t), Bph21(t)	*O. minuta*	Rahman et al. (2009)
bph11(t), bph12(t)	*O. officinalis*	Hirabayashi et al. (1999)

zigzag leafhopper (ZLH), small brown planthopper (SBPH), and green rice leafhopper (GRH), which cause yield losses in rice to a variable extent. The plant hoppers also transmit viral diseases of rice as vectors. Among the plant hoppers, the brown plant hopper (*Nilaparvata lugens*) causes direct damage by sucking plant sap and produces typical symptoms of infestation called *hopper burn*. The brown planthopper is a vector of viruses such as grassy stunt and ragged stunt, and causes devastating damage to rice production worldwide. Growing resistant varieties is the most effective and environment-friendly strategy for protecting the crop from BPH. A total of 21 genes for BPH resistance and a few QTLs have been identified from cultivated and wild species of *Oryza* using RFLP, RAPD, and SSR markers (Brar et al. 2009). Of these 21 resistance genes, 15 are mapped to different chromosomal locations, 8 are tightly linked with molecular markers, and 11 resistance loci have been identified from wild species (Table 1.1). Du et al. (2009) characterized *Bph14*, a gene conferring resistance to BPH, using a map-based cloning strategy. This gene encodes a coiled-coil, nucleotide binding, and leucin-rich repeat (CC-NB-LRR) protein. Expression of *Bph14* activates a salicylic acid signaling pathway and induces callose deposition in phloem cells and trypsin inhibitor production after BPH infestation, thus reducing feeding, growth rate, and longevity of the insects. A number of genes for resistance to other planthoppers have been identified: 11 dominant and 3 recessive genes for GLH, 7 genes for WBPH, 6 for GRH, and 3 for ZLH (Brar et al. 2009). Characterization of resistance genes provides information on the molecular mechanisms of rice defense against insect pests, which will facilitate the development of resistant rice varieties.

1.5 Rice Proteomics

The availability of genome-wide DNA data and cDNA-based microarrays for rice has been proved to be the turning point in understanding the biological systems of the plant. The genome annotation projects that have determined gene structure and function have been undertaken to accurately identify protein-coding genes and their products. Over the past decade considerable advances have been made in rice proteomic research (Helmy et al. 2011). The genome annotation work that involves experimental and computational methods is largely dependent on expression (transcriptional) evidence (Brent 2008), and is highly reliant on *de novo* annotations of protein-coding genes performed using gene prediction programs (Koonin and Galperin 2002; Brent 2008).

The gene/protein prediction tools have been useful in the annotation process, but their prediction accuracy varies greatly depending on the tool or algorithm used and the genome complexity of the organism (Brent 2008; Guigo et al. 2006). Proteins can be identified by different proteomic techniques such as two-dimensional polyacrylamide gel electrophoresis (2D-PAGE), mass spectrometry (MS), and protein microarray approaches (Rakwal and Agrawal 2003). An alternative technology, termed multidimensional protein identification technology (MudPIT), which involves the generation of peptides from a complex protein mixture followed by separation on a strong cation exchange phase in the first dimension and by reverse phase chromatography in the second dimension (Washburn et al. 2001; Wolters et al. 2001), has also been used coupled with 2D-PAGE to analyze protein expression in rice leaf, root, and seed with improved coverage (Koller et al. 2002).

The technique of proteome analysis with 2D-PAGE has the power to monitor global changes that occur in the protein expression of tissues and organisms, as well as under stressed conditions (Komatsu et al. 2003). Komatsu and colleagues constructed rice proteome catalogs by 2D-PAGE of proteins extracted from different tissues of rice and also from rice plants grown under stressed conditions. Image analysis of gels revealed a total of 10,589 protein spots. The N-terminal and internal amino acid sequences were determined by electroblotting separated proteins and by using a protein sequencer or MS of enzyme digested proteins, respectively. Finally, catalogs of rice proteins with information on amino acid sequences and sequence homologies was constructed by Komatsu et al. (2003). The steps involved in rice proteome analysis are given in Figure 1.5. The information generated from the proteome analysis can be used to develop proteome databases for predicting function of unknown proteins. The large-scale peptide sequence information generated from MS-based approaches can also be used for refinement of genome annotation through proteogenomics (Helmy et al. 2011).

The MS-based proteomics approach enables identification and validation of protein-coding genes in an independent and unambiguous way based on translation-level expression evidence. This process of proteogenomics detects naturally occurring proteins based on MS/MS-based approaches (proteomics) and subsequently the proteins are systematically mapped back to the genome sequence (genomics) (Armengaud 2010; Castellana and Bafna 2010). The proteogenomics approach not only validates predicted gene models at a translational level (Jaffe et al. 2004; Wang et al. 2005), but also leads to identification of new gene models (Castellana et al. 2008), determination of start and termination sites of proteins (Tanner et al. 2007), verification of hypothetical and putative genes/proteins (Tanner et al. 2007; Ansong et al. 2008), and verification of splice isoforms at the protein level (Power et al. 2009).

In the case of rice, two proteome databases, one based on 2D-PAGE (Rice Proteome Database; Komatsu 2005) and another based on shotgun proteogenomics (OryzaPG-DB; Helmy et al. 2011), have been reported to date. These databases can be useful resources and data-serving tools for confirming the gene models (Itoh et al. 2007).

1.5.1 Functional Analysis of Rice Proteome

Proteomics has evolved from the need to be able to globally determine the expression patterns of proteins and their respective posttranslational modification in different tissues (Zivy and Vienne 2000), as each cell in a given organism varies in its protein content although they all have the same genome (Collins 2001). The functional analysis of the rice proteome involves detection of variation in protein expression in different specific tissues such as leaf, seed, embryo, internodes, and pollen. The up- or downregulated proteins are

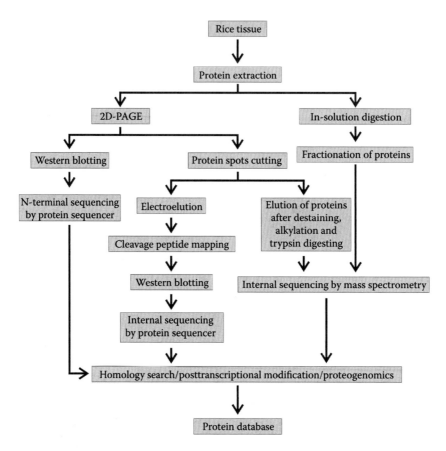

FIGURE 1.5
Steps involved in rice proteomic analysis.

then selected and their sequences or structures are analyzed by comparison with the proteome database or by MS and Edman sequencing, and are sorted into functional categories (Schoof et al. 2002; Komatsu and Tanaka 2005). Plants, being nonmotile, have to be plastic in their response to environmental changes as well as to abiotic and biotic surroundings to grow and develop optimally. Plant hormones play a vital role in signal transduction and in regulation of developmental pathways of the plant (Davies 1995). In the postgenomic era, several studies have been conducted to understand rice proteomics in different growth stages and tissues, and in regulation under different effects in response to external stimuli (hormones, chemicals, etc.).

1.5.1.1 Rice Proteome in Different Organs and Developmental Stages

Koller et al. (2002) identified 2528 unique proteins which included 1022, 1350, and 877 different proteins from leaf, root, and seed, respectively, by using a combination of 2D-PAGE and mudPIT approaches. Functional analysis of these identified proteins categorized them into various functional classes (Koller et al. 2002). The majority of the identified proteins from leaf, root, and seed were classified as being involved in metabolic processes (20.8%), developmental processes (8%), and different cellular activities such as protein synthesis, protein degradation, translation initiation and elongation factors, and signal transduction.

Proteomic analysis of rice seed or embryo revealed differences in expression of proteins involved in metabolic pathways, response to stress, creation of nutrient reservoirs, and other major biological processes (Yang et al. 2007a; Wang et al. 2008b). Among the proteins with altered expression, downregulated genes mainly encoded storage, seed maturation, and aging regulated proteins, while upregulated genes were involved in glycolysis. Differential expression of proteins in rice leaves was also observed in relation to growth stages of the plant (Zhou et al. 2005; Shao et al. 2008). Proteomic analysis of the uppermost internodes of rice at the milky stage revealed high physiological and stress-resistant activities of the internodes (Yang et al. 2006). Similarly, the proteome of mature pollen grains of *japonica* rice had a role in pollen germination and tube growth (Dai et al. 2007).

1.5.1.2 Rice Proteome under Biotic and Abiotic Stresses

1.5.1.2.1 Abiotic-Stress-Responsive Proteome

The response of rice plants to different abiotic and biotic stresses through regulation of the tissue-specific proteome has been surveyed in various studies. Most of these studies have been conducted to understand proteomics of plants growing under important abiotic stresses such as drought, salinity, cold, high temperature, and heavy metal toxicity.

Under rainfed and upland conditions, drought represents one of the most important limitations, mainly affecting spikelet fertility due to deficiency in anther dehiscence. Drought-tolerant cultivars can maintain the tissue water content and recover quickly when the drought period is over. Proteome analysis was undertaken of leaves from 3-week-old rice plants of two lowland and upland *indica* cultivars (Salekdeh et al. 2002). The proteins that increased most in abundance in response to drought were S-like RNase homolog, actin depolymerizing factor, and RuBisCO activase, whereas isoflavone reductase-like proteins decreased significantly. The anther proteins from well-watered and drought-stressed rice cultivars were also compared by Liu and Bennett (2011). They reported 93 proteins with altered expression under drought.

Proteomics analysis of rice leaf sheath in response to salt (NaCl)-induced stress had been performed by Abbasi and Komatsu (2004), and eight proteins were identified having significant and reproducible changes in abundance. They identified five proteins among the eight as photosystem II oxygen-evolving complex protein, fructose bisphosphate aldolases, oxygen-evolving enhancer protein 2 (OEE2), and superoxide dismutase (SOD). Yan et al. (2005) surveyed the proteomes of rice roots and apoplasts under salt stress and identified salt-stress-responsive proteins in rice roots. It was also suggested that apoplastic proteins may play an important role in salt stress response signal pathways. This was also established by Zhang et al. (2009b), who studied the apoplastic proteome of NaCl-treated rice seedlings and reported 10 differentially expressed abundant proteins, which include some important biotic and abiotic stress-related proteins. Among them, an apoplastic protein with extracellular domain-like cysteine-rich motifs (DUF26), *O. sativa* root meander curling (OsRMC), has shown drastically increased abundance in response to salt stress. Dooki et al. (2006) studied proteome patterns of young panicles from rice plants grown in NaCl-containing culture solution at an early reproductive phase and reported differential expression of 13 proteins involved in several salt-responsive mechanisms. The proteome analysis of rice leaf lamina responding to salt stress showed upregulation of 32 stress-responsive proteins including RuBisCO activase, ferritin, and phosphoglycerate kinase (Parker et al. 2006). Proteins from wild type and OSRK1 transgenic rice roots exposed to 150 mM NaCl were analyzed by Nam et al. (2012), who reported upregulation of proteins such as enzymes related to energy regulation, amino acid metabolism, methylglyoxal detoxification, redox regulation, and protein turnover; proteins that were downregulated included enzymes involved in gibberelic acid (GA_3)-induced root growth such as fructose bisphosphate aldolase and methylmalonate seminaldehyde dehydrogenase. It was also suggested that on challenging salt stress, rice roots rapidly change a broad spectrum of energy metabolism.

Systematic functional proteomics studies to investigate proteins that are responsive to low-temperature stress in rice have been conducted in the past. These studies are useful in understanding molecular mechanisms of cold adaptation/resistance of rice. Cui and colleagues (2005) found 60 leaf proteins to be upregulated in responding to progressive low-temperature stress, and out of these 41 were identified as factors of protein biosynthesis, molecular chaperones, proteases, enzymes of cell wall biosynthesis, antioxidative/ detoxifying enzymes, proteins for energy pathway, and proteins involved in signal transduction. It was suggested that the chloroplast proteome is virtually subjective to cold stress as 43.9% of the identified proteins were predicted to be located in chloroplasts. In a different study, Yan et al. (2006) identified 85 differentially expressed proteins from leaves including many novel cold-responsive proteins. A total of 12 upregulated proteins were identified from rice leaves exposed to low temperature by Lee et al. (2007a), including some novel proteins such as cysteine proteinase, thioredoxin peroxidise, a RING zinc finger protein-like, and a fibrillin-like protein. Earlier, Komatsu et al. (1999) reported several proteins to be phosphorylated on cold stress in rice seedlings as phosphorylation of cellular proteins or activation of protein kinases occurs during the initial stage of cold acclimation (Garbalino et al. 1991). As cold stress affects male reproductive development in rice, proteomics analysis of rice anthers in response to low-temperature treatment was studied by Imin et al. (2004). They reported 70 differentially expressed proteins at the young microspore stage. Among these proteins, 12 novel proteins were induced, 47 were upregulated, and 11 were downregulated by cold treatment. Hashimoto and Komatsu (2007) reported altered expression of 39 proteins from cold-treated rice seedlings. The proteins related to energy metabolism were upregulated and defense-related proteins were downregulated, which suggested that energy production is activated in chilling environments. A total of 27 proteins were found to be upregulated in rice roots on exposure to chilling stress (Lee et al. 2009). A group of novel proteins including acetyl transferase, phosphogluconate dehydrogenase, NADP-specific isocitrate dehydrogenase, and fructokinase, in addition to a few other proteins involved in energy production and metabolism, vesicular trafficking, and detoxification, were identified in this study.

Heat stress response in rice is complex as this involves upregulation, downregulation, or both of numerous proteins involved in several metabolic processes. The heat-responsive proteins mainly include enzymes of protein biosynthesis, protein degradation, energy metabolism, and redox homeostasis (for review, see Zou et al. 2011). Lin et al. (2005) detected more than 70 differentially expressed polypeptides in response to high-temperature treatments during caryopsis, which included 21 proteins involved in carbohydrate metabolism, 14 in protein synthesis and sorting, and 9 in stress response. Rice leaf proteome in response to exposure at 42°C was studied by Lee et al. (2007b), and a total of 48 proteins belonging to two major functional groups, heat shock proteins (HSPs) and proteins for energy metabolism, were identified. Similar findings were reported by Gammulla et al. (2010). These proteomics studies suggested that high-temperature stress disrupts energy and carbohydrate metabolism pathways and plants generate HSPs for protection against cellular damage under heat stress. Tolerant cultivars have greater protection against cell damage through accumulation of high-temperature responsive proteins (Han et al. 2009; Jagadish et al. 2010).

Heavy metals such as cadmium, lead, mercury, and arsenic (metalloid) have detrimental effects on plant growth as these metals act as inhibitors of many enzymes with metal cofactors involved in various growth and developmental processes. As heavy-metal-contaminated soils impose threats to rice production, a number of experiments on tissue-specific proteomics were conducted to elucidate plants' response to enhanced concentrations of heavy metals such as cadmium (Aina et al. 2007; Lee et al. 2010), arsenic (Ahsan et al. 2008), and a combination of cadmium, cobalt, copper, lithium, mercury, strontium, and zinc (Hajduch et al. 2001). Altered expression of proteins involved in signal transduction in defensive reaction from rice roots under aluminum stress was also noted (Yang et al. 2007b). Zhang et al. (2009c) investigated the response of rice to copper stress and reported 16 differentially expressed genes in germinating seed embryos. Elevated concentrations of these heavy metals mainly cause upregulation of enzymes cooperating with glutathione (GSH) in reactive oxygen species (ROS) scavenging, which include glutathione-S-transferase (GST) and Cu/Zn–SOD, which protect cellular metabolism against enhanced oxidative stress, since heavy metals act as efficient catalyzers of ROS formation (Kosová et al. 2011).

1.5.1.2.2 Biotic-Stress-Responsive Proteome

Proteomics approaches have been applied in rice for better understanding of the regulatory mechanisms of the plants against various biotic stresses including diseases and insects. It was observed that the identified proteins were mainly involved in carbohydrate and energy metabolism, defense mechanisms, and signaling. The studies on biotic stress-responsive proteomes have been reviewed in Wang et al. (2011).

A number of studies were conducted on elicitors (a group of stimuli such as polysaccharides, small proteins, and chemicals, which elicit plant defense responses) such as chitosan (Agrawal et al. 2002), chitooligisaccharide (Chen et al. 2007a), probenazole (Lin et al. 2008), and CSB I, a small glycoprotein from *Magnepothe oryzae* (Liao et al. 2009). These elicitors trigger host–defense responses by activating and suppressing defense signaling and metabolic pathways (De Wit et al. 2009). The proteomics studies showed upregulation of defense regulated proteins, proteins involved in lignin biosynthesis, flavonoid-type phytoalexins, putative zinc finger proteins, SODs GST, and enzymes for signal transduction.

Rice proteomes in response to infections of different pathogens have been investigated focusing on fungal pathogens such as *M. oryzae* (Kim et al. 2003, 2004, 2009), *Rhizoctonia solani* (Lee et al. 2006), pathogenic bacterium *Xanthomonus oryzae* (Mahmood et al. 2006), and *Rice yellow mottle virus* (Ventelon-Debout et al. 2004). Only one proteomics study has been conducted to understand the mechanism of response to BPH infestation (Wei et al. 2009).

The proteins found to be differentially expressed responding to biotic stresses were analyzed functionally to reveal underlying regulatory mechanisms in biotic stressed rice plants. The proteins included defense-related proteins (chitinase, β-1,3-glucanase, PBZ1, OsPR10, RLK, etc.), ROS-related proteins (SOD, APX, and OsIRL), stress-related proteins (HSP and chaperonin), and Jacalin-like domain proteins. Plasma membrane proteins are also found to be involved in early defense response to biotic stresses (Chen et al. 2007a, 2007b).

1.5.1.3 Hormone Response Proteome

Plant hormones greatly affect growth and development and signal transduction pathways and play an important role in regulation of seed dormancy and germination, stem regulation, leaf growth, flowering, and fruiting. Plants perceive environmental stimuli as information and then respond via signaling pathways, which is important in plant defense responses (Rakwal and Agrawal 2003).

Several proteins in rice leaf sheath, root, and cultured cell suspension have been reported to be regulated by the exogenous application of GA_3 (Tanaka et al. 2004). The proteins that were regulated by GA_3 application in leaf sheath can broadly be categorized into proteins involved in primary metabolism, transcription regulation, and signaling pathways. In roots, GA_3-regulated proteins were mostly those involved in defense reactions, while proteins from cultured cell suspension were mostly those involved in metabolism, energy, cell growth, protein destination, transcription, defense, signal transduction, transporter, and other hypothetical proteins. Similarly, ABA has been found to induce enhanced salt tolerance in treated rice seedlings, and ABA-regulated proteins were mainly involved in energy metabolism, defense, and primary cellular metabolism (Li et al. 2010b). Auxin plays a vital role in apical dominance and lateral root initiation in plants. The effect of auxin on root formation in rice has been studied through proteomic analysis of cell suspension culture (Oguchi et al. 2004). Seven proteins including NADPH-dependent oxidoreductase and methylmalonate-semialdehyde dehydrogenase (MMSDH) were upregulated in auxin-treated root tissues, suggesting that these proteins play an important role in root formation in rice.

Exogenous application of brassinosteroids (naturally occurring plant steroids) and its effects on rice proteome have been reported (Konishi and Komatsu 2003). In the brassinosteroid-treated lamina joint and root, differential expression of proteins related to photosynthesis and defense responses, respectively, had been found. Like brassinosteroids, jasmonic acid (JA) has diverse roles and functions, including a potential involvement in plant defense and signaling pathways. Expression analysis of proteins in leaf and stem tissues from JA-treated rice seedlings indicated that JA affects defense-related gene expression in rice (Rakwal and Komatsu 2000).

1.6 Rice Metabolomics

Metabolomics is one of the newest omics technologies for measuring small molecular metabolite components of cells. Metabolomics implies a qualitative and quantitative approach that enables the parallel assessment of the levels of a broad range of metabolites. The nutritional status of crops is ultimately dependent on their metabolic composition, which is very important for human health (Demmig-Adams and Adams 2002). In recent years, the spectacular technical advances of the postgenomic era have led to understanding of biological functions of unknown genes of rice, mainly through genomics and proteomics approaches. Metabolomics also provides the possibility of identifying gene functions directly connected to rice quality, yield, and defense mechanisms. The cost of metabolomics analysis, though lower than that of transcript profiling (Kopka et al. 2004), currently limits its exploitation (Fernie and Schauer 2008). In addition to this, our knowledge of the plant metabolome usually lags behind that of yield and stress-resistance traits (Fernie et al. 2006). Considering the usefulness of metabolomics studies for understanding the metabolic network and its interaction with the developmental phenotype, it has been targeted for integration in genomics-assisted breeding (Fernie and Schauer 2008). In rice, a comprehensive review of metabolomics has discussed current technological developments and the possibilities of the application of metabolomics in rice improvement (Oikawa et al. 2008). Integration of the results from metabolomics with those of phenomics (discussed below) is a powerful strategy for crop improvement. Metabolomics has been used by investigators in combination with other genomics approaches for some crops such as *Lycopersicon*, *Medicago* and opium poppy, to gain new insights on gene annotation (Fridman et al. 2004; Achnine et al. 2005; Hagel et al. 2008), which resulted in the identification of numerous candidate genes including several in which expression correlates strongly with the levels of metabolites with important nutritional or organoleptic properties (Fernie and Schauer 2008).

Metabolomics tools have been used in molecular breeding of genetically modified (GM) rice in order to improve nutritional quality: for example, development of rice cultivars with high β-carotene content (golden rice; Ye et al. 2000) and high tryptophan content, which was achieved by producing transgenic rice expressing a feedback-insensitive anthranilate synthase α-subunit gene (*OASA1*) (Tozawa et al. 2001; Wakasa et al. 2006). The safety issues related to GM rice can be addressed by using metabolomics techniques as well as other omics tools in combination with allergenicity and toxicity tests to survey a wide range of metabolites (Baker et al. 2006; Catchpole et al. 2005; Dixon et al. 2006). High-throughput metabolome and proteome analysis of transgenic rice plants overexpressing *YK1* gene, a homologue of the HC-toxin reductase gene, and having dihydroflavonol-4-reductase activity, was performed to understand the correlation between gene overexpression and the concentration of metabolites (Takahashi et al. 2005, 2006).

Most of the studies regarding metabolomic natural variation have been conducted in *Arabidopsis*, and few studies have been carried out in other crop species (for review, see Fernie and Schauer 2008). The studies of *Arabidopsis* have demonstrated that integration of transcriptomics with metabolomics analyses is a promising strategy for functional identification of metabolomics-related genes (Hirai et al. 2004, 2005; Watanabe et al. 2008). These studies have facilitated the development of databases based on transcriptomics and expression correlation data (Obayashi et al. 2007), which can be accessed to identify candidate genes for unknown biological processes and for secondary metabolism of *Arabidopsis* (Yonekura-Sakakibara et al. 2007). In the case of rice, integration of metabolomics data with such databases (for example, Rice Expression Database; http://cdna02.dna.affrc.go.jp/RED/) may prove to be useful in functional genomics study. Recently, metabolome data have been used by Lisec et al. (2008) in identifying metabolic QTLs in *Arabidopsis*. Similarly, metabolic QTL/eQTL analysis in rice should be adopted for further improvement of rice quality and taste. Moreover, the power of association mapping in identifying genes underlying phenotypic variation can be applied to establish association between genomic regions/genes and the metabolic variation with a higher mapping resolution than QTL analysis. Overall, advances in metabolomics research in rice and application of postgenomics tools will facilitate metabolomics-assisted breeding, which is certainly a viable option for rice improvement.

1.7 Rice Phenomics

After completion of the genomic sequence of rice, the next challenge is to harness the wealth of genomic information for agricultural applications, and to comprehensively link it to phenotype in a given environment. Computation- and curation-based annotation of the rice genome predicted the locations of genes, including exons, introns, and their putative functions (Itoh et al. 2007; Ouyang et al. 2007). But the earlier annotation has been found to have several limitations, and needs further validation of the present rice gene model, and identification of additional genes by computational and experimental approaches has been proposed (Chern et al. 2007). It was indicated that phenomics or large-scale phenotyping is the natural complement to genome sequencing as a route to rapid advances in biology (Houle et al. 2010). Phenomic-level data are necessary to understand which genomic variants affect phenotypes, and to understand complex plant characteristics. Recent progress in DNA sequencing and phenotyping technologies, in concert with advances in large-scale data management, has broadened the scope of using phenomics in combination with other omics approaches in rice. Phenotyping is widely considered as laborious, slow, and a technically challenging part of a breeding program. Moreover, phenotyping tools commonly require destructive sampling at particular phonological stages of the crop. The labor-intensive and costly nature of conventional field phenotyping, which is considered a bottleneck, can now be addressed by combining novel technologies such as noninvasive imaging, spectroscopy, image analysis, robotics, and high-performance computing (Furbank and Tester 2011).

The word *phenomics*, first used by Michael Soulé (1967), has been defined in many ways. The most appropriate definition of phenomics is the acquisition of high-dimensional phenotypic data on a genome-wide scale (Bilder et al. 2009). Thus plant phenomics is the study of plant growth, performance, and composition. Phenotyping tools can be used in screening large collections of crop germplasm for valuable traits. The screening could be high-throughput, fully automated, and-low resolution, followed by higher-resolution, lower-throughput measurements, and it must be reproducible and of physiological relevance. This is also called forward phenomics. On the other hand, reverse phenomics is the detailed dissection of traits to reveal mechanistic understanding of plant growth and development. This approach involves reduction of a physiological trait to biochemical pathways and ultimately to a gene or genes (Furbank and Tester 2011). Phenomics enables us to draw causal links between genotypes and environmental factors and phenotypes, commonly known as the G–P map (Houle et al. 2010). Recently, plant phenomics projects with the goal of understanding the G–P map by combining genomic data with data on quantitative variation in phenotypes have been initiated (http://plantgenomics.com). The International Plant Phenomics Network (IPPN) was constituted aiming to develop, integrate, and provide novel technologies to analyze plant phenotypes using high-throughput approaches. This international initiative will expand our knowledge on the way environments affect plant structure and function, and will develop new concepts on plants' response to the environment.

Application of various modern tools (Table 1.2) for phenotyping abiotic and biotic stress tolerance in crop plants and trait-based physiological crop breeding has been reviewed comprehensively (Houle et al. 2010; Furbank and Tester 2011). There are some reports on systematic phenomics applications in rice (see Chern et al. 2011 and references thereof). These studies have highlighted phenotypic profiling of genome-wide insertional (Chern et al. 2007; Miyao et al. 2007) or chemical/induced mutant collections and collection of data for a database of the qualitative and quantitative phenotypic traits, which would be a good resource for rice functional genomics study such as association mapping.

1.8 Conclusion

To conclude, the progress in rice genomics research needs to be supplemented by developments in rice proteomics, metabolomics, and phenomics. New knowledge and tools are

TABLE 1.2

Phenomics Approaches for High-Throughput Phenotyping in Plants

Phenomics Tool	Features	Application(s)
High-Throughput Phenotyping of Plant Growth Dynamics		
Noninvasive imaging technology	Image analysis and mathematical treatment of imaging data to extract growth dynamics, morphological characters, and specially described photosynthetic parameters.	Screening of *Arabidopsis* for drought tolerance (Granier et al. 2006). Detection of QTLs linked to biomass in *Arabidopsis* (Meyer et al. 2010).
Three-dimensional plant models	Use of mathematical models known as *L-systems*. Simulation of plant development with a series of generative rules for plant organs.	Modeling of plants using L-systems (Prusinkiewicz et al. 1996, 2002). Measurement of plant shape using generative models (Cui et al. 2010).
Plant organ shape analysis using geometric descriptors	Geometric descriptors of plant organs developed for leaf and root shape analysis. Accurate measurement of plant shape without the need to store large images for comparison.	Plant shape analysis (Costa and Cesar 2001). High-throughput three-dimensional root phenotyping with image analysis software platform has been developed in rice (Clark et al. 2011).
Phenotyping for Abiotic and Biotic Stress Tolerance		
Carbon isotope discrimination (CID)	A reproducible descriptor of transpiration efficiency and stomatal conductance. Sampling for ^{13}C isotope composition can be performed at the end of crop growth.	CID successfully used to breed commercial wheat varieties with greater water use efficiency and yield (Condon et al. 2004).
Infrared thermography	Thermal images of leaf or canopy are recorded in stressed as well as control plants. It enables low-cost, high-throughput field phenotyping for drought tolerance.	Selection of wheat and barley seedlings with higher stomatal conductance under osmotic stress (Sirault et al. 2009).
Chlorophyll fluorescence analysis	A surrogate measurement of photosynthesis under stresses. Large number of plants can be screened at the same developmental stage using a fluorometer. Other photosynthetic parameters can be derived from chlorophyll fluorescence data.	Nonphotochemical quenching (NPQ) has been used as an indicator of stress in crop species (Baker 2008; Jansen et al. 2010). Chlorophyll fluorescence data has been used as a tool for monitoring the progress of fungal growth on leaves (Scholes and Rolfe 2009).
Growth analysis using digital images	This technique uses multiple viewing angles to extract mathematical relationship between the images and plant characteristics. Digital imaging can be used as a high-throughput technique to quantify lesions or chlorotic areas on diseased leaves.	Analysis of plant biomass and leaf area using digital images (Rajendran et al. 2009).

changing the strategies for rice improvement research and will thus increase the throughput of the assays at a reduced cost. Rice genomics research has greatly benefited from new tools and techniques of bioinformatics. Bioinformatics has been proved to be useful for integration and structured interrogation of datasets in the postgenomics era. Similarly to genomics, the rice proteomics research has contributed to better understanding of physiological mechanisms underlying plants' developmental processes and stress response. Moreover, proteomics research could also significantly contribute to identification and detailed characterization of key proteins underlying plant tolerance of a given stress, which can be used as protein biomarkers for that stress. In the near future, rice metabolomics study is expected to be more systematic, with the availability of high-throughput and high-sensitivity analytical technologies and methods. Like other omics studies, phenomics also has the promise to accelerate progress in rice improvement by introducing the recent advances made in computing, robotics, and image analysis technologies. It is obvious that a systematic integration of other omics approaches with the genomics study will enhance rice production to feed the growing world population under the adverse effects of climate change.

References

Abbasi, F. M. and Komatsu, S. 2004. A proteomic approach to analyze salt-responsive proteins in rice sheath. *Proteomics* 4: 2072–81.

Abdurakhmonov, I. Y. and Abdukarimov, A. 2008. Application of association mapping to understanding the genetic diversity of Plant Germplasm Resources. *International Journal of Plant Genomics*, doi:10.1155/2008/574927.

Achnine, L., Huhman, D. V., Farag, M. A., Sumner, L. W., Blount, J. W., and Dixon, R. A. 2005. Genomics-based selection and functional characterization of triterpene glycosyltransferases from the model legume *Medicago truncatula*. *The Plant Journal* 41: 875–87.

Agrawal, G. K., Rakwal, R., Tamogami, S., Yonekura, M., Kubo, A., and Saji, H. 2002. Chitosan activates defense/stress response(s) in the leaves of *Oryza sativa* seedlings. *Plant Physiology and Biochemistry* 40: 1061–69.

Ahsan, N., Lee, D. G., Alam, I., Kim, P. J., Lee, J. J., Ahn, Y. O., Kwak, S. S., et al. 2008. Comparative proteomic study of arsenic-induced differentially expressed proteins in rice roots reveals glutathione plays a central role during As stress. *Proteomics* 8: 3561–76.

Aina, R., Labra, M., Fumagalli, P., Vannini, C., Marsoni, M., Cucchi, U., Bracale, M., et al. 2007. Thiol-peptide level and proteomic changes in response to cadmium toxicity in *Oryza sativa* L. roots. *Environmental and Experimental Botany* 59: 381–92.

Albar, L., Bangratz-Reyser, M., Hébrard, E., Ndjiondjop, M.-N., Jones, M., and Ghesquière, A. 2006. Mutations in the eIF(iso)4G translation initiation factor confer high resistance of rice to Rice yellow mottle virus. *The Plant Journal* 47: 417–26.

Andaya, V. C. and Mackill, D. J. 2003. Mapping of QTLs associated with cold tolerance during the vegetative stage in rice. *Journal of Experimental Botany* 54: 2579–85.

Andaya, V. C. and Tai, T. H. 2006. Fine mapping of *qCTS12* locus, a major QTL for seedling cold tolerance in rice. *Theoretical and Applied Genetics* 113: 467–75.

Andersen, J. R. and Lübberstedt, T. 2003. Functional markers in plants. *Trends in Plant Science* 8: 554–60.

Ansong, C., Purvine, S. O., Adkins, J. N., Lipton, M. S., and Smith, R. D. 2008. Proteogenomics: Needs and roles to be filled by proteomics in genome annotation. *Briefings in Functional Genomics and Proteomics* 7: 50–62.

Arite, T., Iwata, H., Ohshima, K., and Maekawa, M. 2007. *DWARF10*, an *RMS1/MAX4/DAD1* ortholog, controls lateral bud outgrowth in rice. *The Plant Journal* 51: 1019–29.

Armengaud, J. 2010. Proteogenomics and systems biology: Quest for the ultimate missing parts. *Expert Review of Proteomics* 7: 65–77.

Ashikari, M. and Matsuoka, M. 2006. Identification, isolation and pyramiding of quantitative trait loci for rice breeding. *Trends in Plant Science* 11: 344–50.

Ashikari, M., Sakakibara, H., Lin, S., Yamamoto, T., Takashi, T., Nishimura, A., Angeles, E. R., Qian, Q., Kitano, H., and Matsuoka, M. 2005. Cytokinin oxidase regulates rice grain production. *Science* 309: 741–45.

Babu, P. R., Sekhar, A. C., Ithal, N., Markandeya, G., and Reddy, A. R. 2002. Annotation and BAC/PAC localization of nonredundant ESTs from drought-stressed seedlings of an *indica* rice. *Journal of Genetics* 81: 25–44.

Baker, N. R. 2008. Chlorophyll fluorescence: A probe of photosynthesis *in vivo*. *Annual Review of Plant Biology* 59: 89–113.

Baker, J. M., Hawkins, N. D., Ward, J. L., Lovegrove, A., Napier, J. A., and Shewry, P. R. 2006. A metabolomic study of substantial equivalence of field-grown genetically modified wheat. *Plant Biotechnology Journal* 4: 381–92.

Ballini, E., Morel, J. B., Droc, G., Price, A., Courtois, B., Notteghem, J. L., and Tharreau, D. 2008. A genome-wide meta-analysis of rice blast resistance genes and quantitative trait loci provides new insights into partial and complete resistance. *Molecular Plant Microbe Interactions* 21: 859–68.

Bart, R., Chern, M., Park, C. J., and Bartley, L. 2006. A novel system for gene silencing using siRNAs in rice leaf and stem-derived protoplasts. *Plant Methods* 2: 13.

Bilder, R. M., Sabb, F. W., Cannon, T. D., London, E. D., Jentsch, J. D., Parker, D. S., Poldrack, R. A., Evans, C., and Freimer, N. B. 2009. Phenomics: The systematic study of phenotypes on a genome-wide scale. *Neuroscience* 164: 30–42.

Bradbury, L. M. T., Henry, R. J., Jin, Q., and Waters, D. L. E. 2005. A perfect marker for fragrance genotyping in rice. *Molecular Breeding* 16: 279–83.

Brar, D. S., Virk, P. S., Jena, K. K., and Khush, G. S. 2009. Breeding for resistance to planthoppers in rice. In *Planthoppers: New Threats to the Sustainability of Intensive Rice Production Systems in Asia*, eds, K. L. Heong and B. Hardy, pp. 401–28. Los Baños (Philippines): International Rice Research Institute.

Brent, M. R. 2008. Steady progress and recent breakthroughs in the accuracy of automated genome annotation. *Nature Reviews Genetics* 9: 62–73.

Bryan, G. T., Wu, K.-S, Farrall, L., Jia, Y., Hershey, H. P., McAdams, S. A., Faulk, K. N., et al. 2000. A single amino acid difference distinguishes resistant and susceptible alleles of the rice blast resistance gene *Pi-ta*. *The Plant Cell* 12: 2033–46.

Buckler, E. S. and Thornsberry, J. M. 2002. Plant molecular diversity and applications to genomics. *Current Opinion in Plant Biology* 5: 107–11.

Castellana, N. E. and Bafna, V. 2010. Proteogenomics to discover the full content of genomes: A computational perspective. *Journal of Proteomics* 73: 2124–35.

Castellana, N. E., Payne, S. H., Shen, Z., Stanke, M., Bafna, V., and Briggs, S. P. 2008. Discovery and revision of Arabidopsis genes by proteogenomics. *Proceedings of the National Academy of Sciences USA* 105: 21034–38.

Catchpole, G. S., Beckmann, M., Enot, D. P., Mondhe, M., Zywicki, B., Taylor, J., Hardy, N., et al. 2005. Hierarchical metabolomics demonstrates substantial compositional similarity between genetically modified and conventional potato crops. *Proceedings of the National Academy of Sciences USA* 102: 14458–62.

Chen, F., Li, Q., and He, Z. H. 2007a. Proteomic analysis of rice plasma membrane-associated proteins in response to chitooligosaccharide elicitors. *Journal of Integrative Plant Biology* 49: 1–5.

Chen, F., Yuan, Y., Li, Q., and He, Z. 2007b. Proteomic analysis of rice plasma membrane reveals proteins involved in early defence response to bacterial blight. *Proteomics* 7: 1529–39.

Chen, H. Q., Chen, Z. X., Ni, S., Zuo, S. M., Pan, X. B., and Zhu, X. D. 2008. Pyramiding three genes with resistance to Blast by marker-assisted selection to improve rice blast resistance of Jin 23B. *Chinese Journal of Rice Science* 22: 23–27.

Chen, S., Tao, L., Zeng, L., Vega-Sánchez, M. E., Umemura, K., and Wang, G. L. 2006. A highly efficient transient protoplast system for analyzing defence gene expression and protein–protein interactions in rice. *Molecular Plant Pathology* 7: 417–27.

Cheng, C. Y., Motohashi, R., Tsuchimoto, S., Fukuta, Y., Ohtsubo, H., and Ohtsubo, E. 2003. Polyphyletic origin of cultivated rice: Based on the interspersion pattern of SINEs. *Molecular Biology and Evolution* 20: 67–75.

Chern, C.-G., Fan, M.-J., Yu, S.-M., Hour, A. L., Lu, P. C., Lin, Y. C., Wei, F. J., et al. 2007. A rice phenomics study—Phenotype scoring and seed propagation of a T-DNA insertion-induced rice mutant population. *Plant Molecular Biology* 65: 427–38.

Chern, C.-G., Fan, M.-J., Huang, S.-C., et al. 2011. Methods for rice phenomics studies. *Methods in Molecular Biology* 678: 129–38.

Chin, H. G., Choe, M. S., Lee, S.-H., Park, S. H., Koo, J. C., Kim, N. Y., Lee, J. J., et al. 1999. Molecular analysis of rice plants harbouring an Ac/Ds transposable element-mediated gene trapping system. *The Plant Journal* 19: 615–23.

Chin, J. H., Lu, X., Haefele, S. M., Gamuyao, R., Ismail, A., Wissuwa, M., and Heuer, S. 2010. Development and application of gene-based markers for the major rice QTL *Phosphorus uptake 1*. *Theoretical and Applied Genetics* 120: 1073–86.

Chin, J. H., Gamuyao, R., Dalid, C., Bustamam, M., Prasetiyono, J., Moeljopawiro, S., Wissuwa, M., and Heuer, S. 2011. Developing rice with high yield under phosphorus deficiency: *Pup1* sequence to application. *Plant Physiology* 156: 1202–16.

Chono, M., Honda, I., Zeniya, H., Yoneyama, K., Saisho, D., Takeda, K., Takatsuto, S., Hoshino, T., and Watanabe Y. 2003. A semidwarf phenotype of barley *uzu* results from a nucleotide substitution in the gene encoding a putative brassinosteroid receptor. *Plant Physiology* 133: 1209–19.

Chu, Z., Ouyang, Y., Zhang, J., Yang, H., and Wang, S. 2004. Genome-wide analysis of defence responsive genes in bacterial blight resistance of rice mediated by a recessive *R* gene, *xa13*. *Molecular Genetics and Genomics* 271: 111–20.

Clark, R. T., MacCurdy, R. B., Jung, J. K., Shaff, J. E., Mccouch, S. R., Aneshansley, D. J., and Kochian, L. V. 2011. Three-dimensional root phenotyping with a novel imaging and software platform. *Plant Physiology* 156: 455–65.

Collard, B. C. Y. and Mackill, D. J. 2007. Marker-assisted selection: An approach for precision plant breeding in the twenty-first century. *Philosophical Transactions of the Royal Society B: Biological Science* 17: 1–16.

Collard, B. C. Y., Cruz, C. M. V., McNally, K. L., Virk, P. S., and Mackill, D. J. 2008. Rice molecular breeding laboratories in the genomics era: Current status and future considerations. *International Journal of Plant Genomics* 2008: 524–847.

Collins, F. S. 2001. Contemplating the end of the beginning. *Genome Research* 11: 641—43.

Comai, L., Young, K., Till, B. J., Reynolds, S. H., Greene, E. A., Codomo, C. A., Enns, L. C., et al. 2004. Efficient discovery of DNA polymorphisms in natural populations by EcoTILLING. *The Plant Journal* 37: 778–86.

Condon, A. G., Richards, R. A., Rebetzke, G. J., and Farquhar, G. D. 2004. Breeding for high water-use efficiency. *Journal of Experimental Botany* 55: 2447–60.

Cooper, B., Clarke, J. D., Budworth, P., Kreps, J., Hutchison, D., Park, S., Guimil, S., et al. 2003. A network of rice genes associated with stress response and seed development. *Proceedings of the National Academy of Sciences USA* 100: 4945–50.

Costa, L. and Cesar, R. M. 2001. *Shape Analysis and Classification: Theory and Practice*, Boca Raton, FL: CRC Press.

Cui, M.-L., Copsy, L., Green, A. A., Bangham, J. A., and Coen, E. 2010. Quantitative control of organ shape by combinatorial gene activity. *PLoS Biology* 8: e1000538.

Cui, S., Huang, F., Wang, J., Ma, X., Cheng, Y., and Liu, J. 2005. A proteomic analysis of cold stress responses in rice seedlings. *Proteomics* 5: 3162–72.

Dai, S. J., Chen, T. T., Chong, K., Xue, Y. B., Liu, S. Q., and Wang, T. 2007. Proteomics identification of differentially expressed proteins associated with pollen germination and tube growth reveals characteristics of germinated *Oryza sativa* pollen. *Molecular and Cellular Proteomics* 6: 207–30.

Dai, X., Wang, Y., Yang, A., and Zhang, W. H. 2012. *OsMYB2P-1*, an R2R3 MYB transcription factor, is involved in the regulation of phosphate-starvation responses and root architecture in rice. *Plant Physiology* 159: 169–83.

Dardick, C. and Ronald, P. 2006. Plant and animal pathogen recognition receptors signal through non-RD kinases. *PLoS Pathology* 2: e2.

Dardick, C., Chen, J., Richter, T., Ouyang, S., and Ronald, P. 2007. The rice kinase database. A phylogenomic database for the rice kinome. *Plant Physiology* 143: 579–86.

Davies, P. J. (ed.) 1995. *Plant Hormones: Physiology, Biochemistry, and Molecular Biology*. Dordrecht, Netherlands: Kluwer.

de Koning, D. J. and Haley, C. S. 2005. Genetical genomics in humans and model organisms. *Trends in Genetics* 21: 377–81.

De Wit, P. J., Mehrabi, R., Van den Burg, H. A., and Stergiopoulos, I. 2009. Fungal effector proteins: Past, present and future. *Molecular Plant Pathology* 10: 735–47.

Demmig-Adams, B. and Adams, W. W. 2002. Antioxidants in photosynthesis and human nutrition. *Science* 298: 2149–53.

Dixon, R. A., Gang, D. R., Charlton, A. J., Fiehn, O., Kuiper, H. A., Reynolds, T. L., Tjeerdema, R. S., et al. 2006. Applications of metabolomics in agriculture. *Journal of Agricultural and Food Chemistry* 54: 8984–94.

Doi, K., Sobrizal, K., Ikeda, K., Sanchez, P. L., and Kurakazu, T. 2002. Developing and evaluating rice chromosome segment substitution lines. In *IRRI Conference, International Rice Research Institute*, Beijing, China, 275–87.

Dooki, A. D., Mayer-Posner, F. J., Askari, H., Zaiee, A. A., and Salekdeh, G. H. 2006. Proteomic responses of rice young panicles to salinity. *Proteomics* 6: 6498–507.

Du, B., Zhang, W., Liu, B., Hu, J., Wei, Z., Shi, Z. Y., He, R. F., et al. 2009. Identification and characterization of *Bph14*, a gene conferring resistance to brown planthopper in rice. *Proceedings of the National Academy of Sciences USA* 106: 22163–68.

Dubouzet, J. G., Sakuma, Y., Ito, Y., Kasuga, M., Dubouzet, E. G., Miura, S., Seki, M., Shinozaki, K., and Yamaguchi-Shinozaki, K. 2003. *OsDREB* genes in rice, *Oryza sativa* L., encode transcription activators that function in drought-, high-salt- and cold-responsive gene expression. *The Plant Journal* 33: 751–63.

Fan, C. H., Xing, Y. Z., Mao, H. L., Lu, T. T., Han, B., Xu, C. G., Li, X. H., and Zhang, Q. F. 2006. *GS3*, a major QTL for grain length and weight and minor QTL for grain width and thickness in rice encodes a putative trans-membrane protein. *Theoretical and Applied Genetics* 112: 1164–71.

Fernie, A. R. and Schauer, M. 2008. Metabolomics-assisted breeding: A viable option for crop improvement? *Trends in Plant Science* 25: 39–48.

Fernie, A. R., Tadmor, Y., and Zamir, D. 2006. Natural genetic variation for improving crop quality. *Current Opinion in Plant Biology* 9: 196–202.

Flint-Garcia, S. A., Thornsberry J. M., and Buckler, E. S., IV. 2003. Structure of linkage disequilibrium in plants. *Annual Review of Plant Biology* 54: 357–74.

Fridman, E., Carrari, F., Liu, Y. S., Fernie, A., and Zamir, D. 2004. Zooming in on a quantitative trait for tomato yield using interspecific introgressions. *Science* 305: 1786—89.

Fukao, T., Xu, K., Ronald, P. C., and Bailey-Serres, J. 2006. A variable cluster of ethylene response factor–like genes regulates metabolic and developmental acclimation responses to submergence in rice. *The Plant Cell* 18: 2021–34.

Furbank, R. T. and Tester, M. 2011. Phenomics—Technologies to relieve the phenotyping bottleneck. *Trends in Plant Sciences* 16: 635–44.

Gammulla, C., Pascovici, D., Atwell, B., and Haynes, P. A. 2010. Differential metabolic response of cultured rice (*Oryza sativa*) cells exposed to high- and low-temperature stress. *Proteomics* 10: 3001–19.

Gamuyao, R., Chin, J. H., Tanaka-Pariasca, J., Pesaresi, P., Catausan, S., Dalid, C., Slamet-Loedin, I., Tecson-Mendoza, E. M., Wissuwa, M., and Heuer, S. 2012. The protein kinase Pstol1 from traditional rice confers tolerance of phosphorus deficiency. *Nature* 488: 535–41.

Gao, X., Liang, W., Yin, C., Ji, S., Wang, H., Su, X., Guo, C., Kong, H., Xue, H., and Zhang, D. 2010. The SEPALLATA-like gene *OsMADS34* is required for rice inflorescence and spikelet development. *Plant Physiology* 153: 728–40.

Gao, Z., Qian, Q., Liu, X., Yan, M. Q., Dong, G., Liu, J., and Han, B. 2009. *Dwarf 88*, a novel putative esterase gene affecting architecture of rice plant. *Plant Molecular Biology* 71: 265–76.

Garbalino, J. E., Hurkman, W. J., Tanaka, C. K., and Dupont, F. M. 1991. *In vitro* and *in vivo* phosphorylation of polypeptides in plasma membrane and tonoplast-enriched fractions from barley roots. *Plant Physiology* 95: 1219–28.

Garris, A. J., Tai, T. H., Coburn, J., Kresovich, S., and McCouch, S. 2005. Genetic structure and diversity in *Oryza sativa* L. *Genetics* 169: 1631–38.

Goff, S. A. 1999. Rice as a model for cereal genomics. *Current Opinion in Plant Biology* 2: 86–89.

Goff, S. A., Ricke, D., Lan, T. H., Presting, G., Wang, R., Dunn, M., Glazebrook, J., et al. 2002. A draft sequence of the rice genome (*Oryza sativa* L. ssp. *japonica*). *Science* 296: 92–100.

Granier, C., Aguirrezabal, L., Chenu, K., Cookson, S. J., Dauzat, M., Hamard, P., Thioux, J. J., et al. 2006. PHENOPSIS, an automated platform for reproducible phenotyping of plant responses to soil water deficit in *Arabidopsis thaliana* permitted the identification of an accession with low sensitivity to soil water deficit. *New Phytologist* 169: 623–35.

Gu, K., Yang, B., Tian, D., Wu, L. F., Wang, D. J., Sreekala, C., Yang, F., et al. 2005. *R* gene expression induced by a type-III effector triggers disease resistance in rice. *Nature* 435: 1122–25.

Guigo, R., Flicek, P., Abril, J. F., Reymond, A., Lagarde, J., Denoeud, F., Antonarakis, S., et al. 2006. EGASP: The human ENCoDE genome annotation assessment project. *Genome Biology* 7: S2.

Gupta, P. K. and Rustgi, S. 2004. Molecular markers from the transcribed/expressed region of the genome in higher plants. *Functional and Integrative Genomics* 4: 139–62.

Gupta, P. K., Rustgi, S., and Kulwal, P. L. 2005. Linkage disequilibrium and association studies in higher plants: Present status and future prospects. *Plant Molecular Biology* 57: 461–85.

Gupta, S. K., Rai, A. K., Kanwar, S. S., Chand, D., Singh, N. K., and Sharma, T. R. 2012. The single functional blast resistance gene *Pi54* activates a complex defence mechanism in rice. *Journal of Experimental Botany* 63: 757–72.

Hagel, J. M., Weljie, A. M., Vogel, H. J., and Facchini, P. H. 2008. Quantitative 1H nuclear magnetic resonance metabolite profiling as a functional genomics platform to investigate alkaloid biosynthesis in opium poppy. *Plant Physiology* 147: 1805–21.

Hajduch, M., Rakwal, R., Agrawal, G. K., Yonekura, M., and Pretova, A. 2001. High-resolution two-dimensional electrophoresis separation of proteins from metal-stressed rice (*Oryza sativa* L.) leaves: Drastic reductions/fragmentation of ribulose-1,5-bisphosphate carboxylase/oxygenase and induction of stress-related proteins. *Electrophoresis* 22: 2824–31.

Han, F., Chen, H., Li, X., Yang, M., Liu, G., and Shen, S. 2009. A comparative proteomic analysis of rice seedlings under various high-temperature stresses. *Biochimica et Biophysica Acta* 1794: 1625–34.

Harris, S. B. 2002. Virtual rice: Japan sets up the rice simulator project to create an *in silico* rice plant. *EMBO Report* 3: 511–13.

Hasegawa, P. M., Bressan, R. A., Zhu, J. K., and Bohnert, H. J. 2000. Plant cellular and molecular responses to high salinity. *Annual Review in Plant Physiology and Plant Molecular Biology* 51: 463–99.

Hashimoto, M. and Komatsu, S. 2007. Proteomic analysis of rice seedlings during cold stress. *Proteomics* 7: 1293–302.

Helmy, M., Tomita, M., and Ishihama, Y. 2011. OryzaPG-DB: Rice proteome database based on shotgun proteogenomics. *BMC Plant Biology* 11: 63.

Hiei, Y., Ohta, S., Komari, T., and Kumashiro, T. 1994. Efficient transformation of rice (*Oryza sativa* L.) mediated by *Agrobacterium* and sequence analysis. *The Plant Journal* 6: 271–82.

Hieter, P. and Boguski, M. 1997. Functional genomics: It's all how you read it. *Science* 278: 601–2.

Hinsinger, P., Betencourt, E., Bernard, L., Brauman, A., Plassard, C., Shen, J., Tang, X., and Zhang, F. 2011. P for two, sharing a scarce resource: Soil phosphorus acquisition in the rhizosphere of intercropped species. *Plant Physiology* 156: 1078–86.

Hirabayashi, H., Kaji, R., Angeles, E. R., Ogawa, T., Brar, D. S., and Khush, G. S. 1999. RFLP analysis of a new gene for resistance to brown planthopper derived from *O. officinalis* on rice chromosome 4. *Breeding Science* 48(Suppl. 1): 48.

Hirai, M. Y., Yano, M., and Goodenowe, D. B. 2004. Integration of transcriptomics and metabolomics for understanding of global responses to nutritional stresses in *Arabidopsis thaliana*. *Proceedings of the National Academy of Sciences USA* 101: 10205–10.

Hirai, M. Y., Klein, M., Fujikawa, Y., Yano, M., Goodenowe, D. B., Yamazaki, Y., Kanaya, S., et al. 2005. Elucidation of gene-to-gene and metabolite to–gene networks in *Arabidopsis* by integration of metabolomics and transcriptomics. *Journal of Biological Chemistry* 280: 25590–95.

Hittalmani, S., Parco, A., Mew, T. V., Zeigler, R. S., and Huang, N. 2000. Fine mapping and DNA marker-assisted pyramiding of the three major genes for blast resistance in rice. *Theoretical and Applied Genetics* 100: 1121–28.

Hospital, F. 2001. Size of donor chromosome segments around introgressed loci and reduction of linkage drag in marker-assisted backcross programs. *Genetics* 158: 1363–79.

Hospital, F. and Charcosset, A. 1997. Marker-assisted introgression of quantitative trait loci. *Genetics* 147: 1469–85.

Houle, D., Govindaraju, D. R., and Omholt, S. 2010. Phenomics: The next challenge. *Nature Reviews Genetics* 11: 855–67.

Hu, H., You, J., Fang, Y., Zhu, X., Qi, Z., and Xiong, L. 2008. Characterization of transcription factor gene *SNAC2* conferring cold and salt tolerance in rice. *Plant Molecular Biology* 67: 169–81.

Huang, N., Parco, A., Mew, T., Magpantay, G., McCouch, S., Guiderdoni, E., Xu, J., Subudhi, P., Angeles, E. R., and Khush, G. S. 1997. RFLP mapping of isozymes, RAPD and QTLs for grain shape, brown planthopper resistance in a doubled haploid rice population. *Molecular Breeding* 3: 105–13.

Huang, Z., He, G., Shu, L., Li, X., and Zhang, Q. 2001. Identification and mapping of two brown planthopper resistance genes in rice. *Theoretical and Applied Genetics* 102: 929–34.

Huang, X., Qian, Q., Liu, Z., Sun, H., He, S., Luo, D., Xia, G., Chu, C., Li, J., and Fu, X. 2009a. Natural variation at the *DEP1* locus enhances grain yield in rice. *Nature Genetics* 41: 494–97.

Huang, X. H., Feng, Q., Qian, Q., Zhao, Q., Wang, L., Wang, A., Guan, J., et al. 2009b. High-throughput genotyping by whole-genome resequencing. *Genome Research* 19: 1068–76.

Huang, X., Zhao, Y., Wei, X., Li, C., Wang, A., Zhao, Q., Li, W., et al. 2011. Genome-wide association study of flowering time and grain yield traits in a worldwide collection of rice germplasm. *Nature Genetics* 44: 32–9.

Ikeda, A., Ueguchi-Tanaka, M., Sonoda, Y., Kitano, H., Koshioka, M., Futsuhara, Y., Matsuoka, M., and Yamaguchi, J. 2001. Slender rice, a constitutive gibberellins response mutant, is caused by a null mutation of the *SLR1* gene, an ortholog of the height-regulating gene *GAI/RGA/RHT/D8*. *The Plant Cell* 13: 999–1010.

Imin, N., Kerim, T., Rolfe, B. G., and Weinman, J. J. 2004. Effect of early cold stress on the maturation of rice anthers. *Proteomics* 4: 1873–82.

Inoue, T., Zhong, H. S., Miyao, A., Monna, L., Zhong, H. S., Sasaki, T., and Minobe, Y. 1994. Sequence-tagged sites (STSs) as standard landmarkers in the rice genome. *Theoretical and Applied Genetics* 89: 728–34.

Ishii, T., Brar, D. S., Multani, D. S., and Khush, G. S. 1994. Molecular tagging of genes for brown planthopper resistance and earliness introgressed from *Oryza australiensis* into cultivated rice, *O. sativa. Genome* 37: 217–21.

Ishikawa, S., Maekawa, M., Arite, T., Onishi, K., Takamure, I., and Kyozuka, J. 2005. Suppression of tiller bud activity in tillering dwarf mutants of rice. *Plant and Cell Physiology* 46: 79–86.

Itoh, T., Tanaka, T., Barrero, R. A., Yamasaki, C., Fujii, Y., Hilton, P. B., Antonio, B. A., et al. 2007. Curated genome annotation of *Oryza sativa* ssp. *japonica* and comparative genome analysis with *Arabidopsis thaliana. Genome Research* 17: 175–83.

Iyer, A. S. and McCouch, S. R. 2004. The rice bacterial blight resistance gene *xa5* encodes a novel form of disease resistance. *Molecular Plant Microbe Interaction* 17: 1348–54.

Izawa, T. and Shimamoto, K. 1996. Becoming a model plant: The importance of rice to plant science. *Trends in Plant Science* 1: 95–99.

Jaffe, J. D., Berg, H. C., and Church, G. M. 2004. Proteogenomic mapping as a complementary method to perform genome annotation. *Proteomics* 4: 59–77.

Jagadish, S., Muthurajan, R., Oane, R., Wheeler, T. R., Heuer, S., Bennett, J., and Craufurd, P. Q. 2010. Physiological and proteomic approaches to address heat tolerance during anthesis in rice (*Oryza sativa* L.). *Journal of Experimental Botany* 61: 143–56.

Jansen, M. A., Martret, B. L., and Koornneef, M. 2010. Variations in constitutive and inducible UVB tolerance; dissecting photosystem II protection in *Arabidopsis thaliana* accessions. *Physiologia Plantarum* 138: 22–34.

Jansen, R. C. and Nap, J. P. 2001. Genetical genomics: The added value from segregation. *Trends in Genetics* 17: 388–91.

Jena, K. K. and Mackill, D. J. 2008. Molecular markers and their use in marker-assisted selection in rice. *Crop Science* 48: 1266–76.

Jena, K. K., Jeung, J. U., Lee, J. H., Choi, H. C., and Brar, D. S. 2006. High-resolution mapping of a new brown planthopper (BPH) resistance gene, *Bph18(t)*, and marker-assisted selection for BPH resistance in rice (*Oryza sativa* L.). *Theoretical and Applied Genetics* 112: 288–97.

Jeon, J. S., Jang, S., Lee, S., Nam, J., Kim, C., Chung, Y.-Y., Kim S.-R., Lee, Y. H., Cho, Y.-G., and An, G. 2000. *Leafy hull sterile1* is a homeotic mutation in a rice MADS box gene affecting rice flower development. *The Plant Cell* 12: 871–84.

Jiao, Y. and Deng, X. W. 2007. A genome-wide transcriptional activity survey of rice transposable element-related genes. *Genome Biology* 8: R28.

Johnson, J. M., Edwards, S., Shoemaker, D., and Schadt, E. E. 2005. Dark matter in the genome: Evidence of widespread transcription detected by microarray tiling experiments. *Trends in Genetics* 21: 93–102.

Jung, K.-H., Han, M.-J., Lee, Y.-S., Kim, Y. W., Hwang, I., Kim, M. J., Kim, Y. K., Nahm, B. H., and An, G. 2005. *Rice Undeveloped tapetum1* is a major regulator of early tapetum development. *The Plant Cell* 17: 2705–22.

Jung, K.-H., An, G., and Ronald, P. C. 2008. Towards a better bowl of rice: Assigning function to tens of thousands of rice genes. *Nature Reviews Genetics* 9: 91–101.

Juretic, N., Bureau, T. E., and Bruskiewich, R. M. 2004. Transposable element annotation of the rice genome. *Bioinformatics* 20: 155–60.

Kellogg, E. A. 2001. Evolutionary history of grasses. *Plant Physiology* 125: 1198–205.

Khush, G. S. 1997. Origin, dispersal, cultivation and variation of rice. *Plant Molecular Biology* 35: 25–34.

Khush, G. S. 1999. Green revolution: Preparing for the 21st century. *Genome* 42: 646–55.

Khush, G. S. 2001. Green revolution: The way forward. *Nature Review Genetics* 2: 815–22.

Kikuchi, S., Satoh, K., Nagata, T., Kawagashira, N., Doi, K., Kishimoto, N., Yazaki, J., et al. 2003. Collection, mapping, and annotation of over 28,000 cDNA clones from japonica rice. *Science* 301: 376–79.

Kim, S. T., Cho, K. S., Yu, S., Kim, S. G., Hong, J. C., Han, C., Bae, D. W., Nam, M. H., and Kang, K. Y. 2003. Proteomic analysis of differentially expressed proteins induced by rice blast fungus and elicitor in suspension-cultured rice cells. *Proteomics* 3: 2368–78.

Kim, S. T., Kim, S. G., Hwang, D. H., Kang, S. Y., Kim, H. J., Lee, B. H., Lee, J. J., and Kang, K. Y. 2004. Proteomic analysis of pathogen-responsive proteins from rice leaves induced by rice blast fungus, *Magnaporthe grisea*. *Proteomics* 4: 3569–78.

Kim, S. T., Kang, Y. H., Wang, Y., Wu, J., Park, Z. Y., Rakwal, R., Agrawal, G. K., Lee, S. Y., and Kang, K. Y. 2009. Secretome analysis of differentially induced proteins in rice suspension-cultured cells triggered by rice blast fungus and elicitor. *Proteomics* 9: 1302–13.

Kinoshita, T. 1994. Report of the committee on gene symbolization, nomenclature and linkage groups. *Rice Genetics Newsletter* 11: 8–22.

Kinoshita, T. 1995. Report of committee on gene symbolization, nomenclature and linkage groups. *Rice Genetics Newsletter* 12: 9–153.

Kliebenstein, D. 2009. Quantitative genomics: Analyzing intraspecific variation using global gene expression polymorphisms or eQTLs. *Annual Review of Plant Biology* 60: 93–114.

Koller, A., Washburn, M. P., Markus Lange, B., Andon, N. L., Deciu, C., Haynes, P. A., Hays, L., et al. 2002. Proteomic survey of metabolic pathways in rice. *Proceedings of the National Academy of Sciences USA* 99: 11969–74.

Komari, T., Hiei, Y., Ishida, Y., Kumashiro, T., and Kubo, T. 1998. Advances in cereal gene transfer. *Current Opinion in Plant Biology* 1: 161–65.

Komatsu, S. 2005. Rice proteome database: A step toward functional analysis of the rice genome. *Plant Molecular Biology* 59: 179–90.

Komatsu, S. and Tanaka, N. 2005. Rice proteome analysis: A step toward functional analysis of the rice genome. *Proteomics* 5: 938–49.

Komatsu, S., Karibe, H., Hamada, T., and Rakwal, R. 1999. Phosphorylation upon cold stress in rice (*Oryza sativa* L.) seedlings. *Theoretical and Applied Genetics* 98: 1304–10.

Komatsu, S., Konishi, H., Shen, S., and Yang, G. 2003. Rice proteomics: A step toward functional analysis of the rice genome. *Molecular and Cellular Proteomics* 2: 2–10.

Konishi, H. and Komatsu, S. 2003. A proteomics approach to investigating promotive effects of brassinolide on lamina inclination and root growth in rice seedlings. *Biological and Pharmaceuticle Bulletin* 26: 401–8.

Koonin, E. V. and Galperin, M. Y. 2002. *Sequence-Evolution-Function: Computational Approaches in Comparative Genomics*, 1st edn. Boston: Kluwer.

Kopka, J., Fernie, A., Weckwerth, W., Gibon, Y., and Stitt, M. 2004. Metabolite profiling in plant biology: Platforms and destinations. *Genome Biology* 5: 109.

Kosová, K., Vítámvás, P., Prášil, T., and Renaut, J. 2011. Plant proteome changes under abiotic stress—Contribution of proteomics studies to understanding plant stress response. *Journal of Proteomics* 74: 1301–22.

Kou, Y. and Wang, S. 2010. Broad-spectrum and durability: Understanding of quantitative disease resistance. *Current Opinion in Plant Biology* 13: 181–85.

Kovach, M. J., Megan, T. S., and McCouch, S. R. 2007. New insights into the history of rice domestication. *Trends in Genetics* 23: 478–87.

Kumar, S. P., Dalal, V., Singh, N. K., and Sharma, T. R. 2007. Comparative analysis of the 100 kb region containing the *Pi-k(h)* locus between *indica* and *japonica* rice lines. *Genomics, Proteomics and Bioinformatics* 5: 35–44.

Lang, N. T., Subudhi, P. K., Virmani, S. S., Brar, D. S., Khush, G. S., Li, Z., Huang, N. et al. 1999. Development of PCR-based markers for thermosensitive genetic male sterility gene *tms3(t)* in rice (*Oryza sativa* L.). *Hereditas* 31: 121–27.

Latha, R., Rubia, L., Bennett, J., and Swaminathan, M. S. 2004. Allele mining for stress tolerance genes in *Oryza* species and related germplasm. *Molecular Biotechnology* 27: 101–8.

Lee, D.-G., Ahsan, N., Lee, S.-H., Lee, J. J., Bahk, J. D., Kang, K. Y., and Lee, B. H. 2007a. An approach to identify cold-induced low-abundant proteins in rice leaf. *Comptes Rendus Biologies* 330: 215–25.

Lee, D.-G., Ahsan, N., Lee, S.-H., Kang, K. Y., Bahk, J. D., Lee, I. J., and Lee, B. H. 2007b. A proteomic approach in analyzing heat-responsive proteins in rice leaves. *Proteomics* 7: 3369–83.

Lee, D. G., Ahsan, N., Lee, S. H., Lee, J. J., Bahk, J. D., Kang, K. Y., and Lee, B. H. 2009. Chilling stress-induced proteomic changes in rice roots. *Journal of Plant Physiology* 166: 1–11.

Lee, J., Bricker, T. M., Lefevre, M., Pinson, S. R. M., and Oard, J. H. 2006. Proteomic and genetic approaches to identifying defence-related proteins in rice challenged with the fungal pathogen *Rhizoctonia solani*. *Molecular Plant Pathology* 7: 405–16.

Lee, K., Bae, D. W., Kim, S. H., Han, H. J., Liu, X., Park, H. C., Lim, C. O., Lee, S. Y., and Chung, W. S. 2010. Comparative proteomic analysis of the short-term responses of rice roots and leaves to cadmium. *Journal of Plant Physiology* 167: 161–68.

Leung, H., Wu, C., Baraoidan, M., Bordeos, A., Ramos, M., Madamba, S., Cabauatan, P., et al. 2001. Deletion mutants for functional genomics: Progress in phenotyping, sequence assignment and database development. In *Rice Genetics IV. Proceedings of the Fourth International Rice Genetics Symposium*, eds., G. S. Khush, D. S. Brar and B. Hardy, pp. 239–51. New Delhi (India): Science, Los Banos (Philippines): International Rice Research Institute.

Li, H., Liang, W., Jia, R., Yin, C., Zong, J., Kong, H., and Zhang, D. 2010a. The AGL6-like gene *OsMADS6* regulates floral organ and meristem identities in rice. *Cell Research* 20: 299–313.

Li, X. J., Yang, M. F., Chen, H., Qu, L. Q., Chen, F., and Shen, S. H. 2010b. Abscisic acid pretreatment enhances salt tolerance of rice seedlings: Proteomic evidence. *Biochimica et Biophysica Acta* 1804: 929–40.

Li, L., Wang, X., Stolc, V., Li, X., Zhang, D., Su, N., Tongprasit, W., et al. 2006. Genome-wide transcription analyses in rice using tiling microarrays. *Nature Genetics* 38:124–29.

Li, L., Wang, X., Sasidharan, R., Stolc, V., Deng, W., He, H., Korbel, J., et al. 2007. Global identification and characterization of transcriptionally active regions in the rice genome. *PLoS ONE* 2: e294.

Li, X., Qian, Q., Fu, Z., Wang, Y., Xiong, G., Zeng, D., Wang, X., et al. 2003. Control of tillering in rice. *Nature* 422: 618–21.

Li, X., Wang, X., He, K., Ma, Y., Su, N., He, H., Stolc, V., et al. 2008. High-resolution mapping of epigenetic modifications of the rice genome uncovers interplay between DNA methylation, histone methylation, and gene expression. *The Plant Cell* 20: 259–76.

Liao, M., Li, Y., and Wang, Z. 2009. Identification of elicitor-responsive proteins in rice leaves by a proteomic approach. *Proteomics* 9: 2809–19.

Lin, H., Wang, R., Qian, Q., Yan, M., Meng, X., Fu, Z., Yan, C., et al. 2009. *DWARF27*, an iron-containing protein required for the biosynthesis of strigolactones, regulates rice tiller bud outgrowth. *The Plant Cell* 21: 1512–25.

Lin, S. K., Chang, M. C., Tsai, Y. G., and Lur, H. S. 2005. Proteomic analysis of the expression of proteins related to rice quality during caryopsis development and the effect of high temperature on expression. *Proteomics* 5: 2140–56.

Lin, X. H., Zhang, D. P., Xie, Y. F., Gao, H. P., and Zhang, Q. 1996. Identifying and mapping a new gene for bacterial blight resistance in rice based on RFLP markers. *Phytopathology* 86: 1156–59.

Lin, Y. Z., Chen, H. Y., Kao, R., Chang, S. P., Chang, S. J., and Lai, E. M. 2008. Proteomic analysis of rice defence response induced by probenazole. *Phytochemistry* 69: 715–28.

Lipshutz, R. J., Fodor, S. P., Gingeras, T. R., and Lockhart, D. J. 1999. High density synthetic oligonucleotide arrays. *Nature Genetics* 21: 20–24.

Lisec, J., Meyer, R. C., Steinfath, M., Redestig, H., Becher, M., Witucka-Wall, H., Fiehn, O., et al. 2008. Identification of metabolic and biomass QTL in *Arabidopsis thaliana* in a parallel analysis of RIL and IL populations. *The Plant Journal* 53: 960–72.

Liu, J. X. and Bennett, J. 2011. Reversible and irreversible drought-induced changes in the anther proteome of rice (*Oryza sativa* L.) genotypes IR64 and Moroberekan. *Molecular Plant* 4: 59–69.

Liu, G. Q., Yan, H., Fu, Q., Qian, Q., Zhang, Z. T., Zhai, W. X., and Zhu, L. H. 2001. Mapping of a new gene for brown planthopper resistance in cultivated rice introgressed from *Oryza eichingeri*. *Chinese Science Bulletin* 46: 738–42.

Londo, J. P., Chiang, Y. C., Hung, K. H., Chiang, T. Y., and Schaal, B. A. 2006. Phylogeography of Asian wild rice, *Oryza rufipogon*, reveals multiple independent domestications of cultivated rice, *Oryza sativa*. *Proceedings of the National Academy of Sciences USA* 103: 9578–83.

Lukowitz, W., Gillmor, C. S., and Scheible, W. R. 2000. Positional cloning in Arabidopsis. Why it feels good to have a genome initiative working for you. *Plant Physiology* 123: 795–805.

Ma, Q., Dai, X., Xu, Y., Guo, J., Liu, Y., Chen, N., Xiao, J., et al. 2009. Enhanced tolerance to chilling stress in *OsMYB3R-2* transgenic rice is mediated by alteration in cell cycle and ectopic expression of stress genes. *Plant Physiology* 150: 244–56.

Mahmood, T., Jan, A., and Kakishima, M. 2006. Proteomic analysis of bacterial-blight defence-responsive proteins in rice leaf blades. *Proteomics* 6: 6053–65.

Mather, K. A., Caicedo, A. L., Polato, N. R., Olsen, K. M., McCouch, S., and Purugganan, M. D. 2007. The extent of linkage disequilibrium in rice (*Oryza sativa* L.). *Genetics* 177: 2223–32.

McCouch, S. R., Chen, X., Panaud, O., Temnykh, S., Xu, Y., Cho, Y. G., Huang, N., Ishii, T., and Blair, M. 1997. Microsatellite marker development, mapping and applications in rice genetics and breeding. *Plant Molecular Biology* 35: 89–99.

McCouch, S. R., McNally, K. L., Wang, W., and Sackville Hamilton, R. 2012. Genomics of gene banks: A case study in rice. *American Journal of Botany* 99: 407–23.

Metzker, M. L. 2010. Sequencing technologies: The next generation. *Nature Reviews Genetics* 11: 31–46.

Meyer, R. C., Kusterer, B., Lisec, J., Steinfath, M., Becher, M., Scharr, H., Melchinger, A. E., et al. 2010. QTL analysis of early stage heterosis for biomass in Arabidopsis. *Theoretical and Applied Genetics* 120: 227–37.

Miki, D., Itoh, R., and Shimamoto, K. 2005. RNA silencing of single and multiple members in a gene family of rice. *Plant Physiology* 138: 1903–13.

Miyao, A., Iwasaki, Y., Kitano, H., Itoh, J., Maekawa, M., Murata, K., Yatou, O., Nagato, Y., and Hirochika, H. 2007. A large-scale collection of phenotypic data describing an insertional mutant population to facilitate functional analysis of rice genes. *Plant Molecular Biology* 63: 625–35.

Mohan, M., Nair, S., Bentur, J. S., Rao, U. P., and Bennett, J. 1994. RFLP and RAPD mapping of the rice *Gm2* gene that confers resistance to biotype 1 of gall midge (*Orseolia oryzae*). *Theoretical and Applied Genetics* 87: 782–88.

Moncada, P., Martinez, C. P., Borrero, J., Châtel, M., Gauch, H., Guimaraes, E. P., Tohmé, J., and McCouch, S. R. 2001. Quantitative trait loci for yield and yield components in an *Oryza sativa* × *Oryza rufipogon* BC_2F_2 population evaluated in an upland environment. *Theoretical and Applied Genetics* 102: 41–52.

Morinaka, Y., Sakamoto, T., Inukai, Y., Agetsuma, M., Kitano, H., Ashikari, M., and Matsuoka, M. 2006. Morphological alteration caused by brassinosteroid insensitivity increases the biomass and grain production of rice. *Plant Physiology* 141: 924–31.

Nakano, M., Nobuta, K., Vemaraju, K., Tej, S. S., Skogen, J. W., and Meyers, B. C. 2006. Plant MPSS databases: Signature based transcriptional resources for analyses of mRNA and small RNA. *Nucleic Acids Research* 34: D731–35.

Nakashima, K., Tran, L. S., Van Nguyen, D., Fujita, M., Maruyama, K., Todaka, D., Ito, Y., Hayashi, N., Shinozaki, K., and Yamaguchi-Shinozaki, K. 2007. Functional analysis of a NAC-type transcription factor *OsNAC6* involved in abiotic and biotic stress-responsive gene expression in rice. *The Plant Journal* 51: 617–30.

Nam, M. H., Huh, S. M., Kim, K. M., Park, W. J., Seo, J. B., Cho, K., Kim, D. Y., Kim, B. G., and Yoon, I. S. 2012. Comparative proteomic analysis of early salt stress-responsive proteins in roots of SnRK2 transgenic rice. *Proteome Science* 10: 25.

Nobuta, K., Venu, R. C., Lu, C., Beló, A., Vemaraju, K., Kulkarni, K., Wang, W., et al. 2007. An expression atlas of rice mRNAs and small RNAs. *Nature Biotechnology* 25: 473–77.

Obayashi, T., Kinoshita, K., Nakai, K., Shibaoka, M., Hayashi, S., Saeki, M., Shibata, D., Saito, K., and Ohta, H. 2007. ATTED-II: A database of co-expressed genes and *cis* elements for identifying co-regulated gene groups in *Arabidopsis*. *Nucleic Acids Research* 35: D863–69.

Oguchi, K., Tanaka, N., Komatsu, S., and Akao, S. 2004. Methylmalonate-semialdehyde dehydrogenase is induced in auxin-stimulated and zinc-stimulated root formation in rice. *Plant Cell Reports* 22: 848–58.

Oikawa, A., Matsuda, F., Kusano, M., Okazaki, Y., and Saito, K. 2008. Rice metabolomics. *Rice* 1: 63–71.

Oka, H. I. 1988. *Origin of Cultivated Rice*. Tokyo: Japan Scientific Societies Press and Elsevier.

Ouyang, S., Zhu, W., Hamilton, J., Lin, H., Campbell, M., Childs, K., Thibaud-Nissen, F., et al. 2007. The TIGR rice genome annotation resource: Improvements and new features. *Nucleic Acids Research* 35: D883–87.

Parker, R., Flowers, T. J., Moore, A. L., and Harpham, N. V. J. 2006. An accurate and reproducible method for proteome profiling of the effects of salt stress in the rice leaf lamina. *Journal of Experimental Botany* 57: 1109–18.

Peng, S., Huang, J., Sheehy, J. E., Laza, R. C., Visperas, R. M., Zhong, X., Centeno, G. S., Khush, G. S., and Cassman, K. G. 2004. Rice yields decline with higher night temperature from global warming. *Proceedings of the National Academy of Sciences* 101: 9971–75.

Pinson, S. R. M., Capdevielle, F. M., and Oard, J. H. 2005. Confirming QTLs and finding additional loci conditioning sheath blight resistance in rice using recombinant inbred lines. *Crop Science* 45: 503–10.

Potokina, E., Druka, A., Luo, Z., Moscou, M., Wise, R., Waugh, R., and Kearsey, M. J. 2008. Tissue-dependent limited pleiotropy affects gene expression in barley. *The Plant Journal* 56: 287–96.

Power, K. A., McRedmond, J. P., de Stefani, A., Gallagher, W. M., and Gaora, P. O. 2009. High-throughput proteomics detection of novel splice isoforms in human platelets. *PLoS One* 4: e5001.

Provart, N. and McCourt, P. 2004. Systems approaches to understanding cell signalling and gene regulation. *Current Opinion Plant Biology* 7: 605–9.

Prusinkiewicz, P., Hammel, M., Hanan, J., and Mech, R. 1996. L-systems: From the theory to visual models of plants. In *Proceedings of the 2nd CSIRO Symposium on Computational Challenges in Life Sciences*, ed., M. T. Michalewicz. Australia: CSIRO.

Prusinkiewicz, P., Muendermann, L., Karwowski, R., and Lane, B. 2002. The use of positional information in the modelling of plants. In *Proceedings of SIGGRAPH 2001*, Los Angeles, CA, pp. 289–300.

Rabbani, M. A., Maruyama, K., Abe, H., Khan, M. A., Katsura, K., Ito, Y., Yoshiwara, K., Seki, M., Shinozaki, K., and Yamaguchi-Shinozaki, K. 2003. Monitoring expression profiles of rice genes under cold, drought, and high-salinity stresses and abscisic acid application using cDNA microarray and RNA gel-blot analyses. *Plant Physiology* 133: 1755–67.

Rahman, M. L., Jiang, W., Chu, S. H., Qiao, Y., Ham, T. H., Woo, M. O., Lee, J., et al. 2009. High-resolution mapping of two brown planthopper resistance genes, *Bph20(t)* and *Bph21(t)*, originating from *Oryza minuta*. *Theoretical and Applied Genetics* 119: 1237–46.

Rai, A. K., Kumar, S. P., Gupta, S. K., Gautam, N., Singh, N. K., and Sharma, T. R. 2011. Functional complementation of rice blast resistance gene *Pi-kh* (*Pi54*) conferring resistance to diverse strains of *Magnaporthe oryzae*. *Journal of Plant Biochemistry and Biotechnology* 20: 55–65.

Rajendran, K., Tester, M., and Roy, S. J. 2009. Quantifying the three main components of salinity tolerance in cereals. *Plant Cell and Environment* 32: 237–49.

Rakwal, R. and Agrawal, G. K. 2003. Rice proteomics: Current status and future perspectives. *Electrophoresis* 24: 3378–89.

Rakwal, R. and Komatsu, S. 2000. Role of jasmonate in the rice (*Oryza sativa* L.) self-defence mechanism using proteome analysis. *Electrophoresis* 21: 2492–00.

Ramsay, G. 1998. DNA chips: State-of-the-art. *Nature Biotechnology* 16: 40–4.

Ren, Z. H., Gao, J. P., Li, L. G., Cai, X. L., Huang, W., Chao, D. Y., Zhu, M. Z., et al. 2005. A rice quantitative trait locus for salt tolerance encodes a sodium transporter. *Nature Genetics* 37: 1141–46.

Renganayaki, K., Feitz, A. K., and Sadasivam, S. 2002. Mapping and progress toward map-based cloning of brown planthopper biotype-4 resistance gene introgressed from *Oryza officinalis* into cultivated rice, *O. sativa*. *Crop Science* 42: 2112–17.

Ross-Ibarra, J., Morrell, P. L., and Gaut, B. S. 2007. Plant domestication, a unique opportunity to identify the genetic basis of adaptation. *Proceedings of the National Academy of Sciences USA* 104: 8641–48.

Roy, S., Banerjee, A., Senapati, B. K., and Sarkar, G. 2012. Comparative analysis of agro-morphology, grain quality and aroma traits of traditional and Basmati type genotypes of rice, *Oryza sativa* L. *Plant Breeding* 131: 486–92.

Rudd, S., Schoof, H., and Meyer, K. 2005. Plant markers: A database of predicted molecular markers from plants. *Nucleic Acids Research* 33: D628–32.

Saha, S., Sparks, A. B., Rago, C., Akmaev, V., Wang, C. J., Vogelstein, B., Kinzler, K. W., and Velculescu, V. E. 2002. Using the transcriptome to annotate the genome. *Nature Biotechnology* 20: 508–12.

Sahi, C., Agarwal, M., Reddy, M. K., Sopory, S. K., and Grover, A. 2003. Isolation and expression analysis of salt stress-associated ESTs from contrasting rice cultivars using a PCR-based subtraction method. *Theoretical and Applied Genetics* 106: 620–28.

Sahi, C., Singh, A., Blumwald, E., and Grover, A. 2006. Beyond osmolytes and transporters: Novel plant salt-stress tolerance-related genes from transcriptional profiling data. *Physiologia Plantarum* 127: 1–9.

Saijo, Y., Hata, S., Kyozuka, J., Shimamoto, K., and Izui, K. 2000. Overexpression of a single Ca^{2+} dependant protein kinase confers cold and salt/drought tolerance on rice plants. *The Plant Journal* 23: 319–27.

Saito, K., Hayano-Saito, Y., Maruyama-Funatsuki, W., Sato, Y., and Kato, A. 2004. Physical mapping and putative candidate gene identification of a quantitative trait locus *Ctb1* for cold tolerance at booting stage of rice. *Theoretical and Applied Genetics* 109: 515–22.

Sakamoto, T. and Matsuoka, M. 2008. Indentifying end exploiting grain yield genes in rice. *Current Opinion in Plant Biology* 11: 209–14.

Sakamoto, T., Morinaka, Y., Ishiyama, K., Kobayashi, M., Itoh, H., Kayano, T., Iwahori, S., Matsuoka, M., and Tanaka, H. 2003. Genetic manipulation of gibberellin metabolism in transgenic rice. *Nature Biotechnology* 21: 909–13.

Sakamoto, T., Morinaka, Y., Ohnishi, T., Sunohara, H., Fujioka, S., Ueguchi-Tanaka, M., Mizutani, M., et al. 2006. Erect leaves caused by brassinosteroid deficiency increase biomass production and grain yield in rice. *Nature Biotechnology* 24: 105–9.

Salekdeh, Gh. H., Siopongco, J., Wade, L. J., Ghareyazie, B., and Bennett, J. 2002. A proteomic approach to analyzing drought- and salt-responsiveness in rice. *Field Crops Research* 76: 199–219.

Sasaki, H. and Itoh, T. 2010. Massive gene losses in Asian cultivated rice unveiled by comparative genome analysis. *BMC Genomics* 11: 121.

Sasaki, A., Gomi, K., Ueguchi-Tanaka, M., Ishiyama, K., Kobayashi, M., Jeong, D. H., An, G., et al. 2003. Accumulation of phosphorylated repressor for gibberellin signaling in an F-box mutant. *Science* 299: 1896–98.

Sasaki, T., Matsumoto, T., Yamamoto, K., Sakata, K., Baba, T., Katayose, Y., Wu, J., et al. 2002. The genome sequence and structure of rice chromosome 1. *Nature* 420: 312–16.

Schmidt, R. 2000. Synteny: Recent advances and future prospects. *Current Opinion in Plant Biology* 3: 97–102.

Scholes, J. D. and Rolfe, S. A. 2009. Chlorophyll fluorescence imaging as a tool for understanding the impact of fungal diseases on plant performance: A phenomics perspective. *Functional Plant Biology* 36: 880–92.

Schoof, H., Zaccaria, P., Gundlach, H., Lemcke, K., Rudd, S., Kolesov, G., Arnold, R., Mewes, H. W., and Mayer, K. F. X. 2002. MIPS *Arabidopsis thaliana* Database (MAtDB): An integrated biological knowledge resource based on the first complete plant genome. *Nucleic Acids Research* 30: 91–3.

Schulze, T. G. and McMahon, F. J. 2002. Genetic association mapping at the crossroads: Which test and why? Overview and practical guidelines. *American Journal of Medical Genetics B* 114: 1–11.

Schwab, R., Ossowski, S., Riester, M., Warthmann, N., and Weigel, D. 2006. Highly specific gene silencing by artificial microRNAs in *Arabidopsis*. *The Plant Cell* 18: 1121–33.

Second, G. 1982. Origin of the genetic diversity of cultivated rice (*Oryza* spp.): Study of the polymorphism scored at 40 isozyme loci. *Japanese Journal of Genetics* 57: 25–57.

Septiningsih, E. M., Prasetiyono, J., Lubis, E., Tai, T. H., Tjubaryat, T., Moeljopawiro, S., and McCouch, S. R. 2003. Identification of quantitative trait loci for yield and yield components in an advanced backcross population derived from the *Oryza sativa* variety IR64 and the wild relative *O. rufipogon*. *Theoretical and Applied Genetics* 107: 1419–32.

Shang, J., Tao, Y., Chen, X., Zou, Y., Lei, C., Wang, J., Li, X., Zhao, X., Zhang, M., Lu, Z., Xu, J., et al. 2009. Identification of a new rice blast resistance gene, *Pid3*, by genomewide comparison of paired nucleotide-binding site–leucine-rich repeat genes and their pseudogene alleles between the two sequenced rice genomes. *Genetics* 182: 1303–11.

Shao, C. H., Liu, G. R., Wang, J. Y., Yue, C. and Lin, W. 2008. Differential proteomic analysis of leaf development at rice (*Oryza sativa*) seedling stage. *Agricultural Science in China* 7: 1153–60.

Sharma, T. R., Shanker, P., Singh, B. K., Jana, T. K., Madhav, M. S., Gaikwad, K., Singh, N. K., Plaha, P., and Rathour, R. 2005. Molecular mapping of rice blast resistance gene *Pi-kh* in rice variety Tetep. *Journal of Plant Biochemistry and Biotechnology* 14: 127–33.

Sharma, T. R., Rai, A. K., Gupta, S. K., and Singh, N. K. 2010. Broad spectrum blast resistance gene *Pi-kh* designated as *Pi-54*. *Journal of Plant Biochemistry and Biotechnology* 19: 987–89.

Sharma, T. R., Rai, A. K., Gupta, S. K., Vijayan, J., Devanna, B. N., and Ray, S. 2012. Rice blast management through host-plant resistance: Retrospect and prospects. *Agricultural Research* 1: 37–52.

Shiu, S. H., Karlowski, W. M., Pan, R., Tzeng, Y. H., Mayer, K. F., and Li, W. H. 2004. Comparative analysis of the receptor-like kinase family in Arabidopsis and rice. *The Plant Cell* 16: 1220–34.

Sirault, X. R. R., James, R. A., and Furbank, R. T. 2009. A new screening method for osmotic component of salinity tolerance in cereals using infrared thermography. *Functional Plant Biology* 36: 970–77.

Song, W., Wang, G., Chen, L., Kim, H., Pi, L., Holsten, T., Gardner, J. et al. 1995. A receptor kinase-like protein encoded by the rice disease resistance gene, *Xa21*. *Science* 270: 1804–6.

Song, X., Huang, W., Shi, M., Zhu, M. Z., and Lin, H. X. 2007. A QTL for rice grain width and weight encodes a previously unknown RING-type E3 ubiquitin ligase. *Nature Genetics* 39: 623–30.

Song, Y., You, J., and Xiong, L. 2009. Characterization of *OsIAA1* gene, a member of rice Aux/IAA family involved in auxin and brassinosteroid hormone responses and plant morphogenesis. *Plant Molecular Biology* 70: 297–309.

Soulé, M. 1967. Phenetics of natural populations. I: Phenetic relationships of insular populations of the side-blotched lizard. *Evolution* 21: 584–91.

Su, C., Wang, Y., Hsieh, T. Lu, C. A., Tseng, T. H., and Yu, S. M. 2010. A novel *MYBS3*-Dependent pathway confers cold tolerance in rice. *Plant Physiology* 153: 145–58.

Suh, J. P., Jeung, J. U., Lee, I., Choi, Y. H., Yea, J. D., Virk, P. S., Mackill, D. J., and Jena, K.K. 2010. Identification and analysis of QTLs controlling cold tolerance at the reproductive stage and validation of effective QTLs in cold-tolerant genotypes of rice (*Oryza sativa* L.). *Theoretical and Applied Genetics* 120: 985–95.

Suh, J. P., Lee, C. K., Lee, J. H., Kim, J. J., Kim, S. M., Cho, Y. C., Park, S. H., Shin, J. C., Kim, Y. G., and Jena, K.K. 2012. Identification of quantitative trait loci for seedling cold tolerance using RILs derived from a cross between *japonica* and *tropical japonica* rice cultivars. *Euphytica* 184: 101–8.

Sun, X. L., Cao, Y. L., Yang, Z. F., Xu, C. G., Li, X. H., Wang, S. P., and Zhang, Q. F. 2004. *Xa26*, a gene conferring resistance to *Xanthomonas oryzae* PV.*Oryzae* in rice, encodes an LRR receptor Kinase-like protein. *The Plant Journal* 37: 517–27.

Takahashi, H., Hotta, Y., Hayashi, M., Kawai-Yamada, M., Komatsu, S., and Uchimiya, H. 2005. High throughput metabolome and proteome analysis of transgenic rice plants (*Oryza sativa* L.). *Plant Biotechnology Journal* 22: 47–50.

Takahashi, H., Hayashi, M., Goto, F., Sato, S., Soga, S., Nishioka, T., Tomita, M., Kawai-Yamada, M., and Uchimiya, H. 2006. Evaluation of metabolic alteration in transgenic rice overexpressing dihydroflavonol-4-reductase. *Annals of Botany* 98: 819–25.

Takahashi, N., Hamamura, K., Tsunoda, S., Sakamoto, S., and Sato, Y. 1997. Differentiation of eco-types in cultivated rice. In *Science of the Rice Plant*, eds, T. Matsuo, Y. Futsuhara, F. Kikuchi, et al., pp. 112–60. Tokyo: Food and Agriculture Policy Research Centre.

Takai, T., Fukuta, Y., Shiraiwa, T., and Horie, T. 2005. Time-related mapping of quantitative trait loci controlling grain-filling in rice (*Oryza sativa* L.). *Journal of Experimental Botany* 56: 2107–18.

Takeda, T., Suwa, Y., Suzuki, M., Kitano, H., Ueguchi-Tanaka, M., Ashikari, M., Matsuoka, M., and Ueguchi, C. 2003. The *OsTB1* gene negatively regulates lateral branching in rice. *The Plant Journal* 33: 513–20.

Tan, L. B., Li, X. R., Liu, F. X., Sun, X. Y., Li, C. G., Zhu, Z. F., Fu, Y. C., et al. 2008. Control of a key transition from prostrate to erect growth in rice domestication. *Nature Genetics* 40: 1360–64.

Tanaka, N., Konishi, H., Khan, M., and Komatsu, S. 2004. Proteome analysis of rice tissues by two-dimensional electrophoresis: An approach to the investigation of gibberellin regulated proteins. *Molecular Genetics and Genomics* 270: 485–96.

Tanaka, T., Antonio, B. A., Kikuchi, S., Matsumoto, T., Nagamura, Y., Numa, H., Sakai, H., et al. 2008. The rice annotation project database (RAP-DB): 2008 update. *Nucleic Acids Research* 36: D1028–33.

Tanksley, S. D. and Nelson, J. C. 1996. Advanced backcross QTL analysis: A method for the simultaneous discovery and transfer of valuable QTLs from unadapted germplasm into elite breeding lines. *Theoretical and Applied Genetics* 92: 191–203.

Tanner, S., Shen, Z., Ng, J., Florea, L., Guigo, R., Briggs, S. P., and Bafna, V. 2007. Improving gene annotation using peptide mass spectrometry. *Genome Research* 17: 231–39.

The International Rice Genome Sequencing Project (IRGSP). 2005. The map-based sequence of the rice genome. *Nature* 436: 793–800.

Thomson, M. J., de Ocampo, M., Egdane J., Akhlasur Rahman, M., Godwin Sajise, A., Adorada, D. L., Tumimbang-Raiz, E., et al. 2010. Characterizing the *Saltol* quantitative trait locus for salinity tolerance in rice. *Rice* 3: 148–60.

Tian, C., Wan, P., Sun, S., Li, J., and Chen, M. 2004. Genome-wide analysis of the GRAS gene family in rice and *Arabidopsis*. *Plant Molecular Biology* 54: 519–32.

Till, B. J., Reynolds, S. H., Greene, E. A., Codomo, C. A., Enns, L. C., Johnson, J. E., Burtner, C., et al. 2003. Large-scale discovery of induced point mutations with high-throughput TILLING. *Genome Research* 13: 524–30.

Till, B. J., Cooper, J., Tai, T. H., Colowit, P., Greene, E. A., Henikoff, S., and Comai, L. 2007. Discovery of chemically induced mutations in rice by TILLING. *BMC Plant Biology* 7: 19.

Todaka, D., Nakashima, K., Shinozaki, K., and Yamaguchi-Shinozaki, K. 2012. Toward understanding transcriptional regulatory networks in abiotic stress responses and tolerance in rice. *Rice* 5: 6.

Tozawa, Y., Hasegawa, H., Terakawa, T., and Wakasa, K. 2001. Characterization of rice anthranilate synthase alpha-subunit genes *OASA1* and *OASA2*. Tryptophan accumulation in transgenic rice expressing a feedback-insensitive mutant of *OASA1*. *Plant Physiology* 126: 1493–506.

Tyagi, A. K. and Khurana, J. P. 2003. Plant molecular biology and biotechnology research in the post-recombinant DNA era. *Advance in Biochemical Engineering/Biotechnology* 84: 91–121.

Ueguchi-Tanaka, M., Ashikari, M., Nakajima, M., Itoh, H., Katoh, E., Kobayashi, M., Chow, T. Y., et al. 2005. Gibberellin insensitive *dwarf1* encodes a soluble receptor for gibberellin. *Nature* 437: 693–98.

Vance, C. P., Uhde-Stone, C., and Allan, D. L. 2003. Phosphorus acquisition and use: Critical adaptation by plants for securing a non-renewable resource. *New Phytologist* 157: 423–47.

Vanderplank, J. E. 1982. *Host-Pathogen Interactions in Plant Disease*. New York: Academic Press.

Varshney, R. K., Graner, A., and Sorrells, M. E. 2005. Genomics-assisted breeding for crop improvement. *Trends in Plant Science* 10: 621–30.

Varshney, R. K., Hoisington, D. A., and Tyagi, A. K. 2006. Advances in cereal genomics and applications in crop breeding. *Trends in Biotechnology* 24: 490–99.

Varshney, R. K., Nayak, S. N., May, G. D., and Jackson, S. A. 2009. Next-generation sequencing technologies and their implications for crop genetics and breeding. *Trend in Biotechnology* 27: 522–30.

Varshney, R. K., Bansal, K. C., Aggarwal, P. K., Datta, S. K., and Craufurd, P. Q. 2011. Agricultural biotechnology for crop improvement in a variable climate: Hope or hype? *Trends in Plant Science* 16: 363–71.

Vaughan, D. A., Morishima, H., and Kadowaki, K. 2003. Diversity in the *Oryza* genus. *Current Opinion in Plant Biology* 6: 139–46.

Velculescu, V. E., Zhang, L., Vogelstein, B., and Kinzler, K. W. 1995. Serial analysis of gene expression. *Science* 270: 484–87.

Ventelon-Debout, M., Delalande, F., and Brizard, J. P. 2004. Proteome analysis of cultivar specific deregulations of *Oryza sativa indica* and *O. sativa japonica* cellular suspensions undergoing rice yellow mottle virus infection. *Proteomics* 4: 216–25.

Wakasa, K., Hasegawa, H., Nemoto, H., Matsuda, F., Miyazawa, H., Tozawa, Y., Morino, K., et al. 2006. High-level tryptophan accumulation in seeds of transgenic rice and its limited effects on agronomic traits and seed metabolite profile. *Journal of Experimental Botany* 57: 3069–78.

Wang, E., Wang, J., Zhu, X., Hao, W., Wang, L., Li, Q., Zhang, L., et al. 2008a. Control of rice grain-filling and yield by a gene with a potential signature of domestication. *Nature Genetics* 40: 1370–74.

Wang, W. W., Meng, B., Ge, X. M., Song, S., Yang, Y., Yu, X., Wang, L., Hu, S., Liu, S., and Yu, J. 2008b. Proteomic profiling of rice embryos from a hybrid rice cultivar and its parental lines. *Proteomics* 8: 4808–21.

Wang, J., Yu, H., Xie, W., Xing, Y., Yu, S., Xu, C., Li, X., Xiao, J., and Zhang, Q. 2010. A global analysis of QTLs for expression variations in rice shoots at the early seedling stage. *The Plant Journal* 63: 1063–74.

Wang, R., Prince, J. T., and Marcotte, E. M. 2005. Mass spectrometry of the *M. smegmatis* proteome: Protein expression levels correlate with function, operons, and codon bias. *Genome Research* 15: 1118–26.

Wang, S., Sim, T. B., Kim, Y. S., and Chang, Y. T. 2004. Tools for target identification and validation. *Current Opinion in Chemical Biology* 8: 371–77.

Wang, Y., Kim, S. G., Kim, S. T., Agrawal, G. K., Rakwal, R., and Kang, K. Y. 2011. Biotic stress-responsive rice proteome: An overview. *Journal of Plant Biology* 54: 219–26.

Wang, Z., Jia, Y., Rutger, J. N., and Xia, Y. 2007. Rapid survey for presence of a blast resistance gene *Pi-ta* in rice cultivars using the dominant DNA markers derived from portions of the *Pi-ta* gene. *Plant Breeding* 126: 36–42.

Wang, Z.-X, Yano, M., Yamanouchi, U., Iwamoto, M., Monna, L., Hayasaka, H., Katayose, Y., and Sasaki, T. 1999. The *Pib* gene for rice blast resistance belongs to the nucleotide binding and leucine-rich repeat class of plant disease resistance genes. *The Plant Journal* 19: 55–64.

Washburn, M. P., Wolters, D., and Yates, J. R. 2001. Large-scale analysis of the yeast proteome by multidimensional protein identification technology. *Nature Biotechnology* 19: 242–47.

Watanabe, M., Kusano, M., Oikawa, A., Fukushima, A., Noji, M, Saito, K., et al. 2008. Physiological roles of the beta-substituted alanine synthase gene family in *Arabidopsis. Plant Physiology* 146: 310–20.

Wei, Z., Hu, W., Lin, Q., Cheng, X., Tong, M., Zhu, L., Chen, R., and He, G. 2009. Understanding rice plant resistance to the brown planthopper (*Nilaparvata lugens*): A proteomic approach. *Proteomics* 9: 2798–808.

West, M. A. L., Kim, K., Kliebenstein, D. J., van Leeuwen, H., Michelmore, R. W., Doerge, R. W., and St.Clair, D. A. 2007. Global eQTL mapping reveals the complex genetic architecture of transcript level variation in *Arabidopsis. Genetics* 175: 1441–50.

Wolters, D. A., Washburn, M. P., and Yates, J. R. 2001. An automated multidimensional protein identification technology for shotgun proteomics. *Analytical Chemistry* 73: 5683–90.

Wu, C., Li, X., Yuan, W., Chen, G., Kilian, A., Li, J., Xu, C., et al. 2003. Development of enhancer trap lines for functional analysis of the rice genome. *The Plant Journal* 35: 418–27.

Wu, J. L., Wu, C., Lei, C., Baraoidan, M., Bordeos, A., Madamba, M. R., Ramos-Pamplona, M., et al. 2005. Chemical- and irradiation-induced mutants of *indica* rice IR64 for forward and reverse genetics. *Plant Molecular Biology* 59: 85–97.

Xia, C., Chen, H., and Zhu, X. 2012. Identification, mapping, isolation of the genes resisting to bacterial blight and breeding application in rice. *Molecular Plant Breeding* 3: 121–31.

Xiang, Y., Cao, Y. L., Xu, C. G., Li, X. H., and Wang, S. P. 2006. *Xa3*, conferring resistance for rice bacterial blight and encoding a receptor kinase-like protein, is the same as *Xa26. Theoretical and Applied Genetics* 113: 1347–55.

Xiang, Y., Tang, N., Du, H., Ye, H., and Xiong, L. 2008. Characterization of *OsbZIP23* as a key player of the basic leucine zipper transcription factor family for conferring abscisic acid sensitivity and salinity and drought tolerance in rice. *Plant Physiology* 148: 1938–52.

Xiao, J. H., Li, J. M., Grandillo, S., Ahn, S. N., Yuan, L. P., and Tanksley, S. D. 1998. Identification of trait-improving quantitative trait loci alleles from a wild rice relative, *Oryza rufipogon. Genetics* 150: 899–909.

Xu, K., Xia, X., Fukao, T., Canlas, P., Maghirang-Rodriguez, R., Heuer, S., Ismail, A. M., Bailey-Serres, J., Ronald, P. C., and Mackill, D.J. 2006. *Sub1A* is an ethylene response factor-like gene that confers submergence tolerance to rice. *Nature* 442: 705–8.

Xue, W., Xing, Y., Weng, X., Zhao, Y., Tang, W., Wang, L., Zhou, H., et al. 2008. Natural variation in *Ghd7* is an important regulator of heading date and yield potential in rice. *Nature Genetics* 40: 761–67.

Xue, Y., Li, J., and Xu, Z. 2003. Recent highlights of the China rice functional genomics program. *Trends in Genetics* 19: 390–94.

Yamakawa, H., Hirose, T., Kuroda, M., and Yamaguchi, T. 2007. Comprehensive expression profiling of rice grain filling-related genes under high temperature using DNA microarray. *Plant Physiology* 144: 258–77.

Yamamoto, K. and Sasaki, T. 1997. Large-scale EST sequencing in rice. *Plant Molecular Biology* 35: 135–44.

Yamamuro, C., Ihara, Y., Wu, X., Noguchi, T., Fujioka, S., Takatsuto, S., Ashikari, M., Kitano, H., and Matsuoka, M. 2000. Loss of function of a rice *brassinosteroid insensitive1* homolog prevents internode elongation and bending of the lamina joint. *The Plant Cell* 12: 1591–605.

Yan, S., Tang, Z., Su, W., and Sun, W. 2005. Proteomic analysis of salt stress-responsive proteins in rice root. *Proteomics* 5: 235–44.

Yan, S. P., Zhang, Q. Y., Tang, Z. C., Su, W. A., and Sun, W. N. 2006. Comparative proteomic analysis provides new insights into chilling stress responses in rice. *Molecular and Cellular Proteomics* 5: 484–96.

Yang, H., Ren, X., Weng, Q., Zhu, L., and He, G. 2002. Molecular mapping and genetic analysis of a rice brown planthopper (*Nilaparvata lugens* Stål) resistance gene. *Hereditas* 136: 39–43.

Yang, P. F., Liang, Y., Shen, S. H., and Kuang, T. Y. 2006. Proteome analysis of rice uppermost internodes at the milky stage. *Proteomics* 6: 3330–38.

Yang, P. F., Li, X. J., Wang, X. Q., Chen, H., Chen, F., and Shen, S. H. 2007a. Proteomic analysis of rice (*Oryza sativa*) seeds during germination. *Proteomics* 7: 3358–68.

Yang, Q. S., Wang, Y. Q., Zhang, J. J., Shi, W. P., Qian, C. M., and Peng, X. X. 2007b. Identification of aluminum-responsive proteins in rice roots by a proteomic approach: Cysteine synthase as a key player in Al response. *Proteomics* 7: 737–49.

Ye, X., Al-Babili, S., Kloti, A., Zhang, J., Lucca, P., Beyer, P., and Potrykus, I. 2000. Engineering the pro-vitamin A (beta-carotene) biosynthetic pathway into (carotenoid-free) rice endosperm. *Science* 287: 303–05.

Yonekura-Sakakibara, K., Tohge, T., Niida, R., and Saito, K. 2007. Identification of a Flavonol 7-O-rhamnosyltransferase gene determining flavonoid pattern in Arabidopsis by transcriptome coexpression analysis and reverse genetics. *Journal of Biological Chemistry* 282: 14932–41.

Yoshimura, S., Yamanouchi, U., Katayose, Y., Toki, S., Wang, Z. W., Kono, I., Kurata, N., Yano, M., Iwata, N., and Sasaki, T. 1998. Expression of *Xa1*, a bacterial blight resistance gene in rice, is induced by bacterial inoculation. *Proceedings of the National Academy of Sciences USA* 95: 1663–68.

Yu, J., Hu, S., Wang, J., Wong, G. K., Li, S., Liu, B., Deng, Y., et al. 2002. A draft sequence of the rice genome (*Oryza sativa* L. ssp. *indica*). *Science* 296: 79–92.

Yu, J., Wang, J., Lin, W., Li, S., Li, H., Zhou, J., Ni, P., et al. 2005. The Genomes of *Oryza sativa*: A history of duplications. *PLoS Biology* 3: e38.

Yuan, X., Xiao, S., and Taylor, T. N. 2005. Lichen-like symbiosis 600 million years ago. *Science* 308: 1017–20.

Zhang, G., Xu, Q., Zhu, X., Qian, Q., and Xue, H.-W. 2009a. *SHALLOT-LIKE1* Is a KANADI transcription factor that modulates rice leaf rolling by regulating leaf abaxial cell development. *The Plant Cell* 21: 719–35.

Zhang, L., Tian, L.-H., Zhao, J.-F., Song, Y., Zhang, C. J., and Guo, Y. 2009b. Identification of an apo-plastic protein involved in the initial phase of salt stress response in rice root by two-dimensional electrophoresis. *Plant Physiology* 149: 916–28.

Zhang, H. X., Lian, C. L., and Shen, Z. G. 2009c. Proteomic identification of small, copper-responsive proteins in germinating embryos of *Oryza sativa*. *Annals of Botany* 103: 923–30.

Zhang, X., Shiu, S., Cal, A., and Borevitz, J. O. 2008. Global analysis of genetic, epigenetic and transcriptional polymorphisms in *Arabidopsis thaliana* using whole genome tiling arrays. *PLoS Genetics* 4: 12.

Zhang, Y. C., Gong, S. F., Li, Q. H., Sang, Y., and Yang, H. Q. 2006. Functional and signaling mechanism analysis of rice *CRYPTOCHROME 1*. *The Plant Journal* 46: 971–83.

Zhao, K., Tung, C. W., Eizenga, G. C., Wright, M. H., Ali, M. L., Price, A. H., Norton, G. J., et al. 2011. Genome-wide association mapping reveals a rich genetic architecture of complex traits in *Oryza sativa*. *Nature Communications* 2: 467.

Zhou, C., Wang, J., Cao, M., Zhao, K., Shao, J., Lei, T., Yin, J., Hill, G. G., Xu, N., and Liu, S. 2005. Proteomic changes in rice leaves during development of field-grown rice plants. *Proteomics* 5: 961–72.

Zhou, J., Jiao, F., Wu, Z., Li, Y., Wang, X., He, X., Zhong, W., and Wu, P. 2008. *OsPHR2* is involved in phosphate-starvation signalling and excessive phosphate accumulation in shoots of plants. *Plant Physiology* 146: 1673–86.

Zhou, L., Zeng, Y. W., Zheng, W. W., Tang, B., Yang, S. M., Zhang, H. L., Li, J. J., and Li, Z. C. 2010. Fine mapping a QTL *qCTB7* for cold tolerance at the booting stage on rice chromosome 7 using a near-isogenic line. *Theoretical and Applied Genetics* 121: 895–905.

Zhou, P. H., Tan, Y. F., He, Y. Q., Xu, C. G., and Zhang, Q. 2003. Simultaneous improvement for four quality traits of Zhenshan 97, an elite parent of hybrid rice, by molecular marker-assisted selection. *Theoretical and Applied Genetics* 106: 326–31.

Zhu, T., Budworth, P., Chen, W., Provart, N., Chang, H.-S., Guimil, S., Zou, G., and Wang, X. 2003. Transcriptional control of nutrient partitioning during rice grain filling. *Plant Biotechnology Journal* 1: 59–70.

Zivy, M. and Vienne, D. 2000. Proteomics: A link between genomics, genetics and physiology. *Plant Molecular Biology* 44: 575–80.

Zou, J., Liu, C., and Chen, X. 2011. Proteomics of rice in response to heat stress and advances in genetic engineering for heat tolerance in rice. *Plant Cell Reports* 30: 2155–65.

Zou, J., Zhang, S., Zhang, W., Li, G., Chen, Z., Zhai, W., Zhao, X., et al. 2006. The rice *HIGH-TILLERING DWARF1* encoding an ortholog of *Arabidopsis MAX3* is required for negative regulation of the outgrowth of axillary buds. *The Plant Journal* 48: 687–98.

2

Practical Omics Approaches for Drought Tolerance in Rice

Prashant Vikram, PhD; B.P. Mallikarjuna Swamy, PhD; and Arvind Kumar, PhD

CONTENTS

2.1 Drought Stress: Present Scenario and Future Threats 47
2.2 Traditional Donors: Potential Source of Drought Tolerance Genes 48
2.3 Conventional Approaches in Breeding Rice for Drought Tolerance 49
2.4 Omics Approaches for Drought Tolerance 50
    2.4.1 Transcriptomics 50
    2.4.2 Proteomics 51
    2.4.3 Metabolomics 53
    2.4.4 Epigenomics 54
    2.4.5 Genomics 55
    2.4.6 Comparative Genomics 58
2.5 Transgenic Approach: Present Status and Future Prospects 58
2.6 Practical Marker Applications for Drought Tolerance in Rice 61
    2.6.1 Important Considerations in Molecular Breeding for Drought 61
    2.6.2 Drought Molecular Breeding Strategies: MABC and MAQP 62
    2.6.3 Future Perspectives 63
References 66

2.1 Drought Stress: Present Scenario and Future Threats

The rapidly increasing global population and enhanced usage of water for social and economic development have led to an alarming situation for sustainable crop production (Wallace et al. 2003). The agricultural sector is the largest consumer of freshwater resources, around 70%, and this is higher (90%) in developing regions of the world (Margat et al. 2005). The present global climate change trend, increasing population pressure, and shrinking water resources have created threats to crop production in different parts of the world, especially Asia, where two-thirds of the global population reside (Khush 2005).

Rice is one of the most important staple food crops of the world, accounting for more than 20% of global calorie intake. In Asia, where 90% of the world's rice is grown, it is estimated that rice supplies 35%–60% of the total calorie intake (Khush 1997). Rice is probably the most diversely cultivated crop, grown under irrigation, on rainfed sloping upland, on rainfed plain upland, on rainfed lowland, and in deepwater conditions. Rice is a semi-aquatic plant that requires two to three times more water than other food crops such as wheat or maize (Barker et al. 1999). Irrigated rice is by far the most common rice ecosystem, which occupies 55% of the total 158 Mha of cultivated rice area; rainfed lowland rice occupies 54 Mha (34%), rainfed upland rice occupies 14 Mha (9%), and flood-prone areas occupy 11 Mha (7%) (Bouman et al. 2007). Rainfed rice accounts for 40% of the total rice area in Asia. South Asia alone holds 37% of the world's rice area, 50% of which is rainfed (Dawe et al. 2010). The upland ecosystem represents 12% of global rice area and it is the lowest yielding rice ecosystem (Khush 1997). Rainfed rice is also important in sub-Saharan Africa, where it accounts for 84% of the total rice area (Gauchan and Pandey 2012).

The current scenario of global climate change, unpredictable rainfall patterns, and uneven distribution of rainwater leads to severe drought spells in rainfed areas. Although water covers almost 70% of our planet, freshwater resources are limited. It has been estimated that the amount of crop water consumption will be increased by 70%–90% in 2050 and will reach 12,050–13,500 km^3, from the present 7,130 km^3 (de Fraiture et al. 2007). Another estimate predicted that in the next 25 years approximately 15–20 Mha of irrigated rice will suffer water scarcity (Rice Today 2009). It is highly likely that rainfed rice-growing areas will undergo severe spells of drought stress, resulting in a high decline in yield.

Alternative strategies are required to combat the forthcoming drought stress in rainfed environments. Apart from managing water resources, genetic improvement of rice cultivars could be a good option to overcome the problem of drought stress. The development of rice cultivars that can withstand severe drought stress is needed for sustaining rice production in rainfed environments. Farmers in rainfed areas in South Asia are growing rice varieties originally bred for irrigated environments because of their high genetic yield potential (Vikram et al. 2011). This is because the drought-tolerant traditional varieties and landraces grown earlier in such areas had lower yield potential under normal irrigated situations and did not possess very high-quality traits. Genomic regions conferring tolerance to drought in drought-tolerant traditional varieties/landraces could be suitable candidates for genomics-assisted breeding to improve the yield of high-yielding but drought-susceptible rice varieties.

2.2 Traditional Donors: Potential Source of Drought Tolerance Genes

Genetic improvement of existing rice varieties for drought stress could be carried out through mining genes and alleles conferring tolerance of drought. Sources of these genes and alleles could be landraces, wild species, or improved drought-tolerant breeding lines. Drought-tolerant landraces such as Dular, Nagin-22, Dhagaddeshi, Aus 276, Kali Aus, Birsa gora, Vandana, and several commonly used wild accessions could be considered as traditional donors. A number of varieties have been developed through selection from landraces. For example, Nagina-22, a well-known drought-tolerant rice cultivar, was developed through selection from landrace "Rajbhog" grown in the foothills of Nepal (Vikram et al. 2011). Laloo-14 is another drought-tolerant Indian rice variety bred by selection from landraces and released for cultivation in rainfed areas of central India (Madhya Pradesh). Landraces have immense genetic potential in rice improvement for rainfed areas. The genetic diversity harbored in landraces enables them to adapt to a wide range of agroecological niches. Vanniarajan et al. (2012) listed some drought-tolerant Indian landraces. Singh et al. (2005) explored some landraces in the Vindhya hills such as Argenta, Bejhari, Jalhaur, Karghi, Karanfool, Lal Basmati, Lamchoor, Safed Sarkari, and Samasar Dhan, which are well known to farmers as drought-tolerant cultivars. Puri et al. (2010) characterized

TABLE 2.1

Wild *Oryza* Species Characterized for Drought Tolerance

Species	Trait(s)	References
Oryza rufipogon	Water loss, osmotic potential, electrolytical leakage, MDA content, soluble sugar content, and leaf temperature	Zhang et al. (2006)
O. longistaminata	Greater stomatal conductance	Liu et al. (2004)
O. australiensis	Fewer tillers, increased leaf area, carbohydrate in culm and rhizomes before stress, and faster remobilization during stress	Chaturvedi et al. (1996)
O. nivara	Drought avoidance	Khush (1997)
O. breviligulata	Drought avoidance	Khush (1997)
O. longistaminata	Drought avoidance	Khush (1997)
O. meridionalis	Drought avoidance	Khush (1997)
O. rhizomatis	Drought avoidance	Khush (1997)
O. australiensis	Drought avoidance	Khush (1997)

landrace Kataush from the Tarai region of Nepal for drought tolerance. Identification and characterization of drought-tolerant landraces in the centers of diversity of rice would be very useful for further improvement of rice cultivars for rainfed environments.

Wild species of rice are often used for rice genetic improvement. The two cultivated species of rice are *Oryza sativa* and *O. glaberrima*. It is believed that *O. sativa* evolved from *O. rufipogon* and *O. glaberrima* from *O. longistaminata* (Khush 1997). Also, it is likely that landraces derived their drought tolerance from these wild species. *O. rufipogon* and *O. longistaminata* were reported to show high stomatal conductance (Liu et al. 2004); *O. rufipogon* performed well for drought-related traits such as osmotic potential, electrolyte leakage, soluble sugar content, and leaf temperature (Zhang et al. 2006). In a physiological study carried out with three wild species, *O. australiensis* was found to be more tolerant of drought. It has more drought-adaptive traits such as fewer tillers, reduced leaf area, higher accumulation of carbohydrate in culm and rhizomes before stress, and faster remobilization during stress (Chaturvedi et al. 1996). A number of wild species reviewed for drought avoidance are presented in Table 2.1. *O. glaberrima*, mainly cultivated in West Africa, was used to develop New Rice for Africa (NERICA) and some of these varieties were found to be drought tolerant. Wild species could also be helpful in enhancing the genetic yield potential of existing rice varieties.

Many high-yielding but drought-susceptible varieties bred for the irrigated ecosystem such as Swarna, MTU1010, IR64, IR36, IR72, BR11, and TDK1 are grown on millions of hectares in the rainfed ecosystem (Vikram et al. 2011). Efforts are being made to improve such genotypes so as to develop rice varieties that combine high yield potential with good yield under drought and that also possess high-quality traits.

2.3 Conventional Approaches in Breeding Rice for Drought Tolerance

Drought is a highly complex trait affected by both the amount of water and the duration and timing of rainfall. The rice plant employs several mechanisms (escape, avoidance, tolerance) to cope with drought stress. These mechanisms vary by cultivar: for example, thickening of cuticles (to reduce evapotranspiration), design of root systems (to enable the plant to absorb water from deeper soil layers), osmotic adjustment (to enable plants to maintain their turgor pressure during droughts) and, finally, the plant's ability to recover from drought if it is tolerant of desiccation (Nguyen et al. 1997). Efforts have also been made to work on drought-related traits that can be used for breeding purposes. Constitutive root traits (rooting depth, root thickness, branching angle, and root distribution pattern), osmotic adjustment, and several other secondary traits were earlier reported to be effective in breeding for drought tolerance (Kato et al. 2006; Nguyen et al. 1997; Babu et al. 2001). However, limited success has been achieved in improving yield under drought through the use of these traits.

Direct selection for high grain yield under drought is an alternative criterion. This trait has moderate to high heritability and therefore can be used as a suitable selection criterion (Kumar et al. 2007, 2008). Studies conducted at the International Rice Research Institute (IRRI) in recent years have demonstrated the effectiveness of this trait (Atlin and Lafitte 2002; Bernier et al. 2008; Venuprasad et al. 2007; Kumar et al. 2008). Several varieties combining high yield potential and good yield under drought have been developed by employing direct selection for grain yield under drought for cultivation in India, the Philippines, Bangladesh, and Nepal (Kumar et al. 2012).

2.4 Omics Approaches for Drought Tolerance

Omics approaches can be grouped into different categories: (1) transcriptomics, (2) proteomics, (3) metabolomics, (4) epigenomics, (5) genomics, and (6) comparative genomics.

2.4.1 Transcriptomics

Transcriptomics is the study of the transcriptome of an individual. The transcriptome represents the total set of transcripts in a given organism. In-depth studies of quantitative trait loci (QTLs) and genes conferring drought tolerance in rice have revealed the role of transcription factors (TFs). TFs involved in dehydration response can be classified into two broad groups: (a) ABA dependent and (b) ABA independent. ABA-dependent TFs in rice further include two families: (1) basic leucine zipper (bZIP) and (2) NAM, ATAF, CUC2 (NAC). The bZIP class of TFs has ABA-responsive element-binding protein/factors (*AREB/ ABF*) with a consensus sequence ACGTGG/TC, which plays an important role in ABA signaling during dehydration (Yang et al. 2010). *ABF3*-mediated drought tolerance was found to be better than *DREB1A/CBF3*-mediated drought tolerance (Oh et al. 2005). The other TF in this group, NAC, has a highly conserved DNA-binding NAC domain (Palaniswamy et al. 2006; Guo et al. 2008). The *SNAC1* gene in rice is a typical example of this family. Hu et al. (2006) reported that *SNAC1* transgenic rice plants showed higher spikelet fertility and higher seed setting rate under severe drought stress. The ABA-independent class also has two TF families in this group: (1) zinc fingers and (2) AP2/ERFs. Zinc fingers have zinc ions that coordinate with motifs to stabilize the protein folds. Examples of zinc fingers in rice are *ZFP252*, *Zat10/STZ*, and *WRKY* genes (Xu et al. 2008; Xiao et al. 2009; Wu et al. 2009). *STZ* was reported to increase spikelet fertility and grain yield under drought (Xiao et al.

2009). The APETALA 2/ethylene response factor (AP2/ERF) group of TFs has flowering pattern proteins AP2 and C-repeats (CRTs), which are dehydration responsive, and a conserved *cis*-element, A/GCCGAC, in promoters of target genes. The *AtDREB1A/CBF3* gene was overexpressed in rice with drought-inducible promoter *OsHVA22P* and the transgenic rice significantly improved spikelet fertility and yield per plant. The *ZAT10:OsHVA22P* cassette also performed well under managed drought conditions in the field. Some of the TFs and genes encoding them are summarized in Figure 2.1. Although a number of TFs have been reported in rice responding to drought stress, their successful deployment in developing transgenic plants in order to obtain substantial yield gains in field conditions is still awaited.

2.4.2 Proteomics

The set of proteins that are expressed in a particular cell or organism at a particular time is called the *proteome*. The large-scale characterization of an entire protein complement of a cell line, tissue, or organism is called proteomics (Wasinger et al. 1995). It is important to note that the proteome is dynamic and it depends on the immediate environment where it was studied. Therefore, a given genome could have an infinite number of proteomes. Unlike genomics, there is little progress in proteomic applications in rice. Proteome maps are the prerequisite for gel-based proteomic analysis. Several workers developed proteome maps in rice. Koller et al. (2002) developed a proteome map after systematic analysis of rice roots, leaves, and seeds. Such maps were also developed using rice samples at different developmental stages (Komatsu et al. 1999; Nozu et al. 2006; Tanaka et al. 2005). The Rice Proteome Database is already available (Komatsu et al. 2004; National Institute

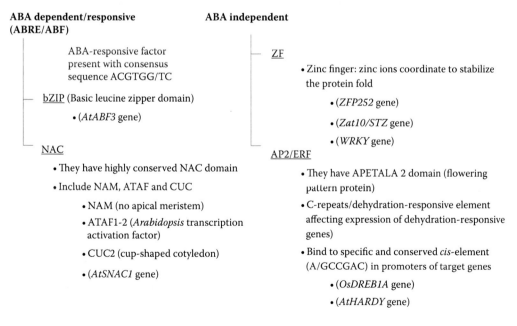

FIGURE 2.1
Drought-responsive transcription factors in rice.

of Agrobiological Sciences). This information could be used in comparing the proteomic analysis of drought-stressed and control plants in such a way as to identify drought-related proteins.

Several workers have carried out proteomic analysis in rice to identify drought-related proteins. Salekdeh et al. (2002) studied the proteome of rice leaves under drought stress. Two cultivars, CT9993 (drought-tolerant) and IR62266 (drought-susceptible), were drought-stressed and subjected to proteome analysis. A total of 27 protein spots showing a differential response pattern were identified in the two genotypes. Using MALDI-MS or ESI-MS/MS, 16 drought-responsive proteins were identified. Among upregulated proteins were an S-like RNase homolog, an actin-depolymerizing factor (ADF), and Rubisco activase, whereas an isoflavone reductase-like protein was found to be downregulated under drought stress. A differential response of cytosolic Cu–Zn superoxide dismutase (SOD) was also observed. The upregulation of cytosolic Cu–Zn SOD under water-deficit conditions was observed in several other studies (Hajheidari et al. 2005; Plomion et al. 2006; Gazanchian et al. 2007; Ke et al. 2009). Babu et al. (2004) isolated late embryogenesis abundant (LEA) proteins from barley and analyzed their role in rice. Ali and Komatsu (2006) conducted a comparative proteome analysis between drought-tolerant cultivar Zonghua 8 and drought-susceptible variety Nipponbare. The tissue targeted in the experiment was the leaf sheath, which was drought-stressed. The downregulation of small and large subunits of Rubisco enzyme and upregulation of SOD, light-harvesting complex chain II (LHC), PSII oxygen-evolving complex protein (OEC), 2-cys peroxiredoxin, and ADF were reported under drought. Rabello et al. (2008) identified 22 proteins putatively associated with drought tolerance in rice through the subtractive DNA hybridization approach. Ke et al. (2009) revealed the upregulation of *LEA*-like protein and Cu–Zn SOD and downregulation of Rieske Fe–S precursor protein under drought-stress conditions. Salekdeh et al. (2002) also reported the downregulation of Rieske Fe-S precursor protein. Ke et al. (2009) reported the alteration in phosphorylation patterns of proteins under drought stress. Such proteins included NAD-malate dehydrogenase, a G-protein β subunit, an abscisic acid- and stress-inducible protein, an S-like ribonuclease, and a poorly defined ethylene-inducible protein. The dephosphorylation of r40C1 and germin-like protein was also observed in their study. Liu and Bennett (2010) studied alterations in proteomes of anthers under drought stress using drought-tolerant rice cultivar Moroberekkan and drought-susceptible genotype IR64. Eight drought-induced proteins were identified in the study, including glyceraldehyde-3-phosphate dehydrogenase, β expansin, and actin-binding proteins. An interesting finding in their study was reversion of most of the induced proteins in IR64 compared with Moroberekkan. Jian-Hua et al. (2010) carried out proteomic analysis with drought-tolerant pyramided lines and identified some differentially displayed proteins. Muthurajan et al. (2010) reported 31 drought-responsive proteins in an experiment conducted under controlled drought.

Overexpression of LHC proteins under drought-stress conditions was also revealed in another study (Li et al. 2002). The higher accumulation of LHCs under abiotic stresses is also reported in other crops (Peng et al. 2009). The amount of ADF protein was enhanced in all organs of rice plants under drought stress (Salekdeh et al. 2002; Ali and Komatsu 2006). A differential response of ADF proteins under salt stress is reported (Yan et al. 2005). It could be easily concluded that the proteins that play an important role in drought stress response are cytosolic Cu–Zn SOD, LEA, small and large subunits of RuBisCO, LHC, OEC, ADF, and Rieske Fe–S precursor protein. A list of up- and downregulated proteins appears in Table 2.2.

TABLE 2.2

List of Up- and Downregulated Proteins under Drought in Rice

Protein Name	Tissue	Regulation	References
Isoflavone reductase-like protein	Leaf	Downregulated	Salekdeh et al. (2002)
r40C1 protein	Leaf	Downregulated	Ke et al. (2009)
Rieske Fe-S precursor protein	Leaf	Downregulated	Ke et al. (2009)
Rieske Fe-S protein	Leaf	Downregulated	Salekdeh et al. (2002)
Rubisco large subunit	Leaf sheath	Downregulated	Ali and Komatsu (2006)
Rubisco small subunit	Leaf sheath	Downregulated	Ali and Komatsu (2006)
2-cys peroxiredoxin	Leaf sheath	Upregulated	Ali and Komatsu (2006)
Actin-binding proteins	Anther	Upregulated	Liu and Bennett et al. (2010)
Actin-depolymerizing factor	Leaf sheath	Upregulated	Ali and Komatsu (2006)
Actin-depolymerizing factor	Leaf	Upregulated	Salekdeh et al. (2002)
Chloroplast ATPase	Leaf sheath	Upregulated	Ali and Komatsu (2006)
Cu–Zn superoxide dismutase, chloroplast	Leaf	Upregulated	Ke et al. (2009)
Cu–Zn superoxide dismutase, cytosolic	Leaf	Upregulated	Salekdeh et al. (2002)
Fructose bisphosphate aldolase	Leaf	Upregulated	Salekdeh et al. (2002)
Glutathione dehydroascorbate reductase	Leaf	Upregulated	Salekdeh et al. (2002)
Glyceraldehyde-3-phosphate dehydrogenase	Anther	Upregulated	Liu and Bennett et al. (2010)
Heterotrimeric G protein β subunit	Leaf	Upregulated	Ke et al. (2009)
Late embryogenesis abundant (LEA)	Leaf	Upregulated	Ke et al. (2009)
Light harvesting complex chain II	Leaf sheath	Upregulated	Ali and Komatsu (2006)
NAD-malate dehydrogenase	Leaf	Upregulated	Ke et al. (2009)
Nucleoside diphosphate kinase	Leaf	Upregulated	Salekdeh et al. (2002)
Oxygen-evolving enhancer protein 2	Leaf sheath	Upregulated	Ali and Komatsu (2006)
PSII oxygen-evolving complex protein	Leaf sheath	Upregulated	Ali and Komatsu (2006)
Rubisco activase	Leaf	Upregulated	Salekdeh et al. (2002)
Serine hydroxylmethyltransferase I	Leaf sheath	Upregulated	Ali and Komatsu (2006)
Superoxide dismutase	Leaf sheath	Upregulated	Ali and Komatsu (2006)
Triose phosphate isomerase	Leaf	Upregulated	Salekdeh et al. (2002)

2.4.3 Metabolomics

The term *metabolome* was first used in the late 1990s (Oliver et al. 1998), referring to the change in the relative concentration of metabolites as a result of the deletion or overexpression of a gene. Comprehensively, metabolome refers to the entire set of metabolites present in a given biological sample. These include metabolic intermediates, hormones, signaling molecules, and secondary metabolites. Metabolomics is the study of the metabolome, which provides an integrated view of the functional status of an organism. Metabolomic studies are done with different analytical techniques (chromatographic techniques, spectroscopic techniques, or the two combined) such as HPLC-MS, GC-MS, CE-UV, and HPTLC (chromatographic) and LC-MS-NMR, GC-IT-MS-MS, and LC-MSMDF (combined) developed in recent years (Abou-Donia et al. 2007; Gotti et al. 2006; Llop et al. 2010). Metabolomic analysis has potential in stress genomics. Differences in metabolite content of stressed and nonstressed plants could be used as an indicator in screening rice genotypes for different abiotic stresses (Genga et al. 2011). Metabolomic changes in plants that undergo stress could be due to an adjustment of concentration of metabolites to restore homeostasis and synthesis/accumulation of compounds that mediate the tolerance mechanism (Genga et al. 2011). Such metabolomic changes could be worked out using the above-mentioned analytical techniques. Metabolites associated with stress phenomena are sugars (i.e., sucrose, glucose, fructose, and trehalose), polyols, betaines, and amino acids such as proline (Chen and Murata 2008; Szabados and Savouré 2010; Shulaev et al. 2008; Smirnoff 1998). Metabolites also involve the compounds playing an important role in stabilizing the photosystem II complex, protecting the structure of enzymes and proteins, maintaining membrane integrity, scavenging reactive oxygen species, acting as chelating agents, redesigning lipids, and sending signals to molecules (Alcázar et al. 2010; Valluru and Van den Ende 2008).

Metabolomics could be widely used in abiotic stresses such as drought. For example, abscisic acid (ABA) is synthesized in plants in response to drought stress. Based on this, stress-signaling pathways have been categorized as ABA-dependent and ABA-independent ones (Yamaguchi-Shinozaki and Shinozaki 2006). Transcription factors were also categorized accordingly. In another example, two rice cultivars (Arborio and Nipponbare) were given osmotic stress and studied via analytical techniques such as nuclear magnetic resonance (NMR) by Fumagalli et al. (2009). The authors suggested differential regulation of sugar and glutamine–glutamate metabolism in response to abiotic stress. Shu et al. (2011) performed a comprehensive genomic, proteomic, and metabolomic analysis of rice seedlings in response to drought stress and concluded that energy consumption was enhanced from a substance stored under drought stress. In particular, there was an increase in the transfer of energy from carbohydrates and fatty acids to amino acids. Also, the expression of enzymes involved in anabolic pathways corresponding to some amino acids increased. Similar studies in other crops such as tomato, castor, and cotton were carried out to work out changes in metabolomes in response to drought stress (Semel et al. 2007; Babita et al. 2010; Levi et al. 2011). The information available from other crops and rice could be helpful in future drought-related metabolomic studies.

2.4.4 Epigenomics

It is well known that plants show acclimation responses resulting in the retention of stress memory for a short span of time (Thomashow 1999). Short-duration stress memories depend on the half-life of stress-induced proteins, RNAs, and metabolites. These stress memories could be longer if there were reprogramming in phenology and morphology. Longer stress memories could be attributed to epigenetic processes (Chinnusamy and Zhu 2009). Heritable (and stable) DNA methylation, histone modifications, and effects of small RNAs are well-known epigenetic processes. Epigenomics involves the study of the molecular mechanisms underlying epigenetics using techniques such as chromatin immunoprecipitation, ChiP-sequencing (this method combines ChiP with next-generation sequencing), methylated-DNA immunoprecipitation, and shotgun bisulfite sequencing. Epigenetic mechanisms have been reported to be associated with drought in *Arabidopsis* and rice (Chinnusamy and Zhu 2009). Wang et al. (2011) studied the DNA methylation pattern between two rice genotypes: DK151 (a drought-tolerant IR64 derivative) and IR64. The study concluded that DNA methylation alteration sites are specific to genotypes, developmental stages, and tissues. They also found that changes in DNA methylation due to drought stress may or may not be reversible. Histone modification such as deacetylation is a well-known phenomenon. Histone deacetylases (HDACs) are the enzymes that mediate histone deacetylation in response to biotic and abiotic stresses. Fu et al. (2007) reported several HDACs in rice. Though efforts have been made in epigenomics, little progress is achieved. The role of epigenetic mechanisms in stress tolerance has long been debated; however, recent developments and reports clearly indicate that epigenomics is an important area. This approach could be helpful in explaining the complexity of drought stress responses in rice.

2.4.5 Genomics

Genomics is the study of the structure and function of genes with the ultimate aim of using them in crop improvement programs. The resources involved in structural and functional genomics studies are QTLs, mutant libraries, full-length cDNAs, expression profiles, and sequence data sets (Jiang et al. 2011). Several genes and QTLs have been cloned/identified for drought tolerance in rice, as reviewed in some recent publications (Vikram et al. 2012a). Quantitative trait loci are one of the most important genomic resources not only for practical breeding applications but also for functional genomics studies. Numerous QTLs were identified for drought tolerance in rice but are seldom used in drought breeding programs due to the phenotyping constraints. Trait selection and precise phenotyping play a crucial role in functional genomics studies on drought. Phenotypic screening for drought tolerance has been carried out in the laboratory using polyethylene glycol (PEG) or in field conditions following physiological or grain-yield-related traits (Kato et al. 2008; Venuprasad et al. 2007; Kumar et al. 2008). Recently, several QTLs were identified for high grain yield under drought. Since drought grain yield QTLs can be directly used for enhancing grain yield under drought stress, these have advantages over the ones identified for secondary traits. Such QTLs were identified for rainfed uplands as well as rainfed lowlands (Bernier et al. 2007; Venuprasad et al. 2009; Vikram et al. 2011). These two drought-prone ecosystems are entirely different and QTLs/genes associated with drought tolerance also express differently. A QTL identified for the upland drought ecosystem may or may not perform well under lowland drought stress. Also, drought QTLs are usually specific to particular genetic backgrounds as well as environments. Therefore, the most suitable QTL will be the one that can perform well in different genetic backgrounds, ecosystems, and environments.

Interestingly, some drought grain yield QTLs are performing well across different genetic backgrounds, ecosystems, and environments. The identification of such QTLs in rice has been possible through the study of a large number of mapping populations and by following a bulked segregant analysis (BSA) approach in genotyping multiple populations simultaneously to reduce genotyping costs as well as effort (Vikram et al. 2012b). Bulked segregant analysis involves genotyping of DNA bulks and parents so that fewer samples are genotyped (see Figure 2.2). Several drought grain-yield-related genomic regions have been identified in rice and are summarized in Figure 2.3. Among them, $qDTY_{1.1}$ showed a consistent effect in approximately 10 different mapping populations (Kumar et al. 2007; Vikram et al. 2011; Ghimire et al. 2012; Venuprasad et al. 2012) under both upland and lowland drought (Venuprasad et al. 2012). This QTL was also responsible for increased yield in irrigated conditions (Vikram et al. 2011). Another QTL identified for aerobic adaptation as well as lowland drought stress is $qDTY_{3.1}$, contributing a genetic variance of around 32% (Venuprasad et al. 2009). Chromosome 3 also harbors an important region, $qDTY_{3.2}$, contributing to a significant increase in yield under drought stress. $qDTY_{3.2}$ was first reported for high grain yield under lowland drought stress (Vikram et al. 2011). Later, an effect under upland drought stress was also observed. Interestingly, markers within this QTL region showed a digenic interaction with another drought grain yield QTL, $qDTY_{12.1}$

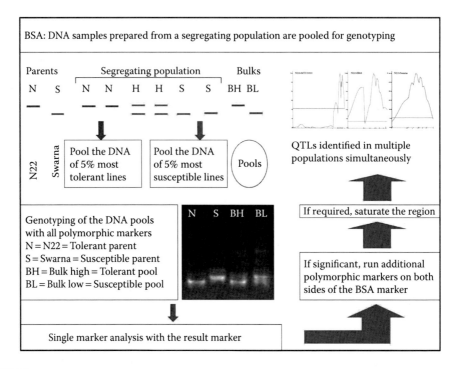

FIGURE 2.2
Bulked segregant analysis for high grain yield under drought in rice.

(Dixit et al. 2012). There was a significant increase in yield due to an additive interaction between the two loci. $qDTY_{3.2}$ also showed a significant additive interaction effect with $qDTY_{1.1}$ in two N22-derived populations (Prashant Vikram, CIMMYT, personal communication). This region was also found to be associated with days to flowering and plant height in an N22/Swarna RIL population. The pleiotropic and interaction effects of $qDTY_{3.2}$ render it an interesting candidate in further genomic analysis for drought tolerance.

The largest effect drought grain yield QTL identified so far is $qDTY_{12.1}$, explaining a genetic variance of around 50%. $qDTY_{12.1}$ was identified in a population derived from

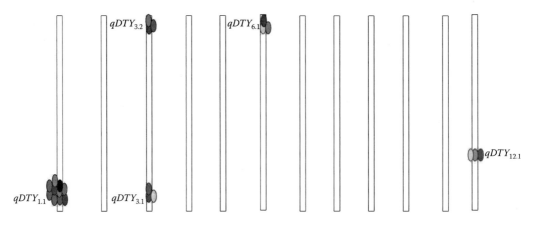

FIGURE 2.3
Genomic regions harboring consistent drought grain yield QTLs.

two upland-adapted cultivars—Vandana and Way Rarem—and the QTL was originally believed to be specified for upland adaptation (Bernier et al. 2007). However, recently the effect of this QTL was found to be significant under lowland drought stress and in multiple environments (Mishra et al. 2013).

It is worth mentioning that donor and recipient backgrounds play a crucial role in QTL studies on drought grain yield. The drought grain yield QTLs identified so far are usually from populations that derive their drought tolerance from either landraces or traditional donors. For example, the positive allele for $qDTY_{1.1}$ was contributed by Nagina 22 (or N22), which is an aus variety developed through selection from landrace Rajbhog grown in the foothills of Nepal (Vikram et al. 2011). Apo, which is a drought-tolerant variety of the Philippines, also contributed a positive allele for the effect of $qDTY_{1.1}$ (Venuprasad et al. 2012). Apo was developed using an Indonesian traditional variety, Benong. $qDTY_{1.1}$ was contributed by an Indian upland variety, Vandana, which was developed through the cross of landrace Kalakeri and improved breeding line C22. Another Indian landrace, Dhagaddeshi, is also reported to be one of the donors for $qDTY_{1.1}$ (Ghimire et al. 2012). Ghimire et al. (2012) also reported that Apo, Dhagaddeshi, and N22 shared a common allele for the peak marker of $qDTY_{1.1}$. In addition to landraces and traditional varieties, $qDTY_{1.1}$ was contributed by breeding lines such as CT9993 (Kumar et al. 2007). Quantitative trait loci for some physiological traits have been identified in the $qDTY_{1.1}$ region, which was also contributed by a landrace. Nootripathu, which is a landrace of Tamil Nadu Province of India, was used as a drought-tolerant donor to identify the QTLs associated with the physiological traits conferring drought tolerance (Gomez et al. 2010). Similarly, $qDTY_{3.2}$ was contributed by several traditional donors, such as N22 and Vandana (Vikram et al. 2011; Dixit et al. 2012), and breeding lines IR77298-5-6-18 and IR74371-46-1-1 (Mishra et al. unpublished). It is highly likely that the positive alleles contributed by N22 and Vandana derived their drought tolerance from landraces Rajbhog and Kalakeri, respectively (Vikram et al. 2011). The donor parent and the recipient parent to be used in population development are equally important. It was long believed that contrasting parents should be used in population development for QTL identification. The higher the contrast between two parents is, the higher the QTL effect will be. However, unless a drought QTL has somewhat similar effects in different genetic backgrounds, it will not be useful in practical breeding applications. For a study on $qDTY_{1.1}$, in which three populations with three different recipient backgrounds using a common donor (N22) were used, it was revealed that the additive effect of the QTL was inversely proportional to the drought tolerance of the recipient parent. The additive effect was highest in a Swarna background, followed by IR64, and lowest in MTU1010 (Vikram et al. 2011). Swarna was the most drought-susceptible, followed by IR64, and MTU1010 was the most tolerant of drought among three varieties (Swarna, IR64, and MTU1010). Therefore, a mapping population should involve the target variety as a recipient parent so that QTLs identified could directly enter MAB programs following one or more backcrosses.

It is noteworthy that one of the drought QTL donor parents, N22, which contributed yield-enhancing alleles under drought stress, also served as a potential resource material for identifying differentially expressed genes (DEGs) under drought stress. N22 has been used extensively in drought functional genomics studies. Several drought-responsive expressed sequence tags (ESTs) and candidate genes for drought tolerance in rice were identified from N22 (Gorantla et al. 2007). Comparative analysis of the expression profiles of N22 and IR64 revealed DEGs under drought (Lenka et al. 2011). Some of them lie in the $qDTY_{1.1}$ region, including 4,5-DOPA dioxygenase extradiol (LOC_Os01g65690), glycosyl transferase (LOC_Os01g65780), amino acid transporters (LOC_Os01g66010), the

MADS-box family gene (LOC_Os01g66290), and serine/threonine protein kinase (LOC_Os01g66860). Amino acid transporters and protein kinases are the most likely candidate genes (Vikram et al. 2011).

QTL identification and expression analysis can be coupled to have a robust estimate of the genes conferring drought tolerance in landraces and traditional donors. A similar approach has been applied in rice for salt tolerance (Pandit et al. 2010) and in wheat for drought tolerance (Kadam et al. 2012). Quantitative trait loci can be identified in RIL/BIL populations and NILs can be developed in a fast-track mode using marker-assisted backcrossing (MABC). Further, the NILs can enter a product delivery system in order to use them as base materials for the expression profiling studies. With the available genomic resources, a significant yield increase can be achieved in drought-prone environments and information pertaining to the physiological mechanisms and genes related to drought tolerance can be obtained.

2.4.6 Comparative Genomics

A comparison of the structure and function of genes across the species unveils the evolutionary history of a crop species. Genetic information in different crop species could be compared for better understanding of the conserved genomic regions in plants. The phylogenetic relationship of members of grasses (or Poaceae family) is well understood now. Several syntenic regions have been identified among different crop species. For example, two-thirds of the genes on chromosome 11 have been found to be distributed on six homologous groups of wheat with a high level of rearrangements in the grass genomes (Singh et al. 2004). Such syntenic gene(s) or genomic regions could be analyzed in detail to investigate their role in plant growth, development, and adaptation across species. One of the most fascinating examples is the syntenic relationship of a major drought grain yield QTL, $qDTY_{1.1}$. This QTL region showed similarity with a QTL in maize on chromosome 3, in wheat on chromosome 4B, and in barley on chromosome 6H (Swamy et al. 2011). It is interesting to note here that grain yield QTL $qDTY_{1.1}$ in rice harbors the green revolution gene *sd1* (Swamy et al. 2011). A similar QTL in wheat (*qGY.4B.1*) that increases grain yield under drought stress harbors another green revolution gene, *Rht1* (Kadam et al. 2012). Both QTLs (*qDTY_{1.1}* and *qGY.4B.1*) were associated with plant height (Vikram et al. 2011; Kadam et al. 2012). Swamy et al. (2011) also reported that the nature of a rice drought grain yield QTL is homologous with that of maize.

A draft genome sequence of some cereal crops, including rice, wheat, and maize, is already available. In addition, different types of sequence data sets of different crops especially related to expressed genes/proteins could serve as a resource for comparative genomic analysis. Next-generation sequencing technology offers a cost-efficient alternative with enormous potential for comparative genomics. An approach involving comparative genomics followed by expression analysis in different species simultaneously could be very useful in targeting the conserved genomic regions associated with drought tolerance in cereal crops. Since rice is highly susceptible to drought compared with other cereal species, such strategies would be quite advantageous.

2.5 Transgenic Approach: Present Status and Future Prospects

To understand the complexity of drought tolerance in rice, transgenic technology is another alternative, and the most common strategy is to overexpress drought-responsive genes.

Several genes have been identified related to signal transduction/regulation, posttranslational modification, and production of metabolites (Yang et al. 2010). These genes were overexpressed in transgenic rice using constitutive and drought-inducible promoters.

Genes conferring drought tolerance have been classified recently into three broad groups: (1) genes associated with metabolites and osmoprotectants, (2) genes related to posttranscriptional/translational modifications, and (3) regulatory genes (Yang et al. 2010). The first group associated with structural or functional genes and their mode of action was comparatively simple. Therefore, initial transgenic studies were based on these genes. Such genes encoded products such as *LEA* proteins, heat shock proteins, and carbohydrates such as proline and phytohormone ABA (Xu et al. 1996; Xiao et al. 2007, 2009; Sato and Yokoya 2008; Zhu et al. 1998). Genes related to posttranslational regulation have been categorized into two broad groups based on their respective functions. One group is related to farnesylation, which involves a gene such as *OsCDPK7* (see Figure 2.4). This gene plays a significant role in the expression of the *LEA* gene (Saijo et al. 2000). The other group relates to protein phosphorylation. One of the protein kinases, *OsCIPK03*, is reported to be an example of this group (Yang et al. 2010). Regulatory elements play a key role in drought response in rice. Transcription factors usually regulate the expression of multiple downstream target genes under drought stress. These elements interact with specific *cis*-elements in the promoter regions of target genes. The whole system involving TFs and *cis*-elements of promoter regions of target genes is considered as a unit called a *regulon*. The major regulons studied for drought responses in rice are *DREB1/CBF*, *DREB2*, *AREB*, and *NAC*. For example, in a recent gene expression profiling using rice near isogenic lines under drought stress, the role of the *NAC* family was emphasized (Nuruzzaman et al. 2012).

Although a number of regulatory elements have been identified and characterized in rice, the development of drought-tolerant transgenic rice is still a challenge. Major concerns in the process are the choice of promoters and a better understanding of molecular, physiological, and metabolic aspects of the target trait. Several studies carried out in the past found constitutive promoters to be poor performers compared with specific promoters. For example, overexpression of TF *DREB1A* was studied using the constitutive CaMV35S promoter as well as an inducible promoter from *rd29A*. The transgenic plants with the inducible promoter performed better under drought than the ones with constitutive promoters (Kasuga et al. 1999). Overexpression of *AtDREB1A* with drought-inducible promoter *OsHVA22P* provided good results under water stress. Expression of genes in specific tissues and at a particular developmental stage in response to drought stress is crucial. Also, the long- and short-term effects of genes and their effects at the cellular level as well as whole-plant phenotype under stress are important. Evaluation of transgenic plants is usually done in a greenhouse under controlled conditions and is seldom done in the field. Transgenics need to undergo evaluation for biosafety regulations, which also delays the process.

As far as the success of transgenic technology is concerned, there are two schools of thought. Some workers believe that transgenic technology is still an attractive option for drought tolerance in rice (Bhatnagar-Mathur et al. 2008; Yang et al. 2010), whereas others have the opposite view (Pray et al. 2011). Bhatnagar-Mathur et al. (2008) suggested a few points to consider for enhancing the efficiency of the transgenic approach: (1) targeting multiple-gene regulation should be preferred over single-gene targeting, (2) the biological cost of the production of different metabolites and their effect on stress tolerance as well as a proper assessment of yield under stress, and (3) physiological responses to drought stress must be assessed and promoters could be chosen accordingly. It is noteworthy that

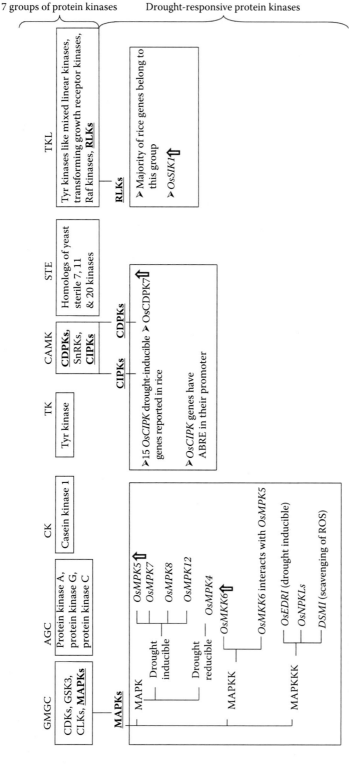

CDKS, calmodulin-dependent kinases; GSK3, glycogen synthase kinase 3; CLKs, CDC-like kinases; MAPK, mitogen-activated protein kinases; CDPKs, calmodulin/calcium-dependent protein kinases; SnRKs, sucrose non-fermenting-related protein kinases; CIPKS, calcineurin B-like protein interacting protein kinases; RLKs, receptor-like kinases.

FIGURE 2.4
Protein kinases responsive to drought in rice.

transgenic research has continued for the past 20 years in different parts of the world, but the only available commercial product is Bt rice meant for insect resistance released in China recently (Tian et al. 2012). Researchers working on transgenics are discouraged because of these constraints, particularly the expression patterns of the transferred genes. This has reduced the enthusiasm of researchers to put effort into developing transgenics for complex traits such as drought tolerance in rice (Pray et al. 2011).

There is no doubt that drought is a complex trait, and a comprehensive knowledge of the molecular, physiological, and biochemical basis of drought tolerance at both the cellular and morphological levels is required. A transgenic approach coupled with conventional and molecular breeding approaches is needed in order to achieve substantial gains in yield. Major drought QTLs could be fine-mapped, followed by cloning and transgenic confirmation of the candidate genes in those regions. The role of several regulons is well established in drought but their deployment in molecular breeding needs to be strategized. Overall, an integrated strategy involving conventional and molecular breeding as well as a transgenic approach is required to enhance the grain yield of existing rice varieties in rainfed environments.

2.6 Practical Marker Applications for Drought Tolerance in Rice

2.6.1 Important Considerations in Molecular Breeding for Drought

The tremendous efforts devoted by plant breeders to developing drought-tolerant rice cultivars through physiological and genetic approaches have met limited success. Slow progress is due to (1) the lack of effective selection criteria and (2) the complexity of the trait. Trait selection is one of the most important considerations in drought molecular breeding. Practically, "grain yield under drought" is a preferred selection criterion showing moderate to high heritability and has been used successfully in drought breeding programs in the past few years at IRRI (Venuprasad et al. 2007; Kumar et al. 2008; Vikram et al. 2011, 2012a; Swamy and Kumar 2011). The second reason for slow growth is complexity of the trait, which is still a bottleneck for conventional as well as molecular plant breeders. Some considerations in the application of drought grain yield QTLs/markers to practical breeding programs are (1) unwanted linkages, (2) additive interactions, and (3) the size of the QTL regions. Unwanted linkages and interaction effects have hampered drought breeding programs (Vikram et al. IRRI, unpublished). For example, drought-tolerant landraces and rice cultivars are usually tall with lower genetic yield potential. After using such lines in drought breeding, this trait appears as linkage drag. Also, some unknown genomic interaction effects are difficult to handle through a conventional plant breeding program. It is interesting to see that QTLs for grain yield under drought usually colocate with QTLs for days to flowering, plant height, biomass increase, harvest index, and several other traits (Bernier et al. 2007; Venuprasad et al. 2009; Vikram et al. 2011; Ghimire et al. 2012). Not only drought grain yield QTLs but also the ones identified for physiological traits colocate with several other traits (Lanceras et al. 2004). This colocation of QTLs in molecular breeding is comparable to the correlation of these colocating traits with drought-tolerance traits in a conventional approach. This linkage can be broken through a marker-assisted approach coupled with phenotypic selection. Marker-assisted and phenotypic selection could be modified in various ways for the purpose. One of the most fascinating examples is $qDTY_{1.1}$, a major and consistent locus for grain yield under drought, which colocated for QTLs related to plant height. Colocation of $qDTY_{1.1}$ for plant height and grain yield under drought could be due either to pleiotropy of the *sd1* gene or to its tight linkage with other drought-tolerance QTLs in the same region (Venuprasad et al. 2012). Bernier et al. (2007) report the association of this QTL region with plant height and flowering under drought stress, but not with grain yield. Babu et al. (2003) also indicated a pleiotropic effect of the *sd1* gene for a QTL effect associated with this region. In contrast, Kumar et al. (2007) reported that this region is associated with grain yield under drought and is linked with the *sd1* gene because the QTL was not significant for plant height. Several other reports also suggest a linkage effect of this QTL region with the *sd1* gene (Jiang et al. 2004; Khowaja et al. 2009). A definite test of linkage versus pleiotropism using *sd1* gene-based markers could resolve the issue. Further studies undertaken at IRRI in recent years indicated tight linkage of a plant height QTL and drought grain yield QTL in this genomic region. Not only the linkage effects but also interaction effects were observed with drought QTLs. Several digenic interactions were also reported to be associated with drought QTLs (Lanceras et al. 2004; Dixit et al. 2012). The epistatic interaction of two drought QTLs, $qDTY_{12.1}$ and $qDTY_{3.2}$, has been reported to provide an additional yield advantage over a single QTL under drought stress (Dixit et al. 2012). Some other digenic interactions between the QTLs $qDTY_{1.1}$, $qDTY_{3.2}$, $qDTY_{3.1}$, and $qDTY_{6.1}$ have been observed in some populations at IRRI (Vikram et al. IRRI, unpublished). Understanding of the linkage and interaction effects of QTLs could help in explaining the complexity of drought tolerance.

Fine-mapping of drought QTLs up to 1 Mb is advocated to be a preferred strategy for introgression purposes (Kamoshita et al. 2008). However, recent findings of linkage and interaction effects of drought QTLs clearly suggest that the focus should be on elimination of the unwanted linkages, of plant height and days to flowering, with QTLs for grain yield under drought. Therefore, for practical breeding purposes, large-effect drought grain yield QTLs could be identified in populations involving target varieties. Digenic interactions could be explained through high-density genotyping in a segregating population to determine additive interactions. Interacting loci providing an additional advantage could be introgressed if there is no detrimental effect under nonstress situations. Unwanted linkages could be eliminated through backcrossing and selecting for plant type at the $BCnF_2$ stage while selecting QTL homozygotes. In the process, the recurrent parent genome could be recovered and, if required, further backcrossing could be done. Lines with a QTL region phenotypically identical to that of the recipient parent could be tested under multiple ecosystems and environments (see Figure 2.5). Fine mapping of QTL regions can be carried out along with their physiological and biochemical characterization so that product delivery and trait discovery can proceed simultaneously.

2.6.2 Drought Molecular Breeding Strategies: MABC and MAQP

MAB is a well-known fast-track alternative for rice genetic improvement (Collard and Mackill 2008; Ye and Smith 2010). Availability of the major and consistent drought grain yield QTLs (genomic regions) in rice has enabled fast-track breeding for high grain yield under drought in the past few years (Swamy and Kumar 2011; Vikram et al. 2012b). These QTLs could be applied through different approaches such as (1) MABC, (2) marker-assisted QTL pyramiding (MAQP), and (3) marker-assisted selection (MAS). Marker-assisted backcrossing is advantageous only when a major QTL or gene governs a trait of interest. For complex traits such as drought, several genes/genomic regions are likely involved in the stress responses. Therefore, pyramiding of QTLs/genes could be more

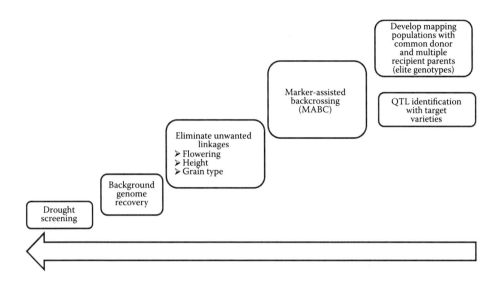

FIGURE 2.5
MAB strategy for rice drought grain yield QTLs.

useful than single-QTL/-gene introgressions. A substantial gain of 1 t/ha or more under rainfed conditions could be successfully achieved through MAQP. Swamy and Kumar (2011) reported a yield advantage of 1.2–2.0 t/ha in pyramided IR64 lines. Several other studies at IRRI also confirmed that MAQP is advantageous over MABC. Marker-assisted QTL pyramiding is therefore a preferred strategy for genetic improvement of rainfed rice varieties.

The development of drought-tolerant rice varieties through MABC/MAQP can be carried out in a more efficient way using single nucleotide polymorphism (SNP) genotyping platforms. Consistency of the drought yield (DTY) QTLs across donor and recipient backgrounds offered an opportunity to develop SNP markers for wide applications. SNPs can be used through sequencing of DTY QTL regions (see Figure 2.6) and a fast-track MAB program can be carried out using such platforms (see Figure 2.7).

2.6.3 Future Perspectives

Apart from MABC and MAQP, MAS can also be an alternative approach. Genomic regions associated with drought tolerance can be accumulated into single recombination events (irrespective of genetic backgrounds) and desired alleles/allele combinations could be used to advance generations. A somewhat similar approach has been advocated through the use of MAGIC (multiparent advanced generation intercross) populations. The development of mapping populations with multiple parents (of diverse origin) and their rigorous phenotyping and genotyping with high-density SNPs could help in pinpointing the major genes and genomic interactions governing a trait of interest.

Genomic selection through training and reference populations offers another approach that can help plant breeders in an efficient way. Plant breeders can easily incorporate marker applications in their breeding populations, which in turn can be used as training and reference populations for genomic selection.

Breeding and mapping populations usually benefit from the genes/alleles from a single parent or the unknown genomic interactions. Breeders usually prefer improved breeding

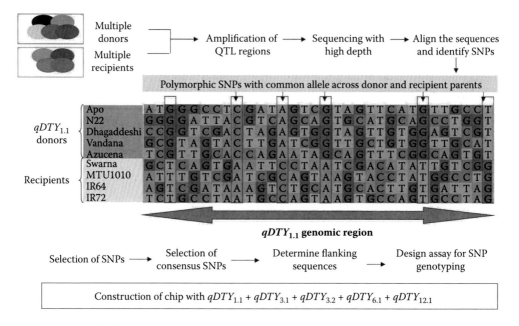

FIGURE 2.6   (See color insert)
SNP marker design for drought grain yield QTLs.

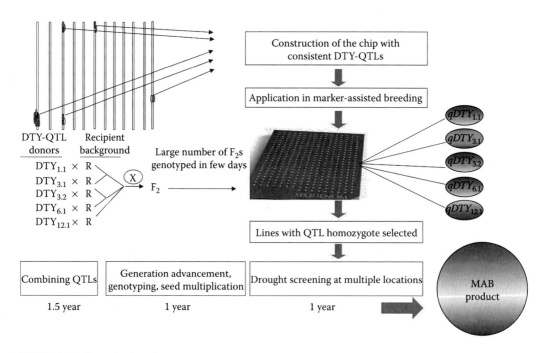

FIGURE 2.7   (See color insert)
Fast-track MAB program using SNP genotyping platforms.

lines for genetic enhancement because they do not want to compromise with quality traits. However, landraces and wild species have enormous potential that can help in improving rice yield in rainfed environments. This can be achieved using genome-wide association analysis (GWAS). GWAS has been carried out for various traits, including drought stress, but a focused study on drought tolerance in rice is yet to be done.

The different conventional and molecular breeding approaches mainly target Mendelian genes, though epigenetic mechanisms are also considered by some workers (Chinnusamy and Zhu 2009). The two approaches can be integrated so as to harness the possible benefits of both. A MAB strategy that can accommodate the possible gains through Mendelian (QTLs/genes) as well as epigenetic mechanisms is proposed in Figure 2.8. Selection of lines from the stress treatment and their use in generation advancement could enable plant breeders to harvest the possible benefits of epigenetic mechanisms.

The development and application of high-throughput phenomics technologies could offer a better alternative method for precision phenotyping support to breeders. Little progress has been made in proteomics, metabolomics, and epigenomics compared with genomic approaches, but knowledge obtained from these approaches is useful in understanding the genetic as well as physiological basis of drought tolerance. Practically, MABC and MAQP were successfully applied in developing high-yielding varieties under rainfed environments. However, SNP marker applications, MAS, and GWAS are likely to be applied in the near future. Efforts aimed at coupling of omics approaches with practical plant-breeding applications can bring significant progress in developing rice cultivars suitable for rainfed environments.

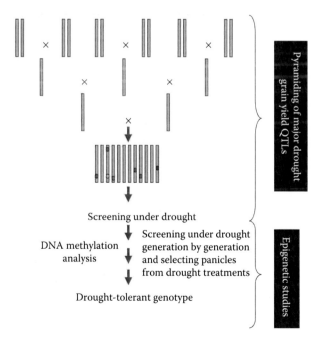

FIGURE 2.8
A possible drought breeding strategy through coupling of MAB with epigenetic phenomena for drought tolerance in rice.

References

Abou-Donia, A.H., Toaima, S.M., Hammoda, H.M., and Shawky, E. 2007. New rapid validated HPTLC method for the determination of lycorine in Amaryllidaceae plants extracts. *Chromatographia*. 65:497–500.

Alcázar, R., Altabella, T., Marco, F., Bortolotti, C., Reymond, M., Koncz, C., Carrasco, P., and Tiburcio, A.F. 2010. Polyamines: Molecules with regulatory functions in plant abiotic stress tolerance. *Planta*. 231(6):1237–1249.

Ali, G.M. and Komatsu, S. 2006. Proteomic analysis of rice leaf sheath during drought stress. *Journal of Proteome Research*. 5:396–403.

Atlin, G.N. and Lafitte, H.R. 2002. Marker-assisted breeding versus direct selection for drought tolerance in rice. In Saxena, N.P. and O'Toole, J.C. (eds), *Field Screening for Drought Tolerance in Crop Plants with Special Emphasis on Rice: Proceedings of the International Workshop on Field Screening for Drought Tolerance in Rice*, December 11–14, 2000, International Crop Research Institute for Semi-Arid Tropics, Patancheru, India. New York: Rockefeller Foundation.

Babita, M., Maheswari, M., Rao, L.M., Shankerb, A.K., and Rao, G.D. 2010. Osmotic adjustment, drought tolerance and yield in castor (*Ricinus communis* L.) hybrids. *Environmental and Experimental Botany*. 69(3):243–249.

Babu, R.C., Shashidhar, H.E., Lilley, J.M., Thanh, N.D., Ray, J.D., Sadasivam, S., Sarkarung, S., O'Toole, J.C., and Nguyen, H.T. 2001. Variation in root penetration ability, osmotic adjustment and dehydration tolerance among accessions of rice adapted to rainfed lowland and upland ecosystems. *Plant Breeding*. 120:233–238.

Babu, R.C., Nguyen, B.D., Chamarerk, V., Shanmugasundaram, P., Chezhian, P., Jeyaprakash, P., Ganesh, S.K. et al. 2003. Genetic analysis of drought resistance in rice by molecular markers: Association between secondary traits and field performance. *Crop Science*. 43:1457–1469.

Babu, R.C., Jingxian, Z., Blum, A., David, H.T.H., Wue, R., and Nguyen, H.T. 2004. *HVA1*, a *LEA* gene from barley confers dehydration tolerance in transgenic rice (*Oryza sativa* L.) via cell membrane protection. *Plant Science*. 166:855–862.

Barker, R., Dawe, D., Tuong, T.P., Bhuiyan, S.I., and Guerra, L.C. 1999. The outlook for water resources in the year 2020: Challenges for research on water management in rice production. In *Assessment and Orientation Towards the 21st Century, Proceedings of 19th Session of the International Rice Commission*, pp. 96–109, September 7–9, 1998, Cairo, Egypt. Rome: FAO.

Bernier, J., Kumar, A., Venuprasad, R., Spaner, D., and Atlin, G.N. 2007. A large-effect QTL for grain yield under reproductive-stage drought stress in upland rice. *Crop Science*. 47:507–516.

Bernier, J., Atlin, G.N., Serraj, R., Kumar, A., and Spaner, D. 2008. Breeding upland rice for drought resistance. *Journal of Food Science and Agriculture*. 88:927–939.

Bhatnagar-Mathur, P., Vadez, V., and Sharma, K.K. 2008. Transgenic approaches for abiotic stress tolerance in plants: retrospect and prospects. *Plant Cell Reporter*. 27:411–424.

Bouman, B., Barker, R., Humphreys, E., Tuong, T.P., Atlin, G., Bennett, J., Dawe, D. et al. 2007. Rice: Feeding the billions. In Molden, D. (ed.), *Water for Food, Water for Life: A Comprehensive Assessment of Water Management in Agriculture*, pp. 515–549. London: Earthscan; Colombo, Sri Lanka: IWMI.

Chaturvedi, G.S., Ram, P.C., Singh, A.K., Ram, P., Ingram, K.T., Singh, B.B., Singh, R.K., and Singh, V.P. 1996. Carbohydrate status of rainfed lowland rices in relation to submergence, drought and shade tolerance. In *Physiology of Stress Tolerance in Rice: Proceedings of the International Conference on Stress Physiology of Rice*, pp. 103–122, February 28 to March 5, 1994, Lucknow: India.

Chen, T.H. and Murata, N. 2008. Glycinebetaine: An effective protectant against abiotic stress in plants. *Trends in Plant Science*. 13(9):499–505.

Chinnusamy, V. and Zhu, J.-K. 2009. Epigenetic regulation of stress responses in plants. *Current Opinion in Plant Biology*. 12:1–7.

Collard, B.C.Y. and Mackill, D.J. 2008. Marker-assisted selection: An approach for precision plant breeding in the twenty first century. *Philosophical Transactions in Royal Biological Society*. 363:557–572.

Dawe, D., Pandey, S., and Nelson, A. 2010. Emerging trends and spatial patterns of rice production. In Pandey, S., Byerlee, D., Dawe, D., Dobermann, A., Mohanty, S., Rozelle, S., and Hardy, B. (eds), *Rice in the Global Economy: Strategic Research and Policy Issues for Food Security*, pp. 15–35. Los Baños, Philippines: International Rice Research Institute.

de Fraiture, C., Wichelns, D., Rockstrom, J., Kemp-Benedict, E., Eriyagama, N., Gordon, L., Hanjra, M. et al. 2007. Looking ahead to 2050: Scenarios of alternative investment approaches. In Molden, D. (ed.), *Water for Food, Water for Life: A Comprehensive Assessment of Water Management in Agriculture*, pp. 91–145. London: Earthscan; Colombo, Sri Lanka: IWMI.

Dixit, S., Swamy, B.P.M., Vikram, P., Bernier, J., Sta Cruz, M.T., Amante, M., Atri, D., and Kumar, A. 2012. Increased drought tolerance and wider adaptability of $qDTY_{12.1}$ conferred by its interaction with $qDTY_{2.3}$ and $qDTY_{3.2}$. *Molecular Breeding*. 30:1767–1779.

Fu, W., Wu, K., and Duan, J. 2007. Sequence and expression analysis of histone deacetylases in rice. *Biochemical and Biophysical Research Communications*. 356:843–850.

Fumagalli, E., Baldoni, E., Abbruscato, P., Piffanelli, P., Genga, A., Lamanna, R., and Consonni, R. 2009. NMR techniques coupled with multivariate statistical analysis: Tools to analyse *Oryza sativa* metabolic content under stress conditions. *Journal of Agronomy and Crop Science*. 195:77–88.

Gauchan, D. and Pandey, S. 2012. Synthesis of key results and implications. In Pandey, S., Gauchan, D., Malabayabas, M., Bool-Emerick, M., and Hardy, B. (eds), *Patterns of Adoption of Improved Rice Varieties and Farm-Level Impacts in Stress-Prone Rainfed Areas in South Asia*, p. 3. Los Baños, Philippines: International Rice Research Institute.

Gazanchian, A., Hajheidari, M., Sima, N. K., and Salekdeh, G. H. 2007. Proteome response of *Elymus elongatum* to severe water stress and recovery. *Journal of Experimental Botany*. 58:291–300.

Genga, A., Mattana, M., Coraggio, I., Locatelli, F., Piffanelli, P., and Consonni, R. 2011. Plant metabolomics: A characterisation of plant responses to abiotic stresses. In Shanker, A. and Venkateswarlu, B. (eds), *Abiotic Stress in Plants—Mechanisms and Adaptations*, pp. 309–350. Rijeka: InTech.

Ghimire, K.H., Quiatchon, L.A., Vikram, P., Swamy, B.P.M., Dixit, S., Ahmed, H.U., Hernandez, J.E., Borromeo, T.H., and Kumar, A. 2012. Identification and mapping of a QTL ($qDTY_{1.1}$) with a consistent effect on grain yield under drought. *Field Crops Research*. 131:88–96.

Gomez, S.M., Boopathi, N.M., Kumar, S.S., Ramasubramanian, T., Chengsong, Z., Jeyaprakash, P., Senthil, A., and Babu, R.C. 2010. Molecular mapping and location of QTLs for drought-resistance traits in indica rice (*Oryza sativa* L.) lines adapted to target environments. *Acta Physiologiae Plantarum*. 32:355–364.

Gorantla, M., Babu, P.R., Lachagari, V.B.R., Reddy, A.M.M., Wusirika, R., Bennetzen, J.L., and Reddy, A.R. 2007. Identification of stress-responsive genes in an indica rice (*Oryza sativa* L.) using ESTs generated from drought stressed seedlings. *Journal of Experimental Botany*. 58:253–265.

Gotti, R., Fiori, J., Bartolini, M., and Cavrini, V. 2006. Analysis of Amaryllidaceae alkaloids from Narcissus by GC-MS and capillary electrophoresis. *Journal of Pharmaceutical and Biomedical Analysis*. 42:17–24.

Guo, A.-Y., Chen, X., Gao, G., Zhang, H., Zhu, Q.-H., Liu, X.-C., Zhong, Y.-F., Gu, X., He, K., and Luo, J. 2008. PlantTFDB: A comprehensive plant transcription factor database. *Nucleic Acids Research*. 36:D966–D969.

Hajheidari, M., Abdollahian-Noghabi, M., Askari, H., Heidari, M., Sadeghian, S.Y., Ober, E.S., and Salekdeh, G.H. 2005. Proteome analysis of sugar beet leaves under drought stress. *Proteomics*. 5:950–960.

Hu, H., Dai, M., Yao, J., Xiao, B., Li, X., Zhang, Q., and Xiong, L. 2006. Overexpressing a NAM, ATAF, and CUC (NAC) transcription factor enhances drought resistance and salt tolerance in rice. *Proceedings of National Academy of Sciences USA*. 103:12987–12992.

Hua, X., Bin-Ying, F.U., Hua-Xue, X.U., and Yang-Sheng, L.I. 2010. Proteomic analysis of PEG-simulated drought stress responsive proteins of rice leaves using a pyramiding rice line at the seedling stage. *Botanical Studies*. 51:137–145.

Jiang, Y., Cai, Z., Xie, W., Long, T., Yu, H., and Zhang, Q. 2011. Rice functional genomics research: Progress and implications for crop genetic improvement. *Biotechnology Advances.* 30:1059–1070.

Jiang, Y.Z., Bagali, P., Gao, Y.M., Lafitte, R., Fu, B.Y., Xu, J.L., Maghirang, R., Domingo, R., Hittalmani, S., Mackill, D., and Li, Z.K. 2004. High resolution mapping of a genomic region harbouring several QTLs and sd1 in rice. In Poland, D., Sawkins, M., Ribaut, J.M., and Hoisington, D. (eds), *Proceedings of a Workshop on Resilient Crops for Water-Limited Environments,* Cuernavaca, Mexico, 24–28 May. Mexico: CIMMYT.

Jian-Hua, X., Bin-Ying, F.U., Hua-Xue, X.U., and Yang-Sheng, L.I. 2010. Proteomic analysis of PEG-simulated drought stress responsive proteins of rice leaves using a pyramiding rice line at the seedling stage. *Botanical Studies.* 51:137–145.

Kadam, S., Singh, K., Shukla, S., Goel, S., Vikram, P., Pawar, V., Gaikwad, K., Chopra, R.K., and Singh, N.K. 2012. Genomic associations for drought tolerance on the short arm of wheat chromosome 4B. *Functional and Integrative Genomics.* 12:447–464.

Kamoshita, A., Babu, R.C., Boopathi, N.M., and Fukai, S. 2008. Phenotypic and genotypic analysis of drought-resistance traits for development of rice cultivars adapted to rainfed environments. *Field Crops Research.* 109:1–23.

Kasuga, M., Liu, Q., Miura, S., Yamaguchi-Shinozaki, K., and Shinozaki, K. 1999. Improving plant drought, salt, and freezing tolerance by gene transfer of a single stress inducible transcription factor. *Nature Biotechnology.* 17:287–291.

Kato, Y., Abe, J., Kamoshita, A., and Yamagishi, J. 2006. Genotypic variation in root growth angle in rice (*Oryza sativa* L.) and its association with deep root development in upland fields with different water regimes. *Plant and Soil.* 287:117–129.

Kato, Y., Hirotsu, S., Nemoto, K., and Yamagishi, J. 2008. Identification of QTLs controlling rice drought tolerance at seedling stage in hydroponic culture. *Euphytica.* 160:423–430.

Ke, Y., Han, G., He, H., and Li, J. 2009. Differential regulation of proteins and phosphoproteins in rice under drought stress. *Biochemistry and Biophysics Research Communications.* 379:133–138.

Khowaja, F.S., Norton, G.J., Courtois, B., and Price, A.H. 2009. Improved resolution in the position of drought-related QTLs in a single mapping population of rice by meta-analysis. *BMC Genomics.* 10:276.

Khush, G.S. 1997. Origin, dispersal, cultivation and variation of rice. *Plant Molecular Biology.* 35:25–34.

Khush, G.S. 2005. What it will take to feed 5.0 billion rice consumers in 2030. *Plant Molecular Biology.* 59:1–6.

Koller, A., Washburn, M.P., Lange, B.M., Andon, N.L., Deciu, C., Haynes, P.A., Hays, L. et al. 2002. Proteomic survey of metabolic pathways in rice. *Proceedings of National Academy of Sciences USA.* 99:11969–11974.

Komatsu, S., Muhammad, A., and Rakwal, R. 1999. Separation and characterization of proteins from green and etiolated shoots of rice (*Oryza sativa* L.): Towards a rice proteome. *Electrophoresis.* 20:630–636.

Komatsu, S., Kojima, K., Suzuki, K., Ozaki, K., and Higo, K. 2004. Rice proteome database based on two-dimensional polyacrylamide gel electrophoresis: Its status in 2003. *Nucleic Acids Research.* 32:D388–D392.

Kumar, A., Bernier, J., Verulkar, S., Lafitte, H.R., and Atlin, G.N. 2008. Breeding for drought tolerance: Direct selection for yield, response to selection and use of drought-tolerant donors in upland and lowland-adapted populations. *Field Crops Research.* 107:221–231.

Kumar, A., Verulkar, S.B., Mandal, N.P., Variar, M., Shukla, V.D., Dwivedi, J.L., Singh, B.N. et al. 2012. High- yielding, drought-tolerant, stable rice genotypes for the shallow rainfed lowland drought prone ecosystem. *Field Crops Research.* 133:37–47.

Kumar, R., Venuprasad, R., and Atlin, G. 2007. Genetic analysis of rainfed lowland rice drought tolerance under naturally occurring stress in Eastern India: Heritability and QTL effects. *Field Crops Research.* 103:42–52.

Lanceras, J.C., Pantuwan, G., Jongdee, B., and Toojinda, T. 2004. Quantitative trait loci associated with drought tolerance at reproductive stage in rice. *Plant Physiology.* 135:384–399.

Lenka, S.K., Katiyar, A., Chinnusamy, V., and Bansal, K.C. 2011. Comparative analysis of drought-responsive transcriptome in Indica rice genotypes with contrasting drought tolerance. *Plant Biotechnology Journal*. 9:315–327.

Levi, A., Paterson, A.H., Cakmak, I., and Saranga, Y. 2011. Metabolite and mineral analyses of cotton near-isogenic lines introgressed with QTLs for productivity and drought related traits. *Physiologia Plantarum*. 141:265–275.

Li, X.P., Muller-Moule, P., Gilmore, A.M., and Niyogi, K.K. 2002. PsbS-dependent enhancement of feedback de-excitation protects photosystem II from photoinhibition. *Proceedings of National Academy of Sciences USA*. 99:15222–15227.

Liu, J.X. and Bennett, J. 2010. Reversible and irreversible drought-induced changes in the anther proteome of rice (*Oryza sativa* L.) genotypes IR64 and Moroberekan. *Molecular Plant*. 4(1):59–69.

Liu, L., Lafitte, R., and Guan, D. 2004. Wild *Oryza* species as potential sources of drought-adaptive traits. *Euphytica*. 138:149–161.

Llop, A., Pocurull, E., and Borull, F. 2010. Automated determination of aliphatic primary amines in wastewater by simultaneous derivatization and headspace solid-phase microextraction followed by gas chromatography-tandem mass spectrometry. *Journal of Chromatography*. 1217:575–581.

Margat, J., Frenken, K., and Faurès, J.M. 2005. Key water resources statistics in aquastat, IWG-Env, International Work Session on Water Statistics, June 20–22, 2005, Vienna.

Mishra, K.K., Vikram, P., Yadaw, R.B., Swamy, B.P.M., Sta Cruz, M.T., Maturan, P., Marker, S., and Kumar A. 2013. $qDTY_{12.1}$: A locus with consistent effect on grain yield under drought in rice. *BMC Genetics*. 14:12.

Muthurajan, R., Shobbar, Z.-S, Jagadish, S.V.K., Bruskiewich, R., Ismail, A., Leung, H., and Bennett, J. 2010. Physiological and proteomic responses of rice peduncles to drought stress. *Molecular Biotechnology*. 48:173–182.

National Institute of Agrobiological Sciences. Rice proteome database. http://gene64.dna.affrc.go.jp/RPD/ (accessed October 2012).

Nguyen, H.T., Babu, R.C., and Blum, A. 1997. Breeding for drought resistance in rice: Physiology and molecular genetics considerations. *Crop Science*. 37:1426–1434.

Nozu, Y., Tsugita, A., and Kamijo, K. 2006. Proteomic analysis of rice leaf, stem and root tissues during growth course. *Proteomics*. 6:3665–3670.

Nuruzzaman, M., Sharoni, A., Satoh, K., Moumeni, A., Venuprasad, R., Serraj, R., Kumar, A., Leung, H., Attia, K., and Kikuchi, S. 2012. Comprehensive gene expression analysis of the *NAC* gene family under normal growth conditions, hormone treatment, and drought stress conditions in rice using near-isogenic lines (NILs) generated from crossing Aday Selection (drought tolerant) and IR64. *Molecular Genetics and Genomics*. 287(5):389–410.

Oh, S.J., Song, S.I., Kim, Y.S., Jang, H.J., Kim, S.Y., Kim, M., Kim, Y.K., Nahm, B.H., and Kim, J.K. 2005. Arabidopsis CBF3/DREB1A and ABF3 in transgenic rice increased tolerance to abiotic stress without stunting growth. *Plant Physiology*. 138:341–351.

Oliver, S.G., Winson, M.K., Kell, D.B., and Baganz, F. 1998. Systematic functional analysis of the yeast genome. *Trends in Biotechnology*. 16:373–378.

Palaniswamy, S.K., James, S., Sun, H., Lamb, R.S., Davuluri, R.V., and Grotewold, E. 2006. AGRIS and AtRegNet: A platform to link *cis* regulatory elements and transcription factors into regulatory networks. *Plant Physiology*. 140:818–829.

Pandit, A., Rai, V., Bal, S., Sinha, S., Kumar, V., Chauhan, M., Gautam, R.K. et al. 2010. Combining QTL mapping and transcriptome profiling of bulked RILs for identification of functional polymorphism for salt tolerance genes in rice (*Oryza sativa* L.). *Molecular Genetics and Genomics*. 284:121–136.

Peng, Z., Wang, M., Li, F., Lv, H., Li, C., and Xia, G. 2009. A proteomic study of the response to salinity and drought stress in an introgression strain of bread wheat. *Molecular Cell Proteomics*. 8:2676–2686.

Plomion, C., Lalanne, C., Claverol, S., Meddour, H., Kohler, A., Bogeat-Triboulot, M.B., Barre, A. et al. 2006. Mapping the proteome of poplar and application to the discovery of drought-stress responsive proteins. *Proteomics*. 6:6509–6527.

Pray, C., Nagrajan, L., Li, L., Huang, J., Hu, R., Selvaraj, K.N., Napasintuwong, O., and Babu, R.C. 2011. Potential impact of biotechnology on adaption of agriculture to climate change: The case of drought tolerant rice breeding in Asia. *Sustainability.* 3:1723–1741.

Puri, R.R., Khadka, K., and Paudyal, A. 2010. Separating climate resilient crops through screening of drought tolerant rice land races in Nepal. *Aronomy Journal of Nepal.* 1:80–84.

Rabello, A.R., Guimarães, C.M., Rangel, P.H.N., da Silva, F.R., Seixas, D., de Souza, E.D., Brasileiro, A.C.M., Spehar, C.R., Ferreira, M.E., and Mehta, Â. 2008. Identification of drought-responsive genes in roots of upland rice (*Oryza sativa* L). *BMC Genomics.* 9:485.

Ribaut, J.M., and Hoisington, D. (eds), *Proceedings of a Workshop on Resilient Crops for Water-Limited Environments*, pp. 215, May 24–28, 2004, Cuernavaca, Mexico. Mexico, DF: CIMMYT.

Rice Today. 2009. Magzine published by international rice research institute. 8(3):16.

Saijo, Y., Hata, S., Kyozuka, J., Shimamoto, K., and Izui, K. 2000. Overexpression of a single Ca2+-dependent protein kinase confers both cold and salt/drought tolerance on rice plants. *Plant Journal.* 23:319–327.

Salekdeh, G.H., Siopongco, J., Wade, L.J., Ghareyazie, B., and Bennett, J. 2002. Proteomic analysis of rice leaves during drought stress and recovery. *Proteomics.* 2(9):1131–1145.

Sato, Y. and Yokoya, S. 2008. Enhanced tolerance to drought stress in transgenic rice plants overexpressing a small heat-shock protein, sHSP17.7. *Plant Cell Reporter.* 27:329–334.

Semel, Y., Schauer, N., Roessner, U., Zamir, D., and Fernie, A.R. 2007. Metabolite analysis for the comparison of irrigated and non-irrigated field grown tomato of varying genotype. *Metabolomics.* 3:289–295.

Shu, L., Lou, Q., Ma, C., Ding, W., Zhou, J., Wu, J., Feng, F. et al. 2011. Genetic, proteomic and metabolomics analysis of regulation of energy storage in rice seedlings in response to drought. *Proteomics.* 11(21):4122–4138.

Shulaev, V., Cortes, D., Miller, G., and Mittler, R. 2008. Metabolomics for plant stress response. *Physiologia Plantarum.* 132:199–208.

Singh, N.K., Raghuvanshi, S., Srivastava, S.K., Gaur, A., Pal, A.K., Dalal, V., Singh, A. et al. 2004. Sequence analysis of the long arm of rice chromosome 11 for rice-wheat synteny. *Functional and Integrative Genomics.* 4:102–117.

Singh, S.P., Mallick, S.S., and Singh, A.K. 2005. Collection of rice land races from Vindhyachal hills. *Agriculture Science Digest.* 25(3):174–177.

Smirnoff, N. 1998. Plant resistance to environmental stress. *Current Opinion in Biotechnology.* 9:214–219.

Swamy, B.P.M. and Kumar, A. 2011. Sustainable rice yield in water short drought prone environments: Conventional and molecular approaches. In Lee, T.S. (ed.), *Irrigation Systems and Practices in Challenging Environments*, pp. 149–168. InTech, Croatia.

Swamy, B.P.M., Vikram, P., Dixit, S., Ahmed, H.U., and Kumar, A. 2011. Meta-analysis of grain yield QTL identified during agricultural drought in grasses showed consensus. *BMC Genomics.* 12:319.

Szabados, L. and Savouré, A. 2010. Proline: A multifunctional amino acid. *Trends in Plant Science.* 15:89–97.

Tanaka, N., Mitsui, S., Nobori, H., Yanagi, K., and Komatsu, S. 2005. Expression and function of proteins during development of the basal region in rice seedlings. *Molecular and Cellular Proteomics.* 4:796–808.

Thomashow, M.F. 1999. Plant cold acclimation: Freezing tolerance genes and regulatory mechanisms. *Annual Review of Plant Physiology and Plant Molecular Biology.* 50:571–599.

Tian, J.-C., Chen, Y., Li, Z.-L., Li, K., Chen, M., Peng, Y.-F., Hu, C., Shelton, A.M., and Ye, G.-Y. 2012. Transgenic Cry1Ab rice does not impact ecological fitness and predation of a generalist spider. *PLoS ONE.* 7(4):e35164.

Valluru, R. and Van den Ende, W. 2008. Plant fructans in stress environments: Emerging concepts and future prospects. *Journal of Experimental Botany.* 59:2905–2916.

Vanniarajan, C., Vinod, K.K., and Pereira, A. 2012. Molecular evaluation of genetic diversity and association studies in rice (*Oryza sativa* L.). *Journal of Genetics.* 91:9–19.

Venuprasad, R., Lafitte, H.R., Atlin, G.N. 2007. Response to direct selection for grain yield under drought stress in rice. *Crop Science.* 47:285–293.

Venuprasad, R., Dalid, C.O., Del Valle, M., Zhao, D., Espiritu, M., Sta Cruz, M.T., Amante, M., Kumar, A., and Atlin, G.N. 2009. Indentification and characterization of large-effect quantitative trait loci for grain yield under lowland drought stress in rice using bulk-segregant analysis. *Theoretical and Applied Genetics*. 120:177–190.

Venuprasad, R., Bool, M.E., Quiatchon, L., Sta Cruz, M.T., Amante, M., and Atlin, G.N. 2012. A large-effect QTL for rice grain yield under upland drought stress on chromosome 1. *Molecular Breeding*. 30:535–547.

Vikram, P., Swamy, B.P.M., Dixit, S., Sta Cruz, M.T., Ahmed, H.U., Singh, A.K., and Kumar, A. 2011. $qDTY_{1.1}$, a major QTL for rice grain yield under reproductive-stage drought stress with a consistent effect in multiple elite genetic backgrounds. *BMC Genetics* 12:89.

Vikram, P., Kumar, A., Singh, A.K., and Singh, N.K. 2012a. Rice: Genomics-assisted breeding for drought tolerance In Tuteja, S.S,. Gill, A.F., and Tiburico, R. (eds.), *Improving Crop Resistance to Abiotic Stress*, pp. 715–731. Germany: Wiley-VCH Verlag GmbH & Co. KGaA.

Vikram, P., Swamy, B.P.M., Dixit, S., Ahmed, H.U., Sta Cruz, M.T., Singh, A.K., Ye, G., and Kumar, A. 2012b. Bulk segregant analysis: An effective approach for mapping drought grain yield QTLs in rice. *Field Crops Research*. 134:185–192.

Wallace, J.S., Acreman, M.C., and Sullivan, C.A. 2003. The sharing of water between society and ecosystems: From conflict to catchment-based co-management. *Philosophical Transactions of the Royal Society B*. 358:2011–2026.

Wang, W.-S., Pan, Y.-J., Zhao, X.-Q., Dwivedi, D., Zhu, L.-H., Ali, J., Fu, B.-Y., and Li, Z.-K. 2011. Drought-induced site-specific DNA methylation and its association with drought tolerance in rice (*Oryza sativa* L.). *Journal of Experimental Botany*. 62(6):1951–1960.

Wasinger, V.C., Cordwell, S.J., Cerpa-Poljak, A., Yan, J.X., Gooley, A.A., Wilkins, M.R., Duncan, M.W., Harris, R., Williams, K.L., and Humphery-Smith, I. 1995. Progress with gene-product mapping of the mollicutes: Mycoplasma genitalium. *Electrophoresis*. 16(7):1090–1094.

Wu, X., Shiroto, Y., Kishitani, S., Ito, Y., and Toriyama, K. 2009. Enhanced heat and drought tolerance in transgenic rice seedlings overexpressing OsWRKY11 under the control of HSP101 promoter. *Plant Cell Reporter*. 28:21–30.

Xiao, B., Huang, Y., Tang, N., and Xiong, L. 2007. Over-expression of a LEA gene in rice improves drought resistance under the field conditions. *Theoretical and Applied Genetics*. 115:35–46.

Xiao, B.Z., Chen, X., Xiang, C.B., Tang, N., Zhang, Q.F., and Xiong, L.Z. 2009. Evaluation of seven function-known candidate genes for their effects on improving drought resistance of transgenic rice under field conditions. *Molecular Plant*. 2:73–83.

Xu, D., Duan, X., Wang, B., Hong, B., Ho, T., and Wu, R. 1996. Expression of a late embryogenesis abundant protein gene, HVA1, from barley confers tolerance to water deficit and salt stress in transgenic rice. *Plant Physiology*. 110:249–257.

Xu, D.Q., Huang, J., Guo, S.Q., Yang, X., Bao, Y.M., Tang, H.J., and Zhang, H.S. 2008. Overexpression of a TFIIIA-type zinc finger protein gene ZFP252 enhances drought and salt tolerance in rice (*Oryza sativa* L.). *FEBS Letters*. 582:1037–1043.

Yamaguchi-Shinozaki, K. and Shinozaki, K. 2006. Transcriptional regulatory networks in cellular responses and tolerance to dehydration and cold stresses. *Annual Review of Plant Biology*. 57:781–803.

Yan, S., Tang, Z., Su, W., and Sun, W. 2005. Proteomic analysis of salt stress-responsive proteins in rice root. *Proteomics*. 5:235–244.

Yang, S., Vanderbeld, B., Jiangxin, W., and Huang, Y. 2010. Narrowing down the targets: towards successful genetic engineering of drought-tolerant crops. *Molecular Plant*. 3(3):469–490.

Ye, G. and Smith, K.F. 2010. Marker-assisted gene pyramiding for cultivar development. *Plant Breeding Reviews*. 33:219–256.

Zhang, X., Zhou, S., Fu, Y., Su, Z., Wang, X., and Sun, C. 2006. Identification of a drought tolerant introgression line derived from Dongxiang common wild rice (*O. rufipogon* Griff.). *Plant Molecular Biology*. 62:247–259.

Zhu, B., Su, J., Chang, M., Verma, D.P.S., Fan, Y.-L., and Wu, R. 1998. Overexpression of a D1-pyrroline-5-carboxylate synthetase gene and analysis of tolerance to water- and salt-stress in transgenic rice. *Plant Science*. 139:41–48.

3

Omics Approaches in Maize Improvement

Pawan K. Agrawal, PhD; Navinder Saini, PhD;
B. Kalyana Babu, PhD; and Jagdish C. Bhatt, PhD

CONTENTS

3.1 Introduction...73
3.2 Genomics...74
    3.2.1 Genetic and Physical Maps..74
    3.2.2 Mapping and Tagging of Genes and QTLs for Agronomic Traits...................75
    3.2.3 Mapping and Tagging of Genes for Quality Traits.....................................76
    3.2.4 Association Mapping and Nested Association Mapping in Maize.................78
    3.2.5 Marker-Assisted Selection for Maize Improvement...................................78
    3.2.6 Success Stories of Commercialization of MAS-Derived QPM Hybrids...........79
        3.2.6.1 MAS for Development of QPM Maize...79
        3.2.6.2 Vivek QPM 9...79
        3.2.6.3 Vivek QPM 21...80
    3.2.7 Genome Sequencing...81
    3.2.8 Functional Genomics..82
    3.2.9 Transgenics for Genomic Studies...83
3.3 Maize Proteomics..84
    3.3.1 Maize Proteomics for Abiotic Stress Tolerance...85
    3.3.2 Maize Proteomics for Biotic Stress Tolerance...85
    3.3.3 Proteomics for Maize Seed Science...86
    3.3.4 Proteome Map of Maize Endosperm...87
3.4 Metabolomics in Maize...87
    3.4.1 Maize Metabolomics for Abiotic Stress Tolerance.....................................88
    3.4.2 Characterization of Metabolism...89
3.5 Conclusion...90
References...90

3.1 Introduction

Maize (*Zea mays* L.) is an important food and feed crop globally. It is the third most important food crop after rice and wheat, in terms of both area and production, and India is the fifth largest producer of maize in the world, contributing 3% of the total global production. Due to economic necessity, most of the people in developing countries are overly dependent on maize as the staple food. It provides 50% of the dietary protein for humans and can comprise 70% of the protein intake for people in developing countries (Deutscher 1978). In Africa and some Asian countries, almost 90% of maize grown is for human consumption and may account for 80%–90% of the energy intake. Together with rice and wheat, it provides at least 30% of the food calories to more than 4.5 billion people in 94 developing countries. Maize has historically been used as a model species for genetics, developmental biology, physiology, and, more recently, genomic studies. The genetic studies on *Zea mays* L. started with Edward East's 1908 report on inbreeding depression and hybrid vigor, and the 1940s saw a cytogenetic breakthrough, for example, transposable elements (TEs) by Barbara McClintock (Walbot 2008). The accumulated cytogenetic and genetic data and more recently vast sequence information derived from genomic studies in maize have provided a wealth of information on the structure, function, and evolution of the maize genome.

3.2 Genomics

Although the traditional plant breeding based on phenotypic selection is very effective, it has suffered from several limitations for complex traits. The advances in omics technologies—that is, comprehensive and integrated genomic, transcriptomic, and proteomic analyses—can elucidate genetic architecture of plant genomes and the relationships between genotype and phenotype. The rapid advances in DNA sequencing technology have made whole-genome sequencing (WGS) both technically and economically feasible. More than 25 economically important plants' genomes have been sequenced (Hamilton and Buell 2012). The next-generation sequencing (NGS) technologies are used not only for WGS but also to allow applications related to target region deep sequencing, epigenetics, transcriptome sequencing (RNA-seq), megagenomics, and genotyping. Affordable high-throughput NGS technology platforms including Roche/454 (http://www.454.com), Illumina (http://www.illumina.com), and AB SOLiD (http://www.appliedbiosystems.com) enabled the genomics community to conduct a number of analyses (Lister and Ecker 2009; Thudi et al. 2012). High-throughput NGS also offers a unique opportunity for WGS via either *de novo* assemblies or mapping to a reference genome (Tenaillon et al. 2012). Maize is an excellent model system for addressing evolutionary dynamics of TEs within and between species, particularly for large, complex, repetitive genomes such as that of maize. The application of omics techniques will greatly improve our ability to assign functions to the gene products.

3.2.1 Genetic and Physical Maps

The first full genetic maps were a part of the seminal monograph on maize genetics published by Emerson, Beadle, and Fraser in 1935. The cytogenetic map was based on the pachytene stage chromosome karyotype developed by Barbara McClintock and on the chromosome's fractional arm length unit, referred to as centiMcClintocks (cMC). After the development of DNA-based markers, the first-generation molecular map based on RFLP markers was developed (Helentjaris et al. 1986; Coe et al. 1987; Burr et al. 1988). These were subsequently saturated with various types of PCR-based SSR markers (Lee et al. 2002; Fu et al. 2006). During the past decade, the hegemony of medium-throughput SSRs, declared as "markers of choice," was broken by single nucleotide polymorphism (SNP) markers (Mammadov et al. 2010). More than 130,000 gene-based SNPs have been identified (Mammadov et al. 2010) and many of these were used for the construction of genetic maps (Shi et al. 2012; Sa et al. 2012). Several types of mapping populations have been used in maize to construct the genetic maps, including F_2 (Coe et al. 1987; Beavis and Grant

1991), immortalized F_2 (Gardiner et al. 1993; Davis et al. 1999), and recombinant inbred line (RIL) populations (Burr et al. 1988; Causse et al. 1996; Taramino and Tingey 1996). By using the multiple crosses, composite maps have also been constructed (Causse et al. 1996). Functional maps using the expressed sequence tagged sites (ESTs) markers have been developed which in turn will be useful to usher in use of a candidate gene approach for discovering the sequence underpinning a quantitative trait locus (QTL) (Falque et al. 2005). More than 1800 molecular maps have been developed using different mapping populations in maize, and are documented in the Maize Genetics and Genomics Database (MaizeGDB, http://www.maizegdb.org). Efforts have been made to combine many physically and genetically mapped probes and genes onto a single consensus map (Schaeffer et al. 2006). A high-resolution physical map based on fingerprinted contigs (1902 genetic markers and 24,006 overgo/BAC-end sequence [BES] markers) (FPCs) of cultivar B73 was constructed (Wei et al. 2007).

3.2.2 Mapping and Tagging of Genes and QTLs for Agronomic Traits

Molecular tools to enhance breeding efficiency and effectiveness have become integral to many maize research programs worldwide. Discovery of genes and their closely linked markers can provide useful information in setting up molecular breeding strategies (Prasanna et al. 2010). Molecular marker-facilitated QTL mapping for yield-related traits in maize was first reported by Stuber et al. (1987). Since then, there have been large numbers of studies for identifying associated major genes (gene tagging) and QTL mapping (Table 3.1). QTLs have been mapped for several important traits such as plant height (Zhang et al. 2007; Teng et al. 2012), popping ability (Babu et al. 2006; Li et al. 2008), biotic stresses such as downy mildew resistance (George et al. 2003; Nair et al. 2005; Sabry et al. 2006), sugarcane mosaic virus (SCMV) resistance (Zhang et al. 2003), maize dwarf mosaic virus (MDMV) resistance (Liu et al. 2006; Shi et al. 2012), common smut resistance (Ding et al. 2008), head smut resistance (Li et al. 2012), gray leaf spot (GLS) (Zhang et al. 2012), *Fusarium moniliforme* ear rot resistance (Zhang et al. 2006), banded leaf and sheath blight (BLSB) resistance (Zhao et al. 2006a,b; Garg et al. 2009), and other biotic stresses; drought stress tolerance (Xiao et al. 2005; Tuberosa et al. 2002; Prasanna et al. 2010); and waterlogging tolerance (Qiu et al. 2007). Maize oil is highly valued for both animal feed and human food. Many studies have been conducted to identify QTL associated with oil content in maize kernels (Goldman et al. 1994; Berke and Rocheford 1995; Mangolin et al. 2004; Laurie et al. 2004; Song et al. 2004; Wassom et al. 2008; Zhang et al. 2008b; Yang et al. 2009) and provitamin A (Zhou et al. 2012). Approximately 50 QTLs for oil content, accounting for over 50% of the phenotypic variation, were identified in a large random mating population (IHO (70) × ILO (70); Laurie et al. 2004). Maize provides a large proportion of the daily intake of energy and other nutrients, including micronutrients for poor populations in many areas of sub-Saharan Africa that have limited access to animal foods (Allen 1993). Biofortification of maize is the best approach for alleviating the micronutrient deficiencies. Significant genetic variation for micronutrient concentrations exists in maize (Pfeiffer and McClafferty 2007). Several QTLs have been reported in maize controlling the mineral accumulation, especially Zn, Fe, Cu, and Mn (Agrawal et al. 2011; Lungaho et al. 2011; Simic et al. 2011; Qin et al. 2012; Figure 3.1). By using the intermated B73 × Mo17 recombinant inbred lines (IBMRILs) whose genome sequences were available, the QTLs affecting the total number of leaves before flowering and the number of leaves below the uppermost ear (NLBE) were mapped and characterized (Lauter et al. 2008). The availability of a physical map of B73 and high genetic resolution of the IBMRIL population resulted in

TABLE 3.1

List of QTLs Linked to Agronomically Important Traits

S. No.	Trait	QTL	References
1	SCMV resistance	Scm1 and Scm2	Duble et al. (2008)
2	Ear rot resistance	ER1	Xiang et al. (2012)
3	Fatty acid composition	pal9	Yang et al. (2010)
4	Plant height	qPH3.1	Teng et al. (2012)
5	Popping expansion volume	qPEV1-1	Li et al. (2008)
6	Gibberella stalk rot	qRfg1 and qRfg2	Yang et al. (2010)
7	Maize rough dwarf disease	QTL	Shi et al. (2012)
8	Kernel oil content	qKO1-1, QTL	Li et al. (2010), Yang et al. (2012)
9	Southern corn leaf blight	rhm1	Zhao et al. (2012)
10	Northern leaf blight	qNLB1.06 and qNLB1.02	Chung et al. (2010)
11	Fusarium ear rot		Li et al. (2011b)
12	Oil content	Oil 3.77	Zhang et al. (2008b)
13	Starch content	QTL	Dong et al. (2006)
14	Downy mildews resistance	QTL	George et al. (2003), Nair et al. (2005), Agrama et al. (1999)
15	Maize head smut	qHS2.09	Weng et al. (2012)
		qHSR1	Chen et al. (2008)
		QTL	Ding et al. (2008)
16	Gray leaf spot	qRgls1	Zhang et al. (2012)
17	Drought tolerance	mQTL	Almeida et al. (2013)
18	Early flowering time	*vgt1* and *vgt2*	Chardon et al. (2005)
19	N-remobilization and postsilking N-uptake	QTLs	Coque et al. (2008)
20	Zinc and iron concentration	ZnK, ZnC, FeK and FeC	Qin et al. (2012)
21	Maize test weight	QTL	Ding et al. (2011)
22	Tassel number	QTL	Gao et al. (2007)
23	Leaf area	QTL	Liu et al. (2010)
24	Leaf angle	QTL	Lu et al. (2007)
25	Leaf orientation	QTL	Lu et al. (2007)
26	Root architecture	QTL	Giuliani et al. (2005)
		Root-yield-1.06	Landi et al. (2010)
		Root-ABA1	Landi et al. (2007)
27	Controlling masculinization of ear tips	QTL	Holland and Coles (2012)

identification of QTLs being localized to smaller physical regions than in nonintermated maize populations (Balint-Kurti and Johal 2008) (Table 3.1).

3.2.3 Mapping and Tagging of Genes for Quality Traits

Gene tagging for quality traits has received great attention to improve the quality of maize for human consumption (Gupta et al. 2009). Normal maize protein is known to have a biological

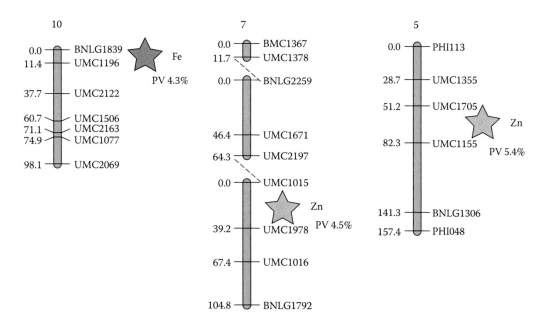

FIGURE 3.1

QTLs for Fe and Zn content in maize kernels. (From Agrawal, P.K., Jaiswal, S.K., Bhatt, J.C., Hossain, F., Nepolean, T., Guleria, S.K., Prasanna, B.M., and Gupta, H.S., 11th Asian Maize Conference held at Nanning, China, Book of extended summary, p. 332, 2011.)

value of 40% that of milk and, therefore, needs protein supplementation from legumes and animal products. The essential amino acids such as lysine and tryptophan are found in reduced quantities. The International Maize and Wheat Improvement Center (Centro Internacional de Mejoramiento de Maíz y Trigo; CIMMYT) developed quality protein maize (QPM), which harbors *opaque2* (*o2*), a recessive single gene along with modifiers, which resulted in nearly twice the amount of lysine and tryptophan, and this makes the protein of QPM equivalent to 90% of milk protein. Another mutant, that is, *opaque-16* in combination with *o2* increased the lysine content by 30% (Yang et al. 2005). Several modifiers have been identified that influenced the lysine content in maize (Holding et al. 2010; Wu and Messing 2011).

Maize is a model cereal crop for developing strategies for showing the promise for the development of provitamin A biofortified maize (Wurtzel et al. 2012). Maize exhibits a broad range of genetic variation for provitamin A content (pVAC) (Kurilich and Juvik 1999; Egesel et al. 2003; Chander et al. 2008). Major genes with large genetic effects (i.e., *crtRB1* and *LcyE*) have been identified and confirmed across different populations (Harjes et al. 2008; Yan et al. 2010). The mapping of provitamins A and total carotenoids in maize grain has been reported by Stevenson et al. (2006), where near-isogenic populations for a mutation in the *phytoene synthase* (*y1*) gene show a huge range in concentrations of β-carotene, α-carotene, β-cryptoxanthin, lutein, zeaxanthin, and total carotenoids. Babu et al. (2012) studied the polymorphism for provitamin A-related genes LcyE5′TE, LcyE3′Indel, and CrtRB1-3′TE. CrtRB1-3′TE had a twofold–tenfold effect on enhancing β-carotene and total pVAC. In higher plants, mineral accumulation appears to be controlled by a complex polygenic phenomenon. Identification of candidate genes involved in micronutrient accumulation using a bioinformatics tool could be one of the potential options. Chauhan (2006) used this approach for the scanning of available maize genome sequences, resulting in the identification of 33 genes predicted to be involved in iron and zinc transport in maize.

Fifteen genes belong to the yellow stripe (*YS*) family, nine to the zinc-regulated transporter/iron-regulated transporter protein (*ZIP*) family, six to the natural resistance-associated macrophage protein (*NRAMP*) family, two to the ferritin family, and one to the Fe^{+3}-chelate reductase oxidase (*FRO*) family. Members of each gene family possessed characteristic signature sequences and transmembrane domains of functionally characterized genes.

3.2.4 Association Mapping and Nested Association Mapping in Maize

Association mapping—the alternate of linkage mapping, also called linkage disequilibrium (LD)—refers to the analysis of statistical associations between genotypes, usually individual SNPs or SNP haplotypes, determined in a collection of individuals, and the traits (phenotypes) of the same individuals (Pritchard et al. 2000). It provides a valuable first step in many cases and a method to validate marker–gene associations, found through independent analyses. Association mapping in a population of diverse lines allows sampling of greater genetic diversity and also provides higher mapping resolution because more recombination events have occurred during historical diversification than during production of a biparental mapping population (Yu and Buckler 2006; Myles et al. 2009). The association mapping can be done either by candidate gene association or by whole genome scan (genome-wide association study) (Rafalski 2010). The candidate gene approach has been used along with association mapping to identify loci involved in the accumulation of carbohydrates and abscisic acid (ABA) metabolites during stress in maize (Setter et al. 2010). The major flowering-time QTL, *Vegetative to generative transition 1* (*Vgt1*), was identified by association mapping studies (Salvi et al. 2007). A significant association of nine candidate genes, for example, *indeterminate spikelet1* and inflorescence branching, *ramosa1* and *ramosa2* and ear structure, *sugary1* and seed oil content, and *terminal ear1* and ear length, were detected in the maize ancestor Balsas teosinte (*Zea mays* ssp. *parviglumis*) (Weber et al. 2007, 2008) in candidate-gene-based association mapping.

With the availability of dense genome-wide genotyping in maize, maize genetic diversity has been used to understand the molecular basis of phenotypic variation and to improve agricultural efficiency and sustainability (McMullen et al. 2009). This has been named *nested association mapping* (NAM), and has been developed based on 25 maize populations, each of which comprised 200 RILs derived by crossing 25 diverse inbred lines to a common inbred line, B73. With a dense coverage (2.6 cM) of common-parent-specific (CPS) markers, the genome information for the 5000 RILs can be inferred based on the parental genome information, leading to genome-wide high-resolution mapping by capturing a total of 136,000 recombination events. The power of NAM with 5000 RILs helped to precisely identify QTLs and genes. With the many ongoing genome sequencing projects, NAM will greatly facilitate the dissection of complex traits in many species in which a similar strategy can readily be applied. Genome-wide NAM, using 1.6 million SNPs, identified multiple candidate genes related to plant defense, including receptor-like kinase genes similar to those involved in basal defense (Poland et al. 2011). This population was also used to study the plant genetic architecture. Tian et al. (2011) reported that the leaf traits are dominated by small effects, with little epistasis, environmental interaction, or pleiotropy, and that the *liguleless* genes have contributed to the more upright leaves.

3.2.5 Marker-Assisted Selection for Maize Improvement

A large number of genes and QTLs have been identified in maize and provide an abundance of DNA marker–trait associations. DNA markers have enormous potential to

TABLE 3.2

Successful Examples of Marker-Assisted Selection for Maize Improvement

Trait	Gene/QTLs	References
Corn borer resistance	QTLs on chromosomes 7, 9 and 10	Willcox et al. (2002)
Earliness and yield	QTLs on chromosomes 5, 8 and 10	Bouchez et al. (2002)
Quality-protein maize	*Opaque 2* gene on chromosome 7	Dreher et al. (2003), Morris et al. (2003), Babu et al. (2005), Gupta et al. (2009), Agrawal and Gupta (2010)
Maize streak virus	QTL on chromosomes 1	Abalo et al. (2009)
Maize head smut		Li et al. (2012), Min et al. (2012)
Northern leaf blight	Genes on chromosome 2, 7, 8	Min et al. (2012)

improve the efficiency and precision of conventional plant breeding via marker-assisted selection (MAS) (Collard et al. 2008). Despite numerous reports of genes and QTLs tagging/mapping in maize (http://www.maizegdb.org/qtl.php), little has been published on the implementation of MAS-based breeding programs. Successful MAS applications have been reported for introgression breeding in maize, including introgressions of transgenes (Ragot et al. 1995) and conversions involving simple (Ho et al. 2002; Morris et al. 2003) or complex traits (Bouchez et al. 2002; Willcox et al. 2002; Johnson 2004; Niebur et al. 2004; Eathington 2005; Crosbie et al. 2006; Ragot and Lee 2007) (Table 3.2). The gene *RsrR* for maize head smut has been transferred into two elite maize inbred lines, "Huangzao4" and "Qi319," through traditional hybridization and marker-assisted selection (Li et al. 2012). Dreher et al. (2003) and Abalo et al. (2009) compared the MAS with conventional selection and observed MAS to be a cost-effective alternative to conventional selection. Economic analyses by Morris et al. (2003) regarding selection of quality-protein maize using theoretical breeding schemes determined that MAS was faster, but more costly, than phenotypic selection. Asea et al. (2012) analyzed the genetic gain and cost efficiency of MAS of maize for improved resistance to multiple foliar pathogens—Northern corn leaf blight (NCLB) caused by *Exserohilum turcicum*, GLS caused by *Cercospora zeae-maydis*, and maize streak caused by maize streak Mastrevirus (MSV)—by transferring six target rQTLs and found them highly effective in improving the level of host resistance to multiple foliar pathogens of maize, and potentially cost-effective.

3.2.6 Success Stories of Commercialization of MAS-Derived QPM Hybrids

3.2.6.1 MAS for Development of QPM Maize

It was already known that the *opaque 2* gene along with associated modifiers resulted in the development of QPM, with protein quality as good as 90% of that of milk protein. Since DNA markers within the exons of this gene were available, it was a viable proposition to use MAS for the introgression of the gene along with the necessary modifiers for the development of QPM maize cultivars. Utilizing this method, many normal maize inbreds and hybrids have been converted into QPM versions and were released for commercial cultivation (Babu et al. 2005; Gupta et al. 2009).

3.2.6.2 Vivek QPM 9

Gupta et al. (2009) successfully converted a promising maize hybrid viz. Vivek Maize Hybrid 9 into a QPM version using MAS. The resulting QPM hybrid, Vivek QPM 9,

FIGURE 3.2   **(See color insert)**
Cobs of Vivek QPM 9.

matures in 85–90 days and yields up to 6 tons maize grain per hectare (Figure 3.2). Vivek QPM 9 was evaluated during *Kharif* 2005 and 2007, under the All India Coordinated Research Project (AICRP) on Maize at seven locations in zone I and thirteen locations in zone IV in India. The performance of the Vivek QPM 9 (5843 kg/ha in zone I and 5435 kg/ha in zone IV) was on a par with Vivek Maize Hybrid 9 (5931 kg/ha and 5404 kg/ha in zone IV) in both the zones over a period of years. Besides, this hybrid showed 41% higher tryptophan in the endosperm than Vivek Maize Hybrid 9. In addition, the new QPM hybrid showed an equal level of resistance to *turcicum* blight—the most important disease of maize crop in the hills. Based on the performance for 2 years in zone I and zone IV, Vivek QPM 9 was released in 2008 for commercial cultivation in zone I and zone IV in India and for the organic conditions in the hills of Uttarakhand (Gupta et al. 2009).

3.2.6.3 Vivek QPM 21

Vivek QPM 21 (QPM version of Vivek Maize Hybrid 21) shows >70% enhancement in tryptophan over the original hybrid, Vivek Maize Hybrid 21. The tryptophan content of Vivek QPM 21 is 0.85 whereas it is 0.49 for Vivek Maize Hybrid 21. Vivek QPM 21 was also tested in the All India Coordinated Trial of *Kharif* 2007, 2008, and 2009, in which it performed equally well in respect of grain yield and other agronomic traits compared to non-QPM national check, Vivek Maize Hybrid 17, and Vivek Maize Hybrid 21. Vivek Maize Hybrid 21 was released for commercial cultivation in zones I, II, and IV in 2006. The parents of this hybrid have been converted into the QPM version using DNA markers, and this hybrid was reconstituted by crossing VQL 1 and VQL 17. This QPM hybrid shows more than 70% enhancement in tryptophan over the original hybrid. In the state trials of Uttarakhand under organic conditions this hybrid gave more than 2.4% higher yield over Vivek Maize Hybrid 21 with an average grain yield of 56.31 q/ha (Tables 3.3 and 3.4). Vivek QPM 21 was released for the state of Uttarakhand, India in the year 2012

TABLE 3.3

Performance (kg) of Vivek QPM 21 under AICRP on Maize

Genotype	2008	2009	Average
FQH 38	6015	6571	6293
Vivek MH 21	5929	6429	6179
Vivek MH 17	5735	5853	5794

TABLE 3.4

Performance (kg) of Vivek QPM 21 under State Varietal Trial, Uttarakhand

Genotype	2008	2010	Average
FQH 38	3936	7327	5631
FQH 55	3823	6820	5321
Vivek MH 21	3406	7592	5499
Vivek MH 23	3399	6499	4949
Vivek MH 9	2811	6227	4519

for commercial cultivation by the State Varietal Release Committee, Uttarakhand, for the hill conditions (Figure 3.3).

3.2.7 Genome Sequencing

On September 20, 2002, the National Science Foundation (NSF), USA, announced the launch of the Maize Genome Sequencing Project; a high-quality assembled complete

FIGURE 3.3 (See color insert)
Cobs of Vivek QPM 21.

genome sequence of cv. B73 was published in 2009 (http://www.maizegdb.org/sequencing_project.php). The other genome sequences that are nearing completion are of Mo17 genotype (http://www.phytozome.net/maize and Palomero Toluqueno). Being a model crop plant sequenced after rice, WGS of maize will facilitate the identification and functional analysis of genes not only in maize but also in other cereals. The maize genome is genetically diploid and consists of 10 chromosomes with an estimated size ranging from 2.3 to 2.7 Gb (Schnable et al. 2009). The maize genome consists mostly of a nongenic, repetitive fraction punctuated by islands of unique or low-copy DNA that harbor single genes or small groups of genes. The repetitive elements contribute significantly to the wide range of diversity within the species and include TEs, ribosomal DNA (rDNA), and high-copy short-tandem repeats mostly present at the telomeres, centromeres, and heterochromatin knobs.

3.2.8 Functional Genomics

Functional genomics allows large-scale gene function analysis as compared to the gene-by-gene approach that has normally been used to understand gene function. The information coming from the sequencing projects provides many inputs about the genes to be analyzed. Gene expression is highly influenced by the environment. Various high-throughput methods are available for detection and quantification of gene expression including the EST library (Adams et al. 1991), serial analysis of gene expression (SAGE) (Velculescu et al. 1995), and microarray (Duggan et al. 1999). The sequencing of the maize genome (Schnable et al. 2009) programs has provided a framework for the identification and functional characterization of genes and genetic networks for crop improvement and basic research. The draft genome of the B73 inbred line by sequencing nearly 17,000 assembled bacterial artificial chromosome (BAC) sequences anchored to the maize physical map is available through the ever-improving genome browsers at http://www.maizesequence.org and MaizeGDB (http://www.maizegdb.org). The availability of global transcriptome profiling technologies, such as DNA microarrays, together with the genome sequence offers the opportunity to understand patterns of transcription in the context of plant growth and development. Functional genomics approaches are being increasingly applied to investigate many aspects of maize biology. Sekhon et al. (2011) describe an atlas of global gene expression covering developmental steps, especially organ- and paralog-specific expression patterns of lignin biosynthetic pathway genes in vegetative organs during the life cycle of a maize plant. They used a NimbleGen microarray containing 80,301 probe sets to profile transcription patterns in 60 distinct tissues representing 11 major organ systems of inbred line B73. Mohanty et al. (2009) developed streamlined methods to generate maize FP-tagged lines using these regulatory elements, allowing analysis of tissue-specific expression and localized function. There are 364,385 maize ESTs in GenBank (2009, http://www.maizecdna.org), the majority of which are drawn from 10 inbred lines. ESTs or full-length cDNA served as the central resource in the study of individual and global gene expression through high-density microarrays and analysis of complex traits (Schadt et al. 2003). The transcriptome was characterized for various traits and organ specificity, for example, gene expression of water-deficit-stressed maize tissues such as female reproductive tissues (Zinselmeier et al. 2002), placenta, and endosperm in developing maize kernels (Yu and Setter 2003; Andjelkovic and Thompson 2006), leaves and roots of three-leaf-old seedlings (Zheng et al. 2004; Jia et al. 2006), developing immature ear and tassel (Zhuang et al. 2007), primary roots (Bassani et al. 2004; Poroyko et al. 2007; Spollen et al. 2008), and leaves and roots of three-leaf-old seedlings of maize with a short anthesis–silking interval (Li et al. 2010; Lu et al.

2011); salt-stressed tissues (Wang et al. 2003) and aluminum toxicity-stressed tissues such as roots (Maron et al. 2008). Functional annotation of full-length cDNA in gene prediction has sharply improved knowledge about the transcriptome of maize (Jia et al. 2006; Soderlund et al. 2009). Jia et al. (2006) constructed a full-length cDNA library of maize line Han 21, and a full-length cDNA macroarray was prepared and the expression analysis resulted in the identification of 79 genes upregulated by stress treatments and 329 downregulated genes in osmotically stressed maize seedlings. Alexandrov et al. (2009) generated 36,565 full-length cDNAs using different tissues and treatments from diverse hybrids, of which 10,084 were determined to be high-quality unique clones. Soderlund et al. (2009) generated 27,455 full-length cDNAs from maize inbred B73 (http://www.maizecdna.org).

Maize is a typical plant with respect to the proportion of its genome that is composed of active TEs. Large populations of maize have been created containing highly active Mutator transposons to saturate the genome with insertional mutations, and these have been used as transposon tagging resources and functional genomics (Candela and Hake 2008). McCarty et al. (2005) developed the Mutator system by creating the unique UniformMu population and database, which facilitate the high-throughput molecular analysis of Mu-tagged mutants and gene knockouts in a uniform inbred background. The public database contains pedigree and phenotypic data (www.plantgdb.org) for over 2000 independent seed mutants selected from a population of 31,548 F_2 lines and integrated with analyses of 34,255 MuTAIL sequences.

3.2.9 Transgenics for Genomic Studies

The function of a gene can also be identified by either gain-of-function or loss-of-function approaches in plants. Gain-of-function is achieved by increasing gene expression of transgenes under the control of a strong promoter (Nakazawa et al. 2003; Weigel et al. 2000). In this approach, phenotypes of gain-of-function mutants that overexpress a member of a gene family can be observed without interference from other family members, which allows the characterization of functionally redundant genes (Ito and Meyerowitz 2000; Nakazawa et al. 2001). RNAinterference (RNAi) is a powerful tool for functional genomics in a number of species (McGinnis 2010). In some cases, the RNAi construct was able to cause a reduction in the steady-state RNA levels of not only the target gene, but also another closely related gene (McGinnis et al. 2007). Maize is susceptible to *Aspergillus flavus* infection and subsequent contamination with aflatoxins, the potent carcinogenic secondary metabolites of the fungus. Using the RNAi technology for pathogenesis-related protein 10 (*PR10*) gene, the expression of *PR10* was reduced by 65% to more than 99% in transgenic callus lines (Chen et al. 2009). In maize, male fertility gene *Ms45* and several anther-expressed genes of unknown function were used to evaluate the efficacy of generating male-sterile plants by transforming with constitutively expressing inverted repeats (IRs) of the *Ms45* promoter. These sterile plants lacked MS45 mRNA due to transcriptional inactivity of the target promoter (Cigan et al. 2005). Cinnamyl alcohol dehydrogenase (CAD) is a key enzyme involved in the last step of monolignol biosynthesis. Transgenic CAD-RNAi plants show a different degree of enzymatic reduction and produced higher levels of ethanol compared to wild-type (Fornaléa et al. 2012). Similarly, RNAi technology has been used for many other studies, for example, nitric oxide-activated calcium/calmodulin-dependent protein kinase (Ma et al. 2012), production of high-amylose through silencing of sbe2a genes (Guan et al. 2011), and maize storage protein mutants γ-zein (Wu and Messing 2010).

The other recent method of analyzing functional genomics is virus-induced gene silencing (VIGS), especially for species not amenable to stable genetic transformation. This technique allows the use of VIGS as a high-throughput method that will exploit the potential of genome and transcriptome projects further (Becker and Lange 2009). The VIGS system based on the brome mosaic virus (BMV) was used to identify maize genes that are functionally involved in the interaction with *Ustilago maydis* (Linde et al. 2011). Shi et al. (2012) determined that *ZmTrm2*, a gene encoding thioredoxin *m* during SCMV infection and silencing in maize by VIGS, significantly enhanced systemic SCMV infection.

3.3 Maize Proteomics

Proteomics has been defined as "the systematic analysis of the protein population in a tissue, cell, or sub-cellular compartment" and is often associated with two-dimensional electrophoresis (2-DE). The concept of "proteome" (for PROTEin complement expressed by a genOME; Wilkins et al. 1996) has emerged recently as a consequence of questions raised by several genome and postgenome projects. The first large-scale proteomic plant work was published on *Arabidopsis*. The DNA/RNA methodologies have allowed innumerable outstanding breakthroughs in genomics; however they have limitations, particularly in physiology, because the protein structure and function cannot be completely described from the gene sequence. After synthesis, the proteins usually undergo posttranslational modifications, such as phosphorylation, which can alter their activity and location in the cell. Proteomics addresses analytical questions about the abundance and distribution of proteins in the organism, the expression profiles of different tissues, and the identification and localization of individual proteins of interest. Present proteomics research aims both at identifying new proteins in relation to their function and ultimately at unraveling how their expression is controlled within regulatory networks. The classical proteomics studies involve 2-DE, which is often combined with protein identification by mass spectrometry (MS) analysis. Nevertheless, the identification of proteins using this approach is limited by the sensitivity of the protein complement to extract preparation, running conditions, and gel composition. Protein array technology is a highly parallel approach to classical proteomics and the concept enables the connection of recombinant proteins to clones identified. The possible applications of plant proteomics are as follows.

1. Identification of proteomes and comparative proteomics with the aim to study differential protein expression. The most widespread techniques are 2-DE, using immobilized pH gradient strips in the first dimension and sodium dodecyl sulfate polyacrylamide gel electrophoresis (SDS-PAGE) in the second dimension for soluble and peripheral membrane proteins. Chromatography and organic solvent fractionation are the most appropriate techniques for the identification of membrane proteins.

2. Identification of protein–protein interactions and multi-subunit complexes using nondenaturing gel electrophoresis (e.g., blue-native gel electrophoresis) and different types of chromatography, and immune affinity purification techniques.

3. Analysis of posttranslational modifications, including phosphorylation, lipid modification, glycosylation, processing, and proteolysis.

4. Structural determination of protein complexes, using limited proteolysis, possibly combined with crosslinking or isotope exchange.

Several classical works have been published on maize proteomics at tissue level and also at organelle level. Several aspects of the proteomics of maize studied so far will be discussed further.

3.3.1 Maize Proteomics for Abiotic Stress Tolerance

Drought is one of the most important environmental stress factors limiting maize plant growth and productivity. In particular, global climate change that causes rising temperature and altered soil moisture has resulted in more severe drought in recent years and made water resources more limiting. To understand the mechanisms of plant adaptation to drought, it is a crucial step to identify the key genes that can be used for engineering transgenic crops with improved drought tolerance without reducing yield or biomass. It is generally believed that roots first perceive a dehydration stress signal when the water deficit reaches a certain level (Comstock 2002). However, the mechanisms of roots' response to drought remain unclear. To identify the proteins associated with drought, study of ABA is helpful. Proteome analysis has been employed to study alterations in protein expression in response to drought and ABA (Hu et al. 2010a) in leaves of maize seedlings. Hu et al. (2011) studied differential expression of proteins in maize roots in response to ABA and drought. To clarify the role of ABA in protein expression of root response to drought, root protein patterns were monitored using a proteomic approach in maize ABA-deficient mutant vp5 and its wild-type Vp5 exposed to drought, and 2-DE was used to identify drought-responsive protein spots in maize roots. Hu et al. conducted experiments under different conditions that express proteins under pathways such as ABA-dependent, ABA-independent, and both. In the ABA-dependent pathway they observed upregulation of an anionic peroxidase and two putative uncharacterized proteins by drought; but in both ABA-dependent and ABA-independent pathways there is an upregulation of glycine-rich RNA binding protein, pathogenesis-related protein, an enolase, a serine/threonine-protein kinase receptor, and a cytosolic ascorbate peroxidase by drought; a nuclear transport factor, a nucleoside diphosphate kinase, a putative uncharacterized protein, and a peroxiredoxin-5 were upregulated by drought in an ABA-independent way. Late embryogenesis abundant (LEA) proteins constitute a complex set of proteins that participate in several plant stress responses. These LEA proteins accumulate naturally in seed or pollen grains, and in plant vegetative tissues during exposure to abiotic stress. During the last stages of seed development, high levels of LEA proteins accumulate as desiccation proceeds. This makes this plant structure especially suitable for LEA proteomic analysis. Amara et al. (2012) have identified 20 unfolded maize embryo proteins, 13 of which belong to the LEA family on the basis of the unusual heat stability and acid solubility characteristic of unfolded proteins. These LEA Pfam group motifs have yielded an estimation of two under group 1, nine under group 2, and 14 in the group 3 LEA proteins in the maize genome. On this basis, they recovered all group 1 members, one member of group 2, and six members of group 3 as components of the maize embryo proteome. The two group 1 proteins identified, Emb5 and Emb564, were detected in high abundance in mature embryos. In contrast, Rab17 was the only member from group 2 found in the maize embryo.

3.3.2 Maize Proteomics for Biotic Stress Tolerance

Less work has been done regarding proteomics for biotic stress tolerance of maize as compared to abiotic stresses. A few proteomic studies on maize diseases have been reported. Maize rough dwarf disease (MRDD) is a viral disease that occurs worldwide, and three pathogens result in MRDD. They are maize rough dwarf virus (MRDV), Mal de Río Cuarto virus (MRCV), and rice black-streaked dwarf virus (RBSDV) (Distéfano et al. 2002; Dovas et al. 2004; Harpaz 1959). Among them, MRDV and MRCV are the primary and important MRDD pathogens in Europe and South America, respectively. The virus RBSDV is considered as the causal agent of MRDD in China and is transmitted in a persistent manner by the insect vector *Laodelphax striatellus* (Wang et al. 2003; Zhang et al. 2001). Kunpeng et al. (2011) studied extensively the proteome profile of maize leaf tissue at the flowering stage after long-term treatment with RBSDV infection. They identified 91 differentially accumulated proteins that belong to multiple metabolic/biochemical pathways. Further analysis of these identified proteins showed that MRDD resulted in dramatic changes in the glycolysis and starch metabolism, and eventually in significant differences in morphology and development between virus-infected and normal plants. Even then, MRDD occurrence increased the demands for G-proteins, antioxidant enzymes, lipoxygenases, and UDP-glucosyl transferase BX9, which may play important roles in response of maize against virus infection. The results also indicated that MRDD is a complicated disease controlled by multigenes participating in different pathways. By detecting proteome changes in composition and concentration, the data generated present cellular status more directly. These responsive proteins reflect changes of biochemical and cellular pathways in the host. Thus proteomic analysis, in combination with transcriptome analysis, can promote our understanding of plant growth and metabolic responses to pathogen infection. Chivasa et al. (2005) analyzed the pathogen-induced changes in the maize extracellular matrix proteins and showed that the extracellular matrix may play complex roles in the defensive processes, especially in signal modulation during pathogen-induced defense responses. However, compared to microarray analysis, few proteomic studies targeting the pathogen-induced protein expression patterns have been reported so far (Šamaj and Thelen 2007). The host resistance strategy for eliminating aflatoxins from corn has been advanced by the discovery of natural resistance traits such as proteins. This progress was aided by the development of a rapid laboratory-based kernel screening assay (KSA) used to separate resistant from susceptible seed, and for investigating kernel resistance. Several proteins associated with resistance (RAPs) have been identified using 1-D PAGE. However, proteomics is now being used to further the discovery of RAPs. This methodology has led to the identification of stress-related RAPs as well as other antifungal proteins. Campo et al. (2004) used a proteomic approach to identify a number of proteins, including pathogenesis-specific proteins and antioxidative enzymes, which are differentially accumulated in response to fungal infection in maize embryos.

3.3.3 Proteomics for Maize Seed Science

Seed viability is an important agronomic trait defined as the potential to produce viable seedlings. Wang et al. (2009) reported that controlled deterioration treatment (CDT) results in the loss of maize seed viability, accompanied by a change in the maize embryo proteome. There are as yet no reports on proteomic characterization of specific proteins associated with seed viability in maize. Recently, Wu et al. (2011) identified specific proteins related to maize seed viability. They used seeds of Zhengdan 958 (one of the high-yield maize hybrids in China) for comparative proteomic analysis. Interestingly, prominent small heat shock proteins, LEA proteins, and antioxidant enzymes were highly upregulated, while two proteases were highly downregulated in R embryos compared to W embryos. One of the LEA proteins was EMB564, which declined in abundance during artificial aging of seeds. These results suggested an association of LEA protein EMB564 with maize seed viability. It would be of interest to use these small proteins to develop quick tests for seed quality.

3.3.4 Proteome Map of Maize Endosperm

During maize seed development, several physiological steps involving complex and inter-related processes can be recognized. From 4 to 10 days after pollination, cell division and differentiation take place, and then storage compounds are synthesized from 12 days after pollination to maturity. At around 25 days after pollination, the relative water content of the endosperm begins to decrease and the dry down process initiates. While morpho-logical steps of this development are well known, until now the underlying physiological and molecular mechanisms have been largely unknown. These complex processes can be resolved by studying the proteome of the endosperm at several key stages, which would give insight on the functional gene products and how their expression is modulated. Mechin et al. (2004) have established a proteome reference map for maize endosperm by 2D gel electrophoresis and protein identification with LC–MS/MS analysis. Among the 632 protein spots processed, 496 were identified by matching against the NCBInr and ZMtuc-tus databases (using the SEQUEST software). Forty-two percent of the proteins were iden-tified against maize sequences, 23% against rice sequences, and 21% against *Arabidopsis* sequences. Identified proteins were not only cytoplasmic but also nuclear, mitochondrial, or amyloplastic. Metabolic processes, protein destination, protein synthesis, cell rescue, defense, cell death, and aging are the most abundant functional categories, comprising almost half of the 632 proteins analyzed in Mechin et al.'s study. The proteome map consti-tutes a powerful tool for physiological studies and is a step for investigating maize endo-sperm development.

3.4 Metabolomics in Maize

After the establishment of technologies for genomics, transcriptomics, and proteomics, the remaining functional genomics challenge is metabolomics. Metabolites are the end products of cellular regulatory processes, and are the ultimate response of biological sys-tems to genetic or environmental changes. Metabolomics is the term coined for essentially comprehensive, nonbiased, high-throughput analyses of complex metabolite mixtures typical of plant extracts. Metabolomics not only assists in the establishment of a deeper understanding of the complex nature of plant metabolic networks and their responses to environmental and genetic change, but also will provide unique insights into the funda-mental nature of plant phenotypes in relation to development, physiology, tissue identity, resistance, biodiversity, and so forth. The cascading effect begins with a modified genome, leading to modified proteins and, consequently, a change in the pattern and concentration of metabolites. Quantitative and qualitative measurements of large numbers of cellular metabolites provide a broad view of the biochemical status of an organism, which can then be used to monitor and assess gene function. The First International Congress on Plant Metabolomics was held in Wageningen, the Netherlands, in April 2002, with the primary goal of bringing together the players who are already active in this field and those who soon plan to be. Metabolomics is driven primarily by recent advances in mass spectrometry (MS) technology and by the goals of functional genomics efforts. In the early

1970s gas chromatography–mass spectrometry (GC–MS) technologies were used to analyze steroids, acids, and neutral and acidic urinary drug metabolites. Metabolic profiling research remained stable in the 1980s with approximately 10–15 publications a year. Workers at the Max Planck Institute in Germany pioneered this approach for plants. Since then, the number of academic and commercial groups using and entering this field has grown exponentially. Plant metabolomics is still a field in its infancy, but the opportunities are almost endless. Metabolomics offers the unbiased ability to characterize and differentiate genotypes and phenotypes based on metabolite levels. The following is just a subset of the possible applications of metabolomics in maize crop improvement.

The combination of nuclear magnetic resonance (NMR) spectroscopy, chemometric methods, and principal component analysis (PCA) is a useful tool for the discrimination of maize silks in respect to their chemical composition, including rapid authentication of the raw material of current pharmacological interest (Fiehn 2002; Halket et al. 2005; Shulaev 2006). Marcelo et al. (2012) determined metabolic fingerprint and pattern recognition of silk extracts from seven maize landraces cultivated in southern Brazil by NMR spectroscopy and chemometric methods. Aqueous extract from maize silks is used by traditional medicine for the treatment of several ailments, mainly related to the urinary system. PCA of the NMR data set showed clear discrimination among the maize varieties by PC1 and PC2, pointing out three distinct metabolic profiles. Target compound analysis showed significant differences ($p < 0.05$) in the content of protocatechuic acid, gallic acid, *t*-cinnamic acid, and anthocyanins, corroborating the discrimination of the genotypes in this study as revealed by PCA analysis. Metabolomics is also useful for predicting the complex heterotic traits in maize, as has been supported by some studies. Christian et al. (2012) crossed 285 diverse Dent inbred lines from worldwide sources with two testers and predicted their combining abilities for seven biomass- and bioenergy-related traits using 130 metabolites. Whole metabolic prediction models were built by fitting effects for all metabolites. Prediction accuracies ranged from 0.60 to 0.80 for metabolites, allowing a reliable screening of large collections of diverse inbred lines for their potential to create superior hybrids. Piccioni et al. (2009) studied NMR metabolic profiling of transgenic maize with the *Cry1A(b)* gene and nontransgenic maize using one- and two-dimensional NMR techniques. About 40 water-soluble metabolites in the maize seed extracts were identified. In particular, ethanol, lactic acid, citric acid, lysine, arginine, glycine-betaine, raffinose, trehalose, α-galactose, and adenine were identified for the first time in the ^{1}H NMR spectrum of maize seed extracts. However, a higher concentration of ethanol, citric acid, glycine-betaine, trehalose, and another compound not yet completely identified was observed in the transgenic extracts than in the nontransgenic samples.

3.4.1 Maize Metabolomics for Abiotic Stress Tolerance

Under drought situations, maize ovule abortion appears to be related to the flux of carbohydrates to the young ear around flowering, and concurrent photosynthesis is required to maintain this above threshold levels (Zinselmeier et al. 1995). Drought has additionally been shown to reduce invertase activities in the ovaries, which would also likely result in a reduced flux of hexose sugars, altered hormone balance, and ovary abortion (Zinselmeier et al. 2000; Chourey et al. 2010). The tested organs of leaf blade, leaf sheath, ear, husk, and silks differed very strongly in their metabolite abundance, with blade tissue being the most diverse in terms of metabolites while containing considerably lower metabolite levels than the other tissues. Subjecting the metabolite data to a PCA separated all tested tissue regarding metabolite composition, independently from genotype or treatment situation, reflecting that major differences exist in metabolite composition within the different tissues. PC1 revealed that blade tissue shows the most contrasting metabolic profile in comparison to other tissues. PC2 separates different treatments. The data of the heat map reflected those of the PCA analysis in demonstrating that leaf blade material definitely exhibits large metabolic changes following drought stress. For example, there were dramatic increases in tryptophan, phenylalanine, and histidine as well as in proline. By contrast, the levels of pyruvic acid decreased (in four of the six hybrids), as did the levels of quinic acid (in five of six hybrids) following drought stress. Similarly, high salinity, caused by either natural (e.g., climatic changes) or anthropic factors (e.g., agriculture), is a widespread environmental stressor that can affect development and growth of salt-sensitive plants, leading to water deficit, the inhibition of intake of essential ions, and metabolic disorders. The application of an NMR-based metabolic profiling approach to the investigation of saline-induced stress in maize plants was studied by Gavaghan et al. (2011). The maize seedlings were grown in 0, 50, or 150 mM saline solution. Plants were harvested after 2, 4, and 6 days ($n = 5$ per class and time point) and H NMR spectroscopy was performed separately on shoot and root extracts. Spectral data were analyzed and interpreted using multivariate statistical analyses. A distinct effect of time/growth was observed for the control group, with relatively higher concentrations of acetoacetate at day 2 and increased levels of alanine at days 4 and 6 in the root extracts, whereas concentration of alanine was positively correlated with the shoot extracts harvested at day 2 and *trans*-aconitic acid increased at days 4 and 6. A clear dose-dependent effect, superimposed on the growth effect, was observed for saline-treated shoot and root extracts. This was correlated with increased levels of alanine, glutamate, asparagine, glycine-betaine, and sucrose, and decreased levels of malic acid, *trans*-aconitic, acid and glucose in shoots. Correlation with salt-load shown in roots included elevated levels of alanine, γ-amino-*N*-butyric acid, malic acid, succinate, and sucrose and depleted levels of acetoacetate and glucose. The metabolic effect of high salinity was predominantly consistent with osmotic stress as reported for other plant species and was found to be stronger in the shoots than in the roots. Using multivariate data analysis it is possible to investigate the effects of more than one environmental stressor simultaneously.

3.4.2 Characterization of Metabolism

To study the role of sucrose synthase (SuSy) in the regulation of the carbon (C) partitioning in central metabolism in maize root tips, and the role of SuSy in C partitioning, metabolic fluxes were analyzed in maize root tips of a double mutant of SuSy genes, sh1 sus1 and the corresponding wild type, W22 (Alonsoa et al. 2007). [U-14C]-glucose pulse labeling experiments permitted the quantification of unidirectional fluxes into sucrose, starch, and cell wall polysaccharides. These results clearly established that, in maize root tips, starch is produced from ADP-Glc synthesized in the plastid and not in the cytosol by sucrose synthase. Unexpectedly, the flux of cell wall synthesis was increased in the double mutant. This observation indicates that, in maize root tips, SH1 and SUS1 are not specific providers for cellulose biosynthesis. Improving N use efficiency (NUE) is particularly relevant to most crops requiring large amounts of N fertilizers to obtain maximum yield, since NUE has been estimated on average to be far less than 50% (Raun and Johnson 1999). N use efficiency can be defined as the grain or biomass yields per unit of available N in the soil, which includes the residual mineral nutrients, the organic N present in the soil, and that provided by fertilization. Metabolomic, transcriptomic, and, to a lesser extent, proteomic studies have been conducted for the high-throughput phenotyping necessary for large-scale physiological, molecular, and quantitative genetic studies, aimed at identifying the function of a particular gene or set of genes involved in the control of complex physiological traits such as NUE (Meyer et al. 2007; Lisec et al. 2008; Kusano et al. 2011). In order to improve knowledge of the physiological and molecular responses of maize leaves to long-term N deprivation, a metabolomic analysis was conducted in parallel with a proteomic and a transcriptomic study at two key stages of plant development (Amiour et al. 2012). It was found that a number of key plant biological functions were either up- or downregulated when N was limiting, including major alterations to photosynthesis, C metabolism, and, to a lesser extent, downstream metabolic pathways. It was also found that the impact of the N deficiency stress resembled the response of plants to a number of other biotic and abiotic stresses, in terms of transcript, protein, and metabolite accumulation. The genetic and metabolic alterations were different during the N assimilation and the grain-filling period, indicating that plant development is an important component for identifying the key elements involved in the control of plant NUE. Pavlik et al. (2010) analyzed metabolome of maize plants growing under different nitrogen nutrition conditions; sequential extraction of fresh biomass was used and isolated fractions were characterized and evaluated using IR spectra. The results showed an increased induction of oxalic acid in plants after 4 g nitrogen application. Degradation of this acid induced oxidative stress, therefore a strong correlation among contents of oxalic acid, flavonoids, and compounds with amide nitrogen (glutathione) was observed in plants growing under 4 g nitrogen nutrition.

3.5 Conclusion

The latest knowledge in omics improves our understanding of DNA variation, transcription patterns, and protein expression profiling. This knowledge and related technologies in maize will make maize a model crop among cereals and other grasses (except rice) by giving new insights for orthologous and paralogous maize genes. This information can be used for other plant species. The most significant breakthrough in agricultural biotechnology will come from the developed and affordable NGS methods to study the structure of genomes and the genetic mechanisms behind economically important traits. The genomics-assisted breeding was effectively used to develop QPM hybrids to overcome the bottlenecks of conventional breeding practices. The integration of multiple omics tools and strategies such as high-throughput genome-scale genotyping platforms including whole-genome resequencing, proteomics, and metabolomics facilitated the discovery of key genes involved in biological processes and aided the design of new plant types in maize, and the development of strategies for future crop breeding programs for the betterment of humankind.

References

Abalo, G., R. Edema, J. Derera, and P. Tongoona. 2009. A comparative analysis of conventional and marker-assisted selection methods in breeding maize streak virus resistance in maize. *Crop Sci* 49:509–20.

Adams, M. D., J. M. Kelley, J. D. Gocayne, M. Dubnick, M. H. Polymeropoulos, H. Xiao, C. R. Merril, et al. 1991. Complementary DNA sequencing: Expressed sequence tags and human genome project. *Science* 252 (5013):1651–56.

Agrama, H. A., M. E. Moussa, M. E. Naser, M. A. Tarek, and A. H. Ibrahim. 1999. Mapping of QTL for downy mildew resistance in maize. *Theor Appl Genet* 99 (3–4):519–23.

Agrawal, P. K. and H. S. Gupta. 2010. Enhancement of protein quality of maize using biotechnological options. *Anim Nutr Feed Techn* 10 (spl):79–91.

Agrawal, P. K., S. K. Jaiswal, J. C. Bhatt, F. Hossain, T. Nepolean, S. K. Guleria, B. M. Prasanna, and H. S. Gupta. 2011. Genetic vatability and QTL mapping of kernel iron and zinc concentration in maize (*Zea mays* L.) genotypes. 11th Asian Maize Conference held at Nanning, China from 7th–11th November 2011, Book of extended summary, p. 332.

Alexandrov, N. N., V. V. Brover, S. Freidin, M. E. Troukhan, T. V. Tatarinova, H. Zhang, T. J. Swaller, et al. 2009. Insights into corn genes derived from large-scale cDNA sequencing. *Plant Mol Biol* 69 (1–2):179–94.

Allen, L. H. 1993. The nutrition CRSP: What is marginal malnutrition, and does it affect human function? *Nutr Rev* 51 (9):255–67.

Almeida, G. D., D. Makumbi, C. Magorokosho, S. Nair, A. Borem, J. M. Ribaut, M. Banziger, B. M. Prasanna, J. Crossa, and R. Babu. 2013. QTL mapping in three tropical maize populations reveals a set of constitutive and adaptive genomic regions for drought tolerance. *Theor Appl Genet*. 126:583–600.

Alonsoa, A. P., P. Raymonda, M. Hernoulda, C. R. Mourob, A. de Graafc, P. Choudhary, M. Lahayeb, Y. Shachar-Hille, D. Rolina, and M. D. Noubhani. 2007. A metabolic flux analysis to study the role of sucrose synthase in the regulation of the carbon partitioning in central metabolism in maize root tips. *Metab Eng* 9:419–32.

Amara, I., A. Odena, E. Oliveira, A. Moreno, K. Masmoudi, M. Pagès, and A. Goday. 2012. Insights into maize LEA proteins: From proteomics to functional approaches. *Plant Cell Physiol* 53 (2):312–29.

Amiour, N., S. Imbaud, G. Clement, N. Agier, M. Zivy, B. Valot, T. Balliau, P. Armengaud, I. Quillere, R. Canas, T. Tercet-Laforgue, and B. Hirel. 2012. The use of metabolomics integrated with transcriptomic and proteomic studies for identifying key steps involved in the control of nitrogen metabolism in crops such as maize. *J Exp Bot* 63:5017–33.

Andjelkovic, V. and R. Thompson. 2006. Changes in gene expression in maize kernel in response to water and salt stress. *Plant Cell Rep* 25 (1):71–79.

Asea, G., B. S. Vivek, P. E. Lipps, and R. C. Pratt. 2012. Genetic gain and cost efficiency of marker-assisted selection of maize for improved resistance to multiple foliar pathogens. *Mol Breeding* 29:515–27.

Babu, R., S. K. Nair, A. Kumar, S. Venkatesh, J. C. Sekhar, N. N. Singh, G. Srinivasan, and H. S. Gupta. 2005. Two-generation marker-aided backcrossing for rapid conversion of normal maize lines to quality protein maize (QPM). *Theor Appl Genet* 111 (5):888–97.

Babu, R., S. K. Nair, A. Kumar, H. S. Rao, P. Verma, A. Gahalain, I. S. Singh, and H. S. Gupta. 2006. Mapping QTLs for popping ability in a popcorn × flint corn cross. *Theor Appl Genet* 112 (7):1392–99.

Babu, R., N. P. Rojas, S. Gao, J. Yan, and K. Pixley. 2012. Validation of the effects of molecular marker polymorphisms in LcyE and CrtRB1 on provitamin A concentrations for 26 tropical maize populations. *Theor Appl Genet* 126:389–99.

Balint-Kurti, P. J. and G. S. Johal. 2008. Maize disease resistance. In Bennetzen, J. L. and Hake, S. C. (eds), *Maize Genetics Handbook: Its Biology*, pp. 251–70. Springer, New York.

Bassani, M., P. M. Neumann, and S. Gepstein. 2004. Differential expression profiles of growth-related genes in the elongation zone of maize primary roots. *Plant Mol Biol* 56 (3):367–80.

Beavis, W. D. and D. Grant. 1991. A linkage map based on information from four F2 opulations of maize (*Zea mays* L.). *Theor Appl Genet* 82:636–44.

Becker, A. and M. Lange. 2009. VIGS—Genomics goes functional. *Trends Plant Sci* 15 (1):1–4.

Berke, T. and T. R. Rocheford. 1995. Quantitative trait loci for flowering, plant and ear height, and kernel traits in maize. *Crop Sci* 35:1542–49.

Bouchez, A., F. Hospital, M. Causse, A. Gallais, and A. Charcosset. 2002. Marker-assisted intro-gression of favorable alleles at quantitative trait loci between maize elite lines. *Genetics* 162 (4):1945–59.

Burr, B., F. A. Burr, K. H. Thompson, M. C. Albertson, and C. W. Stuber. 1988. Gene mapping with recombinant inbreds in maize. *Genetics* 118:519–26.

Campo, S., M. Carrascal, M. Coca, J. Abián, and B. S. Segundo. 2004. The defense response of germi-nating maize embryos against fungal infection: A proteomics approach. *Proteomics* 4:383–96.

Candela, H. and S. Hake. 2008. The art and design of genetic screens: Maize. *Nat Rev Genet* 9 (3):192–203.

Causse, M., S. Santoni, C. Damerval, A. Maurice, A. Charcosset, J. Deatrick, and D. Vienne. 1996. A composite map of expressed sequences in maize. *Genome* 39 (2):418–32.

Chander, S., Y. Q. Guo, X. H. Yang, J. Zhang, X. Q. Lu, J. B. Yan, T. M. Song, T. R. Rocheford, and J. S. Li. 2008. Using molecular markers to identify two major loci controlling carotenoid contents in maize grain. *Theor Appl Genet* 116 (2):223–33.

Chardon, F., D. Hourcade, V. Combes, and A. Charcosset. 2005. Mapping of a spontaneous mutation for early flowering time in maize highlights contrasting allelic series at two-linked QTL on chromosome 8. *Theor Appl Genet* 112 (1):1–11.

Chauhan, R. S. 2006. Bioinformatics approach toward identification of candidate genes for zinc and iron transporters in maize. *Curr Sci* 91:510–15.

Chen, X., W. C. Li, and F. L. Fu. 2009. Bioinformatic prediction of microRNAs and their target genes in maize. *Yi Chuan* 31 (11):1149–57.

Chen, Y., Q. Chao, G. Tan, J. Zhao, M. Zhang, Q. Ji, and M. Xu. 2008. Identification and fine-map-ping of a major QTL conferring resistance against head smut in maize. *Theor Appl Genet* 117 (8):1241–52.

Chivasa, S., W. J. Simon, X. L. Yu, N. Yalpani, and A. R. Slabas. 2005. Pathogen elicitor-induced changes in the maize extracellular matrix proteome. *Proteomics* 5:4894–904.

Chourey, P. S., Q. B. Li, and D. Kumar. 2010. Sugar-hormone crosstalk in seed development: Two redundant pathways of IAA biosynthesis are regulated differentially in the invertase-deficient minature1 (mn1) seed mutant in maize. *Mol Plant* 3:1026–36.

Christian, R., A. C. Eysenberg, C. Grieder, J. Lisec, F. Technow, R. Sulpice, T. Altmann, M. Stitt, L. Willmitzer, and A. E. Melchinger. 2012. Genomic and metabolic prediction of complex heter-otic traits in hybrid maize. *Nat Genet* 44:217–20.

Chung, C. L., J. M. Longfellow, E. K. Walsh, Z. Kerdieh, G. Van Esbroeck, P. Balint-Kurti, and R. J. Nelson. 2010. Resistance loci affecting distinct stages of fungal pathogenesis: Use of introgres-sion lines for QTL mapping and characterization in the maize—Setosphaeria turcica pathosys-tem. *BMC Plant Biol* 10:103.

Cigan, A. M., E. Unger-Wallace, and K. Haug-Collet. 2005. Transcriptional gene silencing as a tool for uncovering gene function in maize. *Plant J* 43 (6):929–40.

Coe, E. H., D. A. Hoisington, and M. G. Neuffer. 1987. Linkage map of corn (maize) (*Zea mays* L.). *Maize Genet Coop Newsl* 61:116–47.

Collard, B. C. Y., C. M. Cruz, K. L. McNally, P. S. Virk, and D. J. Mackill. 2008. Rice molecular breeding laboratories in the genomics era: Current status and future considerations. *Int J Plant Genom* 524847.

Comstock, J. P. 2002. Hydraulic and chemical signalling in the control of stomatal conductance and transpiration. *J Exp Bot* 53:195–00.

Coque, M., A. Martin, J. B. Veyrieras, B. Hirel, and A. Gallais. 2008. Genetic variation for N-remobilization and postsilking N-uptake in a set of maize recombinant inbred lines. 3. QTL detection and coincidences. *Theor Appl Genet* 117 (5):729–47.

Crosbie, T. M., S. R. Eathington, G. R. Johnson, M. Edwards, R. Reiter, S. Stark, R. G. Mohanty, et al. 2006. Plant breeding: Past, present, and future. In Lamkey, K. R. and Lee, M. (eds), *Plant Breeding: The Arnel R.Hallauer International Symposium*, pp. 3–50. Blackwell, Ames, IA.

Davis, G. L., M. D. McMullen, C. Baysdorfer, T. Musket, D. Grant, M. Staebell, G. Xu, et al. 1999. A maize map standard with sequenced core markers, grass genome reference points and 932 expressed sequence tagged sites (ESTs) in a 1736-locus map. *Genetics* 152 (3):1137–72.

Deutscher, D. 1978. The current status of breeding for protein quality in corn. *Adv Exp Med Biol* 105:281-300.

Ding, J. Q., X. M. Wang, S. Chander, and J. S. Li. 2008. Identification of QTL for maize resistance to common smut by using recombinant inbred lines developed from the Chinese hybrid Yuyu22. *J Appl Genet* 49 (2):147–54.

Ding, J. Q., J. L. Ma, C. R. Zhang, H. F. Dong, Z. Y. Xi, Z. L. Xia, and J. Y. Wu. 2011. QTL mapping for test weight by using F(2:3) population in maize. *J Genet* 90 (1):75–80.

Distéfano, A. J., L. R. Conci, M. Hidalgo, F. A. Guzman, H. E. Hopp, and M. del Vas. 2002. Sequence analysis of genome segments S4 and S8 of Mal de Río Cuarto virus (MRCV): Evidence that the virus should be a separate Fijivirus species. *Arch Virol* 147:1699–709.

Dong, Y. B., Y. L. Li, and S. Z. Niu. 2006. [QTL analysis of starch content in maize kernels using the trisomic inheritance of the endosperm model]. *Yi Chuan* 28 (11):1401–6.

Dovas, C. I., K. Eythymiou, and N. I. Katis. 2004. First report of maize rough dwarf virus (MRDV) on maize crops in Greece. *Plant Pathol* 53:238.

Dreher, K., M. Khairallah, J. Ribaut, and M. Morris. 2003. Money matters (I): Costs of field and laboratory procedures associated with conventional and marker-assisted maize breeding at CIMMYT. *Mol Breed* 11:221–34.

Duble, C. M., A. E. Melchinger, L. Kuntze, A. Stork, and T. Lubberstedt. 2008. Molecular mapping and gene action of Scm1 and Scm2, two major QTL contributing to SCMV resistance in maize. *Plant Breed* 119:299–303.

Duggan, D. J., M. Bittner, Y. Chen, P. Meltzer, and J. M. Trent. 1999. Expression profiling using cDNA microarrays. *Nat Genet* 21 (1 Suppl):10–14.

Eathington, S.R. 2005. Practical applications of molecular technology in the development of commercial maize hybrids. In *Proceedings of the 60th Annual Corn and Sorghum Seed Research Conference*, 7–9 December 2005, Chicago [CD-ROM]. American Seed Trade Association, Washington, DC.

Egesel, C.O., J. C. Wong, R. J. Lambert, and T. R. Rocheford. 2003. Combining ability of maize inbreds for carotenoids and tocopherols. *Crop Sci* 43:818–23.

Emerson, R. A., G. W. Beadle, and A. C. Fraser. 1935. A summary of linkage studies in maize. *Cornell University Agricultural Experiment Station Memoirs* 180:1–83.

Falque, M., L. Decousset, D. Dervins, A. M. Jacob, J. Joets, J. P. Martinant, X. Raffoux, N. Ribiere, C. Ridel, D. Samson, A. Charcosset, and A. Murigneux. 2005. Linkage mapping of 1454 new maize candidate gene Loci. *Genetics* 170 (4):1957–66.

Fiehn, O. 2002. Metabolomics—The link between genotypes and phenotypes. *Plant Mol Biol* 48:155–71.

Fornaléa, S., M. Capellades, A. Encina, K. Wang, S. Irar, C. Lapierre, K. Ruel, et al. 2012. Altered lignin biosynthesis improves cellulosic bioethanol production in transgenic maize plants down-regulated for cinnamyl alcohol dehydrogenase. *Mol Plant* 5 (4):817–30.

Fu, Y., T. J. Wen, Y. I. Ronin, H. D. Chen, L. Guo, D. I. Mester, Y. Yang., et al. 2006. Genetic dissection of intermated recombinant inbred lines using a new genetic map of maize. *Genetics* 174 (3):1671–83.

Gao, S. B., M. J. Zhao, H. Lan, and Z. M. Zhang. 2007. [Identification of QTL associated with tassel branch number and total tassel length in maize]. *Yi Chuan* 29 (8):1013–17.

Gardiner, J. M., E. H. Coe, S. Melia-Hancock, D. A. Hoisington, and S. Chao. 1993. Development of a core RFLP map in maize using an immortalized F2 population. *Genetics* 134 (3):917–30.

Garg, A., B. M. Prasanna, and S. V. S. Chauhan. 2009. Simple sequence repeat (SSR) polymorphism in the tropical Asian maize inbred lines differing in resistance to banded leaf and sheath blight (*Rhizoctonia solani* f. sp. sasakii). *Indian J Genet Plant Breed* 67:238–42.

Gavaghan, C. L., J. V. Li, S. T. Hadfield, S. Hole, J. K. Nicholson, I. D. Wilson, P. W. Howe, P. D. Stanley, and E. Holmes. 2011. Application of NMR-based metabolomics to the investigation of salt stress in maize (*Zea mays*). *Phytochem Anal* 22 (3):214–24.

George, M. L., B. M. Prasanna, R. S. Rathore, T. A. Setty, F. Kasim, M. Azrai, S. Vasal, et al. 2003. Identification of QTLs conferring resistance to downy mildews of maize in Asia. *Theor Appl Genet* 107 (3):544–51.

Giuliani, S., M. C. Sanguineti, R. Tuberosa, M. Bellotti, S. Salvi, and P. Landi. 2005. Root-ABA1, a major constitutive QTL, affects maize root architecture and leaf ABA concentration at different water regimes. *J Exp Bot* 56 (422):3061–70.

Goldman, I. L., T. R. Rocheford, and J. W. Dudley. 1994. Molecular markers associated with maize kernel oil concentration in an Illinois high protein 9 Illinois low protein cross. *Crop Sci* 34:908–15.

Guan, S., P. Wang, H. Liu, G. Liu, Y. Ma, and L. Zhao. 2011. Production of high-amylose maize lines using RNA interference in sbe2a. *Afr J Biotechnol* 10:15229–37.

Gupta, H. S., P. K. Aggarwal, V. Mahajan, G. S. Bisht, A. Kumar, P. Verma, A. Srivastava, et al. 2009. Quality protein maize for nutritional security: Rapid development of short duration hybrids through molecular marker assisted breeding. *Curr Sci* 96:230–37.

Halket, J. M., D. Waterman, A. M. Przyborowska, R. K. Patel, P. D. Fraser, and P. M. Bramley. 2005. Chemical derivatization and mass spectral libraries in metabolic profiling by GC/MS and LC/MS/MS. *J Exp Bot* 56:219–43.

Hamilton, J. P. and C. R. Buell. 2012. Advances in plant genome sequencing. *Plant J* 70 (1):177–90.

Harjes, C. E., T. R. Rocheford, L. Bai, T. P. Brutnell, C. B. Kandianis, S. G. Sowinski, A. E. Stapleton, et al. 2008. Natural genetic variation in lycopene epsilon cyclase tapped for maize biofortification. *Science* 319 (5861):330–33.

Harpaz, I. 1959. Needle transmission of a new maize virus. *Nature* 184:77–78.

Helentjaris, T., D. F. Weber, and S. Wright. 1986. Use of monosomics to map cloned DNA fragments in maize. *Proc Natl Acad Sci USA* 83 (16):6035–39.

Ho, C., R. McCouch, and E. Smith. 2002. Improvement of hybrid yield by advanced backcross QTL analysis in elite maize. *Theor Appl Genet* 105 (2–3):440–48.

Holding, D. R., B. G. Hunter, J. P. Klingler, S. Wu, X. Guo, B. C. Gibbon, R. Wu, J. M. Schulze, R. Jung, and B. A. Larkins. 2010. Characterization of opaque2 modifier QTLs and candidate genes in recombinant inbred lines derived from the K0326Y quality protein maize inbred. *Theor Appl Genet* 122 (4):783–94.

Holland, J. B. and N. D. Coles. 2012. QTL controlling masculinization of ear tips in a maize (Zea mays L.) intraspecific cross. *G3 (Bethesda)* 1 (5):337–41.

Hu, X., Y. Li., C. Li, H. Yang, W. Wang, and M. Lu. 2010. Characterization of small heat shock proteins associated with maize tolerance to combined drought and heat stress. *J Plant Growth Regul* 29:455–64.

Hu, X., M. Lu, C. Li, T. Liu, W. Wang, J. Wu, F. Tai, X. Li, and J. Zhang. 2011. Differential expression of proteins in maize roots in response to abscisic acid and drought. *Acta Physiol Plant* 33:2437–46.

Ito, T. and E. M. Meyerowitz. 2000. Overexpression of a gene encoding a cytochrome P450, CYP78A9, induces large and seedless fruit in arabidopsis. *Plant Cell* 12 (9):1541–50.

Jia, J., J. Fu, J. Zheng, X. Zhou, J. Huai, J. Wang, M. Wang, et al. 2006. Annotation and expression profile analysis of 2073 full-length cDNAs from stress-induced maize (*Zea mays* L.) seedlings. *Plant J* 48 (5):710–27.

Johnson, G. R. 2004. Marker assisted selection. In Janick, J. (ed.), *Plant Breeding Reviews*, Vol. 24, pp. 293–310. Wiley.

Kurilich, A. C. and J. A. Juvik. 1999. Quantification of carotenoid and tocopherol antioxidants in *Zea mays*. *J Agric Food Chem* 47:1948–55.

Kusano, M., A. Fukushima, H. Redestig, and K. Saito. 2011. Metabolomic approaches toward understanding nitrogen metabolism in plants. *J Exp Bot* 62:1432–53.

Landi, P., M. C. Sanguineti, C. Liu, Y. Li, T. Y. Wang, S. Giuliani, M. Bellotti, S. Salvi, and R. Tuberosa. 2007. Root-ABA1 QTL affects root lodging, grain yield, and other agronomic traits in maize grown under well-watered and water-stressed conditions. *J Exp Bot* 58 (2):319–26.

Landi, P., S. Giuliani, S. Salvi, M. Ferri, R. Tuberosa, and M. C. Sanguineti. 2010. Characterization of root-yield-1.06, a major constitutive QTL for root and agronomic traits in maize across water regimes. *J Exp Bot* 61 (13):3553–62.

Laurie, C. C., S. D. Chasalow, J. R. LeDeaux, R. McCarroll, D. Bush, B. Hauge, C. Lai, D. Clark, T. R. Rocheford, and J. W. Dudley. 2004. The genetic architecture of response to long-term artificial selection for oil concentration in the maize kernel. *Genetics* 168 (4):2141–55.

Lauter, N., M. J. Moscou, J. Habiger, and S. P. Moose. 2008. Quantitative genetic dissection of shoot architecture traits in maize: Towards a functional genomics approach. *Plant Genome* 1:99–110.

Lee, M., N. Sharopova, W. D. Beavis, D. Grant, M. Katt, D. Blair, and A. Hallauer. 2002. Expanding the genetic map of maize with the intermated B73 x Mo17 (IBM) population. *Plant Mol Biol* 48 (5–6):453–61.

Li, K., C. Xu, and J. Zhang. 2011a. Proteome profile of maize (Zea Mays L.) leaf tissue at the flowering stage after long-term adjustment to rice black-streaked dwarf virus infection. *Gene* 485:106–13.

Li, Z. M., J. Q. Ding, R. X. Wang, J. F. Chen, X. D. Sun, W. Chen, W. B. Song, et al. 2011b. A new QTL for resistance to Fusarium ear rot in maize. *J Appl Genet* 52 (4):403–6.

Li, L., H. Li, J. Li, S. Xu, X. Yang, and J. Yan. 2010. A genome-wide survey of maize lipid-related genes: Candidate genes mining, digital gene expression profiling and co-location with QTL for maize kernel oil. *Sci China Life Sci* 53 (6):690–700.

Li, W. H., X. D. Xu, G. Li, L. Q. Guo, S. W. Wu, Y. Jiang, H. Y. Dong, et al. 2012. Characterization and molecular mapping of RsrR, a resistant gene to maize head smut. *Euphytica* 187:303–11.

Li, Y. L., Y. B. Dong, D. Q. Cui, Y. Z. Wang, Y. Y. Liu, M. G. Wei, and X. H. Li. 2008. The genetic relationship between popping expansion volume and two yield components in popcorn using unconditional and conditional QTL analysis. *Euphytica* 162:345–51.

Linde, K., C. Kastner, J. Kumlehn, R. Kahmann, and G. Doehlemann. 2011. Systemic virus-induced gene silencing allows functional characterization of maize genes during biotrophic interaction with *Ustilago maydis*. *New Phytol* 189:471–83.

Lisec, J., R. C. Meyer, and M. Steinfath, et al. 2008. Identification of metabolic and biomass QTL in *Arabidopsis thaliana* in a parallel analysis of RIL and IL populations. *Plant J* 53:960–72.

Lister, R. and J. R. Ecker. 2009. Finding the fifth base: Genome-wide sequencing of cytosine methylation. *Genome Res* 19 (6):959–66.

Liu, J. C., Q. Chu, H. G. Cai, G. H. Mi, and F. J. Chen. 2010. [SSR linkage map construction and QTL mapping for leaf area in maize]. *Yi Chuan* 32 (6):625–31.

Liu, X. H., Z. B. Tan, and T. Z. Rong. 2009. Molecular mapping of a major QTL conferring resistance to SCMV based on immortal RIL population in maize. *Euphytica* 167:229–35.

Liu, Y., T. Lamkemeyer, A. Jakob, G. Mi, F. Zhang, A. Nordheim, and F. Hochholdinger. 2006. Comparative proteome analyses of maize (Zea mays L.) primary roots prior to lateral root initiation reveal differential protein expression in the lateral root initiation mutant rum1. *Proteomics* 6 (15):4300–8.

Lu, H. F., H. T. Dong, C. B. Sun, D. J. Qing, N. Li, Z. K. Wu, Z. Q. Wang, and Y. Z. Li. 2011. The panorama of physiological responses and gene expression of whole plant of maize inbred line YQ7-96 at the three-leaf stage under water deficit and re-watering. *Theor Appl Genet* 123 (6):943–58.

Lu, M., F. Zhou, C. X. Xie, M. S. Li, Y. B. Xu, W. Marilyn, and S. H. Zhang. 2007. [Construction of a SSR linkage map and mapping of quantitative trait loci (QTL) for leaf angle and leaf orientation with an elite maize hybrid]. *Yi Chuan* 29 (9):1131–38.

Lungaho, M. G., A. M. Mwaniki, S. J. Szalma, J. J. Hart, M. A. Rutzke, L. V. Kochian, R. P. Glahn, and O. A. Hoekenga. 2011. Genetic and physiological analysis of iron biofortification in maize kernels. *PLoS One* 6 (6):20429.

Ma, F., R. Lu, H. Liu, B. Shi, J. Zhang, M. Tan, A. Zhang, and M. Jiang. 2012. Nitric oxide-activated calcium/calmodulin-dependent protein kinase regulates the abscisic acid-induced antioxidant defence in maize. *J Exp Bot* 63 (13):4835–47.

Mammadov, J. A., W. Chen, R. Ren, R. Pai, W. Marchione, F. Yalcin, H. Witsenboer, T. W. Greene, S. A. Thompson, and S. P. Kumpatla. 2010. Development of highly polymorphic SNP markers from the complexity reduced portion of maize [Zea mays L.] genome for use in marker-assisted breeding. *Theor Appl Genet* 121 (3):577–88.

Mangolin, C. A., C. L. de Souza Jr., A. A. F. Garcia, A. F. Garcia, S. T. Sibov, and A. P. de Souza. 2004. Mapping QTLs for kernel oil content in a tropical maize population. *Euphytica* 137:251–59.

Marcelo, M., S. Kuhnen, P. M. M. Lemos, S. K. de Oliveira, D. A. da Silva, M. M. Tomazzoli, A. C. V. Souza, et al. 2012. Metabolomics and chemometrics as tools for chemo(bio)diversity analysis—Maize landraces and propolis. In Varmuza, K. (ed.), *Chemometrics in Practical Applications*, InTech.

Maron, L. G., M. Kirst, C. Mao, M. J. Milner, M. Menossi, and L. V. Kochian. 2008. Transcriptional profiling of aluminum toxicity and tolerance responses in maize roots. *New Phytol* 179:116–28.

McCarty, D. R., A. M. Settles, M. Suzuki, B. C. Tan, S. Latshaw, T. Porch, K. Robin, et al. 2005. Steady-state transposon mutagenesis in inbred maize. *Plant J* 44 (1):52–61.

McGinnis, K. M. 2010. RNAi for functional genomics in plants. *Brief Funct Genomics* 9 (2):111–17.

McGinnis, K., N. Murphy, A. R. Carlson, A. Akula, C. Akula, H. Basinger, M. Carlson, et al. 2007. Assessing the efficiency of RNA interference for maize functional genomics. *Plant Physiol* 143 (4):1441–51.

McMullen, M. D., S. Kresovich, H. S. Villeda, P. Bradbury, H. Li, Q. Sun, S. Flint-Garcia, et al. 2009. Genetic properties of the maize nested association mapping population. *Science* 325 (5941):737–40.

Mechin, V., T. Balliau, C. Sophie, D. Marlene, O. Langella, L. Negroni, J. L. Prioul, T. Claudine, Z. Michel, and C. Damerval. 2004. A two-dimensional proteome map of maize endosperm. *Phytochemistry* 65:1609–18.

Meyer, R. C., M. Steinfath, J. Lisec, et al. 2007. The metabolic signature related to high plant growth rate in *Arabidopsis thaliana*. *Proc Natl Acad Sci U S A* 104:4759–64.

Min, J., Z. ChunYu, H. Khalid, L. Nan, S. Quan, M. Qing, W. Suwen, and L. Feng. 2012. Pyramiding resistance genes to northern leaf blight and head smut in maize. *Int J Agr Biol* 14:430–34.

Mohanty, A., A. Luo, S. DeBlasio, X. Ling, Y. Yang, D. E. Tuthill, K. E. Williams. D, et al. 2009. Advancing cell biology and functional genomics in maize using fluorescent protein-tagged lines. *Plant Physiol* 149 (2):601–5.

Morris, M., K. Dreher, J.-M. Ribaut, and M. Khairallah. 2003. Money matters (II): Costs of maize inbred line conversion schemes at CIMMYT using conventional and marker-assisted selection. *Mol. Breed* 11:235–47.

Myles, S., J. Peiffer, P. J. Brown, E. S. Ersoz, Z. Zhang, D. E. Costich, and E. S. Buckler. 2009. Association mapping: Critical considerations shift from genotyping to experimental design. *Plant Cell* 21 (8):2194–202.

Nair, S. K., B. M. Prasanna, A. Garg, R. S. Rathore, T. A. Setty, and N. N. Singh. 2005. Identification and validation of QTLs conferring resistance to sorghum downy mildew (Peronosclerospora sorghi) and Rajasthan downy mildew (P. heteropogoni) in maize. *Theor Appl Genet* 110 (8):1384–92.

Nakazawa, M., N. Yabe, T. Ichikawa, Y. Y. Yamamoto, T. Yoshizumi, K. Hasunuma, and M. Matsui. 2001. DFL1, an auxin-responsive GH3 gene homologue, negatively regulates shoot cell elongation and lateral root formation, and positively regulates the light response of hypocotyl length. *Plant J* 25 (2):213–21.

Nakazawa, M., T. Ichikawa, A. Ishikawa, H. Kobayashi, Y. Tsuhara, M. Kawashima, K. Suzuki, S. Muto, and M. Matsui. 2003. Activation tagging, a novel tool to dissect the functions of a gene family. *Plant J* 34 (5):741–50.

Niebur, W. S., J. A. Rafalski, O. S. Smith, and M. Cooper. 2004. Applications of genomics technologies to enhance rate of genetic progress for yield of maize within a commercial breeding program. In Fischer, T. (ed.), *New Directions for a Diverse Planet: Proceedings of the 4th International Crop Science Congress*, 26 September–1 October. Brisbane, QLD, Australia.

Pavlík, M., D. Pavlíková, and S. Vašíčková. 2010. Infrared spectroscopy-based metabolomic analysis of maize growing under different nitrogen nutrition. *Plant Soil Environ* 56 (11):533–40.

Pfeiffer, W. H. and B. McClafferty. 2007. HarvestPlus: Breeding crops for better nutrition. *Crop Sci* 47 (S3):S88–105.

Piccioni, F., D. Capitani, L. Zolla, and L. Mannina. 2009. NMR metabolic profiling of transgenic maize with the Cry1Ab gene. *J Agric Food Chem* 57 (14):6041–49.

Poland, J. A., P. J. Bradbury, E. S. Buckler, and R. J. Nelson. 2011. Genome-wide nested association mapping of quantitative resistance to northern leaf blight in maize. *Proc Natl Acad Sci USA* 108 (17):6893–98.

Poroyko, V., W. G. Spollen, L. G. Hejlek, A. G. Hernandez, M. E. LeNoble, G. Davis, H. T. Nguyen, G. K. Springer, R. E. Sharp, and H. J. Bohnert. 2007. Comparing regional transcript profiles from maize primary roots under well-watered and low water potential conditions. *J Exp Bot* 58 (2):279–89.

Prasanna, B. M., K. Pixley, M. L. Warburton, and C. X. Xie. 2010. Molecular marker-assisted breeding options for maize improvement in Asia. *Mol Breed* 26:339–56.

Pritchard, J. K., M. Stephens, and P. Donnelly. 2000. Inference of population structure using multilocus genotype data. *Genetics* 155 (2):945–59.

Qin, H., Y. Cai, Z. Liu, G. Wang, J. Wang, Y. Guo, and H. Wang. 2012. Identification of QTL for zinc and iron concentration in maize kernel and cob. *Euphytica* 187:1–14.

Qiu, F., Y. Zheng, Z. Zhang, and S. Xu. 2007. Mapping of QTL associated with waterlogging tolerance during the seedling stage in maize. *Ann Bot* 99 (6):1067–81.

Rafalski, J. A. 2010. Association genetics in crop improvement. *Curr Opin Plant Biol* 13 (2):174–80.

Ragot, M. and M. Lee. 2007. Marker-assisted selection in maize: Current status, potential, limitations and perspectives from the private and public sectors. In Guimarães, E. P. et al. (eds), *Marker-Assisted Selection, Current Status and Future Perspectives in Crops, Livestock, Forestry, and Fish*, pp. 117–50. FAO, Rome.

Ragot, M., M. Biasiolli, M. F. Delbut, A. Dell'Orco, L. Malgarini, P. Thevenin, J. Vernoy, J. Vivant, R. Zimmermann, and G. Gay. 1995. Marker-assisted backcrossing: A practical example. In Berville, A. and Tersac, M. (eds), *Les Colloques, No. 72*, pp. 45–56. INRA, Paris.

Raun, W. R. and G. V. Johnson. 1999. Improving nitrogen use efficiency for cereal production. *Agron J* 91:357–63.

Sa, K. J., J. Y. Park, K. C. Park, and J. K. Lee. 2012. Analysis of genetic mapping in a waxy/dent maize RIL population using SSR and SNP markers. *Genes Genom* 34:157–64.

Sabry, A., D. Jeffers, S. K. Vasal, R. Frederiksen, and C. Magill. 2006. A region of maize chromosome 2 affects response to downy mildew pathogens. *Theor Appl Genet* 113 (2):321–30.

Salvi, S., G. Sponza, M. Morgante, D. Tomes, X. Niu, K. A. Fengler, R. Meeley, et al. 2007. Conserved noncoding genomic sequences associated with a flowering-time quantitative trait locus in maize. *Proc Natl Acad Sci U S A* 104 (27):11376–81.

Šamaj, J. and J. J. Thelen. 2007. *Plant Proteomics*. Springer-Verlag, Berlin Heidelberg.

Schadt, E. E., S. A. Monks, T. A. Drake, A. J. Lusis, N. Che, V. Colinayo, T. G. Ruff, et al. 2003. Genetics of gene expression surveyed in maize, mouse and man. *Nature* 422 (6929):297–302.

Schaeffer, M., P. Byrne, and E. H. Coe. 2006. Consensus quantitative trait maps in maize: A database strategy. *Maydica* 51 (2):357.

Schnable, P. S., D. Ware, R. S. Fulton, J. C. Stein, F. Wei, S. Pasternak, C. Liang, et al. 2009. The B73 maize genome: Complexity, diversity, and dynamics. *Science* 326 (5956):1112–15.

Sekhon, R. S., H. Lin, K. L. Childs, C. N. Hansey, C. R. Buell, N. de Leon, and S. M. Kaeppler. 2011. Genome-wide atlas of transcription during maize development. *Plant J* 66 (4):553–63.

Setter, T. L., J. Yan, M. Warburton, J. M. Ribaut, Y. Xu, M. Sawkins, E. S. Buckler, Z. Zhang, and M. A. Gore. 2010. Genetic association mapping identifies single nucleotide polymorphisms in genes that affect abscisic acid levels in maize floral tissues during drought. *J Exp Bot* 62 (2):701–16.

Shi, L. Y., Z. F. Hao, J. F. Weng, C. X. Xie, C. L. Liu, D. G. Zhang, M. S. Li, L. Bai, X. H. Li, and S. H. Zhang. 2012. Identification of a major quantitative trait locus for resistance to maize rough dwarf virus in a Chinese maize inbred line X178 using a linkage map based on 514 gene-derived single nucleotide polymorphisms. *Mol Breed* 30:1–11.

Shulaev, V. 2006. Metabolomics technology and bioinformatics. *Brief Bioinform* 7:128–39.

Simic, D., S. Mladenovic Drinic, Z. Zdunic, A. Jambrovic, T. Ledencan, J. Brkic, A. Brkic, and I. Brkic. 2011. Quantitative trait loci for biofortification traits in maize grain. *J Hered* 103 (1):47–54.

Soderlund, C., A. Descour, D. Kudrna, M. Bomhoff, L. Boyd, J. Currie, A. Angelova, et al. 2009. Sequencing, mapping, and analysis of 27,455 maize full-length cDNAs. *PLoS Genet* 5 (11):e1000740.

Song, R., G. Segal, and J. Messing. 2004. Expression of the sorghum 10-member kafirin gene cluster in maize endosperm. *Nucleic Acids Res* 32 (22):e189.

Spollen, W. G., W. Tao, B. Valliyodan, K. Chen, L. G. Hejlek, J. J. Kim, M. E. Lenoble, et al. 2008. Spatial distribution of transcript changes in the maize primary root elongation zone at low water potential. *BMC Plant Biol* 8:32.

Stevenson, D. M., R. E. Muck, K. J. Shinners, and P. J. Weimer. 2006. Use of real time PCR to determine population profiles of individual species of lactic acid bacteria in alfalfa silage and stored corn stover. *Appl Microbiol Biotechnol* 71(3):329–38.

Stuber, C. W., M. D. Edwards, and J. F. Wendel. 1987. Molecular markerfacilitated investigations of quantitative traits loci in maize. II: Factors influencing yield and its component traits. *Crop Sci* 27:639–48.

Taramino, G. and S. Tingey. 1996. Simple sequence repeats for germplasm analysis and mapping in maize. *Genome* 39 (2):277–87.

Tenaillon, O., A. Rodriguez-Verdugo, R. L. Gaut, P. McDonald, A. F. Bennett, A. D. Long, and B. S. Gaut. 2012. The molecular diversity of adaptive convergence. *Science* 335 (6067):457–61.

Teng, F., L. Zhai, R. Liu, W. Bai, L. Wang, D. Huo, Y. Tao, Y. Zheng, and Z. Zhang. 2012. ZmGA3ox2, a candidate gene for a major QTL, qPH3.1, for plant height in maize. *Plant J.* 73 (3):405–16.

Thudi, M., Y. Li, S. A. Jackson, G. D. May, and R. K. Varshney. 2012. Current state-of-art of sequencing technologies for plant genomics research. *Brief Funct Genom* 11 (1):3–11.

Tian, F., P. J. Bradbury, P. J. Brown, H. Hung, Q. Sun, S. Flint-Garcia, T. R. Rocheford, M. D. McMullen, J. B. Holland, and E. S. Buckler. 2011. Genome-wide association study of leaf architecture in the maize nested association mapping population. *Nat Genet* 43 (2):159–62.

Tuberosa, R., S. Salvi, M. C. Sanguineti, P. Landi, M. Maccaferri, and S. Conti. 2002. Mapping QTLs regulating morpho-physiological traits and yield: Case studies, shortcomings and perspectives in drought-stressed maize. *Ann Bot* 89:941–63.

Velculescu, V. E., L. Zhang, B. Vogelstein, and K. W. Kinzler. 1995. Serial analysis of gene expression. *Science* 270 (5235):484–87.

Walbot, V. 2008. Maize genome in motion. *Genome Biol* 9 (4):303.

Wang, H., S. Miyazaki, K. Kawai, M. Deyholos, D. W. Galbraith, and H. J. Bohnert. 2003. Temporal progression of gene expression responses to salt shock in maize roots. *Plant Mol Biol* 52 (4):873–91.

Wang, J. H., S. N. Chen, B. Bai, J. H. Tang, and W. Wang. 2009. Effects of artificial aging on seed viability and proteome of maize seeds. *J Henan Agric Univ* 43:132–35.

Wang, Z. H., S. G. Fang, J. L. Xu, L. Y. Sun, D. W. Li, and J. L. Yu. 2003. Sequence analysis of the complete genome of rice black-streaked dwarf virus isolated from maize with rough dwarf disease. *Virus Genes* 27:163–68.

Wassom, J. J., V. Mikkelineni, M. O. Bohn, and T.R. Rocheford. 2008. QTL for fatty acid composition of maize kernel oil in Illinois high oil 9 B73 backcross-derived lines. *Crop Sci* 48:69–78.

Weber, A., R. M. Clark, L. Vaughn, J. Sanchez-Gonzalez Jde, J. Yu, B. S. Yandell, P. Bradbury, and J. Doebley. 2007. Major regulatory genes in maize contribute to standing variation in teosinte (*Zea mays ssp.* parviglumis). *Genetics* 177 (4):2349–59.

Weber, A. L., W. H. Briggs, J. Rucker, B. M. Baltazar, J. de Jesus Sanchez-Gonzalez, P. Feng, E. S. Buckler, and J. Doebley. 2008. The genetic architecture of complex traits in teosinte (*Zea mays* ssp. parviglumis): New evidence from association mapping. *Genetics* 180 (2):1221–32.

Wei, F., E. Coe, W. Nelson, A. K. Bharti, F. Engler, E. Butler, H. Kim, et al. 2007. Physical and genetic structure of the maize genome reflects its complex evolutionary history. *PLoS Genet* 3 (7):e123.

Weigel, D., J. H. Ahn, M. A. Blazquez, J. O. Borevitz, S. K. Christensen, C. Fankhauser, C. Ferrandiz, et al. 2000. Activation tagging in Arabidopsis. *Plant Physiol* 122 (4):1003–13.

Weng, J., X. Liu, Z. Wang, J. Wang, L. Zhang, Z. Hao, C. Xie, et al. 2012. Molecular mapping of the major resistance quantitative trait locus qHS2.09 with simple sequence repeat and single nucleotide polymorphism markers in maize. *Phytopathology* 102 (7):692–99.

Wilkins, M. R., J. C. Sanchez, A. A. Gooley, R. D. Appel, I. H. Smith, D. F. Hochstrasser, and K. L. Williams. 1996. Progress with proteome projects: Why all proteins expressed by a genome should be identified and how to do it. *Biotechnol Genet Eng Rev* 13:19–50.

Willcox, M. C., M. M. Khairallah, D. Bergvinson, J. Crossa, J. A. Deutsch, G. O. Edmeades, D. Gonzalez-de-Leon, et al. 2002. Selection for resistance to southwestern corn borer using marker-assisted and conventional backcrossing. *Crop Sci* 42:1516–28.

Wu, Y. and J. Messing. 2010. RNA interference-mediated change in protein body morphology and seed opacity through loss of different zein proteins. *Plant Physiol* 153 (1):337–47.

Wu, Y. and J. Messing. 2011. Novel genetic selection system for quantitative trait loci of quality protein maize. *Genetics* 188 (4):1019–22.

Wurtzel, E. T., A. Cuttriss, and R. Vallabhaneni. 2012. Maize provitamin a carotenoids, current resources, and future metabolic engineering challenges. *Front Plant Sci* 3:29.

Wu, X., H. Liu, W. Wang, S. Chen, X. Hu, and C. Li. 2011. Proteomic analysis of seed viability in maize. *Acta Physiol Plant* 33:181–91.

Xiang, K., L. M. Reid, Z. M. Zhang, X. Y. Zhu, and G. T. Pan. 2012. Characterization of correlation between grain moisture and ear rot resistance in maize by QTL meta-analysis. *Euphytica* 183:185–95.

Xiao, Y. N., X. H. Li, M. L. George, M. S. Li, S. H. Zhang, and Y. L. Zheng. 2005. Quantitative trait locus analysis of drought tolerance and yield in maize in China. *Plant Mol Biol Rep* 23:155–65.

Yan, J., C. B. Kandianis, C. E. Harjes, L. Bai, E. H. Kim, X. Yang, D. J. Skinner, et al. 2010. Rare genetic variation at Zea mays crtRB1 increases beta-carotene in maize grain. *Nat Genet* 42 (4):322–27.

Yang, Q., G. Yin, Y. Guo, D. Zhang, S. Chen, and M. Xu. 2010. A major QTL for resistance to Gibberella stalk rot in maize. *Theor Appl Genet* 121 (4):673–87.

Yang, W., Y. Zheng, W. Zheng, and R. Feng. 2005. Molecular genetic mapping of a high-lysine mutant gene (opaque-16) and the double recessive effect with opaque-2 in maize. *Mol Breed* 15 (3):257–69.

Yang, X., Y. Guo, J. Yan, J. Zhang, T. Song, T. Rocheford, and J. S. Li. 2009. Major and minor QTL and epistasis contribute to fatty acid compositions and oil concentration in high-oil maize. *Theor Appl Genet* 120 (3):665–78.

Yang, X., H. Ma, P. Zhang, J. Yan, Y. Guo, T. Song, and J. Li. 2012. Characterization of QTL for oil content in maize kernel. *Theor Appl Genet* 125 (6):1169–79.

Yu, J. and E. S. Buckler. 2006. Genetic association mapping and genome organization of maize. *Curr Opin Biotechnol* 17 (2):155–60.

Yu, L. X. and T. L. Setter. 2003. Comparative transcriptional profiling of placenta and endosperm in developing maize kernels in response to water deficit. *Plant Physiol* 131 (2):568–82.

Zhang, C., Y. Liu, L. Liu, Z. Lou, H. Zhang, H. Miao, X. Hu, Y. Pang, and B. Qiu. 2008a. Rice black streaked dwarf virus P9-1, an α-helical protein, self-interacts and forms viroplasms in vivo. *J Gen Virol* 89:1770–76.

Zhang, H., J. Chen, J. Lei, and M. J. Adams. 2001. Sequence analysis shows that a dwarfing disease on rice, wheat and maize in China is caused by *rice black-streaked dwarf virus*. *Eur J Plant Pathol* 107:563–67.

Zhang, J., X. Q. Lu, X. F. Song, J. B. Yan, T. M. Song, J. R. Dai, T. Rocheford, and J. S. Li. 2008b. Mapping quantitative trait loci for oil, starch, and protein concentrations in grain with high-oil maize by SSR markers. *Euphytica* 162:335–44.

Zhang, F., X. Q. Wan, and G. T. Pan. 2006. QTL mapping ofFusarium moniliforme ear rot resistance in maize. 1. Map construction with microsatellite and AFLP markers. *J Appl Genet* 47 (1):9–15.

Zhang, H. W., Z. B. Tan, R. J. Chen, J. S. Li, and G. Chen. 2003. [Maize starch biosynthesis and its genetic manipulation]. *Yi Chuan* 25 (4):455–60.

Zhang, Y., L. Xu, X. Fan, J. Tan, W. Chen, and M. Xu. 2012. QTL mapping of resistance to gray leaf spot in maize. *Theor Appl Genet* 125 (8):1797–808.

Zhang, Z. M., M. J. Zhao, T. Z. Rong, and G. T. Pan. 2007. SSR linkage map construction and QTL identification for plant height and ear height in maize (*Zea mays* L). *Acta Agronomica Sinica* 33:341–34.

Zhao, M. J., S. B. Gao, Z. M. Zhang, T. Z. Rong, and G. T. Pan. 2006a. Initial identification of quantitative trait loci controlling resistance to banded leaf and sheath blight at elongating and heading date in maize]. *Fen zi xi bao sheng wu xue bao* = *Journal of molecular cell biology/Zhongguo xi bao sheng wu xue xue hui zhu ban* 39 (2):139.

Zhao, M., Z. Zhang, S. Zhang, W. Li, D. P. Jeffers, T. Rong, and G. Pan. 2006b. Quantitative trait loci for resistance to banded leaf and sheath blight in maize. *Crop Sci* 46 (3):1039–45.

Zhao, Y., X. Lu, C. Liu, H. Guan, M. Zhang, Z. Li, H. Cai, and J. Lai. 2012. Identification and fine mapping of rhm1 locus for resistance to Southern corn leaf blight in maize. *J Integr Plant Biol* 54 (5):321–29.

Zheng, J., J. Zhao, Y. Tao, J. Wang, Y. Liu, J. Fu, Y. Jin, et al. 2004. Isolation and analysis of water stress induced genes in maize seedlings by subtractive PCR and cDNA macroarray. *Plant Mol Biol* 55 (6):807–23.

Zhou, M. L., Q. Zhang, M. Zhou, L. P. Qi, X. B. Yang, K. X. Zhang, J. F. Pang, et al. 2012. Aldehyde dehydrogenase protein superfamily in maize. *Funct Integr Genomics* 12 (4):683–91.

Zhuang, Y., G. Ren, G. Yue, Z. Li, X. Qu, G. Hou, Y. Zhu, and J. Zhang. 2007. Effects of water-deficit stress on the transcriptomes of developing immature ear and tassel in maize. *Plant Cell Rep* 26:2137–47.

Zinselmeier, C., M. J. Lauer, and J. S. Boyer. 1995. Reversing drought-induced losses in grain yield: Sucrose maintains embryo growth in maize. *Crop Sci* 35:1390–400.

Zinselmeier, C., J. E. Habben, M. E. Westgate, and J. S. Boyer. 2000. Carbohydrate metabolism in setting and aborting maize ovaries. In Westgate, M. E. and Boote, K. (eds), *Physiology and Modeling Kernel Set in Maize*, pp. 1–14. CSSA Special publication number 29. Crop Science Society of America, Madison.

Zinselmeier, C., Y. Sun, T. Helentjaris, M. Beatty, S. Yang, H. Smith, and J. Habben. 2002. The use of gene expression profiling to dissect the stress sensitivity of reproductive development in maize. *Field Crops Res* 75:111–21.

4

Omics Approaches in Pulses

Abhishek Bohra, PhD; Uday Chand Jha, PhD; Balwant Singh, PhD; Khela Ram Soren, PhD; Indra Prakash Singh, PhD; Sushil Kumar Chaturvedi, PhD; Nagaswamy Nadarajan, PhD; and Debmalya Barh, PhD

CONTENTS

4.1 Introduction 102
4.2 Genome Organization, Taxonomy/Center of Origin, and Production Constraints 102
    4.2.1 Chickpea 102
    4.2.2 Pigeonpea 103
    4.2.3 Cowpea 104
    4.2.4 Lentil 105
4.3 Availability of Genomic Tools and Technologies in Pulses 106
    4.3.1 Mapping Populations 107
        4.3.1.1 Chickpea 107
        4.3.1.2 Pigeonpea 107
        4.3.1.3 Cowpea 108
        4.3.1.4 Lentil 108
    4.3.2 BAC Libraries, BESs and Physical Maps 110
        4.3.2.1 Chickpea 110
        4.3.2.2 Pigeonpea 111
        4.3.2.3 Cowpea 111
        4.3.2.4 Lentil 111
    4.3.3 Molecular Markers 111
        4.3.3.1 Chickpea 111
        4.3.3.2 Pigeonpea 112
        4.3.3.3 Cowpea 112
        4.3.3.4 Lentil 112
    4.3.4 Genetic Linkage Maps 113
        4.3.4.1 Chickpea 113
        4.3.4.2 Pigeonpea 113
        4.3.4.3 Cowpea 114
        4.3.4.4 Lentil 114
    4.3.5 Trait Mapping/Tagging 115
        4.3.5.1 Chickpea 115
        4.3.5.2 Pigeonpea 115
        4.3.5.3 Cowpea 117
        4.3.5.4 Lentil 117
    4.3.6 Transcriptomics and Comparative Analyses 118
        4.3.6.1 Chickpea 118
        4.3.6.2 Pigeonpea 118

        4.3.6.3 Cowpea 119
        4.3.6.4 Lentil 119
4.4   Whole-Genome Sequencing in Pulses 120
4.5   MAB in Pulses 121
4.6   Transgenic Approaches in Pulse Crops 122
4.7   Conclusion 126
References 126

4.1 Introduction

Pulses play a crucial role in combating food insecurity and protein calorie malnutrition, especially for the billions of people in the developing countries. A total of 11 pulse crops are recognized by the Food and Agricultural Organization (FAO), cultivated worldwide in an area of 76 Mha and producing approximately 67.6 Mt with an average yield of 890 kg/ha. Nearly 25% of production comes from India, from a total area of 25 Mha. Nevertheless, most of these pulses are grown mainly as rainfed crops in areas extremely prone to pest/disease incidence and adverse climatic conditions. The conditions are exacerbated by minimal input supply and poor crop management, leading to fluctuating pulse production and stagnated productivity. Genetic improvement of pulse crops through conventional breeding has contributed significantly, with development and release of a number of high-yielding varieties. However, there has not been a substantial increase in the production of pulses in recent decades, and in the current scenario of drastically changing global climate, temperature extremes, irregularity in rainfall, and increasing population pressure, a quantum leap in pulse yields per hectare is needed in order to ensure nutritional security. In this context molecular tools and technologies promise to be a great supplement to the traditional pulse breeding. A few years ago pulses were considered as "orphan crops" because of the meager importance given to the generation of genomic resources in the less-studied pulses. However over the past 5–10 years, with the current state-of-art sequencing and genotyping platforms, dramatic progress has been noticed in the area of pulse genomics, culminating in the availability of drafts of whole-genome sequences for low-profile crops such as pigeonpea. Likewise, whole-genome sequencing is in progress for chickpea, cowpea, lentil, and others. Here we discuss the latest advances in pulse genomics, particularly for chickpea, pigeonpea, cowpea, and lentil, and the future prospects of genomics-enabled breeding for accelerated genetic improvement of pulses.

4.2 Genome Organization, Taxonomy/Center of Origin, and Production Constraints

4.2.1 Chickpea

Chickpea (*Cicer arietinum* L.) is a cool-season legume crop—the second most important legume crop after dry beans worldwide (FAOSTAT 2010). Annually a total of 10.9 Mt of chickpea is harvested from 12 Mha area worldwide, with an average productivity of 910 kg/ha (FAOSTAT 2010). Among the various chickpea-growing countries, India stands at the top, contributing more than 65% of global chickpea production (Gaur et al. 2012a).

In India total chickpea area and production are around 8.17 Mha and 7.48 Mt, respectively, with a mean productivity of 915 kg/ha (DAC 2010). Other chickpea-growing countries include Myanmar, Iran, Mexico, Canada, and the USA (Gaur et al. 2012a). Being a legume crop, its higher protein content (varying between 20% and 30%) makes it an excellent source of amino acids in otherwise protein-deficient cereal-based vegetarian diets. It also contains greater quantities of carbohydrate, and soluble vitamins and minerals such as calcium, magnesium, potassium, phosphorus, iron, and zinc, than other legumes (Ibrikci et al. 2003).

Chickpea is a self-pollinated and diploid crop with a genome size of 740 Mbps organized into eight pairs of chromosomes, that is, $2n = 2x = 16$ (Arumuganathan and Earle 1991). A number of studies have been conducted in chickpea to elucidate the structure and numbers of the different chromosomes (Iyengar 1939; Mercy et al. 1974; Sharma and Gupta 1986; Venora et al. 1991). The morphological differences observed in the chickpea chromosomes using techniques such as Feulgen staining and *in situ* hybridization provided the basis for the development of flow cytogenetics in chickpea (Vlácilová et al. 2002). This flow cytometric analysis represented the first attempt at assigning a particular linkage group (LG 8) to a specific chromosome (Chromosome H). Utility of flow cytometry and DNA markers in integrating the genetic and physical map was further demonstrated in chickpea by Zatloukalová and colleagues (2011).

Approximately 70% of legume species are in the subfamily Papilionoideae (Cannon et al. 2009), which has been divided into four clades: (i) genistoid, (ii) aeschynomenoid/dalbergioid, (iii) hologalegina, and (iv) phaseoloid/millettoid. *Cicer* belongs to the inverted repeat loss subclade (IRLC) within hologalegina along with *Pisum, Trifolium, Medicago, Lens, and Lathyrus* genera. The IRLC is characterized by lack of one copy of the large inverted repeat in the chloroplast genome and most of the plants forming this subtribe are equipped with temperate adaptation (Gepts et al. 2005). The genus *Cicer* comprises 43 species further grouped as annual (9 species) or perennial (33 species), while the remaining one species is designated as unspecified according to van der Maesen (1987). Among the known chickpea species, *C. arietinum* L. is the only domesticated one and shows close resemblance to one of the wild ancestors, *C. reticulatum* (Ladizinsky and Adler 1976). Due to the presence of the endemic progenitor (*C. reticulatum*) in the south-eastern Turkey and adjacent Syrian region, chickpea is believed to have originated here with its domestication starting approximately 11,000 years ago (Zohary and Hopf 1973).

Although it is one of the foremost pulses in terms of global production and area, its productivity has shown more or less stagnated trends. The fluctuations in chickpea production may be attributed to susceptibility to a variety of biotic pathogens and abiotic constraints. Several kinds of diseases have been documented in chickpea; however, only a few such as *Fusarium* wilt (FW) caused by *Fusarium oxysporum* f. sp. *ciceri, Ascochyta* blight (AB) caused by *Ascochyta rabiei* (Pass.) Labr, dry root rot caused by *Rhizoctonia bataticola*, and collar rot caused by *Sclerotium rolfsii* are of global interest. Additionally some of the foliar diseases such as botrytis gray mold (*Botrytis cineria* Pres.) are mainly confined to the cool and humid chickpea-growing areas. Among the insect pests causing substantial yield losses in chickpea, gram pod borer (*Helicoverpa armigera* Hubner) is the most devastating one. The global yield losses due to pod borer damage are estimated to be approximately US$ 500 million. Of the abiotic constraints, drought remains the most challenging yield reducer, leading to 40%–45% yield loss in chickpea (Ahmad et al. 2005).

4.2.2 Pigeonpea

Pigeonpea (*Cajanus cajan* L. Millsp.) ranks sixth in global grain legume production and worldwide it is cultivated in about a 4.70 Mha area with an annual production of 3.69 Mt and a mean productivity of 783 kg/ha (FAOSTAT 2010). India ranks first in annual pigeonpea production with 2.46 Mt, followed by Myanmar (0.2 Mt) and Malawi (0.18 Mt) (FAOSTAT 2010). In India pigeonpea is the second most important pulse crop after chickpea. It is generally grown as a sole crop or is intercropped with cereals such as sorghum, pearlmillet, and maize. Like other pulses, with high protein content (up to 25%) and digestibility it ensures an adequate supply of essential amino acids, especially lysine, to vegetarians.

Genus *Cajanus* comprising 32 species belongs to the tribe *Phaseoleae* and clade millettioid within the subfamily *Papilionoideae* (Ratnaparkhe and Gupta, 2007). All the members of the millettioid clade including soybean, common bean, and cowpea are better adapted to tropical conditions and therefore are also referred to as warm season legumes. Based on the vast genetic diversity and occurrence of several wild relatives, the Indian subcontinent is considered as the primary center of origin for pigeonpea (De 1974), especially given the presence of the most proximal wild form *C. Cajanifolius* in this region (van der Maesen 1990). Nevertheless, some scientists contest this view by postulating Africa as the center of origin due to the existence of a single endemic wild species, *C. kerstingii*.

Pigeonpea is a diploid crop with a gametic chromosome number of $n = x = 11$. To understand the patterns of genome packaging into chromosomes, several karyotype analyses have been reported in pigeonpea (Akinola et al. 1972). All of these studies unanimously reported a similar number of chromosomes in cultivated as well as in wild relatives except for *C. kertsingii*. *C. kertsingii*, native to Africa, was found to carry a somatic chromosome number of $2n = 32$ (Gill and Hussaini 1986). Moreover, a detailed karyotype analysis using pigeonpea cultivars and some of the wild species demonstrated a greater similarity in chromosome morphologies between cultivated pigeonpea (*C. cajan*) and *C. cajanifolius* while differences were observed with the other related species (Ohri and Singh 2002). In terms of the size of the entire genome, based on K-mer statistics pigeonpea genome is estimated to be 833.07 Mb (Varshney et al. 2011), of which 605.78 Mb (Varshney et al. 2011) and 510 Mb (Singh et al. 2011) were captured in two separate whole-genome sequencing projects.

Like other pulse crops, the huge gap existing between the potential and actual yields of pigeonpea is due to a number of biotic and abiotic constraints affecting production severely. The various fungal and viral diseases include FW caused by *Fusarium udum* Butler, sterility mosaic disease (SMD) caused by pigeonpea sterility mosaic virus (PPSMV), and phytophthora blight caused by the fungus *Phytophthora drechsleri*. In addition, infestations from pests such as the pod borer (*Helicoverpa armigera* Hubner), *Maruca* (*Maruca vitrata* Geyer), pod-sucking bugs (*Clavigralla horrida* Germar), and podfly (*Melanagromyza chalcosoma* Spencer) result in substantial yield losses in pigeonpea (Shanower et al. 1999). Among the potential abiotic yield reducers, waterlogging has profound impacts on crop growth and development. Waterlogged conditions, which often become frequent in the rainy season, also predispose susceptibility to *Phytophthora* blight. Soil salinity and aluminum toxicity are other major problems encountered by pigeonpea. Besides, pigeonpea production is limited by other factors such as unpredictability and irregularity of rainfall and poor soil health prevailing in the pigeonpea-growing areas, coupled with cultivation of lower-yielding and low-responsive traditional cultivars.

4.2.3 Cowpea

Cowpea (*Vigna unguiculata* L. Walpers) is a warm season legume commonly grown as an annual crop in the tropical and subtropical regions of the world (Singh et al. 1997). It is also known as southern pea, black-eyed pea, crowder pea, or lobia. It is cultivated in ~11 Mha area distributed in parts of Africa, Asia, Europe, and Central and South America, producing a total of 5.63 Mt of dried cowpeas (FAOSTAT 2010). Interestingly, about 98% of the global harvested area and 95% of the global production are accounted for by African countries, especially Nigeria, Niger, and Burkina Faso. Cowpea is used for diverse purposes such as human consumption, animal feed, cover crop, and as seeds. Its higher tolerance of high temperature and drought helps in ensuring high returns to small-scale farmers in Africa and other parts of the developing world even in prolonged dry spells. By virtue of its ability to fix atmospheric nitrogen and its short life cycle, it offers greater suitability for crop rotation, and its higher protein content makes it one of the best sources of amino acids and vitamins in vegetarian diets.

Cowpea belongs to the tribe Phaseoleae of Millettioid clade under the subfamily Papilionoideae. Four subspecies—more precisely cultigroups—are recognized in *V. unguiculata*, which are *unguiculata*, *biflora* (or *cylindrica*), *sesquipedalis*, and *textilis* (Singh 2005). Owing to the presence of wild ancestors along with other evidence, Southern Africa is considered to be the place where cowpea originated before diffusing to India, West Asia, and Europe. The domestication of cowpea occurred around 4000 years ago, making it one of the pioneer crops that underwent human cultivation. However, the place of domestication—western or north-eastern Africa—has been always a matter of contention for the cowpea community (Coulibaly et al. 2002). Based on similarities to the cultivated cowpea in terms of morphological attributes and growth habit, *V. unguiculata dekindtiana*, a wild and weedy form in West Africa, is considered the most probable progenitor of cowpea (Lush and Evans 1981).

Cowpea is a self-pollinating and diploid ($2n = 2x = 22$) species with an estimated genome size of 620 Mb, which is one of the smallest legume genomes (Timko et al. 2008). There have been a few attempts to discover the number and morphology of cowpea chromosomes through karyotype analysis. All of these attempts reported that the entire cowpea genome is packaged into 11 pairs of chromosomes. A high degree of cytological similarity among the various *V. unguiculata* subspecies is responsible for the frequent crossing among the *V. unguiculata* subspecies (Venora and Padulosi 1997).

Cowpea has been reported to be adversely affected by a wide range of fungal diseases such as seed decay and root and stem rot caused by *Pythium aphanidermatum* (Edson); wilt caused by *Fusurium oxysporum* f. sp. *tracheiphilum*; charcoal rot, seedling damping-off, and ashy stem blight caused by *Macrophomina phaseolina*; damping off and stem canker caused by *Rhizoctonia solani* and *R. bataticola*; brown blotch and anthracnose caused by *Colletotrichum truncatum* Schw. and *C. capsici*; cercospora leaf spot by *Cercospora canescens*; cowpea leaf smut disease by *Protomycopsis phaseoli*; and sclerotium rot by *Sclerotium rolfsii*. Bacterial diseases including bacterial blight and leaf spot (*Xanthomonas axonopodis* pv. *vignicola*) and viral diseases including cowpea yellow mosaic virus (CYMV), cowpea aphid borne mosaic virus (CAMV), and blackeye cowpea mosaic virus (BCMV) have gained attention because of their devastating nature. Also, a number of pests attack cowpea—the problem is much more severe in Africa than in other parts, where substantial yield reductions result from aphids (*Aphis craccivora*), thrips and leaf hopper (*Empoasca kerri*), pod bugs (*Riptortus pedestris*), and spotted pod borer (*Maruca testulalis*). Most importantly, cowpea production is also subjected to parasitism by weeds such as striga (*Striga gesnerioides*) and alectra (*Alectra vogelii*) (Adegbite and Amusa 2008).

4.2.4 Lentil

Lentil (*Lens culinaris* Medik.) is an annual cool season legume grown in developed (Canada, US, Australia, etc.) as well as in developing (India, Nepal, Bangladesh, etc.) countries

(Akibode and Maredia 2011). With a global production of 4.6 Mt and an area of 4.1 Mha, lentil stands fourth among the pulse crops (FAO 2010). In terms of global lentil production and harvested area, Canada ranks first with 1.9 Mt production and 1.3 Mha area, followed by India (0.9 Mt production from 1.3 Mha area) and Turkey (0.4 Mt and 0.2 Mha). In the Mediterranean and subtropical regions, lentil is used as a winter crop in rotation with small cereal grains including wheat and barley, whereas in temperate regions it is primarily grown as a summer legume (Erskine et al. 2011).

Viewed from the standpoint of domestication, lentil is possibly the most ancient food crop, having originated 8000–9000 years ago in the Near East and Asia Minor (Ladizinsky 1979); its subsequent dispersal led to its cultivation in other parts including Europe, the Indian subcontinent, Africa, North America, and Australia. Owing to the genomic harmony and morphological similarity with wild relative *L. orientalis*, the latter is considered the wild progenitor of the cultivated lentil. Genomic harmony is evident from the usual meiotic chromosome pairing and existence of comparatively weaker crossability barriers, yielding fully/partially fertile interspecific hybrids between *L. culinaris* and *L. orientalis*. It is postulated that *L. orientalis* also originated in the Middle East Fertile Crescent (Zohary and Hopf 1973).

Lentil is a self-pollinated and diploid crop with a DNA content of 4.2 pg/1C or 4063 Mbp/1C packaged into seven pairs of chromosomes ($2n = 2x = 14$) (Arumuganathan and Earle 1991). Construction of a standard karyotype for lentil has always been seen as an interesting researchable area and consequently several conflicting reports were published related to total chromosome lengths and chromosome structures. However, the most acceptable report demonstrated seven pairs of chromosomes in the lentil genome with their categorization as acrocentrics (three chromosomes), metacentric (one chromosome) and submetacentrics (the remaining three chromosomes) (Sindhu et al. 1983).

The most notable reason for static lentil yields is vulnerability to drought and heat/water stresses occurring at flowering and pod-filling, especially in low-rainfall areas, whereas in spring-sown regions a major proportion of crop is lost due to terminal drought (Erskine et al. 2011), necessitating identification of trait-specific donors for winter hardiness and cold tolerance. Moreover, though of regional importance, excess water, waterlogging, and salinity also cause sizable losses in lentil production (Erskine and Saxena 1993). In relation to lentil yields, a wide range of fungal diseases are of worldwide importance, including vascular wilt, rust, and AB caused by *Fusarium oxysporum* f.sp. *lentis*, *Uromyces viciae-fabae*, and *Ascochyta fabae* f.sp. *lentis*, respectively. Other noteworthy constraints include anthracnose caused by *Colletotrichum truncatum*, collar rot (*Sclerotiun rolfsii*), root rot (*Rhizoctonia solani*), and sclerotinia white mold (*Sclerotinia sclerotiorum*). which adversely affect lentil production (Muehlbauer et al. 2006). In addition, infestation of a weedy root parasite, crenate broomrape (*Orobanche crenata*), also has adverse impacts on lentil production in the Mediterranean area and western Asia (de la Vega et al. 2011).

4.3 Availability of Genomic Tools and Technologies in Pulses

Remarkable progress has been noted in the generation of molecular tools for pulse crops in the recent past; this includes development of a variety of genomic resources such as mapping populations, bacterial artificial chromosome (BAC) libraries, BAC-end sequences (BESs), large-scale DNA markers, physical maps and linkage maps, transcriptome

TABLE 4.1

Genomics Resources in Chickpea, Pigeonpea, Cowpea, and Lentil

Genomic Resources	Chickpea	Pigeonpea	Cowpea	Lentil
Mapping populations	+	+	+	+
BAC libraries	+	+	+	−
BAC-end sequences	+	+	+	−
ESTs	+	+	+	+
Genomic SSRs	+	+	+	+
Genic or EST–SSRs	+	+	+	+
SNPs	+	+	+	−
DArTs	+	+	−	−
SFPs	−	+	+	−
Transcriptome assemblies	+	+	+	+
Genetic maps	+	+	+	+
Physical map	+	In progress	+	−
Whole-genome sequence	+	+	In progress	In progress

assemblies, and, above all, drafts of whole-genome sequence. A list of all the genomic resources recently made available for various pulse crops is given in Tables 4.1 and 4.2. Recent developments in the field of pulse genomics are discussed in the following section.

4.3.1 Mapping Populations

4.3.1.1 Chickpea

A mapping population segregating for markers/traits with a desirable size sets a prerequisite for conducting any linkage/trait mapping analysis. Several types of biparental populations such as F_2, recombinant inbred line (RIL), backcross (BC), and double haploid (DH) are being used for the construction of genetic maps. Compared to F_2, which is an ephemeral population, DH and RIL populations ensure adequate supply of homozygous seeds, facilitating multilocation/multienvironmental/multiyear testing. Additionally, DH/RILs are relatively flexible in dominant/codominant nature of marker applied as the two marker systems follow the same segregation pattern of 1:1 in these populations. Therefore, the availability of intraspecific DH/RIL population provides an ideal platform for undertaking any quantitative trait locus (QTL) or trait-mapping study. Various interspecific mapping populations derived from *C. arietinum* × *C. reticulatum*, *C. arietinum* × *C. echinospermum* (Gaur and Slinkard 1990; Simon and Muehlbauer 1997) and intraspecific populations were generated in chickpea, focusing on tagging/mapping of important characters (Tar'an et al. 2007; Cho et al. 2002; Udupa and Baum 2003; Flandez-Galvez et al. 2003b) (Table 4.3).

4.3.1.2 Pigeonpea

Several F_2 mapping populations segregating for economically important traits such as FW, SMD, and waterlogging were developed aiming at molecular tagging/mapping of the respective traits (Varshney et al. 2010a). To start with, two reference populations were developed, for broad-based (ICP 28 [*C. cajan*] × ICPW 94 [*C. scarabaeoides*]) and narrow-based (Asha × UPAS120) pigeonpea. Since the cytoplasmic genetic male sterility (CGMS)-derived hybrid system in pigeonpea has been the top priority of breeders as a potent option to

TABLE 4.2

Recent Developments in Chickpea, Pigeonpea, Cowpea, and Lentil Genomics

Genomic Resources	Crop	References
BAC libraries and BESs	Chickpea	Thudi et al. 2011
	Pigeonpea	Bohra et al. 2011
	Cowpea	http://www.comparative-legumes.org/pages/resources
Large-scale SSR/SNP markers	Chickpea	Thudi et al. 2011; Hiremath et al. 2012; Gaur et al. 2012b
	Pigeonpea	Bohra et al. 2011; Dubey et al. 2011; Kassa et al. 2012
	Cowpea	Muchero et al. 2009a; Lucas et al. 2011
High-Throughput Genotyping Platforms		
DArT arrays	Chickpea	Varshney et al. 2010b
	Pigeonpea	Yang et al. 2011
GoldenGate/KASPar assays	Chickpea	Hiremath et al. 2012; Gaur et al. 2012b
	Pigeonpea	Kassa et al. 2012
	Cowpea	Muchero et al. 2009a; Lucas et al. 2011
First genetic maps	Pigeonpea	Yang et al. 2011; Bohra et al. 2011, 2012
High-density genetic maps	Chickpea	Thudi et al. 2011; Hiremath et al. 2012; Gaur et al. 2012b
	Cowpea	Muchero et al. 2009a; Lucas et al. 2011
Physical maps	Chickpea	Zhang et al. 2010
	Cowpea	http://phymap.ucdavis.edu/cowpea

harness heterosis or hybrid vigor, attempts were made to develop F_2 mapping populations segregating for fertility restoration (*Rf*) (Varshney et al. 2010a; Bohra et al. 2012) (Table 4.3).

4.3.1.3 Cowpea

With the growing demands for undertaking marker-assisted selection (MAS) in cowpea breeding, several F_2 as well as RIL populations are now available to facilitate linkage/ association studies and trait mapping in cowpea (Table 4.3). These populations were generated targeting important attributes such as *Striga* resistance, aphid resistance, nematode resistance, *Macrophomina* resistance, and various yield-related traits (Fatokun et al. 1993a; Ouédraogo et al. 2001; Muchero et al. 2009a; 2011) (Table 4.3).

4.3.1.4 Lentil

To begin with, interspecific or intersubspecific F_2 mapping populations were built in lentil due to less DNA polymorphism in the primary gene pool than in other pulses (Zamir and Ladizinsky 1984; Weeden et al. 1992; Eujayl et al. 1997, 1998). Most of these populations were derived from the intersubspecific cross (*L. culinaris* ssp. *orientalis* × *L. culinaris* Med) (Table 4.3). Later, emphasis was given to developing intraspecific RIL mapping populations (Kahraman et al. 2004; Tullu et al. 2008; Saha et al. 2010). In this context, the RIL population, viz. ILL5588 × ILL7537, was the first mapping population based on cultivated cross (Rubeena et al. 2003) (Table 4.3). Based on these mapping populations several DNA

TABLE 4.3

Some of the Important Mapping Populations in Chickpea, Pigeonpea, Cowpea, and Lentil

Mapping Population	Type of Population	Segregating Traits[a]	References
Chickpea			
C. arietinum × *C. reticulatum* (PI 489777)	RIL	FW	Winter et al. 2000
ICCL81001 Cr5-9	RIL	FW	Cobos et al. 2009
CA2139 × JG62	RIL	FW	Halila et al. 2009
PI 359075(1) × FLIP84-92C(2)	RIL	AB	Cho et al. 2004
C. arietinum × *C. echinospermum*	F_2	AB	Collard et al. 2003
C. arietinum × *C. arietinum*	F_2	AB	Flandez-Galvez et al. 2003a,b
C. arietinum × *C. reticulatum*	RIL	AB	Rakshit et al. 2003
ILC 1272 × ILC 3279	RIL	AB	Udupa and Baum 2003
ICCV 96029 × CDC Frontier	F_2	AB	Tar'an et al. 2007
CDC Frontier × ICC96029	F_2	AB	Anbessa et al. 2009
CDC Corinne × ICC96029	F_2	AB	Anbessa et al. 2009
CDC Luna × ICC96029	F_2	AB	Anbessa et al. 2009
Amit × ICC 96029	F_2	AB	Anbessa et al. 2009
C. arietinum × *C. arietinum*	F_2	AB	Bhardwaj et al. 2011
C. arietinum × *C. reticulatum*	F_2	AB	Arymanesh et al. 2010
C. arietinum × *C. arietinum*	$F_{2:3}$	AB	Taleei et al. 2010
ICCV 2 × JG 62	RIL	Gray mold	Anuradha et al. 2011
C. arietinum × *C. arietinum*	RIL	Seed weight, days to flower	Cobos et al. 2007
ICCL81001 × Cr 5-9	RIL	Seed coat thickness, seed coat reticulation	Cobos et al. 2009
ICC 4958 × ICC 1882	RIL	Drought- and yield-related traits	Varshney et al. 2012
ICC 283 × ICC 8261	RIL	Drought- and yield-related traits	Varshney et al. 2012
Pigeonpea			
ICP 28 × ICPW 94	F_2	Pod borer	Bohra et al. 2011; Yang et al. 2011
ICPB 2049 × ICPL 99050	F_2	FW	Bohra et al. 2012
ICPL 8863 × ICPL 20097	F_2	SMD	Gnanesh et al. 2011
ICPA 2043 × ICPR 2671	F_2	Rf	Bohra et al. 2012
ICPA 2043 × ICPR 3467	F_2	Rf	Bohra et al. 2012
ICPA 2039 × ICPR 2447	F_2	Rf	Bohra et al. 2012
ICPA 2039 × ICPR 2438	F_2	Rf	Bohra et al. 2012
TTB7 × ICP 7035	F_2	SMD	Gnanesh et al. 2011
Cowpea			
524B × 219-01	RIL	Seed weight, pod fiber layer thickness	Andargie et al. 2011
"Gorom" × "Tvx3236"	F_2	SR	Ouédraogo et al. 2002
IT81D-994 × Tvx 3236	F_2	SR	Ouédraogo et al. 2002
CB46 × IT93K-503-1	RIL	AR, grain weight, nematode resistance, SR, seedling cold tolerance	Muchero et al. 2009a

(continued)

TABLE 4.3 (Continued)

Some of the Important Mapping Populations in Chickpea, Pigeonpea, Cowpea, and Lentil

Mapping Population	Type of Population	Segregating Traits[a]	References
Dan Ila × TVu-7778	RIL	Drought tolerance, yield, bacterial blight resistance	Muchero et al. 2009a
Yacine × 58-77	RIL	FT, individual grain weight	Muchero et al. 2009a
TVu14676 × IT84S-2246-4	RIL	SR, nematode resistance	Muchero et al. 2009a
CB27 × 24-125B-1	RIL	AR, grain weight, cowpea weevil resistance, FTR	Muchero et al. 2009a
IT93K-503-1 × CB46	RIL	MR	Muchero et al. 2009a
524B × 219-01	RIL	Seed weight, pod shattering, pod fiber layer thickness	Andargie et al. 2011
Lentil			
Eston × Indian head	F_2	AB	Chowdhury et al. 2001
Precoz × WA8949041	RIL	—	Tanyolac et al. 2010
ILL-6002 × ILL-5888	RIL	SB	Saha et al. 2010
Lupa (*L. culinaris*) × Boiss (*L. orientalis*)	F_2	Plant structure, growth habit, and yield	Fratini et al. 2007
ILL7537 × ILL6002	F_2	AB	Rubeena et al. 2006
ILL5588 × ILL7537	F_2	AB	Rubeena et al. 2006
Eston × PI320937	RIL	Earliness and plant height	Tullu et al. 2008
ILL 5588 × L 692-16-1(s)	RIL	FW	Hamwieh et al. 2005
L. culinaris × *L. orientalis*	RIL	—	Tahir and Muehlbauer 1994
CDC Robin × 964a-46	RIL	AB and anthracnose	Tar'an et al. 2003

[a] AB, *Ascochyta* blight; SB, *Stemphylium* blight; FW, *Fusarium* wilt; AR, aphid resistance; SR, *Striga* resistance; Rf, fertility resoration; SMD, sterility mosaic disease.

markers associated with traits such as resistance against AB, FW, *Anthracnose*, cold winter hardiness, and rust were identified in lentil.

4.3.2 BAC Libraries, BESs and Physical Maps

4.3.2.1 Chickpea

Bacterial artificial chromosome libraries are considered as a preferred class of large insert DNA libraries routinely used as seed points for various genetics applications such as genetic marker discovery, map-based gene/QTL cloning, integration of physical with genetic maps, and BAC-based whole-genome sequencing (Farrar and Donnison 2007). An array of BAC-based resources in chickpea was developed including BAC libraries, BESs, and physical map. The development of the first BAC library was reported from genotype FLIP84-92C using enzyme *Hin*dIII (Rajesh et al. 2004). The library contained 23,780 clones with average insert size of 100 kb. Subsequently a series of BAC and binary BAC (BIBAC) libraries was reported from the chickpea genotype Hadas, collectively offering more than 17× genome coverage (Lichitenzveiz et al. 2005; Zhang et al. 2010). End-sequencing of BAC clones was also undertaken, providing large sets of high-quality BESs (Nayak et al. 2010; Thudi et al. 2011). Moreover, in order to build a genome-wide BAC/BIBAC-based physical map, fluorescent-based high information content fingerprinting (HICF) was performed for 67,584 clones belonging to genotype Hadas assembled into a total of 1945 contigs with 11× genome coverage (Zhang et al. 2010).

4.3.2.2 Pigeonpea

Aiming at the development of genomic tools in pigeonpea, a collaborative attempt known as the pigeonpea genomics initiative (PGI) was initiated in 2006 with funding from the Indian Council of Agricultural Research (ICAR), National Science Foundation (NSF) (USA) and the Consultative Group on International Agricultural Research (CGIAR)–Generation Challenge Program (GCP) (Mexico). Consequently two BAC libraries were reported using enzymes *Hind*III and *Bam*HI, consisting of 34,560 clones each. The libraries collectively represented ~11× of the pigeonpea genome (Bohra et al. 2011). The construction of BAC-libraries was accompanied by end-sequencing of 50,000 BAC clones producing a total of 88,860 high-quality BESs. The construction of a genome-wide physical map in pigeonpea is in progress (Varshney et al. 2013a).

4.3.2.3 Cowpea

Two BAC libraries derived from genotype IT97K-499-35 (a *Striga*-resistant and dual-purpose African genotype) using enzymes *Hind*III and *Mbo*I were established. Both the *Hind*III and *Mbo*I libraries comprised 36,864 clones each covering ~9× and ~8× of cowpea genome, respectively. Moreover 33,696 BACs underwent BAC-end sequencing yielding a set of 30,611 BESs (Ma et al. 2009). Similarly, another set of 50,120 BESs was generated and was further surveyed for discovery of BAC-associated microsatellite markers (Nayak et al. 2008). Apart from these two BAC libraries, one more library containing 23,780 clones was reported for genotype FLIP84-92C with threefold depth. A total of 60,000 BAC clones from genotype IT97K-499-35 were chosen for fingerprinting. Consequently a physical map was constructed with 43,717 BACs and 10× coverage (Ma et al. 2009). Fingerprinted contigs (FPC) software allowed assembly of clones into 790 contigs and 2535 singletons. The developed physical map offers a viable tool for computation of minimal tiling path (MTP), which relies on selection of minimally overlapping BAC clones to facilitate whole-genome sequencing in cowpea. The assembly is available online at http://phymap.ucdavis.edu:8080/cowpea.

4.3.2.4 Lentil

To the best of our knowledge, no report is available on the development of BAC libraries in lentil so far (Varshney et al. 2009a).

4.3.3 Molecular Markers

4.3.3.1 Chickpea

With the discovery of molecular markers, viz. restriction fragment length polymorphism (RFLP), by Botstein and colleagues in 1980, a number of DNA-based markers were used in chickpea, demonstrating their extensive utility for various kinds of genetics and genomics applications. These markers include RFLP (Udupa et al. 1993; van Rheenen et al. 1992), randomly amplified polymorphic DNA (RAPD) (Udupa et al. 1993; van Rheenen et al. 1992; Ahmed et al. 2010), and amplified fragment length polymorphism (AFLP) (Sudupak et al. 2004; Shan et al. 2005; Talebi et al. 2008). However, due to the preference of simple sequence repeat (SSR) markers over other marker systems, these emerged as the class of choice for undertaking diversity/linkage/association analyses and trait mapping (Huttel et al. 1999; Upadhyaya et al. 2008; Imtiaz et al. 2008). Instances of discovery of large-scale SSR markers through *in silico* mining of BESs and expressed sequence tags (ESTs) were also reported in chickpea (Nayak et al. 2010; Thudi et al. 2011; Jhanwar et al. 2012). In addition, a plethora of gene-derived functional markers (FMs) such as EST–SSRs, cleaved amplified polymorphic sequence (CAPS), derived CAPS (dCAPS), and conserved intron-spanning primer (CISP) makers have been used in chickpea (Gujaria et al. 2011). The recent emphasis on the development of a high-throughput marker system led to the discovery of several sequence-based genetic markers in chickpea which include mainly diversity arrays technology (DArT) and single-nucleotide polymorphism (SNP) markers (Thudi et al. 2011; Gaur et al. 2012b; Hiremath et al. 2012). Amenability of SNPs to high-throughput genotyping opened new avenues for the establishment of Illumina GoldenGate and competitive allele-specific PCR (KASPar) assays in chickpea (Gaur et al. 2012b; Hiremath et al. 2012).

4.3.3.2 Pigeonpea

The first attempt of using markers for distinguishing different pigeonpea accessions was made by Ladizinsky and Hamel (1980). Subsequently several DNA markers belonging to the different categories such as RFLPs (Nadimpalli et al. 1993; Sivaramakrishnan et al. 1997, 2002), RAPDs (Ratnaparkhe et al. 1995), AFLPs (Panguluri et al. 2006), SSRs (Odeny et al. 2007; Saxena et al. 2010), and DArTs (Yang et al. 2006, 2011) have been employed in pigeonpea. In pigeonpea SSRs have been generated through the traditional genomic library method as well as through *in silico* mining of BESs or ESTs. The first set of large-scale SSR markers comprising >3000 SSRs was developed from BESs and used for linkage mapping and hybridity testing (Bohra et al. 2011, 2012; Gnanesh et al. 2011). Concerning high-throughput genotyping, a DArT array comprising 5376 clones was established in pigeonpea. Similarly, in order to develop microarray-based genotyping systems, a total of 5692 single-feature polymorphism (SFP) markers were identified using the Affymetrix soybean genome arrays (Saxena et al. 2011). Further association of some of the SFP markers was predicted with gene(s)/QTLs imparting drought tolerance; the candidate genes thus identified could be harnessed for future genetic analysis. Recently the whole-genome sequence of pigeonpea genotype has allowed easy access to a plethora of SSR and SNP markers (Varshney et al. 2011; Singh et al. 2011).

4.3.3.3 Cowpea

DNA markers have been extensively used for diversity assessment in cowpea including RFLP (Fatokun et al. 1993b), AFLP (Fatokun et al. 1997), RAPD (Mignouna et al. 1998), and SSR (Li et al. 2001). Recently more than 900 SSR primer pairs were designed and successfully used for linkage analysis in cowpea (Andargie et al. 2011). Information about a large number of SSR markers is available at the cowpea genomics knowledge base (CGKB) (http://cowpeagenomics.med.virginia.edu/CGKB) website. A large set of SNP markers (~10,000) was also identified through mining of cDNA-derived EST sequences, and SNP genotyping of over 700 individuals belonging to seven different RILs was performed using Illumina 1536 GoldenGate assay (Muchero et al. 2009a). The utility of Illumina 1536 GolenGate assay was further demonstrated via SNP genotyping of six additional RIL populations (Lucas et al. 2011).

4.3.3.4 Lentil

The first investigation on molecular markers in lentil was reported by Havey and Muehlbauer (1989). They used RFLP for analyzing genetic diversity among domesticated and wild lentils and for linkage analyses. Later RAPD (Ford et al. 1997), AFLP (Závodná et al. 2000), and inter simple sequence repeat (ISSR) markers (Durán and de la Vega 2004) were widely employed to estimate genetic variability within the *Lens* genus. However, lentil still maintains a low-profile status in terms of availability of SSR markers. The lentil-specific SSR reported by Závodná and colleagues (2000) was successfully used for differentiating closely related cultivars in lentil. Similarly, a new set of microsatellites comprising 14 SSRs was developed from a genomic library of a cultivated lentil accession ILL 5588 and was used for molecular characterization of lentil core collection (Hamwieh et al. 2009). Apart from this, some resistance gene analogs (RGAs) targeting nucleotide-binding sites (NBSs) of resistance genes were generated for *Lens* species using degenerate primer pairs and subsequently used for molecular mapping/linkage analyses (Rubeena et al. 2003).

4.3.4 Genetic Linkage Maps

4.3.4.1 Chickpea

The first genetic linkage map of chickpea was constructed for an F_2 population derived from an interspecific cross (*C. arietinum* × *C. reticulatum*). The map was based on morphological and isozyme markers and covered a map distance of 200 cM. Following this some DNA-based markers such as RFLP and RAPD were integrated along with morphological and isozyme markers covering collectively a length of 550 cM of the chickpea genome (Simon and Muehlbauer 1997). An advancement on this genetic map was seen in the form of 303 mapped markers belonging to various categories such as RAPD, ISSR, AFLP, sequence-tagged microsatellite site (STMS), and sequence-characterized amplified region (SCAR) mapped at a distance of 2078 cM (Winter et al. 2000). Pfaff and Kahl (2003) provided further refinement comprising 296 markers with a length of 2500 cM. The initial intraspecific genetic linkage mapping in chickpea was based on a RIL population derived from the cross ICCV 2 × JG 62 (Cho et al. 2002). The map had 80 loci mapped onto 14 linkage groups (LGs) with a coverage of 297.5 cM. Afterwards, several cultivated maps were built in chickpea (Flandez-Galvez et al. 2003a; Cobos et al. 2005). Efforts were also directed to synthesize a consensus or composite genetic map in chickpea. Segregation data from multiple mapping populations were used and merging of these different genetic maps was accomplished with the help of common markers using appropriate softwares (Radhika et al. 2007; Millan et al. 2010; Gaur et al. 2011). In chickpea, consensus maps are available for both interspecific and intraspecific crosses (Millan et al. 2010). In recent years, a series of highly saturated genetic maps have been reported for the interspecific population (*C. arietinum* [ICC 4958] × *C. reticulatum* [PI 489777]) in chickpea using DArTs, SSRs, and SNPs (Thudi et al. 2011; Gaur et al. 2012b; Hiremath et al. 2012).

4.3.4.2 Pigeonpea

Extremely low level of DNA polymorphism coupled with less availability of molecular markers hampered the development of genetic maps in pigeonpea. Nevertheless, to ensure adequate supply of DNA polymorphisms, DArT system for high-throughput genotyping was established and the first genetic map in pigeonpea was constructed for the reference mapping population ICP 28 (*C. cajan*) × ICPW 94 (*C. scarabaeoides*) using polymorphic DArT markers. The DArT-based paternal and maternal genetic maps consisted of 172 and 122 loci covering map lengths of 451 and 270 cM, respectively (Yang et al. 2011). However, due to the absence of a sufficient number of bridging markers/codominant SSR markers, the two separate genetic maps could not be merged; this challenged the practical utility of DArT-based genetic maps. The first SSR-based genetic map was reported for the same reference population by Bohra et al. (2011). The F_2 population was genotyped with 378 BES–SSRs markers and finally a total of 239 SSR loci were mapped at a distance of 930 cM. Widespread uses of the large-scale BES–SSRs in linkage analyses were further demonstrated by construction of SSR-based intraspecific genetic maps for six different F_2 populations. Additionally, occurrence of common SSR markers among six different cultivated genetic maps led to the synthesis of one integrated or consensus genetic map for pigeonpea (Bohra et al. 2012). The consensus genetic map had 339 SSR loci mapped at a distance of 1059 cM (Bohra et al. 2012).

4.3.4.3 Cowpea

The first linkage map of cowpea was derived from a cross IT84S-2246-4 × TVu 96392 with 717 cM genome coverage using RAPD and AFLP markers (Fatokun et al. 1993a). Following this a cultivated genetic map of cowpea derived from a RIL population (IT84S-2049 × 524B) was constructed with 181 loci. All the loci were mapped on 12 LGs covering a distance of 972 cM with mean intermarker distance of 6.4 cM (Menéndez et al. 1997). Further attempts to integrate 242 DNA markers, specifically AFLPs and RGAs, into the IT84S-2049 × 524B-based genetic map resulted in an improved version with 2670 cM, with one marker per 6.43 cM and 11 LGs (Ouédraogo et al. 2002). Another map with 1620.1 cM was developed using 134 AFLP and 5 SSR markers were mapped onto 11 LGs. The map was based on a cross between Sanzi (resistant to flower thrips FTh) and VITA 7 (susceptible to FTh) consisting of 145 F_{10} RILs (Omo-Ikerodah et al. 2008). An SSR-based genetic map was reported by Andargie and colleagues (2011) for a RIL population (524B × 219-01) with mapping of 202 SSRs onto 11 LGs and spanning 677 cM. The most comprehensive genetic linkage map in cowpea was reported by Muchero and colleagues (2009a). This represented the first instance of implementation of Illumina 1536 GoldenGate assay for high-throughput SNP genotyping providing six RIL-based individual genetic maps with loci ranging between 288 and 436. The map lengths were obtained in the range of approximately 600 cM. Eventually, a gene-based consensus genetic map was synthesized by merging six different genetic maps. The integrated map consisted of 928 SNP loci mapped onto 11 LGs and covering a map length of 680 cM. Furthermore, in order to saturate the SNP-based consensus genetic map, markers from six additional genetic maps were incorporated and a 1100 loci high-density consensus map (680 cM) was constructed from a total of 13 RIL populations (Lucas et al. 2011).

4.3.4.4 Lentil

The first genetic map for *Lens* was developed by Zamir and Ladizinsky (1984) using morphological and isozyme markers. Subsequently a series of genetic maps was reported using these types of markers. However, due to insufficiency in the number of such markers coupled with scanty DNA polymorphism, all the preliminary lentil genetic maps were of rudimentary type which could cover only a fraction of the entire *Lens* genome. Therefore aiming at expanded genome coverage, the first DNA markers-based genetic map in lentil was built for an intersubspecific F_2 population (*L. culinaris* × *L. orientalis*) comprising over 60 individuals. The map had 34 marker loci including 20 RFLPs, 8 isozyme, and 6 morphological markers and spanned a map length of 333 cM (Havey and Muehlbauer 1989). Following the developments in linkage mapping, numerous genetic linkage maps were developed for cultivated lentil as well. Therefore, the first genetic map for cultivated lentil was constructed for an intraspecific F_2 population "ILL 5588 × ILL 7537" (Rubeena et al. 2003). A total of 784 cM map length was obtained, which could ensure map positions to 114 markers (100 RAPDs, 11 ISSRs, and 3 RGAs) with 9 LGs. To date the most comprehensive genetic map in lentil, comprising 283 marker loci, was developed for a RIL population. The genetic map coverage obtained was 751 cM with a mean intermarker distance of 2.65 cM (Hamwieh et al. 2005).

4.3.5 Trait Mapping/Tagging

4.3.5.1 Chickpea

Mapping or tagging of the gene(s)/QTLs underlying resistance to disease such as FW and AB, and abiotic stress tolerance such as drought resistance, has received the most attention from chickpea molecular breeders/geneticists (Table 4.4). The first resistance gene *Hg* against FW was tagged using RAPD markers (Mayer et al. 1997). Molecular tagging of the other resistance genes was conducted against different races, viz. *foc4* race (Ratnaparkhe et al. 1998a,b; Tullu et al. 1998, 1999; Tekeoglu et al. 2000; Winter et al. 2000) and *foc5* race (Ratnaparkhe et al. 1998b; Tekeoglu et al. 2000; Winter et al. 2000; Benko-lseppon et al. 2003). One of the resistance genes for *foc3* race was also found to be associated with STMS marker (Rajesh et al. 2004). The same gene (*foc-3*) was flanking with STMS markers TA96 and TA27 and sequence-tagged site (STS) marker CS27A using RILs derived from WR-315 (resistant) × C-104 (susceptible). Similarly gene(s)/QTLs for AB resistance were also identified and Flandez-Galvez et al. (2003b) reported 6 QTLs conferring AB resistance. Interestingly, most of the underlying QTLs were mapped close to RGAs. Two QTLs associated with seedling resistance were also detected on LG 4 of an interspecific F_2 population (*C. arietinum* × *C. echinospermum*). Quantitative trait locus mapping conducted on an F_2 population, viz. ICCV 96029 × CDC Frontier, revealed a total of three QTLs on LG 3, LG 4, and LG 6 governing more than 50% variation on the phenotype (Tar'an et al. 2007). Three QTLs for BGM resistance were also recovered from an intraspecific RIL population (ICCV 2 × JG 62) and all these QTLs collectively explained more than 40% variation in BGM resistance (Anuradha et al. 2011). Considering drought tolerance in chickpea, plenty of major and minor QTLs were discovered for various root- and yield-related traits on LG 4 from two intraspecific RIL mapping populations, viz. ICC 4958 × ICC 1882 and ICC 283 × ICC 8261 (Gaur et al. 2012a). Efforts are under way to deploy all these QTL(s)/marker(s) in routine chickpea breeding programs.

4.3.5.2 Pigeonpea

The history of trait mapping has not been very encouraging in the case of pigeonpea. Some preliminary efforts using bulked segregants analysis (BSA) have been made for tagging important traits such as FW and SMD resistance (Table 4.4). These studies include RAPD-based molecular tagging of FW resistance and plant type using F_2 populations, viz. GS1 × ICPL 87119 (250 individuals) and T 44-4 × TDI 2004-1 (84 individuals), respectively (Kotresh et al. 2006; Dhanasekar et al. 2010). Similarly, AFLP-based molecular tagging for SMD resistance was reported using an F_2 population derived from a cross between TTB 7 (susceptible to SMD) and BRG 3 (resistant to SMD) (Ganapathy et al. 2009). Moreover, availability of trait-specific mapping populations and the genetic maps also offered new perspectives for undertaking QTL analysis to discover gene(s)/QTLs linked to the respective traits. For instance, genotyping and phenotyping data accumulated on the two F_2

TABLE 4.4

Molecular Mapping/Tagging of Some Economically Important Traits

Crop Name	Traits of Interest[a]	Type of Associated DNA Markers	References
Chickpea	FW	RAPD, STMS/SSR	Rajesh et al. 2004; Cobos et al. 2005; Gowda et al. 2009; Rubio et al. 2003; Sharma et al. 2004; Castro et al. 2010
	AB	RAPD, STMS/SSR/ SCAR	Cobos et al. 2006; Millan et al. 2003; Tar'an et al. 2007; Iruela et al. 2006
	BGM	SSR	Anuradha et al. 2011; Madrid et al. 2008
	Double podding and other morphological traits	STMS	Cho et al. 2002; Rajesh et al. 2002
	100 seed weight, flower pigmentation	STMS	Cho et al. 2002
	Drought tolerance-related traits	SSR	Chandra et al. 2004
Pigeonpea	FW	RAPD	Kotresh et al. 2006
	SMD	SSR, AFLP	Ganapathy et al. 2009; Gnanesh et al. 2011
	Rf	SSR	Bohra et al. 2012
	Plant type	RAPD	Dhanasekar et al. 2010
Cowpea	FW	SNP	Pottorff et al. 2012
	SR	SCAR, AFLP	Ouédraogo et al. 2001, 2002
	MR	SNP	Muchero et al. 2011
	Drought stress-induced premature senescence	AFLP	Muchero et al. 2009b
	Leaf morphology	SNP	Pottorff et al. 2012
	Pod length	SSR	Kongjaimun et al. 2012
	Seed weight, pod shattering/ pod fiber layer thickness	SSR	Andargie et al. 2011
	Aphid resistance	RFLP	Myers et al. 1996
Lentil	AB	AFLP, RAPD, SCAR	Chowdhury et al. 2001; Rubeena et al. 2006; Tullu et al. 2006
	Anthracnose	RAPD	Tullu et al. 2003; Tar'an et al. 2003
	FW	RAPD, AFLP, SSR	Hamwieh et al. 2005
	Radiation–frost tolerance or cold winter hardiness	RAPD, ISSR	Eujayl et al. 1999; Kahraman et al. 2004
	Plant structure, growth habit, and yield	RAPDs, AFLPs, ISSRs, SSRs	Fratini et al. 2007
	SB	SRAP	Saha et al. 2010

[a] AB, *Ascochyta* blight; SB, *Stemphylium* blight; FW, *Fusarium* wilt; AR, aphid resistance; SR, *Striga* resistance; Rf, fertility resoration; SMD, sterility mosaic disease; MR, *Macrophomina* resistance; BGM, botrytis gray mold.

populations TTB 7 × ICP 7035 and ICP 8863 × ICPL 20097 revealed occurrence of a total of six QTLs associated with SMD resistance. Of the six identified QTLs, one major QTL explained PV up to 24% (Gnanesh et al. 2011). Similarly, restoration of fertility in A_4 cytoplasm was also subjected to QTL analysis utilizing the marker segregation and trait phenotyping data from three different mapping populations, viz. ICPA 2039 × ICPR 2447, ICPA 2043 × ICPR 2671, and ICPA 2043 × ICPR 3467. Four QTLs experiencing major effects on the phenotype were identified for restoring fertility in A_4-based CGMS systems in pigeonpea (Bohra et al. 2012). Discovery of the DNA markers associated with FW/SMD resistance, plant type, and Rf would help greatly in MAS for the recovery of desired genotype in a much faster, more precise, and environmentally insensitive manner. However, to ensure the robustness of the above-identified DNA markers, marker/QTL validation in different genetic backgrounds would be required before proceeding for investments using these marker(s)/QTL(s) in pigeonpea breeding program.

4.3.5.3 Cowpea

Owing to its immense importance in sub-Saharan and savannah subsistence agriculture, considerable efforts have been dedicated to mapping/tagging of agronomically important traits in cowpea (Table 4.4). Most of the diseases of economic importance, including FW resistance, *Striga* resistance, flower and foliar thrips, root knot nematode, and bacterial and ashy blight, have been molecularly mapped and the underlying gene(s)/QTLs and DNA markers strongly associated with the traits are identified for facilitating marker-assisted breeding (MAB) for accelerated cowpea improvement. For instance, several small and large effect QTLs (*Mac*1-9) imparting resistance to *Macrophomina phaseolina* and explaining PV in the range of 6% (*Mac*-4) to 40% (*Mac*-2) were discovered from RIL population (IT 93K-503-1 × CB 46). Colocalization of these QTLs with drought-induced senescence was also found to exist. The same population contributed to other drought-related QTLs (*Dro* 1-10) experiencing up to 24% effects on phenotype (Muchero et al. 2009b). Likewise RFLP markers associated with QTLs with major phenotypic effects on seed weight were recovered from an F_2 population (IT 2246-4 × TVNI 963) in cowpea. Interestingly, the cowpea genomic region was orthologous with seed weight-QTLs based on another mungbean F_2 population (VC 3890 × TC l966) (Fatokun et al. 1992). As it is one of the oldest domesticated crops, emphasis was also given to mapping of domestication-related traits, and in the context QTL analysis on RIL population 524B × 219-01 revealed occurrence of several QTLs for seed weight and pod fiber layer thickness, explaining PV up to 19% (Andargie et al. 2011). In another instance, a single major QTL named hastate leaf shape (*Hls*) with more than 70% of PV was obtained in the RIL population (Sanzi × Vita 7). Further, an SNP marker 1_0349 cosegregated with hastate/sub-globose leaf shape. To precisely locate the QTL, it was also placed onto a consensus genetic map and physical map (Pottorff et al. 2012). A mapping population generated through crossing JP 81610 (*V. unguiculata* (L.) Walp. subsp. *unguiculata* Sesquipedalis Group) and TVnu457 (*V. unguiculata* subsp. *unguiculata* var. *spontanea*) contributed a total of seven QTLs for pod length (Kongjaimun et al. 2012).

4.3.5.4 Lentil

In lentil, classical genetics-based linkage mapping methods have led to the establishment of linkage relationships among several traits using them as morphological markers, which include cotyledon color (*I*), glabrous pod (*Glp*), pod indehiscence (*Pi*), anthocyanin production in the stem (*Gs*), dwarfing genes (*Df1* and *Df2*), seed coat spotting (*Scp*), and cotyledon color (*Yc*). (de la Vega et al. 2011). The availability of limited DNA markers in lentil encouraged tagging of traits using BSA relying on the identification of tightly linked DNA marker in a cost-effective and time-saving manner (Michelmore et al. 1991). The method is best suited for tagging of major genes governing oligogenic traits with large phenotypic effects. Several resistance genes such as *ral2* (Chowdhury et al. 2001), *AbR1* for AB resistance (Ford et al. 1999), *Fw* for FW resistance (Eujayl et al. 1998), *LCt-2* against anthracnose (Tullu et al. 2003), and *Frt* imparting radiation–frost tolerance (Eujayl et al.

1999) have successfully been tagged. QTL analyses for dissecting quantitative traits (QTs) governed by several genes were reported for plant height, flowering, and pod dehiscence (Durán et al. 2004). A list of some of the traits that have been mapped/tagged in lentil is given in Table 4.4. Rubeena and colleagues (2003) investigated AB resistance for composite interval mapping (CIM) analysis and reported three associated QTLs with PV more than 50%. Major QTLs experiencing about 30% effects were discovered for earliness and plant height from population Eston × PI 320937 (Tullu et al. 2008). Similarly an F_6-derived RIL population (WA 8649090 × Precoz) provided one major QTL contributing more than 20% to the winter hardiness tolerance (Kahraman et al. 2010).

4.3.6 Transcriptomics and Comparative Analyses

4.3.6.1 Chickpea

A comparatively small number of ESTs (44,157) are available in the public domain (dbEST release 120,701, http://www.ncbi.nlm.nih.gov/dbEST/dbEST_summary.html), warranting an acceleration in the development of gene-based or functional genomics tools. The first comprehensive set comprising 20,162 ESTs was generated using single-pass Sanger sequencing from the water-stressed root tissues (Varshney et al. 2009b). Exciting breakthroughs have been made using GS-FLX 454 technology-based deep transcriptome sequencing, rapidly generating large-scale transcript reads in chickpea (Hiremath et al. 2011; Jhanwar et al. 2012). By virtue of the higher bp lengths of GS-FLX 454 reads, this technology is considered an ideal tool for transcriptome as well as whole-genome sequencing of any nonmodel organism lacking a reference genome. However, the effectiveness of shorter reads in characterization of the transcriptome was also shown by constructing a *de novo* transcriptome assembly with more than 100 million Illumina reads (Garg et al. 2011), and additionally to enable the extensive search in generated data of chickpea genotype ICC 4958, a web resource—the chickpea transcriptome database (CTDB)—has also been developed (http://www.nipgr.res.in/ctdb.html) (Garg et al. 2011). All the large-scale ESTs acted as impending resource for discovery of gene-based DNA markers including EST–SSRs, SNPs, conserved orthologous sequence (COS) markers, and intron-spanning region (ISR) markers. A vast number of ESTs are deposited in the public domain for model legumes such as *Medicago*, *Lotus* as well as for other legumes such as soybean and cowpea, providing greater scope for the identification of conserved regions across genera, viz. microsyntenic and macrosyntenic blocks, as well as allowing the easy flow of molecular information across different legume genera.

4.3.6.2 Pigeonpea

In pigeonpea transcriptomics tools were generated using Sanger as well as next-generation sequencing (NGS) platforms (Raju et al. 2010; Dubey et al. 2011; Dutta et al. 2011; Kudapa et al. 2012). The first Sanger-based assembly comprising 9888 ESTs was developed from FW- and SMD-responsive pigeonpea genotypes (Raju et al. 2010). Currently a set of 25,576 ESTs for pigeonpea is found in the National Center for Biotechnology Information (NCBI) EST database. A total of four trancriptome assemblies have been reported for pigeonpea, with transcript assembly contigs (TACs) ranging from 4,557 to 48,726 (Kudapa et al. 2012). The most recent assembly, designated as *Cajanus cajan* transcriptome assembly version 2 (CcTA v2), includes ESTs generated through various sequencing platforms, viz. Illumina GA IIx (128.9 million short reads), FLX/454 reads (2.19 million longer reads), and Sanger sequencing (18,353 ESTs). Furthermore, comparative analysis of the CcTA v2 with genome sequence of soybean (*Glycine max*) provided a set of 128 ISR markers, of which 116 ISRs were experimentally validated in a panel of pigeonpea genotypes consisting of seven cultivated genotypes and one wild genotype (Kudapa et al. 2012). The assembly CcTA v2 can be accessed through the legume information system (LIS) website (http://cajca.comparative-legumes. org/data/lista_cajca-201012.tgz). Concerning functional or genic DNA markers, a set of 84 EST–SSRs was identified from Sanger ESTs (Raju et al. 2010), while more than 550 EST–SSRs were developed from ESTs generated through 454/FLX sequencing (Dutta et al. 2011).

4.3.6.3 Cowpea

Cowpea enjoys accessibility to a large number of ESTs (187,487) deposited in NCBI. Likewise a large number of EST-derived SNP markers were identified and subsequently mapped in cowpea (Muchero et al. 2009a). All these EST-based resources can be easily retrieved from a web-based program, *HarvEST: Cowpea* version 1.27 (http://www.harvest-web.org). Furthermore, to gain valuable insights into the gene content, genome evolution, and phylogenic relationships within the Leguminosae family, methylation filtration (MF) technology relying on reduced representation of genome was used to sequence the genespace of cowpea genome. By virtue of separation of plant genomes into hypermethylated repetitive and undermethylated generich fractions, MF has already been proved to be of widespread use in sequencing of genespace in various crop plants such as maize (Palmer et al. 2003). The MF-based sequencing of cowpea gene-space accumulated a huge number of gene-space sequence reads (GSRs)—approximately 298,848, with an average length of 610 bp (Timko et al. 2008). Additionally BLAST comparisons with the other legumes' Unigenes database revealed an average similarity of 74.43% to other legume species, being highest with *Phaseolus vulgaris* (88.6%) while the lowest was observed with *Lotus japonicas* (61%). A web-based database, Cowpea Genespace Sequences Knowledge base (CGKB) (http://cowpeagenomics.med.virginia.edu/CGKB), developed under the Cowpea Genomics Initiative (CGI), provides details about these GSR sequences and annotation in a user-friendly manner (Chen et al. 2007).

4.3.6.4 Lentil

In the case of crops that suffer from a dearth of genomic resources, functional genomics provides a potential option to leverage the genomic resource repertoire, facilitating flow of genomic information from resource-rich model plant species to the less-studied crops. Expressed sequence tags of *M. truncatula* (closely related model legume belonging to the clade hologalegina) and *A. thaliana* (model species from family brassicaceae) were chosen for the design of some perfect markers targeting conserved regions. As a result, a total of 100 conserved primers (CPs) were designed, of which 22 CPs showed successful amplification. Using these CPs, an extensive diversity analysis conducted on a panel of 308 lentil accessions (including 175 wild and 133 domesticated lentils) provided additional insights into domestication and distribution of the *Lens* genus (Alo et al. 2011). Transcriptomic resources allow detailed comparative mapping and assessment of phylogenetic relationships among different genera. For instance, 79 intron-targeted amplified polymorphic (ITAP) markers and 18 genomic SSRs were used for construction of the first gene-based genetic map for lentil and affirmed the existence of a syntenic relationship between *Medicago truncatula* and *Lens culinaris* ssp. *Culinaris* (Phan et al. 2006, 2007). Extensive macrosynteny between genera *Lens* and *Medicago* and easy crossgenus transferability of FMs permitted successful integration of 15 *Medicago*-specific EST–SSRs into an intraspecific genetic map of lentil (Gupta et al. 2011). In another attempt to develop *Lens*-specific FMs, transcriptome was sequenced using Roche 454 GS-FLX titanium technology, leading to generation of a massive amount of high-quality ESTs (847,824 ESTs). Subsequent clustering and assembly provided a set of 84,074 unigenes (15,359 contigs and 68,715 singletons), and a comprehensive set of 2393 EST–SSR primers was designed, of which 192 markers were confirmed among 13 *Lens* genotypes (Kaur et al. 2011).

4.4 Whole-Genome Sequencing in Pulses

Next-generation sequencing or second-generation sequencing and, more recently, third-generation sequencing (TGS) technologies have emerged as a highly efficient method offering generation of whole-genome sequence in a much faster, less cumbersome (compared to the BAC by BAC approach), and cost-effective manner. These NGS-based technologies have been exploited successfully not only for resequencing but also for *de novo* whole-genome sequencing of many crops such as apple, cucumber and *Brassica rapa* (Imelfort and Edwards 2009). Among the pulses, pigeonpea became the first to possess a draft of the whole-genome sequence, witnessing a paradigm shift in its status from being an "orphan legume" to a "resource-rich crop." This club comprising sequenced crops was recently joined by chickpea also. In the case of pigeonpea, two of the NGS-based technologies, viz. 454 GS-FLX sequencing (Singh et al. 2011) and Illumina GA HiSeq 2000 sequencing (Varshney et al. 2011), were selected to generate two separate drafts of whole-genome sequences of a popular pigeonpea variety, Asha (ICPL 87119). The 454 GS-FLX sequencing could capture a total of 510 Mb of the pigeonpea genome with ~10× coverage, while the other attempt assembled 605.78 Mb of pigeonpea genome with ~163× coverage. Existence of more than 45,000 protein coding genes was predicted within the pigeonpea genome. In addition several disease- and drought-responsive genes were found to occur in the genome, which can be targeted as candidate genes for future studies. Availability of the whole-genome sequence permits easy access to a large number of DNA markers such as SSR and SNP markers. *In silico* mining of pigeonpea genome sequence permitted the discovery of 309,052 SSRs and 28,104 SNPs (across 12 pigeonpea genotypes) (Varshney et al. 2011). Of the total SSRs found in the genome, primer designing was performed for 23,410 SSRs, and a list of all these SSRs can be accessed from the International Initiative for Pigeonpea Genomics (IIPG) website (http://www.icrisat.org/gt-bt/iipg/Home.html). Similarly a set of hypervariable SSR markers designated as hypervariable "Arhar" simple sequence repeat (HASSR), consisting of 437 SSRs, was developed for future applications in diversity analyses and linkage and association studies (Singh et al. 2011). Most importantly, from the standpoint of evolution valuable insights were gained from the whole-genome sequence such as absence of the whole-genome duplication (WGD) event.

Similarly in chickpea the whole-genome sequence of CDC Frontier (A Kabuli chickpea variety) was reported recently by Varshney et al. (2013b) using Illumina sequencing. This draft of whole-genome shotgun (WGS) sequence provided ~738 Mb of chickpea genome with more than 200× coverage. Additionally resequencing of 90 more genotypes was performed to investigate genetic diversity existing at whole-genome level. Large proportions of DNA markers (more than 75,000 SSRs and SNPs) were revealed through scanning of this whole-genome sequence along with providing acess to 28,269 genes. The whole-genome sequence also provided valuable insights into chickpea domestication and evolution. Likewise, whole-genome sequencing of cowpea and lentil is under way and will likely be completed soon.

4.5 MAB in Pulses

Downstream applications of genomics tools in pulses breeding promise to be a potential add-on to the conventional breeding in order to speed up the progress of pulse genetic improvement. Though, unlike other crops, molecular breeding in pulses is in the infancy stage, the future seems very encouraging. A holistic approach of using these genomics tools and technologies for pulse genetic improvement is demonstrated in Figure 4.1. In chickpea, marker-assisted backcrossing (MABC) using SSR markers has already been initiated to address problems of biotic (AB and FW) as well as abiotic (drought) stresses. The chickpea genotypes, viz. ILC 3279 and Vijay/WR 315, were selected as donors for introgression of AB and FW resistance, respectively, into the elite chickpea genotypes including C 214 (recipient for AB resistance) and Pusa 256/JG 74/Phule G12/Annigeri-1 (recipients for FW resistance) (Varshney et al. 2012). For drought tolerance the hotspot genomic regions harboring several major as well as minor QTLs were targeted for marker-assisted introgression. The region was identified from the RIL populations, viz. ICC 4958 × ICC 1882 and

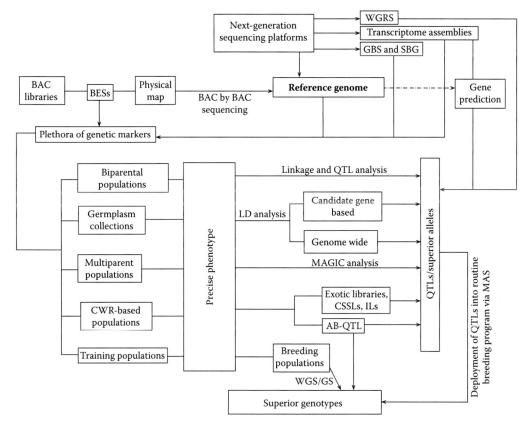

FIGURE 4.1

A holistic approach for integrating modern genomics technologies into breeding for genetic improvement of pulses. *Notes:* WRGS, whole genome resequencing; GBS, genotyping by sequencing; SBG, sequence-based genotyping; CWRs, crop wild relatives; LD, linkage disequilibrium; CSSLs, chromosome segment substitution lines; ILs, introgression lines; AB-QTLs, advanced backcross QTLs; MAGIC, multiparent advanced generation intercross; GS, genomic selection; WGS, whole genome selection.

ICC 283 × ICC 8261, and flanked by several SSR markers. However, the MABC approach may not be appropriate for capturing the phenotypic variations attributed to several minor effects QTLs, therefore some novel molecular breeding methods such as marker-assisted recurrent selection (MARS) have been introduced recently in chickpea to develop chickpea genotypes with enhanced drought tolerance (Gaur et al. 2012a). Marker-assisted recurrent selection offers greater opportunities for recovery of such mosaic genotypes endowed with several superior alleles, hence facilitating precise transfer of such complex traits.

In pigeonpea the limited number of markers/QTLs has discouraged initial investment. Nevertheless some of the methods such as advanced backcross QTL (AB-QTL) are of immense importance because AB-QTL allows not only detection, but introgression of QTLs also into the background of elite cultivars. Advanced backcross QTL provides several possibilities for harnessing high levels of cryptic genetic variations existing among the wild pools (Gibson and Dworkin 2004), especially in crops such as pigeonpea that suffer from the domestication-led extremely narrow genetic base of the primary gene pool. In view of the above, *C. cajanifolius* and *C. acutifolius* are taken as donors for generating the backcross populations for AB-QTL analysis in pigeonpea (Varshney et al. 2013a).

To establish MAS as an integral component of cowpea breeding, serious efforts are under way to develop *Striga*-resistant genotypes in cowpea using molecular tools and technologies. Two tightly linked SCAR markers, 61R and MahSE2, exhibited more than 75% marker efficiencies while selecting desirable genotypes (http://www.iita.org).

4.6 Transgenic Approaches in Pulse Crops

Although the transgenic approach was first applied in 1997 to develop transgenic *Vigna aconitifolia* (Eapen et al. 1987), till now the approach has not succeeded much in pulse crops due to inefficacy of the transformation system so far used (Popelka et al. 2004; Dita et al. 2006). In general, *Agrobacterium*-mediated gene transfer, particle gun bombardment, and electroporation methods of transgenesis were practiced so far to generate transgenic pulses and the *Agrobacterium*-mediated transformation system is found most successful as compared to other two methods. During the early years, antibiotic marker genes were mostly used to test the efficacy and feasibility of the transformation (Eapen et al. 1987). However, in later years, several genes of economic importance have been used in developing various transgenic major pulse crops: such genes are *Lea* (drought resistance), *bar* (herbicide-resistant), *cry* and α-amylase inhibitor (insect resistance), *Als* (weed resistance), and *S2 albumin* (quality and storage improvement). Although transgenic pulses can be developed using explants such as protoplast, embryo axis, cotyledonary node, apical meristem, hypocotyl, plumule, and seed, the most preferred explants for transformation are cotyledonary nodes (Kumar et al. 2004) and young embryonic axes (Krishnamurthy et al. 2000), due to their better efficacy in generating viable transgenic plants in pulses. Table 4.5 describes the techniques, transgenes, and explants so far used to develop various pulse crops with the expected traits.

Development of transgenic pulse crops is limited due to low efficacy of currently available or used transgenic technology. For example, transgenic *V. aconitifolia* can be developed using isolated protoplast and *Agrobacterium*-mediated gene transfer or electroporation (Eapen et al. 1987; Kohler et al. 1987). However, for other pulses these methods do not work. Similarly, axillary meristem and electroporation-based transformation has had some success in lentil, cowpea, and pea (Chowrira et al. 1998); for other pulses electroporation to

TABLE 4.5

Transgenic Approaches in Pulse Crops

Plant Species	Explants Used	Technique Used	Transferred Gene	Expected Trait	References
Phaseolus vulgaris (common bean)	Embryo axis	Particle gun bombardment	*uidA, coat protein, bar*	Herbicide resistance	Russell et al. 1993
	Embryo axis	Particle gun bombardment	*nptII, S2 albumin, uidA*	Bactericidal, storage improvement	Aragao et al. 1996
	Seeds	*Agrobacterium*-mediated gene transfer	*Lea*	Drought resistance	Liu et al. 2005
	Embryo axis	*Agrobacterium*-mediated gene transfer	*ahas/als, nptII*	Herbicide-resistant	Nifantova et al. 2011
Phaseolus acutifolius (tepary bean)	Callus	*Agrobacterium*-mediated gene transfer	*nptII, uidA,*	Mexican bean weevil resistance	Zambre et al. 2005
Vigna unguiculata (cowpea)	Cotyledonary node	*Agrobacterium*-mediated gene transfer	*nptII, uidA*	Herbicide resistance	Chaudhury et al. 2006
	Cotyledonary node	*Agrobacterium*-mediated gene transfer	*bar*	Herbicide resistance	Popelka et al. 2006
	Embryo axis	Particle gun bombardment	*uidA, bar*	Herbicide resistance	Ikea et al. 2003
	Seeds	*Agrobacterium*-mediated gene transfer	*α-amylase inhibitor-1*	Storage pest, bruchid beetle resistance	Solleti et al. 2008
	Embryo axis	*Agrobacterium*-mediated gene transfer	*Atahas*	Imazapyr tolerance	Citadin et al. 2013
Vigna mungo (Black gram, black lentil)	Cotyledonary nodes, embryonic axes, apical meristem	*Agrobacterium*-mediated gene transfer	*uidA, nptII*		Saini et al. 2003; Saini and Jaiwal 2005; Saini and Jaiwal 2007
Vigna aconitifolia (Moth bean)	Protoplast	*Agrobacterium*-mediated gene transfer	*nptII*		Eapen et al. 1987
Vigna radiate (mung bean)	Hypocotyl	Particle gun bombardment	*cryIAc, uidA, nptII*	Insect resistance	Kamble et al. 2003
	Hypocotyl, cotyledonary node, leaf	*Agrobacterium*-mediated gene transfer	*nptII, uidA*		Pal et al. 1991; Jaiwal et al. 2001; Tazeen and Mirza 2004

(continued)

TABLE 4.5 (Continued)

Transgenic Approaches in Pulse Crops

Plant Species	Explants Used	Technique Used	Transferred Gene	Expected Trait	References
	Cotyledonary node	*Agrobacterium*-mediated gene transfer	*Bar, α-amylase inhibitor-1*	Herbicide resistance	Sonia et al. 2007
	Cotyledonary node	*Agrobacterium*-mediated gene transfer	*BjNPR1 of mustard*	Rot pathogen, *Rhizoctonia solani* resistance	Vijayan and Kirti 2012
Vigna angularis (azuki bean)	Epicotyl	*Agrobacterium*-mediated gene transfer	*Bar, hpt*	Herbicide resistance	Khalafalla et al. 2005
	Cotyledonary node	*Agrobacterium*-mediated gene transfer	*alphaAI-2*	Azuki bean weevil resistance	Nishizawa et al. 2007
Vicia narbonensis (narbon bean)	Embryogenic callus	*Agrobacterium*-mediated gene transfer	*Hpt, 2S albumin*	Bactericidal, storage improvement	Pickardt et al. 1991; Pickardt et al. 1998
	Cotyledonary node	*Agrobacterium*-mediated gene transfer	*VfAAP1 from Vicia faba*	Increased storage protein content	Rolletschek et al. 2005
	Cotyledonary node	*Agrobacterium*-mediated gene transfer	*Phosphoenolpyruvate carboxylase*	Improved nutrient	Radchuk et al. 2007
Vigna sequidepalis (Chinese long bean)	Cotyledonary node	*Agrobacterium*-mediated gene transfer	*nptII, uidA*		Ignacimuthu 2000
Lens culinaris (lentil)	Cotyledonary node, decapitated embryo	*Agrobacterium*-mediated gene transfer	*nptII, uidA*		Sarkar et al. 2003
	Cotyledonary node	Particle gun bombardment	*Als*	Weed resistance	Gulati et al. 2002
Cajanus cajan (pigeonpea)	apical meristem, Embryonic axis, cotyledonary nodes, leaf, seed, plumule, nodal segments	*Agrobacterium*-mediated gene transfer	*HN gene of PPRV, hpt, rice chitinase, cryIAb, uidA, nptII*	Insect resistance	Prasad et al. 2004; Kumar et al. 2004; Surekha et al. 2005; Sharma et al. 2006; Surekha et al. 2007

Species	Explant	Method	Gene	Trait	Reference
Pisum sativum (pea)	Embryo axis	Particle gun bombardment	*hpt, uidA*		Thu et al. 2003
	Embryo axis, cotyledonary nodes	*Agrobacterium*-mediated gene transfer	*Hemagglutinin protein of Rinderpest virus*	Edible vaccine	Satyavathi et al. 2003
	Embryo, cotyledons	*Agrobacterium*-mediated gene transfer	*uidA, bar*	Herbicide resistance	Schroeder et al. 1993; Grant et al. 1995; Krejci et al. 2007
	Embryo explants	*Agrobacterium*-mediated gene transfer	*PGIP*	Fungi resistance	Richter et al. 2007
	Axillary meristems	Micro injection	*PEMV virus coat protein*	Pea enation mosaic virus resistance	Chowrira et al. 1998
	Embryo explants	*Agrobacterium*-mediated gene transfer	*PGIP, VST-stilbene synthase, PChit30*	Fungi resistance	Hassan et al. 2012
	Embryo explants	*Agrobacterium*-mediated gene transfer	*Chitinase, glucanase*	Fungi resistance	Amian et al. 2011
Cicer arietinum (chickpea)	Embryonic axis	*Agrobacterium*-mediated gene transfer	*nptII, uidA*		Krishnamurthy et al. 2000
	Embryonic axis	*Agrobacterium*-mediated gene transfer	*cryIAc*	Pod borer resistance	Kar et al. 1997
	Cotyledon	*Agrobacterium*-mediated gene transfer	*α-amylase inhibitor-1*	Seed weevil and insect resistance	Sarmah et al. 2004
	Plumule	*Agrobacterium*-mediated gene transfer	*bar*		Senthil et al. 2004
	Epicotyl	Particle gun bombardment	*cryIAc, nptII, uidA*	Insect resistance	Indurker et al. 2007
	Embryonic axes	*Agrobacterium*-mediated gene transfer	*BtCry*	Insect resistance	Mehrotra et al. 2011
	Embryonic cotyledon	*Agrobacterium*-mediated gene transfer	*Leaf agglutinin gene from Allium sativum*	Aphis resistance	Chakraborti et al. 2009

axillary meristem is not very successful. Therefore, in most cases the transgenic pulse crops are developed using *Agrobacterium*-mediated gene transfer to cotyledonary nodes, which gives better success as compared to other transformation methods and explants used. But again, the conventional *Agrobacterium*-mediated transformation is not successful when compared with other crops such as cereals or cruciferous plants. Therefore, genetically modified *Agrobacterium* with engineered more aggressive *vir* genes and an additional *L-cysteine* gene can improve transfection efficacy in pulses (Olhoft and Somers 2001; Olhoft et al. 2001). There are also concerns about transgene marker-based selection of transgenic clones in culture for pulses. Therefore, the use of selection markers also needs to be standardized and efforts to make marker-free transgenic pulses need to be taken into account (Darbani et al. 2007). Similarly, it is well known that transplastomic plants are preferable due to their ability to prevent horizontal gene transfer. However, in pulses such transplastomic approaches are yet to be established (Eapen 2008). Transgenic approaches have been successfully applied to develop various abiotic stress (drought, salinity, acidity, etc.) tolerance pulses and also high-yielding, improved quality, and nutraceutical-producing varieties using specific transgenes to generate such traits in many crops (Dita et al. 2006). But in pulses, so far transgenic approaches have been taken only to develop biotic stress (insect, pest, virus, bacteria, etc.) tolerance. Therefore, development of various abiotic stress tolerance pulses with high-yielding and pharmaceutical and edible value-added transgenic cultivars is still open. Such varieties can be developed based on a transgenic approach in combination with metabolic engineering using various omics strategies.

4.7 Conclusion

In summary, recent developments in genomics technologies have made a remarkable impact on pulse research. As a result large sets of genomics resources have been made available for further deployment in pulse breeding programs. In this regard, the availability of whole-genome sequences for the major pulses will pave the way for the discovery of a number of candidate gene(s) and DNA markers, as in the case of pigeonpea. Access to thousands of DNA markers, preferably SSRs and SNPs, would be a great support for undertaking genome-wide association studies (GWASs) and WGS or genomics selection (GS) aiming at pulse genetic improvement. Furthermore, reference genomes of a particular pulse crop would open new avenues for facilitating *de novo* resequencing of landraces and wild relatives, which would harness the genome-wide genetic diversity in a much broader and more practical way. The output of all these initial investments would be reflected in MAS-bred superior pulse cultivars with improved productivity and enhanced tolerance to various biotic and abiotic stresses.

References

Adegbite AA and Amusa NA (2008) The major economic field diseases of cowpea in the humid agro-ecologies of South-Western Nigeria. *African Journal of Biotechnology*, 7: 4706–4712.

Ahmad F, Gaur P, and Croser J (2005) Chickpea (*Cicer ariatinum* L.). In: *Genetic Resources, Chromosome Engineering and Crop Improvement: Grains Legumes*, Singh R, and Jauhar P (eds), pp. 185–214. CRC, Boca Raton, FL.

Ahmed F, Khan AI, Awan FS, Sadia B, Sadaquat HA, and Bahadur S (2010) Genetic diversity of chickpea (*Cicer arietinum* L.) germplasm in Pakistan as revealed by RAPD analysis. *Genetic and Molecular Research*, 9: 1414–1420.

Akibode CS and Maredia M (2011) Global and regional trends in production, trade and consumption of food legume crops. Report submitted to the Standing Panel on Impact Assessment (SPIA) of the CGIAR Science Council, FAO, Rome.

Akinola JO, Pritchard AJ, and Whiteman PC (1972) Chromosome number in pigeonpea (*Cajanus cajan* (L.) Millsp.). *Journal of the Australian Institute of Agricultural Science*, 38: 305.

Alo F, Furman BJ, Akhunov E, Dvorak J, and Gepts P (2011) Leveraging genomic resources of model species for the assessment of phylogeny in wild and domesticated lentil. *Journal of Heredity*, 102: 315–329.

Amian AA, Papenbrock J, Jacobsen HJ, and Hassan F (2011) Enhancing transgenic pea (*Pisum sativum* L.) resistance against fungal diseases through stacking of two antifungal genes (chitinase and glucanase). *GM Crops*, 2: 104–109.

Anbessa Y, Taran B, Warkentin TD, Tullu A, and Vandenberg A (2009) Genetic analyses and conservation of QTL for ascochyta blight resistance in chickpea (*Cicer arietinum* L.). *Theoretical and Applied Genetics*, 4: 757–764.

Andargie M, Pasquet RS, Gowda BS, Muluvi GM, and Timko MP (2011) Construction of a SSR-based genetic map and identification of QTL for domestication traits using recombinant inbred lines from a cross between wild and cultivated cowpea (*V. unguiculata* (L.) Walp). *Molecular Breeding*, 28: 413–420.

Anuradha C, Gaur PM, Pande S, Gali KK, Ganesh M, Kumar J, and Varshney RK (2011) Mapping QTL for resistance to botrytis grey mould in chickpea. *Euphytica*, 182: 1–9.

Aragao FJL, Barros LMG, Brasileiro ACM, Ribeiro SG, Smith FD, Sanford JC, Faria JC, et al. (1996) Inheritance of foreign genes in transgenic bean (*Phaseolus vulgaris* L) co-transformed via particle bombardment. *Theoretical and Applied Genetics*, 93: 142–151.

Arumuganathan K and Earle D (1991) Nuclear DNA content of some important plant species. *Plant Molecular Biology Reporter*, 9: 208–219.

Arymanesh N, Nelson MN, Yan G, Clarke HJ, and Siddique KHM (2010) Mapping a major gene for growth habit and QTLs for ascochyta blight resistance and flowering time in a population between chickpea and (*Cicer reticulatum*). *Euphytica*, 3: 307–319.

Benko-Iseppon AM, Winter P, Huttel B, Staginnus C, Muehlbauer FJ, and Kahl G (2003) Molecular markers closely linked to *Fusarium* resistance genes in chickpea show significant alignments to pathogenesis-related genes located on *Arabidopsis* chromosomes1 and 5. *Theoritical and Applied Genetics*, 107: 379–386.

Bhardwaj R, Sandhu JS, Varshney RK, Gaur PM, Kaur L, and Vikal Y (2011) Synteny relationship among the linkage groups of chickpea (*Cicer arietinum* L.). *Journal of Food Legumes*, 24: 91–95.

Bohra A, Dubey A, Saxena RK, Penmetsa RV, Poornima KN, Kumar N, Farmer AD, et al. (2011) Analysis of BAC-end sequences (BESs) and development of BES-SSR markers for genetic mapping and hybrid purity assessment in pigeonpea. *BMC Plant Biology*, 11: 56.

Bohra A, Saxena RK, Gnanesh BN, Saxena KB, Byregowda M, Rathore A, KaviKishor PB, Cook DR, and Varshney RK (2012) An intra-specific consensus genetic map of pigeonpea [*Cajanus cajan* (L.) Millspaugh] derived from six mapping populations. *Theoretical and Applied Genetics*, 125: 1325–1338.

Botstein D, White RL, Skolnick M, and Davis RW (1980) Construction of a genetic map in man using restriction fragment length polymorphisms. *American Journal of Human Genetics*, 32: 314–331.

Cannon SB, May GD, and Jackson SA (2009) Three sequenced legume genomes and many crop species: Rich opportunities for translational genomics. *Plant Physiology*, 151: 970–977.

Castro P, Piston F, Madrid E, Millan T, Gil J, and Rubio J (2010) Development of chickpea near-isogenic lines for Fusarium wilt. *Theoretical and Applied Genetics*, 121: 1519–1526.

Chakraborti D, Sarkar A, Mondal HA, and Das S (2009) Tissue specific expression of potent insecticidal, Allium sativum leaf agglutinin (ASAL) in important pulse crop, chickpea (*Cicer arietinum* L.) to resist the phloem feeding Aphis craccivora. *Transgenic Research*, 18: 529–544.

Chandra S, Buhariwalla HK, Kashiwagi J, Harikrishna S, Sridevi KR, Krishnamurthy L, Serraj R, and Crouch JH (2004) Identifying QTL-linked markers in marker-deficient crops. In: *International Crop Science Congress*, 26 September–1 October, Brisbane, Australia.

Chaudhury D, Madanpotra S, Jaiwal R, Saini R, Kumar PA, and Jaiwal PK (2006) *Agrobacterium tumefaciens* mediated high frequency genetic transformation of an Indian cowpea (*Vigna unguiculata* L. Walp) cultivar and transmission of transgenes into progeny. *Plant Science*, 172: 692–700.

Chen X, Laudeman TW, Rushton PJ, Spraggins TA, and Timko MP (2007) CGKB: An annotation knowledge base for cowpea (*Vigna unguiculata*L.) methylation filtered genomic genespace sequences. *BMC Bioinformatics*, 8: 129.

Cho S, Kumar J, Shultz JK, Anupama K, Tefera F, and Muehlbauer FJ (2002) Mapping genes for double podding and other morphological trait in chickpea. *Euphytica*, 125: 285–292.

Cho S, Chen W, and Muehlbauer FJ (2004) Pathotype-specific factors inchickpea (*Cicer arietinum* L.) for quantitative resistance to ascochyta blight. *Theoretical and Applied Genetics*, 109: 733–739.

Chowdhury MA, Andrahennadi CP, Slinkard AE, and Vandenberg A (2001) RAPD and SCAR markers for resistance to ascochyta blight in lentil. *Euphytica*, 118: 331–337.

Chowrira GM, Cavileer TD, Gupta SK, Lurquin PF, and Berger PH (1998) Coat protein mediated resistance to pea enationmosaic virus in transgenic *Pisum sativum* L. *Transgenic Research*, 7: 265–271.

Citadin CT, Cruz AR, and Aragão FJ (2013) Development of transgenic imazapyr-tolerant cowpea (*Vigna unguiculata*). *Plant Cell Reports*, 32: 537–543.

Cobos MJ, Fernandez MJ, Rubio J, Kharrat M, Moreno MT, Gil J, and Millan T (2005) Linkage map of chickpea (*Cicer arietinum* L.) based on populations from Kabuli Desi cross; location of genes for resistance to fusarium wilt race 0. *Theoretical and Applied Genetics*, 110: 1347–1353.

Cobos MJ, Rubio J, Strange RN, Moreno ML, Gil J, and Millan T (2006) A new QTL for Ascochyta blight resistance in an RIL population derived from an interspecific cross in chickpea. *Euphytica*, 149: 105–111.

Cobos MJ, Rubio J, Fernandez M, Garza R, Moreno MT, and Millan T (2007) Genetic analysis of seed size, yield and days to flowering in a chickpea recombinant inbred line population derived from a Kabuli × Desi cross. *Annals of Applied Biology*, 151: 33–42.

Cobos MJ, Winter P, Kharrat M, Cubero JI, Gil J, Millan T, and Rubio J (2009) Genetic analysis of agronomic traits in a wide cross of chickpea. *Field Crops Research*, 111: 130–136.

Collard BCY, Pang ECK, Ades PK, and Taylor PWJ (2003) Preliminary investigation of QTLs associated with seedling resistance to ascochyta blight from *Cicer echinospermum*, a wild relative of chickpea. *Theoretical and Applied Genetics*, 107: 719–729.

Coulibaly S, Pasquet RS, Papa R, and Gepts P (2002) AFLP analysis of the phenetic organization and genetic diversity of *Vigna unguiculata* L. Walp. reveals extensive gene flow between wild and domesticated types. *Theoretical and Applied Genetics*, 104: 358–366.

DAC (2010) Directorate of Economics and Statistics, DAC, Govt. of India.

Darbani B, Eimanifar A, Stewart Jr. SN, and Chamargo WN (2007) Methods to produce marker free transgenic plants. *Journal of Biotechnology*, 2: 83–90.

De DN (1974) Pigeonpea. In: *Evolutionary Studies in World Crops: Diversity and Change in the Indian Subcontinent*, Hutchinson J (ed.), pp. 79–87. Cambridge University Press, London.

de la Vega MP, Fratini RM, and Muehlbauer FJ (2011) Lentil. In: *Genetics and Genomics of Cool Season Grain Legumes*, Kole C (ed.), pp. 98–150. Science Publishers, Enfield.

Dhanasekar P, Dhumal KH, and Reddy KS (2010) Identification of RAPD marker linked to plant type gene in pigeonpea. *Indian Journal of Biotechnology*, 9: 58–63.

Dita MA, Rispail N, Prats E, Rubiales D, and Singh KB (2006) Biotechnology approaches to overcome biotic and abiotic stress constraints in legumes. *Euphytica*, 147: 1–24.

Dubey A, Farmer A, Schlueter J, Cannon SB, Abernathy B, Tuteja, R, Woodward, J et al. (2011) Defining the transcriptome assembly and its use for genome dynamics and transcriptome profiling studies in pigeonpea (*Cajanus cajan* L.). *DNA Research*, 18: 153–164.

Durán Y and de la Vega MP (2004) Assessment of genetic variation and species relationships in a collection of *Lens* using RAPD and ISSR. *Spanish Journal of Agricultural Research*, 4: 538–544.

Duran Y, Fratini R, Garcia P, and De La Vega MP (2004) An intersubspecific genetic map of Lens. *Theoretical and Applied Genetics*, 108: 1265–1273.

Dutta S, Kumawat G, Singh BP, Gupta DK, Singh S, Dogra V, Gaikwad K, et al. (2011) Development of genic-SSR markers by deep transcriptome sequencing in pigeonpea [*Cajanus cajan* (L.) Millspaugh]. *BMC Plant Biology*, 11: 17.

Eapen S (2008) Advances in development of transgenic pulse crops. *Biotechnology Advances*, 26: 162–168.

Eapen S, Kohler F, Gerdemann M, and Schieder O (1987) Cultivar dependence of transformation rates in mothbean after co-cultivation of protoplasts with *Agrobacterium tumefaciens*. *Theoritical and Applied Genetics*, 75: 207–210.

Erskine W and Saxena MC (1993) Problems and prospects of stress resistance breeding in lentil. In: *Breeding for Stress Tolerance in Cool-Season Food Legumes*, Singh KB and Saxena MC (eds), pp. 51–62. John Wiley and Sons, Chichester.

Erskine W, Sarker A, and Kumar S (2011) Crops that feed the world 3. Investing in lentil improvement toward a food secure world. *Food Security*, 3: 127–139.

Eujayl I, Baum M, Erskine W, Pehu E, and Muehlbauer FJ (1997) The use of RAPD markers for lentil genetic mapping and the evaluation of distorted F$_2$ segregation. *Euphytica*, 96: 405–412.

Eujayl I, Baum M, Powell W, Erskine W, and Pehu E (1998) A genetic linkage map of lentil (*Lens* sp.) based on RAPD and AFLP markers using recombinant inbred lines. *Theoretical and Applied Genetics*, 97: 83–89.

Eujayl I, Erskine W, Baum M, and Pehu E (1999) Inheritance and linkage analysis of frost injury in lentil. *Crop Science*, 39: 639–642.

FAOSTAT (2010) http://faostat.fao.org/.

Farrar K and Donnison IS (2007) Construction and screening of BAC libraries made from *Brachypodium* genomic DNA. *Nature Protocols*, 2: 1661–1674.

Fatokun CA, Danesh D, Menancio-Hautea D, and Young ND (1993a) A linkage map for cowpea [*Vigna unguiculata* (L.) Walp.] based on DNA markers. In: *A Compilation of Linkage and Restriction Maps of Genetically Studied Organisms, Genetic Maps 1992*, O'Brien, JS (ed.), pp. 6.256–6.258. Cold Spring Harbor Laboratory Press, Cold Spring Harbor, New York.

Fatokun CA, Danesh D, and Young ND (1993b) Molecular taxonomic relationships in the genus *Vigna* based on RFLP analysis. *Theoretical and Applied Genetics*, 86: 97–104.

Fatokun CA, Menancio-Hautea DI, Danesh D, and Young ND (1992) Evidence for orthologous seed weight genes in cowpea and mung bean based on RFLP mapping. *Genetics*, 132: 841–846.

Fatokun CA, Mignouna HD, Knox MR, and Ellis THN (1997) AFLP variation among cowpea varieties. In: *Agronomy Abstracts*, p. 156. ASA, Madison.

Flandez-Galvez H, Ford R, Pang ECK, and Taylor PWJ (2003a) An intraspecific linkage map of the chickpea (*Cicer arietinum* L.) genome based on sequence tagged based microsatellite site and resistance gene along marker. *Theoretical and Applied Genetics*, 106: 1447–1456.

Flandez-Galvez H, Ades PK, Ford R, Pang ECK, and Taylor PWJ (2003b) QTL analysis for ascochyta blight resistance in an intraspecific population of chickpea (*Cicer arietinum* L.). *Theoretical and Applied Genetics*, 107: 1257–1265.

Ford R, Pang ECK, and Taylor PWJ (1997) Diversity analysis and species identification in *Lens* using PCR generated markers. *Euphytica*, 96: 247–255.

Ford R, Pang ECK, and Taylor PWJ (1999) Genetics of resistance to ascochyta blight (A. *lentis*) of lentil and the identification of closely linked RAPD markers. *Theoretical and Applied Genetics*, 98: 93–98.

Fratini R, Durán Y, García P, and Pérez de la Vega M (2007) Identification of quantitative trait loci (QTL) for plant structure, growth habit and yield in lentil. *Spanish Journal of Agricultural Research*, 5: 348–356.

Ganapathy KN, Byre Gowda M, Venkatesha SC, Ramachandra R, Gnanesh BN, and Girish G (2009) Identification of AFLP markers linked to sterility mosaic disease in pigeonpea (*Cajanus cajan*) (L.) Millsp. *International Journal of Integrative Biology*, 7: 145–149.

Garg R, Patel RK, Tyagi AK, and Jain M (2011) *De novo* assembly of chickpea transcriptome using short reads for gene discovery and marker identification. *DNA Research*, 18: 53–63.

Gaur PM and Slinkard AE (1990) Genetic control and linkage relations of additional isozyme markers in chickpea. *Theoretical and Applied Genetics*, 80: 648–656.

Gaur PM, Sethy NK, Choudhary S, Shokeen B, Gupta V, and Bhatia S (2011) Advancing the STMS genomic resources fordefining new locations on the intra-specific genetic linkage map of chickpea (*Cicer arietinum* L.). *BMC Genomics*, 12: 117.

Gaur PM, Jukanti AK, and Varshney R (2012a) Impact of genomic technologies on chickpea breeding strategies. *Agronomy*, 2: 199–221.

Gaur R, Azam S, Jeena G, Khan AW, Choudhary S, Jain M, Yadav G, Tayagi AK, Chattopadhyay D, and Bhatia S (2012b) High throughput SNP discovery and genotyping for constructing a saturated Linkage map of chickpea (*Cicer arietinum* L.). *DNA Research*, 19: 357–373.

Gepts P, Beavis WD, Brummer EC, Shoemaker RC, Stalker HT, Weeden NF, and Young ND (2005) Legumes as a model plant family: Genomics for food and feed report of the cross-legume advances through genomics conference. *Plant Physiology*, 137: 1228–1235.

Gibson G and Dworkin I (2004) Uncovering cryptic genetic variation. *Nature Reviews Genetics*, 5: 681–690.

Gill LS and Hussaini SWH (1986) Cytological observations in Leguminosae from Southern Negeria. *Willdenowia*, 15: 521.

Gnanesh BN, Bohra A, Sharma M, Byregowda M, Pande S, Wesley V, Saxena RK, Saxena KB, Kavi Kishor PB, and Varshney RK (2011) Genetic mapping and quantitative trait locus analysis of resistance to sterility mosaic disease in pigeonpea [*Cajanus cajan* (L.) Millsp.]. *Field Crops Research*, 123: 53–61.

Gowda SJM, Radhika P, Kadoo NY, Mhase LB, and Gupta V (2009) Molecular mapping of wilt resistance genes in chickpea. *Molecular Breeding*, 24: 177–183.

Grant JE, Cooper PA, McAra AE, and Frew TJ (1995) Transformation of pea (*Pisum sativum*) using immature cotyledons. *Plant Cell Reports*, 15: 254–258.

Gujaria N, Kumar A, Dauthal P, Dubey A, Hiremath PJ, Bhanu Prakash A, Farmer A, et al. (2011) Development and use of genic molecular markers (GMMs) for construction of a transcript map of chickpea (*Cicer arietinum* L.). *Theoritical and Applied Genetics*, 122: 1577–1589.

Gulati A, Schryer P, and McHughen A (2002) Production of fertile transgenic lentil (*Lens culinaris* Medik) plants using particle bombardment. *In Vitro Cellular Developmental Biology Plant*, 38: 316–324.

Gupta D, Taylor PWJ, Inder P, Phan HTT, Ellwood SR, Mathur PN, Sarker A et al. (2011) Integration of EST- SSR markers of *Medicago truncatula* into intraspecific linkage map of lentil and identification of QTL conferring resistance to ascochyta blight at seedling and pod stages. *Molecular Breeding*, 30: 429–439.

Halila I, Cobos MJ, Rubio J, Millan T, Kharrat M, Marrakchi M, and Gil J (2009) Tagging and mapping a second resistance gene for *Fusarium* wilt race 0 in chickpea. *European Journal of Plant Pathology*, 124: 87–94.

Hamwieh A, Udapa SM, Choumane W, Sarker A, Dreyer F, Jung C, and Baum M (2005) A genetic linkage map of lentil based on microsatellite and AFLP markers and localization of fusarium vascular wilt resistance. *Theoretical and Applied Genetics*, 110: 669–677.

Hamwieh A, Udupa SM, Sarker A, Jung C, and Baum M (2009) Development of new microsatellite markers and their application in the analysis of genetic diversity in lentils. *Breeding Science*, 59: 77–86.

Hassan F, Noorian MS, and Jacobsen HJ (2012) Effect of antifungal genes expressed in transgenic pea (*Pisum sativum* L.) on root colonization with Glomus intraradices. *GM Crops Food*, 3: 301–309.

Havey MJ and Muehlbauer FJ (1989) Linkages between restriction fragment length, isozyme, and morphological markers in lentil. *Theoretical and Applied Genetics*, 77: 395–401.

Hiremath PJ, Farmer A, Cannon SB, Woodward J, Kudapa H, Tuteja R, Kumar A, et al. (2011) Large scale transcriptome analysis in chickpea (*Cicer arietinum* L.), an orphan legume crop of the semi-arid tropics of Asia and Africa. *Plant Biotechnology Journal*, 9: 922–931.

Hiremath PJ, Kumar A, Penmetsa RV, Farmer A, Schlueter JA, Chamarthi SK, Whaley AM, et al. (2012) Large-scale development of cost-effective SNP marker assays for diversity assessment and genetic mapping in chickpea and comparative mapping in legumes. *Plant Biotechnology Journal*, 10: 716–732.

Huttel B, Winter P, Weising K, Choumane W, Weig F, and Kahl G (1999) Sequence tagged microsatellite markers for chickpea (*Cicer arietinum* L.). *Genome*, 42: 210–217.

Ibrikci H, Knewtson S, and Grusak MA (2003) Chickpea leaves as vegetable green for human: Evaluation of mineral composition. *Journal of the Science and Food and Agriculture*, 83: 945–950.

Ignacimuthu S (2000) *Agrobacterium* mediated transformation of *Vigna sequipedalis Koern* (*asparagus bean*). *Indian Jounal of Experimental Biology*, 38: 493–498.

Ikea J, Ingelbrecht I, Uwaifo A, and Thottapilly G (2003) Stable transformation in cowpea (*Vigna unguiculata* walp) using particle gun method. *African Jounal of Biotechnology*, 2: 211–222.

Imelfort M and Edwards D (2009) De novo sequencing of plant genomes using second generation technologies. *Briefings in Bioinformatics*, 10: 609–618.

Imtiaz M, Materne M, Hobson K, vanGinkel M, and Malhotra RS (2008) Molecular genetic diversity and linked resistance to ascochyta blight in Australian chickpea breeding materials and their wild relatives. *Australian Journal of Agricultural Research*, 59: 554–560.

Indurker S, Misra HS, and Eapen S (2007) Genetic transformation of chickpea (*Cicer arietinum* L) with insecticidal crystal protein gene using particle gun bombardment. *Plant Cell Reports*, 26: 755–763.

Iruela M, Rubio J, Barro F, Cubero JI, Millan T, and Gill J (2006) Detection of two quantitative trait loci for resistance to ascochyta blight in an intra-specific cross of chickpea(*Cicer arietinum* L.): Development of SCAR markers associated with resistance. *Theoritical and Applied Genetics*, 112: 278–287.

Iyengar NK (1939) Cytological investigations on the genus *Cicer*. *Annals of Botany*, 3: 271–305.

Jaiwal PK, Kumari R, Ignacimuthu S, Potrykus I, and Sautter C (2001) *Agrobacterium* mediated transformation of mungbean (*Vigna radiata* L Wilczek)—A recalcitrant grain legume. *Plant Science*, 161: 239–247.

Jhanwar S, Priya P, Garg R, Parida SK, Tyagi AK, and Jain M (2012) Transcriptome sequencing of wild chickpea as rich resource for marker development. *Plant Biotechnology Journal*, 10: 690–702.

Kahraman A, Kusmenoglu I, Aydin N, Aydogun A, Erskine W, and Muehlbauer FJ (2004) Genetics of winter hardiness in 10 lentil recombinant inbred line populations. *Crop Science*, 44: 5–12.

Kahraman A, Demirel U, Ozden M, and Muehlbauer FJ (2010) Mapping of QTLs for leaf area and the association with winter hardiness in fall-sown lentil. *African Journal of Biotechnology*, 9: 8515–8519.

Kamble S, Misra HS, Mahajan SK, and Eapen S (2003) A protocol for efficient biolistic transformation of mothbean (*Vigna aconitifolia* L. Marechal). *Plant Molecular Biology Reporter*, 21: 457a–j.

Kar S, Basu D, Das S, Ramakrishanan NA, Mukherjee P, Nayak P, et al. (1997) Expression of cryIAc gene of Bacillus thuringiensis in transgenic chickpea plants inhibits development of pod borer (*Heliothis armigera*) larvae. *Transgenic Research*, 6: 177–185.

Kassa MT, Penmetsa RV, Carrasquilla-Garcia N, Sarma BK, Datta S, Upadhyaya HD, Varshney RK, von Wettberg EJB, and Cook DR (2012) Genetic patterns of domestication in pigeonpea (*Cajanus cajan* (L.) Millsp.) and wild *Cajanus* relatives. *PLoS ONE*, 7 (6): e39563.

Kaur S, Cogan NO, Pembleton LW, Shinozuka M, Savin KW, Materne M, and Forster JW (2011) Transcriptome sequencing of lentil based on second-generation technology permits large-scale unigene assembly and SSR marker discovery. *BMC Genomics*, 12: 265.

Khalafalla MM, El-shemy HA, Mizanur RS, Teraishi M, and Ishimoto M (2005) Recovery of herbicide-resistant Azuki bean (*Vigna angularis* (wild) ohwi and ohashi). *African Journal of Biotechnology*, 4: 61–67.

Kohler F, Golz C, Eapen S, Kohn H, and Schieder O (1987). Stable transformation of mothbean (*Vigna aconitifolia*) via direct gene transfer. *Plant Cell Reports*, 6: 313–317.

Kongjaimun A, Kaga A, Tomooka N, Somta P, Vaughan DA, and Srinives P (2012) The genetics of domestication of yardlong bean, *Vigna unguiculata* (L.) Walp. ssp.*unguiculata* cv.-gr. *Sesquipedalis*. *Annals of Botany*, 109: 1185–1200.

Kotresh H, Fakrudin B, Punnuri S, Rajkumar B, Thudi M, Paramesh H, Lohithswa H, and Kuruvinashetti M (2006) Identification of two RAPD markers genetically linked to a recessive allele of a *Fusarium* wilt resistance gene in pigeonpea (*Cajanus cajan* (L.) Millsp.). *Euphytica*, 149: 113–120.

Krejci P, Matuskova P, Hanacek P, Reinohl V, and Prochazka S (2007) The transformation of pea (*Pisum sativum* L): Applicable methods of *Agrobacterium tumefaciens* mediated gene transfer. *Acta Physiologiae Plantarum*, 29: 157–163.

Krishnamurthy KV, Suhasini K, Sagare AP, Meixner M, de Kathen A, Pickardt T, and Schieder O (2000) *Agrobacterium* mediated transformation of chickpea (*Cicer arietinum* L.) embryo axes. *Plant Cell Reports*, 19: 235–240.

Kudapa H, Bharti AK, Cannon SB, Farmer AD, Mulaosmanovic B, Kramer R, Bohra A, et al. (2012) A comprehensive transcriptome assembly of pigeonpea (*Cajanaus cajan* L.) using Sanger and Second-generation sequencing platforms. *Molecular Plant*, 5: 1020–1028.

Kumar SM, Kumar BK, Sharma KK, and Devi P (2004) Genetic transformation of pigeonpea with rice chitinase gene. *Plant Breeding*, 123: 485–489.

Ladizinsky G (1979) The origin of lentil and its wild gene pool. *Euphytica*, 28: 179–187.

Ladizinsky G and Adler A (1976) Genetic relationships among the annual species of *Cicer* L. *Theoretical and Applied Genetics*, 48: 197–203.

Li CD, Fatokun CA, Ubi B, Singh BB, and Scoles GJ (2001) Determining genetic similarities and relationships among cowpea breeding lines and cultivars by microsatellite markers. *Crop Science*, 41: 189–197.

Lichitenzveiz J, Scheuring C, Dodge J, Abbo S, and Zhang HB (2005) Construction of BAC and BIBAC libraries and their application for generation of SSR markers for genome analysis of chickpea (*Cicer arietinum* L.). *Theoretical and Applied Genetics*, 110: 492–510.

Liu ZC, Park BJ, Kanno A, and Kameya T (2005) The novel use of a combination of sonication and vaccum infiltration in *Agrobacterium* mediated transformation of kidney bean (*Phaseolus vulgaris*) with lea gene. *Molecular Breeding*, 16: 189–197.

Lucas MR, Diop NN, Wanamaker S, Ehlers JD, Roberts PA, and Close TJ (2011) Cowpea soybean synteny clarified through an improved genetic map. *The Plant Genome*, 4: 215–218.

Lush WM and Evans IT (1981) Domestication and improvement of cowpea. *Euphytica*, 30: 579–587.

Ma Y, Wu J, You FM, Bhat PR, Ehlers JT, Roberts PL, Close TJ, Gu YQ, and Luo MC (2009) Genomic resources of cowpea (*Vigna unguiculata*). In: *Plant and Animal Genome Conference XVII*, p. 393. January 10–14, Town and Country Convention Center, San Diego, CA.

Madrid E, Rubiales D, Moral A, Moreno MT, Millan T, Gil J, and Rubio J (2008) Mechanism and molecular markers associated with rust resistance in a chickpea inter-specific cross (*Cicer arietinum* × *Cicer reticulatum*). *European Journal of Plant Pathology*, 121: 43–53.

Mayer MS, Tullu A, Simon CJ, Kumar J, Kaiser WJ, Kraft JM, and Muehlbauer FJ (1997) Development of a DNA marker for fusarium wilt resistance in chickpea. *Crop Science*, 7: 1625–1629.

Mehrotra M, Sanyal I, and Amla DV (2011) High-efficiency *Agrobacterium*-mediated transformation of chickpea (*Cicer arietinum* L.) and regeneration of insect-resistant transgenic plants. *Plant Cell Reports*, 30: 1603–1616.

Menéndez CM, Hall AE, and Gepts P (1997) A genetic linkage map of cowpea (*Vigna unguiculata*) developed from a cross between two inbred, domesticated lines. *Theoretical and Applied Genetics*, 95: 1210–1217.

Mercy ST, Kakar SK, and Chowdhry JB (1974) Cytological studies in three species of the genus *Cicer*. *Cytologia*, 39: 383–390.

Michelmore RW, Paran I, and Kesseli RV (1991) Identification of markers linked to disease-resistance genes by bulked segregant analysis: A rapid method to detect markers in specific genomic regions by using segregating populations. *Proceedings of the National Academy of Sciences USA*, 88: 9828–9832.

Mignouna HD, Ng NQ, Ikca J, and Thottappilly G (1998) Genetic diversity in cowpea as revealed by random amplified polymorphic DNA. *Journal of Genetics and Breeding*, 52: 151–159.

Millan T, Rubio J, Irula M, Daly K, Cubero JI, and Gil J (2003) Marker associated with Ascochyta blight resistance in chickpea and their potential in marker assisted selection. *Field Crops Research*, 84: 373–384.

Millan T, Winter P, Jungling R, Gil J, Rubio J, Cho S, Cobos MJ, et al. (2010) A consensus genetic map of chickpea (*Cicer arietinum* L.) based on 10 mapping populations. *Euphytica*, 175: 175–189.

Muchero W, Diop NN, Bhat PR, Fenton RD, Wanamaker S, Pottorff M, Hearne S, et al. (2009a) A consensus genetic map of cowpea [*Vigna unguiculata* (L) Walp.] and synteny based on EST-derived SNPs. *Proceedings of the National Academy of Sciences USA*, 106: 18159–18164.

Muchero W, Ehlers JD, Close TJ, and Roberts PA (2009b) Mapping QTL for drought stress-induced premature senescence and maturity in cowpea [*Vigna unguiculata* (L) Walp]. *Theoretical and Applied Genetics*, 118: 849–863.

Muchero W, Ehlers JD, Close TJ, and Roberts PA (2011) Genic SNP markers and legume synteny reveal candidate genes underlying QTL for Macrophomina phaseolina resistance and maturity in cowpea [Vigna unguiculata (L) Walp.]. *BMC Genomics*, 12: 8.

Muehlbauer FJ, Cho S, Sarker A, McPhee KE, Coyne CJ, Rajesh PN, and Ford R (2006) Application of biotechnology in breeding lentil for resistance to biotic and abiotic stress. *Euphytica*, 147: 149–165.

Myers GO, Fatokun CA, and Young ND (1996) RFLP mapping of an aphid resistance gene in cowpea (*Vigna unguiculata* L. Walp). *Euphytica*, 91: 181–187.

Nadimpalli RG, Jarret JL, Pathak SC, and Kochert G (1993) Phylogenetic relationships of pigeonpea (*Cajanus cajan*) based on nuclear restriction fragment length polymorphism. *Genome*, 36: 216–223.

Nayak S, Mamo B, Varghese N, Penmetsa RV, Farmer A, Woodward J, Gao J, et al. (2008) BAC-end Associated microsatellites in chickpea (*Cicer arietinum*) and cowpea (*Vigna unguiculata*): A resource for genetic and physical mapping. In: *Plant and Animal Genome Conference XVI*, p. 382. January 12–16, Town and Country Convention Center San Diego, CA.

Nayak SN, Zhu H, Varghese N, Datta S, Choi HK, Horres R, Jungling R, et al. (2010) Integration of novel SSR and gene-based SNP marker loci in the chickpea genetic map and establishment of new anchor points with *Medicago truncatula* genome. *Theoretical and Applied Genetics*, 120: 1415–1441.

Nifantova SN, Komarnickiy IK, and Kuchuk NV (2011) Obtaining of transgenic French bean plants (*Phaseolus vulgaris* L.) resistant to the herbicide pursuit by *Agrobacterium*-mediated transformation. *Tsitol Genet*, 45: 41–45.

Nishizawa K, Teraishi M, Utsumi S, and Ishimoto M (2007) Assessment of the importance of alpha-amylase inhibitor-2 in bruchid resistance of wild common bean. *Theoritical and Applied Genetics*. 114: 755–764.

Odeny DA, Jayashree B, Ferguson M, Hoisington D, Cry LJ, and Gebhardt C (2007) Development, characterization and utilization of microsatellite markers in pigeonpea. *Plant Breeding*, 126: 130–136.

Ohri D and Singh SP (2002) Karyotypic and genome size variation in *Cajanus cajan* (L.) Millsp (pigeonpea) and some wild relatives. *Genetic Resources and Crop Evolution*, 49: 1–10.

Olhoft PM and Somers DA (2001) L-cysteine increases *Agrobacterium*-mediated T-DNA delivery into soybean cotyledonary-node cells. *Plant Cell Reports*, 20: 706–711.

Olhoft PM, Lin K, Galbraith J, Nielsen NC, and Somers DA (2001) The role of thiol compounds in increasing *Agrobacterium*-mediated transformation of soybean cotyledonary node cells. *Plant Cell Reports*, 20: 731–737.

Omo-Ikerodah EE, Fawole I, and Fatokun CA (2008) Genetic mapping of quantitative trait loci (QTLs) with effects on resistance to flower bud thrips (*Megalurothrips sjostedti*) identified in recombinant inbred lines of cowpea. *African Journal of Biotechnology*, 7: 263–270.

Ouédraogo JT, Maheshwari V, Berner D, St-Pierre C-A, Belzile F, and Timko MP (2001) Identification of AFLP markers linked to resistance of cowpea (*Vigna unguiculata* L.) to parasitism by *Striga gesnerioides*. *Theoretical and Applied Genetics*, 102: 1029–1036.

Ouédraogo JT, Gowda BS, Jean M, Close TJ, Ehlers JD, Hall AE, Gillespie AG, et al. (2002) An improved genetic linkage map for cowpea (*Vigna unguiculata* L.) combining AFLP, RFLP, RAPD, biochemical markers, and biological resistance traits. *Genome*, 45: 175–188.

Pal M, Ghosh U, Chandra M, Pal A, and Biswas BB (1991) Transformation and regeneration of mungbean (*Vigna radiata*). *Indian Journal of Biochemistry and Biophysics*, 28: 449–455.

Palmer LE, Rabinowicz PD, O'Shaughnessy AL, Balija VS, Nascimento LU, Dike S, de la Bastide M, Martienssen RA, and McCombie WR (2003) Maize genome sequencing by methylation filtration. *Science*, 302: 2115–2117.

Pfaff T and Kahl G (2003) Mapping of gene-specific markers on the genetic map of chickpea (*Cicer arietinum* L.). *Molecular Genetics and Genomics*, 269: 243–251.

Phan HTT, Ellwood SR, Ford R, Thomas S, and Oliver RP (2006) Differences in syntenic complexity between *Medicago truncatula* with *Lens culinaris* and *Lupinus albus*. *Functional Plant Biology*, 33: 775–782.

Phan HTT, Ellwood SR, Hane JK, Ford R, Materne M, and Oliver RP (2007) Extensive macrosynteny between *Medicago truncatula* and *Lens culinaris* ssp. *culinaris*. *Theoretical and Applied Genetics*, 114: 549–558.

Pickardt T, Meixner M, Schade V, and Shieder O (1991) Transformation of Vicia narbonensis via *Agrobacterium tumefaciens* mediated gene transfer. *Plant Cell Reporst*, 9: 535–538.

Pickardt T, Saalbach I, Waddell D, Meixner M, Muntz K, and Schieder O (1998) Seed specific expression of the 2S albumin gene from Brazilnut (*Bertholletia excelsa*) into transgenic *Vicia narbonensis*. *Molecular Breeding*, 1: 295–301.

Popelka JC, Terryn N, and Higgins TJV (2004) Gene technology for grain legumes: Can it contribute to the food challenge in developing countries? *Plant Science*, 167: 195–206.

Popelka JC, Gollasch S, Moore A, Molvig L, and Higgins TJV (2006) Genetic transformation of cowpea (*Vigna unguiculata* L.) and stable transmission of the transgene to progeny. *Plant Cell Reports*, 25: 304–312.

Pottorff M, Ehlers JD, Fatokun C, Roberts PA, and Close TJ (2012) Leaf morphology in Cowpea [*Vigna unguiculata* (L.) Walp]: QTL analysis, physical mapping and identifying a candidate gene using synteny with model legume species. *BMC Genomics*, 13: 234.

Prasad V, Satyavathi VV, Sanjaya Valli KM, Khandelwal A, Shaila MS, and Sita GL (2004) Expression of biologically active hemagglutinin-neuraminadase protein of peste des petits ruminants virus in transgenic pigeonpea.[*Cajanus cajan* (L) Millsp.]. *Plant Science*, 166: 199–205.

Radchuk R, Radchuk V, Götz KP, Weichert H, Richter A, Emery RJ, Weschke W, and Weber H (2007) Ectopic expression of phosphoenolpyruvate carboxylase in *Vicia narbonensis* seeds: Effects of improved nutrient status on seed maturation and transcriptional regulatory networks. *Plant Journal*, 51: 819–839.

Radhika P, Gowda SJM, Kadoo NY, Mhase LB, Jamadagni BM, Sainani MN, Chandra S, and Gupta VS (2007) Development of an integrated intra-specific map of chickpea (*Cicer arietinum* L.) using two recombinant inbred line populations. *Theoretical and Applied Genetics*, 115: 209–216.

Rajesh PN, Tullu A, Gil J, Gupta, VS, Ranjekar PK, and Muehlbauer FJ (2002) Identification of an STMS marker for double-podding gene in chickpea. *Theoretical and Applied Genetics*, 105: 604–607.

Rajesh PN, Coyne C, Meksem K, Sharma KD, Gupta V, and Muehlbauer FJ (2004) Construction of a HindIII bacterial artificial chromosome library and its use in identification of clones associated with disease resistance in chickpea. *Theoretical and Applied Genetics*, 108: 663–669.

Raju NL, Gnanesh BN, Lekha P, Jayashree B, Pande S, Hiremath PJ, Byregowda M, Singh NK, and Varshney RK (2010) The first set of EST resource for gene discovery and marker development in pigeonpea (*Cajanus cajan* L.). *BMC Plant Biology*, 10: 45.

Rakshit S, Winter M, Tekeoglu M, Munoz JJ, Paff T, and Benko-Iseppon AM (2003) DAF maker tightly linked to a major locus for *Ascochyta* blight resistance in chickpea (*Cicer arietinum* L.). *Euphytica*, 32: 23–30.

Ratnaparkhe MB and Gupta VS (2007) Pigeonpea. In: *Genome Mapping and Molecular Breeding in Plants: Pulses, Sugar and Tuber Crops*, Kole C (ed.), vol. 3, pp. 133–142. Springer, Berlin.

Ratnaparkhe MB, Gupta VS, Ven Murthy MR, and Ranjekar PK (1995) Genetic fingerprinting of pigeonpea (*Cajanus cajan* (L.) Millsp) and its wild relatives using RAPD markers. *Theoretical and Applied Genetics*, 91: 893–898.

Ratnaparkhe MB, Santra DK, Tullu A, and Muehlbauer FJ (1998a) Inheritance of inter-simple-sequence-repeat polymorphisms and linkage with a fusarium wilt resistance gene in chickpea. *Theoretical and Applied Genetics*, 96: 348–353.

Ratnaparkhe MB, Tekeoglu M, and Muehlbauer FJ (1998b) Inter-simple sequence-repeat (ISSR) polymorphisms are useful for finding markers associated with disease resistance gene clusters. *Theoretical and Applied Genetics*, 97: 515–519.

Richter A, Jacobson HJ, de Kathen A, de Lorenzo G, Briviba K, Hain R, Ramsay G, and Kiesecker H (2007) Transgenic peas (Pisum sativum) expressing polygalacturonase inhibiting protein from raspberry (Rubus idaeus) and stilbene synthase from grape (*Vitis vinifera*). *Plant Cell Reports*, 25: 1166–1173.

Rolletschek H, Hosein F, Miranda M, Heim U, Götz KP, Schlereth A, Borisjuk L, Saalbach I, Wobus U, and Weber H (2005) Ectopic expression of an amino acid transporter (VfAAP1) in seeds of *Vicia narbonensis* and pea increases storage proteins. *Plant Physiology*, 137: 1236–1249.

Rubeena, Taylor PWJ, Ades PK, and Ford R (2006) QTL mapping of resistance in lentil (*Lens culinaris* ssp. *culinaris*) to Ascochyta blight (*Ascochyta lentis*). *Plant Breeding*, 125: 506–512.

Rubeena T, Taylor PWJ, and Ford R (2003) Molecular mapping the lentil (*Lens culinaris* ssp. *culinaris*) genome. *Theoretical and Applied Genetics*, 107: 910–916.

Rubio J, Moussa E, Kharrat M, Moreno MT, Millan T, and Gil J (2003) Two genes and linked RAPD markers involved in resistance to *Fusarium oxysporum* f. sp. *ciceris* race 0 in chickpea. *Plant Breeding*, 122: 188–191.

Russell DR, Wallace K, Bathe J, Martinell B, and McCabe D (1993) Stable transformation of *Phaseolus vulgaris* via electric-discharge mediated particle acceleration. *Plant Cell Reports*, 12: 165–169.

Saha GC, Sarker A, Chen W, Vandemark GJ, and Muehlbauer FJ (2010) Inheritance and linkage map positions of genes conferring resistance to *Stemphylium blight* in lentil. *Crop Science*, 50: 1831–1839.

Saini R and Jaiwal PK (2005) Transformation of a recalcitrant grain legume, *Vigna mungo* L. Hepper using *Agrobacterium tumefaciens*-mediated gene transfer to shoot apical meristem cultures. *Plant Cell Report*, 24: 164–171.

Saini R and Jaiwal PK (2007) *Agrobacterium tumefaciens*-mediated transformation of blackgram: An assessment of factors influencing the efficiency of uidA gene transfer. *Journal of Plant Biology*, 51: 69–74.

Saini R, Jaiwal S, and Jaiwal PK (2003) Stable genetic transformation of *Vigna mungo* L. Hepper via *Agrobacterium tumefaciens*. *Plant Cell Reports*, 21: 851–859.

Sarkar RH, Biswas A, Mustafa BM, Mahbub S, and Haque MI (2003) *Agrobacterium* mediated transformation of Lentil (*Lens culinaris* Medik). *Plant Tissue Culture*, 13: 1–12.

Sarmah BK, Moore A, Tate W, Morvig L, Morton RL, Rees RP, Chiaiese P, et al. (2004) Transgenic chickpea seeds expressing high levels of a bean α-amylase inhibitor. *Molecular Breeding*, 14: 73–82.

Satyavathi VV, Prasad V, Khandelwal A, Shaila MS, and Sita GL (2003) Expression of hemagglutinin protein of Rinderpest virus in transgenic pigeon pea [*Cajanus cajan* (L.) Millsp.] plants. *Plant Cell Reports*, 21: 651–658.

Saxena RK, Prathima C, Saxena KB, Hoisington DA, Singh NK, and Varshney RK (2010) Novel SSR markers for polymorphism detection in pigeonpea (*Cajanus* spp.). *Plant Breeding*, 129: 142–148.

Saxena RK, Cui X, Thakur V, Walter B, Close TJ, and Varshney RK (2011) Single feature polymorphisms (SFPs) for drought tolerance in pigeonpea (*Cajanus* spp.). *Functional and Integrated Genomics*, 11: 651–657.

Schroeder HE, Schotz AH, Wardley-Richardson T, Spencer D, and Higgins TJV (1993) Transformation and regeneration of two cultivars of pea *Pisum sativum* L. *Plant Physiology*, 101: 751–757.

Senthil G, Williamson B, Dinkin RD, and Ramsay G (2004) An efficient transformation system for chickpea. *Plant Cell Reports*, 23: 297–303.

Shan F, Clarke HC, Plummer JA, Yan G, and Siddique KHM (2005) Geographical patterns of genetic variation in the world collections of wild annual Cicer characterized by amplified fragment length polymorphisms. *Theoretical and Applied Genetics*, 110: 381–391.

Shanower TG, Romeis J, and Minja EM (1999) Insect pests of pigeonpea and their management. *Annual Review of Entomology*, 44: 77–96.

Sharma KD, Winter P, Kahl G, and Muehlbauer FJ (2004) Molecular mapping of *Fusarium oxysporum* f. sp. *ciceris* race 3 resistance gene in chickpea. *Theoretical and Applied Genetics*, 108: 1243–1248.

Sharma PC and Gupta PK (1986) Cytogenetics of legume genera *Cicer* L and *Lens* L. In: *Genetic and Crop Improvement*, Gupta PK and Bahl JR (eds), Ratogi and company, Meerut, India.

Sharma KK, Lavanya M, and Anjaiah V (2006) *Agrobacterium*-mediated production of transgenic pigeonpea (*Cajanus cajan* L Millsp) expressing synthetic Bt cryIAb gene. *In Vitro Cellular Developmental Biology Plant*, 42: 165–173.

Simon CJ and Muehlbauer FJ (1997) Construction of a chickpea linkage map and its comparison with maps of pea and lentil. *Journal of Heredity*, 88: 115–119.

Sindhu JS, Slinkard AE, and Scoles JE (1983) Studies on variation in *Lens* I. Karyotype. *Lentil Experimental News Service*, 10: 14.

Singh BB (2005) Cowpea *Vigna unguiculata* (L.) Walp. In: *Genetic Resources, Chromosome Engineering and Crop Improvement*, Singh RJ and Jauhar PP (eds), pp. 117–162. CRC, Boca Raton, FL.

Singh BB, Cambliss OI, and Sharma B (1997) Recent advances in cowpea breeding. In: *Advances in Cowpea Research*, Singh BB, Mohan Raj DR, Dashiell K, and Jackai LEN (eds.), pp. 30–49. IITA, Ibadan.

Singh NK, Gupta DK, Jayaswal PK, Mahato AK, Dutta S, Singh S, Bhutani S, et al. (2011) The first draft of the pigeonpea genome sequence. *Journal of Plant Biochemistry and Biotechnology*, 21: 98–112.

Singh NK, Gupta DK, Jayaswal PK, Mahato AK, Dutta S, Singh S, Bhutani S, and Dogra V (1997).

Sivaramakrishnan S, Seetha K, Rao AN, and Singh L (1997) RFLP analysis of cytoplasmic male sterile lines in Pigeonpea (*Cajanus cajan* L. Millsp.). *Euphytica*, 126: 293–299.

Sivaramakrishnan S, Seetha K, and Reddy LJ (2002) Diversity in selected wild and cultivated species of pigeonpea using RFLP of mtDNA. *Euphytica*, 125: 21–28.

Solleti SK, Bakshi S, Purkayastha J, Panda SK, and Sahoo L (2008) Transgenic cowpea (*Vigna unguiculata*) seeds expressing a bean alpha-amylase inhibitor 1 confer resistance to storage pests, bruchid beetles. *Plant Cell Reports*, 27: 1841–1850.

Sonia, SR, Singh RP, and Jaiwal PK (2007). *Agrobacterium tumefaciens* mediated transfer of *Phaseolus vulgaris* alpha-amylase inhibitor-1 gene into mungbean *Vigna radiata* (L.) Wilczek using bar as selectable marker. *Plant Cell Reports*, 26: 187–198.

Sudupak MA, Akkaya MS, and Kence A (2004) Genetic relationships among perennial and annual *Cicer* species growing in Turkey assessed by AFLP fingerprinting. *Theoretical and Applied Genetics*, 108: 937–944.

Surekha Ch, Arundhanti A, and Seshagiri Rao G (2007) Diffrential response of Cajanus cajan varieties to transformation with different strains of *Agrobacterium*. *Journal of Biological Sciences*, 7: 176–181.

Surekha C, Beena MR, Arundhati A, Singh PK, Tuli R, Dutta-Gupta A, and Kirti PB (2005). *Agrobacterium*-mediated genetic transformation of pigeon pea (*Cajanus cajan* (L.) Millsp.) using embryonal segments and development of transgenic plants for resistance against Spodoptera. *Plant Science*, 169: 1074–1080.

Tahir M and Muehlbauer FJ (1994) Gene mapping in lentil with recombinant inbred lines. *Journal of Heredity*, 85: 306–310.

Talebi R, Naji AM, and Fayaz F (2008) Geographical patterns of genetic diversity in cultivated chickpea (*Cicer arietinum* L.) characterized by amplified fragment length polymorphism. *Plant Soil Environment*, 54: 447–452.

Taleei A, Kanouni H, and Baum M (2010) Genetical studies of ascochyta blight resistance in chickpea. *International Journal of Bio-Science and Bio-Technology*, 2: 19–28.

Tanyolac B, Ozatay S, Kahraman A, and Muehlbauer FJ (2010) Linkage mapping of lentil (*Lens culinaris* L.) genome using recombinant inbred lines revealed by AFLP, ISSR, RAPD and some morphologic markers. *Journal of Agricultural Biotechnology and Sustainable Development*, 2: 001–006.

Tar'an B, Buchwaldt L, Tullu A, Banniza S, Warkentin TD, and Vandenberg A (2003) Using molecular markers to pyramid genes for resistance to ascochyta blight and anthracnose in lentil (*Lens culinaris* Medik). *Euphytica*, 134: 223–230.

Tar'an B, Warkentin TD, Tullu A, and Vandenberg A (2007) Genetic mapping of ascochyta blight resistance in chickpea (*Cicer arietinum* L.) using a simple sequence repeat linkage map. *Genome*, 50: 26–34.

Tazeen S and Mirza B (2004) Factors affecting *Agrobacterium tumefaciens* mediated genetic transformation of *Vigna radiata* L. Wilczek. *Pakistan Journal of Botany*, 36: 887–896.

Tekeoglu M, Tullu A, Kaiser WJ, and Muehlbauer FJ (2000) Inheritance and linkage of two genes that confer resistance to Fusarium wilt in chickpea. *Crop Science*, 40: 1247–1251.

Thu TT, Mai TTT, Dewaele E, Farsi S, Tadesse Y, Angenon G, and Jacobs M (2003) In vitro regeneration and transformation of pigeonpea [*Cajanus cajan* (L.) Millsp]. *Molecular Breeding*, 11: 159–168.

Thudi M, Bohra A, Nayak SN, Varghese N, Shah TM, Penmesta RV, Thirunavukkarasu N, et al. (2011) Novel SSR markers from BAC-end sequences, DArT arrays and a comprehensive genetic map with 1,291 marker loci for chickpea (*Cicer arietinum* L.). *PLoS ONE*, 6: e27275.

Timko MP, Rushton PJ, Laudeman TW, Bokowiec MT, Chipumuro E, Cheung F, Town CD, and Chen X (2008) Sequencing and analysis of the gene-rich space of cowpea. *BMC Genomics*, 9: 103.

Tullu A, Buchwaldt T, Warkentin T, Tar'an B, and Vandenberg A (2003) Genetics of resistance to anthracnose and identification of AFLP and RAPD markers linked to the resistance gene in PI 320937 germplasm of lentil (*Lens culinaris Medikus*). *Theoretical and Applied Genetics*, 106: 428–434.

Tullu A, Muehlbauer FJ, Simon CJ, Mayer MS, Kumar J, Kaiser WJ, and Kraft JM (1998) Inheritance and linkage of a gene for resistance to race 4 of Fusarium wilt and RAPD markers in chickpea. *Euphytica*, 102: 227–232.

Tullu A, Kaiser WJ, Kraft JM, and Muehlbauer FJ (1999) A second gene for resistance to race 4 of Fusarium wilt in chickpea and linkage with a RAPD marker. *Euphytica*, 109: 43–50.

Tullu A, Tar'an B, Breitkreutz C, Buchwaldt L, Banniza S, Warkentin TD, and Vandenberg A (2006) A quantitative trait locus for resistance to ascochyta blight (*Ascochyta lentis*) maps close to a gene for resistance to anthracnose (*Colletotrichum truncatum*). *Canadian Journal of Plant Pathology*, 28: 588–595.

Tullu A, Tar'an B, Warkentin T, and Vandenberg A (2008) Construction of an intraspecific linkage map and QTL analysis for earliness and plant height in lentil. *Crop Science*, 48: 2254–2264.

Udupa SM and Baum M (2003) Genetic dissection of pathotype-specific resistance to ascochyta blight resistance in chickpea (*Cicer arietinum* L.) using microsatellite markers. *Theoretical and Applied Genetics*, 106: 1196–1202.

Udupa S, Sharma A, Sharma A, and Pai R (1993) Narrow genetic variability in *Cicer arietinum* as revealed by RFLP analysis. *Journal of Plant Biochemistry and Biotechnology*, 2: 83–86.

Upadhyaya HD, Dwivedi SL, Baum M, Varshney RK, Udupa SM, Gowda CLL, Hoisington D, and Singh S (2008) Genetic structure, diversity, and allelic richness in composite collection and reference set in chickpea (*Cicer arietinum* L.). *BMC Plant Biology*, 8: 106.

van der Maesen LJG (1987) Origin, history and taxonomy of chickpea. In: *The Chickpea*, Saxena MC and Singh KB (eds), pp. 11–34. CAB International, Cambridge.

van der Maesen LJG (1990) Pigeonpea: Origin, history, evolution and taxonomy. In: *The Pigeonpea*, Nene YL, Hall SD, and Sheila VK (eds), pp. 15–46. CAB International, Wallingford.

van Rheenen HA, Reddy MV, Kumar J, and Haware M (1992) Breeding for resistance to soil-borne diseases in chickpea. In: *Disease Resistance in Chickpea*, Singh KB and Saxena MC (eds), pp. 55–70. ICARDA, Aleppo, Syria.

Varshney RK, Hiremath PJ, Lekha P, Kashiwagi J, Balaji J, Deokar AA, Vadez V, et al. (2009a) A comprehensive resource of drought- and salinity- responsive ESTs for gene discovery and marker development in chickpea (*Cicer arietinum* L.). *BMC Genomics*, 10: 523.

Varshney RK, Close TJ, Singh NK, Hoisington DA, and Cook DR (2009b) Orphan legume crops enter the genomics era. *Current Opinion in Plant Biology*, 12: 1–9.

Varshney RK, Penmetsa RV, Dutta S, Kulwal PL, Saxena RK, Datta S, Sharma TR, et al. (2010a) Pigeonpea genomics initiative (PGI): An international effort to improve crop productivity of pigeonpea (*Cajanus cajan* L.). *Molecular Breeding*, 26: 393–408.

Varshney RK, Glaszmann JC, Leung H, and Ribaut JM (2010b) More genomic resources for less-studied crops. *Trends in Biotechnology*, 28: 452–460.

Varshney RK, Chen W, Li Y, Bharti AK, Saxena RK, Schlueter JA, Donoghue MTA, et al. (2011) Draft genome sequence of pigeonpea (*Cajanus cajan*), an orphan legume crop of resource-poor farmers. *Nature Biotechnology*, 30: 83–89.

Varshney RK, Murali Mohan S, Gaur PM, Gangarao NVPR, Pandey MK, Bohra A, Sawargaonkar SL, et al. (2013a) Achievements and prospects of genomics-assisted breeding in three legume crops of the semi-arid tropics. *Biotechnology Advances* (in press).

Varshney RK, Song C, Saxena RK, Azam S, Yu S, Sharpe AG, Cannon S, et al. (2013b) Draft genome sequence of chickpea (*Cicer arietinum*) provides a resource for trait improvement. *Nature Biotechnology*, 31: 240–246.

Venora G and Padulosi S (1997) Karyotypic analysis of wild taxa of *Vigna unguiculata* (L.) Walpers. *Caryologia*, 50: 125–138.

Venora G, Conicella C, Errico A, and Saccardo F (1991) Karyotyping in plants by an image analysis system. *Journal of Genetics and Breeding*, 45: 233–240.

Vijayan S and Kirti PB (2012) Mungbean plants expressing BjNPR1 exhibit enhanced resistance against the seedling rot pathogen, Rhizoctonia solani. *Transgenic Research*, 21: 193–200.

Vlácilová K, Ohri D, Vrána J, Číhalíková J, Kubaláková M, Kahl G, and Doležel J (2002) Development of flow cytogenetics and physical genome mapping in chickpea (*Cicer arietinum* L.). *Chromosomes Research*, 10: 695–706.

Weeden NF, Muehlbauer FJ, and Ladizinsky G (1992) Extensive conservation of linkage relationships between pea and lentil genetic maps. *Journal of Heredity*, 83: 123–129.

Winter P, Benko-Iseppon M, Huttel B, Ratnaparkhe MB, Tullu A, Sonnante G, Pfaff T, et al. (2000) A linkage map of the chickpea (*Cicer arietinum* L.) genome based on recombinant inbred lines from a *C. arietinum* × C. reticulatum cross: Localization of resistance genes for fusarium wilt races 4 and 5. *Theoretical and Applied Genetics*, 101: 1155–1163.

Yang S, Pang W, Harper J, Carling J, Wenzl P, Huttner E, Zong X, and Kilian A (2006) Low level of genetic diversity in cultivated pigeonpea compared to its wild relatives is revealed by diversity arrays technology (DArT). *Theoretical and Applied Genetics*, 113: 585–595.

Yang SY, Saxena RK, Kulwal PL, Ash GJ, Dubey A, Harper JD, Upadhyaya HD, Gothalwal R, Kilian A, and Varshney RK (2011) The first genetic map of pigeonpea based on diversity arrays technology (DArT) markers. *Journal of Genetics*, 90: 103–109.

Zambre M, Goossens A, Cardona C, van Montagu M, Terryn N, and Angenon G (2005) A reproducible genetic transformation system for cultivated *Phaseolus acutifolius* (tepary bean) and its use to assess the role of arcelins in resistance to the Mexican bean weevil. *Theoritical and Applied Genetics*, 110: 914–924.

Zamir D and Ladizinsky G (1984) Genetics of allozyme variants and linkage groups in lentil. *Euphytica*, 33: 329–336.

Zatloukalová P, Hřibová E, Kubaláková M, Suchánková P, Simková H, Adoración C, Kahl G, Millán T, and Doležel J (2011) Integration of genetic and physical maps of the chickpea (*Cicer arietinum* L.) genome using flow-sorted chromosomes. *Chromosome Research*, 19: 729–739.

Závodná M, Kraic J, Paglia G, Gregova E, and Morgante M (2000) Differentiation between closely related lentil (*Lens culinaris* Medik.) cultivars using DNA markers. *Seed Science and Technology*, 28: 217–219.

Zhang X, Schering CF, Zhang M, Dong JJ, Zhang Y, Huang JJ, Lee, MK, et al. (2010) A BAC/BIBAC-based physical map of chickpea (*Cicer arietinum* L). *BMC Genomics*, 11: 501.

Zohary D and Hopf M (1973) Domestication of pulses in the old world. *Science* 182: 887–894.

5

Genetic Engineering in Potato Improvement

Elena Rakosy-Tican, PhD

CONTENTS

5.1   Introduction 139
5.2   Improvement of Potato by Transgenesis 140
    5.2.1   The Use of Reporter Gene gfp to Improve the Efficiency of
        Transformation and Allow the Transfer of Other Genes Directly
        Generating Marker-Free Plants 140
    5.2.2   Genetic Transformation of the Wild Potato Species *Solanum chacoense* 143
    5.2.3   Recent Advances in Improving Potato Quality by Gene Transfer 145
    5.2.4   Potato Plants as Bioreactors 150
    5.2.5   Recent Advances in Improving Resistance to Biotic Stress 151
        5.2.5.1   Breeding Resistance to Late Blight by a Transgenic Approach 151
        5.2.5.2   Breeding Resistance to Viruses 152
        5.2.5.3   Breeding Resistance to Colorado Potato Beetle 153
        5.2.5.4   Breeding Resistance to Other Diseases and Pests 153
    5.2.6   Advances in Improving Resistance to Abiotic Stress 154
5.3   Improving Potato Crop by Cisgenesis 154
5.4   A Few Remarks on Public Acceptance of GM Potatoes 155
5.5   Funding 155
References 155

5.1 Introduction

The potato is one of the most productive and nutritious vegetables, currently ranking third in worldwide crop production. It produces not only high carbohydrate (starch), but also higher quality protein than any vegetable except soybeans. Potatoes are also an important source of animal feed, industrial starch, ethanol, and consequently biofuels. In recent years, human health-promoting compounds, mainly antioxidants (vitamin C, phenolics, and carotenoids) and anticancer agents (gallic acid or α-chaconine), have been studied in relation to healthier food (Lester 2006; Stushnoff et al. 2008; Reddivari et al. 2010). Although the potato has such importance for agriculture, industrial applications, and human health, genetics and genomic studies are less advanced because of its autotetraploidy ($2n = 4x = 48$), heterozygosity, and inbreeding depression, which make genetic study and manipulation quite difficult. However, the potato genome was recently sequenced by using a dihaploid produced by *in vitro* anther culture of *Solanum phureja*, a development that will soon open more possibilities for biotechnological improvement (Huang et al. 2011). These results of the Potato Genome Sequencing Consortium (PGSC), based on homozygous doubled and monoploid potato clones, assembled 86% of the genome (844 Mb). It is assumed that the potato genome contains 39,031 protein-coding genes.

Classical potato breeding has attempted to improve mainly the quantity and quality of tubers, but it is only by genetic transformation and other biotechnological approaches that a faster transfer of traits such as quality of tuber composition and resistance to biotic and abiotic stresses might be possible. Moreover, there are many attempts and results on the transfer and integration of economically important genes in potato crops, and some previous reviews have presented the state of the art in this tuberous crop (Dietze et al. 1995; Kumar 1995; Solomon-Blackburn and Barker 2001; Christou et al. 2006; Mullins et al. 2006; Millam 2007). A simple Google Scholar search for the period 2000–2012, for "transgenic potato" and "review," resulted in 4700 hits, revealing a wealth of information on this subject. Of course, in this context it is almost impossible to prepare an exhaustive review covering all the achievements of potato genetic modifications.

The aim of this chapter is to present the recent advances in potato genetic transformation and the perspectives of their application, pointing out also the progress in our laboratory.

5.2 Improvement of Potato by Transgenesis

The potato as a food security crop has been an important target for both classical and biotechnological improvement. *Agrobacterium tumefaciens*-mediated transformation, which works well with many cultivars, was mainly applied to potato crops (see Rakosy-Tican et al. 2007) and a few wild species of the genus *Solanum*. Although the potato was one of the first crops to be transformed by *Agrobacterium*, the efficiency of the gene transfer varies depending on the genotype, with cultivar Désirée being a kind of model variety (Stiekema et al. 1988; Sheerman and Bevan 1988; Rakosy-Tican et al. 2007). To date many important traits have been targeted by genetic transformation, and the most recent achievements of transgenesis will be presented in this section.

5.2.1 The Use of Reporter Gene gfp to Improve the Efficiency of Transformation and Allow the Transfer of Other Genes Directly Generating Marker-Free Plants

Reporter genes that allow selection and visual detection of transgene expression, such as *uidA* and *luc*, have been used extensively to monitor transformation in plants. The most commonly exploited reported *uidA* gene with its product β-glucuronidase (GUS) (Jefferson et al. 1987) has a crucial drawback, that is, the destructive nature of histochemical assay, and, as such, cannot be used to reveal *in vivo* transgene expression. The other reporter, the *luc* gene, can be monitored *in vivo* but it needs an exogenous substrate, luciferin, and costly equipment (Ow et al. 1986). In contrast, the *gfp* gene, encoding the small green fluorescent protein (GFP) isolated from the jellyfish *Aequorea victoria* (Chalfie et al. 1994), does not need a substrate, is expressed and can be monitored *in vivo* without any cofactor, and is visible under the microscope or even macroscopically under fluorescent illumination (Molinier et al. 2000; Rakosy-Tican et al. 2007) (Figure 5.1). Its expression is visible in whole plants or plant organs, permitting screening of primary transformants. The discrimination of homozygous and hemizygous plants by the intensity of green fluorescence in tobacco and other species has been reported (Molinier et al. 2000; Ghorbel et al. 1999; Zhang et al. 2001), but not in the case of the potato (Rakosy-Tican et al. 2007). High

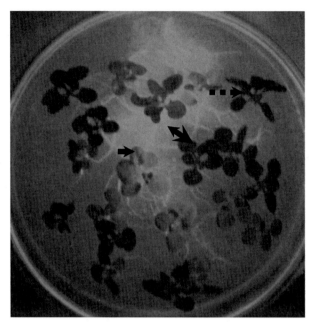

FIGURE 5.1
Gfp reporter gene expression in seedlings of *Nicotiana tabacum* L. cv. Samsun NN under fluorescent illumination (UV lamp B-100AP) showing bright green fluorescence in homozygous (*simple arrow*), intermediate fluorescence in hemizygous state (*double arrow*), or native red fluorescence of chlorophyll in untransformed shoots (*dashed arrow*). Transgene segregation following Mendelian dominant gene model is supported even after many years of *in vitro* subcultures.

stability of the constitutive expression of *gfp* driven by CaMV 35S promoter, after many generations of *in vitro* subculture and self-fertilization, has been observed in tobacco, as well as maintenance of Mendelian dominant gene segregation (Figure 5.1). GFP has become useful for studying tissue-specific promoter expression, protein traffic, cell compartmentalization, and organelle-specific proteins, virus infection, or even virus-induced gene silencing. In our studies, we proved its utility in better monitoring transformation efficiency (Rakosy-Tican et al. 2007 and references therein). GFP has been also used as a marker for somatic hybridization in *Citrus* (Olivares-Fuster et al. 2002) and to obtain green fluorescent flowers (Mercuri et al. 2001; Rakosy-Tican, unpublished data). However, in the potato, *gfp* has been only used in stable chloroplast transformation (Sidorov et al. 1999) before the application in our group for refinement of transformation of dihaploid lines, commercial cultivars, and some wild potato relatives, as a prerequisite for the transfer of other genes, eventually without using selection for antibiotic resistance. Although the use of *nptII* marker gene for selection of stable transgenic cells and regenerated shoots was considered reliable for different target tissues such as leaf disks, internodes, and microtubers, in combination with the new reporter *gfp* it was shown that it allows a large number of escapes, nontransgenic shoots, or chimeras. The use of both genes and selection based on kanamycin and *gfp* visual screening allowed us to improve the transformation for different cultivars and dihaplod lines of the potato (Rakosy-Tican et al. 2007) as well as for the wild *Solanum* species: *S. chacoense* and *S. microdontum* (Rakosy-Tican et al. 2004). Moreover, the protocol once established for a genotype can be further used to generate marker-free transgenic plants with very high efficiency (Rakosy-Tican et al. 2010) or to transfer other genes such as *msh2*, a gene coding for the key protein involved in DNA mismatch repair (MMR) (Rakosy-Tican et al. 2004).

Our results showed that *gfp* is a useful tool to monitor the efficiency of transformation, the effects of antibiotics on organogenesis, the somatic hybrid cells' viability and vigorous growth as reliable selection criteria (*S. tuberosum* + *S. chacoense*) (Figure 5.2), transgenic plant selection, or the occurrence of escapes and chimeras in three potato dihaploid genotypes and four tetraploid cultivars tested. The cultivar Désirée was used widely as a model for potato transformation, but commercial varieties or dihaploid genotypes used in our previous research were not subjected to *Agrobacterium*-mediated gene transfer. Developing an efficient technique for genetic transformation of tetraploid commercial varieties of potato opens the way for direct transfer and integration of genes improving such qualities as resistance to stress or tuber traits. Another study reported for 16 potato commercial cultivars, based on *nptII* marker gene and antisense gene for granule-bound starch synthase (Heeres et al. 2002), low transformation frequencies of 0.02–0.35 transgenic shoots per explant (internodes), while Rommens et al. (2004) showed regeneration of 0.3–0.6 transgenic shoots per explant. In comparison, in our experiments the efficiency of transformation was high for the potato, with some genotypes such as dihaploid 178/10 regenerating 13 green fluorescent shoots per leaf explant or the cultivar Baltica with a mean of two transgenic shoots per internode (Rakosy-Tican et al. 2007). Moreover, efficient genetic transformation of specific dihaploid lines might be an advantage for further tetraploidization and hence the increase in transgene stability in the next generations. Our results reinforced the previous assumption (Domìnguez et al. 2004) that selection with kanamycin does not prevent the regeneration of shoots that are not transformed by *nptII* gene. But, in contrast with *Citrus* (Domìnguez et al. 2004), our molecular analysis on the potato (polymerase chain reaction [PCR] and reverse transcription PCR [RT-PCR]) did not prove any case of *nptII* silencing. Fluorescent microscopy analysis of *gfp* expression

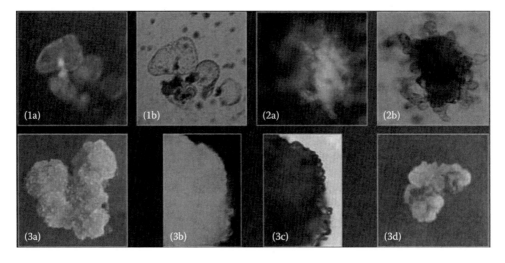

FIGURE 5.2   (See color insert)
The usefulness of *gfp* for monitoring hybrid cell regeneration in somatic hybrids *Solanum tuberosum* (2x) 178/10 + *S. chacoense* (2x): (1a) hybrid cell expressing gfp (UV); (1b) the same cell under light microscopy (LM); (2a) cell colony with GFP (UV); (2b) the same colony under LM; (3a) hybrid callus with vigorous growth; (3b) part of same callus with GFP (UV); (3c) same as (3b) under LM; (3d) *S. chacoense* callus without gfp (control, nontransgenic) at the same age as (3a).

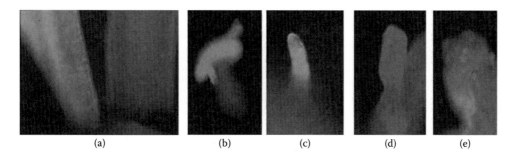

(a) (b) (c) (d) (e)

FIGURE 5.3 (See color insert)

Examples of gfp expression in putative transgenic plants regenerating on kanamycin containing media: (a) potato dihaploid genotype 178/10, *left* transgenic leaf and *right* control; (b) very young shoot with GFP cv. Baltica; (c) young shoot with GFP cv. Desiree; (d) a putative transgenic shoot without GFP cv. Delikat; (e) young shoot of *Solanum chacoense* showing chimeric tissues.

in young regenerated shoots is shown in Figure 5.3 and the PCR amplification products of the *nptII* gene in transgenic potato clones or RT-PCR analysis of *gfp* is exemplified in Figure 5.4. A kanamycin concentration of 50 mg/L, used in our transformation experiments, and *Agrobacterium* cocultivation showed genotype-dependent effects on organogenesis reducing and delaying shoot and callus regeneration (2–4 weeks depending on the genotype) (Figure 5.5). Cefotaxime, used mainly at 250 mg/L, also influences organogenesis with stimulation of shoot regeneration, as shown in *S. chacoense* (Rakosy-Tican et al. 2011). For the species that generate many escapes and chimeras, other authors have recommended the use of a screenable strategy instead of lethal selection (Christou and McCabe 1992). *Gfp* proved to be the best for transgenic shoot selection in previous experiments with other plant species (Ghorbel et al. 1999; Zhang et al. 2001). We have also shown that by using both genes, *gfp* and *nptII*, the real picture of genotype-dependent transformation efficiency is revealed. The effects of culture media, antibiotics, and *Agrobacterium* strains on shoot regeneration may also be better seen and may allow optimization of each particular genotype transformation (Table 5.1). Such a transformation protocol could be then used for efficient transfer of other genes (Rakosy-Tican et al. 2004), eventually eliminating the necessity to use antibiotic resistance genes (Visser 2003; De Vetten et al. 2003; Rakosy-Tican et al. 2010).

5.2.2 Genetic Transformation of the Wild Potato Species *Solanum chacoense*

All living species have developed during evolution systems to repair DNA mismatches caused by replication or recombination. MMR in plants was less investigated in comparison with bacteria. In bacteria two genes are involved in MMR, *mutS* and *mutL* (Marsischky et al. 1996; Modrich and Lahue 1996). In yeasts there are six homologs of *mutS* (*msh1–msh6*) and four homologous genes of *mutL* (*mlh1–mlh3* and *pms1*). In maize and *Arabidopsis* a new gene *msh7* was described. The MMR system involves many proteins, but only MUTS interacts with the heteroduplex DNA, and MUTL works together with MUTS to fix the DNA mismatch. Other proteins are involved in processing the base mispairs. One of the key proteins in the MMR system of plants is MSH2, which forms complexes with other proteins such as MSH6 and MSH3 in order to bind and repair the DNA. Any deficiency in the MMR system will increase DNA instability and recombination (Culligan and Hays 1997). In order to increase recombination efficiency between cultivated and

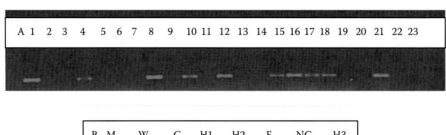

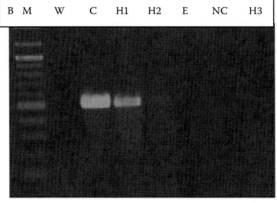

FIGURE 5.4

Examples of transgenic potato clones analyzed with specific primers: (A) PCR profile of *nptII* gene: (1) shoots expressing *gfp* of cv. Agave; (2) putative transgenic shoot without GFP potato cv. Désirée; (3) *Solanum chacoense* control; (4) shoots expressing *gfp* of cv. Désirée; (5) *S. chacoense* HL (high leptine) control; (6) shoots expressing *gfp* of cv. Baltica; (7) cv. Baltica control; (8) shoots expressing *gfp* of *S. chacoense* HL; (9) shoots expressing *gfp* of potato dihaploid line 178/10; (10) shoots expressing *gfp* of cv. Baltica; (11) potato cv. Delikat control; (12) chimerical shoot of cv. Delikat; (13) putative transgenic shoot without GFP of *S. chacoense* HL; (14) putative transgenic shoot without GFP cv. Baltica; (15) chimerical shoot of cv. Delikat; (16) shoots expressing *gfp* of cv. Delikat; (17) chimerical shoot of cv. Delikat; (18) chimerical shoot of cv. Delikat; (19) potato cv. Désirée control; (20) empty; (21) positive control; (22) empty; (23) negative control. (B) RT-PCR of *gfp*: M, 100 bp marker (New England Biolabs); (C) positive control; W, water; E, empty; H1, H2, H3, different chimerical transgenic clones of cv. Delikat; NC, negative control.

wild *Solanum* species somatic hybrids, we attempted to use MSH2-deficient transgenic plants. The gene *atmsh2* from *Arabidopsis thaliana* either in antisense orientation (*As*) or in negative complementary sequence (*Apa*) was transformed into *Agrobacterium tumefaciens* plasmids (Rakosy-Tican et al. 2004). Different accessions of the wild species *S. chacoense* were transformed by using *A. tumefaciens*, after the improvement of the methodology by using the above-mentioned combination of reporter gene *gfp* and marker gene *nptII* (Figures 5.6 and 5.7). The most efficient protocol for genetic transformation of *S. chacoense* was then applied to transfer the genes *atmsh2-as* or *atmsh2-apa*. The plants that proved to be transgenic (Table 5.2) were further used to isolate mesophyll protoplasts and to electrofuse them with potato cultivars Delikat or Désirée. The somatic hybrids of one accession of *S. chacoense* with very high content of repellent glycoalkaloids toward Colorado potato beetle (CPB), leptines, carrying *msh2-as* or *msh2-apa* were compared to the wild type nontransgenic parent for efficiency of somatic hybridization and resistance to CPB. Moreover, random amplified polymorphic DNA (RAPD) markers known to be linked to leptine biosynthesis, OPT-16, OPT-20 and OPQ-2 (Boularte-Medina et al. 2002), were also evaluated. Somatic hybrid plants with resistance to CPB that show the specific RAPD marker OPT-20 were identified. The results showed that the number of hybrids is higher

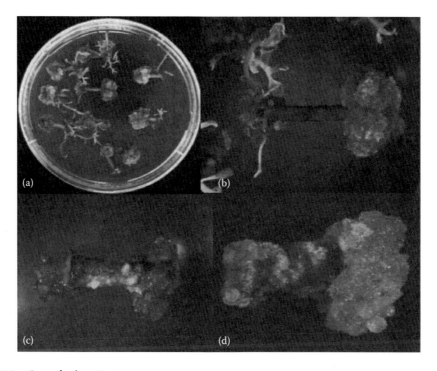

FIGURE 5.5 (See color insert)
The effect of *Agrobacterium* strain and kanamycin selection on delaying organogenesis in cultivar Baltica, after 8 weeks of internodes culture: (a) the general view of regeneration in the control; (b) a detail of a, an internode regenerating in a polar manner calus and shoots; (c) only callus develops after cocultivation with *Agrobacterium tumefaciens* LBA4404; (d) callus formation after cocultivation with *A. tumefaciens* EHA105; the delay in shoot regeneration is to be observed in agroinfected internodes.

for transgenic wild parent with deficiency in MMR, and that some of the somatic hybrids show mutations affecting chlorophyll biosynthesis or plant growth (data not illustrated here). Further research on meiosis and increase in homologous recombination will be conducted at both molecular and cytogenetic levels.

5.2.3 Recent Advances in Improving Potato Quality by Gene Transfer

A number of tuber quality traits have been targeted by the transgenic approach, mainly in the past decade. One of the important traits of the tubers is that they contain toxic glycoalkaloids. Such compounds with bitter taste and potential toxicity to humans are present in *Solanaceae* family and hence in the potato cultivars. Some of the cultivars, particularly the ones involving a wild species cross in their pedigree, have higher glycoalkaloid content than others. Occasionally higher concentrations of glycoalkaloids can be found in commercially cultivated varieties as a result of unfavorable climatic conditions, such as drought and extreme temperatures, a situation that raises questions in relation to global climate change. Since glycoalkaloid biosynthesis is enhanced by exposure to light (sun or artificial) (Lafta and Lorenzen 2000), careless handling and storage after harvest can cause high glycoalkaloid concentrations in tubers. Moreover, it is known that in tubers the highest concentration accumulates in or immediately under the peel. The total glycoalkaloid content should not exceed 20 mg/100 g FW, higher values being poisonous for consumers.

TABLE 5.1

Plant Regeneration in Potato Dihaploid Genotypes after Coculture with *Agrobacterium tumefaciens* LBA4404 Carrying *gfp* and *nptII* Genes

Dihaploid Genotype	Type of Explant	Cocultivation	n	Total Number of Regenerated Shoots	Mean Shoot Length (cm ± SE)
178/10	Stem segments	None (control)	10	360	ND
		Agro C + K, 5 weeks[a]	10	0	ND
	Leaf segments	None (control)	10	262	ND
		Agro C + K 5 weeks[a]	10	345	ND
224/1	Stem segments	None (control)	18	25	21.56 ± 9.6
		Agro C + K, 5 weeks[a]	15	2	2 ± 0
		Agro C + K[b]	17	0	ND
	Leaf segments	None (control)	25	44	12.07 ± 0.3
		Agro C + K, 5 weeks[a]	27	17	8.11 ± 0.8
		Agro C + K[b]	26	7	7.85 ± 1.12
227/5	Stem segments	None (control)	14	0	ND
		Agro C + K, 5 weeks[a]	14	9	10.4 ± 4.4
		Agro C + K[b]	14	24	13.86 ± 8.5
	Leaf segments	None (control)	20	0	ND
		Agro C + K, 5 weeks[a]	20	8	9.5 ± 0.5
		Agro C + K[b]	18	3	11.66 ± 2.27

Note: The effect of kanamycin on plant regeneration varies for each genotype and type of explant.

[a] C + K 5 weeks: 250 mg/L cefotaxime during all steps of culture and 50 mg/L kanamycin during only the first 5 weeks of culture.

[b] C + K: both antibiotics were applied during all steps of culture.

Toxic symptoms in humans include both neurological and gastrointestinal disorders (vomiting, stomach pain, increased heart rate, and even hallucinations). Although breeders analyze all new cultivars and eliminate the ones with high values of toxic glycoalkaloids, there is always a balance between reducing tuber glycoalkaloid content and maintaining high content in the plant as a way to combat diseases and pests. Further improvement of the potato and better knowledge on these compounds are desirable, and genetic transformation has been also applied toward this goal. One useful approach is downregulation of genes coding for enzymes involved in glycoalkaloid biosynthesis. For instance, the *Sgt*1 gene (encoding sterol alkaloid glycosyltransferase) was inhibited, resulting in almost complete inhibition of α-solanine, but the biosynthesis of α-chaconine was increased instead (McCue et al. 2005). In another report, the gene GmSTM1 (for type I sterol methyltransferase) isolated from soybean was transformed into potatoes in order to study the sterol biosynthesis in relation to toxic glycoalkaloid production (Arnqvist et al. 2003). The results revealed that by downregulation of toxic glycoalkaloid biosynthesis, free nonalkylated sterols in the tubers are reduced and this sustains the involvement of cholesterol as a precursor in alkaloid biosynthesis.

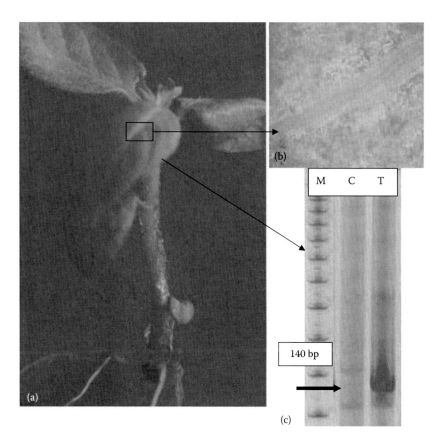

FIGURE 5.6
Genetic transformation of *Solanum chacoense* with *gfp* and *nptII* genes (cocultivation with *Agrobacterium tumefaciens* with the plasmid pHB2892): (a) putative transgenic shoot; (b) microscopic view of *gfp* expression; (c) PCR analysis of *nptII* amplified product (140 bp), M—100 bp marker, C—control, T—transgenic shoot.

Genetically modified (GM) potatoes have been developed with improved resistance against viruses or with other desirable properties. The tuber peels of late blight-resistant *S. tuberosum* cv. Désirée transgenic lines, have been found to contain twice as many glycoalkaloids compared to their conventional counterparts. However, the concentrations in the tuber flesh were nearly the same as or lower than in the conventional variety (Bianco et al. 2003). No significant increase in total glycoalkaloid content was detected in potatoes resistant to CPB or virus Y (Rogan et al. 2000). In other studies the α-solanine and

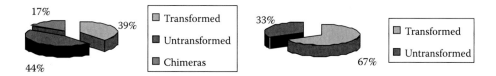

FIGURE 5.7
The efficiency of transformation for two different accession of *Solanum chacoense*, based on the epifluorescence microscopy assay of *gfp* expression: *left,* accession 2095 (not analyzed for leptines); *right,* accession HL (leptine producer).

TABLE 5.2

Results Obtained in Transforming Two Genotypes of *Solanum chacoense* with *msh2* Genes in Antisense (As) or Negative Complementary Gene (Apa) Constructs

	S. chc Apa	S. chc As	S. chc HL Apa	S. chc HL As
Total no. of plants	30	76	14	15
No. of plants forming roots on kanamycin + media	12	37	2	9
No. of plants positive for the specific gene *msh2* (PCR)	1	ND	3	5

Note: The high leptine producer (HL) genotype is compared with another genotype not characterized for leptine biosynthesis (S. chc 2095).

α-chaconine quantities determined by HPLC were significantly lower in transgenic plants than in control potatoes (Stobiecki et al. 2003; Zuk et al. 2003; Matthews et al. 2005).

Moreover, recently the beneficial effect of α-chaconine and gallic acid on human health through the inhibition of prostate cancer cells has been demonstrated *in vitro* (Reddivari et al. 2010).

Another important trait for human health is protein and rare amino acid content in the tubers. Indian researchers (Chakraborty et al. 2000) have developed a transgenic potato much richer in protein and amino acids than conventional varieties. The novel potato incorporates a gene for albumin isolated from an edible crop known as amaranth. This species with broad leaves and small seeds was a staple of the Aztecs and earlier American cultures, and by 1970 was becoming a crop in the US. The new transgenic variety has between 35% and 60% more protein than a conventional potato and higher levels of amino acids, especially lysine, tyrosine, and others, which are present in low quantities in conventional potatoes. In 2007, Hungarian researchers reported elevation of cysteine and glutathione in potato tubers by using a transgenic marker-free approach (Stiller et al. 2007).

The creation of "Golden rice" has driven the attention of scientists to other important crops for biofortification and alleviation of malnutrition. As it is a very important staple food, the potato was hence the next target for increasing its carotenoid content. The yellow and orange flesh potatoes contain higher values of carotenoids, which are well recognized as health-promoting compounds (Millam 2007). Transgenesis was first applied for improving carotenoid content in potato tubers through downregulation of a gene coding for zeaxanthin epoxidase (Romer et al. 2002). Further on, the mechanisms of carotenogenesis and its changes during storage were reported and manipulated by using the gene *crtB*, from *Erwinia uredovora*, encoding phytoene synthase (Morris et al. 2004; Ducreux et al. 2005). The real "Golden potato" has been reported recently by tuber-specific over-expression of a bacterial minipathway (Diretto et al. 2007). The minipathway was made of three genes of bacterial origin (*Erwinia*) encoding phytoene synthase (*CrtB*), phytoene desaturase (*CrtI*), and lycopene beta-cyclase (*CrtY*), under the control of tuber-specific or constitutive promoters. The best results were obtained for tuber-specific promoter-driven expression of the three genes and the total tuber carotenoid content increased 20-fold and β-carotene (provitamin A) 3600-fold. These values are the best ever reported for the potato or other crops, including "Golden rice 2," and can account for 50% of the recommended daily allowance of Vitamin A (Diretto et al. 2007, 2010). The increase of total carotenoid content was also reported (Ducreux et al. 2005), as well as tuber accumulation of astaxantin, a carotenoid with high economic value (Gerjets and Sandmann 2006).

Another compound with great value for human health is vitamin E or tocopherol, which is synthesized mainly in photosynthetic tissues of the plants. The production of this vitamin in potato tubers will be an added value for consumers. By the transfer into the potato of two different genes, isolated from the model plant *Arabidopsis thaliana*, At-HPPD (p-hydroxyphenylpyruvate dioxygenase) or At-HPT (homogentisate phytyltransferase), higher values of vitamin E in tubers were recorded. The group showed that although the synthesis of vitamin E was increased there are still metabolic constraints that do not allow better values (Crowell et al. 2008). Further research is needed in order to obtain biofortification of potato tubers with vitamin E.

The carbohydrate metabolism of the potato can be changed by either classical breeding or gene transfer (Blennow et al. 2003; Dale and Hampson 1995). The ratio of amylose to amylopectin in starch is important for its properties and applications, and can be modified. A waxy potato starch with low amylose content has been developed, and its different physicochemical properties have been examined (McPherson and Jane 1999). High amylopectin starch was also produced in a transgenic potato by downregulation of a gene that controls amylose biosynthesis, *GBSS* (Visser et al. 1991).

Inulin is a carbohydrate with high solubility in water that is naturally synthesized and accumulated in some plants, such as chicory (*Chichorium intybus*) and Jerusalem artichoke (*Helianthus tuberosus*). This compound has gained special interest in recent years by proving its health effects in humans, where it promotes gut microflora and mineral absorption in the intestine, reduces blood lipids and prevents colon cancer (Hellwege et al. 2000 and references therein). A better source of high molecular weight inulin proved to be the globe artichoke (*Cynara scolymus*), containing molecules with a chain length of up to 200. Two genes coding the enzymes known for their role in inulin biosynthesis, 1-*FFT* (fructan:fructan 1-fructosyltransferase) and 1-*SST* (sucrose:sucrose 1-fructosyltransferase), isolated from globe artichoke were transformed into the potato under the control of the constitutive promoter 35S-CaMV. The transgenic potato plants accumulated an inulin with high molecular weight, very similar to the one synthesized by globe artichoke, in the tubers (Hellwege et al. 2000). But inulin synthesis was at the expense of starch, and further research is under way to increase the carbohydrate sink of the potato.

The potato and its waste can also be a source of ethanol or biofuels. The European Commission has identified biofuels as an environmentally friendly way to ensure the security and diversity of energy supply for European Union (EU) transport (Directive 2003/30/EC). Among the different pathways available for producing bioethanol and biodiesel, the use of agriculture-derived feedstock appears to be the most feasible, ready-to-market option. The core limitation for this strategy is the availability of arable land. Reducing the losses caused by diseases, pests and abiotic stress, as a result of recent climate change, might improve yield and hence the use of arable land. In this context the use of transgenic potatoes as a source of biofuels seems feasible both economically and ethically, including consumer acceptance. Economically, potatoes and particularly waste potatoes produced 86.9 GJ/ha in 2010, with 100–110 L ethanol/t; if only 10% of waste potatoes is to be converted in bioethanol the annual production will reach 150 million liters (Ghobadian et al. 2008). Although the potato is not one of the high biomass species growing on marginal land, like *Miscanthus* or *Arundo donax*, its waste can be a good source for local production of bioethanol (Liimatainen et al. 2004). Biotechnological approaches have addressed the modification of starch content, but there are no commercialized products resulting from such research (Smith 2008). The bioethical issue has to be addressed better by the scientific community from the very beginning, but one can see no reasons for consumers to oppose the use of transgenic plants as a source of bioenergy.

5.2.4 Potato Plants as Bioreactors

Transgenic plants in general, and the potato, in particular, have been investigated as delivery systems for edible vaccine against various diseases. Edible plant vaccines have some potential advantages: a less expensive source of antigen, ease of administration, mass production and easy transport, heat-stable vaccines (do not need refrigeration), and systemic and mucosal immunity and safety, since plants do not contain animal or human viruses. In the past decade many advances have been made in the field of plant-derived vaccines (Rybick 2010). Furthermore, plants have been shown to be capable of expressing a multicomponent vaccine that when orally delivered induces a T-helper cell subset 1 response and enables passive immunization. The amount of recombinant antigens in transgenic plants ranged from 0.002% to 0.8% in total soluble protein, depending on the promoters and plants used for transformation. Increased antigen expression levels up to 4.1% total soluble protein have been obtained through transformation of the chloroplast genome (reviewed by Guan et al. 2010). Plant-derived vaccines have been proved to be valuable commodities to the world's health system. However, before their application, optimization of immunization strategies and antigen stability has to be investigated. Throughout the past decade, edible plant vaccine made notable progress. For the first time in the world, in 2006 the USDA approved the use of a plant vaccine, produced by tobacco cells in bioreactor for combating chicken Newcastle disease (http://www.aphis.usda.gov/newsroom/content/2006/01/ndvaccine.shtml). The pharmaceutical proteins can be produced in leaves (tobacco, soybean, lettuce, and *Medicago*), in the seeds of legumes or cereals, in fruits or legumes (tomato, banana, potato), or in textile- and oil-producing species such as *Linum*, cotton, and rapeseed.

The potato was one of the first plants to be transformed toward the production of edible vaccines and other therapeutic proteins and is still an important plant for such studies. Transgenic potato tubers expressing a bacterial antigen stimulated both systemic and mucosal immune responses when they were provided as food. These results sustained from the beginning the proof of concept for the use of plants to produce vaccines for both animals and humans, introduced by the same authors already in 1992 (Mason and Arntzen 1995). Transgenic potato tubers expressing alien proteins have been used for the first clinical trial. Different vaccines have been successfully produced in the potato, such as coat protein (CP) of rotavirus VP6 for acute coloenteritis, *E. coli* endotoxin, CP of Norwalk virus, rabbit hemorrhagic disease virus (RHDV) structural protein VP60, B toxin of *Vibrio cholerae*, hepatitis B surface antigen (HBsAg), papiloma virus proteins E7 and L1, viral proteins of Newcastle disease, to name but a few (see Pribylova et al. 2006; Rybick 2010; Thanavala and Lugade 2010; De Muynck et al. 2010). By using the structural protein VP1 produced in transgenic potatoes, a virus-specific antibody response to foot and mouth disease has been reported (Carrillo et al. 2001). Biologically active salmon interferon has been also produced in potato tubers and has proved its immunogenic activity in human clinical trials (Fukuzawa et al. 2010).

Since the efficiency of the production of alien proteins in plants constrains its application, in the potato a novel approach has been reported, that is, inactivation by RNAi of patatin biosynthesis (Kim et al. 2008). The *in vitro* and *ex vitro* analyzed transgenic potato plants showed a reduction by 99% of patatins in the tubers without changes in tuber growth, yield, or mean weight. The reduction of patatins is an advantage for rapid purification of other tuber-produced glycoproteins.

5.2.5 Recent Advances in Improving Resistance to Biotic Stress

The costs and losses in potato production caused by various diseases and pests can be very high, sometimes compromising the entire crop, as in the case of CPB in Eastern Europe, or of the newly discovered strain of potato virus Y, PVY[NTN], which attacks leaves and tubers, not to mention the well-known case of late blight. The utility of transferring genes involved in traits such as resistance to different biotic stresses is of great interest in potato improvement.

5.2.5.1 Breeding Resistance to Late Blight by a Transgenic Approach

Late blight caused by the oomycete *Phytophthora infestans* (*Pi*) is the most devastating disease of potatoes and other crops worldwide. All the efforts of classical breeding and modern marker-assisted selection failed to develop potato cultivars resistant to this pathogen. There are many reasons for this lack of success in breeding resistant potatoes. One major cause is late blight virulence and adaptability that made the use of R-dominant genes nondurable. For many years breeders were using the crossable wild species *S. demissum* to transfer the 11 R genes to potato cultivars. But it was soon clear that the pathogen was developing new strains that resist R gene and induce hypersensitive response (Park et al. 2009b). In the past decade a lot of new genes have been discovered in the wild relatives of the potato; some have already been sequenced and cloned (Vleeshouwers et al. 2008). The gene *Rpi-ber1* was identified and characterized from *S. berthaultii* situated on chromosome X (Ewing et al. 2000) and further on *Rpi-ber2* closely linked to the first one on the long arm of the same chromosome was found (Park et al. 2009a). From the wild species *Solanum pinnatisectum* the gene *Rpi-pnt1* was described on chromosome VII (Kuhl et al. 2001) and the genes *Rpi-mcq1* from *Solanum mochiquense* and *Rpi-phu1* from *Solanum phureja* were found on chromosome IX (Smilde et al. 2005; Śliwka et al. 2006). Today quite a large number of potato *Rpi* genes and quantitative trait loci are known for resistance to *P. infestans* (Pi-QTL) (Park et al. 2009b). The gene *Rpi-sto1* was cloned from *Solanum stoloniferum*, and *Rpi-pta1* from the species *S. papita,* both being functionally equivalent to the gene *Rpi-blb1*, characterized in *S. bulbocastanum* (*blb*). This wild Mexican diploid potato species ($2n = 2x = 24$) is highly resistant to all known races of *P. infestans*, even under intense disease pressure. *Blb* is a typical example of 1EBN (endosperm balance number) species and hence cannot be crossed directly with the potato. Limited success has been obtained by crossing through bridging with other species. Double-bridge hybrids involving *S. acaule* and *S. phureja* produced the clones ABPT that were used to release the first potato cultivar with resistance from *S. bulbocastanum*, Biogold (Van Rijn BV), in 2004 in Europe (Hermsen and Ramanna 1973; Huang 2005; Bradshaw et al. 2006). Those clones were the source for characterizing the resistance gene *R-abpt* located at a major late blight resistance locus on linkage group IV, which also harbors the *Rpi-blb-3-, -R-2-,* and *R-2*-like genes, all containing sequences characteristic of leucine zipper nucleotide binding site leucine-rich repeat (LZ-NBS-LRR) (Lokossou et al. 2009). Four somatic hybrid clones with resistance from *blb* proved useful to map a dominant resistance gene to late blight *RB* to chromosome VIII (Helgeson et al. 1998; Naess et al. 2000) and to characterize and sequence *RB* gene (Song et al. 2003). By genetic transformation *RB* was introduced in the potato, proving its capacity to confer resistance to a broad spectrum of late blight isolates (Kramer et al. 2009). The introgressions in BC1 and BC2 of *blb* DNA were followed by RAPD and RFLP markers (Naess et al. 2001). To date there are four different NBS-LRR resistance genes identified in and cloned from *blb*: *Rpi-blb1* (van der Vossen et al. 2003), also known as *RB*, and *Rpi-bt1* on chromosome VIII (Song et al. 2003; Oosumi et al. 2009); *Rpi-blb2* on linkage group VI (van der Vossen et al. 2005); and *Rpi-blb3* on linkage group IV (Lokossou et al. 2009). Nevertheless, other new factors involved in late blight resistance are suggested in recent literature (Lokossou et al. 2010). These data demonstrate the complexity of the determination of late blight immunity in *blb* populations. Moreover, recent screening of different *blb* accessions, for the presence of sequenced genes *Rpi-blb1*, *Rpi-blb2*, and *Rpi-blb3*, by gene-specific markers identified any accession possessing all these three genes. The most frequent gene was *Rpi-blb1*, sometimes combined with one of the other two genes (Thieme—personal communication). The complexity of genetic determination of resistance to late blight and continuous coevolution of pathogen and host make this trait difficult for improvement by transgenesis, but gene pyramiding through cisgenesis and gene stacking by somatic hybridization might contribute to resolving this problem (EU project with the acronym DuRPh, started in 2006, Haverkort et al. 2009; internet report 2010, Wageningen UR n.d.; Rakosy-Tican et al. 2013).

Currently, the genetically modified potato Fortuna, which was produced and tested for 6 years by BASF-Plant Science, is assessed by EFSA for its late blight resistance and safety. Fortuna is modified with two stacked genes: *Rpi-blb1* and *Rpi-blb2* (http://www.basf.com/plantscience). Our resultant BC progeny of the somatic hybrids between *S. bulbocastanum* and two potato cultivars (Delikat and Rasant) carrying two other *blb* genes, that is, *Rpi-blb1* and *Rpi-blb3*, proved to be resistant in detached leaf assay and in a field having good fertility and production. They represent another source of potato resistant to foliage blight, which does not raise concerns regarding genetic modification and is more similar to classical hybrids (Rakosy-Tican et al. 2013).

5.2.5.2 *Breeding Resistance to Viruses*

Viruses are considered as the most important parasites of the potato. From 40 virus species known to infect cultivated potatoes, some are more dangerous, that is, the potato leaf roll virus (PRLV), the potato potyviruses Y and A (PVY and PVA), the carlaviruses (PVM and PVS), and the PVX, a potexvirus. Although there is variation between different potato-cultivating areas in the losses caused by virus infection and propagation via seed tubers, the best solution to reduce such loss is to develop resistant cultivars (Valkonen 2007). For many years the transgenic approach was using viral CP genes to induce virus resistance in plants. The first reports were on tobacco, where TMV CP genes were used (Powell Abel et al. 1986), followed by many other viruses in other plant species (Lomonossoff 1995), including resistance to PVY and PVX in potatoes (Lawson et al. 1990). Later, sense and antisense resistance to PRLV was engineered in the potato (Kawchuk et al. 1991). The antisense gene of the PVY CP was also proved to induce virus resistance (Lindbo et al. 1993; Smith et al. 1995). A better strategy was developed later by using RNA-induced silencing (Missiou et al. 2004). PVY is a single-stranded RNA virus, which replicates in the host cell by double-stranded (ds) RNA intermediates. There are four different isolates of PVY that cause great losses in potato crops: PVY^0, PVY^N, PVY^{NTN}, and PVY^NW. Disease symptoms vary depending on the virus isolate and potato cultivar, but also on some environmental factors such as temperature. The use of posttranscriptional gene silencing (PTGS) to induce PVY resistance was developed based on the natural mechanism of plant cells to combat viruses and transposons by mRNA silencing (Voinnet 2001). This strategy involves CP hairpin constructs where two CP sequences in antisense orientation are separated by an intron. Those structures generate dsRNA with intron loop that are recognized by Dicer RNase and produce siRNAs (small interfering RNAs). Being complementary to PVY CP gene, these siRNAs will induce PTGS and hence virus resistance in transgenic potatoes (Missiou et al. 2004). The same biotechnology proved efficient in engineering resistance to PVY in other potato cultivars in a marker-free transformation system, by using either a shooter mutant of *Agrobacterium tumefaciens* (Bukovinszki et al. 2007) or a first step based on the reporter gene *gfp*, in our experiments (Rakosy-Tican et al. 2010, see Section 5.2.1). We also proposed that such transgenic cultivars might be better accepted by consumers for bioethanol production.

5.2.5.3 Breeding Resistance to Colorado Potato Beetle

Besides late blight and viruses there are other diseases and pests that cause great losses in potato production. The colorado potato beetle (CPB) is notorious not only for the loss it causes, mainly in northern latitudes, but also for its adaptability. Worldwide measures against CPB involve the application of toxic chemical insecticides, which have important drawbacks such as the development, under severe selection pressure, of resistance to more than 50 insecticides used worldwide (Whalon et al. 2007), pollution, and negative effects on humans. Genetic engineering has developed the use of genes from *Bacillus thuringiensis* (*Bt*) to combat insect pests in different crops. Consequently, the transgenic *Bt* potato was also produced (Adang et al. 1993; Perlak et al. 1993). In 1995 Monsanto first launched, through its subsidiary, NatureMark, *Bt cry3A* potatoes, also known by the Newleaf® trade name (Grafius and Douches 2008). In the past decade better results have been obtained by assaying the resistance induced by the genes for *Bt* prototoxins (*cry3A* or *cry1Ia1*), or by natural resistance mechanisms such as the biosynthesis of leptine glycoalkaloids or the presence of glandular trichomes in a field no-choice test (Cooper et al. 2007). These results reveal that *Bt* genes provide a high degree of resistance to CPB while the natural mechanisms are less efficient in a no-choice situation. A similar comparison has been done for both CPB and leaf hopper resistance (Ghidiu et al. 2011). In this field evaluation, the *Bt* potato with the genes *cry3A* significantly reduced feeding by CPB since glandular trichomes prove efficient against leafhopper, but none of these mechanisms was effective for both pests. Combining engineered *Bt* resistance with natural mechanisms such as high leptines and glandular thrichomes did not improve the resistance to CPB, as each measure worked better alone (Coombs et al. 2003). Another approach was to combine avidin, a protein having insecticidal properties through its potential to sequester biotin, an essential protein for insect development and growth, with other engineered (*Bt*) or natural (leptine) host resistances (Cooper et al. 2006). The best results were obtained when avidin was applied together with *Bt-cry3A*, showing the toxic effects on different instars of CPB. Nevertheless, the *Bt* potato is the only single technology assuring 100% protection against insect pests in the potato crop, but consumer acceptance of GM plants still limits the application of this biotechnological tool.

5.2.5.4 Breeding Resistance to Other Diseases and Pests

The nematodes are also important pests of the potato, particularly in regions such as the UK and tropical countries. Control of nematode pests has proved very difficult, and different strategies have been evaluated including the transgenic approach. Recently good results have been obtained by transgenesis in controlling the cyst nematode *Globodera pallida*. The trait that was triggered by genetic transformation was suppressing root invasion. A construct carrying the gene for the peptide nAChRbp, which disrupts chemoreception of the host plant by the nematode parasite, driven by a root tip specific promoter isolated from *Arabidopsis thaliana*, MDK4–20 was transformed into potato crops. The resistant plants proved to be safe for the soil organisms, including other beneficial nematode species (Green et al. 2012).

In order to combat other diseases and pests, general mechanisms involved in pathogenesis have been also considered, such as the use of small proteins involved in general defense against bacteria and fungi. The overexpression of the gene *snakin*-1 (SN1) in the potato proved to be effective against both the pathogen *Rhizoctonia solani* and the fungus *Erwinia carotovora* (Almasia et al. 2008).

5.2.6 Advances in Improving Resistance to Abiotic Stress

Climate change has opened new challenges for potato improvement as for many other crops. Because of intensive management, potato yield was not under pressure of abiotic stresses such as drought, high temperatures, frost, or salty soils. As a consequence, potato production is a heavy user of groundwater for irrigation during the summer. Since water use by the industry is becoming an issue for the public and legislatures, better understanding and manipulation of resistance to abiotic stress should be addressed. Genetic transformation is one way toward this goal. Recent research suggests that it is now possible to use transgenic approaches to improve abiotic stress tolerance with fewer genes, as previously anticipated (Zhang et al. 2000). Far beyond the initial attempts to insert single genes, engineering of the regulatory genetic machinery involving transcription factors has emerged as a new tool for controlling the expression of many stress-responsive genes. One such strategy is to express heterologous C-repeat/dehydration-responsive element binding factor 1 (*CBF1*) genes in plants (Thomashow 1999). These genes are known as master switches for the expression of COR genes, increasing stress tolerance without cold stimulation. It was first reported that transgenic tomato expressing the heterologous *Arabidopsis* *CBF1* gene shows drought resistance (Hsieh et al. 2002). Further on, the same strategy was used in the potato by overexpressing *CBF1* gene driven to cytosol by an abscisic acid (ABA) inducible promoter. A number of transgenic clones were produced and evaluated in greenhouse and field tests, some of which show resistance to water deficit and are expected to have both cold and salt stress tolerance (Douches and Kirk—fieldcrop.msu. edu). Transcription factors such as *CBF1* that control COR genes are one example of a transgenic approach that improves environmental stresses in plants.

5.3 Improving Potato Crop by Cisgenesis

Kaare M. Nielsen from Tromsø University Norway (2003) proposed a distinction between plant genetic modification with alien genes, which can be considered transgenesis, and with a plant's own genes, which can be defined as cisgenesis (Nielsen 2003; Rommens et al. 2004; Schouten et al. 2006a,b). Cisgenesis was also proposed and applied to potato crops (Jacobsen and Schouten 2007). This technology has great potential for the potato, a crop that benefits from a great wealth of resistance genes present in more than 226 wild relatives growing in South or Central America, which is the center of origin of *S. tuberosum* subsp. *tuberosum*. Cisgenes are natural genes isolated from the same species or related interbreeding species. The transfer of the cisgene takes place only in one step, contrary to introgression of transgenes where more steps are needed to transfer more genes, including sometimes marker or reporter genes. As for the transgenes, the integration into the recipient genome cannot yet be targeted (Jacobsen and Schouten 2007). A very good example of potato cisgenesis is the isolation and transfer of resistance genes to *Phytophthtora infestans*, such as *Rpi-blb1* (RB), *Rpi-blb2*, or *Rpi-blb3*, from the immune wild species *S. bulbocastanum* into the potato crop genome (Jacobsen and Schouten 2009).

The intragenic approach, which means using only constructs made of plant's own sequences, was successfully applied to silence by tuber-driven expression the genes involved in browning processes or starch degradation (Rommens et al. 2007).

The achievements of cisgenesis will only be of practical utility if EU and member state legislation accepts the transfer of plant genes from interbreeding species as nongenetically modified organism (GMO).

5.4 A Few Remarks on Public Acceptance of GM Potatoes

Social and bioethical concerns in relation to potato transgenesis are following the general trend of public concerns and hot debate that affects practical application of very promising results. The scientific community is trying hard to improve the technology for better acceptance. Issues such as marker-free transgenic potato crops, better targeting of the transgenes, and their stable integration in the genome or tissue-specific expression are new achievements in the field and together with cisgenesis and combinatorial biotechnology, meaning the use of more biotechnological tools for improvement and sustainability of potato or other crops, will, it is hoped, contribute to a better acceptance of genetically improved potato crops.

5.5 Funding

The financial support of a grant of the Romanian Authority for Scientific Research, CNCS-UEFISCDI project number PNII-ID-PCE-2011-3-0586, is acknowledged.

References

Adang MJ, Brody MS, Cardineau G, Eagan N, Roush RT, Shewmaker C, Jones A, Oakes J, McBride K (1993) The reconstruction and expression of a *Bacillus thuringiensis cryIIIA* gene in protoplasts and potato plants. *Plant Mol Biol* 21: 1131–1145.

Almasia NI, Bazzini AA, Hopp HE, Vasquez-Rovere C (2008) Overexpression of *snakin*-1 gene enhances resistance to Rhizoctonia solani and Erwinia carotovora in transgenic potato plants. *Mol Plant Pathol* 9(3): 329–338.

Arnqvist L, Dutta PC, Jonsson L, Sitbon F (2003) Reduction of cholesterol and glycoalkaloid levels in transgenic potato plants by overexpression of a Type 1 Sterol Methyltransferase cDNA. *Plant Physiol* 131: 1792–1799.

Bianco G, Schmitt-Kopplin P, Crescenzi A, Comes S, Kettrup A, Cataldi TR (2003) Evaluation of gly-coalkaloids in tubers of genetically modified virus Y-resistant potato plants (var. Desiree) by non-aqueous capillary electrophoresis coupled with electrospray ionization mass spectrometry (NACE-ESI-MS). *Anal Bioanal Chem* 375: 799–804.

Blennow A, Hansen M, Schulz A, Jorgensen K, Donald AM, Sanderson J (2003) The molecular depo-sition of transgenically modified starch in the starch granule as imaged by functional micros-copy. *J Struct Biol* 143: 229–241.

Boularte-Medina T, Fogelman E, Chani E, Miller AR, Levin I, Levy D, Veilleux RE (2002) Identification of molecular markers associated with leptines in reciprocal backcross families of diploid potato. *Theor Appl Genet* 105: 1010–1018.

Bradshaw JE, Bryan GJ, Ramsay G (2006) Genetic resources (including wild and cultivated *Solanum* species) and progress in their utilisation in potato breeding. *Potato Res* 49: 49–65.

Bukovinszki A, Diveki Z, Csanyi M, Palkovics L, Balazs E (2007) Engineering resistance to PVY in different potato cultivars in a marker-free transformation system using a 'shooter mutant' *A. tumefaciens. Plant Cell Rep* 26: 459–465.

Carrillo C, Wigdorovitz A, Trono K, Dus Santos MJ, Castañón S, Sadir AM, Ordas R, Escribano JM, Borca MV (2001) Induction of a virus-specific antibody response to foot and mouth disease virus using the structural protein VP1 expressed in transgenic potato plants. *Viral Immunol* 14(1): 49–57.

Chakraborty S, Chakraborty N, Datta A (2000) Increased nutritive value of transgenic potato by express-ing a nonallergenic seed albumin gene from *Amaranthus hypochondriacus. PNAS* 97(7): 3724–3729.

Chalfie M, Tu Y, Euskirchen G, Ward WW, Prasher DC (1994) Green fluorescent protein as a marker for gene expression. *Science* 263: 802–805.

Christou P, McCabe DE (1992) Prediction of germ-line transformation events in chimeric R_0 trans-genic soybean plantlets using tissue-specific expression patterns. *Plant J* 2: 283–290.

Christou P, Capell T, Kohli A, Gatehouse JA, Gatehouse AMR (2006) Recent developments and future prospects in insect pest control in transgenic crops. *Trends Plant Sci* 11(6): 302–308.

Coombs JJ, Douches DS, Li W, Grafius EJ, Pett WL (2003) Field evaluation of natural, engineered and combined resistance mechanisms in potato for control of Colorado potato beetle. *J Amer Soc Hort Sci* 128(2): 219–224.

Cooper SG, Douches DS, Grafius EJ (2006) Insecticidal activity of avidin combined with genetically engineered and traditional host plant resistance against Colorado potato beetle (Coleoptera: Chrysomelidae) Larvae. *J Econ Entomol* 99(2): 527–536.

Cooper SG, Douches DS, Coombs JJ, Grafius EJ (2007) Evaluation of natural and engineered resis-tance mechanisms in potato against Colorado potato beetle in a no-choice field study. *J Econ Entomol* 100(2): 573–579.

Crowell EF, McGrath JM, Douches DS (2008) Accumulation of vitamin E in potato (Solanum tuberosum) tubers. *Transgenic Res* 17: 205–217.

Culligan KM, Hays JB (1997) DNA mismatch repair in plants—An *Arabidopsis thaliana* gene that pre-dicts a protein belonging to the MSH2 subfamily of eukaryotic MutS homologs. *Plant Physiol* 115: 833–839.

Dale PJ, Hampson KK (1995) An assessment of morphogenic and transformation efficiency in a range of varieties of potato (*Solanum tuberosum* L.). *Euphytica* 85: 101–108.

De Muynck B, Navarre C, Boutry M (2010) Production of antibodies in plants: Status after twenty years. *Plant Biotechnol J* 8: 529–563.

De Vetten N, Wolters AM, Raemakers K, Van Der Meer I, Ter Stege R, Heeres E, Heeres P, Visser R (2003) A transformation method for obtaining marker-free plants of a cross-pollinating and vegetatively propagated crop. *Nat Biotechnol* 21: 439–442.

Dietze J, Blau A, Willmitzer L (1995) *Agrobacterium*-mediated transformation of potato (*Solanum tuberosum*). In: Potrykus I, Spangenberg G (eds), *Gene Transfer to Plants*. Berlin: Springer-Verlag, pp. 24–29.

Diretto G, Al-Babili S, Tavazza R, Papacchioli V, Beyer P, Giuliano G (2007) Metabolic engineering of potato carotenoid content through tuber-specific overexpression of a bacterial minipathway. *PLoS ONE* 2: e350.

Diretto G, Al-Babili S, Tavazza R, Scossa F, Papacchioli V, Migliore M, Beyer P, Giuliano G (2010) Transcriptional-metabolic networks in b-carotene-enriched potato tubers: The long and winding road to the golden phenotype. *Plant Physiol* 154: 899–912.

Domìnguez A, Cervera M, Pérez RM, Romero J, Fagoaga C, Cubero J, López MM, Juz JA, Navaro L, Peña L (2004) Characterisation of regenerants obtained under selective conditions after *Agrobacterium*-mediated transformation of citrus explants reveals production of silenced and chimeric plants at unexpected high frequencies. *Mol Breed* 14: 171–183.

Douches DS, Kirk WW, Engineered and conventional approaches to develop potatoes for sub-optimal irrigation conditions. Project GREEEN No.: GR03-011. fieldcrop.msu.edu/uploads/documents/GR03-011.pdf (last accessed December 2012).

Ducreux LJ, Morris WL, Hedley PE, Shepherd T, Davies HV, Millam S, Taylor MA (2005) Metabolic engineering of high carotenoid potato tubers containing enhanced levels of beta-carotene and lutein. *J Exp Bot* 56: 81–89.

Ewing EE, Simko I, Smart CD, Bonierbale MW, Mizubuti ESG, May GD, Fry WE (2000) Genetic mapping from field of qualitative and quantitative resistance to *Phytophthora infestans* in a population derived from *Solanum tuberosum* and *Solanum berthaultii*. *Mol Breed* 6: 25–36.

Fukuzawa N, Tabayashi N, Okinaka Y, Furusawa R, Furuta K, Kagaya U, Matsumura T (2010) Production of biologically active Atlantic salmon interferon in transgenic potato and rice plants. *J Biosci Bioeng* 110(2): 201–207.

Gerjets T, Sandmann G (2006) Ketocarotenoid formation in transgenic potato. *J Exp Bot* 57: 3639–3645.

Ghidiu GM, Douches DS, Felcher KJ, Coombs JJ (2011) Comparing host plant resistance, engineered resistance and insecticide treatment for control of Colorado potato beetle and potato leafhopper in potatoes. *Int J Agron* 2011: 4. doi:10.1155/2011/390409.

Ghobadian B, Rahimi H, Tavakkoli Hashjin T, Khatamifar M (2008) Production of bioethanol and sunflower methyl ester and investigation of fuel blend properties. *J Agric Sci Technol* 10: 225–232.

Ghorbel R, Juárez J, Navarro L, Peña L (1999) Green fluorescent protein as a screenable marker to increase the efficiency of generating transgenic woody fruit plants. *Theor Appl Genet* 99: 350–358.

Grafius EJ, Douches DS (2008) The present and future role of insect-resistant genetically modified potato cultivars in IPM. In: Romeis J, Shelton AM, Kennedy GG (eds), *Integration of Insect-Resistant Genetically Modified Crops within IPM Programs*. New York, NY: Springer, pp. 195–221.

Green J, Wang D, Lilley CJ, Urwin PE, Atkinson HJ (2012) Transgenic potatoes for potato cyst nematode control can replace pesticide use without impact on soil quality. *PLoS ONE* 7(2): e30973.

Guan ZJ, Guo B, Huo YL, Guan ZP, Wei YH (2010) Overview of expression of hepatitis B surface antigen in transgenic plants. *Vaccine* 28(46): 7351–7362.

Haverkort J, Struik PC, Visser RGF, Jacobsen E (2009) Applied biotechnology to combat late blight in potato caused by *Phytophthora infestans*. *Potato Res* 52(3): 249–264.

Heeres P, Schippers-Rozenboom M, Jacobsen E, Visser RGF (2002) Transformation of a large number of potato varieties: Genotype-dependent variation in efficiency and somaclonal variability. *Euphytica* 124: 13–22.

Helgeson JP, Pohlman JD, Austin S, Haberlach GT, Wielgus SM, Ronis D, Zambolim L, et al. (1998) Somatic hybrids between *Solanum bulbocastanum* and potato: A new source of resistance to late blight. *Theor Appl Genet* 96: 738–742.

Hellwege EM, Czapla S, Jahnke A, Willmitzer L, Heyer AG (2000) Transgenic potato (*Solanum tuberosum*) tubers synthesize the full spectrum of inulin molecules naturally occurring in globe artichoke (*Cynara scolymus*) roots. *PNAS* 97(15): 8699–8704.

Hermsen JGT, Ramanna MS (1973) Double-bridge hybrids of *Solanum bulbocastanum* and cultivars of *Solanum tuberosum*. *Euphytica* 22: 457–466.

Hsieh T-H, Lee J-T, Charng YY, Chan M-T (2002) Tomato plants ectopically expressing *Arabidopsis CBF1* show enhanced resistance to water deficit stress. *Plant Physiol* 130: 618–626.

Huang S (2005) The discovery and characterization of the major late blight resistance complex in potato: Genomic structure, functional diversity and implications. PhD Thesis Wageningen University, Wageningen.

Jacobsen E, Schouten HJ (2007) Cisgenesis strongly improves introgression breeding and induced translocation breeding of plants. *Trends Biotechnol* 25: 219–223.

Jacobsen E, Schouten HJ (2009) Cisgenesis: An important sub-invention for traditional plant breeding companies. *Euphytica* 170: 235–247.

Jefferson RA, Ravanagh TA, Bevan MW (1987) GUS fusion: β-glucuronidase as a sensitive and versatile gene fusion marker in higher plants. *EMBO J* 6: 3901–3907.

Kawchuk LM, Martin RR, McPherson J (1991) Sense and antisense RNA-mediated resistance to potato leafroll virus in Russet Burbank potato plants. *Mol Plant Microbe Inter* 4: 247–253.

Kim Y-S, Lee Y-H, Kim H-S, Kim M-S, Hahn K-W, Ko J-H, Joung H, Jeon J-H (2008) Development of patatin knockdown potato tubers using RNA interference (RNAi) technology, for the production of human-therapeutic glycoproteins. *BMC Biotechnol* 8: 36.

Kramer LC, Choudoir MJ, Wielgus SM, Bhaskar PB, Jiang J (2009) Correlation between transcript abundance of the RB gene and the level of the RB-mediated late blight resistance in potato. *Mol Plant Microbe Interact* 22: 447–455.

Kuhl JC, Hanneman RE, Havey MJ (2001) Characterization and mapping of Rpi1, a late blight resistance locus from diploid (1EBN) Mexican Solanum pinnatisectum. *Mol Genet Genomics* 265: 977–985.

Kumar A (1995) *Agrobacterium*-mediated transformation of potato genotypes. In: Gartland KMA, Davey MR (eds), *Methods in Molecular Biology*, vol 44. New York, Totowa: Humana Press, pp. 121–128.

Lafta AM, Lorenzen JH (2000) Influence of high temperature and reduced irradiance on glycoalkaloid levels in potato leaves. *J Am Soc Hortic Sci* 125: 563–566.

Lawson C, Kaniewski W, Haley L, Rozman R, Newell C, Sanders P, Tumer NE (1990) Engineered resistance to mixed virus infection in a commercial potato cultivar: Resistance to potato virus X and potato virus Y in transgenic Russet Burbank. *Bio Tech* 8: 127–134.

Lester GE (2006) Environmental regulation of human health nutrients (ascorbic acid, beta-carotene, and folic acid) in fruits and vegetables. *Hort Sci* 41(1): 59–65.

Liimatainen H, Kuokkanen T, Kääriäinen J (2004) Development of bio-ethanol production from waste potatoes. In: Pongrácz E (ed.), *Proceedings of the Waste Minimization and Resources Use Optimization Conference*. June 10th 2004, (University of Oulu, Finland) Oulu: Oulu University Press, pp. 123–129.

Lindbo JA, Silva-Rosales L, Proebsting WM, Dougherty WG (1993) Induction of a highly specific antiviral state in transgenic plants: Implications for regulation of gene expression and virus resistance. *Plant Cell* 5: 1749–1759.

Lokossou AA, Park T-H, van Arkel G, Arens M, Ruyter-Spira C, Morales J, Whisson SC, et al. (2009) Exploiting knowledge of R/Avr genes to rapidly clone a new LZ-NBS-LRR family of late blight resistance genes from potato linkage group IV. *MPMI* 22(6): 630–641.

Lokossou AA, Rietman H, Wang M, Krenek P, van der Schoot H, Henken B, Hoekstra R, Vleeshouwers VGAA, van der Vossen EAG (2010) Diversity, distribution, and evolution of Solanum bulbocastanum late blight resistance genes. *Mol Plant Microbe Inter* 23(9): 1206–1216.

Lomonossoff GP (1995) Pathogen-derived resistance to plant viruses. *Ann Rev Plant Pathol* 33: 323–343.

Marsischky GT, Filosi N, Kane MF, Kolodner R (1996) Redundancy of *Saccharomyces cerevisiae* MSH3 and MSH6 in MSH2-dependent mismatch repair. *Genes Dev* 10: 407–420.

Mason HS, Arntzen CJ (1995) Transgenic plants as vaccine production systems. *TIBTECH* 13: 388–392.

Matthews D, Jones H, Gans P, Coates S, Smith LM (2005) Toxic secondary metabolite production in genetically modified potatoes in response to stress. *J Agric Food Chem* 53: 7766–7776.

McCue KF, Shepherd LVT, Allen PV, Maccree MM, Rockhold DR, Corsini DL, Davies HV, Belknap WR (2005) Metabolic compensation of steroidal glycoalkaloid biosynthesis in transgenic potato tubers: Using reverse genetics to confirm the in vivo enzyme function of a steroidal alkaloid galactosyltransferase. *Plant Sci* 168: 267–273.

McPherson AE, Jane J (1999) Comparison of waxy potato with other root and tuber starches. *Carbohyd Polym* 40: 57–70.

Mercuri A, Sacchetti A, De Benedetti A, Schiva T, Alberti S (2001) Green fluorescent flowers. *Plant Sci* 161: 961–968.

Millam S (2007) Developments in transgenic biology and the genetic engineering of useful traits. In: Vreugdenhil D (ed.), *Potato Biology and Biotechnology: Advances and Perspectives*. Oxford: Elsevier, pp. 669–686.

Missiou A, Kalantidis K, Boutla A, Tzortzakaki S, Tabler M, Tsagris M (2004) Generation of transgenic potato plants highly resistant to potato virus Y (PVY) through RNA silencing. *Molec Breeding* 14: 185–197.

Modrich P, Lahue R (1996) Mismatch repair in replication, fidelity, genetic recombination and cancer biology. *Annu Rev Biochem* 65: 101–133.

Molinier J, Himber C, Hahne G (2000) Use of green fluorescent protein for detection of transformed shoots and homozygous offspring. *Plant Cell Rept* 19: 219–223.

Morris WL, Ducreux L, Griffiths DW, Stewart D, Davies HV, Taylor MA (2004) Carotenogenesis during tuber development and storage in potato. *J Exp Bot* 55: 975–982.

Mullins E, Milbourne D, Petti C, Doyle-Prestwich BM, Meade C (2006) Potato in the age of biotechnology. *Trends Plant Sci* 11(5): 254–260.

Naess SK, Bradeen JM, Wielgus SM, Haberlach GT, McGrath JM, Helgeson JP (2000) Resistance to late blight in *Solanum bulbocastanum* is mapped to chromosome 8. *Theor Appl Genet* 101: 697–704.

Naess SK, Bradeen JM, Wielgus SM, Haberlach GT, McGrath JM, Helgeson JP (2001) Analysis of the introgression of *Solanum bulbocastanum* DNA into potato breeding lines. *Mol Genet Genomics* 265: 694–704.

Nielsen KM (2003) Transgenic organisms: Time for conceptual diversification? *Nat Biotechnol* 2: 227–228.

Olivares-Fuster O, Pena L, Duran-Vila N, Navarro L (2002) Green fluorescent protein as a visual marker in somatic hybridization. *Ann Bot* 89: 491–497.

Oosumi T, Rockhold D, Maccree M, Deahl K, McCue K, Belknap W (2009) Gene Rpi-bt1 from *Solanum bulbocastanum* confers resistance to late blight in transgenic potatoes. *Amer J Potato Res* 86: 456–465.

Ow DW, Wood KV, DeLuca M, de Wet JR, Helinski DR, Howell SH (1986) Transient and stable expression of the firefly luciferase gene in plant cell and transgenic plants. *Science* 234: 856–859.

Park TH, Foster S, Brigneti G, Jones JDG (2009a) Two distinct potato late blight resistance genes from *Solanum berthaultii* are located on chromosome 10. *Euphytica* 165: 269–278.

Park TH, Vleeshouwers VGAA, Jacobsen E, Van der Vossen E, Visser RGF (2009b) Molecular breeding for resistance to *Phytophthora infestans* (Mont.) de Bary in potato (*Solanum tuberosum* L.): A perspective of cisgenesis. *Plant Breeding* 128: 109–117.

Perlak FJ, Stone TB, Muskopf YM, Petersen LJ, Parker GB, McPherson SA, Wyman J, et al. (1993) Genetically improved potatoes: Protection from damage by Colorado potato beetles. *Plant Mol Biol* 22: 313–321.

Powell Abel P, Nelson RS, De B, Hoffmann N, Rogers SG, Fraley RT, Beachy RN (1986) Delay of disease development in transgenic plants that express the tobacco mosaic virus coat protein gene. *Science* 232: 738–743.

Pribylova R, Pavlik I, Bartos M (2006) Genetically modified potato plants in nutrition and prevention of diseases in humans and animals: A review. *Veterinarni Medicina* 51(5): 212–223.

Rakosy-Tican E, Aurori CM, Dijkstra C, Maior MC (2010) Generating marker free transgenic potato cultivars with an hairpin construct of PVY coat protein. *Rom Biotech Lett* 15(1suppl): 63–71.

Rakosy-Tican E, Aurori CM, Aurori A (2011) The effects of cefotaxime and silver thiosulphate on in vitro culture of *Solanum chacoense*. *Rom Biotech Lett* 16(4): 6369–6377.

Rakosy-Tican L, Aurori A, Aurori CM, Ispas G, Famelaer I (2004) Transformation of wild *Solanum* species resistant to late blight by using reporter gene *gfp* and *msh2* genes. *Plant Breeding Seed Sci* (*Warszawa*) 50: 119–128.

Rakosy-Tican L, Aurori CM, Dijkstra C, Thieme R, Aurori A, Davey MR (2007) The usefulness of reporter gene *gfp* for monitoring *Agrobacterium*-mediated transformation of potato dihaploid and tetraploid genotypes. *Plant Cell Rep* 26: 661–671.

Rakosy-Tican E, Thieme R, Nachtigall M, Hammann T, Schubert J (2013) Somatic hybridization between potato commercial varieties and the incongruent species *Solanum bulbocastanum*, II: Stacking two late blight resistance genes *Rpi-blb1* and *Rpi-blb3* into potato gene pool. *Plant Cell Rep* (in press).

Reddivari L, Vanamala J, Safe SH, Miller Jr. JC (2010) The bioactive compounds in potato extracts decrease survival and induce apoptosis in LNCaP and PC3 prostate cancer cells. *Nutr Cancer* 62(5): 601–610.

Rogan GJ, Bookout JT, Duncann DR, Fuchs RL, Lavrik PB, Love SL, Mueth M, et al. (2000) Compositional analysis of tubers from insect and virus resistant potato plants. *J Agric Food Chem* 48: 5936–5945.

Romer S, Lubeck J, Kauder F, Steiger S, Adomat C, Sandmann G (2002) Genetic engineering of a zeaxanthin-rich potato by antisense inactivation and co-suppression of carotenoid epoxidation. *Metab Eng* 4: 263–272.

Rommens CM, Humara JM, Ye J, Yan H, Richael C, Zhang L, Perry R, Swords K (2004) Crop improvement through modification of the plant's own genome. *Plant Physiol* 135: 421–431.

Rommens CM, Harings MA, Swords K, Davies HV, Belknap WR (2007) The intragenic approach as a new extension to traditional plant breeding. *Trends Plant Sci* 12: 397–403.

Rybick EP (2010) Plant-made vaccines for humans and animals. *Plant Biotechnol J* 8: 620–637.

Schouten HJ, Krens FA, Jacobsen E (2006a) Do cisgenic plants warrant less stringent oversight? *Nat Biotechnol* 24: 753.

Schouten HJ, Krens FA, Jacobsen E (2006b) Cisgenic plants are similar to traditionally bred plants. *EMBO Reports* 7: 750–753.

Sheerman S, Bevan MW (1988) A rapid transformation method for *Solanum tuberosum* using binary *Agrobacterium tumefaciens* vectors. *Plant Cell Rep* 7: 13–16.

Sidorov VA, Kasten D, Pang SZ, Hajdukiewicz PTJ, Staub JM, Nehra NS (1999) Stable chloroplast transformation in potato: Use of green fluorescent protein as a plastid marker. *Plant J* 19: 209–216.

Śliwka, J, Jakuczun H, Lebecka R, Marczewski W, Gebhardt C, Zimnoch-Guzowska E (2006) The novel, major locus *Rpi-phu1* for late blight resistance maps to potato chromosome IX and is not correlated with long vegetation period. *Theor Appl Genet* 113: 685–695.

Smilde WD, Brigneti G, Jagger L, Perkins S, Jones JD (2005) *Solanum mochiquense* chromosome IX carries a novel late blight resistance gene *Rpi-moc1*. *Theor Appl Genet* 110: 252–258.

Smith AM (2008) Prospects for increasing starch and sucrose yields for bioethanol production. *Plant J* 54: 546–558.

Smith HA, Powers H, Swaney SL, Brown C, Dougherty WG (1995) Transgenic potato virus Y resistance in potato: Evidence for an RNA-mediated cellular response. *Phytopathol* 85: 864–870.

Solomon-Blackburn RM, Barker H (2001) Breeding virus resistant potatoes (*Solanum tuberosum*): A review of traditional and molecular approaches. *Heredity* 86(1): 17–35.

Song J, Bradeen JM, Naess SK, Raasch JA, Wielgus SM, Haberlach GT, Liu J, et al. (2003) Gene *RB* cloned from *Solanum bulbocastanum* confers broad spectrum resistance to potato late blight. *Proc Natl Acad Sci USA* 100: 9128–9133.

Stiekema WJ, Heidekamp F, Louwerse JD, Verhoeven HA, Dijkhuis P (1988) Introduction of foreign genes into potato cultivars Bintje and DÃÅesiree using an *Agrobacterium tumefaciens* binary vector. *Plant Cell Rep* 7: 47–50.

Stiller I, Dancs G, Hesse H, Hoefgen R, Banfalvi Z (2007) Improving the nutritive value of tubers: Elevation of cysteine and glutathione contents in the potato cultivar White Lady by marker-free transformation. *J Biotech* 128: 335–343.

Stobiecki M, Matysiak-Kata I, Franski R, Skaa J, Szopa J (2003) Monitoring changes in anthocyanin and steroid alkaloid glycoside content in lines of transgenic potato plants using liquid chromatography/mass spectrometry. *Phytochemistry* 62: 959–969.

Stushnoff C, Holm D, Thompson MD, Jiang W, Thompson HJ, Joyce NI, Wilson P (2008) Antioxidant properties of cultivars and selections from the Colorado potato breeding program. *Am J Pot Res* 85: 267–276.

Thanavala Y, Lugade AA (2010) Oral transgenic plant-based vaccine for hepatitis B. *Immunol Res* 46(1–3): 4–11.

The Potato Genome Consortium (2011) Genome sequence and analysis of the tuber crop potato. *Nature* 475:189–195.

Thomashow MF (1999) Plant cold acclimation: Freezing tolerance genes and regulatory mechanisms. *Annu Rev Plant Physiol Plant Mol Bio* 50: 571–599.

Valkonen JPT (2007) Viruses: Economical losses and biotechnological potential. In: Vreugdenhil D (ed.), *Potato Biology and Biotechnology Advances and Perspectives*. Oxford: Elsevier.

van der Vossen EAG, Sikkema A, te Lintel Hekkert B, Gros J, Stevens P, Muskens M, Wouters D, Pereira A, Stiekema W, Allefs S (2003) An ancient *R* gene from the wild potato species *Solanum bulbocastanum* confers broad-spectrum resistance to *Phytophthora infestans* in cultivated potato and tomato. *Plant J* 36: 867–882.

van der Vossen EAG, Gros J, Sikkema A, Muskens M, Wouters D, Wolters P, Pereira A, Allefs S (2005) The *Rpi-blb2* gene from *Solanum bulbocastanum* is an *Mi-1* gene homolog conferring broad spectrum late blight resistance in potato. *Plant J* 44: 208–222.

Visser R (2003) A transformation method for obtaining marker-free plants of a cross-pollinating and vegetatively propagated crop. *Nat Biotechnol* 21: 439–442.

Visser RGF, Somhorst I, Kuipers GJ, Ruys NJ, Feenstra WJ, Jacobsen E (1991) Inhibition of the expression of the gene for granule-bound starch synthase in potato by antisense contructs. *Mol Gen Genet* 225: 289–296.

Vleeshouwers VGAA, Rietman H, Krenek P, Champouret N, Young C, Oh SK, Wang MQ, et al. (2008) Effector genomics accelerates discovery and functional profiling of potato disease resistance and *Phytophthora Infestans* avirulence genes. *PLoS ONE* 3(8): e2875.

Voinnet O (2001) RNA silencing as a plant immune system against viruses. *Trends Genet* 17: 449–459.

Wageningen UR (n.d.) Durable resistance against *Phytophthora* through cisgenic marker-free modification DuRPh. Contact AJ Haverkort (research project). www.durph.wur.nl (last accessed November 2012).

Whalon ME, Mota-Sanchez D, Hollingworth RM (2007) Global database for insect resistance. Entomol Res 37(suppl 1): A11–A73.

Zhang CL, Chen DF, McCormac AC, Scott NW, Elliot MC, Slater A (2001) Use of the GFP reporter as a vital marker for *Agrobacterium*-mediated transformation of suger beet (*Beta vulgaris* L.). *Mol Biotechnol* 17: 109–117.

Zhang J, Klueva NY, Wang Z, Wu R, Ho TD, Nguyen HT (2000) Genetic engineering for abiotic stress resistance in crop plants. *In Vitro Cell Dev Biol Plant* 36: 108–114.

Zuk M, Prescha A, Kepczynski J, Szopa J (2003) ADP ribosylation factor regulates metabolism and antioxidant capacity of transgenic potato tubers. *J Agric Food Chem* 51: 288–294.

www.basf.com/plantscience. A new weapon against *Phytophthora*—Fortuna, the potato variety for your peace of mind (last accessed December 2012).

6

Omics Applications in Brassica Species

Xiaonan Li, PhD; Nirala Ramchiary, PhD; Vignesh Dhandapani, PhD;
Su Ryun Choi, PhD; and Yong Pyo Lim, PhD

CONTENTS

6.1 Introduction 164
6.2 Markers to Genome Sequencing: Dissecting *Brassica* Genomes 164
    6.2.1 Markers to Genetic Maps: Understanding *Brassica* Genomes 164
    6.2.2 Comparative Genome Study: A Step Ahead for Understanding
        Structural Conservation and Diversification between Species 167
    6.2.3 Multinational *Brassica* Genome Project: Start of *Brassica* Omics 167
6.3 Transcriptomics: Creation of Genomic Resources and Marker Development 168
    6.3.1 Expressed Sequence Tag and cDNA Library 169
    6.3.2 Array-Based Resource 169
        6.3.2.1 Expression Microarrays 169
        6.3.2.2 Diversity Arrays Technology 170
    6.3.3 mRNA-Seq and Transcriptome Mapping 170
6.4 *Brassica* Genome Resource Databases 171
6.5 Genetics and Genomics of Important Traits: Dissecting the Functional Loci in
    Brassica Genomes 171
    6.5.1 Conventional Quantitative Trait Loci Mapping and Gene Identification 172
        6.5.1.1 Morphological Traits 172
        6.5.1.2 Flowering-Related Traits 172
        6.5.1.3 Glucosinolates 173
    6.5.2 Whole Genome Resequencing and QTL Mapping for Finding Genetic
        Loci for Important Traits 174
    6.5.3 Association Mapping Approach 175
    6.5.4 Determining Functional Loci through Transcriptome
        and Proteome Analysis 177
        6.5.4.1 Plant Growth and Development 177
        6.5.4.2 Oil Content and Seed Metabolites 179
        6.5.4.3 Male Sterility 179
        6.5.4.4 Stress Tolerance 180
6.6 Conclusion and Perspectives: Integrated Omics Should Be Adopted for
    Breeding *Brassica* Crops 180
References 181

6.1 Introduction

Brassica crops belonging to the mustard family (Brassicaceae) are economically important crops grown worldwide to produce vegetables, oilseeds, mustard condiments, and fodder, which are used in the form of leaves, roots, stems, buds, inflorescences, and seeds (Beilstein et al. 2006). Of the many species belonging to the Brassicaceae, six species, including three diploids (*Brassica rapa*, AA, $n = 10$; *B. oleracea*, CC, $n = 9$; and *B. nigra*, BB, $n = 8$) and three amphidiploids (*B. juncea*, AABB, $n = 18$; *B. napus*, AACC, $n = 19$; and *B. carinata*, BBCC, $n = 17$) derived by natural hybridization of the diploids cultivated commercially, are becoming an integral part of global agriculture and horticulture (U 1935; Prakash and Hinata 1980; Prakash and Chopra 1990). Leafy *B. rapa* (Chinese cabbage and pakchoi) and *B. oleracea* (cabbage, cauliflower, broccoli, turnip, and rutabaga) are cultivated as vegetables (FAOSTAT 2009). For extraction of oil, *B. napus*, *B. rapa* oil type, and *B. juncea* are extensively cultivated in India, China, Europe, the USA, Russia, and Canada.

For several decades, conventional breeding of cultivated *Brassica* species has developed cultivars with high yield, desired morphology, nutritional components, and biotic and abiotic stress phenotypes. However, the time-consuming process of phenotypic selection and environmental influence on the majority of agronomic and quality traits led to the development of other alternatives. With the development of molecular markers, several linkage maps have been developed in cultivated *Brassica* species and molecular markers linked to trait loci have been identified. However, the limitations of conventional molecular markers are that most of the molecular markers are anonymous, so the need to find causal genes creating phenotypic differences encourages the development of new, advanced high-throughput and low-cost technologies that can sequence DNAs, RNAs, and proteins, which directly relate to the phenotypes. In this chapter, an attempt has been made to review the earlier genetics studies and the recent development of omics technology that is applied for improvement of cultivated *Brassica* species.

6.2 Markers to Genome Sequencing: Dissecting *Brassica* Genomes

6.2.1 Markers to Genetic Maps: Understanding *Brassica* Genomes

The most important application of different molecular marker systems is to develop linkage maps for gene mapping and association studies between genotype and phenotype. Genetic linkage maps constructed with a proper marker system and mapping population are essential to understand the genome structure and evolution, mapping quantitative trait loci (QTL) and genes for agronomically important traits, and map-based cloning of trait-specific candidate genes (Snowdon and Friedt 2004). Development of linkage maps in *Brassica* species is as old as the development of molecular markers, and some of the details are elaborated below. The works by many researchers that are cited below highlight some of the important research related to molecular markers' development in *Brassica* species, their localization in the genome by genetic mapping, and their importance toward understanding structure and evolution of *Brassica* genomes subsequently, which contribute greatly to whole-genome sequencing. However, for the detailed studies, readers are advised to refer to *Genetics and Genomics of the Brassicaceae* (Schmidt and Bancroft 2011).

Hybridization-based DNA markers, such as restriction fragment length polymorphism (Botstein et al. 1980), are one of the earliest marker types used in *B. rapa* (Song et al. 1991; Chyi et al. 1992), *B. oleracea* (Slocum et al. 1990; Kianian and Quiros 1992; Landry et al. 1992), and other *Brassica* species (Truco and Quiros 1994; Lagercrantz and Lydiate 1995, 1996; Cheung et al. 1997) for genetic studies. Later, with the evolution of marker technique, this type was replaced by two PCR-based markers, random amplified polymorphic DNA (Williams et al. 1990) and amplified fragment length polymorphism (AFLP) (Vos et al. 1995), used mainly for *Brassica* genetic diversity and genetic mapping studies (Demeke et al. 1992; Thormann et al. 1994; Tanhuanpaa et al. 1996; Pradhan et al. 2003; Zhao et al. 2005). However, because of their random nature and difficulty in genotyping, these were in turn replaced by new advanced marker systems.

Simple sequences repeats (SSRs), also called microsatellites, are the most popular PCR-based DNA markers, and are developed by flanking short repeat motifs found in the genes or in random genomics sequences. SSRs are short tandem repeats of 1–6 nucleotides bases distribution, which are found to be different from species to species (Lowe et al. 2004; Suwabe et al. 2004; Iniguez-Luy et al. 2008). Due to advanced features such as a codominant and multiallele nature, genome-wide distribution, high level of polymorphisms, and locus specificity, SSRs markers have been developed on a large scale (Saal et al. 2001; Lowe et al. 2002, 2004; Suwabe et al. 2002) and have become an indispensable marker system for genetic map construction, comparative mapping, and establishment of genetic relationship studies among *Brassica* species (Sebastian et al. 2000; Suwabe et al. 2004, 2008; Tonguc and Griffiths 2004; Choi et al. 2007). SSRs were identified and developed from the genomic (Plieske and Struss 2001; Suwabe et al. 2002), bacterial artificial chromosome (BAC)-end sequences (Choi et al. 2007) or from expressed sequence tags (ESTs) or genomic sequences *in silico* (Burgess et al. 2006; Ramchiary et al. 2011). The genic microsatellite markers that are developed from ESTs have an advantage over genomics SSRs because the former targets the functional part of the genome which may be directly or indirectly related to the phenotypic change (Ramchiary et al. 2011). A large set of *Brassica* SSRs is available at the *Brassica* community exchange platform (http://www.brassica.info/resource/markers/ssr-exchange.php) (Table 6.1). Further, Panjabi et al. (2008) developed intron polymorphism (IP) markers by flanking the intron of genes and mapping in the *B. juncea* genome.

Single nucleotide polymorphism (SNP) markers are observed due to the point mutations in the DNA sequences and are promising for haplotype map (HapMap) construction, population association studies, and other genetic studies (Gupta et al. 2001, 2005; The International HapMap Consortium 2003). SNPs that occur in the genic regions may cause single amino acid substitutions, premature termination or open frame shift, thereby influencing the gene functions changing trait phenotypes. While SNPs that occur in the intergenic spaces (intron) may not be likely to cause amino acid change directly, they may induce alternative splicing variations (Pagani and Baralle 2004; Kralovicova et al. 2005; Elsharawy et al. 2006). Initially, because of the limited sequence information and high cost of development and genotyping, SNP markers were developed only for a few genes in *Brassica* crops (Hu et al. 2006b). SNPs were identified in *FAE* gene that is involved in erucic acid biosynthesis between high and low erucic acid lines in *B. juncea* (Gupta et al. 2004) and *B. napus* (Rahman et al. 2008), respectively. Two SNP markers developed corresponding to *fad2* and *fad3c* genes are being applied for marker-assisted selection (MAS) for oleic and linolenic acid contents in *B. napus* (Hu et al. 2006b). Additionally, an SNP marker linked to *B. rapa* seed coat color genes was developed by Rahman et al. (2007).

With the development of high-throughput technology such as next-generation sequencing (NGS) technology and genotyping methods such as high-resolution melting (Vossen

TABLE 6.1

Genomic Resources of *Brassica* Species

Databases	Type of Data	Web URL
Annotated SNP database (autoSNP)	SNPs between *Brassica* cultivars	http://autosnpdb.appliedbioinformatics.com.au
Brassica database (BRAD)	Whole-genome sequence of *B. rapa* and the predicted genes	http://brassicadb.org/brad
Brassica Ensembl (BrassEnsembl)	Whole-genome sequence of *B. rapa*, the predicted genes, BACs, ESTs, and unigenes	http://www.brassica.info/BrassEnsembl/index.html
Brassica EST Database (BrED)	ESTs, unigenes, SSRs, and SNPs	http://brassest.cnu.ac.kr
Brassica Genome Gateway	*B. rapa* and *B. oleracea* BAC libraries and *B. napus* SNPs	http://brassica.nbi.ac.uk
Brassica oleracea Database (Bolbase)	Genome and gene information of *B. oleracea*	http://www.ocri-genomics.org/bolbase
Crop genetic database (CropStoreDB)	*Brassica* populations, maps, traits, and QTLs	http://www.cropstoredb.org
Intergrated Marker System for Oilseed Rape Breeding (IMSORB)	*B. napus* markers and BAC clones	http://brassica.nbi.ac.uk/IMSORB
National Centre for Biotechnology Information (NCBI)	ESTs, mRNAs, and genomic DNAs	http://www.ncbi.nlm.nih.gov
The Multinational *Brassica* Genome Project (MBGP)	Current status of *Brassica* sequencing, BACs, ESTs, GSSs, markers, and mapping populations	http://www.brassica.info

et al. 2009), GOOD assay (Sauer et al. 2000), or GeneChips (Gut 2004; Giancola et al. 2006), large-scale genome-wide SNP development becomes feasible and an integral part of genetics and genomics studies of crop plants including *Brassica* species (Gholami et al. 2012). In *B. oleracea*, since the Genome Sequencing Project was initiated in 2009, a large number of SSR and SNP markers were developed from the whole-genome shotgun sequences and firstly used for high-density linkage map construction (Wang et al. 2012b). In *B. rapa*, 21,311 SNPs were characterized by resequencing 1398 sequence-tagged sites based on 557 BAC sequences in eight genotypes (Park et al. 2010). Compared to the diploid *Brassica* species, SNP identification in tetraploid *Brassica* species, that is, *B. napus and B. juncea*, is more complex due to the presence of two progenitor genomes and of duplicated paralogous genes and genomic segments (Westermeier et al. 2009; Durstewitz et al. 2010). In *B. napus*, genome-wide SNP discovery on a large scale was first reported by Trick et al. (2009b) using Solexa transcriptome sequencing technology. Approximately 20 million ESTs were generated and 23,330–41,593 putative SNPs were detected between the two *B. napus* cultivars, Tapidor and Ningyou. Then Westermeier et al. (2009) developed and evaluated 87 SNPs from 18 candidate gene sequences among the six rapeseed varieties with a frequency of 1SNP/247bp. Furthermore, TraitGenetics GmbH group identified 604 SNPs with an average of 1SNP/42bp in 100 amplicons from ESTs in *B. napus* species (Durstewitz et al. 2010). All these SNP markers will have potential for further genetic mapping or association analysis. Information on all the above SNP markers is available now for *Brassica* research communities at the *Brassica* information exchange website (http://www.brassica.info/resource/markers/snp.php). At present, many high-density linkage maps with SSRs, SNPs, IPs, or a combination of all those markers are available especially in *B. juncea*, *B. napus*, *B. rapa*, and *B. oleracea* (Kim et al. 2006, 2009; Gao et al. 2007; Panjabi et al. 2008;

Li et al. 2010; Wang et al. 2011b, 2012b). These maps are developed either based on a single population or as integrated maps from multiple populations to saturate the genome, and at present serve as reference genetic maps for the respective *Brassica* species (Udall et al. 2005; Kaczmarek et al. 2009; Ramchiary et al. 2011; Wang et al. 2011b).

6.2.2 Comparative Genome Study: A Step Ahead for Understanding Structural Conservation and Diversification between Species

The advantage that molecular markers of a *Brassica* species are transferrable to other *Brassica* species helps to accomplish comparative mapping studies between cultivated *Brassica* species and *Arabidopsis thaliana*. Further, since *A. thaliana* is the closest model plant relative of *Brassica* species, and there is conservation of nucleotide sequences between them, many comparative mapping studies have been done with *A. thaliana* (Lagercrantz and Lydiate 1996; Lagercrantz 1998; Panjabi et al. 2008; Parkin et al. 2005). Using the observation of comparative mapping studies between *A. thaliana* and the ancestral karyotypes, 24 crucifer genomic blocks (A–X) have been proposed by Schranz et al. (2006), and are now widely accepted by scientific communities. By comparative genetic mapping between the *Arabidopsis* and *Brassica* species, the presence of segmental duplications and the genome rearrangements of *Brassica* genomes were proposed and confirmed at the micro or macro level (Lagercrantz and Lydiate 1996; Lagercrantz 1998; Osborn et al. 1997, 2003; O'Neill and Bancroft 2000; Rana et al. 2004; Parkin et al. 2005; Panjabi et al. 2008; Kim et al. 2009). Panjabi et al. (2008) made a comprehensive comparative mapping study between *B. juncea* and other *Brassica* species and reported in detail the conservation, diversification, and homologous relationships between the A, B, and C subgenomes of *Brassica* genomes as well as the identification of genomic blocks. Subsequently, in *B. rapa* also identification of 24 crucifer blocks and their confirmation by comparison with previously reported blocks in A genomes of *B. napus* (Parkin et al. 2005) and *B. juncea* (Panjabi et al. 2008) were done (Li et al. 2010; Ramchiary et al. 2011). These maps are being used to identify the potential candidate genes in *A. thaliana* as whole-genome sequence, and functional genomics information of most of the genes is available on this species. This is possible since the homologous ancestral crucifer blocks of *Brassica* species are already identified by comparative mapping with *A. thaliana*. However, the observation of the complicated nature of *Brassica* genomes, such as the presence of duplication and divergence compared to the *A. thaliana*, bigger genome size and diverse plant morphophytes, indicated the need for separate study at the genomic level as *Brassica* species are very important crops worldwide, hence the Multinational *Brassica* Genome Sequencing Project Consortium (MBGP) was formed in the early 2000s.

6.2.3 Multinational *Brassica* Genome Project: Start of *Brassica* Omics

As the first step of whole *Brassica* genome sequencing by MBGP, *B. rapa* ssp. *pekinensis* cv. Chiifu-401-42, a Chinese cabbage inbred line, which has a comparatively small genome size among the cultivated *Brassica* species (Johnston et al. 2005; Lim et al. 2006), was selected as the model for *Brassica* A genome sequencing in 2003. Many research groups from Korea, China, the UK, Canada, the US, Australia, and Japan are involved and committed to the *B. rapa* genome sequencing project. The BAC-to-BAC sequencing approach was adopted by MBGP (Yang et al. 2005). Till now, five large-insert BAC libraries constructed with different restriction enzymes (KBrB, KBrE, KBrH, KBrS1, and KBrS2) were available, providing 53-fold genome coverage by Korea *B. rapa* genome project (KBGP; Yang et al. 2005). A total of 260,637 BAC-end sequences have been generated from 146,688 clones. Comparison of BAC-end sequences to the *A. thaliana* chromosome showed high level of sequence similarity. SSRs derived from BAC-end sequences (BAC–SSRs) were then designated and used for the construction of *B. rapa* reference genetic linkage maps using a doubled haploid population derived from F_1 by crossing between two diverse Chinese cabbage inbred lines, Chiifu and Kenshin. Versions I and II anchored a total of 188 seed BACs (Choi et al. 2007; Kim et al. 2009). Further, this version II reference genetic map was updated and integrated with the addition of BAC-anchored SSR markers (Li et al. 2010; Ramchiary et al. 2011).

Subsequently Mun et al. (2008) published the first genome-wide BAC-based physical map of *B. rapa*, in which they accommodated 242 anchored contigs of the 1428 contigs identified in the 10 linkage groups of *B. rapa*. Thereafter, the Korean *B. rapa* Genome Project (KBGP) group and coauthors published the almost complete sequence of *B. rapa* chromosome A3 (31.9 Mbp) organized into nine contigs (Mun et al. 2010).

NGS technology accelerated the sequencing process. The *Brassica rapa* Genome Sequencing Project Consortium accomplished the draft genome sequence of *B. rapa* cultivar Chiifu with 72× coverage of paired short reads generated by Illumina GA II technology in 2011 (Wang et al. 2011d). The assembled sequence was of 283.8 Mbp of the expected 529 Mbp size and a total of 41,174 protein coding genes were annotated with a mean of 5.03 exons per gene. The rest of the genome could not be covered due to the presence of high-repetitive DNAs. The important finding of the draft genome sequencing was that the *B. rapa* genome contains an average of three paralogous blocks, as reported earlier by genetic mapping studies (Parkin et al. 2005; Panjabi et al. 2008; Kim et al. 2009). The paralogous blocks are found to retain various numbers of genes and their size is different. The paralogous blocks are termed as least, medium, and most fractioned genome based on the highest, medium, and least retention of genes compared to *A. thaliana*, respectively. The phenomenon of retention of more genes in a particular paralogous block is termed *genomic dominance*. The *de novo* assembled genome sequences, predicted gene annotations, and orthologous genes to *A. thaliana*, as well as genetic markers and linkage maps were provided online (http://brassicadb.org/brad).

In 2009, the *B. oleracea* C-Genome Sequencing Project was launched by a whole-genome shotgun strategy using next-generation sequencing technology. The genome assembly of nine pseudochromosomes, gene prediction, and functional annotation are now complete (http://www.ocri-genomics.org/bolbase/index.html).

The present whole-genome sequence analysis (Mun et al. 2010; Wang et al. 2011d) comprehensively provides evidence for genome triplication and chromosome segment rearrangement, as well as gene-preferred retention, which is discovered by preliminary molecular data (Kim et al. 2006). Understanding the genome structure by syntenic comparative genomics with *A. thaliana* would facilitate the identification and isolation of candidate genes, further contributing to the valuable traits in *Brassica* crops.

6.3 Transcriptomics: Creation of Genomic Resources and Marker Development

The successful completion of whole-genome sequencing is no doubt a breakthrough for *Brassica* crops in understanding the genome structures or uncovering the genetic loci hidden in genomic sequences responsible for controlling the important traits. However, to elucidate and decode the functions and regulatory networks of the genes involved in different growing and development stage traits, metabolism, nutrient compounds, and abiotic and biotic stresses, functional analysis of gene expression must be done. The recent advances in transcriptomic analysis or gene expression profiling in different tissues or under different conditions open up new functional genomics areas. Several transcriptomic resources for different *Brassica* species resulted from a collective effort of the *Brassica* research community in the form of ESTs, complementary DNA (cDNA) libraries, microarrays, a new messenger RNA sequencing (mRNA-seq) dataset, and transcriptome maps in order to dissect the relationship between structure, expressions, and functions of genes.

6.3.1 Expressed Sequence Tag and cDNA Library

ESTs and cDNA sequences are the most important functional genomics resources developed from transcriptomes. ESTs are short and vary from 100 to 1000 base pairs, approximately. ESTs are used to identify genes in different tissues and developmental stages that are developed from cDNA libraries or direct RNA sequencing. With the advancement of high-throughput technologies such as NGS, there is rapid and enormous growth in the EST information being deposited in public databases. This greatly facilitates discovery of genes, identification of SNPs and SSRs, gene function annotation, identification of microRNA, the finding of possible alternative splicing, and some proteome analysis by simple computational methods.

Tens of thousands of ESTs generated from cDNA libraries were developed by numerous laboratories focusing on *Brassica* crops. ESTs from several ongoing *Brassica* functional genomics projects could be obtained from the National Center of Biotechnology Information database (Table 6.1), which at present contains 643,944 *B. napus*, 213,605 *B. rapa*, 179,213 *B. oleracea*, 1810 *B. nigra*, and 5518 *B. juncea* ESTs. Apart from this, several *Brassica* EST resources are accessible from other public databases. *Brassica* genomics resources (http://brassicagenomics.ca) contain over 500,000 ESTs from four *Brassica* species (*B. napus*, *B. rapa*, *B. oleracea*, and *B. carinata*) that were generated by the Plant Biotechnology Institute, National Research Council of Canada, and Agriculture and Agri-Food Canada. The *Brassica* EST Database (BrED) at Chungnam National University, Korea (http://brassest.cnu.ac.kr) currently hosts over 52,079 unigenes and their predicted 6,000 SSRs, for which some of the publicly available *B. rapa* ESTs were used along with ESTs developed in-house. The same database contains ~1,300 possible SNP and insertion/deletion (Indel) variations identified from *B. rapa* and *B. napus* EST sequences. The Korea *Brassica* Genome Project (KBrGP) has released 127 K ESTs developed from 33 cDNA libraries prepared from different tissues of *B. rapa* ssp. *pekinensis* cv Chiifu (seedling, cotyledon, young plant, mature green leaf, floral bud, anther, silique, root, etc.) and abiotic stress treatment (cold, heat, and salt treatment) (Lee et al. 2008).

6.3.2 Array-Based Resource

6.3.2.1 Expression Microarrays

A microarrays-based approach for gene expression profiling has been widely used for transcriptome analysis in plants. Microarrays consist of a few thousands of cloned DNA or oligonucleotides spotted onto a solid matrix, as a high-throughput method to examine genome-wide gene expression (Rabbani et al. 2003; Rensink et al. 2005; Lee and Yun 2006). Initially, an *Arabidopsis* ATH1 GeneChip or other low-density *Arabidopsis* microarray chip was used for gene expression study in *Brassica* species (Hammond et al. 2005; Carlsson et al. 2007; Fei et al. 2007; Gaeta et al. 2009; Nishizawa et al. 2012). Up to now, several expression microarrays have been developed in *Brassica* using different technology platforms. John Innes Centre (JIC), United Kingdom, in collaboration with J. Craig Venter Institute, USA and Cogenics, developed a 60-mer oligo *Brassica* microarray containing 94,588 probes using Agilent Technologies (Trick et al. 2009a). Using this chip, gene expression was performed in *B. napus*, *B. oleracea*, and *B. rapa* lines. The EST assembly data, gene ontology distribution of ESTs associated with the array, and the species-specific genes estimations are available at the *Brassica* Gateway website (http://brassica.bbsrc. ac.uk/array_info.html). Two *Brassica* microarrays, namely KBGP-24K and KBGP-50K, were developed by KBrGP using NimbleGen platform. The KBGP-24K microarray with 24,963 unigenes consists of 360,312 probes duplicated in two blocks. This microarray was used for genome-wide transcriptome analysis in response to salt, cold, and drought stresses (Lee et al. 2008). The KBGP-50K covered approximately 30,000 unigenes and 17,500 genes predicted from the seed BAC sequences. A 135K *Brassica* Exon array platform was constructed by Christopher et al. (2010), containing 2.4 million 25-base oligonucleotide probes representing 135,201 gene models. In addition, a number of studies reported use of microarrays for specific functional analysis, such as in the seed development stage (Xiang et al. 2008; Jiang and Deyholos 2010), and for identification of genes responsible for the *Brassica* male sterility trait (Dong et al. 2013). A 135K *Brassica* Exon array platform was conducted by Christopher et al. (2010), containing 2.4 million 25-base oligonucleotide probes representing 135,201 gene models.

6.3.2.2 Diversity Arrays Technology

DArT is a novel microarray-based genotyping method developed firstly using rice genome (Jaccoud et al. 2001). Being a marker system, DArT has an advantage over other conventional marker types. DArT can detect the DNA variation in whole-genome profiling without relying on sequence information and it is a dominant marker system, detecting the presence or absence of marker alleles. Since it was developed, DArT has been applied in various crops for genetic mapping and diversity analysis, such as barley (Wenzl et al. 2004), cassava (Xia et al. 2005), *A. thaliana* (Wittenberg et al. 2005), and pigeonpea (Yang et al. 2006a). The application of DArT markers to *Brassica* species is recent and has been reported only during the past 2 years. Raman et al. (2012b) developed DArT markers for application in genetic studies of *Brassica* species. In their study, 1547 polymorphic DArT markers were successfully identified and used to detect the genetic diversity among *B. napus*, *B. rapa*, *B. juncea*, and *B. carinata* germplasm collections. Moreover, these DArT markers were also mapped in a rapeseed DH population and integrated with previously mapped markers.

6.3.3 mRNA-Seq and Transcriptome Mapping

Messenger RNA sequencing, also called whole transcriptome shotgun sequencing, which is a newly developed method, provides new strategies for gene expression analysis that directly target the coding genes at transcriptome level without sequencing the whole genome. In *B. rapa*, the first application of mRNA-seq was performed on Chinese cabbage for the heading trait. Sequence reads were generated from the rosette and heading leaves and aligned to the reference Chiifu genome (Wang et al. 2012a). Genome-wide gene expression analysis demonstrated a number of transcription factors, endogenous hormones biosynthesis, and protein kinases-related gene families involved in head leaf development in Chinese cabbage. Considering the complex genome of polyploid *B. napus* species and the difficulty in separating and assembling the related sequences from two ancestral genomes, transcriptome sequencing is a good option not only for high-throughput marker discovery, but also for the measurement of transcript abundance (Trick et al. 2009b; Higgins et al. 2012). Bancroft et al. (2011) sequenced *B. napus* leaf transcriptomes of TNDH mapping population, and the ancestral cultivars of the parental lines. In total 23,037 transcriptome-generated SNPs were mapped to saturate the TNDH reference genetic linkage map of *B. napus* (Shi et al. 2009).

6.4 *Brassica* Genome Resource Databases

The *Brassica* database (BRAD) was developed with genetic and genomic data of *Brassica* species and updated with the whole-genome sequence of *B. rapa*. The integrated genome browser in the BRAD database helps to visualize the genes, SSRs, miRNAs, mRNAs, transposons, and genetic markers for selected regions of *B. rapa*. Moreover, the synteny viewer of the BRAD database helps one to find the *B. rapa* synteny genes in *T. parvula*, *A. thaliana*, and *A. lyrata* genomes. Rothamsted Research station and EBI developed the BrassEnsembl browser, which is similar to the BRAD genome browser. It displays the *Brassica* unigenes, markers, and *Arabidopsis* coding DNA sequence (CDS) for requested regions with configurable options. The Biotechnology and Biological Sciences Research Council UK and JIC developed and are maintaining the *Brassica* Genome Gateway with large numbers of BACs, unigenes, and SNPs for *B. napus* and *B. rapa*. The database provides multiple gene predictions for each BAC clone of *B. rapa* (Table 6.1). Additionally, corresponding *B. napus* BAC end sequences, SSRs, similar *A. thaliana* gene models, genetic markers, ORF, and restriction sites are viewable for selected *B. rapa* BAC sequences. More databases related to *Brassica* genomic research are listed in Table 6.1.

AutoSNPdb is the database developed with identified SNPs between the genes of different cultivars of *Brassica* species. This database holds 203,036 SNPs and 40,213 Indels from 169,721 unigenes of *Brassica* species. Similarly the BrED presents 1,300 SNPs and Indels from publicly available EST sequences of *B. rapa* and *B. napus*. Moreover, the same database provides 6,228 simple sequence repeats with predicted primers and their annotated functional information. *B. rapa* tissue-specific EST database contains the functions and gene ontology information of tissue-specific ESTs, and allows users to browse ESTs by tissues such as flower, root, pollen, and silique. These *Brassica* resources play a major role in analyzing and sharing the generated genomic and transcriptomic sequences.

6.5 Genetics and Genomics of Important Traits: Dissecting the Functional Loci in *Brassica* Genomes

The presence of high-level genetic diversity, reflected in the diverse morphological variation among the *Brassica* species, has helped the exploitation of those materials in breeding programs. The important traits exploited in breeding programs using natural variability are growing habit, local area consumer demand, morphological, yield related, disease resistant, and nutritional compound. Conventional breeding mostly relies on subjective observation and screening on the phenotype from generation to generation. But most of the economically important traits are complex, are inherited quantitatively, and are highly influenced by the environmental conditions, which makes breeding programs for selection difficult as well as time-consuming. Construction of molecular maps is of great significance for increasing breeding efficiency as it permits analysis of QTL identification with linked markers, thereby helping marker- or gene-assisted selection of desired trait loci. Significant applications of molecular genetics mapping are not only to study the genome structure in *Brassica* crops (Lagercrantz et al. 1996; Osborn et al. 2003; Parkin et al. 2005; Panjabi et al. 2008), but also to manipulate qualitative or quantitative trait loci of economically important crops (Snowdon and Friedt 2004). MAS or gene-assisted selection could greatly accelerate the breeding process by selecting elite lines with target and desired traits in early stages. Whole-genome sequence and a large number of SNP resources, integrated with genetic and physical maps, have been promising in the dissection of functional genetic loci in plant genomes including *Brassica* species that are associated with genetic variations observed within and between species. The following paragraphs highlight various studies of mapping QTL/gene(s) in *Brassica* species for flowering time, morphological traits, and yield component traits.

6.5.1 Conventional Quantitative Trait Loci Mapping and Gene Identification

6.5.1.1 Morphological Traits

The presence of wide genetic diversity reflecting morphological diversity in *Brassica* species provides a good resource for genetic dissection of plant growth and development. Mapping and comparison of QTL maps based on three F_2 mapping populations having one common parental line, "rapid cycling brassica," detected a total of 86 QTL controlling eight curd-related traits in *B. oleracea* (Lan and Paterson 2000). Sebastian et al. (2002) detected 32 QTL for 27 morphological and developmental traits, including leaf, flowering, axillary bud and stem characters, using a multiple-marker regression approach. In *B. rapa*, Song et al. (1995) mapped three plant height QTL each on linkage groups 4A, 9A, and 5A. A dwarf gene, *DWF2*, was mapped at the end of linkage group R06 (Muangprom and Osborn 2004) in a region having homology to the top of *A. thaliana* chromosome 2 containing *RGA* gene. In addition, Lou et al. (2007) identified a total of three plant heights QTL, each on R02, R03, and R07 in the RCCC $F_{2/3}$ population derived from a cross between a rapid cycling line RC-144 and a vegetable-type Chinese cabbage line CC-156. For the other morphological traits, 48 QTL determining 28 phenotypic traits related to flowering (days to bud and flower), plant height, leaf (pubescence, length, lobes, and petiole characteristics) and stem (stem length and index) traits were detected (Song et al. 1995). Yu et al. (2003) identified 50 QTL based on a Recombinant Inbred population, including 5 for plant growth habit, 6 for plant height, 5 for plant diameter, 7 for leaf length, 4 for leaf width, 6 for leaf length/leaf width ratio, 7 for petiole length, 4 for petiole width, and 6 for bolting character. A total of 22 morphological traits including flowering time, seed-related traits, growth-related traits, leaf-related traits, and turnip-related traits QTL were identified from multiple segregating populations derived from parental crosses involving the three main groups of *B. rapa*, that is, the oleiferous, leafy, and turnip types (Lou et al. 2007).

6.5.1.2 Flowering-Related Traits

Flowering time is a very important developmental trait, with wide variation in flowering among *Brassica* crops. *Brassica* cultivars are differentiated as biennial or annual based on their flowering habits and their requirement for vernalization under low-temperature treatment, which may be controlled by different regulatory networks.

Flowering-related traits were the earliest traits investigated in QTL mapping studies on *Brassica* crops (Kennard et al. 1994; Osborn et al. 1997; Bohuon et al. 1998; Lagercrantz et al. 2002; Osterberg et al. 2002). Four *B. rapa* homologous genes (*BrFLC1, BrFLC2, BrFLC3,* and *BrFLC5*) to *Arabidopsis* MADS-box flowering time regulator *FLC* were cloned and mapped in *B. rapa* (Schranz et al. 2002), which were colocated on several flowering time QTL regions (Teutonico and Osborn 1995; Osborn et al. 1997; Kole et al. 2001). These four copies of flowering time genes *BrFLC1, BrFLC2, BrFLC3,* and *BrFLC5* were mapped to chromosomes R10, R2, and R3, respectively, and the chromosomal regions harboring these genes are found to be collinear with *A. thaliana* (Yang et al. 2006b; Wang et al. 2011d). *BrFLC2* was identified as a candidate gene for a major flowering-time QTL and the vernalization response in different mapping populations of *B. rapa* (Li et al. 2009, 2012; Zhao et al. 2010b). *BrFLC1* was colocated with a QTL region detected in a backcross population derived from a cross between Per and R500 (Kole et al. 2001; Schranz et al. 2002), and a F$_2$ population derived by crossing yellow sarson and *B. rapa* cv. Osome (Li et al. 2009). Further, candidate gene sequence variation association studies revealed that a splicing site variation (A/G) in *BrFLC1* correlated with flowering time variation in vegetable-type *B. rapa*, while an Indel of *BrFLC2* correlated to this in oil-type *B. rapa* (Yuan et al. 2009; Wu et al. 2012). Lou et al. (2007) detected QTL for flowering time in four different populations developed from wide cross between *B. rapa* cultivars. Three QTL for flowering time traits were detected colocalizing with *LFY* (*LEAFY*), *FT* (*FLOWERING LOCUS T*), and *VRN2* (*VERNALIZATION2*) homologs. In *B. oleracea*, five *FLC* copies were cloned and characterized (*BoFLC1, BoFLC2, BoFLC3, BoFLC4,* and *BoFLC5*), while only *BoFLC2* was suggested to affect flowering time (Schranz et al. 2002; Lin et al. 2005; Okazaki et al. 2007). Comparative mapping analysis showed that three flowering time QTL regions on O2, O3, and O9 in *B. oleracea* were homologous to each other and with chromosome regions of *B. nigra* (Lagercrantz and Lydiate 1996), which was also homologous to *Arabidopsis* chromosome 5, a region containing a number of flowering-related genes (Bohuon et al. 1998; Rae et al. 1999). Axelsson et al. (2001) further performed detailed comparative QTL mapping in *B. nigra, B.oleracea, B. rapa* and *B. juncea*, proving that multiple flowering time QTLs in different *Brassica* species are controlled by paralogous genes, which are also homologous to *Arabidopsis CO* genes.

The uniform and desired flowering time is a major breeding goal in rapeseed, because to some extent this trait influences the yield components. Flowering-related gene *FLC, FT,* and *CO* homologs in *B. napus* have been identified, respectively (Robert et al. 1998; Wang et al. 2009; Schranz et al. 2002). Several QTL (more than 30 loci) were detected in different mapping populations that were derived by crossing different flowering habit parental lines under multiple environments (Long et al. 2007; Shi et al. 2009), onto which *BnFLC* paralogs were also comapped with some major QTL regions (Udall et al. 2006; Long et al. 2007). One major QTL region on chromosome N10 corresponding to vernalization response was colocated with *BnFLC1* paralog (Osborn et al. 1997; Tadage et al. 2001). Wang et al. (2009) detected two major QTL clusters for flowering time on A2 and C6, where three out of six *FT* homologs (*BnA2.FT, BnC6.FT.a,* and *BnC6.FT.b*) were mapped. Another candidate gene, *FRIGIDA* (*FRI*) homologs (*BnaA.FRI.a*), was identified and mapped to a QTL region on chromosome A3 for flowering under multiple environments (Wang et al. 2011c). Recently, Raman et al. (2012a) reported that flowering time is controlled by at least 20 loci in rapeseed and predicted a few candidate genes, such as *APETALA1, LEAFY, GA3OXIDASE,* and *FLOWERING LOCUS T,* to be involved.

6.5.1.3 *Glucosinolates*

Although increasing yield is the most important breeding goal in *Brassica* species, recently breeding high and desirable nutritional components for human health has been gaining equally important attention. *Brassica* species are viewed as a group of crops having anti-oxidant, anticarcinogenic, and anticancer properties, thanks to the special compound glu-cosinolates (GSLs), vitamin C, and other phenolic compounds.

Among various nutritional components in *Brassica* crops, the biosynthesis and effect of GSLs is highly studied. Several studies have provided evidence that increased consump-tion of cruciferous vegetables could significantly reduce the risk of cancer, due to the pres-ence of GSLs, particularly of breakdown products from sulforapahnin (Block et al. 1992; Mithen et al. 2000; Talalay and Fahey 2001). To date, most of the *A. thaliana* genes respon-sible for the glucosinolate biosynthesis pathway have been identified, including *GSL-Elong*, *GSL-ALK*, *GSL-OHP*, and *GSL-OH* (Wittstock and Halkier 2000; Halkier and Gershenzon 2006; Hirai et al. 2007; Sonderby et al. 2010; Wittstock and Burow 2010). The ortholog genes in *B. rapa* have been analyzed by comparative studies with *A. thaliana* based on sequencing of ESTs, BAC libraries, and the present assembled genome (Zang et al. 2009; Wang et al. 2011a). Candidate genes *BoGSL-Elong* and *BoGSL-ALK* have been cloned and character-ized in *B. oleracea*, which controlled the side chain elongation and side chain desaturation in glucosinolate synthetic steps, respectively (Li and Quiros 2002, 2003). Lou et al. (2008) reported QTL mapping for glucosinolate content in *B. rapa* leaves using two DH popula-tions: DH38, derived from a cross between yellow sarson R500 and a pak choi variety HK Naibaicai; and DH30, from a cross between R500 and Kairyou Hakata, a Japanese vegetable turnip variety. A total of 16 loci controlling aliphatic glucosinolate accumula-tion were found, three loci each controlling total indolic glucosinolate concentration and aromatic glucosinolate concentrations. The results obtained from comparative analysis between *A. thaliana* and *B. rapa*, together with candidate gene mapping in *B. rapa*, indicated the possibility of involvement of several genes in the glucosinolate biosynthesis pathway.

However, for livestock animals, the presence of GSL breakdown products in oil cake, which contains 30%–40% protein after extraction of oils, when fed to them leads to goiter and abnormal growth and development. Therefore, canola quality cultivars with zero erucic acid and low GSL, *B. napus*, *B. rapa*, and *B. juncea*, have been developed in Canada. Seed glucosinolate content is a complex quantitative trait, which makes it difficult for breeding programs due to linkage drag for high glucosinolate content. Four QTL for seed glucosinolate accumulation located on *B. napus* linkage group A9, C2, C7, and C9 have been identified in different studies (Howell et al. 2003; Zhao and Meng 2003; Quijada et al. 2006; Basunanda et al. 2007). Hasan et al. (2008) developed gene-linked SSR markers for oilseed rape seed glucosinolate content by genome comparison between *Arabidopsis* and *Brassica* together with allele–trait association analysis. Recently, Feng et al. (2012) identi-fied 105 metabolite QTL governing glucosinolate concentrations in both leaves and seeds of *B. napus* and constructed an advanced metabolic network regulating glucosinolate syn-thesis. In *B. juncea*, Ramchiary et al. (2007a) and Bisht et al. (2009) through QTL mapping and candidate gene approach tagged five genetic loci responsible for the development of low-GSL traits. The high and low gene-specific markers (SSR, SNPs, and Indel) are being used for marker-assisted transfer of low-GSL traits in *B. napus* and *B. juncea*.

6.5.2 Whole Genome Resequencing and QTL Mapping for Finding Genetic Loci for Important Traits

Recently, a comprehensive QTL mapping study in *B. rapa* by Li et al. (2012) identified a total of 95 QTL for plant height, bolting/flowering time, leaf shape, and silique-related traits in different crucifer blocks of *B. rapa* genome. Comparative mapping identified five key evo-lutionarily conserved crucifer blocks (R, J, F, E, and W) harboring QTL for morphological and yield component traits between A, B, and C subgenomes of *Brassica* species. By the integration of genome-wide scanning of QTL regions and whole genome resequencing of two parental line DNAs, a few candidate genes and gene-specific SNPs were identified in the detected QTL region on *B. rapa* (Figure 6.1). The candidate gene SNPs were validated experimentally by SNP genotyping in parental lines and mapping population, which further showed comapping in the specific chromosomal region containing trait QTL. The identified candidate genes belong to the gene families cytokinin oxidase/dehydrogenase, gibberellic acid-stimulated *Arabidopsis*, gibberellic acid oxidase (GA2OX, GA20OXs), and growth-related factor, which are involved in the regulation of plant growth and development, flowering, silique traits, seed number, size, and development. This finding revealed that QTL mapping when combined with the whole-genome sequencing/resequencing of parental lines could accelerate the identification of candidate genes and gene sequence variations causing phenotypic differences between the parental lines and segregating population. The identified gene sequence information could be applied for breeding other cultivated *Brassica* species, as conservation of QTL was observed between A, B, and C subgenomes in the above study for leaf-, flowering-, and yield-related traits.

6.5.3 Association Mapping Approach

Association mapping (AM), also known as linkage disequilibrium mapping, as a complementary approach to the conventional QTL mapping, has recently been gaining more attention for genetic studies of complex traits because there is no need to construct a mapping population and it could be applied in natural populations that can harvest natural allelic varaiations, unlike biparental lines in traditional mapping. AM can be employed with either genome-wide association or candidate gene association relying on high-throughput whole-genome scan and SNP discovery by genome resequencing (Zhu et al. 2008). AM has been applied on various plants based on diverse germplasms and large-scale marker resources, mostly on *Arabidopsis* (Ehrenreich et al. 2009; Weigel and Mott 2009; Sterken et al. 2012), maize (Kump et al. 2011; Tian et al. 2011), and rice (Agrama et al. 2007; Huang et al. 2010).

In *Brassica* species, genome-wide association mapping (GWAS) is still in its infancy, although individual genes/marker–traits association has been reported in few studies. Recently, various collections of *B. napus* inbred lines, as well as *B. rapa* core collections, were used to assess the population structure and linkage disequilibrium using molecular markers, which provided useful materials and information for further complex traits association study (Zhao et al. 2010a; Bus et al. 2011; Xiao et al. 2012). Complex agronomical traits (flowering time, leaf shape, plant height, and grain yield) and metabolites (glucosinolate, tocopherols, carotenoids, etc.), as well as protein and oil contents, were given more emphasis in *Brassica* crops for AM studies using SNPs or other markers (Zou et al. 2010; Pino Del Carpio et al. 2011; Wurschum et al. 2012a,b). Zhao et al. (2007) tested the association of morphological traits, phytate content, and phosphate levels in diverse *B. rapa* accessions with AFLP markers. Four subpopulations were estimated by the model-based approach of *structure*, and 27 markers were identified associating with eight traits. A candidate gene AM was performed for rapeseed flowering variation by Wang et al. (2011c). Association analysis of the candidate gene *BnaA.FRI.a*, comapping with a major flowering time QTL region, showed that one SNP in a putative functional site and one haplotype block are associated with flowering time variation. In *B. napus*, Wurschum et al. (2012a) used both multiple-line cross QTL mapping and a joint linkage AM approach for QTL detection of diverse traits with 253 SNPs. Honsdorf et al. (2010) performed an experiment

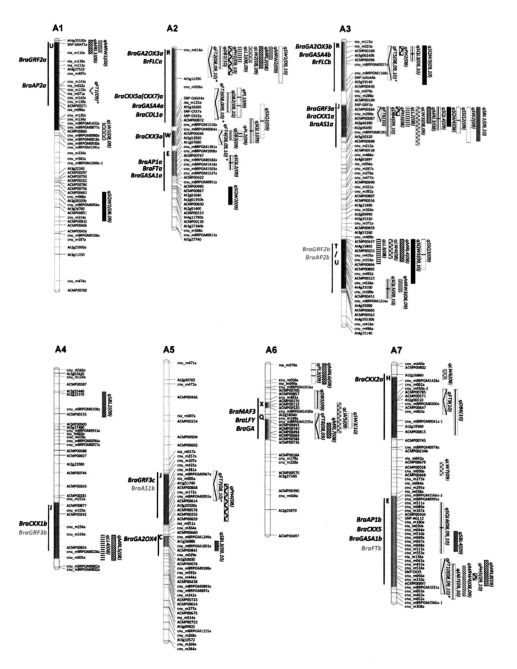

FIGURE 6.1

Distribution of quantitative trait loci (QTL) for morphological and yield component traits in *B. rapa* genome. QTL names are indicated by abbreviations of trait names as follows: flowering time (FT), bolting time (DB), flowering time after vernalization (FT*), plant height (PH), leaf length (LL), leaf width (LW), midrib length (MRL), midrib width (MRW), petiole length (PL), silique length (SQL), silique width (SQW), silique beak length (SBL), seeds per silique (SSQ), and seed weight (SW). The numbers in parentheses indicate the year of QTL detection. The crucifer building blocks in each linkage group of *B. rapa*, which are homologous to 5 chromosomes of *A. thaliana*, are indicated by different colors. Putative candidate genes identified within the QTL blocks are shown in bold *black* letters on the *left* of each linkage group; those outside QTL intervals are shown in bold *gray* letters. (Reprinted from Li, X., Ramchiary, N., Dhandapani, V. et al., *DNA Res*, 20, 1–16, 2012, with permission of Oxford University Press.)

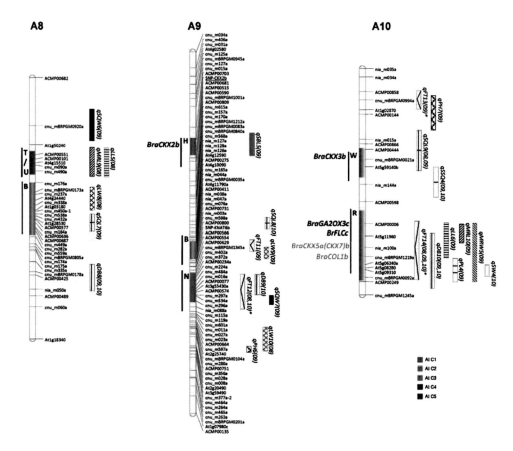

FIGURE 6.1 (Continued)

on whole-genome association analysis of 14 traits in an 84 canola winter rapeseed natural population. All the results showed that AM is feasible in *Brassica* species.

Interestingly, an associative transcriptomics approach was developed using mRNA-Seq to correlate both the gene sequence and gene expression variation with the trait variation in *B. napus* (Harper et al. 2012). Genome-wide association study using both SNP marker and expression variation-based (GEM) markers generated from transcriptome sequences of a large collection of *B. napus* crop types identified a transcription factor gene family as a candidate controlling seed glucosinolate content. These results showed that transcriptome sequencing in genetic mapping and trait association studies provided a comprehensive way to identify genes and genetic regulatory network, even in more complex polyploids *Brassica* species.

6.5.4 Determining Functional Loci through Transcriptome and Proteome Analysis

6.5.4.1 Plant Growth and Development

Numerous genes involved in the process of plant growth and development are regulated and interact in network. Transcript data are important for functional gene discovery to detect gene expression levels, and the developmental regulatory pathway in different tissues and the growing stage (Soeda et al. 2005; Whittle et al. 2010). During seed development, plants accumulate starch, oil, and proteins in differing amounts, which serve a crucial role in human and animal diets as well as in feedstocks for the chemical industry. In this regard, the recent use of transcriptomics and proteomics approaches has begun to provide global views of the structural, regulatory, and enzymatic proteins involved in the seed-developing process in agronomic *Brassica* oil-type crops (Li et al. 2005; Soeda et al. 2005; Agrawal and Thelen 2006; Agrawal et al. 2008; Xiang et al. 2008; Meyer et al. 2012). Li et al. (2005) used an *Arabidopsis* microarray to compare gene expression between normal germinated rapeseed seeds and seeds in which germination was inhibited by polyethylene glycol or abscisic acid (ABA) analog PBI429. Forty genes were identified to associate with the late seed development by both treatments, which were functionally involved in carbohydrate metabolism, cell wall-related process, detoxification of reactive oxygen, and triacylglycerol breakdown. The results suggested a greater importance of ABA signaling than of GA signaling in nongerminating *B. napus* seeds. A transcriptome analysis of secondary-wall-enriched seed coat of *B. napus* canola has been performed using a 90,000 element microarray to identify functional genes (Jiang and Deyholos 2010). The identified genes functional to seed coat development are associated with cell wall metabolism, flavonoid synthesis, and various cell wall transcript factors. Xiang et al. (2008) identified 10,642 unigenes for *Brassica* seed cDNA array preparation, especially for identifying genetic elements contributing to seed development processes. Hudson et al. (2007) used an *A. thaliana* Affymetrix ATH1 array to determine the mechanism of priming on the germination efficiency of cauliflower (*B. oleracea*) seeds.

Moreover, RNA sequencing has been successfully used in transcriptome analysis of *Brassica* crop growth and development (Korbes et al. 2012; Sun et al. 2012; Wang et al. 2012a). Putative precursors and a few novel miRNA affecting seed development were identified from small RNA and RNA-seq libraries by comparing the mature seeds and developing seeds (Korbes et al. 2012). These candidate miRNAs and their target genes would be useful for future functional validation. In polyploid *B. juncea* species, transcriptomes of tumorous stem mustard at different stem developmental stages were sequenced using Illumina short-read technology with a tag-based digital gene expression system (Sun et al. 2012). The gene expression profiles of three swollen stem samples were compared with those of two nonswollen stem samples, and finally 1042 genes with significantly different expression levels were screened out.

Proteomics or comparative proteomics provides a method to trace and compare genome-wide gene expression leading to qualitative and quantitative differences in several development stages, tissues, and even species. Kong et al. (2010) adopted a comparative proteomics approach for *B. rapa* and *B. oleracea* seedlings and used two-dimensional electrophoresis (2-DE) and MALDI-TOF-MS to identify the species-specific proteins. They further developed PCR-based markers based on the A or C genome-specific proteins to distinguish these two species. Comparative proteomics was also applied to three vegetative organs of *B. napus*, leaf, stem, and shoot, using 2-DE to find common and tissue-specific proteomes (Albertin et al. 2009). A 2D reference map was established allowing the identification of 93, 385, and 266 proteins in leaf, stem, and root proteomes, respectively (Albertin et al. 2009). In addition, protein phosphorylation has been shown before to have an important role in seed development (Agrawal and Thelen 2006). To better characterize the extent of protein phosphorylation during seed maturation, recently Meyer et al. (2012) performed a large-scale mass spectrometry-based phosphoproteomic study on seed development of rapeseed, *Arabidopsis*, and soybean; 2001 phosphopeptides were identified, containing 1026 phosphorylation sites, from which 956 proteins were identified.

An integrated approach that combines ESTs, protein profiling, and microarray analyses was taken to look into the gene expression landscape in the endosperm of the oilseed crop *B. napus* by Huang et al. (2009). Their results provided insight into several prominent metabolic pathways in the endosperm that have implications for improving the productivity and quality of canola, including starch and sugar metabolism-related genes, developmental stage favored genes, and a transcription factor *LEAFY COTYLEDON (LEC1)*.

6.5.4.2 Oil Content and Seed Metabolites

The major fatty acid (FA) constituents of *Brassica* oil are oleic acid (C18:1), linoleic acid (C18:2), linolenic acid (C18:3), arachidic acid (C20:0), eicosenoic acid (C20:1), and erucid acid (C22:1). Low erucic acid in rapeseed improves the quality of the oil. Several studies of QTL mapping for oil content and seed fatty acids have been done in *B. napus* (Burns et al. 2003; Qiu et al. 2006; Zhao et al. 2006) and in *B. juncea* (Gupta et al. 2004; Ramchiary et al. 2007b, Yadava et al. 2012). Major QTL for seed erucic acid content are closely linked to two genes involved in the synthesis of FA, *FAE1.1* and *FAE1.2* in both *B. napus* and *B. juncea* (Ecke et al. 1995; Gupta et al. 2004; Qiu et al. 2006). Ramchiary et al. (2007b) identified seven QTL for oil content, two of which mapped to the locations of two *FAE1* genes. This suggested that the erucic acid content is a major determinant of seed oil content in *Brassica* species (Ecke et al. 1995; Burns et al. 2003).

High oleic acid content in oilseed *Brassica* species is a desirable trait. A few major loci corresponding to *FAD2* and *FAD3* genes, which code for the enzymes involved in the desaturation of linoleic to linolenic acid, have been found to control high oleic acid and low linolenic acid content in rapeseed and *B. juncea* (mustard) (Barret et al. 1999; Schierholt et al. 2001; Falentin et al. 2007; Jagannath et al. 2011). To explore more related genes associated with the oleic acid content, a microarray technology was employed to investigate the difference of gene expression between two lines with high and low oleic acid contents (Guan et al. 2012). A set of 562 differentially expressed genes were identified and classified into 23 functional categories. Their results suggested that many candidate genes for high oleic acid content coordinated together in metabolic networks in addition to inactivation of the *FAD2* and *FAD3* genes.

Seed oil content is highly affected by the environment during the seed maturation process. The QTL method and microarray analysis were combined to investigate the candidate genes related to the seed oil content under high-temperature conditions. Zhu et al. (2012) developed two near-isogenic lines and analyzed the gene expression at a QTL region controlling oil content. Another 454 pyrosequencing was performed to explore the regulatory mechanism of elevated photosynthesis in the silique wall (Hua et al. 2012). Their result highlighted that the silique wall photosynthesis-related genes have an important effect on seed oil content in terms of maternal effects. Recently, a *B. napus* candidate gene *BnGRF2* (growth-regulation factor 2-like gene) was elucidated with a role in seed yield and seed oil production, which was induced by cell number and plant photosynthesis, using a transgenic approach and transcriptome analysis (Liu et al. 2012).

6.5.4.3 Male Sterility

B. rapa 300-K oligomeric probe microarrays (Br300K) and *A. thaliana* flower-specific cDNA microarrays were used to detect the gene expression differences between cytoplasmic male sterility (CMS) line and its maintainer line of *B. rapa* and *B. napus*, respectively, in order to investigate the regulatory mechanism of CMS (Carlsson et al. 2007; Dong et al. 2013). In the

B. oleracea male sterility study, Kang et al. (2008) used an *A. thaliana* whole-genome microarray to analyze gene expression of male sterile and fertile buds. Kinase genes and cell wall modification genes are also found to be significantly downregulated in male sterile line. Chen et al. (2009) analyzed the flower bud expression profiles of the *B. napus* S45A (male sterile) and S45B (male fertile) lines using an *A. thaliana* ATH1 oligonucleotide array to elucidate the mechanism of the recessive genic male sterility (*Bnms1*). The results suggested that *Bnms1* might be directly related to lipid/fatty acid metabolism, and the lack of lipid/fatty acids metabolism might be responsible for the male sterility of the *Bnms1* mutant.

6.5.4.4 Stress Tolerance

Plant growth and development are to a large extent affected by environmental stress, which affects *Brassica* plant production. Transcriptome profiling analysis is a promising strategy to detect the whole-genome level gene expression of plants in abiotic or biotic stress conditions (Hammond et al. 2005; Lee et al. 2008; Ahmed et al. 2012). A proteomic approach is also widely applied to analyze plant responses to biotic and abiotic stresses (Peck et al. 2001; Kim et al. 2003) by typical proteomics techniques, such as 2-DE and mass spectrometry (MS).

For nutrition deficit studies, *B. oleracea* transcriptional response to phosphorus stress has been analyzed by using *A. thaliana* ATH1-121501 GeneChip array. Transcriptional response of *B. oleracea* to mineral nutrient phosphorus stress was quantified and 99 genes were identified as significantly regulated in this stress condition (Hammond et al. 2005). Proteomic and metabolic changes of phosphorus (Pi) deficiency in oilseed rape were conducted by comparative proteome analysis by Yao et al. (2011). Another experiment designed for sulfur (S) and chromium (Cr) deficit in *B. juncea* was performed to explore gene regulation by transcriptome profiling (Schiavon et al. 2012).

To investigate the response of rapeseed to drought stress, Mohammadi et al. (2012) analyzed the difference of protein expression profile among rapeseed drought-sensitive and drought-tolerant lines, together with their F1 hybrid, by a proteomics approach. Their results indicate that V-type H(+) ATPase, plasma-membrane associated cation-binding protein, HSP 90, and elongation factor EF-2 have a role in the drought tolerance of rapeseed, and the heat shock protein 70 and tubulin beta-2 might induce drought sensitivity.

Fernandez-Garcia et al. (2011) analyzed the effect of salt stress on the contents of xylem sap in *B. oleracea* by MS. By combining metabolome and proteome analysis, many organic compounds and enzymes involved in xylem differentiation and lignifications under salt stress were clarified, which is helpful in elucidating the mechanisms in plant development and signaling under salt stress.

It is notable that all of the identified functional genes involved in plant development, particularly stress tolerance, will be of value in future transgenic and reverse genetic manipulation for breeding engineering.

6.6 Conclusion and Perspectives: Integrated Omics Should Be Adopted for Breeding *Brassica* Crops

Although significant progress has been made in genetics and genomics of *Brassica* crop plants, only a handful of genes governing important major traits are identified. The majority of the important agricultural traits such as morphology, nutritional components, and biotic and abiotic stresses are governed by many genes or a gene network with small effects. Further research is necessary for development of important tools for omics-based *Brassica* breeding. The presence of duplicated and repetitive DNAs complicates the proper alignment and identification of actual causal genes out of many paralogs, as genome sequence information of the five cultivated *Brassica* genomes other than *B. rapa* is still not available. The recent advances in the development of low-cost, high-throughput technologies to analyze whole-genome DNAs, RNAs, protein, and metabolites encourages the combined use of those technologies with traditional genetics for breeding *Brassica* crops. This combined use would foster rapid progress in identification of genes and gene network involved in plant growth and development, biosynthesis of particular metabolites, high seed yield, oil content, heterosis, and biotic and abiotic stresses. Further, since the sequences of *Brassica* species are highly conserved with other *Brassica* species, DNA and genomics information of extensively studied *B. rapa* and *B. napus* could be applied in other *Brassica* species. Detailed phenotypic (phenomics) study of a large collection of natural populations combined with transcriptome, proteome, and metabolome studies under different environmental conditions would help to identify the root cause of phenotypic differences observed in different *Brassica* species and *Brassica* lines. Further, present low-cost next-generation technology could be used for transcriptome and genome sequencing so that a large number of SNP markers could be used for GWAS and the development of a *Brassica* HapMap. In the near future, application of integrated omics would be of great help in predicting phenotype based on genotype, which would be very important in the development of breeding *Brassica* cultivars with desired nutrition, biotic/abiotic stress, and yield components.

References

Agrama, H.A., Eizenga, G.C., and Yan, W. 2007. Association mapping of yield and its components in ricecultivars. *Mol Breeding* 19: 341–356.

Agrawal, G.K. and Thelen, J.J. 2006. Large scale identification and quantitative profiling of phosphoproteins expressed during seed filling in oilseed rape. *Mol Cell Proteomics* 5(11): 2044–2059.

Agrawal, G.K., Hajduch, M., Graham, K., and Thelen, J.J. 2008. In-depth investigation of the soybean seed-filling proteome and comparison with a parallel study of rapeseed. *Plant Physiol* 148(1): 504–518.

Ahmed, N.U., Park, J.I., Seo, M.S., et al. 2012. Identification and expression analysis of chitinase genes related to biotic stress resistance in *Brassica. Mol Biol Rep* 39(4): 3649–3657.

Albertin, W., Langella, O., Joets, J., et al. 2009. Comparative proteomics of leaf, stem, and root tissues of synthetic *Brassica napus. Proteomics* 9(3): 793–799.

Axelsson, T., Shavorskaya, O., and Lagercrantz, U. 2001. Multiple flowering time QTLs within several *Brassica* species could be the result of duplicated copies of one ancestral gene. *Genome* 44: 856–864.

Bancroft, I., Morgan, C., Fraser, F., et al. 2011. Dissecting the genome of the polyploid crop oilseed rape by transcriptome sequencing. *Nat Biotechnol* 29(8): 762–766.

Barret, P., Delourme, R., Brunel, D., Jourdren, C., Horvais, R., and Renard, M. 1999. Low linolenic acid level in rapeseed can be easily assessed through the detection of two single base substitution in fad3 genes. In: *Proceeding of the 10th International Rapeseed Congress.* Canberra, Australia, pp. 26–29.

Basunanda, P., Spiller, T.H., Hasan, M., et al. 2007. Marker-assisted increase of genetic diversity in a double-low seed quality winter oilseedrape genetic background. *Plant Breed* 126: 581–587.

Beilstein, M.A., Al-Shehbaz, I.A., and Kellogg, E.A. 2006. Brassicaceae phylogeny and trichome evolution. *Am J Bot* 93: 607–619.

Bisht, N.C., Gupta,V., Ramchiary, N., et al. 2009. Fine mapping of loci involved with glucosinolate biosynthesis in oilseed mustard (*Brassica juncea*) using genomic information from allied species. *Theor Appl Genet* 118(3): 413–421.

Block, G., Patterson, B., and Subar, A. 1992. Fruit, vegetables and cancer preventiion: A review of the epidemiological evidence. *Nutr Cancer* 18: 1–29.

Bohuon, E.J.R., Ramsay, L.D., Craft, J.A., et al. 1998. The association of flowering time quantitative trait loci with duplicated regions and candidate loci in *Brassica oleracea*. *Genetics* 150: 393–401.

Botstein, D., White, R.L., Skolnick, M., and Davis, R.W. 1980. Construction of a genetic linkage map in man using restriction fragment length polymorphisms. *Am J Hum Genet* 32: 311–331.

Burgess, B., Mountford, H., Hopkins, C.J., et al. 2006. Identification and characterization of single sequence repeat (SSR) markers derived in silico from *Brassica oleracea* genome shotgun sequences. *Mol Ecol Notes* 6: 1191–1194.

Burns, M.J., Barnes, S.R., Bowman, J.G., Clarke, M.H., Werner, C.P., and Kearsey, M.J. 2003. QTL analysis of an intervarietal set of substitution lines in *Brassica napus*: (i) Seed oil content and fatty acid composition. *Heredity* 90(1): 39–48.

Bus, A., Korber, N., Snowdon, R.J., and Stich, B. 2011. Patterns of molecular variation in a species-wide germplasm set of *Brassica napus*. *Theor Appl Genet* 123(8): 1413–1423.

Carlsson, J., Lagercrantz, U., Sundstrom, J., et al. 2007. Microarray analysis reveals altered expression of a large number of nuclear genes in developing cytoplasmic male sterile *Brassica napus* flowers. *Plant J* 49(3): 452–462.

Chen, Y., Lei, S., Zhou, Z., et al. 2009. Analysis of gene expression profile in pollen development of recessive genic male sterile *Brassica napus* L. line S45A. *Plant Cell Rep* 28(9): 1363–1372.

Cheung, W.Y., Champagne, G., Hubert, N., and Landry, B.S. 1997. Comparison of the genetic maps of *Brassica napus* and *Brassica oleracea*. *Theor Appl Genet* 94: 569–582.

Choi, S.R., Teakle, G.R., Plaha, P., et al. 2007. The reference genetic linkage map for the multinational *Brassica rapa* genome sequencing project. *Theor Appl Genet* 115: 777–792.

Christopher, G.L., Graham, N.S., Lochlainn, S.O., et al. 2010. A *Brassica* exon array for whole-transcript gene expression profiling. *Plos One* 5(9): e12812.

Chyi, Y.S., Hoenecke, M.E., and Sernyk, J.L. 1992. A genetic map of restriction fragment length polymorphism loci for *Brassica rapa* (syn. *campestris*). *Genome* 25: 746–757.

Demeke, T., Adams, R.P., and Chibbar, R. 1992. Potential taxonomic use of random amplified polymorphic DNA (RAPD): A case study in *Brassica*. *Theor Appl Genet* 84: 990–994.

Dong, X., Kim, W.K., Lim, Y.P., Kim, Y.K., and Hur, Y. 2013. Ogura-CMS in Chinese cabbage (*Brassica rapa* ssp. *pekinensis*) causes delayed expression of many nuclear genes. *Plant Sci* 199–200: 7–17.

Durstewitz, G., Polley, A., Plieske, J., et al. 2010. SNP discovery by amplicon sequencing and multiplex SNP genotyping in the allopolyploid species *Brassica napus*. *Genome* 53(11): 948–956.

Ecke, W., Uzunova, M., and Weissleder, K. 1995. Mapping the genome of rapeseed (*Brassica napus* L.). II Localization of genes controlling erucic acid synthesis and seed oil content. *Theor Appl Genet* 91: 972–977.

Ehrenreich, I.M., Hanzawa, Y., Chou, L., et al. 2009. Candidate gene association mapping of *Arabidopsis* flowering time. *Genetics* 183(1): 325–335.

Elsharawy, A., Manaster, C., Teuber, M., et al. 2006. SNPSplicer: Systematic analysis of SNP-dependent splicing in genotyped cDNAs. *Hum Mutat* 27: 1129–1134.

Falentin, C., Brégeon, M., Lucas, M.O., et al. 2007. Identification of *fad2* mutations and development of allele-specific markers for high oleic acid content in rapeseed (*Brassica napus* L.). In: *Proceedings of the 12th International Rapeseed Congress*. Wuhan, China, pp. 117–119.

FAOSTAT data. 2009. http://faostat.fao.org/site/567/default.aspx#ancor.

Fei, H., Tsang, E., and Cutler, A.J. 2007. Gene expression during seed maturation in *Brassica napus* in relation to the induction of secondary dormancy. *Genomics* 89: 419–428.

Feng, J., Long, Y., Shi, L., et al. 2012. Characterization of metabolite quantitative trait loci and metabolic networks that control glucosinolate concentration in the seeds and leaves of *Brassica napus*. *New Phytol* 193(1): 96–108.

Fernandez-Garcia, N., Hernandez, M., Casado-Vela, J., et al. 2011. Changes to the proteome and targeted metabolites of xylem sap in *Brassica oleracea* in response to salt stress. *Plant Cell Environ* 34(5): 821–836.

Gaeta, R.T., Yoo, S.Y., Pires, J.C., et al. 2009. Analysis of gene expression in resynthesized *Brassica napus* Allopolyploids using arabidopsis 70mer oligo microarrays. *PLoS One* 4(3): e4760.

Gao, M., Li, G., Yang, B., Qiu, D., Farnham, M., and Quiros, C. 2007. High-density *Brassica oleracea* linkage map: Identification of useful new linkages. *Theor Appl Genet* 115(2): 277–287.

Gholami, M., Bekele, W.A., Schondelmaier, J., and Snowdon, R.J. 2012. A tailed PCR procedure for cost-effective, two-order multiplex sequencing of candidate genes in polyploid plants. *Plant Biotechnol J* 10(6): 635–645.

Giancola, S., McKhann, H.I., Berard, A., et al. 2006. Utilization of the three high-throughput SNP genotyping methods, the GOOD assay, Amplifluor and TaqMan, in diploid and polyploid plants. *Theor Appl Genet* 112(6): 1115–1124.

Guan, M., Li, X., and Guan, C. 2012. Microarray analysis of differentially expressed genes between *Brassica napus* strains with high- and low-oleic acid contents. *Plant Cell Rep* 31(5): 929–943.

Gupta, P.K., Roy, J.K., and Prasad, M. 2001. Single nucleotide polymorphisms: A new paradigm for molecular marker technology and DNA polymorphism detection with emphasis on their use in plants. *Curr Sci* 80: 524–534.

Gupta, V., Mukhopadhyay, A., Arumugam, N., Sodhi, Y.S., Pental, D., and Pradhan, A.K. 2004. Molecular tagging of erucic acid trait in oilseed mustard (*Brassica juncea*) by QTL mapping and single nucleotide polymorphisms in *FAE1* gene. *Theor Appl Genet* 108(4): 743–749.

Gupta, P.K., Rustgi, S., and Kulwai, P.I. 2005. Linkage disequilibrium and association studies in higher plants: Present status and future prospects. *Plant Mol Biol* 57: 461–485.

Gut, I.G. 2004. An overview of genotyping and single nucleotide polymorphisms (SNP). In: Rapley, R. and Harbron, S. (eds), *Molecular Analysis and Genome Discovery*. Wiley, Chichester, pp. 43–64.

Halkier, B.A. and Gershenzon, J. 2006. Biology and biochemistry of glucosinolates. *Annu Rev Plant Biol* 57: 303–333.

Hammond, J.P., Broadley, M.R., Craigon, D.J., et al. 2005. Using genomic DNA-based probe-selection to improve the sensitivity of high-density oligonucleotide arrays when applied to heterologous species. *Plant Methods* 1(1): 10.

Harper, A.L., Trick, M., Higgins, J. et al. 2012. Associative transcriptomics of traits in the polyploid crop species *Brassica napus*. *Nat Biotechnol* 30: 798–802.

Hasan, V., Friedt, W., Pons-Kühnemann, J., Freitag, N.M., Link, K., and Snowdon, R.J. 2008. Association ofgene-linked SSR markers to seed glucosinolate content in oilseed rape (*Brassica napus* ssp.*napus*). *Theor Appl Genet* 116: 1035–1049.

Higgins, J., Magusin, A., Trick, M., Fraser, F., and Bancroft, I. 2012. Use of mRNA-seq to discriminate contributions to the transcriptome from the constituent genomes of the polyploid crop species *Brassica napus*. *BMC Genomics* 13: 247.

Hirai, M.Y., Sugiyama, K., Sawada, Y. et al. 2007. Omics-based identification of *Arabidopsis* Myb transcription factors regulating aliphatic glucosinolate biosynthesis. *Proc Natl Acad Sci USA* 104: 6478–6483.

Honsdorf, N., Becker, H.C., Ecke, W., et al. 2010. Association mapping for phenological, morphological, and quality traits in canola quality winter rapeseed (*Brassica napus* L.). *Genome* 53(11): 899–907.

Howell, P.M., Sharpe, A.G., and Lydiate, D.J. 2003. Homoelogous loci control the accumulation of seedglucosinolates in oilseed rape (*Brassica napus*). *Genome* 46: 454–460.

Hu, S.W., Fan, Y.F., Zhao, H.X., et al. 2006a. Analysis of MS2Bnap genomic DNA homologous to MS2 gene from *Arabidopsis thaliana* in two dominant digenic male sterile accessions of oilseed rape (*Brassica napus* L.). *Theor Appl Genet* 113: 397–406.

Hu, X., Sullivan-Gilbert, M., Gupta, M., and Thompson, S.A. 2006b. Mapping of the loci controlling oleic and linolenic acid contents and development of *fad2* and *fad3* allele-specific markers in canola (*Brassica napus* L.). *Theor Appl Genet* 113(3): 497–507.

Hua, W., Li, R.J., Zhan, G.M., et al. 2012. Maternal control of seed oil content in *Brassica napus*: The role of silique wall photosynthesis. *Plant J* 69(3): 432–444.

Huang, Y., Chen, L., Wang, L., et al. 2009. Probing the endosperm gene expression landscape in *Brassica napus*. *BMC Genomics* 10: 256.

Huang, X., Wei, X., Sang, T., et al. 2010. Genome-wide association studies of 14 agronomic traits in rice landraces. *Nature Genetics* 42: 961–967.

Hudson, M.E., Bruggink, T., Chang, S.H., et al. 2007. Analysis of gene expression during *Brassica* seed germination using a cross-species microarray platform. *Crop Sci* 47: S-96–S-112.

Iniguez-Luy, F.L., Voort, A.V., and Osborn, T.C. 2008. Development of a set of public SSR markers derived from genomic sequence of a rapid cycling *Brassica oleracea* L. genotype. *Theor Appl Genet* 117: 977–985.

Jaccoud, D., Peng, K., Feinstein, D., and Kilian, A. 2001. Diversity arrays: A solid state technology for sequence information independent genotyping. *Nucl Acids Res* 29: e25.

Jagannath, A., Sodhi, Y.S., Gupta, V., et al. 2011. Eliminating expression of erucic acid-encoding loci allows the identification of "hidden" QTL contributing to oil quality fractions and oil content in *Brassica juncea* (Indian mustard). *Theor Appl Genet* 122(6): 1091–1103.

Jiang, Y. and Deyholos, M.K. 2010. Transcriptome analysis of secondary-wall-enriched seed coat tissues of canola (*Brassica napus* L.). *Plant Cell Rep* 29(4): 327–342.

Johnston, J.S., Pepper, A.E., Hall, A.E., et al. 2005. Evolutin of genome size in Brassicaceae. *Ann Bot* 95: 229–235.

Kaczmarek, M., Koczyk, G., Ziolkowski, P.A., and Sadowski, J. 2009. Comparative analysis of the *Brassica oleracea* genetic map and the *A. thaliana* genome. *Genome* 52: 620–633.

Kang, J., Zhang, G., Bonnema, G., Fang, Z., and Wang, X. 2008. Global analysis of gene expression in flower buds of Ms-cd1 *Brassica oleracea* conferring male sterility by using an *Arabidopsis* microarray. *Plant Mol Biol* 66(1–2): 177–192.

Kennard, W.C., Slocum, M.K., Figdore, S.S., and Osborn, T.C. 1994. Genetic analysis of morphological variation in *Brassica oleracea* using molecular markers. *Theor Appl Genet* 87: 721–732.

Kianian, S.F. and Quiros, C.F. 1992. Generation of a *Brassica oleracea* composite RFLP map: Linkage arrangements among various populations and evolutionary implications. *Theor Appl Genet* 84: 544–554.

Kim, H.N., Chal, J.S., Cho, T.J., and Kim, H.Y. 2003. Salicylic acid and wounding induce defense-related proteins in Chinese cabbage. *Korean J Biol Sci* 7: 313–219.

Kim, J.S., Chung, T.Y., King, G.J., et al. 2006. A sequence-tagged linkage map of *Brassica rapa*. *Genetics* 174(1): 29–39.

Kim, H., Choi, S.R., Bae, J., et al. 2009. Sequenced BAC anchored reference genetic map that reconciles the ten individual chromosomes of *Brassica rapa*. *BMC Genomics* 10: 432.

Kole, C., Quijada, P., Michaels, S.D., Amasino, R.M., and Osborn, T.C. 2001. Evidence for homology of flowering-time genes *VFR2* from *Brassica rapa* and *FLC* from *Arabidopsis thaliana*. *Theor Appl Genet* 102: 425–430.

Kong, F., Ge, C., Fang, X., Snowdon, R.J., and Wang, Y. 2010. Characterization of seedling proteomes and development of markers to distinguish the *Brassica* A and C genomes. *J Genet Genomics* 37(5): 333–340.

Korbes, A.P., Machado, R.D., Guzman, F., et al. 2012. Identifying conserved and novel microRNAs in developing seeds of *Brassica napus* using deep sequencing. *PLoS One* 7(11): e50663.

Kralovicova, J., Christensen, M.B., and Vorechovsky, I. 2005. Biased exon/intron distribution of cryptic and de novo 3'splice sites. *Nucleic Acids Res* 33: 4882–4898.

Kump, K.L., Bradbury, P.J., Wisser, R.J., et al. 2011. Genome-wide association study of quantitative resistance to southern leaf blight in the maize nested association mapping population. *Nature Genetics* 43: 163–168.

Lagercrantz, U. 1998. Comparative mapping between *Arabidopsis thaliana* and *Brassica nigra* indicates that *Brassica* genomes have evolved through extensive genome replication accompanied by chromosome fusions and frequent rearrangements. *Genetics* 150: 1217–1228.

Lagercrantz, U. and Lydiate, D.J. 1995. RFLP mapping in *Brassica nigra* indicates differing recombination rates in male and female meioses. *Genome* 38: 255–264.

Lagercrantz, U. and Lydiate, D.J. 1996. Comparative genome mapping in *Brassica*. *Genetics* 144: 1903–1910.

Lagercrantz, U., Osterberg, M.K., and Lascoux, M. 2002. Sequence variation and haplotype structure at the putative flowering-time locus *COL1* of *Brassica nigra*. *Mol Biol Evol* 19: 1474–1482.

Lan, T.H. and Paterson, A.H. 2000. Comparative mapping of quantitative trait loci sculpting the curd of *Brassica oleracea*. *Genetics* 155(4): 1927–1954.

Landry, B.S., Hubert, N., Crete, R., Chiang, M.S., Lincoln, S.E., and Etoh, T. 1992. A genetic map for *Brassica oleracea* based on RFLP markers detected with expressed DNA sequences and mapping of resistance genes to races 2 of *Plasmodiophora brassicae* (Woronin). *Genome* 35: 409–420.

Lee, S. and Yun, S.C. 2006. The ozone stress transcriptome of pepper (*Capsicum annuum* L.). *Mol Cells* 21: 197–205.

Lee, S.C., Lim, M.H., Kim, J.A., et al. 2008. Transcriptome analysis in *Brassica rapa* under the abiotic stresses using *Brassica* 24K oligo microarray. *Mol Cells* 26(6): 595–605.

Li, G. and Quiros, C.F. 2002. Genetic analysis, expression and molecular characterization of BoGSLELONG, a major gene involved in the aliphatic glucosinolate pathway of *Brassica* species. *Genetics* 162: 1937–1943.

Li, G. and Quiros, C.F. 2003. In planta side-chain glucosinolate modification in *Arabidopsis* byintroduction of dioxygenase *Brassica* homolog BoGSL-ALK. *Theor Appl Genet* 106: 1116–1121.

Li, F., Wu, X., Tsang, E., and Cutler, A.J. 2005. Transcriptional profiling of imbibed *Brassica napus* seed. *Genomics* 86(6): 718–730.

Li, F., Kitashiba, H., Inaba, K., and Nishio, T. 2009. A *Brassica rapa* linkage map of EST-based SNP markers for identification of candidate genes controlling flowering time and leaf morphological traits. *DNA Res* 16(6): 311–323.

Li, X., Ramchiary, N., Choi, S.R., et al. 2010. Development of a high density integrated reference genetic linkage map for the multinational *Brassica rapa* Genome Sequencing Project. *Genome* 53: 939–947.

Li, X., Ramchiary, N., Dhandapani, V., et al. 2012. Quantitative trait loci mapping in *Brassica rapa* revealed the structural and functional conservation of genetic loci governing morphological and yield component traits in the A, B, and C subgenomes of *Brassica* species. *DNA Res* 20(1): 1–16.

Lim, Y.P., Plaha, P., Choi, S.R., et al. 2006. Toward unraveling the structure of *Brassica rapa* genome. *Physiologia Plantarum* 126: 585–591.

Lin, S., Wang, J., Poon, S., Su, C., Wang, S., and Chiou, T. 2005. Differential regulation of *FLOWERING LOCUS C* expression by vernalization in cabbage and *Arabidopsis*. *Plant Physiol* 137: 1037–1048.

Liu, J., Hua, W., Zhan, G.M., et al. 2012. The *BnGRF2* gene (GRF2-like gene from *Brassica napus*) enhances seed oil production through regulating cell number and plant photosynthesis. *J Exp Bot* 63(10): 3727–3740.

Long, Y., Shi, J., Qiu, D., et al. 2007. Flowering time quantitative trait Loci analysis of oilseed brassica in multiple environments and genomewide alignment with *Arabidopsis*. *Genetics* 177(4): 2433–2444.

Lou, P., Zhao, J., Kim, J.S., et al. 2007. Quantitative trait loci for flowering time and morphological traits in multiple populations of *Brassica rapa*. *J Exp Bot* 58(14): 4005–4016.

Lou, P., Zhao, J., He, H., et al. 2008. Quantitative trait loci for glucosinolate accumulation in *Brassica rapa* leaves. *New Phytol* 179(4): 1017–1032.

Lowe, A.J., Jones, A.E., Raybould, A.F., Trick, M., Moule, C.L., and Edwards, K.J. 2002. Transferability and genome specificity of a new set of microsatellite primers among *Brassica* species of the U triangle. *Mol Ecol Notes* 2: 7–11.

Lowe, A.J., Moule, C., Trick, M., and Edwards, K.J. 2004. Efficient large-scale development of microsatellites for marker and mapping applications in *Brassica* crop species. *Theor Appl Genet* 108: 1103–1112.

Meyer, L.J., Gao, J., Xu, D., and Thelen, J.J. 2012. Phosphoproteomic analysis of seed maturation in *Arabidopsis*, rapeseed, and soybean. *Plant Physiol* 159(1): 517–528.

Mithen, R.F., Dekker, M., Verkerk, R., Rabot, S., and Johnson, I.T. 2000. The nutritional significance, biosynthesis and bioavailability of glucosinolates in human foods. *J Sci Food Agric* 80: 967–984.

Mohammadi, P.P., Moieni, A., and Komatsu, S. 2012. Comparative proteome analysis of drought-sensitive and drought-tolerant rapeseed roots and their hybrid F1 line under drought stress. *Amino Acids* 43(5): 2137–2152.

Muangprom, A. and Osborn, T.C. 2004. Characterization of a dwarf gene in *Brassica rapa*, including the identification of a candidate gene. *Theor Appl Genet* 108: 1378–1384.

Mun, J.-H., Kwon, S.-J., Yang, T.-J., et al. 2008. The first generation of a BAC-based physical map of *Brassica rapa*. *BMC Genomics* 9: 280.

Mun, J.H., Kwon, S.J., Seol, Y.J., et al. 2010. Sequence and structure of *Brassica rapa* chromosome A3. *Genome Biol* 11(9): R94.

Nishizawa, T., Tamaoki, M., Kaneko, Y., et al. 2012. High-throughput capture of nucleotide sequence polymorphisms in three *Brassica* species (*Brassicaceae*) using DNA microarrays. *Am J Bot* 99(3): e94–e96.

Okazaki, K., Sakamoto, K., Kikuchi, R., et al. 2007. Mapping and characterization of *FLC* homologs and QTL analysis of flowering time in *Brassica oleracea*. *Theor Appl Genet* 114: 595–608.

O'Neill, C.M. and Bancroft, I. 2000. Comparative physical mapping of segments of the genome of *Brassica oleracea* var. *alboglabra* that are homoeologous to sequenced regions of chromosomes 4 and 5 of *Arabidopsis thaliana*. *Plant J* 23: 233–243.

Osborn, T.C., Kole, C., Parkin, I.A.P., et al. 1997. Comparison of flowering time genes in *Brassica rapa*, *B. napus* and *Arabidopsis thaliana*. *Genetics* 146: 1123–1129.

Osborn, T.C., Butrulle, D.V., Sharp, A.G., et al. 2003. Detection and effects of a homoeologous reciprocal transposition in *Brassica napus*. *Genetics* 165: 1569–1577.

Osterberg, M.K., Shavorskaya, O., Lascoux, M., and Lagercrantz, U. 2002. Naturally occurring indel variation in the *Brassica nigra COL1* gene is associated with variation in flowering time. *Genetics* 161(1): 299–306.

Pagani, F. and Baralle, F.E. 2004. Genomic variants in exons and introns: Identifying the splicing spoilers. *Nat Rev Genet* 5: 389–396.

Panjabi, P., Jagannath, A., Bisht, N., et al. 2008. Comparative mapping of *Brassica juncea* and *Arabidopsis thaliana* using Intron Polymorphism (IP) markers: Homoeologous relationships, diversification and evolution of the A, B and C *Brassica* genomes. *BMC Genomics* 9: 113.

Park, S., Yu, H.-J., Mun, J.-H., and Lee, S.-C. 2010. Genome-wide discovery of DNA polymorphism in *Brassica rapa*. *Mol Genet Genomics* 283: 135–145.

Parkin, I.A.P., Gulden, S.M., Sharpe, A.G., et al. 2005. Segmental structure of the *Brassica napus* genome based on comparative analysis with *Arabidopsis thaliana*. *Genetics* 171: 765–781.

Peck, S.C., Nühse, T.S., Hess, D., Iglesias, A., Meins, F., and Boller, T. 2001. Directed proteomics identifies a plant-specific protein rapidly phosphorylated in response to bacterial and fungal elicitors. *Plant Cell* 13: 1467–1475.

Pino Del Carpio, D., Basnet, R.K., De Vos, R.C., et al. 2011. Comparative methods for association studies: A case study on metabolite variation in a *Brassica rapa* core collection. *PLoS One* 6(5): e19624.

Plieske, J. and Struss, D. 2001. Microsatellite markers for genome analysis in *Brassica*. I development in *Brassica napus* and abundance in *Brassicaceae* species. *Theor Appl Genet* 102: 689–694.

Pradhan, A.K., Gupta, V., Mukhopadhyay, A., Arumugam, N., Sodhi, Y.S., and Pental, D. 2003. A high-density linkage map in *Brassica juncea* (Indian mustard) using AFLP and RFLP markers. *Theor Appl Genet* 106: 607–614.

Prakash, S. and Hinata, K. 1980. Taxonomy, cytogenetics and origin of crop *Brassica*, a review. *Opera Bot* 55: 1–57.

Prakash, S. and Chopra, V.L. 1990. Cytogenetics of crop Brassicas and their allies. In: Tsuchiya, T. and Gupta, P.K. (eds), *Chromosome Engineering of Plants-Genetics and Breeding, II*. Elsevier science Publishers, The Netherlands, pp. 161–180.

Qiu, D., Morgan, C., Shi, J., et al. 2006. A comparative linkage map of oilseed rape and its use for QTL analysis of seed oil and erucic acid content. *Theor Appl Genet* 114(1): 67–80.

Quijada, P.A., Udall, J.A., Lambert, B., and Osborn, T.C. 2006. Quantitative trait analysis of seed yield and other complex traits in hybrid spring rapeseed (*Brassica napus* L.): 1. Identification of genomic regions from winter germplasm. *Theor Appl Genet* 113: 549–561.

Rabbani, M.A., Maruyama, K., Abe, H., et al. 2003. Monitoring expression profiles of rice genes under cold, drought, and high-salinity stresses and abscisic acid application using cDNA microarray and RNA gel-blot analyses. *Plant Physiol* 133: 1755–1767.

Rae, A.M., Howell, E.C., and Kearsey, M.J. 1999. More QTL for flowering time revealed by substitution lines in *Brassica oleracea*. *Heredity* 83: 586–596.

Rahman, M., McVetty, P.B., and Li, G. 2007. Development of SRAP, SNP and multiplexed SCAR molecular markers for the major seed coat color gene in *Brassica rapa* L. *Theor Appl Genet* 115(8): 1101–1107.

Rahman, M., Sun, Z., McVetty, P.B., and Li, G. 2008. High throughput genome-specific and gene-specific molecular markers for erucic acid genes in *Brassica napus* (L.) for marker-assisted selection in plant breeding. *Theor Appl Genet* 117: 895–904.

Raman, H., Raman, R. Eckermann, P., et al. 2012a. Genetic and physical mapping of flowering time loci in canola (*Brassica napus* L.). *Theor Appl Genet* 126(1): 119–132.

Raman, H., Raman, R., Nelson, M.N., et al. 2012b. Diversity array technology markers: Genetic diversity analyses and linkage map construction in rapeseed (*Brassica napus* L.). *DNA Res* 19(1): 51–65.

Ramchiary, N., Bisht, N.C., Gupta, V., et al. 2007a. QTL analysis reveals context-dependent loci for seed glucosinolate trait in the oilseed *Brassica juncea*: Importance of recurrent selection backcross scheme for the identification of 'true' QTL. *Theor Appl Genet* 116(1): 77–85.

Ramchiary, N., Padmaja, K.L., Sharma, S., et al. 2007b. Mapping of yield influencing QTL in *Brassica juncea*: Implications for breeding of a major oilseed crop of dryland areas. *Theor Appl Genet* 115(6): 807–817.

Ramchiary, N., Nguyen, D.V., Li, X., et al. 2011. Genic microsatellite markers in *Brassica rapa*: Development, characterization, mapping, and their utility in other cultivated and wild *Brassica* relatives. *DNA Res* 18: 305–320.

Rana, D., Boogaart, T., O'Neill, C.M., et al. 2004. Conservation of the microstructure of genome segments in *Brassica napus* and its diploid relatives. *Plant J* 40: 725–733.

Rensink, W.A., Iobst, S., Hart, A., Stegalkina, S., Liu, J., and Buell, C.R. 2005. Gene expression profiling of potato responses to cold, heat, and salt stress. *Funct Integr Genomics* 5: 201–207.

Robert, L.S., Robson, F., Sharpe, A., Lydiate, D.J., and Coupland, G. 1998. Conserved structure and function of the *Arabidopsis* flowering time gene *CONSTANS* in *Brassica napus*. *Plant Mol* 37: 763–772.

Saal, B., Plieske, J., Hu, J., Quiros, C.F., and Struss, D. 2001. Microsatellite markers for genome analysis in *Brassica* II assignment of rapeseed microsatellites to the A and C genomes and genetic mapping in *Brassica oleracea* L. *Theor Appl Genet* 102: 695–699.

Sauer, S., Lechner, D., Berlin, K. et al. 2000. Full flexibility genotyping of single nucleotide polymorphisms by the GOOD assay. *Nucleic Acids Res* 28: E100.

Schiavon, M., Galla, G., Wirtz, M., et al. 2012. Transcriptome profiling of genes differentially modulated by sulfur and chromium identifies potential targets for phytoremediation and reveals a complex S-Cr interplay on sulfate transport regulation in *B. juncea*. *J Hazard Mater* 239–240: 192–205.

Schierholt, A., Rcker, B., and Becker, H.C. 2001. Inheritance of high oleic acid mutations in winter oilseed rape (*Brassica napus* L.). *Crop Sci* 41: 1444–1449.

Schmidt, R. and Bancroft, I. (eds) 2011. *Genetics and Genomics of the Brassicaceae*. Springer, New York.

Schranz, M.E., Quijada, P., Sung, S.B., Lukens, L., Amasino, R.M., and Osborn, T.C. 2002. Characterization and effects of the replicated time gene *FLC* in *Brassica rapa*. *Genetics* 162: 1457–1468.

Schranz, M.E., Lysak, M.A., and Mitchell-Olds, T. 2006. The ABC's of comparative genomics in the *Brassicaceae*: Building blocks of crucifer genomes. *Trends Plant Sci* 11: 535–542.

Sebastian, R.L., Howell, E.C., King, G.K., Marshall, D.F., and Kearsey, M.J. 2000. An integrated AFLP and RFLP *Brassica oleracea* linkage map from two morphologically distinct double haploid mapping populations. *Theor Appl Genet* 100: 75–81.

Sebastian, R.L., Kearsey, M.J., and King, G.J. 2002. Identification of quantitative trait loci controlling developmental characteristics of *Brassica oleracea* L. *Theor Appl Genet* 104(4): 601–609.

Shi, J., Li, R., Qiu, D., et al. 2009. Unraveling the complex trait of crop yield with quantitative trait loci mapping in *Brassica napus*. *Genetics* 182(3): 851–861.

Slocum, M.K., Figdore, S.S., Kennard, W.C., Suzuki, J.Y., and Osborn, T.C. 1990. Linkage arrangement of restriction fragment length polymorphism loci in *Brassica oleracea*. *Theor Appl Genet* 80: 57–64.

Snowdon, R.J. and Friedt, W. 2004. Molecular markers in *Brassica* oilseed breeding: Current status and future possibilities. *Plant Breed* 123: 1–8.

Soeda, Y., Konings, M.C.,Vorst, O., et al. 2005. Gene expression programs during *Brassica oleracea* seed maturation, osmopriming, and germination are indicators of progression of the germination process and the stress tolerance level. *Plant Physiol* 137(1): 354–368.

Sonderby, I.E., Burow, M., Rowe, H.C., Kliebenstein, D.J., and Halkier, B.A. 2010. A complex interplay of three R2R3 MYB transcription factors determines the profile of aliphatic glucosinolates in *Arabidopsis*. *Plant Physiol* 153: 348–363.

Song, K.M., Suzuki, J.Y., Slocum, M.K., Williams, P.H., and Osborn, T.C. 1991. A linkage map of *Brassica rapa* base on restriction fragment length polymorphism loci. *Theor Appl Genet* 82: 296–304.

Song, K., Slocum, M.K., and Osborn, T.C. 1995. Molecular marker analysis of genes controlling morphological variation in *Brassica rapa* (syn. *campestris*). *Theor Appl Genet* 90: 1–10.

Sterken, R., Kiekens, R., Boruc, J., et al. 2012. Combined linkage and association mapping reveals CYCD5;1 as a quantitative trait gene for endoreduplication in *Arabidopsis*. *Proc Natl Acad Sci USA* 109(12): 4678–4683.

Sun, Q., Zhou, G., Cai, Y., et al. 2012. Transcriptome analysis of stem development in the tumourous stem mustard *Brassica juncea* var. *tumida* Tsen et Lee by RNA sequencing. *BMC Plant Biol* 12: 53.

Suwabe, K., Iketani, H., Nunome, T., Kage, T., and Hirrai, M. 2002. Isolation and characterization of microsatellites in *Brassica rapa* L. *Theor Appl Genet* 104: 1092–1098.

Suwabe, K., Iketani, H., Nunome, T., Ohyama, A., Hirai, M., and Fukuoka, H. 2004. Characteristics of microsatellites in *Brassica rapa* genome and their potential utilization for comparative genomics in cruiferae. *Breed Res* 54: 85–90.

Suwabe, K., Morgan, C., and Bancroft, I. 2008. Integration of *Brassica* A genome genetic linkage map between *Brassica napus* and *B. rapa*. *Genome* 51: 169–176.

Tadage, M., Sheldon, C.C., Helliwell, C.A., Stoutjesdijk, P., Dennis, E.S., and Peacock, W.J. 2001. Control of flowering time by *FLC* orthologues in *Brassica napus*. *Plant J* 28: 545–553.

Talalay, P. and Fahey, J.W. 2001. Phytochemicals from cruciferous plants protect against cancer by modulating carcinogen metabolism. *J Nutr* 131: 3027S–3033S.

Tanhuanpaa, P.K., Vilkki, J.P., and Vilkki, H.J. 1996. A linkage map of spring turnip rape based on RFLP and RAPD markers. *Agric Food Sci Finl* 5: 209–217.

Teutonico, R.A. and Osborn, T.C. 1995. Mapping loci controlling vernalization requirement in *Brassica rapa*. *Theor Appl Genet* 91: 1279–1283.

The International HapMap Consortium. 2003. The international HapMap project. *Nature* 426: 789–796.

Thormann, C.E., Ferreira, M.E., Carmago, L.E.A., Tivang, J.G., and Osborn, T.C. 1994. Comparison of RFLP and RAPD markers to estimate genetic relationships within and among cruciferous species. *Theor Appl Genet* 88: 973–980.

Tian, F., Bradbury, P.J., Brown, P.J., et al. 2011. Genome-wide association study of leaf architecture in the maizenested association mapping population. *Nat Genet* 43: 159–162.

Tonguc, M. and Griffiths, P.D. 2004. Genetics relationships of *Brassica* vegetables determined using database derived simple sequence repeats. *Euphityca* 137: 193–201.

Trick, M., Cheung, F., Drou, N., et al. 2009a. A newly-developed community microarray resource for transcriptome profiling in *Brassica* species enables the confirmation of *Brassica*-specific expressed sequences. *BMC Plant Biol* 9: 50.

Trick, M., Long, Y., Meng, J., and Bancroft, I. 2009b. Single nucleotide polymorphism (SNP) discovery in the polyploid *Brassica napus* using Solexa transcriptome sequencing. *Plant Biotechnol J* 7(4): 334–346.

Truco, M.J. and Quiros, C.F. 1994. Structure and organization of the B genome based on a linkage map in *Brassica nigra. Theor Appl Genet* 89: 590–598.

U, N. 1935. Genome analysis in *Brassica* with special reference to the experimental formation of *B. napus* and peculiar mode of fertilization. *Jpn J Bot* 7: 389–452.

Udall, J.A., Quijada, P.A., and Osborn, T.C. 2005. Detection of chromosomal rearrangements derived from homeologous recombination in four mapping populations of *Brassica napus* L. *Genetics* 169: 967–979.

Udall, J.A., Quijada, P.A., Lambert, B., and Osborn, T.C. 2006. Quantitative trait analysis of seed yield and other complex traits in hybrid spring rapeseed (*Brassica napus* L.): 2. Identification of alleles from unadapted germplasm. *Theor Appl Genet* 113(4): 597–609.

Vos, P., Hogers, R., Bleeker, M., et al. 1995. AFLP: A new technique for DNA fingerprinting. *Nucl Acids Res* 21: 4407–4414.

Vossen, R.H., Aten, E., Roos, A., and den Dunnen, J.T. 2009. High-resolution melting analysis (HRMA): More than just sequence variant screening. *Hum Mutat* 30(6): 860–866.

Wang, J., Long, Y., Wu, B., et al. 2009. The evolution of *Brassica napus FLOWERING LOCUS T* paralogues in the context of inverted chromosomal duplication blocks. *BMC Evol Biol* 9: 271.

Wang, H., Wu, J., Sun, S., et al. 2011a. Glucosinolate biosynthetic genes in *Brassica rapa. Gene* 487: 135–142.

Wang, J., Lydiate, D.J., Parkin, I.A., et al. 2011b. Integration of linkage maps for the Amphidiploid *Brassica napus* and comparative mapping with *Arabidopsis* and *Brassica rapa. BMC Genomics* 12: 101.

Wang, N., Qian, W., Suppanz, I., et al. 2011c. Flowering time variation in oilseed rape (*Brassica napus* L.) is associated with allelic variation in the *FRIGIDA* homologue *BnaA.FRI.a. J Exp Bot* 62(15): 5641–5658.

Wang, X., Wang, H., Wang, J., et al. 2011d. The genome of the mesopolyploid crop species *Brassica rapa. Nat Genet* 43: 1035–1039.

Wang, F., Li, L., Li, H., et al. 2012a. Transcriptome analysis of rosette and folding leaves in Chinese cabbage using high-throughput RNA sequencing. *Genomics* 99(5): 299–307.

Wang, W., Huang, S., Liu, Y., et al. 2012b. Construction and analysis of a high-density genetic linkage map in cabbage (*Brassica oleracea* L. var. *capitata*). *BMC Genomics* 13(1): 523.

Weigel, D. and Mott, R. 2009. The 1001 genomes project for *Arabidopsis thaliana. Genome Biol* 10(5): 107.

Wenzl, P., Carling, J., Kudrna, D., et al. 2004. Diversity arrays technology (DArT) for whole-genome profiling of barley. *PNAS* 101: 9915–9920.

Westermeier, P., Wenzel, G., and Mohler, V. 2009. Development and evaluation of single-nucleotide polymorphism markers in allotetraploid rapeseed (*Brassica napus* L.). *Theor Appl Genet* 119(7): 1301–1311.

Whittle, C.A., Malik, M.R., Li, R., and Krochko, J.E. 2010. Comparative transcript analyses of the ovule, microspore, and mature pollen in *Brassica napus. Plant Mol Biol* 72(3): 279–299.

Williams, J.G.K., Kubelik, A.R., Livak, K.J., Rafalski, J.A., and Tingey, S. 1990. DNA polymorphisms amplified by arbitrary primers are useful as genetic markers. *Nucl Acids Res* 25: 6531–6535.

Wittenberg, A.H., Van der Lee, T., Cayla, C., Kilian, A., Visser, R.G., and Schouten, H.J. 2005. Validation of the high-throughput marker technology DArT using the model plant *Arabidopsis thaliana. Mol Gen Genomics* 274: 30–39.

Wittstock, U. and Halkier, B.A. 2000. Cytochrome P450 CYP79A2 from *Arabidopsis thaliana* L. Catalyzes the conversion of L-phenylalanine to phenylacetaldoxime in the biosynthesis of benzylglucosinolate. *J Biol Chem* 275(19): 14659–14666.

Wittstock, U. and Burow, M. 2010. Glucosinolate breakdown in *Arabidopsis*: Mechanism, regulation and biological significance. *Arabidopsis Book* 8: e0134.

Wu, J., Wei, K., Cheng, F., et al. 2012. A naturally occurring InDel variation in *BraA.FLC.b* (*BrFLC2*) associated with flowering time variation in *Brassica rapa. BMC Plant Biol* 12: 151.

Wurschum, T., Liu, W., Maurer, H.P., Abel, S., and Reif, J.C. 2012a. Dissecting the genetic architecture of agronomic traits in multiple segregating populations in rapeseed (*Brassica napus* L.). *Theor Appl Genet* 124(1): 153–161.

Wurschum, T., Maurer, H.P., Dreyer, F., and Reif, J.C. 2012b. Effect of inter- and intragenic epistasis on the heritability of oil content in rapeseed (*Brassica napus* L.). *Theor Appl Genet* 126: 435–441.

Xia, L., Peng, K., Yang, S., et al. 2005. DArT for high-throughput genotyping of cassava (Manihot esculenta) and its wild relatives. *Theor Appl Genet* 110: 1092–1098.

Xiang, D., Datla, R., Li, F., et al. 2008. Development of a *Brassica* seed cDNA microarray. *Genome* 51(3): 236–242.

Xiao, Y., Cai, D., Yang, W., et al. 2012. Genetic structure and linkage disequilibrium pattern of a rapeseed (*Brassica napus* L.) association mapping panel revealed by microsatellites. *Theor Appl Genet* 125(3): 437–447.

Yadava, S.K., Arumugam, N., Mukhopadhyay, A., et al. 2012. QTL mapping of yield-associated traits in *Brassica juncea*: Meta-analysis and epistatic interactions using two different crosses between east European and Indian gene pool lines. *Theor Appl Genet* 125(7): 1553–1564.

Yang, T.J., Kim, J.S., Lim, K.B., et al. 2005. The Korea *Brassica* genome project: A glimpse of the *Brassica* genome based on comparative genome analysis with *Arabidopsis*. *Comp Funct Genomics* 6: 138–146.

Yang, S., Pang, W., Ash, G., et al. 2006a. Low level of genetic diversity in cultivated Pigeonpea compared to its wild relatives is revealed by diversity arrays technology. *Theor Appl Genet* 113: 585–595.

Yang, T.J., Kim, J.S., Kwon, S.J., et al. 2006b. Sequence-level analysis of the diploidization process in the triplicated *FLOWERING LOCUS C* region of *Brassica rapa*. *Plant Cell* 18: 1339–1347.

Yao, Y., Sun, H., Xu, F., Zhang, X., and Liu, S. 2011. Comparative proteome analysis of metabolic changes by low phosphorus stress in two *Brassica napus* genotypes. *Planta* 233(3): 523–537.

Yu, S.C., Wang, Y.J., and Zheng, X.Y. 2003. Mapping and analysis QTL controlling some morphological traits in Chinese cabbage (*Brassica campestris* L. ssp. *pekinensis*). *Acta Genet Sin* 30: 1153–1160.

Yuan, Y.X., Wu, J., Sun, R.F., et al. 2009. A naturally occurring splicing site mutation in the *Brassica rapa FLC1* gene is associated with variation in flowering time. *J Exp Bot* 60: 1299–1308.

Zang, Y.X., Kim, H.U., Kim, J.A., et al. 2009. Genome-wide identification of glucosinolate synthesis genes in *Brassica rapa*. *FEBS J* 276(13): 3559–3574.

Zhao, J. and Meng, J. 2003. Detection of loci controlling seed glucosinolate content and their associationwith Sclerotinia resistance in *Brassica napus*. *Plant Breed* 122: 19–23.

Zhao, J., Wang, X., Deng, B., et al. 2005. Genetic relationships within *Brassica rapa* as inferred from AFLP fingerprints. *Theor Appl Genet* 110: 1301–1314.

Zhao, J., Becker, H.C., Zhang, D., Zhang, Y., and Ecke, W. 2006. Conditional QTL mapping of oil content in rapeseed with respect to protein content and traits related to plant development and grain yield. *Theor Appl Genet* 113(1): 33–38.

Zhao, J., Paulo, M.J., Jamar, D., et al. 2007. Association mapping of leaf traits, flowering time, and phytate content in *Brassica rapa*. *Genome* 50(10): 963–973.

Zhao, J., Artemyeva, A., Del Carpio, D.P., et al. 2010a. Design of a *Brassica rapa* core collection for association mapping studies. *Genome* 53(11): 884–898.

Zhao, J., Kulkarni, V., Liu, N., Carpio, D.P.D., Bucher, J., and Bonnema, G. 2010b. BrFLC2 (*FLOWERING LOCUS C*) as a candidate gene for a vernalization response QTL in *Brassica rapa*. *J Exp Bot* 61: 1817–1825.

Zhu, C.S., Gore, M., Buckler, E.S., and Yu, J.M. 2008. Status and prospects of association mapping in plants. *Plant Genome* 2: 121–133.

Zhu, Y., Cao, Z., Xu, F., et al. 2012. Analysis of gene expression profiles of two near-isogenic lines differing at a QTL region affecting oil content at high temperatures during seed maturation in oilseed rape (*Brassica napus* L.). *Theor Appl Genet* 124(3): 515–531.

Zou, J., Jiang, C., Cao, Z., et al. 2010. Association mapping of seed oil content in *Brassica napus* and comparison with quantitative trait loci identified from linkage mapping. *Genome* 53(11): 908–916.

7

Integrating Omics Approaches in Sugarcane Improvement

Rachayya M. Devarumath, PhD; Sachin B. Kalwade, PhD; Pranali A. Kulkarni, PhD; Suman S. Sheelavanthmath, PhD; and Penna Suprasanna, PhD

CONTENTS

7.1 Introduction 191
7.2 Sugarcane Initiatives 195
    7.2.1 Sugarcane Synteny and Colinearity with Other Poaceae Members 195
    7.2.2 Sugarcane Genome 196
        7.2.2.1 Sequencing 197
        7.2.2.2 Gene Discovery 199
7.3 Sugarcane Transcriptome 200
    7.3.1 Sources of Transcriptomes 200
    7.3.2 Functional Integrity of Transcriptomics in Sugarcane 203
    7.3.3 Comparative Transcriptomics and Reverse Genetics 206
    7.3.4 Transcriptome Polymorphism for Functional Studies 207
7.4 Transcriptome Changes in Response to Abiotic and Biotic Stresses 208
7.5 Integration of Metabolomics with Genomics 212
7.6 Sugarcane Proteomics 214
7.7 Sugarcane Molecular Breeding 216
    7.7.1 DNA Markers in Sugarcane 216
    7.7.2 Genetic Linkage Maps in Sugarcane 217
    7.7.3 QTLs in Sugarcane 219
    7.7.4 Molecular Markers for Diagnostics 221
7.8 Genetic Transformation in Sugarcane 221
7.9 Metabolic Engineering of Sugarcane and Biofactory 227
7.10 Genetic Variability through *In Vitro* Culture and Mutagenesis 228
7.11 Conclusions and Future Perspective 229
References 230

7.1 Introduction

Sugarcane belongs to the grass family (*Poaceae*) and is the source of about 75% of the world's sucrose, with a value of more than US$150 billion per year. Apart from its traditional use as a source of sugar, sugarcane is fast becoming a source for ethanol and biomass production, which is an alternative energy source to the nonrenewable resources in many countries, helping to build up their economies. Sugarcane is cultivated in the tropical and subtropical regions of more than 90 countries, with the area under cultivation close to 20 million hectares (FAO; http://faostat.fao.org/, http://www.illovo.co.za/worldofsugar). The Indian sugar industry plays an important role in the global market as the world's second largest producer after Brazil, producing nearly 15% of world sugar and 25% of world sugarcane per annum under a wide range of agroclimatic conditions. Currently, the industry produces around 300–350 million tons (Mt) of cane, 20–22 Mt of white sugar, and 6–8 Mt of jaggery and khandsari to meet the domestic consumption of sweeteners (Solomon 2011). After the petroleum crisis of the mid-1970s, sugarcane often was used as a source of bagasse for boiler fuel and alcohol for motor fuel in many countries. This has accelerated sugarcane production in most of the developing and developed countries globally (Alexander 1986). India is also looking forward through molecular and allied research after the E10 ethanol blending program (Solomon 2011). It has been stated that the developing countries consume roughly 26% of the world's energy (Louime et al. 2012). Worldwide sugarcane researchers are trying to enhance the potential of sugarcane to make it a substitute for the nonrenewable fossil fuels. The progress of sugarcane research emerges from conventional breeding, genome understanding, gene discovery, and molecular breeding. From selection of existing variations in prehistoric times to the current biparental/multiparental crossing and subsequent use of nonconventional techniques, crop improvement has concentrated mostly on improving the yield and sugar content.

Sugarcane is a genetically complex polyploidy grass of family Poaceae, tribe Andropogoneae, and genus *Saccharum*. It comprises six species, namely *S. officinarum* L. ($2n = 80$), two wild species *S. robustum* Brandes and Jeswiet ex Grassl ($2n = 60–80$), and *S. spontaneum* L. ($2n = 40–128$), and three secondary species, *S. barberi* Jeswiet ($2n = 81–124$), *S. sinense* Roxb. ($2n = 111–120$), and *S. edule* Hassk. ($2n = 60, 70, 80$) (Selvi et al. 2006); *S. officinarum*, *S. spontaneum*, and *S. robustum* represent the basic species. *S. officinarum*, however, is believed to have evolved through hybridization of species such as *Erianthus arundinaceus* (Retz.) Jeswiet, *S. spontaneum*, and *S. robustum* (Daniels et al. 1975), whereas *S. barberi* and *S. sinense* are the secondary ones believed to be natural hybrids between *S. officinarum* and *S. spontaneum* (Daniels and Roach 1987). The cultivated sugarcane *Saccharum* spp. are believed to have originated from complex hybridization events (termed *nobilization*) between *S. officinarum*, *S. barberi*, *S. sinense*, and the wild related species *S. spontaneum* (Sreenivasan et al. 1987). Other genera such as *Erianthus* Michx., *Miscanthus* Anderss, *Narenga* Burkiee, and *Sclerostachya* (Hack.) A. Camus are closely related to *Saccharum*. Mukherjee (1957) coined the term "*Saccharum* complex" to encompass all the above species and genera, which constitute an inbreeding group (Daniels et al. 1975). The mutual relationship and actual contribution of these different genera, however, remain unclear due to their high and variable ploidy levels (D'Hont et al. 1996; Nair et al. 1999). Current sugarcane cultivars are estimated to possess 80%–90% of the genome from *S. officinarum* and 10%–20% from *S. spontaneum* (Grivet et al. 1996; Hoarau et al. 2002; D'Hont 2005; Piperidis et al. 2010). Sugarcane's basic monoploid genome ranges between 760 and 930 Mb depending on the cultivar breeding history, which is more than twice the size of the rice genome (389 Mb) and is close to the size of the sorghum genome (730 Mb) (D'Hont and Glaszmann 2001).

To understand the integrated omics (transcriptomics, metabolomics, and proteomics) approaches in sugarcane for important traits such as sucrose concentration, fiber content, higher yield, and tolerance of cold, drought (water deprivation), and other stress conditions, it is important to know about the physiology, genome structure, functional integrity, and colinearity of sugarcane with other more or less similar crops. Sugarcane is a typical glycophyte and hence exhibits stunted growth or no growth under salinity, with its yield falling to 50% or less of its true potential. To meet the demand, the development of new sugarcane varieties for improved productivity, tolerance to biotic and abiotic stresses, nutrient management, and sugar recovery is a challenge that has to be overcome.

Drought and salinity are among the most severe limitations on the yield of sugarcane. Indian sugarcane is affected by salinity and water scarcity, creating osmotic and ionic stress for the plant. This stress induces various biochemical and physiological responses in plants as a survival mechanism (Seki et al. 2003; Patade et al. 2010). Various strategies are developed in plants under stress conditions. In terms of biochemical responses, stress induces accumulation of functioning solutes, such as proline, sugar, sugar alcohol, and betaine in plants (Ingram and Bartels 1996; Rhodes and Hanson 1993). These compounds facilitate adaptation of plants under severe circumstances. Regarding the genetic responses, a variety of genes have been reported to be induced by stress in various plant species. The function of the expressed proteins has been predicted to play a vital role in the adaptive response to stress (Ingram and Bartels 1996; Bray 1997; Seki et al. 2001). Salinity in the root zones of sugarcane decreases the sucrose yield, through its effect on both biomass and juice quality. The complexity and polygenic nature of salinity tolerance has further limited the efforts to develop tolerant crop varieties through conventional breeding practices. In this regard, biotechnological approaches including somaclonal variation, and *in vitro* mutagenesis and selection are being applied for the isolation of agronomically useful mutants (Jain 2005; Suprasanna et al. 2012). Agronomically improved sugarcane varieties endowed with tolerance to biotic and abiotic stresses are highly beneficial, as unfavorable environmental factors can challenge cultivation and crop productivity. Although crops tolerant to biotic and abiotic stresses have been selected by traditional breeding programs, increasing the pace is essential to develop improved varieties. Transcriptome approaches have proved to be an efficient tool in finding functional genes, and are useful for the development of new sugarcane cultivars in association with the other omics such as proteomics and metabolomics (Figure 7.1).

Sugarcane is a complex polyploidy crop, and hence no single technique is found to be best for confirmation of polygenic and phenotypic characteristics. Approaches such as metabolomics and proteomics open up the complex network association through *in silico* methodology and laboratory work for particular traits. The stress-inducible genes in response to abnormal conditions have been classified into two groups: proteins directly protecting against stress, and proteins involved in regulation of gene expression and signal transduction. In the case of drought, the first group includes proteins that probably function by protecting cells from dehydration, such as the enzymes required for the biosynthesis of various osmoprotectants, late embryogenesis abundant proteins, and detoxification enzymes. The second group includes regulatory proteins such as transcription factors and protein kinases (Seki et al. 2003; Shinozaki and Yamaguchi-Shinozaki 2000). Identification of such biochemical markers is of great importance in plant breeding strategy since this could help to improve field performance under stress conditions.

Several sugarcane expressed sequence tag (EST) databases have been generated (Casu et al. 2003; Carson and Botha 2002; Carson et al. 2002a,b; Vettore et al. 2003; Ma et al. 2004; Bower et al. 2005; Gupta et al. 2010). The publicly available sugarcane ESTs were assembled into tentative consensus sequences referred to as the Sugarcane Gene Index, mainly composed by sequences from the Brazilian sugarcane EST project (SUCEST; Vettore et al. 2003). The SUCEST project generated 237,954 ESTs, which were organized into 43,141 putative unique sugarcane transcripts referred to as sugarcane-assembled sequences (SASs). These ESTs were used to develop molecular markers such as microsatellite simple sequence

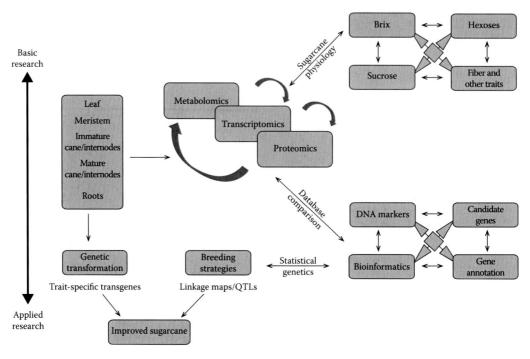

FIGURE 7.1
Diagrammatic representation of the central role of integrated omics (metabolomics, transcriptomics, and proteomics) and their contribution to the development of improved sugarcane through genetic transformation and classical breeding.

repeats (SSRs) and single-nucleotide polymorphisms (SNPs), which were successfully used to produce linkage maps and identify quantitative trait loci (QTLs) for important agronomical traits (Oliveira et al. 2007; Pastina et al. 2012; Somerville et al. 2010).

Apart from the morphological, physiological, and more refined transcriptomics studies, sucrose and other hexose sugars are of great importance as they persist up to the last month in the sink tissue of the sugarcane growth cycle. Metabolite interconversions and retention are under the control of cascades of enzymes and transporter and vacuolar cells in sugarcane. Metabolite profiling provides a deeper insight into complex regulatory processes of the potential metabolites such as fructose, glucose, sucrose, and other derivatives by using high-throughput techniques by which the metabolic phenotypes can be determined. The protein profiling of the stress-inducible sugarcane by two-dimensional (2-D) electrophoresis and its characterization predict the role of different proteins and their involvement in the cell signaling mechanisms that are consistently induced in response to the stresses (Jangpromma et al. 2007). From the integrated omics data one can depict gene–protein–metabolite networks as a functional model. Since sugarcane diversification is a factor in the selection of diversified cultivars that can be used for the introgression of desirable characteristics, molecular markers lead the genome dissection or analysis. Molecular approaches have evolved over the past four to five decades to address these issues and make sugarcane breeding programs faster and more precise. Molecular markers such as restriction fragment length polymorphisms (RFLPs), random-amplified polymorphic DNA (RAPD), amplified fragment length polymorphisms (AFLPs), SSRs, single-strand conformation polymorphisms (SSCPs), intersimple sequence repeats (ISSRs), single-nucleotide polymorphisms (SNPs), target region amplification polymorphisms (TRAPs), and sequence-related amplified polymorphisms (SRAPs) have been used for various applications in sugarcane breeding to assess the genetic variability, to construct genome maps, and to tag genes for economically important traits and for comparative and functional genomics (Selvi and Nair 2010; Swapna and Srivastava 2012). In early genomic research, molecular markers helped to elucidate the genome structure of modern sugarcane cultivars and derive phylogenetic relationships among the *Saccharum* complex. Later much work focused on the construction of genetic maps, the detection of major genes, and the usage of linked markers in selection programs. Expressed sequence tags and microarrays have been used for the identification and functional analysis of an array of candidate genes that are expressed under various biotic and abiotic stress conditions. Molecular markers have great potential in improving the efficiency of the breeding programs, by serving as a valuable selection tool for the breeder. In conventional sugarcane breeding programs one cycle takes, on average, 12 years from hybridization to the release of cultivars. Markers linked to economically important traits would increase the efficiency of indirect and early selection in sugarcane for several traits at one point in time, as well as assisting the precise selection of the genotypes (Selvi and Nair 2010).

There has been a tremendous success in gene transfer from a wide variety of plant and nonplant sources to sugarcane. With the availability of efficient transformation systems, selectable marker genes and genetic engineering tools, it should also be possible to clone and characterize useful genes and improve commercially important traits into elite germplasm that subsequently lead to the development of an ideal plant type of sugarcane (Lakshmanan et al. 2005; Suprasanna et al. 2011). The range of potential applications being developed through transgenic plants in sugarcane include insect resistance, virus resistance, altered sucrose content, lignin modification, sucrose accumulation, and herbicide resistance. Sustained efforts are also being made to engineer sugarcane varieties that can produce high-value compounds such as pharmaceutically important proteins, functional foods and nutraceuticals, biopolymers, precursors and enzymes, and biopigments. This will go a long way in launching sugarcane as a biofactory (Brumbley et al. 2008; Suprasanna et al. 2011). This review chapter outlines sugarcane omics and its role in improvement through biotechnological tools.

7.2 Sugarcane Initiatives

7.2.1 Sugarcane Synteny and Colinearity with Other Poaceae Members

Grass family members have evolved as native grass, weeds, or crops (Clayton and Renvoize 1999). They possess a highly conserved genome organization; less diversification in and among the genes resulted in the evolution of new species independently belonging to the same family with common ancestors (Bennetzen and Freeling 1993, 1997; Ming et al. 1998). Grasses are highly competitive, and some species, such as sugarcane, sorghum, and maize, contain anatomical and physiological adaptations to optimize CO_2 fixation for carbohydrate biosynthesis as C4 pathways, with more or less similar physiobiochemical properties. While at the genetic level use of molecular marker techniques showed evolutionary trends, the composite association of the possible targets can be traced through the multi-network aspect of omics, molecular data, and bioinformatic tools and comparison with model plants (Figure 7.1).

Molecular marker techniques are preferentially used to dissect the genome because of its specificity, robustness, and the fact that it requires less DNA and is not affected by environmental conditions during the developmental stages of the plant. There are numerous polymerase chain reaction (PCR)-based molecular marker techniques such as RAPD (Nair et al. 1999; Kawar et al. 2009), RFLP (Powell et al. 1996), AFLP (Selvi et al. 2005), SSR or microsatellite (Selvi et al. 2003; Maccheroni et al. 2009), sequence tag microsatellite (STMS) (Singh et al. 2005; Babu et al. 2010), ISSR (Shrivastava and Gupta 2008; Kalwade et al. 2012), and TRAP (Alwala et al. 2006a) that have been used to explain genetic diversity among different accessions of sugarcane and revealed the association and evolution among the hybrid cultivars and the wild-type sugarcane. Likewise the comparative mapping of sugarcane with sorghum showed that the sugarcane and sorghum chromosomes are highly syntenic and conserved within the order of DNA sequence in their chromosomes (Glaszmann et al. 1997; Dufour et al. 1997; Guimaraes et al. 1997).

Janoo et al. (2003) studied the alcohol dehydrogenase (Adh) genes of sugarcane and sorghum and found a high colinearity in gene order, gene structure, and nucleotide similarity, close to 95%. These findings were significant in establishing the genetic relationship and synteny of sugarcane with the sorghum genome (Dufour et al. 1997; Glaszmann et al. 1997; Ming et al. 1998). Bowyer and Leegood (1997) had demonstrated the $NADP^+$–malic enzyme (NADP–ME) pathway operation in sugarcane, maize, and sorghum; this finding shows the association of biochemical pathways in grass family members. A highly expressed photosynthesis-related phosphoenolpyruvatecarboxykinase (PEPCK) had already been detected and validated in maize leaf bundle sheath cells (Furumoto et al. 2000; Taiz and Zeiger 1998). The limited availability of sequence information from sugarcane and the sorghum tissue library opens the way for gene prediction and exploring the addition of informative genes to the public database via transcriptome analysis.

7.2.2 Sugarcane Genome

Present-day modern cultivars are interspecific hybrids having genome $2n = 100-300$ and are believed to have originated from a complex hybridization event; it is estimated that about 70%–80% are from *S. officinarum* ($2n = 80$), 10%–20% from *S. spontaneum* ($2n = 40-128$), (Sreenivasan et al. 1987; Oliveira et al. 2007; Butterfield et al. 2001), and few are from recombination of two genomes (up to 10%–25%) (D'Hont et al. 1996). The genetic view of *S. officinarum* and *S. spontaneum* shows basic chromosome $x = 10$ and $x = 8$, respectively (Grivet and Arruda 2002). The interspecific crosses that were a turning point in the evolution of modern cultivars were followed by backcrosses to *S. officinarum* clones to recover types adapted to increase sugar content (Price 1965; Arceneaux 1965). This introgression event multiplies the alleles, and the polyaneuploid nature of the sugarcane genotypes was coined as *nobilization* of sugarcane.

Because of the unusual meiosis and aneuploid nature of the sugarcane genome, there is uneven distribution of homologs with chromosome inherited by a recombination event and in the progenies developed through a crossing event program (Burner and Legendre 1994; Grivet et al. 1996). Hence the coexistence of the chromosome was introduced from the crosses mostly performed within interspecific hybrids, *S. officinarum*, *S. spontaneum*, and *S. barberi*, *Erianthus* used for introgression of the desirable characters such as drought tolerance, high sucrose content, and disease resistance (Grivet and Arruda 2002). The non-replicated genome size of a somatic sugarcane cell (2C) is estimated to be 7440 Mb in *S. officinarum* because of the octaploid nature of its genome; however, a complete nonredundant chromosome set should be eight-fold smaller, ~930 Mb (D'Hont and Glaszmann 2001). The modern cultivars of sugarcane have around 120 chromosomes and genome size of 10,000 Mb (D'Hont 2005), which is about 10 times higher, and comparable with the estimated monoploid genome of *S. officinarum* and *S. spontaneum*: around 930 and 750 Mb, respectively. The sorghum genome comprises ~760 Mb, which is twice that of the rice genome (~390) (Arruda and Silva 2007).

With interest in the wide genome of sugarcane in comparison with the other crop genomes, most of the research is being conducted by geneticists trying to decode the relationships of the complex gene networks. Among the Poaceae species the level of the genome varies from diploid to decaploid (Arumuganathan and Earle 1991; Paterson et al. 2009). The syntenic region of the genome maintains the order of gene, which suggests the origin and conservation of the gene function (Manetti et al. 2012). The expansion of the genome in grass has strong support from the transposable element (TE) intervening between coding genes (McClintock 1956; Janoo et al. 2003). TEs are categorized into two groups: transposons and retrotransposons. The long terminal repeat (LTR) retrotransposons are the most abundant retroelements in the plant. The transposons show a deletion–insertion mechanism on the genome which is associated with the specific transposase protein that acts as a driving signal molecule, while movement of the retrotransposons depends on the active site of the transcription, which provides the substrate for reverse transcription into DNA copy and is reinserted into the genome and increases its copy number after each propagation cycle. Recent studies reveal that apart from the TEs' retraction (Kalendar et al. 2000; Piegu et al. 2006), there is gene remodeling resulting in the generation of new genes (Kazazian 2004; Cordaux et al. 2006); this shows expansion of a new regulatory network for altering gene expression (Kashkush et al. 2003; Muotri et al. 2007; Feschotte 2008). Studies on TEs on barley (Shirasu et al. 2000) and hexaploid wheat (Devos et al. 2005) strongly support and provide close association of TE with genome structure. Hence the grass families, particularly sugarcane, with wide genomes, possess an extent of TEs that can progressively be triggered and analyzed through the functional transcriptomic approach. This can unravel the complexity of important traits of sugarcane such as sucrose accumulation, fiber content, and pathogen-resistant protein studies. Also, the transcriptome modulation under abiotic stress induces the signal for adaptation or resistance to salt, temperature, drought, toxic metal, and nutrient stress.

Recent research has revealed that the mutator-like transposases are the most represented transposons transcript in the sugarcane transcriptome (Manetti et al. 2012). Sugarcane transcriptome analysis and reverse genetics studies confirmed that the 173 amino acid mutator-like transposons provided evidence of at least four (I–IV) classes in the monocots/eudicots (Rossi et al. 2004).

7.2.2.1 Sequencing

The intricacies and size of the genome have been a major restraint in sugarcane genetic improvement. While continued selective breeding for improved sucrose accumulation has been responsible for over half of the yield augmentation in the past 50 years, it has been seen as having reached a plateau due to constraints on the gene pool exploited in the traditional breeding agenda (Mariotti 2002). Present knowledge about sugarcane gene regulation is short of sufficient genomic resources to comprehend transcriptomic variation among sugarcane genotypes or even among different copies of the same gene in the same individual cultivar. The goal of the Sugarcane Genome Sequencing Initiative (SUGESI) was to identify different strategies for sugarcane genome sequencing in the light of new technologies. The utmost efforts in sugarcane-applied genomics are based on categorization of the sugarcane genes discovered in ESTs, especially from the expression patterns evaluated by quantitative real-time PCR (qRT-PCR) and microarray and macroarray assays. An integrated use of these approaches, viz. whole-genome shotgun sequencing (SP80-3280, Q165, LA Purple and SES 208), bacterial artificial chromosome (BAC) end sequencing, R570 BAC selection and sequencing, ChIP-Seq, EST resequencing and high-density genetic mapping, was adopted by SUGESI to obtain a reference sequence for one sugarcane cultivar but mainly focused on BAC cloning. Bacterial artificial chromosomes are often used to sequence the genome of organisms. Short pieces of the organism's DNA are amplified as an insert in a BAC vector and then sequenced. The sequenced parts are rearranged *in silico*, resulting in the genomic sequence of the organism. To date, only two sugarcane BACs have been deposited in the public databases (Jannoo et al. 2007). The initially formulated strategy to target the gene-rich active euchromatin was designed using the BAC cloning and sequencing for cultivar R570, as it is the most intensively characterized to date. It is the only sugarcane cultivar that has a publicly accessible BAC library. Around 4000–5000 BAC sequences of monoploid euchromatic genome of this cultivar were isolated (Souza et al. 2011). Further BAC clone selection was done through EST hybridization which generated the sequence data. The genome coverage of using this library is to be around 1.3×, suggesting that recovery of all alleles would not be efficient. BAC screening was undertaken using a fast and efficient 3D-pool approach which was developed at the French Plant Genomic Resources Center (CNRGV) followed by PCR amplification of genes of interest. The sugarcane genomics community is now looking forward to taking up the resources for sugarcane genome analyses by the development of additional BAC libraries from other cultivars and parental species (*S. officinarum* and *S. spontaneum*) as well as producing a detailed physical map from one reference cultivar, probably R570. This detailed analysis of the sugarcane BACs will make available crucial information for understanding sugarcane genome structure. Further, sequencing efforts stimulate gene and allele discovery and crop improvement in sorghum as they did in rice. Genetic analysis showed close similarity with sorghum and hence sugarcane improvement can be enhanced by the application of genomic tools to explore wild relatives in the *Sorghum* and *Saccharum* genera.

Sugarcane genome sequencing and assembly could be advanced by introducing the technological advances of third-generation sequencers having longer read lengths. These third-generation DNA-sequencing technologies will, it is hoped, provide the breakthrough for amassing this autopolyploid genome. The basic theory of these methods is electrical detection of the DNA sequences, by reading the electron tunneling current measured across the pore transversally to the translocating DNA, which varies with a base passing through the pore between the electrodes (Lee and Thundat 2001, 2004; Zwolak and Di Ventra 2005; Zhang et al. 2006b; Lagerqvist et al. 2006; Zikic et al. 2006). The sequencing of the GT-rich and G-rich templates of sugarcane was through BigDye™ Terminator v3.0 with Applied Biosystems 3130 Genetic Analyzer ABI 3730 DNA Analyzer, ABI PRISM® 3100 and 310 Analyzers (Applied Biosystems, Foster City, CA), and Illumina Genome Analyzer IIx DNA sequencers. The BigDye Terminator has shown accuracy for a wide range of applications including the comparative long-read and *de novo* sequencing of vast genomes. This improved dye terminator method showed enhancements in the new chemistry combinations which imparted even dye mobility characteristics, and optimized and strong signal balance especially for longer read lengths, which benefited the sequencing of these BAC vectors. Mutant populations will aid augmentation of the scope for gene discovery and genetic manipulation in the sugarcane genome. The protocols for EcoTILLING (a variant of TILLING: targeting induced local lesions in genomes) (Cordeiro et al. 2006) and quantitative SNP analysis will be valuable for gene mapping, gene discovery, and association genetics in sugarcane. The accessibility of a sorghum genome sequence will further hasten the potential to apply these techniques in both sorghum and sugarcane. Gene discovery will also be assisted by application of advances in expression profiling tools, as have been applied to other crop species in the Poaceae (McIntosh et al. 2007). In the supplementary data provided by Paterson et al. (2009) and analysis carried out by Wang et al. (2010), sequences of 20 sugarcane BACs were compared to their respective homologous regions in sorghum and 189 and 209 genes could be identified from sorghum and sugarcane, respectively, based on gene-calling algorithms and EST information. Of these, 12 appeared to be sorghum-specific and 19 to be from sugarcane. These comparative data signify the occurrence of gene duplication, which might have led to some differences in the gene family numbers between the two grasses.

7.2.2.2 Gene Discovery

Advancement in modern technologies comprising highly efficient SNP identification and genome mapping, DNA microarray technologies for gene expression analysis, RNA interference (RNAi) technology, and the rapid improvement in data mining tools is expected to have a major influence on future sugarcane crop improvement programs. Alternative strategies are being applied for comprehending differentially expressed genes at reasonable extents. The projects include the cDNA suppression subtraction hybridization (SSH) used to study abiotic stresses such as salt (Patade et al. 2010) and water deficits (Prabu et al. 2011) and in response to fungal pathogens (Borras-Hidalgo et al. 2005). The cDNA-AFLP technique has also been put forward to identify genes associated with the infection of rust fungus (Carmona et al. 2004). Practicing these techniques provided useful candidate genes but little information regarding the overall reprogramming of the gene expression and its function. Recent studies reveal the gene expression in sugarcane to be highly enhanced by the advent of qRT-PCR and semi-qRT-PCR. The use of the gene-specific primers allows the qRT-PCR to work powerfully for the differential analysis of gene expression within the gene families.

Identification and classification of the genes expressed in disease-resistant sugarcane, and those involved in carbohydrate metabolism (Ulian and Burnquist 2000; Grivet et al. 2001; Casu et al. 2005), have appended new information to the sugarcane gene pool. Moreover, various tissue-specific EST libraries of Indian subtropical sugarcane varieties, viz. CoS767 and Co 1148 (Gupta et al. 2010), sugarcane cv. Co740 (Prabu et al. 2011), Co 62175 (Pagariya et al. 2010), and SCGS-infected sugarcane (Kawar et al. 2010) have been characterized and greatly contributed to the gene discovery process. Nearly 25 sequence clusters related to water-deficit stress showed greater relative expression during 9 h dehydration stress in CoS767 variety (Gupta et al. 2010). Further, Prabu et al. (2011), based on sqRT-PCR analysis, showed higher transcript expression of 22-kDa drought-induced protein, transcription factors such as WRKY and ScMYBAS1-3 (Prabu et al. 2011), transcription factor-inducible promoter PScMYBAS1 (Prabu and Prasad 2012), myo-inositol-3-phosphate synthase (MIPS) and ornithine-oxo-acid amino transferase (OAT) at initial stages of stress induction in Co740, with a gradual decrease in advanced stages. The cDNA RAPD-based gene profile expression of Co62175 (Pagariya 2012) led to the identification of 335 early growth stage related, differentially expressed transcript-derived fragments (TDFs), of which 156 up- and 85 downregulated were sequenced. The response of SCGS infection in sugarcane (Kawar et al. 2010) triggered transcriptional mechanism expressing the R2R3-MYB (SoMYB18, Acc No: FJ560976) transcription factor gene (Kulkarni et al. 2011), which could be potent in the regulation of secondary metabolism signal transduction during biotic, abiotic, and other environmental stresses.

Additionally, a cDNA library from sugarcane plants stressed with NaCl (200 mM) for 0.5–18 h (Patade et al. 2011) was constructed to find mRNA species that are differentially expressed in sugarcane in response to salinity stress. To understand the polyethylene glycol (PEG) 8000 (Patade 2009) and salt-induced genes in sugarcane leaves, the SSH approach provided target rare transcripts, such as those contributing to cell signaling and the regulation of gene expression (Patade et al. 2011). Applications of the gene expression profiling in segregating populations and collecting varied genotypes have exhibited disparity in traits such as sucrose accumulation (Casu et al. 2005; Papini-Terzi et al. 2009), putative serine–threonine protein phosphatase-encoding gene (TC142199), gene of unknown function (TC115292) (Casu et al. 2005), and number of genes that showed differential regulation among the low and high BRIX genotypes (Papini-Terzi et al. 2009).

Sugarcane genomics is also focusing on the isolation of novel transcription factors and studies in their functional characterization. One such attempt has been made to isolate MYB transcription factors from sugarcane wild relative species such as *E. arundinaceus*, EaMYB2R, EaMYB18 (Kulkarni and Devarumath 2012; Kulkarni et al. 2013) *S. spontaneum* SsMYB2R (in review), *S. barberi* SbMYB18, *S. officinarum* Q63 SoQMYB18 and Vellai SoVMYB18, Narenga NaMYB18, *E. ciliaris* EcMYB18, *E. elegans* EeMYB18, and *S. robustum* SrMYB18. The isolation and *in silico* characterization of these transcription factors will reveal the function of the domains and motifs involved in the various transcriptional regulatory mechanisms and expression studies. The transformation of these factors in the cultivated sugarcane might revolutionize the development of sugarcane with resistance to diseases and abiotic stresses, and increase metabolism, vigor, and rationing capability.

In addition, a different and promising approach for gene discovery in sugarcane and its improvement is to use quantitative methods to detect allele dosage in sugarcane in combination with techniques such as pyrophosphate sequencing using the Pyrosequencer™ platform (Cordeiro et al. 2006) and mass spectrometry using the Sequenom™ platform (Cordeiro et al. 2006). These methods allow the quantitative detection of frequencies of consensus to alternate SNP bases at any particular SNP locus. Utilizing a group of SNP markers developed to the same EST or gene makes it possible to infer the likely copy number of the EST or gene in the genome. These data will provide information to stockpile the updated characteristics of the constructive alleles from the single genotype of hybrid cane to be figured out through statistical approaches (Cordeiro et al. 2006). Categorization and identification of such QTLs by molecular markers will help to exploit and facilitate the gene accumulation.

7.3 Sugarcane Transcriptome

7.3.1 Sources of Transcriptomes

Transcriptome analysis is a significant approach to screen candidate genes, predict gene function, and discover *cis*-regulatory motifs in the selected experimental plants. The sugarcane genome is not yet completely sequenced because of the wide genome size (10 Gb) of modern cultivars (D'Hont and Glaszmann 2001), but the mRNA as a transcribable part of the genome constitutes most of the important traits, such as sucrose synthesis, transportation, accumulation, pathogenic disease resistance to red rot and yellow leaf, tolerance of drought and salt, and cellulose content for biomass production. Transcriptome can target different tissues of plant starting from root, stem, meristematic and leaf tissues of sugarcane (Azevedo et al. 2011). Transcriptome analysis is the hybridization-based method used in microarray and GeneChips through which large-scale gene expression profiles can be acquired. The information for identification of the genes in the targeted sugarcane can basically be taken from *in silico* methodologies, mainly from probe hybridization array, ESTs, or known genes of other related crops. The rapid accumulation of large-scale data containing gene expression profiles and the ability of related databases to support the availability of such large repositories of data have provided access to large amounts of information in the public domain. These public domain data are an efficient and valuable resource for many secondary uses such as coexpression and comparative analyses (Mochida and Shinozaki 2010). Furthermore, as a next-generation DNA sequencing application, deep sequencing of short fragments of expressed RNAs, including sRNAs, is quickly becoming an efficient tool for use with genome-sequenced species (Harbers and Carninci 2005; de Hoon and Hayashizaki 2008). Currently more than 1000 species' ESTs have been submitted to an EST database (dbEST) which is made available through NCBI (http://www.ncbi.nlm.nih.gov/projects/dbEST). Most of the sugarcane hybrid cultivars have been used for the EST analysis shown in Table 7.1 (data extracted from NCBI dbESTs). Plants represent ~18% of all ESTs in the dbEST, and sugarcane, maize, barley, rice, and wheat contribute half of these sequences (Arruda and Silva 2007). Some of the crops' ESTs are compared in Table 7.2. Previously in the 1990s, gene discovery was limited in sugarcane (Vettore et al. 2003) and confined mostly to sucrose metabolism gene identification (Kumar et al. 1992;

TABLE 7.1

Expressed Sequence Tag Corresponding to
Saccharum Species and Hybrids Lodged in dbEST
(NCBI: Nov. 2012)

Saccharum **Species and Hybrids**	**# of ESTs**
S. arundinaceum	340
S. officinarum	105
Saccharum hybrid cultivar	283,117
Saccharum hybrid cultivar SP80-3280	135,534
Saccharum hybrid cultivar CoS767	25,382
Saccharum hybrid cultivar Co1148	1069
Saccharum hybrid cultivar CoC671	315
Saccharum hybrid cultivar Co740	310
Saccharum hybrid cultivar Co62175	57
Saccharum hybrid cultivar Co86032	27
Saccharum hybrid cultivar (mixed)	73,778
Saccharum hybrid cultivar SP70-1143	24,313
Saccharum hybrid cultivar Q117	9,141
Saccharum hybrid cultivar CP72-2086	7,993
Saccharum hybrid cultivar NCo376	535
Saccharum hybrid cultivar BO91	41
Saccharum hybrid cultivar H50-7209	27
Saccharum hybrid cultivar TCP02-4589	9
All other taxa	49

TABLE 7.2

Comparison of Submitted EST Sequences of Sugarcane with Plant Species

Sr. No.	Species Name	ESTs	ETs	TCs	Singleton ESTs	Singleton ETs	Total Unigen
1	*S. officinarum* (sugarcane)	282,683	499	42,377	78,924	41	121,342
2	*Sorghum bicolor* (sorghum)	208,008	1,426	23,442	22,326	257	46,025
3	*Hordeum vulgare* (barley)	502,606	4,630	43,310	39,494	177	82,981
4	*Zea mays* (maize)	1,689,899	8,746	112,156	202,621	310	315,087
5	*Triticum aestivum* (wheat)	1,053,965	3,750	93,508	128,166	251	221,925
6	*Arabidopsis thaliana*	972,675	131,754	49,009	55,364	8,454	112,827
7	*Oryza sativa* (rice)	1,230,504	134,877	82,184	99,774	20,184	202,142
8	*Glycine max* (soybean)	1,354,268	2,464	73,178	63,866	130	137,174
9	*Panicum virgatum* (switchgrass)	544,707	2	65,074	37,152	0	102,226
10	*Solanum tuberosum* (potato)	236,740	2,819	31,972	30,175	183	62,330
11	*Solanum lycopersicum* (tomato)	299,800	2,989	28,167	24,139	196	52,502
12	*Helianthus annuus* (sunflower)	132,841	1,966	20,130	32,717	269	53,116
13	*Vigna unguiculata* (cowpea)	186,216	149	19,333	13,582	14	32,929
14	*Nicotiana tabacum* (tobacco)	324,058	2,265	46,752	70,022	190	116,964
15	*Citrus sinensis* (orange)	202,400	170	26,081	72,791	26	98,898
16	*Theobroma cacao* (cocoa)	156,925	191	14,724	31,514	24	46,262
17	*Coffea canephora* (coffee)	68,360	88	9,298	13,409	5	22,712
18	*Gossypium* (cotton)	351,954	2,315	50,873	67,043	76	117,992
19	*Vitis vinifera* (grape)	357,238	25,479	34,634	32,779	14,716	82,129
20	*Volvox carteri* F. nagariensis (green algae)	131,863	55	9,284	9,523	6	18,813

Note: Data taken from The Gene Index Project database.

Bugos and Thom 1993; Bucheli et al. 1996). Only 786 DNA sequence entries of all types, derived from various species of the *Saccharum* genus, were listed in a GeneBank release in 2003 (Vettore et al. 2003). Simultaneously and successively the EST resources have been deposited in the databases; primarily ESTs were derived from a major research consortium in Brazil (Vettore et al. 2003) that was credited with more than 80% of the current published data of sugarcane. However, Australia, South Africa (reviewed in Casu et al. 2005), and India have recently taken an initiative in sugarcane EST acquisition through transcriptome study (Gupta et al. 2010). Sugarcane ESTs can be accessed from various databases and these are described by Casu (2010). The largest collection was generated by a consortium of Brazilian sugarcane EST databases (SUCEST) with approximately 238,000 ESTs from 26 libraries made from diverse tissue from several Brazilian varieties (http://sucest-fun.org, Vettore et al. 2001, 2003), organized into 43,141 putative unique sugarcane transcripts containing 26,803 contigs and 16,338 singletons, referred to as sugarcane assembled sequences (SASs). An internal redundancy analysis suggested an SAS containing 33,000 sugarcane genes. The transcriptome analysis of SUCEST libraries was built through virtual northern blotting and array hybridization in tissues (Arruda 2001; Nogueira et al. 2003; Vettore et al. 2001; De Rosa et al. 2005). Australian researchers worked on the Q117 sugarcane cultivar (Casu et al. 2003, 2004) in the USA from CP72-2086 (Ma et al. 2004), while Gupta et al. (2010) worked on CoS 767, Co 1148, and other Co-canes varieties in India. Sugarcane gene information is uploaded in most of the databases (Casu 2010). However, the Gene Index Project at the Dana Faber Institute (DFCI) has started the Sugarcane Gene Index database known as SoGI (http://compbio.dfci.harvard.edu/tgi/cgi-bin/tgi/gimain.pl?gudb=s_officinarum). Additional sugarcane transcript clustering activities were also undertaken by TIGR and PlantGDB (http://www.plantgdb.org). The various crops with their related entries in the SoGI database are shown in Table 7.2. The SoGI entry in 2010 deposited for sugarcane transcriptome has 282,683 ESTs which are partial, single-pass sequences from either end of a cDNA clone, assembled into 42,377 tentative consensuses (TCs) derived as nonhuman transcript assembly, along with the 499 mature transcripts nearer to full-length cDNA. It also has 78,924 singleton ESTs with 41 expressed transcripts (ETs) derived as nonhuman transcript. These are the major resources for obtaining transcript information through which researchers can identify and genetically validate new transcriptome in sugarcane contributing to important traits.

7.3.2 Functional Integrity of Transcriptomics in Sugarcane

Nowadays, the comprehensive and high-throughput analysis technique has been introduced into the research work on sugarcane transcriptome. For the expressional analysis, identification of new genes and sequencing of whole-genome or large-scale cDNA clones can be achieved through the DNA microarray and GeneChip technique. Transcriptome profiling can be done through the different types of DNA microarray systems. SUCAST and SUCAMET catalogs are part of the SUCEST-FUN regulatory network project that aims to study gene expression in sugarcane (Papini-Terzi et al. 2005, 2009; Rocha et al. 2007; Waclawovsky et al. 2010). In comparison with the other widely grown cereal crops such as sorghum, maize, wheat, and rice, as well as Arabidopsis, the transcriptome analyses are substantially lower in sugarcane. However, a wide range of tissues have been targeted in sugarcane for making cDNA libraries: about 70 independent libraries have been developed (Manners and Casu 2011). The sugarcane transcriptomes obtained from various research groups possess a high degree of parallelism with the present sorghum transcriptome (Paterson et al. 1995), followed by maize and barley, with the sugarcane at local and at global scale (Paterson et al. 1995; Souza et al. 2011). Hence, the transcriptome profiling of the sugarcane became easy in terms of obtaining EST sequence information, gene annotation, and assignments, and it yields the potential role of the genes for the important qualitative and quantitative traits in sugarcane as well as for candidate gene marker identification. Another landmark for the transcriptome was that of the sorghum and rice genomes both at physical level (Bower et al. 2005) and at nucleotide level (Paterson et al. 2009), revealing that each of these genomes comprise two qualitatively different components. The euchromatin region comprises gene-rich, poor repeats that contribute 98% of recombination in each genome, compared with 60% of the sorghum and rice genome (250–300 Mb), while the heterochromatin regions were gene-poor with some important genes and a repeat-rich region contributed to 2% recombination at genome level (Paterson et al. 2009). Despite the successive genetic evolutions, comparison of the sugarcane transcriptome with the monocots and dicots proved to be significant (Vettore et al. 2003). Around 71% of the sugarcane protein matches the *Arabidopsis* and 82% matches the rice proteome. Casu et al. (2004), as well as Damaj et al. (2010), demonstrated the role of dirigent protein, which is highly expressed in mature sugarcane culm and diverse monocot. It belongs to the family of cell wall proteins putatively involved in lignan biosynthesis and plant defense. Expressed sequence tags obtained through various targeted experimental methods in comparison with SASs submitted to the SUCEST database have had a greater impact on gene discovery (Vettore et al. 2003). These SASs allowed the targeting of over 30,000 sugarcane genes (Vettore et al. 2003); however, 11% of SASs may correspond to the genes restricted to monocots and 18% of the SASs may represent the protein restricted to sugarcane (Vincentz et al. 2004). Transcript analysis has come into existence with alternative strategies applied in sugarcane such as cDNA suppression hydridization, which has been used to study abiotic stresses such as salt and water deficit (Prabu et al. 2011; Patade et al. 2010). Similarly, cDNA-AFLP technique has been used to identify genes associated with infection by rust fungus *Puccinia melanocephela* (Carmona et al. 2004). Unlike the array-based gene profiling, this technique is also useful for identifying candidate genes for elucidating genome and genes' polymorphism at functional level (Manners and Casu 2011). The qRT-PCR technique also helps in studying the confirmation of highly expressed genes that are enhanced by the exogenous stress imposed by glucose, sucrose, and other compounds (Iskandar et al. 2004, 2011; Casu et al. 2007). In SUCEST, the SASs were functionally categorized and grouped into 18 broad categories according to the role in biological process and pathways. Fifty percent of the SASs were associated with the five broad categories including cellular dynamics, stress response, protein metabolism, bioenergetics, and cellular communication/signal transduction, while 17% of the SASs correspond to genes whose roles were unknown or could not be assigned with confidence. The 43,141 SASs as putative unique sugarcane transcripts in the SUCEST database were analyzed for the conserved protein domains using the ESTScan algorithm and domain search in the Pfam database (Iseli et al. 1999; Bateman et al. 2000). In terms of SASs a tissue was considered enriched if it contained at least three ESTs originating exclusively from a single sugarcane tissue, and hence 1234 were found to be tissue enriched, corresponding to 5716 ESTs, while 1020 SASs contained three or four ESTs. These tissue-enriched SASs occupied about 2.5% of the total ESTs. Most of the ESTs were tissue-enriched SASs rich in prolamine; these amounted to 360 ESTs. The systematic transcriptomics data of sugarcane subjected to further analysis found 40,756 proteins encoding for the 43,141 SASs by ESTScan, and high percentages (87%) of the SASs were not assigned as a protein compared with the public databases while 65 yielded two proteins, which leads to 40,821 amino acid sequences ranging from 6 to 1782 amino acids. This could indicate that the sequence corresponding to the 5' or 3' UTRs tends to be less conserved among organisms (Vettore et al. 2003). Protein domain differences were found in 40,821 protein sequences; 12,921 sequences possessed at least one domain while 1415 possessed different domains. The detailed and extensive sugarcane transcriptomic research conducted by Vettore et al. (2003) suggested that most of the domains found in SUCEST were of the nucleoprotein phenylalanine–glycine (FG) domain, which is predominantly involved in the nuclear pore complex protein assembly for transport of the cargo proteins. The next repeated domains were the leucine-rich repeat (LRR) and Huntington, elongation factor 3-like HEAT repeats related to armadillo/β-catenine-like repeats greatly involved in cytoplasmic protein and intracellular protein transport and protein–protein interactions. Also M repeats of bacterial M protein as well as pentatricopeptide repeats (PPRs) of unknown function were found in plants. Leucine-rich repeat and protein kinases were found to be more frequent domains in sugarcane. The modulation of the transcriptome (ESTs) when subjected to the particular stress treatment allows unraveling of the extent of genes' involvement in important traits such as sucrose accumulation, higher yield, and fiber content in sugarcane. Cytochrome p450s are one of the big family of enzymes involved in the environmental detoxification of toxins; their conserved domains were specific to sugarcane root SASs. Cell type proliferation is commonly associated with the specific transcription factors. Twenty-eight SASs containing transcription factor domain were tissue-enriched, among which AP2-domain was the most represented. These SASs confer ethylene responsiveness (Ecker 1995), while the AP2 domain is specific for temperature-stress calli, seed, root–shoot transition zone, flower, stress, and roots.

Gene regulation unravels the complex phenomenon of the up- and downregulation of genes in sugarcane and establishes the relationship with plant and animal transcription factors (Riechmann et al. 2000). Around 5% of the SASs (1415) revealed functions of RNA metabolism and transcription. This was comparable with the *A. thaliana* transcription factors. About 60% of the transcription factors found in *A. thaliana* or rice were also present in the SUCEST database of sugarcane. One of the interesting findings in sugarcane was the presence of large group of SAS-encoding proteins containing C3H-type zinc-finger domains C–x8–C–x5–C–x3–H, which were not frequently present in *A. thaliana* (Riechmann et al. 2000). However, the main receptor class identified was the serine/threonine receptor kinase family containing the leucine-rich repeat (LRR), which is postulated to mediate signaling by recognition of plant peptides and possibly pathogens. This domain was also found in disease-resistance genes (R-genes; Ellis et al. 2000). There were 274 LRR domains containing SASs in the SUCEST database. Also, G-α-protein and Ras-like domain-mediated signal transduction were predominantly involved in cell signaling of sugarcane, with 5 and 72 SASs of SUCEST, respectively. There was less evidence of phosphotyrosine recognition by SH2 domains, but SH3 domains may be associated with SH2 domains (Morton and Campbell 1994) and occur in three SASs of the SUCEST database. Sugarcane has a well-developed complex defense response to the biotic and abiotic factors. The major responses are mediated by the LRR domain-rich proteins, as there were 274 LRR domains containing SASs in the SUCEST database of which 46 SAS-encoding WRKY transcription factor domains are associated with regulation of plant–pathogen defense mechanisms in sugarcane. Some other genes related to defense such as chitinase, β-1,3-glucanase, chalcone synthase, isoflavone reductases, catalases, and superoxide dismutases have putative orthologs in sugarcane and are highly conserved in the plant kingdom. The new transcript analysis techniques include serial analysis of gene expression (SAGE), which is based on conventional Sanger-based sequencing of contigs of short tagged ligated cDNA-restriction fragments (Velculescu et al. 1995) and is applied to sugarcane (Calsa and Figueira 2007). Calsa and Figueira (2007) targeted the sugarcane leaf tissue of 15-month-old field-grown plants of cultivar SP80-3280 by SAGE. Of the 50,000 clones, 480 clones were sequenced; 9482 valid tags with 5227 unique sequences were extracted and compared with the SASs for transcriptome match and validation, of which 3659 (70%) matched the sugarcane SASs with putative function, while 872 tags (16%) matched without SAS without putative annotation and only 173 (3.3%) did not match any sugarcane ESTs; and 1014 (19.4) matched more than one SAS with distinct annotation (multiple hits). A total of 1557 unique tags (29.8%) were associated with genes coding for chloroplast-targeted proteins as a part of photosynthetic complexes. Photosystem I (PS I) was identified as the most frequent gene product location on the basis of gene ontology (GO), followed by thylakoid complex, plastocyanin, oxygen-evolving protein, chlorophyll a/b-binding (CAB) proteins, and nuclear genes related to photosynthesis and carbon fixation in sugarcane leaves; this revealed alternative C4 metabolism. Other sequencing projects contributed to unraveling the sugarcane transcriptome. Ma et al. (2004)

sequenced 9216 ESTs from three different cDNA libraries of apex, leaf, and mature inter-nodes. About 57% of these sequences have significant matches with previously character-ized protein while 28% corresponded to the unknown protein function. In recent years the advent of more next-generation high-throughput sequencing platforms (Mezker 2010; Simon et al. 2009) applied to cDNA populations has provided new options for sequencing of cDNA. High-throughput instruments such as Roche GS FLX (454) and paired end Illumina GA II (Solexa) sequencing of PCR products will directly reveal similar allelic complexity in tran-script analysis in polyploid sugarcane. Comparison of mature and immature sugarcane stem internodes transcriptome also opened up the relationship of highly expressed ESTs during sucrose accumulation and stem maturation through microarray (competitive hybridization experiment) and northern blot analysis of selected potentially differentially expressed tran-script in Q117, 74C42, and Q124 sugarcane genotypes (Casu et al. 2004). More advanced tran-scriptome analysis was carried out through Affymetrix GeneChip sugarcane genome array in Q117 sugarcane variety by targeting the meristem, internodes 1–3, 8, and 20 (Casu et al. 2007). The resultant highly expressed 119 transcripts cellulose synthase (CesA) and cellulose synthase-like (Csl) gene families were predominantly expressed in the mature stem followed by the transcript involved in the cell wall metabolism and lignifications in sugarcane. Over 3500 genes involved in the signal transduction, transcription, development, cell cycle, stress response, and interaction with the pathogens were identified in the SUCEST database (Souza et al. 2011). Microarray expression profiling of these transcripts showed a signaling protein in flowers, roots, leaves, lateral buds, and internodes (Papini-Terzi et al. 2005). Unraveling through the complex network of the polyploid sugarcane are small RNAs involved in the regulatory roles in cellular pathways (Broderson et al. 2008; Vazquez et al. 2010). These small RNAs, ~21 nucleotides long, are called microRNAs. These are recruited by RNA-induced silencing complex (RISC; including Argonaute and other proteins), which combines with mRNA to inhibit or degrade the target mRNA and thus translation is interrupted. Zhang et al. (2010) identified 32 potential microRNAs in sugarcane in comparison with the public databases, while Zanca et al. (2010) identified a further 19 sugarcane microRNAs of which six were found in most tissues in both of the ancestral species of *S. officinarum* and *S. sponta-neum*. Progressive increase in transcripts of the miRNA 159 was observed under short-term PEG stress and various members of the MYB transcription factor family were predicted as the potential targets of miR159 (Patade and Suprasanna 2010). Hence these sugarcane tran-scriptomes provide new insights for functional studies associated with sucrose synthesis and accumulation.

7.3.3 Comparative Transcriptomics and Reverse Genetics

As stated above, the grasses are highly competitive, and some species such as sugarcane, sorghum, and maize contain anatomical and physiological adaptations to optimize CO_2 fixation for carbohydrate biosynthesis as C4 pathways yield similar physiobiochemical properties. Among the Poaceae species, the level of genome varies from diploid to deca-ploid (Arumuganathan and Earle 1991; Peterson et al. 2009). The syntenic region of the genome maintained the order of genes, which suggests the origin and conservation of the gene function (Manetti et al. 2012). Sorghum is physiologically very close to sugarcane and is believed to have diverged from a common ancestor approximately 8 million years ago, hence sorghum would help to build up a foundation of transcriptome of sugarcane (Jannoo et al. 2007; Paterson et al. 2009; Manners and Casu 2011). A huge amount of data obtained from the sugarcane transcriptome set up the relation with the candidate genes and their functional role through bioinformatic tools at the preliminary level, and this could enhance the functional genomics transcriptome research in sugarcane. Expressed transcript comparison of sugarcane with the major crops such as sorghum, maize, and rice revealed about 97%, 93%, and 86% mean sequence identities, which illustrate the proximity of sugarcane at the transcript level (Paterson et al. 2009). Wang et al. (2010) found the 20 sugarcane BACs that were associated with the sorghum-specific transcripts. Calvino et al. (2008) highlighted the carbohydrate relationship found in sugarcane culm in comparison with the sweet sorghum as a part of the up- and downregulation of the genes during sucrose and other hexose accumulation in the grain; these findings help in the sugarcane orthologous determination.

Sucrose accumulation is one of the major areas of research in sugarcane functional genomics. There has been less information available on account of sucrose accumulation and the transport from the source to the sink tissues of sugarcane. About 50%–60% of sucrose is found in the stem dry weight of sugarcane (Bull and Glasziou 1963), particularly during stem maturation (Moore 1995; Walsh et al. 2005) with the involvement of metabolic and physiological processes (Whittaker and Botha 1997; Botha and Black 2000). The gene expression pattern has been studied in sugarcane and provides strong transcript evidence of sucrose accumulation during stem maturation with immature and mature internodes (Casu et al. 2003, 2004, 2005). Some of the genes such as sucrose synthase, sucrose phosphate synthase (SPS), and soluble acid invertase perform significant modulation during sucrose accumulation in the stem maturation. PST type 2a has been an informative finding in sugar transporter and it may also be involved in the sugar translocation in the phloem companion cells and associated parenchyma in maturing stem (Casu et al. 2003). Collection of a large amount of information provided strong support to develop the gene network and hypothetical models of the major regulatory mechanism of sugarcane pathways. Various groups of researchers have studied expressional analysis to confirm the upregulation of selected genes (Borecky et al. 2006; Jackson et al. 2007; Nogueira et al. 2005; Vicentini et al. 2009). Sugarcane is a highly polyploid crop having polyallelic genetic regulation of major traits such as sucrose accumulation and cellulose and fiber contents. Hence the reverse genetics approach could be as significant to achieve the target as phenotypic effects and biological functions of specific gene sequence obtained by DNA sequencing. Different reverse genetics techniques are available in other crops: (a) site-directed mutagenesis in promoter of the open reading frame region or codon to identify important amino acid residues for protein function of target genes, (b) gene silencing through RNAi mechanism for gene inhibition without mutation by targeting the mRNA at the promoter region or cleavage of mRNA transcript by binding to it to allow Dicer protein to cleave the targeted mRNA. Endogenous gene silencing of sugarcane has been conducted in regenerated transgenic sugarcane plants using both antisense (Groenewald and Botha 2008; Botha et al. 2001) and RNAi-expressed transgene (Osabe et al. 2009). Suppression of the fructose 6-phosphate 1-phosphotransferase (PFP) demonstrated a functional role for PFP in sucrose accumulation in immature internodes and a critical role in glycolytic carbon flow (van der Merwe et al. 2010). As PFP is developmentally specific to sucrose accumulation in the culm region of the sugarcane (Whittaker and Botha 1999), (c) transgene interference allows overexpression of the targeted wild or mutated genes; these can be achieved in sugarcane through genetic transformation and regeneration via either biolistic bombardment (Bower and Birch 1992; Basnayake et al. 2011) or *Agrobacterium*-mediated techniques (Arencibia et al. 1998; Joyce et al. 2010).

7.3.4 Transcriptome Polymorphism for Functional Studies

Functional study of the transcript is a potential tool toward the determination of candidate genes in sugarcane. The study of SNP and insertions/deletions (Indels) could yield stable molecular markers and it may be useful to screen the mapping population. In sugarcane, SSCP was carried out to find out the expression profile of signal transduction components in a population segregating for sucrose content (Felix et al. 2009). Single-base polymorphism study was used with the gel-based assay, such as SSCP and denaturating gradient gel electrophoresis (DGGE), but these techniques have some constraints as they do not pinpoint the location of or the type of polymorphism present in the DNA fragment (Defrancesco and Perkel 2001). The SNP detection carried out through TILLING is a powerful tool for the reverse genetics study (Gilchrist and Haughn 2005). It is based on restriction fragment cutting with mismatch-specific endonuclease to detect induced or natural DNA polymorphisms formed during heteroduplex. Vettore et al. (2003) considered the SASs for SNP with more than four read. In all, 42,936 SNPs were detected for 14,445 SASs (mean of 2.97 SNP per SAS). This value was considerably higher than that of the reported human genome with similar data (Garg et al. 1999; Picoult-Newberg et al. 1999; Deutsch et al. 2001). The integrity of the SNPs obtained by Vettore et al. (2003) was tested for the 51 restriction SNPs on the panel of 55 genotypes including representatives of the *S. officinarum, S. spontaneum*, and modern cultivars. Thirty-one (61%) polymorphisms indicated that the majority of SNPs succeeded in sugarcane. However, in a detailed analysis of five genes of two small family 6-phosphogluconate dehydrogenase (Grivet et al. 2001) and Adh (Grivet et al. 2003), 15.4 SNP and 3.6 Indels were detected (Vettore et al. 2003). SNP analyses of the 178 cDNAs encoding for Adh in the four sugarcane cultivars revealed 37 SNPs in the coding and untranslated region of three Adh genes (Grivet et al. 2003). Most of the INDELs were limited to the 5′ and 3′ untranslated region while SNPs were observed throughout the gene sequence. These variations can be used to fingerprint and identify individual genotypes (Cordeiro et al. 2006). Similarly, the polymorphism at the functional level of transcript can be maximized through expression sequence tag–simple sequence repeat (EST–SSR) study. A survey of the SUCEST database revealed over 2000 SASs containing dinucleotide, trinucleotide, and tetranucleotide SSRs (Pinto et al. 2006) and experimentally it was proved that 23 SSR markers out of the 30 tested SSRs proved scorable for polymorphism in 18 commercial sugarcane clones. EST-derived RFLP markers have been developed to target the sugarcane content (Da Silva and Bressiani 2005; Pinto et al. 2010). Ten SNPs were reported from the SPS gene of SPS family III from 400 nucleotide (nt) amplified region, and poor association with the variation in sucrose content was found (McIntyre et al. 2006). Hence intensive studies are needed to rectify the genetic changes that contribute functionally to sucrose accumulation, high fiber, and high yield. The resultant transcriptome and its experimental validation for the functional marker development can become useful in the screening of segregating populations for desirable traits in sugarcane.

7.4 Transcriptome Changes in Response to Abiotic and Biotic Stresses

Plants gradually evolved their complex defensive strategies to protect themselves against attack by viruses, insects, and other pathogens (Haruta et al. 2001). Various diseases of sugarcane have been reported worldwide to cause yield losses. Sugarcane is challenged by wide ranges of stresses such as viral, bacterial, fungal, nematode, and herbivorous infestation. Bacterial diseases include Gumming disease caused by *Xanthomonas axonopodis* pv. *vasculorum;* leaf scald caused by *Xanthomonas albilineans;* and root stunting disease (RSD) caused by *Leifsonia xyli* subp. *xyli*. Fungal diseases include black rot caused by *Ceratocystis adiposa*; brown spot caused by *Cercospora longipes*; downy mildew caused by *Peronosclerospora sacchari*; red rot caused by *Glomerella tucumanensis*; eye spot caused by *P. sacchari*; and wilt caused by *Fusarium sacchari*. Viral diseases include Fiji disease caused by sugarcane Fiji disease virus; mosaic caused by sugarcane mosaic; and yellow leaf caused by sugarcane yellow leaf virus. Diseases caused by nematodes include lesion by *Pratylenchus* spp. and root-knot by *Meloidogyne* spp. Phytoplasma cause sugarcane grassy shoot (SCGS) diseases.

Sugarcane has a significant defensive system to counter pathogens through its prime barriers such as spines and wax on the leaves and stem region. However, the stem portion possesses visible phenolic shading that can make a barrier to landing and invasion of pathogens. Volatiles that attract predators of the insect herbivores (Birkett et al. 2000) include secondary metabolites (Baldwin 2001) and densely covered trichomes (Fordyce and Agrawal 2001). Despite the defensive system barrier, pathogens invade and cause detectable infestations to the targeted plant tissues such as leaves, stem, and roots, resulting in a major loss of crop yield. Plant–insect interactions are the dynamic system whereby plants form a barrier against insects, but insects also develop resistance, such as detoxification of the toxic compounds (Scott and Wen 2001), avoidance mechanisms (Zangerl and Berenbaum 1990), sequestration of poison (Nishida 2002), and alteration of gene expression patterns in their guts. Successive transcript studies have been conducted throughout the world to show that the pathogen resistance triggers defense signaling with synergistic and antagonistic outcomes (Beckers and Spoel 2006). Some of the growth regulators, mainly jasmonic acid, ethylene, and salicyclic acid, actively and synergistically take part in the specific and general defense response in sugarcane. Current research on methyl jasmonate (MJ) shows that an increase in transcript level under MJ stress alters signal transduction, phenylpropanoid pathway, and up- and downregulation of oxidative stress genes. In some other crops ethylene synthesis increases the response to bacteria, fungi (Broekaert et al. 2006), and insects (Kahl et al. 2000). However, in sugarcane, a putative ethylene receptor and two putative transcription factors are regulated during the association with the nitrogen-fixing endophytic bacteria (Cavalcante et al. 2007). Transcriptome induction during the biotic response can lead to overcoming the biotic attack through the transcript expression in leaf and other tissues of the sugarcane.

Considering the role of the plant hormone MJ, extensive research has come up with significant results about its action in the plant defense mechanism (Bower et al. 2005; Rocha et al. 2007; De Rosa et al. 2005). Sugarcane roots were treated with MJ and 829 ESTs were isolated; these were subjected to cDNA microarray study along with the 4793 sugarcane mature and immature stem tissues, of which 21 ESTs corresponding to transcripts were significantly induced by MJ in roots and 23 were reduced in expression following MJ treatment. The induction of six transcripts identified in the microarray analysis was tested and confirmed using northern blotting. Homologs of genes encoding lipoxygenase and PR-10 proteins were induced 8–24 h after MJ treatment, while the other four selected transcripts were induced at later time points. These results lay a foundation for further studies of induced pest and disease resistance in sugarcane roots (Bower et al. 2005). Other experiments (Wang and Fristensky 2001) were performed with MJ solution treatment to the soil containing plants, and roots were harvested after 1, 3, and 10 days. The highest induction of transcript that encodes dirigent protein was found, which is involved in lignin framework and fungal attack protection. MJ treatment to the leaves and changes in the gene expression profiling have been evaluated using a cDNA microarray of 1545 genes (Rocha et al. 2007) as well as the influence of transcription factors such as MYB, NAC, and Aux/IAA, and protein kinases were found to be up- and downregulated under the MJ response. In the case of the symbiotic association of endophytic bacteria with the sugarcane, it is believed that the sugarcane produces an active defense response (Vinagre et al. 2006); 1545 genes in sugarcane revealed that *Gluconacetobacter diazotrophicus* and *Herbaspirillum seropedicae* endophytic bacteria activate a distinct class of defense genes such as R-protein-encoding gene, salicyclic acid biosynthesis genes, and transcription factor (Rocha et al. 2007). In the differential transcript expression study against different stress compounds, a huge volume of transcriptome data was generated that can be used against the biotic diseases. Some success has been achieved by targeting fungus such as smut, bacteria such as RSD bacteria, and pests such as lepidopterous leaf rollers and leaf miners, root and shoot borers, with a view to the development of improved sugarcane cultivars. The heterologous expression system for defense-related proteins such as sugarcane Cry protein (Braga et al. 2003) or the soyabean proteinase inhibitors (PIs) encoding genes has led to increased resistance against the sugarcane borer *D. saccharalis*. However, a major sugarcane pest and filamentous pathogenic fungus, *Trichoderma reesei*, can be controlled through the cysteine protease inhibitor (Soares-Costa et al. 2002). A sugarcane EST database has been developed containing hundreds of transcripts deposited in response to insect herbivores (Falco et al. 2001) and diazotrophic endophytes (Lambais 2001; Nogueira et al. 2001). Through genetic transformation, biotic disease or stress can be overcome by introducing novel genes to develop new varieties. Numerous attempts have been made to achieve this under the influence of a strong promoter. The biolistic method of genetic transformation developed the resistance to the Fiji disease virus (FDV) in Q124 callus-raised plants. The genetic construct, particularly the transgene, was under the ubiquitin promoter and Nos termination sequence (McQualter et al. 2001). Hence the identification and characterization of genes associated with the biotic stress may lead to the development of healthy and resistant sugarcane cultivars for higher yield through the genetic transformation. The major breakthrough is increasing the transgene expression for controlling the biotic agents through the use of a strong promoter so that it can produce resistant protein to the threshold sufficient to kill the biotic agents.

Besides the biotic stress attack, abiotic factors are also important to study the transcriptome changes in stress conditions. Various abiotic factors such as temperature (including freezing, chilling, and high temperature), water-deficit stress, and nutrient stress have been studied in model plants as well as in tropical plants such as sugarcane. Sugarcane possesses C4 mechanism and is grown in almost all the tropical and subtropical areas of the globe (Inman-Bamber and Smith 2005; Zhang et al. 2006c). It is amenable to high-temperature growth, but sugarcane is also cultivated in cold regions, some of which are prone to freezing temperatures. The study of transcripts under all possible abiotic stresses should be encouraged so that in the future tolerant sugarcane cultivars can be developed so as to overcome extreme conditions. Among the abiotic stresses in sugarcane, water deficit is a major stress in India and other countries because of the poor irrigation facilities and electricity shortages. These factors necessitate the development of drought-resistant varieties in the future (Kates and Parris 2003; Riera et al. 2005). In the transcriptome aspect, a number of studies have been conducted using cDNA microarray methodologies and periodic stress over hours or days, which are preferred to transcriptome studies. Rocha et al. (2007) exposed plants to 24, 72, and 129 h of water deprivation and studied the expression of an array of genes through a cDNA microarray study in sugarcane (Menossi et al. 2008). Prabu et al. (2011) indicated the upregulated genes identified in the Co740 sugarcane variety under water-deficit conditions. PCR-based cDNA suppression subtractive hybridization and dot blot technique were applied to the selected 158 clones. Sequencing and annotation of ESTs with the public database found that most EST-encoded proteins are involved in cellular organization, protein metabolism, signal transduction, and transcription. Further, semiquantitative reverse transcriptase PCR found that the WRKY-like transcription factor and 22 kDa drought-induced protein, abscissic acid (ABA)-inducible gene, HVA22, and MIPS were expressed in the stressed plants in comparison with controls. Hence this transcript strongly supports drought-responsive genes and may lead to overcoming drought in susceptible and moderately resistant sugarcane through the genetic transformation approach. In a similar way, the stress conditions generate reactive oxygen species (ROSs) that can be neutralized by catalase activity through the conversion of hydrogen peroxide to water and oxygen (Chagas et al. 2008). In association with an increase in expression of a gene encoding a peroxidase, drought tolerance in sugarcane has been determined by Rodrigues et al. (2009). In temperature stresses, a sugarcane transcript modulates efficiently and markedly under different conditions of freezing, chilling, and higher temperatures.

Sugarcane is mainly meant for the sucrose content; initially the potential photosynthetic enzymes such as rubisco and sucrose accumulation enzymes such as stromal and cytosolic fructose-1,6-bisphosphatase (FBPase), sucrose-phosphate synthase, and sucrose synthase were studied under cold stress applied to leaves (Holaday et al. 1992; Hurry et al. 1995) and the relation of photosynthetic enzymes and sucrose accumulation was established. Under short-term exposure to low temperature, changes in enzyme activation state seem to play a more important role in compensating plant photosynthesis (Holaday et al. 1992). However, species that do not tolerate low temperature may not have the same ability to acclimatize by upregulation of enzymes (Briiggemann et al. 1992; Holaday et al. 1992). Du et al. (1999) investigated the mechanism of photosynthetic changes in sugarcane leaves in response to chilling temperature (10°C) using three species (*S. sinense* R. cv. Yomitanzan, *Saccharum* sp. NiF$_4$, and *S. officinarum* L. cv. Badira). There was a substantial accumulation of aspartate and the level of alanine was increased in leaves during the chilling treatment, hence it is suggested that NADP-MDH and PPDK are key enzymes that may determine cold sensitivity in the photosynthesis of sugarcane (Du et al. 1999). The first report on the use of cDNA array to discover sugarcane genes modulated by cold stress was conducted by Nogueira et al. (2003), and some more evidence in relation to cold sensitivity (Du et al. 1999) has been studied. On exposure to low temperature (4°C), the expression of 25 genes was repressed while 34 genes were upregulated. One of the genes, SsNAC23, belongs to the NAC family of transcriptional factors that are involved in biotic and abiotic stresses (herbivory) and development (Olsen et al. 2005). At high temperatures such as 40°C, the synthesis of proline, glycinebetaiine, and soluble sugars was enhanced (Wahid 2007; Wahid and Close 2007), while carotenoids and flavonoids (Wahid 2007) were also detected. These data may reinforce the experimental approach that can lead to the transcript evaluation and validation for temperature stress. Higher concentration of Cd also increases the toxicity and antioxidant response in sugarcane when it is exposed through the irrigation system (Sereno et al. 2007; Fornazier et al. 2002). Zn, which interferes with the normal mitosis, leads to inhibition of DNA synthesis (Jain et al. 2010) as well as photosynthetic pigment content in leaf tissue, which can be related to disturbance in photosynthesis and interference with the sucrose accumulation in sugarcane. The genes associated with the metal ions also have led the transcriptome research in sugarcane to overcome the problem of soil and water pollution. Phytohormones have also been used to study transcriptome analysis. Abscissic acid was sprayed on sugarcane leaves and the gene expression was evaluated using cDNA array (Rocha et al. 2007); it was found that the transcripts were matched with the two Ser/Thr kinases genes. These genes were upregulated and some drought-responsive genes and MJ biosynthetic genes were also elevated (Rocha et al. 2007) with the complex network of ethylene, gibberellin, and salicylic acid biosynthesis. ABA shows an antagonistic response between the ABA and auxin pathways. However, there is some evidence that the protein similar to auxin-responsive protein GH3 was repressed by ABA treatment (Hagen et al. 1984). This seems to be a target for the genetic manipulation for expression of the genes involved in sucrose synthesis, transport, accumulation in culm, and stress response to biotic and abiotic factors in sugarcane.

7.5 Integration of Metabolomics with Genomics

Plants represent an excellent biochemical factory producing an array of small, simple and complex molecules that are beneficial to plant survival and as food and medicine to humankind (Oksman-Caldentey and Inzé 2004). Metabolomics or metabolome research has emerged as a postgenomic revolution. The term was coined first in the field of micro-biology (Tweeddale et al. 1998). Nowadays it is a more powerful tool in drug discovery and biomarker development for organ-specific toxicity (Harrigan and Goodacre 2003) and also in disease prediction in humans. The importance of metabolomics has increased tremendously because of intermediates of metabolism and integration with other omics data, such as transcriptomic data, which can be analyzed *in silico* to depict gene–protein metabolic networks (Thimm et al. 2004). Plant metabolites have enormous chemical diversity, estimated at ~200,000, even more than microorganisms and animals (Dixon and Strack 2003; Trethewey 2004). A single plant species, *A. thaliana*, might produce ~5000 metabolites (Bino et al. 2004)—significantly more than a microorganism (~1500) or animal (~2500). A complex genome system such as sugarcane, maize, or other Poaceae members may generate thousands of primary and secondary metabolites. Primary metabolites are directly involved in normal growth, development, and reproduction, while secondary metabolites usually perform ecological functions and have a different structure from primary metabolites (Oksman-Caldentey and Saito 2005). Metabolomics also has the potential not only to provide deeper insight into complex regulatory processes but also to determine phenotypes directly. Data-mining tools such as principal component analysis have enabled the assignment of *metabolic phenotypes* using large data sets. Hence, metabolite profiling has become a tool to significantly extend and enhance the power of existing functional genomics approaches.

Sugarcane metabolomics is still in its infancy because of the difficulties encountered in the determination of stable functional molecules, markers for complex traits, and so forth. Compressive plant chemical profiling and its validation are the key components of metabolomics (Fiehn 2002; Bino et al. 2004) and can explore stress biology or sugar metabolism. Investigation of the metabolome (the total complement of metabolites in a specific tissue) to understand metabolic regulation has arisen in recent times. This has become possible by technological advances facilitating simultaneous measurement of multiple metabolite levels (up to 160 compounds in the potato) (Roessner et al. 2000) in a single tissue within an hour. Techniques such as mass specrophotometry (MS) combined with a number of chromatographic methods such as gas chromatography–mass spectrometry (GC–MS), liquid chromatography–mass spectrometry (LC–MS) (Fiehn et al. 2000; Roessner et al. 2001), high-performance liquid chromatography–mass spectrometry (HPLC–MS) (Von Roepenack-Lahaye et al. 2004; Yamazaki et al. 2003a,b), and, more recently, capillary electrophoresis–mass spectrophotometry (CE–MS) (Sato et al. 2004) have been used for the determination of metabolites. More advanced Fourier-transform ion cyclotron resonance (FT-ICR) mass spectrometry and time-of-flight (TOF) mass spectrometry are also used for large fingerprinting purposes (Aharoni et al. 2002; Brown et al. 2005). Nuclear magnetic resonance (NMR) spectroscopy analysis coupled with liquid chromatography is also available for metabolite analysis (Griffin 2003; Kikuchi et al. 2004). Most usable powerful instruments and their specificity to the variety of metabolites are shown in Table 7.3. Some carbohydrates quantify radio-labeled sugars by inline isotope after HPLC separation (Black and Botha 2000). MetaCrop is a database that summarizes diverse information about metabolic pathways in crop plants and allows the automatic export of information for the creation of detailed metabolic models (http://metacrop.ipk-gatersleben.de).

Crops such as *Zea mays, Hordeum vulgare, A. thaliana, Triticum aestivum, Oryza sativa, Solanum tuberosum, Brassica napus, Beta vulgaris,* and *Medicago trucatula* have been publicly accessed by MetaCrop with a total of 62 pathways, 566 reactions, and 22 compartments (Schreiber et al. 2011). More data of relevance to sugarcane can be traced with *Z. mays, H. vulgare,* and *A. thaliana.* Other databases for understanding metabolic pathways include Plant Cyc (http://www.plantcyc.org). MapMan (Thimm et al. 2004), KaPPA-view (Tokimatsu et al. 2005; Sakurai et al. 2011), Arabidopsis Reactome (http://www.arabidopsisreactome.org/about.html), KEGG (Kanehisa et al. 2008), and KEGG PLANT (http://www.genome.jp/kegg/pathway.html) (Okazaki and Saito 2012). Bosch et al. (2003) explained metabolite profiling in sugarcane by targeting the third and seventh internodes of 12-month-old N19 and US6656-15 sugarcane varieties. The extracts were analyzed using *GC–MS* (Thermo Finnigan TRACE GCMS) *analysis* with ribitol as internal standard. Carbohydrates such as sucrose, glucose, fructose, inositol, raffinose, xylose, and maltose, and low-abundance metabolites such as amino and organic acids, sugar phosphates, sugars, and sugar alcohols

TABLE 7.3

Metabolite Detection Using a Variety of Analytical Techniques for Metabolomics

Metabolite Class	Typical Metabolites	Instruments
Amino acids and their derivatives	Twenty amino acids, β-alanine, GABA, etc.	CE–MS, GC–MS, LC–MS, NMR
Amines	Polyamines, etc.	CE–MS, GC–MS
Alkaloids	Polar alkaloids (e.g., pyrrolizidine alkaloids)	LC–MS, NMR
Fatty acids and their derivatives	Saturated/unsaturated aliphatic monocarboxylic acids and their derivatives	GC–MS
Isoprenoids	Terpenoids and their derivatives	GC–MS (nonpolar), GC–MS (polar)
Nucleic acids and their derivatives	Purines, pyramidine, mono-, di-, and triphosphate nucleosides	GC–MS (partially)
Organic acids in central metabolism	TCA cycle intermidiates, etc.	CE–MS, GC–MS, LC–MS, NMR, LC–MS (partially)
Pigments	Carotenoids, chlorophylls, anthocynine, etc.	HPLC, LC–MS
Polar lipids	Phospholipids, mono-, di-, and triacylglycerols	LC–MS
Sugar and its derivatives	Mono-, di-, and trisaccharides, sugar alcohols and sugar mono- and diphosphates	GC–MS, CE–MS (sugar phosphates), CE–DAD (partially)
Volatiles	Phenylpropanoid volatiles, aliphatic alcohols, aldehydes, ketones, etc.	GC–MS
Other secondary metabolites	Polar phenylpropanoids (e.g., flavonoids), etc.	LC–MS

Note: CE–DAD, capillary electrophoresis-diode array detection; GABA, γ-aminobutyric acid; TCA, tricarboxylic acid cycle.

were estimated. A higher amount of amino acids and organic acids in immature inter-nodes indicated the metabolic scenario in sugarcane. Similarly, the trehalose and its phosphate derivative trehalose-6-phosphate have come up as signaling molecules involved in carbon partitioning in plants (Eastmond et al. 2003). The authors also worked on carbohy-drate profiling to determine the individual fructose, glucose, and sucrose by using HPLC in support of the selection of sugarcane genotypes with higher sucrose yields (Kalwade et al. 2011). High- and low-sucrose F1 sugarcane genotypes were tested at 6, 7, 9, 11, and 13 months for the xylitol, mannitol, trehalose, arabinose, lactose, glucose, fructose, maltose, and sucrose contents using HPLC and high-performance anion exchange chromatography with pulsed amperometric detection (HPAEC-PAD) methods and established a significant similarity with the soluble solid content (Brix). The RNAs from bulk of high- and low-sugar sugarcane genotypes were hybridized to cDNA microarray. Twenty-four ESTs were differentially expressed, of which omega-3 fatty acid desaturase (FAD8) was expressed in mature leaves of higher sugar content genotypes; however, two proteins of the inositol metabolism, O-methyltransferase and 1,4,5-triphosphate phosphatase, were expressed in mature leaves of low sugar content genotypes, hence these findings provided a better rela-tionship of the concentrations of the detected metabolites with associated ESTs to depict the gene–protein–metabolite network. Some of the metabolomic research in *A. thaliana* was of great support to the study of metabolomics in plant science. Two ecotypes of the *A. thaliana* Col-2 and C24 and their mutants were subjected to GC–MS-based metabolite profiling, indicating that these ecotypes diverged from each other more than their mutant diverged from its parent ecotypes. Such ecotype divergences happen irrespective of the single gene or point mutation that can support the multigenic aspect of some genes, which can control vital pathways such as carbohydrate metabolism, sucrose transport, and the lignin synthesis in sugarcane. *Arabidopsis pal1* and *pal2* mutants lacking the function of two phenylalanine ammonia lyase genes showed no clear phenotypic alteration but the detailed carbohydrate and amino acid profiling in combination with the transcript showed the specific function of *PAL1* and *PAL2* (Casacuberta and Santiago 2003).

7.6 Sugarcane Proteomics

Proteomic analysis offers an added approach to potentially increase the understanding of the molecular mechanisms of any plant subjected to stress conditions. Proteomics is a powerful and advanced technique applied in sugarcane crop research. The technique follows a basic principle to segregate complex protein mixtures to produce visible results. Changes in protein profiles in response to different environmental effects and diseases can be resolved with this method. This technique is highly in demand as the gene func-tion and its physiological outcome can be analyzed effectively. For understanding or determining the function of any gene, identification of the encoded proteins is essen-tial (Abbasi and Komatsu 2004). Many proteomic approaches offer powerful tools that have been implemented for systemic study of proteins isolated under stress conditions in plants (Agrawal and Rakwal 2008; Cho et al. 2010). The most popular strategies used in proteomics are two-dimensional electrophoresis (2-DE) and matrix-assisted laser desorp-tion/ionization-time of flight mass spectrometry (MALDI-TOF-MS), which examine the differential and comparative expression of proteins expressed in plants (Agrawal et al. 2009; Komatsu and Yano 2006; Ding et al. 2011; Falvo et al. 2011; Kim et al. 2011). Ramagopal

(1990) pointed out that the 2-D approach presents a unique prospect to clone and exemplify a large amount of proteins expressed by the sugarcane genome. 2-DE with immobilized pH gradients facilitates the separation of intricate mixtures of proteins according to isoelectric point, molecular mass, and relative abundance (Görg et al. 2004). Several current proteomic studies have been executed on various crops under different abiotic stresses, such as osmotic stress in rice and rapeseed (Zang and Komatsu 2007; Toorchi et al. 2009), drought in rice, wheat, and sugar beet (Jorge et al. 2006; Salekdeh et al. 2002; Vincent et al. 2005; Ali and Komatsu 2006; Hajheidari et al. 2007; Caruso et al. 2009; Xu et al. 2009), high salinity in rice (Parker et al. 2006; Chen et al. 2009; Jellouli et al. 2010), and low temperature in winter rye (Amme et al. 2006; Gao et al. 2009). Sugarcane proteomic analyses have examined general protein polymorphisms (Ramagopal 1990), changes in protein expression after dedifferentiation of leaf tissue in callus culture (Ramagopal 1994), and drought-stress-responsive proteins (Jangpromma et al. 2010; Sugihario et al. 2002). Nevertheless, application of proteomics knowledge in sugarcane needs to be extended and meticulously explored (Amalraj et al. 2010; Sugihario et al. 2002; Jangpromma et al. 2010). Using these proteomic tools, many researchers are investigating abiotic stress-related proteins isolated from sugarcane in comparative studies based on 2-DE and MALDI-TOF-MS.

The 2-DE system was used in sugarcane mainly to study protein variation during leaf dedifferentiation (Ramagopal 1994) and to identify a drought-inducible protein localized in the bundle sheath cells (Sugihario et al. 2002). Studies carried out in sugarcane revealed many proteins to be upregulated when subjected to water-deficit stress. These differentially expressed proteins are identified by 2-DE and MALDI-TOF-MS. Zhou et al. (2012) studied the identification of proteins from sugarcane under osmotic stress and obtained drought-inducible 22 kDa protein and ribulosebisphosphate carboxylase small chain proteins (Rubisco small subunit) which were upregulated while ATP synthase delta chain and isoflavone reductase-like (IRL) proteins were downregulated. Thus the drought-inducible proteins and energy metabolism and antioxidant defense-related proteins play a vital role under osmotic stress conditions in sugarcane. Further, the properties of these proteins were identified by using different bioinformatic tools and analyses, suggesting that the proteins were more of an acidic, unstable nature, transmembrane and enriched with hydrophobic amino acids such as leucine and alanine, whereas the drought-inducible 22 kDa protein was a hydrophilic and nontransmembrane protein enriched with glutamic acid having involvement in adaptation to drought stress via different signaling cascades. Zhou et al. (2012) further stated that the tryptic digested peptides extracted from gels were analyzed by MALDI-TOF-MS and the obtained peptide mass fingerprints (PMFs) to be compared with those present in the SWISS-PROT, NCBI, and MSDB databases using the MASCOT program (http://www.matrixscience.com). Menossi et al. (2008) studied sugarcane plants of three drought-tolerant (RB92579, RB867515, and SP79-1011) and three drought-sensitive varieties (RB72454, RB855536, and RB855113) and evaluated the proteome. Leaf, stem, and root proteome mixtures were subjected to 2-DE/MALDI-TOF-MS to obtain the differentially expressed peptides (DEPs). Identification of these DEPs will certainly help in functionally relevant biological data interpretation and linkage to physiological profiles.

Another proteomic approach was used by Freddy (2011) to study protein differential expression against bacterial infection by *X. albilineans*, which causes leaf scald disease in sugarcane. The aim of the study was to identify and characterize the differentially expressed proteins in sugarcane plants infected by leaf scald disease. Comprehending the different conditions related to resistance and susceptibility in different sugarcane clones will possibly allow the monitoring of the reaction of sugarcane to leaf scald in time and elucidate molecular aspects of the expression of resistance. The information generated by the proteome analysis could be integrated with sugarcane genomic data to improve the understanding of the nature of sugarcane and the abiotic and biotic forms of interaction. In addition, it may be possible to utilize some of the proteome information acquired to develop molecular candidate markers that could be applied in marker-assisted selection.

Proteins that are differentially expressed in the duration of these stress conditions may be used in the selection and development of new sugarcane varieties with improved drought or salinity tolerance, considering the growing demand for varied renewable energy sources. However, further investigation to identify more functional proteins and validation of their functions to ascertain their involvement in producing resistant varieties is desired.

7.7 Sugarcane Molecular Breeding

7.7.1 DNA Markers in Sugarcane

DNA markers have contributed greatly to fingerprinting of elite genetic stocks for assessing genetic diversity, increasing the efficiency of trait selection, and diagnostics. Various molecular marker systems including use of biochemical markers such as isozymes were reported for genetic diversity studies in sugarcane (Glaszmann et al. 1989). Later the initial work on DNA-based markers used as nuclear ribosomal DNA marker (rDNA) (Glaszmann et al. 1990) and nuclear RFLP analysis of germplasm diversity (Lu et al. 1994; Burnquist et al. 1992; Jannoo et al. 1999) exhibited a high level of heterozygosity among cultivated clones. Restriction studies using mitochondria and chloroplast genes (D'Hont et al. 1993; Sobral et al. 1994) indicated that *S. spontaneum* had a distinctly different restriction pattern. The most widely used markers are amplified fragment length polymorphism markers (Bess et al. 1998; Lima et al. 2002; Selvi et al. 2005, 2006) and SSR markers—either genomic or EST (Cordeiro et al. 1999, 2000, 2001, 2003; Devarumath et al. 2012; Hemaprabha et al. 2006; Oliveira et al. 2007; Pan 2006; Pan et al. 2004; Parida et al. 2009, 2010; Pinto et al. 2006; Selvi et al. 2003; Swapna et al. 2011; Singh et al. 2011). RAPD markers have been used to assess the genetic diversity in elite and exotic sugarcane germplasm (Kawar et al. 2009; Srivastava and Gupta 2006; Nair et al. 1999, 2002) and to construct genetic maps (Mudge et al. 1996). The RAPD technique is also used for detecting genetic change in tissue culture raised plants (Taylor et al. 1995; Zucchi et al. 2002) and for detecting genetic fidelity in meristem raised TC plantlets (Devarumath et al. 2007; Doule et al. 2008; Tawar et al. 2008). The potential of ISSR markers for molecular profiling was assessed in sugarcane using 42 varieties from subtropical India (Srivastava and Gupta 2008). Virupakshi and Naik (2008) used organellar genome ISSR markers (cplSSR and mtISSR) to analyze red rot disease-resistant, moderately resistant, and susceptible elite sugarcane (*Saccharum* spp. hybrid) genotypes. The results indicated that these markers may be used as a new tool for the identification of the disease-resistant varieties. In sugarcane, AFLP markers have been used to study diversity among tropical and subtropical Indian sugarcane cultivars (Selvi et al. 2005), and *Saccharum* complex and *Erianthus* (Selvi et al. 2006; Bess et al. 1998). Single strand conformation polymorphism from genomics as well as ESTs has been used in sugarcane. Here amplified products are converted into single strands and electrophoresed. Variations arising out of conformational changes in the single strand are seen in this case (Swapna et al. 2011). The report revealed different degrees of polymorphism among the sugarcane clones depending on the plant material and the marker system studied. A limited level of variability is also revealed among cultivated clones (Harvey et al. 1994; Nair et al. 2002). Overall the species-level clones exhibit a high degree of variability with molecular markers.

The SSR markers have been used to fingerprint 180 sugarcane varieties and the data are stored in a database that could provide a source of information to identify varieties of unknown or disputed origin and as additional information in plant breeding rights applications and for quality assurance for delivery of new cultivars to the industry (Piperidis et al. 2004). Target region amplification polymorphism has also been used to characterize the germplasm from the genera *Saccharum*, *Miscanthus*, and *Erianthus* with the help of six primers designed using sucrose- and cold tolerance-related EST sequences (Alwala et al. 2006a,b). Sequence related amplified polymorphism has also been used in sugarcane for mapping studies (Alwala et al. 2009). A conserved intron scanning primers (CISP) marker system is based on conserved sequences used in sugarcane (Khan et al. 2011). Recently SNP-based markers have been replacing other marker types in many species, because, in general, SNPs are common in the genome, both within and between genes, and they can be cheap to assay. Significant resources have been devoted to the development of SNPs as high-throughput markers and also to SNP discovery. Extensive SNP discovery projects have been undertaken for high-throughput use in marker-assisted breeding, for population studies in different crop plants, such as maize (Ching et al. 2002), rice (McNally et al. 2006), and also in sugarcane (Bundock et al. 2009, 2012), and diversity array technology (DArT) marker is the most advanced markers system used in sugarcane (Heller-Uszynska et al. 2011). Recently Devarumath et al. (2013) used SNP and TRAP markers for assessment of genetic diversity in sugarcane. SNPs and DArT markers demonstrated that the markers can effectively dis-cover and score a large number of polymorphisms. These markers are expected to have a major impact in the future for sugarcane improvement.

7.7.2 Genetic Linkage Maps in Sugarcane

The sugarcane genome has limited classical genetic studies while genetically simpler crops have made remarkable gains (Barnes and Bester 2000). The genetic complexity is due to the aneuploidy, heterozygous nature, and chromosome numbers in various homo(eo) logy groups due to its aneuploidy (Hoarau et al. 2001). The elevated ploidy levels, cyto-genetic complexity of interspecific hybrids, and the difficulty of controlled hybridization have further complicated genetic dissection studies (Hogarth 1987). However, the advent of a diverse array of molecular marker systems such as AFLPs, RFLPs, and SSRs revealed several alleles for a single locus. These markers may be single dose (SD) and present as a single copy, double dose (DD) and present as two copies, or multidose and present as multiple copies. In recent times, the efficiency of developing genetic linkage maps in sugarcane has increased and these markers have eventually been used in gene tagging, QTL mapping, and map-based cloning (Cunff et al. 2008). The initial difficulty in mapping polyploids using molecular markers was due to the inability to identify the genotypes of marker phenotypes where a large number of genotypes for each marker phenotype are possible in a segregating population (Wu et al. 1992). Efforts in unraveling the sugarcane genome now seem to be promising, with the development of theoretical aspects of genetic mapping in polyploids by Wu et al. (1992) using single-dose fragments (SDFs).

Genetic linkage maps have been developed for sugarcane cultivars as well as for their ancestral species using the full-sib (F1) individuals (pseudotest cross strategy) and RAPD, RFLP, AFLP, SSRs, SRAP, TRAP, and EST-SSR markers. Initial efforts by da Silva et al. (1993)

in developing genetic maps of sugarcane using RFLP marker on haploid progeny of SES 208 and cross population of ADP 068 map consisted of 216 SD loci comprising 44 linkage groups (LGs), and the estimated genome coverage was 86% with at least one marker at a distance of every 25 cM. This was followed by Al-Janabi et al. (1993) using an RAPD map with 208 SD loci comprising 41 LGs and the predicted genome coverage was 85% with at least one marker every 30 cM. A further effort by da Silva et al. (1993) extended the RFLP map to 319 loci through the development of a method to map DD and triple-dose (TD) markers. Later these two maps were combined by da Silva et al. (1995) into a single genetic linkage consensus map comprising 527 markers randomly distributed across the genome with coverage of 94% with 64 LGs and bearing an average interval of 6 cM between the markers. The results also suggested this clone was an auto-octaploidy. A linkage map was constructed for *S. officinarum* using RAPDs SD marker (Mudge et al. 1996) and RFLPs (Ming et al. 1998), and for *S. robustum* with RFLPs (Ming et al. 1998; Hoarau et al. 2001). The RFLP maps by Ming were developed from interspecific crosses of *S. officinarum* × *S. spontaneum*, providing a total of four maps, two for *S. officinarum* (clones Green German and Muntok Java) and a further two for *S. spontaneum* (clones IND81-146 and PIN84-1). On Green German (2n = 97–117), 418 markers were mapped, of which 270 were distributed over 72 LGs and covered 2304 cM; with Muntok Java (2n = 140), 355 markers were mapped and 206 markers were also distributed over 72 LGs covering 1443 cM. Of the 385 markers mapped on IND81-146 (2n = 52–56), 248 were distributed over 69 groups covering 2063 cM, and with the 297 markers on PIN84-1 (2n = 96), 182 were also assembled in 69 LGs over 1103 cM (Ming et al. 1998).

For the first time, maps of cultivated genotype SP70-1006 were initiated on the selfed progeny (D'Hont et al. 1994). Further, this map was transferred and developed on cultivar R570 by Grivet et al. (1996) using RFLP probes from maize and sugarcane; 480 markers identified 96 LGs. Similarly the self population of R570 was used for genome mapping using AFLP markers (Hoarau et al. 2001). Using 37 primer combinations, 939 SD AFLP markers on R570, of which 887 were assigned into 120 cosegregation or LGs with an average distance of 6.5 cM between two markers. The same population had been used previously to construct a map using RFLP markers by the same group (Grivet et al. 1996). This RFLP map consisted of 408 loci on 96 cosegregation groups (CGs) arranged in 10 putative homologous groups (HGs). The mapping effort by this group led to the identification of a major rust resistance gene *Bru1* (Daugrois et al. 1996; Asnaghi et al. 2000, 2004) and efforts are under way to isolate it by map-based cloning (Cunff et al. 2008).

AFLP and SSR markers in a F_1 population of IJ 76-514 × Q 165 are used for development of a genetic map for the sugarcane cultivar Q 165 (Aitken et al. 2005). A total of 112 (40 AFLP and 72 SSR) primers were used to generate 967 SD and 123 DD markers. Out of a total of 136 markers, 127 markers were grouped to eight homology groups, which correspond to the basic chromosome number of *S. spontaneum*, and 910 markers were distributed across 116 LGs covering a total map length of 9058.3 cM. Similarly, Aitken et al. (2007) developed a linkage map for IJ76-514 using simplex and duplex markers in a segregation population of Q165 × IJ76-514. They used the same set of primers and revealed 595 markers. Of these, 240 were simplex markers (segregated in a 1:1 ratio); 178 of these were distributed in 47 LGs with 62 unlinked markers—this shows *S. officinarum* with the lower number of markers with an additional 234 duplex markers and 80 biparental simplex markers generated a total of 123 LGs. Using multiallele SSR markers, repulsion phase linkage and alignment with Q165 linkage maps were identified with 10 HGs from 105 LGs. Repulsion phase linkage analysis indicated that IJ76-514 is neither a complete polyploid nor an allopolyploid.

Oliveira et al. (2007) used EST markers for commercial linkage mapping in a commercial cross SP80-180 × Sp80-4966. The data analysis showed that 72.5% SD markers were generated: 198 CGs, of which 120 were grouped into 14 HGs. On the basis of BLAST studies, putative functions were assigned to 113 EST-SSR and 6 EST-RFLP markers. Thus this map is populated with functional associated markers. Alwala et al. (2008) constructed linkage maps and genome analysis in the progeny of an interspecific cross in sugarcane using TRAP, SRAP, and AFLP markers. The map comprised 146 linked markers with 49 LGs and 121 linked markers in 45 LGs for *S. officinarum* and *S. spontaneum*, respectively.

Recently, LCP 85-384, considered one of the most successful sugarcane cultivars of the Louisiana sugar industry (Andru et al. 2011), was used to construct a molecular genetic linkage map using AFLP, SSR, and TRAP markers using the selfed progeny and based on polymorphism-derived 64, 19, 12 primer pairs, respectively. Of 1111 polymorphic markers detected, 773 simplex (segregated in a 3:1 ratio) and 182 duplex (segregate in a 77:4 ratio) markers were used to construct the map. Of the 955 markers, 718 simplex and 66 duplex markers were assigned to 108 CGs with a cumulative map length of 5617 cM and a density of 7.16 cM per marker. Fifty-five simplex and 116 duplex markers remained unlinked. With an estimated genome size of 12,313 cM for LCP 85-384, the map covered approximately 45.6% of the genome. Forty-four of the 108 CGs were assigned into 9 HGs based on information from locus-specific SSR and duplex markers, and repulsion phase linkages detected between CGs.

DArT marker technique has recently been used by Heller-Uszynska et al. (2011) in molecular analysis of a cross from Q165 × IJ 76-514 even though linkage map construction is yet to be reported.

7.7.3 QTLs in Sugarcane

QTLs, or groups of genes or alleles controlling complex traits, can be used for locating genomic regions associated with quantitative traits and validation of the identified markers as a selection tool and integrating them in the breeding programs. The prerequisites for QTL detection are identifying markers linked to trait to generate/have a mapping population that displays the genetic variation for the trait of interest followed by screening them with molecular markers having sufficient phenotypic data on the population and being able to establish genetic LGs. The identification of QTLs requires knowledge of the genome of the species under investigation, and this can be acquired by utilizing existing genetic maps, comparative mapping, and association mapping.

There are inherent difficulties in the detection of QTLs in the complex polyploid genome of sugarcane. The majority of QTL analyses and the statistical methods used to identify QTLs have been based on diploid crops that have homologous pairs of chromosomes arranged in sets and, therefore, have predictable meiotic outcomes, single copy molecular markers, and progeny that segregate in expected patterns. Since sugarcane is a highly heterozygous crop, F_1 progenies from two highly heterozygous parents or a selfed population were used as a mapping population for the identification of markers and QTLs. Most of the research used segregating population to map QTLs and involved F_1 from crosses between two cultivars (Kang et al. 1983; Milligan et al. 1990) or cultivated varieties and wild species (Ming et al. 1998).

Most of the agronomic traits in sugarcane are controlled by polygenes and the genetic analysis of such traits has been difficult due to polyploidy. It is known that in sugarcane, alleles that occur in low frequency in a population are the ones that will show segregation and can be used for mapping (Wu et al. 1992). Despite these difficulties, comparative mapping has been particularly useful for QTL analysis in sugarcane. Synteny with diploid relatives, in particular sorghum, has been beneficial in constructing genetic maps and for identifying QTLs for homologous agronomic traits in sugarcane (Paterson et al. 1995; Dufour et al. 1996, 1997; Grivet et al. 1996; Guimaraes et al. 1997; Ming et al. 1998, 2002; Bowers et al. 2003; Jordan et al. 2004). In sugarcane the first published QTL analysis by Sills et al. (1995) used RAPD marker to evaluate mapping population of 44 F_1 progenies from a cross between *S. officinarum* and *S. robustum*. This analysis included QTLs for stalk number, tasseled stalks, and stalks with smut. Similarly, Mudge et al. (1996) linked a RAPD marker identified for eye spot resistance in a segregating population of 84 F_1 progenies from a *S. officinarum* × *S. spontaneum* cross, and Guimaraes et al. (1997) used a RFLP marker linked with short-day flowering in a mapping population of 100 F_1 individuals. Association DNA markers were used with a disease-resistant population for a study of inheritance of rust resistance in a population of 142 F_1 from rust-resistant cultivar R570. Data showed a 3:1 segregation, suggesting the presence of a rust resistance gene in sugarcane cultivar R570 (Daugrois et al. 1996).

Asnaghi et al. (2000) saturated the region surrounding the major rust gene utilizing comparative mapping between sugarcane, sorghum, maize, and rice. They used RFLPs on 88 selfed mapping progeny of R570, adding 217 SD markers to the genetic map of which 66% mapped to the rust-resistant gene. Rossi et al. (2003) used the Brazil-based SUCEST EST database to identify candidate genes for pathogen resistance and construct a map of their distribution on the sugarcane genome, resulting in 88 resistant gene analogs (RGAs) out of a total of 261,609 EST sequences analyzed. Of these, 50 RGAs were located on a reference map constructed on sugarcane cultivar R570 (Grivet et al. 1996; Hoarau et al. 2001). Using SD markers, 148 RFLP and 55 SSR loci were mapped on 112 selfed progeny from R570, and Hoarau et al. (2001) used this population to construct an AFLP genetic map. Later Asnaghi et al. (2004) used bulked segregation analysis (BSA) and AFLPs to map the resistance gene (*Bru*1) using 658 individuals derived from selfing of clone R570; these clones were established from using R570 as the male parent and four susceptible cultivars as the female parent. These individuals were analyzed for segregation of rust resistance.

The earliest report for mapping of sugar content and sugar yield-related traits on sugarcane was carried out using a candidate gene approach by Ming et al. (2001, 2002). Using linkage analysis in two different interspecific crosses, *S. officinarum* (Green German) × *S. spontaneum* (IND81-146) 264 F_1 plants and another cross from *S. spontaneum* (PIN 84-1) × *S. officinarum* (Muntock Java), 239 F_1 plants were used for sugar-related mapping studies. They identified 14 marker traits and associated 8 from *S. officinarum* (Green German) and 6 from *S. spontaneum* (IND81-146), showing 38.6% phenotypic variation for sugar and 36% variation, respectively. The 22 QTLs from PIN × MJ contributed to 68.3% variation with 18 *S. officinarum* (MJ) QTLs alone, contributing 45.7% of variation and the four *S. spontaneum* (PIN) QTLs showing 33.4% variation.

The first extensive QTL mapping study in cultivated sugarcane R570 was developed by Hoarau et al. (2002) on a population in a replicated trial over two crop cycles using 1000 AFLP markers. Forty-five quantitative trait alleles (QTAs) were mapped linking to four traits such as stalk length (SL), stalk diameter (SD), number of millable canes (NMS), and brix (BR). However, the QTL effects were inconsistent between two crop cycles. The QTAs had a small individual effect and accounted for between 3% and 7% of the phenotypic variation. In another cross, Q117 × MQ77-340, QTL analysis identified 23 marker trait associations (MTAs) from commercial cane sugar (CCS), Pol, and BR (Reffay et al. 2005).

QTL affecting sucrose content in Australian cultivar Q165 were studied by Aitken et al. (2006) using the same population and markers used by Aitken et al. (2005) in their earlier publication. BR, Pol, and CCS were included as sugar traits. For QTLs detection was based on SD and multidose (MD) markers. A total of 37 QTLs for BR and Pol were identified, of which 8 were specific to the early period and 9 specific to mature cane. Each QTL explained 3%–9% phenotypic variation for the two traits. Out of 37, 30 QTLs were mapped into 12 genomic regions, which spread over 6 HGs. The same population was further used for yield-related trait mapping by the group (Aitken et al. 2008).

Jordan et al. (2004) used RFLP and radio-labeled amplified fragments (RAFs) in the population of Q117 × 74C42 with 108 individuals for yield-related traits. In this study they used some sorghum-related probes (Paterson et al. 1995), observed benefits of comparative mapping between sorghum and sugarcane, and suggested that useful markers for sugarcane can be identified from sorghum QTLs associated with related traits. Alwala et al. (2008) showed that sugar-related BR and Pol traits were mapped using the progeny of a *S. officinarum* (La Striped) × *S. spontaneum* cross (SES 147B) on the basis of AFLP, SRAP, and TRAP markers. A total of 41 QTLs (30 from *S. officinarum* and 11 from *S. spontaneum*) were identified from two traits. Individual QTLs explained phenotypic variation of 15.2%–21.6%. Nine digenic interactions were also identified. Pinto et al. (2010), in their mapping studies using RFLP-ESTs, could associate sucrose synthase-derived markers with a QTL having a negative effect on cane yield and positive effect on Pol in two crop cycles. Thus a different marker system has been used in locating different agronomy-related trait QTLs in different sugarcane populations.

7.7.4 Molecular Markers for Diagnostics

The development of new and improved molecular assays to detect various sugarcane pathogens is progressing in different laboratories worldwide (Braithwaite and Smith 2001). Molecular diagnostic tests, based mostly on nucleic acids, are highly sensitive and relatively easy to use compared with the traditional detection methods such as histology, electron microscopy, sap transmission onto indicator plants, or the isolation and culture of the causal agents. Due to their high sensitivity, molecular diagnostic methods are capable of detecting pathogens even in asymptomatic plants with an extremely low pathogen titer. Molecular tests for diseases such as ratoon stunting (Fegan et al. 1998; Pan et al. 1998), Fiji disease (Smith and van de Velde 1994; Smith et al. 1994), mosaic (Smith and van de Velde 1994), striate mosaic (Thompson et al. 1998), sugarcane yellow leaf virus (Moonan and Mirkov 2002; Chatenet et al. 2001), smut (Albert and Schenck 1996), sorghum mosaic (Yang and Mirkov 1997), and SCBV (Braithwaite et al. 1995) have been developed and are being applied to screen quarantine germplasm. With advances in genome sequencing and the continued refinement of various techniques in recombinant DNA technology, development of more reliable, faster, and more cost-effective molecular diagnostic tests for all the important sugarcane pathogens can be expected in the near future.

7.8 Genetic Transformation in Sugarcane

Besides being an economically important crop for food and energy, there are several other factors that make sugarcane a suitable candidate for improvement through genetic transformation. Sugarcane improvement by classical breeding is difficult because of its complex polyploidy nature, variable fertility, genotype versus environment interactions, and long duration (10–15 years) for development of a new variety. Thus traditional back-crossing to recover elite genotypes with desired agronomic traits is a very difficult task in sugarcane. The availability of tissue culture regeneration systems from various explants makes this crop a suitable candidate for genetic manipulation. In addition, the gene transfer techniques are well established in sugarcane and the vegetative propagative nature of sugarcane can easily pass the transgene to progenies and maintain the same without loss. Tremendous progress has been made in sugarcane genetic engineering and several genes targeted toward its improvement have been introduced. Genes for disease/pest resistance, for drought tolerance, and for quality improvement such as sugar accumulation have been introduced into sugarcane (Table 7.3).

The success of transgenic sugarcane plant production depends on three major aspects: the technology used for transformation, the target tissue/explants, and the tissue culture regeneration system used. Somatic cells with good embryogenic potential are ideal targets for integration of transgenes since each has the potential to become an individual plant. Various explant types (axillary buds, apical meristems, immature inflorescences, and leaf segments) have been used successfully to regenerate full plants in sugarcane, indicating that a wide range of totipotent target tissues are available for genetic transformation. Various transformation methods such as electroporation, PEG, particle bombardment, and *Agrobacterium* have been tried in sugarcane with varying degrees of success in the production of fertile transgenic plants. Microprojectile bombardment and *Agrobacterium*-mediated gene delivery are the most adopted methodologies for sugarcane transformation. The improvements in the microprojectile method and its simplicity made this technology unavoidable in sugarcane genetic engineering. Embryogenic callus can be used for transformation via the microprojectile method to develop transgenic plants. Despite being the most useful, robust, and routinely applied method, the biolistic DNA method often leads to a complex transgene integration pattern, which can cause problems in subsequent analysis. On the contrary, *Agrobacterium*-mediated transformation gained more usage due to its simplicity and efficiency in producing single-copy integration of transgenes. Several genes (for disease/pest resistance, salt and drought tolerance, and sugar accumulation) targeting improvement have been introduced into sugarcane (Altpeter and Oraby 2010; Hotta et al. 2011; Lakshmanan et al. 2005; Srikanth et al. 2011 2011; Suprasanna et al. 2011) (Table 7.4).

The first experiments in sugarcane genetic transformation were reported by Chen et al. (1987) when a kanamycin-resistance gene was introduced into protoplasts through electroporation and PEG treatment. However, the PEG-mediated transformation did not receive much attention because its low transformation efficiency (one per 10^6) of treated protoplast and poor reproducibility was reported in the sugarcane cultivar F164. The electroporation method of sugarcane transformation was found to be more suitable than the PEG method. Stably transformed calli were observed in different sugarcane cultivars using the electroporated protoplast method (Chowdhury and Vasil 1992; Rathus and Birch 1992). Rathus and Birch (1992) reported one transformation event per 10^2–10^4 electrophorated protoplasts for sugarcane cultivars Q63 and Q96. This frequency is sufficient for regeneration of useful transgenic plants, but the production of the transgenic plants is less due to lack of regeneration from protoplasts. Later, Arencibia et al. (1995) used embryogenic cells and Arencibia et al. (1992) used meristamatic tissues of *in vitro* grown plants to develop transgenic plants through the electroporation method.

The early attempts at transformation in sugarcane using *Agrobacterium*, with or without virulent gene inductions and other treatments that enhance infection, met with little success (Birch and Maretzki 1993). First reports by Arencibia et al. (1998) demonstrated the production of morphologically normal transgenic plants. Similarly, Enriquez-Obregon et al. (1998) developed herbicide-resistant sugarcane plants by *Agrobacterium*-mediated transformation using different strains of *Agrobacterium tumefaciens* and changes in the steps in the protocol for cocultivation and plantlet regeneration. For the first time, visual

TABLE 7.4

List of the Traits Introduced to Sugarcane for Genetic Improvement

Trait	Gene	Method of Gene Transfer	References
Reporter and Selection Systems			
Neopmycin phosphotransferase	*npt*-II	Microprojectile	Bower and Birch 1992
β-Glucuronidase	*uid*-A	Microprojectile	Bower and Birch 1992
Hygromycin phosphotransferase	*Hpt*	*Agrobacterium*	Arencibia et al. 1998
Green fluorescent protein	*Gfp*	*Agrobacterium*	Elliott et al. 1998
Phosphinothricin acetyl transferase	*Bar*	*Agrobacterium*	Manickavasagam et al. 2004
Phosphomannose isomerase	*manA*	Microprojectile	Jain et al. 2007
Herbicide resistance			
Bialophos	*Bar*	Microprojectile	Gallo-Meagher and Irvine 1996
Phosphinothricine			
Glufosinate ammonium	*Bar*	*Agrobacterium*	Enriquez-Obregon et al. 1998
	Pat	Microprojectile	Leibbrandt and Snyman 2003
Disease Resistance			
SCMV	*SCMV-CP*	Microprojectile	Joyce et al. 1998a
Sugarcane leaf scald	*alb*D	Microprojectile	Zhang et al. 1999
SrMV	*SrMV-CP*	Microprojectile	Ingelbrecht et al. 1999
Puccinia melanocephala	*Glucanase, chitanase, and ap24*	*Agrobacterium*	Enriquez et al. 2000
SCYLV	*SCYLV-CP*	Microprojectile	Gilbert et al. 2005, 2009
Fiji leaf gall	*FDVS9 ORF 1*	Microprojectile	McQualter et al. 2004
Pest Resistance			
Sugarcane stem borer	*cry*1A	Microprojectile	Arencibia et al. 1999
Sugarcane stem borer	*cry*1Ab	Microprojectile	Braga et al. 2003
Sugarcane stem borer	*cry*1Ab	Microprojectile	Arvinth et al. 2010
Sugarcane stem borer	*cry*1Aa3	*Agrobacterium*	Kalunke et al. 2009
Proceras venosatus	Modified *cry*1Ac	Microprojectile	Weng et al. 2010
Sugarcane canegrub	*Gna*	Microprojectile	Legaspi and Mirkov 2000
Mexican rice borer	*Gna*	Microprojectile	Sétamou et al. 2002
Ceratovacuna lanigera	*Gna*	*Agrobacterium*	Zhangsun et al. 2007
Scirpophaga excerptalis	*Aprotinin*	Microprojectile	Christy et al. 2009
Metabolic Engineering/Alternative Products			
Sucrose accumulation	*Antisense soluble acid invertase*	Microprojectile	Ma et al. 2004
Fructooligosaccharide	*IsdA*	*Agrobacterium*	Enriquez et al. 2000
Polyphenol oxidase	*Ppo*	Microprojectile	Vickers et al. 2005
Polyhydroxybutyrate	*phaA, phaB, phaC*	Microprojectile	Brumbley et al. 2007
p-Hydroxybenzoic acid	*hch*1 and *cp*1	Microprojectile	McQualter et al. 2004
Mannose	*manA*	Microprojectile	Jain et al. 2007
Isomaltulose	*SI*	Microprojectile	Wu and Birch 2007
Proline overproduction	*P5CS*	Microprojectile	Molinari et al. 2007
Tripsin inhibitors		Microprojectile	Falco and Silver-Filho 2003

selection of transgene using green fluorescent protein (GFP) as a visual marker has also been reported (Elliott et al. 1998, 1999). Manickavasagam et al. (2004) produced herbicide BASTA-resistant sugarcane with close to 50% transformation efficiency by infecting axillary buds with *Agrobacterium*.

Microprojectile-mediated transformation is a widely used method for sugarcane transformation (Birch 1997; Moore 1999). Developed in the 1960s for the inoculation of intact plants with infectious viral particles (MacKenzie et al. 1966), particle bombardment was used by Hawaiian researchers to produce transgenic plants of a *Saccharum* species (Maretzki et al. 1990). Investigations on microprojectile-mediated transformation by Franks and Birch (1991) in Australia led to the development of the first transgenic sugarcane plants from a commercial cultivar in 1992 (Bower and Birch 1992). Subsequently, microprojectile-mediated transformation of several commercially cultivated sugarcane genotypes was reported from a number of laboratories worldwide (Birch and Maretzki 1993; Bower et al. 1996; Birch 1997; Irvine and Mirkov 1997; Joyce et al. 1998a,b; Moore 1999; Nutt et al. 1999; Weng et al. 2006; Arvinth et al. 2010).

There were two principal technologies driving the rapid development of microprojectile-mediated methods for sugarcane transformation: (1) the availability of the equipment required for microprojectile-mediated transformation and (2) the ability to produce somatic embryogenic systems from a range of sugarcane genotypes. Among the different tissues tested, embryogenic callus appears to be the preferred target due to its high transformability and regenerability. Nonetheless, regenerable cell suspension (Franks and Birch 1991; Chowdhury and Vasil 1992), apical meristems (Gambley et al. 1993), and immature leaf whorls and inflorescences (Elliott et al. 1999; Lakshmanan et al. 2003) have also been successfully used for microprojectile-mediated sugarcane transformation. The applicability to a wide range of target tissues and genotypes, and the simplicity of operation make the microprojectile approach the preferred method for sugarcane transformation.

Transgenic plant production requires selectable marker genes that enable the selection of transformed cells, tissue, and plants. The most routinely used are those that exhibit resistance to antibiotics or herbicides. Since there are perceived risks in the deployment of transgenic plants containing these markers, alternate selection systems referred to as positive selection and marker-free systems have become useful (Suprasanna and Ganapathi 2010). In sugarcane, Jain et al. (2007) used mannose for the selection of embryogenic callus and found that increased mannose improved the overall transformation efficiency by reducing the number of selection escapes. High-level constitutive transgene expression in all tissues at all stages of development is always aimed for in transgenic plant production. The expression of a transgene can be effectively targeted to specific tissues or to particular stages of plant development by including promoters to get the desired pattern of regulation. The cauliflower mosaic virus (CaMV) 35S promoter is one of the most commonly used in dicots but has poor activity in sugarcane transgenic studies. Maize ubiquitin 1 promoter showed better expression of GUS gene than the CaMV 35S, EMU, and rice actin 1 promoters (Rathus et al. 1993; Gallo-Meagher and Irvine 1996). Sugarcane Ubi promoters Ubi-4 and Ubi-9 showed higher levels of expression in both rice and sugarcane calli (Wei et al. 1999; Wei 2001). However, it was observed that the Ubi-9 promoter poses a problem due to posttranscriptional silencing (Wei et al. 2003). Yang et al. (2003) isolated 11 promoters from the genomic libraries of sugarcane elongation factor 1α and sugarcane proline-rich protein and studied their expression with GUS gene in sugarcane callus. The expression level is more or less similar to that of the Ubi-1 promoter. However, the promoter of sugarcane UDP glucose dehydrogenase gene is expressed only in intermodal regions and requires the intron for its strong expression (van der Merwe et al. 2003). Rice ubiquitin promoter RUBQ2 has increased transgene expression by about 1.6-fold over the maize Ubi-1 promoter in sugarcane (Liu et al. 2003). Increased GFP expression was noted by a promoter isolated from banana streak virus (Schenk et al. 2001). Mudge et al. (2009) isolated eight promoters from cultivar Q117, of which at least three are associated with expressed alleles. All of the isolated promoter variants were tested for their ability to drive reporter gene expression in sugarcane. The genetic manipulation of sugarcane to enhance phenotypes such as sugar levels and borer resistance requires not only appropriate genes but also promoters. A sugarcane stem-specific promoter, UQ67P, has been isolated and shown to be able to drive reporter gene expression in stem tissue (Hansom et al. 1999).

Significant progress has been made toward the development of transgenic sugarcane resistant to viral disease. The transfer of coat protein genes (SCMV-CP) conferred resistance against the sugarcane mosaic virus (SCMV) (Joyce et al. 1998a,b), sorghum mosaic virus (SrMV) (Ingelbrecht et al. 1999), and sugarcane yellow mosaic virus (Rangel et al. 2003). Genes from bacteria such as *Bacillus thuringiensis* (*Bt*) and *Bacillus sphaericus*, protease inhibitors, plant lectins, ribosome-inactivating proteins, secondary plant metabolites, and small RNA viruses have been used alone or in combination with conventional host plant resistance to develop crop cultivars that suffer less damage from insect pests (Hilder and Boulter 1999). Arencibia et al. (1997) reported the first sugarcane transgenics expressing a truncated *cry*1A(b) gene coding the active region of Bt δ-endotoxin for resistance to *Diatraea saccharalis*. Screening of the transgenic lines resulted in the selection of five events exhibiting significant protection against the borer in spite of very low expression of the toxin (0.59–1.35 ng/mg of soluble leaf protein). The synthetic *cry*1A(c) gene with altered GC content produced transgenic lines with resistance to sugarcane stem borer *P. venosatus*, which is a major sugarcane borer pest in China (Weng et al. 2006). In a subsequent study, Weng et al. (2010) further enhanced the GC content of the *cry*1Ac gene and obtained a five-fold higher toxin expression level than that obtained in their earlier study. Transgenic sugarcane for borer resistance using the *Cry* 1Aa3 gene was also reported (Kalunke et al. 2009). Various traits and genetic manipulation of transgenes are mentioned in Table 7.4. Recently, Arvinth et al. (2010) assessed the efficacy of native *Cry*1Aa, *Cry*1Ab, and *Cry*1Ac against *C. infuscatellus* in *in vitro* bioassays through the diet-surface contamination method and observed that *Cry*1Ab is the most toxic among the three compounds (Table 7.4).

Pathogenesis-related (PR) proteins are induced in plants in response to attack by microbial pathogens or insect pests (Muthukrishnan et al. 2001). The antifungal properties of chitinases (PR protein gene) and β-1,3-glucanases have been widely reported (Collinge et al. 2001; Muthukrishnan et al. 2001). Enriquez et al. (2000) have introduced β-1,3-glucanase, chitinase, and ap24 genes against the *Puccinia melanocephala* pathogen. Plants with albicidin detoxification capacity equivalent to 1–10 ng of AlbD enzyme per mg of leaf protein did not develop chlorotic disease symptoms in inoculated leaves, whereas all untransformed control plants developed severe symptoms (Zhang et al. 1999). The albD gene for albicidin detoxification has been studied to explore engineered inactivation of pathogenesis factors as a control strategy against bacterial diseases. Expression in transgenic sugarcane of a novel gene for albicidin detoxification conferred a high level of resistance to chlorotic symptom induction, and multiplication and systemic invasion by *X. albilineans*. Transgenic sugarcane cultivar Q124 with significantly enhanced resistance to Fiji disease was produced by a transgene encoding a translatable version of FDV segment 9 ORF 1 (McQualter et al. 2004).

PIs of both plant and animal origin have been expressed in sugarcane to impart resistance to borers. Transgenic sugarcane plants with proteinase inhibitor gene and snowdrop lectin gene exhibited resistance against cane grubs (Allsopp and Suasa-ard

2000; Nutt et al. 2001). Falco and Silva-Filho (2003) developed sugarcane transgenic lines expressing soybean Kunitz trypsin inhibitor (SKTI) and soybean Bowman-Birk inhibitor (SBBI) under the control of the maize Ubi-1 promoter. The transgenic lines were evaluated against *D. saccharalis* in insect feeding trials. The larvae that fed on leaves with transgenic lines expressing SBBI did not show any significant change in mortality, whereas those that fed on leaves from SKTI-expressing plants showed a slightly higher mortality. Transgenic sugarcane plants with snowdrop lectin gene (*Galanthus nivalis* agglutinin, *gna*) negatively affected the development and reproduction of the Mexican rice borer (Sétamou et al. 2002). Legaspi and Mirkov (2000) also observed considerable growth inhibition of sugarcane stalk borers when they were fed on transgenic sugarcane engineered with lectin genes. A bioassay of transgenic sugarcane plants engineered with the snowdrop lectin (gna) gene demonstrated a significant resistance to the woolly aphid (*Ceratovacuna lanigera* Zehnther) (Zhangsun et al. 2007). Christy et al. (2009) transferred aprotinin genes to sugarcane cultivars. The *in vivo* bioassay studies showed that the larvae of the top borer *Scirpophaga excerptalis* Walker (Lepidoptera: Pyralidae) that fed on transgenics showed a significant reduction in weight, which indicated impairment of their development.

Sucrose content is a highly desirable trait in sugarcane. Sugarcane cultivars differ in their capacity to accumulate sucrose, and breeding programs routinely perform crosses to identify genotypes that are able to produce more sucrose. In this regard, transgenic approaches to manipulate native genes that influence metabolism may have significant application. One of the key enzymes in sucrose synthesis metabolism is *SPS*, which catalyzes the synthesis of sucrose-6-phosphate from UDP-glucose and fructose-6-phosphate. Transgenic plants that overexpress *SPS* have been attempted in sugarcane but were less successful (Grof et al. 1996). Ma et al. (2000) reported that two-fold increases in sucrose concentration were noted in the liquid culture of sugarcane engineered with soluble acid. However, up to a 70% reduction of soluble acid invertase activity in the immature internodes of transgenic sugarcane plants had no significant impact on sucrose concentration (Botha et al. 2001). The transgenic plants expressing sense constructs of enzyme polyphenol oxidase (PPO) acting on phenolic compounds produced dark-colored polymers when sugarcane (*Saccharum* spp.) was crushed to release the juice.

Most of the field trials of transgenic sugarcane are concerned with transgene expression. Few research studies including comprehensive field trials and agronomic data have been carried out. Leibbrandt and Snyman (2003) conducted the field trials of herbicide-resistant transgenic sugarcane. Herbicide resistance was expressed even after the third vegetative propagation. Agronomic traits such as height, diameter, the number of stalks, fiber content, disease resistance, and yield of transgenic clones were not significantly different from those of untransformed sugarcane plants. However the field trials conducted by Arencibia et al. (1999) in insect-resistant transgenic sugarcane produced from electroporation revealed some morphological, physiological, and phytopathological variations. Further these variations were analyzed and confirmed at the molecular level by using AFLP. Vickers et al. (2005) have reported that significant reduction in the yield performance was noted in the transgenic sugarcane compared with tissue culture-derived plants and normal field-grown plants. Butterfield et al. (2002) conducted different kinds of field trials. They crossed transgenic sugarcane with multiple copies of herbicide-resistant (*bar*) and SrMV-resistant genes with nontransgenic sugarcane varieties and analyzed the progenies. Depending on the linkage relationships between transgenes in the parent plants, distinct segregation patterns were observed in the progeny.

7.9 Metabolic Engineering of Sugarcane and Biofactory

Development of plant-based expression systems has resulted in the production of recombinant proteins of pharmaceutical and industrial significance. These include viral proteins, vaccines, antimicrobial peptides, antibodies, pharmaceuticals, bioactive compounds, and industrial compounds. In the era of successful molecular farming, the challenge of confining transgenes and transgene products is crucial to the development of plants as biofactory systems for the production of value-added chemicals and pharmaceuticals, especially in sugarcane for biofuel, bioplastics, and alternative sugars. Sugarcane has advantages for such containment as a secure platform (Wang et al. 2005). It is a C4 plant that is capable of highly efficient carbon fixation, accumulates and stores large amounts of carbon as sucrose, plus cellulose and hemicelluloses, and has a very good storage system (stem) and a large biomass. Generally, commercial sugarcane cultivars are propagated by using setts; moreover, normal flowering is almost never observed under field conditions, so there will be little chance of pollen drift or production of viable seed. In the case of sugarcane, the principal food product is the refined crystal, which is essentially free of protein, as compared to a whole fruit or vegetable. The various traits of interest can be manipulated in sugarcane for genetic improvement. Some of these are listed in Table 7.4.

There have been successful attempts to develop sugarcane as a biofactory. Enriquez et al. (2000) expressed *Acetobacter diazotrophicus* levansucase (lsdA) as this enzyme acts on sucrose yielding high amounts of 1-ketose, which is a fructooligosaccharide essential for healthy nutrition of humans and animals. Brumbley et al. (2004) developed transgenic sugarcane producing polyhydroxybutyrate (PHB), a biodegradable thermoplastic compound. PHB accumulated up to 1.2% of the total weight in leaves without any growth penalties. Another industrially important compound, para-hydroxybenzoic acid, was also produced in transgenic sugarcane by expressing bacterial enzymes (chorismate pyruvate-lyase and 4-hydroxycinnamoyl CoA hydratase) (McQualter et al. 2004). The pHBA overexpressing line accumulated 7.3% dry weight in the leaf and 1.5% dry weight in the stem tissue. Wang et al. (2005) demonstrated the expression of human granulocyte macrophage colony stimulating factor (GM-CSF, 0.02% of the total soluble protein) in transgenic sugarcane plants. This study could provide a highly secure system for the production of pharmaceutical proteins. Petrasovits et al. (2007) used a single gene from apple (*Malus domestica*) to engineer sugarcane to make the alternative sugar sorbitol. The best line contained sorbitol at 12% dry weight in leaves and 1% dry weight in stems. Hamerli and Birch (2011) developed sugarcane plants that produced the sucrose isomers trehalulose and isomaltulose through expression of a vacuole-targeted trehalulose synthase modified from the gene in *Pseudomonas mesoacidophila* MX-45. Trehalulose concentration in juice increased with internode maturity, reaching about 600 mM, with near-complete conversion of sucrose in the most mature internodes.

Sugarcanes (*Saccharum* hybrids) are attractive candidates for metabolic engineering aimed at sustainable production of value-added biomaterials and feedstocks, particularly those derived from sucrose (a-D-glucopyranosyl-1,2-D-fructofuranose), the major storage product in sugarcane (Birch 2007). Successful examples of establishing biorefineries at sugar mills to produce biofuel and bioplastics and of engineering sugarcane to make new bioplastics and alternative sugars demonstrate that this crop has the potential to contribute to the bioeconomy (Brumbley et al. 2007).

Although microprojectile bombardment has been the main method of sugarcane transformation, other methods for transgene incorporation into sugarcane are needed.

For example, genetic transformation mediated by *Agrobacterium tumefaciens* should be studied and applied to sugarcane more frequently as it has the potential to become more efficient with appropriate manipulation and *in vitro* culture conditions, selection of the best age of the calli, type and stage of embryogenic culture, and improvement in the virulence of *A. tumefaciens*. Compared to other methods, transformation mediated by *A. tumefaciens* is a simple and low-cost method. Furthermore, it can transfer relatively long segments of DNA with little rearrangement and at a relatively low number of integrated copies. Another important transformation technique is the transformation of chloroplasts. Chloroplast genetic engineering offers a number of unique advantages compared to conventional transgenic technologies, including high protein expression levels (De Cosa et al. 2001), integration into the plastome via homologous recombination without position effects or gene silencing (Daniell et al. 2001), and the expression of several transgenes in a single transcriptional unit due to the chloroplast's prokaryotic origin (Bock 2001).

Over the next few years, the first generation of commercial sugarcane transgenics, with herbicide and insect resistance, will be available as a result of improved transformation efficiency, the design of new transformation vectors, and new transgene design rules to avoid silencing. The development of strategies for incorporating polygenic traits, the hyperexpression of transgenes, and the containment of transgenes within the transgenic plants (Daniell 1999; Daniell et al. 2001; Maliga 2004), together with the possibility of engineering native genes without significant genetic rearrangements (Beetham et al. 1999), are valuable innovations that could be utilized for the improvement of sugarcane in the near future.

For the successful release of transgenic sugarcane, various scientific, legislative, and public perception issues must also be addressed. Transformation systems that do not incorporate any nontransgene DNA into the plant, and utilize nonantibiotic selection and plant gene-based selection strategies would be a good start toward overcoming regulatory and public perception issues. In addition, the ability to control transgene expression through induction, developmental control, or tissue specificity will provide a platform for the production of a range of new compounds in sugarcane at commercially useful levels (Lakshmanan et al. 2005).

7.10 Genetic Variability through *In Vitro* Culture and Mutagenesis

Somaclonal variation has emerged as an important *in vitro* culture tool for crop improvement. This system has been adopted for improving the quality and production of sugarcane and somaclones for yield, sugar recovery, disease resistance, drought tolerance, and maturity have been isolated. Somaclones for eye spot disease, Fiji disease, downy mildew disease, smut, and rust resistance; improvement in yield; and sucrose content have been isolated and adopted in improvement programs. It is hence imperative that genetic variability in tissue culture-derived plants be assessed to select appropriate material for the breeding programs. Sugarcane was among the first plants where somaclonal variation was reported (Heinz and Mee 1969; Larkin and Scowcroft 1981). Culture-induced variation is observed, particularly when plants are produced via a callus stage upon exposure to high levels of auxin (3–5 mg/l 2,4-D) and long culture periods (Irvine et al. 1991; Zucchi et al. 2002).

Physical and chemical mutagens have been applied to *in vitro* cultures so as to enhance the frequency of genetic variation and obtain beneficial modifications in cultivars (Patade et al. 2008). Physical (gamma rays and ion beams) and chemical (ethyl methanesulfonate [EMS], sodium azide, and sodium nitrite) mutagens have been used successfully.

In vitro selection at the cellular level has been successful in isolating mutants for desirable traits (Suprasanna et al. 2008a) and imposing *in vitro* selection pressure by incorporating fungal pathotoxins or fungal culture filtrates has been practiced for selecting disease resistance (Rai et al. 2011) and incorporation of sodium chloride, PEG, and mannitol for selecting salt or drought tolerance (Suprasanna et al. 2008b). In sugarcane, somaclonal variant lines resistant to eye spot disease caused by *Helminthosporium sacchari* were selected (Larkin and Scowcroft 1983). Various researchers have used this strategy to select embryogenic cells tolerant to the causal agent of red rot (Ali et al. 2008; Sengar et al. 2009).

Assessment of genetic fidelity among micropropagated plants is important, especially in polyploid plants like sugarcane. Besides morphological analysis, molecular markers have been used for detecting genetic change in tissue culture raised plants (Lal et al. 2008; Suprasanna et al. 2006, 2010), for detecting genetic fidelity in meristem raised tissue culture plantlets (Devarumath et al. 2007; Tawar et al. 2008; Doule et al. 2008), and for characterizing salt- and drought-tolerant radiation-induced variants (Patade et al. 2006). Field testing of tissue-cultured progeny has been conducted by several researchers and clones for improved traits have been obtained (Geetha and Padmanabhan 2002; Sandhu et al. 2008). Screening for mutations in mutagenized populations is possible through genome-wide chips or other high-throughput genotyping technologies and to understand the molecular mechanisms involved in trait development.

7.11 Conclusions and Future Perspective

Gene discovery through different *omics* approaches has been significant for sugarcane improvement programs, and functional information on the genes is unraveling mechanisms of plant adaptation and responses to the biotic and abiotic environment. EST-SSRs) are being used successfully to infer genetic relationships and genetic diversity. The SUCEST sequencing program has been significant for worldwide sugarcane breeding programs, and functional analysis based on cDNA arrays has become a sought-after approach for uncovering pathways of plant responses to stressful environments. Increasing sucrose content is a major research goal since an increase in sucrose yield is more valuable. A better understanding of how sugarcane plants cope with cold and drought stresses could aid in the development of cultivars better suited to particular areas.

Development of new markers and their integration in genetic maps will undoubtedly boost programs to speed up the breeding and improvement efforts. Genetic manipulation of sugarcane has become successful owing to the optimization of different conditions for transformation techniques and cultivars, followed by field trials. The current decade has witnessed the development of techniques for silencing of genes or their overexpression to study their function and to produce new and novel phenotypes that are otherwise not possible through conventional means. Metabolomics studies coupled with gene expression studies are certainly the potential tools of current and future sugarcane research. Transcriptomic analysis of transgenic plants with altered genes of interest will unravel gene regulatory networks associated with important agronomic traits. Cultivars tolerant to biotic and abiotic stresses are important to expand cultivation in environmentally challenged regions. The prospect of employing stress-related genes as markers for breeding or for genetic manipulation will certainly reduce the impact of the sugarcane crop on the environment. A better perception of how sugarcane plants manage with stresses could help in the development of cultivars suitable to particular areas.

References

Abbasi, F.M. and Komatsu, S. 2004. A proteomic approach to analyse salt responsive proteins in rice leaf sheath. *Proteomics* 4:2072–2081.

Agrawal, G.K. and Rakwal, R. 2008. *Plant Proteomics: Technologies, Strategies, and Applications*. Wiley: Hoboken.

Agrawal, G.K., Jwa, N.S., and Rakwal, R. 2009. Rice proteomics: Ending phase I and the beginning of phase II. *Proteomics* 9:935–963.

Aharoni, A., de Vos, C.H., Verhoeven, H.A., et al. 2002. Non targeted metabolome analysis by use of Fourier transform ion cyclotron mass spectrometry. *OMICS*. 6:217–234.

Aitken, K.S., Jackson, P.A., and McIntyre, C.L. 2005. A combination of AFLP and SSR markers provides extensive map coverage and identification of homo(eo)logous linkage groups in a sugarcane cultivar. *Theor Appl Genet* 110:789–801.

Aitken, K.S., Jackson, P.A., and McIntyre, C.L. 2006. Quantitative trait loci identified for sugar related traits in sugarcane (*Saccharum* spp.) cultivar *Saccharum officinarum* population. *Theor Appl Genet* 112:1306–1311.

Aitken, K.S., Jackson, P.A., and McIntyre, C.L. 2007. Construction of genetic linkage map for *Saccharum officinarum* incorporating both simplex and duplex markers to increase genome coverage. *Genome* 50:742–756.

Aitken, K.S., Hermann, S., Karno, K., Bonnett, G.D., McIntyre, L.C., and Jackson, P.A. 2008. Genetic control of yield related stalk traits in sugarcane. *Theor Appl Genet* 117:1191–1203.

Albert, H.H. and Schenck, S. 1996. PCR amplification from a homolog of the bE mating-type gene as a sensitive assay for the presence of *Ustilago scitaminea* DNA. *Plant Dis* 80:452–457.

Alexander, A.G. 1986. Rapid implementation options and opportunities for energy cane biomass producers. Energy from biomass and wastes. FAO (preprint): Washington, DC, April 7–10. http://www.fao.org/docrep/003/s8850e/S8850E04.htm.

Ali, G.M. and Komatsu, S. 2006. Proteomic analysis of rice leaf sheath during drought stress. *J Proteome Res* 5:396–403.

Ali, A., Naz, S., Siddiqui, F.A., and Iqbal, J. 2008. An efficient protocol for large scale production of sugarcane through micropropagation. *Pak J Bot* 40(1):139–149.

Al-Janabi, S.M., Honeycutt, R.J., McClelland, M., and Sobral, B.W.S. 1993. A genetic linkage map of *Saccharum spontaneum* L. SES 208. *Genetics* 134:1249–1260.

Allsopp, P.G. and Suasa-ard, W. 2000. Sugarcane pest management strategies in the new millennium. In: *Proceedings of the International Society Sugar Cane Technology, Sugarcane Entomol Workshop*, p. 4. Khon Kaen.

Altpeter, F. and Oraby, H. 2010. Sugarcane. In: Kempken, F. and Jung, C. (eds), *Genetic Modification of Plants. Biotechnology in Agriculture and Forestry*, vol. 64, pp. 453–472. Springer: New York.

Alwala, S., Suman, A., Arro, J.A., Veremis, J.C., and Kimberg, C.A. 2006a. Target region amplification polymorphism (TRAP) for assessing genetic diversity in sugarcane germplasm collections. *Crop Sci* 46:448–455.

Alwala, S., Kimberg, C.A., Gravois, K.A., and Bischoff, K.P. 2006b. TRAP: A new tool for sugarcane breeding; comparison with AFLP and co-efficient of parentage. *Sugarcane Int* 24(6):11–21.

Alwala, S., Collins, A., Kimbeng, J., Veremis, C., and Gravois, K.A. 2008. Linkage mapping and genome analysis in a Saccharum interspecific cross using AFLP, SRAP and TRAP markers. *Euphytica* 164:37–51.

Alwala, S., Collins, A., Kimbeng, J., Veremis, C., and Gravois, K.A. 2009. Identification of molecular markers associated with sugar-related traits in a Saccharum interspecific cross. *Euphytica* 167:127–142.

Amalraj, R.S., Selvaraj, N., Veluswamy, G.K., et al. 2010. Sugarcane proteomics: Establishment of a protein extraction method for 2-DE in stalk tissues and initiation of sugarcane proteome reference map. *Electrophoresis* 31:1959–1974.

Amme, S., Maltros, A., Schlesier, B., and Mock, H.P. 2006. Proteome analysis of cold stress response in *Arabidopsis thaliana* using DIGEtechnology. *J Exp Bot* 57:1537–1546.

Andru, S., Pan, Y.-B., Thongthawee, S., Burner, D.M., and Kimbeng, C.A. 2011. Genetic analysis of the sugarcane (*Saccharum spp.*) cultivar 'LCP 85–384'. I. Linkage mapping using AFLP, SSR, and TRAP markers. *Theor Appl Genet* 123:77–93.

Arceneaux, G. 1965. Cultivated sugarcane if the world and their botanical derivation. *Proc Int Soc Sugar Cane Technol* 12:844–854.

Arencibia, A., Molina, P., Gutierrez, C., et al. 1992. Regeneration of transgenic sugarcane (*Saccharum officinarum* L.) plants from intact meristematic tissues transformed by electroporation. *Biotechnol Aplicada* 9:156–165.

Arencibia, A., Carmona, E., Cornide, M.T., Castiglione, S., O'Relly, J., Cinea, A., Oramas, P., and Sala, F. 1999. Somaclonal variation in insect resistant transgenic sugarcane (*Saccharum* hybrid) plants produced by cell electroporation. *Transgenic Res* 8:349–360.

Arencibia, A., Molina, P., De la Riva, G., Selman-Houssein, G., et al. 1995. Production of transgenic sugarcane (*Saccharum officinarum* L.) plants by intact cell electroporation. *Plant Cell Rep* 14:305–309.

Arencibia, A., Vazquez, R.I., Prieto, D., et al. 1997. Transgenic sugarcane plants resistant to stem borer attack. *Mol Breed* 3:247–255.

Arencibia, A.D., Carmona, E.R., Tellez, P., et al. 1998. An efficient protocol for sugarcane (*Saccharum* spp. L) transformation mediated by *Agrobacterium tumefaciens*. *Transgenic Res* 7:213–222.

Arruda, P. 2001. Sugarcane transcriptome: A landmark in plant genomics in the tropics. *Genet Mol Biol* 24:1–296.

Arruda, P. and Silva, T.R. 2007. Transcriptome analysis of the sugarcane genome for crop improvement. In: Varshney, R.K. and Tuberosa, R. (eds), *Genomic Assisted Crop Improvement*, vol. 2, pp. 483–494. Springer: Dordrecht.

Arumuganathan, K. and Earle, E.D. 1991. Nuclear DNA content of some important plant species. *Plant Mol Biol Rep* 9:208–219.

Arvinth, S., Arun, S., Selvakesavan, R.K., et al. 2010. Genetic transformation and pyramiding of aprotinin expressing sugarcane with cry1Ab for shoot borer (*Chilo infuscatellus*) resistance. *Plant Cell Rep* 29(4):383–395.

Asnaghi, C., Paulet, F., Kaye, C., et al. 2000. Application of synténie across Poaceae to determine the map location of a rust resistance gene of sugarcane. *Theor Appl Genet* 101:962–969.

Asnaghi, C., Roques, D., Ruffel, S., et al. 2004. Targeted mapping of a sugarcane rust resistance gene (*Bru*1) using bulked segregant analysis and AFLP markers. *Theor Appl Genet* 108:759–764.

Azevedo, R.A., Carvalho, R.F., Cia, M.C., and Gratão, L.P. 2011. Sugarcane under pressure: An overview of biochemical and physiological studies of abiotic stress. *Tropical Plant Biol* 4:42–51.

Babu, C., Kodalingam, K., Natarajan, U.S., Govindaraj, P., and Shanthi, R.M. 2010. Assessment of genetic diversity detected by co-ancestry of crosses and STMS markers among sugarcane genotypes (*Saccharum* spp. hybrids). *Sugar Cane Int* 28(5):206–215.

Baldwin, I.T. 2001. An ecologically motivated analysis of plant-herbivore interactions in native tobacco. *Plant Physiol* 127:1449–1458.

Barnes, J.M. and Bester, A.E. 2000. Genetic mapping in sugarcane: Prospects and progress in the South African sugar industry. *Proc S Afr Sugar Tech Assoc* 74.

Basnayake, S.W.V., Moyle, R., and Birch, G. 2011. Embryogenic callus proliferation and regeneration conditions for genetic transformation of diverse sugarcane cultivars. *Plant Cell Rep* 30(3):439–448.

Beckers, G.J.M. and Spoel, S.H. 2006. Fine-tuning plant defence signalling: Salicylate versus jasmonate. *Plant Biol* 8(1):1–10.

Beetham, P.R., Kipp, P.B., Sawycky, X.L., Arnzen, C.J., and May, G.D. 1999. A tool for functional plant genomics: Chimeric RNA/DNA oligonucleotides cause in vivo gene-specific mutations. *Proc Natl Acad Sci USA* 96:8774–8778.

Bennetzen, J.L. and Freeling, M. 1993. Grasses as a single genetic system: Genome composition, collinearity and compatibility. *Trends Genet* 9:259–261.

Bennetzen, J.L. and Freeling, M. 1997. The unified grass genome: Synergy in synteny. *Genome Res* 7:301–306.

Bess, P., Taylor, G., Carroll, B., et al. 1998. Assessing genetic diversity in a sugarcane germplasm collection using an automated AFLP analysis. *Genetica* 104:143–153.

Bino, R.J., Hall, R.D., Fiehn, O., et al. 2004. Potential of metabolomics as a functional genomics tool. *Trends Plant Sci* 9:418–425.

Birch, R.G. 1997. Transgenic sugarcane: Opportunities and limitations. In: Keating, B.A. and Wilson, J.R. (eds), *Intensive Sugarcane Production: Meeting the Challenge Beyond 2000*, pp. 125–140. CAB International: Wallingford.

Birch, R.G. 2007. Metabolic engineering in sugarcane: Assisting the transition to a bio-based economy. In: Verpoorte, R., Alfermann, A.W., and Johnson, T.S. (eds), *Applications of Plant Metabolic Engineering*, pp. 249–281. Springer: Berlin.

Birch, R.G. and Maretzki, A. 1993. Transformation of sugarcane. In: Bajaj, Y.P.S. (ed.), *Plant Protoplasts and Genetic Engineering IV. Biotechnology in Agriculture and Forestry*, vol. 23, pp. 348–360. Springer-Verlag: Heidelberg.

Birkett, M.A., Campbell, C.A.M., Chamberlain, K., et al. 2000. New roles for cis-jasmone as an insect semichemical and plant defence. *Proc Natl Acad Sci USA* 97:9329–9334.

Black, K. and Botha, F.C. 2000. Sucrose phosphate synthase and sucrose synthase activity during maturation of intermodal tissue in sugarcane. *Aus J Plant Physiol* 27:81–85.

Borecky, J., Nogueira, F.T.S., de Oliveira, K.A.P., Maia, I.G., Vercesi, A.E. et al. 2006. The plant energy-dissipating mitochondrial systems: Depicting the genomic structure and the expression profiles of the gene families of uncoupling protein and alternative oxidase in monocots and dicots. *J Exp Bot* 57:849–864.

Bock, R. 2001. Transgenic plastids in basic research and plant biotechnology. *J Mol Biol* 312:425–438.

Borras-Hidalgo, O., Thomma, B.P.H.J., Carmona, E., et al. 2005. Identification of genes induced in disease-resistant somaclones upon inoculation with *Ustilago scitaminea* or *Bipolaris sacchari*. *Plant Physiol Biochem* 43:1115–1121.

Bosch, S., Rohwer, J.M., and Botha, F.C. 2003. The sugarcane metabolome. *Proc S Afr Sugar Technol Assoc* 7:129–133.

Botha, F.C. and Black, K.G. 2000. Sucrose phosphate synthase and sucrose synthase activity during maturation of intermodal tissue in sugarcane. *Aus J Plant Physiol* 27:81–85.

Botha, F., Sawyer, B., and Birch, R. 2001. Sucrose metabolism in the culm of transgenic sugarcane with reduced soluble acid invertase activity. In: *Proceedings International Society Sugarcane Technologists XXIV Congress*, pp. 588–591. ASSCT, Mackay.

Bower, R. and Birch, R.G. 1992. Transgenic sugarcane plant via microprojectile bombardment. *Plant J* 2:409–416.

Bower, R., Elliott, A.R., Potier, B.A.M., and Birch, R.G. 1996. High-efficiency, microprojectile-mediated cotransformation of sugarcane, using visible or selectable markers. *Mol Breed* 2:239–249.

Bower, N.I., Casu, R.E., Maclean, D.J., et al. 2005. Transcriptional response of sugarcane roots to methyl jasmonate. *Plant Sci* 168(3):761–772.

Bowers, J.E., Abbey, C., Anderson, S., Chang, C., Draye, X., Hoppe, A.H., et al. 2003. A high density genetic recombination map of sequence-tagged sites for *Sorghum*, as a framework for comparative structural and evolutionary genomics of tropical grains and grasses. *Genetics* 165:367–386.

Bowyer, J.R. and Leegood, R.C. 1997. Photosynthesis. In: Day, P.M. and Habone, J.B. (eds), *Plant Biochemistry*, pp. 49–110. Academic Press: San Diego.

Braga, D.P.V., Arrigoni, E.D.B., Silva-Filho, M.C., and Ulian, E.C. 2003. Expression of the Cry1Ab protein in genetically modified sugarcane for the control of *Diatraea saccharalis* (Lepidoptera: Crambidae). *J New Seeds* 5(2–3):209–221.

Braithwaite, K.S., Egeskov, N.M., and Smith, G.R. 1995. Detection of sugarcane bacilliform virus using the polymerase chain reaction. *Plant Disease* 79:792–796.

Braithwaite, K.S. and Smith, G.R. 2001. Molecular-based diagnosis of sugarcane virus diseaseas. In: Rao, G.P., Ford, R.E., Tosic, M., and Teakle, D.S. (eds), *Sugarcane Pathology, Volume II: Viruses and Phytoplasma Diseases*, pp. 175–192. Oxford and IBH Publishing: New Delhi.

Bray, E.A. 1997. Plant responses to water deficit. *Trends Plant Sci* 2:48–54.

Briiggemann, W., van der Kooij, T.A.W., and van Hasselt, P.R. 1992. Long-term chilling of young tomato plants under low light and subsequent recovery II. Chlorophyll fluorescence, carbon metabolism and activity of ribulose-1,5-bisphosphate carboxylase oxygenase. *Planta* 176:179–187.

Broderson, P., Sakvarelidze-Achard, L., Bruun-Rasmussen, M., Dunoyer, P., Yamamoto, Y.Y., et al. 2008. Widespread translational inhibition by plant miRNAs and siRNAs. *Science* 320:1185–1190.

Broekaert, W.F., Delaur´e, S.L., De Bolle, M.F.C., and Cammue, B.P.A. 2006. The role of ethylene in host-pathogen interactions. *Annu Rev Phytopathol* 44:393–416.

Brown, S.C., Kruppa, G., and Dasseux, J.L. 2005. Metabolomics applications of FT-ICR mass spectrometry. *Mass Spectrom Rev* 24:223–231.

Brumbley, S.M., Purnell, M.P., Petrasovits, L.A., Nielsen, L.K., and Twine, P.H. 2007. Developing the sugarcane biofactory for high value biomaterials. *Int Sugar J* 109:5–15.

Brumbley, S.M., Purnell, M.P., Petrasovits, L.A., O'Shea, M.G., and Nielsen, L.K. 2004. Development of sugarcane as a biofactory for biopolymers. In: *Plant & Animal Genomes XII Conference*, San Diego, CA. W117. http://63.141.253.172/pag/12/abstracts/W27_PAG12_117.html.

Brumbley, S.M., Snyman, S.J., Gnanasambandam, A., Joyce, P.A., Hermann, S.R., da Silva, J.A.G., et al. 2008. Sugarcane. In: Kole, C. and Hall, T.C. (eds), Compendium of Transgenic Crop Plants: Transgenic Sugar, Tuber and Fiber Crops, pp. 1–58. Blackwell: Oxford.

Bucheli, C.S., Dry, I.B., and Robinson, S.P. 1996. Isolation of a full-length cDNA encoding polyphenol oxidase from sugarcane, a C4 grass. *Plant Mol Biol* 31:1233–1238.

Bugos, R.C. and Thom, M. 1993. Glucose transporter cDNAs from sugarcane. *Plant Physiol* 103:1469–1470.

Bull, T.A. and Glasziou, K.T. 1963. The evolutionary significance of sugar accumulation in *Saccharum*. *Aust J Biol Sci* 16:737–742.

Bundock, P., Casu, R., and Henry, R. 2012. Enrichment of genomic DNA for polymorphism detection in a non-model highly polyploid crop plant. Plant *Biotechnol J* 10:657–667.

Bundock, P.C., Eliott, F.G., Ablett, G., et al. 2009. Targeted single nucleotide polymorphism (SNP) discovery in a highly polyploid plant species using 454 sequencing. *Plant Biotechnol J* 7(4):347–354.

Burner, D.M. and Legendre, B.L. 1994. Cytogenetic and fertility characteristics of elite sugarcane clones. *Sugar Cane* 1:6–10.

Burnquist, W.L., Sorrells, M.E., and Tanksley, S. 1992. Characterization of genetic variability in *Saccharum* germplasm by means of restriction fragment length polymorphism (RFLP) analysis. *Proc Int Soc Sugarcane Technol* 21:355–365.

Butterfield, M.K., D'Hont, A., and Berding, N. 2001. The sugarcane genome: A synthesis of current understanding, and lesson for breeding and biotechnology. *Proc S Afr Sug Technol Ass* 75:1–5.

Butterfield, M.K., Irvine, J.E., Valdez Garza, M., and Mirkov, T.E. 2002. Inheritance and segregation of virus and herbicide resistance transgenes in sugarcane. *Theor Appl Genet* 104:797–803.

Calsa, T. and Figueira, A. 2007. Serial analysis of gene expression in sugarcane (*Saccharum* spp.) leaves revealed alternative C4 metabolism and putative antisense transcript. *Plant Mol Biol* 63:745–762.

Calvino, M., Bruggmann, R., and Messing, J. 2008. Screen of genes linked to high-sugar content in stems by comparative genomics. *Rice* 1:166–176.

Carmona, E., Vargas, D., Borroto, C.J., et al. 2004. cDNA-AFLP analysis of differential gene expression during the interaction between sugarcane and *Puccinia melanocephala*. *Plant Breed* 123:499–501.

Carson, D.L. and Botha, F.C. 2002. Gene expressed in sugarcane maturing intermodal tissue. *Plant Cell Rep* 20:1075–1081.

Carson, D., Huckett, B., and Botha, F. 2002a. Differential gene expression in sugarcane leaf and intermodal tissues of varying maturity. *S Afr J Bot* 68:434–442.

Carson, D.L., Huckett, B.I., and Botha, F.C. 2002b. Sugarcane ESTs differentially expressed in immature and maturing internodal tissue. *Plant Sci* 162:289–300.

Caruso, G., Cavaliere, C., Foglia, P., et al. 2009. Analysis of drought responsive proteins in wheat (*Triticum durum*) by 2D-PAGE and MALDI-TOF mass spectrometry. *Plant Sci* 177:570–576.

Casacuberta, J.M. and Santiago, N. 2003. Plant LTRretrotransposons and MITEs: Control of transposition and impact on the evolution of plant genes and genomes. *Gene* 311(1–2):1–11.

Casu, R.E. 2010. Role of bioinformatics as a tool for sugarcane research. In: Henry, R. and Kole, C. (eds), *Genetics, Genomics and Breeding of Sugarcane*, pp. 229–248. CRC Press, Enfield.

Casu, R.E., Grof, C.P.L., Rae, A.L., et al. 2003. Identification of a novel sugar transporter homologue strongly expressed in maturing stem vascular tissues of sugarcane by expressed sequence tag and microarray analysis. *Plant Mol Biol* 52:371–386.

Casu, R.E., Dimmock, C.M., Chapman, S.C., et al. 2004. Identification of differentially expressed transcripts from maturing stem of sugarcane by *in silico* analysis of stem expressed sequence tags and gene expression profiling. *Plant Mol Biol* 54:503–517.

Casu, R.E., Manners, J.M., Bonnett, G.D., et al. 2005. Genomics approaches for the identification of genes determining important traits in sugarcane. *Field Crops Res* 92:137–147.

Casu, R.E., Jarmey, J.M., Bonnett, G.D., and Manners, J.M. 2007. Identification of transcripts associated with cell wall metabolism and development in the stem of sugarcane by Affymetrix GeneChip Sugarcane Genome Array expression profiling. *Funct Integr Genomics* 7:153–167.

Cavalcante, J.J.V., Vargas, C., Nogueira, E.M., et al. 2007. Members of the ethylene signalling pathway are regulated in sugarcane during the association with nitrogen-fixing endophytic bacteria. *J Exp Bot* 58(3):673–686.

Chagas, R.M., Silveira, J.A.G., Ribeiro, R.V., Vitorello, V.A., and Carrer, H. 2008. Photochemical damage and comparative performance of superoxide dismutase and ascorbate peroxidase in sugarcane leaves exposed to paraquat-induced oxidative stress. *Pestic Biochem Physiol* 90:181–188.

Chatenet, M., Delage, C., Ripolles, M., et al. 2001. Detection of sugarcane yellow leaf virus in quarantine and production of virus-free sugarcane by apical meristem culture. *Plant Dis* 85:1177–1180.

Chen, W.H., Gartland, K.M.A., Davey, M.R., et al. 1987. Transformation of sugarcane protoplasts by direct uptake of a selectable chimaeric gene. *Plant Cell Rep* 6:297–301.

Chen, S., Gollop, N., and Heuer, B. 2009. Proteomic analysis of salt-stressed tomato (*Solanum lycopersicum*) seedlings: Effect of genotype and exogenous application of glycinebetaine. *J Exp Bot* 7:2005–2019.

Ching, A., Caldwell, K.S., Jung, M., et al. 2002. SNP frequency, haplotype structure and linkage disequilibrium in elite maize inbred lines. *BMC Genet* 3:19.

Cho, W.K., Chen, X.Y., Rim, Y., et al. 2010. Extended latex proteome analysis deciphers additional roles of the lettuce laticifer. *Plant Biotechnol Rep* 4:311–319.

Chowdhury, M.K.U. and Vasil, I. 1992. Stably transformed herbicide resistant callus of sugarcane via microprojectile bombardment of cell suspension cultures and electroporation of protoplasts. *Plant Cell Rep* 11:494–498.

Christy, L.A., Aravith, S., Saravanakumar, M., et al. 2009. Engineering sugarcane cultivars with bovine pancreatic trypsin inhibitor (aprotinin) gene for protection against top borer (*Scirpophaga excerptalis* Walker). *Plant Cell Rep* 28:175–184.

Clayton, W.D. and Renvoize, S.A. 1999. *Genera Graminum: Grasses of the World*. Kew Publishing: London.

Collinge, D.B., Borch, J., Madriz-Ordeñana, K., and Newman, M.A. 2001. The responses of plants to pathogens. In: Hawkesford, M.J. and Buchner, P. (eds), *Molecular Tools for the Assessment of Plant Adaptation to the Environment*, pp. 131–158. Kluwer: Dordrecht.

Cordaux, R., Udit, S., Batzer, M.A., and Feschotte, C. 2006. Birth of a chimeric primate gene by capture of the transposase gene from a mobile element. *Proc Natl Acad Sci USA* 103:8101–8106.

Cordeiro, G.M., Maguire, T.L., Edwards, K.J., and Henry, R.J. 1999. Optimisation of a microsatellite enrichment technique in Saccharum spp. *Plant Mol Biol Rep* 17:225–229.

Cordeiro, G.M., Taylor, G.O., and Henry, R.J. 2000. Characterization of microsatellite markers from sugarcane (*Saccharum* spp.) a highly polyploidy species. *Plant Sci* 155:161–168.

Cordeiro, G.M., Casu, R., McIntyre, C.L., Manners, J.M., and Henry, R.J. 2001. Microsatellite markers from sugarcane (*Saccharum* spp.): ESTs cross-transferable to Erianthus and sorghum. *Plant Sci* 160:1115–1123.

Cordeiro, G.M., Pan, Y.B., and Henry, R.J. 2003. Sugarcane microsatellites for the assessment of genetic diversity in sugarcane germplasm. *Plant Sci* 165:181–189.

Cordeiro, G.M., Eliott, F., McIntyre, C.L., Casu, R.E., and Henry, R.J. 2006. Characterisation of single nucleotide polymorphisms in sugarcane ESTs. *Theor Appl Genet* 113:331–343.

Cunff, L.L., Garsmeur, O., Raboin, L.M., Pauquet, J., Telismart, H., Selvi. A., et al. 2008. Diploid/polyploidy syntenic shuttle mapping and haplotype-specific chromosome walking toward a rust resistance gene (Bru1) in highly polyploid sugarcane (2n 12 × 115). *Genetics* 180:649–660.

D'Hont, A. 2005. Unraveling the genome structure of polyploids using FISH and GISH; examples of sugarcane and banana. *Cytogenet Genome Res* 109(1–3):27–33.

D'Hont, A. and Glaszmann, J.C. 2001. Sugarcane genome analysis with molecular markers, a first decade of research. *Proc Int Soc Sugarcane Technol* 24:556–559.

D'Hont, A., Lu, Y.H., Feldmann, P., and Glaszmann, J.C. 1993. Cytoplasmic diversity in sugarcane revealed by heterologous probes. *Sugar Cane* 1:12–15.

D'Hont, A., Lu, Y.H., Le'on, D.G.D., et al. 1994. A molecular approach to unravelling the genetics of sugarcane, a complex polyploid of Andropogonaea tribe. *Genome* 37:222–230.

D'Hont, A., Grivet, L., Feldmann, P., et al. 1996. Characterization of the double genome structure of modern sugarcane cultivars (*Saccharum* spp.) by molecular cytogenetics. *Mol Gen Genet* 250:405–413.

Da Silva, J. A. and Bressiani, J. A. 2005. Sucrose synthase molecular marker associated with sugar content in elite sugarcane progeny. *Genet Mol Biol* 28:294–298.

Da Silva, J., Honeycutt, R.J., Burnquist, W., Al-Janabi, S.M., Sorrells, M.E., Tanksley, S.D., and Sobral, B.W.S. (1995) *Saccharum spontaneum* L. 'SES 208' genetic linkage map combining RFLP and PCR-based markers. *Mol Breeding* 1:165–179.

Da Silva, J.A.G., Sorrells, M.E., Burnquist, W., and Tanksley, S.D. 1993. RFLP linkage map and genome analysis of Saccharum spontaneum. *Genome* 36:782–791.

Damaj, M.B., Beremand, P.D., Buenrostro-Nava, M.T., et al. 2010. Isolating promoters of multigene family members from the polyploidy sugarcane genome by PCR-based walking in BAC DNA. *Genome* 53:840–847.

Daniell, H. 1999. Environmentally friendly approaches to genetic engineering. *In Vitro Cell Dev Biol Plant* 35:361–36.

Daniell, H., Streatfield, S.J., and Wycoff, K. 2001. Medical molecular farming: Production of antibodies, biopharmaceuticals and edible vaccines in plants. *Trend Plant Sci* 6:219–225.

Daniels, J. and Roach, B. 1987. Taxonomy and evolution. In: Heinz, D.J. (ed.), *Sugarcane Improvement through Breeding*, pp. 7–84. Elsevier: Amsterdam.

Daniels, J., Smith, P., Paton, N., and Williams, C. 1975. The origin of the genus *Saccharum*. *Sugarcane Breed Newsl* 36:24–39.

Daugrois, J.H., Grivet, L., Roques, D., et al. 1996. A putative major gene for rust resistance linked with an RFLP marker in sugarcane cultivar R 570. *Theor Appl Genet* 92:1059–1064.

De Cosa, B., Moar, W., Lee, S.B., Miller, M., and Daniell, H. 2001. Overexpression of the Bt cry2Aa2 operon in chloroplasts leads to formation of insecticidal crystals. *Nat Biotechnol* 19:71–74.

Defrancesco, L. and Perkel, J.M. 2001. In search of genomic variation: A wealth of technologies exists to find elusive genetic polymorphisms. *Scientist* 15:24.

de Hoon, M. and Hayashizaki, Y. 2008. Deep cap analysis gene expression (CAGE): Genome-wide identification of promoters, quantification of their expression, and network inference. *Biotechniques* 44:627–632.

De Rosa Jr., V.E., Nogueira, F.T.S., Menossi, M., Ulian, E.C., and Arruda, P. 2005. Identification of methyl jasmonate-responsive genes in sugarcane using cDNA arrays. *Braz J Plant Physiol* 17:173–180.

Deutsch, S., Iseli, C., Bucher, P., Antonarakis, S.E., and Scott, H.S. 2001. A cSNP map and database for human Chromosome 21. *Genome Res* 11:300–307.

Devarumath, R.M., Doule, R.B., Kawar, P.G., Naikebawane, S.B., and Nerkar, Y.S. 2007. Field performance and RAPD analysis to evaluate genetic fidelity of tissue culture raised plants vis-à-vis conventional setts derived plants of sugarcane. *Sugar Tech* 9(1):17–22.

Devarumath, R.M., Kalwade, S.B., Bundock, P., Eliott, F.G., and Henry, R. 2013. Independent target region amplification polymorphism (TRAP) and single nucleotide polymorphism (SNP) marker utility in genetic evaluation of sugarcane genotypes. *Plant Breeding* (forthcoming), doi:10.1111/pbr.12092.

Devarumath, R.M., Kalwade, S.B., Kawar, P.G., and Sushir, K.V. 2012. Assessment of genetic diversity in sugarcane germplasm using ISSR and SSR markers. *Sugar Tech* 14(4):334–344.

Devos, K.M., Ma, J., Pontaroli, A.C., Pratt, L.H., and Bennetzen, J.L. 2005. Analysis and mapping of randomly chosen bacterial artificial chromosome clones from hexaploid bread wheat. *Proc Natl Acad Sci USA* 102:19243–19248.

Ding, C., You, J., Liu, Z., Rehmani, M.I.A., Wang, S., Li, G., Wang, Q., and Ding, Y. 2011. Proteomic analysis of low nitrogen stress-responsive proteins in roots of rice. *Plant Mol Biol Rep* 29:618–625.

Dixon, R.A. and Strack, D. 2003. Phytochemistry meets genome analysis, and beyond. *Phytochemistry* 62:815–816.

Doule, R.B., Kawar, P.G., Nerkar, Y.S., and Devarumath, R.M. 2008. Field performance of promising somaclonal variants and RAPD analysis to assess genetic variation in sugarcane (*Saccharum* spp.). *Indian J Genet Plant Breed* 68(3):301–306.

Du, Y.C., Nose, A., and Wasanow, K. 1999. Effects of chilling temperature on photosynthetic rates, photosynthetic enzyme activities and metabolite levels in leaves of three sugarcane species. *Plant Cell Environ* 22:317–324.

Dufour, P., Deu, M., Grivet, L., D'Hont, A., Paulet, F., et al. 1997. Construction of a composite sorghum genome map and comparison with sugarcane, a related complex polyploid. *Theor Appl Genet* 94:409–418.

Dufour, P., Grivet, L., D'Hont, A., Deer, M., Tronche, G., Glaszmann, J.C., and Hamon, P. 1996. Comparative genetic mapping between duplicated segments on maize chromosome 3 and 8 and homologous regions in sorghum and sugarcane. *Theor Appl Genet* 92:1024–1030.

Eastmond, P.J., Li, Y., and Graham, I.A. 2003. Is trehalose-6-phosphate a regulator of sugar metabolism in plants? *J Exp Bot* 54(382):537– 553.

Ecker, J.R. 1995. The ethylene signal transduction pathway in plants. *Science* 268:667–675.

Elliott, A.R., Campbell, J.A., Bretell, R.I.S., and Grof, C.P.L. 1998. *Agrobacterium* mediated transformation of sugarcane using GFP as a screenable marker. *Aust J Plant Physiol* 25:739–743.

Elliott, A.R., Campbell, J.A., Dugdale, B., Brettell, R.I.S., and Grof, C.P.L. 1999. Green-fluorescent protein facilitates rapid in vivo detection of genetically transformed plant cells. *Plant Cell Rep* 18:707–714.

Ellis, J., Dodds, P., and Pryor, T. 2000. Structure, function and evolution of plant disease resistance genes. *Curr Opin Plant Biol* 3:278–284.

Enriquez, G.A., Trujillo, L.E., Menendez, C., et al. 2000. Sugarcane (*Saccharum* hybrid) genetic transformation mediated by *Agrobacterium tumefaciens*: Production of transgenic plants expressing proteins with agronomic and industrial value. In: Arencibia, A.D. (ed.), *Plant Genetic Engineering: Towards the Third Millennium*, pp. 76–81. Elsevier Science: Amsterdam.

Enriquez-Obregon, G.A., Vazquez, P.R.I., Prieto, S.D.L., Riva-Gustavo, A.D.L., and Selman, H.G. 1998. Herbicide resistant sugarcane (*Saccharum officinarum* L.) plants by *Agrobacterium*-mediated transformation. *Planta* 206:20–27.

Falco, M.C. and Silva-Filho, M.C. 2003. Expression of soybean proteinase inhibitors in transgenic sugarcane plants: Effects on natural defense against *Diatraea saccharalis*. *Plant Physiol Biochem* 41:761–766.

Falco, M.C., Marbach, P.A.S., Pompermayer, P., Lopes, F.C.C., and Silva-Filho, F.C. 2001. Mechanisms of sugarcane response to herbivory. *Genet Mol Biol* 24(1–4):113–122.

Falvo, S., Acquadro, A., Albo, A.G., America, T., and Lanteri, S. 2011. Proteomic analysis of PEG-fractionated UV-C stress-response proteins in globe artichoke. *Plant Mol Biol Rep* 30:111–122.

FAO. 2008. FAOSTAT. Food and Agriculture Organization of the United Nations. Website: http://faostat.fao.org/site/291/default.aspx.

Fegan, M., Croft, B.J., Teakle, D.S., Hayward, A.C., and Smith, G.R. 1998. Sensitive and specific detection of *Clavibacter xyli* subsp. xyli, causal agent of ratoon stunting disease of sugarcane, with a polymerase chain reaction-based assay. *Plant Pathol* 47:495–504.

Felix, J.M., Papini-Terzi, F.S., Rocha, F.R., et al. 2009. Expression profile of signal transduction components in a sugarcane population segregating for sugar content. *Trop Plant Biol* 2:98–109.

Feschotte, C. 2008. Transposable elements and the evolution of regulatory networks. *Nat Rev Genet* 9:397–405.

Fiehn, O. 2002. Metabolomics: The link between genotypes and phenotypes. *Plant Mol Biol* 48:155–171.

Fiehn, O., Kopka, J., Doermann, P., et al. 2000. Metabolite profiling for plant functional genomics. *Nat Biotechnol* 18:1157–1161.

Fordyce, J.A. and Agrawal, A.A. 2001. The role of plant trichomes and caterpillar group size on growth and defense of the pipevine swallowtail *Battus philenor*. *J Animal Ecol* 70:997–1005.

Fornazier, R.F., Ferreira, R.R., Vitoria, A.P., Molina, S.M.G., Leas, P.J., and Azevedo, R.A. 2002. Effect of cadmium on antioxidant enzyme activities in sugarcane. *Biol Plant* 45:91–97.

Franks, T. and Birch, R.G. 1991. Gene transfer into intact sugarcane cells using microprojectile bombardment. *Aust J Plant Physiol* 18:471–480.

Freddy, F.G.O. 2011. Comparative proteome and qPCR analysis of the sugarcane reaction to leaf scald caused by *Xanthomonas albilineans*. Ph.D. Thesis, Louisiana State University.

Furumoto, T., Hata, S., and Izui, K. 2000. Isolation and characterization of cDNAs for differentially accumulated transcripts between mesophyll cells and bundle sheath strands of maize leaves. *Plant Cell Physiol* 41:1200–1209.

Gallo-Meagher, M. and Irvine, J.E. 1996. Herbicide resistant transgenic sugarcane plants containing the bar gene. *Crop Sci* 36:1367–1374.

Gambley, R.L., Ford, R., and Smith, G.R. 1993. Microprojectile transformation of sugarcane meristems and regeneration of shoots expressing b-glucuronidase. *Plant Cell Rep* 12:343–346.

Gao, F., Zhou, Y., Zhu, W., et al. 2009. Proteomic analysis of cold stress-responsive proteins in *Thellungiella rosette* leaves. *Planta* 230:1033–1046.

Garg, K., Green, P., and Nickerson, D.A. 1999. Identification of candidate coding region single nucleotide polymorphisms in 165 human genes using assembled expressed sequence tags. *Genome Res* 9:1087–1092.

Geetha, S. and Padmanabhan, D. 2002. Evaluation of tissue culture raised sugarcane for yield and quality. *Sugar Tech* 4:179–180.

Gilbert, R.A., Gallo-Meagher, M., Comstock, J.C., Miller, J.D., Jain, M., and Abouzid, A. 2005. Agronomic evaluation of sugarcane lines transformed for resistance to sugarcane mosaic virus strain E. *Crop Science* 45:2060–2067.

Gilbert, R.A., Glynn, N.C., Comstock, J.C., and Davis, M.J. 2009. Agronomic performance and genetic characterization of sugarcane transformed for resistance to sugarcane yellow leaf virus. *Field Crop Res* 111:39–46.

Gilchrist, E.J. and Haughn, G.W. 2005. TILLING without a plough: A new method with applications for reverse genetics. *Curr Opin Plant Biol* 8:211–215.

Glaszmann, J.C., Fautret, A., Noyer, J.L., Feldmann, P., and Lanaud, C. 1989. Biochemical genetic markers in sugarcane. *Theor Appl Genet* 78:537–543.

Glaszmann, J.C., Lu, Y.H., and Lanaud, C. 1990. Variation of nuclear ribosomal DNA in sugarcane. *J Genet Breed* 44:191–198.

Glaszmann, J.C., Dufour, P., Grivet, L., et al. 1997. Comparative genome analysis between several tropical grasses. *Euphytica* 96:13–21.

Görg, A., Weiss, W., and Dunn, M.J. 2004. Review: Current two-dimensional electrophoresis technology for proteomics. *Proteomics* 4:3665–3685.

Griffin, J.L. 2003. Metabonomics: NMR spectroscopy and pattern recognition analysis of body fluids and tissues for characterisation of xenobiotic toxicity and disease diagnosis. *Curr Opin Chem Biol* 7:648–654.

Grivet, L. and Arruda, P. 2002. Sugarcane genomics: Depicting the comle genome organization of resistance gene analogs in soyabean. *Genome* 43:86–93.

Grivet, L., D' Hont, A., Roques, D., et al. 1996. RFLP mapping in cultivated Sugarcane (*Saccharum* spp.): Genome organization in a highly polyploid and aneuploid interspecific hybrid. *Genetics* 142:987–1000.

Grivet, L., Glaszmann, J.C., and Arruda, P. 2001. Sequence polymorphism from EST data in sugarcane: A fine analysis of 6-phosphogluconate dehydrogenase genes. *Genet Mol Biol* 24:161–167.

Grivet, L., Glaszmann, J.C., Vincentz, M., da Silva, F.R., and Arruda, P. 2003. ESTs as a source for sequence polymorphism discovery in sugarcane: Example of the *Adh* genes. *Theor Appl Genet* 106:190–197.

Groenewald, J-H. and Botha, F.C. 2008. Down-regulation of pyrophosphate: Fructose 6-phosphate 1-phosphotransferase (PFP) activity in sugarcane enhances sucrose accumulation in immature internodes. *Transgenic Res* 17:85–92.

Grof, C.P.L., Glassop, D., Quick, W.P., Sonnewald, U., and Campbell, J. A. 1996. Molecular manipulation of sucrose phosphate synthase in sugarcane. In: Wilson, J.R., Hogarth, D.M., Campbell, J.A., and Garside, A.L. (eds), *Sugarcane: Research Towards Efficient and Sustainable Production*, pp. 124–126. CSIRO Division of Tropical Crops and Pastures: Brisbane.

Guimaraes, C.T., Sills, G.R., and Sobral, B.W.S. 1997. Comparative mapping of Andropogoneae: *Saccharum* L. (sugarcane) and its relation to sorghum and maize. *Proc Natl Acad Sci USA* 94:14261–14266.

Gupta, V., Raghuvanshi, S., Gupta, A., et al. 2010. The water-deficit stress- and red-rot-related genes in sugarcane. *Funct Integr Genomics* 10:207–214.

Hagen, G., Kleinschmidt, A., and Guilfoyle, T. 1984. Auxinregulated gene expression in intact soybean hypocotyl and excised hypocotyl sections. *Planta* 162(2):147–153.

Hajheidari, M., Eivazi, A., Buchanan, B.B., et al. 2007. Proteomics uncovers a role for redox in drought tolerance in wheat. *J Proteome Res* 6:1451–1460.

Hamerli, D. and Birch, R.G. 2011. Transgenic expression of trehalulose synthase results in high concentrations of the sucrose isomer trehalulose in mature stems of field-grown sugarcane. *Plant Biotechnol J* 9(1):32–37.

Hansom, S., Bower, R., Zhang, L., et al. 1999. Regulation of transgene expression in sugarcane. *Proc Int Soc Sugar Cane Technol* 23:278–289.

Harbers, M. and Carninci, P. 2005. Tag-based approaches for transcriptome research and genome annotation. *Nat Methods* 2:495–502.

Harrigan, G.G. and Goodacre, R. (eds) 2003. *Metabolic Profiling: Its Role in Biomarker Discovery and Gene Function Analysis*, vol. XIV. Kluwer Academic Publishers.

Haruta, H., Kato, A., and Todokoro, K. 2001. Isolation of a novel interleukin-1-inducible nuclear protein bearing ankyrin-repeat motifs. *J Biol Chem* 276:12485–12488.

Harvey, H., Huckett, B.I., and Botha, F.C. 1994. Use of polymerase chain reaction and random amplification of polymorphic DNAs for the determination of genetic distances between 21 sugarcane varieties. *Proc South Afr Sugar Technol Assoc* 68:36–40.

Heinz, D.J. and Mee, G.W.P. 1969. Plant differentiation from callus tissue of Saccharum species. *Crop Sci* 9:346–348.

Heller-Uszynska, K., Uszynska, G., and Huttner, E. 2011. Diversity array technology effectively reveals DNA polymorphism in a large and complex genome of sugarcane. *Mol Breed* 28:37–55.

Hemaprabha, G., Natarajan, U.S., Balasundaram, N., and Singh, N.K. 2006. STMS based genetic divergence among common parents and its use in identifying productive cross combinations for varietal evolution in sugarcane (*Saccharum* sp.). *Sugarcane Int* 24(6):22–27.

Hilder, V.A. and Boulter, D. 1999. Genetic engineering of crop plants for insect resistance: A critical review. *Crop Prot* 18:177–191.

Hoarau, J.Y., Offmann, B., and D'Hont, A. 2001. Genetic dissection of a modern sugarcane cultivar (*Saccharum* spp.) I. Genome mapping with AFLP markers. *Theor Appl Genet* 103:84–97.

Hoarau, J.Y., Grivet, L., and Offmann, B. 2002. Genetic dissection of a modern sugarcane cultivar (*Saccharum* spp.) II. Detection of QTLs for yield components. *Theor Appl Genet* 105:1027–1037.

Hogarth, D.M. 1987. Genetics of sugarcane. In: Heinz, D.J. (ed.) *Sugarcane Improvement through Breeding*, pp. 255–271. Elsevier: New York.

Holaday, A.S., Martindale, W., Alred, R., Brooks, A.L., and Leegood, R.C. 1992. Changes in activities of enzymes of carbon metabolism in leaves during exposure of plants to low temperature. *Plant Physiol* 98(3):1105–1114.

Hotta, C.T., Lembke, C.G., Domingues, D.S., et al. 2011. The biotechnology roadmap for sugarcane improvement. *Trop Plant Biol* 3:75–87.

Hurry, V.M., Strand, A., Tobiaeson, M., Gardestrom, P., and Oquist, G. 1995. Cold hardening of spring and winter wheat and rape results in differential effects on growth, carbon metabolism, and carbohydrate content. *Plant Physiol* 109(2):697–706.

Ingelbrecht, I.L., Irvine, J.E., and Mirkov, T.E. 1999. Posttranscriptional gene silencing in transgenic sugarcane. Dissection of homology dependent virus resistance in a monocot that has a complex polyploid genome. *Plant Physiol* 119:1187–1198.

Ingram, J. and Bartels, D. 1996. The molecular basis of dehydration tolerance in plants. *Annu Rev Plant Physiol Plant Mol Biol* 47:377–403.

Inman-Bamber, N.G. and Smith, D.M. 2005. Water relations in sugarcane and response to water deficits. *Field Crops Res* 92:185–202.

Irvine, J.E., Benda, G.T.A., Legendre, B.L., and Machado, G.R. 1991. The frequency of marker changes in sugarcane plants regenerated from callus culture. II. Evidence for vegetative and genetic transmission, epigenic effects and chimeral disruption. *Plant Cell Tiss Organ Cult* 26:115–125.

Irvine, J.E. and Mirkov, T.E. 1997. The development of genetic transformation of sugarcane in Texas. *Sugar J* 60:25–29.

Iskandar, H.M., Casu, R.E., Fletcher, A.T., Schmidt, S., Xu, J., Maclean, D. J., Manners J.M., and Bonnett G.D. 2011. Identification of drought-response genes and a study of their expression during sucrose accumulation and water deficit in sugarcane culms. *BMC Plant Biol* 11:12.

Iskandar, H.M., Simpson, R.S., Casu, R.E., Bonnett, G.D., Maclean, D J., and Manners, J.M. 2004. Comparison of reference genes for quantitative real-time polymerase chain reaction analysis of gene expression in sugarcane. *Plant Mol Biol Rep* 22:325–337.

Jackson, M.A., Rae, A.L., Casu, R.E., et al. 2007. A bioinformatic approach to the identification of a conserved domain in a sugarcane legumain that directs GFP to the lytic vacuole. *Funct Plant Biol* 34:633–644.

Jain, S.M. 2005. Major mutation assisted plant breeding programs supported by FAO/IAEA. *Plant Cell Tiss Org Culture* 82:113–123.

Jain, M., Chengalrayan, K., Abouzid, A., and Gallo, M. 2007. Prospecting the utility of a PMI/mannose selection system for the recovery of transgenic sugarcane (*Saccharum* spp. hybrid) plants. *Plant Cell Rep* 26:581–590.

Jain, R., Srivastava, S., Solomon, S., Shrivastava, A.K., and Chandra, A. 2010. Impact of excess zinc on growth parameters, cell division, nutrient accumulation, photosynthetic pigments and oxidative stress of sugarcane (*Saccharum* spp). *Acta Physiol Plant* 32:979–986.

Jangpromma, N., Kitthaisong, S., Daduang, S., Jaisil, P., and Thammasirirak, S. 2007. 18 kDa Protein accumulation in sugarcane leaves under drought stress. *KMITL Sci Tech J* 7(S1):44–54.

Jangpromma, N., Kitthaisong, S., Lomthaisong, K., et al. 2010. A proteomics analysis of drought stress responsive proteins as biomarker for drought-tolerant sugarcane cultivars. *Am J Biochem Biotechnol* 6:89–102.

Jannoo, N., Grivet, L., D'Hont, A., and Arruda, P. 2003. Genomic sequencing in sugarcane: First insight into the physical organization of the genome and microsynteny with other grasses. Abstracts of PAG XI Conference, San Diego, Abstract W182. URL: http://www.intl-pag. org//11/abstracts/W26_W182_XI.html.

Jannoo, N., Grivet, L., Chantret, N., et al. 2007. Orthologous comparison in a gene-rich region among grasses reveals stability in the sugarcane polyploidy genome. *Plant J* 50:574–585.

Jannoo, N., Grivet, L., D'Hont, A., and Arruda, P. 2003. Genomic sequencing in sugarcane: First insight into the physical organization of the genome and microsynteny with other grasses. Abstracts of PAG XI Conference, San Diego, Abstract W182. http://www.intl-pag.org//11/ abstracts/ W26_W182_XI.html.

Jannoo, N., Grivet, L., Seguin, M., Paulet, F., Domaingue, R., Rao, P.S., Dookun, A., D'Hont, A., and Glaszmann, J.C. 1999. Molecular investigation of the genetic base of sugarcane varieties. *Theor Appl* Genet 99:171–184.

Jellouli, N., Ben, J.H., Daldoul, S., Chenennaoui, S., Ghorbel, A., Ben, S.A., and Gargouri, A. 2010. Proteomic and transcriptomic analysis of grapevine pr10 expression during salt stress and functional characterization in yeast. *Plant Mol Biol Rep* 28:1–8.

Jordan, D.R., Casu, R.E., Besse, P., Carroll, B.C., Berding, N., and McIntyre, C.L. 2004. Markers associated with stalk number and suckering in sugarcane colocate with tillering and rhizomatousness QTLs in sorghum. *Genome* 47:988–993.

Jorge, I., Navarro, R.M., Lenz, C., Ariza, D., and Jorrín, J. 2006. Variation in the holm oak leaf proteome at different plant developmental stages, between provenances and in response to drought stress. *Proteomics* 6:207–214.

Joyce, P.A., McQualter, R.B., Bernad, M.J., and Smith, G.R. 1998a. Engineering for resistance to SCMV in sugarcane. *Acta Hort* 461:385–391.

Joyce, P.A., McQualter, R.B., Handley, J.A., Dale, J.L., Harding, R.M., and Smith, G.R. 1998b. Transgenic sugarcane resistant to sugarcane mosaic virus. In: Hogarth, D.M. (ed.), *Proc. Aust. Soc. Sugarcane Technol* 20:204–210.

Joyce, P., Kuwahata, M., Turner, N., and Lakshmanan, P. 2010. Selection systems and co-cultivation medium are important determinants of *Agrobacterium*-mediated transformation of sugarcane. *Plant Cell Rep* 29:173–183.

Kahl, J., Siemens, D.H., Aerts, R.J., et al. 2000. Herbivore-induced ethylene suppresses a direct defense but not a putative indirect defense against an adapted herbivore. *Planta* 210(2):336–342.

Kalendar, R., Tanskanen, J., Immonen, S., Nevo, E., and Schulman, A.H. 2000. Genome evolution of wild barley (*Hordeum spontaneum*) by BARE-1 retrotransposon dynamics in response to sharp microclimatic divergence. *Proc Natl Acad Sci USA* 97:6603–6607.

Kalunke, R.M., Kolge, A.M. Babu, K.H., and Prasad, D.T. 2009. *Agrobacterium* mediated transformation of sugarcane for borer resistance using Cry 1Aa3 gene and one-step regeneration of transgenic plants. *Sugar Tech* 11(4):355–359.

Kalwade, S.B., Devarumath, R.M., Kawar, P.G., and Sushir, K.V. 2012. Genetic profiling of sugarcane genotypes using Inter Simple Sequence Repeat (ISSR) markers. *Electron J Plant Breeding* 3(1):621–628.

Kalwade, S.B., Takle, S.P., Kawar, P.G., Devarumath R.M., Babu, K.H., and Patil, S.V. 2011. Estimation of carbohydrates by HPLC for development of molecular marker tools. In: Proc. of 10th Conv. of STAI and DSTA, pp. 26–35.

Kanehisa, M., Araki, M., Goto, S., Hattori, M., et al. 2008. KEGG for linking genome to life and the environment. *Nucleic Acids Res* 36:D480-D484.

Kang, M.S., Miller, J.D., and Tai, P.Y.P. 1983. Genetic and phenotypic path analyses and heritability in sugarcane. *Crop Science* 23:643–647.

Kashkush, K., Feldman, M., and Levy, A.A. 2003. Transcriptional activation of retro transposons alters the expression of adjacent genes in wheat. *Nat Genet* 33:102–106.

Kates, R.W. and Parris, T.M. 2003. Long-term trends and a sustainability transition. *Proc Natl Acad Sci USA* 100(14):8062–8067.

Kawar, P.G., Devarumath, R.M., and Nerkar, Y. 2009. Use of RAPD marker for assessment of genetic diversity in sugarcane cultivars. *Indian J Biotechnol* 8:67–71.

Kawar, P.G., Pagariya, M.C., Dixit, G.B., and Theertha Prasad, D. 2010a. Identification and Isolation of SCGS phytoplasma-specific fragments by riboprofiling and development of specific diagnostic tool. *J Plant Biochem Biotechnol* 19:185–194.

Kazazian, H.H. 2004. Mobile elements: Drivers of genome evolution. *Science* 303:1626–1632.

Khan, M.S., Yadav, S., Srivastava, S., et al. 2011. Development and utilization of conserved intron-scanning primers in sugarcane. *Aust J Bot* 59(1):38–45.

Kikuchi, J., Shinozaki, K., and Hirayama, T. 2004. Stable isotope labeling of *Arabidopsis thaliana* for an NMR-based metabolomics approach. *Plant Cell Physiol* 45:1099–1104.

Kim, S.G., Wang, Y., Wu, J., Kang, K.Y., and Kim, S.T. 2011. Physiological and proteomic analysis of young rice leaves grown under nitrogen starvation conditions. *Plant Biotechnol Rep* 5:309–315.

Komatsu, S. and Yano, H. 2006. Update and challenges on proteomics in rice. *Proteomics* 6:4057–4068.

Kulkarni, P.A. and Devarumath, R.M. 2012. *In silico* novel MYB transcription factor [EaMYB2R], from *Erianthus arundinaceus*. *Online J Bioinform* 13(1):167–183.

Kulkarni, P.A., Pagariya, M.C., Devarumath, R.M., et al. 2011. Isolation, molecular characterization and expression studies of SoMYB18 from Phytoplasma infected *Saccharum officinarum* hybrid. Abstract (P082) in Plant and Animal Genome Conference XIX, January 15–19, San Diego, CA.

Kulkarni, P.A., Prabu, G.R., and Devarumath, R.M. 2013. Isolation and in *silico* depiction of novel R2-R3 MYB transcription factors from sugarcane. *Adv Bio Tech* 12(11):1–7.

Kumar, A.S., Moore, P.H., and Maretzki, A. 1992. Amplification and cloning of sugarcane sucrose synthase cDNA by anchored PCR. *PCR Methods Appl* 2:70–75.

Lagerqvist, J., Zwolak, M., and Di Ventra, M. 2006. Fast DNA sequencing via transverse electronic transport. *Nano Lett* 6:779–782.

Lakshmanan, P., Geijskes, R.J., Elliott, A.R., et al. 2003. Direct regeneration tissue culture and transformation systems for sugarcane and other monocot species. *Proc Int Soc Sugar Cane Technol Mol Biol* 4:25.

Lakshmanan, P., Geijskes, R.J., Aitken, K.S., et al. 2005. Sugarcane biotechnology: The challenges and opportunities. *In Vitro Cell Dev Biol Plant* 41:345–363.

Lal, M., Singh, R., Srivastava, S., et al. 2008. RAPD marker based analysis of micropropagated plantlets of sugarcane for early evaluation of genetic fidelity. *Sugar Tech* 10:99–103.

Lambais, M.R. 2001. *In silico* differential display of defense related expressed sequence tags from sugarcane tissues infected with diazotrophic endophytes. *Genet Mol Biol* 24(1–4):103–111.

Larkin, P.J. and Scowcroft, W.R. 1981. Somaclonal variation: A novel source of variability from cell cultures for plant improvement. *Theor Appl Genet* 60:197–214.

Larkin, P.J. and Scowcroft, W. 1983. Somaclonal variation and eyespot toxin tolerance in sugarcane. *Plant Cell Tiss Org Culture* 2:111–121.

Lee, J.W. and Thundat, T.G. 2001. DNA and RNA sequencing by nanoscale reading through programmable electrophoresis and nanoelectrode-gated tunneling and dielectric detection. United States Patent 6905586 app no 10/055881.

Lee, J.W. and Thundat, T.G. 2004. Separation and counting of single molecules through nanofluidics, programmable electrophoresis, and nanoelectrode-gated molecular detection. US Patent Appl Publ 23 pp US 20040124084 A1.

Legaspi, J.C. and Mirkov, T.E. 2000. Evaluation of transgenic sugarcane against stalk borers. In: Allsopp, P.G. and Suasaard, W. (eds), *Proc Int Soc SugCane Technol SugCane Entomol Workshop*, Khon Kaen, Thailand, vol. 4, pp. 68–71.

Leibbrandt, N.B. and Snyman, S.J. 2003. Stability of gene expression and agronomic performance of a transgenic herbicide-resistant sugarcane line in South Africa. *Crop Sci* 43:671–678.

Lima, M.L.A., Garcia, K.M., Oliveira, K.M., et al. 2002. Analysis of genetic similarity detected by AFLP and co-efficient of parentage among genotypes of sugarcane (*Saccharum* spp.). *Theor Appl Genet* 104:30–38.

Liu, D.W., Oard, S.V., and Oard, J.H. 2003. High transgene expression levels in sugarcane (*Saccharum officinarum* L.) driven by the rice ubiquitin promoter RUBQ2. *Plant Sci* 165:743–750.

Louime, C., Marshall, R.W., and Vasanthaiah, H.K.N. 2012. Genomics and potential bioenergy applications in developing world. In: Nelson, K.E. and John-Nelson, B. (eds), *Genomics Applications for the Developing Worlds, Advances in Microbial Ecology*, pp. 263–272.

Lu, Y.H., D'Hont, A., Walker, D.J.T., et al. 1994. Relationship among ancestral species of sugarcane revealed with RFLP using single copy maize nuclear probes. *Euphytica* 78:7–18.

Ma, H., Albert, H.H., Paull, R., and Moore, P.H. 2000. Metabolic engineering of invertase activities in different subcellular compartments affects sucrose accumulation in sugarcane cells. *Aust J Plant Physiol* 27:1021–1030.

Ma, H.M., Schulze, S., Lee, S., et al. 2004. An EST survey of the sugarcane transcriptome. *Theor Appl Genet* 108:851–863.

Maccheroni, W., Jordao, H., De Gaspari, R., De Moura, G.L., and Matsuoka, S. 2009. Development of a dependable microsatellite-based fingerprinting system for sugarcane. *Sugar Cane Int* 27:47–52.

MacKenzie, D.R., Anderson, P.M., and Wernham, C.C. 1966. A mobile air blast inoculator for pot experiments with maize dwarf mosaic virus. *Plant Disease Rep* 50:363–367.

Maliga, P. 2004. Plastid transformation in higher plants. *Annu Rev Plant Biol* 55:289–313.

Manetti, M.E., Rossi, M., Cruz, G.M.D., et al. 2012. Mutator system derivatives isolated from sugarcane genome sequence. *Trop Plant Biol* 5(3):233–243.

Manickavasagam, M., Ganapathi, A., Anbazhagan, V.R., et al. 2004. *Agrobacterium* mediated genetic transformation and development of herbicide resistant sugarcane (*Saccharum* species hybrids) using axillary buds. *Plant Cell Rep* 23:134–143.

Manners, J. and Casu, R. 2011. Transcriptome analysis and functional genomics of sugarcane. *Trop Plant Biol* 4:9–21.

Maretzki, A., Sun, S.S., Nagai, C., Bidney, D., Houtchens, K.A., and Dela Cruz, A. 1990. Development of a transformation system for sugarcane. *VII Int. Congr. Plant Tiss. Cell Cult.*, Amsterdam, p. 68.

Mariotti, J.A. 2002. Selection for sugar cane yield and quality components in subtropical climates. *Sugar Cane Int* March/April:22–26.

McClintock, B. 1956. Controlling elements and the gene. *Cold Spring Harb Symp Quant Biol* 21:197–216.

McIntosh, S.R., Watson, L., Bundock, P.C., et al. 2007. SAGE of the most abundant transcripts in the developing wheat Caryopsis. *Plant Biotechnol J* 5:69–83.

McIntyre, C.L., Jackson, M., Cordeiro, G.M., et al. 2006. The identification and characterization of allele of sucrose phosphate synthase gene family III in sugarcane. *Mol Breeding* 18:39–50.

McNally, K., Bruskiewich, R., Mackill, D., Buell, C., Leach, J., and Leung, H. 2006. Sequencing multiple and diverse rice varieties. Connecting whole-genome variation with phenotypes. *Plant Physiol* 141:26–31.

McQualter, R.B., Harding, R.M., Dale, J.L., and Smith, G.R. 2001. Virus derived transgenes confer resistance to Fiji disease in transgenic sugarcane plants. *Int Soc Sugar Cane Technol* 2:584–585.

McQualter, R.B., Dale, J.L., Harding, R.H., McMahon, J.A., and Smith, G.R. 2004. Production and evaluation of transgenic sugarcane containing a Fiji disease virus (FDV) genome segment S9-derived synthetic resistance gene. *Aust J Agric Res* 55:139–145.

Menossi, M., Silva-Filho, M.C., Vincentz, M., Van-Sluys, M.A., and Souza, G.M. 2008. Sugarcane functional genomics: Gene discovery for agronomic trait development. *Int J Plant Genom* 2008:458732.

Mezker, M.L. 2010. Sequencing technologies: The next generation. *Nat Rev Genet* 11:31–45.

Milligan, S.B., Gravois, K.A., Bischoff, K.P., and Martin, F.A. 1990. Crop effects on broad-sense heritabilities and genetic variances of sugarcane yield components. *Crop Sci* 30:344–349.

Ming, R., Liu, S.C., Lin, Y.R., et al. 1998. Alignment of the *Sorghum* and *Saccharum* chromosomes: Comparative genome organization and evolution of a polysomic polyploid genus and its diploid cousin. *Genetics* 150:1663–1682.

Ming, R., Liu, S.-C., Moore, P.H., Irvine, J.E., and Paterson, A.H. 2001. QTL analysis in a complex autopolyploid: Genetic control of sugar content in sugarcane. *Genome Res* 11:2075–2084.

Ming, R., Wang, Y.W., Draye, X., et al. 2002. Molecular dissection of complex traits in autopolyploids: Mapping QTLs affecting sugar yield and related traits in sugarcane. *Theor Appl Genet* 105:332–345.

Mochida, K. and Shinozaki, K. 2010. Genomics and bioinformatics resources for crop improvement. *Plant Cell Physiol* 51(4):497–523.

Molinari, H.B.C., Marur, C.J., Daros, E., et al. 2007. Evaluation of the stress-inducible production of praline in transgenic sugarcane (*Saccharum* spp.): Osmotic adjustment, chlorophyll fluorescence and oxidative stress. *Physiol Plant* 130:218–229.

Moonan, F. and Mirkov, T.E. 2002. Analyses of genotypic diversity among North, South, and Central American isolates of *Sugarcane Yellow Leaf Virus*: Evidence for Colombian origins and for intraspecific spatial phylogenetic variation. *J Virol* 70(3):1339–1348.

Moore, P. 1995. Temporal and spatial regulation of sucrose accumulation in the sugarcane stem. *Aust J Plant Physiol* 22:661–679.

Moore, P.H. 1999. Progress and development in sugarcane biotechnology. *Proc Int Soc Sugar Cane Technol* 23:241–258.

Moore, P.H. 1999. Progress and development in sugarcane biotechnology. In: Singh, V. and Kumar, V. (eds), *Proceedings of International Society Sugar Cane Technology*, vol. 23, pp. 241–258. New Delhi.

Morton, C.J. and Campbell, I.D. 1994. SH3 domains. Molecular 'velcro.' *Curr Biol* 4:615–617.

Mudge, J., Andersen, W.R., Kehrer, R.L., and Fairbanks, D.J. 1996. A RAPD genetic map of *Saccharum officinarum*. *Crop Sci* 36:1362–1366.

Mudge, S.R., Osabe, K., Casu, R.E., et al. 2009. Efficient silencing of reporter transgenes coupled to known function promoters in sugarcane, a highly polyploidy crop species. *Planta* 229:549–558.

Mukherjee, S. 1957. Origin and distribution of *Saccharum*. *Bot Gaz* 19:55–61.

Muotri, A.R., Marchetto, M.C.N., Coufal, N.G., and Gage, F.H. 2007. The necessary junk: New functions for transposable elements. *Hum Mol Genet* 16:159–167.

Muthukrishnan, S., Kramer, K.J., Zhang, H., et al. 2001. Use of chitinases as biopesticides. In: Uragami, T., Kurita, K., and Fukamizo, T. (eds), *Chitin and Chitosan: Chitin and Chitosan in Life Science*. Kodansha Scientific.

Nair, N.V., Nair, S., Sreenivasan, T.V., and Mohan, M. 1999. Analysis of genetic diversity and phylogeny in Saccharum and related genera using RAPD markers. *Genet Resour Crop Evol* 46:73–79.

Nair, V.N., Selvi, A., Sreenivasan, T.V., and Pushpalatha, K.N. 2002. Molecular diversity in Indian sugarcane cultivars as revealed by randomly amplified DNA polymorphism. *Euphytica* 127:219–225.

Nishida, R. 2002. Sequestration of defensive substances from plant by Lepidoptera. *Annu Revi Entomol* 47:57–92.

Nogueira, E.D., Vinagre, F., Masuda, H.P., et al. 2001. Expression of sugarcane genes induced by inoculation with *Gluconacetobacter diazotrophicus* and *Herbaspirillum rubrisubalbicans*. *Genet Mol Biol* 24(1–4):199–206.

Nogueira, F.T.S., de Rosa Jr., V.E., Menossi, M., Ulian, E.C., and Arruda, P. 2003. RNA expression profiles and data mining of sugarcane response to low temperature. *Plant Physiol* 132:1811–1824.

Nogueira, F.T.S., Schlogl, P.S., Camargo, S.R., Fernandez, J.H., Rosa, V.E., Pompermayer, P., and Arruda, P. 2005. *SsNAC23*, a member of the NAC domain protein family, is associated with cold, herbivory and water stress in sugarcane. *Plant Sci* 169:93–106.

Nutt, K.A., Allsopp, P.G., Geijskes, R.J., McKeon, M.G., and Smith, G.R. 2001. Canegrub resistant sugarcane. In: Proc. Int. Soc. Sugar Cane Technol., Brisbane, 24:584–585.

Nutt, K.A., Allsopp, P.G., McGhie, T.K., et al. 1999. Transgenic sugarcane with increased resistance to canegrubs. *Proc Aust Soc Sugarcane Technol* 21:171–176.

Okazaki, Y. and Saito, K. 2012. Recent advances of metabolomics in plant biotechnology. *Plant Biotechnol Rep* 6:1–15.

Oksman-Caldentey, K.M. and Inzé, D. 2004. Plant cell factories in the postgenomic era: New ways to produce designer secondary metabolites. *Trends Plant Sci* 9:433–440.

Oksman-Caldentey, K.M. and Saito, K. 2005. Intergating genomics and metabolomics for engineering plant metabolic pathways. *Curr Opin Biotech* 16:174–179.

Oliveira, K.M., Pinto, L.R., Margarido, G.R.A., et al. 2007. Functional integrated genetic linkage map based on EST-markers for sugarcane (*Saccharum* spp.) commercial cross. *Mol Breeding* 20(3):189–208.

Olsen, A.N., Ernst, H.A., Leggio, L.L., and Skriver, K. 2005. NAC transcription factors: Structurally distinct, functionally diverse. *Trends Plant Sci* 10(2):79–87.

Osabe, K., Mudge, S., Graham, M., and Birch, R. 2009. RNAi mediated downregulation of PDS gene expression in sugarcane (*Saccharum*), a highly polyploid crop. *Trop Plant Biol* 2:143–148.

Pagariya, M.C. 2012. Candidate genes as molecular markers for evaluating and validating sugarcane germplasm for salinity stress. Ph.D. Thesis, Shivaji University Kolhapur, Maharashtra, India.

Pagariya, M.C., Harikrishnan, M., Kulkarni, P.A., Devarumath, R.M., and Kawar, P.G. 2010. Physio-biochemical analysis and transcript profiling of *Saccharum officinarum* L. submitted to salt stress. *Acta Physiol Plant* 33:1411–1424.

Pan, Y.B. 2006. Highly polymorphic microsatellite DNA markers for sugarcane germplasm evaluation and variety identity testing. *Sugar Tech* 8(4):246–256.

Pan, Y.B., Grisham, M.P., Burner, D.M., and Damann, K.E. 1998. A polymerase chain reaction protocol for the detection of *Clavibacter xyli* subsp. xyli, the causal bacterium of sugarcane ratoon stunting disease. *Plant Dis* 82:285–290.

Pan, Y.B., Burner, D.M., Legendre, B.L., Grisham, M.P., and White, W.H. 2004. An assessment of the genetic diversity within a collection of Saccharum spontaneum with RAPD PCR. *Genet Res Crop Evol* 51:895–903.

Papini-Terzi, F.S., Rocha, F.R., Vencio, R.Z., et al. 2005. Transcription profiling of signal transduction-related genes in sugarcane tissues. *DNA Res* 12:27–38.

Papini-Terzi, F.S., Rocha, F.R., Vencio, R.Z.N., et al. 2009. Sugarcane genes associated with sucrose content. *BMC Genomics* 10:120.

Parida, S.K., Kalia, S.K., and Kaul, S. 2009. Informative genomic microsatellite markers for efficient genotyping applications in sugarcane. *Theor Appl Genet* 118:327–338.

Parida, S.K., Pandit, A., and Gaikwad, K. 2010. Functionally relevant microsatellites in sugarcane unigenes. *BMC Plant Biol* 10:251.

Parker, R., Flowers, T.J., Moore, A.L., and Harpham, N.V.J. 2006. An accurate and reproducible method for proteome profiling of the effects of salt stress in the rice leaf. *J Exp Bot* 57:1109–1118.

Pastina, M.M., Malosetti, M., Gazaffi, R., Mollinari, M., Margarido, G. R.A., Oliveira, K. M., et al. 2012. A mixed model QTL analysis for sugarcane multiple-harvest-location trial data. *Theor Appl Genet* 124:835–849.

Patade, V.Y. 2009. Studies on salt stress responses of sugarcane (*Saccharum officinarum* L.) using physiological and molecular approaches. PhD. Thesis, University of Pune, Pune, Maharashtra, India.

Patade, V.Y. and Suprasanna, P. 2010. Short-term salt and PEG stresses regulate expression of microRNA, miR159 in sugarcane *J Crop Sci Biotech* 13:177–182.

Patade, V.Y., Suprasanna, P., Kulkarni, U.G., and Bapat, V.A. 2006. Molecular profiling using RAPD technique of abiotic stress (salt and drought) tolerant regenerants of sugarcane Cv. Coc-671. *Sugar Tech* 8:63–68.

Patade, V.Y., Suprasanna, P., and Bapat, V.A. 2008. Gamma irradiation of embryogenic callus cultures and in vitro selection for salt tolerance in sugarcane (Saccharum offcinarum L.). *Agr Sci China* 7(9):101–105.

Patade, V.Y., Rai, A.N., and Suprassana, P. 2010. Expression of a sugarcane shaggy-like kinase (SuSK) gene identified through cDNA subtractive hybridisation in sugarcane (*Saccharum officinarum* L.). *Protoplasma* 248:613–621.

Patade, V.Y., Rai, A.N., and Suprasanna, P. 2011. Expression analysis of sugarcane shaggy-like kinase (*SuSK*) gene identified through cDNA subtractive hybridization in sugarcane (*Saccharum officinarum* L.). *Protoplasma* 248(3):613–621.

Paterson, A.H., Bowers, J.E., Bruggmann, R., Dubchak, I., Grimwood, J., Gundlach, H., Habere, G., Hellsten, U., et al. 2009. The *Sorghum bicolor* genome and the diversification of grasses. *Nature* 457:551–556.

Paterson, A.H., Schertz, K.F., Lin, Y.R., Liu, S.C., and Chang, Y.L. 1995. The weediness of wild plants: Molecular analysis of genes influencing dispersal and persistence of Johnsongrass, Sorghum-Halepense (L) Pers. *Proc Natl Acad Sci USA* 92:6127–6131.

Paterson, A.H., Bowers, J.E., Bruggmann, R., et al. 2009. The Sorghum bicolor genome and the diversification of grasses. *Nature* 457:551–548.

Petrasovits, L.A., Purnell, M.P., Nielsen, L.K., and Brumbley, S.M. 2007. Production of polyhydroxybutyrate in sugarcane. *Plant Biotechnol J* 5:162–172.

Picoult-Newberg, L., Ideker, T.E., Pohl, M.G., et al. 1999. Mining SNPs from EST databases. *Genome Res* 9:167–174.

Piegu, B., Guyot, R., Picault, N., et al. 2006. Doubling genome size without polyploidization: Dynamics of retrotransposition-driven genomic expansions in *Oryza australiensis*, a wild relative of rice. *Genome Res* 16:1262–1269.

Pinto, L.R., Oliveira, K.M., Marconi, T., et al. 2006. Characterization of novel sugarcane expressed sequence tag microsatellites and their comparison with genomic SSRs. *Plant Breed* 125:378–384.

Pinto, L.R., Garcia, A.A.F., Pastina, M.M., et al. 2010. Analysis of genomic and functional RFLP derived markers associated with sucrose content, fiber, and yield QTLs in sugarcane (*Saccharum* spp.) commercial cross. *Euphytica* 172:313–327.

Piperidis, G., Piperidis, N., and D'Hont, A. 2010. Molecular cytogenetic investigation of chromosome composition and transmission in sugarcane. *Mol Genet Genomics* 284:65–73.

Piperidis, G., Rattey, A.R., Taylor, G.O., and Cox, M.C. 2004. DNA markers: A tool for identifying sugarcane varieties. In: Hogarth, D.M. (ed.), *Proceedings of Australia Society Sugarcane Technology*, vol. 26.

Powell, W., Machrar, G.C., and Proven, J. 1996. Polymorphism revealed by simple sequence repeats. *Trends Plant Sci* 7:215–222.

Prabu, G. and Prasad, D.T. 2012. Functional characterization of sugarcane MYB transcription factor gene promoter (PScMYBAS1) in response to abiotic stresses and hormones. *Plant Cell Rep* 31(4):661–669.

Prabu, G., Kawar, P.G., Pagariya, M.C., and TheerthaPrasad, D. 2011. Identification of water deficit stress upregulated genes in sugarcane. *Plant Mol Biol Rep* 29(2):291–304.

Price, S. 1965. Interspecific hybridization in sugarcane breeding. *Proc Int Soc Sugar Cane Technol* 12:1021–1026.

Rai, M.K., Kalia, R.K., Singh, R., Gangola, M.P., and Dhawan, A.K. 2011. Developing stress tolerant plants through in vitro selection: An overview of the recent progress. *Environ Exp Bot* 71(1):89–98.

Ramagopal, S. 1990. Protein polymorphism in sugarcane revealed by two-dimensional gel analysis. *Theoret Appl Gen* 79:297–304.

Ramagopal, S. 1994. Protein variation accompanies leaf dedifferentiation in sugarcane (*Saccharum officinarum*) and is influenced by genotype. *Plant Cell Rep* 13:692–696.

Rangel, P., Gomez, L., Victoria, J.I., and Angel, F. 2003. Transgenic plants of CC 84-75 resistant to the virus associated with the sugarcane yellow leaf syndrome. In: *Proceedings of International Society Sugar Cane Technology Molecular Biology*. Workshop, Montpellier, vol. 4, p. 30.

Rathus, C. and Birch, R.G. 1992. Stable transformation of callus from electroporated sugarcane protoplasts. *Plant Sci* 82:81–89.

Rathus, C., Bower, R., and Birch, R.G. 1993. Effects of promoter, intron and enhancer elements on transient gene expression in sugarcane and carrot protoplasts. *Plant Mol Biol* 23:613–618.

Reffay, N., Jackson, P.A., Aitken, K.S., Hoarau, J.Y., D'Hont, A., Besse, P., and McIntyre, C.L. 2005. Characterization of genome regions incorporated from an important wild relative into Australian sugarcane. *Mol Breeding* 15:367–381.

Rhodes, D. and Hanson A.D. 1993. Quaternary ammonium and tertiary sulfonium compounds in higher plants. *Annu Rev Plant Physiol Plant Mol Biol* 44:357–384.

Riechmann, J.L., Heard, J., Martin, G., Reuber, L., Jiang, C., Keddie, J., et al. 2000. *Arabidopsis* transcription factors: Genome-wide comparative analysis among eukaryotes. *Science* 290:2105–2110.

Riera, M., Valon, C., Fenzi, F., Giraudat, J., and Leung, J. 2005. The genetics of adaptive responses to drought stress: Abscisic acid-dependent and abscisic acid-independent signalling components. *Physiol Plantarum* 123(2):111–119.

Rocha, F.R., Papini-Terzi, F.S., Nishiyama Jr., M.Y., et al. 2007. Signal transduction-related responses to phytohormones and environmental challenges in sugarcane. *BMC Genomics* 8:71.

Rodrigues, F.A., Laia, M.L., and Zingaretti, S.M. 2009. Analysis of gene expression profiles under water stress in tolerant and sensitive sugarcane plants. *Plant Sci* 176:286–302.

Roessner, U., Luedemann, A., Brust, D., et al. 2001. Metabolic profiling allows comprehensive phenotyping of genetically or environmentally modified plant systems. *Plant Cell* 13:11–29.

Roessner, U., Wagner, C., Kopka, J., Trethewey, R.N., and Willmitzer, L. 2000. Simultaneous analysis of metabolites in potato tuber by gas chromatography–mass spectrometry. *Plant J* 23:131–142.

Rossi, M., Araujo, P.G., Paulet, F., et al. 2003. Genomic distribution and characterization of EST-derived resistance gene analogs (RGAs) in sugarcane. *Mol Genet Genomics* 269:406–419.

Rossi, M., Araujo, P.G., de Jesus, E.M., Varani, A.M., and Van Sluys, M.A. 2004. Comparative analysis of Mutator-like transposases in sugarcane. *Mol Genet Genomics* 272:194–203.

Sakurai, N., Ara, T., Ogata, Y., et al. 2011. Kappa-view4: A metabolic pathway database for representation and analysis of correlation networks of gene co-expression and metabolite co-accumulation and omics data. *Nucleic Acid Res* 39:D677–D684.

Salekdeh, G.H., Siopongco, J., Wade, L.J., Ghareyazie, B., and Bennett, J. 2002. Proteomic analysis of rice leaves during drought stress and recovery. *Proteomics* 2:1131–1145.

Sandhu, S., Gosal, S., Thind, K., et al. 2008. Field performance of micropropagated plants and potential of seed cane for stalk yield and quality in sugarcane. *Sugar Tech* 11:34–38.

Sato, S., Soga, T., Nishioka, T., and Tomita, M. 2004. Simultaneous determination of the main metabolites in rice leaves using capillary electrophoresis mass spectrometry and capillary electrophoresis diode array detection. *Plant J* 40:151–163.

Schenk, P.M., Remans, T., Sagi, L., et al. 2001. Promoters for pregenomic RNA of banana streak badnavirus are active for transgene expression in monocot and dicot plants. *Plant Mol Biol* 47:399–412.

Schreiber, F., Colmsee, C., Czauderna, T., et al. 2011. MetaCrop 2.0: Managing and exploring information about crop plant metabolism. *Nucleic Acids Res* 40(D1):D1173–D1177.

Scott, J.G. and Wen, Z. 2001. Cytochromes P450 of insects: The tip of the iceberg. *Pest Manag Sci* 57:958–967.

Seki, M., Kamei, A., Yamaguchi-Shinozaki, K., and Shinozaki, K. 2003. Molecular responses to drought, salinity and frost: Common and different paths for plant protection. *Curr Opin Biotech* 14:194–199.

Seki, M., Narusaka, M., Abe, H., Kasuga, M., Yamaguchi-Shinozaki, K. et al. 2001. Monitoring the expression pattern of 1300 *Arabidopsis* genes under drought and cold stresses by using a full-length cDNA microarray. *Plant Cell* 13:61–72.

Selvi, A. and Nair, N.V. 2010. Molecular breeding in sugarcane. *Int J Agric Environ Biotechnol* 3:115-127.

Selvi, A., Nair, N.V., Balasundaram, N., and Mohapatra, T. 2003. Evaluation of maize microsatellite markers for genetic diversity analysis and fingerprinting in sugarcane. *Genome* 46(3):394–403.

Selvi, A., Nair, N.V., Noyer, J.L., et al. 2005. Genomic constitution and genetic relationship among the tropical and subtropical Indian sugarcane cultivars revealed by AFLP. *Crop Sci* 45:1750–1757.

Selvi, A., Nair, N.V., and Noyer, J.L. 2006. AFLP analysis of the phonetic organization and genetic diversity in the sugarcane complex, Saccharum and Erianthus. *Genet Resour Crop Evol* 53:831–842.

Sengar, A.S., Thind, S., Kumar, B., Pallavi, M., and Gosal, S.S. 2009. In vitro selection at cellular level for red rot resistance in sugarcane (*Saccharum* sp.). *Plant Growth Regul* 58(2):201–209.

Sereno, M.L., Almeida, R.S., Nishimura, D.S., and Figueira, A. 2007. Response of sugarcane to increasing concentrations of copper and cadmium and expression of metallothionein genes. *J Plant Physiol* 164:1499–1515.

Sétamou, M., Bernal, J.S., Legaspi, J.C., Mirkov, T.E., and Legaspi, B.C. 2002. Evaluation of lectin-expressing transgenic sugarcane against stalkborers (Lepidoptera: Pyralidae): Effects on life history parameters. *J Econ Entomol* 95:469–477.

Shinozaki, K. and Yamaguchi-Shinozaki, K. 2000. Molecular responses to dehydration and low temperature: Differences and cross-talk between two stress signaling pathways. *Curr Opin Plant Biol* 3:217–223.

Shirasu, K., Schulman, A.H., Lahaye, T., and Schulze-Lefert, P. 2000. A contiguous 66-kb barley DNA sequence provides evidence for reversible genome expansion. *Genome Res* 10:908–915.

Shrivastava, S. and Gupta, P. 2008. Inter simple sequence repeat profile as a genetic marker system in sugarcane. *Sugar Tech* 10(1):48–52.

Sills, G., Bridges, W., Al-Janabi, S., and Sobral, B.W.S. 1995. Genetic analysis of agronomic traits in a cross between sugarcane (*Saccharum officinarum* L.) and its presumed progenitor (*S. robustum* Brandes, Jesw. Ex. Grassl). *Mol Breeding* 1:355–363.

Simon, S.A., Zhai, J., Nandety, R.S., McCormick, K.P., Zeng, J., et al. 2009. Short-read sequencing technologies for transcriptional analyses. *Annu Rev Plant Biol* 60:305–333.

Singh, R.K., Singh, P., Misra, P., Singh, S.P., and Singh, S.B. 2005. STMS marker for tagging high sugar genes in sugarcane. *Sugar Tech* 7(2–3):74–76.

Singh, R.K., Singh, R.B., Singh, S.P., and Sharma, M.L. 2011. Identification of sugarcane microsatellites associated to sugar content in sugarcane and transferability to other cereal genomes. *Euphytica* 182(3):335–354.

Smith, G.R. and van de Velde, R. 1994. Detection of sugarcane mosaic virus and Fiji disease virus in diseased sugarcane using the polymerase chain reaction. *Plant Dis* 78:557–561.

Smith, G.R., Clarke, M.L., Van de Velde, R., and Dale, J.L. 1994. Chemiluminescent detection of Fiji disease virus in sugarcane with biotinylated DNA probes. *Arch Virol* 136:325–334.

Soares-Costa, A., Beltramini, L.M., Thiemann, O.H., and Henrique-Silva, H. 2002. A sugarcane cystatin: Recombinant expression, purification, and antifungal activity. *Biochem Biophys Res Commun* 296(5):1194–1199.

Sobral, B.W.S., Braga, D.P.V., Lahood, E.S., and Keim, P. 1994. Phylogenetic analysis of chloroplast restriction enzyme site mutations in the Saccharinaea, Griseb. subtribe of the Andropogonaea Dumort tribe. *Theor Appl Genet* 87:843–853.

Solomon, S. 2011. The Indian sugar industry: An overview. *Sugar Tech* 13(4):255–265.

Somerville, C., Young, H., Taylor, C., et al. 2010. Feedstocks for lignocellulosic biofuels. *Science* 329:790–792.

Souza, G.M., Berges, H., and Bocs, S. 2011. The sugarcane genome challenge: Strategies for sequencing a highly complex genome. *Trop Plant Biol* 4:145–156.

Sreenivasan, T.V., Ahloowalia, B.S., and Heinz, D.J. 1987. Cytogenetics. In: Heinz, D.J. (ed.), *Sugarcane Improvement Through Breeding*, pp. 211–253. Elsevier: New York.

Srikanth, J., Subramonian, N., and Premachandran, M.N. 2011. Advances in transgenic research for insect resistance in sugarcane. *Trop Plant Biol* 4:52–61.

Srivastava, S. and Gupta, P. 2008. Inter simple sequence repeat profile as a genetic marker system in sugarcane. *Sugar Tech* 10(1):48–52.

Srivastava, S. and Gupta, P.S. 2006. Low level of genetic diversity in sugarcane varieties of India as assessed by RAPD markers. In: *Proceedings of International Symposium on Technologies to Improve Sugar Productivity in Developing Countries, Session 7: Molecular Biology, Biotechnology and Tissue Culture in Sugar Crops*, pp. 574–578. Guilin, P.R. China.

Sugihario, B., Ermawati, N., Mori, H., et al. 2002. Identification and characterization of a gene encoding drought-inducible protein localized in the bundle sheath cell of sugarcane. *Plant Cell Physiol* 43:350–354.

Suprasanna, P., Desai, N.S., Sapna, G., and Bapat. V.A. 2006. Monitoring genetic fidelity in plants derived through direct somatic embryogenesis in sugarcane by RAPD analysis. *J New Seeds* 8(3): 1–9.

Suprasanna, P. and Ganapathi, T.R. 2010. Engineering the plant genome: Prospects of selection systems using non-antibiotic marker genes. *GM Crop* 1(3):1–9.

Suprasanna, P., Rupali, C., Desai, N.S., and Bapat, V.A. 2008a. Partial desiccation improves plant regeneration response of gamma-irradiated embryogenic callus in sugarcane (*Saccharum* Spp.). *Plant Cell Tissue Org Cult* 92:101–105.

Suprasanna, P., Patade, V.Y., Vaidya, E.R., and Patil, V.D. 2008b. Radiation induced in vitro muta-genesis, selection for salinity tolerance and characterization in sugarcane. In: Shu, Q.Y. (ed.), *Induced Plant Mutations in the Genomics Era. Proceedings of International Symposium on Induced Mutations in Plants*, pp. 159–162. Vienna, IAEA, Austria. Food and Agriculture Organization of the United Nations: Rome.

Suprasanna, P., Manjunatha, B.R., Patade, V.Y., Desai, N.S., and Bapat, V.A. 2010. Profiling of culture-induced variation in sugarcane plants regenerated via direct and indirect somatic embryogen-esis by using transposon-mediated polymorphism. *Sugar Tech* 12(1):26–30.

Suprasanna, P., Patade, V., Desai, N.S., et al. 2011. Biotechnological developments in sugarcane improvement: An overview. *Sugar Tech* 13:322–335.

Suprasanna, P., Sonawane, B.V., and Patade, V.Y. 2012. In vitro mutagenesis and selection in plant tissue cultures and their prospects for crop improvement. *Biorem Biodiv Bioavail* 6:6–14.

Swapna, M., Sivaraju, K., Sharma, R.K., Singh, N.K., and Mohapatra, T. 2011. Single-strand confor-mational polymorphism of EST-SSRs: A potential tool for diversity analysis and varietal iden-tification in sugarcane. *Plant Mol Biol Rep* 29:505–513.

Swapna, M. and Srivastava, S. 2012. Molecular marker application for improving sugar content in sugarcane. *Springer Briefs in Plant Science*. DOI: 10.1007/978-1-4614-2257-0_1. pp. 1–49.

Taiz, L. and Zeiger, E. 1998. *Plant Physiology*, pp. 214–215. Sinauer Associates: Sunderland.

Tawar, P.N., Sawant, R.A., Dalvi, S.G., et al. 2008. An assessment of somaclonal variation in micro-propagated plants of sugarcane by RAPD markers. *Sugar Tech* 10(2):124–127.

Taylor, P.W.J., Geijskes, J.R., Ko, H.L., et al. 1995. Sensitivity of random amplified polymorphic DNA analysis to detect genetic change in sugarcane during tissue culture. *Theor Appl Genet* 90:1169–1173.

Thimm, O., Blasing, O., Gibon, Y., et al. 2004. A user-driven tool to display genomics data sets onto diagrams of metabolic pathways and other biological processes. *Plant J* 37:914–939.

Thompson, N., Choi, Y., and Randles, J.W. 1998. Sugarcane striate mosaic disease: Development of a diagnostic test. In: *Proc. 7th Int Congr Plant Path*, Edinburgh (Abstract 3.3.21).

Tokimatsu, T., Sakurai, N., Suzuki, H., Ohta, H., Nishitani, K., Koyama, T., et al. 2005. KaPPA-view: A web-based analysis tool for integration of transcript and metabolite data on plant metabolic pathway maps. *Plant Physiol* 138:1289–1300.

Toorchi, M., Yukawa, K., Nouri, M.Z., and Komatsu, S. 2009. Proteomics approach for identifying osmotic-stress-related proteins in soybean roots. *Peptides* 30:2108–2117.

Trethewey, R. 2004. Metabolite profiling as an aid to metabolic engineering in plants. *Curr Opin Plant Biol* 7:196–201.

Tweeddale, H., Notley-McRobb, L., and Ferenci, T. 1998. Effect of slow growth on metabolism of *Escherichia coli*, as revealed by global metabolite pool ('Metabolome') analysis. *J Bacteriol* 180:5109–5116.

Ulian, E.C. and Burnquist, W.L. 2000. Sucest: O projeto genoma da cana-de-açúcar. *STAB* 18(4):24–25.

van der Merwe, M.J., Groenewald, J.H., and Botha, F.C. 2003. Isolation and evaluation of a develop-mentally regulated sugarcane promoter. *Proc South African Sugar Cane Technol* 77:146–169.

van der Merwe, M.J., Groenewald, J.-H., Stitt, M., Kossmann, J., and Botha, F.C. 2010. Down-regulation of pyrophosphate: Fructose 6-phosphate 1-phosphotransferase activity in sugar-cane culms enhances sucrose accumulation due to elevated hexose-phosphate levels. *Planta* 231:595–608.

Vazquez, F., Legrand, S., and Windels, D. 2010. The biosynthetic pathways and biological scopes of plant small RNAs. *Trends Plant Sci* 15:337–345.

Velculescu, V.E., Zhang, L., Vogelstein, B., and Kinzler, K.W. 1995. Serial analysis of gene expression. *Science* 270:484–487.

Vettore, A., da Silva, F., Kemper, E., and Arruda, P. 2001. The libraries that made SUCEST. *Genet Mol Biol* 24:1–7.

Vettore, A.L., da Silva, F.R., Kemper, E.L., et al. 2003. Analysis and functional annotation of an expressed sequence tag collection for tropical crop sugarcane. *Genome Res* 13:2725–2735.

Vicentini, R., Felix, J.M., Dornelas, M.C., and Menossi, M. 2009. Characterisation of a sugarcane (*Saccharum* spp.) gene homolog to the brassinosteroid insensitive1-associated receptor kinase 1 that is associated to sugar content. *Plant Cell Rep* 28:481–491.

Vickers, J.E., Grof, C.P.L., Bonnett, G.D., Jackson, P.A., and Morgan, T.E. 2005. Effects of tissue culture, biolistic transformation, and introduction of PPO and SPS gene constructs a performance of sugarcane clones in the field. *Aust J Agric Res* 56:57–68.

Vinagre, F., Vargas, C., Schwarcz, K., et al. 2006. SHR5: A novel plant receptor kinase involved in plant-N2-fixing endophytic bacteria association. *J Exp Bot* 57(3):559–569.

Vincent, D., Lapierre, C., Pollet, B., et al. 2005. Water deficits affect caffeate O-methyltransferase, lignification, and related enzymes in maize leaves: A proteomic investigation. *Plant Physiol* 137:949–960.

Vincentz, M., Cara, F.A.A., Okura, V.K., da Silva, F.R., Pedrosa, G.L., Hemerly, A.S., et al. 2004. Evaluation of monocot and eudicot divergence using the sugarcane transcriptome. *Plant Physiol* 134:951–959.

Virupakshi, S. and Naik, G.S. 2008. ISSR analysis of chloroplast and mitochondrial genome can indicate the diversity in sugarcane genotypes for red rot resistance. *Sugar Tech* 10(1):65–70.

Von Roepenack-Lahaye, E., Degenkolb, T., Zerjeski, M., et al. 2004. Profiling of Arabidopsis secondary metabolites by capillary liquid chromatography coupled to electrospray ionization quadrupole time-of-flight mass spectrometry. *Plant Physiol* 134:548–559.

Waclawovsky, J., Sato, P.M., Lembke, C.G., Moore, P.H., and Souza, G.M. 2010. Sugarcane for bioenergy production: An assessment of yield and regulation of sucrose content. *Plant Biotechnol J* 8:1–14.

Wahid, A. 2007. Physiological implications of metabolite biosynthesis for net assimilation and heat-stress tolerance of sugarcane (*Saccharum officinarum*) sprouts. *J Plant Res* 120:219–228.

Wahid, A. and Close, T.J. 2007. Expression of dehydrins under heat stress and their relationship with water relations of sugarcane leaves. *Biol Plant* 51:104–109.

Walsh, K.B., Sky, R.B., and Brown, S.M. 2005. The anatomy of the pathway of sucrose unloading within the sugarcane stem. *Funct Plant Biol* 32:367–374.

Wang, Y. and Fristensky, B. 2001. Transgenic canola lines expressing pea defense gene DRR206 have resistance to aggressive blackleg isolates and to *Rhizoctonia solani*. *Mol Breeding* 8(3):263–271.

Wang, M.L., Goldstein, C., Su, W., Moore, P.H., and Albert, H.H. 2005. Production of biologically active GM-CSF in sugarcane: A secure biofactory. *Transgenic Res* 14(2):167–178.

Wang, J., Roe, B., Macmil, S., et al. 2010. Micro collinearity between autopolyploid sugarcane and diploid sorghum genomes. *BMC Genomics* 11:261.

Wei, H. 2001. *Isolation and Characterization of Two Sugarcane Polyubiquitin Gene Promoters and Matrix Attachment Regions.* University of Hawaii: Manoa.

Wei, H., Albert, H.H., and Moore, P.H. 1999. Differential expression of sugarcane polyubiquitin genes and isolation of promoters from two highly expressed members of the gene family. *J Plant Physiol* 155:513–519.

Wei, H., Moore, P.H., and Albert, H.H. 2003. Comparative expression analysis of two sugarcane polyubiquitin promoters and flanking sequences in transgenic plants. *J Plant Physiol* 160:1241–1251.

Weng, L.X., Deng, H.H., Xu, J.L., et al. 2006. Regeneration of sugarcane elite breeding lines and engineering of strong stem borer resistance. *Pest Manage Sci* 62:178–187.

Weng, L.X., Deng, H.H., Xu, J.L., et al. 2010. Transgenic sugarcane plants expressing high levels of modified cry1Ac provide effective control against stem borers in field trials. *Transgenic Res* 20(4):759–772.

Whittaker, A. and Botha, F. C. 1997. Carbon particinung during sucrose accumulation in sugarcane internodal tissue. *Plant Physiol* 115:1651–1659.

Whittaker, A. and Botha, F.C. 1997. Carbon partitioning during sucrose accumulation in sugarcane stem regulated by the difference between the activities of soluble acid invertase and sucrose phosphate synthase. *Plant Physiol* 115:609–616.

Whittaker, A. and Botha, F.C. 1999. Pyrophosphate: Fructose 6-phosphate 1-phosphotransferase activity patterns in relation to sucrose storage across sugarcane varieties. *Physiol Plantarum* 107:379–386.

Wu, L.G. and Birch, R.G. 2007. Doubled sugar content in sugarcane plants modified to produce a sucrose isomer. *Plant Biotechnol J* 51:109–117.

Wu, K.K., Burnquist, W., Sorrells, M.E., et al. 1992. The detection and estimation of linkage in polyploids using single-dose restriction fragments. *Theor Appl Genet* 83:294–300.

Xu, G., Li, C.Y., and Yao, Y.A. 2009. Proteomics analysis of drought stress responsive proteins in *Hippophae rhamnoides* L. *Plant Mol Biol Rep* 27:153–161.

Yamazaki, M., Nakajima, J., Yamanashi, M., et al. 2003a. Metabolomics and differential gene expression in anthocyanin chemo-varietal forms of *Perilla frutescens*. *Phytochemistry* 62:987–995.

Yamazaki, Y., Urano, A., Sudo, H., et al. 2003b. Metabolite profiling of alkaloids and strictosidine synthase activity in camptothecin producing plants. *Phytochemistry* 62:461–470.

Yang, M.Z., Bower, R., Burow, M.D., Paterson, A.H., and Mirkov, T.E. 2003. A rapid and direct approach to identify promoters that confer high levels of gene expression in monocots. *Crop Sci* 43:1805–1814.

Yang, Z.N. and Mirkov, T.E. 1997. Sequence and relationships of sugarcane mosaic and sorghum mosaic strains and development of RT-PCR-based RFLPs for strain discrimination. *Phytopathology* 87:932–939.

Zanca, A.S., Vicentini, R., Ortiz-Morea, F.A., Del Bem, L.E.V., da Silva, M.J., et al. 2010. Identification and expression analysis of microRNAs and targets in the biofuel crop sugarcane. *BMC Plant Biol* 10:260.

Zang, X. and Komatsu, S. 2007. A proteomics approach for identifying osmotic-stress related proteins in rice. *Phytochemistry* 68:426–437.

Zangerl, A.R. and Berenbaum, M.R. 1990. Furanocoumarin induction in wild parsnip: Genetics and population variation. *Ecology* 71:1933–1940.

Zhang, L., Xu, J., and Birch, R.G. 1999. Engineered detoxification confers resistance against a pathogenic bacterium. *Nat Biotechnol* 17:1021–1024.

Zhang, B.H., Pan, X.P., and Anderson, T.A. 2006a. Identification of 188 conserved maize microRNAs and their targets. *FEBS Letters* 580(15):3753–3762.

Zhang, X.G., Krstic, P.S., Zikic, R., Wells, J.C., and Fuentes-Cabrera, M. 2006b. Firstprinciples transversal DNA conductance deconstructed. *Biophys J* 91:L04–L06.

Zhang, S.-Z., Yang, B.-P., Feng, C.-L., et al. 2006c. Expression of the *Grifola frondosa* trehalose synthase gene and improvement of drought-tolerance in sugarcane (*Saccharum officinarum* L.). *J Integr Plant Biol* 48:453–459.

Zhang, Z., Yu, J., Li, D., Zhang, Z., Liu, F., Zhou, X., et al. 2010. PMRD: Plant microRNA database. *Nucl Acids* Res 38:D806–D813.

Zhangsun, D.T., Luo, S.L., Chen, R.K., and Tang, K.X. 2007. Improved *Agrobacterium*-mediated genetic transformation of GNA transgenic sugarcane. *Biologia* (*Bratislava*) 62(4):386–393.

Zikic, R., Krstić, P.S., Zhang, X.G., et al. 2006. Characterization of the tunneling conductance across DNA bases. *Phys Rev EStat Nonlin Soft Matter Phys* 74(1):011919.

Zucchi, M.I., Arizono, H., Morais, V.I., Fungaro, M.H.P., and Vieira, M.L.C. 2002. Genetic instability of sugarcane plants derived from meristem cultures. *Genet Mol Biol* 25(1):91–96.

Zwolak, M. and Di Ventra, M. 2005. Electronic signature of DNA nucleotides via transverse transport. *Nano Lett* 5:421–424.

8

Recent Advances in Temperate Fruit Crops: An Omics Perspective

Md. Abdur Rahim, PhD and Livio Trainotti, PhD

CONTENTS

8.1   Background...251
8.2   Genomics..252
8.3   Transcriptomics...258
    8.3.1   Expressed Sequence Tags (ESTs)...258
    8.3.2   Microarray 259
    8.3.3   RNA-seq 264
8.4   Proteomics..266
8.5   Metabolomics 270
References...273

8.1 Background

Fruits can be defined as fertilized and matured ovaries that bear seeds to maintain generations. They are rich in different types of vitamins (C: ascorbic acid; B1: thiamine; B3: niacin; B6: pyridoxine; and B9: folic acid), minerals, and dietary fibers (Craig and Beck, 1999; Wargovich, 2000). During ripening, many fruits accumulate different types of bioactive chemicals such as flavonoids (including anthocyanins, flavones, and flavonols) and carotenoids which protect humans against cancer, cardiovascular diseases, and other chronic diseases (Rao and Rao, 2007). In addition, anthocyanins increase antioxidant levels in serum (Mazza et al., 2002), regulate cholesterol distribution (Xia et al., 2007), and protect human red blood cells (RBCs) from oxidative damage (Tedesco et al., 2001). Fleshy fruits such as tomato, apple, peach, grape, pear, kiwifruit, watermelon, papaya, and mango are economically important, either fresh or as ingredients of foods such as jams, cookies, muffins, yoghurt, ice cream, different types of cakes, and drinks such as fruit juices, wine, brandy, and so on. Therefore, the fruit development and ripening process is of foremost importance for fruit crop researchers. Fruit ripening is a complex process which is modulated by genetic, biochemical, developmental, and environmental processes (Klee and Giovannoni, 2011), leading to the accumulation of sugars, flavors, pigments (Singh et al., 2010), and changes in the fruit texture. The fruit-ripening control is mainly a combination of different molecular pathways with environmental factors, and a crucial contribution to its elucidation has been given by using ripening-affected mutants (Klee and Giovannoni, 2011). These studies allowed the identification of key regulators of the ripening process, thus opening the way for their use in genetic improvement programs. The genetic improvement of these fruit crops through traditional breeding programs is a long-term process, and therefore molecular and genetic approaches are crucial to further improve fruit crops. Recently, the cost of genome sequencing has decreased, and sequencing technology is improving rapidly; as a result, the number of sequenced plant genomes is increasing. Following the completion of the genome sequencing and annotation projects of several fruit crops, there is a definite capability to identify the functions of genes and the possibility to utilize them to improve fruit quality, working on qualitative as well as quantitative traits. To date, tomato, apple, pear, peach, Chinese plum, grape, strawberry, papaya, date palm, melon, watermelon, sweet orange, clementine, and banana are the sequenced fruit genomes. Furthermore, other genome sequencing projects are in progress (sweet cherry, raspberry, and kiwi). Besides the genome sequencing of different fruit crops, expressed sequence tags (ESTs), a number of anchored molecular markers, and genetic and physical maps were developed, and these tools might be useful for the improvement of these crops. Nowadays, global analysis of transcriptomes (transcriptomics) could easily be done, either by microarray or by RNA sequencing, in order to understand the expression of genes in a particular organ or tissue under variable conditions, thus initiating functional studies at gene level. Moreover, proteomics allow us to identify and start the functional characterization of proteins, understanding posttranscriptional modifications and protein–protein interactions (van Wijk, 2001). Proteomic studies could also be performed at the cellular and sub-cellular tissue levels, as well as organ level, to understand changes in protein levels at various developmental stages. Lastly, metabolomics is the study of metabolites, which are the end products of cellular regulatory processes (Fiehn, 2002); metabolic profiles could be indicators of cellular processes of an organism that more specifically act as bridges between genotypes and phenotypes. In this chapter we focus on the "omics" perspective of major temperate fruit crops, citing omics studies in other plant systems.

8.2 Genomics

Genomics is the study of genomes through the determination of most, if not all, of their DNA sequences and the development of fine maps. The genetic map is the linear arrangement of genes on chromosomes, obtained following the inheritance of traits through generations, and distances among loci are given as recombination frequencies, while in physical maps the position and distances of genes on the chromosomes are obtained through DNA sequencing and expressed in base pairs (bps). The development of a genetic map for a species is important in order to know the genetic system and as a tool to identify the genes that control traits. Linking genes to traits leads to the development of genetic linkage maps (Table 8.1). Genetic maps are an important tool for mapping qualitative and quantitative trait loci (QTL), and comparative mapping of species (Dirlewanger et al., 2004). Currently, several genetic linkage maps are available for different fruit species, focusing on molecular markers linked to important agronomic traits, which could help breeders in improving resistance in future against biotic and abiotic stresses, as well as fruit quality, through a molecular breeding approach. Moreover, sequenced genomes are also an important source of information to assist in finding genes and understanding their functions, and improving disease diagnostics and forensic research in medical science, as well as agricultural science and systems biology. The first sequenced plant genome was that of *Arabidopsis thaliana* (The Arabidopsis Genome Initiative, 2000), a milestone in genomics. To date, 13 fruit crop genomes have been completely sequenced and the related papers

TABLE 8.1

Genetic Linkage Maps Available in Different Fruit Crops

Fruit Crops	Reference Map	Markers	Coverage (cM)	References
Apple	Rome Beauty × White Angel	253	950	Hemmat et al., 1994
	Fiesta × Discovery	840	1371	Liebhard et al., 2003
	Telamon × Braeburn	Telamon: 259 Braeburn: 264	1039	Kenis and Keulemans, 2005
Pear	Bartlett × Housui	226	949	Yamamoto et al., 2002
Peach	Ferjalou Jalousia® × Fantasia (J × F)	270	712	Dirlewanger et al., 1998
	Almond Texas × Peach Earlygold (T × E)	141	538	Lambert et al., 2004
	Ferjalou Jalousia® × Fantasia (J × F)	181	621.2	Dirlewanger et al., 2006
Grape	Syrah × Pinot Noir	994	1245	Troggio et al., 2007
Strawberry	*F. vesca × F. nubicola*	182	424	Sargent et al., 2006
Tomato	*S. lycopersicum × S. pimpinellifolium*	250	1002.4	Sharma et al., 2008
Melon	*Védrantais* × PI 414723	318	1411	Perin et al., 2002
Papaya	SunUp × AU9	707	1069.9	Chen et al., 2007
Banana	Borneo × Pisang Lilin	489	1197	Hippolyte et al., 2010
Sweet Cherry	Emperor Francis (partial map)	80	503.3	Stockinger et al., 1996
Kiwifruit	Diploid *A. chinensis*	644	2078	Fraser et al., 2009

published (Figure 8.1). Moreover, clementine genome sequencing has also been completed, but the related publications are not yet published (Table 8.2); finally, more are on the way, such as sweet cherry (*Prunus avium*), European plum (*Prunus domestica*), and red raspberry (*Rubus idaeus*) (http://genomevolution.org/wiki/index.php/Sequenced_plant_genomes).

Tomato is the model plant for fleshy fruit development and climacteric ripening (The Tomato Genome Consortium 2012). The sequencing initiative International Solanaceae Genome Project was undertaken by ten countries in November 2003. Finally, genome sequencing was completed using an inbred cultivar, Heinz 1706, by the Tomato Genome Consortium and the genome article was published in *Nature* (The Tomato Genome Consortium 2012). As regards the fruit, interesting findings are the reduced numbers of genes belonging to cytochrome P450 subfamilies and the limited expression of the remaining ones in the fruit, thus avoiding the accumulation of toxic alkaloids, and the expansion and neofunctionalization of genes controlling fruit characteristics, such as fleshiness, pigment production, and softening.

Despite the fact that the "fruit" model system is tomato, the first sequenced "fruit" genome was the grapevine PN40024 genotype, derived by successive selfings of *Pinot noir* to reduce the heterozygosity of this cultivar and thus facilitate the assembly of a high-quality genome obtained through a whole-genome shotgun (WGS) approach. Its genome sequencing was started in December 2005 by the French–Italian Public Consortium for Grapevine Genome Characterization and the first draft genome sequence, with an 8x coverage, was published in Summer 2007 (Jaillon et al., 2007). Compared to previously sequenced genomes, expansions of gene families with roles in aromatic features and the absence of recent genome duplications were reported. A 12x coverage of this genome is now available (http://www.genoscope.cns.fr/externe/GenomeBrowser/Vitis/). The grapevine genome race saw the participation of another consortium, which reported the sequence of the heterozygous variety *Pinot noir*. Velasco et al. (1997) succeeded in obtaining the sequence of a heterozygous genome by means of Sanger shotgun sequencing and

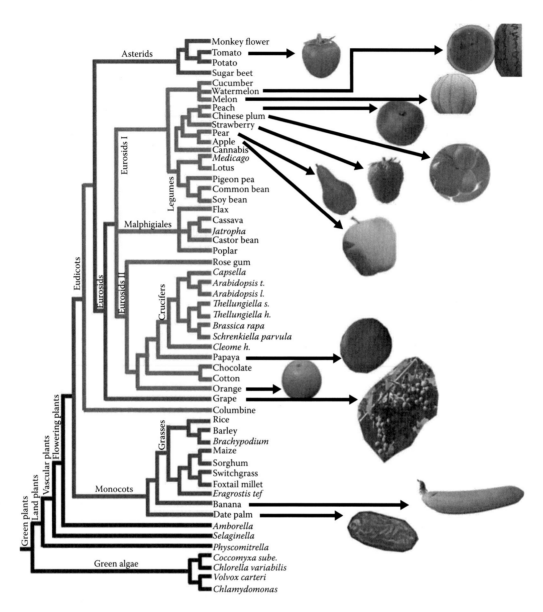

FIGURE 8.1   (See color insert)
Phylogenetic tree of the plant genomes up to 14 April 2013. Pictures on the right side of the tree represent the sequenced fruit genomes. (Modified from http://genomevolution.org/wiki/index.php/Sequenced_plant_genomes, accessed on 14th April 2013.)

highly efficient sequencing by synthesis (SBS), together with dedicated assembly programs, thus being able to obtain more than 2 million SNPs, of which more than 1.7 million were mapped in 86.7% of anchored genes.

The apple (*Malus × domestica*) genome sequencing was started on 1st September 2008 with the collaboration of 18 research institutions financially supported by the Autonomous Province of Trento, Italy, WSU Agricultural Research Center, USDA-NRI USA, and New Zealand Foundation for Research Science and Technology, and it was published two years later (Velasco et al., 2010). It was reported to have more than 57,000 genes, which, although

TABLE 8.2

Sequenced Fruit Crop Genomes on Publicly Available Databases

	Arabidopsis[a]	Tomato	Apple	Pear	Grapes	Peach	Chinese Plum	Strawberry
Genome size (Mb)	135	950	881.3	512.0	487	265	280	240
ESTs	>1.2 million	301,822	325,053		381,609	89,166	4,589	60,012
Protein-coding transcripts	27,416	42,257	95,216	42,812	30,434	28,689	25,854	34,809
Genome sequenced (year)	2000	May 2012	October 2010	November 2012	August 2007	April 2010	May 2012	December 2010
Chromosome number (2n)	$n = x = 5$	$n = x = 12$	$n = 17$	$n = 17$	$n = x = 19$	$n = x = 8$	$n = x = 8$	$n = x = 7$
Life cycle	Annual	Annual	Perennial	Perennial	Perennial	Perennial	Perennial	Perennial
Fruit type	Silique	Berry	Pome	Pome	Berry	Drupe	Drupe	Receptacle
Means of propagation	Seeds	Seeds	Vegetative	Vegetative	Vegetative	Seeds & vegetative	Seeds & vegetative	Seeds & vegetative
Database	http://www.arabidopsis.org	http://www.solgenomics.net	http://www.applegenome.org, http://www.rosaceae.org	http://peargenome.njau.edu.cn	http://www.genoscope.cns.fr/vitis, http://www.vitisgenome.it, http://www.appliedgenomics.org	http://www.phytozome.net, http://www.rosaceae.org	http://prunusmumegenome.bjfu.edu.cn	http://www.strawberrygenome.org, http://www.rosaceae.or
Reference	The Arabidopsis Genome Initiative, 2000	The Tomato Genome Consortium, 2012	Velasco et al., 2010	Wu et al., 2013	The French–Italian Public Consortium for Grapevine Genome Characterization, 2007	Verde et al., 2013	Zhang et al., 2012b	Shulaev et al., 2010g

(continued)

TABLE 8.2 (Continued)

Sequenced Fruit Crop Genomes on Publicly Available Databases

	Papaya	Melon	Watermelon	Date Palm	Banana	Sweet Orange	Clementine
Genome size (Mb)	372	375	430	380	472.2	367	296
ESTs	85,961		129,067	7.8 million	43,800	3.8 million	604,455
Protein-coding transcripts	13,311	27,427	24,444	30,684–40,378	36,542	25,376	25,385
Genome sequenced (year)	December 2007	April 2012	November 2012	May 2011	June 2012	November 2012	—
Chromosome number (2n)	$n = x = 9$	$n = x = 12$	$n = x = 11$	$n = x = 9$	$n = x = 11$	$n = x = 9$	$n = x = 9$
Life cycle	Perennial	Annual	Annual	Perennial	Annual	Perennial	Perennial
Fruit type	Berry	Berry	Pepo	Berry	Berry	Hesperidium	Hesperidium
Means of propagation	Seeds	Seeds	Seeds	Seeds & vegetative	Vegetative	Seeds & vegetative	Seeds & vegetative
Database	http://www.papayabase.org	https://melonomics.net	http://www.iwgi.org	http://qatar-weill.cornell.edu/research/datepalmGenome/	http://www.musagenomics.org	http://citrus.hzau.edu.cn/orange/	http://www.phytozome.net/clementine.php
Reference	Ming et al., 2008	Garcia-Mas et al., 2012	Guo et al., 2012	Al-Dous et al., 2011	D'Hont et al., 2012	Xu et al., 2013	—

[a] This is not a fruit species but it was included as a model plant.

probably an overestimation (Wu et al., 2013), is the maximum among other sequenced plant genomes, with a set of 992 genes involved in disease resistance, an expansion of the StMADS11-like subclade, belonging to the type II MADS-box gene family and probably responsible for the formation of the typical fruit of the Pyreae tribe, the pome, and an expansion of gene families related to sorbitol metabolism, that is the form in which photosynthesis-derived carbohydrates are transported in Rosaceae. The sequencing project also developed a very large set of SNP markers (around 3 million, with an average polymorphism rate within the domestic cultivars of 4.8 SNPs/kb, 5.7 and 9.6 SNPs/kb between Golden Delicious and the wild relatives *M. sieversii* and *M. sylvestris*, respectively.) that will be very useful in genetic studies and for molecular breeding. This achievement is also a milestone for other pome fruit species such as pear, whose genome has recently been released by an international pear genome consortium headed by the Chinese group at Nanjing Agricultural University and including other Chinese, American, Japanese, and European laboratories (Wu et al., 2013). Even if more than half of the pear genome consists of repetitive sequences, the very high coverage of its sequence (194x) and the high-density genetic maps comprising 2005 SNP markers allowed the anchoring of 75.5% of the sequence to all 17 chromosomes. The genome contains 42,812 protein-coding genes, and, besides the repetitive sequences, is very similar to that of its close relative apple, from which it has been estimated to have diverged since ~5.4 to 21.5 MYA. As mentioned above, the comparative analyses of the two genomes allowed a re-estimation of apple gene number down to 45,293, which is very close to that of pear and to those of other genomes of similar sizes. Particular emphasis has been given to describing genes critical for self-incompatibility, lignified stone cells formation (a unique feature of pear fruit), sorbitol metabolism, and volatile compounds of fruit, all expanded in the pear genome.

Peach (*Prunus persica*) is the model for drupe fruit tree genomics within the Rosaceae family (Georgi et al., 2002). Its genome sequencing project was launched by the Joint Genome Institute (JGI) at the Plant and Animal Genome XV Meeting in 2007, thus aggregating several laboratories in the international consortium International Peach Genome Initiative (IPGI), which released a draft genome sequence of peach cultivar Lovell on 1st April 2010 (Verde et al., 2013). Approximately 227 Mbp, corresponding to 99.3% of the peach genome, were aligned in 8 pseudomolecules using a network of *Prunus* linkage maps, and containing 27,852 protein-coding genes.

The strawberry genome was published in December 2010 (Shulaev et al., 2010) and was a milestone for the model for nonclimacteric fruit ripening. Although it is the genome of the woodland strawberry, it will be an invaluable tool for the octoploid-cultivated strawberry. As regards fruit biology, it is worth mentioning that, among the genes overrepresented in the fruit transcriptome, 92 were unique to strawberry clusters and 84 of these were previously unidentified genes, thus posing new challenges for the study of the biology of this false fruit.

Following genome sequencing, the discovery of molecular markers along with the chromosomes becomes much easier, and could be used for future improvements to genetic maps that are useful for the improvement of quality traits. In apple the first molecular marker linkage map was constructed for *Rome Beauty* × *White Angel* (Hemmat et al., 1994). There were 253 markers in *White Angel* on 24 linkage groups, and 156 markers in *Rome Beauty* with 21 linkage groups. Later, Maliepaard et al. (1998) constructed another linkage map from apple, with a population of 152 individuals coming from the cross of cultivars *Prima* and *Fiesta* using RFLP, RAPD, isozyme, AFLP, SCAR, and microsatellite markers, and they reduced the linkage groups to 17. They also provided the position of some useful genes such as those for resistance to scab (*Vf*) and rosy leaf curling aphid (*Sd1*), fruit acidity (*Ma*), and self-incompatibility (*S*). Recently, Han et al. (2011), using SSR markers, were able to unambiguously position 470 contigs on the 17 apple linkage groups, thus anchoring the physical map onto the genetic map of the apple genome, yielding an integrated map spanning ~421 Mb. Lastly, Antanaviciute et al. (2012) constructed a comprehensive saturated linkage map (SNP-based) spanning 1282.2 cM employing the International RosBREED SNP Consortium (IRSC) Malus array by incorporating data for 2272 (out of the total 7867 on the array) SNP markers onto the map of the rootstock M432 progeny. In peach, a genetic linkage map was constructed with an F_2 population for Ferjalou Jalousia × Fantasia (*J × F*). They reported six traits that follows Mendelian inheritance (Dirlewanger et al., 2006). Furthermore, they mapped a new trait that follows Mendelian inheritance, called *trees bearing aborting fruits*, which is linked with the trait *flat shape of the fruit*. Ogundiwin et al. (2009) reported a fruit quality map for the *Prunus* genus containing 133 genes for fruit texture, pigmentation, flavor, and chilling injury resistance.

Comparative genomics is an emerging and powerful discipline of biological science that could assess common structures and functions of genes in the genomes of different species. Comparative genomics shows the degree of genome conversion between species to enable the transfer of genetic information (Celton et al., 2009). Furthermore, it could allow the use of trait introgression from wild relatives to increase the genetic diversity of cultivated crop species. The genome sequences of fruit crop species allow the development of millions of SNP markers, thus permitting breeders to develop better cultivars faster and to revisit the rich trait reservoir present in wild relatives, thereby empowering biodiversity-based breeding.

8.3 Transcriptomics

The total set of transcripts in the cell of a species is known as a transcriptome. Transcriptomics is a high-throughput approach for global gene expression analysis of various cell types, tissues, and organs at different developmental stages and after different treatments or in different genotypes. Initially, expressed sequence tags (ESTs) were used to generate transcriptomics data. Presently, several transcriptomics tools are available to predict genes and their putative functions as well as to measure the expression level of genes. Usually, a high-density DNA hybridization-based technique (microarray) needs sequence information to be applied to a given species, while sequencing-based technology (RNA sequencing, RNA-seq) can also be used on systems for which no previous sequence information data are available.

8.3.1 Expressed Sequence Tags (ESTs)

ESTs are short fragments of cDNAs, and have proved to be a powerful tool for the discovery of new genes. By using EST resources it is possible to predict the genetic basis of many divergent traits in fruit crops (Crowhurst et al., 2008), discover genes and markers for gene mapping, and estimate genetic relationships (Bell et al., 2008) for the improvement of cultivars. The EST datasets are an important resource for the development of simple sequence repeat (SSR) markers (Wang et al., 2012). In addition, EST data for a species can give the relative gene expression in different tissues (Ewing et al., 1999). To date, the largest numbers of ESTs available among fruit crops are for grapevine (Figure 8.2). The available EST resources for different fruit crops are presented in Table 8.3. The expression analysis of ESTs by da Silva et al. (2005) was aimed at identifying the role of different genes involved with berry ripening in grapevine.

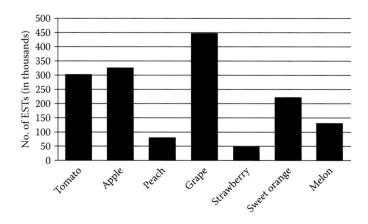

FIGURE 8.2
Overview of number of ESTs available for major fruit crops in the NCBI EST database. (From http://www.ncbi.
nlm.nih.gov/nucest, accessed on 2nd November 2012.)

In tomato, 120,892 ESTs were developed from 26 different cDNA libraries, reaching the number of 27,274 unigenes (Van der Hoeven et al., 2002). Gasic et al. (2009) reported over 300,000 ESTs in 43 cDNA libraries of apple (cv. Royal Gala) covering 34 various tissues and treatments. Sometimes molecular markers (microsatellites and simple sequence repeats) generated from ESTs can be useful tools for comparative mapping among species and evolutionary studies. In peach, a transcript map was built by mapping 1236 out of the available 9984 ESTs against a BAC library (Horn et al., 2005). Moreover, 41,519 ESTs were found from cDNA libraries from 4 different fruit and storage conditions, allowing 10,830 unigenes to be obtained (Vizoso et al., 2009) and insights to be gained into chilling-induced disorders. The first cDNA library of *Prunus mume* was developed by Li et al. (2010) using floral and fruit tissues. They took 8656 high-quality ESTs and assembled them in 4473 unigenes, of which 2981 were singletons. Furthermore, they also studied markers for polymorphism within the *Prunus* species using 14 EST-SSRs, finding 65% of marker polymorphisms in plum, and only 46% in peach. As regards strawberry, 47,952 ESTs for *Fragaria vesca* and 10,988 ESTs for *Fragaria × ananassa* are available on the NCBI EST-database that makes up 7096 unigenes in *F. × ananassa* (Bombarely et al., 2010). Although not strictly related to fruit biology, Rivarola et al. (2011) developed 41,430 ESTs in *Fragaria vesca* from plants treated with different abiotic stress conditions (water, temperature, and osmotic).

8.3.2 Microarray

Microarray is a hybridization-based large-scale gene expression profiling technique. After collection of EST data, the macroarray (nylon filter-based) (Trainotti et al., 2003; Srivastava et al., 2010) technique was used for gene expression profiling, but shortly after this the emergence of a technique with high-density probes arranged on a glass slide (microarray) became the most popular for global gene expression profiling in different organisms. Currently several microarray platforms are available from different vendors (Affymetrix, Agilent, Illumina, and Nimblegen, just to cite a few). The amount of microarray transcriptomics data has increased rapidly for different fruit crops in recent years, and, even though the new RNA-seq seems the technology that better fits transcriptomics projects, it will probably also still be a valuable research tool in the near future due to its

TABLE 8.3

Available EST Resources for Different Fruit Species in Publicly Available Databases

Common Name	Botanical Name	Transcript Assemblies			Gene Indices		
		ESTs + cDNAs	Assemblies	Singletons	ESTs + cDNAs	TC sequences	Singletons
Thale-cress[a]	*Arabidopsis thaliana*	682,040	27,983	120,385	1,104,399	49,009	63,818
Tomato	*Solanum lycopersicum*	251,814	21,523	32,268	302,789	28,167	24,335
Cultivated apple	*Malus × domestica*	250,228	26,757	50,847	—	—	—
Peach	*Prunus persica*	65,979	6,596	21,845	78,633	9,633	16,454
Grape	*Vitis vinifera*	313,050	21,627	40,880	382,717	34,634	47,495
Sour cherry	*Prunus cerasus*	1,252	115	824	—	—	—
American plum	*Prunus armeniaca*	15,032	2,372	3,571	—	—	—
Summer grape	*Vitis aestivalis*	2,093	247	1,174	—	—	—
Frost grape	*Vitis riparia*	1,949	516	529	—	—	—
Calloose grape	*Vitis shuttleworthii*	10,704	1,692	4,261	—	—	—
American blueberry	*Vaccinium corymbosum*	4,893	614	3,194	—	—	—
Woodland strawberry	*Fragaria vesca*	45,085	4,825	8,624	—	—	—
Sweet orange	*Citrus sinensis*	93,451	11,061	18,222	202,570	26,081	72,817
Sour orange	*Citrus aurantium*	5,048	829	2,584	—	—	—
Mandarin orange	*Citrus reticulata*	3,619	688	1,386	—	—	—
Clementine	*Citrus clementina*	61,091	5,222	43,167	117,964	32,287	10,231
Rough lemon	*Citrus macrophylla*	1,049	77	89	—	—	—
Cleopatra mandarin	*Citrus reshni*	2,840	188	2,220	—	—	—
Satsuma orange	*Citrus unshiu*	4,492	428	2,307	—	—	—
Grapefruit	*Citrus × paradisi*	8,000	1,087	3,819	—	—	—
Common melon	*Cucumis melo*	5,643	829	2,505	—	—	—
Pineapple	*Ananas comosus*	5,672	709	2,923	—	—	—
Avocado	*Persea americana*	8,758	1,109	5,591	—	—	—
Papaya	*Carica papaya*	—	—	—	74,675	9,890	31,827

Source: http://plantta.jcvi.org/cgi-bin/plantta_release.pl; http://compbio.dfci.harvard.edu/cgi-bin/tgi/gimain.pl.

[a] This is not a fruit species but it was included as a model plant.

TABLE 8.4

Transcriptomics Data Available in Tomato

Transcriptomic Data	References
Carpel development	Pascual et al., 2009
Maturity and ripening-stage specific	Srivastava et al., 2010
Fruit ripening	Karlova et al., 2011
Root response to a nitrogen-enriched soil	Ruzicka et al., 2010
Abscission-related transcriptome	Meir et al., 2010
Nematode resistance	Schaff et al., 2007
Auxin crosstalk	Swarup et al., 2002
Chromoplast gene expression	Kahlau and Bock, 2008
Ethylene and fruit development	Alba et al., 2005

lower cost and greater ease of data analysis. Here we summarize some recent important transcriptomics data on major fruit crops.

Tomato is the model plant for the study of fleshy fruit, and microarray transcriptomics is well established in respect of fruit development and ripening (climacteric), biotic and abiotic stresses, plant hormone, and secondary metabolites (Table 8.4). The process of development and ripening of the fruit has a very complex molecular basis, upon which are involved many transcription factors (TFs) as well as hormones. The plant hormone ethylene is crucial for tomato fruit ripening, as well as for other climacteric fruits (Chen et al., 2004). For example, in the tomato mutant *Never ripe* (*Nr*), fruits do not undergo ripening and abscission, even if treated with exogenous ethylene, as a result of a change of a single amino acid in the NR ethylene receptor (Klee and Giovannoni, 2011). This mutant demonstrates the significance of ethylene in climacteric fruit ripening. As in other plant systems, microarray analyses had tremendous success in unveiling the basic regulatory circuits responsible for the peculiarities of the different developmental stages in tomato. Tomato has five distinct fruit developmental stages: organogenesis, expansion, maturation, ripening, and senescence (Alba et al., 2005). Extensive microarray analyses in wild type and *Nr* fruit showed that ethylene is crucial for multiple aspects of development both prior to and during fruit ripening in tomato. Importantly, it showed that a complex network of TFs and signal transduction elements is impacted by ethylene, highlighting a complex regulatory network responsible for the regulation of fruit ripening at multiple levels. Microarray analyses were used also for the functional characterization of *APETALA2a* (*AP2a*; Karlova et al., 2011), a TF that regulates ripening of tomato fruit via regulation of ethylene biosynthesis and acts together with *COLORLESS NONRIPENING* (*CNR*) in a negative feedback loop. *AP2a* RNAi fruits showed alteration in expression of genes involved in several metabolic pathways, such as those for phenylpropanoid and carotenoid synthesis, as well as in hormone synthesis and perception. Moreover, genes involved in chromoplast differentiation and other ripening-associated processes were also differentially expressed. Nakano et al. (2012) identified the *MACROCALYX* (a MADS-box TF) gene as a key player in tomato fruit pedicel abscission zone development by microarray analyses of *JOINTLESS* fruit abscission zones.

Among fruits of the Rosaceae species, apple has the largest accumulation of microarray data from fruit development to ripening, as well as some microarray data available for floral development (Lee et al., 2007). Janssen et al. (2008) reported global gene expression in apple from floral development to ripening using eight different developmental stages. Highly expressed genes were grouped into the four most important expression patterns, as follows: floral buds, actively dividing cells, high starch containing and expanding cells, and cells during ripening. In apple, ethylene biosynthesis takes place before the enhanced production of volatiles. Ethylene controls aroma biosynthesis (Schaffer et al., 2007) and ester biosynthesis (Defilippi et al., 2005). Exogenous ethylene application results in an increase in aroma production and softening of the flesh (Schaffer et al., 2007). Furthermore, the treatment-induced genes were also induced during normal fruit ripening (Janssen et al., 2008). Ethylene controls aroma production at the transcriptional level, but not all genes involved in aroma synthesis are regulated by ethylene; the hormone often controls the first and always the last gene of the different pathways (Schaffer et al., 2007). Microarray has not only monitored fruit development, but also other aspects of plant biology relevant to the fruit, such as the activation of its abscission zone. This is very important in the agronomical practice of thinning, in which hormone-like chemicals are used to induce the naturally occurring fruit drop in order to better regulate fruit load, thus aiming to optimize fruit quality. In 2011, two papers described the effect of benzyladenine (BA; Botton et al., 2011) and naphthaleneacetic acid (NAA; Zhu et al., 2011) on fruitlet abscission. Botton et al. proposed a model of starvation-induced abscission sustained almost entirely by microarray data in which if starvation is perceived by the seed, the abscission program is started. More than 700 genes differentially expressed in the NAA-treated fruit abscission zone largely overlapped with those differentially expressed in shading-induced abscission, thus showing that the auxin effect was largely due to the downregulation of genes involved in photosynthesis, cell cycle, and membrane/cellular trafficking, while genes responsible for abscisic acid (ABA), ethylene biosynthesis and signaling, cell wall degradation, and programmed cell death were upregulated (Zhu et al., 2011). For pear, another pome type fruit, the major changes in expression profiles occur in two phases, such as the preclimacteric and the entrance into the climacteric period. In the first phase, transcripts encoding kinases and phosphatases were induced, while another set of genes was activated at the onset of the climacteric period (Fonseca et al., 2004).

A number of microarray transcriptomics are also available for peach fruit. The first transcriptome profiling of peach was done by Trainotti et al. (2003) by nylon macroarray. Thereafter, a glass microarray platform (μPEACH1.0) with approximately 4800 oligonucleotide probes, corresponding to a set of ESTs expressed during the last stages of fruit development, was prepared (ESTree Consortium, 2005). As with other climacteric fruit, ethylene plays an important role in peach ripening. Nonetheless, Trainotti et al. (2007) reported on auxin signaling in peach mesocarp during ripening and found that the auxin receptors, auxin response factors (ARFs), and Aux/IAA had enhanced expression during ethylene-induced ripening. Moreover, they showed an intense crosstalk between the ethylene- and auxin-regulated pathways as well as a distinctive role for auxin. A transcriptome analysis was conducted with nectarine by Ziliotto et al. (2008) on fruit treated with 1-methylcyclopropene (1-1-MCP). They compared treated fruit with untreated fruit: only 9 genes in the treated fruit, but 90 genes in untreated fruit were differentially expressed compared to harvest time. Further, the role of 1-1-MCP was clarified by the comparison of expression profiles in treated and nontreated fruit after 24 hours of incubation, when 106 genes were differentially expressed. In the case of treated fruit, genes that were previously found to be involved in the fruit-ripening process (e.g., tissue softening, color, and sugar metabolism) were differentially expressed in an opposite manner than in untreated fruit. The same platform was used by Bonghi et al. (2011) to identify the genes responsible for the seed–mesocarp relationship by transcription profiling during various developmental stages and ripening. They identified some markers for each developmental stage of seed and mesocarp. Furthermore, they translated the *Arabidopsis* hormonometer platform to peach and analyzed the microarray data, thus identifying candidates for hormone signals in the seed and mesocarp and showing that auxin, cytokinins (CK), and gibberellic acid (GA) were candidates for early fruit development, with ABA and ethylene for later stages. Comparative transcript profiling was performed between peach cv. Fantasia and apricot cv. Goldrich during fruit development by microarray by Manganaris et al. (2011). They found that 70% of genes showed a very similar expression model in both species, indicating quite a similar transcriptome in apricot and peach fruit, but also found genes with species-specific expression, such as those encoding some Aux/IAA and heat-shock proteins. Chilling injury (CI) and diseases arising in peach fruit that undergo postharvest cold storage are among the main problems for the peach industry (Figure 8.3). The ChillPeach cDNA microarray platform contains 4261 UniGene-derived probes that have been selected from an EST collection prepared from mesocarp tissues of individuals contrasting for CI susceptibility (Ogundiwin et al., 2009). It has been shown that at least 516 genes were upregulated by the CI treatment in peach mesocarp, thus showing the usefulness of both the microarray platform and the mapping population. This research also highlighted that not all of the CI-modulated genes in peach behave as their *Arabidopsis* orthologs, with some behaving in the opposite way, probably because of the specific features of the mesocarp tissue. Pirona et al. (2012) reported the development of a new array platform, particularly enriched with genes related to flavor synthesis. The comparison of the ripe fruit transcriptomes of two cultivars contrasting for their aroma properties highlighted the differential expression of 12 transcripts involved in the metabolism of phenylpropanoids, esters, lactones, and norisoprenoids during peach fruit ripening.

Grape (*Vitis vinifera*) is a nonclimacteric fruit (Chervin et al., 2004), and several authors have described the grape berry development and ripening process. Grimplet et al. (2007) analyzed grape berry transcriptomes in different berry tissues such as skin, pulp, and seed, and also observed the impact of water-deficit stress on tissue-specific expression. They found that the seed transcriptome differs significantly from those of the pulp and the skin, as the latters do

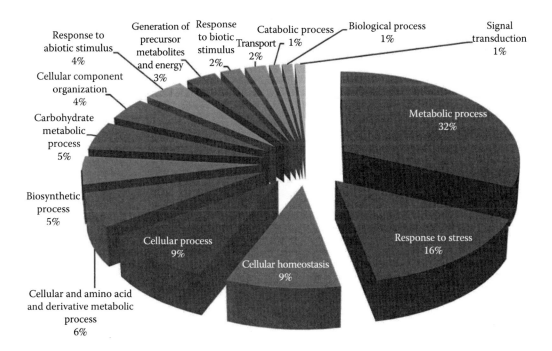

FIGURE 8.3
Functional categories of differentially accumulated proteins in peach fruit during normal fruit softening and under postharvest conditions that led to fruit chilling injury. (Modified from Nilo, R., Saffie, C., Lilley, K., Baeza-Yates, R., Cambiazo, V., Campos-Vargas, R., González, M. et al., *BMC Genomics*, 11, 1, 43, 2010.)

from each other. Genes related to flavonoid biosynthesis, pathogen resistance, and cell wall modification had higher expression levels in exocarp than in other tissues. In contrast, the maximum transcript accumulation of genes involved in cell wall functions and transport processes was in mesocarp tissues, whereas in seeds, genes related to seed storage proteins, phenylpropanoid biosynthetic enzymes, and late embryogenesis abundant proteins were prominent. Under water-deficit conditions, genes related to phenylpropanoid biosynthesis, aroma, lignin, proanthocyanin, and ethylene metabolisms were abundant at transcript level in both the skin and pulp tissues. Grimplet et al. argued that the internal ethylene modulates some metabolisms associated with berry development and ripening. Wheeler et al. (2009) reported that the ABA level during véraison has an impact on berry ripening time in grapes. Lund et al. (2008) performed affymetrix microarray analyses using 32 individual berries of *Vitis vinifera* cv. Cabernet Sauvignon and reported that VvNCED2 (ABA biosynthesis enzyme) and VvGCR2 (a putative ortholog to a reported ABA receptor) were well correlated with ripening initiation.

The first fruit microarray ever used examined strawberry gene expression and allowed the identification of a novel strawberry *alcohol acyltransferase* (*SAAT*) gene that is involved in flavor biogenesis during ripening (Aharoni et al., 2000). Aharoni and O'Connell (2002) carried out a large-scale sequencing and gene expression profiling by microarray using a set of 1701 cDNA clones to elucidate fruit-ripening process in strawberry (*F.* × *ananassa*) in two different organs: receptacle and achene. Of the 441 cDNAs differentially expressed, 259 and 182 were specific for achene and receptacle, respectively. They provide an overview of the transcriptomics of strawberry fruit development and maturation, such as signal transduction and major regulatory and metabolic pathways. They showed that ABA signal transduction pathway is involved during achene maturation. The transcript level of several cell wall-associated genes (cinnamyl alcohol dehydrogenases and cinnamoyl-CoA reductases) was abundant during the ripening stage of the lignification process of receptacle. A cDNA which is homologous to the tobacco and *Arabidopsis* ethylene response factors (ERFs) reveals that ethylene plays a role in late achene development. A transcriptome analysis was performed by Bombarely et al. (2010) using receptacles from ripe fruits of cultivated (*F.* × *ananassa*, cv. Camarosa) and wild (*F. vesca*) strawberry. They found that genes related to stress response (mostly heat-shock proteins) were more abundant among the genes upregulated in *F. vesca* (13.1%). On the other hand, genes involved in the "regulation of cellular processes" were highly represented among the genes upregulated in *F.* × *ananassa*, and most of these genes encode for proteins related to signaling processes, some of them related to hormone action (specially auxin).

As regards the pepo fruit type, in watermelon Wechter et al. (2008) reported that metabolisms related to fruit ripening deal with the vascular system and ethylene, thus supporting the importance of the gaseous hormone for the ripening of nonclimacteric fruits also. In melon (*Cucumis melo* L.), Mascarell-Creus et al. (2009) showed global gene expression profiles during fruit ripening, during which deregulation in ethylene signaling, sugar metabolism, and cell wall-loosening enzymes occurs.

8.3.3 RNA-seq

A recent advancement in transcriptomics is RNA sequencing (RNA-seq): that is, the usage of a high-throughput sequencing technique (deep sequencing) for cDNA-derived templates, which will be a better alternative to microarrays for gene expression profiling (Marioni et al., 2008). Using this novel technique, the detection of transcript levels is much more accurate (Wang et al., 2009), also allowing the detection of short RNAs (sRNAs). Weber et al. (2007) used this novel deep sequencing technique in *Arabidopsis* and found

17,449 gene loci with larger coverage of the transcriptome. Currently, few RNA-seq-based transcriptomics data are available for a limited number of fruit crops, but the gap will be bridged very quickly in the near future.

In tomato, Moxon et al. (2008) performed high-throughput pyrosequencing for the identification of both conserved and nonconserved micro RNAs (miRNAs) and other sRNAs in fruit and leaf and found several conserved miRNAs showing tissue-specific expression, indicating that miRNAs might play a role in fleshy fruit development. They further showed four novel nonconserved miRNAs, and one of the validated targets belonged to the CTR family related to fruit ripening. They suggested that fruit development and ripening might be under miRNA regulation. Also, Zuo et al. (2012) showed the significance of miRNAs in fruit ripening and ethylene response.

Xia et al. (2012) studied miRNAs and their targets and expression in apple. As previously reported by Yu et al. (2011), the most important sRNA targets are TFs. They showed that, besides conserved canonical miRNAs, apple has evolved a variety of specific miRNAs with distinct expression patterns and these miRNAs target several genes related to various functions. They also reported an additional short *MdTAS3* family, target of miR390, and thus apple ARFs might play a more complicated role in the auxin signaling pathway. Moreover, they found two important distinct regulatory networks in apple. These are direct miRNAs targeting a large number of *MYB*s, and miR828-activated and *MYB*-derived siRNA-cascaded targeting of 77 genes primarily outside the *MYB* family, which has not yet been reported in other species (Xia et al., 2012).

In peach, Zhu et al. (2012) reported on RNA sequencing and identified 47 peach-specific and 47 known miRNAs or families. Three of the miRNAs collectively target 49 MYBs, 19 of which were involved in phenylpropanoid metabolism, suggesting that they were involved in stone hardening and fruit pigmentation. These results indicated that miRNAs might regulate the fruit development and ripening of peach. Barakat et al. (2012) and Colaiacovo et al. (2012) found hundreds of conserved and nonconserved miRNAs and their target genes in peach. Barakat and coworkers also identified a number of cold-responsive (CR) miRNAs and their predicted gene targets present in CR-QTLs and BD-QTLs (BD: bloom date). Colaiacovo's work places more emphasis on isomiRs, that is, the population of variants of miRNAs coming from the same precursors, which potentially extend the species' miRNome. These novel miRNA datasets open the avenue for other studies more specifically related to fruit biology.

Zenoni et al. (2010), starting from the cDNAs of three fruit developmental stages, generated over 59 million short sequence reads by massively parallel sequencing (RNA-seq) in *Vitis vinifera* cv. Corvina and identified 17,324 genes expressed at berry development, of which 6695 were expressed in a stage-specific manner. They suggested differences in expression for genes in several functional categories and a significant transcriptional complexity. The integration of microarray and RNA-seq analyses gave a comprehensive transcriptome map in grape, during both the reproductive and vegetative phases (Fasoli et al., 2012). The 54 samples were grouped according to maturity rather than organ identity, with vegetative/green samples separated from mature/woody ones.

It has to be mentioned that many RNA-seq reports are related to general plant biology rather than solely to fruit biology, but fruit-specific RNA-seq approaches are going to be very common in the near future for nonmodel fruit crops as well. An example of this scenario is presented by a paper on Chinese berry, in which the analysis of mRNA sequence data made it possible to build an interactive pathway that highlighted an increase in energy-related metabolism, catalytic activity, and anthocyanin biosynthesis correlating with color change during the fruit-ripening process (Feng et al., 2012).

8.4 Proteomics

Proteome analysis is a useful approach for identifying the biochemical and physiological processes of fruit ripening (Prinsi et al., 2011), and comparative proteomics is also an important tool to better understand fruit softening and postharvest ripening (Nilo et al., 2010). To date, few data are available at the proteomic level in fruit crops and the available databases for plant proteomics resources are generally still missing fruit data (Table 8.5). However, this field is rapidly growing, and here we present some recent developments on proteomics in fruit development and ripening.

The first tomato fruit proteomics investigation was reported by Rose et al. (2004) who, describing the technical cues of plant proteomics, showed comparisons among ripening stages and with pepper. Rocco et al. (2006) compared two ecotypes during ripening and found some common differentially expressed proteins involved in carbon metabolism, redox status control, energy generation, defense, stress response, and cellular signaling. The first exhaustive report of a proteomics approach describing tomato during fruit growth and ripening was undertaken by Faurobert et al. (2007). They extracted total proteins from pericarp tissues at six developmental stages and separated them by two-dimensional gel electrophoresis (2-DE). They found major variations in the protein levels at different stages. During early fruit growth, protein spots were mainly related to amino acid metabolism or protein syntheses which were upregulated at the cell division stage and downregulated at later stages, suggesting that they were involved in the cell division processes. Likewise, proteins linked to photosynthesis and cell wall formation increased rapidly during the cell expansion stage. Most of the proteins related to C compounds and carbohydrate metabolism or oxidative processes were expressed at the highest rates during the maturity stage, indicating their role in the fruit-ripening processes, during which abundant proteins related to stress responses and fruit senescence were also expressed. Besides proteomics, transcriptomics and metabolomics were employed on tomato pericarp by Osorio et al. (2011), using three well-known mutants, namely *nonripening (nor)*, *ripening inhibitor (rin)*, and *Never ripe (Nr)*, and wild type at various ripening stages (39, 42, and 52 days after pollination, DAP). They identified 158 proteins that were differentially expressed at these three stages (Table 8.6). Among these stages, 52 DAP had the highest number of differentially expressed proteins in mutants (123 for *nor*, 71 for *rin*, and 29 for *Nr*) and 17 proteins were common among the three. The common proteins were related to stress response (pathogenesis-related, heat shock, glutathione S-transferase like, catalase), cell wall metabolism (polygalacturonase), hormone biosynthesis (1-aminocyclopropane-1-carboxylate oxidase), and secondary metabolism (flavonoid glucosyltransferase). Moreover, they also identified proteins (Table 8.6) whose cognate transcripts were differentially

TABLE 8.5

Some Proteome Databases Available for Plant Oganisms

Database	References	Web link
PPDB	Sun et al., 2009	http://ppdb.tc.cornell.edu
UniProt	Schneider et al., 2005	http://www.uniprot.org/program/Plants
GabiPD	Usadel et al., 2012	http://www.gabipd.org/
pep2pro	Hirsch-Hoffmann et al., 2012	http://www.pep2pro.ethz.ch
ProMEX	Hummel et al., 2007	http://promex.pph.univie.ac.at/promex/
ARAMEMNON	Schwacke et al., 2003	http://aramemnon.botanik.uni-koeln.de/

TABLE 8.6

Differentially Expressed mRNA Abundance and Related Protein Level in Tomato Mutants (*nor, rin and Nr*) at 39, 42, and 52 DAP

	mRNA Levels			Protein Levels		
Stages	*nor*	*rin*	*Nr*	*nor*	*rin*	*Nr*
39 DAP	Sn-1↓, and Sn-2↓	Sn-1↓, and Sn-2↓	No change	Sn-1↓, and Sn-2↓	Sn-1↓, and Sn-2↓	Protein levels were likely Up- or downregulated
42 DAP	SGRP-1↑	SGRP-1↑	PG 2↑, pectinesterase↑, Cys proteinase 2↑, GA 20 oxidase↑, and ACC oxidase↑	SGRP-1↑	SGRP-1↑	PG 2↑, pectinesterase↑, Cys proteinase 2↑, GA 20 oxidase↑, and ACC oxidase↓
52 DAP	PDC↓, GST-like↓, MSR↓, PG 2↓, HSP 83↓, ADH 2↓, Sn-1↑, and Sn-2↑	PDC↓, GST-like↓, MSR↓, PG 2↑, HSP 83↓, ADH 2↓, AQP↓, sHSP↑ PEP carboxylase 2↑, and ASR4↑	HSP 83↓, PDC↓, ADH 2↓, OEE protein 2↓, Sn-1↑, Sn-2↑, pectinesterase 1↑, PR protein P2↑, lipoxygenase A ↑, lipoxygenase↑, sHSP↑, PG 2↑, Sh-Rnase↓, Sn-1↑ and Sn-2↑	PDC↓, GST-like↓, MSR↓, PG 2↓, HSP 83↓, ADH 2↓, Sn-1↑, and Sn-2↑	PDC↓, GST-like↓, MSR↓, PG 2↓, HSP 83↓, ADH 2↓, AQP↓, sHSP↓ PEP carboxylase 2↑, and ASR4↑,	HSP 83↓, PDC ↓, ADH 2↓, OEE protein 2↓, Sn-1↑, Sn-2↑, pectinesterase 1↑, PR protein P2↑, lipoxygenase A ↑, lipoxygenase↑, sHSP↓, PG2↓, and Sh-Rnase↑ Sn-1↑ and Sn-2↑

Source: Modified from Osorio, S., Alba, R., Damasceno, C. M., Lopez-Casado, G., Lohse, M., Zanor, M. I., Tohge, T. et al., *Plant Physiology*, 157, 1, 405–425, 2011.

Note: ↑ and ↓ represent up- and downregulations, respectively; ADH, alcohol dehydrogenase; AQP, aquaporin-like protein; Cys proteinase, cysteine proteinase; GA *20-oxidase*, gibberellin 20 oxidase; MSR, met sulfoxide reductase; PDC, pyruvate decarboxylase; *PEP carboxylase*, phosphoenolpyruvate carboxylase; PG, poly-galacturonase; OEE protein, oxygen-evolving enhancer protein; PR protein, pathogenesis-related; sHSP, small heat-shock protein; HSP, heat-shock protein.

expressed by microarray (TOM1array), and a substantial degree of posttranscriptional regulatory activity was found during early ripening, while a substantial correlation between transcripts and proteins was found in later stages. Importantly, a strong correlation was found between ripening-related genes and specific metabolites, such as organic acids, sugars, and cell wall-derived molecules. The role of the *RIN* gene in the generation of aroma compounds has been revealed through a proteomics approach which compared WT and mutant fruits (Qin et al., 2012). RIN directly interacts with the promoters of several genes of the lipoxygenase (LOX) pathway, and its mutation in *rin* plants leads to a defect in aroma production.

To date, very little is known about apple proteomics. The first systematic protein repertoire of apple pseudocarp tissues was reported by Guarino et al. (2007) using MALDI-TOF-MS and μLC-ESI-IT-MS/MS techniques and identified 44 protein spots. These identified proteins were involved in energy production, ripening, and stress response. They also reported two proteins, Mal d1 and Mal d2, that are major apple allergens occurring in Annurca flesh. The senescence process of fruits, including apple, is associated with vibrant alterations in the mitochondrial proteome (Qin et al., 2009). Variations in mitochondrial protein levels were related to the TCA cycle, the electron transport chain, and stress response during fruit senescence. After analyzing the mitochondrial proteome of fruit stored at high and low oxygen concentrations, the latter a common practice in postharvest apple storage, it was predicted that reactive oxygen species (ROS) participate in the regulation of fruit senescence by changing the expression of specific mitochondrial proteins. The impacts of proteomes of fruits with contrasting firmness characteristics have been compared in order to gain insight into pulp texture (Marondedze and Thomas, 2012). Besides the expected differences related to ethylene metabolism, which was downregulated in the firmer genotypes, a GTP-binding protein and cytoskeleton proteins were more abundant in the firmer fruits.

As described in apple, Qin et al. (2009b) reported the oxidative damage occurring to mitochondrial proteins in peach and the impact that ROS might have in this senescence program (Qin et al., 2009a). Postharvest chilling injury of peach fruit may occur during the long cold storage periods necessary to ship fruit to distant markets. Amelioration of fruit quality is obtained with a heat treatment that can enhance self-defense potentiality. This is because of the accumulation of proteins related to stress responses, such as HSP, cysteine proteases, dehydrin, and repression of a polyphenol oxidase (Lara et al., 2009). Similarly, Zhang et al. (2011b) found 30 proteins modulated by heat treatment including HSPs, glycolytic enzymes, ascorbate peroxidase, and others involved in fruit ripening. Nilo et al. (2010) reported 43 protein spots corresponding to 18% of the total number analyzed, showing significant changes after the CI treatment. Of these, 39 have been identified by mass spectrometry (MS) and some, such as endopolygalacturonase, catalase, isocitrate dehydrogenase, pectin methylesterase, and dehydrins, were important for distinguishing between healthy and chill-injured fruit. The proteome contribution to the contrasting mesocarp textures of two peach cultivars such as Oro A (nonmelting flesh, NMF) and Bolero (melting flesh, MF) aided understanding of the biochemical and physiological processes which distinguish the shift of fruit from the nonclimacteric (unripe) to the climacteric (ripe) stage (Prinsi et al., 2011). They recognized 53 protein spots by 2-DE which were significantly changed in both unripe and ripe stages and NMF and MF. Of these, 30 were identified by LC-ESI-MS/MS, such as enzymes related to primary metabolism, ethylene biosynthesis, and proteins responsible for secondary metabolism and stress response. The ethylene biosynthesis enzyme 1-aminocyclopropane-1-carboxylic acid oxidase (ACO) was the protein with the highest change in relative abundance during the fruit shift from the preclimacteric to the climacteric stage, suggesting that ACO forms could also be modulated during the ripening process by posttranslational modifications. Moreover, ROS scavenger enzymes were more abundant in NMF compared to MF fruits, suggesting that a higher oxidative stress, associated with senescence, occurred in the latter. The effect of the genotype impact on the fruit proteome during ripening has been extended by comparing three peach and two nectarine MF varieties, one of the latter ones being a clingstone with the other four freestone (Nilo et al., 2012). The low genetic variability of peach is also reflected at the proteome level, which is conserved among varieties, with changes being mainly quantitative and not qualitative. Lastly, Zhang et al. (2012a) reported the effect of 1-MCP and ethephon on the protein profiles of peach fruit (MF cv. Huiyulu) during ripening using 2-DE and MALDI-TOF/TOF techniques. They found that differently expressed proteins are related to various metabolic pathways including energy and metabolism, cell structure, protein fate, stress response and defense, and ripening and senescence. According to them, ethephon induces the expressions of ACO, abscisic acid stress ripening-like protein

(ASR), and cell structure-related proteins, while 1-MCP inhibits the degradation of starch, weakens the metabolism of the glycolytic pathway, strongly depresses the expression of ACO and ASR, and induces the expression of chaperonin 60 and HSPs. Mesocarp and endocarp formation during the early development of peach fruit was investigated by Hu et al. (2011), who showed that 68 proteins, most of which were related to primary or secondary metabolism, followed differential expression, suggesting that enzymes involved in the lignin and flavonoid pathways obeyed the diverse spatiotemporal expression, the two being in competition.

The tremendous economic impact of grapevine cultivation has attracted a lot of attention to the elucidation of its biology, thus making this species an attractive model system. This interest resulted in one of the first plant proteomics papers being about the comparison of the ripe berry proteome of six different cultivars (Sarry et al., 2004). This survey highlighted as relevant those berry proteins having a role in energy and primary metabolisms, defense, and stress response and a considerable accumulation of dehydrin, invertase, and a putative TF, in addition to pathogenesis-related proteins such as chitinase and thaumatin-like proteins. An extensive survey of proteome changes during Nebbiolo berry development (starting one month after flowering to completely ripe stage, with a sampling interval of 10 days) highlighted that, with the ripening process, there is a general decrease of glycolysis and an increase of pathogenesis-related proteins. Moreover, contrary to what was observed in climacteric fruit, oxidative stress decreases during ripening, while extensive cytoskeleton rearrangement also takes place in this period in grape berries (Giribaldi et al., 2007). In order to understand the proteome contribution of berry skin to wine quality, this tissue was assayed at five developmental stages from véraison to full maturation in the Barbera variety (Negri et al., 2008). Contrary to what had been previously observed in the pulp (Giribaldi et al., 2007), many enzymes involved in primary metabolism were induced during ripening, including those of the last five steps of the glycolytic pathway. Moreover, many proteins related to stress responses were remarkably highly differentially regulated, indicating that this tissue is important in mediating berry protection. Most, if not all, of the proteome changes observed after véraison are under ABA control, and the hormone regulation is similar to that observed in vegetative tissues undergoing ABA-mediated stress responses (Giribaldi et al., 2010). Proteomics were added to transcriptomics and metabolomics to have a comprehensive approach toward the ripening and postharvest withering of cultivar Corvina berries (Zamboni et al., 2010). The holistic approach allowed identification of stage-specific biomarkers belonging to the three different classes of biopolymers (i.e., mRNAs, proteins, and metabolites) and the development of hypothesis-free and hypothesis-driven strategies to integrate the different kinds of omics data into biologically significant networks. The proteomic approach undertaken in cultivar Muscat Hamburg confirmed previous findings and highlighted the switch of the carbohydrate metabolism from malate during the green growing phases to hexoses during ripening (Martínez-Esteso et al., 2011). The withering processes of cv. Corvina berries were also investigated by Di Carli et al. (2011) who, beside confirming previous findings, demonstrated an active modulation of metabolic pathways throughout the slow dehydration process, including de novo protein synthesis in response to the stress condition and further evolution of physiological processes originated during ripening, thus giving scientific evidence to traditional winemaking practices.

Proteomic approaches have also been undertaken in other fruit crops during fruit development and ripening, such as strawberry (Bianco et al., 2009; Park et al., 2006), papaya (Nogueira et al., 2011; Huerta-Ocampo et al., 2012), sweet orange (Pan et al., 2009; Muccilli et al., 2009) and others, but are not discussed here due to space constraints.

8.5 Metabolomics

Fruit metabolomics is the analytical process of identifying and quantifying metabolites present in the fruit to gain knowledge of the metabolic processes occurring in the organ. Plants are the unique source of different metabolites, and over 200,000 metabolites belong to the plant kingdom (Schauer and Fernie, 2006). It is the endpoint of the omics that is closest to the phenotype (Dettmer et al., 2006) and because of that it is very attractive. The application of this area of research is also vast, including drug discovery for human health, toxicology, food quality, and functional genomics. Tomato, grape, and strawberry are the best-studied fruit systems in respect of plant metabolites (Hanhineva and Aharoni, 2009), but metabolomics is also a rapidly growing field in nonmodel crop species (several databases related to fruit metabolomics are presented in Table 8.7). Besides primary metabolism molecules and hormones, the most investigated compounds and related biosynthetic pathways are those belonging to carotenoids (Namitha et al., 2011; Günther et al., 2011), flavonoids (Winkel-Shirley, 2001), anthocyanins (Takos et al., 2006; Boss et al., 1996), and other phenolics (Cheng and Breen, 1991).

In tomato, Schauer et al. (2005) described the first comprehensive comparative analysis of the metabolite composition of leaf and fruit cells, comparing them with those of five wild species. They found a tremendous variation in over 60 quantifiable metabolites in both the leaves and fruits of the wild species. Also, in the same year, the detection of more than 300 volatile metabolites from the fruits of 94 tomato genotypes was described (Tikunov et al., 2005), thus paving the way for the forthcoming fruit metabolomics. This massive approach needs the development of common metabolite databases dedicated to liquid chromatography–mass spectrometry (LC–MS), and this has been achieved, for example, with the Metabolome Tomato Database (MoTo DB), a metabolite database specifically developed for tomato (Moco et al., 2006). As a proof of concept, the use of this tool revealed that tomato flavonoids and α-tomatine are found in the peel, whereas other alkaloids and some specific phenylpropanoids are mainly present in the flesh of tomato fruits. As it is the synthesis site of many aroma compounds, the peel has been one of the most favored tissues investigated by this approach. Mintz-Oron et al. (2008) identified 100 peel-enriched primary and secondary metabolites including glycoalkaloids and amyrin-type pentacyclic triterpenoids as well as polar metabolites related to cuticle and cell wall metabolism and protection against photooxidative stress, alkaloids, organic acids, and sugars. Furthermore, they found the upregulation of 574 transcripts in the fruit skin compared to the flesh, and, among those, many were involved in the biosynthesis of wax, cutin, and phenylpropanoids. Metabolomics has also been used to characterize the effect of the deregulation of gene expression in transgenic models, as in the case of *DE-ETIOLATED1* (*DET1*). Fruit-specific downregulation of *DET1* causes a quantitative increase in carotenoid, tocopherol, phenylpropanoids, flavonoids, and anthocyanidins and a qualitative difference within the phenolics (Enfissi et al., 2010). In recent years the integration of the omics tools became increasingly apparent, as in the above-cited study by Osorio et al. (2011), who reported that ripening-associated transcripts had a strong correlation with specific metabolite groups such as organic acids, sugars, and cell wall-related metabolites. As an example, regarding the hormonal regulation on the metabolome, it has been reported that the downregulation of the *SlLoxB* gene not only affected the methyl jasmonate levels, thus lowering also those of polyamines, but also those of many amino acids and other metabolites, thus supporting the important role of jasmonates in fruit quality (Kausch et al., 2012). Flavor is one of the important quality characteristics of fruits during ripening and this is mainly due to volatile ester compounds (Beekwilder et al., 2004). In tomato, 20–30 chemicals control the flavor, including

TABLE 8.7

Metabolomic Resources Available for Different Fruit Crops

Database	Description	References	Web Link
MoTo DB	LC–MS-based metabolome database for tomato	Moco et al., 2006	http://www.ab.wur.nl/moto/
KOMICS	LC-FT-ICR-MS-based metabolome database for tomato cultivar Microtom	Iijima et al., 2008	http://www.kazusa.or.jp/komics/en/
METACROP	Summarizes diverse information about metabolic pathways in crop plants and allows automatic export of information for the creation of detailed metabolic models	—	http://metacrop.ipk-gatersleben.de/
MassBase	Mass chromatogram-based database to develop pipeline software for extracting metabolite peaks and annotating metabolites	—	http://webs2.kazusa.or.jp/massbase/
KEEG Pathway	Large-scale datasets in genomics, transcriptomics, proteomics, and metabolomics	Ogata et al., 1999	http://www.genome.jp/kegg/pathway.html#metabolism
PNM 7.0 (grape and papaya)	Plant metabolic network database involved in annotating genomes, metabolic pathway databases	Seaver et al., 2012	http://www.plantcyc.org/about/general_information.faces
MetaCyc	Highly curated, nonredundant reference database of small-molecules metabolism for both primary and secondary metabolisms	Caspi et al., 2008	http://biocyc.org/download.shtml
LycoCyc	Tomato metabolic pathway database	—	http://solcyc.solgenomics.net/LYCO/NEW-IMAGE?type=OVERVIEW&force=t
GMD	Golm Metabolome Database provides information about mass spectral libraries; metabolite profiling experiments	Kopka et al., 2005	http://csbdb.mpimp-golm.mpg.de/gmd.html
PRIMe	Platform for RIKEN Metabolomics for the measurements and analysis tool for metabolites based on NMR, GC/MS, LC/MS, and CE/MS	Akiyama et al., 2008	http://prime.psc.riken.jp/?action = standard_index
Plant MetGenMAP	Plant metabolites	Joung et al., 2009	http://bioinfo.bti.cornell.edu/tool/MetGenMAP
KNApSAcK	Database for searching metabolites from MS peaks, molecular weights, and molecular formulas, and species	Shinbo et al., 2006	http://kanaya.aist-nara.ac.jp/KNApSAcK/
GDR	Cyc pathways of apple, peach, and strawberry	—	http://www.rosaceae.org/node/312449

chemicals derived from amino acids, fatty acids, and carotenoids (Klee and Giovannoni, 2011). Tomato fruit liking has been investigated with a combination of targeted metabolomics and the exploitation of natural variation providing novel insights into flavor chemistry, the interactions between taste and retronasal olfaction, and how aroma volatiles make contributions to perceived sweetness independent of sugar concentration, thus suggesting novel strategies to increase the perception of sweetness without adding sugar (Tieman et al., 2012).

In apple, the integration of the omics data is not yet as developed as for tomato and grape. So far, only papers describing the analyses of metabolomics data and related to postharvest storage are available. Prestorage irradiation with UV-light altered the apple fruit metabolic pathways associated with ethylene biosynthesis, acid metabolism, flavonoid biosynthesis, and fruit texture, as demonstrated by alterations in more than 200 compounds, of which 78 could be identified (Rudell et al., 2008). The same group also investigated the effect of superficial scald development, associating several compounds with the disease before the appearance of the symptoms (Rudell et al., 2009). In the progress of developing metabolomics for apple, Aprea et al. (2011) set up protocols, software, and a library for compound identification. Lee et al. (2012) reported that, during hypoxic controlled atmosphere storage, amino acids and volatile metabolites were affected by 1-MCP treatment while carbohydrates and organic acids were not. The 1-MCP treatment-induced flesh-browning symptoms and metabolic changes were associated with the development of the disease. Lastly, the genotype effect of six commercially grown apple cultivars was assessed by GC–MS. Apple metabolites directly correlated with different fruit quality traits, for example antioxidant activity, total phenolics, and total anthocyanins, thus paving the way for the elucidation of these quality traits (Cuthbertson et al., 2012).

Peach is a stone fruit producing a fleshy mesocarp, and several biochemical changes occur from fruit development to ripening. Lombardo et al. (2011) carried out an extensive GC–MS metabolomics profiling of peach fruit (cv. Dixiland) from early developmental stages to postharvest ripening. They showed the metabolism of sugars and sugar alcohols, organic acid (fatty acid, caffeoylquinic acid, and putrescine), and N_2-containing compounds. At harvest, the most abundant sugar is glucose, followed by sucrose and fructose. But, while the two hexoses showed maximum concentration at the S3 stage, sucrose steadily grew up to S4 and remained constant during postharvest. The alcohol sugar sorbitol had an accumulation profile similar to the two hexoses, and those profiles were also confirmed at the molecular level with the expression of mRNAs related to their metabolisms. The expression profiles of catabolic and anabolic sucrose-related genes suggest the cycling of sugar during postharvest. A step decrease in amino acid concentration has been observed at harvest, suggesting that they might be substrates for the respiratory burst associated with ripening. The fraction of volatile compounds has been assessed by HS–SPME–GC–MS with a nontargeted approach using developing and ripening fruit of two contrasting genotypes, ripe fruit differing for genetic background and location, and fruits after postharvest (Sánchez et al., 2012). Alcohols, ketones, aldehydes, esters, lactones, carboxylic acids, phenolics, and terpenoids for a total of 110 compounds were identified and quantified in peach, showing a high degree of coregulation within groups, thus unveiling the good orchestration of the synthesis of aroma and flavor compounds during peach ripening.

Metabolomics in grape took advantage of the techniques developed to trace chemical compounds in wine. Those techniques were developed as analytical tools against fraud, but are also used for better characterization of the peculiarities of different wines (Son et al., 2008). Thus (1)H NMR spectroscopy, associated with pattern recognition methods, was used to characterize grape metabolomes, though mainly at a qualitative level, and showed that environmental vineyard conditions can affect the chemical composition of grapes and their wines (Son et al., 2009). In a holistic approach (i.e., mRNAs, proteins, and metabolites), more than 100 metabolites have been identified and quantified by HPLC–MS. As mentioned above, this extensive profiling, besides giving in-depth details about grape berry biology, allowed the identification of stage-specific biomarkers belonging to the three different classes of biopolymers (Zamboni et al., 2010). Lastly, Fortes et al. (2011) provided information about the main metabolic profile changes of grapes during ripening, associating them with parallel changes of the transcriptome over two vintages in Trincadeira cultivar. They found many novel ripening-related genes which were involved in carbohydrate metabolism, amino acid metabolisms, and epigenetic factors as well as signaling pathways of phytohormones including receptors, TFs, and kinases.

Strawberry is a model for the ripening of nonclimacteric fruit and is rich in secondary metabolites. Fait et al. (2008) analyzed primary and secondary metabolites using GC–MS and LC–MS both in achenes and receptacles during development and ripening of the edible part usually called "fruit." They found that receptacle content in sugars and sugar derivatives increased with ripening, while the achene had lower levels of carbon and nitrogen-rich compounds, but for those metabolites related to storage. Proanthocyanidins and flavonol derivatives were abundant at early development, whereas anthocyanins were abundant during red ripening in the receptacle. On the other hand, ellagitannin and flavonoid levels were high at early and late developmental stages, respectively. The volatile components of the strawberry fruit metabolome were investigated with an untargeted approach by GC–MS in order to gain insight into flavor formation (Zhang et al., 2011a). This analysis showed the high degree of connection among individual metabolites and metabolic pathways and the correlation between primary and secondary metabolisms.

The recent developments in fruit metabolomics could be an opportunity for fruit crop breeders to improve specific quality traits by exploring fruit genetic resources as well as potential marker development for marker-assisted molecular plant breeding. As was recently shown for tomato (DiLeo et al., 2011), the integration of the omics data is not trivial and the choice of different statistical approaches can have different strengths in the extraction of biologically meaningful data.

References

Aharoni, A., L. C. P. Keizer, H. J. Bouwmeester, Z. Sun, M. Alvarez-Huerta, H. A. Verhoeven, J. Blaas, et al. 2000. Identification of the SAAT gene involved in strawberry flavor biogenesis by use of DNA microarrays. *The Plant Cell* 12 (5): 647–661.

Aharoni, A. and A. P. O'Connell. 2002. Gene expression analysis of strawberry achene and receptacle maturation using DNA microarrays. *Journal of Experimental Botany* 53 (377): 2073–2087.

Akiyama, K., E. Chikayama, H. Yuasa, Y. Shimada, T. Tohge, K. Shinozaki, M. Y. Hirai, T. Sakurai, J. Kikuchi, and K. Saito. 2008. PRIMe: A web site that assembles tools for metabolomics and transcriptomics. *In Silico Biology* 8 (3): 339–345.

Alba, R., P. Payton, Z. Fei, R. McQuinn, P. Debbie, G. B. Martin, S. D. Tanksley, and J. J. Giovannoni. 2005. Transcriptome and selected metabolite analyses reveal multiple points of ethylene control during tomato fruit development. *The Plant Cell* 17 (11): 2954–2965.

Al-Dous, E. K., B. George, M. E. Al-Mahmoud, M. Y. Al-Jaber, H. Wang, Y. M. Salameh, E. K. Al-Azwani, et al. 2011. De novo genome sequencing and comparative genomics of date palm (*Phoenix dactylifera*). *Nature Biotechnology* 29 (6): 521–527.

Antanaviciute, L., F. Fernández-Fernández, J. Jansen, E. Banchi, K. M. Evans, R. Viola, R. Velasco, J. M. Dunwell, M. Troggio, and D. J. Sargent. 2012. Development of a dense SNP-based linkage map of an apple rootstock progeny using the *Malus infinium* whole genome genotyping array. *BMC Genomics* 13 (1): 203.

Aprea, E., H. Gika, S. Carlin, G. Theodoridis, U. Vrhovsek, and F. Mattivi. 2011. Metabolite profiling on apple volatile content based on solid phase microextraction and gas-chromatography time of flight mass spectrometry. *Journal of Chromatography A* 1218 (28): 4517–4524.

Barakat, A., A. Sriram, J. Park, T. Zhebentyayeva, D. Main, and A. Abbott. 2012. Genome wide identification of chilling responsive microRNAs in *Prunus persica*. *BMC Genomics* 13 (1): 481.

Beekwilder, J., M. Alvarez-Huerta, E. Neef, F. W. A. Verstappen, H. J. Bouwmeester, and A. Aharoni. 2004. Functional characterization of enzymes forming volatile esters from strawberry and banana. *Plant Physiology* 135 (4): 1865–1878.

Bell, D. J., L. J. Rowland, J. J. Polashock, and F. A. Drummond. 2008. Suitability of EST-PCR markers developed in highbush blueberry for genetic fingerprinting and relationship studies in lowbush blueberry and related species. *Journal of the American Society for Horticultural Science* 133 (5): 701–707.

Bianco, L., L. Lopez, A. G. Scalone, M. Di Carli, A. Desiderio, E. Benvenuto, and G. Perrotta. 2009. Strawberry proteome characterization and its regulation during fruit ripening and in different genotypes. *Journal of Proteomics* 72 (4): 586–607.

Bombarely, A., C. Merchante, F. Csukasi, E. Cruz-Rus, J. L. Caballero, N. Medina-Escobar, R. Blanco-Portales, et al. 2010. Generation and analysis of ESTs from strawberry (*Fragaria xananassa*) fruits and evaluation of their utility in genetic and molecular studies. *BMC Genomics* 11 (1): 503.

Bonghi, C., L. Trainotti, A. Botton, A. Tadiello, A. Rasori, F. Ziliotto, V. Zaffalon, G. Casadoro, and A. Ramina. 2011. A microarray approach to identify genes involved in seed-pericarp cross-talk and development in peach. *BMC Plant Biology* 11 (1): 107.

Boss, P. K., C. Davies, and S. P. Robinson. 1996. Expression of anthocyanin biosynthesis pathway genes in red and white grapes. *Plant Molecular Biology* 32 (3): 565–569.

Botton, A., G. Eccher, C. Forcato, A. Ferrarini, M. Begheldo, M. Zermiani, S. Moscatello, et al. 2011. Signaling pathways mediating the induction of apple fruitlet abscission. *Plant Physiology* 155 (1): 185–208.

Caspi, R., H. Foerster, C. A. Fulcher, P. Kaipa, M. Krummenacker, M. Latendresse, S. Paley, et al. 2008. The MetaCyc database of metabolic pathways and enzymes and the BioCyc collection of Pathway/Genome databases. *Nucleic Acids Research* 36 (suppl 1): D623–D631.

Celton, J. M., D. Chagné, S. D. Tustin, S. Terakami, C. Nishitani, T. Yamamoto, and S. E. Gardiner. 2009. Update on comparative genome mapping between *Malus* and *Pyrus*. *BMC Research Notes* 2 (1): 182.

Chen, C., Q. Yu, S. Hou, Y. Li, M. Eustice, R. L. Skelton, O. Veatch, et al. 2007. Construction of a sequence-tagged high-density genetic map of papaya for comparative structural and evolutionary genomics in Brassicales. *Genetics* 177 (4): 2481–2491.

Chen, G., L. Alexander, and D. Grierson. 2004. Constitutive expression of EIL-like transcription factor partially restores ripening in the ethylene-insensitive *nr* tomato mutant. *Journal of Experimental Botany* 55 (402): 1491–1497.

Cheng, G. W. and P. J. Breen. 1991. Activity of phenylalanine ammonia-lyase (PAL) and concentrations of anthocyanins and phenolics in developing strawberry fruit. *Journal of the American Society for Horticultural Science* 116 (5): 865–869.

Chervin, C., A. El-Kereamy, J. P. Roustan, A. Latché, J. Lamon, and M. Bouzayen. 2004. Ethylene seems required for the berry development and ripening in grape, a non-climacteric fruit. *Plant Science* 167 (6): 1301–1305.

Colaiacovo, M., L. B. I. Centomani, C. Crosatti, L. Giusti, L. Orrù, G. Tacconi, A. Lamontanara, L. Cattivelli, and P. Faccioli. 2012. A survey of microRNA length variants contributing to miR-Nome complexity in peach (*Prunus persica* L.). *Frontiers in Plant Science* 3: 165.

Craig, W. and L. Beck. 1999. Phytochemicals: Health protective effects. *Canadian Journal of Dietetic Practice and Research* 60 (2): 78–84.

Crowhurst, R. N., A. P. Gleave, E. A. MacRae, C. Ampomah-Dwamena, R. G. Atkinson, L. L. Beuning, S. M. Bulley, et al. 2008. Analysis of expressed sequence tags from *Actinidia*: Applications of a cross species EST database for gene discovery in the areas of flavor, health, color and ripening. *BMC Genomics* 9 (1): 351.

Cuthbertson, D. J., P. K. Andrews, J. P. Reganold, N. M. Davies, and B. M. Lange. 2012. Utility of metabolomics toward assessing the metabolic basis of quality traits in apple fruit with an emphasis on antioxidants. *Journal of Agricultural and Food Chemistry* 60 (65): 8552–8560.

da Silva, F.G., A. Iandolino, F. Al-Kayal, M. C. Bohlmann, M. A. Cushman, H. Lim, A. Ergul, et al. 2005. Characterizing the grape transcriptome. Analysis of expressed sequence tags from multiple *Vitis* species and development of a compendium of gene expression during berry development. *Plant Physiology* 139 (2): 574–597.

Defilippi, B. G., A. A. Kader, and A. M. Dandekar. 2005. Apple aroma: Alcohol acyltransferase, a rate limiting step for ester biosynthesis, is regulated by ethylene. *Plant Science* 168 (5): 1199–1210.

Dettmer, K., P. A. Aronov, and B. D. Hammock. 2006. Mass spectrometry-based metabolomics. *Mass Spectrometry Reviews* 26 (1): 51–78.

D'Hont, A., F. Denoeud, J. M. Aury, F. C. Baurens, F. Carreel, O. Garsmeur, B. Noel, et al. 2012. The banana (*Musa acuminata*) genome and the evolution of monocotyledonous plants. *Nature* 488: 213–217.

Di Carli, M., A. Zamboni, M. E. Pè, M. Pezzotti, K. S. Lilley, E. Benvenuto, and A. Desiderio. 2011. Two-dimensional differential in gel electrophoresis (2D-DIGE) analysis of grape berry proteome during postharvest withering. *Journal of Proteome Research* 10 (2): 429–446.

DiLeo, M. V., G. D. Strahan, M. den Bakker, and O. A. Hoekenga. 2011. Weighted correlation network analysis (WGCNA) applied to the tomato fruit metabolome. *PLoS ONE* 6 (10): e26683.

Dirlewanger, E., E. Graziano, T. Joobeur, F. Garriga-Calderé, P. Cosson, W. Howad, and P. Arús. 2004. Comparative mapping and marker-assisted selection in Rosaceae fruit crops. *Proceedings of the National Academy of Sciences of the United States of America* 101 (26): 9891–9896.

Dirlewanger, E., P. Cosson, K. Boudehri, C. Renaud, G. Capdeville, Y. Tauzin, F. Laigret, and A. Moing. 2006. Development of a second-generation genetic linkage map for peach [*Prunus persica* (L.) Batsch] and characterization of morphological traits affecting flower and fruit. *Tree Genetics & Genomes* 3 (1): 1–13.

Dirlewanger, E., V. Pronier, C. Parvery, C. Rothan, A. Guye, and R. Monet. 1998. Genetic linkage map of peach [*Prunus persica* (L.) Batsch] using morphological and molecular markers. *Theoretical and Applied Genetics* 97 (5): 888–895.

Enfissi, E. M. A., F. Barneche, I. Ahmed, C. Lichtlé, C. Gerrish, R. P. McQuinn, J. J. Giovannoni, et al. 2010. Integrative transcript and metabolite analysis of nutritionally enhanced DE-ETIOLATED1 downregulated tomato fruit. *The Plant Cell Online* 22 (4): 1190–1215.

ESTree Consortium. 2005. Development of an oligo-based microarray (µPEACH 1.0) for genomics studies in peach fruit. *Acta Horticulturae* 682: 263–268.

Ewing, R. M., A. B. Kahla, O. Poirot, F. Lopez, S. Audic, and J. M. Claverie. 1999. Large-scale statistical analyses of rice ESTs reveal correlated patterns of gene expression. *Genome Research* 9 (10): 950–959.

Fait, A., K. Hanhineva, R. Beleggia, N. Dai, I. Rogachev, V. J. Nikiforova, A. R. Fernie, and A. Aharoni. 2008. Reconfiguration of the achene and receptacle metabolic networks during strawberry fruit development. *Plant Physiology* 148 (2): 730–750.

Fasoli, M., S. Dal Santo, S. Zenoni, G. B. Tornielli, L. Farina, A. Zamboni, A. Porceddu, et al. 2012. The grapevine expression atlas reveals a deep transcriptome shift driving the entire plant into a maturation program. *The Plant Cell* 24 (9): 3489–3505.

Faurobert, M., C. Mihr, N. Bertin, T. Pawlowski, L. Negroni, N. Sommerer, and M. Causse. 2007. Major proteome variations associated with cherry tomato pericarp development and ripening. *Plant Physiology* 143 (3): 1327–1346.

Feng, C., M. Chen, C. Xu, L. Bai, X. Yin, X. Li, A. C. Allan, I. B. Ferguson, and K. Chen. 2012. Transcriptomic analysis of Chinese bayberry (*Myrica rubra*) fruit development and ripening using RNA-seq. *BMC Genomics* 13 (1): 19.

Fiehn, O. 2002. Metabolomics–the link between genotypes and phenotypes. *Plant Molecular Biology* 48 (1): 155–171.

Fonseca, S., L. Hackler, Á. Zvara, S. Ferreira, A. Baldé, D. Dudits, M. S. Pais, and L. G. Puskás. 2004. Monitoring gene expression along pear fruit development, ripening and senescence using cDNA microarrays. *Plant Science* 167 (3): 457–469.

Fortes, A. M., P. Agudelo-Romero, M. S. Silva, K. Ali, L. Sousa, F. Maltese, Y. H. Choi, et al. 2011. Transcript and metabolite analysis in Trincadeira cultivar reveals novel information regarding the dynamics of grape ripening. *BMC Plant Biology* 11 (1): 149.

Fraser, L. G., G. K. Tsang, P. M. Datson, H. N. De Silva, C. F. Harvey, G. P. Gill, R. N. Crowhurst, and M. A. McNeilage. 2009. A gene-rich linkage map in the dioecious species *Actinidia chinensis* (kiwifruit) reveals putative X/Y sex-determining chromosomes. *BMC Genomics* 10 (1): 102.

Garcia-Mas, J., A. Benjak, W. Sanseverino, M. Bourgeois, G. Mir, V. M. González, E. Hénaff, et al. 2012. The genome of melon (*Cucumis melo* L.). *Proceedings of the National Academy of Sciences* 109 (29): 11872–11877.

Gasic, K., Y. Han, S. Kertbundit, V. Shulaev, A. F. Iezzoni, E. W. Stover, R. L. Bell, M. E. Wisniewski, and S. S. Korban. 2009. Characteristics and transferability of new apple EST-derived SSRs to other rosaceae species. *Molecular Breeding* 23 (3): 397–411.

Georgi, L., Y. Wang, D. Yvergniaux, T. Ormsbee, M. Inigo, G. Reighard, and A. Abbott. 2002. Construction of a BAC library and its application to the identification of simple sequence repeats in peach [*Prunus persica* (L.) Batsch]. *Theoretical and Applied Genetics* 105 (8): 1151–1158.

Giribaldi, M., I. Perugini, F. X. Sauvage, and A. Schubert. 2007. Analysis of protein changes during grape berry ripening by 2 DE and MALDI TOF. *Proteomics* 7 (17): 3154–3170.

Giribaldi, M., L. Gény, S. Delrot, and A. Schubert. 2010. Proteomic analysis of the effects of ABA treatments on ripening *Vitis vinifera* berries. *Journal of Experimental Botany* 61 (9): 2447–24458.

Grimplet, J., L. G. Deluc, R. L. Tillett, M. D. Wheatley, K. A. Schlauch, G. R. Cramer, and J. C. Cushman. 2007. Tissue-specific mRNA expression profiling in grape berry tissues. *BMC Genomics* 8 (1): 187.

Guarino, C., S. Arena, L. De Simone, C. D'Ambrosio, S. Santoro, M. Rocco, A. Scaloni, and M. Marra. 2007. Proteomic analysis of the major soluble components in Annurca apple flesh. *Molecular Nutrition & Food Research* 51 (2): 255–262.

Günther, C. S., C. Chervin, K. B. Marsh, R. D. Newcomb, and E. J. F. Souleyre. 2011. Characterisation of two alcohol acyltransferases from kiwifruit (*Actinidia spp.*) reveals distinct substrate preferences. *Phytochemistry* 72 (8): 700–710.

Guo, S., J. Zhang, H. Sun, J. Salse, W. J. Lucas, H. Zhang, Y. Zheng, et al. 2012. The draft genome of watermelon (*Citrullus lanatus*) and resequencing of 20 diverse accessions. *Nature Genetics* 45 (1): 51–58.

Han, Y., D. Zheng, S. Vimolmangkang, M. A. Khan, J. E. Beever, and S. S. Korban. 2011. Integration of physical and genetic maps in apple confirms whole-genome and segmental duplications in the apple genome. *Journal of Experimental Botany* 62 (14): 5117–5130.

Hanhineva, K. and A. Aharoni. 2009. Metabolomics in fruit development. *Molecular Techniques in Crop Improvement*, Springer Science; Berlin, Germany, 675–693.

Hemmat, M., N. F. Weedon, A. G. Manganaris, and D. M. Lawson. 1994. Molecular marker linkage map for apple. *Journal of Heredity* 85 (1): 4–11.

Hippolyte, I., F. Bakry, M. Seguin, L. Gardes, R. Rivallan, A. M. Risterucci, C. Jenny, et al. 2010. A saturated SSR/DArT linkage map of *Musa acuminata* addressing genome rearrangements among bananas. *BMC Plant Biology* 10 (1): 65.

Hirsch-Hoffmann, M., W. Gruissem, and K. Baerenfaller. 2012. pep2pro: The high-throughput proteomics data processing, analysis, and visualization tool. *Frontiers in Plant Science* 3: 123.

Horn, R., A. C. Lecouls, A. Callahan, A. Dandekar, L. Garay, P. McCord, W. Howad, et al. 2005. Candidate gene database and transcript map for peach, a model species for fruit trees. *Theoretical and Applied Genetics* 110 (8): 1419–14128.

Hu, H., Y. Liu, G. L. Shi, Y. P. Liu, R. J. Wu, A. Z. Yang, Y. M. Wang, B. G. Hua, and Y. N. Wang. 2011. Proteomic analysis of peach endocarp and mesocarp during early fruit development. *Physiologia Plantarum* 142 (4): 390–406.

Huerta-Ocampo, J. Á., J. A. Osuna-Castro, G. J. Lino-López, A. Barrera-Pacheco, G. Mendoza-Hernández, A. De León-Rodríguez, and A. P. Barba de la Rosa. 2012. Proteomic analysis of differentially accumulated proteins during ripening and in response to 1-MCP in papaya fruit. *Journal of Proteomics.* 75 (7): 2160–2169.

Hummel, J., M. Niemann, S. Wienkoop, W. Schulze, D. Steinhauser, J. Selbig, D. Walther, and W. Weckwerth. 2007. ProMEX: A mass spectral reference database for proteins and protein phosphorylation sites. *BMC Bioinformatics* 8 (1): 216.

Iijima, Y., Y. Nakamura, Y. Ogata, K. Tanaka, N. Sakurai, K. Suda, T. Suzuki, et al. 2008. Metabolite annotations based on the integration of mass spectral information. *The Plant Journal* 54 (5): 949–962.

Jaillon, O., J. M. Aury, B. Noel, A. Policriti, C. Clepet, A. Casagrande, N. Choisne, S. Aubourg, et al. 2007. The grapevine genome sequence suggests ancestral hexaploidization in major angiosperm phyla. *Nature* 449 (7161): 463–467.

Janssen, B. J., K. Thodey, R. J. Schaffer, R. Alba, L. Balakrishnan, R. Bishop, J. H. Bowen, et al. 2008. Global gene expression analysis of apple fruit development from the floral bud to ripe fruit. *BMC Plant Biology* 8 (1): 16.

Joung, J. G., A. M. Corbett, S. M. Fellman, D. M. Tieman, H. J. Klee, J. J. Giovannoni, and Z. Fei. 2009. Plant MetGenMAP: An integrative analysis system for plant systems biology. *Plant Physiology* 151 (4): 1758–1768.

Kahlau, S., and R. Bock. 2008. Plastid transcriptomics and translatomics of tomato fruit development and chloroplast-to-chromoplast differentiation: Chromoplast gene expression largely serves the production of a single protein. *The Plant Cell* 20 (4): 856–874.

Karlova, R., F. M. Rosin, J. Busscher-Lange, V. Parapunova, P. T. Do, A. R. Fernie, P. D. Fraser, C. Baxter, G. C. Angenent, and R. A. de Maagd. 2011. Transcriptome and metabolite profiling show that APETALA2a is a major regulator of tomato fruit ripening. *The Plant Cell* 23 (3): 923–941.

Kausch, K. D., A. P. Sobolev, R. K. Goyal, T. Fatima, R. Laila-Beevi, R. A. Saftner, A. K. Handa, and A. K. Mattoo. 2012. Methyl jasmonate deficiency alters cellular metabolome, including the aminome of tomato (*Solanum lycopersicum* L.) fruit. *Amino Acids* 42 (2–3): 843–856.

Kenis, K., and J. Keulemans. 2005. Genetic linkage maps of two apple cultivars (*Malus* × *domestica* Borkh.) based on AFLP and microsatellite markers. *Molecular Breeding* 15 (2): 205–219.

Klee, H. J., and J. J. Giovannoni. 2011. Genetics and control of tomato fruit ripening and quality attributes. *Annual Review of Genetics* 45: 41–59.

Kopka, J., N. Schauer, S. Krueger, C. Birkemeyer, B. Usadel, E. Bergmüller, P. Dörmann, et al. 2005. GMD@ CSB. DB: The golm metabolome database. *Bioinformatics* 21 (8): 1635–1638.

Lambert, P., L.S. Hagen, P. Arus, and J.M. Audergon. 2004. Genetic linkage maps of two apricot cultivars (*Prunus armeniaca* L.) compared with the almond *Texas* × peach *Earlygold* reference map for *Prunus*. *Theoretical and Applied Genetics* 108 (6): 1120–1130.

Lara, M. V., J. Borsani, C. O. Budde, M. A. Lauxmann, V. A. Lombardo, R. Murray, C. S. Andreo, and M. F. Drincovich. 2009. Biochemical and proteomic analysis of 'Dixiland' peach fruit (*Prunus persica*) upon heat treatment. *Journal of Experimental Botany* 60 (15): 4315–4333.

Lee, J., D. R. Rudell, P. J. Davies, and C. B. Watkins. 2012. Metabolic changes in 1-methylcyclopropene (1-MCP)-treated 'Empire'apple fruit during storage. *Metabolomics* 8 (4): 742–753.

Lee, Y. P., G. H. Yu, Y. S. Seo, S. E. Han, Y. O. Choi, D. Kim, I. G. Mok, W. T. Kim, and S. K. Sung. 2007. Microarray analysis of apple gene expression engaged in early fruit development. *Plant Cell Reports* 26 (7): 917–926.

Li, X., L. Shangguan, C. Song, C. Wang, Z. Gao, H. Yu, and J. Fang. 2010. Analysis of expressed sequence tags from *Prunus mume* flower and fruit and development of simple sequence repeat markers. *BMC Genetics* 11 (1): 66.

Liebhard, R., B. Koller, L. Gianfranceschi, and C. Gessler. 2003. Creating a saturated reference map for the apple (*Malus* × *domestica* Borkh.) genome. *Theoretical and Applied Genetics* 106 (8): 1497–1508.

Lombardo, V. A., S. Osorio, J. Borsani, M. A. Lauxmann, C. A. Bustamante, C. O. Budde, C. S. Andreo, M. V. Lara, A. R. Fernie, and M. F. Drincovich. 2011. Metabolic profiling during peach fruit development and ripening reveals the metabolic networks that underpin each developmental stage. *Plant Physiology* 157 (4): 1696–1710.

Lund, S. T., F. Y. Peng, T. Nayar, K. E. Reid, and J. Schlosser. 2008. Gene expression analyses in individual grape (*Vitis vinifera* L.) berries during ripening initiation reveal that pigmentation intensity is a valid indicator of developmental staging within the cluster. *Plant Molecular Biology* 68 (3): 301–315.

Maliepaard, C., F. H Alston, G. Van Arkel, L. M. Brown, E. Chevreau, F. Dunemann, K. M. Evans, et al. 1998. Aligning male and female linkage maps of apple (*Malus pumila* Mill.) using multi-allelic markers. *Theoretical and Applied Genetics* 97 (1): 60–73.

Manganaris, G. A., A. Rasori, D. Bassi, F. Geuna, A. Ramina, P. Tonutti, and C. Bonghi. 2011. Comparative transcript profiling of apricot (*Prunus armeniaca* L.) fruit development and on-tree ripening. *Tree Genetics & Genomes* 7 (3): 609–616.

Marioni, J. C., C. E. Mason, S. M. Mane, M. Stephens, and Y. Gilad. 2008. RNA-seq: An assessment of technical reproducibility and comparison with gene expression arrays. *Genome Research* 18 (9): 1509–1517.

Marondedze, C. and L. A. Thomas. 2012. Apple hypanthium firmness: New insights from comparative proteomics. *Applied Biochemistry and Biotechnology* 168 (2): 306–326.

Martínez-Esteso, M. J., S. Sellés-Marchart, D. Lijavetzky, M. A. Pedreño, and R. Bru-Martínez. 2011. A DIGE-based quantitative proteomic analysis of grape berry flesh development and ripening reveals key events in sugar and organic acid metabolism. *Journal of Experimental Botany* 62 (8): 2521–2569.

Mascarell-Creus, A., J. Cañizares, J. Vilarrasa-Blasi, S. Mora-García, J. Blanca, D. Gonzalez-Ibeas, M. Saladié, et al. 2009. An oligo-based microarray offers novel transcriptomic approaches for the analysis of pathogen resistance and fruit quality traits in melon (*Cucumis melo* L.). *BMC Genomics* 10 (1): 467.

Mazza, G., C. D. Kay, T. Cottrell, and B. J. Holub. 2002. Absorption of anthocyanins from blueberries and serum antioxidant status in human subjects. *Journal of Agricultural and Food Chemistry* 50 (26): 7731–7737.

Meir, S., S. Philosoph-Hadas, S. Sundaresan, K. S. V. Selvaraj, S. Burd, R. Ophir, B. Kochanek, M. S. Reid, C. Z. Jiang, and A. Lers. 2010. Microarray analysis of the abscission-related transcriptome in the tomato flower abscission zone in response to auxin depletion. *Plant Physiology* 154 (4): 1929–1956.

Ming, R., S. Hou, Y. Feng, Q. Yu, A. Dionne-Laporte, J. H. Saw, P. Senin, et al. 2008. The draft genome of the transgenic tropical fruit tree papaya (*Carica papaya* Linnaeus). *Nature* 452 (7190): 991–996.

Mintz-Oron, S., T. Mandel, I. Rogachev, L. Feldberg, O. Lotan, M. Yativ, Z. Wang, R. Jetter, I. Venger, and A. Adato. 2008. Gene expression and metabolism in tomato fruit surface tissues. *Plant Physiology* 147 (2): 823–851.

Moco, S., R. J. Bino, O. Vorst, H. A. Verhoeven, J. De Groot, T. A. Van Beek, J. Vervoort, and C. H. R. De Vos. 2006. A liquid chromatography-mass spectrometry-based metabolome database for tomato. *Plant Physiology* 141 (4): 1205–1218.

Moxon, S., R. Jing, G. Szittya, F. Schwach, R. L. R. Pilcher, V. Moulton, and T. Dalmay. 2008. Deep sequencing of tomato short RNAs identifies microRNAs targeting genes involved in fruit ripening. *Genome Research* 18 (10): 1602–1609.

Muccilli, V., C. Licciardello, D. Fontanini, M. P. Russo, V. Cunsolo, R. Saletti, G. Reforgiato Recupero, and S. Foti. 2009. Proteome analysis of *Citrus sinensis* L. (Osbeck) flesh at ripening time. *Journal of Proteomics* 73 (1): 134–152.

Nakano, T., J. Kimbara, M. Fujisawa, M. Kitagawa, N. Ihashi, H. Maeda, T. Kasumi, and Y. Ito. 2012. MACROCALYX and JOINTLESS interact in the transcriptional regulation of tomato fruit abscission zone development. *Plant Physiology* 158 (1): 439–450.

Namitha, K. K., S. N. Archana, and P. S. Negi. 2011. Expression of carotenoid biosynthetic pathway genes and changes in carotenoids during ripening in tomato (*Lycopersicon esculentum*). *Food & Function* 2 (3–4): 168–173.

Negri, A. S., B. Prinsi, M. Rossoni, O. Failla, A. Scienza, M. Cocucci, and L. Espen. 2008. Proteome changes in the skin of the grape cultivar Barbera among different stages of ripening. *BMC Genomics* 9 (1): 378.

Nilo, R., C. Saffie, K. Lilley, R. Baeza-Yates, V. Cambiazo, R. Campos-Vargas, M. González, et al. 2010. Proteomic analysis of peach fruit mesocarp softening and chilling injury using difference gel electrophoresis (DIGE). *BMC Genomics* 11 (1): 43.

Nogueira, S. B., C. A. Labate, F. C. Gozzo, E. J. Pilau, F. M. Lajolo, and J. R. Oliveira do Nascimento. 2011. Proteomic analysis of papaya fruit ripening using 2DE-DIGE. *Journal of Proteomics* 75 (4): 1428–1439.

Ogata, H., S. Goto, K. Sato, W. Fujibuchi, H. Bono, and M. Kanehisa. 1999. KEGG: Kyoto encyclopedia of genes and genomes. *Nucleic Acids Research* 27 (1): 29–34.

Ogundiwin, E. A., C. P. Peace, T. M. Gradziel, D. E. Parfitt, F. A. Bliss, and C. H. Crisosto. 2009. A fruit quality gene map of *Prunus. BMC Genomics* 10 (1): 587.

Osorio, S., R. Alba, C. M. Damasceno, G. Lopez-Casado, M. Lohse, M. I. Zanor, T. Tohge, et al. 2011. Systems biology of tomato fruit development: Combined transcript, protein, and metabolite analysis of tomato transcription factor (*nor, rin*) and ethylene receptor (*nr*) mutants reveals novel regulatory interactions. *Plant Physiology* 157 (1): 405–425.

Pan, Z., Q. Liu, Z. Yun, R. Guan, W. Zeng, Q. Xu, and X. Deng. 2009. Comparative proteomics of a lycopene accumulating mutant reveals the important role of oxidative stress on carotenogenesis in sweet orange (*Citrus sinensis* [L.] Osbeck). *Proteomics* 9 (24): 5455–5470.

Park, S., J. D. Cohen, and J. P. Slovin. 2006. Strawberry fruit protein with a novel indole-acyl modification. *Planta* 224 (5): 1015–1022.

Pascual, L., J. M. Blanca, J. Cañizares, and F. Nuez. 2009. Transcriptomic analysis of tomato carpel development reveals alterations in ethylene and gibberellin synthesis during pat3/pat4 parthenocarpic fruit set. *BMC Plant Biology* 9 (1): 67.

Perin, C., L. Hagen, V. De Conto, N. Katzir, Y. Danin-Poleg, V. Portnoy, S. Baudracco-Arnas, J. Chadoeuf, C. Dogimont, and M. Pitrat. 2002. A reference map of *Cucumis melo* based on two recombinant inbred line populations. *Theoretical and Applied Genetics* 104 (6–7): 1017–1034.

Pirona, R., A. Vecchietti, B. Lazzari, A. Caprera, R. Malinverni, C. Consolandi, M. Severgnini, et al. 2012. Expression profiling of genes involved in the formation of aroma in two peach genotypes. *Plant Biology* 15 (3): 443–451.

Prinsi, B., A. S. Negri, C. Fedeli, S. Morgutti, N. Negrini, M. Cocucci, and L. Espen. 2011. Peach fruit ripening: A proteomic comparative analysis of the mesocarp of two cultivars with different flesh firmness at two ripening stages. *Phytochemistry* 72 (10): 1251–1262.

Qin, G., Q. Wang, J. Liu, B. Li, and S. Tian. 2009. Proteomic analysis of changes in mitochondrial protein expression during fruit senescence. *Proteomics* 9 (17): 4241–4253.

Qin, G., X. Meng, Q. Wang, and S. Tian. 2009. Oxidative damage of mitochondrial proteins contributes to fruit senescence: A redox proteomics analysis. *Journal of Proteome Research* 8 (5): 2449–2462.

Qin, G., Y. Wang, B. Cao, W. Wang, and S. Tian. 2012. Unraveling the regulatory network of the MADS box transcription factor RIN in fruit ripening. *The Plant Journal* 70 (2): 243–255.

Rao, A.V. and L. G. Rao. 2007. Carotenoids and human health. *Pharmacological Research* 55 (3): 207–216.

Rivarola, M., A. P. Chan, D. E. Liebke, A. Melake-Berhan, H. Quan, F. Cheung, S. Ouyang, K. M. Folta, J. P. Slovin, and P. D. Rabinowicz. 2011. Abiotic stress-related expressed sequence tags from the diploid strawberry *Fragaria vescaf. semperflorens. The Plant Genome* 4 (1): 12–23.

Rocco, M., C. D'Ambrosio, S. Arena, M. Faurobert, A. Scaloni, and M. Marra. 2006. Proteomic analysis of tomato fruits from two ecotypes during ripening. *Proteomics* 6 (13): 3781–3791.

Rose, J. K. C., S. Bashir, J. J. Giovannoni, M. M. Jahn, and R. S. Saravanan. 2004. Tackling the plant proteome: Practical approaches, hurdles and experimental tools. *The Plant Journal* 39 (5): 715–733.

Rudell, D. R., J. P. Mattheis, and E. A. Curry. 2008. Prestorage ultraviolet– white light irradiation alters apple peel metabolome. *Journal of Agricultural and Food Chemistry* 56 (3): 1138–1147.

Rudell, D. R., J. P. Mattheis, and M. L. A. T. M. Hertog. 2009. Metabolomic change precedes apple superficial scald symptoms. *Journal of Agricultural and Food Chemistry* 57 (18): 8459–8466.

Ruzicka, D. R., F. H. Barrios-Masias, N. T. Hausmann, L. E. Jackson, and D. P. Schachtman. 2010. Tomato root transcriptome response to a nitrogen-enriched soil patch. *BMC Plant Biology* 10: 75.

Sánchez, G., C. Besada, M. I. Badenes, A. J. Monforte, and A. Granell. 2012. A Non-Targeted Approach Unravels the Volatile Network in Peach Fruit. *PLoS ONE* 7 (6): e38992.

Sargent, D. J., J. Clarke, D. W. Simpson, K. R. Tobutt, P. Arus, A. Monfort, S. Vilanova, et al. 2006. An enhanced microsatellite map of diploid *Fragaria*. *Theoretical and Applied Genetics* 112 (7): 1349–1359.

Sarry, J. E., N. Sommerer, F. X. Sauvage, A. Bergoin, M. Rossignol, G. Albagnac, and C. Romieu. 2004. Grape berry biochemistry revisited upon proteomic analysis of the mesocarp. *Proteomics* 4 (1): 201–215.

Schaff, J. E., D. M. Nielsen, C. P. Smith, E. H. Scholl, and D. M. K. Bird. 2007. Comprehensive transcriptome profiling in tomato reveals a role for glycosyltransferase in Mi-mediated nematode resistance. *Plant Physiology* 144 (2): 1079–1092.

Schaffer, R. J., E. N. Friel, E. J. F. Souleyre, K. Bolitho, K. Thodey, S. Ledger, J. H. Bowen, et al. 2007. A genomics approach reveals that aroma production in apple is controlled by ethylene predominantly at the final step in each biosynthetic pathway. *Plant Physiology* 144 (4): 1899–1912.

Schauer, N. and A. R. Fernie. 2006. Plant metabolomics: Towards biological function and mechanism. *Trends in Plant Science* 11 (10): 508–516.

Schauer, N., D. Zamir, and A. R. Fernie. 2005. Metabolic profiling of leaves and fruit of wild species tomato: A survey of the *Solanum lycopersicum* complex. *Journal of Experimental Botany* 56 (410): 297–307.

Schneider, M., A. Bairoch, C. H. Wu, and R. Apweiler. 2005. Plant protein annotation in the UniProt knowledgebase. *Plant Physiology* 138 (1): 59–66.

Schwacke, R., A. Schneider, E. Van Der Graaff, K. Fischer, E. Catoni, M. Desimone, W. B. Frommer, U. I. Flügge, and R. Kunze. 2003. ARAMEMNON, a novel database for arabidopsis integral membrane proteins. *Plant Physiology* 131 (1): 16–26.

Seaver, S. M. D., C. S. Henry, and A. D. Hanson. 2012. Frontiers in metabolic reconstruction and modeling of plant genomes. *Journal of Experimental Botany* 63 (6): 2247–2258.

Sharma, A., L. Zhang, D. Niño-Liu, H. Ashrafi, and M. R. Foolad. 2008. A *Solanum lycopersicum* × *Solanum pimpinellifolium* linkage map of tomato displaying genomic locations of R-genes, RGAs, and candidate resistance/defense-response ESTs. *International Journal of Plant Genomics* 2008: 926090.

Shinbo, Y., Y. Nakamura, M. Altaf-Ul-Amin, H. Asahi, K. Kurokawa, M. Arita, K. Saito, D. Ohta, D. Shibata, and S. Kanaya. 2006. KNApSAcK: A comprehensive species-metabolite relationship database. *Plant Metabolomics* 57: 165–181.

Shulaev, V., D. J. Sargent, R. N. Crowhurst, T. C. Mockler, O. Folkerts, A. L. Delcher, P. Jaiswal, et al. 2010. The genome of woodland strawberry (*Fragaria vesca*). *Nature Genetics* 43 (2): 109–116.

Singh, R., S. Rastogi, and U. N. Dwivedi. 2010. Phenylpropanoid metabolism in ripening fruits. *Comprehensive Reviews in Food Science and Food Safety* 9 (4): 398–416.

Son, H. S., G. S. Hwang, K. M. Kim, H. J. Ahn, W. M. Park, F. Van Den Berg, Y. S. Hong, and C. H. Lee. 2009. Metabolomic studies on geographical grapes and their wines using 1H NMR analysis coupled with multivariate statistics. *Journal of Agricultural and Food Chemistry* 57 (4): 1481–1490.

Son, H.S., K. M. Kim, F. Van den Berg, G. S. Hwang, W. M. Park, C. H. Lee, and Y. S. Hong. 2008. 1H nuclear magnetic resonance-based metabolomic characterization of wines by grape varieties and production areas. *Journal of Agricultural and Food Chemistry* 56 (17): 8007–8016.

Srivastava, A., A. K. Gupta, T. Datsenka, A. K. Mattoo, and A. K. Handa. 2010. Maturity and ripening-stage specific modulation of tomato (*Solanum lycopersicum*) fruit transcriptome. *GM Crops* 1 (4): 237–249.

Stockinger, E. J., C. A. Mulinix, C. M. Long, T. S. Brettin, and A. F. Lezzoni. 1996. A linkage map of sweet cherry based on RAPD analysis of a microspore-derived callus culture population. *Journal of Heredity* 87 (3): 214–218.

Sun, Q., B. Zybailov, W. Majeran, G. Friso, P. D. Olinares, and K. J. Van Wijk. 2009. PPDB, the plant proteomics database at cornell. *Nucleic Acids Research* 37 (suppl 1): D969–D974.

Swarup, R., G. Parry, N. Graham, T. Allen, and M. Bennett. 2002. Auxin cross-talk: Integration of signalling pathways to control plant development. *Plant Molecular Biology* 49 (3–4): 409–424.

Takos, A. M., F. W. Jaffé, S. R. Jacob, J. Bogs, S. P. Robinson, and A. R. Walker. 2006. Light-induced expression of a MYB gene regulates anthocyanin biosynthesis in red apples. *Plant Physiology* 142 (3): 1216–1232.

Tedesco, I., G. Luigi Russo, F. Nazzaro, M. Russo, and R. Palumbo. 2001. Antioxidant effect of red wine anthocyanins in normal and catalase-inactive human erythrocytes. *The Journal of Nutritional Biochemistry* 12 (9): 505–511.

The Arabidopsis Genome Initiative. 2000. Analysis of the genome sequence of the flowering plant *Arabidopsis thaliana*. *Nature* 408 (6814): 796–715.

The French–Italian Public Consortium for Grapevine Genome Characterization. 2007. The grapevine genome sequence suggests ancestral hexaploidization in major angiosperm phyla. *Nature* 449 (7161): 463–467.

The Tomato Genome Consortium. 2012. The tomato genome sequence provides insights into fleshy fruit evolution. *Nature* 485: 635–641.

Tikunov, Y., A. Lommen, C. H. R. de Vos, H. A. Verhoeven, R. J. Bino, R. D. Hall, and A. G. Bovy. 2005. A novel approach for nontargeted data analysis for metabolomics. Large-scale profiling of tomato fruit volatiles. *Plant Physiology* 139 (3): 1125–1137.

Tieman, D., P. Bliss, L. M. McIntyre, A. Blandon-Ubeda, D. Bies, A. Z. Odabasi, G. R. Rodríguez, et al. 2012. The chemical interactions underlying tomato flavor preferences. *Current Biology* 22 (11): 1035–1039.

Trainotti, L., A. Tadiello, and G. Casadoro. 2007. The involvement of auxin in the ripening of climacteric fruits comes of age: The hormone plays a role of its own and has an intense interplay with ethylene in ripening peaches. *Journal of Experimental Botany* 58 (12): 3299–3308.

Trainotti, L., D. Zanin, and G. Casadoro. 2003. A cell wall oriented genomic approach reveals a new and unexpected complexity of the softening in peaches. *Journal of Experimental Botany* 54 (389): 1821–1832.

Troggio, M., G. Malacarne, G. Coppola, C. Segala, D. A. Cartwright, M. Pindo, M. Stefanini, et al. 2007. A dense single-nucleotide polymorphism-based genetic linkage map of grapevine (*Vitis vinifera* L.) anchoring pinot noir bacterial artificial chromosome contigs. *Genetics* 176 (4): 2637–2650.

Usadel, B., R. Schwacke, A. Nagel, and B. Kersten. 2012. GabiPD–the GABI primary database integrates plant proteomic data with gene-centric information. *Frontiers in Plant Science* 3: 154.

Van der Hoeven, R., C. Ronning, J. Giovannoni, G. Martin, and S. Tanksley. 2002. Deductions about the number, organization, and evolution of genes in the tomato genome based on analysis of a large expressed sequence tag collection and selective genomic sequencing. *The Plant Cell* 14 (7): 1441–1456.

van Wijk, K. J. 2001. Challenges and prospects of plant proteomics. *Plant Physiology* 126 (2): 501–508.

Velasco, R., A. Zharkikh, J. Affourtit, A. Dhingra, A. Cestaro, A. Kalyanaraman, P. Fontana, et al. 2010. The genome of the domesticated apple (*Malus domestica* Borkh.). *Nature Genetics* 42 (10): 833–839.

Velasco, R., A. Zharkikh, M. Troggio, D. A. Cartwright, A. Cestaro, D. Pruss, M. Pindo, et al. 2007. A high quality draft consensus sequence of the genome of a heterozygous grapevine variety. *PloS One* 2 (12): e1326.

Verde, I., Abbott, A.G., Scalabrin, S., Jung, S., Shu, S., Marroni, F., Zhebentyayeva, et al. 2013. The high-quality draft genome of peach (*Prunus persica*) identifies unique patterns of genetic diversity, domestication and genome evolution. *Nature Genetics* 45: 487–494.

Vizoso, P., L. A. Meisel, A. Tittarelli, M. Latorre, J. Saba, R. Caroca, J. Maldonado, et al. 2009. Comparative EST transcript profiling of peach fruits under different post-harvest conditions reveals candidate genes associated with peach fruit quality. *BMC Genomics* 10 (1): 423.

Wang, X. C., L. Guo, L. F. Shangguan, C. Wang, G. Yang, S. C. Qu, and J. G. Fang. 2012. Analysis of expressed sequence tags from grapevine flower and fruit and development of simple sequence repeat markers. *Molecular Biology Reports* 39 (6):6825–6834.

Wang, Z., M. Gerstein, and M. Snyder. 2009. RNA-seq: A revolutionary tool for transcriptomics. *Nature Reviews Genetics* 10 (1): 57–63.

Wargovich, M. J. 2000. Anticancer properties of fruits and vegetables: The role of oxidative stress and antioxidants in plant and human health. *HortScience* 35 (4): 573–575.

Weber, A. P. M., K. L. Weber, K. Carr, C. Wilkerson, and J. B. Ohlrogge. 2007. Sampling the arabidopsis transcriptome with massively parallel pyrosequencing. *Plant Physiology* 144 (1): 32–42.

Wechter, W. P., A. Levi, K. R. Harris, A. R. Davis, Z. Fei, N. Katzir, J. J. Giovannoni, et al. 2008. Gene expression in developing watermelon fruit. *BMC Genomics* 9 (1): 275.

Wheeler, S., B. Loveys, C. Ford, and C. Davies. 2009. The relationship between the expression of abscisic acid biosynthesis genes, accumulation of abscisic acid and the promotion of *Vitis vinifera* L. berry ripening by abscisic acid. *Australian Journal of Grape and Wine Research* 15 (3): 195–204.

Winkel-Shirley, B. 2001. Flavonoid biosynthesis. A colorful model for genetics, biochemistry, cell biology, and biotechnology. *Plant Physiology* 126 (2): 485–493.

Wu, J. Z. Wang, Z. Shi, S. Zhang, R. Ming, S. Zhu, M. A. Khan, et al. 2013. The genome of the pear (*Pyrus bretschneideri* Rehd.). *Genome Research* 23 (2): 396–408.

Xia, M., W. Ling, H. Zhu, Q. Wang, J. Ma, M. Hou, Z. Tang, L. Li, and Q. Ye. 2007. Anthocyanin prevents CD40-activated proinflammatory signaling in endothelial cells by regulating cholesterol distribution. *Arteriosclerosis, Thrombosis, and Vascular Biology* 27 (3): 519–524.

Xia, R., H. Zhu, Y. An, E. P. Beers, and Z. Liu. 2012. Apple miRNAs and tasiRNAs with novel regulatory networks. *Genome Biology* 13 (6): R47.

Xu, Q., L. L. Chen, X. Ruan, D. Chen, A. Zhu, C. Chen, D. Bertrand, et al. 2013. The draft genome of sweet orange (*Citrus sinensis*). *Nature Genetics* 45: 59–66.

Yu, H., C. Song, Q. Jia, C. Wang, F. Li, K. K. Nicholas, X. Zhang, and J. Fang. 2011. Computational identification of microRNAs in apple expressed sequence tags and validation of their precise sequences by miR RACE. *Physiologia Plantarum* 141 (1): 56–70.

Zamboni, A., M. D. Carli, F. Guzzo, M. Stocchero, S. Zenoni, A. Ferrarini, P. Tononi, et al. 2010. Identification of putative stage-specific grapevine berry biomarkers and omics data integration into networks. *Plant Physiology* 154: 1439–1459.

Zenoni, S., A. Ferrarini, E. Giacomelli, L. Xumerle, M. Fasoli, G. Malerba, D. Bellin, M. Pezzotti, and M. Delledonne. 2010. Characterization of transcriptional complexity during berry development in *Vitis vinifera* using RNA-seq. *Plant Physiology* 152 (4): 1787–1795.

Zhang, L., L. Jiang, Y. Shi, L. Haibo, R. Kang, and Z. Yu. 2012. Post-harvest 1-methylcyclopropene and ethephon treatments differently modify protein profiles of peach fruit during ripening. *Food Research International* 48 (2): 609-619.

Zhang, Q., W. Chen, L. Sun, F. Zhao, B. Huang, W. Yang, Y. Tao, et al. 2012. The genome of *Prunus mume*. *Nature Communications* 3: 1318.

Zhang, J., X. Wang, O. Yu, J. Tang, X. Gu, X. Wan, and C. Fang. 2011a. Metabolic profiling of strawberry (*Fragaria × ananassa* Duch.) during fruit development and maturation. *Journal of Experimental Botany* 62: 1103–1118.

Zhang, L., Z. Yu, L. Jiang, J. Jiang, H. Luo, and L. Fu. 2011b. Effect of post-harvest heat treatment on proteome change of peach fruit during ripening. *Journal of Proteomics* 74 (7): 1135–1149.

Ziliotto, F., M. Begheldo, A. Rasori, C. Bonghi, and P. Tonutti. 2008. Transcriptome profiling of ripening nectarine (*Prunus persica* L. Batsch) fruit treated with 1-MCP. *Journal of Experimental Botany* 59 (10): 2781–2791.

Zhu, H., C. D. Dardick, E. P. Beers, A. M. Callanhan, R. Xia, and R. Yuan. 2011. Transcriptomics of shading-induced and NAA-induced abscission in apple (*Malus domestica*) reveals a shared pathway involving reduced photosynthesis, alterations in carbohydrate transport and signaling and hormone crosstalk. *BMC Plant Biology* 11 (1): 138.

Zhu, H., R. Xia, B. Zhao, Y. An, C. D. Dardick, A. M. Callahan, and Z. Liu. 2012. Unique expression, processing regulation, and regulatory network of peach (*Prunus persica*) miRNAs. *BMC Plant Biology* 12 (1): 149.

Zuo, J., B. Zhu, D. Fu, Y. Zhu, Y. Ma, L. Chi, Z. Ju, Y. Wang, B. Zhai, and Y. Luo. 2012. Sculpting the maturation, softening and ethylene pathway: The influences of microRNAs on tomato fruits. *BMC Genomics* 13 (1): 7.

9

Omics Approaches in Tropical Fruit Crops

Kundapura V. Ravishankar, PhD; Kanupriya, PhD; Ajitha-kumar Rekha, PhD; Anuradha Upadhyay, PhD; Chinmaiyan Vasugi, PhD; N. Vijayakumari, PhD; Pooja Kishnani, PhD; and Makki R. Dinesh, PhD

CONTENTS

9.1 Introduction..286
9.2 Banana...287
    9.2.1 Introduction...287
    9.2.2 Genome Information...287
        9.2.2.1 Markers...287
        9.2.2.2 Mapping..288
        9.2.2.3 Whole Genome Sequencing..288
        9.2.2.4 Transcriptomes..289
    9.2.3 Proteomics and Metabolomics...289
    9.2.4 Conclusion and Future Perspectives..290
9.3 Citrus..290
    9.3.1 Introduction...290
    9.3.2 Genomic Information..291
        9.3.2.1 Markers...291
        9.3.2.2 Mapping..291
        9.3.2.3 Whole Genome Sequencing..292
        9.3.2.4 Transcriptome Analysis...292
    9.3.3 Proteomics and Metabolomics...293
    9.3.4 Conclusion and Future Perspectives..293
9.4 Grape...294
    9.4.1 Introduction...294
    9.4.2 Genome Information...294
        9.4.2.1 Markers...294
        9.4.2.2 Mapping..295
        9.4.2.3 Whole Genome Information...295
        9.4.2.4 Transcriptomics..296
    9.4.3 Proteomics..296
    9.4.4 Conclusion and Future Perspectives..297
9.5 Guava..297
    9.5.1 Introduction...297
    9.5.2 Genomic Information..298
        9.5.2.1 Markers...298
        9.5.2.2 Mapping..298
    9.5.3 Conclusion and Future Perspectives..299

9.6 Mango..299
     9.6.1 Introduction...299
     9.6.2 Genome Information...300
             9.6.2.1 Markers..300
             9.6.2.2 Mapping 301
             9.6.2.3 Transcriptome...301
     9.6.3 Metabolomics and Proteomics...301
     9.6.4 Conclusion and Future Strategies 302
9.7 Papaya..302
     9.7.1 Introduction...302
     9.7.2 Genomic Information...303
             9.7.2.1 Markers..303
             9.7.2.2 Mapping 304
             9.7.2.3 Whole Genome Sequencing...304
             9.7.2.4 Transcriptomic Studies...305
     9.7.3 Metabolomics and Proteomic Studies 305
     9.7.4 Conclusion and Future Perspectives...306
9.8 Pomegranate 307
     9.8.1 Introduction...307
     9.8.2 Genome Information...307
             9.8.2.1 Markers..307
             9.8.2.2 Mapping 307
     9.8.3 Metabolomics 308
     9.8.4 Conclusion and Future Perspectives...308
9.9 General Conclusion 309
References...309

9.1 Introduction

With recent advances in techniques such as pyrosequencing, mass spectrometry, high-performance liquid chromatography (HPLC), and bioinformatics, a great deal of knowledge and data are generated in biology. At present, the "omics" sciences include genomics, proteomics, and metabolomics. Genomic study includes sequencing entire sets of DNA molecules inside the cell and studying the expression profile of genes under various conditions (transcriptomics). Expression profiling of mRNA in a cell population has been extensively used to understand the underlying mechanism of different processes related to the different stages of development of organs such as shoots, roots, flowers, and fruits, and tissue response to different biotic and abiotic stresses. The identified genes and their alleles have become markers for use in plant breeding programs. The development of linkage maps and whole genome sequencing have become an integral part of crop improvement programs. Quantitative trait loci (QTLs), the DNA regions on chromosomes which affect or control polygenic traits, can easily be located using linkage maps and the phenotyping of mapping populations. Therefore, DNA markers, mapping, whole genome sequencing, and trancriptome analysis have become part of crop genetics.

Next, the active molecules in biological processes are proteins, which are responsible for a number of processes within the cell. The complete set of proteins in a cell can be referred to as its proteome, and the study of protein structure and function in a cell or organ at different stages or under different abiotic and biotic stress situations is known as proteomics. Proteomics needs extensive genomic data and bioinformatic tools to make accurate predictions. The metabolome is the complete set of low molecular weight compounds in a cell or tissue. These compounds are the substrates and by-products of enzymatic reactions and have a direct effect on the phenotype of the cell or organ. Thus, metabolomics study examines the profile of these compounds at a specified time under specific environmental conditions. In crop science, especially in fruit crops, proteomic and metabolomic studies are in their early stages. The omics such as genomics, proteomics, and metabolomics will help extensively in explaining the genotype and phenotype of an organism.

9.2 Banana

9.2.1 Introduction

Cultivated bananas belong to the family Musaceae of the order Zingiberales. It has two genera: *Musa* and *Ensete*. Two important species of the genus *Musa*, *M. acuminata* (2n = 22; referred as the A genome) and *M. balbisiana* (2n = 22; referred as the B genome), have contributed to the world's banana cultivars. Being one of the oldest and most highly evolved fruits it has a number of cultivated types selected by humans involving diploid AA, AB genomic groups and triploid AAA, AAB, and ABB genomic groups as per the classifications made by Simmonds (1962), although AAA Cavendish and AAB Plantains are the most widely grown globally. *Musa* is further divided into four sections: *Eumusa*, *Rhodhochlymus*, *Australimusa*, and *Callimusa*. *Eumusa* and *Rhodochlymus* have a chromosome number of n = 11, whereas *Australimusa* and *Callimusa* have n = 10 chromosomes. The related genus *Ensete* has n = 9 chromosomes. Members of the families Zingiberaceae and Strelitziaceae are closely related to Musaceae (Sharrock, 1997).

9.2.2 Genome Information

9.2.2.1 *Markers*

All available markers, including random amplification of polymorphic DNA (RAPD), amplified fragment length polymorphism (AFLP), inter-simple sequence repeat (ISSR), methylation-sensitive amplified polymorphism (MSAP), simple sequence repeat (SSR), sequence-tagged microsatellite (STMS), minisatellite and microsatellite markers, chloroplast- and mitochondrial-specific markers, have been used to study various aspects such as phylogenetic relationships, diversity, evolution, parentage analysis, differentiation of A and B genomes, and the development of trait-specific markers in *Musa* (Carreel et al., 2002; Rekha et al., 2001; Ravishankar et al., 2012; Creste et al., 2006; Mark et al., 2009; Miller et al., 2010). Of late, single-nucleotide polymorphism (SNP) markers are found to be more useful as they have an impact on protein function and thus on phenotype. The studies have revealed the close relationships among the cultivars and species and have helped to differentiate the cultivars, species, and different genomic groups. The chloroplast- and mitochondria-specific markers helped the breeders to understand the evolution and involvement of male and female parents among natural hybrids. Several sequence-characterized amplified region (SCAR) and resistance gene analog (RGA) markers were identified in relation to the important diseases like Fusarium wilt and leaf Sigatoka (Uma et al., 2011).

9.2.2.2 Mapping

In order to develop a genetic linkage map, an F1 progeny of 180 individuals was obtained from a cross between two genetically distant accessions of *M. acuminata*: Borneo and Pisang Lilin (P. Lilin). Based on the gametic recombination of each parent, two parental maps composed of SSR and Diversity Arrays Technology (DArT) markers were developed. From this study, a synthetic map with 11 linkage groups containing 489 markers (167 SSRs and 322 DArTs) covering 1197 cM was constructed. This is the first saturated map proposed as a "reference *Musa* map" for further analyses. The authors also proposed two complete parental maps with interpretations of structural rearrangements localized on the linkage groups. The structural heterozygosity in P. Lilin was hypothesized to result from a duplication, likely accompanied by an inversion on another chromosome (Hippolyte et al., 2010).

The first linkage map for the fungal plant pathogen *Mycosphaerella fijiensis*, which causes a serious leaf spot disease in banana, was developed. This genetic linkage map consists of 298 AFLP and 16 SSR markers with 23 linkage groups. The map can be utilized for anchoring contigs in the genome sequence, evolutionary studies in comparison with other fungi, to identify QTLs associated with aggressiveness or oxidative stress resistance and with the available genome sequence of banana (Manzo-Sánchez et al., 2008).

There are some important problems in the cultivation of bananas which cannot be tackled by conventional breeding techniques. This makes it important to study and understand the genome. Attempts were made to map various traits which included banana streak virus (BSV), wilt resistance, nematode resistance, and the development of a core map. Several populations were involved in the development of the map, the important ones being *M. balbisiana* × *M. acuminata* (AAAA), SF 265 (AA) × *M. acuminata* ssp. *banksii* (AA), *M. acuminata* ssp. *truncata* × *M. acuminata* ssp. *madang*. Restriction fragment length polymorphism (RFLP) and SSR markers were employed to develop the BSV-related map. RAPD and AFLP were used to develop the map related to wilt resistance. M-53 was widely in use to develop bacterial artificial chromosome (BAC) libraries and the core map of the banana. Several institutes and research groups have been involved, of which CIRAD, France, is the leading institute.

9.2.2.3 Whole Genome Sequencing

A decision to sequence the whole genome of *Musa* was made in the year 2001. It was a difficult task as the development of mapping populations was very tough. There were no segregating populations involving the BB genome. It was in the year 2009 that serious efforts were first made to sequence the entire banana genome. By then, sequencing was comparatively easy and faster, as high-throughput sequencers and data analyses were available. In August 2012, the *Musa* genome was published, which enlightened the banana workers, especially breeders and geneticists. The *Musa* genome size is larger than that of rice but smaller than that of wheat and maize. The DNA content ranges from 1.22 to 1.33 pg, with about 534–615 Mbp for *M. acuminata*. The genome of *M. balbisiana* is smaller in size, at 1.03–1.16 pg. About 30% of a genome largely consists of repetitive DNAs, transposable elements, satellite DNA, simple sequence repeats, and tandem repeats, causing variation in genome size. The presence of transposable elements and simple sequence repeats was revealed by *in situ* hybridization techniques (Heslop-Harrison & Schwarzacher, 2007).

Recently, whole genome sequence data for the 523-megabase genome of a double haploid *Musa acuminata* genotype were described. Three rounds of whole genome duplications were detected in *Musa* lineage, followed by gene loss and chromosome rearrangements. The genome of DH-Pahang, a doubled haploid *M. acuminata* genotype ($2n = 22$) of the subspecies *malaccensis* that contributed one of the three *acuminata* genomes of Cavendish, was sequenced.

A total of 27.5 million Roche/454 single reads and 2.1 million Sanger reads were used, representing a coverage of 20.53 of the 523-megabase (Mb) DH-Pahang genome. In addition, Illumina data were used to correct sequence errors. The assembly consisted of 24,425 contigs and 7513 scaffolds with a total length of 472.2 Mb, which represented 90% of the estimated DH-Pahang genome size. A total of 70% of the assembly (332 Mb) was assigned along the 11 *Musa* linkage groups of the Pahang genetic map (D'Hont et al., 2012).

An earlier *Musa* and rice synteny map was developed using *Musa* BAC clones and RFLP probes. Comparison of genomic regions from *M. acumninata* and *M. balbisiana* revealed a highly conserved genome structure, and indicated that these genomes diverged circa 4.6 Mya. A comparative analysis was done to improve the basic understanding of monocot genome evolution between distantly related monocot species like rice and *Musa* (Lescot et al., 2008).

9.2.2.4 Transcriptomes

In order to investigate the transcriptional changes induced by the Fusarium wilt-causing fungus *Fusarium oxysporum* f.sp. *cubense* (Foc) in banana roots, a cDNA library from infected banana roots with Foc Tropical Race 4 (Foc TR 4) was generated. A total of 21,622 (85.94%) unique sequences were annotated and 11,611 were assigned to specific metabolic pathways using the Kyoto Encyclopedia of Genes and Genomes database. The expression of genes in the phenylalanine metabolism, phenylpropanoid biosynthesis, and α-linolenic acid metabolism pathways was affected by Foc TR4 infection (Li et al., 2012).

The response of the banana (*Musa* spp.) leaf transcriptome to drought stress using a genomic DNA (gDNA)-based probe selection strategy was employed using microarray techniques to detect differentially expressed *Musa* transcripts. A cross-hybridization of *Musa* gDNA to the rice GeneChip® genome array, ~33,700 gene-specific probe sets revealed a high degree of homology. A total of 2910 *Musa* gene homologs with a greater than two-fold difference in expression levels were identified. These drought-responsive transcripts included many functional genes associated with plant biotic and abiotic stress responses, as well as a range of regulatory genes known to be involved in coordinating abiotic stress responses. These abiotic stress response genes included members of the ERF, DREB, MYB, bZIP, and bHLH transcription factors. Fifty-two of these drought-sensitive *Musa* transcripts were homologous to genes underlying QTLs for drought and cold tolerance in rice, including in two instances of QTLs associated with a single underlying gene. The list of drought-responsive transcripts also included genes identified in publicly available comparative transcriptomics experiments (Davey et al., 2009).

9.2.3 Proteomics and Metabolomics

Studies related to proteomics and metabolomics in *Musa* are scanty. Proteomics is used as a tool to discover drought-tolerant varieties by using *in vitro* multiplied plants. Five varieties representing different genomic constitutions such as AAAh, AAA, AAB, AABp, and ABB were selected. The ABB varieties showed less stress-induced growth reduction.

An analysis of stress-induced proteins showed that 112 proteins were significantly more abundant in stressed plants. This also clearly indicated that there is a new balance in the stressed plants and that the respiration, metabolism of reactive oxygen species (ROS) and several dehydrogenases involved in nicotinamide adenine dinucleotide (NAD)/NADH homeostasis play an important role (Vanhove et al., 2012).

In another study, investigations were done on the influence of sucrose-mediated osmotic stress in a dehydration-tolerant variety. The maintenance of an osmoprotective intracellular sucrose concentration, the enhanced expression of particular genes of the energy-conserving glycolysis, and the conservation of cell wall integrity are essential to maintain homeostasis, to acclimate, and to survive dehydration. Comparison of the dehydration-tolerant variety with a dehydration-sensitive variety helped in distinguishing several genotype-specific proteins (isoforms), and associating the dehydration-tolerant variety with the proteins involved in energy metabolism (such as phosphoglycerate kinase, phosphoglucomutase, and uridine diphosphate (UDP)-glucose pyrophosphorylase) and the proteins associated with stress adaptation (e.g., OSR40-like proteins, abscisic stress ripening protein-like protein) (Carpentier et al., 2007).

The use of advanced mass spectrometry techniques and the *Musa* mRNA database in combination with the Uniprot viridiplantae database allowed the identification of 1131 proteins. In this first in-depth exploration of the banana fruit proteome studies, several already-known allergens, such as *Musa a 1*, pectinesterase, and superoxide dismutase, and some potentially new allergens were identified. Additionally, several enzymes involved in the degradation of starch granules and strictly correlated to the ripening stage were also recognized (Esteve et al., 2012).

9.2.4 Conclusion and Future Perspectives

The banana whole genome sequence has been published and is a stepping stone for the improvement of banana, which is a challenge. It helps in identifying the genes responsible for important agronomic characters, such as fruit quality, disease, and pest resistance, as they affect banana production. The genomic studies are of immense use in understanding the evolution of plants, and in particular monocotyledons. Since banana is known to be an important source of starch it is essential to evolve varieties/cultivars resistant to various biotic and abiotic stresses. The various markers, transcriptomes, and related proteins would help in gene identification and better understanding of the plant–host interactions.

The future of banana genome analysis involves establishing linkage maps, especially for the important traits like disease and drought tolerance, and developing disease resistance among important cultivars of the global banana industry and locally preferred types. In the present scenario of climate change and water scarcity, it is necessary to develop or identify varieties that can sustain extreme temperatures and water deficiency.

9.3 Citrus

9.3.1 Introduction

Citrus is one of the world's most important fruit crops, with a total world production of 115.65 million tons (FAO, 2011; Soost et al., 1996). The most significant citrus-producing regions are in the Americas (Brazil, USA, Argentina, and Mexico, primarily), the Mediterranean basin (Southern Europe, Southwest Asia, and North Africa), Asia (including China, India, and Japan), and South Africa. Citrus industries in many production areas generate substantial regional revenue. Brazil, USA, China, Mexico, and Spain are the five largest citrus producers in the world (FAO, 2006). Despite the economic significance of the genus, *Citrus* breeding programs have had minimal success because of various biological factors including sterility (Soost and Cameron, 1975), self and cross incompatibility, widespread nucellar embryony (Frost and Soost, 1968; Soost and Cameron, 1975), and long juvenile periods resulting in a large plant size at maturity. With globalization of *Citrus* production and increased human travel throughout the world, devastating citrus diseases have been rapidly spreading, thus threatening the viability and the very future of citrus production globally. However, the genetic challenges and the lack of understanding of the fundamental mechanisms underlying the critical traits described above present tremendous impediments to the progress needed to incorporate the needed genes. In this context, the development and application of genomic information generated through studies of *Citrus* omics act as powerful tools.

9.3.2 Genomic Information

9.3.2.1 *Markers*

Following isozymes markers (Torres et al., 1978), several kinds of nuclear marker have since been developed and used for *Citrus* genetic studies. Dominant markers such as RAPD (Higashi et al., 2000), ISSR (Fang et al., 1997; Fang et al., 1998), AFLP (Liang et al., 2006), SCAR (Nicolosi et al., 2000), and cleaved amplified polymorphic sequences (CAPS) from expressed sequence tags (ESTs; Omura et al., 2006) have been useful for large-scale characterization of genomes for which previous genomic sequence information was not available. In the last 15 years, a broad international collaboration in the citrus community developed sets of SSR markers (Luro et al., 2007; Froelicher et al., 2008; Chen et al., 2008; Ollitrault et al., 2008; Cuenca et al., 2009; Aleza et al., 2010; Ferrante et al., 2010; Bernet et al., 2010). The implementation of large EST databases has allowed the development of many more SSR markers.

9.3.2.2 *Mapping*

Molecular screening of hybrid progeny based on established linkage relationships could be an invaluable tool to expedite the selection process. *Citrus* and the closely related genera are partially sexually compatible in varying degrees; they are primarily diploid with a few known triploids and occasional tetraploid forms ($2n = 2x = 18$), and they possess fairly small genomes (e.g., sweet orange has been said to be around 367 Mb, or approximately three times that of *Arabidopsis* (Arumuganathan and Earle, 1991). *Citrus* species should be amenable to many of the commonly used techniques and approaches related to genomic research, including genetic and physical mapping, full genome sequencing, and functional genomics studies aimed at unraveling the complexities of the key traits of interest. Because of the valuable characteristics within some of the related genera that are absent from *Citrus*, particularly cold tolerance and resistance to citrus canker (caused by *Xanthomonas axonopodis* pv. *citri*) from kumquat (*Fortunella*) and multiple stress-tolerant and disease-resistance traits from *P. trifoliata* (including freeze avoidance and resistance to citrus tristeza virus (CTV), *Phytophthora* and citrus nematode [*Tylenchulus semipenetrans*]), many of the genetic mapping projects, and some of the physical mapping as well, have focused on *P. trifoliata* through intergeneric hybrids with *Citrus*.

The main challenge for a comprehensive and meaningful description of the genomes is the integration of the DNA marker-based genetic maps with physical maps. Large genomic DNA insert-containing libraries are required for physical mapping, positional cloning, and genome sequencing of complex genomes.

9.3.2.3 Whole Genome Sequencing

For the generation of high-resolution physical maps, the construction of BAC libraries containing clones with large DNA fragments appears to be indispensable. Studies in this direction in relation to CTV disease resistance were led by Mamata (2006). Fine mapping of loci involved in polyembryony was studied by Hong et al. (2001), Nakano et al. (2008), and Kepiro and Roose (2009).

Two full-length annotated genome assemblies, based on Sanger sequencing, were produced and made available to researchers: first one derived from the haploid Clementine plant and second one from Ridge Pineapple, a clone of the sweet orange. To explore transcriptomic responses of citrus species and hybrids to a wide range of conditions, several EST databases and microarray platforms have been freely shared for utilization of genome sequence information (Gmitter et al., 2012). Citrus researchers (Chen et al., 2006; Tao et al., 2007; Gmitter et al., 2012) actively engaged in national and international collaborations integrated their efforts for genetic improvement programs with fast evolving genomic tools. Work from Gmitter et al. (2012) helped in better understanding of the control of gene expression in citrus polyploids, which are being used in the development of seedless triploid selections or as rootstocks. The rapidly advancing technologies described by him aim to overcome the risk of devastating diseases that threaten to limit production. Further work is in progress (in Dr. F.G. Gmitter's laboratory at the University of Florida, Citrus Research and Education Centre, Lake Alfred, Riverside, USA, and by different scientists in Brazil, France, and Japan) to expand and improve citrus genome sequence resources and tools, in order to enable the application of sequence-derived knowledge in improving citrus plants and study their interactions with biotic and abiotic factors.

Research on citrus genomics in China was originally set up for EST exploration and utilization, gene cloning, genomic sequencing, and genetic map construction. The first cDNA library published in China was constructed from Cara Cara navel orange pulp by Tao et al. (2006).

9.3.2.4 Transcriptome Analysis

The large-scale sequencing and analysis of the *Citrus* EST database is a fundamental part of genomics research to enable gene discovery and annotation. Katz et al. (2007) classified the proteins from citrus juice cells as per the putative function and assigned the function according to the known biosynthetic pathway through the *Citrus* EST database. Functional evaluation of the transcriptome was studied by Shimizu et al. (2009) and Delseny et al. (2010). In addition to these efforts, initial attempts to identify micro RNAs in *Citrus* were reported by Xu et al. (2009).

Accumulation of the nucleotide sequences for expressed genes with functional annotation enables expression profiling in citrus tissues. These efforts have been made by developing DNA microarrays. Different types of microarray, spotted array, *in situ* synthesized oligonucleotide for short (25-mer oligo) or long (60-mer oligo) probes have been developed.

Applications for expression profiling toward gene mining with these microarrays have been reported (Fujii et al., 2008; Xu et al., 2010).

Ollitrault et al. (2001) analyzed 96 SSR markers of allotetraploidy products between *Citrus deliciosa* and six other species at CIRAD France. They did not reveal any inconsistency with total addition of parental genomes. Evaluations of phenome, proteome, and transciptome were conducted using a field trial planted in a complete randomized design with three trees (replications) grafted to Carrizo citrange rootstock for six genotypes. Leaf volatile compounds of the six allotetraploid hybrids sharing Willow leaf mandarin as their common parent were analyzed by gas chromatography–mass spectrography (GC–MS), and the systematic dominance of mandarin traits was observed (Gancel et al., 2003). The leaf proteomes of two allotetraploid somatic hybrids combining *C. deliciosa* with *C. aurantifolia* and *Fortunella margarita* were analyzed by 2D electrophoresis (Gancel et al., 2006). Recently, genome-wide gene expression analysis on fruit pulp of a *Citrus* interspecific somatic allotetraploid between *C. reticulata* cv. Willowleaf mandarin + *C. limon* cv. Eureka lemon was done using a Citrus 20 K cDNA microarray (Bassene et al., 2010). The authors observed a global downregulation of the allotetraploid hybrid transcriptome compared to a theoretical midparent. The potential implication of nonadditive gene expression in the phenotype of citrus somatic hybrids was illustrated by analyzing the carotenoid and abscisic acid (ABA) contents and the expression of the genes of the carotenoid/ABA biosynthesis pathway (Bassene et al., 2009).

Data mining of nucleotide sequences obtained from expressed genes or genome sequence assemblies provides various types of putative polymorphic regions that could be useful for molecular marker development. SSR regions have been mined from EST sequences and utilized for marker development and linkage mapping (Luro et al., 2008; Roose et al., 2006).

9.3.3 Proteomics and Metabolomics

Proteomics has become a powerful tool in plant research in the last few years. The development of state-of-the-art liquid chromatography–mass spectrometry (LC–MS)/MS technology, fine separation techniques, genomic and EST databases for a variety of species, and powerful bioinformatics tools enables the understanding and assessment of protein function, their relative abundance, the modifications affecting enzyme activity, their interaction with other proteins, and localization. Proteomics research has been conducted in several plant species, mainly using two-dimensional electrophoresis (2DE) gels. The most successful studies are those that use separation of subcellular compartments such as mitochondria (Heazlewood et al., 2004; Lister et al., 2004), chloroplast (Giacomelli et al., 2006), endoplasmic reticulum (Maltman et al., 2002), peroxisomes (Fukao et al., 2002), and cell walls (Slabas et al., 2004), since they contain a limited number of proteins which help in protein identification. Katz et al. (2007) classified the proteins from citrus juice cells as per the putative function and assigned the function according to the known biosynthetic pathway.

9.3.4 Conclusion and Future Perspectives

The current status of citrus genome, proteome, and metabolomics research, including the development of fundamental tools, the applications currently under way, and envisaged solutions by the citrus experts to the intriguing problems facing the citrus industries of the world are reviewed. Future analyses of the currently available genomes include exploration of citrus genome evolution and duplication phylogeny and the origins of the sweet orange (the most widely grown citrus type in the world), identification of SSR and SNP sets, comparisons of gene content and genome structure, and in the longer term will come studies of gene and allelic content as they relate to citrus biochemistry, metabolism, physiology, disease resistance, stress tolerance, fruit quality, and productivity. The foundational reference genome from the haploid Clementine will serve well for accessing and utilizing the newly produced genome sequences. Deep sequencing of citrus transcriptomes is already underway in gene expression studies of host responses to pathogens, with a major emphasis currently on Huanglongbing (Citrus greening or HLB, a most serious disease ravaging citrus production on a nearly global basis).

The tagging of genes conferring resistance/tolerance should be a high priority for the programs envisaging applications to breeding. Understanding the current knowledge of citrus genomics, proteomics, and metabolomics, and integrating this with conventional breeding would allow access to and combination of genes with precision to manage biotic and abiotic factors.

9.4 Grape

9.4.1 Introduction

Grape is the most important fruit crop in the world. The majority of the commercially grown grape varieties belong to the European species *Vitis vinifera*, family Vitaceae. Many wild species, viz. *V. rupestris*, *V. berlendieri*, *V. riparia*, and their hybrids are used as rootstocks. *V. vinifera* is a diploid plant with $2n = 38$ chromosomes and a relatively small genome size of 500 Mb.

9.4.2 Genome Information

9.4.2.1 Markers

Grapevine is a vegetative propagated, heterozygous plant. Germplasm conservation of grape is in the form of *ex situ* field collection. The management of such collection requires accurate identification to reduce redundancy, maintain true-to-typeness, and to streamline the addition of new accessions. Several molecular markers such as RFLP, RAPD, microsatellites, and retrotransposon-based markers have been developed in grape and used for characterization and identification. However, microsatellite or SSR markers are found to be the most suitable and extensively used to analyze the genetic diversity, phylogeny, and species interrelationships in grape. First reported by Thomas et al. (1993), a large number of microsatellite markers have been developed through international collaboration in the form of the Vitis Microsatellite Consortium, as well as several research groups (Bowers et al., 1996, 1999; Merdinoglu et al., 2005; Di Gaspero et al., 2005). Microsatellite markers are successfully used for pedigree analysis, variety identification, genetic relationship analysis, analysis of clonal variation, and so forth. In recent years, emphasis has been on the analysis of large collections (Laucou et al., 2011; Santana et al., 2010; Cipriani et al., 2010; Ibanez et al., 2009; Le Cunff et al., 2008) for genetic diversity analysis and identification of duplicates, misnomer, and synonymy for germplasm management. These studies have helped in unraveling the history of the evolution and domestication of grape cultivation (This et al., 2006).

AFLP markers have been sparingly used for genetic diversity analysis (Cervera et al., 2000; Martinez et al., 2004; Upadhyay et al., 2007), but have been used more frequently for detecting clonal variation (Vignani et al., 2002; Blaich et al., 2007; Moncada and Hinrichsen, 2007; Upadhyay et al., 2011). SNP marker discovery and use in grape are facilitated by sequencing of the grape genome and the availability of high-throughput sequencing techniques. Identification of SNPs has been reported by several workers (Salmaso et al., 2004; Lijavetzky et al., 2007; Myles et al., 2010) and used for the genotyping of grape genotypes. A set of 48 stable SNP markers is proposed as a standard set for grapevine genotyping (Cabezas et al., 2011). Myles et al. (2010) developed and used SNP arrays to distinguish *V. vinifera* and its wild *Vitis* species, and also for assessing the population structure and level of linkage disequilibrium in grape. Based on SNP analysis, the genetic structure and the domestication history of grape were confirmed (Myles et al., 2011), and the pedigree relationship between the present set of commercial cultivars was analyzed.

9.4.2.2 Mapping

Microsatellite, AFLP, RAPD, and SNP markers have been extensively used for the construction of saturated linkage maps in grape. There are more than 25 reports of molecular linkage maps for *Vitis vinifera* and other *Vitis* species. An integrated reference map for cultivated grapevine (*Vitis vinifera* L.) from three crosses, based on 283 SSR, 350 AFLP, and 501 SNP-based markers, have been developed by Vezzulli et al. (2009). A dense SNP-based genetic linkage map of grapevine which also anchors the Pinot Noir physical map has been constructed and made publicly available (Troggio et al., 2007). The framework map and the integrated map from the International Grape Genome Programs are made available at the National Center for Biotechnology Information (NCBI; http://www.ncbi.nlm.nih.gov/mapview).

QTLs have been identified for disease resistance (Akkurt et al., 2007; Blasi et al., 2011; Fischer et al., 2004; Lowe and Walker 2006; Riaz et al., 2006, 2011; Xu et al., 2008; Marguerit et al., 2009; Zhang et al., 2009; Welter et al., 2007), fruit quality (Doligez et al., 2006; Fournier-Level et al., 2009), phenology-related traits (Costantini et al., 2008), seedlessness (Doligez et al., 2002; Mejia et al., 2007; Gebauer et al., 2009), flower sex (Dalbó et al., 2000), berry characters (Cabezas et al., 2006), aroma compounds (Doligez et al., 2006; Duchêne et al., 2009), berry color (Fournier-Level et al., 2009), fruit yield component (Fanizza et al., 2005), and developmental stages (Duchêne et al., 2012). Marker-assisted selection for transfer of QTL has been used for seedlessness, pierce disease resistance, and powdery mildew (Pauquet et al., 2001).

9.4.2.3 Whole Genome Information

The full genome sequence of the grapevine was completed in 2007 (Jaillon et al., 2007; Velasco et al., 2007). Jaillon et al. (2007) reported the sequencing of self-bred genotype PN40024 derived from cultivar Pinot Noir. The 8X draft sequence contained a set of 30,434 protein-coding genes. Forty-one percent of the genome was found to be composed of repetitive/transposable elements. The genes involved in wine characteristics have higher copy number. Analysis of paralogous regions suggested that the grape haploid genome is derived from three ancestral genomes. In another study, Velasco et al. (2007) sequenced the genome of Pinot Noir clone ENTAV115, the size of which was estimated to be 506 Mb. A consensus sequence of the genome and a set of mapped SNP markers were generated. The candidate genes and genomic regions involved in different traits like wine quality and disease resistance/susceptibility were identified. Under the aegis of the International

Grape Genome Program (IGGP), a 12X assembly of the genome sequence was released at NCBI with the V0 version of gene annotation in 2009, and was later upgraded to V1 version. The assembly data were prepared by a French–Italian public consortium. The 12X assembly is composed of 2059 scaffolds with a total length of 485 Mb.

9.4.2.4 Transcriptomics

The expression profiling or transcriptomics which analyzes the expression levels of mRNA in a cell population has been extensively used to understand the underlying mechanism of different processes related to different stages of grape berry development and grapevine response to different biotic and abiotic stresses. Large numbers of ESTs have been generated by several workers (Peng et al., 2007; Terrier et al., 2001). Da Silva et al. (2005) developed a compendium of gene expression during berry development in multiple *Vitis* species. These resources were used by Tillett et al. (2011) to identify tissue-specific and abiotic stress-responsive gene expression in wine grapes. VitisNet, a system of grapevine molecular networks, was developed (Grimplet et al., 2009a), which allowed integration of different resources. Zamboni et al. (2010) integrated transcriptomic, proteomic, and metabolomics data into networks and identified putative stage-specific biomarkers for berry development. In a recent study, Fasoli et al. (2012) analyzed transcriptomes of about 54 samples and developed a genome-wide transcriptome atlas of grapevine. The analysis revealed fundamental transcriptomic reprogramming during the maturation process.

The transcriptome of different tissues like leaf (Ablett et al., 2000; Kobayashi et al., 2009; Liu et al., 2012), berry (Ablett et al., 2000), berry skin (Ali et al., 2011; Kobayashi et al., 2009; Lijavetzky et al., 2012; Waters et al., 2005, 2006), and berry flesh (Lijavetzky et al., 2012), as well as different stages of berry growth like berry development (Costenaro-da-Silva et al., 2010; Fortes et al., 2011; Terrier et al., 2005; Zenoni et al., 2010) and berry ripening (Davies and Robinson, 2000; Guillaumie et al., 2011; Kobayashi et al., 2009; Pilati et al., 2007), has been used to analyze the differential gene expressions and to identify major genes expressed during different stages. Genes and regulatory pathways involved in grapevine response to different abiotic stresses like water deficit (Castellarin et al., 2007; Cramer et al., 2007; Grimplet et al., 2007), salinity (Cramer et al., 2007; Daldoul et al., 2012; Tattersall et al., 2007), osmotic stress (Tattersall et al., 2007), heat stress (Liu et al., 2012), UV-B radiation (Pontin et al., 2010), and chilling requirements (Mathiason et al., 2009; Tattersall et al., 2007) have been studied through transcriptome analysis, mainly using microarrays. The influence of source–sink ratio (Pastore et al., 2011) on berry characteristics and postharvest water loss (Rizzini et al., 2009) helped in identifying the genes affecting the quality of grape and wine.

9.4.3 Proteomics

Proteomics or the large-scale analysis of proteins is gaining importance in the study of biological systems. The availability of the grape genome sequence and techniques for high-throughput protein identification have provided the impetus for proteome analysis, which allows the study of the manifestation of gene expression. The grape proteome has been analyzed to study the effect of physical factors like the herbicide flumioxazin (Castro et al., 2005), ABA treatment (Giribaldi et al., 2010), water deficit (Vincent et al., 2007; Grimplet et al., 2009b), salt stress (Vincent et al., 2007; Jellouli et al., 2008), photoperiod-induced growth (Victor et al., 2010), and gibberellic acid (GA3; Wang et al., 2012). Pathogen-related genes and proteins have been identified by analysis of the interaction of plants with phytoplasma (Margaria and Palmano, 2011), *Plasmopara viticola* (Milli et al., 2012), Esca proper (Pasquier et al., 2012; Spagnolo et al., 2011), and *Xylella fastidiosa* (Yang et al., 2011). The protein changes occurring during berry ripening (Deytieux et al., 2007; Giribaldi et al., 2007; Martínez-Esteso et al., 2011a; Negri et al., 2008; Zhang et al., 2008), postharvest withering (Di Carli et al., 2010), and different stages of berry development (Martínez-Esteso et al., 2011b; Negri et al., 2011) have been studied to improve the understanding of the underlying cellular processes during berry development which affect wine quality.

9.4.4 Conclusion and Future Perspectives

Grapevine is considered the model plant for a fruit crop. Its relatively small genome size as well as its economic importance attracts the attention of researchers the world over. The sequencing of the grape genome is the most important accomplishment and has facilitated the integration of different omics to investigate the different processes involved in different stages of development, fruit and wine quality, and response to biotic and abiotic stresses. The information generated through these studies has significantly increased the knowledge on grapevine biology.

Further understanding of the factors affecting fruit and wine quality and response to biotic and abiotic stresses will help in evolving strategies for the development of grape genotypes with improved characteristics that are suitable for varied climate conditions. Integration of data generated through various omics approaches will pave the way for elucidation of the networks and pathways associated with the different processes.

9.5 Guava

9.5.1 Introduction

Guava (*Psidium guajava* L. $2n = 2x = 22$; 496 Mbp) belongs to the genus *Psidium* of the family Myrtaceae. This genus has nine species (USDA, Natural resource conservation service) and all are native to the tropical regions of northern South America (Risterucci et al., 2005). They are now naturalized throughout the tropics and subtropics of Asia, Africa, subtropical regions of North America, and Australia. Mexico, Brazil, India, and Thailand are the largest producers of guava. The fruits, which are a rich source of dietary fiber, vitamins A and C, and folic acid, are eaten fresh or processed into juices and jams. The leaves and bark have pharmacological actions and have been used in folk medicine for gastrointestinal tract ailments, gum problems, fever, and cough (Kamath et al., 2008).

Conventional breeding in guava by selection, intervarietal and interspecific hybridization has been carried out in the past to give rise to superior varieties but the application of molecular tools for breeding has not received much attention. Until recently, molecular markers and linkage maps were not available for this crop. The EU Project on "Improvement of guava: Linkage mapping and QTL analysis as a basis for marker assisted selection" provided a breakthrough, and genomic resources were generated. The NCBI, which serves as a repository of computerized databases and analysis tools for biology-related information, currently (December, 2012) contains a mere 83 nucleotide, 22 GSS and 31 protein sequences for guava.

9.5.2 Genomic Information

9.5.2.1 Markers

Valdés-Infante et al. (2003) were the first to develop genomic resources in guava by using AFLP markers to characterize the guava accessions of Cuba. They were also the first workers to develop the linkage map in the population Enana Cubana roja × N6. This linkage map was later enhanced by Rodriguez et al. (2007) and contained 220 AFLP markers. Several QTLs for vegetative traits and fruit characteristics were integrated into this map to make it more informative (Rodriguez et al., 2007). The SSR markers were initially developed by Risterucci et al. (2005) using a (GA)n and (GT)n microsatellite-enriched library, and 23 nuclear SSR loci were characterized in the guava species (*Psidium guajava* L.). These SSR markers have been put to use by several workers for assessing the genetic diversity of guava in different countries. The guava accession of Mexico was analyzed by Sánchez-Teyer et al. (2010) using 6 SSR loci and 5 AFLP primer combinations in an automatic sequencer. In this analysis, a total of 57 individual accessions from 10 different regions of Mexico were included. The AFLP- and SSR-based dendrogram grouped the guava accessions into two main groups with at least five different subclusters. In another study, microsatellite markers were used to characterize the diversity of guava germplasm in the United States (Viji et al., 2010). Seven polymerase chain reaction (PCR) primers amplifying 14 previously developed microsatellite loci were used to identify and characterize 13 guava accessions. The diversity in nine Indian pink and white flesh cultivars of guava was characterized using 23 microsatellite markers by Kanupriya et al. (2011). The pink cultivars and white cultivars separated into two distinct groups. The EU project (2005–2009) on "Improvement of guava: Linkage mapping and QTL analysis as a basis for marker assisted selection" provided the necessary impetus for development of hundreds of microsatellite (SSR) markers in this crop. The Guavamap project webpage contains a list of 344 SSR primer pairs. These primers are available in the public domain and can be used for the development of additional linkage maps and for the identification of markers and QTLs for traits of interest.

9.5.2.2 Mapping

Under the Guavamap project, individual and combined linkage maps for three mapping populations (Enana × N6, Enana × Suprema Roja, and Enana × Belic L-207) based on AFLP and SSR markers were constructed (Ritter et al., 2010a). For each population, AFLP primer combinations in the range of 100–120 were analyzed, which generated 700–1100 segregating AFLP fragments. In addition, 50–200 SSR markers were also analyzed for the same three mapping populations. The distribution pattern of the markers indicated that Enana was slightly less heterozygous than the other parents used for the mapping population and all the parents shared a large part of the gene pool. Based on the fragments specific to either parent, linkage groups were constructed and 11 linkage groups corresponding to 11 chromosomes of the haploid guava were obtained. A combined parental linkage map of each mapping population contained 850, 427, and 408 markers, respectively, having lengths of 1500–2200 cM each. The individual linkage groups had 35–100 markers each, covering lengths of 150–240 cM. A partial alignment of the maps was performed using common, codominant SSRs, which are supposed to map identical locations in different genetic backgrounds.

The QTL analysis (Ritter et al., 2010b) for vegetative traits like plant height, petiole length, and leaf length and width, along with traits affecting yield like fruit number, fruit weight, length and width, total yield, external pulp thickness, seed number and weight, and quality traits like vitamin C content, acidity, and total soluble solids (TSS) was performed. The mapping population had sufficient variation for these traits to allow an efficient QTL analysis. A total of 2–13 QTLs/traits were detected in the three populations and over 100 QTLs were detected in all. Individual QTLs explained between 5% and over 40% of the total variance. Total variance explained by the sum of all detected QTLs varied between 20% and over 50% between traits and populations. Some QTLs colocated with or closely linked to SSR markers are being used for marker-assisted breeding in different genetic backgrounds. These are the major genomic resources present in guava. Besides these, a few isolated genomic studies are indicated in GenBank accessions, like molecular cloning and expression of genes such as 1-aminocyclopropane-1-carboxylate oxidase (ACO1), ACO2, phytoene synthase, and fatty acid hydroperoxide lyase.

9.5.3 Conclusion and Future Perspectives

At present, the genomic resources in guava are limited to linkage maps and QTLs, while resources like gene sequences are scarce. EST sequencing using next-generation sequencing technology will generate a set of expressed genes in a particular tissue, giving an insight into the active genes under those conditions. These can be used for novel gene discovery, as well as mapping. Association mapping in guava could also prove to be a useful tool for analyzing the presence of traits of interest to breeders in germplasm collections of guava.

9.6 Mango

9.6.1 Introduction

Mango (*Mangifera indica* L.), popularly known as the "King of Fruits," is believed to have originated in eastern India (Knight, 1980). The genus *Mangifera* belongs to the order Sapindales in the family Anacardiaceae, which is a family of mainly tropical species with 73 genera (c. 850 species), with a few representatives in temperate regions. It consists of 69 species and is mostly restricted to tropical Asia. The highest diversity occurs in Malaysia, particularly in peninsular Malaya, Borneo, and Sumatra, representing the heart of the distribution range of the genus. The natural occurrence of all the *Mangifera* species extends as far north as 27° latitude and as far east as the Caroline Islands (Bompard and Schnell, 1997). Mukherjee (1958) reported that mango is an allopolyploid based on the high chromosome number, high number of nuclear chromosomes, secondary association of bivalent, regular pairing, absence of multivalent formation, and good pollen fertility. The *Mangifera* species, viz., *M. indica*, *M. caloneura*, *M. sylvatica*, *M. foetida*, *M. caesia*, *M. odorata*, and *M. zeylanica*, have been studied, and found to have chromosome numbers of $2n = 40$ and $n = 20$ (Mukherjee, 1950, 1957, 1963). Chromosome numbers and ploidy status of other *Mangifera* species have yet to be studied (Bompard and Schnell, 1997). The mango genome consists of 4.39×10^8 bp (Arumuganathan and Earle, 1991) and has 20 chromosomes, most of which are small. The common mango was closely related to *M. sylvatica* Roxb., *M. laurina* Bl., *M. oblongifolia* Hook. f., *Mangifera macrocarpa* Bl., *M. foetida*, and *M. Odorata* (Eiadthong et al., 2000).

Mango is commercially grown in over 103 countries of the world, but nowhere is it so greatly valued as in India. The fruit is a rich source of vitamins A and C. There is good demand for mango fruits and their processed products on both the national and international markets. It is necessary that one knows the desirable characteristics of a variety before taking up conventional or molecular breeding. There is very limited work on molecular breeding and the mapping of the genotypes in mango compared to other fruit crops like papaya, banana, and guava.

9.6.2 Genome Information

9.6.2.1 Markers

Various biotechnological methods to understand and improve the crop genetics are being applied, which reduce the time required for breeding. Markers can be used as a tag to isolate unique types using marker-assisted selection. Even genetic distance studies among genotypes can be taken up using these markers. Using the molecular approach, one can genetically manipulate a plant to have desirable characteristics. However, molecular studies in mango are very few.

Initially, RAPD markers were used to study the genetic diversity of mango (Schnell et al., 1995; Ravishankar et al., 2000; Ravishankar et al., 2004; Kumar et al., 2001). Later, other more reproducible and reliable markers like AFLP and SSR were employed.

AFLP markers are quite suitable for cultivar identification, estimation of genetic relationships, and mapping of QTLs in mango (Kashkush et al., 2001). Kashkush et al. also constructed a genetic linkage map, which consisted of 13 linkage groups and covered 161.5 cM defined by 34 AFLP markers. The information generated by AFLP analysis regarding genetic relatedness and diversity existing in the mango gene pool is useful for breeding improved mango varieties (Yamanaka et al., 2006). Fang et al. (1999) constructed a genetic fingerprinting of two mango cultivars, viz., Keitt and Tommy, employing AFLP. A study using 31 F1 progenies from crosses between Alphonso and Palmer led to the construction of maps for each cultivar that were useful for analyzing correlations of traits like fruit size, shape, and color (Phumichai et al., 2000). The phylogenetic relationships among 14 *Mangifera* L. species, including three economically important species, were analyzed by comparing 217 AFLP markers (Eiadthong et al., 2000).

SSR markers are widely used as a versatile tool in plant breeding programs as well as in evolutionary studies because of their ability to show diversity among cultivars (Viruel et al., 2005; Schnell et al., 2006; Ravishankar et al., 2011). SSRs, also known as microsatellites, are an efficient type of molecular marker based on tandem repeats of short (2–6 nucleotides) DNA sequences (Charters et al., 1996), and target highly variable and numerous loci. Among the 64 Florida cultivars evaluated in the parentage analysis by Schnell et al. (2006), the genetic background was found to be based on as few as four Indian cultivars and the polyembryonic cultivar Turpentine. Two Indian cultivars, Mulgoba and Sandersha, are in the background of most Florida types, with Amini, Bombay, Cambodiana, Long, Julie, Turpentine, and Nam doc Mai making lesser contributions. Microsatellites were successfully used for genetic diversity analysis of the indigenous type Appemidi. The analysis of the 211 bands generated by the 14 SSR markers showed the unambiguous discrimination of the 43 mango genotypes. The dendrogram resulted in the grouping of accessions into two major clusters, viz., cluster I with highly acidic types and cluster II with less acidic and high TSS group (Vasugi et al., 2012). The work related to SNP in mango is very recent. Khan and Kamran Azim (2012) used SNP for mango cultivar identification and phylogenetics. They characterized rpl20-rps12 and atp-rbcL intergenic spacers of chloroplast DNA from 19 mango cultivars to analyze the suitability of these regions for assessing intraspecific variation.

9.6.2.2 Mapping

Mango is a highly heterozygous crop, and hence we cannot predict progeny performance. It is therefore important to study and understand the genome. Very little work has been done on mapping. The first genetic linkage map in mango was reported by Kashkush et al. (2001), utilizing AFLP markers and 29 progeny from a cross of Tommy Atkins × Keitt in Israel. They were able to map 34 AFLP loci and produced a crude linkage map that identified 13 of the 20 linkage groups covering 160 cM. A second map has been produced using 60 progeny from a cross of Keitt × Tommy Atkins in China using AFLP markers. Eighty-one markers with the correct segregation ratios were identified and 39 of these were used to identify 15 linkage groups. The average distance between two adjacent markers was 14.74 cM. Improvement of the mango recombination map requires the development of more codominant molecular markers.

9.6.2.3 Transcriptome

In mango, very few studies have addressed gene expression during fruit development and ripening. In mango, variations in qualities dependent on the pre-harvest environment have also been reported (Hoffman et al., 1997), which indicate that the genes, other than those related to flavor, might play a regulatory role. Under such circumstances, transcriptome analysis through its development and ripening would be an ideal approach. In Alphonso, more than 90% of the total aroma compounds in fresh fruit pulp are mono and sesquiterpene hydrocarbons (Engel and Tressl, 1983; Idestein and Schreier, 1985). In addition, the color-conferring compounds like anthocynins and isoflavonoids are also products of the same terpenoid biosynthesis pathways (Bohlmann et al., 1998). The study of this pathway is the most appropriate start for transcriptomic studies. At molecular level, only very few genes have been studied, such as peroxisomal thiolase mRNA (Bojorquez and Gomez-Lim, 1995), alternative oxidase (AOX), and uncoupling proteins (UCP; Considine et al., 2001). AOX and UCP play an important role in postclimacteric senescence processes. Vasanthaiah et al. (2006) indicated that oxidative stress is the probable cause for the spongy tissue in Alphonso. High monodehydroascorbate reductase (MDHAR) transcript levels indicate the high antioxidant levels in mango fruit and consequently ensure the high quality of nutrition (Pandit et al., 2007).

9.6.3 Metabolomics and Proteomics

Metabolomics is the analysis of the complete metabolome (all the metabolites that one organism produces) at a particular stage (Fiehn, 2002). It provides the phenotypical response at the metabolic level in a particular environmental circumstance. It helps to monitor the phenotypic variability of one genotype in response to environmental changes in drought (Fumagalli et al., 2009), nutrient availability (Hirai et al., 2004, 2005), pollutants (Jones et al., 2007; Bundy et al., 2008), salinity (Fumagalli et al., 2009), temperature (Michaud and Denlinger, 2007), and biotic interactions (Choi et al., 2006), among other ecological factors. These studies are especially applicable in plants because metabolomic studies enable the simultaneous analysis of primary compounds together with secondary compounds, which have a defensive and protective function. The unique volatile metabolites associated with microorganisms that cause food spoilage were studied in mango (Ibrahim et al., 2011).

Protein profile analysis of mango fruit during ripening could help identify key steps in the metabolic control of mango fruit quality. A first comparative proteomic investigation between the preclimacteric and climacteric stages of mango fruits of the Keitt cultivar to identify the most variable proteins within the pulp resulted in the identification of differentially abundant protein species (Andrade et al., 2012).

9.6.4 Conclusion and Future Strategies

The use of molecular markers in mango breeding would greatly assist in solving a number of difficult challenges for breeders such as the development of complex family structures for recombination mapping and for recurrent selection. The more challenging problem is the development of large mapping populations and the accumulation of phenotypic data to identify QTL regions associated with traits of interest. The other problem in perennials like mango is the long juvenile period of 5–7 years. Once the populations are made and evaluated, the application of marker-assisted selection would be straightforward. Mapping quantitative traits is difficult because the genotype is never unambiguously inferred from the phenotype. Gene expression studies in mangoes are limited. It is desirable to identify the genes uniquely expressed for particular traits. At the molecular level, no studies have been carried out to isolate genes involved in pest and disease resistance and physiological disorders.

The breeding programs aim to develop new cultivars that are suitable for modern high-tech horticultural production systems. Mango has a comparatively small haploid genome size about three times as large as *Arabidopsis* (Arumuganathan and Earle, 1991). Therefore, there is an urgent need for sequencing the whole genome. This information should be exploited using modern molecular techniques to develop superior inter/intraspecific cultivars of desirable traits, viz., regular and precocious bearing, dwarf types, good quality with better longevity, free from physiological disorders, pests and diseases.

9.7 Papaya

9.7.1 Introduction

Papaya (*Carica papaya* L., $2n = 2x = 18$, 372 Mbp) is a tropical fruit tree, first domesticated in the eastern lowlands of the central Americas (Nakasone and Paull, 1998) and now cultivated around the tropical and warmer subtropical areas of the world (Villegas, 1997). Papaya belongs to the family Caricaceae and, until recently, this family was thought to comprise about 31 species in four genera. However, a recent taxonomic revision proposed that some species formerly assigned to *Carica* were more appropriately classified in the genus *Vasconcella* (Badillo, 2002). The species in this genus are called wild papayas and are distantly related to *Carica papaya*. They possess several desirable traits such as resistance to pathogens, cold tolerance, and high sugar content of fruits, which can be used for the improvement of cultivated papaya if the pre- and postzygotic barriers are overcome (Drew et al., 1998).

The annual worldwide production of papaya is estimated at 6.9 mt (FAOSTAT, 2007). It is largely used for the consumption of fresh fruit and for use in jams, jellies, pies, juices, and as dried and crystallized fruit (Villegas, 1997). Nutritionally, papaya is an excellent source of calcium and vitamins A, C, folate, and riboflavin (Nakasone and Paull, 1998). The green unripe fruits produce an enzyme called papain which has industrial and medicinal use. Papaya has now naturalized in many areas of the world. The undomesticated papaya was a spindly plant with almost inedible fruits, but during domestication the species has undergone considerable changes in fruit size, fruit flesh color, mating system, and growth habit (Manshardt and Moore, 2003).

9.7.2 Genomic Information

9.7.2.1 Markers

Microsatellite markers are the most widely used markers in major crops (Gupta and Varshney, 2000). This class of markers is highly reliable, codominant in inheritance, relatively simple and cheap to use, and generally highly polymorphic in nature. Several microsatellite markers have been developed for papaya (Sharon et al., 1992; Parasnis et al., 1999; Santos et al., 2003; Collard and Mackill, 2008; Eustice et al., 2008; Oliveira et al., 2008, 2010a). Reliable and highly polymorphic markers have been used for the construction of genetic maps (Chen et al., 2007), sexual differentiation (Parasnis et al., 1999; Santos et al., 2003), and to assess genetic diversity (Pérez et al., 2007; Oliveira et al, 2010a).

The developed SSR markers are being used to identify plants presenting high levels of homozygosity in segregating progenies for the development of new inbred papaya lines, which are needed in order to avoid F1 segregation. Crossing between these inbreds will allow development of hybrids with vigor superior to the parents (Oliveira et al., 2010b). SSR markers are also being used for genetic characterization of papaya plants derived from the first backcross generation (Ramos et al., 2011).

In papaya, fruit flesh color is controlled by a single gene as segregating populations show a 3:1 inheritance of yellow:red fruit flesh color. Using a candidate gene and map-based cloning approach a chromoplast-specific lycopene β cyclase CpCYC-b gene was identified. A 2 bp insertion found in the CpCYC-b coding region has been identified as resulting in a frame shift mutation and a premature stop codon. Using this information, a simple PCR-based screening test has been developed based on the CPFC1 marker to identify individuals in a segregating population with the desired flesh color (Blas et al., 2010).

Sex inheritance in papaya is determined by a dominant M allele for males, a different dominant Mh allele for hermaphrodites, and a recessive m allele for females. The MhMh, MhM, and MM genotypes are lethal, thus leading to a 2:1 (hermaphrodite:female) segregation ratio after hermaphrodite plants (Mhm) have self pollinated (Ming et al., 2007); thus, all-hermaphrodite progeny is not possible. Since female plants have less commercial value than hermaphrodites, identification of hermaphrodites at an early stage is desirable. Several male-hermaphrodite-specific markers have been developed by RAPD or AFLP, which were converted to SCAR markers, for example T12 and W11 (Deputy et al., 2002), napF (Parasnis et al., 2000), and papaya sex determination marker (PSDM; Urasaki et al., 2002) to select for hermaphrodites from female papaya seedlings. Recently, a loop-mediated isothermal amplification (LAMP) reaction has been developed for quick identification of sex in the papaya field (Hsu et al., 2012). LAMP is a novel technique that requires a special primer set design which recognizes a total of six regions of the targeted DNA sequence (Notomi et al., 2000). The advantage is visual judgment based on turbidity or fluorescence of the reaction mixture without the need for specific equipment (Hsu et al., 2011).

9.7.2.2 Mapping

The first genetic map of papaya consisted of three morphological markers: sex form, flower color, and stem color (Hofmeyr, 1939). When RAPD markers were developed, the second genetic map of papaya became available with 62 RAPD markers (Sondur et al., 1996). A high density linkage map was constructed with 1498 AFLP markers, papaya ringspot coat protein marker, morphological sex type, and fruit flesh color markers using 54 F2 plants derived from the cross Kapoho × SunUp. These markers were mapped to 12 major linkage groups covering a total length of 3294.2 cM. This map revealed severe suppression of recombination around the sex determination locus, with 225 markers cosegregating with sex types (Ma et al., 2003). Microsatellite markers derived from BAC end sequences and whole genome shotgun sequences were used to analyze and map 54 F2 plants derived from the varieties AU9 and SunUp. This map validated the suppression of recombination at the male-specific region of the Y chromosome (MSY) mapped on LG1 and at potential centromeric regions of other linkage groups. A total of 707 markers, including 706 microsatellite loci and the morphological marker fruit flesh color, were mapped into nine major and three minor linkage groups. The resulting map spanned 1069.9 cM with an average distance of 1.5 cM between adjacent markers (Chen et al., 2007). AFLP, SSR, and morphological markers were used by Blas et al. (2009) to construct a high-density genetic linkage map of papaya, previously developed using an F2 mapping population derived from the intraspecific cross AU9 × SunUp. The comprehensive genetic map spanned 945.2 cM and covered 9 major and 5 minor linkage groups containing 712 SSR, 277 AFLP, and 1 morphological markers. Yu et al. (2009) constructed a BAC-based physical map of papaya and integrated it with the genetic map and genome sequence. This integrated map facilitated the draft genome assembly, and 72.4% of the genome was aligned to the genetic map. This map is proving to be a valuable resource for comparative genomics and map-based cloning of agronomically and economically important genes and for sex chromosome research.

QTL mapping has been done to identify the multiple genes controlling papaya fruit size and shape. A large variation is present in the germplasm for these traits. The weight of the fruit may vary from 0.2 to 10 kg, while the shape of the fruit is linked to sex, ranging from spherical to ovate, cylindrical, or pyriform. An F2 mapping population produced from a cross between the Thai variety Khaek Dum, bearing 1.2 kg, red-fleshed fruit was crossed with the Hawaiian Solo type bearing 0.2 kg, yellow-fleshed fruits. Fourteen QTLs with phenotypic effects ranging from 5% to 23% were identified across six linkage groups (Blas et al., 2009). The 1-LOD interval surrounding each QTL was searched for candidate genes. Five candidate genes (without QTL association) that showed homology to one of three previously identified tomato loci (ovate, sun, or fw 2.2) affecting the tomato fruit size and shape were identified in the papaya genome.

9.7.2.3 Whole Genome Sequencing

Papaya was selected for genome sequencing since it is considered to be an excellent model crop to study the evolution of sex chromosomes and for tropical fruit tree genomics. At 372 Mbp, the papaya genome is three times the size of *Arabidopsis*, with which it shared a common ancestor about 72 million years ago (Wikstrom et al., 2001). The transgenic female SunUp variety was chosen for sequencing using whole genome shotgun (WGS) with Sanger sequencers (Ming et al., 2008). Some interesting features emerged from the genome sequencing, such as that papaya has about 25% fewer genes than *Arabidopsis*, with the reduction across most gene families and biosynthetic pathways. The lower gene number is believed to be due to the absence of recent genome-wide duplication of the papaya genome, unlike *Arabidopsis*. Papaya contains fewer (only 25%) resistance genes than *Arabidopsis*, which signifies that it may have evolved some alternative system of defense. The genes involved in fruit ripening, circadian rhythm, and light signaling are less abundant in papaya compared to other sequenced genomes, while the genes in the MADS-box family and genes involved in the development of volatiles in fruit are tremendously amplified (Ming et al., 2012).

9.7.2.4 Transcriptomic Studies

Prior to sequencing the whole genome, a normalized subtractive cDNA library was constructed using pooled RNA from roots, leaves, seeds, calli, the three sex forms of flowers, and three ripening stages of papaya fruit. Over 50,000 EST sequences were generated yielding 16,432 unigenes for genome annotation (Ming et al., 2008).

Papaya produces climacteric fruits with soft and sweet pulp rich in health-promoting phytochemicals. Despite its importance, limited research has been done on transcriptional modifications during fruit growth, ripening, and its control. In some of the initial studies, cloning and characterization of aminocyclopropanecarboxylate (ACC) oxidase cDNA (CP-ACO1, L-76283 by Lin et al. (1997) and CP-ACO-2 by Chen et al. (2003)) from papaya fruit were reported. Devitt et al. (2006) generated a large number of ESTs from randomly selected clones of two independent fruit cDNA libraries derived from yellow- and red-fleshed papaya varieties. Of the 1171 ESTs, the most abundant sequences encoded were chitinase, 1-aminocyclopropane-1-carboxylic acid (ACC) oxidase, catalase, and methionine synthase. The identified ESTs showed significant similarity to genes associated with fruit softening, aroma, and color biosynthesis. In another study, Fabi et al. (2010) identified ripening-related genes affected by ethylene using cDNA-AFLP. The transcript profiling revealed 71 differentially expressed genes between noninduced and ethylene-induced fruits. The genes involved in ethylene biosynthesis, regulation of transcription, and stress response or plant defense were reported (heat shock proteins, polygalacturonase-inhibiting protein, and acyl-CoA oxidases). Most of the transcription factors were negatively affected by ethylene except for a 14-3-3 protein, an AP2 domain-containing factor, a salt-tolerant zinc finger protein, and a suppressor of PhyA-105 1. Fruit quality-related genes, like genes for cell wall structure or metabolism, volatiles or pigment precursors, and vitamin biosynthesis, were also found. In a recent study by Fabi et al. (2012), analysis of ripe papaya transcriptome was done by using a cross-species (XSpecies) microarray technique using RNA from unripe and ripe papaya to probe the Affymetrix *Arabidopsis* GeneChip ATH 1-121501. The profile of ripening-related gene expression in papaya showed a total of 414 ripening-related genes which indicated that transcription factors (TFs) of the MADS-box, NAC, and AP2/ERF gene families were involved in the control of ripening. Similarities between papaya and tomato were found with respect to expression of genes encoding proteins involved in primary metabolism, regulation of transcription, biotic and abiotic stresses, and cell wall metabolism.

9.7.3 Metabolomics and Proteomic Studies

The quality and volatile attributes of attached and detached Pluk Mai Lei papaya during fruit ripening were investigated by Fuggate et al. (2010), who found that the levels of methanol and ethanol increased steadily during ripening, with esters formed from ethyl alcohol predominating from half-ripe through senescence phases. The alcohol dehydrogenase activity increased tremendously during early ripening, while alcohol acetyltransferase was active throughout ripening. No difference was found in the profile of volatiles of attached and detached fruits. Papaya is a rich source of antioxidants and vitamin C. The carotenoid development of pre- and postharvest, red-fleshed papaya fruits was investigated by Schweiggert et al. (2011) using HPLC with diode-array detection coupled to mass spectrometry. Esterified xanthophylls such as β-cryptoxanthin laurate and caprate were found to be the most abundant pigments during carotenoid biosysnthesis. Fruit maturation led to an increase of carotenoids, particularly β-cryptoxanthin laurate, and total lycopene contents disproportionately increased. Total carotenoid contents of fully ripe papaya ranged from 5414 to 6214 μg/100 g of FW, while corresponding biosynthetic precursors such as phytoene, phytofluene, and ζ-carotene were only detected in trace amounts. A metabolic study was conducted to investigate fruit flesh gelling, a postharvest disorder in papaya fruits. The juice of affected and nonaffected fruits was compared using nuclear magnetic resonance (NMR). Also, a comparison was made with fruits which initiated the degradation process. Analysis of the data by principal component analysis (PCA) revealed that all three types of fruits displayed a specific profile. Healthy fruits were characterized by the presence of malic acid and the absence of ethanol. Both fruits with flesh gelling and those that had started to degrade contained no detectable levels of malic acid and variable levels of ethanol. Furthermore, increased levels of lactic acid, acetic acid, and succinic acid were encountered. In comparison with fruits that had started to degrade, the fruits with flesh gelling showed higher levels of asparagines in particular (Schripsema et al., 2010). Rivera et al. (2010) reported the phenolic and carotenoid profiles of papaya fruit under low temperature storage. No significant difference in carotenoid content was found in relation to storage temperature. Sancho et al. (2011) conducted qualitative and quantitative analysis of the major phytochemicals found in papaya flesh and skin during four stages of ripening, using HPLC–MS. Phenolic compounds identified in the fruit skin decreased with ripening, while the carotenoids, along with vitamin C, increased in flesh with ripening.

Proteins play a central role in biological processes like ripening of fruits, and differential proteomics can help in the identification of proteins affected during the process. A comparative analysis of climacteric and pre-climacteric papayas using 2DE-DIGE identified 27 proteins which were classified into six main categories related to the metabolic changes occurring during ripening (Nogueira et al., 2012). Proteins from the cell wall (α-galactosidase and invertase), ethylene biosynthesis (methionine synthase), climacteric respiratory burst, stress response, synthesis of carotenoid precursors (hydroxymethylbutenyl 4-diphosphate synthase, GcpE), and chromoplast differentiation (fibrillin) were identified. In another study, to understand the physiological and biochemical modifications in fruit after 6 and 18 days of harvest and the effect of 1-MCP (1-methylcyclopropene), a nontoxic ripening retardant, 2DE analysis was done. The protein profile showed 27 spots which were identified by nano-LC-ESI/MS/MS. Some spots corresponding to cell wall-degrading enzymes related to fruit ripening were identified, while others corresponding to oxidative damage protection, protein folding, cell growth, and survival were found to be induced by 1-MCP (Huerta et al., 2012).

9.7.4 Conclusion and Future Perspectives

Papaya has great nutritional, industrial, and medicinal value. The whole genome sequencing of papaya along with integrated genetic and physical maps, EST database, metabolomic and proteomic studies, and other genomic resources will provide a breakthrough in papaya improvement and in the exploration of its nutritional and medicinal applications. The papaya industry will benefit from the sequencing of the MSY region and its X chromosome counterpart, which will aid in the identification of sex-determining genes controlling the male, female, and hermaphrodite sex forms in papaya.

9.8 Pomegranate

9.8.1 Introduction

Pomegranate (*Punica granatum* L.; $2n = 2x = 18$) is a diploid, perennial, woody plant in the Lytheraceae family. The genome size of a pomegranate has been ascertained to be 704 Mb, about six times the size of *Arabidopsis thaliana* (Bennett and Leitch, 2010). The pomegranate is believed to have originated in Iran (Simmonds, 1976; Levin, 1994) and from there it diversified to other regions like Mediterranean countries, India, China, and Afghanistan through ancient trade routes. It is one of the oldest known edible fruits (Damania, 2005) and is highly prized for its nutritional and medicinal properties (Noda et al., 2002). In recent years, the demand for pomegranate has increased due to the presence of abundant quantities of antioxidants and other phytochemicals in the fruit and peel. Although some effort has been made in the development of superior cultivars by conventional breeding of this crop, the availability of genomic resources is very poor.

9.8.2 Genome Information

9.8.2.1 Markers

In the past, several studies have been undertaken to ascertain the genetic variability present in this ancient fruit crop. Most of the studies were initially restricted to dominant markers like RAPD (Zamani et al., 2007; Dorgac et al., 2008; Sarkhosh et al., 2006; Talebi et al., 2003), ISSR (Talebi et al., 2003), AFLP (Jbir et al., 2008; Rahimi et al., 2003), and directed amplification of minisatellite DNA (DAMD; Narzary et al., 2009). In 2007, Koohi-Dehkordi et al. published the first report describing the isolation of 15 microsatellites in pomegranate. Since then, 178 SSR markers have been isolated by different workers in this crop, and all of them have been developed by the microsatellite-enriched small insert library preparation method (Hasnaoui et al., 2010; Soriano et al., 2011; Currò et al., 2010; Pirseyedi et al., 2010; Ebrahimi et al., 2010).

9.8.2.2 Mapping

A genetic linkage map of the pomegranate was constructed using 62 individuals of F1 progeny derived from crossing a commercial pomegranate variety from Iran, Malas Danesiyah Esfahani (MDE), and Bihaste Danesefid Ravar (BDR), a genotype that carries a unique soft-seed character, using AFLP markers. Twenty-six marker–primer combinations were used which yielded 216 useful markers, of which 144 were scored in MDE and 130 in BDE. In the MDE map, 67 AFLP markers covered 435.3 cM of the genome and included seven major groups and five minor groups, with a mean map distance of

6.5 cM between adjacent markers. In the other parent, BDR, 66 AFLP markers produced seven major and three minor groups which covered 344.2 cM of the genome. The mean map distance was 5.21 cM. A consensus map was constructed by using 67 AFLP markers which covered 335.5 cM and eight marker groups (Sarkhosh et al., 2012). Association analysis for important morphological traits in 202 pomegranate genotypes using 30 microsatellites was undertaken by Basaki et al. (2011). Morphological traits such as sunburn sensitivity, hull cracking, fruit height, fruit diameter, fruit shape, calyx shape, calyx type, fruit taste, and several other commercially important traits were considered. From 30 primers used, 7 pairs were found polymorphic and produced 23 alleles in 202 pomegranate genotypes.

9.8.3 Metabolomics

Since pomegranate has a lot of medicinal value, a few studies have addressed antioxidant and anti-cancer properties of pomegranate juice using a metabolomic approach. Pomegranate (*Punica granatum* L.) fruits are widely consumed as juice (PJ). The potent antioxidant and anti-atherosclerotic activities of PJ are attributed to its polyphenols, including punicalagin, the major fruit ellagitannin, and ellagic acid (EA). Punicalagin is the major antioxidant polyphenol ingredient in PJ. Punicalagin, EA, a standardized total pomegranate tannin (TPT) extract, and PJ were evaluated for *in vitro* antiproliferative, apoptotic, and antioxidant activities. In an experiment on cancer cells, the superior bioactivity of PJ compared to its purified polyphenols illustrated the multifactorial effects and chemical synergy of the action of multiple compounds compared to single purified active ingredients (Seeram et al., 2005). In another study, the antioxidant activity of pomegranate juices was evaluated by four different methods (ABTS, DPPH, DMPD, and FRAP) and compared to those of red wine and a green tea infusion. Commercial pomegranate juices showed an antioxidant activity, 18–20 Trolox equivalent antioxidant capacity (TEAC), three times higher than those of red wine and green tea (6–8 TEAC). The activity was higher in commercial juices extracted from whole pomegranates than in experimental juices obtained from the arils only (12–14 TEAC). HPLC-DAD and HPLC–MS analyses of the juices revealed that commercial juices contained the pomegranate tannin punicalagin (1500–1900 mg/L), while only traces of this compound were detected in the experimental juice obtained from arils in the laboratory. This shows that pomegranate industrial processing extracts some of the hydrolyzable tannins present in the fruit rind. This could account for the high antioxidant activity of commercial juices compared to the experimental ones. In addition, anthocyanins, ellagic acid derivatives, and hydrolyzable tannins were detected and quantified in the pomegranate juices (Gil et al., 2000).

9.8.4 Conclusion and Future Perspectives

In pomegranate, the development of genomic resources has started very recently. There are two major problems with pomegranate: first, the presence of low diversity in germplasm, and second, the availability of few markers, especially microsatellite markers. There is an urgent need to develop genomic resources like SSR and SNP markers. Since it is the lone cultivated member of the family Lytheraceae, whole genome sequencing would help us to better understand this crop and also to develop varieties/hybrids with disease resistance, especially against devastating bacterial blight.

9.9 General Conclusion

The whole genome sequence has been published for banana, grape, papaya, and citrus. It greatly helps in identifying genes responsible for important agronomic characters, such as fruit quality, disease, and pest resistance. The genomic studies are of immense use in understanding the evolution of plants. The various markers, transcriptomes, and related proteins would help in gene identification and better understanding of the plant–host interactions. In these crops, tagging of genes conferring resistance/tolerance needs to be given a high priority. Understanding of the current knowledge of genomics, proteomics, and metabolomics, and integrating this with conventional breeding would allow access to and combination of genes with precision to manage biotic and abiotic factors. Grapevine is considered as the model plant for a fruit crop, with its relatively small genome size as well as its economic importance. The sequencing of the grape genome has facilitated the integration of different omics to investigate different processes involved in different stages of development, fruit and wine quality, and response to biotic and abiotic stresses. Understanding the factors affecting fruit and wine quality and response to biotic and abiotic stresses will help in evolving strategies for the development of grape genotypes with improved characteristics and suitability for varied climatic conditions.

The whole genome sequencing of papaya along with integrated genetic and physical maps, EST database, metabolomic and proteomics studies, and other genomic resources will provide a breakthrough in papaya improvement and the exploration of its nutritional and medicinal applications. In mango, guava, and pomegranate, development of genomic resources activity has started very recently. There is an urgent need to develop genomic resources like SSR and SNP markers. These genomic resources will contribute to the development of high-density linkage maps and the identification of QTLs for various traits. Further, this information would help in whole genome sequencing in these crops.

References

Ablett, E., Seaton, G., Scott, K., Shelton, D., Graham, M.W., Baverstock, P., Lee, L.S., and Henry, R. (2000). Analysis of grape ESTs: Global gene expression patterns in leaf and berry. *Plant Science* 159(1): 87–95.

Akkurt, M., Welter, L., Maul, E., Töpfer, R., and Zyprian, E. (2007). Development of SCAR markers linked to powdery mildew (*Uncinulanecator*) resistance in grapevine (*Vitis vinifera* L. and *Vitis* sp.). *Molecular Breeding* 19(2): 103–111.

Aleza, P., Juárez, J., Cuenca, J., Ollitrault, P., and Navarro, L. (2010). Recovery of citrus triploid hybrids by embryo rescue and flow cytometry from 2x × 2x sexual hybridisation and its application to extensive breeding programs. *Plant Cell Report* 29: 1023–1034.

Ali, M., Howard, S., Chen, S., Wang, Y., Yu, O., Kovacs, L., and Qiu, W. (2011). Berry skin development in Norton grape: Distinct patterns of transcriptional regulation and flavonoid biosynthesis. *BMC Plant Biology* 11(1): 7.

Andrade, J.M., Toledo, T.T., Nogueira, S.B., Cordenunsi, B.R., Lajolo, F.M., and Nascimento, J.R.O. (2012). 2D-DIGE analysis of mango (*Mangifera indica* L.) fruit reveals major proteomic changes associated with ripening. *Journal of Proteomics* 75: 3331–3341.

Arumuganathan, K. and Earle, E.D. (1991). Nuclear DNA content of some important plant species. *Plant Molecular Biology Reporter* 9: 208–218.

Badillo, V.M. (2002). Carica L. vs. Vasconcella St. Hil. (Caricaceae) con la Rehabilitacion de este Ultimo. *Ernstia* 10: 74–79.

Basaki, T., Choukan, R., Nekouei, S.M.K., Mardi, M., Majidi, E., Faraji, S., and Zeinolabedini, M. (2011). Association analysis for morphological traits in pomegranate (*punica geranatum* l.) using microsatellite markers. *Middle-East Journal Scientific Research* 9 (3): 410–417.

Bassene, J.B., Froelicher, Y., Dhuique-Mayer, C., Mouhaya. W., Ferrer. R.M., Ancillo, G., Morillon, R., Navarro, L., and Ollitrault, P. (2009). Nonadditive phenotypic and transcriptomic inheritance in a *Citrus* allotetraploid somatic hybrid between *C. reticulata* and *C. limon*: The case of pulp carotenoid biosynthesis pathway. *Plant Cell Reporter* 28: 1689–1697.

Bassene, J.B., Froelicher, Y., Dubois, C., Ferrer, R.M., Navarro, L., Ollitrault, P., and Ancillo, G. (2010). Non-additive gene regulation in a *Citrus* allotetraploid somatic hybrid between *C. reticulata* Blanco and *C. limon* (L.) Burm. *Heredity* 105: 299–308.

Bennett, M.D. and Leitch, I.J. (2010). Angiosperm DNA C-values database (release 7.0, December 2010). http://www.kew.org/cvalues/.

Bernet, G.P., Fernandez-Ribacoba, J., Carbonell, E.A., and Asins, M.J. (2010). Comparative genome-wide segregation analysis and map construction using a reciprocal cross design to facilitate *Citrus* germplasm utilization. *Molecular Breeding* 25(4): 659–673.

Blaich, R., Konradi, J., Rühl, E., and Forneck, A. (2007). Assessing genetic variation among Pinot noir (*Vitis vinifera* L.) clones with AFLP markers. *American Journal of Enology and Viticulture* 58(4): 526–529.

Blas, A.L., Yu, Q., Chen, C., Veatch, O., Moore, P.H., Paull, R.E., and Ming, R. (2009). Enrichment of a papaya high-density genetic map with AFLP markers. *Genome* 52:716–725.

Blas, A.L., Ming, R., Liu, Z., Veatch, O.J., Paull, R.E., Moore, P.H., and Yu, Q. (2010). Cloning of the papaya chromoplast-specific lycopene b-cyclase, CpCYC-b, controlling fruit flesh color reveals conserved microsynteny and a recombination hot spot. *Plant Physiology* 152: 2013–2022.

Blasi, P., Blanc, S., Wiedemann-Merdinoglu, S., Prado, E., Rühl, E., Mestre, P., and Merdinoglu, D. (2011). Construction of a reference linkage map of *Vitis amurensis* and genetic mapping of Rpv8, a locus conferring resistance to grapevine downy mildew. *Theoretical and Applied Genetics* 123(1): 43–53.

Bohlmann, J., Meyer-Gauen, G., and Croteau, R. (1998). Plant terpenoid synthases: Molecular biology and phylogenetic analysis. *Proceedings National Academy of Science* 95: 4126–4133.

Bojorquez, G. and Gomez-Lim, M.A. (1995). Peroxisomal thiolasem RNA is induced during mango fruit ripening. *Plant Molecular Biology* 28: 811–20.

Bompard, J.M. and Schnell, R.J. (1997). Taxomomy and systematics. In: Litz, R.E. (ed.), *The Mango, Botany, Production and Uses*, pp. 21–48. CABI International, New York.

Bowers, J.E., Dangl, G.S., and Meredith, C.P. (1999). Development and characterization of additional microsatellite DNA markers for grape. *American Journal of Enology and Viticulture* 50(3): 243–246.

Bowers, J.E., Dangl, G.S., Vignani, R., and Meredith, C.P. (1996). Isolation and characterization of new polymorphic simple sequence repeat loci in grape, *Vitis vinifera* L. *Genome* 39(4): 628–633.

Bundy, J.G., Sidhu, J.K., Rana, F., Spurgeon, D.J., Svendsen, C., Wren, J.F., Stürzenbaum, S.R., Morgan, A.J., and Kille, P. (2008). "Systems toxicology" approach identifies coordinated metabolic responses to copper in a terrestrial non-model invertebrate, the earthworm Lumbricus rubellus. *BMC Biology* 6: 25.

Cabezas, J.A., Cervera, M.T., Ruiz-Garcia, L., Carreno, J., and Martinez-Zapater, J.M. (2006). A genetic analysis of seed and berry weight in grapevine. *Genome* 49(12): 1572–1585.

Cabezas, J.A., Ibanez, J., Lijavetzky, D., Velez, D., Bravo, G., Rodriguez, V., Carreno, I., et al. (2011). A 48 SNP set for grapevine cultivar identification. *BMC Plant Biology* 11(153): 12.

Carpentier, S., Witters, E., Laukens, K., Van Onckelen, H., Swennen, R., and Panis, B. (2007). Banana (Musa spp.) as a model to study the meristem proteome: Acclimation to osmotic stress. *Proteomics* 7: 92–105.

Carreel, F., Gonzalez de Leon, D., Lagoda, P., Lanaud, C., Jenny, C., Horry, J.P, and Tezenas du Montcel, H. (2002). Ascertaining maternal and paternal lineage within Musa by chloroplast and mitochondrial DNA RFLP analyses. *Genome*, 45(4): 679–692.

Castellarin, S.D., Pfeiffer, A., Sivilotti, P., Degan, M., Peterlunger, E., and Di Gaspero, G. (2007). Transcriptional regulation of anthocyanin biosynthesis in ripening fruits of grapevine under seasonal water deficit. *Plant, Cell and Environment* 30(11): 1381–1399.

Castro, A.J., Carapito, C., Zorn, N., Magne, C., Leize, E., Van Dorsselaer, A. and Clement, C. (2005). Proteomic analysis of grapevine (*Vitis vinifera* L.) tissues subjected to herbicide stress. *Journal of Experimental Botany* 56(421): 2783–2795.

Cervera, M.T., Cabezas, J.A., Sanchez-Escribano, E., Cenis, J.L., and Martinez-Zapater, J.M. (2000). Characterization of genetic variation within table grape varieties (*Vitis vinifera* L.) based on AFLP markers. *Vitis* 39(3): 109–114.

Charters, Y.M., Robertson, A., Wilkinson, M.J., and Ramsay, G. (1996). PCR analysis of oilseed rape cultivars (*Brassicanapus* L. ssp. *oleifera*) using 5'-anchored simple sequence repeat (SSR) primers. *Theoretical and Applied Genetics* 92: 442–447.

Chen, Y.T., Lee, Y.R., Yang, C.Y., Wang, Y.T., Yang, S., Shaw, J. (2003). A novel papaya ACC oxidase gene (CP-ACO2) associated with late stage fruit ripening and leaf senescence. *Plant Science* 164: 531–540.

Chen, C., Zhou, P., Choi, Y.A., Huang, S., and Gmitter, F.G. (2006). Mining and characterizing microsatellites from citrus ESTs. *Theoretical and Applied Genetics* 112: 1248–1257.

Chen, C., Yu, Q., Hou, S., Li, Y., Eustice, M., Skelton, R.L., Veatch, O., et al. (2007). Construction of a sequence-tagged high density genetic map of papaya for comparative structural and evolutionary genomics in Brassicales. *Genetics* 177: 2481–2491.

Chen, C., Bowman, K.D., Choi, Y.A., Dang, P.M., Rao, M.N., Huang, S., Soneji, J.R., McCollum, T.G., Gmitter, F.G. Jr. (2008). EST-SSR genetic maps for *Citrus sinensis* and *Poncirus trifoliata*. *Tree Genetics & Genomes* 4: 1–10.

Choi, Y.H, Kim, H.K., Linthorst, H.J.M., Hollander, J.G., Lefeber, A.W.M., Erkelens, C., Nuzillard, J.M., and Verpoorte, R. (2006). NMR metabolomics to revisit the tobacco mosaic virus infection in Nicotiana tabacum leaves. *Journal of Natural Products* 69: 742–748.

Cipriani, G., Spadotto, A., Jurman, I., Gaspero, G.D., Crespan, M., Meneghetti, S., Frare, E., et al. (2010). The SSR-based molecular profile of 1005 grapevine (*Vitis vinifera* L.) accessions uncovers new synonymy and parentages, and reveals a large admixture amongst varieties of different geographic origin. *Theoretical and Applied Genetics* 121(8): 1569–1585.

Collard, B.C.Y. and Mackill, D.J. (2008). Marker-assisted selection: An approach for precision plant breeding in the twenty-first century. *Philosophical Transactions of The Royal Society B* 363: 557–572.

Considine, M.J., Daley, D.O., and Whelan, J. (2001). The expression of alternative oxidase and uncoupling protein during fruit ripening in mango. *Plant Physiology* 126: 1619–1629.

Costantini, L., Battilana, J., Lamaj, F., Fanizza, G., and Grando, M.S. (2008). Berry and phenology-related traits in grapevine (*Vitis vinifera* L.): From quantitative trait loci to underlying genes. *BMC Plant Biology* 8(1): 38.

Costenaro-da-Silva, D., Passaia, G., Henriques, J.A.P., Margis, R., Pasquali, G., and Revers, L.F. (2010). Identification and expression analysis of genes associated with the early berry development in the seedless grapevine (*Vitis vinifera* L.) cultivar Sultanine. *Plant Science* 179(5): 510–519.

Cramer, G., Ergül, A., Grimplet, J., Tillett, R., Tattersall, E.R., Bohlman, M., Vincent, D., et al. (2007). Water and salinity stress in grapevines: Early and late changes in transcript and metabolite profiles. *Functional and Integrative Genomics* 7(2): 111–134.

Creste, S., Benatti, T.R., Orsi, M.R., Risterucci, A.M., and Figueira, A. (2006). Isolation and characterization of microsatellite loci from a commercial cultivar of *Musa acuminata*. *Molecular Ecology Notes*. 6: 303–306.

Cuenca, J., Navarro, L., and Ollitrault, P. (2009). Origin of 2n gametes in *C. reticulata* cv Fortune mandarin. Paper presented at the International Conference on Polyploidy, Hybridization and Biodiversity, May 17–20, Saint Malo, France.

Currò, S., Caruso, M., Distefano, G., Gentile, A., and La Malfa, S. (2010). New microsatellite loci for pomegranate, *Punica granatum* (Lythraceae). *AJB Primer Notes Protocols Plant Sci*ence e58–e60.

Dalbó, M.A., Ye, G.N., Weeden, N.F., Steinkellner, H., Sefc, K.M., and Reisch, B.I. (2000). A gene controlling sex in grapevines placed on a molecular marker-based genetic map. *Genome* 43(2): 333–340.

Daldoul, S., Mliki, A., and Höfer, M.U. (2012). Suppressive subtractive hybridization method analysis and its application to salt stress in grapevine (*Vitis vinifera* L.). *Russian Journal of Genetics* 48(2): 179–185.

Damania, A.B. (2005). The pomegranate: Its origin, folklore, and efficacious medicinal properties. In: Nene, Y.L. (Ed.), *Agriculture Heritage of Asia: Proceedings of the International Conference*, pp. 175–183, Asian Agri History Foundation, Secunderabad, India.

da Silva, F.G., Iandolino, A., Al-Kayal, F., Bohlmann, M.C., Cushman, M.A., Lim, H., Ergul, A., et al. (2005). Characterizing the grape transcriptome. Analysis of expressed sequence tags from multiple *Vitis* species and development of a compendium of gene expression during berry development. *Plant Physiology* 139(2): 574–597.

Davey, M.W., Graham, N.S., Vanholme, B., Swennen, R., May, S.T., and Keulemans, J. (2009). Heterologous oligonucleotide microarrays for transcriptomics in a non-model species; a proof-of-concept study of drought stress in *Musa*. *BMC Genomics 2009* 10: 436.

Davies, C. and Robinson, S.P. (2000). Differential screening indicates a dramatic change in mRNA profiles during grape berry ripening. Cloning and characterization of cDNAs encoding putative cell wall and stress response proteins. *Plant Physiology* 122(3): 803–812.

Delseny, M., Han, B., and Hsing, Y.L. (2010). High throughput DNA sequencing: The new sequencing revolution. *Plant Sci* 179: 407–422.

Deputy, J.C., Ming, R., Ma, H., Liu, Z., Fitch, M.M., Wang, M., Manshardt, R., and Stiles, J.I. (2002). Molecular markers for sex determination in papaya (*Carica papaya* L.). *Theoretical and Applied Genetics* 106(1): 107–111.

Devitt, L.C., Sawbridge, T., Holton, T.A., Mitchelson, K., and Dietzgen, R.G. (2006). Discovery of genes associated with fruit ripening in Carica papaya using expressed sequence tags. *Plant Science* 170: 356–363.

Deytieux, C., Geny, L., Lapaillerie, D., Claverol, S., Bonneu, M., and Donèche, B. (2007). Proteome analysis of grape skins during ripening. *Journal of Experimental Botany* 58(7): 1851–1862.

D'Hont, A., Denoeud, F., Aury, J.-M., Baurens, F.-C., Carreel, F., Garsmeur, O., Noel, B., et.al. (2012). The banana (*Musa acuminata*) genome and the evolution of monocotyledonous plants. *Nature* 488:213–217.

Di Carli, M., Zamboni, A., Pè, M.E., Pezzotti, M., Lilley, K.S., Benvenuto, E., and Desiderio, A. (2010). Two-dimensional differential in gel electrophoresis (2D-DIGE) analysis of grape berry proteome during postharvest withering. *Journal of Proteome Research* 10(2): 429–446.

Di Gaspero, G., Cipriani, G., Marrazzo, M.T., Andreetta, D., Prado Castro, M.J., Peterlunger, E., and Testolin, R. (2005). Isolation of (AC)n-microsatellites in *Vitis vinifera* L. and analysis of genetic background in grapevines under marker assisted selection. *Molecular Breeding* 15: 11–20.

Doligez, A., Bouquet, A., Danglot, Y., Lahogue, F., Riaz, S., Meredith, C., Edwards, K., and This, P. (2002). Genetic mapping of grapevine (*Vitis vinifera* L.) applied to the detection of QTLs for seedlessness and berry weight. *Theoretical and Applied Genetics* 105(5): 780–795.

Doligez, A., Audiot, E., Baumes, R., and This, P. (2006). QTLs for muscat flavor and monoterpenic odorant content in grapevine (*Vitis vinifera* L.). *Molecular Breeding* 18: 109–125.

Dorgac, C., Ozgen, M., Simsek, O., Kakar, Y.A., Kiyga, Y., Celebi, S., Gunduz, K., and Serce, S. (2008) Molecular and morphological diversity among pomegranate (*Punica granatum*) cultivars in eastern mediteranean region of Turkey. *African J Biotechol*. 7: 1294–1301.

Drew, R.A., O'Brien, C.M., and Magdalita, P.M. (1998). Development of interspecific *Carica* hybrids. *Proceedings of the International Symposium on Biotechnology of Tropical and Subtropical Species*, part II, Brisbane, Queensland, Australia, 29 September to 3 October 1997, 285–291.

Duchêne, E., Butterlin, G., Claudel, P., Dumas, V., Jaegli, N., and Merdinoglu, D. (2009). A grapevine (*Vitis vinifera* L.) deoxy-d-xylulose synthase gene colocates with a major quantitative trait loci for terpenol content. *Theoretical and Applied Genetics* 118(3): 541–552.

Duchêne, E., Butterlin, G., Dumas, V., and Merdinoglu, D. (2012). Towards the adaptation of grapevine varieties to climate change: QTLs and candidate genes for developmental stages. *Theoretical and Applied Genetics* 124(4): 623–635.

Ebrahimi, S., Sayed-Tabatabaei, B.E., and Sharifnabi, B. (2010). Microsatellite isolation and characterization in pomegranate (*Punica granatum* L.). *Iranian J Biotechnol.* 8(3): 159–163.

Eiadthong, W., Yonemori, K., Kanzaki, S., and Sugiura, A. (2000). Amplified fragment length polymorphism analysis for studying genetic relationships among *Mangifera* species in Thailand. *Journal of the American Society for Horticultural Science*, 125: 160–164.

Engel, K. and Tressl, R. (1983). Studies on volatile components of two mango varieties. *Journal of Agriculture and Food Chemistry.* 31: 796–801.

Esteve, C., D'Amato, A., Marina, M.L., García, M.C., Citterio, A., and Righetti, P.G. (2012). In-depth proteomic analysis of banana (*Musa* spp.) fruit with combinatorial peptide ligand libraries. *Electrophoresis* 34: 207–214.

Eustice, M., Yu, Q., Lai, C.W., Hou, S., Thimmapuram, J., Liu, L., Alam, M., Moore, P.H., Presting, G.G., and Ming, R. (2008). Development and application of microsatellite markers for genomic analysis of papaya. *Tree Genetics & Genomes* 4: 333–341.

Fabi, J.P., Mendes, L.R.B.C., Lajolo, F.M., and Nascimento, J.R.O. (2010) Transcript profiling of papaya fruit reveals differentially expressed genes associated with fruit ripening. *Plant Science* 179: 225–233.

Fabi, J.P., Seymour, G.B., Graham, N.S., Broadley, M.R., May, S.T., Lajolo, F.M., Cordenunsi, B.R., and Oliveira do Nascimento, J.R. (2012). Analysis of ripening-related gene expression in papaya using an *Arabidopsis*-based microarray. *BMC Plant Biology* 12: 242.

Fang, D.Q., Roose, M.L., Krueger, R.R., and Federici, C.T. (1997) Finger printing trifoliate orange germplasm accessions with isozymes, RFLPs and inter-simple sequence repeat markers. *Theoretical and Applied Genetics* 95: 211–219.

Fang, D.Q., Fedrici, C.T., Roose, M.L. (1998). A high-resolution linkage map of citrus Tristeza virus resistance gene region in *pomcitrus trifoliate* (L) Raf. *Genetics* 150: 883–890.

Fanizza, G., Lamaj, F., Costantini, L., Chaabane, R., and Grando, M.S. (2005). QTL analysis for fruit yield components in table grapes (*Vitis vinifera*). *Theoretical and Applied Genetics* 111(4): 658–664.

FAO (2006). Crop production statistics. Food and Agriculture Organization of the United Nations. http://faostat.fao.org/site/567/default.aspx#ancor.

Fasoli, M., Dal Santo, S., Zenoni, S., Tornielli, G.B., Farina, L., Zamboni, A., Porceddu, A., et al. (2012). The grapevine expression atlas reveals a deep transcriptome shift driving the entire plant into a maturation program. *The Plant Cell Online* 24(9): 3489–3505.

Ferrante, S.P., Lucretti, S., Reale, S., De Patrizio, A., Abbate, L., Tusa, N., and Scarano, M.T. (2010). Assessment of the origin of new citrus tetraploid hybrids (2n = 4x) by means of SSR markers and PCR based dosage effects. *Euphytica* 173: 223–233.

Fiehn, O. (2002). Metabolomics: The link between genotypes and phenotypes. *Plant Molecular Biology* 48: 155–171.

Fischer, B.M., Salakhutdinov, I., Akkurt, M., Eibach, R., Edwards, K.J., Töpfer, R., and Zyprian, E.M. (2004). Quantitative trait locus analysis of fungal disease resistance factors on a molecular map of grapevine. *Theoretical and Applied Genetics* 108(3): 501–515.

Food and Agriculture Organization of United Nations Annual Statistics (2012). Citrus fruit fresh and processed annual statistics 2012, Rome, Italy.

Fortes, A., Agudelo-Romero, P., Silva, M., Ali, K., Sousa, L., Maltese, F., Choi, Y., et al. (2011). Transcript and metabolite analysis in Trincadeira cultivar reveals novel information regarding the dynamics of grape ripening. *BMC Plant Biology* 11(1): 149.

Fournier-Level, A., Cunff, L.L., Gomez, C., Doligez, A., Ageorges, A., Roux, C., Bertrand, Y., Souquet, J. M., Cheynier, V., and This, P. (2009). Quantitative genetic bases of anthocyanin variation in grape (*Vitis vinifera* L. ssp. *sativa*) berry: A quantitative trait locus to quantitative trait nucleotide integrated study. *Genetics* 183(3): 1127–1139.

Froelicher, Y., Dambier, D., Costantino, G., Lotfy, S., Didout, C., Beaumont, V., Brottier, P., Risterucci, A.-M., Luro, F., and Ollitrault, P. (2008) Characterization of microsatellite markers in *Citrus reticulata* Blanco. *Molecular Ecology Resources* 8(1): 119–122.

Frost, H.B. and Soost, R.K. (1968): Seed reproduction: Development of gametes and embryos. In: W. Reuther, H.J. Batchelor, and L.D. Bachelor (eds.). *The Citrus Industry.* Vol. I. University of California Press, Barkeley.

Fuggate, P., Aree, C.W., Noichinda, S., and Kanlayanarat, S. (2010). Quality and volatile attributes of attached and detached "Pluk Mai Lie" papaya during fruit ripening. *Scientia Horticulturae* 126: 120–129.

Fujii, H., Shimada, T., Sugiyama, A., Endo, T., Nishikawa, F., Nakano, M., Ikoma, Y., Shimizu, T., and Omura, M. (2008). Profiling gibberellins (GA3)-responsive genes in mature mandarin fruit using a citrus 22K oligoarray. *Scientia Horticulture* 116: 291–298.

Fukao, Y., Hayashi, M., and Nishimura, M. (2002). Proteomic analysis of leaf peroxisomal proteins in greening cotyledons of Arabidopsis thaliana. *Plant Cell Physiology* 43: 689–696.

Fumagalli, E., Baldoni, E., Abbruscato, P., Piffanelli, P., Genga, A., Lamanna, R., and Consonni, R. (2009). NMR techniques coupled with multivariate statistical analysis: Tools to analyse Oryza sativa metabolic content under stress conditions. *Journal of Agronomy and Crop Science* 195: 77–88.

Gancel, A.L., Ollitrault, P., Froelicher, Y., Tomi, F., Jacquemond, C., Luro, F., and Brioullet, J.M. (2003). Leaf volatile compounds of seven Citrus somatic tetraploid hybrids sharing willow leaf mandarin (*Citrus deliciosa* Ten) as their common parent. *Journal of Agriculture and Food Chemistry* 51: 6006–6013.

Gancel, A.L.., Grimplet, J., Sauvage, F.X., Ollitrault, P., and Brillouet, J.M. (2006). Predominant expression of diploid mandarin leaf proteome in two citrus mandarin derived somatic allotetraploid hybrids. *J Agric Food Chem* 54(17): 6212–6218.

Gebauer, M., Mejia, N., Munoz, L., Hewstone, N., and Hinrichsen, P. (2009). A genetic linkage map of seedless table grapes (*Vitis vinifera* L.) developed for the analysis of seedlessness and fruit quality QTLs. *Acta Horticulturae (ISHS)* 827: 369–376.

Giacomelli, L., Rudella, A., and van Wijk, K.J. (2006). High light response of the thylakoid proteome in arabidopsis wild type and the ascorbate-deficient mutant *vtc2-2*. A comparative proteomics study. *Plant Physiology* 141: 685–701.

Gil, M.I., Tomás-Barberán, F.A., Hess-Pierce, B., Holcroft, D.M., and Kader, A.A. (2000). Antioxidant activity of pomegranate juice and its relationship with phenolic composition and processing. *Journal of Agriculture and Food Chemistry* 48: 4581–4589.

Giribaldi, M., Perugini, I., Sauvage, F.-X., and Schubert, A. (2007). Analysis of protein changes during grape berry ripening by 2-DE and MALDI-TOF. *Proteomics* 7(17): 3154–3170.

Giribaldi, M., Geny, L., Delrot, S., and Schubert, A. (2010). Proteomic analysis of the effects of ABA treatments on ripening *Vitis vinifera* berries. *Journal of Experimental Botany* 61(9): 2447–2458.

Gmitter, G. Jr., Chen, C., Machado, M.A., Alves de Souza, A., Ollitrault, P., Froehlicher, Y., and Shimizu, T. (2012). Citrus genetics. *Tree Genetics & Genomes* 8: 611–626.

Grimplet, J., Deluc, L., Tillett, R., Wheatley, M., Schlauch, K., Cramer, G., and Cushman, J. (2007). Tissue-specific mRNA expression profiling in grape berry tissues. *BMC Genomics* 8(1): 187.

Grimplet, J., Cramer, G.R., Dickerson, J.A., Mathiason, K., Van Hemert, J., and Fennell, A.Y. (2009a). VitisNet: "Omics" integration through grapevine molecular networks. *PLoS ONE* 4(12): e8365.

Grimplet, J., Wheatley, M.D., Jouira, H.B., Deluc, L.G., Cramer, G.R., and Cushman, J.C. (2009b). Proteomic and selected metabolite analysis of grape berry tissues under well-watered and water-deficit stress conditions. *Proteomics* 9(9): 2503–2528.

Guillaumie, S., Fouquet, R., Kappel, C., Camps, C., Terrier, N., Moncomble, D., Dunlevy, J., Davies, C., Boss, P., and Delrot, S. (2011). Transcriptional analysis of late ripening stages of grapevine berry. *BMC Plant Biology* 11(1): 165.

Gupta, P.K. and Varshney, R.K. (2000). The development and use of microsatellite markers for genetic analysis and plant breeding with emphasis on bread wheat. *Euphytica* 113: 163–185.

Hasnaoui, N., Buonamici, A., Sebastiani, F., Mars, M., Trifi, M., and Vendramin, G.G. (2010). Development and characterization of SSR markers for pomegranate (*Punica granatum* L.) using an enriched library. *Conservation Genetic Resources* 2(1): 283–285. 10.1007/s12686-010-9191-8.

Heazlewood, J.L., Tonti-Filippini, J.S., Gout, A.M., Day, D.A., Whelan, J., and Millar, A.H. (2004). Experimental analysis of the Arabidopsis mitochondrial proteome highlights signaling and regulatory components, provides assessment of targeting prediction programs, and indicates plant-specific mitochondrial proteins. *Plant Cell* 16: 241–256.

Heslop-Harrison, J.S. and Schwarzacher, T. (2007). Domestication, genomics and the future for banana: Review. *Annals of Botany* 100: 1073–1084.

Higashi, H., Hironaga, T., Sennenbara, T., Kunitake, H., and Komatsu, H. (2000). Phylogenetic classification of *Fortunella* species using RAPD method (in Japanese). *Journal of Japanese Society Horticultural Science* 2: 288.

Hippolyte, I., Bakry, F., Seguin, M., Gardes, L., Rivallan, R., Risterucci, A.M., Jenny, C., et al. (2010). A saturated SSR/DArT linkage map of Musa acuminata addressing genome rearrangements among bananas. *BMC Plant Biology* 10(1): 65.

Hirai, M.Y., Yano, M., Goodenowe, D.B., Kanaya, S., Kimura, T., Awazuhara, M., Arita, M., Fujiwara, T., and Saito, K. (2004). Integration of transcriptomics and metabolomics for understanding of global responses to nutritional stresses in Arabidopsis thaliana. *Proceedings of National Academy of Science USA* 101: 10205–10210.

Hirai, M.Y., Klein, M., Fujikawa, Y., Yano, M., Goodenowe, D.B., Yamazaki, Y., Kanaya, S., et al. (2005). Elucidation of gene-to-gene and metabolite-to-gene networks in Arabidopsis by integration of metabolomics and transcriptomics. *Journal of Biological Chemistry* 280: 25590–25595.

Hoffman, P.J., Smith, L.G., Joyce, D.C., Johnson, G.I., and Meiburg, G.F. (1997). Bagging of mango (*Mangifera indica* cv. "Keitt") fruit influences fruit quality and mineral composition. *Postharvest Biology and Technology*, 12: 83–91.

Hofmeyr, J.D.J. (1939). Sex-linked inheritance in *Carica Papaya* L. *South African Journal Science* 36: 283–285.

Hong, Q.B., Xiang, S.Q., Chen, K.L., and Chen, L.G. (2001). Two complementary dominant genes controlling apomixis in genus Citrus and Poncirus. *Acta Genetica Sinica* 28: 1062–1067.

Hsu, T.H., Adiputra, Y.T., Ohta, H., and Gwo, J.C. (2011). Species and sex identification of Formosa landlocked salmon using loop-mediated isothermal amplification. *Molecular Ecology Resources* 11: 802–807.

Hsu, T.H., Gwo, J.C., and Lin, K.H. (2012). Rapid sex identification of papaya (Carica papaya) using multiplex loop-mediated isothermal amplification (mLAMP). *Planta* 236(4): 1239–1246.

Huerta-Ocampo, J.Á., Osuna-Castro, J.A., Lino-López, G.J., Barrera-Pacheco, A., Mendoza-Hernández, G., De León-Rodríguez, A., and Barba de la Rosa, A.P. (2012). Proteomic analysis of differentially accumulated proteins during ripening and in response to 1-MCP in papaya fruit. *Journal Proteomics* 75(7): 2160–2169.

Ibanez, J., Vargas, A.M., Palancar, M., Borrego, J., and deAndres, M.T. (2009). Genetic relationships among table grape varieties. *American Journal of Enology and Viticulture* 60: 35–42.

Ibrahim, A.D., Sani, A., Manga, S.B., Aliero, A.A., Joseph, R.U., Yakubu, S.E., and Ibafidon, H. (2011). Microorganisms associated with volatile metabolites production in soft rot disease of sweet pepper fruits. (Tattase). *International Journal of Biotechnology and Biochemistry* 7(6): 25–35.

Idestein, H. and Schreier P. (1985). Volatile constituents of Alphonso mango (*Mangifera indica*). *Phytochemistry* 24(10): 2313–2316.

Jaillon, O., Aury, J.M., Noel, B., Policriti, A., Clepet, C., Casagrande, A., Choisne, N., et al. (2007). The grapevine genome sequence suggests ancestral hexaploidization in major angiosperm phyla. *Nature* 449(7161): 463–468.

Jbir, R., Hasnaoui, N., Mars, M., Marrakchi, M., and Trifi, M. (2008). Characterization of Tunisian pomegranate (*Punica granatum* L.) cultivars using amplified fragment length polymorphism analysis. *Scientia Horticulturae* 115: 231–237.

Jellouli, N., Ben Jouira, H., Skouri, H., Ghorbel, A., Gourgouri, A., and Mliki, A. (2008). Proteomic analysis of Tunisian grapevine cultivar Razegui under salt stress. *Journal of Plant Physiology* 165(5): 471–481.

Jones, O.A.H., Walker, L.A., Nicholson, J.K., Shore, R.F., and Griffin, J.L. (2007). Cellular acidosis in rodents exposed to cadmium is caused by adaptation of the tissue rather than an early effect of toxicity. *Comparative Biochemistry and Physiology* D 2: 316–321.

Kamath, J.V., Rahul, N., Ashok Kumar, C.K., and Laxmi, S.M. (2008). *Psidium guajava* L. A Review. *International Journal of Green Pharmacy* 2(1): 9–12.

Kanupriya, Madhavi Latha, P., Aswath, C., Reddy, L., Padmakar, B., Vasugi, C., and Dinesh, M.R. (2011). Cultivar identification and genetic fingerprinting of guava (*Psidium guajava*) using microsatellite markers. *International Journal Fruit Science* 11: 184–196.

Kashkush, K., Jinggui, F., Tomer, E., Hillel, J., and Lavi, U. (2001). Cultivar identification and genetic map of mango (*Mangifera indica*). *Euphytica* 122: 129–136.

Katz, E., Fon, M., Lee, Y.J., Phinney, B.S., Sadka, A., and Blumwald, E., (2007). The citrus fruit proteome: Insights into citrus fruit metabolism. *Planta* 226(4): 989–1005.

Kepiro, J. and Roose, M. (2009). AFLP markers closely linked to a major gene essential for nucellar embryony (apomixis) in *Citrus maxima* × *Poncirus trifoliata*. *Tree Genetics & Genomes* 6: 1–11.

Khan, I.A. and Kamran Azim, M. (2012). Variations in intergenic spacer rpl20-rps12 of mango (*Mangifera indica*) chloroplast DNA: Implications for cultivar identification and phylogenetic analysis. *Plant Systematics and Evolution* 292(3): 249–255.

Knight, R.J.R. (1980). Origin and world importance of tropical and subtropical fruit crops. In: Nagy, S. and Shaw, P.E. (eds) *Tropical and Subtropical Fruits*. AVI, Westport, CT, USA.

Kobayashi, H., Fujita, K., Suzuki, S., and Takayanagi, T. (2009). Molecular characterization of Japanese indigenous grape cultivar "Koshu" (*Vitis vinifera*) leaf and berry skin during grape development. *Plant Biotechnology Reports* 3(3): 225–241.

Koohi-Dehkordi, M., Sayed Tabatabaei, B.E., Yamchi, A., and Daneshshahraki, A. (2007). Microsatellites markers in pomegranate. *Acta Hort.* 760: 179–183.

Kumar, N.V.H., Narayanaswamy, P., Prasad, D.T., Mukunda, G.K., and Sondur, S.N. (2001). Estimation of genetic diversity of commercial mango (*Mangifera indica* L.) cultivars using RAPD markers. *Journal of Horticultural Science and Biotechnology* 76(5): 529–533.

Laucou, V., Lacombe, T., Dechesne, F., Siret, R., Bruno, J.P., Dessup, M., Dessup, T., et al. (2011). High throughput analysis of grape genetic diversity as a tool for germplasm collection management. *Theoretical and Applied Genetics* 122(6): 1233–1245.

Le Cunff, L., Fournier-Level, A., Laucou, V., Vezzulli, S., Lacombe, T., Adam-Blondon, A.-F., Boursiquot, J.-M., and This, P. (2008). Construction of nested genetic core collections to optimize the exploitation of natural diversity in *Vitis vinifera* L. subsp. sativa. *BMC Plant Biology* 8(1): 31.

Lescot, M., Piffanelli, P., Ciampi, A.Y., Ruiz, M., Blanc, G., Leebens-Mack, J., da Silva, F.R., et al. (2008). Insights into the Musa genome: Syntenic relationships to rice and between Musa species. *BMC Genomics* 9(1): 58.

Levin, G.M. (1994). Pomegranate (*Punica granatum* L.) plant genetic resources in Turkmenistan. *Plant Genetics Resources Newsletter* 97: 31–36.

Li, C.Y., Deng, G.M., Yang, J., Viljoen, A., Jin, Y., Kuang, R.B., Zuo, C.W., et al. (2012). Transcriptome profiling of resistant and susceptible Cavendish banana roots following inoculation with Fusarium oxysporum f. sp. cubense tropical race 4. *BMC Genomics* 13(1): 374.

Liang, X.G., Guo, Q., He, Q., and Li, X. (2006). AFLP analysis and the taxonomy of Citrus. Acta Horticulturae 760: XXVII International Horticultural Congress—IHC2006: II International Symposiumon Plant Genetic Resources of Horticultural Crops. Seoul, Korea.

Lijavetzky, D., Cabezas, J., Ibanez, A., Rodriguez, V., and Martinez-Zapater, J. (2007). High throughput SNP discovery and genotyping in grapevine (*Vitis vinifera* L.) by combining a re-sequencing approach and SNPlex technology. *BMC Genomics* 8(1): 424.

Lijavetzky, D., Carbonell-Bejerano, P., Grimplet, J., Bravo, G., Flores, P., Fenoll, J., Hellín, P., Oliveros, J. C., and Martínez-Zapater, J.M. (2012). Berry flesh and skin ripening features in *Vitis vinifera* as assessed by transcriptional profiling. *PLoS ONE* 7(6): e39547.

Lin, C.T., Lin, M.T., and Shaw, J.F. (1997). Cloning and characterization of a cDNA for 1-aminicyclopropene-1-1carboxylate oxidase from papaya fruit. *Journal of Agriculture and Food Chemistry* 45: 526–530.

Lister, R., Chew, O., Lee, M.-N., Heazlewood, J.L., Clifton, R., Parker, K.L., Millar, A.H., and Whelan, J. (2004). A transcriptomic and proteomic characterization of the Arabidopsis mitochondrial protein import apparatus and its response to mitochondrial dysfunction. *Plant Physiology* 134: 777–789.

Liu, G.-T., Wang, J.-F., Cramer, G., Dai, Z.-W., Duan, W., Xu, H.-G., Wu, B.-H., Fan, P.-G., Wang, L.-J., and Li, S.H. (2012). Transcriptomic analysis of grape (*Vitis vinifera* L.) leaves during and after recovery from heat stress. *BMC Plant Biology* 12(1): 174.

Lowe, K.M. and Walker, M.A. (2006). Genetic linkage map of the interspecific grape rootstock cross Ramsey (*Vitis champinii*) × RipariaGloire (*Vitis riparia*). *Theoretical and Applied Genetics* 112(8): 1582–1592.

Luro, F., Constantino, G., Billot, C., Froehlicher, Y., Morillon, R., Ollitrault, Terol J., Talon, M., Gmitter, F.G. Jr, and Chen C. (2007). Genetic maps of Clementine mandarin and intergeneric *Clementine* × Poncirus using genomic and EST microsatellite markers. Plant & Animal Genome XIV Conference (P487), San Diego, CA.

Luro, F., Costantino, G., Terol, J., Argout, X., Allario, T., Wincker, P., Talon, M., Ollitrault, P., and Morillon, R. (2008). Transferability of the ESTSSRs developed on Nules clementine (*Citrus clementina* Hort ex Tan) to other Citrus species and their effectiveness for genetic mapping. *BMC Genomics* 9: 287.

Ma, H., Moore, P.H., Liu, Z., Kim, M.S., Yu, Q., Fitch, M.M.M., Sekioka, T., Paterson, A.H., and Ming, R. (2004). High density linkage mapping revealed suppression of recombination at the sex determination locus in papaya. *Genetics* 166: 419–436.

Maltman, D.J., Simon, W.J., Wheeler, C.H., Dunn, M.J., Wait, R., and Slabas, A.R. (2002). Proteomic analysis of the endoplasmic reticulum from developing and germinating seed of castor (*Ricinus communis*). *Electrophoresis* 23: 626–639.

Mamta, R. (2006). Refinement of the citrus tristeza virus resistance gene (Ctv) positional map in *Poncirus trifolia* and generation of transgenic grapefruit (*Citrus paradisi*). Plant lines with candidate resistance genes in this region. *Plant Molecular Biology* 61(3): 399–414.

Manshardt, R.M. and Moore, P.H. (2003). Natural history of papaya and the Caricaceae. Abstract #792 in 'Plant Biology 2003', Proceedings of the American Society of Plant Biologists meeting held 25–30 July 2003, Honolulu.

Manzo-Sánchez, G., Zapater, M.F., Luna-Martínez, F., Conde-Ferráea, L., Carlier, J., James-Kay, A., and Simpson, J. (2008). Construction of a genetic linkage map of the fungal pathogen of banana *Mycosphaerellafijiensis*, causal agent of black leaf streak disease. *Current Genetics* 53: 299–311.

Margaria, P. and Palmano, S. (2011). Response of the *Vitis vinifera* L. cv. "Nebbiolo" proteome to Flavescence dorée phytoplasma infection. *Proteomics* 11(2): 212–224.

Marguerit, E., Boury, C., Manicki, A., Donnart, M., Butterlin, G., Nemorin, A., Wiedemann-Merdinoglu, S., Merdinoglu, D., Ollat, N., and Decroocq, S. (2009). Genetic dissection of sex determinism, inflorescence morphology and downy mildew resistance in grapevine. *Theoretical and Applied Genetics* 118(7): 1261–1278.

Mark, W.D., Graham, N.S., Vanholme, B., Swennen, R., May, S.T., and Keulemans, J. (2009). Heterologous oligonucleotide microarrays for transcriptomics in a non-model species; a proof-of-concept study of drought stress in *Musa*. *BMC Genomics* 10: 436.

Martinez, L., Cavagnaro, P., Masuelli, R., and Rodriguez, J. (2004). Evaluation of diversity among Argentine grapevine (*Vitis vinifera* L.) varieties using morphological data and AFLP markers. *Electronic Journal of Biotechnology* 3.

Martinez-Esteso, M.J., Casado-Vela, J., Selles-Marchart, S., Elortza, F., Pedreno, M.A., and Bru-Martinez, R. (2011a). iTRAQ-based profiling of grape berry exocarp proteins during ripening using a parallel mass spectrometric method. *Molecular BioSystems* 7(3): 749–765.

Martínez-Esteso, M.J., Sellés-Marchart, S., Lijavetzky, D., Pedreño, M.A., and Bru-Martínez, R. (2011b). A DIGE-based quantitative proteomic analysis of grape berry flesh development and ripening reveals key events in sugar and organic acid metabolism. *Journal of Experimental Botany* 62(8): 2521–2569.

Mathiason, K., He, D., Grimplet, J., Venkateswari, J., Galbraith, D., Or, E., and Fennell, A. (2009). Transcript profiling in *Vitis riparia* during chilling requirement fulfillment reveals coordination of gene expression patterns with optimized bud break. *Functional and Integrative Genomics* 9(1): 81–96.

Mejia, N., Gebauer, M., Munoz, L., Hewstone, N., Munoz, C., and Hinrichsen, P. (2007). Identification of QTLs for seedlessness, berry size, and ripening date in a seedless × seedless table grape progeny. *American Journal of Enology and Viticulture* 58(4): 499–507.

Merdinoglu, D., Butterlin, G., Bevilacqua, L., Chiquet, V., Adam-Blondon, A. F., and Decroocq, S. (2005). Development and characterization of a large set of microsatellite markers in grapevine (*Vitis vinifera* L.) suitable for multiplex PCR. *Molecular Breeding* 15: 349–366.

Michaud, M.R. and Denlinger, D.L. (2007). Shifts in the carbohydrate, polyol, and amino acid pools during rapid cold-hardening and diapause-associated cold-hardening in flesh flies (Sarcophaga crassipalpis): A metabolomic comparison. *Journal of Comparative Physiology B* 177: 753–763.

Miller, et al. (2010). Characterization of novel microsatellite markers in *Musa acuminata* subsp. *burmannicoides*, var. Calcutta 4. *BMC Research Notes* 2010 3: 148.

Ming, R., Yu, Q.Y., and Moore, P.H. (2007). Sex determination in papaya. *Seminars in Cell Development Biology* 18: 401–408.

Ming, R., Hou, S., Feng, Y., Yu, Q., Dionne-Laporte, A., Saw, JH, Senin, P, et al. (2008). The draft genome of the transgenic tropical fruit tree papaya (Carica papaya Linnaeus). *Nature* 452: 991–996.

Ming, R., Mu, Q., and Moore, P.H. (2012). Papaya genome and genomics. In: Schnell, R.J. and Priyadarshan, P.M. (eds) *Genomics of Tree Crops*, pp. 241–261, Springer, New York.

Moncada, X. and Hinrichsen, P. (2007). Limited genetic diversity among clones of red wine cultivar Carmenere as revealed by microsatellite and AFLP markers. *Vitis* 46: 174–180.

Mukherjee, S.K. (1950). Mango: Its allopolyploid nature. *Nature* 166: 196–197.

Mukherjee, S.K. (1957). Cytology of some Malayan species of *Mangifera*. *Cytologia* 22: 239–241.

Mukherjee, S.K. (1958). The origin of mango. *Indian Journal of Horticulture* 15: 129–134.

Mukherjee, S.K. (1963). Cytology and breeding of mango. *Punjab Horticulture Journal* 3: 107–115 (observed in India. *Economic Botany* 7).

Myles, S., Boyko, A.R., Owens, C.L., Brown, P.J., Grassi, F., Aradhya, M.K., Prins, B. et al. (2011). Genetic structure and domestication history of the grape. *Proceedings of the National Academy of Sciences*. DOI: 10.1073/pnas.1009363108.

Myles, S., Chia, J.-M., Hurwitz, B., Simon, C., Zhong, G.Y., Buckler, E., and Ware, D. (2010). Rapid genomic characterization of the GenusVitis. *PLoS ONE* 5: e8219.

Nakano, M., Shimizu, T., Kuniga, T., Nesumi, H., and Omura, M. (2008). Mapping and haplotyping of the flanking region of the polyembryony locus in Citrus unshiu Marcow. *Journal of Japanese Society Horticultural Science* 77: 109–114.

Nakasone, H.Y. and Paull, R.E. (eds) (1998). Papaya. *Tropical Fruit*, CABI, New York.

Narzary, D., Mahar, K.S., Rana, T.S., and Ranade, S.A. (2009). Analysis of genetic diversity among wild pomegranate in western Himalayas using PCR methods. *Sci Horti.* 121: 237–242.

Negri, A., Prinsi, B., Rossoni, M., Failla, O., Scienza, A., Cocucci, M., and Espen, L. (2008). Proteome changes in the skin of the grape cultivar Barbera among different stages of ripening. *BMC Genomics* 9(1): 378.

Negri, A., Robotti, E., Prinsi, B., Espen, L., and Marengo, E. (2011). Proteins involved in biotic and abiotic stress responses as the most significant biomarkers in the ripening of Pinot Noir skins. *Functional and Integrative Genomics* 11(2): 341–355.

Nicolosi, E., Deng, Z.N., Gentile, A., La Malfa, S., Continella, G., and Tribulato, E. (2000). Citrus phylogeny and genetic origin of important species as investigated by molecular markers. *Theoretical and Applied Genetics* 100: 1155–1166.

Noda, Y., Kaneyuki, T., Mori, A., and Packer, L. (2002). Antioxidant activities of pomegranate fruit extract and its anthocyanidins: Delphinidin, cyanidin, and pelargonidin. *Journal of Agricultural and Food Chemistry* 50(1): 166–171.

Nogueira, S.B., Labate, C.A., Gozzo, F.C., Pilau, E.J., Lajolo, F.M., and Nascimento, J.R.O. (2012). Proteomic analysis of papaya fruit ripening using 2DE-DIGE. *Journal of Proteomics* 75: 1428–1439.

Notomi, T., Okayama, H., Masubuchi, H., Yonekawa, T., Watanabe, K., Amino, N., and Hase, T. (2000). Loop-mediated isothermal amplification of DNA. *Nucleic Acids Research* 28: e63.

Oliveira, E.J., Dantas, J.L.L., Castellen, M.S., and Machado, M.D. (2008). Identificação de microssaté-lites para o mamoeiro por meio da exploração do banco de dados de DNA. *Revista Brasileira De Fruticultura* 30: 625–628.

Oliveira, E.J., Amorim, V.B.O., Matos, E.L.S., Costa, J.L., Castellen, M.S., Pádua, J.G., and Dantas, J.L.L. (2010a). Polymorphism of microsatellite markers in papaya (*Carica papaya* L). *Plant Molecular Biology Reporter* 28: 519–530.

Oliveira, E.J., Santos Silva, A., Carvalho, F.M., Santos, L.F., Costa, J.L., Amorim, V.B.O., and Dantas, J.L.L. (2010b). Polymorphic microsatellite marker set for Carica papaya L. and its use in molec-ular-assisted selection. *Euphytica* 173: 279–287.

Oliveira, E.J., Fraife Filho, G.A., Freitas, J.P.X., de Dantas, J.L.L., and Resende, M.D.V. de. (2012). Plant selection in F2 segregating populations of papaya from commercial hybrids. *Crop Breeding Applied Biotechnology*, 12(3): 191–198.

Ollitrault, P., Dambier, D., Froelicher, Y., Luro, F., and Cottin, R. (2001). La diversite des agrumes: Structuration et exploitation par hybridation somatique. *Comptes-rendus de l'Academie d' agri-culture* 86(8): 197–222.

Omura, M., Shimada, T., Endo, T., Fujii, H., and Shimizu, T. (2006). Paper presented at the Plant & Animal Genome XIII Conference.Development of SNPs assay for identification of Citrus culti-vars., January 14–18.

Pandit, S.S., Mitra, S., Giri, A.P., Pujari, K.H., Bhimarao, P.P., Jambhale, N.D., and Gupta, V.S. (2007). Genetic diversity analysis of mango cultivars using inter simple sequence repeat markers. *Current Science* 93(8): 1135–1141.

Parasnis, A., Ramakrishna, W., Chowdari, K., Gupta, V., and Ranjekar, P. (1999). Microsatellite (GATA) n reveals sex-specific differences in papaya. *Theoretical and Applied Genetics* 99: 1047–1052.

Parasnis, A.S., Gupta, V.S., Tamhankar, S.A., and Ranjekar, P.K. (2000). A highly reliable sex diagnos-tic PCR assay for mass screening of papaya seedlings. *Molecular Breeding* 6(3): 337–344.

Pasquier, G., Lapaillerie, D., Vilain, S., Dupuy, J.-W., Lomenech, A.-M., Claverol, S., Gény, L., Bonneu, M., Teissedre, P.-L., and Donèche, B. (2012). Impact of foliar symptoms of "Esca proper" on pro-teins related to defense and oxidative stress of grape skins during ripening. *Proteomics* 13: 108–118.

Pastore, C., Zenoni, S., Tornielli, G.B., Allegro, G., Dal Santo, S., Valentini, G., Intrieri, C., Pezzotti, M., and Filippetti, I. (2011). Increasing the source/sink ratio in *Vitis vinifera* (cvSangiovese) induces extensive transcriptome reprogramming and modifies berry ripening. *BMC Genomics* 12(1): 631.

Pauquet, J., Bouquet, A., This, P., and Adam-Blondon, A.F. (2001). Establishment of a local map of AFLP markers around the powdery mildew resistance gene *Run*1 in grapevine and assessment of their usefulness for marker assisted selection. *Theoretical and Applied Genetics* 103(8): 1201–1210.

Peng, F.Y., Reid, K.E., Liao, N., Schlosser, J., Lijavetzky, D., Holt, R., Martínez Zapater, J. M., et al. (2007). Generation of ESTs in *Vitis vinifera* wine grape (Cabernet Sauvignon) and table grape (Muscat Hamburg) and discovery of new candidate genes with potential roles in berry devel-opment. *Gene* 402(1–2): 40–50.

Pérez, J.O., d'Eeckenbrugge, G.C., Risterucci, A., Dambier, D., and Ollitrault, P. (2007). Papaya genetic diversity assessed with microsatellite markers in germplasm from the Caribbean region. *Acta Horticulturae* 740: 93–101.

Phumichai, J.C., Babprasert, C., Chunwongse, C., and Sukonsawan, S. (2000). Molecular mapping of mango cultivars "Alphonso" and "Palmer." *Acta Horticulturae* 509: 193–207.

Pilati, S., Perazzolli, M., Malossini, A., Cestaro, A., Dematte, L., Fontana, P., Dal Ri, A., Viola, R., Velasco, R., and Moser, C. (2007). Genome-wide transcriptional analysis of grapevine berry ripening reveals a set of genes similarly modulated during three seasons and the occurrence of an oxidative burst at veraison. *BMC Genomics* 8(1): 428.

Pirseyedi, S.M., Valizadehghan, S., Mardi, M., Ghaffari, M.R., Mahmoodi, P., Zahravi, P., Zeinalabedini, Z., and Nekoui, S.M.K. (2010). Isolation and characterization of novel microsatellite markers in pomegranate (*Punica granatum* L.). *International Journal of Molecular Science* 11: 2010–2016.

Pontin, M., Piccoli, P., Francisco, R., Bottini, R., Martinez-Zapater, J., and Lijavetzky, D. (2010). Transcriptome changes in grapevine (*Vitis vinifera* L.) cv. Malbec leaves induced by ultraviolet-B radiation. *BMC Plant Biology* 10(1): 224.

Rahimi, T., Sayed Tabatabaei, B.E., Sharifnabi, B., and Ghobadi, C. (2006). Genetic relationships between Iranian pomegranate (*Punica granatum L.*) cultivars, using Amplified Fragment Length Polymorphism (AFLP) marker. *Iranian J Agric Sci* 36: 1373–1379.

Ramos, H.C.C., Pereira, M.G., Silva, F.F., Gonçalves, L.S.A., Pinto, F.O., de Souza Filho, G.A., and Pereira, T.S.N. (2011). Genetic characterization of papaya plants (*Carica papaya L.*) derived from the first backcross generation. *Genetics and Molecular Research*. 10(1): 393–403.

Ravishankar, K.V., Lalitha, A., and Dinesh, M.R. (2000). Assessment of genetic relatedness among a few Indian mango varieties using RAPD marker. *The Journal of Horticultural Science and Biotechnology* 75: 198–201.

Ravishankar, K.V., Chandrashekara, P., Sreedhar, S.A., Dinesh, M.R., Lalitha, A., and Saiprasad, G.V.S. (2004). Diverse genetic bases of Indian polyembryonic and monoembryonic mango (*Mangifera indica* L) Cultivars. *Current Science* 87: 870–871.

Ravishankar, K.V., Mani, B.H., Dinesh, M.R., and Lalitha, A. (2011). Development of new microsatellite markers from Mango (*Mangifera indica*) and cross-species amplification. *American Journal of Botany* 98(4): e96–e98.

Ravishankar, K.V., Vidhya, L., Cyriac, A., Rekha, A., Goel, R., Singh, N.K., and Sharma, T.R. (2012). Development of SSR markers based on a survey of genomic sequences and their molecular analysis in banana (*Musa* spp.) *Journal of Horticultural Science and Biotechnology* 87: 84–88.

Rekha, A., Ravishankar, K.V., Anand, L., and Hiremath, S.C. (2001). Genetic and genomic diversity in banana (*Musa* species and cultivars) based on D² analysis and RAPD markers. *Infomusa* 10: 29–34.

Riaz, S., Krivanek, A.F., Xu, K., and Walker, M.A. (2006). Refined mapping of the Pierce's disease resistance locus, *PdR1*, and Sex on an extended genetic map of *Vitis rupestris* × *V. arizonica*. *Theoretical and Applied Genetics* 113(7): 1317–1329.

Riaz, S., Tenscher, A.C., Ramming, D.W., and Walker, M.A. (2011). Using a limited mapping strategy to identify major QTLs for resistance to grapevine powdery mildew (*Erysiphenecator*) and their use in marker-assisted breeding. *Theoretical and Applied Genetics* 122(6): 1059–1073.

Risterucci, A.M., Duval, M.F., Rohde, W., and Billotte, N. (2005). Isolation and characterization of microsatellite loci from Psidium guajava L. *Molecular Ecology Notes* 5(4): 745–748.

Ritter, E., Herran, A., Valdés-Infante, J., Rodríguez-Medina, N.N., Briceño, A., Fermin, G., Sanchez-Teyer, F., et al. (2010a). Comparative linkage mapping in three guava mapping populations and construction of an integrated reference map in guava. *Acta Horticulturae* (ISHS) 849: 175–182.

Ritter, E., Rodríguez-Medina, N.N., Velásquez, B., Rivero, D., Rodríguez, J.A., Martínez, F., and Valdés-Infante, J. (2010b.) Qtl (quantitative trait loci) analysis in guava. *Acta Horticulturae* (ISHS) 849: 193–202.

Rivera-Pastrana, D.M., Yahia, E. M., and González-Aguilar, G.A. (2010). Phenolic and carotenoid profiles of papaya fruit (*Carica papaya L.*) and their contents under low temperature storage. *Journal of Science Food Agriculture* 90: 2358–2365.

Rizzini, F.M., Bonghi, C., and Tonutti, P. (2009). Postharvest water loss induces marked changes in transcript profiling in skins of wine grape berries. *Postharvest Biology and Technology* 52(3): 247–253.

Rodriguez, N., Valdés-Infante, J., Becker, D., Velazquez, B., Gonzalez, G., Sourd, D., Rodriguez, J., Billotte, N., Risterucci, A.M., Ritter, E., and Rohde, W. (2007). Characterization of guava accessions by SSR markers, extension of the molecular linkage map, and mapping of QTLs for vegetative and reproductive characters. *Acta Hort.* 735: 201–216.

Roose, M., Wanamaker, S., Lyon, M., Close, T., Davies, C., and Mei, G. (2006). Paper presented at the plant & animal genome XIV Conference. *Gene Chip for Citrus*, January 14–18, San Diego, CA.

Salmaso, M., Faes, G., Segala, C., Stefanini, M., Salakhutdinov, L., Zyprian, E., Toepfer, R., Grando, M. S., and Velasco, R. (2004). Genome diversity and gene haplotypes in the grapevine (*Vitis vinifera* L.), as revealed by single nucleotide polymorphisms. *Molecular Breeding* 14(4): 385–395.

Sánchez-Teyer, L.F., Barraza-Morales, A., Keb, L., Barredo, F., Quiroz-Moreno, A., O'Connor-Sánchez, A., and Padilla-Ramírez, J.S. (2010). Assessment of genetic diversity of Mexican guava germplasm using dna molecular markers. *Acta Horticulturae* (ISHS) 849: 133–138.

Sancho, L.E., Yahia, E.M., and Aguilar, G.A. (2011). Identification and quantification of phenols, carotenoids, and vitamin C from papaya (Carica papaya L., cv. Maradol) fruit determined by HPLC-DAD-MS/MS-ESI. *Food Research International* 44: 1284–1291.

Santana, J.C., Heuertz, M., Arranz, C., Rubio, J.A., Martinez-Zapater, J.M., and Hidalgo, E. (2010). Genetic structure, origins, and relationships of grapevine cultivars from the Castilian Plateau of Spain. *American Journal of Enology and Viticulture* 61: 214–224.

Santos, S.C., Ruggiero, C., Silva, C.L.S.P., and Lemos, E.G.M. (2003). A microsatellite library for *Carica papaya* L. cv. Sunrise "Solo". *Revista Brasileira de Fruticultura* 25: 263–267.

Sarkhosh, A., Zamani, Z., Fatahi, R., and Ebadi, A. (2006). RAPD markers reveal polymorphism among some Iranian pomegranate (*Punica granatum* L.) genotypes. *Scientia Horticulturae* 111: 24–29.

Sarkhosh, A., Zamani, Z., Fatahi, R., Wiedow, C., Chagné, D., and Gardiner, S.E. (2012). A pomegranate (*Punica granatum* L.) linkage map based on AFLP markers. *Journal of Horticulture Science and Biotechnology* 87(1): 1–6.

Schnell, R.J., Ronning, C.M., and Knight, R.J. Jr. (1995). Identification of cultivars and validation of genetic relationships in *Mangifera indica* L. using RAPD markers. *Theoretical Applied Genetics* 90: 269–274.

Schnell, R.J., Brown, J.S., Olano, C.T., Meerow, A.W., Campbell, R.J., and Kuhn, D.N. (2006). Mango genetic diversity analysis and pedigree inferences for Florida cultivars using microsatellite markers. *Journal of American Society of Horticultural Science* 131(2): 214–224.

Schripsema, J., Vianna, M.D., Rodrigues, P.A.B., de Oliveira, J.G., and Franco, R.W.A. (2010). Metabolomic investigation of fruit flesh gelling of papaya fruit (*carica papaya* l. "golden") by nuclear magnetic resonance and principle component analysis. *Acta Horticulturae* 851: 505–512.

Schweiggert, R.M., Steingass, C.B., Mora, E., Esquivel, P., and Carle, R. (2011). Carotenogenesis and physico-chemical characteristics during maturation of red fleshed papaya fruit (*Carica papaya* L.). *Food Research International* 44: 1373–1380.

Seeram, N.P., Adamsa, L.S., Henning, S.M., Niua, Y., Zhang, Y., Nairb, M.G., and Hebera, D. (2005). In vitro antiproliferative, apoptotic and antioxidant activities of punicalagin, ellagic acid and a total pomegranate tannin extract are enhanced in combination with other polyphenols as found in pomegranate juice. *Journal of Nutritional Biochemistry* 16: 360–367.

Sharon, D., Hillel, J., Vainstein, A., and Lavi, U. (1992). Application of DNA fingerprints for identification and genetic analysis of *Carica papaya* and other *Carica* species. *Euphytica* 62: 119–126.

Sharrock, S. (1997). The banana and its relatives. Focus Paper III, in Annual report, INIBAP.

Simmonds, N.W. (1976). *Evolution of Crop Plants*, Longman, London, UK.

Simmonds, N.W. (1962). *The Evolution of Bananas*, Longman, London, UK.

Shimizu, T., Fujii, H., Nishikawa, F., Shimada, T., Kotoda, N., Yano, K., and Endo, T. (2009). Data mining of citrus sequence data sets to develop microarrays for expression and genomic analysis. *Plant & Animal Genome XVII Conference (p414)*, San Diego, CA.

Slabas, A.R., Ndimba, B., Simon, W.J., and Chivasa, S. (2004). Proteomic analysis of the Arabidopsis cell wall reveals unexpected proteins with new cellular locations. *Biochemical Society Transactions* 32: 524–28.

Sondur, S.N., Manshardt, R.M., and Stiles, J.I. (1996). A genetic linkage map of papaya based on randomly amplified polymorphic DNA markers. *Theoretical and Applied Genetics* 93: 547–553.

Soost, R.K. and Cameron, J.W. (1975). Citrus. In: J. Janick and J.N. Moore (eds.), *Advances in Fruit Breeding*. Purdue University Press, West Lafayette, pp. 507–540.

Soost, R.K., Roose, M., Janick, J., and Moore, J.N. (1996). Citrus. In *Fruit Breeding, Vol 1: Tree and Tropical Fruits*. John Wiley, New York.

Soriano, J.M., Zuriaga, E., Rubio, P., Lla´cer, G., Infante, R., and Badenes, M.L. (2011). Development and characterization of microsatellite markers in pomegranate (*Punica granatum* L.). *Mol Breeding* 27: 119–128.

Spagnolo, A., Magnin-Robert, M., Alayi, T.D., Cilindre, C., Mercier, L., Schaeffer-Reiss, C., Van Dorsselaer, A., Clément, C., and Fontaine, F. (2011). Physiological changes in green stems of *Vitis vinifera* L. cv. Chardonnay in response to Esca proper and apoplexy revealed by proteomic and transcriptomic analyses. *Journal of Proteome Research* 11(1): 461–475.

Talebi, B.M., Sharifi, N.B., and Bahar, M. (2003). Analysis of genetic diversity in pomegranate culti-vars of Iran, using random amplified polymorphic DNA (RAPD) markers. In: *Proceedings of the Third National Congress of Biotechnology*, volume 2, pp. 343–345, Iran.

Tao, N.-G, Xu, J., Chen, Y.-J., and Deng, X.-X. (2006) Construction and characterization of a cDNA library from the pulp of Cara Cara navel orange (*Citrus sinensis* Osbeck). *Journal of Integrative Plant Biology* 48: 315–319.

Tao, N.-G., Hu, Z.-Y., Liu, Q., Xu, J., Cheng, Y.-J., Guo, L.-L., Guo, W.-W., and Deng, X.-X. (2007). Expression of phytoene synthase gene (*Psy*) is enhanced during fruit ripening of Cara Cara navel orange (*Citrus sinensis* Osbeck). *Plant Cell Reports* 26: 837–843.

Tattersall, E.R., Grimplet, J., DeLuc, L., Wheatley, M., Vincent, D., Osborne, C., Ergül, A., et al. (2007). Transcript abundance profiles reveal larger and more complex responses of grapevine to chilling compared to osmotic and salinity stress. *Functional and Integrative Genomics* 7(4): 317–333.

Terrier, N., Ageorges, A., Abbal, P., and Romieu, C. (2001). Generation of ESTs from grape berry at various developmental stages. *Journal of Plant Physiology* 158(12): 1575–1583.

Terrier, N., Glissant, D., Grimplet, J., Barrieu, F., Abbal, P., Couture, C., Ageorges, A., et al. (2005). Isogene specific oligo arrays reveal multifaceted changes in gene expression during grape berry (*Vitis vinifera* L.) development. *Planta* 222(5): 832–847.

This, P., Lacombe, T., and Thomas, M.R. (2006). Historical origins and genetic diversity of wine grapes. *Trends in Genetics* 22(9): 511–519.

Thomas, M.R., Matsumoto, S., Cain, P., and Scott, N.S. (1993). *Theoretical and Applied Genetics* 86: 173–180.

Tillett, R., Ergul, A., Albion, R., Schlauch, K., Cramer, G., and Cushman, J. (2011). Identification of tissue-specific, abiotic stress-responsive gene expression patterns in wine grape (*Vitis vinifera* L.) based on curation and mining of large-scale EST data sets. *BMC Plant Biology* 11(1): 86.

Torres, A.M., Soost, R.K., and Diedenhofen, U. (1978). Leaf isozymes as genetic markers inCitrus. *American Journal of Botany* 65(8): 869–881.

Troggio, M., Malacarne, G., Coppola, G., Segala, C., Cartwright, D.A., Pindo, M., Stefanini, M., et al. (2007). A dense single-nucleotide polymorphism-based genetic linkage map of grapevine (*Vitis vinifera* L.) anchoring pinot noir bacterial artificial chromosome contigs. *Genetics* 176(4): 2637–2650.

Uma, S., Saraswathi, M.S., Rekha, A., and Ravishankar, K.V. (2011). Banana and Plantain. In: *Advances in Horticultural Biotechnology—Molecular Markers and Marker Assisted Selection Vol. III: Fruit Crops, Plantation Crops and Spices*. Vedam Books, New Delhi, India.

Upadhyay, A., Saboji, M.D., Reddy, S., Deokar, K., and Karibasappa, G.S. (2007). AFLP and SSR marker analysis of grape rootstocks in Indian grape germplasm. *Scientia Horticulturae* 112: 176–183.

Upadhyay, A., Aher, L.B., and Karibasappa, G.S. (2011). Detection of variation among clonal selec-tions of grapevine (*Vitis vinifera* L.) Kishmish Chernyi by AFLP analysis. *Journal of Horticultural Science and Biotechnology* 86(3): 230–234.

Urasaki, N., Tokumoto, M., Tarora, K., Ban, Y., Kayano, T., Tanaka, H., Oku, H., Chinen, I., and Terauchi, R. (2002). A male and hermaphrodite specific RAPD marker for papaya (Carica papaya L.). *Theoretical and Applied Genetics* 104: 281–285.

USDA, Natural Resource Conservation Service, Classification for kingdom Plantae down to family *Myrtaceae* (http://plants.usda.gov/java/Classification).

Valdés-Infante, D., Becker, D., Rodríguez, N., Velázquez, B., González, G., Sourd, D., Rodríguez, J., Ritter, E., and Rohde, W. (2003). Molecular characterization of Cuban accessions of guava (*Psidium guajava* L.), establishment of a first molecular linkage map and mapping of QTLs for vegetative characters. *Journal of Genetics & Breeding* 57: 349–358.

Vanhove, A.C., Vermaelen, W., Panis, B., Swennen, R., and Carpentier, S.C. (2012). Screening the banana biodiversity for drought tolerance: Can an in vitro growth model and proteomics be used as a tool to discover tolerant varieties and understand homeostasis. *Frontiers in Plant Science* 3: 176.

Vasanthaiah, H.K.N., Ravishankar, K.V., Shivashankara, K.S., Anand, L., Narayanaswamy, P., Mukunda, G.K., and Prasad, T.G. (2006). Cloning and characterization of differentially expressed genes of internal breakdown in mango fruit (*Mangifera indica*). *Journal of Plant Physiology* 163: 671–679.

Vasugi, C., Dinesh, M.R., Sekar, K., Shivashankara, K.S., Padmakar, B., and Ravishankar, K.V. (2012). Genetic diversity in unique indigenous mango accessions (Appemidi) of the Western Ghats for certain fruit characteristics *Current Science* 103(2): 199–207.

Velasco, R., Zharkikh, A., Troggio, M., Cartwright, D.A., Cestaro, A., Pruss, D., Pindo, M., et al. (2007). A high quality draft consensus sequence of the genome of a heterozygous grapevine variety. *PLoS ONE* 2(12): e1326.

Vezzulli, S., Troggio, M., Coppola, G., Jermakow, A., Malacarne, G., Facci, M., Carthwright, D., et al. (2009). Physical anchoring and integrated genetic mapping among five elite cultivars of *Vitis vinifera* L. *Acta Horticulturae* 827: 35–40.

Victor, K., Fennell, A., and Grimplet, J. (2010). Proteomic analysis of shoot tissue during photoperiod induced growth cessation in V. riparia Michx. grapevines. *Proteome Science* 8(1): 44.

Vignani, R., Scali, M., Masi, E., and Cresti, M. (2002). Genomic variability in *Vitis vinifera* L. Sangiovese assessed by microsatellite and non-radioactive AFLP test. *Electronic Journal of Biotechnology* 5(1): 3–4. Doi: 10.2225/vol5-issue1-fulltext-2.

Viji, G., Harris, D.L., Yadav, A.K., and Zee, F.T. (2010). Use of microsatellite markers to characterize genetic diversity of selected accessions of guava (*Psidium guajava*) in the united states. *Acta Horticulturae* 859: 169–176.

Villegas, V.N. (1997). *Carica papaya* L. In: *Edible Fruits and Nuts,* Verheij, E.W.M. and Coronel, R.E. (eds), volume 2. Wageningen University, The Netherlands.

Vincent, D., Ergül, A., Bohlman, M.C., Tattersall, E.A.R., Tillett, R.L., Wheatley, M.D., Woolsey, R., et al. (2007). Proteomic analysis reveals differences between *Vitis vinifera* L. cv. Chardonnay and cv. Cabernet Sauvignon and their responses to water deficit and salinity. *Journal of Experimental Botany* 58(7): 1873–1892.

Viruel, M.A., Escribano, P., Barbieri, M., Ferri, M., and Hormaza, J.I. (2005). Fingerprinting, embryo type and geographic differentiation in mango (*Mangifera indica* L., Anacardiaceae) with microsatellites. *Molecular Breeding* 15: 383–393.

Wang, Z., Zhao, F., Zhao, X., Ge, H., Chai, L., Chen, S., Perl, A., and Ma, H. (2012). Proteomic analysis of berry-sizing effect of GA3 on seedless *Vitis vinifera* L. *Proteomics* 12(1): 86–94.

Waters, D.E., Holton, T., Ablett, E., Lee, L.S., and Henry, R.J. (2005). cDNA microarray analysis of developing grape (*Vitis vinifera* cv. Shiraz) berry skin. *Functional and Integrative Genomics* 5(1): 40–58.

Waters, D.L.E., Holton, T.A., Ablett, E.M., Slade Lee, L., and Henry, R.J. (2006). The ripening wine grape berry skin transcriptome. *Plant Science* 171(1): 132–138.

Welter, L., Göktürk-Baydar, N., Akkurt, M., Maul, E., Eibach, R., Töpfer, R., and Zyprian, E. (2007). Genetic mapping and localization of quantitative trait loci affecting fungal disease resistance and leaf morphology in grapevine (*Vitis vinifera* L). *Molecular Breeding* 20(4): 359–374.

Wikstrom, N., Savolainen, V., and Chase, M.W. (2001). Evolution of the angiosperms: Calibrating the family tree. *Proceedings of the Royal Society of London* 268: 2211–2220.

Xu, Q., Yu, K., Zhu, A., Ye, J., Liu, Q., Zhang, J., and Deng., X. (2009). Comparative transcripts profiling reveals new insight into molecular processes regualting lycopene accumulation in sweet orange (*Citrus sinensis*) red-flesh mutant. *BMC Genomics* 10: 540.

Yamanaka, N., Hasran, M., Xu, D.H., Tsunematsu, H., Idris, S., and Ban, T. (2006). Genetic relationship and diversity of four Mangifera species revealed through AFLP analysis. *Biomed Life Science of Earth and Environment* 53(5): 949–954.

Yang, L., Lin, H., Takahashi, Y., Chen, F., Walker, M.A., and Civerolo, E.L. (2011). Proteomic analysis of grapevine stem in response to Xylellafastidiosa inoculation. *Physiological and Molecular Plant Pathology* 75(3): 90–99.

Yu, Q., Tong, E., Skelton, R.L., Bowers, J.E., Jones, M.R., Murray, J.E., Hou, S., et al. (2009). A physical map of the papaya genome with integrated genetic map and genome sequence. *BMC Genomics* 10: 371.

Zamani, Z., Sarkhosh, A., Fatahi, R., and Ebadi, A. (2007). Genetic relationships among pomegranate genotypes studied by fruit characteristics and RAPD markers. *Journal of Horticultural Science and Biotechnology* 82: 11–18.

Zamboni, A., Di Carli, M., Guzzo, F., Stocchero, M., Zenoni, S., Ferrarini, A., Tononi, P., et al. (2010). Identification of putative stage-specific grapevine berry biomarkers and omics data integration into networks. *Plant Physiology* 154(3): 1439–1459.

Zenoni, S., Ferrarini, A., Giacomelli, E., Xumerle, L., Fasoli, M., Malerba, G., Bellin, D., Pezzotti, M., and Delledonne, M. (2010). Characterization of transcriptional complexity during berry development in *Vitis vinifera* using RNA-seq. *Plant Physiology* 152(4): 1787–1795.

Zhang, J.K., Hausmann, L., Eibach, R., Welter, L.J., Töpfer, R., and Zyprian, E.M. (2009). A framework map from grapevine V3125 (*Vitis vinifera* Schiavagrossa × Riesling) × rootstock cultivar Börner (*Vitis riparia* × *Vitis cinerea*) to localize genetic determinants of phylloxera root resistance. *Theoretical and Applied Genetics* 119(6): 1039–1051.

Zhang, J., Ma, H., Feng, J., Zeng, L., Wang, Z., and Chen, S. (2008). Grape berry plasma membrane proteome analysis and its differential expression during ripening. *Journal of Experimental Botany* 59(11): 2979–2990.

10

Catharanthus roseus: *The Metabolome That Represents a Unique Reservoir of Medicinally Important Alkaloids under Precise Genomic Regulation*

Ashutosh K. Shukla, PhD and Suman P.S. Khanuja, PhD

CONTENTS

10.1 Preamble...326
10.2 The Plant 326
    10.2.1 History, Origin, Distribution, and Habitat...327
    10.2.2 Nomenclature and Taxonomy...327
    10.2.3 Botany..329
    10.2.4 Agriculture 330
        10.2.4.1 Soil and Climatic Conditions 330
        10.2.4.2 Manures and Fertilizers..330
        10.2.4.3 Weeding and Irrigation...330
        10.2.4.4 Propagation...330
        10.2.4.5 Growth Regulators...331
        10.2.4.6 Harvesting and Yield..331
        10.2.4.7 Pests, Diseases, and Control Measures 332
    10.2.5 Natural and Induced Variation...333
    10.2.6 Phytochemistry 333
        10.2.6.1 Occurrence and Distribution of Alkaloids............................334
        10.2.6.2 Extraction and Detection Methods...334
        10.2.6.3 Biological Function of Alkaloids..337
        10.2.6.4 Pharmacology and Human Usage...338
        10.2.6.5 Essential Oils and Nonalkaloidal Constituents 339
10.3 OMICS: Application Tools and Studies 340
    10.3.1 Genomics Studies..341
    10.3.2 Transcriptomics Studies...341
        10.3.2.1 ESTs...341
        10.3.2.2 cDNA-AFLP 341
        10.3.2.3 Deep Transcriptome Sequencing..342
        10.3.2.4 Genes Encoding the Enzymes Involved in TIA Biosynthesis 342
    10.3.3 Proteomics Studies 344
        10.3.3.1 Enzymology..345
        10.3.3.2 Cellular and Subcellular Compartmentation 355
        10.3.3.3 Regulation of TIA Biosynthesis 357

10.3.4 Metabolomics Studies 362
10.3.5 Ultimate Alkaloidomics: The Metabolic Reprogramming of *C. roseus*
        TIA Biosynthetic Machinery 363
10.4 Some Recent Advances 364
10.5 Future Prospects and the Gaps to Be Filled 365
List of Abbreviations and Symbols 367
Acknowledgments 369
References 369

10.1 Preamble

Alkaloids are one of the major groups of secondary metabolites and have provided several leads for the development of modern drugs. They comprise pharmacologically active, nitrogen-containing heterocyclic compounds and their general classification is based on their chemical structures, with terpenoid indole alkaloids (TIAs) being one of the most well characterized classes of alkaloids. The *Catharanthus roseus* plant produces more than 130 TIAs and no other single plant species is reported to produce such a wide array of complex alkaloids. It is also one of the most extensively investigated medicinal plants. It derives its significance particularly from its expensive and scarce shoot-specific bisindole alkaloids, vinblastine and vincristine, which are established anticancer agents and perennially in high demand. To date, the plant remains the most important source of these bisindole alkaloids (produced in *planta* by the condensation of monomeric TIAs, vindoline, and catharanthine) despite some reports of their total synthesis. The bisindole alkaloids and vindoline are produced only in the aerial green parts of the plant and require differentiated plant tissue for biosynthesis, whereas catharanthine is produced throughout the plant and could also be produced in undifferentiated cell cultures. As a consequence of decades of research on the chemistry, biochemistry, and molecular biology of TIA biosynthesis and the recent application of omics technologies, *C. roseus* has emerged as a model nonmodel plant species to study plant alkaloidomics.

However, there are some inherent challenges associated with research on this important medicinal plant, which make it interesting to researchers, for example, its recalcitrance to genetic transformation, the complexity of the TIA structures, and the complex subcellular compartmentalization of TIA biosynthesis. Although some success has been achieved recently in overcoming these challenges at genomic, transcriptomic, proteomic, and metabolomic levels, a lot remains to be done to understand and exploit the regulatory switches that make this genome so dynamic in expression. This expression is not only organ/tissue specific, but also temporally precise regarding developmental stage. This chapter will provide an overview of the recent progress in research on *C. roseus* and also highlight future prospects and the gaps to be filled.

10.2 The Plant

The Madagascar periwinkle, *C. roseus* (L.) G. Don, is one of the most extensively investigated medicinal plants. The first comprehensive book covering the phytochemical,

chemical, and pharmacological research on *C. roseus* alkaloids was published by Taylor and Farnsworth (1975). Since then, a tremendous amount of literature has been generated on this plant and the trend is steadily increasing (Verpoorte et al. 1997). Most research endeavors on this plant have focused on its pharmacologically important alkaloids. The roots of the plant accumulate ajmalicine and serpentine, which are important components of medicines for controlling high blood pressure and other types of cardiovascular maladies. The shoot (leaves and stem) of the plant is a source of bisindole alkaloids, vincaleukoblastine (vinblastine), and vincaleurocristine (vincristine), which are indispensable parts of most anticancer chemotherapies. Low yields of the bisindole alkaloids coupled with their high market value due to huge demand and the rapid life cycle of the plant have prompted funding of several research projects worldwide. These projects aim to increase the alkaloid productivity of the plant by studying the purpose of alkaloid biosynthesis by the plant and the regulation of the secondary metabolism pathways.

10.2.1 History, Origin, Distribution, and Habitat

The plant was known for its medicinal values even in 50 BC (Virmani et al. 1978). However, it is popularly called Madagascar periwinkle because it is believed that it is indigenous to central Madagascar. *C. roseus* was introduced to many parts of the world in the eighteenth century. It is believed to have been introduced into India by Portuguese missionaries in the middle of the eighteenth century via Goa (Mishra and Kumar 2000). In India, the plant was initially grown in cemeteries on account of its perpetually flowering nature. Later it spread all over the tropical and subtropical parts of the country and could be seen growing wild over the plains, lower foothills, and terai areas in North India and the hills of South India. Currently, *C. roseus* has a pantropical distribution (Stearn 1975) and has become naturalized throughout the tropical and subtropical parts of the world.

10.2.2 Nomenclature and Taxonomy

Initially there was a lot of ambiguity in the botanical literature regarding the correct nomenclature of the genus of the plant. Several names such as *Ammocallis rosea*, *Lochnera rosea*, and, most commonly, *Vinca rosea* were in use (Mishra and Kumar 2000; Stearn 1975). The genus *Vinca* was established by Linnaeus in 1753 in his *Species Plantarum* (1: 209), where he distinguished two species *V. major* and *V. minor*. To these two originally European species he added the tropical *Vinca rosea* (now *Catharanthus roseus*) in 1759, although it did not fit his generic description as regards the stamens. Ludwig Reichenbach was the first to recognize *V. rosea* as being generically distinct from the genus *Vinca* and in 1828 he proposed for it the generic name *Lochnera*. However, the name did not receive nomenclatural validity as he failed to provide any generic description for the genus *Lochnera*. In 1835, George Don assigned the name *Catharanthus* to the genus typified by *V. rosea* in his *General system of gardening and botany* (4: 95). The name *Catharanthus* is derived from the Greek words katharos (pure) and anthos (flower), referring to the neatness, beauty, and elegance of the flower. The differences between the two genera *Vinca* and *Catharanthus* have now been well established (Stearn 1966). Although the name periwinkle is attributed to both genera, they differ markedly in their habitat, floral characters, distribution, biochemistry, and cytology. The corolla tube of the *Vincas* gradually widens from below upward (funnel-shaped), filaments bend forward and then backward like a knee (Linnaeus described them as inflexa, retroflexa), and the anthers, which are each crowned with a short hairy flap-like appendage, arch over the stigma. On the contrary, the corolla tube in the members of *Catharanthus* is cylindrical and the filaments are so short that anthers appear to be directly borne on the corolla tube, except in the case of *C. lanceus* where the filaments are long. Also, the anthers in *Catharanthus* species lack terminal appendages and converge conically above the stigma. *Vincas* are essentially plants occurring in Europe, Western and Central Asia and naturalized in North America, whereas *Catharanthus* belongs to the Old World, chiefly Madagascar and India (Virmani et al. 1978). Stomata occur only on the dorsal surface of the leaves of *V. major*, *V. minor*, and *V. difformis*, but they occur on both the surfaces of *V. herbacea* and in all the species of *Catharanthus* (Virmani et al. 1978). The *Vincas* have chromosome number $2n = 46$ and $2n = 92$, whereas all the species of *Catharanthus* have $2n = 16$. During the course of its spread and naturalization over the tropics and subtropics, *C. roseus* acquired many vernacular names, as mentioned in Table 10.1.

The taxonomic key to the genus *Catharanthus* (Mishra and Kumar 2000) classifies it as— Family: Apocynaceae; Subfamily: Plumeroideae; Tribe: Plumerieae; Subtribe: Alstoniiae; Genus: *Catharanthus*. The genus *Catharanthus* comprises eight species of small shrubs and herbs, six of which are predominantly indigenous to Madagascar (Mishra and Kumar 2000; Stearn 1975). Table 10.2 gives information about the designation, origins, and distribution of the eight species.

TABLE 10.1

Vernacular Names of *Catharanthus roseus*

S. No.	Country	Vernacular Names
1	India	Sadabahar, Sadaphul, Nayantara, Rattanjot, Billaganneru, Gul Feringhi, Ainskati, Sudukadu mallikai, Cape-periwinkle, Nityakalyani, Baramasi, Church-yard blossom, Dead-man's flower
2	West Indies	Old maid, Ramgoat rose, Cayenne jasmine, Magdalena, Vicaria
3	Phillipines	Chichirica
4	Japan	Nichinchi
5	England	Madagascar Periwinkle
6	Indonesia	Indische maagdepalm, Soldatenbloem, Kembang sari tijna, Kembang tembaga

Source: Mishra, P. and Kumar, S., *J. Med. Arom. Plant Sci.*, 22, 306–337, 2000; Virmani, O.P., Srivastava, G.N., and Singh, P., *Indian Drugs*, 15, 231–252, 1978.

TABLE 10.2

The Species of the Genus *Catharanthus*

S. No.	Name of the Species	Origin
1	*C. roseus* (L.) G. Don	Madagascar, now naturalized throughout the tropics
2	*C. ovalis* Markgraf	Madagascar
3	*C. trichophyllus* (Baker) Pichon	Madagascar
4	*C. longifolius* (Pichon) Pichon	Madagascar
5	*C. coriaceus* Markgraf	Madagascar
6	*C. lanceus* (Bojer ex A. DC.) Pichon	Madagascar
7	*C. scitulus* (Pichon) Pichon	Madagascar
8	*C. pusillus* (Murray) G. Don	India, Sri Lanka

According to the Integrated Taxonomic Information System (ITIS), the Taxonomic Serial Number (TSN) for *Catharanthus roseus* (L.) G. Don is 30,168 and its currently accepted taxonomic hierarchy is as follows:

Kingdom: Plantae, plants

Subkingdom: Tracheobionta, vascular plants

Superdivision: Spermatophyta, seed plants

Division: Magnoliophyta, flowering plants

Class: Magnoliopsida, dicotyledons

Subclass: Asteridae

Order: Gentianales

Family: Apocynaceae, dogbane family

Genus: *Catharanthus* G. Don, periwinkle

Species: *Catharanthus roseus* (L.) G. Don, Madagascar periwinkle

10.2.3 Botany

C. roseus is a fast-growing, erect, evergreen, everblooming, perennial herb or small shrub, 32–131 cm high and woody at the base. The branching starts from the base itself. The stem diameter ranges from 1.1 to 8.9 cm and the internodal distance varies from 0.5 to 6.8 cm. The stem and leaves contain milky latex. The plant is pubescent when young and glabrous when mature. The leaves are petiolate, opposite or alternate, thick and leathery, 3–8 cm long and 1.5–5.0 cm broad. Leaf shape is oblong or obovate, the base is acute while the apex is rounded to mucronate and the margin is entire. Veins are prominent on the lower surface. The upper surface of the leaf is deep green in color while the lower surface has a lighter green shade. The petiole length ranges from 0.22 to 1.80 cm and the petiole diameter is 0.10–0.43 cm. The stomata are 7–30 μm long and 7–30 μm wide. Flowers occur in terminal racemes, colored violet (var. roseus), white (var. albus) or white with a red eye (var. ocellatus) and range from 2.0 to 5.1 cm in diameter. The calyx lobes are linear to subulate and pubescent. The calyx is five parted and the sepal length range is 0.09–0.70 cm. The corolla tube is cylindrical, 1.2–3.0 cm long and finely pubescent. The flower is normally five petalled. There are five stamens attached to the middle of the corolla tube or just below the mouth. The filaments are very short and not geniculate. The anthers are dorsifixed to the filament and their shape ranges from sagittate to narrow lanceolate. The pollen is ellipsoid or subglobose, smooth, and 10–60 μm long. The anthers are free from the stigma. The style is 0.5–3.0 cm long and slender. There are two distinct carpels and 10–30 ovules in two series in each carpel. The fruit consists of two long, cylindric, pointed follicles (mericarps) 0.6–3.5 cm long and diverging or parallel. The follicle diameter is 0.15–0.40 cm. There are numerous (3–35) blackish seeds in each follicle. The seeds are 0.05–0.70 cm in size and oblong or cylindrical in shape. The hilum is located in a longitudinal depression on one side. The seed surface is minutely reticulate. The root is 5–25 cm long and its diameter is 0.5–6.0 cm. Mishra et al. (2000) have given detailed descriptors for *C. roseus*. According to Janaki Ammal and Bezbaruah (1963), the tetraploid plants grow more vigorously with bigger flowers. The floral morphology of *C. roseus* is conducive for self as well as insect or cross-pollination.

10.2.4 Agriculture

The agriculture of *C. roseus* has been well worked out and is briefly described here.

10.2.4.1 Soil and Climatic Conditions

The plant does well in a wide variety of soils and climate. Although the plant is highly heat and drought tolerant (Gupta 1977), it prefers tropical and subtropical climatic conditions (Mishra and Kumar 2000). In India it grows well in Tamil Nadu, Karnataka, Gujarat, Madhya Pradesh, and Assam (Virmani et al. 1978). The plant also grows in the subtropical areas of North India, although the low temperature in winter is deleterious for its growth, flowering, and seed formation. The plant grows on practically all types of soil, but highly alkaline and waterlogged soils are unsuitable. Light, sandy loam soil rich in humus or treated with raw liquid cow dung provides an aerated substratum to the crop and is preferred for its commercial cultivation (Mishra and Kumar 2000; Virmani et al. 1978). Harvesting of the roots is also easier in such soils. In South India, the plant is widely grown on red laterite soils and occasionally on black cotton soils (Gupta 1977). A well-distributed rainfall of 100 cm and above is highly beneficial for a rainfed crop. The plant is mildly salt tolerant.

10.2.4.2 Manures and Fertilizers

Application of farmyard manure (FYM) at the rate of about 15 t/ha is suitable for growth of roots and aerial parts. Green manuring the field is also highly beneficial and can replace application of FYM. In case of nonavailability of organic manure, an application of a basal dose of 30 kg K_2O, 30 kg P_2O_5, and 20 kg urea per hectare is also advised. Top dressing of 20 kg nitrogen in two equal split doses during the growing season at a duration of 30–45 days is highly beneficial to the crop (Virmani et al. 1978).

10.2.4.3 Weeding and Irrigation

The crop may require at most two weedings. The first weeding may be done 2 months after seed sowing/transplanting, and the second may be done 2 months later. Fluchoraline at 0.75 kg/ha or alachor at 1 kg/ha provides effective control over a wide range of weeds in the plant (Mishra and Kumar 2000). In areas where the rainfall is distributed evenly throughout the year, the plants do not require any irrigation. However, in areas where the rainfall is restricted to a particular period, 4–5 irrigations are needed to get optimum yield. Some degree of drought can be withstood by the plant but under severe water stress, at leaf water potential below −1.9 Mpa, there was a slowing down of growth associated with some increase in alkaloid content (Saenz et al. 1993). Talha et al. (1975) studied the effect of soil moisture deficit (SMD) on growth and alkaloidal content of *C. roseus* at four different SMD levels (25%, 50%, 75%, and 95%). They found that 25%–50% SMD level was the best to increase the dry weight of the leaf, stem, and root. However, a further increase in SMD from 50% to 75% caused reduction in herb yield, albeit with substantial increase in alkaloid content.

10.2.4.4 Propagation

Freshly collected seeds are the most preferred propagules for *C. roseus*. Seeds can be directly sown in the field or seedlings transplanted into the field from pregrown nurseries.

In direct sowing, seeds are sown in well-prepared pulverized fields at the commencement of the rainy season in rows 45 cm apart, and later the seedlings obtained are thinned out to maintain a distance of at least 30 cm between plants. About 2.5 kg of seeds are sufficient for a hectare of land. Alternatively, seedlings are raised in a nursery to economize on seed utilization. About 500 g of seeds is enough to provide seedlings for one hectare. Seeds are sown in nursery beds, about 2 months ahead of the proposed transplanting of the seedlings, which should synchronize with the onset of the rainy season. Seeds normally germinate in about 10 days, and 60-day-old seedlings are suitable for transplanting. A spacing of 45 × 30 cm accommodates about 74,000 seedlings per hectare. Variable plant spacings of 45 × 15 cm, 45 × 20 cm, and 45 × 30 cm are reported to have no effect on root yield, but higher leaf yield is obtained at a plant spacing of 45 × 20 cm (Mishra and Kumar 2000). A plant spacing of 30 × 20 cm has also been found to be optimal. Kulkarni et al. (1987) found that dense planting of tetraploids led to higher yields. Vegetative propagation of the plant is also possible through cuttings planted in the field in a manner similar to the seedlings.

10.2.4.5 Growth Regulators

Colchicine-induced tetraploid plants of *C. roseus* have thicker stems, thicker leaves, and larger flowers than diploid plants (Virmani et al. 1978). *C. roseus* plants treated for seven weeks with weekly doses of 100 mg gibberellic acid (GA) and harvested biweekly showed a significant increase in plant height and dry weight of stems. The leaf shape changed from obovate to lanceolate but there was no effect on flowering (Masoud et al. 1968). The total alkaloid content in the leaves and roots decreased at the final harvest. However, vinblastine levels increased in the leaves of GA-treated plants (Virmani et al. 1978). Nitrogen fertilizer application and long day conditions cause an increase in alkaloid content of the roots under normal conditions but fail to compensate for the negative effect of GA on the total alkaloid content in the roots (Virmani et al. 1978). Phytohormone 2,4-dichlorophenoxyacetic acid (2,4-D) acts as a defoliator, whereas indole acetic acid (IAA) and GA application results in an increase in leaf number and plant height (Virmani et al. 1978). This effect was more pronounced in the postfruiting period. In autotetraploids, exogenous application of boron caused reduction in leaf width and follicle lengths (Mishra and Kumar 2000). Addition of growth retardants such as (2-chloroethyl) trimethylammonium chloride (CCC) resulted in the production of lateral branches and an increase in alkaloid production in *C. pusillus* (Basu 1992). An increase in the duration of the photosynthetic period increases the yield of root alkaloids and the weight of aerial parts (Virmani et al. 1978).

10.2.4.6 Harvesting and Yield

Harvesting should be done 12 months after sowing. The crop is cut about 7.5 cm above the ground and dried for stems, leaves, and seeds. The field is then irrigated to attain the desired moisture level, followed by ploughing and collection of roots. The roots are washed well and dried in shade. The alkaloid content of the roots is at a maximum when the plants flower. Leaves can be harvested twice before the final harvest. The first stripping is done six months after planting and the second three months after the first cutting. Finally, the whole plant is harvested after another 2–3 months. The leaves are also dried in shade for around 15 days until they become crisp, losing about 75% of their weight. The leaves kept for drying should be spread uniformly and turned periodically to avoid any fungal attack and fermentation. The dried leaves are stored in a cool dry place. Under irrigated conditions, a crop planted in one hectare on average yields about 3.6 tons of air-dried leaves,

1.5 tons of stems, and 1.5 tons of roots. The yield under rainfed conditions is around 2 tons of leaves, 1 ton of stems, and 0.75 tons of roots per hectare (Virmani et al. 1978). Under field conditions, plants show natural variation in the biosynthesis and accumulation of secondary metabolites due to genotypic differences as well as environmental and seasonal changes (Choudhury and Gupta 2002).

10.2.4.7 Pests, Diseases, and Control Measures

Although the *C. roseus* plants are hardy and fairly resistant to pests and diseases, several pathogens of viral, fungal, bacterial, and mycoplasmal origins have been encountered in this plant. Thrips and aphids have been identified as the common pests of this plant. *C. roseus* plants are susceptible to a disease similar to the spike disease of *Santalum album* (Sandalwood) (Virmani et al. 1978). It is characterized by a bushy appearance of the plant in which the leaves become smaller and internodes shorten. In later stages excessive branching is induced. The diseased plants exhibit hyperplasia or increased vegetative activity. In all the stages of the disease, the flowers are green and develop phyllody. Green rosette disease has occasionally been observed on these plants (Garga 1958). In this disease the leaf and flower size is reduced but there is no discoloration or distortion of any floral part. There is no fruiting in such plants. The fungal disease twig blight/top-rot or die back is reported to have been caused by *Phytophthora nicotianae*, *Pythium debaryanum*, *P. butleri*, *P. aphanidermatum*, and *Colletotrichum dermatium*, whereas leaf spot is caused by *Alternaria tenussima*, *A. alternata*, *Rhizoctonia solani*, *Ophiobolus catharanthicola*, *Haplosporella marathwadensis*, *Glomerilla cingulata*, and *Myrothecium roridum* (Mishra and Kumar 2000). The *Myrothecium* leaf spot appears as lesions, which spread to become papery and greenish brown with zonations of translucent areas appearing in concentric rings. Dark-colored sporodochia of the fungus surrounded by cottony mycelial growth develop on translucent areas under highly humid conditions. The spots are subcircular to oval, elongated or irregularly scattered, or coalescing to form bigger patches usually 0.3–1.9 cm in diameter. In full-blown cases of the disease almost the entire lamina gets affected. Even the petiole is affected causing premature leaf fall and a naked appearance of the plant. The disease finally extends to the stem through the petiole causing symptoms of die back. The fruiting bodies of the fungus have also been observed in plenty on the petiole and stem (Virmani et al. 1978).

The fungal diseases foot-rot and wilt are caused by *Sclerotium rolfsii* and *Fusarium solani*. Seedlings are highly susceptible to fusarium wilt while the effect on adult plants is gradual. Leaves are the first casualty in this disease, becoming flaccid and yellow before falling off. Wilting occurs in acropetal succession. The *vein clearing* phenomenon, as observed in cotton wilt, is a characteristic symptom in *C. roseus*. As the disease advances, characteristic wrinkles appear on the bark of the stem extending from the base upward. The roots of the affected plant subsequently show signs of decay and rot. The entire cortex of the tap root peels off and disintegrates. The younger lateral branches of the root are affected later. The cortices of the stem as well as roots of the diseased plant are dominated by fungal hyphae, which also extend to the wood and accumulate in the vessels (Virmani et al. 1978). Some of the fungal diseases can be controlled by foliar sprays of the fungicide Captafol Foltaf 80 WP, Carbendazim Bavistin 50 WP, and Benomyl 50 WP (Kalra et al. 1991).

A mycoplasmal/bacterial disease with chlorotic, wilting, and vein-clearing symptoms together with several anatomical deviations is reported to be transmitted by a sharp shooter insect, *Onconteopia nigrans* (Mishra and Kumar 2000).

Viral diseases of the plant have also been reported (Mishra and Kumar 2000; Samad et al. 2008). Leaf mosaic disease, caused by a viral pathogen, results in irregular yellow patches and malformation, accompanied by necrosis and wilting. Another viral disease causes extreme shortening of nodes and internodes, and reduced fruit and seed formation. Amelioration of the diseased plants could be achieved by regular sprays of oxytetracycline, GA, tetracycline, and 0.5% aqueous solution of nicotine sulfate.

10.2.5 Natural and Induced Variation

The natural populations of *C. roseus* have been observed to harbor considerable genetic variability. The commercial crop has wide morphological and chemical variations, with different ranges of heritability due to a fair amount of outcrossing existing in the species (Kulkarni et al. 1984). Though the plant is basically self pollinated, Krishnan et al. (1979) reported frequent outcrossing (13.6%), whereas Kulkarni (1999) reported that outcrossing between white × white-flowered plants ranged from 28.3% to 65.3%, between white × pink-flowered plants it ranged from 12.2% to 15.1%, and in the *die back* disease-resistant variety Nirmal it was 11.4%. The natural variability has been exploited to develop horticultural and drug types. The horticultural cultivars have been developed on the basis of differences in heights and combinations of different shades of petal and corona colors (Mishra and Kumar 2000). Variation arising from polyploidy, cross breeding, and induced mutagenesis has been variously employed in *C. roseus*. Autotetraploids of different genetic backgrounds, when compared with corresponding diploids, were found to be relatively more tolerant to die back and the collar- and root-rot diseases (Kulkarni and Ravindra 1987). In certain strains the autotetraploids were observed to possess broader leaves with lower length/width ratio, reduced pollen fertility, larger pollens, shorter follicles, heavier seeds, lower leaf dry matter, lower vinblastine content, and higher total alkaloid content (Mishra and Kumar 2000). The higher ploidy level derivatives were found to be usually highly sterile with low seed germinability (Janaki Ammal and Bezbaruah 1963). Kulkarni et al. (1987) studied the performance of diploids and induced autotetraploids of *C. roseus* under different levels of nitrogen and plant spacing.

Experimental hybrids have been described between *C. roseus* and *C. trichophyllus* (Sevestre-Rigouzzo et al. 1993). Levy et al. (1983) observed significant heterosis for leaf and root yields in crosses involving three pure lines. However, heterosis was not observed for ajmalicine content of the roots.

Kulkarni et al. (1999) have shown large differences in the morphology as well as leaf and root alkaloid contents in three induced mutants of *C. roseus*. Nirmal, which is a superior *C. roseus* variety having a high level of field resistance to die back disease, has been developed and released by CIMAP. Prabal is another variety developed at CIMAP (Dwivedi et al. 2001). Starting with Nirmal, Kulkarni et al. (2003) (U.S. Patent No. 6,548,746) have developed Dhawal, a better, distinct, and high alkaloid-producing periwinkle plant, through mutation breeding. It combines the characters of improved alkaloid yield and higher herbage with tolerance to the die back disease. Dhawal can be distinguished morphologically by its distinctly undulating and wavy leaf margin. It is currently one of the best varieties available for advanced molecular biology work and is also stable for commercial cultivation, maintaining the stability of the improved characters.

10.2.6 Phytochemistry

Researchers focused their attention on *C. roseus* in the 1950s when they learned of a tea Jamaicans were drinking to treat diabetes. The reported folkloric usage of the plant as an oral hypoglycemic agent prompted its phytochemical examination by two independent groups: one at the University of Western Ontario (Noble, Beer, and Cutts) and the other at

Eli Lilly and Company, Indianapolis (Svoboda's group). Folkloric usage provided a guiding force for selecting the plants for screening as it was realized that a randomized selection from the 250,000 known higher plants would have in all probability been fruitless, especially when taxonomists had projected an existence of some 500,000 unidentified species (Svoboda and Blake 1975). However, neither group could substantiate the hypoglycemic activity of the plant. The observation by Noble et al. (1958) of a toxic depletion of white cells and bone marrow depression produced in rats by certain fractions of these extracts eventually led to the isolation of vincaleukoblastine sulfate (its generic name adopted by the United States Adopted Names Committee is vinblastine). On the other hand, Svoboda and Johnson observed in certain extracts and fractions a reproducible oncolytic activity primarily against P1534 leukemia, a transplanted acute lymphocytic leukemia, in DBA/2 mice. They isolated leurosine, a new dimeric alkaloid closely related in chemical structure to vinblastine, and were subsequently successful in isolating vinblastine sulfate itself (Johnson et al. 1959; Svoboda 1958; Svoboda et al. 1959). Svoboda's group devised a new technique of selective or differential extraction for this purpose. Soon it was realized that neither leurosine nor vinblastine, nor any therapeutic combination thereof, was responsible for the high percentage of indefinite survivors produced by certain fractions of the extract. This led Svoboda's group to the isolation of leurosidine and leurocristine (its generic name adopted by the United States Adopted Names Committee is vincristine), using another innovative method, the gradient pH technique (Svoboda 1961). Leurosivine and rovidine are the other antineoplastic alkaloids of *C. roseus* but their activities are much lower as compared to vinblastine, leurosine, leurosidine, and leurocristine. Of these six antineoplastic agents, only vincristine and vinblastine have found extensive use in the management of human neoplasms.

10.2.6.1 Occurrence and Distribution of Alkaloids

Terpenoid indole alkaloids (TIAs), found mainly in the plants of the families Apocynaceae, Loganiaceae, and Rubiaceae, constitute one of the largest groups of alkaloids with more than 3000 members having rich structural diversity. The *Catharanthus* alkaloids currently comprise a group of about 130 TIAs (Van der Heijden et al. 2004). Wide differences have been noted in the compositions of the alkaloids isolated from the underground (root) and aerial tissues (shoot/leaves) of the plant. The alkaloid content of the various plant tissues varies considerably as found by different investigators working in different parts of the world (Table 10.3). These variations might be due to different agroclimatic regions, genetically different genotypes/cultivars, and alkaloid extraction procedures. Some of the alkaloids of *C. roseus* are classified according to their major (but not necessarily exclusive) source tissue (Table 10.4).

10.2.6.2 Extraction and Detection Methods

While the isolation of any alkaloid is an individual problem, several standard techniques have been in existence for the preliminary extraction from the crude drug and for the subsequent separation and purification of the specific alkaloid. They depend upon typical alkaloid characteristics—soluble in polar solvents (like water) at low pH and in nonpolar solvents (like chloroform and diethyl ether, which are usually immiscible with water) at higher pH (Verpoorte 1980). In plants such as *C. roseus*, which produce numerous alkaloids, the main problem boils down to the resolution of complex alkaloidal mixture, which is contaminated with nonalkaloidal materials. Using the alkaloidal properties, separation of alkaloids from nonalkaloids can be achieved by means of solvent–solvent extractions.

TABLE 10.3

Variation in Alkaloid Content of Various Parts/Tissues of
C. roseus

Plant Part/Tissue	Percent Alkaloid Content (Dry Weight Basis)
Root	0.125–2.60
Root bark	2.50–9.00
Leaf	0.32–2.56
Stem	0.07–0.46
Fruit	~0.40
Seed	~0.18
Flower	0.005–0.84
Pericarp	~1.14

Source: Virmani, O.P., Srivastava, G.N., and Singh, P., *Indian Drugs*, 15, 231–252, 1978; Mishra, P. and Kumar, S., *J. Med. Arom. Plant Sci.*, 22, 306–337, 2000; Kulkarni, R.N., Baskaran, K., Chandrashekara, R.S., Khanuja, S.P.S., Darokar, M.P., Shasany, A.K., Uniyal, G.C., Gupta, M.M., and Kumar, S., *'Dhawal'*, a high alkaloid producing periwinkle plant. U. S. Patent No. 6,548,746, 2003 and references cited therein; Shukla, A.K., Shasany, A.K., Gupta, M.M., and Khanuja, S.P.S., *J. Exp. Bot.*, 57, 3921–3932, 2006.

Most approaches have failed to take full advantage of the relative basicities of the alkaloids during extraction, treating the problem in two steps: isolation and purification. These are, however, inherently related, and the initiation of purification procedures as early as possible during extraction can often be advantageous. The utilization of the "selective" or "differential" extraction technique coupled with the gradient pH technique resulted in the initial isolation of some 64 alkaloids from mature *C. roseus* plants by Svoboda's group (Svoboda and Blake 1975).

During the extraction procedure, artifact formation must be avoided. The temperature and pH of the extraction procedure must also be optimized as these parameters have also been found to affect alkaloid recovery from *C. roseus* leaves (Shukla et al. 1997). Supercritical fluid extraction has also been used for vindoline (Song et al. 1992) and vinblastine (Choi et al. 2002). Once the extraction and purification are complete, separation and quantitation have to be performed. The chromatographic techniques such as thin layer chromatography (TLC), gas–liquid chromatography (GLC), and high-performance liquid chromatography (HPLC) are widely used nowadays in the analysis of alkaloids. TLC is the least sensitive (10–100 ng) of the three chromatographic techniques and is complicated for quantitative purposes, but it allows more samples to be analyzed per unit of time. Besides, it also allows the analysis of less-clean samples, identification by means of selective spray reagents, visibility of all applied compounds after development, and ease of scale-up for preparative work. GLC is the most sensitive (10^{-12}–10^{-13} g) chromatographic technique and often allows the unequivocal identification of compounds when used in the combination such as gas chromatography–mass spectrometry (GC–MS). It allows easy quantitation but is not suitable for compounds that are less volatile or unstable at high temperatures. Derivatization with acetyl or trimethylsilyl groups may sometimes solve this problem. HPLC offers an enormous separation power because of the high number of theoretical plates which can be obtained with HPLC columns, and also because of the many combinations of stationary and mobile phases, which offer more possibilities for solving difficult separation problems. It is easy for quantitative analysis and can also be used for the analysis of unstable

TABLE 10.4

Some Alkaloids Isolated from Different Source Tissues of *C. roseus*

Leaf	Root	Leaf and Root Both	Others[a]
Vinaspine, vincarodine, perividine, lochnericine, lochneridine, lochnerinine, lochrovicine, lochrovidine, lochrovine, catharosine, desacetylvindoline, vincolidine, vincoline, vindoline, vindolinine, vindorosine, perimivine, carosine, catharicine, catharine, deacetylvinblastine, isoleurosine (4′-deoxyvinblastine), vincristine, neoleurocristine, neoleurosidine, pleurosine, rovidine, vinaphamine, vinblastine, vincathicine, vincamicine, vindolicine, vindolidine (vindorosine), pericyclivine, catharanthamine, N-demethylvinblastine, deacetoxyvinblastine, leurocolombine, leurosidine N_b-oxide, 5-oxoleurosine, pseudovinblastinediol, vinamidine (catharinine), vincadioline, anhydrovinblastine, deacetoxyleurosine	Alstonine, ammorosine, cathindine, cavincidine, akuammicine, lochnerivine, maandrosine, ammocalline, pericalline, virosine, leurosivine, vinosidine, serpentine, lochnerine, akuammine	Catharanthine, cavincine, dihydrositsirikine, isositsirikine, sitsirikine, perivine, perosine, mitraphylline, leurosidine (vinrosidine), leurosine, carosidine, ajmalicine[c], reserpine, cleavimine	Vinsedine[b], vinsedine[b], 19-acetoxy-11-hydroxytabersonine, 19-acetoxy-11-methoxytabersonine, N-acetylvincoside, cathenamine, coronaridine, corynantheine, corynantheine aldehyde, 4,21-dehydrogeissoschizine, 19-*epi*-vindolinine, 19-*epi*-ajmalicine, gelssoschizine, horhammericine, horhammerinine, isovincoside, 11-methoxytabersonine, preakuammicine, stemmadenine, strictosidine lactam, tabersonine, vallesiachotamine, vincoside, yohimbine, roseadine, leurosinine, roseamine, strictosidine, isovallesiachotamine, 3-*epi*-ajmalicine, 3-*epi*-19-*epi*-ajmalicine, tetrahydroalstonine, akuammigine, pseudoindoxylajmalicine, 7-hydroxyindoleneineajmalicine, antirhine, akuammiline, deacetylakuammiline, 10-hydroxy-deacetylakuammiline, pleiocarpamine, xylosyloxyakuammicine, dihydrocondylcarpine, vinervine, 19-hydroxytabersonine, 19-hydroxy-11-methoxytabersonine, minovincinine, vindolinine N-oxide, 19-*epi*-vindolinine N-oxide, N,N-dimethyltryptamine, (16R)-19,20-(E)-isositsirikine, (16R)-19,20-(Z)-isositsirikine, 21-hydroxycyclolochnerine, tubotaiwine, strychnan glycoside, vingramine[b], methylvingramine[b]

Source: Svoboda, G.H. and Blake, D.A., *The Catharanthus Alkaloids*, Marcel Dekker, New York, 1975; Virmani, O.P., Srivastava, G.N., and Singh, P., *Indian Drugs*, 15, 231–252, 1978; Blasko, G. and Cordell, G.A., *The Alkaloids*, Academic Press, San Diego, 1990; Mishra, P. and Kumar, S., *J. Med. Arom. Plant Sci.*, 22, 306–337, 2000 and references cited therein; Jossang, A., Fodor, P., and Bodo, B., *J. Org. Chem.*, 63, 7162–7167, 1998.

a Includes alkaloids isolated from biosynthesis experiments and *in vitro* cultures.
b Isolated from seeds.
c Other names for ajmalicine are vinceine, vincaine, raubasine, deltayohimbine, and tetrahydroserpentine.

alkaloids. Sensitivity differs according to the detector used but is in general in the range 1–100 ng.

   C. roseus alkaloids have been analyzed using TLC (Farnsworth et al. 1964), GC–MS (Ylinen et al. 1990), LC–MS (Auriola et al. 1990), flow-injection electrospray ionization mass spectrometry (Favretto et al. 2001), HPLC (Naaranlahti et al. 1987; Volkov and Grodnitskaya 1994; Singh et al. 2000; Tikhomiroff and Jolicoeur 2002), and enzyme-linked immunosorbent assay (ELISA) (Cibotti et al. 1990). Specific radioimmunoassays (RIAs) have also been developed for the bisindole alkaloids: vinblastine (Hirata et al. 1989) and vincristine (Huhtikangas et al. 1987), as well as their precursors vindoline (Westekemper et al. 1980) and catharanthine (Deus-Neumann et al. 1987). However, for most investigators, HPLC remains the most convenient method of analysis of *C. roseus* alkaloids, and C_{18} reversed-phase columns are the most popular. Now it has also been made possible to detect many TIAs and their precursors simultaneously in single runs (Singh et al. 2004). Gupta et al. (2005) have described the use of a Chromolith column, which gives better resolution than the C_{18} column, for the simultaneous detection of vincristine, vinblastine, catharanthine, and vindoline. Photodiode array and fluorescence detection methods are mostly used, but electrochemical detection methods have also been coupled with HPLC to analyze *C. roseus* alkaloids (Naaranlahti et al. 1989b). A capillary electrophoresis–mass spectrometry (CE–MS) approach for the simultaneous determination of vinblastine, vindoline, and catharanthine has been developed by Chen et al. (2011), and an ultraperformance liquid chromatography/mass spectrometry method to simultaneously quantify vindoline, catharanthine, serpentine, and ajmalicine has been reported by He et al. (2011). Since only the final drugs (vincristine and vinblastine) and their precursors (vindoline and catharanthine) are commercially available and most other alkaloids and biosynthetic intermediates are not marketed, they have to be obtained from the plant by investigators using preparative or semipreparative methods (Naaranlahti et al. 1990), for use as reference compounds. Lopez et al. (2011) have synthesized two molecularly imprinted polymers (MIPs) for catharanthine and vindoline for specific extraction of these natural TIAs from *C. roseus* extracts by solid-phase extraction (SPE).

10.2.6.3 *Biological Function of Alkaloids*

TIAs, like most other secondary metabolites, have ecochemical and defensive roles for the plant. The biological activity inherent in the *in planta* role of these metabolites is exploited for human usage to alleviate various disorders and diseases. Regulation of plant secondary metabolite biosynthesis cannot be seen as separate from the role of these products for the plant. However, it is difficult to pinpoint the exact *in planta* role of secondary metabolites like the TIAs. In the case of *C. roseus*, an antifeedent activity against *Spodoptera* larvae has been reported for vinblastine and catharanthine (Meisner et al. 1981). Antifeedant activity against *Spodoptera* caterpillars for *C. roseus* leaf extracts has been described as well (Chockalingam et al. 1989; Meisner et al. 1981). A nematocidal effect has been reported for serpentine (Verpoorte et al. 1997). Luijendijk et al. (1996a) investigated the involvement of strictosidine in the antimicrobial and antifeedent activities of *C. roseus* leaves. Strictosidine and its deglucosylation product, specifically formed by strictosidine β-D-glucosidase (SGD), were shown to be active against several microorganisms. On the contrary, neither the intact glucoside nor the aglycone product was found to exhibit antifeedent activity against *Spodoptera exigua* larvae, as was found for intact *C. roseus* leaves and leaf extracts. Besides alkaloids, other compounds are also responsible for the antifeedent activity of *C. roseus* leaves, as demonstrated by Singh et al.

(2003), who found that the n-hexane fraction of the acetone extract from *C. roseus* leaves contains α-amyrin acetate and oleanolic acid, which are insect growth regulators active against tobacco caterpillar (*S. litura* F.) and gram pod borer (*Helicoverpa armigera* Hub.). Strictosidine in combination with SGD shows strong antifungal activity (Luijendijk et al. 1996a; Verpoorte et al. 1997). Minimum inhibitory concentrations were found to be about 0.33 mM against some *Fusarium* species and *Cladosporium cucumerinum*, 1 mM against *Trichoderma viride*, and below 0.008 mM against *Phytophthora infestans*. Strictosidine itself is also active against *P. infestans*. Strictosidine and SGD levels are particularly high in young leaves of *C. roseus* (260 µg/g fresh weight, which is ca. 0.5 mM and 750 pkat/g, respectively). Based on the fact that strictosidine is found in the vacuole, whereas the highly specific SGD is localized elsewhere, it seems that this combination of substrate and enzyme plays a role in the plant's defense in the case of wounding (Luijendijk et al. 1996a; Verpoorte et al. 1997), and that strictosidine is a phytoanticipin. Thus, it is highly probable that the alkaloids have important ecochemical functions in the defense of the plant against pathogenic organisms and herbivores.

10.2.6.4 Pharmacology and Human Usage

C. roseus has a history of folkloric uses across the world. The folkloric usage of different parts of the plant for the treatment of various maladies has been reviewed in detail by previous workers (Mishra and Kumar 2000; Virmani et al. 1978). A number of activities like antidiabetic, antimalarial, antihelminthic, antidiuretic, diuretic, antibacterial, and hypotensive have been associated with *C. roseus*. However, the importance of this plant in modern medicine has been realized only after the chance discovery of antineoplastic alkaloids contained in its leaves. The pharmacological significance of ajmalicine, the major alkaloid from the roots, stems from its use in the treatment of hypertension and obstructive circulatory diseases (Verpoorte et al. 1991). It is reported to improve cerebral circulation. Serpentine, which is also present in *C. roseus* roots, is used as a hypotensive (Mishra and Kumar 2000).

The discovery of antineoplastic bisindole alkaloids (vincristine and vinblastine) from *C. roseus* represents one of the most important introductions of plant products into the cancer chemotherapeutic armamentarium (Neuss and Neuss 1990). These alkaloids are useful in the treatment of both malignant and nonmalignant diseases. Vincristine and vinblastine are used to treat a variety of thrombocytopenic disorders such as idiopathic thrombocytopenic purpura, thrombotic thrombocytopenic purpura, and chemotherapy-induced microangiopathic hemolytic anemia. Although vincristine and vinblastine are useful in platelet and platelet-associated disorders, it is in treating malignancy that they are truly an indispensable part of the pharmacopoeia (Neuss and Neuss 1990). Vincristine sulfate and vinblastine sulfate are sold under the trade names ONCOVIN and VELBAN/VELBE, respectively, by Lilly. Although both these drugs produce a wide range of biochemical effects in cells and tissues, the principal mechanisms of cytotoxicity relate to their interactions with tubulin and disruption of microtubule function, particularly of microtubules comprising the mitotic spindle apparatus, leading to metaphase arrest. They bind rapidly and reversibly to binding sites on tubulin that are distinct from those of the taxanes, colchicines, podophyllotoxin, and guanosine triphosphate (GTP). Structurally, vincristine and vinblastine are identical except for a single substitution on the vindoline nucleus, where vincristine and vinblastine possess formyl and methyl groups, respectively. Although this minor difference does not fundamentally alter the mechanism of action and tubulin-binding properties of these agents, the antitumor and toxicologic profiles of vincristine and vinblastine differ significantly. The principal use of vinblastine is in the treatment of Hodgkin's disease, while the chief use of vincristine is in the treatment of acute lymphocytic leukemia in children. While the dose-limiting toxicity for vinblastine is leukopenia and bone marrow depression, that for vincristine is neurotoxicity (Weiss et al. 1974). Children tolerate vincristine therapy better than adults. Vinblastine is also highly effective in the treatment of testicular neoplasms. Other types of lymphomas may also respond to treatment with combination regimens that include vinblastine. Vinblastine is also used in the treatment of Kaposi's sarcoma, mycosis fungoides, and carcinoma of the breast (McCormack 1990). Vincristine is also a component of regimens with established value in the management of Hodgkin's disease and other lymphomas, and pediatric tumors such as Wilm's tumor and embryonal rhabdosarcoma (McCormack 1990). First available in the 1960s, the *Catharanthus* alkaloids are now included as part of virtually every successful combination chemotherapy program, both because of their unique action and because of their unique toxicities (Neuss and Neuss 1990). They are exceedingly important in both curative and palliative regimens (Table 10.5). Combination chemotherapeutic regimens are designed based on the belief that greater cumulative effects can be seen by using agents with differing sites (within the cell) and times (within the cell cycle) of action. Cumulative toxicities may be less than single toxicities because of the varying toxicities of different agents.

The method of administration and the dosage levels used for the bisindoles (vincristine and vinblastine), like many other antineoplastic agents, are closely tied to their clinical toxicity. Careful titration of the dose to achieve minimal toxicity consistent with therapeutic benefit is necessary for successful clinical use. The bisindole alkaloid drugs are administered intravenously once a week. To reduce the toxic side effects of these drugs on "innocent bystander" tissues, new approaches for drug targeting have been employed. Monoclonal antibodies that are reactive with tumor-associated antigens have been evaluated as vehicles for delivering the toxic bisindole alkaloids to tissues that present these antigens (Pearce 1990). The therapeutic index of vincristine can be enhanced significantly through the use of a liposomal delivery system, which also results in a slight decrease in drug toxicity (Waterhouse et al. 2005).

The acquisition of resistance by tumor cells to various chemotherapeutic drugs, a phenomenon known as multiple drug resistance (MDR), represents a further hurdle to be overcome in cancer chemotherapy. MDR is manifested by the reduced intracellular drug accumulation resulting from increased drug efflux by P-glycoprotein (P-gp) or P-170, which is an ATP-dependent efflux pump encoded by the *mdr-1* gene. The over-expression of *mdr-1* gene transcript P-gp is responsible for MDR and is induced by *Catharanthus* alkaloids (vinblastine and vincristine). Due to the likely role of P-gp in clinical drug resistance, many investigations have focused on strategies to inhibit the function or expression of this protein. Indole-3-carbinol, a glucobrassicin metabolite of cruciferous vegetables, resulted in about 80% reversal of the *Catharanthus* alkaloid-induced P-gp expression in comparison with 65%–70% reversal by verapamil, a well-known MDR reversing agent (Arora and Shukla 2003).

10.2.6.5 Essential Oils and Nonalkaloidal Constituents

The essential oil of the *C. roseus* leaf is reported to contain citronellyl acetate, cadinene (a sesquiterpene hydrocarbon), and 2-heptanol (Virmani et al. 1978). In addition to the TIAs, several nonalkaloidal compounds have also been isolated from *C. roseus* (Table 10.6).

TABLE 10.5

Curative and Palliative Regimens Containing *C. roseus* Alkaloids

Acronym	Drugs	Disease
Curative		
MOPP	Nitrogen mustard, vincristine, procarbazine, prednisone	Hodgkin's lymphoma
CHOP	Cyclophosphamide, daunomycin, vincristine, prednisone	Non-Hodgkin's lymphoma
ABVD	Daunomycin, bleomycin, vinblastine, dacarbazine	Hodgkin's lymphoma
PVB	Cisplatinum, vinblastine, bleomycin	Testicular cancer
MACOP-B	Methotrexate, daunomycin, cyclophosphamide, vincristine, prednisone, bleomycin	Lymphoma
MVAC	Methotrexate, vinblastine, daunomycin, cyclophosphamide	Bladder cancer
Palliative		
COP	Cyclophosphamide, vincristine, prednisone	Lymphoma
POC	Procarbazine, vincristine, cyclophosphamide	Melanoma, small-cell lung cancer
VATH	Vinblastine, daunomycin, Thio-TEPA, Halotestin	Breast cancer
VP	Vinblastine/vindesine, cisplatinum	Non-small-cell lung cancer

Source: Neuss, N. and Neuss, M.N., *The Alkaloids*, Academic Press, San Diego, 1990.

TABLE 10.6

Some Nonalkaloidal Constituents from *C. roseus*

Plant Organ	Nonalkaloidal Compound
Whole plant	Ursolic acid, secologanic acid, deoxyloganin, loganin, sweroside, dehydrologanin
Leaf	Formic acid, stearic acid, palmitic acid, lochnerol, lochnerallol, *O*-pyrocatechuic acid, tannins, volatile oil, and a mannoside
Leaves and stem	Roseoside-A, C13 glycoside, adenosine
Root	L(+) bornesitol, D-camphor, choline
Flower	Hirsutidin, petunidin, malvidin, kaempferol, quercetin, β-sitosterol
Seed	Loganic acid, oil

Source: Virmani, O.P., Srivastava, G.N., and Singh, P., *Indian Drugs*, 15, 231–252, 1978; Mishra, P. and Kumar, S., *J. Med. Arom. Plant Sci.*, 22, 306–337, 2000 and references cited therein.

10.3 OMICS: Application Tools and Studies

With the advent of the omics era, new tools have increasingly become available for the dissection of the TIA biosynthetic pathway in *C. roseus*. This section discusses the application of omics-based tools on the plant.

10.3.1 Genomics Studies

Genetical genomics has been used to study the plant, whereby an integrated genetic linkage map of *C. roseus* based on different types of molecular (Random Amplification of Polymorphic DNA [RAPD], Inter-Simple Sequence Repeat [ISSR], etc.) and morphological markers has been constructed (Gupta et al. 2007). In another attempt, a large array of sequence-tagged microsatellite site (STMS) markers and gene-targeted markers (GTMs) were generated for the construction of a framework linkage map of *C. roseus* (Shokeen et al. 2011). This has been followed by characterization of variation and quantitative trait loci (QTLs) related to TIA yield in a recombinant inbred line mapping population of *C. roseus* (Sharma et al. 2012). Transferability of expressed sequence tags (ESTs)-based SSR markers from *C. roseus* to other medicinal plants has also been studied (Mishra et al. 2011). The isolation of microsatellites of *C. roseus* has been preferentially done using enriched libraries (Bhatia and Shokeen 2009).

Recently, cytogenetic characterization of *C. roseus* and estimation of its genome size have also been carried out. The C-value of this species was estimated to be 1C = 0.76 pg, corresponding to 738 Mbp (Guimarães et al. 2012). A loop-mediated isothermal amplification (LAMP)-based method has also been developed for authentication of *C. roseus* (Chaudhary et al. 2012). Taken together, these will serve as a foundation for future genomics studies and molecular breeding in plants.

10.3.2 Transcriptomics Studies

A highly efficient method of poly $(A)^+$ mRNA isolation (which is a basic prerequisite for transcriptomics studies) from medicinal and aromatic plant (MAP) tissues (rich in secondary metabolites) in general and *C. roseus* in particular has been developed for facilitating downstream transcriptomics studies (Shukla et al. 2005).

10.3.2.1 ESTs

Since whole-genome sequencing is not immediately feasible for most of the medicinal plants like *C. roseus* due to prohibitive cost, ESTs and transcriptome datasets remain the only source of nucleotide sequence data for such plants. First reports of ESTs of *C. roseus* came in 2006, and ESTs were reported from aerial and underground organs of the plant (Murata et al. 2006; Shukla et al. 2006). Transcriptome analysis in *C. roseus* leaves and roots for comparative TIA profiles has provided the real-time information on how the genetic expression of the TIA pathway is partitioned between root and shoot tissues, while also integrating the ontogeny of expression from early seedling stage to mature plant development (Shukla et al. 2006). Later, leaf epidermis-specific ESTs were isolated from *C. roseus* and it was observed that the leaf epidermome (complement of proteins expressed in the leaf epidermis) contained most of the TIA biosynthetic genes (Murata et al. 2008). Presently (as on 1 January, 2013), the dbEST (NCBI) contains 20,168 ESTs of *C. roseus*. The *C. roseus* ESTs have been used for the mining and characterization of short sequence repeat (SSR) markers (Joshi et al. 2011).

10.3.2.2 cDNA-AFLP

A cDNA-amplified fragment-length polymorphism (cDNA-AFLP) approach has been employed to differentiate the leaf and root transcriptomes on a numerical basis (Shukla et al. 2012). Earlier, genome-wide transcript profiling by cDNA-AFLP combined with metabolic profiling of elicited *C. roseus* cell cultures has yielded a collection of known and previously undescribed transcript tags and metabolites associated with TIAs (Rischer et al. 2006). Gene-to-gene and gene-to-metabolite networks have been elucidated by studying correlations between the expression profiles of gene tags and the accumulation profiles of metabolite peaks. These networks also revealed that the different branches of TIA biosynthesis and various other metabolic pathways are subject to differing hormonal regulation.

10.3.2.3 Deep Transcriptome Sequencing

Deep transcriptome sequencing provides a direct tool for accessing the gene content in organisms possessing huge genomes (i.e., >100 Mb). Any tissue can be used to generate cDNAs from mRNA populations and sequenced to generate ESTs that can be assembled into a nonredundant set of sequences (contigs and singleton) to represent the transcriptome. The nonredundant sequences are then annotated for putative function using a combination of bioinformatic means such as sequence searches on protein databases, motif/domain elucidation, and biochemical pathway mapping coupled with subcellular localization predictions. Transcript abundance data provide an elaborate measure of gene expression profiles on a per tissue basis. The predicted function, coupled with expression frequency, ultimately facilitates identification of candidate genes pertinent to a pathway of interest as well as nonpathway targets (like those belonging to primary/intermediary metabolism), whose expression is consistent with synthesis of compounds. The Medicinal Plant Consortium is a NIH-supported multi-institutional project (GM092521) focused on providing transcriptomic and metabolomic resources for 14 key medicinal plants to the global research community for the advancement of drug production and development (http://medicinalplantgenomics.msu.edu/). *C. roseus* is included among these 14 plants. The final transcriptome assemblies and expression data for the 14 medicinal plants have been completed. The Medicinal Plant Genomics Website is maintained and hosted by the Buell Lab in the Plant Biology Department at Michigan State University.

The 1000 plants (oneKP or 1KP) initiative is another public–private partnership generating large-scale gene sequence information for 1000 different species of plants (http://www.onekp.com/index.html). *C. roseus* is included in the list of these 1000 plants also for deep transcriptome sequencing from its flower, floral buds, leaves, and stem.

10.3.2.4 Genes Encoding the Enzymes Involved in TIA Biosynthesis

Genes encoding many of the enzymes involved in TIA biosynthesis have been isolated/cloned (Table 10.7). A *C. roseus hmgr* cDNA has been cloned (Maldonado-Mendoza et al. 1992). The existence of differentially regulated *hmgr* isogenes in *C. roseus* has been substantiated (Maldonado-Mendoza et al. 1994), but the exact number of *hmgr* genes has still to be determined (Verpoorte et al. 1997). The promoter and five prime untranslated region (5′ UTR) of the *hds* gene, whose product is involved in the 2-C-methyl-D-erythritol-4-phosphate (MEP) pathway, has also recently been isolated (GenBank Accession JN217103, Ginis et al. 2012). The *C. roseus g10h* gene promoter sequence (GenBank Accession EF363554) has been isolated by a polymerase chain reaction (PCR)-based genome walking method (Suttipanta et al. 2007). The *g10h* gene product has been revealed to have a dual function in the biosynthesis of terpenoids and phenylpropanoids (Sung et al. 2011). The promoter of the *C. roseus cpr* gene has also been studied (GenBank Accession Y09417, Cardoso et al. 1997).

TABLE 10.7

Genes Encoding Enzymes Involved in the TIA Biosynthetic Pathway of *C. roseus*

S. No.	Gene	Accession No.	References
1	Hydroxymethylglutaryl-CoA synthase (*hmgs*)	JF739871	Unpublished
2	Hydroxymethylglutaryl-CoA reductase (*hmgr*)	M96068	Maldonado-Mendoza et al. 1992
		AY623812[a]	Unpublished
3	Mevalonate kinase (*mvk*)	HM462019	Simkin et al. 2011
4	5-Phosphomevalonate kinase (*pmk*)	HM462020	Simkin et al. 2011
5	Mevalonate 5-diphosphate decarboxylase (*mvd*)	HM462021	Simkin et al. 2011
6	1-Deoxy-D-xylulose-5-phosphate synthase (*dxps*)	AJ011840	Chahed et al. 2000
		DQ848672	Unpublished
7	1-Deoxy-D-xylulose-5-phosphate reductoisomerase (*dxr*)	AF250235	Veau et al. 2000
8	4-Diphosphocytidyl-2*C*-methyl-D-erythritol synthase (*cms*)	FJ177510	Unpublished
9	4-Diphosphocytidyl-2-*C*-methyl-D-erythritol kinase (*cmk*)	DQ848671	Unpublished
10	2-*C*-methyl-D-erythritol 2,4-cyclodiphosphate synthase (*mecs*)	AF250236	Veau et al. 2000
11	1-Hydroxy-2-methyl-2-butenyl 4-diphosphate synthase (*hds*)	AY184810	Oudin et al. 2007b
12	1-Hydroxy-2-methyl-2-butenyl 4-diphosphate reductase (*hdr*)	DQ848676	Unpublished
13	Isopentenyl diphosphate isomerase (*idi*)	EU135981	Guirimand et al. 2012a
14	Geranyl pyrophosphate synthase (*gpps*)	EU622902	Unpublished
15	Geraniol-10-hydroxylase (*g10h*)	AJ251269	Collu et al. 2001
16	10-Hydroxygeraniol oxidoreductase (*10hgo*)	AY352047	Unpublished
17	S-adenosyl-L-methionine:loganic acid methyltransferase (*lamt*)	EU057974	Murata et al. 2008
18	Cytochrome P450 reductase (*cpr*)	X69791	Meijer et al. 1993a
19	3-Deoxy-D-arabino-heptulosonate-7-phosphate synthase (*dhs*)	DQ859024	Ramani et al. 2010
20	Anthranilate synthase (*as*)	AJ250008	Unpublished
21	Tryptophan decarboxylase (*tdc*)	M25151	De Luca et al. 1989
		X67662[a]	Ouwerkerk et al. 1999a
22	Secologanin synthase (*sls*)	L10081	Irmler et al. 2000; Vetter et al. 1992
23	Strictosidine synthase (*str*)	X53602	McKnight et al. 1990
		X61932	Pasquali et al. 1992
		Y10182[a]	Pasquali et al. 1999
24	Strictosidine β-D-glucosidase (*sgd*)	AF112888	Geerlings et al. 2000
		EU072423	Unpublished
25	Tabersonine 19-hydroxylase (*t19h*)	HQ901597	Giddings et al. 2011
26	Tabersonine 16-hydroxylase (*t16h*)	AJ238612	Schroder et al. 1999
		FJ647194	Guirimand et al. 2011b
27	16-Hydroxytabersonine-16-*O*-methyltransferase (*16omt*)	EF444544	Levac et al. 2008
28	S-adenosyl-L-methionine:16-methoxy-2,3-dihydro-3-hydroxy-tabersonine-N-methyltransferase (*nmt*)	HM584929[b]	Liscombe et al. 2010
29	Desacetoxyvindoline 4-hydroxylase (*d4h*)	AF008597[a]	Vazquez-Flota et al. 1997
		U71604[b]	
		U71605[b]	
30	Acetyl-CoA: 4-*O*-deacetylvindoline 4-*O*-acetyl-transferase (*dat*)	AF053307[a]	St-Pierre et al. 1998
31	Minovincinine 19-hydroxy-*O*-acetyltransferase (*mat*)	AF253415[a]	Unpublished
32	Peroxidase (*prx*)	AM236087	Costa et al. 2008
		AY924306	Kumar et al. 2007

[a] Gene (DNA) sequence.
[b] Partial mRNA sequence.

De Luca et al. (1989) reported the cloning and sequencing of the full-length *tdc* cDNA, by screening with TDC antibodies in an expression library. The sequence has significant homology with dopa decarboxylase of *Drosophila melanogaster* (39% at the amino acid level) and other mammalian aromatic amino acid decarboxylase genes. The homology was such that a similar secondary structure for these proteins is expected. Sequence similarities are also observed with other amino acid decarboxylases, suggesting an evolutionary link. The *tdc* cDNA from *C. roseus* has been heterologously expressed in tobacco plants (Songstad et al. 1990, 1991), and it increased their levels of tryptamine and tyramine, the product of tyrosine decarboxylation. A fine example of what metabolic engineering can achieve in secondary metabolism has been provided by the transformation of *Brassica napus* with the *C. roseus tdc* cDNA (Chavadej et al. 1994). The seed of this oil-producing crop has limited use as animal feed due in part to the presence of indole glucosinolates, which make the protein meal less palatable. The introduced *tdc* cDNA redirects tryptophan pools away from indole glucosinolate production and into tryptamine. The mature seeds of the transgenic *B. napus* plants contain reduced levels of indole glucosinolates but no tryptamine, achieving a potentially economically useful product.

The cDNA encoding *str* was first isolated from *Rauvolfia serpentina* (Kutchan et al. 1988). McKnight et al. (1990) used a part of the sequence of this gene for a probe to screen a *C. roseus* cDNA library to isolate the periwinkle gene. The *C. roseus* enzyme has been expressed in tobacco (McKnight et al. 1991) and *E. coli* (Roessner et al. 1992). Pasquali et al. (1992) determined the complete mRNA sequence for *C. roseus str*.

Geerlings et al. (2000) isolated a *sgd* cDNA clone from a *C. roseus* cDNA library. Zárate et al. (2001) demonstrated the expression of *C. roseus sgd* cDNA in a transgenic suspension culture of *Nicotiana tabacum*. The promoter of the *C. roseus prx* gene involved in the TIA dimerization step has also been isolated (GenBank Accession AM236088).

Although transgenics hold tremendous promise for the production of secondary metabolites such as TIAs, they have their own limitations. Whitmer et al. (2003) demonstrated the long-term instability of alkaloid production by stably transformed (with *str* and *tdc* genes) cell lines of *C. roseus*. The creation of artificial metabolic sinks in plants by genetic engineering of key branch points may have serious consequences for the metabolic pathways being modified. Transgenic potato tubers expressing the *C. roseus tdc* gene accumulated tryptamine, which resulted in decreased levels of tryptophan, phenylalanine, and phenylalanine-derived phenolic compounds and increased susceptibility to *P. infestans* as compared with nontransformed tubers (Yao et al. 1995).

10.3.3 Proteomics Studies

Proteomics approaches have also been applied to study the biosynthesis of TIAs in *C. roseus*. In two-dimensional gel electrophoresis (2-DE), sample preparation is one of the most critical steps and needs to be optimized for each type of sample. For proteome analysis in *C. roseus*, a sequential solubilization procedure for the solubilization of proteins after precipitation in trichloroacetic acid and acetone has been developed (Jacobs et al. 2001). Jacobs et al. (2005) carried out MALDI-MS/MS-based proteome analysis in the plant toward the identification of novel proteins involved in TIA biosynthesis. In the past, the major limitation for the identification of proteins from *C. roseus* by mass spectrometric analysis has been the weak database resources for proteins and genes, which are now expanding fast. The problem could be overcome by acquiring sequence data (through fragmentation of peptides measured by mass spectrometry or Edman degradation) and not relying solely on peptide masses. Comparison of protein patterns from alkaloid-producing and nonproducing *C. roseus* cells showed the specific occurrence of a 28 kDa polypeptide (CrPS) restricted to cells accumulating TIAs. It was purified by preparative 2-DE, digested with trypsin, microsequenced with the Edman degradation method, and was ultimately found to belong to the alpha/beta hydrolase superfamily (Lemenager et al. 2005).

In TIA biosynthesis in *C. roseus*, it has become customary to divide the pathway into five parts. The first two lead to the biosynthesis of tryptophan and geraniol and are similar to or even a part of the primary metabolic pathway. The third part leads from tryptophan to tryptamine, and the fourth one from geraniol to secologanin. Both these paths are also present in other plants, which do not biosynthesize TIAs. The fifth part leads to the condensation of tryptamine and secologanin to produce strictosidine, which is the focal point for the biosynthesis of a plethora of TIAs. Strictosidine biosynthesis is found in many plant species producing TIAs, mostly belonging to the families Apocynaceae, Loganiaceae, and Rubiaceae. Beyond strictosidine the pathway diverges and even in different parts of the *C. roseus* plant (like root and shoot) different products are formed.

10.3.3.1 Enzymology

The enzymology of TIA biosynthesis in *C. roseus* has previously been reviewed by different research groups (Mishra and Kumar 2000; Misra et al. 1996; Verpoorte et al. 1997). The elaborate biosynthetic pathway for TIA biosynthesis is depicted in Figure 10.1. The presence of different TIAs and transcripts within the aerial and underground tissues of the plant confirms the expression of distinct tissue-specific pathways responsible for the precise tissue and organ-specific compartmentation of TIA biosynthesis (Laflamme et al. 2001; Shukla et al. 2006).

10.3.3.1.1 Biosynthesis of Tryptophan

The indole moiety of TIAs originates from tryptophan, an aromatic amino acid, which is derived from chorismate via anthranilate. Chorismate is a major branching point in plant primary and secondary metabolism where the shikimate pathway branches into different subpathways. The shikimate pathway leading to chorismate is located in the plastids. The largest flux of carbon atoms from chorismate goes into the phenylalanine/tyrosine pathway, among others, leading to lignin and important groups of secondary metabolites such as flavonoids and anthocyanins (Verpoorte et al. 1997). L-Tryptophan is formed from chorismate through a biosynthetic pathway consisting of five enzymatically controlled steps (Radwanski and Last 1995), and in plants it is required for providing precursors for secondary metabolites, such as indole alkaloids, apart from its primary requirement in protein synthesis. The formation of anthranilate and the following four steps are invariant in all organisms studied to date. Anthranilate synthase (AS, EC 4.1.3.27) catalyzes the conversion of chorismate to anthranilate, the first step in this pathway. Although two forms of AS have been reported in 5-methyltryptophan-resistant tobacco cell cultures (a plastidial isoenzyme, which is strongly inhibited by tryptophan, and a cytosolic one, which is not affected by tryptophan), in *C. roseus* the non-tryptophan-regulated form of AS has not been detected. AS has been isolated from *C. roseus* cells and purified to apparent homogeneity (Poulsen et al. 1993). It is a tetramer, consisting of two large (ca. 67 kDa) and two small (ca. 25.5 kDa) subunits. The larger α-subunit is responsible for the conversion of chorismate into anthranilate and the smaller β-subunits are responsible for the generation of the substrate NH_3 from glutamine. The enzyme shows Michaelis–Menten kinetics for the amino-group donor glutamine, chorismate, and the cofactor Mg^{++} (Km values 0.37 mM, 67 μM, and 0.26 mM, respectively). It shows positive cooperativity of chorismate binding

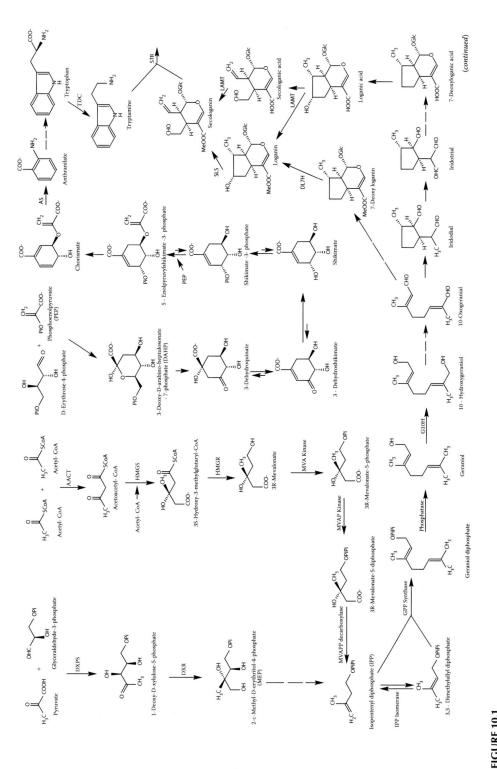

(continued)

FIGURE 10.1
The terpenoid indole alkaloid (TIA) biosynthetic pathway in *C. roseus*.

FIGURE 10.1 (Continued)

at higher levels of tryptophan. The tryptophan binding sites showed positive cooperativity at higher concentrations of chorismate. AS is induced by elicitation and the combination of 2,3-dihydrobenzoic acid (DHBA) and UV light. The next enzyme in the tryptophan pathway is anthranilate-5-phosphoribosyl transferase (EC 2.4.2.18). In some microbial enzymes it is part of the AS enzyme complex, but this transferase activity could not be measured in the purified *C. roseus* enzyme. So far, none of the enzymes leading to tryptophan, after AS, have been studied in *C. roseus* except for a time course study of tryptophan synthase (EC 4.2.1.20) activity during the growth of suspension-cultured and immobilized cells of *C. roseus*. There is not enough evidence at the moment to prove or disprove the existence of a dual (plastidial and cytosolic) pathway for tryptophan biosynthesis in *C. roseus* (Verpoorte et al. 1997).

10.3.3.1.2 *Biosynthesis of Geraniol*

The C_5-unit isopentenyl pyrophosphate (IPP) is the building block of the isoprenoids. It is isomerized to dimethylallyl pyrophosphate (DMAPP), the starter molecule of the isoprenoid pathway. Coupling DMAPP with one or more IPP molecules yields the basic structures that form the backbone of terpenoid biosynthesis. Earlier it was believed that IPP is formed exclusively via the cytosolic mevalonate-dependent pathway, but later it was shown that a plastidial mevalonate-independent pathway also leads to IPP, and in fact it is the latter pathway which is the major route for the biosynthesis of TIAs in *C. roseus* (Contin et al. 1998).

The mevalonate-dependent pathway starts with the formation of mevalonic acid (MVA) from three molecules of acetyl-CoA. Acetoacetyl-CoA thiolase (AACT, acetyl-CoA: acetyl-CoA C-transferase, EC 2.1.3.9) catalyzing the first step, the Claisen-type condensation of two molecules of acetyl-CoA, was partially purified from a cell-suspension culture of *C. roseus* (van der Heijden and Verpoorte 1995). The enzyme consists of several identical subunits with a molecular mass of 41 kDa, which is similar to the value reported for the avian enzyme. The subcellular localization and the substrate specificity (short-, medium-, or long-chain CoA-esters) of this enzyme from *C. roseus* have not yet been studied. A role in sterol biosynthesis is suggested by the fact that during its chromatographic purification the *C. roseus* enzyme always coelutes with HMG-CoA synthase, the next enzyme in the MVA biosynthesis (Van der Heijden et al. 1994b). HMG-CoA synthase (EC 4.1.3.5), catalyzing the formation of HMG-CoA and CoASH from acetyl-CoA and acetoacetyl-CoA, and not yet characterized from plants, was partially purified from *C. roseus*. It is an unstable enzyme that is rapidly inactivated in the presence of a relatively high concentration of salt (>200 mM). The molecular mass of the enzyme is estimated to be about 100 kDa. Etiolated seedlings of *Raphanus sativus* (radish) have been shown to possess a modified pathway whereby HMG-CoA is formed directly from three molecules of acetyl-CoA, without the release of acetoacetyl-CoA from the enzyme involved. Although the major flux of HMG-CoA moves toward the mevalonate pathway, some of it is channeled away by HMG-CoA lyase (EC 4.1.3.4) which forms acetoacetate and acetyl-CoA from HMG-CoA, 3-methylglutaryl-CoA hydratase (EC 4.2.1.18) which converts HMG-CoA into 3-methylglutaconyl-CoA, and certain 3'-nucleotidases which dephosphorylate the 3'-position of HMG-CoA (Van der Heijden et al. 1994a; Verpoorte et al. 1997). In the presence of NADPH, HMG-CoA is reduced to MVA by HMG-CoA reductase (HMGR, *R*-mevalonate: NADP$^+$ oxidoreductase, CoA acylating, EC 1.1.1.34), which is one of the few well-characterized plant enzymes, as it is a major site of regulation of terpenoid biosynthesis. MVA is phosphorylated in two steps to the mono- and diphosphate (MVAP and MVAPP), by the specific ATP-dependent enzymes MVA kinase (ATP-mevalonate-phosphotransferase, EC 2.7.1.36) and

MVAP kinase (EC 2.7.4.2), respectively. MVAPP is converted into IPP by a decarboxylase (EC 4.1.1.33).

The mevalonate-independent pathway was first discovered in bacteria (Rohmer et al. 1993), and later it was also reported in plants (Lichtenthaler et al. 1997a,b) including *C. roseus* (Arigoni et al. 1997; Contin et al. 1998). Other recent studies have also established that this novel pathway is involved in the biosynthesis of geraniol (Eisenreich et al. 1997). The synthesis of IPP via the mevalonate-independent pathway starts with the transketolase-type condensation of pyruvate and glyceraldehydes-3-phosphate to yield 1-deoxy-D-xylulose-5-phosphate (DXP), which is catalyzed by DXP synthase (DXPS). A cDNA clone for this enzyme has been characterized from *C. roseus* suspension cultures (Chahed et al. 2000). The second step of this pathway is the conversion of DXP into 2-C-methyl-D-erythritol-4-phosphate (MEP), a reaction catalyzed by DXP reductoisomerase (DXR). The gene for this enzyme from peppermint has been cloned and heterologously expressed by Lange and Croteau (1999). MEP is the first committed intermediate of the mevalonate-independent pathway, which is therefore called the MEP pathway. Further, less-characterized steps lead to IPP in the MEP pathway. 2-C-Methyl-D-erythritol 2,4-cyclodiphosphate synthase (MECS) is an enzyme, which acts at a step in between MEP and IPP. The cDNAs for DXR and MECS from *C. roseus* have been characterized (Veau et al. 2000). Some other genes of the MEP pathway have also been characterized (Table 10.7).

The compartmental separation between the two different IPP biosynthetic pathways is not absolute. Metabolic cross talk (metabolite exchange) occurs between the cytosolic and plastidial pathways of isoprenoid biosynthesis (Bick and Lange 2003) and its extent depends on the species as well as on the presence and concentration of exogenous precursors (Eisenreich et al. 2001). Cooperation between both pathways has been reported in the formation of chamomile sesquiterpenes (Adam et al. 1999).

Once IPP is formed and isomerized into DMAPP (the actual starter molecule for isoprenoid biosynthesis) by IPP isomerase (EC 5.3.3.2), IPP and DMAPP are coupled in a head-to-tail manner by geranyl diphosphate synthase to yield geranyl diphosphate (GPP), the precursor for the monoterpenes, including iridoids such as secologanin. Geraniol is believed to be formed by the action of phosphatase on GPP (Verpoorte et al. 1997), but further confirmation is required.

10.3.3.1.3 Biosynthesis of Tryptamine

L-Tryptophan is converted into tryptamine by the action of tryptophan decarboxylase (TDC, EC 4.1.1.28). Since its first purification by Noe et al. (1984), this enzyme has been well characterized (Fernandez et al. 1989; Pennings et al. 1989). It is a cytosolic enzyme consisting of two identical subunits (of 54 kDa) and shows Michaelis–Menten kinetics for its substrate tryptophan (Km 75 µM). It also accepts 5-hydroxy-, 5-fluoro-, 4-fluoro, and 5-methyltryptophan as substrates. D-Tryptophan acts as a noncompetitive inhibitor, and tryptamine as a competitive inhibitor (Km 310 µM) (Noe et al. 1984) of this enzyme. The enzyme contains two molecules of pyrroloquinolinequinone (PQQ) and two molecules of pyridoxal phosphate, and Mn^{++} and Mg^{++} ions have a stabilizing effect on it. On the contrary, ATP increases the rate of its inactivation. The monomeric enzyme is probably irreversibly inactivated through conjugation with ubiquitin and subsequently proteolyzed (Fernandez and De Luca 1994).

10.3.3.1.4 Biosynthesis of Secologanin

Geraniol is the substrate for the secologanin biosynthetic pathway. It is hydroxylated on C-10 position by *G10H*, which is regarded as a potential site for regulation, as it catalyzes the first committed step of secologanin biosynthesis. This enzyme is a cytochrome P-450 monooxygenase, which is evident from its properties. It is membrane-bound, is dependent on NADPH and molecular oxygen, and displays a light-reversible inhibition by carbon monoxide. Cytochrome P-450 enzymes are dependent on a membrane-bound reductase (NADPH:cytochrome P-450 reductase, EC 1.6.2.4), a flavoprotein that is involved in the electron transfer from NADPH to the cytochrome P-450 heme group. G10H of *C. roseus* was partially purified from seedlings by Madyastha et al. (1976), and it was later purified to apparent homogeneity from suspension-cultured cells (Meijer 1993; Meijer et al. 1993b). The purified enzyme has a molecular weight of 56 kDa and is localized in provacuolar membranes, unlike many other cytochrome P-450 enzymes that are present in the ER. The NADPH:cytochrome P-450 reductase has been purified by means of affinity chromatography (Meijer et al. 1993a). The molecular weight of the *C. roseus* reductase is 78 kDa; it strictly depends on NADPH, and contains flavin mononucleotide (FMN) and Flavin adenine dinucleotide (FAD) as cofactors.

Soluble oxidoreductases have been isolated from *C. roseus*, which in the presence of NAD$^+$ or NADP$^+$ could oxidize 10-hydroxygeraniol into the dialdehyde 10-oxogeranial (Madyastha and Coscia 1979). 10-Oxogeranial cyclizes to yield iridodial and this is followed by generation of a third aldehyde function to form iridotrial. Of the further steps leading to secologanin, only a few have been studied in more detail. These steps include oxidations, ring closure, glucosylation, and methylation. The carboxylic acid function is methylated by the enzyme *S*-adenosyl-L-methionine:loganic acid methyltransferase (LAMT), which has been partially purified from *C. roseus* seedlings (Madyastha et al. 1973). It catalyzes the transfer of a methyl group from S-adinosyl-L-methionine (SAM) to loganic acid (Km 12.5 mM) or secologanic acid and its activity is maximal just after germination. As only secologanin and not secologanic acid can be used for alkaloid biosynthesis, the methylation of the iridoids is a crucial step in alkaloid biosynthesis (Guarnaccia et al. 1974; Madyastha and Coscia 1979). In the last step, the five-membered ring of loganin is opened to yield secologanin. It has been speculated that the enzyme responsible for this reaction might be one of the targets for 2,4-D in its inhibitory effect on alkaloid accumulation in suspension cultures of *C. roseus* (Verpoorte et al. 1997). Irmler et al. (2000) have identified CYP72A1, a cytochrome P450 enzyme, as secologanin synthase (SLS), which converts loganin into secologanin. According to their hypothesis, in the pathway from geraniol to secologanin, several steps (including possible P450 reactions) are responsible for the conversion of the dialdehyde into 7-deoxyloganin, which is converted into loganin by deoxyloganin 7-hydroxylase (DL7H).

10.3.3.1.5 Biosynthesis of TIAs

The Pictet–Spengler-type condensation of tryptamine and secologanin yields 3-α (S)-strictosidine, which is the central intermediate in TIA biosynthesis. In the next step, a glucosidase splits off the sugar moiety and the reactive dialdehyde formed is further converted through different pathways to a cascade of products, including ajmalicine, catharanthine, tabersonine, and vindoline.

10.3.3.1.5.1 Biosynthesis of Strictosidine: The Gateway to Various TIAs TIA biosynthesis actually starts with the stereospecific coupling of tryptamine and secologanin by strictosidine synthase (STR, EC 4.3.3.2) to form strictosidine, which is the common precursor for all TIAs. The first partial purification of STR from *C. roseus* was reported by Treimer and Zenk (1979) and by Mizukami et al. (1979). The molecular weight of STR is around 38 kDa and it does not require any cofactor. It is highly specific for both substrates and is not inhibited by indole alkaloids such as ajmalicine, vindoline, and catharanthine. The enzyme is present in a series of isoforms in cell cultures, and out of the seven forms detected four were characterized by Pfitzner and Zenk (1989). They differ in pI values (between 4.3 and 4.8) and Km values for tryptamine (from 0.9 to 6.6 mM) but all have maximum activity at pH 6.7. Different isoforms of STR are also present in the leaves of the plant. De Waal et al. (1995) reported the characterization of six isoforms of STR isolated from *C. roseus* cell cultures, all of which were shown to be glycosylated, and had broader pH optima (6–7.5) and much lower (8.2–9.4 μM) Km than the previously reported values (Pfitzner and Zenk 1989). Although it was earlier believed that STR was a cytosolic enzyme, it was later unequivocally shown that it is located in the vacuole (Verpoorte et al. 1997).

10.3.3.1.5.2 Role of Strictosidine β-D-glucosidase   Strictosidine is the precursor for a broad variety of TIAs in different tissues/organs of *C. roseus* itself as well as in many other plant species, all of which share the first part of this pathway. From strictosidine onward, diversification occurs. The basis for this diversity lies in the first place at the dialdehyde, which is formed after deglucosylation of strictosidine and consists of a number of reactive groups (two aldehyde functions, one allylic double bond, and two secondary amine functions). From strictosidine, the first step toward the various types of TIAs is the deglucosylation by SGD (EC 3.2.1.105). The key to the diversification in TIAs must be at the glucosidase or in the steps directly after this enzyme (Verpoorte et al. 1997). Previously, the glucosidase was assumed to be part of a large enzyme complex, the so-called ajmalicine synthase, which catalyzed the formation of ajmalicine out of tryptamine and secologanin, in the presence of NADPH (Scott et al. 1977). This conclusion was later disputed by Hemscheidt and Zenk (1980), who showed that two forms of a highly specific β-D-glucosidase, which could be inhibited by D(+)-gluconic acid-δ lactone (a glucosidase inhibitor), could be chromatographically separated from STR. Stevens (1994) and Luijendijk (1995) have studied SGD extensively. The synthetic substrate *p*-nitrophenyl-β-D-glucoside, used to measure glucosidase activity, is not accepted as a substrate by SGD (Stevens 1994; Hemscheidt and Zenk 1980). SGD has been purified to apparent homogeneity from *C. roseus* cell cultures, which contain a relatively high amount of SGD activity (Luijendijk et al. 1998). Size-exclusion chromatography on Superose 6 HR of crude *C. roseus* protein extracts resulted in an elution of all SGD in the void volume, indicating a molecular mass of over 1500 kDa. It appears that SGD occurs in cell extracts as an aggregated high molecular mass protein complex, which is probably responsible for its stability. Using an SGD activity-specific staining method (Luijendijk et al. 1996b), native polyacrylamide gel electrophoresis (PAGE) of the purified SGD preparation yielded a consistent pattern of three equidistant activity bands (at about 930, 650 [major band], and 240 kDa). As sodium dodecyl sulfate (SDS)- PAGE of purified SGD yielded only one band of 63 kDa, the native enzyme is probably built up of a number of subunits (hydrophobically bound), with the smallest active form being a tetramer (240 kDa). Treatment with proteases such as trypsin results in smaller fragments, but with retention of activity. SGD was found to be highly stable at temperatures up to 50°C and showed broad pH optima, ranging from 6 to approximately 8.5. SGD activity is inhibited (approximately 50%) by Cu^{++} (1 mM) and serpentine (1 mM). SGD is highly specific for its substrate and, apart from strictosidine, only 10-methoxystrictosidine could be converted by it to some extent. The Km for strictosidine is in the range 10–18 μM. Terpenoid precursors (sweroside, loganin, and secologanin) are not accepted as substrates by SGD. The 3-H stereochemistry is essential for SGD activity (Luijendijk et al. 1998). Later, Geerlings et al. (2000) also found the molecular weight of SGD to be 63 kDa under denaturing conditions, but the native PAGE of the purified protein yielded three bands of ~250, 500, and 630 kDa (all three contained SGD activity). Like many other β-glucosidases, SGD was found to be glycosylated but whether this is required for activity is not yet known. In the plant, the highest SGD activity is found in the leaves, although stem, root, and flower have also been reported to possess SGD activity though to a lesser extent (Geerlings et al. 2000). Transcript abundance of *sgd* has been found to increase when the plant is infected with *Spiroplasma citri* (causing Periwinkle lethal yellows), which demonstrates the potential utility of this gene as a host biomarker to increase the fidelity of *S. citri* detection and also in breeding programs to develop stable disease-resistant varieties (Nejat et al. 2012).

Although initially a hemiacetal/aglycone is formed, the final product of the SGD reaction is cathenamine. Earlier workers proposed that 4,21-dehydrocorynantheine aldehyde and 4,21-dehydrogeissoschizine were intermediates in cathenamine formation and the formation of cathenamine from the latter compound was believed to be catalyzed by cathenamine synthase (Stöckigt 1980). However, Stevens (1994) could not detect the presence of 4,21-dehydrogeissoschizine in the SGD-catalyzed formation of cathenamine. It was postulated that the carbinolamine of cathenamine (21-hydroxyajmalicine) was the intermediate rather than 4,21-dehydrogeissoschizine. The formation of the latter compound is probably under enzymatic control and is thus the opening to other biosynthetic pathways. This is in accordance with the results reported by Hemscheidt (1983). The formation of cathenamine from strictosidine occurs in the presence of glucosidase alone and cathenamine synthase is not required for this reaction. Interestingly, SGD from *C. roseus* appears to turnover vincoside, the diastereomer of strictosidine, indicating that the stereoselectivity of the TIA pathway is not maintained by SGD (Yerkes et al. 2008).

10.3.3.1.5.3 Biosynthesis of Ajmalicine The formation of ajmalicine from cathenamine requires a reduction. Hemscheidt (1983) and Stöckigt et al. (1983) described an enzyme cathenamine reductase (CR), which used cathenamine as substrate and NADPH as cofactor, yielding ajmalicine and 19-epi-ajmalicine.

10.3.3.1.5.4 Biosynthesis of Catharanthine Knowledge of the biosynthesis of catharanthine is very limited. Two separate hypotheses propose its genesis from tabersonine and stemmadenine. However, the former hypothesis has not found much support. It is currently believed that the catharanthine biosynthetic pathway goes from strictosidine via 4,21-dehydrogeissoschizine, stemmadenine, and dehydrosecodine (Verpoorte et al. 1997).

10.3.3.1.5.5 Biosynthesis of Vindoline Vindoline, which is found only in the green parts of the *C. roseus* plant and not in the roots or cell-suspension cultures, is biosynthesized from the branch point intermediate tabersonine through the action of six sequential enzymatic steps, first proposed by De Luca et al. (1986) and now almost completely elucidated using a combination of *C. roseus* plants and cell-suspension cultures.

In the first step, tabersonine is hydroxylated at C-16 by tabersonine 16-hydroxylase (T16H), a cytochrome P-450 monooxygenase, to form 16-hydroxytabersonine. St-Pierre and De Luca (1995) reported the first characterization of T16H and detected its activity in total protein extracts from young leaves of *C. roseus* plants. The hydroxylase activity was found to be dependent on NADPH and molecular oxygen, and was inhibited by carbon monoxide, clotrimazole, miconazole, and cytochrome c. In a linear sucrose gradient, T16H activity was localized in the ER fraction. The pH optimum of the enzyme is 7.5–8.0, and the Km for tabersonine and NADPH are, respectively, 11 and 14 μM. In the mature *C. roseus* plants, T16H enzyme activity is abundant in young leaves, but it is 50- to 100-fold lower in flower buds and roots, respectively, and absent in stems and old leaves. Although *C. roseus* cell cultures express T16H activity, the level is only 20% of that in young leaves.

In the second step, the newly introduced hydroxyl moiety is methylated by 16-hydroxytabersonine 16-O-methyltransferase (OMT, EC 2.1.1.94) to yield 16-methoxytabersonine. This enzyme requires SAM as cosubstrate and, as for T16H, it is found in both cell cultures and young leaves of *C. roseus* plants (St-Pierre and De Luca 1995).

The third step involves a hydroxylation mediated by hydration of the 2,3-double bond to form 16-methoxy-2,3-dihydro-3-hydroxytabersonine (Kutchan 1998).

S-adenosyl-L-methionine:16-methoxy-2,3-dihydro-3-hydroxy-tabersonine-N-methyltransferase (NMT, EC 2.1.1.99) catalyzes the fourth step by transferring a methyl group from SAM to N of 16-methoxy-2,3-dihydro-3-hydroxytabersonine to form 16-methoxy-2,3-dihydro-3-hydroxy-N-methyltabersonine (desacetoxyvindoline). This enzyme has high substrate specificity, the reduced 2,3 double bond in the tabersonine skeleton being essential (De Luca et al. 1987). Absence of the 3-hydroxy group resulted in 60% lower N-methylation rate, compared with the natural substrate. Besides, the presence of the 6,7 double bond is a necessary structural requirement for NMT activity. NMT is localized in the thylakoids of the chloroplasts (De Luca and Cutler 1987). The partially purified enzyme was further characterized by Dethier and De Luca (1993), who solubilized it with CHAPS and determined its apparent molecular weight to be 60,000.

The fifth step is again a hydroxylation, catalyzed by desacetoxyvindoline 4-hydroxylase (D4H, EC 1.14.11.11), a 2-oxoglutarate-dependent dioxygenase, which converts desacetoxyvindoline into deacetylvindoline (De Carolis et al. 1990; De Carolis and De Luca 1993, 1994b). The enzyme was purified to apparent homogeneity from *C. roseus* by De Carolis and De Luca (1994a), who found it to be highly substrate specific. The molecular masses of the native and denatured D4H were found to be 45 and 44.7 kDa, respectively, suggesting a monomeric structure. The enzyme has an optimum activity at pH 7.5. Isoelectric focusing under denaturing conditions resolved the purified D4H into three isoforms of pIs 4.6, 4.7, and 4.8. Apart from the cosubstrate, 2-oxoglutarate, the enzyme requires ascorbate, ferrous ions (which reactivate the inactive enzyme in a time-dependent manner), and molecular oxygen for activity. The Km values for 2-oxoglutarate, O_2, and desacetoxyvindoline are, respectively, 45, 45, and 0.03 µM, and those for Fe^{++} and ascorbate are, respectively, 8.5 and 200 µM. Succinate (one of the reaction products) is a competitive inhibitor (Ki 9 mM) for 2-oxoglutarate, but noncompetitive for desacetoxyvindoline and O_2. Deacetylvindoline is a noncompetitive inhibitor (Ki 115 µM) for 2-oxoglutarate, O_2, and desacetoxyvindoline. CO_2 causes 50% inhibition at a concentration of 7.5 mM. Substrate interaction kinetics and product inhibition studies suggested an ordered ter ter mechanism for D4H where 2-oxoglutarate is the first substrate to bind followed by O_2 and desacetoxyvindoline, while deacetylvindoline is the first product to be released followed by CO_2 and succinate. The highest D4H activity is found in the leaves of *C. roseus*, whereas stems and fruits have only about 8% and 5% of the activity found in leaves, respectively. D4H activity is absent in the flowers and roots of the plant (Vazquez-Flota et al. 1997).

The final step in the vindoline biosynthesis is the acetylation of deacetylvindoline to form vindoline, which is catalyzed by acetyl-CoA: 4-O-deacetylvindoline 4-O-acetyltransferase (DAT, EC 2.3.1.107). The enzyme has high selectivity for deacetylvindoline, uses acetyl-CoA as cosubstrate, and also catalyzes the reverse reaction. It is only found in the vindoline-containing parts of the plant—mainly in the leaves, less in the stems, not in the roots (Fahn et al. 1985). After the first characterization of the enzyme in crude enzyme preparations (De Luca et al. 1985; De Luca et al. 1986; Fahn et al. 1985), more detailed studies were made of highly purified DAT (Fahn and Stöckigt 1990; Power et al. 1990), which has a pH optimum of 7.5–9. The enzyme has Km values of 6.5 and 1.3 μM for acetyl-coenzyme A and deacetylvindoline, respectively. Inhibition of DAT by tabersonine (50% inhibition at 45 μM), coenzyme A (50% inhibition at 37 μM), and cations (K^+, Mg^{++}, and Mn^{++}) has been observed. Vindoline causes only 40% inhibition of DAT at a concentration of 500 μM. After SDS-PAGE of purified DAT, two major proteins were observed, having molecular masses of 33 and 21 kDa. On native PAGE gels three bands were seen, and isoelectric focusing-PAGE yielded one diffused protein band (pI 4.7–5.3) in addition to two minor protein bands (pI 5.7 and 6.1) (Power et al. 1990). Acetyl-CoA: DAT might thus be a heterodimeric enzyme occurring in differently charged isoforms. Fahn and Stöckigt (1990) found five discernible forms of DAT, with isoelectric points between 4.3 and 5.4, and each form consisting of two subunits of 20 and 26 kDa, respectively. The Ki for coenzyme A was found to be 8 μM.

The overall pathway from tabersonine to vindoline requires one mole of NADPH, two moles of SAM, and one mole of acetyl CoA per mole of vindoline formed. The first two enzymes of the pathway are found in both cell cultures and leaves of the plant, whereas the last three enzymes are restricted to the leaf tissue. *C. roseus* cell cultures cannot produce the bisindole alkaloids (vinblastine and vincristine), because the last three enzymes required for vindoline biosynthesis are absent from cell cultures (St-Pierre and De Luca 1995).

10.3.3.1.5.6 Biosynthesis of Bisindole Alkaloids (Vinblastine and Vincristine) The enzymatic formation of vinblastine from vindoline and catharanthine in *C. roseus* plants has itself been a point of contention. Although horseradish peroxidase (Goodbody et al. 1988a), hemin and microperoxidase (Verpoorte et al. 1991), and nonspecific peroxidases in *C. roseus* are capable of catalyzing the oxidative coupling of the two monomeric units (vindoline and catharanthine) to 3′,4′-anhydrovinblastine, it has not yet been unequivocally demonstrated that this reaction is mediated by a substrate- and species-specific enzyme *in vivo* in *C. roseus*. Misawa et al. (1988) found that the crude enzyme obtained from suspension cultures of *C. roseus* catalyzed the coupling of vindoline and catharanthine to form bisindole alkaloids, and substantially improved yields could be realized through use of FMN and manganous ion in the reaction mixture. This coupling also occurred nonenzymatically in the presence of FMN (or FAD) and Mn^{++}. It has also been shown to be promoted by irradiation with near-ultraviolet light at low temperature (Asano et al. 2010). Endo et al. (1988) demonstrated that at least five isoenzymes of peroxidase nature from cell-suspension cultures of *C. roseus* were responsible for the coupling of vindoline and catharanthine to form 3′,4′-anhydrovinblastine. Some of these peroxidases, if not all, may be responsible for the biosynthesis of dimeric alkaloids in intact plants of *C. roseus*. It was also postulated that the peroxidases may only be involved in the activation of catharanthine and the actual coupling of activated catharanthine and vindoline could be nonenzymatic, but no experimental proof was offered for this hypothesis. Although 3′,4′-anhydrovinblastine is present in high amounts in *C. roseus* leaves (Goodbody et al. 1988b), it is not the true biosynthetic precursor of vinblastine (Kutney et al. 1988a). The enzymatic coupling mechanism is thought to be similar to the modified Polonovski chemical coupling process whereby catharanthine is oxidized to a reactive species, which is then coupled to vindoline to form a highly unstable dihydropyridinium (iminium) intermediate. The iminium intermediate subsequently undergoes 1,4-reduction, hydroxylation, and reduction to yield vinblastine. On the contrary, 3′,4′-anhydrovinblastine is formed by 1,2-reduction of the iminium intermediate, which can also be performed chemically by sodium borohydride. The iminium ion may also exist as a natural product in the plant (Goodbody et al. 1988b). Although

3′,4′-anhydrovinblastine is not a true intermediate in the biosynthesis of vinblastine, it can be converted to vinblastine by cell-free systems derived from both leaves and suspension cultures. In the presence of $FeCl_3$, 3′,4′-anhydrovinblastine is converted to vinblastine in 50% yield (Verpoorte et al. 1991). Kutney et al. (1988b) have described a "five-step one-pot" chemical synthesis of vinblastine from catharanthine and vindoline, with a yield of 40%. The conversion of vinblastine into vincristine by means of cell cultures of *C. roseus* has been described by Hamada and Nakazawa (1991).

10.3.3.2 Cellular and Subcellular Compartmentation

Alkaloid biosynthesis, which requires the adaptation of cellular activities to perform specialized metabolism without compromising general homeostasis, is accomplished spatially and temporally, by restricting product biosynthesis and accumulation to particular cells and to defined times of plant development. Compartmentation of the TIA biosynthetic pathway in *C. roseus* occurs at cellular as well as subcellular levels.

C. roseus has simple, elliptical mesomorphic leaves (Mersey and Cutler 1986) that are composed of several types of cells. The upper and lower epidermis are composed of thin-walled cells arranged in a single layer, whereas the mesophyll is arranged into a single layer of elongated palisade parenchyma on the adaxial side and a thicker multicellular spongy parenchyma on the abaxial side of the leaf. In addition, unbranched, nonarticulated laticifers are associated with the veins (Yoder and Mahlberg 1976), which are curved and diverge from the midrib at a 35°–45° angle (Mersey and Cutler 1986). Branching from these are smaller veins generally composed of a tracheid and a laticifer. Yoder and Mahlberg (1976) used chemical indicators to identify laticifers and "specialized parenchyma cells" as the sites of alkaloid accumulation in *C. roseus*. Latex could be collected from *C. roseus* fruits and was shown to contain various TIAs (Eilert et al. 1985). Direct observation of *C. roseus* leaves by epifluorescence microscopy showed the random distribution of cells throughout the mesophyll (palisade as well as spongy) that displayed distinctive autofluorescent properties (Mersey and Cutler 1986). Leaf sections and protoplast preparations revealed the presence of larger yellow autofluorescent cells with few chloroplasts, compared with the surrounding red autofluorescent mesophyll cells. These idioblast cells (Mersey and Cutler 1986), which occur in several plant families, are probably morphologically related to laticifers (Fahn 1988), and they may be associated with the biosynthesis and accumulation of secondary products (Platt and Thomson 1992; Postek and Tucker 1983). However, in the most elaborate study on multicellular compartmentation of alkaloid biosynthesis in *C. roseus*, St-Pierre et al. (1999) used *in situ* RNA hybridization and immunocytochemistry to establish the cellular distribution of TIA biosynthesis in the plant. *Tdc* and *str* mRNAs were found to be present in the epidermis of stems, leaves, and flower buds, whereas they appeared in most protoderm and cortical cells around the apical meristem of root tips. In marked contrast, *d4h* and *dat* mRNAs were associated with the laticifer and idioblast cells of leaves, stems, and flower buds but were not found in the root tissue. Some *d4h* transcripts were also found in the upper epidermis of developmentally younger leaf tissue but they disappeared in developmentally older upper epidermis while being retained in idioblasts and laticifers. Expression of the TDC protein was completely restricted to the upper and lower leaf epidermis, as was the case for *tdc* transcript. Similarly, expression of D4H and DAT proteins appeared to be restricted to idioblasts and laticifers, which are also the sites of expression for the respective mRNAs. In contrast to the expression of *d4h* transcripts, which was also observed in the upper epidermis of developmentally younger leaf tissue, the D4H protein was detected only in idioblasts and laticifers. The identical results obtained for the localization of *tdc, d4h,* and *dat* transcripts compared with TDC, D4H, and DAT proteins strongly suggested that at least two different cell types within the same tissue are involved in the biosynthesis of vindoline in *C. roseus.* The leaf base contained the highest levels of all four mRNAs (*tdc, str, d4h,* and *dat*) and of the three enzymes (TDC, D4H, and DAT). The middle portion of the leaf expressed lower levels of this pathway, whereas virtually no expression was observed near the tip of the leaf blade. This basipetal gradient of expression for the genes involved in vindoline biosynthes is correlated with the denser distribution of laticifers characteristic for the basal area compared with those found in the middle section or at the tip of the leaf. This expression pattern suggested activation of the vindoline pathway genes in young immature leaf tissues and rapid downregulation with tissue maturation. It also revealed a spatial separation of the vindoline pathway within particular cells in leaf, stem, and flower tissues. The early vindoline pathway was found to be restricted to the epidermal layers of immature leaves and stems, whereas the late steps committed exclusively to vindoline formation were located in laticifers and idioblasts, which are specialized cells involved in accommodation and sequestration of toxic TIAs (probable sites for the biochemical coupling of vindoline and catharanthine to produce highly toxic bisindoles). The clear differential cell localization of the early and late stages of vindoline biosynthesis in leaves strongly suggests that undetermined poststrictosidine compounds are mobilized from the epidermis to laticifers and idioblasts, where at least the last two reactions of vindoline biosynthesis take place. The distribution of *tdc* and *str* in actively growing shoots and roots of *C. roseus* plants suggested that the meristems of different organs are capable of making tryptamine and monoterpenoid indole alkaloids. The actual presence of different alkaloids in both above-ground and underground tissues suggested that distinct tissue-specific biosynthetic pathways are expressed.

Burlat et al. (2004) have demonstrated coexpression of three mevalonate-independent pathway genes (*dxps, dxr,* and *mecs*) and *g10h* in the internal phloem parenchyma of young aerial organs of *C. roseus,* which adds a new level of complexity to the multicellular nature of TIA biosynthesis. They have predicted the translocation of pathway intermediates from the internal phloem parenchyma to the epidermis and, ultimately, to laticifers and idioblasts during TIA biosynthesis.

At the subcellular level, compartmentation also plays a crucial role in TIA biosynthesis. All the steps do not necessarily occur in the same cell, for example, ajmalicine and serpentine are typical products of the roots, whereas vindoline and the bisindole alkaloids are typical of the leaves. Using immunofluorescence labeling, Brisson et al. (1992) showed that in protoplasts from the leaf, vindoline is located in the vacuoles. With immunogold labeling coupled with cryotechniques, a higher resolution could be obtained. Besides the central vacuole, it also enabled the observation of vesicles in the cytoplasm and to a lesser extent, associated with or in the vicinity of chloroplast, through immunodetection. Although detection of some gold particles in and around the chloroplast may be due to the recognition of biosynthetic intermediates like deacetylvindoline by the antivindoline antibody, the presence of the gold label associated with vesicles in the cytoplasm was indicative of their role in packaging and shuttling of vindoline toward the central vacuole. The biosynthesis of TIAs requires at least three cellular compartments: the plastids for the production of the terpenoid moiety and tryptophan, the cytosol for the decarboxylation of tryptophan, and the vacuole for the coupling of tryptamine with secologanin. Further steps of the alkaloid biosynthesis occur in the cytosol and even in the chloroplasts (like NMT) for certain alkaloids (Verpoorte et al. 1997). Biphasic alkaloid accumulation kinetics, such as those reported for ajmalicine and vindoline in *C. roseus* protoplasts (McCaskill et al. 1988), might point to the occurrence of both a selective uptake mechanism and an ion-trap mechanism (Verpoorte et al. 1997). The fact that the various steps of the biosynthetic pathway of the TIAs occur in different cell compartments implies that transport of intermediates and products is involved, which provides an inbuilt regulatory mechanism for the process.

Recently, Guirimand et al. (2009) have developed a high-efficiency green fluorescent protein (GFP) imaging approach to systematically localize TIA biosynthetic enzymes within *C. roseus* cells following a biolistic-mediated transient transformation. On the basis of this protocol, the subcellular localization of HDS and G10H has next been characterized. HDS has been found to accumulate in the plastids and associated stromules, whereas G10H has been shown to be an ER-anchored protein. It has been shown that isoprenoid biosynthetic enzymes PMK, MVD, and a short isoform of farnesyl diphosphate synthase of *C. roseus* are exclusively localized to peroxisomes (Guirimand et al. 2012b). SGD subcellular localization appears to be unclear. Initially it was believed to be localized to the ER. However, a combination of GFP-imaging, bimolecular fluorescence complementation, and electromobility shift-zymogram experiments revealed that STR was localized to the vacuole, whereas SGD was shown to accumulate as a highly stable supramolecular aggregate within the nucleus (Guirimand et al. 2010). In epidermal cells of aerial organs, vacuole-accumulated strictosidine may either become the precursor of all TIAs after export from the vacuole, or become the substrate for a defense mechanism based on the massive protein cross-linking that occurs due to organelle membrane disruption during biotic/herbivore attacks. This is possible due to physical separation of the vacuolar strictosidine-synthesizing enzyme (STR) and the nucleus-targeted enzyme catalyzing its activation through deglucosylation (SGD). To further build upon this hypothesis, Guirimand et al. (2011a) carried out a cellular and subcellular study of three enzymes catalyzing the synthesis of the two strictosidine precursors (tryptamine and secologanin). RNA *in situ* hybridization demonstrated that the transcript of *lamt*, catalyzing the penultimate step of secologanin synthesis, is specifically localized in the epidermis. A combination of GFP-imaging, bimolecular fluorescence complementation assays, and yeast two-hybrid analysis established that both LAMT and TDC form homodimers in the cytosol, thereby preventing their passive diffusion to the nucleus. SLS is anchored to the ER via an N-terminal helix, thereby allowing the production of secologanin on the cytosolic side of the ER membrane. This makes it mandatory for secologanin and tryptamine to be transported to the vacuole to achieve strictosidine biosynthesis. This in turn demonstrates the significance of transtonoplast translocation events during TIA biosynthesis.

10.3.3.3 Regulation of TIA Biosynthesis

Regulation of TIA biosynthesis in *C. roseus* can be achieved at different levels: product, enzyme, mRNA, and DNA (genes). Concerning regulation, two major possibilities can be distinguished: control by endogenous, developmentally controlled signals, or by exogenous signals. For regulation through endogenous factors, studies on compartmentation and the role of plant growth hormones as well as level of differentiation are relevant aspects. For the latter type of regulation, studies of the plant–insect and plant–microorganism interactions are of interest, including the signal molecules and the signal transduction chains.

Level of differentiation plays an important regulatory role in TIA biosynthesis in *C. roseus*. Studies have already been undertaken using whole plants, callus cultures, and cell-suspension cultures of *C. roseus* to investigate this aspect. Reda (1978) studied the distribution and accumulation of the total alkaloids in *C. roseus* during six different stages of flowering and fruiting. The highest concentration of alkaloids (using perivine as the standard) was found in the roots at the start of flowering, and the minimum concentration was found in the stems during the full fruiting stage. The rate of alkaloid accumulation decreased during fruit maturation, and the most active stage of alkaloid biosynthesis was at the start of flowering in all vegetative organs. Daddona et al. (1976) in the course of studies aimed at the investigation of the formation of the bisindole alkaloids in whole plants of *C. roseus* also examined the catabolism of some of the principal alkaloids, including vindoline and catharanthine, using $^{14}CO_2$ and monitoring the level of incorporation with time. Turnover was found to occur more rapidly in apical cuttings than in intact plants, possibly because of the energy required for general anabolism of the plant and root generation. Babcock and Carew (1962) were the first to study conditions for the growth of callus cultures of *C. roseus* and analytical work commenced when Harris et al. (1964) demonstrated that in the presence of 0.5 mg/l kinetin the callus grew rapidly and probably produced vindoline as well as a number of other alkaloids. Krueger et al. (1982) have described vindoline-producing leaf organ cultures of *C. roseus* whereby a typical 2.5 g fresh weight inoculum produced 29 g fresh weight of leaf material after 35 days without the presence of any dedifferentiated tissue. Endo et al. (1987) examined the production of alkaloids in root and shoot cultures induced from seedlings of *C. roseus*. The pattern of alkaloids in the root cultures was similar to that of the roots from intact plants. Thus, ajmalicine and catharanthine were produced but neither vindoline nor the bisindoles could be detected in the root cultures. Similarly, the pattern of the alkaloid content of the shoot cultures was similar to that of the leaves of the intact plant, showing the presence of catharanthine, ajmalicine, and a significant amount of vindoline. Although, the bisindoles, anhydrovinblastine, leurosine, and catharine were present in the shoot cultures, vinblastine and vincristine could not be detected. The influence of cellular differentiation and elicitation on intermediate (SGD) and late (DAT) steps of TIA biosynthesis in *C. roseus* has been studied (Shukla et al. 2010). The effect of *P. aphanidermatum* homogenate and methyl jasmonate (MeJa) on *in vitro* cultures of *C. roseus* representing increasing levels of differentiation (suspension < callus < shoots) has been analyzed in terms of TIA accumulation and transcript abundance of SGD and DAT. Differentiation was found to be essential for expression of DAT but not SGD, and vindoline biosynthetic potential increased with it. Thus, vindoline biosynthesis seems to be dependent on both organogenesis and light. In shoot cultures it has also been shown to be synchronously activated with morphogenesis through the last biosynthetic step (Campos-Tamayo et al. 2008). Hirata et al. (1987) have successfully induced multiple shoot cultures of *C. roseus* from seedlings in the presence of 1.0 mg/l benzyladenine. Vindoline and catharanthine were the predominating alkaloids in the MSC-B-1 line. Miura et al. (1987) detected vinblastine in callus cultures for the first time, and continued work by the same group led to the first isolation of vinblastine from a multiple shoot culture of *C. roseus* (Miura et al. 1988). Vinblastine content in multiple shoot cultures was greater than that in the callus culture but less than that observed for the parent plant. Studies using suspension cultures of *C. roseus* were first reported by Patterson and Carew (1969). Besides the level of differentiation, the developmental clock of the plant also plays a key role in regulating TIA biosynthesis, as shown in many studies using developing seedlings (De Luca et al. 1986, 1988; Fernandez et al. 1989). Plant growth hormones play an important role in these processes, but little work has been done on the direct relationship between growth hormones and alkaloid accumulation in intact plants. Most of this work has been done using cell cultures of *C. roseus* (Verpoorte et al. 1997). Light was shown to play a major role in TIA biosynthesis, since seedlings grown in the dark had tabersonine as the major compound, whereas those grown in the light had vindoline as the major alkaloid (De Luca et al. 1986; Verpoorte et al. 1997). STR is not under strict control as it could be detected in all tissues through the whole period of germination and development of a seedling, with the maximum occurring simultaneously with that of TDC. The activity of TDC is strongly regulated during the development of the seedling (De Luca et al. 1988; Fernandez et al. 1989). TDC enzyme activity shows a clear transient maximum at Day 5 of the germination, followed after 24 h by some of the enzymes (NMT and DAT, both of which are found only in the hypocotyls and cotyledons) of the vindoline pathway. Light was not found to influence TDC activity, but NMT and DAT activities increased 30% and 10-fold, respectively, upon illumination. The highest TDC protein levels in mature plants were detected in the youngest leaves. Auxins were found to induce TDC activity in radicles of *C. roseus* seedlings. It was also found that auxins delay (by 24 h) but do not suppress the light-mediated induction of DAT and the vindoline biosynthetic pathway remains intact in auxin-treated seedlings (Aerts et al. 1992). The finding that heterotrophic cell cultures of *C. roseus* possess a slight capability for vindoline production (Naaranlahti et al. 1989a) is in accordance with vindoline biosynthesis being found in the green tissues of the plant.

Light is an important parameter, which influences TIA biosynthesis in both plants and cell cultures of *C. roseus*. Under the influence of light, serpentine levels increase in cell cultures, whereas ajmalicine levels decrease (Verpoorte et al. 1997). Hirata et al. (1993) reported that artificial near-ultraviolet light with a peak at 370 nm and the light of natural radiation with wavelengths between 290 and 380 nm stimulated the synthesis of bisindole alkaloids in intact plants of *C. roseus*. The artificial light also specifically stimulated an *in vitro* FMN-mediated, nonenzymatic coupling of vindoline and catharanthine to synthesize an iminium intermediate, the *in vivo* precursor of bisindole alkaloids. These results suggested that near-ultraviolet light is necessary for catharanthine oxidation as a trigger reaction for bisindole alkaloid biosynthesis in the plant. DAT, the last enzyme in vindoline biosynthesis, is induced by light in etiolated seedlings (Aerts and De Luca 1992; De Luca et al. 1986, 1988). Furthermore, T16H is also induced by light in seedlings (St-Pierre and De Luca 1995). D4H activity has also been shown to be light inducible by De Carolis et al. (1990).

The influence of different growth hormones on *C. roseus* plant cell cultures has been extensively studied and described in various reviews (Moreno et al. 1995; Morris 1986; Van der Heijden et al. 1989; Verpoorte et al. 1997). In general, the conclusion is that auxins (specially 2,4-D) inhibit alkaloid production. The role of calcium as a second messenger in the transduction pathway leading to the inhibitory effect of 2,4-D in regulating TIA biosynthesis in *C. roseus* has been studied (Poutrain et al. 2009). Abscisic acid has been found to stimulate the intracellular accumulation of catharanthine and vindoline when added to suspension cultures of *C. roseus* (Smith et al. 1987). It was also shown that an increase in the alkaloid content of mature leaves occurs when watered plants are treated with 1 mM abscisic acid (Saenz et al. 1993). The Ca^{2+}-calmodulin system is believed to be involved in the alkaloid-accumulation-enhancing effect of cytokinins (Verpoorte et al. 1997). Recently, using an inducible RNAi system targeting an element of cytokinin signaling, it has been demonstrated that a relationship exists between cytokinin signaling and the TIA biosynthetic pathways (Amini et al. 2012).

Secondary plant products like TIAs are supposed to be defense compounds directed against phytopathogenic microorganisms (bacteria and fungi) and phytophagous animals. Plant cell-suspension cultures are often severely repressed in their biosynthetic capacity for secondary products. However, if such cell cultures are challenged by a pathogenic organism, or a cell wall preparation or homogenate of such a pathogenic organism, the synthesis of secondary products increases drastically. Substances triggering the chemical defense system are called elicitors and the plant-derived inducible defense compounds are termed phytoalexins (Kutchan et al. 1991). Pasquali et al. (1992) have demonstrated the coordinated regulation of *str* and *tdc* in *C. roseus* by auxin and elicitors. In cell-suspension cultures both the genes are rapidly downregulated by auxin. In contrast, both genes are strongly induced by fungal elicitors such as *P. aphanidermatum* culture filtrate and yeast extract. Induction is a rapid, transcriptional event occurring independent of *de novo* protein synthesis (as it was found not to be affected by cycloheximide). Eilert et al. (1986) demonstrated that 10-day-old subcultures of *C. roseus* cells responded better than older or younger ones, and it was also found that a *P. aphanidermatum* homogenate concentration of 5% and a *Rhodotorula rubra* homogenate concentration of 0.5% effected maximum alkaloid yields. Pauw et al. (2004b) have postulated that activation of the oxidative burst by yeast elicitor in *C. roseus* occurs independently of the activation of the genes involved in alkaloid biosynthesis. However, Xu and Dong (2005) have found that elicitor-induced nitric oxide burst is essential for triggering catharanthine synthesis in *C. roseus* suspension cells. The abiotic elicitor vanadyl sulfate also increases alkaloid production (ajmalicine, catharanthine, and tryptamine) in *C. roseus* cell cultures (Tallevi and DiCosmo 1988). G10H activity was found to be induced by the treatment of *C. roseus* suspension-cultured cells with phenobarbital, and inhibited by treatment with ketoconazole (Contin et al. 1999). The alkaloid accumulation increased after phenobarbital treatment, whereas it decreased after ketoconazole treatment. In contrast, phenobarbital and ketoconazole did not affect the *in vivo* conversion rate of loganin to secologanin. Salicylic acid (0.1 mM) shows a weak inducing effect on *str* and *tdc* steady-state mRNA levels in *C. roseus* cells after 8–24 h of exposure (Pasquali et al. 1992). There are conflicting reports on the effect of sodium nitroprusside (SNP), a donor of NO, on catharanthine formation in *C. roseus* cells. While Zhou et al. (2010) found SNP to cause a dramatic decrease of the catharanthine concentration, Li et al. (2011) reported that SNP stimulates catharanthine formation in *C. roseus* cells. Jasmonate is a well-established endogenous signal compound in the signal transduction chain in plant development and in the response to elicitors (Verpoorte et al. 1997). Aerts et al. (1994) showed that MeJa stimulated the tabersonine–vindoline biosynthetic pathway to a greater extent than the catharanthine biosynthetic pathway in *C. roseus* seedlings. MeJa has been shown to induce ATP biosynthesis deficiency and accumulation of proteins related to secondary metabolism in *C. roseus* hairy roots (Ruiz-May et al. 2011). The jasmonate biosynthetic pathway is an integral part of the elicitor-triggered signal transduction pathway that results in the coordinate expression of the *tdc* and *str* genes, and protein kinases have been shown to act both upstream and downstream of the jasmonates (Menke et al. 1999b). A promoter element involved in jasmonate- and elicitor-responsive gene expression (JERE) was identified in the TIA biosynthetic gene *str* (Menke et al. 1999a). Yeast one-hybrid screening with JERE as bait identified octadecanoid-responsive Catharanthus AP2-domain transcription factors (ORCA1 and ORCA2) of the APETALA2/ethylene responsive factor (AP2/ERF)-domain class (Menke et al. 1999a). The AP2/ERF transcription factors form large families unique to plants, and are characterized by the AP2/ERF DNA-binding domain (Memelink et al. 2001; Vom Endt et al. 2002). *Orca2* gene expression was found to be induced by MeJa and elicitor, whereas *orca1* expressed constitutively (Menke et al. 1999a). Furthermore, the ORCA2 protein transactivated *str* promoter activity by sequence-specific binding to the JERE (Menke et al. 1999a). ORCA3, another member of the AP2/ERF-domain transcription factor family in *C. roseus*, was isolated by T-DNA activation tagging (Van der Fits and Memelink 2000). The expression of the *orca3* gene was induced by MeJa with similar kinetics to *orca2* expression (Van der Fits and Memelink 2001). ORCA3 also binds to the JERE and activates *str* expression. Ectopic expression of ORCA3 in cultured cells from *C. roseus* increased the expression of the TIA biosynthetic genes *tdc*, *str*, *cpr* (which encodes a cytochrome P450-reductase that acts with G10H, a P450 enzyme), and *d4h*. ORCA3 also regulates two genes that encode primary metabolic enzymes (α-subunit of AS and DXPS) that are involved in the formation of TIA precursors. This indicates that ORCA3 is a central regulator of TIA biosynthesis that acts pleiotropically on several steps of the TIA pathway and activates the biosynthesis of TIA precursors. However, ORCA3 does not regulate the genes *g10h* and *dat*. Transgenic cells that overexpress ORCA3 accumulate significantly more tryptophan and tryptamine. However, since no TIAs are detected, the terpenoid branch of the pathway remains limiting for TIA production. This is confirmed by the increase in TIA production when cells that overexpress ORCA3 are fed with the terpenoid precursor loganin (Gantet and Memelink 2002; Van der Fits and Memelink 2000). In the case of ORCA transcription factors, since cycloheximide did not inhibit MeJa-induced target gene expression, it is likely that preexisting ORCA is somehow activated by posttranslational modification and protein–protein interactions. Since phosphorylation has been shown to play a key role in the activation of TIA genes (Menke et al. 1999b), it is possible that jasmonic acid activates ORCAs via phosphorylation, or that ORCAs have to be phosphorylated to be active (Vom Endt et al. 2002).

Although ORCA3 has an important role in regulating TIA biosynthesis, it is not sufficient to regulate the complete pathway, which indicates the involvement of additional transcription factors. An enhancer domain of the *str* promoter was used as a bait in a yeast one-hybrid screening, which resulted in the isolation of CrBPF1, a periwinkle homolog of the MYB-like transcription factor BPF1 from parsley (Van der Fits et al. 2000). The expression of *CrBPF1* is induced by elicitor but not by jasmonate, which is indicative of the fact that elicitor induces *str* expression in *C. roseus* cells via jasmonic acid-dependent and independent pathways. The G-box (5'-CACGTG-3'), a promoter element that is conserved in plants, is located adjacent to the JERE element in the *str* promoter. It is an active *cis*-regulatory element *in planta* (Ouwerkerk and Memelink 1999a). A yeast one-hybrid screen using the G-box as bait resulted in the isolation of G-box binding factors (CrGBF) of the basic leucine zipper class and MYC-type bHLH transcription factors (CrMYC). In *C. roseus* suspension cells, *CrMYC1* mRNA levels are induced by fungal elicitor and jasmonate, suggesting that CrMYC1 may be involved in the regulation of gene expression in response to these signals (Chatel et al. 2003). CrGBF1 and CrGBF2 were shown to repress *str* expression (Sibéril et al. 2001). They also bind *in vitro* to a G-box-like element in the *tdc* promoter (Sibéril et al. 2001), which indicates that CrGBFs could coordinately regulate several TIA biosynthetic genes. However, Pasquali et al. (1999) found that the G-box is not essential for the elicitor responsiveness of the *str* promoter. Recombinant CrMYC2 binds to the G-box of *str in vitro* and *CrMYC2* expression is induced by jasmonate (Gantet and Memelink 2002). The bHLH transcription factor CrMYC2 controls the jasmonate-responsive expression of the ORCA genes that regulate alkaloid biosynthesis in *C. roseus* (Zhang et al. 2011). The nuclear targeting of this transcription factor involves the cooperation of three domains located in the C-terminal region of the protein (Hedhili et al. 2010).

Ouwerkerk et al. (1999a) have studied the UV-B-responsive regions in the promoter of *tdc* from *C. roseus*. The elicitor-responsive promoter regions of the *tdc* gene of *C. roseus* have also been studied in detail (Ouwerkerk and Memelink 1999b). Ouwerkerk et al. (1999b) have described the interaction of nuclear factors GT-1 and 3AF1 with multiple sequences within the promoter of the *tdc* gene of *C. roseus* and demonstrated a functional role for GT-1 in the induction of *tdc* expression by UV light.

Pauw et al. (2004a) have demonstrated that zinc finger proteins act as transcriptional repressors of alkaloid biosynthesis genes in *C. roseus*. They performed a yeast one-hybrid screening with the elicitor-responsive part of the *tdc* promoter to identify three members of the Cys₂/His₂-type (transcription factor IIIA-type) zinc finger protein family from *C. roseus* (ZCT1, ZCT2, and ZCT3), which bind in a sequence-specific manner to the *tdc* and *str* promoters *in vitro* and repress the activity of these promoters in trans-activation assays. In addition, the ZCT proteins can repress the activation of APETALA2/ethylene response-factor domain transcription factors, the ORCAs, on the *str* promoter.

Recently a *C. roseus* WRKY transcription factor (CrWRKY1) that is preferentially expressed in roots and induced by the phytohormones jasmonate, GA, and ethylene has been described (Suttipanta et al. 2011). Its overexpression in *C. roseus* hairy roots upregulated many TIA pathway genes, including TDC, as well as the transcriptional repressors ZCT1 (for zinc-finger *C. roseus* transcription factor 1), ZCT2, and ZCT3, but repressed the transcriptional activators ORCA2, ORCA3, and CrMYC2. Compared with control roots, CrWRKY1 hairy roots accumulated up to threefold higher levels of serpentine, which indicates its key role in determining the root-specific accumulation of serpentine in *C. roseus* plants. A soybean transcription factor GmMYBZ2 has been found to repress catharanthine biosynthesis in the hairy roots of *C. roseus* (Zhou et al. 2011). Besides, tetraploidization has also been shown to increase the contents of TIAs in *C. roseus* (Xing et al. 2011). Bax, a mammalian proapoptotic member of the Bcl-2 family, has also been demonstrated to be a potential regulatory factor for plant secondary metabolite biosynthesis as shown in the *C. roseus* model (Xu and Dong 2007). A novel gene encoding an MDR-like ABC transporter protein (CrMDR1) has been cloned from *C. roseus*, which may be involved in the transport and accumulation of secondary metabolites (Jin et al. 2007). Transcription factor Agamous-like 12 from *Arabidopsis* has been shown to promote tissue-like organization and alkaloid biosynthesis in *C. roseus* suspension cells (Montiel et al. 2007).

Functional analysis of some of the TIA pathway gene promoters has also been accomplished. For example, the DAT gene promoter has been functionally analyzed by transient expression in *C. roseus* cell suspensions by Wang et al. (2010b). Makhzoum et al. (2011) also analyzed the DAT gene promoter using transient *C. roseus* and stable *N. tabacum* transformation systems. Vom Endt et al. (2007) have identified a bipartite jasmonate-responsive promoter element in the *C. roseus* ORCA3 transcription factor gene that interacts specifically with AT-Hook DNA-binding proteins.

Calmodulin isoforms are required for TIA biosynthesis in *C. roseus* cells whereby they act on regulation of expression of genes encoding the enzymes that catalyze early steps of the pathway, such as DXR and G10H (Poutrain et al. 2011). It has been shown that proteins prenylated by type I protein geranylgeranyltransferase act positively on the jasmonate signaling pathway triggering the biosynthesis of TIAs in *C. roseus* (Courdavault et al. 2009).

TIA biosynthesis is regulated in time, by development, and by environmental factors that control the activity of transcriptional factors, such as ORCA, which program the whole process. The use of central transcription factors to modulate plant metabolism avoids the time-consuming step of acquiring knowledge about all the enzymatic steps in a complex metabolic pathway. However, it might require the equally time-consuming step of unraveling complex regulatory networks.

10.3.4 Metabolomics Studies

A number of metabolomic studies have been attempted on *C. roseus*. A comprehensive metabolomic profiling of *C. roseus* infected by 10 types of phytoplasmas was carried out using one-dimensional and two-dimensional NMR spectroscopy followed by principal component analysis (PCA), which identified those metabolites that were present in different levels in phytoplasma-infected *C. roseus* leaves as compared to healthy ones (Choi et al. 2004). Metabolic discrimination of *C. roseus* calli according to their relative locations has also been carried out using ¹H-NMR and PCA (Yang et al. 2009). NMR-based metabolomics has been employed to monitor the metabolic changes in salicylic acid-elicited *C. roseus* cell-suspension cultures (Mustafa et al. 2009).

Alkaloids in aqueous root extracts of *C. roseus* were studied using HPLC–DAD–ESI–MS/MS, which allowed the identification of 19-*S*-vindolinine, vindolinine, ajmalicine and an ajmalicine isomer, tabersonine, catharanthine, serpentine, and a serpentine isomer (Ferreres et al. 2010). Targeted metabolite analysis on different parts of this species (leaves, stems, seeds, and petals) has been achieved to analyze phenolics by HPLC–DAD and organic acids and amino acids by HPLC–UV (Pereira et al. 2009).

The influence of seasonal, ontogenic, and genotypic parameters on the pattern of TIAs biosynthesized in the leaves of *C. roseus* has been elaborately studied for determining the best harvest time and type of tissue for maximal yield of key TIAs (Shukla 2005). An elaborate screening of 64 cultivars of *C. roseus* for the content of vindoline, catharanthine, and serpentine has been carried out by Chung et al. (2011). The distribution and accumulation of vindoline, catharanthine, and vinblastine in the root, stem, leaf, flower, and fruit of *C. roseus* cultivated in China have also been studied (Yu et al. 2010).

The volatile composition of *C. roseus* has been studied by using headspace solid-phase microextraction (HS-SPME) and GC–MS (De Pinho et al. 2009). Efforts are still ongoing to isolate new TIAs from *C. roseus*, and the list of alkaloids known to be produced in *C. roseus* is expanding.

10.3.5 Ultimate Alkaloidomics: The Metabolic Reprogramming of *C. roseus* TIA Biosynthetic Machinery

Recently, there have been efforts, mainly by the research group of Dr Sarah O'Connor of the John Innes Centre, UK, to produce alkaloids with unnatural structures and possible novel bioactivities. This has been achieved through various means.

Study of precursor-directed biosynthesis in *C. roseus* has revealed that the major bottleneck in the production of unnatural alkaloids is the stringent substrate specificity of STR, which catalyzes the biosynthesis of strictosidine from secologanin and tryptamine. The design of enzyme mutants with broadened substrate specificities allows enzymatic production of a greater variety of strictosidine analogs (Runguphan and O'Connor 2009). A rational redesign of the STR binding pocket to selectively accommodate secologanin substrate analogs has been achieved (Chen et al. 2006). Bernhardt et al. (2007) have devised an efficient screening method for rapid identification of STR variants that accept tryptamine analogs not turned over by the wild-type enzyme. Structure-based engineering of STR would ultimately lead to generation of large alkaloid libraries for future pharmacological screenings (Loris et al. 2007).

Targeted suppression of substrate biosynthesis has also been shown to be a viable strategy for programming a plant alkaloid pathway to more effectively produce desirable unnatural products. It has been shown that RNA silencing of tryptamine biosynthesis in *C. roseus* hairy root culture eliminates all production of TIAs, and the chemically silent background thus produced could be exploited to produce unnatural alkaloids (not contaminated by the presence of normal natural alkaloids) by providing unnatural tryptamine analog in the medium (Runguphan et al. 2009). For example, it was found that exogenously providing a tryptamine analog like 5-fluorotryptamine to the *tdc*-silenced hairy root cultures led to the biosynthesis of unnatural, fluorinated alkaloids.

Integration of carbon–halogen bond formation into medicinal plant metabolism is another way of achieving the biosynthesis of unnatural alkaloids. Chlorination biosynthetic machinery of soil bacteria can be successfully introduced into *C. roseus*, whereby the prokaryotic halogenases function within the context of the plant cell to generate chlorinated tryptophan that is then shuttled into TIA metabolism yielding chlorinated TIAs. Thus a new functional group (halide) is introduced into the complex TIA metabolism of *C. roseus*, and is incorporated in a predictable and regioselective manner onto unnatural plant alkaloids (Runguphan et al. 2010). It has been found that *C. roseus* produces chlorinated TIAs after the introduction of bacterial tryptophan halogenase genes into the plant (van Pée 2011). This approach has been further refined by reengineering a tryptophan halogenase to preferentially chlorinate a direct alkaloid precursor (tryptamine) rather than the native substrate tryptophan (Glenn et al. 2011). TDC, which converts tryptophan to tryptamine, accepts 7-chlorotryptophan at only 3% of the efficiency of the native substrate tryptophan, thereby creating a bottleneck in the production of halogenated unnatural alkaloids, which can be circumvented by this approach as tryptamine is a direct precursor to many TIAs.

Chemo-enzymatic approaches leading to diversification of alkaloids with novel improved structures also provide new opportunities for the future (Yang et al. 2010). Thus, despite its genetic and developmental complexity, *C. roseus* seems to be a viable platform for synthetic biology efforts (Runguphan et al. 2010).

10.4 Some Recent Advances

To date, many review articles have been published on various aspects related to TIAs of *C. roseus* such as enzymology (Mishra and Kumar 2000; Misra et al. 1996; Verpoorte et al. 1997), chemical analysis (Hisiger and Jolicoeur 2007), chemistry and biology (O'Connor and Maresh 2006), biosynthesis (El-Sayed and Verpoorte 2007; Hedhili et al. 2007; Loyola-Vargas et al. 2007; Oudin et al. 2007a), transcription factors (Memelink and Gantet 2007), compartmentation (Mahroug et al. 2007; Verma et al. 2012), transportation and sequestration (Roytrakul and Verpoorte 2007), genetic modification (Zarate and Verpoorte 2007), metabolic engineering (Zhou et al. 2009), and bioreactors (Zhao and Verpoorte 2007). Goosens and Rischer (2007) have reviewed the application of functional genomics for gene discovery in alkaloid-producing plants. With the establishment of cellular localization models and the application of various omics technologies, *C. roseus* and *Papaver somniferum* (opium poppy) have emerged as model nonmodel systems/species to study *in planta* alkaloid metabolism (Facchini and De Luca 2008). This emerging area may be aptly termed *alkaloidomics*.

Metabolic engineering through the overexpression of DXPS and G10H or AS (alpha subunit) has been shown to increase TIA accumulation in *C. roseus* hairy roots (Peebles et al. 2011). Similarly, overexpression of G10H and ORCA3 in the hairy roots of *C. roseus* has been shown to improve catharanthine production (Wang et al. 2010a). Goklany et al. (2009) have assessed the limitations to TIA biosynthesis in *C. roseus* hairy root cultures through gene expression profiling and precursor feeding. Increased vincristine production has been reported from *A. tumefaciens* C58–induced shooty teratomas of *C. roseus* (Begum et al. 2009). Improved accumulation of ajmalicine and tetrahydroalstonine has been observed in *C. roseus* cells overexpressing an ABC transporter gene (*CjMDR1*) from *Coptis japonica* (Pomahacová et al. 2009).

Genes that are differentially expressed in leaves of *C. roseus* infected by Candidatus Phytoplasma pyri have been identified (De Luca et al. 2011). Recently, bacterial endophytes (*Staphylococcus sciuri* and *Micrococcus* sp.) have been identified from *C. roseus* that have been found to enhance the *in planta* content of key TIAs like vindoline (Tiwari et al. 2013).

With a view to establishing the role of candidate genes in TIA metabolism identified through *in silico* and omics approaches, it is necessary to determine the subcellular localization of the encoded proteins. Fusion with fluorescent proteins can now be used as a quite effective and reliable tool to investigate this question (Duarte et al. 2010).

Recently, some success has been achieved in the transformation of the recalcitrant *C. roseus* plant. A protocol for genetic transformation of *C. roseus* by *Agrobacterium rhizogenes* A4 has recently been reported (Zhou et al. 2012). Verma and Mathur (2011) have shown *A. tumefaciens*–mediated transgenic plant production via direct shoot bud organogenesis from pre-plasmolyzed leaf explants of *C. roseus*, whereas Wang et al. (2012) have developed an efficient regeneration and transformation system for *C. roseus* using *A. tumefaciens* and hypocotyls as explants. Guirimand et al. (2009) have optimized the transient transformation of *C. roseus* cells by particle bombardment and applied it to the subcellular localization of HDS and G10H. However, it is still a challenge to generate transformed *C. roseus* plants in a robust manner. For metabolic engineering efforts to either improve alkaloid content or provide alternative sources of the bisindole alkaloids, the prerequisite is the isolation and functional characterization of the genes involved. To assay gene function in *C. roseus* in a rapid manner, a virus-induced gene silencing (VIGS) method based on Tobacco Rattle Virus vectors (pTRV) has been recently developed and used to demonstrate *in vivo* silencing of known vindoline biosynthetic pathway genes (Liscombe and O'Connor 2011).

Through an elegant experiment, Roepke et al. (2010) demonstrated how *C. roseus* keeps catharanthine and vindoline spatially separated from each other *in planta*, accounting for the low levels of the bisindole alkaloids in the plant. It was shown that even though the entire production of the monomers (catharanthine and vindoline) occurs in young developing leaves, catharanthine accumulates in leaf wax exudates of leaves, whereas vindoline is found within leaf cells.

10.5 Future Prospects and the Gaps to Be Filled

The antineoplastic bisindole alkaloids and their precursor vindoline are produced only in the aerial parts of the plant and require differentiated plant tissue for biosynthesis, whereas catharanthine is produced throughout the plant and could also be produced in cell-suspension cultures. With the current knowledge it seems highly improbable that the *in planta* biosynthetic potential of the *C. roseus* plant for bisindole alkaloids could be raised beyond a certain threshold limit due to their highly mitotoxic nature, which poses the problem of cellular containment. Besides, large-scale production of bisindole alkaloids in suspension-cultured cells has been unsuccessful due to nonexpression of the vindoline biosynthetic pathway (as one of the enzymes involved in its biosynthesis, N-methyltransferase, NMT is localized in the thylakoid membrane). This leaves the option for a semisynthetic approach for their synthesis from precursors catharanthine and vindoline. Catharanthine (plant- or cell-suspension culture-derived) is easily sourced, but for vindoline the plant leaves are the only source. Hence, there is a need to identify and analyze stages/tissues/genotypes (metabolic blocks) of the plant that accumulate vindoline. In view of these facts the following points are most pertinent for future research on *C. roseus*:

- *Deep transcriptome sequencing data utilization*: Since *C. roseus* is a model nonmodel species, where the genome is yet to be sequenced completely, the best possibility lies in utilization of the deep transcriptome sequencing data to elucidate novel gene functions. A global effort is already under way in this direction under the 1KP project.

- *Strictosidine—the junction point*: This central biosynthetic intermediate gives rise to several different alkaloidal structures in *C. roseus* as well as related species like *R. serpentina*. It must be investigated how a single and simple structure gives rise to so many complex alkaloids. Since strictosidine is the precursor for a broad variety of TIAs in different tissues/organs of *C. roseus* itself as well as in many other plant species, all of which share the first part of this pathway, it is imperative that the diversification in the pathways occurs from strictosidine onward.

- *Tabersonine—the branch point*: Tabersonine, which can be converted to either vindoline or 19-O-acetylhörhammericine, represents a branch point in TIA biosynthesis. Recently, CYP71BJ1 has been discovered that forms a part of the pathway leading to 19-O-acetylhörhammericine (Giddings et al. 2011). Further discovery of such enzymes acting on branch point intermediates will help in elucidating the regulatory controls involved in channeling the metabolic flux for TIA biosynthesis in *C. roseus*.

- *The catharanthine pathway*: The pathway from strictosidine to catharanthine is poorly understood and is yet to be completely elucidated. This part of the bisindole alkaloid biosynthesis has not been targeted as, unlike vindoline, catharanthine does not represent a bottleneck (it is produced in aerial as well as underground parts of the plant, as well as in undifferentiated cell cultures).

- *Metabolic blocks*: Bioprospecting for metabolic blocks (for specific TIAs like vindoline) in the available and collected germplasm must be carried out. Plant breeding and selection should be carried out for specific alkaloid (like vindoline) accumulation.

- *Novel regulatory genes*: The ORCAs do not regulate all the genes of the TIA pathway (although ORCA3 regulates the majority of them), which means that there is a possibility that other regulatory genes (apart from ORCAs) are involved in TIA biosynthesis and they are yet to be elucidated.

- *Biotic/abiotic factors*: The effect of various biotic as well as abiotic factors on TIA biosynthesis and accumulation in the plant needs to be elucidated. The biotic factors could include endophytes, pathogens (like phytoplasma, virus, etc.), and microbes in the rhizosphere, whereas the abiotic factors could include seasonal effects and temperature variations. Besides, the spatial, temporal, and developmental age-related variations in TIA content in the plant must also be defined comprehensively (although some isolated studies have already been done in this direction) for optimal TIA biosynthesis and harvest.

- *Heterologous systems*: Attempts should be made to produce some of the significant TIAs in yeast. This might also help to provide some labeled intermediates for pathway elucidation. Earlier, an attempt was made to express the STR and SGD genes in *Saccharomyces cerevisiae* (Geerlings et al. 2001). The transgenic yeast thus produced was capable of forming strictosidine upon feeding tryptamine and secologanin. The yeast was also able to use the extract of *Symphoricarpus albus* berries as a source for secologanin and carbohydrates, whereby only externally added tryptamine was required to produce indole alkaloids. This model could also be extended in future to produce other TIAs from strictosidine.

- *Biotransformation and production of novel alkaloids*: The *C. roseus* cells could be used for biotransformation of various phytomolecules for value addition. Apart from this, the TIA pathway could be reprogrammed to yield alkaloids with novel structures and activities.

- *Isolation of standards of TIA intermediates*: A chemistry effort is required to produce the standards as well as analytical methods to analyze the TIA intermediates. A crucial consideration for better understanding of the TIA biosynthetic pathway at both the regulatory and structural levels in future research schemes has to be the analysis of a wide range of intermediates and secondary metabolites rather than the final products alone, as it will provide information about the distribution of metabolic flux around key branch points. Although major interest lies in the poststrictosidine pathway leading toward vindoline via tabersonine, measuring the level of the poststrictosidine metabolic flux into all the branching pathways is required if TIA biosynthesis is to be studied in totality.

- *Sequestration of pathway intermediates*: Since the TIA biosynthetic pathway passes through many subcellular compartments, studying the sequestration and transportation of intermediates and their ultimate storage will be an interesting aspect for studying the TIA pathway in totality. Simultaneously, an elaborate proteomics analysis of the subcellular compartments may be undertaken to identify novel proteins.

List of Abbreviations and Symbols

AACT	Acetoacetyl-CoA thiolase
AP2/ERF	APETALA2/ethylene responsive factor
AS	Anthranilate synthase
CCC	(2-Chloroethyl) trimethylammonium chloride
cDNA-AFLP	Complementary deoxyribonucleic acid-amplified fragment length polymorphism
CMK	4-Diphosphocytidyl-2-*C*-methyl-D-erythritol kinase
CMS	4-Diphosphocytidyl-2-*C*-methyl-D-erythritol synthase
CPR	Cytochrome P450 reductase
CR	Cathenamine reductase
DAT	Acetyl-CoA:4-O-deacetylvindoline 4-O-acetyl-transferase
2DE	2-Dimensional gel electrophoresis
D4H	Desacetoxyvindoline 4-hydroxylase
DHBA	2,3-Dihydrobenzoic acid
DHS	3-Deoxy-D-arabino-heptulosonate-7-phosphate synthase

DL7H	Deoxyloganin 7-hydroxylase
2,4-D	2,4-Dichlorophenoxyacetic acid
DMAPP	Dimethylallyl pyrophosphate or dimethylallyl diphosphate
DXP	1-Deoxy-D-xylulose-5-phosphate
DXPS	1-Deoxy-D-xylulose-5-phosphate synthase
DXR	1-Deoxy-D-xylulose-5-phosphate reductoisomerase
EC	Enzyme Commission
ER	Endoplasmic reticulum
EST	Expressed sequence tag
FAD	Flavin adenine dinucleotide
FMN	Flavin mononucleotide
GA	Gibberellic acid
GFP	Green fluorescent protein
G10H	Geraniol-10-hydroxylase
GPP	Geranyl pyrophosphate or geranyl diphosphate
GPPS	Geranyl pyrophosphate synthase
HDR	1-Hydroxy-2-methyl-2-butenyl 4-diphosphate reductase
HDS	1-Hydroxy-2-methyl-2-butenyl 4-diphosphate synthase
10HGO	10-Hydroxygeraniol oxidoreductase
HMGR	Hydroxymethylglutaryl-CoA reductase
HMGS	Hydroxymethylglutaryl-CoA synthase
IAA	Indole acetic acid
IDI	Isopentenyl diphosphate isomerase
IPP	Isopentenyl pyrophosphate or isopentenyl diphosphate
JERE	Jasmonate and elicitor-responsive element
LAMT	*S*-adenosyl-L-methionine:loganic acid methyltransferase
MAT	Minovincinine 19-hydroxy-*O*-acetyltransferase
ME	Methyl jasmonate
MEP	2-*C*-methyl-D-erythritol-4-phosphate
MECS	2-*C*-methyl-D-erythritol 2,4-cyclodiphosphate synthase
MVA	Mevalonic acid
MVAP	Mevalonate phosphate
MVAPP	Mevalonate pyrophosphate or mevalonate diphosphate
MVD	Mevalonate 5-diphosphate decarboxylase
MVK	Mevalonate kinase
NADPH	Nicotinamide adenine dinucleotide phosphate
NEU	N-nitroso-N-ethyl urea
NMT	*S*-adenosyl-L-methionine:16-methoxy-2,3-dihydro-3-hydroxy-tabersonine-*N*-methyltransferase
OMT	16-hydroxytabersonine-16-*O*-methyltransferase
ORCA	Octadecanoid-responsive *Catharanthus* AP2-domain protein
PAGE	Polyacrylamide gel electrophoresis
PCA	Principal component analysis
P-gp	P-glycoprotein

PMK	5-Phosphomevalonate kinase
PRX	Peroxidase
SAM	*S*-adenosyl-L-methionine
SDS	Sodium dodecyl sulfate
SLS	Secologanin synthase
STR	Strictosidine synthase
SGD	Strictosidine β-D-glucosidase
TDC	Tryptophan decarboxylase
TDF	Transcript-derived fragment
T16H	Tabersonine 16-hydroxylase
T19H	Tabersonine 19-hydroxylase
TIA	Terpenoid indole alkaloid

Acknowledgments

The authors are very grateful to Dr. Yogendra N. Shukla, Former Scientist, CSIR-CIMAP, Lucknow, for critically reading this manuscript. They are also thankful to Dr. Richa Pandey, Scientist, CSIR-IICT, Hyderabad, for her help in drawing the chemical structures. The continued support of CSIR, India, at various stages for projects and facilities to support the authors' work is duly acknowledged.

References

Adam, K.P., Thiel, R., Zapp, J. 1999. Incorporation of 1-[1-¹³C] deoxy-D-xylulose in chamomile sesquiterpenes. *Arch. Biochem. Biophys.* 369:127–132.

Aerts, R.J., Alarco, A.-M., De Luca, V. 1992. Auxins induce tryptophan decarboxylase activity in radicles of *Catharanthus* seedlings. *Plant Physiol.* 100:1014–1019.

Aerts, R.J., De Luca, V. 1992. Phytochrome is involved in the light-regulation of vindoline biosynthesis in *Catharanthus*. *Plant Physiol.* 100:1029–1032.

Aerts, R.J., Gisi, D., De Carolis, E., De Luca, V., Baumann, T.W. 1994. Methyl jasmonate vapor increases the developmentally controlled synthesis of alkaloids in *Catharanthus* and *Cinchona* seedlings. *Plant J.* 5:635–643.

Amini, A., Andreu, F., Glévarec, G., Rideau, M., Crèche, J. 2012. Down-regulation of the CrHPT1 histidine phosphotransfer protein prevents cytokinin-mediated up-regulation of CrDXR, and CrG10H transcript levels in periwinkle cell cultures. *Mol. Biol. Rep.* 39:8491–8496.

Arigoni, D., Sagner, S., Latzel, C., Eisenreich, W., Bacher, A., Zenk, M.H. 1997. Terpenoid biosynthesis from 1-deoxy-D-xylulose in higher plants by intramolecular skeletal rearrangement. *Proc. Natl. Acad. Sci. USA* 94:10600–10605.

Arora, A., Shukla, Y. 2003. Modulation of vinca-alkaloid induced P-glycoprotein expression by indole-3-carbinol. *Cancer Lett.* 189:167–173.

Asano, M., Harada, K., Yoshikawa, T., Bamba, T., Hirata, K. 2010. Synthesis of anti-tumor dimeric indole alkaloids in *Catharanthus roseus* was promoted by irradiation with near-ultraviolet light at low temperature. *Biosci. Biotechnol. Biochem.* 74:386–389.

Auriola, S., Naaranlahti, T., Kostiainen, R., Lapinjoki, S.P. 1990. Identification of indole alkaloids of *Catharanthus roseus* with liquid chromatography/mass spectrometry using collison-induced dissociation with the thermospray ion repeller. *Biomed. Environ. Mass Spectrom.* 19:400–404.

Babcock, P.A., Carew, D.P. 1962. Tissue culture of the Apocynaceae. I. Culture requirements and alkaloid analysis. *Lloydia* 25:209–213.

Basu, D. 1992. Effects of growth retardant on growth, flowering and alkaloid production in *Catharanthus pusillus*. *Ind. Forester* 118:659–661.

Begum, F., Nageswara Rao, S.S., Rao, K., Prameela Devi, Y., Giri, A., Giri, C.C. 2009. Increased vincristine production from *Agrobacterium tumefaciens* C58 induced shooty teratomas of *Catharanthus roseus* G. Don. *Nat. Prod. Res.* 23:973–981.

Bernhardt, P., McCoy, E., O'Connor, S.E. 2007. Rapid identification of enzyme variants for reengineered alkaloid biosynthesis in periwinkle. *Chem. Biol.* 14:888–897.

Bhatia, S., Shokeen, B. 2009. Isolation of microsatellites from *Catharanthus roseus* (L.) G. Don using enriched libraries. *Methods Mol. Biol.* 547:289–302.

Bick, J.A., Lange, B.M. 2003. Metabolic cross talk between cytosolic and plastidial pathways of isoprenoid biosynthesis: Unidirectional transport of intermediates across the chloroplast envelope membrane. *Arch. Biochem. Biophys.* 415:146–154.

Blasko, G., Cordell, G.A. 1990. Isolation, structure elucidation, and biosynthesis of the bisindole alkaloids of *Catharanthus*. In Brossi, A. and Suffness, M. (eds), *The Alkaloids*, vol. 37, pp. 1–76. Academic Press, San Diego.

Brisson, L., Charest, P.M., De Luca, V., Ibrahim, R.K. 1992. Immunocytochemical localization of vindoline in mesophyll protoplasts of *Catharanthus roseus*. *Phytochemistry* 31:465–470.

Burlat, V., Oudin, A., Courtois, M., Rideau, M., St-Pierre, B. 2004. Co-expression of three MEP pathway genes and geraniol 10-hydroxylase in internal phloem parenchyma of *Catharanthus roseus* implicates multicellular translocation of intermediates during the biosynthesis of monoterpene indole alkaloids and isoprenoid-derived primary metabolites. *Plant J.* 38:131–141.

Campos-Tamayo, F., Hernández-Domínguez, E., Vázquez-Flota, F. 2008. Vindoline formation in shoot cultures of *Catharanthus roseus* is synchronously activated with morphogenesis through the last biosynthetic step. *Ann. Bot.* 102:409–415.

Cardoso, M.I., Meijer, A.H., Rueb, S., Machado, J.A., Memelink, J., Hoge, J.H. 1997. A promoter region that controls basal and elicitor-inducible expression levels of the NADPH: Cytochrome P450 reductase gene (Cpr) from *Catharanthus roseus* binds nuclear factor GT-1. *Mol. Gen. Genet.* 256:674–681.

Chahed, K., Oudin, A., Guivarch, N., Hamdi, S., Chenieux, J.-C., Rideau, M., Clastre, M. 2000. 1-Deoxy-D-xylulose 5-phosphate synthase from periwinkle: cDNA identification and induced gene expression in terpenoid indole alkaloid-producing cells. *Plant Physiol. Biochem.* 38:559–566.

Chatel, G., Montiel, G., Pre, M., Memelink, J., Thiersault, M., Saint-Pierre, B., Doireau, P., Gantet, P. 2003. CrMYC1, a *Catharanthus roseus* elicitor- and jasmonate-responsive bHLH transcription factor that binds the G-box element of the strictosidine synthase gene promoter. *J. Exp. Bot.* 54:2587–2588.

Chaudhary, A.A., Hemant, Mohsin, M., Ahmad, A. 2012. Application of loop-mediated isothermal amplification (LAMP)-based technology for authentication of *Catharanthus roseus* (L.) G. Don. *Protoplasma* 249:417–422.

Chavadej, S., Brisson, N., McNeil, J.N., De Luca, V. 1994. Redirection of tryptophan leads to production of low indole glucosinolate canola. *Proc. Natl. Acad. Sci. USA* 91:2166–2170.

Chen, Q., Li, N., Zhang, W., Chen, J., Chen, Z. 2011. Simultaneous determination of vinblastine and its monomeric precursors vindoline and catharanthine in *Catharanthus roseus* by capillary electrophoresis-mass spectrometry. *J. Sep. Sci.* 34:2885–2892.

Chen, S., Galan, M.C., Coltharp, C., O'Connor, S.E. 2006. Redesign of a central enzyme in alkaloid biosynthesis. *Chem. Biol.* 13:1137–1141.

Chockalingam, S., Nalina Sundari, M.S., Thenmozhi, S. 1989. Impact of the extract of *Catharanthus roseus* on feeding and enzymatic digestive activities of *Spodoptera litura*. *J. Environ. Biol.* 10:303–307.

Choi, Y.H., Tapias, E.C., Kim, H.K., Lefeber, A.W., Erkelens, C., Verhoeven, J.T., Brzin, J., Zel, J., Verpoorte, R. 2004. Metabolic discrimination of *Catharanthus roseus* leaves infected by phytoplasma using 1H-NMR spectroscopy and multivariate data analysis. *Plant Physiol.* 135:2398–2410.

Choi, Y.H., Yoo, K.P., Kim, J. 2002. Supercritical fluid extraction and liquid chromatography-electrospray mass analysis of vinblastine from *Catharanthus roseus*. *Chem. Pharm. Bull.* 50:1294–1296.

Choudhury, S., Gupta, K. 2002. Influence of seasons on biomass and alkaloid productivity in *Catharanthus roseus*. *J. Med. Arom. Plant Sci.* 24:664–668.

Chung, I.M., Kim, E.H., Li, M., Peebles, C.A., Jung, W.S., Song, H.K., Ahn, J.K., San, K.Y. 2011. Screening 64 cultivars *Catharanthus roseus* for the production of vindoline, catharanthine, and serpentine. *Biotechnol. Prog.* 27:937–943.

Cibotti, M.C., Freier, C., Andrieux, J., Plat, M., Cosson, L., Bohuon, C. 1990. Monoclonal antibodies to bis-indole alkaloids of *Catharanthus roseus* and their use in enzyme-linked immuno-sorbent-assays. *Phytochemistry* 29:2109–2114.

Collu, G., Unver, N., Peltenburg-Looman, A.M.G., van der Heijden, R., Verpoorte, R., Memelink, J. 2001. Geraniol 10-hydroxylase, a cytochrome P450 enzyme involved in terpenoid indole alkaloid biosynthesis. *FEBS Lett.* 508:215–220.

Contin, A., Collu, G., van der Heijden, R., Verpoorte, R. 1999. The effects of phenobarbital and ketoconazole on the alkaloid biosynthesis in *Catharanthus roseus* cell suspension cultures. *Plant Physiol. Biochem.* 37:139–144.

Contin, A., van der Heijden, R., Lefeber, A.W.M., Verpoorte, R. 1998. The iridoid glucoside secologanin is derived from the novel triose phosphate/pyruvate pathway in *Catharanthus roseus* cell culture. *FEBS Lett.* 434:413–416.

Costa, M.M., Hilliou, F., Duarte, P., Pereira, L.G., Almeida, I., Leech, M., Memelink, J., Barceló, A.R., Sottomayor, M. 2008. Molecular cloning and characterization of a vacuolar class III peroxidase involved in the metabolism of anticancer alkaloids in *Catharanthus roseus*. *Plant Physiol.* 146:403–417.

Courdavault, V., Burlat, V., St-Pierre, B., Giglioli-Guivarc'h, N. 2009. Proteins prenylated by type I protein geranylgeranyltransferase act positively on the jasmonate signalling pathway triggering the biosynthesis of monoterpene indole alkaloids in *Catharanthus roseus*. *Plant Cell Rep.* 28:83–93.

Daddona, P.E., Wright, J.L., Hutchinson, C.R. 1976. Alkaloid catabolism and mobilization in *Catharanthus roseus*. *Phytochemistry* 15:941–945.

De Carolis, E., Chan, F., Balsevich, J., De Luca, V. 1990. Isolation and characterization of a 2-oxoglutarate dependant dioxygenase involved in the second-to-last step in vindoline biosynthesis. *Plant Physiol.* 94:1323–1329.

De Carolis, E., De Luca, V. 1993. Purification, characterization and kinetic analysis of a 2-oxoglutarate-dependant dioxygenase involved in vindoline biosynthesis from *Catharanthus roseus*. *J. Biol. Chem.* 268:5504–5511.

De Carolis, E., De Luca, V. 1994a. A novel 2-oxoglutarate-dependant dioxygenase involved in vindoline biosynthesis: Characterization, purification and kinetic properties. *Plant Cell Tiss. Org. Cult.* 38:281–287.

De Carolis, E., De Luca, V. 1994b. 2-Oxoglutarate-dependant dioxygenase and related enzymes: Biochemical characterization. *Phytochemistry* 36:1093–1107.

De Luca, V., Balsevich, J., Kurz, W.G.W. 1985. Acetyl coenzyme A: Deacetylvindoline-O-acetyltransferase, a novel enzyme from *Catharanthus*. *J. Plant Physiol.* 121:417–428.

De Luca, V., Balsevich, J., Tyler, R.T., Eilert, U., Panchuk, B.D., Kurz, W.G.W. 1986. Biosynthesis of indole alkaloids: Developmental regulation of the biosynthetic pathway from tabersonine to vindoline in *Catharanthus roseus*. *J. Plant Physiol.* 125:147–156.

De Luca, V., Balsevich, J., Tyler, R.T., Kurz, W.G.W. 1987. Characterization of a novel N-methyltransferase (NMT) from *Catharanthus roseus* plants. *Plant Cell Rep.* 6:458–461.

De Luca, V., Cutler, A.J. 1987. Subcellular localization of enzymes involved in indole alkaloid biosynthesis in *Catharanthus roseus*. *Plant Physiol.* 85:1099–1102.

De Luca, V., Fernandez, J.A., Campbell, D., Kurz, W.G.W. 1988. Developmental regulation of enzymes of indole alkaloid biosynthesis in *Catharanthus roseus*. *Plant Physiol.* 86:447–450.

De Luca, V., Marineau, C., Brisson, N. 1989. Molecular cloning and analysis of cDNA encoding a plant tryptophan decarboxylase: Comparison with animal DOPA decarboxylases. *Proc. Natl. Acad. Sci. USA* 86:2582–2586.

De Luca, V., Capasso, C., Capasso, A., Pastore, M., Carginale, V. 2011. Gene expression profiling of phytoplasma-infected Madagascar periwinkle leaves using differential display. *Mol. Biol. Rep.* 38:2993–3000.

De Pinho, P.G., Gonçalves, R.F., Valentão, P., Pereira, D.M., Seabra, R.M., Andrade, P.B., Sottomayor, M. 2009. Volatile composition of *Catharanthus roseus* (L.) G. Don using solid-phase microextraction and gas chromatography/mass spectrometry. *J. Pharm. Biomed. Anal.* 49:674–685.

De Waal, A., Meijer, A.H., Verpoorte, R. 1995. Strictosidine synthase from *Catharanthus roseus*: Purification and characterization of multiple forms. *Biochem. J.* 306:571–580.

Dethier, M., De Luca, V. 1993. Partial purification of an *N*-methyltransferase involved in vindoline biosynthesis in *Catharanthus roseus*. *Phytochemistry* 32:673–678.

Deus-Neumann, B., Stockigt, J., Zenk, M.H. 1987. Radioimmunoassay for the quantitative determination of catharanthine. *Planta Med.* 53:184–188.

Duarte, P., Memelink, J., Sottomayor, M. 2010. Fusion with fluorescent proteins for subcellular localization of enzymes involved in plant alkaloid biosynthesis. *Methods Mol. Biol.* 643:275–290.

Dwivedi, S., Singh, M., Singh, A.P., Singh, V., Uniyal, G.C., Khanuja, S.P.S., Kumar, S. 2001. Registration of a new variety Prabal of *Catharanthus roseus*. *J. Med. Arom. Plant Sci.* 23:104–106.

Eilert, U., Nesbitt, L.R., Constabel, F. 1985. Laticifers and latex in fruits of periwinkle, *Catharanthus roseus*. *Can. J. Bot.* 63:1540–1546.

Eilert, U., Constabel, F., Kurz, W.G.W. 1986. Elicitor-stimulation of monoterpene indole alkaloid formation in suspension cultures of *Catharanthus roseus*. *J. Plant Physiol.* 126:11–22.

Eisenreich, W., Rohdich, F., Bacher, A. 2001. Deoxyxylulose phosphate pathway to terpenoids. *Trends Plant Sci.* 6:78–84.

Eisenreich, W., Sagner, S., Zenk, M.H., Bacher, A. 1997. Monoterpenoid essential oils are not of mevalonate origin. *Tetrahedron Lett.* 38:3889–3892.

El-Sayed, M., Verpoorte, R. 2007. *Catharanthus* terpenoid indole alkaloids: Biosynthesis and regulation. *Phytochem. Rev.* 6:277–305.

Endo, T., Goodbody, A., Misawa, M. 1987. Alkaloid production in root and shoot cultures of *Catharanthus roseus*. *Planta Med.* 53:479–482.

Endo, T., Goodbody, A., Vukovic, J., Misawa, M. 1988. Enzymes from *Catharanthus roseus* cell suspension cultures that couple vindoline and catharanthine to form 3′,4′-anhydrovinblastine. *Phytochemistry* 27:2147–2149.

Facchini, P.J., De Luca, V. 2008. Opium poppy and Madagascar periwinkle: Model non-model systems to investigate alkaloid biosynthesis in plants. *Plant J.* 54:763–784.

Fahn, A. 1988. Secretory tissues in vascular plants. *New Phytol.* 108:229–257.

Fahn, W., Gundlach, H., Deuss-Neumann, B., Stöckigt, J. 1985. Late enzymes of vindoline biosynthesis. Acetyl-CoA: 17-*O*-deacetylvindoline 17-*O*-acetyltransferase. *Plant Cell Rep.* 4:333–336.

Fahn, W., Stöckigt, J. 1990. Purification of acetyl-CoA: 17-*O*-deacetylvindoline 17-*O*-acetyltransferase from *Catharanthus roseus* leaves. *Plant Cell Rep.* 8:613–616.

Farnsworth, N.R., Blomster, R.N., Damratoski, D., Meer, W.A., Cammarato, L.V. 1964. Studies on *Catharanthus* alkaloids. VI. Evaluation by means of thin layer chromatography and ceric ammonium sulfate spray reagent. *Lloydia* 27:302–314.

Favretto, D., Piovan, A., Filippini, R., Caniato, R. 2001. Monitoring the production yields of vincristine and vinblastine in *Catharanthus roseus* from somatic embryogenesis. Semiquantitative determination by flow-injection electrospray ionization mass spectrometry. *Rapid Commun. Mass Spectrom.* 15:364–369.

Fernandez, J.A., De Luca, V. 1994. Ubiquitin-mediated degradation of tryptophan decarboxylase from *Catharanthus roseus*. *Phytochemistry* 36:1123–1128.

Fernandez, J.A., Owen, T.G., Kurz, W.G.W., De Luca, V. 1989. Immunological detection and quantitation of tryptophan decarboxylase in developing *Catharanthus roseus* seedlings. *Plant Physiol.* 91:79–84.

Ferreres, F., Pereira, D.M., Valentão, P., Oliveira, J.M., Faria, J., Gaspar, L., Sottomayor, M., Andrade, P.B. 2010. Simple and reproducible HPLC-DAD-ESI-MS/MS analysis of alkaloids in *Catharanthus roseus* roots. *J. Pharm. Biomed. Anal.* 51:65–69.

Gantet, P., Memelink, J. 2002. Transcription factors: Tools to engineer the production of pharmacologically active plant metabolites. *Trends Pharmacol. Sci.* 23:563–569.

Garga, R.P. 1958. Studies on virus diseases of plants in Madhya Pradesh. I. Green rosette of *Vinca*. *Curr. Sci.* 27:493–494.

Geerlings, A., Ibanez, M.M.-L., Memelink, J., van der Heijden, R., Verpoorte, R. 2000. Molecular cloning and analysis of strictosidine β-D-glucosidase, an enzyme in terpenoid indole alkaloid biosynthesis in *Catharanthus roseus*. *J. Biol. Chem.* 275:3051–3056.

Geerlings, A., Redondo, F.J., Contin, A., Memelink, J., van der Heijden, R., Verpoorte, R. 2001. Biotransformation of tryptamine and secologanin into plant terpenoid indole alkaloids by transgenic yeast. *Appl. Microbiol. Biotechnol.* 56:420–424.

Giddings, L.A., Liscombe, D.K., Hamilton, J.P., Childs, K.L., DellaPenna, D., Buell, C.R., O'Connor, S.E. 2011. A stereoselective hydroxylation step of alkaloid biosynthesis by a unique cytochrome P450 in *Catharanthus roseus*. *J. Biol. Chem.* 286:16751–16757.

Ginis, O., Courdavault, V., Melin, C., Lanoue, A., Giglioli-Guivarc'h, N., St-Pierre, B., Courtois, M., Oudin, A. 2012. Molecular cloning and functional characterization of *Catharanthus roseus* hydroxymethylbutenyl 4-diphosphate synthase gene promoter from the methyl erythritol phosphate pathway. *Mol. Biol. Rep.* 39:5433–5447.

Glenn, W.S., Nims, E., O'Connor, S.E. 2011. Reengineering a tryptophan halogenase to preferentially chlorinate a direct alkaloid precursor. *J. Am. Chem. Soc.* 133:19346–19349.

Goklany, S., Loring, R.H., Glick, J., Lee-Parsons, C.W. 2009. Assessing the limitations to terpenoid indole alkaloid biosynthesis in *Catharanthus roseus* hairy root cultures through gene expression profiling and precursor feeding. *Biotechnol. Prog.* 25:1289–1296.

Goodbody, A.E., Endo, T., Vukovic, J., Kutney, J.P., Choi, L.S.L., Misawa, M. 1988a. Enzymic coupling of catharanthine and vindoline to form 3',4'-anhydrovinblastine by horseradish peroxidase. *Planta Med.* 54:136–140.

Goodbody, A.E., Watson, C.D., Chapple, C.C.S., Vukovic, J., Misawa, M. 1988b. Extraction of 3',4'-anhydrovinblastine from *Catharanthus roseus*. *Phytochemistry* 27:1713–1717.

Goosens, A., Rischer, H. 2007. Implementation of functional genomics for gene discovery in alkaloid producing plants. *Phytochem. Rev.* 6:35–49.

Guarnaccia, R., Botta, L., Coscia, C.J. 1974. Biosynthesis of acidic iridoid monoterpene glucosides in *Vinca rosea*. *J. Am. Chem. Soc.* 96:7079–7084.

Guimarães, G., Cardoso, L., Oliveira, H., Santos, C., Duarte, P., Sottomayor, M. 2012. Cytogenetic characterization and genome size of the medicinal plant *Catharanthus roseus* (L.) G. Don. *AoB Plants* 2012:pls002.

Guirimand, G., Burlat, V., Oudin, A., Lanoue, A., St-Pierre, B., Courdavault, V. 2009. Optimization of the transient transformation of *Catharanthus roseus* cells by particle bombardment and its application to the subcellular localization of hydroxymethylbutenyl 4-diphosphate synthase and geraniol 10-hydroxylase. *Plant Cell Rep.* 28:1215–1234.

Guirimand, G., Courdavault, V., Lanoue, A., Mahroug, S., Guihur, A., Blanc, N., Giglioli-Guivarc'h, N., St-Pierre, B., Burlat, V. 2010. Strictosidine activation in Apocynaceae: Towards a "nuclear time bomb"? *BMC Plant Biology* 10:182.

Guirimand, G., Guihur, A., Ginis, O., Poutrain, P., Héricourt, F., Oudin, A., Lanoue, A., St-Pierre, B., Burlat, V., Courdavault, V. 2011a. The subcellular organization of strictosidine biosynthesis in *Catharanthus roseus* epidermis highlights several trans-tonoplast translocations of intermediate metabolites. *FEBS J.* 278:749–763.

Guirimand, G., Guihur, A., Phillips, M.A., Oudin, A., Glévarec, G., Melin, C., Papon, N., Clastre, M., St-Pierre, B., Rodríguez-Concepción, M., Burlat, V., Courdavault, V. 2012a. A single gene encodes isopentenyl diphosphate isomerase isoforms targeted to plastids, mitochondria and peroxisomes in *Catharanthus roseus*. *Plant Mol. Biol.* 79:443–459.

Guirimand, G., Guihur, A., Poutrain, P., Héricourt, F., Mahroug, S., St-Pierre, B., Burlat, V., Courdavault, V. 2011b. Spatial organization of the vindoline biosynthetic pathway in *Catharanthus roseus*. *J. Plant Physiol.* 168:549–557.

Guirimand, G., Simkin, A.J., Papon, N., Besseau, S., Burlat, V., St-Pierre, B., Giglioli-Guivarc'h, N., Clastre, M., Courdavault, V. 2012b. Cycloheximide as a tool to investigate protein import in peroxisomes: A case study of the subcellular localization of isoprenoid biosynthetic enzymes. *J. Plant Physiol.* 169:825–829.

Gupta, M.M., Singh, D.V., Tripathi, A.K., Pandey, R., Verma, R.K., Singh, S., Shasany, A.K., Khanuja, S.P.S. 2005. Simultaneous determination of vincristine, vinblastine, catharanthine and vindoline in leaves of *Catharanthus roseus* by high performance liquid chromatography. *J. Chromatogr. Sci.* 43:450–453.

Gupta, R. 1977. Periwinkle produces anti-cancer drug. *Indian Farming* 27(4):11–13.

Gupta, S., Pandey-Rai, S., Srivastava, S., Naithani, S.C., Prasad, M., Kumar, S. 2007. Construction of genetic linkage map of the medicinal and ornamental plant *Catharanthus roseus*. *J. Genet.* 86:259–268.

Hamada, H., Nakazawa, K. 1991. Biotransformation of vinblastine to vincristine by cell suspension cultures of *Catharanthus roseus*. *Biotechnol. Lett.* 13:805–806.

Harris, A.L., Nylund, H.B., Carew, D.P. 1964. Tissue culture studies of certain members of the Apocynaceae. *Lloydia* 27:322–327.

He, L., Yang, L., Xiong, A., Zhao, S., Wang, Z., Hu, Z. 2011. Simultaneous quantification of four indole alkaloids in *Catharanthus roseus* cell line C20hi by UPLC-MS. *Anal. Sci.* 27:433–438.

Hedhili, S., Courdavault, V., Giglioli-Guivarc'h, N., Gantet, P. 2007. Regulation of the terpene moiety biosynthesis of *Catharanthus roseus* terpene indole alkaloids. *Phytochem. Rev.* 6:341–351.

Hedhili, S., De Mattei, M.V., Coudert, Y., Bourrié, I., Bigot, Y., Gantet, P. 2010. Three non-autonomous signals collaborate for nuclear targeting of CrMYC2, a *Catharanthus roseus* bHLH transcription factor. *BMC Res. Notes* 3:301.

Hemscheidt, T. 1983. Bildung and umsetzung van imminium-cathenamine katalysiert durch spezifische enzyme aus *Catharanthus roseus*. PhD. Thesis, Universität München.

Hemscheidt, T., Zenk, M.H. 1980. Glucosidases involved in indole alkaloid biosynthesis of *Catharanthus roseus* cell cultures. *FEBS Lett.* 110:187–191.

Hirata, K., Asada, M., Yatani, E., Miyamoto, K., Miura, Y. 1993. Effects of near-ultraviolet light on alkaloid production in *Catharanthus roseus* plants. *Planta Med.* 59:46–50.

Hirata, K., Kobayashi, M., Miyamoto, K., Hoshi, T., Okazaki, M., Miura, Y. 1989. Quantitative determination of vinblastine in tissue cultures of *Catharanthus roseus* by radioimmunoassay. *Planta Med.* 55:262–264.

Hirata, K., Yamanaka, A., Kurano, N., Miyamoto, K., Miura, Y. 1987. Production of indole alkaloids in multiple shoot cultures of *Catharanthus roseus* (L.) G. Don. *Agric. Biol. Chem.* 51:1311–1317.

Hisiger, S., Jolicoeur, M. 2007. Analysis of *Catharanthus roseus* alkaloids by HPLC. *Phytochem. Rev.* 6:207–234.

Huhtikangas, A., Lehtola, T., Lapinjoki, S., Lounasmaa, M. 1987. Specific radioimmunoassay for vincristine. *Planta Med.* 53:85–87.

Irmler, S., Schröder, G., St-Pierre, B., Crouch, N.P., Hotze, M., Schmidt, J., Strack, D., Matern, U., Schröder, J. 2000. Indole alkaloid biosynthesis in *Catharanthus roseus*: New enzyme activities and identification of cytochrome P450 CYP72A1 as secologanin synthase. *Plant J.* 24:797–804.

Jacobs, D.I., Gaspari, M., van der Greef, J., van der Heijden, R., Verpoorte, R. 2005. Proteome analysis of the medicinal plant *Catharanthus roseus*. *Planta* 221:690–704.

Jacobs, D.I., van Rijssen, M.S., van der Heijden, R., Verpoorte, R. 2001. Sequential solubilization of proteins precipitated with trichloroacetic acid in acetone from cultured *Catharanthus roseus* cells yields 52% more spots after two-dimensional electrophoresis. *Proteomics* 1:1345–1350.

Janaki Ammal, E.K., Bezbaruah, H.P. 1963. Induced tetraploidy in *Catharanthus roseus* (L.) G. Don. *Proc. Indian Acad. Sci.* (*Section B*) 57:339–342.

Jin, H., Liu, D., Zuo, K., Gong, Y., Miao, Z., Chen, Y., Ren, W., Sun, X., Tang, K. 2007. Molecular cloning and characterization of Crmdr1, a novel MDR-type ABC transporter gene from *Catharanthus roseus*. *DNA Seq.* 18:316–325.

Johnson, I.S., Wright, H.F., Svoboda, G.H. 1959. Experimental basis for clinical evaluation of antitumor principles from *Vinca rosea* Linn. *J. Lab. Clin. Med.* 54:830.

Joshi, R.K., Kar, B., Nayak, S. 2011. Exploiting EST databases for the mining and characterization of short sequence repeat (SSR) markers in *Catharanthus roseus* L. *Bioinformation* 5:378–381.

Jossang, A., Fodor, P., Bodo, B. 1998. A new class of bisindole alkaloids from the seeds of *Catharanthus roseus*: Vingramine and Methylvingramine. *J. Org. Chem.* 63:7162–7167.

Kalra, A., Ravindra, N.S., Chandrashekhar, R.S. 1991. Influence of foliar application of fungicides on dieback disease caused by *Pythium aphanidermatum* and alkaloid yield of periwinkle (*Catharanthus roseus*). *Ind. J. Agric. Sci.* 61:949–951.

Krishnan, R., Naragund, V.R., Vasantha Kumar, T. 1979. Evidences for outbreeding in *Catharanthus roseus*. *Curr. Sci.* 48:80–82.

Krueger, R.J., Carew, D.P., Lui, J.H.C., Staba, E.J. 1982. Initiation, maintenance and alkaloid content of *Catharanthus roseus* leaf organ cultures. *Planta Med.* 45:56–57.

Kulkarni, R.N. 1999. Evidence for phenotypic assortative mating for flower colour in periwinkle. *Plant Breeding* 118:561–564.

Kulkarni, R.N., Baskaran, K., Chandrashekara, R.S., Khanuja, S.P.S., Darokar, M.P., Shasany, A.K., Uniyal, G.C., Gupta, M.M., Kumar, S. 2003. 'Dhawal', a high alkaloid producing periwinkle plant. U. S. Patent No. 6,548,746.

Kulkarni, R.N., Baskaran, K., Chandrashekara, R.S., Kumar, S. 1999. Inheritance of morphological traits of periwinkle mutants with modified contents and yields of leaf and root alkaloids. *Plant Breeding* 118:71–74.

Kulkarni, R.N., Dimri, B.P., Rajagopal, K., Suresh, N., Chandrashekar, R.S. 1984. Variability for quantitative characters in periwinkle (*Catharanthus roseus* (L) G. Don). *Indian Drugs* 22:61–64.

Kulkarni, R.N., Rajagopal, K., Chandrashekar, R.S., Dimri, B.P., Suresh, N., Rajeshwar Rao, B.R. 1987. Performance of diploids and induced autotetraploids of *Catharanthus roseus* under different levels of nitrogen and plant spacings. *Plant Breeding* 98:136–140.

Kulkarni, R.N., Ravindra, N.S. 1987. Resistance to *Pythium aphanidermatum* in diploids and induced autotetraploids of *Catharanthus roseus*. *Planta Med.* 53:356–359.

Kumar, S., Dutta, A., Sinha, A.K., Sen, J. 2007. Cloning, characterization and localization of a novel basic peroxidase gene from *Catharanthus roseus*. *FEBS J.* 274:1290–1303.

Kutchan, T.M. 1998. Molecular genetics of plant alkaloid biosynthesis. In Cordell, G.A. (ed.), *The Alkaloids*, vol. 50, pp. 257–316. Academic Press, San Diego.

Kutchan, T.M., Dittrich, H., Bracher, D., Zenk, M.H. 1991. Enzymology and molecular biology of alkaloid biosynthesis. *Tetrahedron* 47:5945–5954.

Kutchan, T.M., Hampp, N., Lottspeich, F., Beyreuther, K., Zenk, M.H. 1988. The cDNA clone for strictosidine synthase from *Rauvolfia serpentina*: DNA sequence determination and expression in *Escherichia coli*. *FEBS Lett.* 237:40–44.

Kutney, J.P., Choi, L.S.L., Nakano, J., Tsukamoto, H. 1988a. Biomimetic chemical transformation of 3',4'-anhydrovinblastine to vinblastine and related bisindole alkaloids. *Heterocycles* 27:1837–1843.

Kutney, J.P., Choi, L.S.L., Nakano, J., Tsukamoto, H., McHugh, M., Boulet, C.A. 1988b. A highly efficient and commercially important synthesis of the antitumor *Catharanthus* alkaloids vinblastine and leurosidine from catharanthine and vindoline. *Heterocycles* 27:1845–1853.

Laflamme, P., St-Pierre, B., De Luca, V. 2001. Molecular and biochemical analysis of a Madagascar periwinkle root-specific minovincinine-19-hydroxy-*O*-acetyltransferase. *Plant Physiol.* 125:189–198.

Lange, B.M., Croteau, R. 1999. Isoprenoid biosynthesis via mevalonate-independent pathway in plants: Cloning and heterologous expression of 1-Deoxy-D-xylulose 5-phosphate reductoisomerase from peppermint. *Arch. Biochem. Biophys.* 365:170–174.

Lemenager, D., Ouelhazi, L., Mahroug, S., Veau, B., St-Pierre, B., Rideau, M., Aguirreolea, J., Burlat, V., Clastre, M. 2005. Purification, molecular cloning, and cell-specific gene expression of the alkaloid-accumulation associated protein CrPS in *Catharanthus roseus*. *J. Exp. Bot.* 56:1221–1228.

Levac, D., Murata, J., Kim, W.S., De Luca, V. 2008. Application of carborundum abrasion for investigating the leaf epidermis: Molecular cloning of *Catharanthus roseus* 16-hydroxytabersonine-16-*O*-methyltransferase. *Plant J.* 53:225–236.

Levy, A., Milo, J., Ashri, A., Palevitch, D. 1983. Heterosis and correlation analysis of the vegetative components and ajmalicine content in the roots of the medicinal plant—*Catharanthus roseus* (L.) G. Don. *Euphytica* 32:557–564.

Li, M., Peebles, C.A., Shanks, J.V., San, K.Y. 2011. Effect of sodium nitroprusside on growth and terpenoid indole alkaloid production in *Catharanthus roseus* hairy root cultures. *Biotechnol. Prog.* 27:625–630.

Lichtenthaler, H.K., Rohmer, M., Schwender, J. 1997a. Two independent biochemical pathways for isopentenyl diphosphate and isoprenoid biosynthesis in higher plants. *Physiol. Plant.* 101:643–652.

Lichtenthaler, H.K., Schwender, J., Disch, A., Rohmer, M. 1997b. Biosynthesis of isoprenoids in higher plant chloroplasts proceeds via a mevalonate independent pathway. *FEBS Lett.* 400:271–274.

Liscombe, D.K., O'Connor, S.E. 2011. A virus-induced gene silencing approach to understanding alkaloid metabolism in *Catharanthus roseus*. *Phytochemistry* 72:1969–1977.

Liscombe, D.K., Usera, A.R., O'Connor, S.E. 2010. Homolog of tocopherol C methyltransferases catalyzes N methylation in anticancer alkaloid biosynthesis. *Proc. Natl. Acad. Sci. USA* 107:18793–18798.

Lopez, C., Claude, B., Morin, P., Max, J.P., Pena, R., Ribet, J.P. 2011. Synthesis and study of a molecularly imprinted polymer for the specific extraction of indole alkaloids from *Catharanthus roseus* extracts. *Anal. Chim. Acta.* 683:198–205.

Loris, E.A., Panjikar, S., Ruppert, M., Barleben, L., Unger, M., Schübel, H., Stöckigt, J. 2007. Structure-based engineering of strictosidine synthase: Auxiliary for alkaloid libraries. *Chem. Biol.* 14:979–985.

Loyola-Vargas, V.M., Galaz-Avalos, R.M., Ku-Cauich, R. 2007. *Catharanthus* biosynthetic enzymes: The road ahead. *Phytochem. Rev.* 6:307–339.

Luijendijk, T.J.C. 1995. Strictosidine glucosidase in alkaloid biosynthesis. PhD. Thesis, Leiden University.

Luijendijk, T.J.C., Meijden, E.V.D., Verpoorte, R. 1996a. Involvement of strictosidine as a defensive chemical in *Catharanthus roseus*. *J. Chem. Ecol.* 22:1355–1366.

Luijendijk, T.J.C., Stevens, L.H., Verpoorte, R. 1996b. Reaction for the localization of strictosidine glucosidase activity on polyacrylamide gels. *Phytochem. Anal.* 7:16–19.

Luijendijk, T.J.C., Stevens, L.H., Verpoorte, R. 1998. Purification and characterisation of strictosidine β-D-glucosidase from *Catharanthus roseus* cell suspension cultures. *Plant Physiol. Biochem.* 36:419–425.

Madyastha, K.M., Coscia, C.J. 1979. Enzymology of indole alkaloid biosynthesis. *Rec. Adv. Phytochem.* 13:85–129.

Madyastha, K.M., Guarnaccia, R., Baxter, C., Coscia, C.J. 1973. S-Adenosyl-L-methionine: Loganic acid methyltransferase. A carboxyl-alkylating enzyme from *Vinca rosea*. *J. Biol. Chem.* 248:2497–2501.

Madyastha, K.M., Meehan, T.D., Coscia, C.J. 1976. Characterization of a cytochrome P-450 dependant monoterpene hydroxylase from the higher plant *Vinca rosea*. *Biochemistry* 15:1097–1102.

Mahroug, S., Burlat, V., St-Pierre, B. 2007. Cellular and sub-cellular organisation of the monoterpenoid indole alkaloid pathway in *Catharanthus roseus*. *Phytochem. Rev.* 6:363–381.

Makhzoum, A., Petit-Paly, G., St Pierre, B., Bernards, M.A. 2011. Functional analysis of the DAT gene promoter using transient *Catharanthus roseus* and stable *Nicotiana tabacum* transformation systems. *Plant Cell Rep.* 30:1173–1182.

Maldonado-Mendoza, I.E., Burnett, R.J., Lòpez-Meyer, M., Nessler, C.L. 1994. Regulation of 3-hydroxy-3-methylglutaryl-coenzyme A reductase by wounding and methyl jasmonate. *Plant Cell Tiss. Org. Cult.* 38:351–356.

Maldonado-Mendoza, I.E., Burnett, R.J., Nessler, C.L. 1992. Nucleotide sequence of a cDNA encoding 3-hydroxy-3-methylglutaryl coenzyme A reductase from *Catharanthus roseus*. *Plant Physiol.* 100:1613–1614.

Masoud, A.N., Sciuchetti, L.A., Farnsworth, N.R., Blomster, R.N., Meer, W.A. 1968. Effect of gibberellic acid on the growth, alkaloid production, and VLB content of *Catharanthus roseus*. *J. Pharm. Sci.* 57:589–593.

McCaskill, D.G., Martin, D.L., Scott, A.I. 1988. Characterization of alkaloid uptake by *Catharanthus roseus* (L.) G. Don protoplasts. *Plant Physiol.* 87:402–408.

McCormack, J.J. 1990. Pharmacology of antitumor bisindole alkaloids from *Catharanthus*. In Brossi, A. and Suffness, M. (eds), *The Alkaloids*, vol. 37, pp. 205–228. Academic Press, San Diego.

McKnight, T.D., Bergey, D.R., Burnett, R.J., Nessler, C.L. 1991. Expression of enzymatically active and correctly targeted strictosidine synthase in transgenic tobacco plants. *Planta* 185:148–152.

McKnight, T.D., Roessner, C.A., Devagupta, R., Scott, A.I., Nessler, C.L. 1990. Nucleotide sequence of a cDNA encoding the vacuolar protein strictosidine synthase from *Catharanthus roseus*. *Nucleic Acids Res.* 18:4939.

Medicinal Plant Genomics Website. http://medicinalplantgenomics.msu.edu/ (accessed November 13, 2012).

Meijer, A.H. 1993. Cytochrome P-450 and secondary metabolism in *Catharanthus roseus*. PhD. Thesis, Leiden University.

Meijer, A.H., Cardoso, M.I.L., Voskuilen, J.T., de Waal, A., Verpoorte, R., Hoge, J.H.C. 1993a. Isolation and characterization of a cDNA clone from *Catharanthus roseus* encoding NADPH: Cytochrome P-450 reductase, an enzyme essential for reactions catalysed by cytochrome P-450 mono-oxygenases in plants. *Plant J.* 4:47–60.

Meijer, A.H., de Waal, A., Verpoorte, R. 1993b. Purification of the cytochrome P-450 enzyme geraniol-10-hydroxylase from cell cultures of *Catharanthus roseus*. *J. Chromatogr.* 635:237–249.

Meisner, J., Weissenberg, M., Palevitch, D., Aharonson, N. 1981. Phagodeterrency induced by leaves and leaf extracts of *Catharanthus roseus* in the larva of *Spodoptera littoralis*. *J. Econ. Entomol.* 74:131–135.

Memelink, J., Gantet, P. 2007. Transcription factors involved in terpenoid indole alkaloid biosynthesis in *Catharanthus roseus*. *Phytochem. Rev.* 6:353–362.

Memelink, J., Verpoorte, R., Kijne, J.W. 2001. ORCAnisation of jasmonate-responsive gene expression in alkaloid metabolism. *Trends Plant Sci.* 6:212–219.

Menke, F.L.H., Champion, A., Kijne, J.W., Memelink, J. 1999a. A novel jasmonate- and elicitor-responsive element in the periwinkle secondary metabolite gene *Str* interacts with a jasmonate- and elicitor-inducible AP2-domain transcription factor, ORCA2. *EMBO J.* 18:4455–4463.

Menke, F.L.H., Parchmann, S., Mueller, M.J., Kijne, J.W., Memelink, J. 1999b. Involvement of the octadecanoid pathway and protein phosphorylation in fungal elicitor-induced expression of terpenoid indole alkaloid biosynthetic genes in *Catharanthus roseus*. *Plant Physiol.* 119:1289–1296.

Mersey, B.G., Cutler, A.J. 1986. Differential distribution of specific indole alkaloids in leaves of *Catharanthus roseus*. *Can. J. Bot.* 64:1039–1045.

Misawa, M., Endo, T., Goodbody, A., Vukovic, J., Chapplet, C., Choi, L., Kutney, J.P. 1988. Synthesis of dimeric indole alkaloids by cell free extracts from cell suspension cultures of *Catharanthus roseus*. *Phytochemistry* 27:1355–1359.

Mishra, P., Kumar, S. 2000. Emergence of Periwinkle *Catharanthus roseus* as a model system for molecular biology of alkaloids: Phytochemistry, pharmacology, plant biology and in vivo and in vitro cultivation. *J. Med. Arom. Plant Sci.* 22:306–337.

Mishra, P., Singh, M., Dwivedi, S., Kumar, S. 2000. Descriptors of periwinkle *Catharanthus roseus*. *J. Med. Arom. Plant Sci.* 22:268–272.

Mishra, R.K., Gangadhar, B.H., Yu, J.W., Kim, D.H., Park, S.W. 2011. Development and characterization of EST based SSR markers in Madagascar periwinkle (*Catharanthus roseus*) and their transferability in other medicinal plants. *Plant Omics J.* 4:154–162.

Misra, N., Luthra, R., Kumar, S. 1996. Enzymology of indole alkaloid biosynthesis in *Catharanthus roseus*. *Indian J. Biochem. Biophys.* 33:261–273.

Miura, Y., Hirata, K., Kurano, N. 1987. Isolation of vinblastine in callus culture with differentiated roots of *Catharanthus roseus* (L.) G. Don. *Agric. Biol. Chem.* 51:611–614.

Miura, Y., Hirata, K., Kurano, N., Miyamoto, K., Uchida, K. 1988. Formation of vinblastine in multiple shoot culture of *Catharanthus roseus. Planta Med.* 54:18–20.

Mizukami, H., Nordlöv, H., Lee, S.-L., Scott, A.I. 1979. Purification and properties of strictosidine synthetase (an enzyme condensing tryptamine and secologanin) from *Catharanthus roseus* cultured cells. *Biochemistry* 18:3760–3763.

Montiel, G., Breton, C., Thiersault, M., Burlat, V., Jay-Allemand, C., Gantet, P. 2007. Transcription factor Agamous-like 12 from Arabidopsis promotes tissue-like organization and alkaloid biosynthesis in *Catharanthus roseus* suspension cells. *Metab. Eng.* 9:125–132.

Moreno, P.R.H., Van der Heijden, R., Verpoorte, R. 1995. Cell and tissue cultures of *Catharanthus roseus*: A literature survey II. Updating from 1988 to 1993. *Plant Cell Tiss. Org. Cult.* 42:1–25.

Morris, P. 1986. Regulation of product synthesis in cell cultures of *Catharanthus roseus*. II. Comparison of production media. *Planta Med.* 52:121–126.

Murata, J., Bienzle, D., Brandle, J.E., Sensen, C.W., De Luca, V. 2006. Expressed sequence tags from Madagascar periwinkle (*Catharanthus roseus*). *FEBS Lett.* 580:4501–4507.

Murata, J., Roepke, J., Gordon, H., De Luca, V. 2008. The leaf epidermome of *Catharanthus roseus* reveals its biochemical specialization. *Plant Cell* 20:524–542.

Mustafa, N.R., Kim, H.K., Choi, Y.H., Verpoorte, R. 2009. Metabolic changes of salicylic acid-elicited *Catharanthus roseus* cell suspension cultures monitored by NMR-based metabolomics. *Biotechnol. Lett.* 31:1967–1974.

Naaranlahti, T., Lapinjoki, S.P., Huhtikangas, A., Toivonen, L., Kurtén, U., Kauppinen, V., Lounasmaa, M. 1989a. Mass spectral evidence of the occurrence of vindoline in heterotrophic cultures of *Catharanthus roseus* cells. *Planta Med.* 55:155–157.

Naaranlahti, T., Nordstrom, M., Huhtikangas, A., Lounasmaa, M. 1987. Determination of *Catharanthus* alkaloids by reversed-phase high-performance liquid chromatography. *J. Chromatogr.* 410:488–493.

Naaranlahti, T., Nordstrom, M., Lapinjoki, S.P., Huhtikangas, A. 1990. Isolation of *Catharanthus* alkaloids by solid-phase extraction and semipreparative HPLC. *J. Chromatogr. Sci.* 28:173–174.

Naaranlahti, T., Ranta, V.P., Jarho, P., Nordstrom, M., Lapinjoki, S.P. 1989b. Electrochemical detection of indole alkaloids of *Catharanthus roseus* in high performance liquid chromatography. *Analyst* 114:1229–1231.

Nejat, N., Vadamalai, G., Dickinson, M. 2012. Expression patterns of genes involved in the defense and stress response of *Spiroplasma citri* infected Madagascar Periwinkle *Catharanthus roseus. Int. J. Mol. Sci.* 13:2301–2313.

Neuss, N., Neuss, M.N. 1990. Therapeutic use of bisindole alkaloids from *Catharanthus*. In Brossi, A. and Suffness, M. (eds), *The Alkaloids*, vol. 37, pp. 229–240. Academic Press, San Diego.

Noble, R.L., Beer, C.T., Cutts, J.H. 1958. Role of chance observation in chemotherapy: *Vinca rosea. Ann. N. Y. Acad. Sci.* 76:882.

Noe, W., Mollenschott, C., Berlin, J. 1984. Tryptophan decarboxylase from *Catharanthus roseus* cell suspension cultures: Purification, molecular and kinetic data of the homogeneous protein. *Plant Mol. Biol.* 3:281–288.

O'Connor, S.E., Maresh, J.M. 2006. Chemistry and biology of monoterpene indole alkaloid biosynthesis. *Nat. Prod. Rep.* 23:532–547.

One Thousand Plant Transcriptome Project. http://www.onekp.com/(accessed November 13, 2012).

Oudin, A., Courtois, M., Rideau, M., Clastre, M. 2007a. The iridoid pathway in *Catharanthus roseus* alkaloid biosynthesis. *Phytochem. Rev.* 6:259–276.

Oudin, A., Mahroug, S., Courdavault, V., Hervouet, N., Zelwer, C., Rodríguez-Concepción, M., St-Pierre, B., Burlat, V. 2007b. Spatial distribution and hormonal regulation of gene products from methyl erythritol phosphate and monoterpene-secoiridoid pathways in *Catharanthus roseus. Plant Mol. Biol.* 65:13–30.

Ouwerkerk, P.B.F., Hallard, D., Verpoorte, R., Memelink, J. 1999a. Identification of UV-B light-responsive regions in the promoter of the tryptophan decarboxylase gene from *Catharanthus roseus. Plant Mol. Biol.* 41:491–503.

Ouwerkerk, P.B.F., Memelink, J. 1999a. A G-box element from the *Catharanthus roseus* strictosidine synthase (*Str*) gene promoter confers seed-specific expression in transgenic tobacco plants. *Mol. Gen. Genet.* 261:635–643.

Ouwerkerk, P.B.F., Memelink, J. 1999b. Elicitor-responsive promoter regions in the tryptophan decarboxylase gene from *Catharanthus roseus*. *Plant Mol. Biol.* 39:129–136.

Ouwerkerk, P.B.F., Trimborn, T.O., Hilliou, F., Memelink, J. 1999b. Nuclear factors GT-1 and 3AF1 interact with multiple sequences within the promoter of the Tdc gene from Madagascar periwinkle: GT-1 is involved in UV light-induced expression. *Mol. Gen. Genet.* 261:610–622.

Pasquali, G., Erven, A.S.W., Ouwerkerk, P.B.F., Menke, F.L.H., Memelink, J. 1999. The promoter of the strictosidine synthase gene from periwinkle confers elicitor-inducible expression in transgenic tobacco and binds nuclear factors GT-1 and GBF. *Plant Mol. Biol.* 39:1299–1310.

Pasquali, G., Goddijn, O.J.M., de Waal, A., Verpoorte, R., Schilperoort, R.A., Hoge, J.H.C., Memelink, J. 1992. Coordinated regulation of two indole alkaloid biosynthetic genes from *Catharanthus roseus* by auxin and elicitors. *Plant Mol. Biol.* 18:1121–1131.

Patterson, B.D., Carew, D.P. 1969. Growth and alkaloid formation in *Catharanthus roseus* tissue cultures. *Lloydia* 32:131–140.

Pauw, B., Hilliou, F.A.O., Martin, V.S., Chatel, G., de Wolf, C.J.F., Champion, A., Pre, M., van Duijn, B., Kijne, J.W., van der Fits, L., Memelink, J. 2004a. Zinc finger proteins act as transcriptional repressors of alkaloid biosynthesis genes in *Catharanthus roseus*. *J. Biol. Chem.* 279:52940–52948.

Pauw, B., van Duijn, B., Kijne, J.W., Memelink, J. 2004b. Activation of the oxidative burst by yeast elicitor in *Catharanthus roseus* cells occurs independently of the activation of genes involved in alkaloid biosynthesis. *Plant Mol. Biol.* 55:797–805.

Pearce, H.L. 1990. Medicinal chemistry of bisindole alkaloids from *Catharanthus*. In Brossi, A. and Suffness, M. (eds), *The Alkaloids*, vol. 37, pp. 145–204. Academic Press, San Diego.

Peebles, C.A., Sander, G.W., Hughes, E.H., Peacock, R., Shanks, J.V., San, K.Y. 2011. The expression of 1-deoxy-D-xylulose synthase and geraniol-10-hydroxylase or anthranilate synthase increases terpenoid indole alkaloid accumulation in *Catharanthus roseus* hairy roots. *Metab. Eng.* 13:234–240.

Pennings, E.J.M., Groen, B.W., Duine, J.A., Verpoorte, R. 1989. Tryptophan decarboxylase from *Catharanthus roseus* is a pyridoxo-quinoprotein. *FEBS Lett.* 255:97–100.

Pereira, D.M., Ferreres, F., Oliveira, J., Valentão, P., Andrade, P.B., Sottomayor, M. 2009. Targeted metabolite analysis of *Catharanthus roseus* and its biological potential. *Food Chem. Toxicol.* 47:1349–1354.

Pfitzner, U., Zenk, M.H. 1989. Homogeneous strictosidine synthase isoenzymes from cell suspension cultures of *Catharanthus roseus*. *Planta Med.* 55:525–530.

Platt, K.A., Thomson, W.W. 1992. Idioblast oil cells of avocado: Distribution, isolation, ultrastructure, histochemistry, and biochemistry. *Int. J. Plant Sci.* 153:301–310.

Pomahacová, B., Dusek, J., Dusková, J., Yazaki, K., Roytrakul, S., Verpoorte, R. 2009. Improved accumulation of ajmalicine and tetrahydroalstonine in *Catharanthus* cells expressing an ABC transporter. *J. Plant Physiol.* 166:1405–1412.

Postek, M.T., Tucker, S.C. 1983. Ontogeny and ultrastructure of secretory oil cells in *Magnolia grandiflora* L. *Bot. Gaz.* 144:501–512.

Poulsen, C., Bongaerts, R., Verpoorte, R. 1993. Purification and characterization of anthranilate synthase from *Catharanthus roseus*. *Eur. J. Biochem.* 212:431–440.

Poutrain, P., Mazars, C., Thiersault, M., Rideau, M., Pichon, O. 2009. Two distinct intracellular Ca2+-release components act in opposite ways in the regulation of the auxin-dependent MIA biosynthesis in *Catharanthus roseus* cells. *J. Exp. Bot.* 60:1387–1398.

Poutrain, P., Guirimand, G., Mahroug, S., Burlat, V., Melin, C., Ginis, O., Oudin, A., Giglioli-Guivarc'h, N., Pichon, O., Courdavault, V. 2011. Molecular cloning and characterisation of two calmodulin isoforms of the Madagascar periwinkle *Catharanthus roseus*. *Plant Biol. (Stuttg)*. 13:36–41.

Power, R., Kurz, W.G.W., De Luca, V. 1990. Purification and characterization of acetylcoenzyme A: Deacetylvindoline 4-O-acetyltransferase from *Catharanthus roseus*. *Arch. Biochem. Biophys.* 279:370–376.

Radwanski, E.R., Last, R.L. 1995. Tryptophan biosynthesis and metabolism: Biochemical and molecular genetics. *Plant Cell* 7:921–934.

Ramani, S., Patil, N., Jayabaskaran, C. 2010. UV-B induced transcript accumulation of DAHP synthase in suspension-cultured *Catharanthus roseus* cells. *J. Mol. Signal.* 5:13.

Reda, F. 1978. Distribution and accumulation of alkaloids in *Catharanthus roseus* G. Don during development. *Pharmazie* 33:233–234.

Rischer, H., Oresic, M., Seppänen-Laakso, T., Katajamaa, M., Lammertyn, F., Ardiles-Diaz, W., Van Montagu, M.C., Inzé, D., Oksman-Caldentey, K.M., Goossens, A. 2006. Gene-to-metabolite networks for terpenoid indole alkaloid biosynthesis in *Catharanthus roseus* cells. *Proc. Natl. Acad. Sci. USA* 103:5614–5619.

Roepke, J., Salim, V., Wu, M., Thamm, A.M., Murata, J., Ploss, K., Boland, W., De Luca, V. 2010. Vinca drug components accumulate exclusively in leaf exudates of Madagascar periwinkle. *Proc. Natl. Acad. Sci. USA* 107:15287–15292.

Roessner, C.A., Devagupta, R., Hasan, M., Williams, H.J., Scott, A.I. 1992. Purification of an indole biosynthetic enzyme, strictosidine synthase, from a recombinant strain of *Escherichia coli*. *Protein Expr. Purif.* 3:295–300.

Rohmer, M., Knani, M., Simonin, P., Sutter, B., Sahm, H. 1993. Isoprenoid biosynthesis in bacteria: A novel pathway for the early steps leading to isopentenyl diphosphate. *Biochem. J.* 295:517–524.

Roytrakul, S., Verpoorte, R. 2007. Role of vacuolar transporter proteins in plant secondary metabolism: *Catharanthus roseus* cell culture. *Phytochem. Rev.* 6:383–396.

Ruiz-May, E., De-la-Peña, C., Galaz-Ávalos, R.M., Lei, Z., Watson, B.S., Sumner, L.W., Loyola-Vargas, V.M. 2011. Methyl jasmonate induces ATP biosynthesis deficiency and accumulation of proteins related to secondary metabolism in *Catharanthus roseus* (L.) G. hairy roots. *Plant Cell Physiol.* 52:1401–1421.

Runguphan, W., Maresh, J.J., O'Connor, S.E. 2009. Silencing of tryptamine biosynthesis for production of nonnatural alkaloids in plant culture. *Proc. Natl. Acad. Sci. USA* 106:13673–13678.

Runguphan, W., O'Connor, S.E. 2009. Metabolic reprogramming of periwinkle plant culture. *Nat. Chem. Biol.* 5:151–153.

Runguphan, W., Qu, X., O'Connor, S.E. 2010. Integrating carbon-halogen bond formation into medicinal plant metabolism. *Nature* 468:461–464.

Saenz, L., Santamaria, J.M., Villanueva, M.A., Loyola-Vargas, V.M., Oropeza, C. 1993. Changes in the alkaloid content of plants of *Catharanthus roseus* (L) G. Don as a result of water stress and treatment with abscisic acid. *J. Plant Physiol.* 142:244–247.

Samad, A., Ajayakumar, P.V., Gupta, M.K., Shukla, A.K., Darokar, M.P., Somkuwar, B., Alam, M. 2008. Natural infection of periwinkle (*Catharanthus roseus*) with Cucumber mosaic virus, subgroup IB. *Australas. Plant Dis. Notes* 3:30–34.

Schroder, G., Unterbusch, E., Kaltenbach, M., Schmidt, J., Strack, D., De Luca, V., Schroder, J. 1999. Light-induced cytochrome P450-dependant enzyme in indole alkaloid biosynthesis: Tabersonine 16-hydroxylase. *FEBS Lett.* 458:97–102.

Scott, A.I., Lee, S.L., Wan, W. 1977. Indole alkaloid biosynthesis: Partial purification of "ajmalicine synthetase" from *Catharanthus roseus*. *Biochem. Biophys. Res. Commun.* 75:1004–1009.

Sevestre-Rigouzzo, M., Nef-Campa, C., Ghesquière, A., Chrestin, H. 1993. Genetic diversity and alkaloid production in *Catharanthus roseus*, *C. trichophyllus* and their hybrids. *Euphytica* 66:151–159.

Sharma, V., Chaudhary, S., Srivastava, S., Pandey, R., Kumar, S. 2012. Characterization of variation and quantitative trait loci related to terpenoid indole alkaloid yield in a recombinant inbred line mapping population of *Catharanthus roseus*. *J. Genet.* 91:49–69.

Shokeen, B., Choudhary, S., Sethy, N.K., Bhatia, S. 2011. Development of SSR and gene-targeted markers for construction of a framework linkage map of *Catharanthus roseus*. *Ann Bot.* 108:321–336.

Shukla, A.K. 2005. Molecular studies on biosynthesis of shoot alkaloids in *Catharanthus roseus* (L.) G. Don. PhD. Thesis, Department of Biochemistry, University of Lucknow, India.

Shukla, A.K., Shasany, A.K., Gupta, M.M., Khanuja, S.P.S. 2006. Transcriptome analysis in *Catharanthus roseus* leaves and roots for comparative terpenoid indole alkaloid profiles. *J. Exp. Bot.* 57:3921–3932.

Shukla, A.K., Shasany, A.K., Khanuja, S.P.S. 2005. Isolation of poly (A)⁺ mRNA for downstream reactions from some medicinal and aromatic plants. *Indian J. Exp. Biol.* 43:197–201.

Shukla, A.K., Shasany, A.K., Khanuja, S.P.S. 2012. cDNA-AFLP-based numerical comparison of leaf and root organ cDNAs in *Catharanthus roseus. OMICS: A J. Integr. Biol.* 16:397–401.

Shukla, A.K., Shasany, A.K., Verma, R.K., Gupta, M.M., Mathur, A.K., Khanuja, S.P.S. 2010. Influence of cellular differentiation and elicitation on intermediate and late steps of terpenoid indole alkaloid biosynthesis in *Catharanthus roseus. Protoplasma* 242:35–47.

Shukla, Y.N., Rani, A., Kumar, S. 1997. Effect of temperature and pH on the extraction of total alkaloids from *Catharanthus roseus* leaves. *J. Med. Arom. Plant Sci.* 19:430–431.

Sibéril, Y., Benhamron, S., Memelink, J., Giglioli-Guivarc'h, N., Thiersault, M., Boisson, B., Doireau, P., Gantet, P. 2001. *Catharanthus roseus* G-box binding factors 1 and 2 act as repressors of strictosidine synthase gene expression in cell cultures. *Plant Mol. Biol.* 45:477–488.

Simkin, A.J., Guirimand, G., Papon, N., Courdavault, V., Thabet, I., Ginis, O., Bouzid, S., Giglioli-Guivarc'h, N., Clastre, M. 2011. Peroxisomal localisation of the final steps of the mevalonic acid pathway *in planta. Planta* 234:903–914.

Singh, D., Mehta, S.S., Neoliya, N.K., Shukla, Y.N., Mishra, M. 2003. New possible insect growth regulators from *Catharanthus roseus. Curr. Sci.* 84:1184–1186.

Singh, D.V., Maithy, A., Verma, R.K., Gupta, M.M., Kumar, S. 2000. Simultaneous determination of *Catharanthus* alkaloids using reversed phase high performance liquid chromatography. *J. Liq. Chrom. Rel. Technol.* 23:601–607.

Singh, D.V., Pandey-Rai, S., Srivastava, S., Kumar, R.S., Mishra, R., Kumar, S. 2004. Simultaneous quantification of some pharmaceutical *Catharanthus roseus* leaf and root terpenoid indole alkaloids and their precursors in single runs by reversed-phase liquid chromatography. *J. AOAC Int.* 87:1287–1296.

Smith, J.I., Smart, N.J., Kurz, W.G.W., Misawa, M. 1987. Stimulation of indole alkaloid production in cell suspension cultures of *Catharanthus roseus* by abscisic acid. *Planta Med.* 53:470–474.

Song, K.M., Park, S.W., Hong, W.H., Lee, H., Kwak, S.S., Liu, J.R. 1992. Isolation of vindoline from *Catharanthus roseus* by supercritical fluid extraction. *Biotechnol. Prog.* 8:583–586.

Songstad, D.D., De Luca, V., Brisson, N., Kurz, W.G.W., Nessler, C.L. 1990. High levels of tryptamine accumulation in transgenic tobacco expressing tryptophan decarboxylase. *Plant Physiol.* 94:1410–1413.

Songstad, D.D., Kurz, W.G.W., Nessler, C.L. 1991. Tyramine accumulation in *Nicotiana tabacum* transformed with a chimeric tryptophan decarboxylase gene. *Phytochemistry* 30:3245–3246.

St-Pierre, B., De Luca, V. 1995. A cytochrome P-450 monooxygenase catalyzes the first step in the conversion of tabersonine to vindoline in *Catharanthus roseus. Plant Physiol.* 109:131–139.

St-Pierre, B., Laflamme, P., Alarco, A.M., De Luca, V. 1998. The terminal O-acetyltransferase involved in vindoline biosynthesis defines a new class of proteins responsible for coenzyme A-dependant acyl transfer. *Plant J.* 14:703–713.

St-Pierre, B., Vazquez-Flota, F.A., De Luca, V. 1999. Multicellular compartmentation of *Catharanthus roseus* alkaloid biosynthesis predicts intercellular translocation of a pathway intermediate. *Plant Cell* 11:887–900.

Stearn, W.T. 1966. *Catharanthus roseus*, the correct name for the Madagascar periwinkle. *Lloydia* 29:196–200.

Stearn, W.T. 1975. A synopsis of the genus *Catharanthus* (Apocynaceae). In Taylor, W.I. and Farnsworth, N.R. (eds), *The Catharanthus Alkaloids—Botany, Chemistry, Pharmacology, and Clinical Use*, pp. 9–44. Marcel Dekker, New York.

Stevens, L.H. 1994. Formation and conversion of strictosidine in the biosynthesis of monoterpenoid indole and quinoline alkaloids. PhD. Thesis, Leiden University.

Stöckigt, J. 1980. The biosynthesis of heteroyohimbine-type alkaloids. In Phillipson, J.D. and Zenk, M.H. (eds), *Indole and Biogenetically Related Alkaloids*, pp. 113–141. Academic Press, London.

Stöckigt, J., Hemscheidt, T., Höfle, G., Heinstein, P., Formacek, V. 1983. Steric course of hydrogen transfer during enzymatic formation of 3α-heteroyohimbine alkaloids. *Biochemistry* 22:3448–3452.

Sung, P.H., Huang, F.C., Do, Y.Y., Huang, P.L. 2011. Functional expression of geraniol 10-hydroxylase reveals its dual function in the biosynthesis of terpenoid and phenylpropanoid. *J. Agric. Food Chem.* 59:4637–4643.

Suttipanta, N., Pattanaik, S., Gunjan, S., Xie, C.H., Littleton, J., Yuan, L. 2007. Promoter analysis of the *Catharanthus roseus* geraniol 10-hydroxylase gene involved in terpenoid indole alkaloid biosynthesis. *Biochim. Biophys. Acta.* 1769:139–148.

Suttipanta, N., Pattanaik, S., Kulshrestha, M., Patra, B., Singh, S.K., Yuan, L. 2011. The transcription factor CrWRKY1 positively regulates the terpenoid indole alkaloid biosynthesis in *Catharanthus roseus. Plant Physiol.* 157:2081–2093.

Svoboda, G.H. 1958. A note on several new alkaloids from *Vinca rosea* Linn. I: Leurosine, virosine, perivine. *J. Am. Pharm. Assoc. Sci. Ed.* 47:834.

Svoboda, G.H. 1961. Alkaloids of *Vinca rosea* (*Catharanthus roseus*) 1X: Extraction and characterization of leurosidine and leurocristine. *Lloydia* 24:173–178.

Svoboda, G.H., Blake, D.A. 1975. The phytochemistry and pharmacology of *Catharanthus roseus* (L) G. Don. In Taylor, W.I. and Farnsworth, N.R. (eds), *The Catharanthus Alkaloids*, pp. 45–83. Marcel Dekker, New York.

Svoboda, G.H., Neuss, N., Gorman, M. 1959. Alkaloids of *Vinca rosea* Linn. (*Catharanthus roseus* G. Don.) V. Preparation and characterization of alkaloids. *J. Am. Pharm. Assoc. Sci. Ed.* 48:659–666.

Talha, M., Radwan, A.S., Negm, S. 1975. The effect of soil moisture deficit on growth and alkaloidal content of *Catharanthus roseus* G. Don. *Curr. Sci.* 44:614–616.

Tallevi, S.G., DiCosmo, F. 1988. Stimulation of indole alkaloid content in vanadium-treated *Catharanthus roseus* suspension cultures. *Planta Med.* 54:149–152.

Taylor, W.I., Farnsworth, N.R. 1975. *The Catharanthus Alkaloids*. Marcel Dekker, New York.

Tikhomiroff, C., Jolicoeur, M. 2002. Screening of *Catharanthus roseus* secondary metabolites by high-performance liquid chromatography. *J. Chromatogr. A* 955:87–93.

Tiwari, R., Awasthi, A., Mall, M., Shukla, A.K., Satya Srinivas, K.V.N., Syamasundar, K.V., Kalra, A. 2013. Bacterial endophyte-mediated enhancement of *in planta* content of key terpenoid indole alkaloids and growth parameters of *Catharanthus roseus. Ind. Crop Prod.* 43:306–310.

Treimer, J.F., Zenk, M.H. 1979. Purification and properties of strictosidine synthase, the key enzyme in indole alkaloid formation. *Eur. J. Biochem.* 101:225–233.

Van der Fits, L., Memelink, J. 2000. ORCA3, a jasmonate-responsive transcriptional regulator of plant primary and secondary metabolism. *Science* 289:295–297.

Van der Fits, L., Memelink, J. 2001. The jasmonate-inducible AP2/ERF-domain transcription factor ORCA3 activates gene expression via interaction with a jasmonate-responsive promoter element. *Plant J.* 25:43–53.

Van der Fits, L., Zhang, H., Menke, F.L.H., Deneka, M., Memelink, J. 2000. A *Catharanthus roseus* BPF-1 homologue interacts with an elicitor-responsive region of the secondary metabolite biosynthetic gene *Str* and is induced by elicitor via a jasmonate-independent signal transduction pathway. *Plant Mol. Biol.* 44:675–685.

Van der Heijden, R., de Boer-Hlupa, V., Verpoorte, R., Duine, J.A. 1994a. Enzymes involved in the metabolism of 3-hydroxy-3-methylglutaryl-coenzyme A in *Catharanthus roseus. Plant Cell Tiss. Org. Cult.* 38:345–349.

Van der Heijden, R., Jacobs, D.I., Snoeijer, W., Hallard, D., Verpoorte, R. 2004. The *Catharanthus* alkaloids: Pharmacognosy and biotechnology. *Curr. Med. Chem.* 11:607–628.

Van der Heijden, R., Verpoorte, R., 1995. Metabolic enzymes of 3-hydroxy-3-methylglutaryl-coenzyme A in *Catharanthus roseus. Plant Cell Tiss. Org. Cult.* 43:85–88.

Van der Heijden, R., Verpoorte, R., Duine, J.A. 1994b. Biosynthesis of 3S-hydroxy-3-methylglutaryl-coenzyme A in *Catharanthus roseus*: Acetoacetyl-CoA thiolase and HMG-CoA synthase show similar chromatographic behaviour. *Plant Physiol. Biochem.* 32:807–812.

Van der Heijden, R., Verpoorte, R., Ten Hoopen, H.J.G. 1989. Cell and tissue cultures of *Catharanthus roseus* (L) G Don: A literature survey. *Plant Cell Tiss. Org. Cult.* 18:231–280.

van Pée, K.H. 2011. Transformation with tryptophan halogenase genes leads to the production of new chlorinated alkaloid metabolites by a medicinal plant. *Chembiochem* 12:681–683.

Vazquez-Flota, F., De Carolis, E., Alarco, A.-M., De Luca, V. 1997. Molecular cloning and characterization of desacetoxyvindoline-4-hydroxylase, a 2-oxoglutarate dependant-dioxygenase involved in the biosynthesis of vindoline in *Catharanthus roseus* (L.) G. Don. *Plant Mol. Biol.* 34:935–948.

Veau, B., Courtois, M., Oudin, A., Chenieux, J.C., Rideau, M., Clastre, M. 2000. Cloning and expression of cDNAs encoding two enzymes of the MEP pathway in *Catharanthus roseus*. *Biochim. Biophys. Acta.* 1517:159–163.

Verma, P., Mathur, A.K. 2011. *Agrobacterium tumefaciens*-mediated transgenic plant production via direct shoot bud organogenesis from pre-plasmolyzed leaf explants of *Catharanthus roseus*. *Biotechnol. Lett.* 33:1053–1060.

Verma, P., Mathur, A.K., Srivastava, A., Mathur, A. 2012. Emerging trends in research on spatial and temporal organization of terpenoid indole alkaloid pathway in *Catharanthus roseus*: A literature update. *Protoplasma* 249:255–268.

Verpoorte, R. 1980. Isolation and separation methods for indole alkaloids. In Phillipson, J.D. and Zenk, M.H. (eds), *Indole and Biogenetically Related Alkaloids*, pp. 91–112. Academic Press, London.

Verpoorte, R., Van der Heijden, R., Van Gulik, W.M., Ten Hoopen, H.J.G. 1991. Plant biotechnology for the production of alkaloids: Present status and prospects. In Brossi, A. (ed.), *The Alkaloids*, vol. 40, pp. 1–187. Academic Press, San Diego.

Verpoorte, R., Van der Heijden, R., Moreno, P.R.H. 1997. Biosynthesis of terpenoid indole alkaloids in *Catharanthus roseus* cells. In Cordell, G.A. (ed.), *The Alkaloids*, vol. 49, pp. 221–299. Academic Press, San Diego.

Vetter, H.-P., Mangold, U., Schröder, G., Marner, F.-J., Werck-Reichhart, D., Schröder, J. 1992. Molecular analysis and heterologous expression of an inducible cytochrome P-450 protein from periwinkle (*Catharanthus roseus* L.). *Plant Physiol.* 100:998–1007.

Virmani, O.P., Srivastava, G.N., Singh, P. 1978. *Catharanthus roseus*: The tropical periwinkle. *Indian Drugs* 15:231–252.

Volkov, S.K., Grodnitskaya, E.I. 1994. Application of high performance liquid chromatography to the determination of vinblastine in *Catharanthus roseus*. *J. Chromatogr. B. Biomed. Appl.* 660:405–408.

Vom Endt, D., Kijne, J.W., Memelink, J. 2002. Transcription factors controlling plant secondary metabolism: What regulates the regulators? *Phytochemistry* 61:107–114.

Vom Endt, D., Soares e Silva, M., Kijne, J.W., Pasquali, G., Memelink, J. 2007. Identification of a bipartite jasmonate-responsive promoter element in the *Catharanthus roseus* ORCA3 transcription factor gene that interacts specifically with AT-Hook DNA-binding proteins. *Plant Physiol.* 144:1680–1689.

Wang, C.T., Liu, H., Gao, X.S., Zhang, H.X. 2010a. Overexpression of G10H and ORCA3 in the hairy roots of *Catharanthus roseus* improves catharanthine production. *Plant Cell Rep.* 29:887–894.

Wang, Q., Xing, S., Pan, Q., Yuan, F., Zhao, J., Tian, Y., Chen, Y., Wang, G., Tang, K. 2012. Development of efficient *Catharanthus roseus* regeneration and transformation system using *Agrobacterium tumefaciens* and hypocotyls as explants. *BMC Biotechnol.* 12:34.

Wang, Q., Yuan, F., Pan, Q., Li, M., Wang, G., Zhao, J., Tang, K. 2010b. Isolation and functional analysis of the *Catharanthus roseus* deacetylvindoline-4-O-acetyltransferase gene promoter. *Plant Cell Rep.* 29:185–192.

Waterhouse, D.N., Madden, T.D., Cullis, P.R., Bally, M.B., Mayer, L.D., Webb, M.S. 2005. Preparation, characterization, and biological analysis of liposomal formulations of vincristine. *Methods Enzymol.* 391:40–57.

Weiss, H.D., Walker, M.D., Wiernik, P.H. 1974. Neurotoxicity of commonly used antineoplastic agents (second of two parts). *N. Engl. J. Med.* 291:127–133.

Westekemper, P., Wieczorek, U., Gueritte, F., Langlois, N., Langlois, Y., Potier, P., Zenk, M.H. 1980. Radioimmunoassay for the determination of the indole alkaloid vindoline in *Catharanthus*. *Planta Med.* 39:24–37.

Whitmer, S., Canel, C., van der Heijden, R., Verpoorte, R. 2003. Long-term instability of alkaloid production by stably transformed cell lines of *Catharanthus roseus*. *Plant Cell Tiss. Org. Cult.* 74:73–80.

Xing, S.H., Guo, X.B., Wang, Q., Pan, Q.F., Tian, Y.S., Liu, P., Zhao, J.Y., Wang, G.F., Sun, X.F., Tang, K.X. 2011. Induction and flow cytometry identification of tetraploids from seed-derived explants through colchicine treatments in *Catharanthus roseus* (L.) G. Don. *J. Biomed. Biotechnol.* 2011:793198. doi: 10.1155/2011/793198.

Xu, M., Dong, J. 2005. Elicitor-induced nitric oxide burst is essential for triggering catharanthine synthesis in *Catharanthus roseus* suspension cells. *Appl. Microbiol. Biotechnol.* 67:40–44.

Xu, M., Dong, J. 2007. Involvement of nitric oxide signaling in mammalian Bax-induced terpenoid indole alkaloid production of *Catharanthus roseus* cells. *Sci. China C Life Sci.* 50:799–807.

Yang, L., Zou, H., Zhu, H., Ruppert, M., Gong, J., Stöckigt, J. 2010. Improved expression of His(6)-tagged strictosidine synthase cDNA for chemo-enzymatic alkaloid diversification. *Chem. Biodivers.* 7:860–870.

Yang, S.O., Kim, S.H., Kim, Y., Kim, H.S., Chun, Y.J., Choi, H.K. 2009. Metabolic discrimination of *Catharanthus roseus* calli according to their relative locations using ^{1}H-NMR and principal component analysis. *Biosci. Biotechnol. Biochem.* 73:2032–2036.

Yao, K., De Luca, V., Brisson, N. 1995. Creation of a metabolic sink for tryptophan alters the phenylpropanoid pathway and the susceptibility of potato to *Phytophthora infestans*. *Plant Cell* 7:1787–1799.

Yerkes, N., Wu, J.X., McCoy, E., Galan, M.C., Chen, S., O'Connor, S.E. 2008. Substrate specificity and diastereoselectivity of strictosidine glucosidase, a key enzyme in monoterpene indole alkaloid biosynthesis. *Bioorg. Med. Chem. Lett.* 18:3095–3098.

Ylinen, M., Suhonen, P., Naaranlahti, T., Lapinjoki, S.P., Huhtikangas, A. 1990. Gas chromatographic-mass spectrometric analysis of major indole alkaloids of *Catharanthus roseus*. *J. Chromatogr.* 505:429–434.

Yoder, L.R., Mahlberg, P.G. 1976. Reactions of alkaloid and histochemical indicators in laticifers and specialized parenchyma cells of *Catharanthus roseus* (Apocynaceae). *Am. J. Bot.* 63:1167–1173.

Yu, J., Yuan, S., Pang, H., Zhang, X., Jia, X., Tang, Z., Zu, Y. 2010. Distribution and accumulation of vindoline, catharanthine and vinblastine in *Catharanthus roseus* cultivated in China. *Zhongguo Zhong Yao Za Zhi* 35:3093–3096.

Zárate, R., Bonavia, M., Geerlings, A., van der Heijden, R., Verpoorte, R. 2001. Expression of strictosidine β-D-glucosidase cDNA from *Catharanthus roseus*, involved in the monoterpene indole alkaloid pathway, in a transgenic suspension culture of *Nicotiana tabacum*. *Plant Physiol. Biochem.* 39:763–769.

Zarate, R., Verpoorte, R. 2007. Strategies for the genetic modification of the medicinal plant *Catharanthus roseus* (L.) G. Don. *Phytochem. Rev.* 6:475–491.

Zhang, H., Hedhili, S., Montiel, G., Zhang, Y., Chatel, G., Pré, M., Gantet, P., Memelink, J. 2011. The basic helix-loop-helix transcription factor CrMYC2 controls the jasmonate-responsive expression of the ORCA genes that regulate alkaloid biosynthesis in *Catharanthus roseus*. *Plant J.* 67:61–71.

Zhao, J., Verpoorte, R. 2007. Manipulating indole alkaloid production by *Catharanthus roseus* cell cultures in bioreactors: From biochemical processing to metabolic engineering. *Phytochem. Rev.* 6:435–457.

Zhou, M.L., Hou, H.L., Zhu, X.M., Shao, J.R., Wu, Y.M., Tang, Y.X. 2011. Soybean transcription factor GmMYBZ2 represses catharanthine biosynthesis in hairy roots of *Catharanthus roseus*. *Appl. Microbiol. Biotechnol.* 91:1095–1105.

Zhou, M.L., Shao, J.R., Tang, Y.X. 2009. Production and metabolic engineering of terpenoid indole alkaloids in cell cultures of the medicinal plant *Catharanthus roseus* (L.) G. Don (Madagascar periwinkle). *Biotechnol. Appl. Biochem.* 52(Pt 4):313–323.

Zhou, M.L., Zhu, X.M., Shao, J.R., Wu, Y.M., Tang, Y.X. 2010. Transcriptional response of the catharanthine biosynthesis pathway to methyl jasmonate/nitric oxide elicitation in *Catharanthus roseus* hairy root culture. *Appl. Microbiol. Biotechnol.* 88:737–750.

Zhou, M.L., Zhu, X.M., Shao, J.R., Wu, Y.M., Tang, Y.X. 2012. An protocol for genetic transformation of *Catharanthus roseus* by *Agrobacterium rhizogenes* A4. *Appl. Biochem. Biotechnol.* 166:1674–1684.

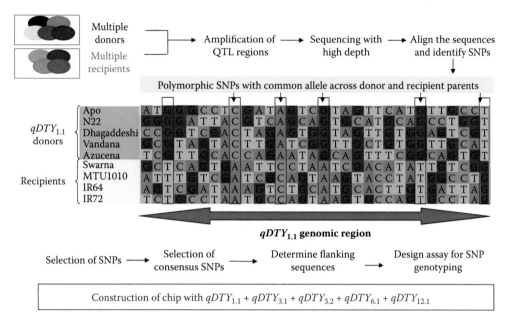

FIGURE 2.6
SNP marker design for drought grain yield QTLs.

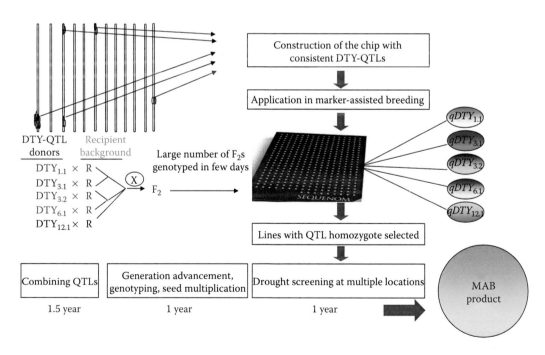

FIGURE 2.7
Fast-track MAB program using SNP genotyping platforms.

FIGURE 3.2
Cobs of Vivek QPM 9.

FIGURE 3.3
Cobs of Vivek QPM 21.

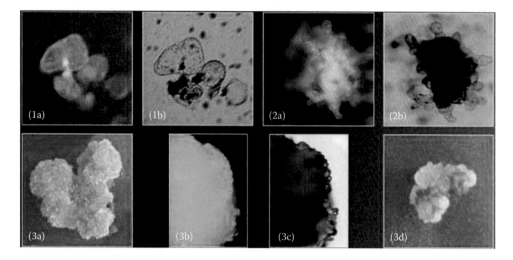

FIGURE 5.2
The usefulness of *gfp* for monitoring hybrid cell regeneration in somatic hybrids *Solanum tuberosum* (2x) 178/10 + *S. chacoense* (2x): (1a) hybrid cell expressing gfp (UV); (1b) the same cell under light microscopy (LM); (2a) cell colony with GFP (UV); (2b) the same colony under LM; (3a) hybrid callus with vigorous growth; (3b) part of same callus with GFP (UV); (3c) same as (3b) under LM; (3d) *S. chacoense* callus without gfp (control, nontransgenic) at the same age as (3a).

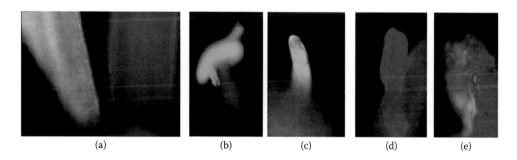

FIGURE 5.3
Examples of gfp expression in putative transgenic plants regenerating on kanamycin containing media: (a) potato dihaploid genotype 178/10, *left* transgenic leaf and *right* control; (b) very young shoot with GFP cv. Baltica; (c) young shoot with GFP cv. Desiree; (d) a putative transgenic shoot without GFP cv. Delikat; (e) young shoot of *Solanum chacoense* showing chimeric tissues.

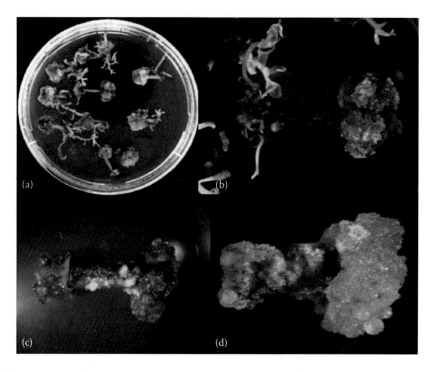

FIGURE 5.5
The effect of *Agrobacterium* strain and kanamycin selection on delaying organogenesis in cultivar Baltica, after 8 weeks of internodes culture: (a) the general view of regeneration in the control; (b) a detail of a, an internode regenerating in a polar manner callus and shoots; (c) only callus develops after cocultivation with *Agrobacterium tumefaciens* LBA4404; (d) callus formation after cocultivation with *A. tumefaciens* EHA105; the delay in shoot regeneration is to be observed in agroinfected internodes.

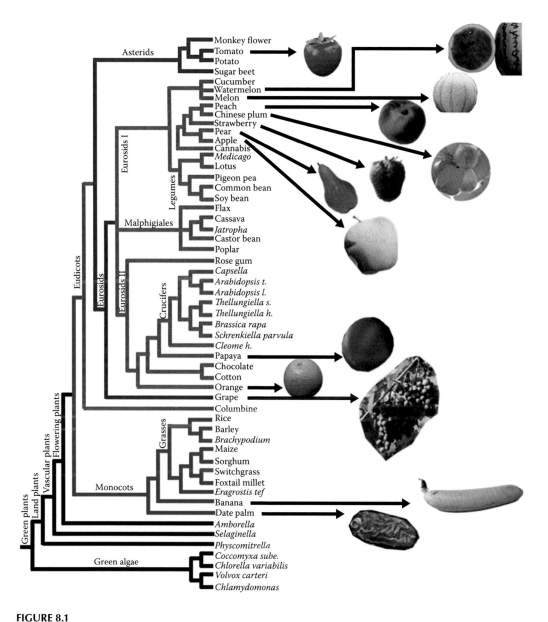

FIGURE 8.1
Phylogenetic tree of the plant genomes up to 14 April 2013. Pictures on the right side of the tree represent the sequenced fruit genomes. (Modified from http://genomevolution.org/wiki/index.php/Sequenced_plant_genomes, accessed on 14th April 2013.)

FIGURE 12.2

Transgenic ornamentals with flower color variation. (a) Color modification of *Torenia hybrida* cv. Summerwave Blue. (b) Color modification of *Lobelia erinus*. (c) Carnations incorporated with cyanidin-based pigments. (d) Transgenic carnation expressing F3′5′H and petunia DFR genes. (e) Cosuppression of the F3′5′H gene in torenia cv. Summerwave Blue produced pink flowers. Overexpression of a torenia F3′H gene generated darker pink flowers. (f) Orange petunia producing pelargonidin pigment. (g) A pale yellow petunia expressing a *Lotus japonica PKR* gene. (h) Morning glory flowers with natural transposons. *Ipomoea purpurea* (CHS-D::Tip100), Ipomoea nil (DFR-B::Tpn1).

FIGURE 12.5
Genetically engineered "blue rose."

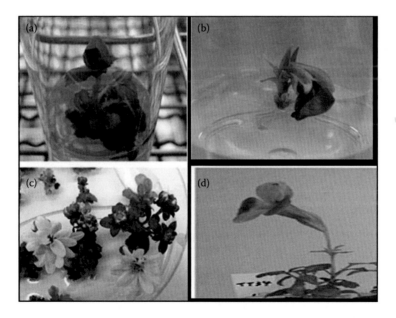

FIGURE 12.6
Early flowering of transgenic ornamentals. (a) Rose, (b) carnation, and (c) chrysanthemum expressing *Arabidopsis FT* gene. (d) Torenia expressing the *FT* gene flowering in low-level, indoor light conditions.

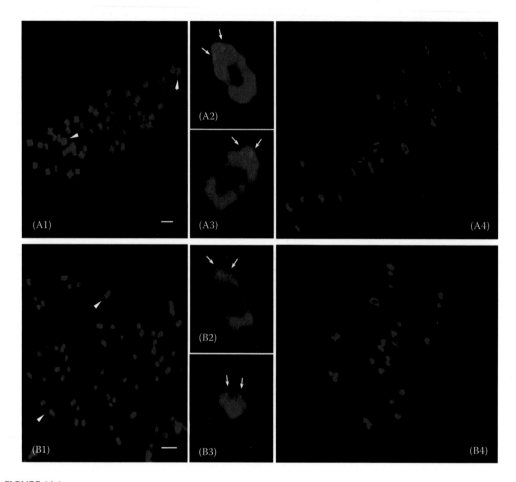

FIGURE 14.1
Eight chromosome-specific BAC clones of cotton identification and chromosomal and subchromosomal location by FISHing to *G. hirsutum* NTs. (From Wang, K. et al., *Theor Appl Genet*, 113, 73–80, 2006c.) **A1** to **H1** show the hybridization of BAC probes (Table 1) to eight chromosomes in mitotic cells to identify chromosome-specific BAC clones, respectively: **(A1)** BAC clone 104O10, **(B1)** 62K03, **(C1)** 98H10, **(D1)** 35J07, **(E1)** 14G14, **(F1)** 59B08, **(G1)** 37F17, and **(H1)** 50D03. Arrowheads point to the dual chromosome-specific FISH signals (*red*). All bars are 5 μm. **A2**, **A3** to **H2**, **H3** show the chromosomal and subchromosomal locations of FISH sites. *Arrows* indicate the dual FISH signals (*red*). All the figures of IVs and bivalents were derived from the integrated metaphase cell images. For example, **A4** and **B4** show the integrated cell images of **A2** and **B3**, respectively.

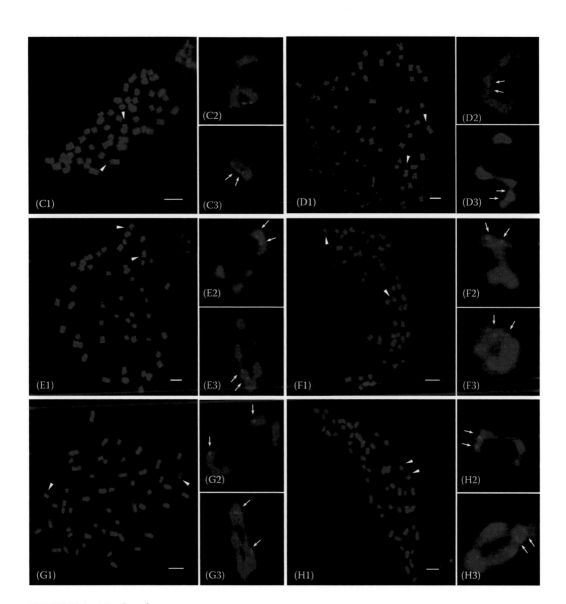

FIGURE 14.1 (Continued)

FIGURE 15.6
Poplar plantation.

FIGURE 15.7
Pulp pine plantation.

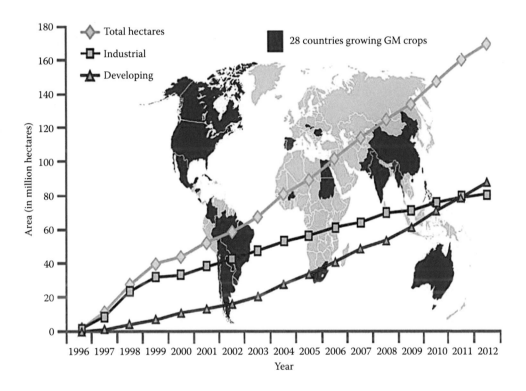

FIGURE 16.3
Global spread of GM crops, 1996–2012. (From James, C., Global Status of Commercialized Biotech/GM Crops: 2012. ISAAA Brief No. 44. ISAAA, Ithaca, NY.)

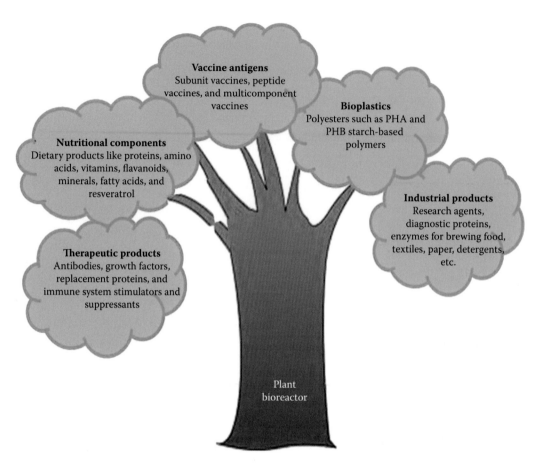

FIGURE 18.1
Plant bioreactor tree showing biotechnological advances.

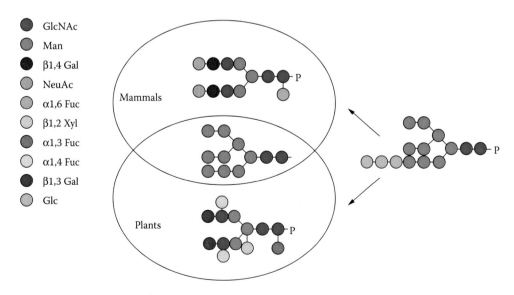

FIGURE 18.3
Mammalian and plant glycan structures.

11

Omics of Secondary Metabolic Pathways in Withania somnifera *Dunal (ashwagandha)*

Neelam S. Sangwan, PhD; Laxmi N. Misra, PhD;
Sandhya Tripathi, PhD; and Amit K. Kushwaha, PhD

CONTENTS

11.1 Introduction 385
11.2 *Withania somnifera*, the Indian Ginseng: An Introduction 386
    11.2.1 Advances in the Chemistry and Biology of *W. somnifera* 386
        11.2.1.1 Advances in Metabolomics 387
    11.2.2 Developments in Genomics, Proteomics and Transcriptomics 394
        11.2.2.1 Genomic Approaches for Gene Discovery in *W. somnifera* 394
        11.2.2.2 Proteomic Advances in *W. somnifera* 396
        11.2.2.3 Advances in Genomic and Transcriptomic Approaches 398
11.3 Conclusions and Perspectives 402
Acknowledgment 403
References 404

11.1 Introduction

The term *genomics*, coined by Tom Roderick, is not conclusive but usually deals with whole genome analysis of organisms. Genomics can be classified into functional and structural genomics. While structural genomics deals with the arrangement of nucleotides in a genome, functional genomics entails their role in the phenotypes, that is, the expression of the genes. Other fields under the regimen of genomics are cognitive genomics, epigenomics, metagenomics, and comparative genomics based on the subject and direction of the study. Transcriptomics is the study of the set of genes expressed in a particular cell, tissue, or organ at a particular stage or time at RNA level. Sometimes it is also included in the area of genomics. It is directly related to the proteome and metabolome, as it defines the array of proteins and metabolites present in a cell at a given time. Proteomics studies the set of proteins expressed/present at a given time in the cell. Proteins are among the most dynamic biomolecules of the cell and changes in the array of proteins result directly in the function and phenotype of the cell at a particular point. Often described as the next step to genomics, proteomics seems to be more complicated as there are fewer proteins than genes. The metabolome is the collection of all the low-molecular-weight compounds other than primary metabolites present in an organism, while their comprehensive analysis is termed as metabolomics (Schauer and Fernie, 2006). Metabolomics is the progression of the omics approaches from DNA, RNA and proteins to molecular prints that cellular processes leave behind.

11.2 *Withania somnifera*, the Indian Ginseng: An Introduction

Withania somnifera, or winter cherry, is an important plant because of its medicinal properties. The angiospermic genus *Withania* is a member of the family Solanaceae and consists of 23 species. The plant, also known as ashwagandha or ginseng, is reputed for its medicinal importance and has been extensively used in Indian traditional medicinal systems like ayurveda and unani. The main pharmacological attributes of the plant are described as physiological and metabolic restoration; protection against arthritis, aging, and cancer; cognitive function improvement in geriatric states; and recovery from neurodegenerative disorders (Singh et al., 1982; Budhiraja et al., 2000; Jayaprakasam and Nair, 2003; Jayaprakasam et al., 2004; Riaz et al., 2004; Mondal et al., 2010; Grover et al., 2012). Due to its pharmacological value, the plant has been an essential constituent of over 100 traditional medicine formulations, making it equivalent to Chinese ginseng (Sangwan et al., 2004). The important medicinal properties have been attributed to the metabolome of different classes of withanolides present exclusively in *W. somnifera,* hence they are associated with the name of the plant. These are derived from steroidal lactones such as withaferin A, withanone, and withanolide A (Table 11.1 and Figure 11.1) and many others found in relatively smaller quantities (Sangwan et al., 2004). Besides other uses of the extract and plant parts, these bioactive phytochemicals are useful for the treatment of several diseases like arthritis, and as an antitumor, anti-inflammatory, anticancer, and antistress medicine (Mondal et al., 2010; Efferth and Greten, 2012). *W. somnifera* is widely distributed throughout the Indian states of Maharashtra, Rajasthan, Uttar Pradesh, Haryana, Tamil Nadu, Madhya Pradesh, and others (Tuli and Sangwan, 2010).

The medicinal properties of *W. somnifera* have rendered this plant a subject of constant scientific investigations and resulted in several studies and observations detailing the reason for the therapeutic marvels of the plant. Growth in the field of omics has led to an increase in omic studies of several important medicinal plants, including *W. somnifera.* Advances in these fields concerning *W. somnifera* are discussed below.

11.2.1 Advances in the Chemistry and Biology of *W. somnifera*

W. somnifera is a prominent medicinal plant of Indian medicinal systems and its chemical constituents have been comprehensively studied. Several types of secondary metabolites such as steroids and alkaloids have been reported. A number of alkaloids, withanolides, and sitoindosides (Table 11.1) have been isolated and reported from various parts of *Withania* species (Misra et al., 2005; Mirjalili et al., 2009, Tuli and Sangwan, 2010; Sabir et al., 2008, 2010, 2011, 2012; Mishra et al., 2012a,b). The withanolides are a group of naturally occurring C_{28}-steroidal lactones built on an intact or rearranged ergostane framework. The basic structure is designated as the withanolide skeleton, defined as 22-hydroxy ergostan-26-oic acid-26, 22-lactone (Figure 11.1). The distribution of withanolides is not confined to solanaceous plants, as these have been isolated from marine organisms and from members of other families such as Taccaceae, Fabaceae, and Lamiaceae, suggesting their wider occurrence and distribution. The first isolated compound of this group was withaferin A (4β,27-dihydroxy-1-oxo-5β,6β-epoxywitha-2,24-dienolide). The novel structure and interesting bioactivity of withaferin A initiated a thorough scientific study of the plant leading to isolation of various compounds with similar structures. Besides the principle bioactive compounds, withanolides, several other constituents—namely, withanamides from fruits, ashwagandhanolides from roots, and alkaloids from leaves—are also reported (Chaurasiya et al., 2008). The tentatively identified alkaloidal composition of the plant includes nicotine, somniferine, somniferinine, withanine,

TABLE 11.1

Secondary Metabolites Isolated/Identified from *W. somnifera*

S. No.	Secondary Metabolites	References
1	Withaferin A	Lavie et al., 1965
2	Withanolide D	Lavie et al., 1968
3	4β-Hydroxy-1-oxo-5β,6β-epoxywitha-2,24-dienolide; 4β,20α,(R)dihydroxy-1-oxo-5β,6β-epoxywith-2-enolide; 4β-hydroxy-1-oxo-5β,6β-epoxywith-2-enolide	Kirson et al., 1970
4	5α,17α-Dihydroxy-1-oxo-6α,7α-epoxy-22R-witha-2,24-dienolide; 5α,17α-dihydroxy-1-oxo-22R-witha-2,6,24-trienolide; 5α,27-dihydroxy-1-oxo-6α, 7α-epoxy-22R-witha-2,24-dienolide; 4β,17α-dihydroxy-1-oxo-5β,6β-epoxy-22R-witha-2,24-dienolide; 17α,27-dihydroxy-1-oxo-22R-witha-2,5,24-trienolide; 7α,27-dihydroxy-1-oxo-22R-witha-2,5,24-trienolide; 7α,27-dihydroxyl-1-oxo-22R-witha-2,5,24-trienolide; 1α,3β,5α-trihydroxy-6α,7α-epoxy-22R-with-24-enolide; 4β-hydroxy-5β,6β-epoxy-1-oxo-22R-witha-2,14,24-trienolide	Kirson et al., 1971
5	Withanolide G, withanolide H, withanolide I, withanolide J, withanolide K, withanolide L, withanolide M	Glotter et al., 1973
6	Withanolide E, withanolide F, withanolide S, withanolide P	Glotter et al., 1977
7	(20R,22R)-14α,20αF-dihydroxy-1-oxowitha-2,5,16,24-tetraenolid	Velde and Lavie, 1982
8	Withanolide C	Bessalle and Lavie, 1992
9	Calystegins	Bekkouche et al., 2001
10	Withanosides I–VII	Masuda et al., 2001
11	Withanamides and 24,25-dihydrowithanolide VI	Jayaprakasham et al., 2004
12	27-Deoxy-16-en-withaferin A; 2,3-dihydro-3β-hydroxy withanone; 27-acetoxy-3-oxo-witha-1,4,24-trienolide; 2,3-dihydro withanone-3β-O-sulfate	Misra et al., 2005
13	Iso-withanone; 6α,7α-epoxy-1α,3β, 5α-trihydroxy-witha-24-enolide	Lal et al., 2006
14	16β-Acetoxy-6α,7α-epoxy-5α-hydroxy-1-oxowitha-2,17(20),24-trienolide; 5,7α-epoxy- 6α,20α-dihydroxy-1-oxowitha-2,24-dienolide	Misra et al., 2008
15	Withanolide Z	Pramanick et al., 2008

pseudowithanine, tropine, pseudotropine, 3α-tigloyloxytropane, choline, cuscohygrine, *dl*-isopelletierine, and the new alkaloids anaferine and anhygrine. In addition, starch, reducing sugars, amino acids, hentriacontane, glycosides, dulcitol, withanicil, chlorogenic acids, calystegines, withanone, condensed tannin, and flavonoids were also identified from several parts of the plant (Tuli and Sangwan, 2010).

11.2.1.1 Advances in Metabolomics

The metabolic chemistry of *W. somnifera* has been extensively studied owing to its medicinal importance. Initially studies were focused on isolation and identification of biologically

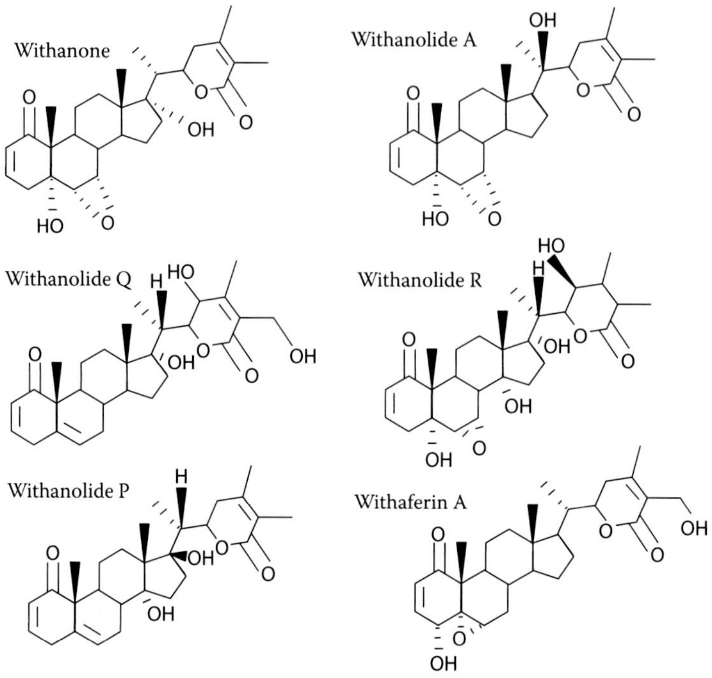

FIGURE 11.1
Structures of different withanolides.

active molecules and it was found to produce a plethora of diverse compounds: namely, withanolides, withanamides, and alkaloids (Table 11.1). This enormous diversity of secondary metabolites could be deemed responsible for the variety of pharmacological activities ascribed to the plant. With the advances in understanding of metabolic chemistry along with developments in the field of metabolomics, it became pertinent to study the metabolome of the plant as a whole in order to understand the qualitative and quantitative variations of the metabolome presented by genotypic and phenotypic variations. Further, the analysis of the total metabolome of a plant is vital to understand the complex biochemistry of the plant in question. Significant advances in analytical techniques like nuclear magnetic resonance (NMR), GC–MS, and LC–MS have presented new opportunities for metabolomics research to simultaneously explore a large number of metabolites quantitatively and qualitatively (Elyashberg et al., 2009). Comprehensive metabolomic study is required not only to corroborate the relation between chemical complexity and pharmacology, but also to decipher biochemical pathways by correlating metabolites to genes. *W. somnifera* is known for the presence of withanolides; therefore understanding its chemistry is extremely important to explore the metabolome and the major pathways involved in its metabolism. Some of the salient features of the withanolides are detailed here.

11.2.1.1.1 Chemical Entity of Withanolides

The term withanolide has been given to a group of compounds characterized by a C_{28} basic steroidal skeleton with a nine-carbon side chain in which C-22 and C-26 are oxidized to form a δ-lactone ring (Figure 11.1). The common names in current use for such structures are withanolides or withasteroids, with a basic skeleton structure of 22-hydroxy ergostan-26-oic acid-26,22-lactone (Ray and Gupta, 1994; Sangwan et al., 2004). Thus, withanolides are a group of naturally existing C_{28} steroidal lactones built on an intact or rearranged ergostane framework. A 1-keto or hydroxy function in ring A is a general feature of this class of compounds (Glotter, 1991). From the biogenetic point of view, the withanolides can be considered to have a cholestane-type structure with an extra methyl group at C-24 and various oxygenated groups or double bonds placed at different sites of the skeleton (Budhiraja et al., 2000).

11.2.1.1.2 Isolation and Analysis of Withanolides

The screening of compounds of natural origin on the pharmaco-chemical basis has been the source of discovery of several therapeutic agents. The approach of random screening of natural compounds and extracts for the discovery of new bioactive molecules has been highly successful. A chemotaxonomic and target-directed approach has also played a crucial role in natural product research. These considerations often result in the discovery of molecules with desired biological characteristics. Even now, drugs from higher plants continue to occupy an important place in modern medicine and as neutraceuticals (Newman and Cragg, 2007; Misra et al., 2013). Further, chemical modifications of this parental structure often open up new ways for therapeutic agents with better potential (Ray and Gupta, 1994). Therefore, it is of immense importance to develop efficient methods for the extraction and isolation of secondary metabolites from plant tissues.

Available methods are used frequently and routinely for the isolation of secondary metabolites, though modifications are made depending on the type of chemical moieties that are to be isolated (Misra et al., 2005, 2008). The most popular method of isolating bioactive molecules includes bioassay-guided extraction, followed by fractionation, purification, and structure elucidation of compounds. Alternatively, compounds can also be isolated following the procedure of solvent extraction, isolation, and structure elucidation of major as well as minor phytochemicals. This method sometimes yields unexpected biologically active phytochemicals from plants and is more likely to produce new results (Misra et al., 2005, 2008). Another method is a modern tool-based approach involving rapid, automated screening systems to identify bioactive compounds. Such high-throughput methods screen large libraries of molecules and their derivatives, using machine-based screening assays reflecting a wide array of metabolomes (Misra et al., 2005, 2008, 2012).

Fresh parts of *W. somnifera* can also be used for the isolation of withanolides, with the advantages of higher recovery and understanding the phytochemical composition of the tissue in a state of total metabolic arrest (Sangwan et al., 2005). Extraction of the fresh herb in gradient methanolic solution results in good yields of withanolide constituent of the tissues used for isolation (Sangwan et al., 2005; Chaurasiya et al., 2008). Due to the highly complex mixtures of withanolides present in the pooled fractions, each pooled fraction needs to be further column chromatographed. In some cases, where amounts are smaller, purification of the withanolides by thin layer chromatography (TLC) in preparatory mode is attempted, to obtain a pure form of the compound. However, TLC-based methods compromise the overall yield of the individual metabolites, as there is some loss involved with the isolation procedure (Misra et al., 2005, 2008). Extracts of fresh as well as dry material are processed for separation by column and TLC using a procedure similar to the one described above. The brown chloroform extract showed prominent spots on the TLC plate at R_f ranging from 0.37 to 0.75 along with many minor spots. But in this case, more polar compounds are not extracted as compared to the methanol extract. Therefore, the separation process becomes a bit simpler (Chaurasiya, 2007). The chloroform extract is adsorbed on silica gel to make a dried slurry and is chromatographed over the column of silica gel. The adsorbed extract is loaded in a glass column using silica gel as the stationary phase and *n*-hexane as the mobile phase. The elution is carried out in *n*-hexane and EtOAc with solvent gradient polarity from *n*-hexane to EtOAc. The polarity is increased by sequentially adding 5%, 25%, 50%, and 75% ethyl acetate in *n*-hexane, then EtOAc, and finally 5% and 10% methanol is added in ethyl acetate. Hundreds of small fractions (150 ml each) are collected and pooled into several major fractions based on their compatibility on the TLC pattern as described above for the isolations from dried plant material.

11.2.1.1.3 Occurrence of Withanolides in Different Parts of W. somnifera

Almost all parts of ashwagandha contain withanolides. Although the types of withanolides are distinct for each part, some structural commonality may exist. Leaves, stems, and berries (including seeds) are the most prominent producers of withanolides and thus have the potential for the generation of withanolides.

Although the roots are the main subject of prescriptions in ayurvedic and folklore systems of medicines, their chemical investigation has been limited to the identification of a few steroids and some alkaloids (Ray and Gupta, 1994). Recent investigations (Misra et al., 2008, 2012) have resulted in the characterization of several withanolidal compounds from the roots (Figure 11.1 and Table 11.1). Two new withanolides, 16β-acetoxy-6α,7α-epoxy-5α-hydroxy-1-oxowitha-2,17(20),24-trienolid and 5α,7α-epoxy-6α,20α-dihydroxy-1-oxowitha-2,24-dienolide, as well as seven known withanolides, 2,3-dihydro-3-hydroxywithaferin A (viscosalactone B, withaferin A, withanolide D, withanolide B, withanolide A, 27-hydroxy withanolide B, and 27-hydroxy withanolide A) were shown to occur in the roots of *W. somnifera*. Among the known compounds, viscosalactone B was reported in the roots of *W. somnifera* for the first time (Misra et al., 2008). The structures of these compounds were elucidated by spectroscopic methods: namely, UV, IR, ¹H NMR, ¹³C NMR, 2D NMR, mass spectra, and chemical transformations (Misra et al., 2008; Elyashberg et al., 2009). Recently, a peculiar dimer sulfide has been isolated from the roots of *W. somnifera* showing inhibition against cancer cell lines (Subbaraju et al., 2006).

Phytochemical studies on the stem bark have shown five new withanolides—namely, withasomnillide, withasomniferanolide, somniferanolide, somniferwithanolide, and somniwithanolide—on the basis of spectroscopic techniques and chemical means (Ahmad and Dough, 2002; Namdeo et al., 2011). The leaves of *W. somnifera* also possess two major compounds, identified as withanone and withaferin A (Lavie et al., 1965; Kirson et al., 1970, 1971; Abraham et al., 1975; Glotter et al., 1977; Rahman et al., 1993). The minor withanolides isolated from this group are 27-deoxy-17-hydroxywithaferin A and withanolides (Dhalla et al., 1961; Singh and Kumar, 1998). Lavie et al. (1965) isolated a unique structure having both 20-H and 20-OH withanolides (Table 11.1). The major withanolides of this group are withaferin A (0.825% dry leaves) and withanolide D (0.047% dry leaves). Furthermore, the other minor withanolides are 27-deoxy withaferin A, 2,3-dihydro withaferin A, withanolide P, withanolide F, dihydro-27-deoxy-withaferin A, and dihydro-withanolide D. Atta-ur-Rahman et al. (1993) also reported several known withanolides along with a new dienone withanolide from ashwagandha. The new withanolide 3α-methoxy-2,3-dihydro-27-deoxywithaferin A was isolated from leaves of *W. somnifera* along with withaferin A, 27-deoxywithaferin A, 2,3-dihydro withaferin A, and 3β-methoxy-2,3-dihydro-withaferin A (Anjaneyulu and Rao, 1997). Recently, identification of very interesting withanolides with unusual chemical features has been reported in *W. somnifera* leaves (Misra et al., 2005). Among the new compounds, a rare 3-O-sulfate group with saturation in ring A and an unusual 1,4-dien-3-one group with a rare C-16 double bond were reported. The structures of all the compounds were elucidated by spectroscopic methods and chemical transformation

(Misra et al., 2005). Furmanowa et al. (2001) isolated withaferin A with immunosuppressive activity from the shoot-tip of the plant. Jayaprakasam and Nair (2003) isolated several cyclooxygenase-2-enzyme (COX-2) inhibitory and antitumor withanolides.

In *W. somnifera*, the fruits also have substantial amounts (10%) of fatty acid oil (60%) rich in linoleic acid (Anonymous, 1976; Bhakre et al., 1993; Pachori et al., 1994) and the seeds also contain withanolides (Kundu, 1976). Ahmad and Dough (2002) also isolated two withanolides along with some minor amounts of coumarins and triterpenoids from the fruits of *W. somnifera*. Our investigations on *W. somnifera* fruits have afforded some novel withanolides of chemical interest. The chloroform extract of the fresh berries of *W. somnifera* was found to have stigmasterol, its glucoside, withanone, and 27-hydroxy withanolide A, along with two new withanolides: namely, iso-withanone and 6α,7α-epoxy-1α,3β,5α-trihydroxy-witha-24-enolide (Lal et al., 2006).

11.2.1.1.4 Structural Characterization of Withanolides

Spectroscopic, chemical, and chromatographic methods are routinely used for the identification of natural products and are also employed for withanolides. In the case of withanolides, there are several functional oxygen groups that are identifiable in the infrared spectrum. In particular, the hydroxyl 2-en-1-one and α,β-unsaturated δ-lactone give characteristic bands quite useful for their identification. The bands can be seen at 3460 cm^{-1} for OH and at 1660–1710 cm^{-1} for α,β-unsaturated δ-lactone and 2-en-1-one functional groups in the IR spectrum of the typical withanolides (Misra et al., 2005, 2008; Lal et al., 2006). NMR spectroscopy is a powerful tool for identifying nuclei based on the interaction of electromagnetic fields with a sample in the applied magnetic field. Using the ^{1}H NMR technique, it is quite possible to recognize the presence of a withanolide skeleton in an extract or fraction or in a pure compound. This technique is also very helpful for determination of the stereochemistry of different chiral centers in withanolides. The most important is the proton at C-22, which was keenly observed by us and was later formulated to predict exact stereoisomers at C-17 by its chemical shift value (Lal et al., 2006). The coupling constant and its splitting pattern are often regarded as the withanolide fingerprint. It appears as a double triplet in the typical withanolide, wherein the C-20 and C-23 are unsubstituted, but it changes to a double doublet when any of these positions is substituted.

Identification of withasteroids in pharmaceutical preparations, extracts, and fractions is a great challenge for analytical chemists. Therefore, by recording the ^{1}H NMR spectra of several withanolides, we have concluded that it could be useful to diagnose the presence of withanolides even in the complex mixture by measuring the ^{1}H NMR spectrum (Misra et al., 2008). This method will enable us to prove the authenticity of the claim about *W. somnifera* in herbal markets. Since recording ^{1}H NMR does not cause any loss in the amount of the sample, and requires a small quantity, little time, and a small amount of solvent, it could be considered as an effective tool for the identification of withasteroids in *W. somnifera* extracts or fractions. As most (>95%) withasteroids contain α,β-unsaturated ketone in ring A, it is imperative to locate prominent signals responsible for this group in the ^{1}H NMR spectrum of the fractions of the herbal preparations. The 2,3-unsaturated ketone in ring A, epoxide in ring B, and δ-lactone in ring E give characteristic signals in ^{1}H NMR spectra that are extremely helpful in the detection of such a system in the molecule. The above characteristic signals can easily ascertain the presence of *W. somnifera* withasteroids and its types even in commercial herbal extracts and preparations.

^{13}C NMR also gives structural information, but its use for the detection of withanolides in a mixture or an extract is not possible, in contrast to the proton NMR. However, the distortionless enhancement by polarization transfer (DEPT) and insensitive nuclei enhanced by polarization transfer (INEPT) techniques add to the knowledge of finding the methyl, methylene, methine, and tetrasubstituted carbons in an organic molecule (Misra et al., 2005, 2008; Lal et al., 2006). In the EI mass of withanolides, a base peak at m/z 125 formed by the fission of the C-20/C-22 bond is considered as the diagnostic tool for withanolides having an α,β-unsaturated-δ-lactone moiety. Similarly, the hydroxyl group at C-20 facilitates the cleavage of the C-17/C-20 bond and gives rise to peaks at m/z 152 and m/z 169. It has also been observed that the base peak at m/z 125 may also be formed from ring A of the withanolide with the 5-hydroxy-2-en-1-one system through a McLafferty rearrangement and C-5/C-6 bond cleavage (Misra et al., 2005, 2008; Lal et al., 2006). Since the mass spectrum contains numerous ions, their identification is rather difficult.

Further, structural analysis of *W. somnifera* has been performed to elucidate the structure of its new components. For example, ashwagandholide, a new dimeric natural product, was isolated by Subbaraju et al. (2006) and its molecular formula was determined as $C_{56}H_{78}O_{12}S$ on the basis of the molecular ion observed at m/z 975.5285. The structure of this compound was elucidated using 2D NMR data and information obtained from comparison of experimental spectra with the structures and spectra of related molecules (Table 11.2). The StrucEluc computer-aided structure elucidation (CASE) program was used in common mode to show that the processing time and the number of generated structures would be unmanageable. Therefore a fragment search using ^{13}C NMR was performed and 5524 fragments were found in the fragment library. The displayed fragments were ranked in decreasing order of the number of carbon atoms. It was confirmed that a fragment is

TABLE 11.2

Description of Deposited Entries for Chemical Entities in *W. somnifera* in PubChem

Substance	Structure	SID	Source
Ajagandha	Not available	13533660	ChemID plus (ZC 80617000)
Ajagandha	Not available	135338547	ChemID plus (0090147436)
Ajagandha	Not available	135360432	ChemID plus (ZC80600000)
Ajagandha	Not available	135339095	ChemID plus (ZC80616750)
Ajagandha	Not available	643255	NIAID (105974)
NSC329512	Not available	458986	DTP/NCI (329512)
Withaferin A	Available	26759748 [CID:16760705]	Calbiochem (681535)
Shwagandholine NIOSH/ ZC7900000	Not available	135344378	ChemID plus (ZC 790000000)
GNF-Pf-105	Available	131334592 [CID:45489105]	GNF/Scripps Winzeler lab (GNF-pf-105)
Withanolides, D05435800	Not available	135358834	ChemID plus (D054348000)
Withanolides, D054358	Not available	53837785	Comparative Toxicogenomics Database (D054358)
6-[(1S)((14S,2R,11R)-11,14-Dihydroxy-2-methyl-3-oxotetra cyclo[8.7.0.0<2,7>.0<11,15>] heptadeca-5,7-dien-14-yl) hydroxymethyl]-3,4-dimethyl-5H-6-hydropyran-2-one	Available	640800 [CID:482272]	NIAID (097803)

a substructure of the structure and its carbon chemical shifts are very close to the values measured for the full structure.

Earlier attempts to comprehensively study the metabolite status were confined to the withanolides (Chaurasiya et al., 2007, 2008). Later, focus was shifted toward understanding of the whole metabolome in a particular tissue at a specific time. In a similar report, metabolic profiles of crude extract of leaves and roots of *W. somnifera* were produced by NMR and chromatographic (HPLC and GC–MS) techniques (Chatterjee et al., 2010). A total of 62 metabolites from leaves and 48 from roots were identified as primary and secondary metabolites. Twenty-nine of these were fatty acids, organic acids, amino acids, sugars, and sterol-based compounds common to the two tissues. Eleven biologically active sterol–lactones were also identified. Twenty-seven of the identified metabolites were quantified. Comparative analysis of the two profiles showed highly significant qualitative and quantitative differences between the leaf and root tissues with respect to the secondary metabolites (Chatterjee et al., 2010). In another report, to study the metabolic changes in growth and development of fruits, profiling was done at different stages by NMR spectroscopy (Sidhu et al., 2011). As a result, 17 metabolites were characterized and quantified from different stages of growth. Quantitative and qualitative analysis of stage-specific metabolite profiles of fruits identified specific stages when fruits could be harvested to obtain particular biomolecules for desirable bioactivity. Concurrently, *W. somnifera* leaves, stems, and roots collected from six different regions in India were metabolically profiled using ^{1}H NMR spectroscopy (Namdeo et al., 2011). This study demonstrated that the leaves have the richest metabolic diversity, harboring amino acids, flavonoids, lipids, organic acids, phenylpropanoids, and sugars along with the main secondary metabolites of the plant, the withanolides. Further, the ratio of two groups of withanolides—4-OH-5,6-epoxy withanolides (withaferin A-like steroids) and 5-OH-6,7-epoxy withanolides (withanolide A-like steroids)—was found to be a key discriminating feature of leaf samples from different origins (Table 11.1; Chaurasiya et al., 2009).

Thereafter, in another attempt at metabolic characterization, the metabolic profile of *W. somnifera* fruits was produced and analyzed by LC–MS (Bolleddula et al., 2012). The study resulted in the identification of 62 metabolites including 32 withanamides, 22 withanolides, 3 steroidal saponins, 2 lignanamides, feruloyl tyramine, methoxy feruloyl tyramine, and a diglucoside of hydroxyl palmitic acid. Six new withanolides, a new hydroxy fatty acid diglucoside, and several known compounds in the extract were also identified (Bolleddula et al., 2012).

11.2.1.1.5 *Metabolome Diversity*

A wide germplasm collection representing metabolome, genome, and proteome variability has been investigated (Chaurasiya et al., 2009; Dhar et al., 2006). The collected plants from different geographic locations were found to be morphologically and chemically different. As a part of systematic studies on phytochemical variability in *W. somnifera* in qualitative as well as quantitative terms, several promising accessions were screened for physiological yield and chemical variability (Dhar et al., 2006). Based on this, *W. somnifera* genotypes are important and valuable as plant resources of phytochemical and pharmaceutical interest. As part of our New Millennium Indian Technology Leadership Initiative (NMITLI) program, we have collected germplasm from many geographical locations in the country, analyzed it for various parameters, and generated a large body of information about genetic and isoenzymatic diversity in the species versus withanolide phytochemical variability (Chaurasiya et al., 2009). This information on metabolites is the key to genetic improvement programs and, for the development of conservation and management strategies for useful accessions and to avoid redundancy and screen elite plant types. Genomic and expressed molecular diversity was assessed through RAPD, and isoenzyme and phytochemical (withanolidal) profiling using the methods developed in our laboratory. The observed differences among accessions were scored and utilized for genetic diversity analysis, chemical constituent/withanolide variability analysis, similarity matrix analysis, and dendrogram construction. The results revealed a rational geographic relationship among the genotypes of W. *somnifera* based on the metabolome, proteome, isozyme, and randomly amplified polymorphism (Chaurasiya et al., 2009; Dhar et al., 2006).

11.2.2 Developments in Genomics, Proteomics and Transcriptomics

11.2.2.1 Genomic Approaches for Gene Discovery in W. somnifera

The main aim of functional genomics is to decode unknown genes and their functions. Though W. *somnifera* has been well characterized for its phytochemicals owing to its pharmaceutical value, extensive studies have not been carried out at genome and transcriptome levels. However, to understand the molecular biology of withanolide biosynthesis, several studies have been done in recent times. A number of genes involved in secondary metabolic pathways have been isolated individually from the plant and characterized for their functions *in vitro* (Sharma et al., 2007; Gupta et al., 2011, 2012a,b; Akhtar et al., 2012; Bhat et al., 2012; Goosens et al., 2003; Table 11.1 and Figure 11.2). Recently, Senthil et al. (2010) reported the generation of expressed sequence tags (ESTs) from leaves and roots of the plant for the first time. ESTs depict a quick gene complement of the organism expressed at a particular time in a particular tissue and could well be utilized to identify and isolate genes representing particular pathways (Figure 11.2 and Table 11.6). The study resulted in 1047 leaf cDNA and 1034 root cDNA clones representing 48.5% and 61.5% unique sequences. Later, the sequences were categorized for their putative functions. Cellular processes, and membrane and catalytic activities were among the most abundant categories within the biological processes, cellular components, and molecular function groups. Several candidate genes involved in the withanolide biosynthetic pathways were identified from these ESTs: namely, HMG CoA synthase, squalene epoxidase, and sesquiterpene cyclase from the MVA pathway and CDP-ME kinase from the MEP pathway (Gupta et al., 2013; Akhtar et al., 2012; Chaurasiya et al., 2012; Figure 11.2 and Table 11.3). The other genes putatively involved in withanolide biosynthesis were also marked such as cytochrome P450s and glycosyltransferases. After sequencing cDNA for root and leaf in the study, 1034 and 1047 ESTs were obtained, respectively, which were clustered to 230 contigs for root and 239 contigs for leaf, with 398 and 539 transcripts, respectively, considered as redundant. Additionally, at an E-value cutoff value of 1×10^{-14}, 1022 root ESTs showed matches above the value and 12 ESTs showed matches below it. On the other hand, in the case of leaf ESTs, 1038 transcripts were above the cutoff value and 9 EST sequences were below it. Further, matches of ESTs with different kingdoms—namely, eukaryote (39 root and 42 leaf), fungus/yeast (10 root and 8 leaf), animal (236 root and 171 leaf), prokaryote (25 root and 12 leaf) including bacteria and archaebacteria, and virus (3 root and 4 leaf)—were also reported (Senthil et al., 2010).

The genes that were found to be highly expressed in the study were universal stress protein with 11 ESTs, zinc finger protein with 9 ESTs, UDP-arabinose-4-epimerase with 9 ESTs, methionine synthase with 9 ESTs, cytochrome P450 with 7 ESTs, and UDP-adipose/xylulose synthase with 7 ESTs. Further, phosphoglucomutase, elongation factor-1-gamma chain, aspartyl protease heat shock protein, and xylulose-6-phosphate-1-dehydrogenase were expressed with 6 ESTs each.

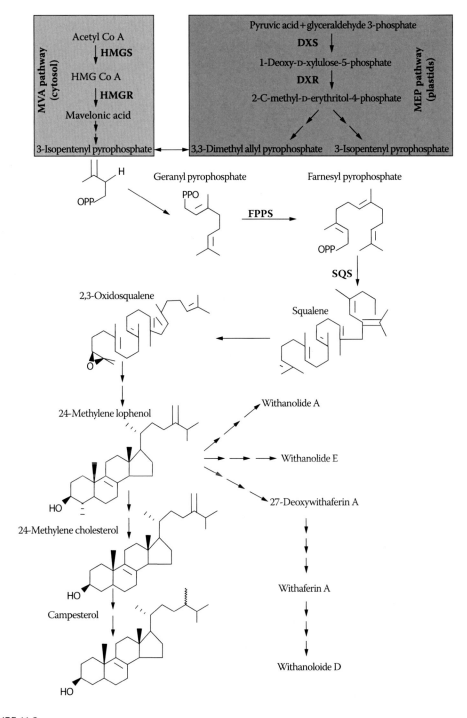

FIGURE 11.2

Proposed pathway for withanolide biosynthesis: putative branches synthesizing different withanolides are elaborated. Enzymes, characterized so far, are also depicted as hydroxymethylglutaryl-CoA synthase (HMGS), hydroxymethylglutaryl-CoA reductase (HMGR), farnesyl diphosphate synthase (FPPS), squalene synthase (SQS), 1-deoxy-D-xylulose 5-phosphate reductoisomerase (DXR), and 1-deoxy-D-xylulose-5-phosphate synthase (DXS). Multiple steps are represented by multiple arrows.

TABLE 11.3

Genes/Proteins Isolated and Characterized from *W. somnifera*

S. No.	Gene/Protein	References
1	3β-Hydroxy sterol glucosyltransferase gene	Sharma et al., 2007
2	3β-Hydroxy-specific sterol glucosyltransferase enzymes	Madina et al., 2007a
3	27β-Hydroxy glucosyltransferase enzyme	Madina et al., 2007b
4	Farnesyl diphosphate synthase (FPPS) gene	Gupta et al., 2011
5	1-Deoxy-ᴅ-xylulose-5-phosphate synthase (DXS) and 1-deoxy-ᴅ-xylulose-5-phosphate reductase (DXR) genes	Gupta et al., 2012
6	Squalene synthase gene	Gupta et al., 2013; Bhat et al., 2012
7	3-Hydroxy-3-methylglutaryl coenzyme A reductase gene	Akhtar et al., 2012

Later in 2011, Pal et al. reported the generation of 495 leaf ESTs and 76 root ESTs (Table 11.6) from the plant. Two of the leaf ESTs were then converted to the full-length sequences and identified as sterol methyltransferase and obstusifoliol 14α-demethylase. These two enzymes are thought to be involved in withanolide biosynthesis. A few other genes, such as 1-deoxy-D-xylulose-5-phosphate reducto-isomerase (DXR), hydroxymethylglutaryl-CoA reductase (HMGR), and glycosyltransferases, were also identified. Besides these reports, 742 ESTs have been submitted to the database from different workers to date (Figure 11.2 and Table 11.3).

11.2.2.2 Proteomic Advances in W. somnifera

Two-dimensional polyacrylamide gel electrophoresis (2D-PAGE) coupled with matrix-assisted laser desorption/ionization/time-of-flight mass spectrometry (MALDI-TOF MS) offers a powerful technique to analyze complex mixtures of proteins. In secondary metabolism, an array of proteins is involved in tandem catalyzing conversion of primary metabolites into compounds harboring medicinal properties. Besides these enzymes, other proteins are engaged in transport and regulatory functions. Thus, the study of proteomes is increasingly essential to analyze secondary metabolism.

Besides technological advances in proteome studies, *W. somnifera* has remained overlooked until recently. In 2011, the first report on proteome analysis of *W. somnifera* root tissues came from Dhar and coworkers, who earlier presented the first comparative proteomic analysis of seed and leaf proteins. Proteins were separated on 2D-PAGE in a pH gradient of 3–10, resulting in 434 spots. A total of 167 spots (82 from seeds and 85 from leaves) were selected for peptide mass fingerprinting by MALDI-TOF-MS. Analysis resulted in the identification of 70 seed proteins and 74 leaf proteins based on the homology with the proteins submitted to the http://www.uniprot.org database (Table 11.4). At the same time, Senthil et al. (2011) reported a comparative proteome of roots grown *in vitro* and *in vivo*. Isolated proteins were analyzed on 2D-PAGE in a 4–7 pH range. Comparative analysis of 2D gels showed a high degree of similarity between protein profiles of roots grown *in vitro* and *in vivo*. Thirty-five protein spots were commonly present while nine were unique to the *in vitro* profile. These individual spots were isolated and subjected to MALDI-TOF-MS analysis after treatment with trypsin. Among the 35 spots analyzed, 22 showed significant matches with proteins in the database and were categorized into general cell metabolism, defense-related proteins, and secondary metabolite production (Table 11.4). The overall data suggested that expression patterns and profiles of the proteins are similar in both *in vitro* and *in vivo* roots.

TABLE 11.4

Major Proteins and Corresponding Genes as Reported in Uniprot Database for *W. somnifera*

Protein Names	Gene Names	Length (AA)
Cycloartenol synthase		758
1-Deoxy-D-xylulose-5-phosphate synthase	DXS dxs	717
Sterol glucosyltransferase	SgtL1	701
NADPH dehydrogenase subunit	ndhF	689
NADPH-cytochrome P450 reductase	CPR1	686
Maturase	matR	581
3-Hydroxy-3-methylglutaryl coenzyme A reductase		575
Squalene epoxidase		531
Glycosyltransferase	GT1	485
1-Deoxy-D-xylulose-5-phosphate reductoisomerase	DXR dxr	475
Ribulose-1,5-biphosphate carboxylase/oxygenase	rbcl	456
ATP synthase subunit beta	atpB	336
Squalene synthase	SQS	413
Squalene synthase		411
Putative progesterone 5-beta-reductase		388
Farnesyl pyrophosphate synthase	FPPS	343
Maturase K	matK	314
Glycosyltransferase	SAP1	310
L-asparaginase		280
MADS domain MPF2-like transcription factor	a206	254
MADS domain MPF2-like transcription factor	b206	249
Putative sterol glucosyltransferase	SgtL2	236
MPF2-like-B	MPF2- like	235
MPF2-like-A	MPF2- like	232
Glycosyltransferase	SAP2	204
MPF1-like-A	MPF1- like	194
MPF1-like-B	MPF1- like	194
Superoxide dismutase [Cu–Zn]		154
Flavolonol glucosyltransferase	FGT	121
Phenolic glycosyltransferase	PGT	102
Putative sulfolipid biosynthesis protein	SQD1	94
Ribosomal protein S16	rps16	39
PsbA	psbA	10

In another study, by Nagappan et al. (2012), comparative protein profiles of the root of Korean ginseng (*Panax ginseng*) and *W. somnifera* were generated with 2D-PAGE. From the *W. somnifera* profile, 35 protein spots were chosen for the MALDI-TOF-MS analysis and 22 proteins were ascribed with their function in the categories of cellular metabolism, defense, and secondary metabolism. Proteomics is an upcoming area for identification of a broad spectrum of genes that are expressed in living systems. The technique is nowadays being applied to detect protein changes *in vitro* and *in vivo*. To better understand the proteins and enzymes involved in the withanolide biosynthetic pathway, one such study was conducted that involved two-dimensional gel electrophoresis (2DE) and mass spectroscopic analysis of *in vitro* grown adventitious root and *in vivo* root samples of

W. somnifera. The resulting protein spots were passed through a homology search using MASCOT software available at http://www.Matrixscience.com to reveal a high level of similarity (Senthil et al., 2011). It was found that similar protein spots were expressed in both *in vitro* and *in vivo* samples and were categorized according to their biological functions into different groups, with general cell metabolism, protein synthesis, RNA synthesis, cell differentiation, defense mechanism, and secondary metabolism being the main ones. Out of all the protein spots, four were found to be involved in secondary metabolite production (Senthil et al., 2011).

It was concluded that there is a similar developmental process for *in vitro* and *in vivo* systems, though *in vitro* roots are developed independently of the shoot organ. It was shown earlier that roots synthesize withanolides independently (Sangwan et al., 2008). The report is suggested to be the first one on the establishment of a 2DE reference proteome of *W. somnifera* root samples (Jadhav et al., 2012). Structural characterization of the flavonoid glycosyltransferase from *W. somnifera* was performed in a study by construction of the three-dimensional model of flavonoid-specific glycosyltransferases (WsFGT) based on the crystal structure of plant UGTs. Multiple amino acid sequence alignment was performed with available protein sequences and 3D models of WsFGT were generated by aligning the target sequence with multiple structure-based alignment of template proteins (Figure 11.3a). The model thus obtained was assessed by various tools and the final refined model (http://pymol.org; http://swissmodel.expasy.org; http://www.rcsb.org/pdb/home/home.do) revealed a GT-B-type fold (Jadhav et al., 2012). Docking studies showed that the sugar donors and acceptors interacted with the active site of WsFGT (Figure 11.3a,b, and c).

11.2.2.3 Advances in Genomic and Transcriptomic Approaches

Utilization of the latest bioinformatic tools and databases may provide a set of information for gene mining and establish the secondary metabolite pathway and genomic and transcriptomic characterization of the plant (Figure 11.2). Using the Entrez search engine from the National Center for Biotechnology Information (http://www.ncbi.nlm. nih.gov), we found that the collection of related protein sequences, epigenetic maps and data sets, conserved domains, gene-oriented clusters of transcript sequences, SNPs, and 3D structure has yet to be reported for *Withania* (Tables 11.4 and 11.5). Further, no records have been reported for the Genome Survey Sequence (GSS) database. Also, the pathways and systems of interacting molecules are not yet known. Gene-centered information has been reported for LRP1 low-density lipoprotein receptor-related protein (*Mus musculus*). (More information about chromosomal locations can be obtained from the gene subsection of NCBI for *Withania*.) *Withania* data were introduced into the MeSH database in 2003 and tree number(s) B01.650.940800.575.100.905.950 was assigned to it. Transcriptome and gene expression data have been submitted by the National Botanical Research Institute (NBRI) with 1 Gbase and 1577 Mbytes of SRA data (Table 11.6). The accession for submission is PRJNA168347. The NCBI NLM catalog comprises three publications and one online book is available (Atal, 1975; Booth and Upton, 2000; Williamson, 2002). Bioinformatics studies have been performed on the differential activities of the two closely related withanolides, withaferin A and withanone, and experimental evidence thus generated has shown that they are two structurally similar withanolides from *W. somnifera*. Alcoholic leaf extract, rich in withanone, showed cytotoxic activity to cancerous cells. Further, two phytochemicals have shown differential activity in normal and cancerous cells *in vitro* and *in vivo*. The bioinformatics tools have been applied on four genes, that is, mortalin, p53, p21, and Nrf2, which were identified by loss-of-function

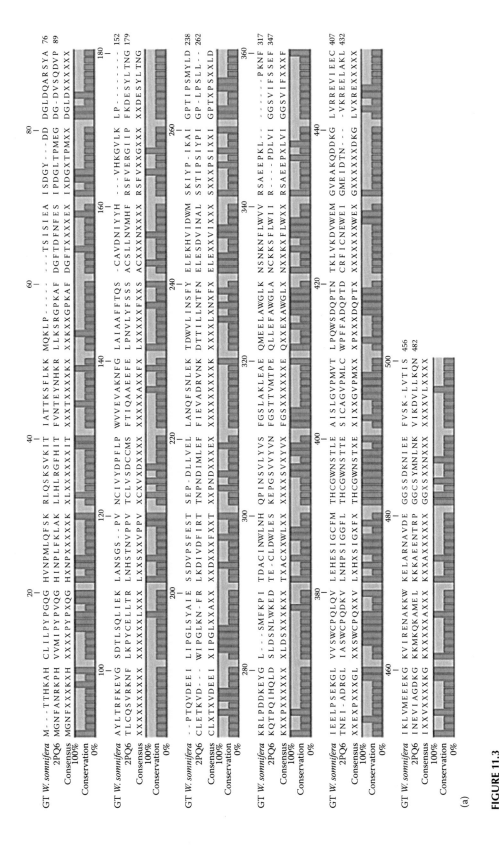

FIGURE 11.3

(a) Sequence alignment of WsFGT from *W. somnifera* with template (PDB code: 2PQ6) using CLC genomics workbench. (b) Homology model of flavonoid glycosyltransferase from *W. somnifera* (black solid-surface model). (c) Superimposition of 3D structure of glycosyltransferase from *W. somnifera* and *M. truncatula* as template (2PQ6: light gray solid surface).

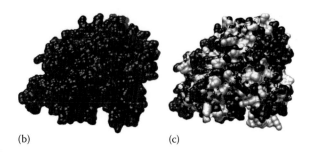

(b)   (c)

FIGURE 11.3   (Continued)

TABLE 11.5

Number of Entries in NCBI Entrez Search
Engine for *W. somnifera*

Entry Type	No. of Entries
EST	742
Pubmed	511
Pubmed Central	327
Nucleotide	282
Protein	186
Pub Chem Bioassay	32
PoP SET	19
UniSTS	11
Biosample	9
NLM catalog	3
Genome	1
Taxonomy	1
Gene	1
Bio Project	1
Pub Chem compound	1
MeSH	1

TABLE 11.6

Summary of ESTs Obtained from cDNA Libraries Submitted in NCBI for *W. somnifera*

ID	Data Type	Study
168134	EST:LIBEST_025134	*W. somnifera* subtracted root-specific cDNA library
168133	EST:LIBEST_025133	*W. somnifera* subtracted leaf-specific cDNA library
167980	EST:LIBEST_024979	*W. somnifera* subtracted root cDNA library in Lambda Zap Express vector
167979	EST:LIBEST_024978	*W. somnifera* subtracted leaf cDNA library in Lambda Zap Express vector
1041435	SRA:SRS344590	Transcriptome analysis of leaf and root tissues of *W. somnifera* (Indian ginseng) using 454 sequencing platform
168672	EST:LIBEST_025678	*W. somnifera* cv. WS-Y-08 root tissue
168314	EST:LIBEST_025315	*W. somnifera* cv. WSR08 leaf cDNA library
168246	EST:LIBEST_025247	*W. somnifera* WSR leaf tissue cDNA library
153009	EST:LIBEST_021275	*W. somnifera* leaf tissue cDNA

screenings. Docking efficacy of withanolide A and withanone was checked against four genes and it was revealed that the two closely related phytochemicals have differential binding properties to the chosen cellular targets and can raise differential molecular responses (Vaishnavi et al., 2012). It was concluded from the study that withaferin A binds strongly to the selected targets and acts as a strong cytotoxic agent both for normal and cancer cells. On the other hand, withanone has weak binding to the targets, has shown milder cytotoxicity toward cancer cells, and is safe for normal cells. This experimental evidence has thus provided important perspectives on the use of withaferin A and withanone for cancer treatment and the development of new anticancer phytochemical cocktails (Vaishnavi et al., 2012).

In another report, Senthil and coworkers (2010) generated and analyzed the first transcriptomes, ESTs in the leaves and roots of *W. somnifera*. Annotations were assigned to the ESTs for the genes with a potential role in photosynthesis (cytochrome-P450), pathogenesis (arginine decarboxylase, chitinase), and withanolide biosynthesis (squalene epoxidase, CDP-ME kinase). CpG islands were also identified on the basis of calculations performed using a 200 bp window moving across the sequence at 1 bp intervals (Senthil et al., 2010). Results were also compared with the Pfam database, which uses HMMER implementation for HMM profiles. The sequences were searched for domains after translation to proteins using http://pfam.sanger.ac.uk. The study revealed the presence of CDP-ME kinase from the MEP pathway and HMG CoA synthase from the MVA pathway (Figure 11.2). Squalene epoxidase, responsible for the formation of oxidosqualene from squalene, was also reported using similar bioinformatics approaches. In the study, differential expression of withanolides in root and leaf tissues was elucidated and comparative transcriptomic analysis in these plant tissues was performed. One more study revealed that reciprocal loss of CArG boxes and auxin response elements (AREs) drives expression divergence of MPF2-like MADS box genes controlling calyx inflation. In this study, Khan and coworkers (2012) performed a bioinformatics analysis to report that the evolution of MPF2-like genes has entailed degenerative mutations in a core promoter CArG box and an auxin response factor (ARF)-binding element in the large first introns in the coding region. Using phylogenetic shadowing, site-directed mutagenesis, and motif swapping they have shown that the loss of a conserved CArG box in MPF2-like-B of *W. somnifera* is responsible for impeding its expression in sepals. Conversely, loss of an ARE in MPF2-like-A relaxed the constraint of expression in sepals, thus suggesting that ARE is an active suppressor of MPF2-like gene expression in sepals, which in contrast is activated through the CArG box (Khan et al., 2012). The observed expression divergence in MPF2-like genes due to reciprocal loss of *cis*-regulatory elements has therefore added to genetic and phenotypic variations in the withanone and enhanced the potential of natural selection for the adaptive evolution of inflated calyx syndrome (ICS). Docking and molecular dynamics simulation studies have been performed to elucidate the binding mechanism of the prospective herbal drug withaferin A onto the structure of DNA polymerase of the herpes simplex virus. Docking simulations suggested high binding affinity of the ligand to the receptor. For enzyme inhibitor interactions, key residue identification, docking results corroboration, and dynamic behavior of the system studied were evaluated by performing long *de novo* molecular dynamics simulations (Khan et al., 2012). These simulations support the hypothesis that withaferin A is a potential ligand to target/inhibit DNA polymerase of the herpes simplex virus. These types of studies will help in the design of selective inhibitors of DNA polymerase with specificity and potent activity in order to strengthen the therapeutic arsenal available against the herpes simplex virus as a dangerous biological warfare agent (Grover et al., 2011a).

Computational approaches have been applied to determine that withanone binds to the TPX2-Aurora A complex in cancer cells. Disruption of the active TPX2-Aurora A association complex was discerned by docking analysis suggesting molecular dynamics simulation resulted in the thermodynamic and structural stability of TPX2-Aurora A in complex with withanone, which further substantiates the binding (Grover et al., 2012). A computational rationale of naturally occurring withanone's ability to alter the kinase signaling pathway in an ATP-independent manner was reported and inactivation of the TPX2-Aurora A complex by withanone was experimentally proved, demonstrating that the TPX2-Aurora A complex is a target of withanone, a potential natural anticancer drug. Similarly, *W. somnifera* comprises a large number of steroidal lactones known as withanolide that show various pharmacological activities. Docking of 26 withaferin and 14 withanolide from *W. somnifera* into the three-dimensional structure of protein kinase G (PknG) of *M. tuberculosis* using GLIDE was described (Natchimuthu and Sekar, 2012). Inhibitor binding positions and affinity were evaluated by Glidescore using scoring functions that identified withanolide E, F, and D and withaferin-diacetate 2 phenoxy ethyl carbonate as potential inhibitors of PknG. In the case of prostate cancer cells, it has been highlighted that the key signaling pathways through which the immunomodulatory effects of ashwagandha are mediated show three distinct pathways—the JAK-STAT pathway, the apoptosis pathway, and the MAPK signaling pathway—as functionally annotated by the DAVID Pathway Viewer (Aalinkeel et al., 2008). Among these, the JAK-STAT pathway appears to be key as it modulates both apoptosis and MAPK signaling. One more study on *Withania* investigated the development of cross-resistance between standard anticancer drugs and withanolides. The molecular determinants of sensitivity and resistance of tumor cells toward withanolides were also elucidated. To depict the possible mode of action of the main active constituent of *W. somnifera*, withanone, in modulating the NF-κB signaling pathway capability, molecular docking and dynamics simulation studies were performed (Grover et al., 2010), which led to the conclusion that withaferin A is a potent anticancer agent ascertained by its potent NF-κB-modulating capability (Figure 11.4). The simulation studies clearly revealed the dynamic structural stability of NF-κB essential modulator (NEMO)/IKKβ in complex with the drug withaferin A (http://hex.loria.fr; http://pymol.org), together with the inhibitory mechanism (Grover et al., 2010).

11.3 Conclusions and Perspectives

Current technologies employed for metabolic profiling such as proteomics, genomics, and transcriptomics are of remarkable value for comprehensive strategies for characterizing and understanding the uniqueness of plants. In addition, they will most probably have an important role to play in safety testing of genetically modified foodstuffs and in metabolomics-assisted breeding. As far as metabolites are concerned, it has been estimated that more than several thousand metabolites are present in plants; their profiles and interplay with several physiological and metabolic situations alter concomitantly to any change happening around them. A large proportion of this enormous diversity derives from compounds of secondary metabolism pathways assumed to play many metabolic roles. However, it has been extremely difficult to analyze the metabolic difference in a real-time manner. However, with the advent of modern omic approaches, such monitoring in interactive and integrative modes is made possible. Although current protocols do not cover the full complement of the

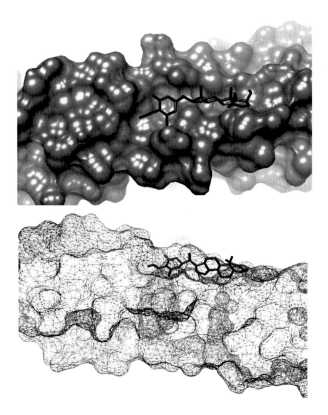

FIGURE 11.4
Different views for interactions of docked withaferin A with NEMO receptor: (*upper*) surface view; (*lower*) mesh view.

plant cell, improvement in the coverage of metabolomics techniques has been made. With the combination of next-generation technologies and analytical precision-based metabolite profiling, it would appear likely that much important information concerning the regulation of primary metabolism will become accessible (Tohge and Fernie, 2009; Schauer et al., 2006). Further development and combination of many analytical techniques will additionally allow a fuller description of the metabolome status of a plant. The presented omic approaches offer the concurrent investigation of complex biological processes at different levels (genes, proteins, and metabolites), corroborating their effect on one another. The relation between genes, proteins, and metabolites could be ascertained in one shot following these approaches. A quick survey of the literature clearly indicates that there is growing interest in the application of omics approaches in the biochemical study of plants, particularly of *W. somnifera* in recent times owing to its unique medicinal importance.

Acknowledgment

The authors are thankful to the New Millennium Indian Technology Leadership Initiative (NMITLI), New Delhi, and Department of Biotechnology (DBT), New Delhi, for providing the financial grant to carry out various studies in the author's laboratory, producing several observations and results compiled and cited in this chapter.

References

Aalinkeel, R., Hu, Z., Nair, B.B., Sykes, D.E., Reynolds, J.L., Mahajan, S.D., and Schwartz, S.A. 2008. Genomic analysis highlights the role of the JAK-STAT signaling in the anti-proliferative effects of dietary flavonoid: 'Ashwagandha' in prostate cancer cells. *Evid Based Complement Alternat Med* 7: 177–187.

Abraham, A., Kirson, I., Lavie, D., and Glotter, E. 1975. The withanolides of *Withania somnifera* chemotypes I and II. *Phytochemistry* 14: 189.

Ahmad, M. and Dough, A. 2002. New withanolides and other constituents from the fruit of *Withania somnifera*. *Arch Pharm Med Chem* 6: 267.

Akhtar, N., Gupta, P., Sangwan, N.S., Sangwan, R.S., and Trivedi, P.K. 2012. Cloning and functional characterization of 3-hydroxy-3-methylglutaryl coenzyme A reductase gene from *Withania somnifera*: An important medicinal plant. *Protoplasma* 250: 613–622.

Anjaneyulu, A.S.R. and Rao, D.S. 1997. A new withanolide from the leaves of *Withania somnifera*. *Indian J Chem* 36B: 161.

Anonymous. 1976. *The Wealth of India*, vol. 8. New Delhi: Publication and Information Directorate, pp. 37–38.

Atal, C.K. 1975. Phamacognosy and phytochemistry of *Withania somnifera* (Linn) Duna (ashwagandha). In *Indian Medicine and Homeopathy*. New Delhi: Central Council of Research.

Bekkouche, K., Daali, Y., Cherkaoui, S., Veuthey, J., and Christen, P. 2001. Calystegine distribution in some solanaceous species. *Phytochemistry* 58: 455–462.

Bessalle, R. and Lavie, D. 1992. Withanolide C, A chlorinated withanolide from *Withania somnifera*. *Phytochemistry* 31: 3648–3651.

Bhakre, H.A., Khotpal, R.R., and Kulkarni, A.S. 1993. Lipid composition of Withania somnifera, Phoenix silvestris and Indigofera enualphylla seeds of central India. *J Food Sci Technol* 30: 382.

Bhat, W.W., Lattoo, S.K., Razdan, S., Dhar, N., Rana, S., Dhar, R.S., Khan, S., and Vishwarkarma, R.A. 2012. Molecular cloning, bacterial expression and promoter analysis of squalene synthase from *Withania somnifera* (L.) Dunal. *Gene* 499: 25–36.

Bolleddula, J., Fitch, W., Vareed, S.K., and Nair, M.G. 2012. Identification of metabolites in *Withania somnifera* fruits by liquid chromatography and high-resolution mass spectrometry. *Rapid Commun Mass Spectrom* 26: 1277–1290.

Booth, S.J. and Upton, R. 2000. *Aswagandha Root, Withania Somnifera: Analytical, Quality Control, and Therapeutic Monograph*. Santa Cruz, CA: American Herbal Pharmacopoeia.

Budhiraja, R.D., Krishan, P., and Sudhir, S. 2000. Biological activity of withanolides. *J Sci Ind Res* 59: 904–911.

Chatterjee, S., Srivastava, S., Khalid, A., Singh, N., Sangwan, R.S., Sidhu, O.P., Roy, R., Khetrapal, C.L., and Tuli, R. 2010. Comprehensive metabolic fingerprinting of *Withania somnifera* leaf and root extracts. *Phytochemistry* 71: 1085–1094.

Chaurasiya, N.D., Gupta, V.K., and Sangwan, R.S. 2007. Leaf ontogenic phase-related dynamics of withaferin a and withanone biogenesis in ashwagandha (*Withania somnifera* Dunal.): An important medicinal herb. *J Plant Biol* 50: 508–513.

Chaurasiya, N.D., Sangwan, R.S., Misra, L.N., Tuli. R., and Sangwan, N.S. 2009. Metabolic clustering of a core collection of Indian ginseng *Withania somnifera* Dunal through DNA, isoenzyme, polypeptide and withanolide profile diversity. *Fitoterapia* 80: 496–505.

Chaurasiya, N.D., Sangwan, N.S., Sabir, F., Misra, L.N., and Sangwan, R.S. 2012. Withanolide biosynthesis recruits both mevalonate and DOXP pathways of isoprenogenesis in Ashwagandha *Withania somnifera* L. (Dunal). *Plant Cell Rep* 31: 1889–1897.

Chaurasiya, N.D., Uniyal, G.C., Lal, P., Misra, L.N., Sangwan, N.S., Tuli, R., and Sangwan, R.S. 2008. Analysis of withanolides in root and leaf of *Withania somnifera* by HPLC with photodiode array and evaporative light scattering detection. *Phytochem Anal* 19: 148–154.

Dhalla, N.S., Sastry, M.S., and Malhotra, C.L. 1961. Chemical studies on the leaves of *Withania somnifera*. *J Pharm Sci* 50: 876.

Dhar, R.S., Gupta, S.B., Singh, P.P., Razdan, S., Bhat, W.W., Rana, S., Lattoo, S.K., and Khan, S. 2011. Identification and characterization of protein composition in *Withania somnifera*: An Indian ginseng. *J Plant Biochem Biotechnol* 21: 77–87.

Dhar, R.S., Verma, V., Suri, K.A., Sangwan, R.S., Satti, N.K., Kumar, A., Tuli, R., and Qazi, G.N. 2006. Phytochemical and genetic analysis in selected chemotypes of *Withania somnifera*. *Phytochemistry* 67: 2269–2276.

Efferth, T. and Greten, H.J. 2012. *In silico* analysis of microarray-based gene expression profiles predicts tumor cell response to withanolides. *Microarrays* 1: 44–63.

Elyashberg, M., Blinov, K., Molodtsov, S., Smurnyy, Y., Williams, A.J., and Churanova, T. 2009. Computer-assisted methods for molecular structure elucidation: Realizing a spectroscopist's dream. *J Cheminform* 1: 3.

Furmanowa, M., Gajdzis-Kuls, D., Ruszkowska, J., Czarnocki, Z., Obidoska, G., Sadowska, A., Rani, R., and Upadhyay, S.N. 2001. In vitro propagation of *Withania somnifera* and isolation of withanolides with immunosuppressive activity. *Planta Med* 67: 146–149.

Glotter, E. 1991. Withanolides and related ergostane-type steroids. *Nat Prod Rep* 8: 415–440.

Glotter, E., Abraham, A., Günzberg, G., and Kirson, I. 1977. Naturally occurring steroidal lactones with a 17α-oriented side chain. Structure of withanolide E and related compounds. *J Chem. Soc* 1: 341–346.

Glotter, E., Kirson, I., Abraham, A., and Lavie, D. 1973. Constituents of *Withania somnifera* Dun-XIII: The withanolides of chemotype III. *Tetrahedron* 29: 1353–1364.

Goossens, A., Hakkinen, S.T., Laakso, I., Seppanen-Laakso, T., Biondi, S., De Sutter, V., Lammertyn, F., Nuutila, A.M., Soderlund, H., Zabeau, M., Inze, D., and Oksman-Caldentey, K.M. 2003. A functional genomics approach toward the understanding of secondary metabolism in plant cells. *Proc Natl Acad Sci* 100: 8595–8600.

Grover, A., Agrawal, V., Shandilya, A., Bisaria, V.S., and Sundar, D. 2011a. Non-nucleosidic inhibition of Herpes simplex virus DNA polymerase: Mechanistic insights into the anti-herpetic mode of action of herbal drug withaferin A. *BMC Bioinformatics* 12: S22.

Grover, A., Shandilya, A., Agrawal, V., Pratik, P., Bhasme, D., Bisaria, V.S., and Sundar, D. 2011b. Hsp90/Cdc37 Chaperone/co-chaperone complex, a novel junction anticancer target elucidated by the mode of action of herbal drug Withaferin A. *BMC Bioinformatics* 12: S30.

Grover, A., Shandilya, A., Punetha, A., Bisaria, V.S., and Sundar, D. 2010. Inhibition of the NEMO/IKKβ association complex formation, a novel mechanism associated with the NF-κB activation suppression by *Withania somnifera*'s key metabolite withaferin A. *BMC Genomics* 11: S25.

Grover, A., Singh, R., Shandilya, A., Priyandoko, D., Agrawal, V., Bisaria, V.S., Wadhwa, R., Kaul, S.C., and Sundar, D. 2012. Ashwagandha derived withanone targets TPX2-Aurora A complex: Computational and experimental evidence to its anticancer activity. *PLoS One* 7(1): e30890.

Gupta, P., Akhtar, N., Tewari, S.K., Sangwan, R.S., and Trivedi, P.K. 2011. Differential expression of farnesyl diphosphate synthase gene from *Withania somnifera* in different chemotypes and in response to elicitors. *Plant Growth Regul* 65: 93–100.

Gupta, N., Sharma, P., Kumar, S.R.J., Vishwakarma, R.K., and Khan, B.M. 2012. Functional characterization and differential expression studies of squalene synthase from *Withania somnifera*. *Mol Biol Rep* 39: 8803–8812.

Gupta, P., Agarwal, A.V., Akhtar, N., Sangwan, R.S., Singh, S.P., and Trivedi, P.K. 2013. Cloning and characterization of 2-C-methyl-D-erythritol- 4-phosphate pathway genes for isoprenoid biosynthesis from Indian ginseng, *Withania somnifera*. *Protoplasma* 250: 285–295.

Jadhav, S.K.R., Patel, K.A., Dholakia, B.B., and Khan, B.H. 2012. Structural characterization of a flavonoid glycosyltransferase from *Withania somnifera*. *Bioinformation* 8: 943–949.

Jayaprakasam, B. and Nair, M.G. 2003. Cyclooxygenase-2 enzyme inhibitory withanolides from *Withania somnifera* leaves. *Tetrahedron* 59: 841–849.

Jayaprakasam, B., Strasburg, G.A., and Nair, M.G. 2004. Potent lipid peroxidation inhibitors from *Withania somnifera* fruits. *Tetrahedron* 60: 3109–3121.

Kawamura, A., Brekman, A., Grigoryev, Y., Hasson, T.H., Takaoka, A., Wolfe, S., and Soll, C.E. 2006. Rediscovery of natural products using genomic tools. *Bioorg Med Chem Lett* 16: 2846–2849.

Khan, M.R., Hu, J., and Ali, G.M. 2012. Reciprocal loss of CArG-boxes and auxin response elements drives expression divergence of MPF2-like MADS-box genes controlling calyx inflation. *PLoS One* 7(8): e42781.

Kirson, I., Glotter, E., Abraham, A., and Lavie, D. 1970. Constituents of *Withania somnifera* Dun: XI. The structure of three new withanolides. *Tetrahedron* 26: 2209–2219.

Kirson, I., Glotter, E., Lavie, D., and Abraham, A. 1971. Constitutents of *Withania somnifera* Dun. Part XII. The withanolides of an Indian chemotype. *J Chem Soc C* 2032–2044.

Kundu, A.B., Mukherjee, A., and Dey, A.K. 1976. A new withanolide from the seeds of *Withania somnifera* Dunal. *Indian J Chem* 14B: 434–435.

Lal, P., Misra, L., Sangwan, R.S., and Tuli, R. 2006. New withanolides from fresh berries of *Withania somnifera*. *Z Naturforsch* 61: 1143–1147.

Lavie, D., Glotter, E., and Shavo, Y. 1965. Constituents of *Withania somnifera* Dun. The structure of withaferin A. *J Chem Soc* 7517.

Lavie, D., Glotter, E., and Shvo, Y. 1965. Constituents of *Withania somnifera* Dun. III. The side chain of withaferin A. *J Org Chem* 30: 1774–1778.

Lavie, D., Kirson, I., and Glotter, E. 1968. Constituents of *Withania somnifera* Dun. Part X. The structure of withanolide D. *Isr J Chem* 6: 671–678.

Madina, B.R., Sharma, L.K., Chaturvedi, P., Sangwan, R.S., and Tuli, R. 2007a. Purification and physico-kinetic characterization of 3ß-hydroxy specific sterol glucosyltransferase form *Withania somnifera* (L.) and its stress response. *Biochim Biophys Acta* 1774: 392–402.

Madina, B.R., Sharma, L.K., Chaturvedi, P., Sangwan, R.S., and Tuli, R. 2007b. Purification and characterization of a novel glucosyltransferase specific to 27ß-hydroxy steroidal lactones form *Withania somnifera* and its role in stress responses. *Biochim Biophys Acta* 1774: 1199–1207.

Masuda, M., Suzui, M., and Weinstein, I. 2001. Effects of epigallocatechin-3-gallate on growth, epidermal growth factor receptor signaling pathways, gene expression, and chemosensitivity in human head and neck squamous cell carcinoma cell lines. *Clin Cancer Res* 7: 4220.

Mirjalili, M.H., Moyano, E., Bonfill, M., Cusido, R.M., and Palazon, J. 2009. Steroidal lactones from *Withania somnifera*, an ancient plant for novel medicine. *Molecules* 14: 2373–2393.

Misra, L., Lal, P., Sangwan, R.S., Sangwan, N.S., Uniyal, G.C., and Tuli, R. 2005. Unusually sulfated and oxygenated steroids from *Withania somnifera*. *Phytochemistry* 66: 2702–2707.

Misra, L., Mishra, P., Pandey, A., Sangwan, R.S., Sangwan, N.S., and Tuli, R. 2008. Withanolides from *Withania somnifera* roots. *Phytochemistry* 69: 1000–1004.

Misra, L.N., Mishra, P., Pandey, A., Sangwan, R.S., and Sangwan, N.S. 2012. 1,4-Dioxane and ergosterol derivatives from *Withania somnifera* roots. *J Asian Nat Prod Res* 14: 39–45.

Misra, L.N. 2013. Traditional phytomedicinal systems, scientific validations and current popularity as neutraceuticals. *Int J Trad Nat Med* 2: 27–75.

Mishra, S., Sangwan, R.S., Bansal, S., and Sangwan, N.S. 2012. Efficient transgenic plant production of *Withania coagulans* (Stocks) Dunal mediated by *Agrobacterium tumefaciens* from leaf explants of in vitro multiple shoot culture. *Protoplasma* 250: 451–458.

Mondal, S., Mandal, C., Sangwan, R., Chandra, C., and Mandal, C. 2010. Withanolide D induces apoptosis in leukemia by targeting the activation of neutral sphingomyelinase-ceramide cascade mediated by synergistic activation of c-Jun N-terminal kinase and p38 mitogen-activated protein kinase. *Molecular Cancer* 9: 239.

Nagappan, A., Karunanithi, N., Sentrayaperumal, S., Park, K.I., Park, H.S., Lee, D.H., Kang, S.R., Kim, J.A., Senthil, K., Natesan, S., Muthurajan, R., and Kim, G.S. 2012. Comparative root protein profiles of Korean ginseng (*Panax ginseng*) and Indian ginseng (*Withania somnifera*). *Am J Chin Med* 40: 203–218.

Namdeo, A.G., Sharma, A., Yadav, K.N., Gawande, R., Mahadik, K.R., Lopez-Gresa, M.P., Kim, H.K., Choi, Y.H., and Verpoorte, R. 2011. Metabolic characterization of *Withania somnifera* from different regions of India using NMR spectroscopy. *Planta Med* 77: 1958–1964.

Natchimuthu, N. and Sekar, A. 2011. Insights from the molecular docking of withanolide derivatives to the target protein PknG from *Mycobacterium tuberculosis*. *Bioinformation* 7: 1–4.

Newman, D.J. and Cragg, G.M. 2007. Natural products as sources of new drugs over the last 25 years. *J Nat Prod* 70: 461–477.

Pachori, N., Rathee, P.S., and Mishra, S.N. 1994. The seed oil of Withania somnifera W.S.-20 variety. *Asian J Chem* 6: 442–444.

Pal, S., Singh, S., Shukla, A.K., Gupta, M.M., Khanuja, P.S., and Shasany, A.K. 2011. Comparative withanolide profiles, gene isolation, and differential gene expression in the leaves and roots of *Withania somnifera*. *J Hort Sci Biotech* 86: 391–397.

Pramanick, S., Roy, A., Ghosh, S., Majumder, H.K., and Mukhopadhyay, S. 2008. Withanolide Z, a new chlorinated withanolide from *Withania somnifera*. *Planta Med* 74: 1745–1748.

Rahman, A., Abbas, S., Shahwar, D., Jamal, S.A., and Choudhary, M.I. 1993. New withanolides from *Withania* spp. *J Nat Prod* 56: 1000–1006.

Ray, A.B. and Gupta, M. 1994. Withasteroids, a growing group of naturally occurring steroidal lactones. *Fortschr Chem Org Naturst* 63: 1–106.

Riaz, N., Malik, A., Rehman, A., Nawaz, S.A., Muhammad, P., and Choudhary, M.I. 2004. Cholinesterase-inhibiting withanolides from *Ajuga bracteosa*. *Chem Biodivers* 1: 1289–1294.

Sabir, F., Kumar, A., Tiwari, P., Pathak, N., Sangwan, R.S., Bhakuni, R.S., and Sangwan, N.S. 2010. Bioconversion of artemisinin to its nonperoxidic derivative deoxyartemisinin through suspension cultures of *Withania somnifera* Dunal. *Z Naturforsch* 65c: 607–612.

Sabir, F., Mishra, S., Sangwan, R.S., Jadaun, J.S., and Sangwan, N.S. 2012a. Qualitative and quantitative variations in withanolides and expression of some pathway genes during different stages of morphogenesis in *Withania somnifera* Dunal. *Protoplasma* 250: 539–549.

Sabir, F., Sangwan, R.S., Kumar, R., and Sangwan, N.S. 2012b. Salt stress induced responses in growth and metabolism in callus cultures and differentiating in vitro shoots of Indian ginseng (*Withania somnifera* Dunal). *J Plant Growth Reg* 31: 537–548.

Sabir, F., Sangwan, R.S., Chaurasiya, N.D., Misra, L.N., and Sangwan, N.S. 2008. In vitro withanolides production by *Withania somnifera* Dunal cultures. *Z Naturforsch* 63c: 409–412.

Sabir, F., Sangwan, R.S., Singh, J., Pathak, N., Mishra, L.N., and Sangwan, N.S. 2011. Biotransformation of withanolides by cell suspension cultures of *Withania somnifera* (Dunal). *Plant Biotechnol Rep* 5: 127–134.

Sangwan, R.S., Chaurasiya, N.D., Lal, P., Mishra, L.N., Tuli, R., and Sangwan, N.S. 2008. Root contained withanolide A is inherently de novo synthesized within roots in Ashwagandha (*Withania somnifera*). *Physiol Plant* 133: 278–287.

Sangwan, R.S., Chaurasiya, N.D., Misra, L.N., Lal, P., Uniyal, G.C., and Sangwan, N.S. 2005. An improved process for isolation of withaferin A from plant materials and products there from. US Patent 7108870.

Sangwan, R.S., Chaurasiya, N.D., Misra, L.N., Lal, P., Uniyal, G.C., Sharma, R., Sangwan, N.S. Suri, K.A., Qazi, G.N., and Tuli, R. 2004. Phytochemical variability in commercial herbal products and preparations of *Withania somnifera* (Ashwagandha). *Curr Sci* 86: 461–465.

Schauer, N. and Fernie, A.R. 2006. Plant metabolomics: Towards biological function and mechanism. *Trends Plant Sci* 11: 508–516.

Schauer, N., Semel, Y., Roessner, U., Gur, A., Balbo, I., Carrari, F., Pleban, T., Perez-Melis, A., Bruedigam, C., Kopka, J., Willmitzer, L., Zamir, D., and Fernie, A.R. 2006. Comprehensive metabolic profiling and phenotyping of interspecific introgression lines for tomato improvement. *Nat Biotech* 24: 447–454.

Senthil, K., Karunanithi, N., Kim, G.S., Nagappan, A., Sundareswaran, S., Natesan, S., and Muthurajan, R. 2011. Proteome analysis of in vitro and *in vivo* root tissue of *Withania somnifera*. *African J Biotech* 10: 16875–16883.

Senthil, K., Wasnik, N.G., Kim, Y.J., and Yang, D.C. 2010. Generation and analysis of expressed sequence tags from leaf and root of *Withania somnifera* (Ashwgandha). *Mol Biol Rep* 37: 893–902.

Sharma, L.K., Bhaskara, R.M., Chaturvedi, P., Sangwan, R.S., and Tuli, R. 2007. Molecular cloning and characterization of one member of 3b-hydroxy sterol glucosyltransferase gene family in *Withania somnifera*. *Arch Biochem Biophys* 460: 48–55.

Sidhu, O.P., Annarao, S., Chatterjee, S., Tuli, R., Roy, R., and Khetrapal, C.L. 2011. Metabolic alterations of *Withania somnifera* (L.) Dunal fruits at different developmental stages by NMR spectroscopy. *Phytochem Anal* 22: 492–502.

Singh, S. and Kumar, S. 1998. *Withania somnifera: The Indian Ginseng Ashwagandha*. Central Institute of Medicinal and Aromatic Plants, Lucknow, India.

Singh, N., Nath, R., Lata, A., Singh, S.P., Kohli, R.P., and Bhargava, K.P. 1982. Withania somnifera (Ashwagandha), a rejuvenating herbal drug which enhances survival during stress (an adaptogen). *Int J Crude Drug Res* 20: 29–35.

Subbaraju, G.V., Vanisree, M., Rao, C.V., Sivaramakrishna, C., Sridhar, P., Jayprakasam, B., and Nair, M.G. 2006. Ashwagandhanolide, a bioactive dimeric thiowithanolide isolated from the roots of *Withania somnifera*. *J Nat Prod* 69: 1790–1792.

Tohge, T. and Fernie, A.R. 2009. Web-based resources for mass-spectrometry-based metabolomics: A user's guide. *Phytochemistry* 70: 450–456.

Tuli, R. and Sangwan, R.S. (eds) 2010. *Ashwagandha (Withania somnifera): A Model Indian Medicinal Plant*. Council of Scientific and Industrial Research (CSIR), New Delhi, India.

Vaishnavi, K., Saxena, N., Shah, N., Singh, R., Manjunath, K., Uthayakumar, M., Kanaujia, S.P., Kaul, S.C., Sekar, K., and Wadhwa, R. 2012. Differential activities of the two closely related withanolides, withaferin A and withanone: Bioinformatics and experimental evidences. *PLoS One* 7: e44419.

Velde, V.V. and Lavie, D. 1982. A Δ^{16}-withanolide in *Withania somnifera* as a possible precursor for α-side-chains. *Phytochemistry* 21: 731–733.

Williamson, E.M. 2002. *Major Herbs of Ayurveda*. Edinburgh: Churchill Livingstone.

12

Genetic Engineering in Ornamental Plants

**Rajesh Kumar Dubey, PhD; Simrat Singh, PhD;
Gurupkar Singh Sidhu, PhD; and Manisha Dubey, PhD**

CONTENTS

12.1 Introduction 410
    12.1.1 Historical Perspective 410
    12.1.2 Biotechnological Developments in Ornamental Crops in India 410
12.2 Prospect of Genetically Engineered Ornamental Crops 411
    12.2.1 Application of Genetic Engineering in Ornamentals 411
        12.2.1.1 Flower Color 412
        12.2.1.2 Plant Architecture 416
        12.2.1.3 Flower Morphology 417
        12.2.1.4 Modification of Floral Scent 417
        12.2.1.5 Extension of Vase Life 418
        12.2.1.6 Disease and Pest Management 421
        12.2.1.7 Abiotic Stress Tolerance 422
12.3 Plant-Specific Applications of Biotechnology 423
    12.3.1 Evolution of Bioluminescent Orchids 423
    12.3.2 Isolation of Somaclonal Variants with Noble Traits 423
    12.3.3 The Transgenic Blue Roses 423
    12.3.4 *In Vitro* Flowering in Transgenic Ornamentals 424
    12.3.5 Transgenic Chrysanthemums against Herbivore Repellence 424
    12.3.6 Transgenic *Oncidium* and *Odontoglossum* Orchid Species for
        Enhanced Vase Life 425
    12.3.7 Induction of Male Sterility in Transgenic Chrysanthemum 425
12.4 Role of Molecular Markers in Ornamental Crops 426
    12.4.1 Morphological Markers 426
    12.4.2 Biochemical Markers 427
    12.4.3 Molecular Markers 428
        12.4.3.1 Properties of an Ideal DNA Marker 429
        12.4.3.2 Restriction Fragment Length Polymorphism 430
        12.4.3.3 Random Amplified Polymorphic DNA 431
        12.4.3.4 Amplified Fragment Length Polymorphism 431
        12.4.3.5 Microsatellites or Simple Sequence Repeat 432
        12.4.3.6 Expressed Sequence Tags 433
12.5 Commercialization of Transgenic Ornamental Plants 433
12.6 Challenges in Commercialization of Transgenic Ornamentals 434
References 434

12.1 Introduction

Ornamental crops are valued for their aesthetic appearance. Any morphological traits exhibiting novel features attract large number of customers and have potential economic value. The demand for novel flower colors for enhanced aesthetic appearance, longer vase life, wide adaptability to soil and climatic factors, and tolerance to biotic and abiotic stresses coupled with increased flower yield are the main breeding objectives for commercially important ornamental crops. With the advent of biotechnological tools, there is vast potential for significant improvement in different aspects of ornamental crops.

In the global market, ornamental horticulture accounts for U.S.$250–400 billion (Chandler and Sanchez, 2012). Among ornamental crops, cut flowers make a significant contribution in terms of monetary gains, contributing one-third of the value of the global ornamental horticulture market. It is estimated that the global cut flower industry shares U.S.$27 billion in annual retail sales (Chandler, 2003). Efforts are being made worldwide to boost the ornamental horticulture industry using biotechnology, and the focus is on the development of new flower colors and novel plant architecture, as these are the main features which determine consumer interest. Besides alterations in plant architecture, efforts are also being made to have year-round flowering in certain ornamental genera to keep them in bloom throughout the year.

12.1.1 Historical Perspective

Ornamental crops, being highly heterozygous, have not been exploited to their full potential due to limited knowledge of their genetic makeup. For a long time, new cultivars have been evolved utilizing conventional breeding principles. These methods are laborious and are limited to crossing between the same or related species. The gene pool available is limited to the genetic background of the parents.

With the advent of genetic engineering, carnation varieties were developed as the world's first transgenic flower crops. Genetically engineered carnation development is credited to Florigene in collaboration with Suntory Ltd, which evolved the Moon series carnation for the markets of Japan, Australia, and North America. So, with the advancement of genetic transformation technologies, efforts are now being directed toward characterizing useful genes in diverse ornamental genera and transferring these genes to other commercially important ornamentals in order to alter their genotype for the expression of the desired trait. To date, more than 30 ornamental species have been transformed, including anthurium, begonia, carnation, chrysanthemum, cyclamen, datura, daylily, gentian, gerbera, gladiolus, hyacinth, iris, lily, lisianthus, orchid, pelargonium, petunia, poinsettia, rose, snapdragon, and torenia (Deroles et al., 2002).

12.1.2 Biotechnological Developments in Ornamental Crops in India

With the establishment of the National Biotechnology Board in 1982, the government of India recognized the potential of biotechnology. This board then became a full-fledged department in 1986. The Department of Biotechnology supported the establishment of seven centers for plant molecular biology throughout the country. To date, there are about 50 public research units in India utilizing the tools of modern biotechnology for agriculture, especially techniques for cell and tissue culture as well as the protocols of gene transfer. The government of India grants an estimated U.S.$ 15 million annually to plant biotechnology research, while the private sector shares about U.S.$ 10 million.

Biotechnology research in India is mainly dominated by tissue cultural techniques, especially micropropagation for supply of quality planting material.

India has got the edge in exporting winter flowers to Western countries, where the energy demand for heating greenhouses increases during severe winters. However, winters are mild in most parts of India, thus providing a favorable climate for producing winter flowers at a comparatively lower cost. India also has a number of other advantages, such as favorable climatic conditions, access to neighboring countries by sea and air, varying agroclimatic zones, availability of cheap labor, advanced tissue culture facilities, and the presence of high genetic diversity in ornamental crops.

12.2 Prospect of Genetically Engineered Ornamental Crops

Genetic transformation in higher plants was first reported by Zambryskpi et al. (1983). Successful transformations have been described for over 120 species in 35 different families (Birch, 1997), and the number is increasing every year. With the advancement in the development of a range of *Agrobacterium*-mediated gene transfers and direct DNA delivery techniques (particle bombardment), along with appropriate tissue culture techniques for regenerating whole plants, a number of transformations have been possible in diverse flora of ornamental plants.

Genetic transformation is superior to mutation breeding because the target trait can be modified directly by incorporating related genes, whereas in mutation breeding the desirable traits can be obtained only when the genes related to the target traits are altered accidently. Genetically engineered plants are presently cultivated on an area of 58.7 million ha, by about 5.5–6.0 million farmers in 16 countries (James, 2002). However, the application and commercialization of these transgenic ornamental crops is limited in spite of the availability of diverse germplasm. This is attributed to the lack of efficient transformation systems for ornamental species and the limited markets for ornamental produce as compared to major food and fiber crops. Substantial interest in the genetic improvement of ornamentals has now been generated due to changing lifestyles and increased disposal income for aesthetic pleasures.

The genetic modification of ornamental plants for both biotic and abiotic stress tolerance, early flowering, and alternation of flower color and plant morphology is of major commercial importance. Recent advances in biotechnology, especially transgenics, have allowed the accelerated modification and improvement of desired characteristics in ornamentals. In orchids the first report on genetic transformation was made by Kuehnle and Sugii (1992), who evolved transgenic *Dendrobium* using particle bombardment. Chia et al. (1994) were the first to genetically engineer an orchid with the luciferase gene isolated from firefly. The tissue to be genetically engineered for molecular breeding must have a totipotency to form new plants. For effective and efficient plant transformation, gene transfer methods and the vectors for gene expression must be compatible with the plant genotype and the tissue under treatment (Swarnapiria, 2009).

12.2.1 Application of Genetic Engineering in Ornamentals

Genetic engineering holds enormous potential for exploiting the available genes in the biological system for the genetic enhancement of ornamentals. Using conventional breeding methods, new varieties have been evolved that have desirable characters, but one limitation associated with traditional breeding is the limited gene pool, which is evident in any given species. In addition, the heterozygous nature of ornamentals accounts for a diverse and complex genetic makeup that limits the delineation of the influence of a particular characteristic that might be under the influence of polygenes. With the discovery of molecular markers in the 1980s, research on the characterization of plant species revealing genetic variation at the DNA level has gained pace and allowed the categorization of genetic diversity in ornamental plants.

The recent advances in genetic engineering make it an alternative technology for use in conjunction with traditional breeding for the improvement of ornamental crops. Certainly, genetic engineering has the capability of increasing the available gene pool for crop improvement. One advantage of such an approach lies in the ability to alter a single trait without altering other genetic traits of the plant. Current efforts are aimed at developing techniques for high-value flower crops, such as carnation, chrysanthemum (Shinoyama et al., 2012a), rose (Katsumoto et al., 2007), petunia, and orchid (Hsiao et al., 2011) to evolve new cultivars with distinct features such as novel flower colors, dwarf plant morphology, novel flower architecture, and enhanced vase life.

12.2.1.1 Flower Color

Flower color is derived from flavonoids, carotenoids, and betalains (Tanaka et al., 2005). Flavonoids and their colored class of compounds, anthocyanins, contribute to a wide range of colors. Flavonoids consist of more than 10 classes of compounds. Anthocyanins confer orange, red, magenta, violet, and blue colors. Aurones and chalcones are yellow pigments, while flavones and flavonols are colorless or very pale yellow.

The flavonoid biosynthetic pathways leading to floral pigment accumulation have been well characterized and the genes encoding relevant enzymes and transcriptional factors have been isolated. Various factors besides anthocyanins determine the final color of the flower. Anthocyanin structure, type and concentration, coexisting compounds, pH of vacuoles, metal ion type and concentration, anthocyanins localization, and shape of surface of cells all contribute to the final color (Yoshida et al., 2009). Of the various known anthocyanins, six are commonly known for flower color: pelargonidin, cyanidin, peonidin, delphinidin, petunidin, and malvidin. Delphinidin and its derivatives account for blue coloring, whereas pelargonidin accounts for red coloring.

12.2.1.1.1 Biosynthesis of Flavonoids

Flavonoids are a class of phenylpropanoids that are classified depending on the structure of the C-ring. Anthocyanidins are the aglycone chromophore form of anthocyanins and are also direct precursors of anthocyanins in their biosynthesis. An increase in the hydroxyl group number confers a color shift to blue, and thus anthocyanins derived from delphinindin, cyanidin, and pelargonidin tend to yield violet/blue, red/magenta, and orange/intense red flower color, respectively (Yoshida et al., 2009). Flowers which do not have delphinidin-based anthocyanins lack blue violet varieties.

The glycosylation, acylation, and methylation of anthocyanins are the primary source of structural diversity and color variation (Figure 12.1). An increase in the number of the hydroxyl group of B-rings imparts a bluer color to the anthocyanins derived from anthocyanidins, while methylation of the 3′ or 5′ hydroxyl group results in reddening. Anthocyanidins are modified with glycosyl or acyl moieties in a species-specific manner by specific glycosyltransferases and acyltransferases. Aromatic acylation of anthocyanins shifts their color toward blue and increases their stability. In gentian, the anthocyanin is gentiodelphin, in

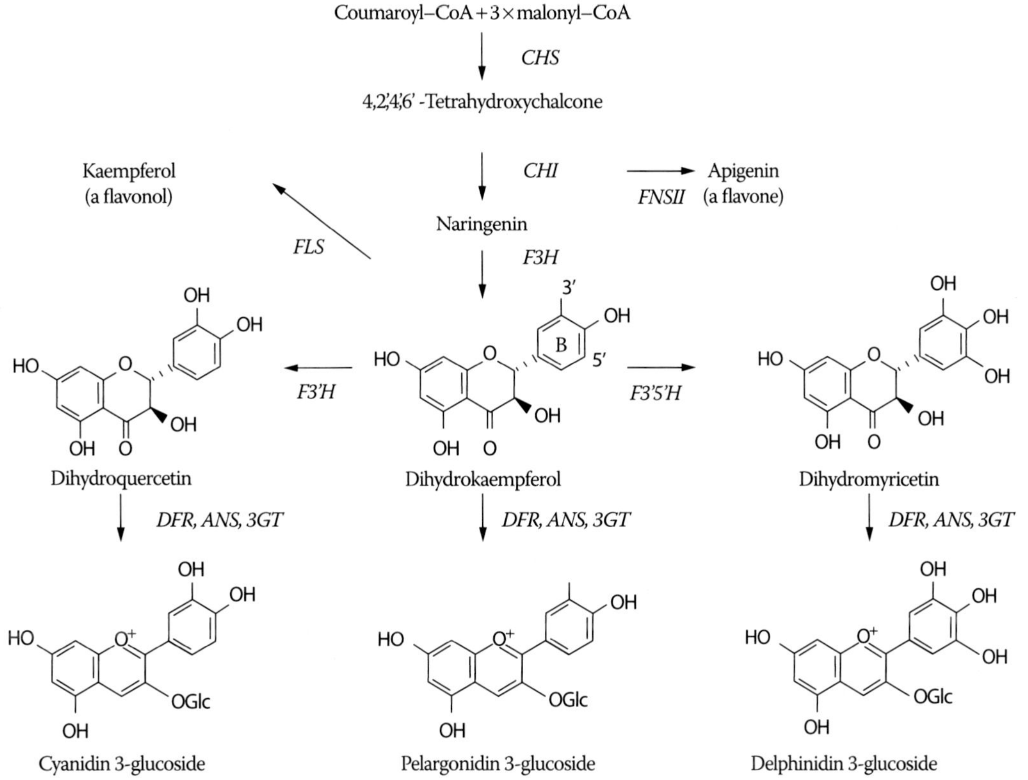

FIGURE 12.1
Anthocyanin biosynthetic pathway for the production of different colors. Flavonoid biosynthetic pathways relevant to flower color. CHS, chalcone synthase; CHI, chalcone isomerase; F3H, flavanone 3-hydroxylase; F3′H, flavonoid 3′-hydroxylase; F3′5′H, flavonoid 3′,5′-hydroxylase; DFR, dihydroflavonol 4-reductase; ANS, anthocyanidin synthase; 3GT, UDP-glucose, anthocyanidin 3-*O*-glucosyltransferase; FNS, flavone synthase; FLS, flavonol synthase.

which caffeoyl moiety at the 3′ position of the gentiodelphin has been shown to contribute to intramolecular staking, resulting in a stable blue color of flower. In general, 3-glucosylation occurs prior to 5-glucosylation, and the two glucosylations are separately catalyzed by UDP-glucosyl-dependent anthocyanidin 3-glucosyl transferase (3GT) and anthocyanidin 5-glucosyl transferase (5GT). The gene encoding an enzyme that catalyzes the transfer of aromatic acyl groups to anthocyanins is described by Tanaka et al. (2009).

The role of metal ions in flower color has been well summarized by Yoshida et al. (2009) The X-ray crystallographic structure of protocyanin in *Centaurea cyanus* (Corn Flower) and commelinin from *Commelina communis* (Asiatic dayflower) showed the involvement of Ca^{2+}, Mg^{2+}, and Fe^{3+} ions in the development of blue color (Shoji et al., 2007). Anthocyanins varied color from red in low pH environments to blue under neutral or alkaline pH environments. Vacuolar pH is generally regulated by vacuolar ATPase and pyrophosphase in plant cells. Various transcriptional factors and their roles in the regulation of vacuolar pH and anthocyanins biosynthesis have been documented by Quattrocchio et al. (2006).

In petunia, mutation of any of the PH1 to PH7 loci results in flowers with a blue color rather than a purple color and increased petal pH homogenates due to inhibition of vacuolar acidification. Tanaka (2009) notably demonstrated that the modification of B-ring hydroxylation by the gene encoding F3***′H and F3′5′H is the key player of flower color modification. Holton et al. (1993) isolated the gene encoding flavonoid 3'5'-hydroxylase, which paved the way for the development of color-modified *D. caryophyllus* and *R. hybrida*. The key genes responsible for the anthocyanin (Nishihara and Nakatsuka, 2011), flavonoid (Togami et al., 2011), and carotenoid (Cazzonelli and Pogson, 2010) biosynthesis and metabolism pathways have been identified, allowing the introduction and modification of flower color in many ways.

However, the biosynthesis and metabolic pathways of anthocyanin may be modified, but the color of the flower is also dependent upon many other factors, such as co-pigmentation and vacuolar pH. An Australian biotech company Florigene has made progress in identifying pH genes and modifying vacuolar pH.

12.2.1.1.2 Modification of Flower Color

Modification of flower color is likely to give a striking first impression to consumers. Some ornamental species, however, have only a narrow color spectrum, while in others specific colors such as shades of blue are lacking. Thus, the introduction of new colors (Table 12.1 and Figure 12.2) and the alteration of different characteristics (Table 12.2) through genetic engineering hold great potential in ornamental research and are likely to have a major impact on the industry. This potential has attracted the attention of several biotechnology companies worldwide (Davies and Schwinn, 1997).

To achieve flower color modification toward blue by overexpression of delphinidin-governing genes, the host must have the following characteristics (Katsumoto et al., 2007):

1. Higher vacuolar pH
2. Accumulation of flavonols that were expected to be copigments
3. No F3'H activity
4. Accumulation of pelargonidin rather than cyanidin

12.2.1.1.2.1 Carnation   In 1997, the first genetically modified blue carnation was introduced to the market (Tanaka et al., 2005). Blue carnations are produced as a result of expression of the petunia *F3'5'H* gene in a red carnation cultivar, accumulating pelargonidin that resulted in flowers containing a mixture of delphinidin and pelargonidin pigments. White carnation cultivars specifically lacking *DFR* (dihydroflavonol 4-reductase) gene were selected for transformation to avoid competition between endogenous enzymes (DFR in this case) and the introduced F3'5'H. Expression of a petunia or pansy *F3'5'H* gene and a petunia *DFR* gene (petunia DFR efficiently reduces dihydromyricetin but not dihydrokaempferol) resulted in almost exclusive accumulation of delphinidin. Since 1996

TABLE 12.1

Transgenic Ornamental Plants with Novel Flower Color

Ornamental Species Transformed	Novel Color Created	References
Cyclamen persicum (Cyclamen)	Purple to red/pink	Boase et al. (2010)
Gentiana triflora (Japenese gentian)	Blue to white	Nakatsuka et al. (2010, 2011)
Lotus japonicas	Light yellow to yellow/ orange	Suzuki et al. (2007)
Phalaenopsis spp.	Pink to light pink	Chen et al. (2011)
Torenia x hybrida (torenia)	Blue to pink	Nakamura et al. (2010)
Tricytris spp. (Toad lily)	Red to white	Kamiishi et al. (2011)

FIGURE 12.2 (See color insert)
Transgenic ornamentals with flower color variation. (a) Color modification of *Torenia hybrida* cv. Summerwave Blue. (b) Color modification of *Lobelia erinus*. (c) Carnations incorporated with cyanidin-based pigments. (d) Transgenic carnation expressing F3′5′H and petunia DFR genes. (e) Cosuppression of the F3′5′H gene in torenia cv. Summerwave Blue produced pink flowers. Overexpression of a torenia F3′H gene generated darker pink flowers. (f) Orange petunia producing pelargonidin pigment. (g) A pale yellow petunia expressing a *Lotus japonica PKR* gene. (h) Morning glory flowers with natural transposons. *Ipomoea purpurea* (CHS-D::Tip100), Ipomoea nil (DFR-B::Tpn1).

these transgenic blue carnations have been sold as Moon® series carnations by Florigene and Suntory Ltd. The blue color of carnation flowers is stable following repeated vegetative propagation. The GE carnations comprise eight varieties developed by Florigene Pty Ltd and Suntory Ltd (Dobres, 2011). Genetically modified carnations were first marketed in Australia in 1997 and are now grown in South America, Australia, and Japan.

12.2.1.1.2.2 Rose Expression of a pansy *F3′5′H* gene in rose petals resulted in significant accumulation of delphinidin. Expression of the pansy *F3′5′H* gene and an iris *DFR* gene in combination with the knockdown of the endogenous *DFR* gene also resulted in the accumulation of delphinidin and a change of flower color to violet (Katsumoto et al., 2007). However, these transgenic roses grew poorly, and have not yet been commercialized.

12.2.1.1.2.3 Chrysanthemum Expression of a pansy *F3′5′H* gene under the control of rose chalcone synthase promoter (Brugliera, 2009) resulted in a transgenic chrysanthemum with a flower color change from pink to violet. The expression of transgenic chrysanthemum has been found to be stable, after vegetative propagation, across two flowering

TABLE 12.2

Genetic Engineering Approaches for Altering Various Characteristics in Ornamental Crops

Genetic Engineering Approaches Used	Trait Modified	Gene(s) Involved	Ornamental Crops/Reference
Flower Color			
Antisense technology	Pink-white	Chs (chalchone synthase)	Chrysanthemum (Florigene) (Courtney-Gutterson et al., 1994)
	Brick red/orange flowers	DFR (dihydroxyflavonol-4-reductase)	Verbena, petunia (Suzuki et al., 2000)
Agrobacterium tumefaciens mediated	Violet/mauve color	F3'5'H (flavonoid 3'5'-hydroxylase) DFR (dihydroflavonoid 4-reductase) from petunia	Carnation (Florigene, Suntory Ltd.)
	White to yellow color	CmCCD4a-RNAi	Chrysanthemum (Ohmiya et al., 2006)
Vase Life			
Antisense technology	Delayed petal senesence	Suppression of gene encoding ACC synthetase	Carnation (Tanaka et al., 2005)
Agrobacterium tumefaciens mediated	Reduced sensitivity to exogenous ethylene	Ethylene receptor mutant gene etr 1-1	*Oncidium* and *Odontoglossum* orchid species (Raffeiner et al., 2009)
Plant Size			
Agrobacterium rhizogenes mediated	Decreased plant height, internodal length, reduced apical dominance	rol C	*C. morifolium* (Dolgov, 2009), *D. Caryophyllus* (Zuker et al., 1999)
Cold Tolerance			
Agrobacterium tumefaciens mediated	Frost tolerance, dwarf plants	AtCBF3 gene	*Petunia exserta and P. integrifolia* (Warner, 2011)
Drought Tolerance			
Agrobacterium tumefaciens mediated	Stress tolerance	DREB 1A	Chrysanthemum (Hong et al., 2006)

seasons. Expression of a *Campamula F3'5'H* gene under the control of chrysanthemum flavanone 3-hydroxylase promoter (Noda et al., 2010) also resulted in similar delphinidin production and petal color changes.

12.2.1.2 Plant Architecture

Many potentially useful genes that are involved in biosynthesis and biochemical pathways associated with flower and plant morphology have been isolated and cloned. Transcriptional factors regulating plant development and biosynthetic or regulatory genes involved in hormonal regulation have also been characterized. However, few of these genes have actually been applied to ornamental crops. More sophisticated regulation of expression of these genes may produce ornamental crops with novel forms having market potential. Here are a few examples to highlight the genetic modifications in ornamentals to alter their morphology:

- The rolC gene from *Agrobacterium rhizogenes* encodes a cytokinin-b-glucosidase. The transgenic plants of petunia cv. Mitchell expressing rolC driven by a CaMV35S promoter exhibited various morphological changes such as reduced plant height, leaf, and flower size, and increased branching (Winefield et al., 1999). Introduction of rolA, B, and C genes into *Rosa hybrida* cv. Moneyway reportedly resulted in improved rooting characteristics (Van der Salm et al., 1997).

- Profuse lateral branching helps in producing floriferous plants ideally suited for garden decoration. The overexpression of a petunia zinc-finger type transcription factor, Lateral-shoot Inducing Factor (LIF), in petunia under the control of a CaMV35S promoter resulted in a dramatic increase in the number of lateral shoots. Secondary branches, which rarely form in wild-type petunias, were common in these transgenic petunias. These plants had a decreased number of enlarged cells in the stem, leaf, and flower. The level of free cytokinins was found to be lower in stem and leaves. Indeed cytokinin regulation is important if the goal is to increase the number of flowers. Also the ratio of cytokinin concentration can be altered in plants by the expression of oncogenes from Ti plasmid isolated from *Agrobacterium tumefaciens* or the Ri plasmid of *A. rhizogenes* (Mol et al., 1995).

- Chemicals such as uniconazole are widely used to produce dwarf plants. Many genes involving gibberellin biosynthesis and signaling have been isolated. Among them, a semidominant mutation allele of GAI, gai-1, greatly reduces gibberellic acid responsiveness during vegetative development. This gene can be used to generate dwarf plants: Suntory Ltd has successfully induced a dwarfing effect by introducing genomic gai-1 sequences that reduce GA response during vegetative growth, which otherwise is controlled by the application of growth retardants to produce dwarf plants.

12.2.1.3 Flower Morphology

Genetically engineering flower architecture involves alternation in flowering behavior with conversion of indeterminate to determinate habit and vice versa by altering the desired portion of floral organs. Genes responsible for determining the identity of flower organs have been identified from the weed *Arabidopsis thaliana*. These genes are known to design flowers with any of the desired floral organs in any position on the flower. For example, it is now possible to generate double varieties, developed by replacing the anthers with petals, by identification of responsible genes. Mutation in a single gene can result in the creation of more prolific varieties by conversion of determinate inflorescence into indeterminate inflorescence.

Many cut flower species have been genetically engineered for determined inflorescence, that is, a single terminal flower is produced. Contrary to this, plants such as *Antirrihinum*, *Arabidopsis*, and *Petunia* have indeterminate inflorescence that can generate an indefinite number of flowers with profuse lateral branching. In these species, mutation in a single gene is sufficient to convert inflorescence from indeterminate to determinate inflorescence.

12.2.1.4 Modification of Floral Scent

Floral fragrance comprises an assortment of volatiles, which are not only involved in plant reproductive processes, but also in plant-to-plant interactions, defense, and abiotic stress responses. It is likely to attract consumers due to its sensory attributes. Floral scent is made up of various compounds. Over 700 of these have been identified in 60 families of plants (Knudsen et al., 1993). They are fatty acid derivatives such as benzenoids, phenylpropanoids, and terpenoids (monoterpenes and sesquiterpenes). The structures of hundreds of these scent compounds have been characterized. Although the number of cloned genes involved in the biosynthesis of floral scent compounds is steadily increasing, knowledge of biochemical and molecular pathways of the biosynthesis of scent compounds is still limited (Dudareva and Pichersky, 2002). Generally, floral scent compounds are synthesized *de novo* in petals. The enzymes involved in the biosynthesis of floral scents are expressed in petal epidermal cells and their expression is regulated at the transcriptional level, dependent on the stage of petal development.

Mature petunia flowers mainly release benzoids. This emission has a circadian rhythm, reaching its maximum at dusk. Volatiles are not stored during periods of low emission, but rather are synthesized *de novo*.

The synthesis of volatile compounds in carnation flower is developmentally regulated, and it has been suggested that their synthesis is membrane associated and that partitioning into the cytosol occurs in accordance with partition coefficients (Schade et al., 2001).

Antirrihinum petal cuticles do not provide any diffusive resistance to volatiles and so permit rapid emission of these compounds (Goodwin et al., 2003). These results imply that expression of a floral scent biosynthetic gene in a transgenic plant can lead to a modification of floral scent. Snapdragon also emits the monoterpenes, myrcene and (E)-bocimene, which are biosynthesized from geranyl pyrophosphate.

The first isolated structural gene encoding a floral scent biosynthetic enzyme was S-linalool synthase (LIS) from *Clarkia breweri*, a plant native to California that emits a strong sweet scent, of which S-Linalool is a major component. S-Linalool is biosynthesized from geranyl pyrophosphate, an intermediate of various terpenoids. The LIS gene is highly expressed in the epidermal cells of petals and in cells of the transmitting tract of the stigma and style, and the level of protein produced is transcriptionally regulated. It has also been shown that specialized scent glands and related organs are nonessential to the production of floral scent (Dudareva et al., 1996), which suggests that floral scent can be modified irrespective of the presence of scent glands.

Other molecular tools for the modification of floral scents have also been reported. Two genes encoding acetyl CoA:benzylalcohol acetyltransferase and benzyl CoA:benzylalcohol benzoyl transferase that are responsible for the production of benzylacetate and benzyl-benzoate, respectively, and are involved in scent biosynthesis have been cloned from *C. breweri* and were subsequently characterized (Dudareva et al., 1998).

These results indicate that successful modification of floral scents could be achieved by optimizing both the expression of *de novo* biosynthetic genes and the availability of their substrate. *Antirrihinum* (Snapdragon flowers) have been known to show coordinated regulation of phenylpropanoid and isoprenoid scent production (Dudareva et al., 2003). Even the petals of *Arabidopsis*, a self-pollinated plant, produce small amounts of terpenes and contain terpene synthase (Chen et al., 2003).

12.2.1.5 Extension of Vase Life

The postharvest life of flowers is influenced primarily by nutrition, stage of harvest, composition of vase solution, microbial colonization, and endogenous production of ethylene, which is a common plant ripening hormone associated with senescence. Endogenous

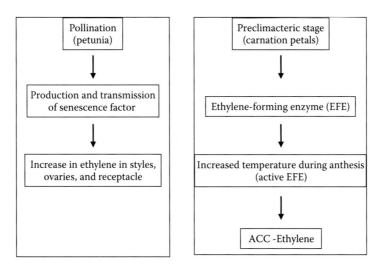

FIGURE 12.3
Ethylene-sensitive flower senescence.

ethylene production triggers flower senescence in carnation alone. Some varieties of rose experience petal drop on exposure to exogenous ethylene during the transport and storage of flowers and fruits.

All cut flowers are susceptible, in varying degrees, to microbial growth in vase water, leading to blockage of vascular tissue (air embolism) which prevents movement of water in the stem (xylem) and results in reduced vase life. Such microbial population build up through all stages of the postharvest treatment; sanitation practices including handling of cut flowers and preparation of vase water can control the problem. Lack of nutrients, primarily sugars in vase solution, can also promote senescence. Again, such deficiencies can be ameliorated through application of nutrient additives, antibiotics, and bactericides to vase water.

To extend vase life, carnations are typically treated with silver thiosulfate (STS) solution which interferes with the perception of ethylene by the flower, thus reducing the sensitivity of endogenous and exogenous ethylene. To safeguard the environment from hazardous pesticides, crops need to be genetically engineered with tolerance to various biotic stresses.

Together, an understanding of the action of ethylene on floral senescence (Figure 12.3) and the characterizations of genes responsible for suppression of ethylene production have provided an opportunity for genetic engineers to address chemical-free extension of vase life in carnation. However, commercialization of such a product is yet to occur due to the increased technical costs of genetically engineering varieties with extended vase life. Also, consumers are reluctant to pay more due to the availability of chemically treated flowers at a comparatively cheaper cost.

Hormones (other than ethylene) in promoting flower senescence:

- Feeding carnation flowers with 6-methyl purine, an inhibitor of cytokinin oxidase/dehydrogenase, resulted in an increased life span of petals, suggesting that ethylene promotes inactivation of cytokinins and facilitates the senescence process (Taverner et al., 2000).

- In ethylene-sensitive flowers such as carnation, abscisic acid (ABA) has been found to accelerate flower senescence.

- Application of auxins (IAA) to flowers has been found to stimulate senescence by hastening the rise in ethylene production.

- Gibberellic acid is known to delay senescence in some cut carnation flowers by acting as an antagonist to ethylene.

- Polyamines (putrescine, spermidine, and spermine) (PAs) have been reported as effective antisenescence agents and have been found to retard chlorophyll loss, membrane deterioration, and increases in RNase, as well as protease activities, all of which help to slow the senescence process.

- Methyl-jasmonates have been found to accelerate senescence in *Petunia hybrida, Dendrobium,* and *Phalaenopsis.*

- Coconut milk, as a source of cytokinins, has been shown to delay the senescence of cut *Allamanda* flowers.

- The introduction of a mutated ethylene receptor gene also reduced ethylene sensitivity in the orchids *Oncidium spp.* and *Odontoglossum spp.* (Raffeiner et al., 2009).

Other strategies have been used to extend the vase life in carnations (Figure 12.4) in order to minimize the application of chemicals. The first involved downregulation of ethylene production in carnation flowers via post transcriptional floral-specific gene silencing of a gene encoding ACC Oxidase (ACO) or ACC Synthase (ACS). These enzymes catalyze the ethylene biosynthesis. Although the effect of exogenous ethylene on carnation flowers during transit did not show many visual symptoms of reduced vase life, the perception still remains that the flower stems are less attractive than chemically treated stems.

FIGURE 12.4
Long-life carnation with downregulated petal ACC synthase (Florigene Ltd). (a) Carnations became senescent 2 weeks after harvest; (b) STS-treated carnation; (c) transgenic carnation.

12.2.1.6 Disease and Pest Management

The ornamental industry is vulnerable to a range of diseases which result in significant annual losses and increased production costs. Increased production costs are a direct result of management practices for disease prevention.

Generally, fungicides and insecticides are in common use against both the agents and vectors of diseases infesting ornamental plants. These are often expensive and toxic to a wide range of beneficial organisms in soil, and also deleterious to health in both humans and animals. Moreover, cost benefit is gained by the application of broad-spectrum fungicides directed at a wide range of pathogens.

Among the various strategies used against pathogens, a large proportion of which are pathogenic fungi, conventional plant breeding approaches have been the best exploited. However, these rely on the existence of adequate resources of resistance genes in the crop itself or in its wild relatives. Genetic engineering has the potential to overcome some of these barriers and lower production costs. However, genetic engineering strategies for disease and insect resistance must take into account the production costs, otherwise the varieties produced will be rendered unprofitable as compared to their chemical-control counterparts. To reduce the dependence on hazardous chemicals, there is an urgent need to breed crops utilizing tools of genetic engineering to develop resistance against fungal, bacterial, viral, and other disease-causing pathogens.

Various approaches of genetic engineering have been used to incorporate disease and insect tolerance into ornamental plants (Table 12.3). There are over 100,000 different species of fungus, more than 8,000 of them capable of causing diseases in plants, as reported by Agrios (1988). All plants are susceptible to fungal attack, and typically one fungus can attack more than one plant species. Almost all fungi spend part of their life cycle in the soil or on plant debris in the soil. Different symptoms may present themselves as a result of infection on different hosts by the same fungus. Commonly observed symptoms in plants infected by fungi are necrotic lesions, rot, wilt, stunting, rust, and mildew.

While plant defense mechanisms are complex, they appear to have common signal pathways. Necrotic lesions represent a self-defense mechanism by plants through localized cell death, to produce an additional barrier to invasion by the pathogen to other cells and tissues. Furthermore, a broad range of genes is activated in a plant in response to the

TABLE 12.3

Genetic Engineering Approaches for Incorporating Disease and Pest Tolerance

Genetic Engineering Approaches Used	Trait Modified	Gene(s) Involved	Flower Crops
Biolistic method	Blackspot	Chitinase gene	*Rosa hybrida* cv. Glad tidings (Marchant, 1998)
Agrobacterium mediated	Powdery mildew	*Ace-AMP1*	*Rosa hybrida* cv. Carefree Beauty (Li et al., 2003)
Biolistic and abiolistic methods	1. *Cymbidium mosaic virus* 2. *Erwinia carotovora*	*CymMV CP pflp*	*Phalaenopsis* spp. (Liao et al., 2004), *Ocidium* spp. (You et al., 2003)
Abiolistic method	*Heliothus viresens*	*cry1AB*	*D. grandiflora* (Van wordragen et al., 1993)
Agrobacterium mediated	*Lepidoptera and aphids*	*N*-methyltransferases genes (CaXMT1, CaMXMT1 and CaDXMT1)	*C. morifolium* cv. Shinba (Kim et al., 2011)

generation of typical symptoms of invasion by a pathogen (Martin et al., 2003). Some of these responses are specific to a certain pathogen or group of pathogens, while others are involved in general stress response. A limited number of such genes have been characterized, some of which hold relevance to plant-pathogen recognition and response to pathogen attack.

Several genes have been identified (Table 12.3) which encode enzymes involved in the synthesis of compounds toxic to disease-causing agents with an inhibitory effect on the growth of fungi. The resistance of plants, and more specifically floral crops, to fungal pathogens has been generally limited to the expression of hydrolytic enzymes or antimicrobial compounds (Punja, 2001).

It has been estimated that, on average, up to 20% of each year's carnation crop is lost due to fusarium wilt disease. Fusarium wilt is caused by the fungus *Fusarium oxysporum* sp. Dianthi (principally race 2). This disease is known to be soil-borne, and the means of controlling its spread rely on various soil treatments. Histological evidence (Baayen, 1998) indicates that the fungus first invaded the carnation plants through the roots. Invasion follows by the colonization and degradation of the vascular system of the root and subsequently of the stem. While the epidermis and cortex of carnation root can be colonized by *F. oxysporum*, the main route to the stem is through the vascular system via root wounds. Carnation plants are infected by fusarium wilt as a result of prolonged cultivation over a period of 2–3 years. An alternative to ward off this disease is growing carnations in a soil-less system or hydroponics, which is relatively expensive but effectively confines the spread of innoclum in other production plots.

Garden roses are susceptible to different diseases in their specific casual organism, such as downy mildew (*Peronospora sparsa*), powdery mildew (*Sphaerotheca pannosa*), and blackspot (*Diplocarpon rosae*). Some of the cultural practices followed to curtail the spread of infection include the selection of disease-free planting material, sanitation, and soil sterilization. The *Rosa hybrida* cv. Glad Tidings was transformed with a basic class I chitinase gene via biolistic gene delivery. Some of the resulting transgenic rose plants exhibited reduced sensitivity to blackspot infection (Marchant, 1998). Li et al. (2003) have reported increased resistance of *R. hybrida* cv. Carefree Beauty to powdery mildew by expression of an antimicrobial protein gene (Ace-AMP1).

Expression of genes coding for caffeine production in transgenic *Dendranthema grandiflora* (Chrysanthemum) was shown to confer resistance to gray mold (Kim et al., 2011). GM virus-resistant lines of the pot plant *Euphorbia pulcherrima* (poinsettia) and Phalaenopsis orchids have also been reported.

The insect resistance genes currently utilized in GM food crops are primarily based on cry endotoxin genes isolated from *Bacillus thurigensis*. These are only effective against a narrow range of pests; however, tolerance to susceptible pests has been demonstrated in transgenic chrysanthemum plants of *D. grandiflorum* carrying the cry1 Ab gene of *Bacillusl thurigensis var. Kurstaki HD-1* (Shinoyama and Mochizuki, 2006). The demonstration of aphid resistance in *D. grandiflora* genetically engineered to produce caffeine (Kim et al., 2011) is a significant recent development.

12.2.1.7 Abiotic Stress Tolerance

As well as genetically modifying flower color and plant shape, higher tolerance against abiotic factors (stresses such as drought, heat, frost, and salinity) is an area of focus for ornamental plant breeding. The credit for this work goes to Ornamental Bioscience GmbH, a joint venture between Selecta Klemm GmbH (Germany) and Mendel Biotechnology

Inc. (USA) that aims to reach this goal by a genetic engineering approach combining the mendel transcription factor technology. *Petunia hybrida* served as the first model system, in which several transcription factors were introduced by *Agrobacterium*-mediated gene transfer. Frost tolerance in *Petunia hybrida* has been reported to have been increased by the transfer of the CBF 3 gene from *Arabidopsis thaliana* (Warner, 2011), and this would potentially increase adaptability for growing petunias in a wider range of environments.

After transformation, transgenic lines are subjected to an initial screening for drought and frost tolerance. Promising lines with enhanced tolerance against one or more of these stress factors have been identified.

12.3 Plant-Specific Applications of Biotechnology

12.3.1 Evolution of Bioluminescent Orchids

The world's first and only genetically modified bioluminescent orchid has been successfully developed by Professor Chia Tet Fatt from the National Institute of Education (NIE). For developing the bioluminescent orchid, tissues from orchids (the *Dendrobium* genus) were transformed with the incorporation of the luciferase gene isolated from firefly. Using a method called *particle bombardment*, biologically active DNA from the firefly gene was delivered into orchid tissues. Transformed cells were identified by their bioluminescence trait. These transformed tissues were propagated to generate progeny of transgenic plants (plants with a foreign gene incorporated). This process was repeated several times, and the bioluminescent trait was found to be stable in all transgenic plants. This confirms that the firefly luciferase gene has been integrated into the orchid.

The bioluminescent trait of the orchid uses its own energy to create light at night that glows as a greenish fluorescence visible to the human eye, for up to five hours at a time. This greenish-white light is emitted from the whole orchid, including the roots, stem, leaves, and petals. The intensity of the light produced varies across the different parts, ranging from 5,000 to 30,000 photons per second.

12.3.2 Isolation of Somaclonal Variants with Noble Traits

Somaclonal variation is an important source of new cultivar development of ornamental crops. So far, at least 63 tropical foliage plants belonging to the Araceae family have been developed from a selection of somaclonal variants, and these cultivars have been reported to differ morphologically and genetically (via AFLP analysis) from their parents. The occurrence of somaclonal mutants may be a result of hidden genetic mutations that have been accumulated during years of asexual propagation of selected cultivars or due to the genetic instability of recent interspecific hybrids. Many ornamental foliage plants selected from somaclonal variants are found to be stable in production with the appearance of novel characters.

12.3.3 The Transgenic Blue Roses

Suntory Holdings Limited has successfully developed the world's first genetically engineered "blue rose." In an effort to achieve blue rose, breeders have crossed rose varieties

FIGURE 12.5 (See color insert)
Genetically engineered "blue rose."

grown all around the world. However, true blue roses, derived from the presence of blue pigment, have not yet come into being. This is due to the fact that genes in rose petals which encode the enzyme that is necessary to create the blue pigment, delphinidin, are not functional (the enzyme is known as flavonoid 3'5'-hydroxylase).

Suntory, together with Florigene Pty Ltd, started the joint development of biotechnology-driven "blue roses" by retrieving the genes necessary to create blue pigments from other plants, such as petunia and pansy, and transferring these into roses. Unlike the roses created using conventional breeding technologies, the roses developed using this process had almost 100% Delphinidin in their petals (Figure 12.5), thus exhibiting different hues of blue color in their petals.

12.3.4 *In Vitro* Flowering in Transgenic Ornamentals

Control of flowering is a very important practice in the ornamental industry. The rice Hd3a protein and *Arabidopsis* FT protein have been shown to correspond to florigen, a phytohormone that promotes flowering. Overexpression of these genes induces early flowering in transgenic plants, including chrysanthemum (Tanaka and Aida, 2010). When the *Arabidopsis FT* gene under a *Cauliflower mosaic virus* 35S promoter was expressed in rose (cultivar Lavande), carnation (cultivar Vega), and chrysanthemum (cultivar Improved Regan), transgenics flowered *in vitro* after transformation (Figure 12.6).

Transgenic petunia, *Nierembergia*, and torenia expressing the *FT* gene also flowered earlier than control plants. Transgenic torenia (cultivar Summerwave Blue) plants (Figure 12.6) expressing the *FT* gene continued flowering under low light (300 lux, room light condition), whereas the control did not bloom. The transgenic plants could therefore be used as indoor flowering plants.

12.3.5 Transgenic Chrysanthemums against Herbivore Repellence

Chrysanthemum plants have been transformed to express three N-methyltransferases that are known to mediate caffeine biosynthetic pathways. The transformed plants produced caffeine at approximately 3 mg/g fresh tissue and showed herbivore repellence.

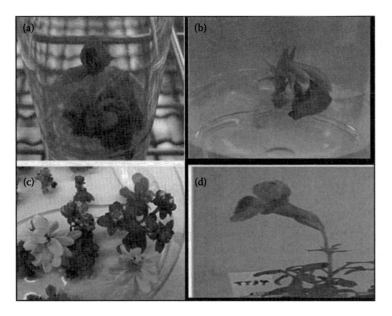

FIGURE 12.6 **(See color insert)**
Early flowering of transgenic ornamentals. (a) Rose, (b) carnation, and (c) chrysanthemum expressing *Arabidopsis*
FT gene. (d) Torenia expressing the *FT* gene flowering in low-level, indoor light conditions.

The results indicate that caffeine-producing chrysanthemum are resistant not only against
Lepidoptera pests, but also against herbivores, caterpillars, and aphids, which cause extensive damage to crops during the vegetative stage of development (Kim et al., 2011). Selected
chrysanthemum plants were transformed by an *Agrobacterium tumefaciens* strain (LBA4404)
harboring the pBIN-NMT777, a multigene expression vector containing the three cof-
fee *N*-methyltransferase genes (CaXMT1, CaMXMT1 and CaDXMT1) and two selection
marker genes (NPT II and HPT).

12.3.6 Transgenic *Oncidium* and *Odontoglossum* Orchid
Species for Enhanced Vase Life

Oncidium and *Odontoglossum* orchid species were transformed with the ethylene receptor
mutant gene *etr1-1*, isolated from *Arabidopsis* under control of a flower-specific promoter to
reduce the sensitivity to exogenous ethylene. Protocorm-like bodies (PLBs) of both orchid
genera were regenerated from leaf tip explants. Leaf tips and PLBs, cultured in liquid and
solid media, were compared as targets for genetic transformation. PLBs of *Oncidium* and
Odontoglossum cultured in a solid medium were successfully transformed with an expression vector containing *nptII* and *gus* genes driven by the cauliflower mosaic virus (CaMV)
35S promoter (Chan, 2003).

12.3.7 Induction of Male Sterility in Transgenic Chrysanthemum

Shinoyama et al. (2012b) reported the induction of male sterility in transgenic chrysanthemums by expression of a mutated ethylene receptor gene, Cm-ETR1/H69A. This
gene also caused delayed tapetum degradation of the anther sac and a reduction in pollen grains. With the expression of the above gene, the transfer of pollen from transgenic

chrysanthemums to its wild plants can be prevented. The expression of the gene was suppressed at low temperature as the mature pollen grains were observed at 10°C–15°C.

12.4 Role of Molecular Markers in Ornamental Crops

More recent developments were boosted by the fast progress made in the area of molecular genetics, and this led to the development of two major areas: the use of molecular markers and approaches to genetically modify plants through genetic engineering. Markers have been used over the years for the classification of plants. Markers are any trait of an organism that can be identified with confidence and relative ease. There are three major types of markers (Table 12.4 and Figure 12.7): morphological markers, which themselves are phenotypic traits or characters; biochemical markers, which include allelic variants of enzymes called isozymes; and DNA (or molecular) markers, which reveal sites of variation in DNA (Jones et al., 1997). Molecular markers represent genetic differences between individual organisms or species. Generally, they do not represent the target genes themselves but act as *signs* or *flags* (Collard et al., 2005). Genetic markers that are located in proximity to genes (i.e., tightly linked) may be referred to as gene *tags*. Molecular markers are now being widely used in genetic diversity analysis (Table 12.5) and as a selection tool in the development of ornamental cultivars (Arus, 2000; Debener, 2001; Meerow, 2005; Riek and Debener, 2010).

12.4.1 Morphological Markers

Morphological markers are scored visually and their inheritance can be followed with the naked eye. The traits included in this group are plant height, disease response, photoperiod, sensitivity, shape or color of flowers, fruits and seeds, and so on. Although they are generally scored quickly, simply, and without laboratory equipment, such markers are not put to much use, due to the following reasons:

- Genotypes can be generally ascertained at whole plant or plant organ level
- Such markers frequently cause major alternations in the phenotype

TABLE 12.4.

Comparison between Morphological, Isozyme, and DNA Markers

Feature	Morphological Markers	Biochemical Molecular Markers	DNA-Based Markers
Feature of the organism scored	Phenotype	Protein	DNA base sequence
Biological meaning of the markers	Consequences of gene action	Genes that are expressed	DNA sequences, may or may not represent genes
Plant material required for detection	Intact plant or plant organ	Little amount of tissue	Little to medium amount of tissue, and no matter what tissue is used
Efforts required for detection	Simple	Moderate	Moderate to difficult
Ease of use	Very easy	Moderately difficult	Moderately difficult to difficult

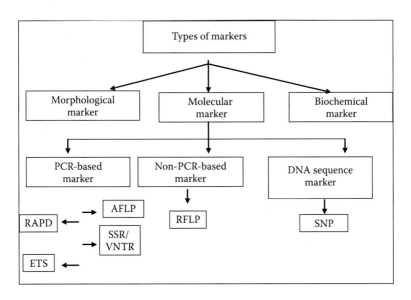

FIGURE 12.7
Various types of markers.

- Dominant, recessive interactions frequently prevent distinguishing all genotypes associated with morphological traits
- Morphological markers mask the effect of linked minor genes, making it nearly impossible to identify desirable linkages
- Limited in number
- Influenced by environment and also specific stage of the analysis

12.4.2 Biochemical Markers

These markers differentiate variations upon protein analysis. Biochemical or isozyme markers have been used for over 60 years for various research purposes in order to delineate phylogenetic relationships, to estimate genetic variability, and to characterize plant genetic resources. Isozymes were defined as structurally different molecular forms of an enzyme with qualitatively the same catalytic function. Isozymes originate through amino acid alterations, which cause changes in net charge, or the spatial structure (conformation)

TABLE 12.5

Number of Publications on Marker Application in Ornamentals
(March 2012)

Ornamental Crop	Number of Papers on Marker Development or Application	Genotyping, Genetic Diversity, Genetic Variability	Genetic Analysis, Mapping
Rose	111	76	35
Lilly	33	26	7
Chrysanthemum	29	24	5
Carnation	17	9	8

of the enzyme molecules and also, therefore, their electrophoretic mobility. After specific staining, the isozyme profile of individual samples can be observed. Isozyme analysis reveals the polymorphism of sequences of certain proteins (indirectly revealing the variation in DNA sequences from which they are translated). These were used to study the incompatibility and genetic variation among ornamental Cacti.

Advantages:

- Allozyme analysis does not require DNA extraction or the availability of sequence information
- They are quick and easy to use
- Simple analytical procedures
- Allozymes are codominant markers that have high reproducibility
- Zymograms (the banding pattern of isozymes) can be readily interpreted in terms of loci and alleles

Disadvantages:

- Relatively low abundance and low level of polymorphism
- Affected by environmental conditions
- Banding profile obtained for a particular allozyme marker may change depending on the type of tissue used for the analysis

12.4.3 Molecular Markers

As a tool to analyze genetic differences between genotypes at the level of the DNA, molecular markers were applied first in the field of human genetics (Botstein et al., 1980), but were quickly adopted by plant geneticists and breeders. A molecular marker is a DNA sequence that is readily detected and whose inheritance can easily be monitored. The uses of molecular markers are based on the naturally occurring DNA polymorphism, which forms the basis for designing strategies to exploit it for applied purposes. A marker must be polymorphic, that is, it must exit in different forms so that the chromosomes carrying the mutant genes can be distinguished from the chromosomes with the normal gene by a marker it also carries. Genetic polymorphism is defined as the simultaneous occurrence of a trait in the same population of two discontinuous variants or genotypes. Unlike protein markers, DNA markers segregate as single genes and they are not affected by the environment. DNA is easily extracted from plant materials and its analysis can be cost and labor effective. The first such DNA markers to be utilized were fragments produced by restriction digestion—the restriction fragment length polymorphism (RFLP)-based genes marker. Consequently, several marker systems have been developed. The major advantages provided by molecular markers are:

- Selectively neutral
- Available in unlimited numbers
- High resolution that may display genotypic differences down to single base pairs

All types of molecular markers have been applied to ornamental crops (Table 12.6; Arus, 2000; Debener, 2001), but due to their high reliability and the high information content

TABLE 12.6

Genetic Diversity Analysis and Identification of Genes Linked to Diseases of Different Ornamental Crops Using Molecular Markers

Crop	Marker Used	References
Gladiolus	Morphological and RAPD markers	Pragya et al., 2010
Ficus species	RAPD and ISSR	Hadia et al., 2008
Chrysanthemum	RAPD	Chatterjee et al., 2005
Pelargonium	RAPD, AFLP, microsatellite	Lesur et al., 2001
Rose	RAPD	Lewis et al., 2004
Gladiolus	RAPD Fusarium Wilt, (*in vitro* selection)	Nasir et al., 2012
Anthurium andreanum	RAPD	Nowbuth et al., 2005

Crop	Disease/Marker Used	Markers Linked	References
DNA Markers for Disease Diagnostics			
Lilium	TBV (Tulip breaking virus)/AFLP	24 linkage groups (Lilium $2n = 2x = 24$) 251 markers covering 1367 cM. Four QTLs for Fusarium resistance mapped to linkage groups 1, 5, 13, and 16, respectively. The resistance gene for TBV was placed on linkage group 9	Van Heusden et al., 2002
Carnation	Bacterial Wilt/RAPD	3 linkage groups, 3 genes are concerned with resistance to bacterial wilt	Onozaki et al., 2003

AFLP and microsatellite markers are used increasingly and are currently the markers of choice. Markers can be used very effectively to distinguish genotypes and therefore provide novel tools for the protection of breeders rights (Lesur et al., 2001). Furthermore, information from marker profiles can be used to analyze genetic relationships between genotypes or to infer phylogenetic relationships between species and closely related genera (Weising et al., 1995). These data provide valuable information about the genetic diversity among relatives of ornamental plants and may subsequently be used to broaden the gene pool of a particular ornamental crop.

12.4.3.1 Properties of an Ideal DNA Marker

An ideal molecular marker must have certain desirable properties:

1. Highly polymorphic nature: it must be polymorphic as it is polymorphism that is measured for genetic diversity studies
2. Codominant inheritance: determination of homozygous and heterozygous states of diploid organisms
3. Frequent occurrence in genome: a marker should be evenly and frequently distributed throughout the genome
4. Selective neutral behaviors: the DNA sequences of any organism are neutral to environmental conditions or management practices
5. Easy access (availability): it should be easy, fast, and cheap to detect
6. Easy and fast assay

7. High reproducibility

8. Easy exchange of data between laboratories

It is extremely difficult to find a molecular marker which meets all of the above criteria. A wide range of molecular techniques are available that detect polymorphism at the DNA level. Depending on the type of study undertaken, a marker system can be identified that would fulfill at least a few of the above characteristics (Weising et al., 1995). Various types of molecular markers are utilized to evaluate DNA polymorphism and are generally classified as hybridization-based markers and polymerase chain reaction (PCR)-based markers. In the former, DNA profiles are visualized by hybridizing the restriction enzyme-digested DNA to a labeled probe, which is a DNA fragment of known origin or sequence. PCR-based markers involve *in vitro* amplification of particular DNA sequences or loci, with the help of specifically or arbitrarily chosen oligonucleotide sequences (primers) and a thermostable DNA polymerase enzyme. The amplified fragments are separated electrophoretically, and banding patterns are detected by different methods such as staining and autoradiography. PCR is a versatile technique invented during the mid-1980s (Saiki et al., 1985). Ever since thermostable DNA polymerase was introduced in 1988, the use of PCR in research and clinical laboratories has increased tremendously. The primer sequences are chosen to allow base-specific binding to the template in reverse orientation. PCR is extremely sensitive and operates at a very high speed. Its application for diverse purposes has opened up a multitude of new possibilities in the field of molecular biology.

12.4.3.2 Restriction Fragment Length Polymorphism

Restriction fragment length polymorphism (RFLP) is a technique in which organisms may be differentiated by the analysis of patterns derived from cleavage of their DNA. If two organisms differ in the distance between sites of cleavage of particular *restriction endonucleases*, the length of the fragments produced will differ when the DNA is digested with a restriction enzyme. The similarity of the patterns generated can be used to differentiate species (and even strains) from one another. This technique is mainly based on a special class of enzymes, that is, restriction endonucleases. They have their origin in the DNA rearrangements that occur due to evolutionary processes, point mutations within the restriction enzyme recognition site sequences, insertions or deletions within the fragments, and unequal crossing over. Size fractionation is achieved by gel electrophoresis and, after transfer to a membrane by Southern blotting, fragments of interest are identified by hybridization with radioactive labeled probes. Different sizes or lengths of restriction fragments are typically produced when different individuals are tested. Such a polymorphism can be used to distinguish plant species, genotypes, and, in some cases, individual plants (Karp, 1997).

Advantages:

- Moderately polymorphic
- High genomic abundance and random distribution
- Codominant markers
- High reproducibility

RFLP markers were used for the first time in the construction of genetic maps by Botstein et al. (1980).

Disadvantages:

- Require large quantities (1–10 µg) of purified, high molecular weight DNA
- The requirement of radioactive isotopes makes the analysis relatively expensive and hazardous. The assay is time-consuming and labor intensive
- Inability to detect single base changes restricts their use in detecting point mutations

12.4.3.3 Random Amplified Polymorphic DNA

Random amplified polymorphic DNA (RAPD) is a PCR-based technology. The method is based on enzymatic amplification of target or random DNA segments with arbitrary primers. In 1990, Welsh and McClelland developed a new PCR-based genetic assay, namely randomly amplified polymorphic DNA (RAPD). This procedure detects nucleotide sequence polymorphisms in DNA by using a single primer of arbitrary nucleotide sequence. In this reaction, a single species of primer anneals to the genomic DNA at two different sites on complementary strands of DNA template. If these priming sites are within an amplifiable range of each other, a discrete DNA product is formed through thermocyclic amplification. On average, each primer directs the amplification of several discrete loci in the genome, making the assay useful for efficient screening of nucleotide sequence polymorphism between individuals.

Advantages:

- Quick and easy to assay
- Low quantities of template DNA required
- No sequence data for primer construction needed
- High genomic abundance and randomly distributed throughout the genome
- Dominant markers

Disadvantages:

- Low reproducibility
- High sensitivity to the reaction conditions
- Require purified, high-molecular-weight DNA
- RAPD markers not locus-specific, and band profiles cannot be interpreted in terms of loci and alleles

12.4.3.4 Amplified Fragment Length Polymorphism

Amplified fragment length polymorphism (AFLP) is essentially intermediate between RFLPs and PCR. AFLP is based on selectively amplifying a subset of restriction fragments from a complex mixture of DNA fragments obtained after digestion of genomic DNA with restriction endonucleases. Polymorphisms are detected from differences in the length of the amplified fragments by polyacrylamide gel electrophoresis (PAGE).

Steps:

1. Restriction of DNA and ligation of oligonucletide adapters
2. Preselective amplification

3. Selective amplification
4. Gel analysis of amplified fragments

Advantages:

- High genomic abundance
- Highly polymorphic
- Considerable reproducibility
- No sequence data for primer construction are required

The use of AFLP in genetic marker technologies has become the dominant tool due to its capability to disclose a high number of polymorphic markers by a single reaction (Vos et al., 1995).

Disadvantages:

- Need for purified, high-molecular-weight DNA
- Dominant nature
- Relatively expensive

12.4.3.5 Microsatellites or Simple Sequence Repeat

Microsatellites are also known as simple sequence repeats (SSRs), and are sections of DNA consisting of tandemly repeating mono-, di-, tri-, tetra-, or penta-nucleotide units that are arranged throughout the genomes of most eukaryotic species. Microsatellite sequences are especially suited to distinguishing closely related genotypes; because of their high degree of variability, they are favored in population studies and for the identification of closely related cultivars. Microsatellite polymorphism can be detected by Southern hybridization or PCR. Microsatellites, like minisatellites, represent tandem repeats, but their repeat motifs are shorter (1–6 base pairs). If nucleotide sequences in the flanking regions of the microsatellite are known, specific primers (generally 20–25 bp) can be designed to amplify the microsatellite by PCR. Microsatellites and their flanking sequences can be identified by constructing a small-insert genomic library, screening the library with a synthetically labeled oligonucleotide repeat, and sequencing the positive clones.

Advantages:

- They are codominant markers with high heritability.
- They are highly reproducible.
- They are highly polymorphic.
- They distinguish closely related genotypes.

Disadvantages:

- They require large amounts of time and labor for production of primer.
- They usually require PAGE.
- They require large amounts of DNA.
- They involve costly primer development.

TABLE 12.7

Number of ESTs Listed under the
NCBI[a] Taxonomy Browser (March
2012)

Genera	ESTs Listed at NCBI
Petunia	50,705
Lilium	3,154
Rosa	9,295
Antirrihinum	25,310
Chrysanthemum	7,109

[a] National Centre for Biotechnology
Information.

12.4.3.6 Expressed Sequence Tags

Expressed Sequence Tags (ESTs) are short (about 200–500 bp) sequenced fragments of randomly collected mRNAs expressed in a tissue or a cell. The mRNAs are collected and reverse-transcribed into cDNA that is then cloned and sequenced. ESTs represent the expression of genes in a tissue or a cell at a certain time or condition, and permit rapid identification of the expressed genes. The concept is to sequence bits of DNA that represent genes expressed in certain cells, tissues, or organs from different organisms and use these tags to fish a gene out of a portion of chromosomal DNA by matching base pairs. EST libraries have been constructed for some important ornamental crop plants (Table 12.7). The potential of developing EST libraries for ornamental crops is enormous, but as yet is largely unrealized due to the difficulty in obtaining funding for such research on ornamental crops.

Molecular markers have been much more widely applied to food and fiber crops than to ornamentals. To a large extent, this probably reflects the levels of funding available to researchers in ornamental crops. Molecular genetic applications are costly and difficult to initiate without considerable expenditure. However, we are likely to see an ever-increasing reliance on genetic marker technology in programs of evaluation and enhancement of ornamental germplasm as the technology becomes more ubiquitous in university departments and research institutions. Traditional breeders of ornamental crops have much to gain by applying these tools to their selection programs. We can also expect that large-scale genetic mapping efforts will be applied to the highest value ornamental crop genera, such as roses, in the near future.

12.5 Commercialization of Transgenic Ornamental Plants

The global flower industry thrives on novelty. Genetic engineering provides a valuable means of expanding the ornamental gene pool, thus promoting the generation of new commercial varieties. There has been extensive research on the genetic transformation of different flowering plant species, and many ornamental species have now been successfully transformed, including those that are most important commercially.

To date, more than 30 ornamental species have been transformed, including anthurium, begonia, carnation, chrysanthemum, cyclamen, datura, daylily, gentian, gerbera, gladiolus, hyacinth, iris, lily, lisianthus, orchid, pelargonium, petunia, poinsettia, rose, snapdragon, and torenia (Deroles et al., 2002). New ornamental plant varieties are being created by breeders in response to consumer demand for new products. In general terms, engineered traits are valuable to either the consumer or the producer. At present only consumer traits appear able to provide a return capable of supporting what is still a relatively expensive molecular breeding tool. The commercialization of genetically engineered flowers is currently confined to novel colored carnations. The production of novel flower color is the first success story in ornamental genetic engineering; however, further products are expected given the level of activity in the field.

Other traits that have received attention include floral scent, floral and plant morphology, senescence of flowers both on the plant and postharvest, and disease resistance. To date, there are only a few ornamental genetically modified products in development and only one, a carnation genetically modified for flower color, is available in the marketplace. There are approximately 8 ha of transgenic carnation in production worldwide, largely in South America. The other breeding programs on color modification or alteration of plant architecture and height remain focused on rose, gerbera, and various pot-plant species.

12.6 Challenges in Commercialization of Transgenic Ornamentals

Of greater importance is the cost of research and development of GM ornamentals. Only those crops or traits which have high market-demand are genetically modified. It is very difficult to conduct trails of GM ornamentals over large area in different countries as the regulatory approval takes several years. This poses a barrier for the GM products currently in development. Royalties may also be required for the developers of GM crops, but negotiations with governments are complicated. In ornamentals, identity preservation is unlikely to be a significant obstacle to commercialization, but unless the existing cost barriers and regulatory uncertainty are addressed, ornamental breeding strategies will continue to shift away from GE approaches. The challenges of commercialization have been well documented by Dobres (2011), Sexton and Zilberman (2011), Strauss (2011), and Chandler and Sanchez (2012).

References

Agrios, G.N. 1988. *Plant Pathology*. 3rd ed. Academic Press, San Diego, p. 803.

Arus, P. 2000. Molecular markers for ornamental breeding. *Acta Horticulturae* 508: 91–98.

Baayen, R.P. 1998. The histology of susceptibility and resistance of carnation to fusarium wilt. *Acta Horticulturae* 216: 119–124.

Birch, R.G. 1997. Plant transformation: Problems and strategies for practical application. *Annual Reviews of Plant Physiology and Plant Molecular Biology* 48: 297–326.

Boase, M.R., Lewis, D.H., Davies, K.M., Marshall, G.B., Patel, D., Schwinn, K.E., and Deroles, S.C. 2010. Isolation and antisense suppression of flavonoid 3'5'-hydroxylase modifies flower pigments and colour in cyclamen. *BMC Plant Biology* 10: 107.

Botstein, D., White, R.L., Skolnick, M., and Davis, R.W. 1980. Construction of a genetic map in man using restriction fragment length polymorphisms. *American Journal of Human Genetics* 32: 314–331.

Brugliera, F. 2009. Genetically modified chrysanthemums. International Patent Publication Number WO/2009/062253.

Cazzonelli, C.I. and Pogson, B.J. 2010. Source to sink: Regulation of carotenoid biosynthesis in plants. *Trends in Plant Science* 15: 266–274.

Chan, M.T. 2003. *Agrobacterium tumefaciens*-mediated transformation of an *Oncidium* orchid. *Plant Cell Reports* 21: 993–998.

Chandler, S.F. 2003. Commercialization of genetically modified ornamental plants. *Journal of Plant Biotechnology* 5: 69–77.

Chandler, S.F. and Sanchez, C. 2012. Genetic modification; the development of transgenic ornamental plant varieties. *Plant Biotechnology Journal* 10(8): 891–903.

Chatterjee, J., Mandal, A.K., Ranade, S.A., and Dutta, S.K. 2005. Estimation of genetic diversity of four chrysanthemum mini cultivars. *Pakistan Journal of Biological Science* 8(4): 546–549.

Chen, F., Tholl, D., D'Auria, J.C., Farooq, A., Pichersky, E., and Gershenzoh, J. 2003. Biosynthesis and emission of terpenoid volatiles from arabidopsis flowers. *Plant Cell* 15: 481–494.

Chen, W.H., Hsu, C.Y., Cheng, H.Y., Chang, H., Chen, H.H., and Ger, M.J. 2011. Down regulation of putative UDP-glucose: Flavonoid 3-Oglucosyltransferase gene alters flower colouring in Phalaenopsis. *Plant Cell Reports* 30: 1007–1017.

Chia, T.F., Chan, Y.S., and Chua, N.H. 1994. The firefly luciferous gene as a non invasive reporter for *Dendrobium* transformation. *Plant Journal* 6: 441–446.

Collard, B.C.Y., Jahufer, M.Z.Z., Brouwer, J.B., and Pang, E.C.K. 2005. An introduction to markers, quantitative trait loci (QTL) mapping and marker-assisted selection for crop improvement: The basic concepts. *Euphytica* 142: 169–196.

Courtney-Gutterson, N., Napoli, C., Lemieux, C., Morgan, A., Firoozababy, E., and Robinson, K.E.P. 1994. Modification of flower colour in Florist's Chrysanthemum: Production of a white-flowering variety through molecular genetics. *Biotechnology* 12: 268–271.

Davies, K.M. and Schwinn, K.E. 1997. Flower colour. *Book Biotechnology of Ornamental Plants*. pp. 259–294.

Debener, T. 2001. Molecular tools for modern ornamental plant breeding and selection. *Acta Horticulturae* 552: 121–127.

Deroles, S.C., Boase, M.R., Lee, C.E., and Peters, T.A. 2002. Gene transfer to plants. In: Breeding for Ornamentals: Classical and Molecular Approaches (Vainstein, A., ed.), pp. 155–196. Kluwer Academic Publishers, Dordrecht.

Dobres, M.S. 2011. Prospects for commercialisation of transgenic ornamentals. In Transgenic Horticultural Crops; Challenges and Opportunities (Mou, B. and Scorza, R., eds), pp. 305–316. CRC, Boca Raton, FL.

Dolgov, S.V. 2009. Early flowering transgenic Chrysanthemum plants. *Acta Horticulturae* 836: 241–246.

Dudareva, N. and Pichersky, E. 2002. Biochemical and molecular genetic aspects of floral scents. *Plant Physiology* 122: 627–633.

Dudareva, N., Cseke, L., Blanc, V.M., and Pichersky, E. 1996. Evolution of floral scent in Clarkia: Novel patterns of Slinalool synthase gene expression in the *C. breweri* flowers. *Plant Cell* 8: 1137–1148.

Dudareva, N., D'Auria, J.C., Nam, K.H., Raguso, R.A., and Pichersky, E. 1998. Acetyl- oA:benzylalcohol acetyltransferase: An enzyme involved in floral scent production in *Clarkia breweri*. *Plant Journal* 14: 297–304.

Dudareva, N., Martin, D., Kish, C.M., Kolosova, N., Gorenstein, N., Faldt, J., Miller, B., and Bohlmann, J. 2003. (E)- b-ocimene synthase and myrcene synthase genes of floral scent biosynthesis in snapdragon: Function and expression of three terpene synthase genes of a new terpene synthase subfamily. *Plant Cell* 15: 1227–1241.

Goodwin, S.M., Kolosova, N., Kish, C.M., Wood, K.V., Dudareva, N., and Jenks, M.A. 2003. Cuticle characterization and volatile emissions of petals in Antirrhinum majus. *Plant Physiology* 117: 435–443.

Hadia, H.A., El-Mokadem, H.E., and El-Tayeb, H.F. 2008. Phylogenetic relationship of four *Ficus* species using random amplified polymorphic DNA (RAPD) and Inter-simple Sequence Repeat (ISSR) markers. *Journal of Applied Sciences Research* 4(5): 507–514.

Holton, T.A., Brugliera, F., Lester, D.R., Tanaka, Y., Hyland, C.D., Menting, J.G., Lu, C.Y., Farcy, E., Stevenson, T.W., and Cornish, E.C. 1993. Cloning and expression of cytochrome P450 genes controlling flower colour. *Nature* 366: 276–279.

Hong, B., Tong, Z., Ma, N., Kasuga, M., Yamaguchi-Shinozuka, K., and Gao, J.P. 2006. Expression of the *Arabidopsis DREB1A* gene in transgenic chrysanthemum enhances tolerance to low temperature. *Journal of Horticultural Science and Biotechnology* 81: 1002–1008.

Hsiao, Y.Y., Pan, Z.J., Hsu, C.C., Yang, Y.P., Hsu, Y.C., Chuang, Y.C., Shih, H.H., Chen, W.H., Tsai, W.C., and Chen, H.H. 2011. Research on orchid biology and biotechnology. *Plant Cell Physiology* 52: 1467–1486.

James, C. 2002. Global status of commercialized transgenic crops: 2002. ISAAA Briefs No. 27. ISAAA, Ithaca.

Jones, N., Ougham, H., and Thomas, H. 1997. Markers and mapping: We are all geneticists now. *New Phytologist* 137: 165–177.

Kamiishi, Y., Otani, M., Takagi, H., Han, D.S., Mori, S., Tatsuzawa, F., Okuhara, H., Kobayashi, H., and Nakano, M. 2011. Flower colour alteration in the liliaceous ornamental *Tricyrtis* sp. by RNA interference-mediated suppression of the chalcone synthase gene. *Molecular Breeding* 30(2): 671–680.

Karp, A. 1997. Reproducibility testing of RAPD, AFLP and SSR markers in plants by a network of European laboratories. *Molecular Breeding* 3: 381–390.

Katsumoto, Y., Mizutani, M., Fukui, Y., Brugliera, F., Holton, T., Karan, M., Nakamura, N., et al. 2007. Engineering of the rose flavonoid biosynthetic pathway successfully generated blue-hued flowers accumulating delphinidin. *Plant Cell Physiology* 48: 1589–1600.

Kim, Y.S., Lim, S., Yoda, H., Choi, C.S., Choi, Y.E., and Sano, H. 2011. Simultaneous activation of salicylate production and fungal resistance in transgenic chrysanthemum producing caffeine. *Plant Signal Behaviour* 6(3): 409–412.

Knudsen, J.T., Tollsten, L., and Bergstrom, G.L. 1993. Floral scents: A checklist of volatile compounds isolated by headspace techniques. *Phytochemistry* 33: 253–280.

Kuehnle, A.R and Sugii, N. 1992. Transformation of Dend- robium orchid using particle gun bombardment of proto corms. *Plant Cell Reports* 11: 484–488.

Lesur, C., Becher, A., Wolff, K., Weising, K., Steinmetz, K., Peltier, D., and Boury, S. 2001. DNA fingerprints for *Pelargonium* cultivar identification. Proceedings of International Symposium on Molecular Markers (Doré et al., eds). *Acta Horticulturae*, p. 546.

Lewis, A., Mary, C., and Morvillo, N. 2004. Investigating the identity of rose varieties utilizing Randomly Amplified Polymorphic DNA (RAPD) Analysis. *Proceedings of Florida State Horticulture Society* 117: 312–316.

Li, X, Gasic, K., Cammue, B., Broekaert, W., and Korban, S.S. 2003. Transgenic rose lines harboring an antimicrobial protein gene, Ace-AMP1, demonstrate enhanced resistance to powdery mildew (*Sphaerotheca pannosa*). *Planta* 218(2): 226–32.

Liao, L., ChunPan, J.I., Chan, Y.L., Hsu, Y.H., Chen, W.H., and Chan, M.T. 2004. Transgene silencing in *Phalaenopsis* expressing the coat protein of Cymbidium Mosaic Virus is a manifestation of RNA-mediated resistance. *Molecular Breeding* 13(3): 229–242.

Marchant, R. 1998. Expression of a chitinase transgene in rose (*Rosa hybrida* L.) reduces development of blackspot disease (*Diplocarpon rosae* Wolf). *Molecular Breeding* 4: 187–194.

Martin, G.B., Bogdmore, A.J., and Sessa, G. 2003. Understanding the function of plant resistance proteins. Annual review. *Plant Biology* 54: 23–61.

Meerow, A.W. 2005. Molecular genetic characterization of new floricultural germplasm. *Acta Horticulturae* 683: 43–63.

Mol, J.N.M., Holton, T.A., and Koes, R.E. 1995. Floriculture: Genetic engineering of commercial traits. *Trends in Biotechnology* 13: 350–355.

Nakamura, N., Mizutani, F., Fukui, M., Ishiguro, Y., Suzuki, K., and Tanaka, Y. 2010. Generation of red flower varieties from blue *Torenia hybrida* by redirection of the flavonoid pathway from delphinidin to pelargonidin. *Plant Biotechnology* 27: 375–383.

Nakatsuka, T., Mishiba, K., Kubota, A., Abe, Y., Yamamura, S., Nakamura, N., Tanaka, Y., and Nishihara, M. 2010. Genetic engineering of novel flower colour by suppression of anthocyanin modification genes in gentian. *Journal of Plant Physiology* 167: 231–237.

Nakatsuka, T., Saito, M., Yamada, E., and Nishihara, M. 2011. Production of picotee type flowers in Japanese gentian by CRES-T. *Plant Biotechnology* 28: 173–180.

Nasir, I.A., Jamal, A., Rahman, Z., and Husnain, T. 2012. Molecular analyses of gladiolus lines with improved resistance against fusarium wilt. *Pakistan Journal of Botany* 44(1): 73–79.

Nishihara, M. and Nakatsuka, T. 2011. Genetic engineering of flavonoid pigments to modify flower colour in floricultural plants. *Biotechnology Letters* 33(3): 433–441.

Noda, N., Aida, R., Sato, S., Ohmiya, A., and Tanaka, Y. 2010. Method for production of chrysanthemum plant having petals containing modified anthocyanin. International Floriculture, Ornamental and Plant Biotechnology Volume V © 2008 Global Science Books, UK. Patent Publication Number WO/2010/122850.

Nowbuth, P., Khittoo, G., Bahorun, T., and Venkatasamy, S. 2005. Assessing genetic diversity of some *Anthuriumandraeanum* Hort. cut-flower cultivars using RAPD Markers. *African Journal of Biotechnology* 4(10): 1189–1194.

Ohmiya, A., Kishimoto, S., Aida, R., Yoshioka, S., and Sumitomo, K. 2006. Carotenoid cleavage dioxygenase (CmCCD4a) contributes to white colour formation in chrysanthemum petals. *Plant Physiology* 142: 1193–1201.

Onozaki, T., Ikeda, H., Tanikawa, N., and Shibata, M. 2003. Identification of random amplified polymorphic DNA markers linked to bacterial wilt resistance in carnations. Proceeding of 21st IS on Classical/Molecular Breeding (Forkmann, G. et al., ed.). *Acta Horticulturae*, p. 612.

Pragya, R., Bhat, K.V., Misra, R.L., and Ranjan, J.K. 2010. Analysis of diversity and relationships among *Gladiolus* cultivars using morphological and RAPD markers. *Indian Journal of Agricultural Sciences* 80(9): 766–772.

Punja, Z.K. 2001. Genetic engineering of plants to enhance resistance to fungal pathogens: A review of progress and future prospects. *Canadian Journal of Plant Pathology* 23: 216–235.

Quattrocchio, F., Verweij, W., Kroon, A., Spelt, C., Mol, J., and Koes, R. 2006. PH4 of Petunia is an R2R3 MYB protein that activates vacuolar acidification through interactions with basic-helix-loop-helix transcription factors of the anthocyanin pathway. *Plant Cell* 18: 1274–1291.

Raffeiner, B., Serek, M., and Winkelmann, T. 2009. *Agrobacterium tumefaciens*-mediated transformation of *Oncidium* and *Odontoglossum* orchid species with the ethylene receptor mutant gene *etr1-1 Plant Cell Tissue and Organ Culture* 98(2): 125–134.

Riek, J. De. and Debener, T. 2010. Present use of molecular markers in ornamental breeding. *Acta Horticulturae* 855: 77–83.

Saiki, R.K., Scharf, S., Faloona, F., Mullis, K.B., Horn, G.T., Erlich, H.A., and Arnheim, N. 1985. Enzymatic amplification of beta-globin genomic sequences and restriction site analysis for diagnosis of sickle cell anemia. *Science* 230: 1350–1354.

Schade, F., Legge, R.L., and Thompson, J.E. 2001. Fragrance volatiles of developing and senesing carnation flower. *Phytochemistry* 56: 703–710.

Sexton, S. and Zilberman, D. 2011. The economic and marketing challenges of horticultural biotechnology. In: *Transgenic Horticultural Crops; Challenges and Opportunities* (Mou, B. and Scorza, R., eds), pp. 175–192. CRC, Boca Raton, FL.

Shinoyama, H. and Mochizuki, A. 2006. Insect resistant Chrysanthemum [*Dendranthema grandiflorum* (Ramat.) Kitamura]. *Acta Horticulturae* 714: 177–184.

Shinoyama, H., Aida, R., Ichikawa, H., Nomura1, Y., and Mochizuki, A. 2012a. Genetic engineering of chrysanthemum (*Chrysanthemum morifolium*): Current progress and perspectives. *Plant Biotechnology* 29: 323–337.

Shinoyama, H., Sano, T., Saito, M., Ezura, H., Aida, R., Nomura, Y., and Kamada, H. 2012b. Induction of male sterility in transgenic chrysanthemums *Chrysanthemum morifolium* Ramat.) by expressing a mutated ethylene receptor gene, *Cm-ETR1/H69A*, and the stability of this sterility at varying growth temperatures. *Molecular breeding* 29: 285–295.

Shoji, K., Miki, N., Nakajima, N., Momonoi, K., Kato, C., and Yoshida, K. 2007. Perianth bottom-specific blue colour development in tulip cv. Murasakizuisho requires ferric ions. *Plant Cell Physiology* 48: 243–251.

Strauss, S.H. 2011. Why are regulatory requirements a major impediment to genetic engineering of horticultural crops? In: *Transgenic Horticultural Crops; Challenges and Opportunities* (Mou, B. and Scorza, R., eds), pp. 249–262. CRC, Boca Raton, FL.

Suzuki, K., Zue, H., Tanaka, Y., Fukui, Y., Fukuchi-Mizutani, M, Murakami, Y., Katsumoto, Y., Tsuda, S., and Kusumi, T. 2000. Flower colour modifications of *Torenia hybrida* by cosuppression of anthocyanin biosynthesis genes. *Molecular Breeding* 6: 239–246.

Suzuki, S., Nishihara, M., Nakatsuka, T., Misawa, N., Ogiwara, I., and Yamamura, S. 2007. Flower colour alteration in *Lotus japonicus* by modification of the carotenoid biosynthetic pathway. *Plant Cell Reports* 26: 951–959.

Swarnapiria, R. 2009. Genetic transformation in ornamentals: A review. *Agricultural Reviews* 30(2): 120–131.

Tanaka, Y. and Aida, R. 2010. Genetic engineering in floriculture. In: *Molecular Technique in Crop Improvement* (Mohan, J.S. and Brars, D.S., eds.), pp. 695–718. Springer, New York.

Tanaka, Y., Brugliera, F., and Chandler, S. 2009. Recent progress of flower colour modification by biotechnology. *International Journal of Molecular Science* 10: 5350–5369.

Tanaka, Y., Katsumoto, Y., Brugliera, F., and Mason, J. 2005. Genetic engineering in floriculture. *Plant Cell Tissue Organ Culture* 80: 1–24.

Taverner, E.A., Letham, D.S., Wang, J., and Cornish, E. 2000. Inhibition of carnation petal inrolling by growth retardants and cytokinins. *Australian Journal of Plant Physiology* 27: 357–362.

Togami, J., Okhuhara, H., Nakamura, N., Ishiguro, K., Hirose, C., Ochiai, M., Fukui, Y., Yamaguchi, M., and Tanaka, Y. 2011. Isolation of cDNAs encoding tetrahydroxychalcone 2-glucosyltransferase activity from carnation, cyclamen, and catharanthus. *Plant Biotechnology* 28: 231–238.

Van der Salm, T.P.M., van der Toorn, C.J.G., Bouwer, R., Hanisch ten Cate, C.H., and Dons, H.J.M. 1997. Production of ROL gene transformed plants of *Rosa hybrida* L. and characterization of their rooting ability. *Molecular Breeding* 3: 39–47.

Van Heusden, A.W., Jongerius, M.C, van Tuyl, J.M., Straathof, Th.P. and Mes, J.J. 2002. Molecular assisted breeding for disease resistance in lily. Proc. XX Eucarpia Symposium on New Ornamentals II (Van Huylenbroeck, J. et al., eds) *Acta Horticulturae*, p. 572.

Van Wordragen, M.F., Honée, G., and Dons, H.J.M. 1993. Insect resistant chrysanthemum calluses by introduction of a *Bacillus thuringiensis* crystal protein gene. *Transgenic Research* 2: 170–180.

Vos, P., Hogers, R., Bleeker, M., Reijans, M., Lee van de, T., Hornes, M., Frijters, A., Pot, J., Peleman, J., Kuiper, M., and Zabeau, M. 1995. AFLP: A new technique for DNA fingerprinting. *Nucleic Acids Research* 23: 4407–4414.

Warner, R. 2011. Genetic approaches to improve cold tolerance of Petunia. *Floriculture International*, June 15–16.

Weising, K., Nybom, H., Wolff, K., and Meyer, W. 1995. *DNA Fingerprinting in Plants and Fungi* (Arbor, A., ed.), pp. 1–3. CRC, Boca Raton.

Welsh, J. and McClelland, M. 1990. Fingerprinting genomes using PCR with arbitrary primers. *Nucleic Acids Research* 18: 7213–7218.

Winefield, C., Lewis, D., Arathoon, S., and Deroles, S. 1999. Alteration of Petunia plant from through the introduction of the *rolC* gene from *Agrobacterium rhizogenes*. *Molecular Breeding* 5: 543–551.

Yoshida, K., Mori, M., and Kondo, T. 2009. Blue flower colour development by anthocyanins: From chemical structure to cell physiology. *Natural Product Reports* 26: 884–915.

You, S.J., Liau, C.H., Huang, H.E., Feng, T.Y., Prasad, V., and Hsiao, H.H. 2003. Sweet pepper ferredoxin-like protein (*pflp*) gene as a novel selection marker for orchid transformation. *Planta* 217: 60–65.

Zambryskpi, H., Genetelloj, J.C., Leemansm, V.O., and Schell, J. 1983. Ti-plasmid vector for the introduction of DNA into plant cells without alteration of their normal regeneration capacity. *EMBO Journal* 2: 2143–2150.

Zuker, A., Ahroni, A., Tzfira, T., Ben-Meir, H., and Vainstein, A. 1999. Wounding by bombardment yields highly efficient *Agrobacterium*-mediated transformation of carnation (*Dianthus caryophyllus* L.). *Molecular Breeding* 5: 367–375.

13

Omics Advances in Tea (Camellia sinensis)

Mainaak Mukhopadhyay, PhD; Bipasa Sarkar, PhD; and Tapan Kumar Mondal, PhD

CONTENTS

13.1 Introduction..439
13.2 Omics Resources...440
13.3 Marker-Based Omics Approaches..442
    13.3.1 Pregenomics Markers...442
        13.3.1.1 Morphological Markers..442
        13.3.1.2 Biochemical Markers 442
        13.3.1.3 Cytological Markers...443
    13.3.2 Postgenomics Markers 443
        13.3.2.1 Randomly Amplified Polymorphic DNA (RAPD)...........................448
        13.3.2.2 Inter-Simple Sequence Repeat (ISSR) 448
        13.3.2.3 Restriction Fragment Length Polymorphism (RFLP)......................448
        13.3.2.4 Simple Sequence Repeats (SSR)...449
        13.3.2.5 Sequence Tagged Microsatellite Site (STMS)............................449
        13.3.2.6 Amplified Fragment Length Polymorphism (AFLP)......................449
        13.3.2.7 Single Nucleotide Polymorphism (SNP)....................................449
        13.3.2.8 Other Marker-Based Applications..449
13.4 Transgenomics..450
13.5 Functional Genomics..450
13.6 Proteomics and Metabolomics...453
13.7 Conclusion 454
References...455

13.1 Introduction

Tea is made from the young leaf of the plant *Camellia sinensis* and serves as a nonalcoholic beverage globally. The tea industry generates employment in more than 52 countries, especially in Asia and Africa. A few of its wild genera such as *C. japonica* and *C. oleracea* are also important for their beautiful ornamental flowers and oil-bearing capability.

Progress in conventional breeding in tea and related genera is slow and hence there is scope for omic approaches for varietal improvement. Since the work of Forrest (1969) on *in vitro* culturing of tea cells, a tremendous amount of work has been completed or initiated on several aspects of functional and structural genomics. This word is discussed in this chapter. The various biotechnological studies on tea are displayed chronologically in Figure 13.1.

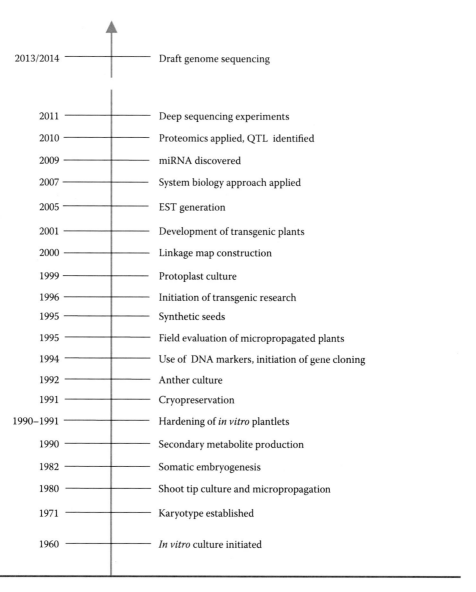

FIGURE 13.1
Time line of landmark discoveries in tea.

13.2 Omics Resources

The existing cultivated taxa of tea include three natural hybrids of *C. sinensis* (L.) O. Kuntze or China type, *C. assamica* (Masters) or Assam type and *C. assamica* subsp. *lasiocalyx* (Planchon ex Watt.) or Cambod or Southern type. They originated in Southeast Asia (Wight 1959) but are cultivated within the latitudinal range 45°N to 34°S, which traverses about 52 countries (Deka et al. 2006; Mondal et al. 2004). This genus has more than 325 species (Mondal 2002a; Mondal et al. 2003) and over 600 popular genotypes of tea are cultivated worldwide, many of which are embodied with unique traits such as high quality, high yield, and tolerance to biotic or abiotic stresses (Mondal et al. 2004). In addition, more than 3000 cultivars of *C. japonica*, a wild species of tea, are cultivated globally due to its excellent floricultural value. Further, several other related genera such as *C. reticulata, C. sasanqua,* and *C. saluensis* are also popular due to their high ornamental value. A few species such as *C. oleifera, C. semiserrata,* and *C. chekiangolomy* are used to produce oil from mature seeds, which is used in the pharmaceutical industry on a limited scale. The diploid genome size of *Camellia* ($2n = 2x = 30$) was found to be 4 giga base pairs (Tanaka et al. 2005) and karyotypes range from 1.28 to 3.44 μm (Bezbaruah 1971). Tea chromosomes are smaller and clustered together at metaphase due to "stickiness." The ratios of long arm:short arm for all 15 pairs of chromosomes are consistent (1–1.91), which advocates a monophyletic origin for all *Camellia* species.

In the absence of a whole genome sequence, genomic resources are essential, especially for woody plants where sequence-driven research such as marker-assisted breeding, association mapping, cloning of genes, and mapping of quantitative **trait** loci (QTLs) is extremely important to develop elite genotypes within a short period of time. In *Camellia*, several genomic resources have been generated (Table 13.1). For example, 48,335 expressed sequence tags (as of 30 September, 2012) are available in the National Center for Biotechnology Information (NCBI) database. These are mainly associated with quality and a few are associated with abiotic and biotic stresses. Lin et al. (2011) constructed a bacterial artificial chromosome (BAC) library of tea and generated 401,280 clones with an average insert size of 135 kb. Among the other genomic resources, while Kamunya et al. (2010) identified 23 minor QTLs, approximately 250 polymorphic simple sequence repeat (SSR) markers have been validated by various groups. Apart from that, a few single nucleotide polymorphisms (SNPs) (Huang et al. 2007) and microRNAs (Das and Mondal 2010; Prabu and Mandal 2010; Mohanpuria and Yadav 2012) have also been identified by *in silico* analysis, although no experimental validation was reported in these studies.

Use of genomic resources may enhance the varietal improvement of this woody plant. Conventional tea breeding has contributed significantly to the genetic improvement of tea over the past several decades, but it is slow and labor intensive. The principle of tea breeding primarily consists of the selection of existing variations for desirable traits or development of hybrid seeds by growing two selected parents in an isolated orchard, which so far have produced several improved cultivars in various tea-growing countries of the world (Deka et al. 2006). Details of breeding objectives, techniques, and achievements both in tea

TABLE 13.1

Different Genetic and Genomic Resources Available in Tea

Name	Quantity	References
Popular cultivars	600	Mondal et al. (1994)
Different species of *Camellia*	325	Mondal et al. (1994)
BAC clones	401,280	Lin et al. (2011)
QTL	23	Kamunya et al. (2010)
Cloned genes	55	NCBI
miRNA	16	Das and Mondal (2010); Prabu and Mandal (2010); Mohanpuria and Yadav (2012)
ESTs	48,335	NCBI
SSR markers	250	Kato et al. (2008); Sharma et al. (2009)
SNPs	5	Huang et al. (2007)

and other *Camellia* species have been reviewed and updated (Wachira 1990; Mondal 2000c, 2008; Ghosh-Hazra 2001). As leaves are the only harvestable plant part, breeding objectives are primarily focused on developing high-yielding cultivars, although a few high-quality cultivars have also been developed. Recently efforts have also been made to understand the physiological and molecular mechanisms of various abiotic stresses such as drought, nutritional stress, and water logging as well as biotic stresses such as blister blight and tea mosquito bug (Deka et al. 2006; Mukherjee et al. 2005; Mukhopadyay et al. 2012, 2013a,b).

13.3 Marker-Based Omics Approaches

Darwin and Mendel established the fundamentals of plant breeding and genetics at the turn of the twentieth century and likewise, the assimilation of biotechnological advances and genomics and molecular markers with conventional breeding practices has created the basis for molecular plant breeding, an interdisciplinary science that transformed twenty-first-century crop improvement (Moose and Mumm 2008). In tea and *Camellia* species, a number of markers have been used, which are discussed below.

13.3.1 Pregenomics Markers

13.3.1.1 Morphological Markers

Morphological variations have been identified, standardized, and recurrently used for several years by tea breeders globally (Kulasegaram 1980; Mondal 2002c, 2008). Barua (1963) portrayed morphology along with anatomy, which were elaborated by Bezbaruah (1971). While tea taxonomists use criteria like leaf and bush architecture and floral biology (Banerjee 1992) for the identification of species (Table 13.2), bush vigor, pruning recovery, height, root mass, root–shoot ratio, dry matter production, and partitioning are used as yield indicators by tea breeders (Banerjee 1992). However, with the advancement of science, attention has shifted toward the development and use of biochemical markers.

13.3.1.2 Biochemical Markers

The presence and quantity of calcium oxalate crystals and catechins in leaf, sclereids in lamina, terpene index, polyphenols, amino acids, chlorophyll, carotenoids (carotene, lutein, violaxanthine, and neoxanthine), and ratio of dihydroxylated to trihydroxylated catechins are some of the successful biochemical markers used for organoleptic efficiency in tea, although readers may refer to excellent reviews for a detailed account (Wachira 1990; Singh 1999; Ghosh-Hazra 2001). On the other hand, caffeine, volatile compounds (Seurei 1996), pigmentation (Banerjee 1992), and leaf pubescence (Wight and Barua 1954) are some of the important quality determinants (Table 13.2). Later, isozyme techniques became popular for studying genetic diversity, parentage identification, and so forth. These were initially standardized by Hairong et al. (1987) and were soon being used in tea by several workers (Xu et al. 1987; Ikeda et al. 1991; Chengyin et al. 1992; Anderson 1994; Singh and Ravindranath 1994; Yang and Sun 1994; Borthakur et al. 1995; Chen 1996; Sen et al. 2000). However, with the advancement of more robust DNA markers, uses of isozymes gradually became limited.

TABLE 13.2

Pregenomics Markers Used in Genetic Improvement of Tea

Criterion	References
Quantity and shape of the sclereids	Barua (1958)
Bush vigor	Barua and Dutta (1959)
Leaf geometry	Banerjee (1987)
Volatile flavor compound	Borse et al. (2002)
Leaf pose, color, serration of the margin, and angle	Eden (1976)
Chlorophyll content/photosynthesis rate	Ghosh-Hazra (2001)
Quantitative changes in chlorophyll-a, chlorophyll-b, and carotenoids	Hazarika and Mahanta (1984)
Epicuticular wax	Kabir et al. (1991)
Dry matter production and partitioning	Magambo and Cannell (1981)
Green leaf catechin and ratio of dihydroxylated to trihydroxylated catechin	Magoma et al. (2000)
Root length	Nagarajah and Ratnasurya (1981)
Caffeine and volatile flavor compounds	Owuor and Obanda (1998)
Leaf, floral biology, and growth morphology	Sealy (1958)
Chloroform test	Sanderson (1964)
Pruning litter weight	Satyanarayan and Sharma (1982)
Anthocyanin pigmentations in young leaves	Satyanarayan and Sharma (1986)
Evenness of the flush, plucking density, and recovery time of pruning	Singh (1999)
Terpene index	Takeo (1981)
Leaf pubescence	Wight and Barua (1954)
Phloem index	Wight (1954)
Isozymes	Xu et al. (1987); Ikeda et al. (1991); Chengyin et al. (1992); Anderson (1994); Singh and Ravindranath (1994); Yang and Sun (1994); Borthakur et al. (1995); Chen (1996); Sen et al. (2000)

13.3.1.3 Cytological Markers

Cytological markers of *Camellia* were elaborately studied in the early 1970s and karyotypic data was accumulated for a number of species of this genus (Beretta et al. 1988; Fukushima et al. 1966; Ackerman 1971; Kondo 1975, 1978a,b, 1979; Mondal 2011), as reviewed by Kondo (1975). Karyotype analysis by grouping chromosome size is difficult in the *Camellia* taxa, since the chromosomes are graded imperceptibly from the largest to the smallest and homologous chromosomes do not appear identical (Kondo 1975, 1978a,b). Karyotypic variability and divergence among the several accessions of different species of *Camellia* were revealed by the C-banding method (Kondo and Parks 1981). But more recently, insightful techniques have shifted the attention of the scientific community from karyotyping to other DNA-based markers.

13.3.2 Postgenomics Markers

Application of DNA-based molecular markers in tea and other wild species is limited to techniques such as genetic fingerprinting, kinness identification, and diversity analysis, which are tabulated in Table 13.3.

TABLE 13.3

Molecular Markers Used in Tea and Other *Camellia* Species

Country	Marker	Objectives	Population Size	References
India	RAPD	Genetic fidelity	17 micropropagated plants	Balasaravanan et al. (2002)
	AFLP	Genetic diversity	49 genotypes	Balasaravanan et al. (2003)
	RAPD	Genetic diversity	14 genotypes	Bera and Saikia (2002)
	RAPD	Genetic fidelity	7 micropropagated plants	Borchetia et al. (2009)
	RFLP	Detection of adulteration in made tea	10 different tea samples	Dhiman and Singh (2003)
	AFLP, SSR and RAPD	QTL identification	42 genotypes	Kamunya et al. (2010)
	AFLP	Genetic diversity	14 genotypes	Karthigeyan et al. (2008)
	AFLP	Genetic diversity	29 genotypes	Mishra and Sen Mandi (2001)
	AFLP	Marker for drought tolerance	29 genotypes	Mishra and Sen-Mandi (2004)
	AFLP	Genetic diversity	29 genotypes	Mishra et al. (2009)
	RAPD	Genetic diversity	27 genotypes	Mondal (2000)
	ISSR	Genetic diversity	25 genotypes	Mondal (2002b)
	RAPD	Genetic fidelity	18 micropropagated plants	Mondal and Chand (2002)
	RAPD	Development of DNA isolation technique	28 genotypes	Mondal et al. (2000a)
	AFLP	Genetic diversity	32 genotypes	Paul et al. (1997)
	AFLP	Genetic diversity	32 genotypes	Rajasekaran (1997)
	RAPD and ISSR	Genetic diversity	10 genotypes	Roy and Chakraborty (2007)
	RAPD and ISSR	Genetic diversity	21 genotypes	Roy and Chakraborty (2009)
	SSR	Development of SSR marker	32 genotypes of tea and 2 genotypes of C. *japonica*	Sharma et al. (2009)
	SSR	Cross-species validation of SSR	18 genotypes	Sharma et al. (2011)
	5S rDNA	Phylogenetic relationship	28 genotypes	Singh and Ahuja (2006)
	RAPD	Authentication of made tea sample	11 genotypes	Singh et al. (1999)
	RAPD	Genetic fidelity	16 vegetative propagated plants	Singh et al. (2004)
	RFLP	Genetic organization	1 tea and 11 non-tea species	Singh et al. (2011)
	ISSR	Genetic fidelity	20 genotypes	Thomas et al. (2006)
Japan	SSR	Identification of cultivars	41 genotypes	Kato et al. (2008)
	*mat*K	Genetic diversity	118 genotypes	Katoh et al. (2003)
	STS-RFLP	Authentication of made tea sample	46 samples	Kaundun and Matsumoto (2003a)

TABLE 13.3 (Continued)

Molecular Markers Used in Tea and Other *Camellia* Species

Country	Marker	Objectives	Population Size	References
	CAPS	Species-specific probe	50 genotypes	Kaundun and Matsumoto (2003b)
	SSR	Genetic diversity	24 genotypes	Kaundun and Matsumoto (2002)
	SSR	Maternal inheritance of cp genome	6 genotypes	Kaundun and Matsumoto (2011)
	RFLP	Genetic analysis	30 genotypes	Matsumoto et al. (1994)
	RFLP	Genetic analysis of PAL gene	472 genotypes	Matsumoto et al. (2002)
	SSR	Development and validation of SSR	157 genotypes	Ohsako et al. (2008)
	e-RAPD	Development of DNA isolation	16 genotypes	Tanaka and Taniguchi (2002)
	RAPD	Identification of true crosses	2 parents and 38 F_1 progenies	Tanaka and Yamaguchi (1996)
	Gene specific PCR-RFLP	Phylogenic relationship	162 genotypes of 49 *Camellia* spp.	Tanikawa et al. (2008)
	SSR	Development of SSR		Ueno et al. (1999)
	SSR	Genetic diversity	518 genotypes of *C. japonica*	Ueno et al. (2000)
	CAPS	Cultivar identification	63 genotypes	Ujihara et al. (2011)
China	RAPD	Genetic analysis	23 species	Chen and Yamaguchi (2002)
	RAPD	Discrimination of genotypes	20 species	Chen and Yamaguchi (2005)
	RAPD	Genetic analysis	15 genotypes	Chen et al. (1998)
	RAPD	Genetic analysis	5 genotypes	Chen et al. (1999)
	RAPD	Molecular phylogeny	24 species	Chen et al. (2002b)
	RAPD and RFLP	Development of technique	7 genotypes	Chen et al. (1997)
	RAPD	Genetic analysis	7 genotypes	Chen et al. (1998)
	RAPD	Genetic analysis	24 species	Chen et al. (2002a)
	RAPD	Genetic diversity	15 genotypes	Chen et al. (2005a)
	RAPD	Genetic diversity	45 genotypes	Fang et al. (2003)
	SSR	Genetic diversity	185 genotypes	Fang et al. (2012)
	ISSR	Genetic diversity	14 genotypes	Hou et al. (2007)
	AFLP	Linkage map	69 F_1 genotypes	Huang et al. (2005)
	RAPD and ISSR	Genetic mapping	94 F_1 genotypes	Huang et al. (2006)
	SNP	Identification of SNP	40 genotypes	Huang et al. (2007)
	RAPD	Genetic diversity	23 genotypes	Hui et al. (2004)
	SSR	Development of SSR markers	24 genotypes of tea and 2 genotypes of *C. japonica*	Hung et al. (2008)
	ISSR	Genetic diversity	181 genotypes	Ji et al. (2011)
	SSR	Development of SSR markers	10 genotypes	Jin et al. (2006)
	SSR	Genetic diversity	42 genotypes	Jin et al. (2007)

(continued)

TABLE 13.3 (Continued)

Molecular Markers Used in Tea and Other *Camellia* Species

Country	Marker	Objectives	Population Size	References
	RAPD	Genetic diversity	2 genotypes	Li et al. (2003)
	RAPD	Genetic diversity	69 genotypes	Li et al. (2005)
	SSR	Development of BAC library for tea genome sequence	1 genotypes	Lin et al. (2011)
	ISSR	Genetic diversity	25 genotypes	Liu et al. (2008)
	ISSR	Genetic diversity	134 genotypes	Liu et al. (2010)
	PAL gene, *rpl32-trnL* spacer	Genetic diversity and erosion	21 genotypes of *C. taliensis*	Liu et al. (2012a)
	ISSR	Genetic diversity	134 genotypes	Liu et al. (2012b)
	RAPD	Genetic diversity	71 genotypes	Luo et al. (2004)
	SSR	Genetic diversity	45 genotypes of 6 different species	Ma et al. (2010)
	SSR	Validation of markers	21 genotypes	Ma et al. (2012)
	RAPD	Genetic diversity	240 genotypes	Shen et al. (2007)
	SRAP	Genetic diversity	25 genotypes	Shen et al. (2009)
	RAPD, AFLP	Genetic diversity	120 genotypes of *C. nitidissima*	Tang et al. (2006)
	RAPD, AFLP	Genetic analysis	48 genotypes	Wachira et al. (2001)
	ISSR	Genetic diversity	114 genotypes of *C. reticulata*	Wang and Ruan (2012)
	RAPD	Genetic diversity	15 genotypes	Wang et al. (2010)
	ISSR	Genetic diversity	84 genotypes of *C. euphlebia*	Wei et al. (2005)
	ISSR	Genetic diversity	250 genotypes of *C. nitidissima*	Wei et al. (2008)
	SSR	Development of SSR markers	25 genotypes of *C. nitidissima*	Wei et al. (2010)
	RAPD	Genetic diversity	27 genotypes	Wen et al. (2002)
	SSR	Development of SSR markers	150 genotypes of *C. chekiangoleosa*	Wen et al. (2012)
	RAPD	Genetic diversity	31 genotypes	Wu et al. (2002a)
	RAPD	Parentage identification	3 hybrids	Wu et al. (2002b)
	AFLP	Genetic diversity	34 genotypes	Xiao et al. (2007)
	SSR	Development of SSR markers	24 genotypes of *C. taliensis*	Yang et al. (2009)
	RAPD	Germplasm evaluation	200 genotypes	Yang et al. (2003)
	ISSR	Parent selection	48 genotypes	Yao et al. (2008)
	SSR	Genetic diversity	450 genotypes	Yao et al. (2012)
	ISSR	Identification of clones	10 genotypes of *C. oleifera*	Zhang et al. (2007)
	AFLP	Genetic diversity	23 genotypes	Zhao et al. (2006)
	SSR	Genetic diversity	40 genotypes	Zhao et al. (2008)

TABLE 13.3 (Continued)

Molecular Markers Used in Tea and Other *Camellia* Species

Country	Marker	Objectives	Population Size	References
Kenya	RAPD, AFLP	Genetic linkage map	90 genotypes of F_1 progenies	Hackett et al. (2000)
	RAPD, ISSR	Assessment of mating system	180 progenies of 6 genotypes	Muoki et al. (2007)
	RAPD	Genetic analysis	38 genotypes	Wachira et al. (1995)
	RAPD	Diagnosis of gene introgression	28 genotypes	Wachira et al. (1997)
	RAPD and AFLP	Genetic diversity	24 genotypes	Wachira et al. (2001)
Italy	SSR	Genetic relationships	132 accessions of *Camellia* spp	Caser et al. (2010)
Taiwan	RAPD, ISSR	Genetic analysis	37 genotypes	Lai et al. (2001)
	RAPD	Genetic diversity	96 genotypes	Luo et al. (2004)
	Intronic sequence of *RPB2* gene	Barcoding of species	25 genotypes of 4 species	Su et al. (2009)
	ITS of nrDNA	Phylogenic relationship	7 species	Vijayan and Tsou (2008)
	Nuclear ITS	Molecular taxonomy	112 species of *Camellia*	Vijayan et al. (2009)
Turkey	AFLP	Genetic diversity	32 genotypes	Kafkas et al. (2009)
Sri Lanka	RAPD	Genetic diversity	39 genotypes	Mewan et al. (2005)
	SSR	Genetic map	3 genotypes	Mewan et al. (2007)
	SSR	Genetic diversity	27 genotypes	Ariyaratne and Mewan (2009)
	RAPD	Genetic diversity	46 genotypes	Goonetilleke et al. (2009)
South Africa	RAPD	Genetic diversity	5 genotypes	Wright et al. (1996)
South Korea	RAPD	Genetic analysis	6 genotypes	Kaundun and Park (2002)
	RAPD	Phylogenic relationship	27 genotypes	Kaundun et al. (2000)
	AFLP	Genetic analysis	37 genotypes	Lee et al. (2003)
	RAPD	Genetic diversity	20 genotypes	Park et al. (2002)
Pakistan	RAPD	Genetic diversity	75 genotypes	Afridi et al. (2011)
	RAPD	Genetic diversity	24 genotypes	Gul et al. (2007)
Portugal	RAPD	Genetic analysis	71 genotypes	Jorge et al. (2003)
U.S.A.	SBA	Evolutionary studies	30 species	Prince and Parks (1997)
	SBA	Phylogenic relationship	19 species	Prince and Parks (2000)
	SBA	Evolutionary studies	35 species	Prince and Parks (2001)
U.K.	SSR	Development of SSR markers	15 genotypes	Freeman et al. (2004)

Note: SBA, sequence-based analysis.

Randomly amplified polymorphic DNA (RAPD) was the first DNA marker used for characterization of tea germplasm (Wachira et al. 1995). Following that, a number of markers by several groups have been used, although advanced areas such as mapping of QTLs and gene introgression by marker-assisted selection are either in their initial stages (Kamunya et al. 2010) or have not been attempted in tea.

13.3.2.1 Randomly Amplified Polymorphic DNA (RAPD)

RAPD markers are the most widely used in *Camellia* species and have been extensively used for the study of genetic diversity in tea (Table 13.3) and related species and their hybrids. A wide range of markers, random 10-oligomer to chloroplast-specific sequences, have been used and tallied with the former monogram on *Camellia* for phylogenetic relationships. Taxonomic classification, as mentioned in Chang's manual for diverse *Camellia* species, was confirmed (Prince and Parks 1997, 2000; Thakor 1997; Tiao and Parks 1997, 2003; Yoshikawa and Parks 2001; George and Adam 2006; Orel et al. 2007).

13.3.2.2 Inter-Simple Sequence Repeat (ISSR)

Several workers in tea-producing countries have used inter-simple sequence repeat (ISSR) markers, which are similar to RAPD markers, mainly for fingerprinting of their own cultivars (Table 13.3). ISSR markers have also been used to characterize the genotypes of other *Camellia* species. For instance, the level and pattern of genetic diversity of 84 individuals from natural populations of *C. euphlebia*, a rare and endangered species of Guangxi province of China, were studied using 100 ISSRs (Wei et al. 2005). The results indicated a relatively low level of genetic diversity at the species level and population level, and a relative degree of differentiation amid populations with low gene flow. Inbreeding and limited gene flow might be the key factors in the genetic structure of *C. euphlebia* and strategies were proposed for genetic conservation. *C. nitidissima* Chi (Theaceae), a popular ornamental species with golden-yellow flowers, have receded greatly due to deforestation in recent decades. Genetic diversity and differentiation of 12 natural populations and one *ex situ* conserved population in China were analyzed using ISSR markers and the study indicated a low level of genetic diversity at both species and population levels but a high degree of differentiation among populations. In contrast, the *ex situ* population contained higher genetic variability and it was proposed that all the wild *C. nitidissima* populations should be protected *in situ* (Wei et al. 2008).

13.3.2.3 Restriction Fragment Length Polymorphism (RFLP)

Restriction fragment length polymorphism (RFLP) markers were used to study genetic variations of Japanese green tea and sorted them into five groups, which distinguished Japanese green tea cultivars from others on the basis of grouping (Matsumoto et al. 1994). Kaundun and Matsumoto (2003b) employed STS-RFLP using the sequence information of three genes (phenyl-ammonia lyase, chalcone synthase, and dihydroflavonol 4-reductase) and demonstrated their utility for authentication of 46 processed green tea samples made out of Yabukita, a popular Japanese cultivar that faces adulteration problems due to price realization.

13.3.2.4 Simple Sequence Repeats (SSR)

SSRs, that is, microsatellites, are well recognized due to hypervariability, PCR scoring simplicity, codominance, and high reproducibility (Wu and Tanksley 1993). Ueno et al. (1999) developed 21 SSR markers in *C. japonica* to genotype 53 ecotypes. Several workers used these to study the genetic diversity of tea (Table 13.3) either by constructing an SSR-rich genomic library or mining the SSRs from the public domain data and validating them among the diverse sets of genotypes (Table 13.3) found to be useful in understanding the genetic structure of a particular population.

13.3.2.5 Sequence Tagged Microsatellite Site (STMS)

Recently, Caser et al. (2010) used sequence tagged microsatellite site (STMS) markers to study the genetic variation with 132 accessions of representing 22 different *Camellia* species. Cross-transferability was obtained that fruitfully amplified polymorphic alleles in all analyzed species and 96 alleles were scored. The distribution of genetic variation, attributed by AMOVA, particularly highlighted genetic overlap among *C. sasanqua* cultivars and those belonging to *C. vernalis*, *C. hiemalis*, and *C. hybrida*. So, it was revealed that STMS markers are suitable for the detection of genetic variability of Camellia genotypes, although nobody has yet used these markers.

13.3.2.6 Amplified Fragment Length Polymorphism (AFLP)

During the late 1990s, amplified fragment length polymorphism (AFLP) markers with five enzyme–primer combinations were used to study the genetic diversity of 32 tea genotypes of Indian and Kenyan origin and revealed 73 unambiguous polymorphic bands (Paul et al. 1997). They concluded that Assam clones from India and Kenya clustered closely, indicating a common ancestry. In the same year, AFLP was used to analyze 42 tea clones of India (Rajasekaran 1997). It was reported that 90% of the UPASI clones are inbred and incongruous for commercial cultivation. Since then, several groups have used AFLP techniques primarily to understand the genetic diversity for the purpose of identification of duplicates or developing the core (Table 13.3).

13.3.2.7 Single Nucleotide Polymorphism (SNP)

Detection of allelic differences or variations in the plant genetic resources (PGRs) is an important application of genomic resources and can be achieved by highly robust DNA-based markers such as SNPs or haplotypes (i.e., a group of SNPs that are linked to a particular trait). Although large-scale SNPs have been discovered in several crops using various sources of sequences, in tea it is in its initial stages. Huang et al. (2007) identified the SNP present in the coding region of polyphenol oxidase from different genotypes of tea. Thus there is tremendous scope for large-scale discovery and identification of SNP markers.

13.3.2.8 Other Marker-Based Applications

The construction of genetic linkage maps is a significant step for mapping of agronomically important genes, which will be very useful for developing improved cultivars. However, the prerequisite for a genetic linkage map is the development of a biparental mapping population from where segregation patterns of the marker can be tagged. This is difficult in a woody species like tea due to its perennial nature with long gestation periods. However, in tea, a pseudo-testcross or semiunstructured population thought to be derived from two known heterogenous parents was scored for RAPD and AFLP markers in order to develop a low-density linkage map (Hackett et al. 2000), but a high-density linkage map is not yet available.

To date, systematic organelle DNA analysis for tea has not been reported, although we made a first attempt to study the choloroplast (cp) DNA of tea (Borthakur et al. 1998). Since then, choloroplast-specific DNA sequences such as intergenic spacer between the *trnL* (UAA) 3' exon and *trn* F (GAA), or *rbcL* and *mat* K, allelic variations of the *rbcL* gene, have been used to classify different species of *Camellia* (Thakor 1997; Prince and Parks 1997, 2000). Further, to facilitate the efficacy of nrITS in elucidating the interspecific relationships of *Camellia*, Vijayan and Tsou (2008) studied seven closely and seven distantly related species. Extensive study of *Camellia*, based on allelic variations of ITS sequences, showed well-resolved interspecies relationships and, thus, the potential of nrITS in deducing phylogenetic relationships was revealed.

Although the detailed limitations of conventional breeding of tea are well documented by Mondal et al. (2004), primary bottlenecks are due to its perennial nature and long gestation period. Therefore, to circumvent the constraints of tea breeding, genetic engineering with transferral of potential genes has become the apt alternative, as discussed below.

13.4 Transgenomics

Perhaps genetic transformation or transgenomics is the technique in varietal improvement in tea with the most potential, mainly for two reasons: (1) it can overcome the barrier of breeding techniques for targeted gene transfer and (2) tea made from transgenic plants will be more acceptable commercially as tea leaves are heated at 120°C during the manufacturing, a temperature that is sufficient to detoxify any foreign protein introduced into the plant through the transgenic approach. For genetic transformation of tea, two different methods were attempted: using either explant cocultivation with *Agrobacterium* Ti/Ri plasmid-based binary/cointegrate vectors or particle gun bombardment (biolistics); the former brought more success. Efforts were made to genetically transform tea using different wild/engineered strains of *Agrobacterium tumefaciens* or *A. rhizogenes* (Table 13.4). Nevertheless, a complete protocol for production of transgenic tea plants was developed for the first time by Mondal et al. (2001b,c). More recently, *Agrobacterium*-mediated genetic transformation has been exploited to reduce caffeine by either suppressing glutathione synthetase (Mohanpuria et al. 2008) or silencing the caffeine synthase gene (Mohanpuria et al. 2011). However, to date, transgenic tea plants are not commercially available for large-scale cultivation anywhere in the world.

13.5 Functional Genomics

In tea, functional genomics was initiated with the isolation of the chalcone synthase gene from a Japanese green tea cultivar Yabukita, revealing its organ-specific and

TABLE 13.4

A Chronological Summary of Transgenic Tea Research

Techniques	Remarks	Gene	References
A. tumefaciens	Antibiotic selection for *Camellia* species was reported	—	Tosca et al. (1996)
A. rhizogenes	First attempt to induce hairy root formation	*rolB*	Zehra et al. (1996)
A. rhizogenes	Transformation technology was exploited to promote roots to facilitate hardening of the micropropagated tea plants	*rolB*	Konwar et al. (1998)
A. tumefaciens	Preliminary study for gene transfer to tea plants	*gus-intron*	Matsumoto and Fukui (1998)
A. tumefaciens	Standardization of somatic embryogenesis and transient expression of *gus* gene	*npt-II*	Mondal et al. (1999)
Particle bombardment	First attempt at standardization of the biolistic-mediated transformation protocol	*npt-II*	Akula and Akula (1999)
A. tumefaciens	Transgenic calli were produced utilizing the phenolic inducer acetosyringone at an effective range (100–500 µM)	*npt-II*	Matsumoto and Fukui (1999)
A. tumefaciens	Detailed study on *Bt* gene transformation was reported	*bt*	Luo and Liang (2000)
A. tumefaciens	Development of selection system for putative transformants	*npt-II*	Mondal et al. (2001d)
A. tumefaciens	Production of transgenic plants from transformed somatic embryos	*npt-II* and *gus-intron*	Mondal et al. (2001b)
A. tumefaciens	Tea leaves with glabrous surface having lower phenol and wax content were identified to be more suitable for infection	*npt-II*	Kumar et al. (2004)
A. tumefaciens and particle bombardment	Green fluorescence protein gene was transferred with organelle target signals	*gfp*	Kato et al. (2004)
A. tumefaciens and particle bombardment	Attempt was made for standardization of the protocol	*npt-II*	Wu et al. (2003)
A. tumefaciens	Attempts to overcome the bactericidal effect of tea leaf polyphenol	*npt-II*	Sandal et al. (2007)
A. tumefaciens	Silencing of glutathione synthetase gene in callus	*gs*	Mohanpuria et al. (2008)
A. rhizogenes	High production of catechin in hairy root culture	*rolB*	John et al. (2009)
A. tumefaciens	Caffeine-free plant production	*cs*	Mohanpuria et al. (2011)

Notes: *rolB*, rooting-locus gene B; *gus-intron*, beta-glucuronidase; *npt-II*, neomycin phosphotransferase II; *gfp*, green florescence protein; *gs*, glutathione synthetase; *cs*, caffeine synthase.

sugar-responsive expression (Takeuchi et al. 1994). Since then a significant amount of work has been done in tea and its wild relatives, targeting isolation of individual genes for various traits (Figure 13.2). Two types of effort have been made in tea: (1) cloning of individual genes associated with particular traits and (2) differential gene expression that leads to the identification of genes that are associated to a trait. Interestingly, the majority of the genes

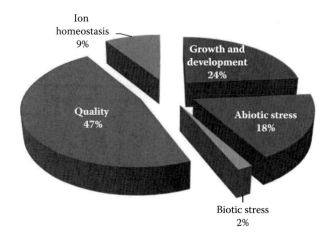

FIGURE 13.2
Different trait-specific genes cloned from *Camellia*.

cloned so far are associated to quality followed by yield, as there is major demand for these two parameters in the tea industry.

Tea leaf contains a high amount of polyphenols (35%), the majority of which are catechins. Therefore, several genes related to catechin biosynthesis pathways such as phenylalanine ammonialyase, chalcone synthase, flavanone 3-hydroxylase, flavonoid 3p, 5p-hydroxylase, dihydroflavonol 4-reductase, anthocyanidin synthase, and anthocyanidin reductase have been cloned. Genes related to caffeine synthesis pathways such as S-adenosylmethionine synthase gene, caffeine synthase, and 7-N-methyltransferase have been isolated. Isolation of these genes also opens up a new era of development of caffeine-free tea plant production through RNAi technology (Kato et al. 2000; Mohanpuria et al. 2011). The β-glucosidases and β-primeverosides are two of the most important endogenous enzymes responsible for tea's floral aroma that have been cloned and characterized (Choi et al. 2007). Theanine, a major amino acid of tea, has a great influence on the taste of green tea. Therefore manipulating theanine biosynthesis will be useful for improving the quality of green tea. However, until now, glutamine synthase is the only gene of theanine biosynthesis that has been cloned (Mohanpuria et al. 2011). Apart from that, several other genes, related to photosynthesis, such as protein D1 and violaxanthin de-epoxidase genes, β-tubulin, and few ion homeostasis genes (ammonium transporter, ATP sulfurylase, etc.) have been cloned. While the majority of the genes have been cloned using rapid amplification cDNA ends, few earlier attempts used other methods including heterologous probes and screening of cDNA libraries.

Understanding the global transcript changes with stress, cell type, tissue, and developmental stage is a robust approach. Although several approaches are available (Mondal and Sutoh 2013), suppression subtractive hybridization (SSH) and cDNA-AFLP have been widely used to study differential gene expression in tea. For example, SSH has been used to understand the molecular regulation of secondary metabolism pathways in young leaves (Park et al. 2004; Chen et al. 2005b), to identify cold-responsive genes (Wang et al. 2009), to identify dormancy-associated genes (Krishnaraj et al. 2011; Yang et al. 2012; Phukon et al. 2012), and to identify drought-responsive gens (Das et al. 2012; Das et al. 2013; Gupta et al. 2012a). Recently, seven cDNA libraries from various organs of tea plants were used to generate 17,458 ESTs, which will be a valuable genomic resource in tea (Taniguchi et al. 2012).

Interestingly, to investigate the molecular mechanism by which light regulates phenylpropanoid metabolism, a light-induced SSH library of tea calli was constructed. Several genes in lignin biosynthesis pathways were identified in the light-induced library (Wang et al. 2012). To isolate upregulated genes in tea leaves under mild infestation of green leafhopper (*Empoasca vitis* Göthe), a SSH library was constructed and several genes were identified (Yang et al. 2011). Interestingly, they found that insect infection triggers the several genes that are involved in quality. Furthermore, through the cDNA-AFLP approach, they identified 108 transcript-derived fragments that were expressed differentially in the leaf of tolerant cultivars (Gupta et al. 2012b). Eungwanichayapant and Popluechai (2009) reported that the expression of phenylalanine ammonia-lyase 1, chalcone synthase, dihydroflavonol 4-reductase, leucoanthocyanidin reductase, and flavanone 3-hydroxylase genes was higher in young shoots than in mature leaves, which is obvious as more of the quality compounds that are the products of these genes are found in young leaves.

In Taiwan, Oriental Beauty, a highly flavored tea made from leaf hopper infestation, is very popular. In order to find the gene responsible for this, the gene expression profile and chemical profile were investigated by DNA microbead array. Interestingly, several stress-responsive genes were found to be upregulated in infested tea leaves (Choi et al. 2007). Darjeeling teas of India are grown at the highest altitudes in the world and are preferred for their flavor, aroma, and quality. Apart from the genetic makeup of the plant, earlier reports suggest that insect infestation, particularly jassids and thrips, triggers the aroma and flavor formation in Darjeeling tea. Several gene and transcription factors were identified by the SSH library of leaves that are highly infested by insects (Gohain et al. 2012).

Recently, next-generation sequencing (RNA-seq) technology has been widely used for gene expression and identification. For the first time, an extensive transcriptome dataset has been generated by an RNA-seq analysis from the young leaf of tea plants. The coverage of the transcriptome is comprehensive enough to discover all known genes of several major metabolic pathways. This transcriptome dataset can serve as an important public information platform for gene expression, genomics, and functional genomic studies in tea (Shi et al. 2011). In the same year, Jiang et al. (2011) did deep transcriptome sequencing of *C. oleifera*, *C. chekiangoleosa*, and *C. brevistyla* using the 454 GS FLX platform. After removal of the adaptor sequence and low-complexity sequences, 182,766, 190,545, and 132,147 reads of *C. oleifera*, *C. chekiangoleosa*, and *C. brevistyla* were generated, respectively. These readings were assembled into 49,909 contigs, which were annotated using BLAST searches. Based on this information, they could estimate the similarity of gene expression in different species, and detect and validate SNPs.

13.6 Proteomics and Metabolomics

Proteomics studies correlate the potential protein modifications to particular phenotypes through techniques such as high-performance liquid chromatography (HPLC), mass spectrometry, sodium dodecyl sulfate polyacrylamide gel electrophoresis (SDS-PAGE), two-dimensional gel electrophoresis, *in silico* protein modeling, and matrix-assisted laser desorption/ionization (MALDI).

Li et al. (2008) first performed a proteomics analysis of tea pollen and then identified the differentially expressed proteins of pollen that are associated with cold stress. Further, changes in protein expression in the embryo of tea in response to desiccation have been investigated and identified by Chen et al. (2011). Proteomic analysis revealed the presence of 23 proteins related to defense response, metabolism, and redox status that were upregulated under desiccation. This finding is particularly important as tea seeds, being recalcitrate, maintain a high rate of metabolism and water content during the seed maturation stage. In China, an albino mutant of tea was identified that has typical white leaf phenotype. In order to understand the protein involved in the albino leaf, proteomics analysis was done, which identified 61 proteins during the three developmental stages of an albino cultivar of tea. Important proteins that play crucial roles in the periodic albinism were identified (Li et al. 2011).

Metabolomics is also a powerful technique and in tea it is particularly relevant as quality is a complex trait, a mixture of several metabolites of various natures. Metabolic characteristics associated with climatic variables were investigated by ^{1}H nuclear magnetic resonance (NMR) spectroscopy in tea grown in different areas of South Korea, thereby allowing for the assessment of quality in green tea production (Lee et al. 2010). Differential metabolites expressed in shade-grown tea plants have been detected by liquid chromatography-mass spectrometry and gas chromatography-mass spectrometry coupled with a multivariate data set (Ku et al. 2010) that helped to understand the effect of shade on tea growth. The dependence of the global green tea metabolome on plucking positions was investigated through NMR analysis coupled with a multivariate statistical data set (Lee et al. 2011) that correlated the plucking shoots with quality.

13.7 Conclusion

Functional and structural genomics are both new developments in biotechnology. These "omics" approaches along with other techniques of microscopy as well as DNA sequencing have great potential for deciphering several fundamental questions of gene expression in model systems such as *Arabidopsis*. Despite their limitations, significant work has been done in tea with conventional breeding, micropropagation, and other cell culture techniques (Mondal et al. 1998, 2000b, 2001a; Mondal 2002c). A limited number of molecular markers have also been developed and utilized for germplasm characterization. However, omics has not been applied on a large scale, perhaps primarily due to the nonavailability of a complete genome sequence in tea. Therefore, generation of large numbers of SSR and SNP markers, their utilization for mapping of agronomically important QTLs, and marker-assisted breeding for gene introgression are currently required. In the genus *Camellia*, developing biparental mapping population is difficult and hence association mapping approaches will be a better alternative, at least in tea. However, such an approach remains unexplored for identifying the agronomically important genes. Although in several aspects of tea molecular biology work needs to be done, priority should be given to the following:

- Undertake *de novo* whole genome sequencing for development of a reference genome.
- Resequence popular trait-specific cultivars to generate large-scale SNP markers and use these for association mapping, gene introgression, high-density linkage map constructions, and so on.

References

Ackerman, W.L. 1971. Genetic and cytological studies with *Camellia* and related genera. Technical Bull. No. 1427, USDA, U.S. Government Printing Office, Washington, DC.

Afridi, S.G., Ahmad, H., Alam, M., Khan, I.A., Hassan, M. 2011. DNA landmarks for genetic diversity assessment in tea genotypes using RAPD markers. *African Journal of Biotechnology*, 10, 15477–15482.

Akula, A., Akula, C. 1999. Somatic embryogenesis in tea (*Camellia sinensis* (L) O Kuntze. In: Jain, S.M., Gupta, P.K., Newton, R.J. (eds), *Somatic Embryogenesis in Woody Plants*, vol. 5, pp. 239–259. Dordrecht: Kluwer Academic Publishers.

Anderson, S. 1994. Isozyme analysis to differentiate between tea clones. *Inligtingsbulletin Institut vir Tropiese en Subtropiese Gewasse*, (266), 15.

Ariyaratne, P.N.K., Mewan, K.M. 2009. Study of genetic relationships of tea (*Camellia sinensis* (L.) O. Kuntze) using SSR markers and pedigree analysis. In: *Proceedings of 9th Agricultural Research Symposium*, vol. 1, pp. 293–298.

Balasaravanan, T., Pius, P.K., Kumar, R.R. 2002. Assessment of genetic fidelity among the in vitro propagated culture lines of *Camellia sinensis* (L.) O kuntze using RAPD markers. In: *Proceedings of the 15th Plantation Crops Symposium, PLACROSYM XV*, pp. 181–184. 10th–13th December, Mysore.

Balasaravanan, T., Pius, P.K., Kumar, R.R., Muraleedharan, N., Shasany, A.K. 2003. Genetic diversity among south Indian tea germplasm (*Camellia sinensis, C. assamica* and *C. assamica* spp. Lasiocalyx) using AFLP markers. *Plant Science*, 165, 365–372.

Banerjee, B. 1987. Can leaf aspect affect herbivory? A case study with tea. *Ecology*, 68, 839–844.

Banerjee, B. 1992. Selection and breeding of tea. In: Willson, K.C., Clifford, M.N. (eds), *Tea: Cultivation to Consumption*, pp. 26–86. London: Chapman and Hall.

Barua, D.N. 1958. Leaf scIereids in the taxonomy of the Thea Camellias. I. Wilson's and related camellias. *Phytomorphology*, 8, 257–264.

Barua, P.K. 1963. Classification of tea plant. *Two and Bud*, 10, 3–11.

Barua, D.N., Dutta, A.C. 1959. Leaf scIereids in taxonomy of *Thea camellias* II. *Camellia sinensis* L. *Phytomorphology*, 9, 372–382.

Bera, B., Saikia, H. 2002. Randomly amplified polymorphic DNA (RAPD) marker analysis in tea (*Camellia sinensis* L) generative clones. In: *Proceedings of the 15th Plantation Crops Symposium, PLACROSYM XV*, pp. 235–238. 10th–13th December, Mysore.

Beretta, D., Vanoli, M., Eccher, T. 1988. The influence of glucose, vitamins and IBA on rooting of *Camellia* shoots *in vitro*. *Acta Horticulturae*, 227, 473–475.

Bezbaruah, H.P. 1971. Cytological investigation in the family Theaceae I. Chromosome numbers in some *Camellia* species and allied genera. *Caryologia*, 24, 421–426.

Borchetia, S., Das, S.C., Handique, P.J., Das, S. 2009. High multiplication frequency and genetic stability for commercialization of the three varieties of micropropagated tea plants (*Camellia* spp.). *Scientia Horticulturae*, 120, 544–550.

Borse, B.B., Jagan Mohan, R.L., Nagalakshmi, S., Krishnamurthy, N. 2002. Fingerprint of black teas from India: Identification of the regio-specific characteristics. *Food Chemistry*, 79, 419–424.

Borthakur, S., Mondal, T.K., Borthakur, A., Deka, P.C. 1995. Variation in peroxidase and esterase isoenzymes in tea leaves. *Two and Bud*, 42, 20–23.

Borthakur, S., Mondal, T.K., Parveen, S.S., Guha, A., Sen, P., Borthakur, A., Deka, P.C. 1998. Isolation of chloroplast DNA from tea, *Camellia* sp. *Indian Journal of Experimental Biology*, 36, 1165–1167.

Caser, M., Torello Marinoni, D., Scariot, V. 2010. Microsatellite-based genetic relationships in the genus Camellia: Potential for improving cultivars. *Genome*, 53, 384–399.

Chen, C. 1996. Analysis on the isozymes of tea plants F1 hybrids. *Journal of Tea Science*, 16, 31–33.

Chen, L., Yamaguchi, L. 2002. Genetic diversity and phylogeny of tea plant (*Camellia sinensis*) and its related species and varieties in the section Thea genus *Camellia* determined by randomly amplified polymorphic DNA analysis. *Journal of Horticultural Science and Biotechnology*, 77, 729–732.

Chen, L., Yamaguchi, S. 2005. RAPD markers for discriminating tea germplasms at the inter-specific level in China. *Plant Breeding*, 124, 404–409.

Chen, L., Chen, D., Gao, Q., Yang, Y., Yu, F. 1997. Isolation and appraisal of genomic DNA from tea plant (*Camellia sinensis* (L.) O.Kuntze). *Journal of Tea Science*, 17, 177–181.

Chen, L., Gao, Q.-K., Chen, D.-M., Xu, C.-J. 2005a. The use of RAPD markers for detecting genetic diversity, relationship and molecular identification of chinese elite tea genetic resources [*Camellia sinensis* (L.) O. Kuntze] preserved in a tea germplasm repository. *Biodiversity Conservation*, 14, 1433–1444.

Chen, L., Zhao, L.P., Gao, Q.K. 2005b. Generation and analysis of expressed sequence tags from the tender shoots cDNA library of tea plant (*Camellia sinensis*). *Plant Science*, 168, 359–363.

Chen, L., Yang, Y., Yu, F., Gao, Q., Chen, D. 1998. Genetic diversity of 15 tea (*Camellia sinensis* (L.) O.Kuntze) cultivars using RAPD markers. *Journal of Tea Science*, 18, 21–27.

Chen, L., Wang, P.-S., Yamaguchi, S. 2002a. Discrimination of wild tea germplasm resources (*Camellia* sp.) using RAPD markers. *Agricultural Science of China*, 1, 1105–1110.

Chen, L., Yamaguchi, S., Wang, P.-S., Xu, M., Song, W.-X., Tong, Q.-Q. 2002b. Genetic polymorphism and molecular phylogeny analysis of section Thea based on RAPD markers. *Journal of Tea Science*, 22, 19–24.

Chen, L., Yu, F., Yang, Y., Gao, Q., Chen, D., Xu, C. 1999. A study on genetic stability of excellent germplasm (*Camellia sinensis* (L.) O. Kuntze) using RAPD markers. *Journal of Tea Science*, 19, 13–16.

Chen, Q., Yang, L., Ahmad, P., Wan, X., Hu, X. 2011. Proteomic profiling and redox status alteration of recalcitrant tea (*Camellia sinensis*) seed in response to desiccation. *Planta*, 233, 583–592.

Chengyin, L., Weihua, L., Mingjun, R. 1992. Relationship between the evolutionary relatives and the variation of esterase isozymes in tea plant. *Journal of Tea Science*, 12, 15–20.

Choi, J-Y., Mizutani, M., Shimizu, B-I., Kinoshita, T., Ogura, M., Tokoro, K., Lin, M-L., Sakata, K. 2007. Chemical profiling and gene expression profiling during manufacturing process of Taiwan oolong tea, "Oriental Beauty". *Bioscience, Biotechnology and Biochemistry*, 71, 1476–1486.

Das, A., Mondal, T.K. 2010. *In silico* analysis of miRNA and their targets in tea. *American Journal of Plant Science*, 1, 77–86.

Das, A., Das, S., Mondal, T.K. 2012. Identification of differentially expressed gene profiles in young roots of tea [*Camellia sinensis* (L.) O. Kuntze] subjected to drought stress using suppression subtractive hybridization. *Plant Molecular Biology Report*, 30, 1088–1101.

Das, A., Saha, D., Mondal, T.K. 2013. An optimized method for extraction of RNA from tea roots for functional genomics analyses. *Indian Journal of Biotechnology*, 12, 129–132.

Deka, A., Deka, P.C., Mondal, T.K. 2006. Tea. In: Parthasarathy, V.A., Chattopadhyay, P.K., Bose, T.K. (eds), *Plantation Crops-I*, pp. 1–148. Calcutta: Naya Udyog.

Dhiman, B., Singh, M. 2003. Molecular detection of cashew husk (*Anacardium occidentale*) adulteration in market samples of dry tea (*Camellia sinensis*). *Planta Medicago*, 69, 882–884.

Eden, T. 1976. *Tea*, p. 236. London: Longman.

Eungwanichayapant, P.D., Popluechai, S. 2009. Accumulation of catechins in tea in relation to accumulation of mRNA from genes involved in catechin biosynthesis. *Plant Physiology and Biotechnology*, 47, 94–97.

Fang, S.-W., Hua, P.-R., Sheng, W.-P., Mei, X., Xing, D.-H., Ping, Z.-Y., Hua, L.-J., et al. 2003. RAPD analysis of tea trees in Yunnan. *Scientia Agriculture Sinica*, 36, 1582–1587.

Fang, W., Cheng, H., Duan, Y., Jiang, X., Li, X. 2012. Genetic diversity and relationship of clonal tea (*Camellia sinensis*) cultivars in China as revealed by SSR markers. *Plant Systamitic and Evolution*, 298, 469–483.

Forrest, GI. 1969. Studies on the polyphenol metabolism of tissue culture derived from the tea plant (*C. sinensis* L.). *Biochemical Journal*, 113, 765–772.

Freeman, S.J., West, C.J., Lea, V., Mayes, S. 2004. Isolation and characterization of highly polymorphic microsatellites in tea (*Camellia sinensis*). *Molecular Ecology and Notes*, 4, 324–326.

Fukushima, E., Iwasa, S., Endo, N., Yoshinari, T. 1966. Cytogenetics studies in *Camellia*. I. Chromosome survey in some *Camellia* species. *Japanese Journal of Horticulture*, 35, 413–421.

George, O., Adam, M. 2006. Investigation into the evolutionary origins of Theaceae and genus *Camellia*. *International Camellia Journal*, 38, 78–89.

Ghosh-Hazra, N. 2001. Advances in selection and breeding of tea: A review. *Journal of Plantation Crops*, 29, 1–17.

Gohain, B., Borchetia, S., Bhorali, P., Agarwal, N., Bhuyan, LP., Rahman, A., Sakata, K., et al. 2012. Understanding Darjeeling tea flavour on a molecular basis. *Plant Molecular Biology*, 78, 577–597.

Goonetilleke, W.A.S.N.S.T., Priyantha, P.G.C., Mewan, K.M., Gunasekare, M.T.K. 2009. Assessment of genetic diversity of tea (*Camellia sinensis* L.O. Kuntze) as revealed by RAPD-PCR markers. *Journal Natural Science Foundation of Sri Lanka*, 37, 147–150.

Gul, S., Ahmad, H., Khan, I.A., Alam, M. 2007. Assessment of diversity of tea genotypes through RAPD markers. *Pakistan Journal of Biological Science*, 10, 2609–2611.

Gupta, S., Bharalee, R., Bhorali, P., Bandyopadhyay, T., Gohain, B., Agarwal, N., Ahmed, P., et al. 2012a. Identification of drought tolerant progenies in tea by gene expression analysis. *Functional and Integrative Genomics*, 12, 543–563.

Gupta, S., Bharalee, R., Bhorali, P., Das, S.K., Bhagawati, P., Bandyopadhyay, T., Gohain, B., et al. 2012b. Molecular analysis of drought tolerance in tea by cDNA-AFLP based transcript profiling. *Molecular Biotechnology*, 53(3), 237–248.

Hackett, C.A., Wachira, F.N., Paul, S., Powell, W., Waugh, R. 2000. Construction of a genetic linkage map for *Camellia sinensis* (tea). *Heredity*, 85, 346–355.

Hairong, X., Qiqing, T., Wanfanz, Z. 1987. Studies on the genetic tendency of tea plant hybrid generation using isozyme technique. In: *Proceedings of International Symposium on Tea Quality and Human Health*, pp. 21–25. 4th–9th November, China.

Hazarika, M., Mahanta, P.K. 1984. Composition changes in chlorophylls and carotenoids during the four flushes of tea in North-East India. *Journal of Science and Food Agriculture*, 35, 298–303.

Hou, Y.J., He, Q., Li, Z.L., Li, P.W., Liang, G.L., Xu, J. 2007. ISSR applied to the germplasms identification of *Camellia sinensis*. *Southwest China Journal of Agricultural Science*, 26, 1272–1276.

Huang, F.P., Liang, Y.R., Lu, J.L., Chen, R.B. 2006. Genetic mapping of first generation of backcross in tea by RAPD and ISSR markers. *Journal of Tea Science*, 26, 171–176.

Huang, J.A., Huang, Y.H., Luo, J.W., Li, J.X., Gong, Z.H., Liu, Z.H. 2007. Identification of single nucleotide polymorphism in polyphenol oxidase gene in tea plant (*Camellia sinensis*). *Journal of Hunan Agricultural University*, 33, 454–458.

Huang, J.A., Li, J.X., Huang, Y.H., Luo, J.W., Gong, Z.H., Liu, Z.H. 2005. Construction of AFLP molecular markers linkage map in tea plant. *Journal of Tea Science*, 25, 7–15.

Hui, L.-X., Lin, L.-C., Peng, S.-Z., Wu, L.-J., Wen, S.-C., Hua, G.-Z., Xuan, C., et al. 2004. Analysis of genetic relationships among "Rucheng Baimao Cha" plants with RAPD method. *Journal of Tea Science*, 24, 33–36.

Hung, C.-Y., Wang, K.-H., Huang, C.-C., Gong, X., Ge, X.-J., Chiang, T.-Y. 2008. Isolation and characterization of 11 microsatellite loci from *Camellia sinensis* in Taiwan using PCR-based isolation of microsatellite arrays (PIMA). *Conservation Genetics*, 9, 779–781.

Ikeda, N., Kawada, M., Takeda, Y. 1991. Isozymic analysis of *Camellia sinensis* and its interspecific hybrids. In: *Proceedings of International Symposium of Tea Science*, p. 98. 26th–28th August, Shizuoka, Japan.

Ji, P.Z., Li, H., Gao, L.Z., Zhang, J., Chen, G.Z.Q., Huang, X.Q. 2011. ISSR diversity and genetic differentiation of ancient tea (*Camellia sinensis* var. assamica) plantations from China: Implications for precious tea germplasm conservation. *Pakistan Journal of Botany*, 43, 281–291.

Jiang, C., Wen, Q., Chen, Y., Xu, L.A., Huang, M.R. 2011. Efficient extraction of RNA from various *Camellia* species rich in secondary metabolites for deep transcriptome sequencing and gene expression analysis. *African Journal of Biotechnology*, 10, 16769–16773.

Jin, J.Q., Cui, H.R., Chen, W.Y., Lu, M.Z., Yao, Y.L., Xin, Y., Gong, X.C. 2006. Data mining for SSR in ESTs and development of EST-SSR marker in tea plant (*Camellia sinensis*). *Journal of Tea Science*, 26, 17–23.

Jin, J.Q., Cui, H.R., Gong, X.C., Chen, W.Y., Xin, Y. 2007. Studies on tea plants (*Camellia sinensis*) germplasms using EST-SSR marker. *Yi Chuan*, 29, 103–108.

John, K.M.M., Joshi, D.S., Mandal, A.K.A., Kumar, S.R., Raj Kumar, R. 2009. *Agrobacterium* mediated hairy root production in tea leaves [*Camellia sinensis* L (O) kuntze]. *Indian Journal of Biotechnology*, 8, 430–434.

Jorge, S., Pedroso, M.C., Neale, D.B., Brown, G. 2003. Genetic differentiation of Portuguese tea plant using RAPD markers. *Horticultural Science*, 38, 1191–1197.

Kabir, S.E., Ghosh-Hajra, N., Chaudhuri, T.C. 1991. Performance of certain clones under the agroclimatic conditions of Darjeeling. *Tea Board of India Technical Bulletin*, 5, 1–8.

Kafkas, S., Ercisxli, S., Doğan, Y., Ertürk, Y., Haznedar, A., Sekban, R. 2009. Polymorphism and genetic relationships among tea genotypes from Turkey revealed by amplified fragment length polymorphism markers. *Journal of American Society of Horticultural Science*, 134, 428–434.

Kamunya, S.M., Wachira, F.N., Pathak, R.S., Korir, R., Sharma, V., Kumar, R., Bhardwaj, P., Chalo, R., Ahuja, P.S., Sharma, R.K. 2010. Genome mapping and testing for quantitative trait loci in tea (*Camellia sinensis* (L.) O. Kuntze). *Tree Genetics and Genomes*, 6, 915–929.

Karthigeyan, S., Rajkumar, S., Sharma, R.K., Gulati, A., Sud, R.K., Ahuja, P.S. 2008. High level of genetic diversity among the selected accessions of tea (*Camellia sinensis*) from abandoned tea gardens in western Himalaya. *Biochemistry and Genetics*, 46, 810–819.

Kato, F., Taniguchi, F., Monobe, M., Ema, K., Hirono, H., Maeda-Yamamoto, M. 2008. Identification of Japanese tea (*Camellia sinensis*) cultivars using SSR marker. *Journal of Japanese Society of Food Science and Technology*, 55, 49–55.

Kato, M., Mizuno, K., Crozier, A., Fujimura, T., Ashihara, A. 2000. Caffeine synthase gene from tea leaves. *Nature*, 406, 956–957.

Kato, M., Uematu, K., Niwa, Y. 2004. Transformation of green fluorescent protein in tea plant. In: *Proceeding of the International Conference on O-Cha (tea) Cultivation and Science*, p. 55. 4–6 November, Shizuoka, Japan.

Katoh, Y., Katoh, M., Takeda, Y., Omori, M. 2003. Genetic diversity within cultivated teas based on nucleotide sequence comparison of ribosomal RNA maturase in chloroplast DNA. *Euphytica*, 134, 287–295.

Katsuo, K. 1969. Anther culture in tea plant (a preliminary report). *Study of Tea*, 4, 31.

Kaundun, S.S., Matsumoto, S. 2002. Heterologous nuclear and chloroplast microsatellite amplification and variation in tea, *Camellia sinensis*. *Genome*, 45, 1041–1048.

Kaundun, S.S., Matsumoto, S. 2003a. Development of CAPS markers based on three key genes of the phenylpropanoid pathway in tea, *Camellia sinensis* (L.) O. Kuntze, and differentiation between assamica and sinensis varieties. *Theoretical and Applied Genetics*, 106, 375–383.

Kaundun, S.S., Matsumoto, S. 2003b. Identification of processed Japanese green tea based on polymorphism generated by STS-RFLP analysis. *Journal of Food Science and Chemistry*, 51, 1765–1770.

Kaundun, S.S., Matsumoto, S. 2011. Molecular evidence for maternal inheritance of the chloroplast genome in tea, *Camellia sinensis* (L.) O. Kuntze. *Journal of Science Food and Agriculture*, 91, 2660–2663.

Kaundun, S.S., Park, Y.-G. 2002. Genetic structure of six Korean tea populations revealed by RAPD-PCR markers. *Crop Science*, 42, 594–601.

Kaundun, S.S., Zhyvoloup, A., Park, Y.-G. 2000. Evaluation of genetic diversity among elite tea (*Camellia sinensis* var. sinensis) accessions using RAPD markers. *Euphytica*, 115, 7–16.

Kondo, K. 1975. Cytological studies in cultivated species of *Camellia*, PhD Thesis, University of North Carolina, Chapel Hill.

Kondo, K. 1978a. Cytological studies in cultivated species of *Camellia*. In: *Encyclopedia of Camellia*, Japan Camellia Society, vol. 2, p. 456. Tokyo: Kodansha Publishing.

Kondo, K. 1978b. Cytological studies in cultivated species of *Camellia*. *Shi-Kaki*, 99, 41–53.

Kondo, K. 1979. Cytological studies in cultivated species of Camellia. V. Intraspecific variation of karyotypes in two species of sect Thea. *Japanese Journal of Breeding*, 29, 205–210.

Kondo, K., Parks, C.R. 1981. Cytological studies in cultivated species of Camellia. VI. Giemsa C-banded karyotypes of seven accessions of *Camellia japonica* L. *sensu lato*. *Japanese Journal of Breeding*, 31, 25–34.

Konwar, B.K., Das, S.C., Bordoloi, B.J., Dutta, R.K. 1998. Hairy root development in tea through *Agrobacterium rhizogenes*-mediated genetic transformation. *Two and Bud*, 45, 19–20.

Krishnaraj, T., Gajjeraman, P., Palanisamy, S., Chandrabose, S.R.S., Mandal, A.K.A. 2011. Identification of differentially expressed genes in dormant (banjhi) bud of tea (*Camellia sinensis* (L.) O. Kuntze) using subtractive hybridization approach. *Plant Physiology and Biochemistry*, 49, 565–571.

Ku, K.M., Choi, J.N., Kim, J., Kim, J.K., Yoo, L.G., Lee, S.J., Hong, Y.S., Lee, C.H. 2010. Metabolomics analysis reveals the compositional differences of shade grown tea (*Camellia sinensis* L.). *Journal of Agricultural and Food Chemistry*, 58, 418–426.

Kulasegaram, S. 1980. Technical developments in tea production. *Tea Quarterly*, 49, 157–183.

Kumar, N., Pandey, S., Bhattacharya, A., Ahuja, P.S. 2004. Do leaf surface characteristics affect *Agrobacterium* infection in tea [*Camellia sinensis* (L.) O Kuntze]? *Journal of Bioscience*, 29, 309–317.

Lai, J.-A., Yang, W.-C., Hsiao, J.-Y. 2001. An assessment of genetic relationships in cultivated tea clones and native wild tea in Taiwan using RAPD and ISSR markers. *Botanical Bulletin Academic Sinica*, 42, 93–100.

Lee, S., Kim, J., Sano, J., Ozaki, Y., Okubo, H. 2003. Phylogenic relationships among tea cultivars based on AFLP analysis. *Journal of Food Agriculture*, 47, 289–299.

Lee, J.E., Lee, B.J., Chung, J.O., Hwang, J.A., Lee, S.J., Lee, C.H., Hong, Y.S. 2010. Geographical and climatic dependencies of green tea (*Camellia sinensis*) metabolites: A ^{1}H NMR-based metabolomics study. *Journal of Agricultural and Food Chemistry*, 58, 10582–10589.

Lee, J.E., Lee, B.J., Hwang, J.A., Ko, K.S., Chung, J.O., Kim, E.H., Lee, S.J., Hong, Y.S. 2011. Metabolic dependence of green tea on plucking positions revisited: A metabolomic study. *Journal of Agriculture and Food Chemistry*, 59, 10579–10585.

Li, C., Zheng, Y., Zhou, J., Xu, J., Ni, D. 2011. Changes of leaf antioxidant system, photosynthesis and ultrastructure in tea plant under the stress of fluorine. *Biologia Plantarum*, 55, 563–566.

Li, J., Chen, J., Zhang, Z., Pan, Y. 2008. Proteome analysis of tea pollen (*Camellia sinensis*) under different storage conditions. *Journal of Agriculture and Food Chemistry*, 56, 7535–7544.

Li, J., Jiang, C.J., Wang, Z.X. 2005. RAPD analysis on genetic diversity of the pre concentrated core germplasms of *Camellia sinensis* in China. *Yi Chuan*, 27, 765–771.

Li, B., Yin, Y., Zhou, Y., Deng, P.Q., Yang, H.W., Li, B., Yin, Y., Zhou, Y., Deng, P.Q., Yang, H.W. 2003. Genetic diversity of two sexual tea cultivars detected by RAPD markers. *Journal of Tea Science*, 23, 46–150.

Lin, J., Kudrna, D., Wing, R.A. 2011. Construction, characterization, and preliminary BAC-end sequence analysis of a bacterial artificial chromosome library of the tea plant (*Camellia sinensis*). *Journal of Biomedical and Biotechnology*, 2, 476723–476731.

Liu, Y., Yang, S.X., Ji, P.Z., Gao, L.Z. 2012a. Phylogeography of *Camellia taliensis* (Theaceae) inferred from chloroplast and nuclear DNA: Insights into evolutionary history and conservation. *BMC Evolutionary Biology*, 12, 92–105.

Liu, B., Sun, X., Wang, Y., Li, Y., Cheng, H., Xiong, C., Wang, P. 2012b. Genetic diversity and molecular discrimination of wild tea plants from Yunnan Province based on inter-simple sequence repeats (ISSR) markers. *African Journal of Biotechnology*, 11, 11566–11574.

Liu, B.Y., Wang, P.S., Ji, P.Z., Xu, M., Cheng, H. 2008. Study on genetic diversity of peculiar sect. *Thea* (L.) Dye in Yunnan by ISSR markers. *Journal of Yunnan Agricultural University*, 23, 302–308.

Liu, B.Y., Li, Y.Y., Tang, Y.C., Wang, L.Y., Cheng, H., Wang, P.S. 2010. Assessment of genetic diversity and relationship of tea germplasm in Yunnan as revealed by ISSR markers. *Acta Agronomica Sinica*, 36, 391–400.

Luo, J.W., Shi, Z.P., Shen, C.W., Liu, C.L., Gong, Z.H., Huang, Y.H., Luo, J.W., et al. 2004. The genetic diversity of tea germplasms [*Camellia sinensis* (L.) O. Kuntze] by RAPD analysis. *Acta Agronomica Sinica*, 30, 266–269.

Luo, Y.Y., Liang, Y.R. 2000. Studies on the construction of *Bt* gene expression vector and its transformation in tea plant. *Journal of Tea Science*, 20, 141–147.

Ma, J.Q., Ma, C.L., Yao, M.Z., Jin, J.Q., Wang, Z.L., Wang, X.C., Chen, L. 2012. Microsatellite markers from tea plant expressed sequence tags (ESTs) and their applicability for cross-species/genera amplification and genetic mapping. *Scientia Horticulture*, 134, 167–175.

Ma, J.Q., Zhou, Y.H., Ma, C.L., Yao, M.Z., Jin, J.Q., Wang, X.C., Chen, L. 2010. Identification and characterization of 74 novel polymorphic EST-SSR markers in the tea plant, *Camellia sinensis* (Theaceae). *American Journal of Botany*, 97, 153–156.

Magambo, M.J.S., Cannell, M.G.R. 1981. Dry matter production and partition in relation to yield of tea. *Experimental Agriculture*, 17, 33–38.

Magoma, G.N., Wachira, F.N., Obanda, M., Imbuga, M., Agong, S.G. 2000. The use of catechins as biochemical markers in diversity studies of tea (*Camellia sinensis*). *Genetic Resources and Crop Evolution*, 47, 107–114.

Matsumoto, S., Fukui, M. 1998. *Agrobacterium tumefaciens* mediated gene transfer in tea plant (*Camellia sinensis*) cells. *Journal of Agricultural Research and Quarterly*, 32, 287–291.

Matsumoto, S., Fukui, M. 1999. Effect of acetosyringone application on *Agrobacterium*-mediated transfer in tea plant (*Camellia sinensis*). *Bulletin of National Research Institute of Vegetable, Ornamental and Tea, Shizuoka, Japan*, 14, 9–15.

Matsumoto, S., Kiriwa, Y., Takeda, Y. 2002. Differentiation of Japanese green tea as revealed by RFLP analysis of phenyl-alanine ammonia-lyase DNA. *Theoretical and Applied Genetics*, 104, 998–1002.

Matsumoto, S., Takeuchi, A., Hayastsu, M., Kondo, S. 1994. Molecular cloning of phenylalanine ammonia-lyase cDNA and classification of varieties and cultivars of tea plants (*Camellia sinensis*) using the tea PAL cDNA probes. *Theoretical and Applied Genetics*, 89, 671–675.

Mewan, K.M., Liyanage, A.C., Everard, J.M.D.T., Gunasekara, M.T.K., Karunanayake, E.H. 2005. Studying genetic relationships among the tea (*Camellia sinensis*) cultivars in Sri Lanka using RAPD markers. *Sri Lanka Journal of Tea Science*, 70(1), 42–53.

Mewan, K.M., Saha, M.C., Konstatin, C., Pang, Y., Abeysinghe, I.S.B., Dixon, R.A. 2007. Construction of genomic and EST-SSR based genetic linkage map of tea (*Camellia sinensis*). In: *Proceedings of the 4th International Conference on O-Cha (Tea) Culture and Science*, p. 52. 2–4 November, Shizuoka, Japan.

Mishra, R.K., Sen-Mandi, S. 2001. DNA fingerprinting and genetic relationship study of tea plants using amplified fragment length polymorphism (AFLP) technique. *Indian Journal of Plant Genetic Resource*, 4, 148–149.

Mishra, R.K., Sen-Mandi, S. 2004. Molecular profiling and development of DNA marker associated with drought tolerance in tea clones growing in Darjeeling. *Current Science*, 87, 60–66.

Mishra, R.K., Chaudhary, S., Ahmad, A., Pradhan, M., Siddiqi, T.O. 2009. Molecular analysis of tea clones (*Camellia sinensis*) using AFLP markers. *International Journal of Integrative Biology*, 5, 130–135.

Mohanpuria, P., Kumar, V., Ahuja, P.S., Yadav, S.K. 2011. *Agrobacterium*-mediated silencing of caffeine synthesis through root transformation in *Camellia sinensis* L. *Molecular Biotechnology*, 48, 235–243.

Mohanpuria, P., Rana, N.K., Yadav, S.K. 2008. Transient RNAi based gene silencing of glutathione synthetase reduces glutathione content in *Camellia sinensis* (L.) O. Kuntze somatic embryos. *Biologia Plantarum*, 52, 381–384.

Mohanpuria, P., Yadav, S.K. 2012. Characterization of novel small RNAs from tea (*Camellia sinensis* L.). *Molecular Biology Reporter*, 39, 3977–3986.

Mondal, T.K. 2002a. Micropropagation of tea (*Camellia sinensis*). In: Jain, S.M. (ed.), *Micropropagation of Woody Plants*, pp. 671–720. Dordrecht: Kluwer.

Mondal, T.K. 2002b. Detection of genetic diversity among the Indian tea (*Camellia sinensis*) germplasm by inter-simple sequence repeats (ISSR). *Euphytica*, 128, 307–315.

Mondal, T.K. 2002c. *Camellia* biotechnology: A bibliographic search. *International Journal of Tea Science*, 1, 28–37.

Mondal, T.K. 2008. Tea. In: Kole, C., Hall, T.C. (eds), *Compendium of Transgenic Crop Plants: Transgenic Plantation Crops, Ornamentals and Turf Grasses*, pp. 99–115. London: Blackwell.

Mondal, T.K. 2011. Camellia. In: Kole, C. (ed.), *Wild Crop Relatives: Genomics and Breeding Resources Plantation and Ornamental Crops*, p. 15–40. Berlin: Springer.

Mondal, T.K., Chand, P.K. 2002. Detection of genetic instability among the micropropagated tea [*Camellia sinensis* (L.) O. Kuntze] by RAPD analysis. *In Vitro Cellular and Developmental Biology-Plant*, 37, 1–5.

Mondal, T.K., Sutoh, K. 2013. Application of next generation sequencing for abiotic stress tolerance of plant. In: Barh, D., Zambare, V., Azevedo, V. (eds), *Applications in Biomedical, Agricultural, and Environmental Sciences*. pp. 347–365 CRC Press (*In press*).

Mondal, T.K., Ahuja, P.S., Chand, P.K. 2000a. Molecular characterization of biodiversity in tea (*Camellia sinensis* (L.) O. Kuntze) germplasm and ex situ conservation through *in vitro* culture. Biodiversity conservation for environment protection, Utkal University, Bhubaneswar, india, pp. 9–10. 16th, 17th March. 2000.

Mondal, T.K., Bhattacharya, A., Laxmikumaran, M., Ahuja, P.S. 2004. Recent advances in tea biotechnology. *Plant, Cell, Tissue and Organ Culture*, 75, 795–856.

Mondal, T.K., Bhattacharya, A., Sood, A., Ahuja, P.S. 2000b. Factor effecting induction and storage of encapsulated tea (*Camellia sinensis* L. O.Kuntze) somatic embryos. *Tea*, 21, 92–100.

Mondal, T.K., Bhattacharya, A., Ahuja, P.S. 2001a. Induction of synchronous secondary embryogenesis of Tea (*Camellia sinensis*). *Journal of Plant Physiology*, 158, 945–951.

Mondal, T.K., Bhattacharya, A., Ahuja, P.S. 2001b. Development of a selection system for *Agrobacterium*-mediated genetic transformation of tea (*Camellia sinensis*). *Journal of Plantation Crops*, 29, 45–48.

Mondal, T.K., Bhattacharya, A., Ahuja, P.S., Chand, P.K. 2001c. Factor effecting *Agrobacterium tumefaciens* mediated transformation of tea (*Camellia sinensis* (L). O. Kuntze). *Plant Cell Report*, 20, 712–720.

Mondal, T.K., Bhattacharya, A., Sharma, M., Ahuja, P.S. 2001d. Induction of in vivo somatic embryogenesis in tea (*Camellia sinensis*) cotyledons. *Current Science*, 81, 101–104.

Mondal, T.K., Bhattacharya, A., Laxmikumaran, M., Ahuja, P.S. 2004. Recent advance in tea biotechnology. *Plant Cell Tissue and Organ Culture*, 75, 795–856.

Mondal, T.K., Bhattacharya, A., Sood, A., Ahuja, P.S. 1998. Micropropagation of tea using thidiazuran. *Plant Growth Regulation*, 26, 57–61.

Mondal, T.K., Bhattachraya, A., Sood, A., Ahuja, P.S. 1999. An efficient protocol for somatic embryogenesis and its use in developing transgenic tea (*Camellia sinensis* (L) O. Kuntze) for field transfer. In: Altman, A., Ziv, M., Izhar, S. (eds), *Plant Biotechnology and In Vitro Biology in 21st Century*, pp. 101–104. Dordrecht: Kluwer Academic Publishers.

Mondal, T.K., Bhattachrya, A., Sood, A., Ahuja, P.S. 2002. Factors affecting germination and conversion frequency of somatic embryos of tea. *Journal of Plant Physiology*, 159, 1317–1321.

Mondal, T.K., Satya, P., Medda, P.S. 2003. India needs national tea germplasm repository. In: *International Conference on Global Advances in Tea Science*, p. 58. 20–22nd November, Calcutta, India.

Moose, S.P., Mumm, R.H. 2008. Molecular plant breeding as the foundation for 21st century crop improvement. *Plant Physiology* 147, 969–977.

Mukherjee, S., Mondal, T.K., Mohan Kumar, P., Jayachandran, R. 2005. Field level control of stem cancer (*Macrophomea theicola*) diseases of tea (*Camellia sinensis* (L). O. Kuntze) by Trichoderma. *Journal of the Agricultural and Horticultural Society of India*, 15, 20–25.

Mukhopadyay, M., Bantawa, P., Das, A., Sarkar, B., Bera, B., Ghosh, P.D., Mondal, T.K. 2012. Changes of growth, photosynthesis and alteration of leaf antioxidative defence system of tea (*Camellia sinensis* (L.) O. Kuntze) seedling under aluminum stress. *Biometals*, 25, 1141–1154.

Mukhopadyay, M., Das, A., Subba, P., Bantawa, P., Sarkar, B., Ghosh, P.D., Mondal, T.K. 2013a. Structural, physiological and biochemical profiling of tea plantlets (*Camellia sinensis* (L.) O. Kuntze) under zinc stress. *Biologia Plantarum*, 2012, 1–7.

Mukhopadyay, M., Ghosh, P.D., Mondal, T.K. 2013b. Effect of boron deficiency on photosynthesis and antioxidant responses of young tea (*Camellia sinensis* (L.) O. Kuntze) plantlets. *Russian Journal of Plant Physiology*, 60, 633–639.

Muoki, R.C., Wachira, F.N., Pathak, R.S., Kamunya, S.M. 2007. Assessment of the mating system of *Camellia sinensis* in biclonal seed orchards based on PCR markers. *Journal Horticulture Science and Biotechnology*, 82, 733–738.

Nagarajah, S., Ratnasurya, R. 1981. Clonal variability in root growth and drought resistance in tea (*Camellia sinensis*). *Plant and Soil*, 60, 153–155.

Ohsako, T., Ohgushi, T., Motosugi, H., Oka, K. 2008. Microsatellite variability within and among local landrace populations of tea, *Camellia sinensis* (L.) O. Kuntze, in Kyoto, Japan. *Genetics Resources and Crop Evolution*, 55, 1047–1053.

Orel, G., Marchant, A.D., Wei, C.F., Curry, A.S. 2007. Molecular investigation and assessment of *C. azalea* (syn. *C.changii*Ye 1985) as potential breeding material. *International Camellia Journal*, 39, 64–75.

Owuor, P.O., Obanda, M. 1998. The use of chemical parameters as criteria for selecting for quality in clonal black tea in Kenya: Achievements, problems and prospects—A review. *Tea*, 19, 49–58.

Park, J.S., Kim, J.B., Hahn, B.S., Kim, K.H., Ha, S.H., Kim, J.B., Kim, Y.H. 2004. EST analysis of genes involved in secondary metabolism in *Camellia sinensis* (tea), using suppression subtractive hybridization. *Plant Science*, 166, 953–961.

Park, Y.G., Kaundun, S.S., Zhyvoloup, A. 2002. Use of the bulked genomic DNA-based RAPD methodology to assess the genetic diversity among abandoned Korean tea plantations. *Genetics Resource and Crop Evolution*, 49, 159–165.

Paul, S., Wachira, F.N., Powell, W., Waugh, R. 1997. Diversity and genetic differentiation among population of Indian and Kenyan tea (*Camellia sinensis* (L.) O. Kuntze) revealed by AFLP markers. *Theoretical and Applied Genetics*, 94, 255–263.

Phukon, M., Namdev, R., Deka, D., Modi, S.K., Sen, P. 2012. Construction of cDNA library and preliminary analysis of expressed sequence tags from tea plant [*Camellia sinensis* (L) O. Kuntze]. *Gene*, 506, 202–206.

Prabu, G.R., Mandal, A.K. 2010. Computational identification of miRNAs and their target genes from expressed sequence tags of tea (*Camellia sinensis*). *Genome, Proteome and Bioinformatics*, 8, 113–121.

Prince, L.M., Parks, C.R. 1997. Evolutionary relationships in the tea subfamily Theoideae based on DNA sequence data. *International Camellia Journal*, 29, 135–144.

Prince, L.M., Parks, C.R. 2000. Estimation of relationships of Theoideae (Theaceae) inferred from DNA data. *International Camellia Journal*, 32, 79–84.

Prince, M.L., Parks, C.R. 2001. Phylogenetic relationships of Theaceae inferred from chloroplast DNA sequence data. *American Journal of Botany*, 88, 2309–2320.

Rajasekaran, P. 1997. Development of molecular markers using AFLP in tea. In: Varghese, J.P. (ed.), *Molecular Approaches to Crop Improvement. Proceedings of National Seminar on Molecular Approaches to Crop Improvement*. 29th–31st December, Kottayam, Kerala, India.

Roy, S.C., Chakraborty, B.N. 2007. Evaluation of genetic diversity in tea of the Darjeeling foot hills, India using RAPD and ISSR markers. *Journal of Hill Research*, 20, 13–19.

Roy, S.C., Chakraborty, B.N. 2009. Cloning and sequencing of chitinase gene specific PCR amplified DNA fragment from tea plant (*Camellia sinensis*) and analyzed the nucleotide sequence using bioinformatics algorithms. *Canadian Journal of Pure and Applied Science*, 3, 795–800.

Sandal, I., Saini, U., Lacroix, B., Bhattacharya, A., Ahuja, P.S., Citovsky, V. 2007. *Agrobacterium*-mediated genetic transformation of tea leaf explants: Effects of counteracting bactericidal of leaf polyphenols without loss of bacterial virulence. *Plant Cell Report*, 26, 169–176.

Sanderson, G.W. 1964. The chemical composition of fresh tea flush as affected by clone and climate. *Tea Quarterly*, 35, 101–109.

Satyanarayan, N., Sharma, V.S. 1982. Biometric basis for yield prediction in tea clonal selection. In: *Proceedings of PLACROSYM IV*, pp. 237–243. 3–5 December, 1981, Mysore, India.

Satyanarayan, N., Sharma V.S. 1986. Tea (*Camellia L.* spp) germplasm in south India. In: Srivastava, H.C., Vatsya, B., Menon, K.K.G. (eds), *Plantation Crops: Opportunity and Constraints*, pp. 173–179. New Delhi: Oxford IBH Publishing.

Sealy, J.R. 1958. *A Revision of the Genus Camellia*, pp. 58–60. London: Royal Horticultural Society.

Sen, P., Bora, U., Roy, B.K., Deka, P.C. 2000. Isozyme characterization in *Camellia* spp. *Crop Research*, 19, 519–524.

Seurei, P. 1996. Tea improvement in Kenya: A review. *Tea*, 17, 76–81.

Sharma, R.K., Bhardwaj, P., Negi, R., Mohapatra, T., Ahuja, P.S. 2009. Identification, characterization and utilization of unigene derived microsatellite markers in tea (*Camellia sinensis* L.). *BMC Plant Biology*, 9, 53–77.

Sharma, H., Kumar, R., Sharma, V., Kumar, V., Bhardwaj, P., Ahuja, P.S., Sharma, R.K. 2011. Identification and cross-species transferability of 112 novel unigene-derived microsatellite markers in tea (*Camellia sinensis*). *American Journal of Botany*, 98, 133–138.

Shen, C.W., Huang, Y.H., Huang, J.A., Luo, J.W., Liu, C.L., Liu, D.H. 2007. RAPD analysis for genetic diversity of typical tea populations in Hunan Province. *Journal of Agricultural Biotechnology*, 15, 855–860.

Shen, C.W., Ning, Z.X., Huang, J.A., Chen, D., Li, J.X. 2009. Genetic diversity of Camellia sinensis germplasm in Guangdong Province based on morphological parameters and SRAP markers. *Ying Yong Sheng Tai Xue Bao*, 20, 1551–1558.

Shi, C.Y., Yang, H., Wei, C.L., Yu, O., Zhang, Z.Z., Jiang, C.J., Sun, J., Li, Y.Y., Chen, Q., Xia, T., Wan, X.C. 2011. Deep sequencing of the *Camellia sinensis* transcriptome revealed candidate genes for major metabolic pathways of tea-specific compounds. *BMC Genomics*, 12, 131–150.

Singh, D., Ahuja, P.S. 2006. 5S rDNA gene diversity in tea (*Camellia sinensis* (L.) O. Kuntze) and its use for variety identification. *Genome*, 49, 91–96.

Singh, H.P., Ravindranath, S.D. 1994. Occurrence and distribution of PPO activity in floral organs of some standard and local cultivars of tea. *Journal of Science Food and Agriculture*, 64, 117–120.

Singh, I.D. 1999. Plant improvement. In: Jain, N.K. (ed.), *Global Advances in Tea*, pp. 427–448. New Delhi: Aravali Book International.

Singh, M., Bandana, Ahuja, P.S. 1999. Isolation and PCR amplification of genomic DNA from market samples of dry tea. *Plant Molecular Biology Report*, 17, 171–178.

Singh, M., Dhiman, B., Sharma, C. 2011. Characterization of a highly repetitive DNA sequence in *Camellia sinensis* (L.) O. Kuntze genome. *Journal of Biotechnology Research*, 3, 78–83.

Singh, M., Saroop, J., Dhiman, B. 2004. Detection of intra-clonal genetic variability in vegetatively propagated tea using RAPD markers. *Biologia Plantarum*, 48, 113–115.

Su, M.H., Hsieh, C.F., Tsou, C.H. 2009. The confirmation of *Camellia formosensis* (Theaceae) as an independent species based on DNA sequence analyses. *Botanical Study*, 50, 477–485.

Takeo, T. 1981. Variations in amounts of linalool and geraniol produced in tea shoots by mechanical injury. *Phytochemistry*, 30, 2149–2151.

Takeuchi, A., Matsumoto, S., Hayatsu, M. 1994. Chalcone synthase from *Camellia sinensis* isolation of the cDNAs and the organ-specific and sugar-responsive expression of the genes. *Plant Cell and Physiology*, 35, 1011–1018.

Tanaka, J., Taniguchi, F. 2002. Emphasized-RAPD (e-RAPD): A simple and efficient technique to make RAPD Bands clearer. *Breeding Science*, 52, 225–229.

Tanaka, J.I., Yamaguchi, S. 1996. Use of RAPD markers for the identification of parentage of tea cultivars. *Bulletin of National Research Institute of Vegetable, Ornamental Plant and Tea*, 9, 31–36.

Tanaka, T., Mizutani, T., Shibata, M., Tanikawa, N., Parks, C.R. 2005. Cytogenetic studies on the origin of *Camellia × vernalis*. V. Estimation of the seed parent of *C. × vernalis* that evolved about 400 years ago by cpDNA analysis. *Journal of Japanese Society of Horticultural Science*, 74, 464–468.

Tang, S., Bin, X., Wang, L., Zhong, Y. 2006. Genetic diversity and population structure of yellow Camellia (*Camellia nitidissima*) in China as revealed by RAPD and AFLP markers. *Biochemical Genetics*, 44, 449–461.

Taniguchi, F., Fukuoka, H., Tanaka, J. 2012. Expressed sequence tags from organ-specific cDNA libraries of tea (*Camellia sinensis*) and polymorphisms and transferability of EST-SSRs across *Camellia* species. *Breeding Science*, 62, 186–195.

Tanikawa, N., Onozaki, T., Nakayama M., Shibata, M. 2008. PCR-RFLP analysis of chloroplast DNA variations in the atpI-atpH spacer region of the genus Camellia. *Journal of Japan Society Horticultural Science*, 77, 408–417.

Thakor, B.H. 1997. A re-examination of the phylogenetic relationships within the genus Camellia. *International Camellia Journal*, 29, 130–134.

Thomas, J., Vijayan, D., Joshi, S.D., Joseph Lopez, S., Raj Kumar, R. 2006. Genetic integrity of somaclonal variants in tea (*Camellia sinensis* (L.) O Kuntze) as revealed by inter simple sequence repeats. *Journal of Biotechnology*, 123, 149–154.

Tiao, J.X., Parks, C.R. 1997. Identification of closely related Camellia hybrid and mutant using molecular markers. *International Camellia Journal*, 29, 111–116.

Tiao, J.X., Parks, C.R. 2003. Molecular analysis of the genus *Camellia*. *International Camellia Journal*, 35, 57–65.

Tosca, A., Pondofi, R., Vasconi, S. 1996. Organogenesis in *Camellia x williamsii*: Cytokinin requirement and susceptibility to antibiotics. *Plant Cell Report*, 15, 541–544.

Ueno, S., Tomaru, N., Yoshimaru, H., Manabe, T., Yamamoto, S. 2000. Genetic structure of *Camellia japonica* L. in an old-growth evergreen forest, Tsushima, Japan. *Molecular Ecology*, 9, 647–656.

Ueno, S., Yoshimaru, H., Tomaru, N., Yamamoto, S. 1999. Development and characterization of microsatellite markers in *Camellia japonica* L. *Molecular Ecology*, 8, 335–336.

Ujihara, T., Taniguchi, F., Tanaka, J., Hayashi, N. 2011. Development of expressed sequence tag (EST)-based cleaved amplified polymorphic sequence (CAPS) markers of tea plant and their application to cultivar identification. *Journal of Agriculture and Food Chemistry*, 59, 1557–1564.

Vijayan, K., Tsou, C.H. 2008. Technical report on the molecular phylogeny of *Camellia* with nr ITS: The need for high quality DNA and PCR amplification with *Pfu*-DNA polymerase. *Botanical Studies*, 49, 177–188.

Vijayan, K., Zhang, W.J., Tsou, C.H. 2009. Molecular taxonomy of *Camellia* (Theaceae) inferred from NRITS sequences. *American Journal of Botany*, 96, 1348–1360.

Wachira, F. 1990. Desirable tea plants: An overview of a search for markers. *Tea*, 11, 42–48.

Wachira, F.N., Waugh, R., Hackett, C.A., Powell, W. 1995. Detection of genetic diversity in tea (*Camellia sinensis*) using RAPD markers. *Genome*, 38, 201–210.

Wachira, F., Powell, W., Waugh, R. 1997. An assessment of genetic diversity among *Camellia sinensis* L. (cultivated tea) and its wild relatives based on randomly amplified polymorphic DNA and organelle specific STS. *Heredity*, 78, 603–611.

Wachira, F., Tanaka, J., Takeda, Y. 2001. Genetic variation and differentiation in tea (*Camellia sisnensis*) germplasm revealed by RAPD and AFLP variation. *Journal of Horticultural Science and Biotechnology*, 76, 557–563.

Wang, B.Y., Ruan, Z.Y. (2012). Genetic diversity and differentiation in *Camellia reticulata* (Theaceae) polyploid complex revealed by ISSR and ploidy. *Genetics and Molecular Research*, 11, 503–511.

Wang, L., Li, X., Zhao, Q., Jing, S., Chen, S., Yuan, H. 2009. Identification of genes induced in response to low-temperature treatment in tea leaves. *Plant Molecular Biology Reporter*, 27, 257–265.

Wang, X.F., Zheng, W.H., Zheng, H.X., Xie, Q.Q., Zheng, H.Y., Tang, H., Tao, Y.L. 2010. Optimization of RAPD-PCR reaction system for genetic relationships analysis of 15 Camellia cultivars. *African Journal of Biotechnology*, 9, 798–804.

Wang, Y.S., Gao, L.P., Wang, Z.R., Liu, Y.J., Sun, M., Zeng, W., Yuan, H. 2012. Light induced expression of genes involved in phenylpropanoid biosynthetic pathways in callus of tea (*Camellia sinensis* (L.) O. Kuntze). *Scientia Horticulturae*, 133, 72–83.

Wei, X., Cao, H.L., Jlang, Y.S., Ye, W.H., Ge, X.J., Li, F. 2008. Population genetic structure of *Camellia nitidissima* (Theaceae) and conservation implications. *Botanical Study*, 49, 147–153.

Wei, J.Q., Chen, Z.Y., Wang, Z.F., Tang, H., Jiang, Y.S., Wei, X., Li, X.Y., Qi, X.X. 2010. Isolation and characterization of polymorphic microsatellite loci in *Camellia nitidissima* chi (Theaceae). *American Journal of Botany*, 97, 89–90.

Wei, X., Wei, J.Q., Cao, H.L., Li, F., Ye, W.H. 2005. Genetic diversity and differentiation of *Camellia euphlebia* (Theaceae) in Guangxi, China. *Annals of Botany Fennici*, 42, 365–370.

Wen, S.C., Wu, L.J., Peng, S.J., Hua, G.Z., Ping, T.H., Zhi, L.F., Huan, H.Y., et al. 2002. Study on genetic polymorphism of tea plants in Anhua Yuntaishan population by RAPD. *Journal of Hunan Agricultural University*, 28, 320–325.

Wen, Q., Xu, L., Gu, Y., Huang, M., Xu, L. 2012. Development of polymorphic microsatellite markers in *Camellia chekiangoleosa* (Theaceae) using 454-ESTs. *American Journal of Botany*, 99, 203–205.

Wight, W. 1954. Morphological basis of quality in tea. *Nature*, 173, 630–631.

Wight, W. 1959. Nomenclature and classification of tea plant. *Nature*, 183, 1726–1728.

Wight, W., Barua, D.N. 1954. Morphological basis of quality in tea. *Nature*, 173, 630–631.

Wright, L.P., Apostolides, Z., Louw, A.I. 1996. DNA fingerprinting of clones. In: Whittle, A.M., Khumalo, F.R.B. (eds), *Proceedings of the 1st Regional Tea Research Seminar*, pp. 44–50. 22nd–23rd March, Blantyre, Malawi.

Wu, H., Sparks, C., Amoah, B., Jones, H.D. 2003. Factors influencing successful *Agrobacterium*-mediated genetic transformation of wheat. *Plant Cell Report*, 21, 659–668.

Wu, K.S., Tanksley, S.D. 1993. Abundance, polymorphism and genetic mapping of microsatellite in rice. *Molecular Genetics*, 241, 225–235.

Wu, L.J., Peng, S.J., Wen, S.C., Lin, L.C., Hua, G.Z., Huan, H.Y., Luo, J.W., et al. 2002a. Studies on genetic relationships of tea cultivars [*Camellia sinensis* (L.) O. Kuntze] by RAPD analysis. *Journal of Tea Science*, 22, 140–146.

Wu, L.J., Peng, S.J., Xian, L.J., Wen, S.C., Huan, H.Y., Hua, G.Z., Luo, J.W., et al. 2002b. Study on the application of RAPD technique to parentage identification of tea plant. *Journal of Hunan Agricultural University*, 28, 502–505.

Xiao, L.Z., Yan, C.Y., Li, J.X., Luo, J.W, He, Y.M., Zhao, C.Y. 2007. AFLP analysis on genetic diversity of Fenghuang-Dancong tea plant germplasm. *Journal of Tea Science*, 27, 280–285.

Xu, H., Ton, Q., Zhuang, W. 1987. Studies on genetic tendency of tea plant hybrid generation using isozyme technique. In: *Proceedings of International Tea Quality Human Health Symposium*, pp. 21–25. China Tea Science Society, Hangzhou, China.

Yang, H., Xie, S., Wang, L., Jing, S., Zhu, X., Li, X., Zeng, W., Yuan, H. 2011. Identification of up-regulated genes in tea leaves under mild infestation of green leafhopper. *Scientia Horticultuare*, 130, 476–481.

Yang, J.-B., Yang, J., Li, H.-T., Zhao, Y., Yang, S.X. 2009. Isolation and characterization of 15 micro-satellite markers from wild tea plant (*Camellia taliensis*) using FIASCO method. *Conservation Genetics*, 10, 1621–1623.

Yang, Y., Sun, T. 1994. Study on the esterase isoenzyme in tea mutagenic breeding. *China Tea*, 16, 4–9.

Yang, Y.J., Wang, X.C., Ma, C.L. 2012. Cloning and bioinformatics analysis of full-length cDNA of actin gene (*CsActin*1) from tea plant (*Camellia sinensis* (L.) O. Kuntze). *Bulletin of Botanical Research*, 32, 69–76.

Yang, Y.J., Yu, F.L., Chen, L., Zeng, J.M., Yang, S.J., Li, S.F., Shu, A.M., et al. 2003. Elite germplasm evaluation and genetic stability of tea plants. *Journal of Tea Science*, 23, 1–8.

Yao, M.Z., Chen, L., Liang, Y.R. 2008. Genetic diversity among tea cultivars from China, Japan and Kenya revealed by ISSR markers and its implication for parental selection in tea breeding pro-gramme. *Plant Breeding*, 127, 166–172.

Yao, M.Z., Ma, C.-L., Qiao, T.-T., Jin, J.-Q., Chen, L. 2012. Diversity distribution and population struc-ture of tea germplasms in China revealed by EST-SSR markers. *Tree Genetics and Genomes*, 8, 205–220.

Yoshikawa, N., Parks, C.R. 2001. Systematic studies of *Camellia Japonica* and closely related species. *International Camellia Journal*, 33, 117–121.

Zehra, M., Banerjee, S., Mathur, A.K., Kukreja, A.K. 1996. Induction of hairy roots in tea (*Camellia sinensis* (L.) using *Agrobacterium rhizogenes*. *Current Science*, 70, 84–86.

Zhang, Y.L., Zhao, L.P., Ma, C.L., Chen, L. 2007. Molecular identification, bioinformatic analysis and prokaryotic expression of the cyclophilin gene full length cDNA from tea plant (*Camellia sinen-sis*). *Journal of Tea Science*, 27, 120–126.

Zhao, C.Y., Zhou, L.H., Luo, J.W., Huang, J.A., Tan, H.P. 2006. AFLP analysis of genetic diversity of tea plant germplasm in Guangdong province. *Journal of Tea Science*, 26, 249–252.

Zhao, L.P., Liu, Z., Chen, L., Yao, M.Z., Wang, X.C. 2008. Generation and characterization of 24 novel EST derived microsatellites from tea plant (*Camellia sinensis*) and cross-species amplification in its closely related species and varieties. *Conservation Genetics*, 9, 1327–1331.

14

Omics Approaches to Improving Fiber Qualities in Cotton

Tianzhen Zhang, PhD; Xiangdong Chen, PhD; and Wangzhen Guo, PhD

CONTENTS

14.1 Introduction 467
14.2 Cotton Genetic and Physical Maps 468
14.3 Fiber QTL Mapping 471
14.4 Integrating Genetics and Genomics Approaches 472
14.5 Transcriptome Analyses of Fiber Development 473
14.6 Proteome Analyses of Fiber Development 480
14.7 Future Prospects 481
     14.7.1 Transcriptomics, Proteomics, and Metabolomics 481
     14.7.2 Integrating Omics 481
     14.7.3 Quantitative Genetics 482
Acknowledgments 482
References 482

14.1 Introduction

Cotton (*Gossypium* spp.) is one of the most important natural fibers and edible oil crops in the world. Upland cotton (*G. hirsutum* L.), with its high yield properties, accounts for about 95% of annual worldwide cotton production; the extra-long staple (ELS) or Pima cotton (*G. barbadense* L.), which has superior quality fiber properties, accounts for the other 5%. The two tetraploid species are allotetraploids ($2n = 4x = 52$) composed of two ancestral genomes designated A-subgenome (hereafter At) *G. herbaceum* and D-subgenome (hereafter Dt) *G. raimondi*, originating from a polyploidy event ~1–2 million years ago (Wendel and Cronn 2003).

Cotton fiber qualities, including length, strength, and fineness, are known to be controlled by genes affecting cell elongation and secondary cell wall (SCW) biosynthesis. Genetic improvements to cotton fiber quality could be accelerated by omics approaches. The international cotton research communities have made considerable progress in structural genomics, such as enhancement of genetic maps, mapping of important economic traits or genes, and molecular-assisted pyramid breeding. Compared to genetic approaches, omics involve relatively new technologies for cotton in functional genomics research. In the present review, we focus on the major advances in cotton genomics, transcriptomics, proteomics, and metabolomics in recent years, and discuss future prospects for cotton omics research.

14.2 Cotton Genetic and Physical Maps

Genome research has demonstrated great promise for the continued and enhanced genetic improvement of crop plants (Zhang et al. 2008). Here, we summarize the major recent advances in cotton structural genomic research, such as genetic and physical maps. The development of a large number of expression sequence tags (ESTs) has provided a good source of polymerase chain reaction (PCR)-based primers for targeting simple sequence repeats (SSRs). Four high-density molecular linkage maps composed of more than two thousand loci from interspecific hybrid (*G. hirsutum* × *G. barbadense*) populations have been reported (Rong et al. 2004; Yu et al. 2011, 2012; Zhao et al. 2013).

The Paterson Group in the USA constructed the first molecular linkage map of the *Gossypium* species from an interspecific *G. hirsutum* Palmeri × *G. barbadense* K101 F_2 population based on RFLPs (Reinisch et al. 1994). The map contained 705 loci that were assembled into 41 linkage groups and spanned 4675 cM. This map was further advanced by Rong et al. (2004), comprising 2584 loci at 1.74 cM intervals and covered all 13 homeologous chromosomes of the allotetraploid cottons. Chee et al. (2004) developed EST-derived markers for construction of a genetic map from the Palmeri × K101 F_2 population.

The Zhang Group in China initially developed a large number of EST-SSR markers and constructed a high-density and gene-rich genetic map containing 3414 loci and covering 3667 cM, with an average intermarker distance of 1.08 cM based on the (TM-1 × Hai7124) BC_1 population (Han et al. 2004, 2006; Song et al. 2005; Guo et al. 2007, 2008; Zhao et al. 2013). The map will serve as a valuable genomic resource for tetrapliod cotton genome assembly, for cloning genes related to superior agronomic traits, and for further comparative genome analysis in *Gossypium*.

Lacape et al. (2003) in France constructed a combined RFLP-SSR-AFLP map based on an interspecific *G. hirsutum* var. VH8-4602 × *G. barbadense* var. Guazuncho2 backcross population of 75 BC_1 plants (Lacape et al. 2003). The map consists of 888 loci that are ordered into 37 linkage groups and span 4400 cM. This map was updated, mostly with new SSR markers, to contain 1160 loci that spanned 5519 cM with an average distance between loci of 4.8 cM (Nguyen et al. 2004). Lacape et al. (2009) report the development of a new interspecific cotton recombinant inbred line (RIL) population of 140 lines deriving from Guazuncho2 × VH8-4602 using AFLP and SSR, and found that 255 loci were an excellent colinearity with the BC_1 map. A consensus BC_1-RIL map based upon 215 individuals (75 BC_1 + 140 RIL) was built, consisting of 1745 loci and spanning 3637 cM.

Recently, the tetraploid cotton genome-wide comprehensive reference map (CRM) was constructed from 28 public cotton genetic maps. The initial CRM contained 7424 markers and represented over 93% of the combined mapping information from the 28 individual maps. The current output is stored and displayed through CottonDB (http://www.cottondb.org), the public cotton genome database (Yu et al. 2010). This not only facilitates the identification and positioning of QTLs and candidate genes, but also provides a basic structure for the genome sequence assembly. By integrating six previously reported high-density maps, a consensus genetic map of tetraploid cotton, consisting of 8245 loci, spanning4070 cM, with an average of 2 loci per cM, was produced (Blenda et al. 2012).

It is imperative to construct physical maps based on bacterial artificial chromosomes (BAC) for genomics research, and advances in molecular cytogenetic techniques will speed up this objective. Fluorescence *in situ* hybridization (FISH) using BAC clones as probes has commonly been applied to chromosome identification (Wang et al. 2006b, 2007a). Based on two BAC libraries of 0-613-2R and TM-1 (Yin et al. 2006; Hu et al. 2009), six linkage groups

(LGs) A01, A02, A03, D02, D03, and D08 to chromosomes 13, 8, 11, 21, 24, and 19 were assigned using BAC-FISH and translocations (Figure 14.1). As a result, all 26 chromosome pairs in tetraploid cotton were identified (Wang et al. 2007a), and 13 homeologous chromosome pairs using a new chromosome nomenclature (A1-13 and D1-13) were established (Wang et al. 2006c). This set of BAC markers enables us to make associations between chromosomes and their genetic linkage groups, and also provides convenient and reliable landmarks for establishing physical linkage with unknown targeted sequences.

Advances in the construction of cotton physical maps based on BAC were reported. An integrated genetic and physical map of homoeologous chromosomes 12 and 26 in Upland cotton were reported (Xu et al. 2008). The Paterson Group reported a draft physical map of a D-genome cotton species (*G. raimondii*): a total of 13662 BAC-end sequences and 2828

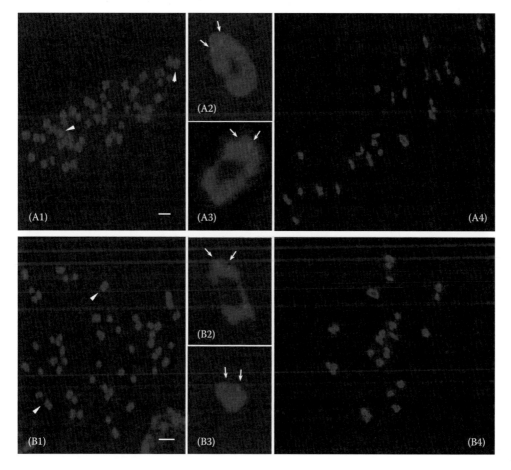

FIGURE 14.1 **(See color insert)**

Eight chromosome-specific BAC clones of cotton identification and chromosomal and subchromosomal location by FISHing to *G. hirsutum* NTs. (From Wang, K. et al., *Theor Appl Genet*, 113, 73–80, 2006c.) **A1** to **H1** show the hybridization of BAC probes (Table 1) to eight chromosomes in mitotic cells to identify chromosome-specific BAC clones, respectively: **(A1)** BAC clone 104O10, **(B1)** 62K03, **(C1)** 98H10, **(D1)** 35J07, **(E1)** 14G14, **(F1)** 59B08, **(G1)** 37F17, and **(H1)** 50D03. Arrowheads point to the dual chromosome-specific FISH signals (*red*). All bars are 5 μm. **A2**, **A3** to **H2**, **H3** show the chromosomal and subchromosomal locations of FISH sites. *Arrows* indicate the dual FISH signals (*red*). All the figures of IVs and bivalents were derived from the integrated metaphase cell images. For example, **A4** and **B4** show the integrated cell images of **A2** and **B3**, respectively.

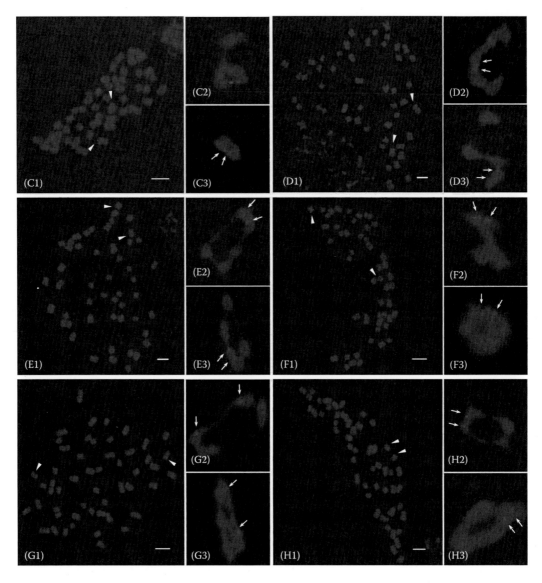

FIGURE 14.1   **(See color insert) (Continued)**

DNA probes were used in genetically anchoring 1585 contigs to a cotton consensus genetic map (Lin et al. 2010), and they released the complete cotton D-genome scaffold sequences through the Joint Genome Institute (JGI; http://www.jgi.doe.gov) (Paterson et al. 2012). Recently, the Chinese Academy of Agricultural Sciences (CAAS) reported that a draft genome of diploid cotton *G. raimondii* has been sequenced and assembled (Wang et al. 2012a). Sequencing of the *G. raimondii* genome will not only provide a major source of candidate genes important for the genetic improvement of cotton quality and productivity, but may also serve as a reference for the assembly of the tetraploid *G. hirsutum* genome. The whole-genome sequence of tetraploid cotton has been entered into a calendar by an international coalition of cotton genomic scientists.

14.3 Fiber QTL Mapping

Molecular linkage map construction has contributed greatly to locating fiber QTL. Important QTLs related to cotton fiber quality have been tagged (Table 14.1). More fiber quality QTL online information databases were constructed between different cotton species (http://www.cottondb.org/wwwroot/table_QTL_names.php). Meta-analysis of polyploid cotton QTL shows unequal contributions of subgenomes to a complex network of genes and gene clusters implicated in lint fiber development (Rong et al. 2007). QTLs for fiber quality and yield more often mapped to Dt than At for intraspecific mapping populations (Shen et al. 2005, 2006, 2007; Wang et al. 2006b, 2007b; Qin et al. 2008a); however, using interspecific mapping populations, more QTLs for fiber quality, particularly for fiber length and strength, were detected on At than Dt chromosomes (He et al. 2007, 2008b; Chen et al. 2012). This indicated the important contribution of the D-subgenome in improving fiber quality traits in *G. hirsutum*. In-depth understanding of distribution and gene expression of At and Dt in tetraploid cottons will help to improve fiber qualities (Jiang et al. 1998; Saha et al. 2006; Yang et al. 2006).

TABLE 14.1

Progress of QTL Mapping for Fiber Quality Traits in Cotton

Parental Materials	References
(*G. hirsutum* L. lines, HS46 × MARCABUCAG8US-1-88) F_2	Shappley et al. 1998
(CAMD-E × Sea Island Seaberry) F_2[a]	Jiang et al. 1998
(D5678ne × Prema) $F_{2:3}$	Ulloa and Meredith 2000
(*G. hirsutum* cv. TM-1 × 3-79 *G. barbadense*) F_2[a]	Kohel et al. 2001
(*G. hirsutum* cv. Siv'on × *G. barbadense* cv. F-177) F_2 and F_3[a]	Paterson et al. 2003
(7235 × TM-1) F_2 and F_3	Zhang et al. 2003
(7235 × TM-1) F_2	Guo et al. 2003
(*G. hirsutum*.cv. Acala-44 × *G. barbadense* cv. Pima S-7) F_2[a]	Mei et al. 2004
(*G.hirsutum* cv. Tamcot 2111 × *G. barbadense* cv. Pima S6) BC_3F_2[a]	Draye et al. 2005; Chee et al. 2005a,b
(Guazuncho 2 × VH8) BC_1, BC_2 and BC_2S_1[a]	Lacape et al. 2005
7235, HS427-10, PD6992, TM-1 × SM3	Shen et al. 2005
(Handan208 × Pima 90) F_2[a]	Lin et al. 2005
(Yumian1 × T586) F_2 and $F_{2:3}$	Zhang et al. 2005
(7235 × TM-1) RILs	Shen et al. 2006
(TM-1 × 3-79) RILs[a]	Frelichowski et al. 2006
(Zhongmiansuo12 × J8891) RILs	Wang et al. 2006b
(7235 × TM-1) RILs	Shen et al. 2007
(Zhongmiansuo12 × J8891) RILs	Wang et al. 2007b
(Handan 208 × Pima 90) $F_{2:3}$[a]	He et al. 2007
(Yumian1 × T586) F_2 and $F_{2:3}$	Wan et al. 2007
Simian3/Sumian12//Zhong4133/8891	Qin et al. 2008a
(Handan 208 × Pima 90) $F_{2:3}$[a]	He et al. 2008a
(7TR-133, 7TR-132, and 7TR-214 × TM-1) F_2 and $F_{2:3}$	Chen et al. 2009
(CRI 8 × Pima 90-53) F_2[a]	Liu et al. 2009
(Guazuncho 2 × VH8) BC_1, BC_2, BC_2S_1 and RILs[a]	Lacape et al. 2010
(TM-1 × Hai7124) BC_1S_1[a]	Chen et al. 2012

[a] Interspecific cross.

In tetraploid cottons, many QTLs for fiber qualities were detected on the Dt that derived from an ancestor with no spinnable fibers. For example, Jiang et al. (1998) suggested that Dt QTL may partly explain the fact that domestication and breeding of tetraploid cottons have resulted in fiber with a higher quality than those achieved by parallel improvement of the At diploid cottons which produce spinnable fibers. The merger of the At and Dt in tetraploid cottons, where each genome has a different evolutionary history, may have offered unique avenues for phenotypic response to selection (Zhang et al. 2008). Using three elite fiber lines of Upland cotton, three pairs of homoeologous QTLs were detected (Shen et al. 2005). Similar homoeologous QTLs for FS on A8/D8 also were reported (Zhang et al. 2005; He et al. 2007).

The position and effect of QTLs for fiber quality were not comparable in the different populations and environments evaluated (Draye et al. 2005; Rong et al. 2007). This suggests that the QTL studies conducted thus far have detected only a small number of loci for fiber growth and development and that additional QTLs remain to be discovered (Chee et al. 2005a,b). Therefore, these traits need to be studied in multiple environments, including different years and locations, because of environmental variation. For example, Wang et al. (2006b, 2007b) tagged a stable fiber length QTL on D2; however, they simultaneously detected five significant QTLs for fiber strength, micronaire, and so on, in four environments in Xiangzamian2 (ZMS 12 × J8891). This result was confirmed through further analyses (He et al. 2007; Qin et al. 2008a).

Zhang et al. (2003) detected eight molecular markers linked with a major FS QTL ($QTLFS_1$) that explained more than 30% of the phenotypic variation in a *G. anomalum* introgression line 7235. This major QTL was stable in comparative mapping of RIL and F_2 populations (Shen et al. 2005, 2006, 2007) and was efficiently used in MAS breeding to improve fiber strength (Guo et al. 2003). In order to fine-map this QTL, three overlapping RILs, developed from a cross between 7235 and TM-1, were backcrossed to TM-1 to produce three large mapping populations. Surprisingly, five tightly linked or clustered QTLs were detected that overlapped our previously identified major QTL region (Chen et al. 2009). These five QTLs act like a major QTL, perhaps representing a single major gene for fiber strength, explaining a total phenotypic variance of 28.8%–59.6%.

QTLs for fiber quality occur genetically in clusters in the cotton genome (Shappley et al. 1998; Ulloa et al. 2000, 2005; Mei et al. 2004; Wang et al. 2006a; Shen et al. 2006, 2007; Rong et al. 2007). Ulloa et al. (2005) suggested the possible existence of highly recombined regions in the cotton genome with abundant putative genes. QTL clusters might exert their multiple functions to compensate for a numerical deficiency, expanding their roles in cotton growth and development (Wang et al. 2006a). In particular, the majority of important QTLs for different traits have been found clustered in the same interval or in neighboring intervals (Shappley et al. 1998; Ulloa et al. 2000, 2005; Mei et al. 2004; Wang et al. 2006a; Shen et al. 2006, 2007; Rong et al. 2007). These results indicate that genes controlling fiber development and yield can be linked, or that they are likely to be pleiotropic, resulting in negative relationships between fiber and yield components that cause complications for plant breeders.

14.4 Integrating Genetics and Genomics Approaches

Genetical genomics is a novel approach linking natural genetic variation to gene expression variation, thereby allowing the identification of genomic loci containing gene expression modulators (eQTLs). Recently, only two reports have focused on the transcriptional changes between *G. hirsutum* and *G. barbadense* fibers using microarrays (Alabady et al. 2008; Al-Ghazi et al. 2009). However, fiber quality genes are less well-characterized, and little is known about the underlying biological causes of these differences in cotton fiber qualities.

Claverie et al. (2012) used quantitative cDNA-AFLP to identify eQTLs and eQTL hotspots among a population of interspecific *G. hirsutum* × *G. barbadense* recombinant inbred lines (RILs) at 10 days post anthesis (DPA) and 22 DPA. The eQTL hotspots were compared to the location of phenotypic QTLs for fiber characteristics among the RILs, and several cases of colocalization were detected. However, little was known about transcript information corresponding to fiber genes or understanding the underlying biological causes of these differences in cotton fiber qualities.

Chen et al. (2012) revealed molecular mechanisms for fiber differential development between *G. barbadense* cv. Hai7124 and *G. hirsutum* acc. TM-1 by genetical genomics. An interspecific backcrossed population from Hai7124 and TM-1 for fiber characteristics was evaluated in four-year environments under field conditions, and 12 QTL and QTL-by-environment interactions were detected by multi-QTL joint analysis. Further analysis of fiber growth and gene expression between TM-1 and Hai7124 showed greater differences at 10 and 25 DPA. In these two important periods for fiber performances, a total of 916 eQTLs were identified significantly ($P < 0.05$) by integrating genome-wide expression profiling with linkage analysis using the same genetic materials. Many positional cis-/trans-acting eQTL and eQTL hotspots were detected across the genome. By comparative mapping of eQTL and fiber QTL, a dataset of candidate genes affecting fiber qualities was generated. Real-time quantitative RT-PCR (qRT-PCR) analysis confirmed the major differential genes regulating fiber cell elongation or SCW synthesis. These data collectively support a molecular mechanism for *G. hirsutum* and *G. barbadense* through differential gene regulation causing difference in fiber qualities. The downregulated expression of abscisic acid (ABA) and ethylene signaling pathway genes, and the high-level and long-term expression of positive regulators including auxin and cell wall enzyme genes for fiber cell elongation at the fiber developmental transition stage may account for superior fiber qualities.

14.5 Transcriptome Analyses of Fiber Development

Cotton fiber is an excellent model for cellular development and elongation, which occurs in four overlapping stages: initiation, elongation, secondary cell wall (SCW) synthesis, and maturation (Basra and Malik 1984; Kim and Triplett 2001). The cotton fiber initiation stage acts as a developmental switch to determine the number of fibers on each ovule, whereas the rate and duration of cell elongation/expansion determine fiber length, and the duration of SCW affects fiber strength and fineness (Smart et al. 1998; Ruan et al. 2001). Therefore, cotton fiber transcriptomics mostly focus on functional identification of crucial genes for improving fiber yield and quality. Since 2003, many specifically or preferentially expressed genes have been identified in cotton fiber (Table 14.2).

Transcription factors play essential roles in cotton fiber initiation. Previous findings illustrate that complex networks (MYB–bHLH–WD40) control the *Arabidopsis* trichome cell fate (Ramsay and Glover 2005). The initiation of cotton fiber cells was found to be developmentally similar to that of *Arabidopsis* trichomes (Guan et al. 2007). Therefore, identification of

TABLE 14.2

Genes Related to Cotton Fiber Development Published since 2003

Stages	Genes	Putative Functions	References
Initiation	GaMYB2 (FIF1)	GaMYB2 is predominantly expressed early in the development of cotton fibers; it rescued the trichome formation of the *Arabidopsis* gl1 mutant.	Wang et al. 2004; Shangguan et al. 2008
	GaRDL1	GaRDL1 contains a homeodomain-binding L1 box involved in activating the RDL1-P3 promoter in *Arabidopsis* trichomes, and a MYB-binding motif, *RDL1* was expressed mainly in developing fiber cells.	Wang et al. 2004
	GhRac	GhRac plays a crucial role in the initiation of cotton fiber cells.	Li et al. 2005a
	GhMyb25	GhMYB25 regulates early fiber development.	Wu et al. 2006; Machado et al. 2009
	GaHOX1	GaHOX1 is predominantly expressed in cotton fiber cells at early developmental stages, and it is a functional homolog of GL2 in plant trichome development.	Guan et al. 2008
	GbML1	GbML1 interacts with GbMYB25 to control cotton fiber development.	Zhang et al. 2010b
	GhRDL1, GhMYB2	They were highly expressed during fiber cell initiation, and activated *Arabidopsis* seed hair development.	Guan et al. 2011
	GhAux1–9	GhAux4–7 were preferentially expressed in 0 DPA ovules, GhAux8 in fiber early elongation stages, and GhIAA16 in fiber initiation and secondary cell wall stage, GhAux9 was specifically expressed in developing fibers.	Han et al. 2012
	GbPDF1	GbPDF1 is involved in cotton fiber initiation via the core cis-element HDZIP2ATATHB2.	Deng et al. 2012
	GhHD-1	A homeodomain leucine zipper gene that mediates epidermal cell differentiation in cotton.	Walford et al. 2012
Initiation and elongation	GhMYB109	GhMYB109 is specifically expressed in initial and elongating fibers, and reveals a largely conserved mechanism of the R2R3 MYB transcription factor in cell fate determination in plants.	Suo et al. 2003; Pu et al. 2008
	SuSy	SuSy plays important roles in fiber initiation and elongation.	Ruan et al. 2003
	GhlecRK	GhlecRK is a lectin-like membrane protein that may play an important role in fiber development.	Zuo et al. 2004
	GhDET2	GhDET2 and BRs play a crucial role in the initiation and elongation of cotton fiber cells.	Luo et al. 2007
	GhFLA1, GhAGP2,3,4	The *FLAs* are essential for the initiation and elongation of cotton fiber development.	Liu et al. 2008; Li et al. 2010a
	GhGA20ox1-3	GhGA20ox1 is expressed preferentially in elongating fibers, while GhGA20ox2,3 transcripts accumulated mainly in ovules; they promote fiber initiation and elongation by regulating GA synthesis	Xiao et al. 2010
	GhSusA1	Overexpression of GhSusA1 increases plant biomass and improves cotton fiber yield and quality.	Jiang et al. 2012
Elongation	GhKCBP	GhKCBP plays a role in interphase cell growth, most likely by interacting with cortical microtubules.	Ji et al. 2003; Preuss et al. 2004
	GhABP	GhABP was activated during the fast fiber elongation period.	Ji et al. 2003

TABLE 14.2 (Continued)

Genes Related to Cotton Fiber Development Published since 2003

Stages	Genes	Putative Functions	References
	GhMAPK	*GhMAPK* was activated during the fast fiber elongation period.	Ji et al. 2003
	GhWBC1	Encoding an ATP-binding cassette transporter of the WBC subfamily, the overexpressed *GhWBC1* interferes with Arabidopsis seed and silique development.	Zhu et al. 2003
	GhAGPs	*GhAGPs* play an important role in the developmental stages of fiber cell elongation.	Feng et al. 2004
	GhGLP1	*GhGLPs* may be important for cell wall expansion.	Kim and Triplett 2004
	GhBRI1	Mutations in the BRI1 genes of several species result in dwarfed plants with reduced sensitivity to BR	Sun et al. 2004
	GhKCR1-2	Encoding 3-ketoacyl-CoA reductases and preferentially expressed during fiber elongation period; *GhKCR1* and *GhKCR2* play an important role in very long-chain fatty acids biosynthesis.	Qin et al. 2005
	GhACT1	Encoding an actin, and involved in fiber elongation but not in fiber initiation.	Li et al. 2005b
	GhPFN1	Encoding profiling, may be involved in the rapid elongation of cotton fibers by promoting actin polymerization.	Wang et al. 2005
	GhBG	Specifically expressed in fiber cells highly abundant in 5-17 DPA, and can lead to a significant increase in cell length and width, and a remarkable decrease in the length/width ratio when transformed into yeast.	Ma et al. 2006
	GhACO1-3	*GhACO1-3* promotes fiber cell elongation.	Shi et al. 2006
	GhKCS13/ CER6	Encoding 3-ketoacyl-CoA synthase, involved in very-long-chain fatty acids biosynthesis, and promote cotton fiber and *Arabidopsis* cell elongation by activating ethylene biosynthesis.	Qin et al. 2007a
	GhAPX1	*GhAPX1* expression is upregulated in response to an increase in cellular H_2O_2 and ethylene, and encodes a functional enzyme that is involved in H_2O_2 homeostasis during fiber development.	Li et al. 2007a; Qin et al. 2008b
	GhTUA9	*GhTUA9* gene specifically expressed in fiber and involved in cell elongation.	Li et al. 2007b
	Gh14-3-3L	Predominantly expressed during early fiber development, and reached the peak of expression in 10 DPA fiber cells, suggesting that it may be involved in regulating fiber elongation.	Shi et al. 2007; Zhang et al. 2010a
	GhEF1As	Translation elongation factor 1A-1, 2, 4, 5, and 9 active at early fiber elongation.	Xu et al. 2007
	GhSMT2-1,2	*GhSMT2-1, 2* play a role in fiber elongation.	Luo et al. 2008
	GhTUBs	Nine *GhTUBs* were highly expressed in elongating fiber cells as compared with fuzzless-lintless mutant ovules, and were induced by gibberellin, ethylene, brassinosteroids, and lignoceric acid; are involved in fiber development.	He et al. 2008b
	GhGS	Expressed differentially between 7235 and TM-1 at 8 DPA fibers, significantly correlated with fiber strength QTL on D7.	He et al. 2008c

(continued)

TABLE 14.2 (Continued)

Genes Related to Cotton Fiber Development Published since 2003

Stages	Genes	Putative Functions	References
	GhUGT1,2	UDP-Glycosyltransferases, involved in plants responding to osmotic stress.	Tai et al. 2008
	GhGID1a,b	GA receptor genes; ectopic expression of the GhGID1a in the rice gid1-3 mutant plants rescued the GA-insensitive dwarf phenotype.	Lorenzo et al. 2008
	GhSLR1a,b	DELLA protein genes; ectopic expression of GhSLR1b in wild type *Arabidopsis* led to reduced growth and upregulated expression of DELLA-responsive genes.	Lorenzo et al. 2008
	GhXTH1-1,2, Ga/GrXTH	*GhXTH1-1* gene was specifically expressed in cotton fiber elongation, and expression levels of *GhXTH1-1* may relate to fiber length.	Michailidis et al. 2009
	GhH6L	Predominantly expressed in the fiber fast elongation period.	Wu et al. 2009
	GhECR1-2	Encoding trans-2-enoyl-CoA reductase (ECR) and upregulated expression during cotton fiber elongation, GhECR1 and 2 were involved in fatty acid elongation during cotton fiber development.	Song et al. 2009
	GhPEL	Encoding a pectate lyase in cell wall loosening by depolymerization of the de-esterified pectin during fiber elongation in cotton.	Wang et al. 2010a
	GhPOX1	*GhPOX1* regulates fiber elongation by ROS signaling.	Mei et al. 2009
	GhPFN2	Overexpression of a profilin (*GhPFN2*) promotes preterminated cell elongation, and results in a marked decrease in the length of mature fibers.	Wang et al. 2010b
	GhATPδ1	*GhATPδ1* (ATP synthase δ1 subunit) is important for the activity of mitochondrial ATP synthase and is probably related to cotton fiber elongation.	Pang et al. 2010a
	GhVIN1	High activity of vacuolar invertase is required for cotton fiber and Arabidopsis root elongation through osmotic dependent and independent pathways.	Wang et al. 2010c
	GhPEPC1,2	The high activity of PEPC is required in cotton fiber elongation.	Li et al. 2010b
	GhXTH1-3	*GhXTH1* is the predominant XTH in elongating fibers and its expression limits cotton fiber elongation.	Lee et al. 2010
	GhAPY1,2	*GhAPY1,2* regulates fiber elongation.	Clark et al. 2010
	Annexin, AnnBj1	Constitutive expression of mustard annexin, AnnBj1 enhances abiotic stress tolerance and fiber quality in cotton under stress	Divya et al. 2010
	GhBCPs	*GhBCPs* may participate in the regulation of fiber development and response to high-salinity and heavy metal stresses in cotton.	Ruan et al. 2011
	GhAnx1	*GhAnx1* is a typical annexin protein, highly expressed in fibers, especially during the elongation stage.	Zhou et al. 2011
	GhMADS11	*GhMADS11* may function in fiber cell elongation.	Li et al. 2011
	GbRL2	*GbRL2* may participate in the development of various organs and may be a target for genetic improvement of fiber.	Zhang et al. 2011
	GhCDKA	*GhCDKA* is an A-type cyclin-dependent kinase, highly expressed at fast elongation stage.	Gao et al. 2012

TABLE 14.2 (Continued)

Genes Related to Cotton Fiber Development Published since 2003

Stages	Genes	Putative Functions	References
Elongation and SCW	*GhGlcAT1*	*GhGlcAT1* may be involved in noncellulose polysaccharides biosynthesis of the cotton cell wall.	Wu and Liu 2005; Wu et al. 2007a
	GhRLK1	*GhRLK1* is expressed in fast elongation and transition stage of cell elongation and SCW; involved in the induction and maintenance of active secondary wall formation.	Li et al. 2005c
	GhCAD6,1	*GhCAD6,1* may play a role in the secondary cell wall stages.	Fan et al. 2009
	GhADF1	*GhADF1* plays key roles in the fiber elongation and SCW stages.	Wang et al. 2009
	GbTLP1	*GbTLP1* is involved in secondary cell wall stages.	Munis et al. 2010

comparable transcription factors in cotton is very important for dissecting fiber initiation mechanisms. Functional analyses have demonstrated that *GaMYB2*, *GaRDL1*, *GaHOX1*, and *GhMYB109* play essential roles in the regulatory networks during cotton fiber initiation (Wang et al. 2004; Shangguan et al. 2008; Guan et al. 2008; Pu et al. 2008).

In addition to the biosynthesis of various phytohormones, fast polarized growth of a cotton fiber cell requires the biosynthesis of plasma membrane and cell wall components, along with cell wall loosening and expansion. Functional analyses demonstrated that several genes related to the cytoskeleton (*GhTUB1*, *GhWBC1*, *GhPFN1*, *GhACT1*, *GhTUA9*, and *GhTUBs*) and four genes related to cell wall biosynthesis or cell expansion (*GhRLK1*, *GhGlcAT1*, *GhPEL*, and *GhATPδ1*) play important roles in fiber elongation (Table 14.2).

Two recent reports show great advances in improvements to fiber quality. Zhang et al. (2011) found that the accumulation of the plant hormone indole-3-acetic acid (IAA) is required for the initiation of fiber cell development. Further targeted expression of the IAA biosynthetic gene iaaM, driven by the promoter of the petunia MADS box gene Floral Binding protein 7 (FBP7), increased IAA levels in the epidermis of cotton ovules at the fiber initiation stage. This substantially increased the number of lint fibers, and fiber fineness was notably improved. This represented a major breakthrough in cotton high fiber yield and superior fiber quality breeding. Jiang et al. (2012) reported the characterization of a novel sucrose synthase (*SusA1*) gene from a superior quality fiber germplasm line 7235 in *G. hirsutum*. Suppression of *GhSusA1* in transgenic cotton reduced fiber quality and decreased the boll size and seed weight. Importantly, overexpression of this gene increased fiber length and strength, with the latter indicated by the enhanced thickening of cell walls during the secondary wall formation stage. These findings identified *GhSusA1* as a key regulator of sink strength in cotton, which is closely associated with productivity, and hence a promising candidate gene that can be developed to increase cotton fiber yield and quality.

Because cotton fiber is an excellent general model for cell elongation, the elongation phase is perhaps the best-studied period of fiber development (Kim and Triplett 2001). Several cDNA libraries derived from ovules, fibers, and other tissues from cultivated tetraploid cotton were constructed (Ji et al. 2003; Arpat et al. 2004; Liu et al. 2006; Shi et al. 2006; Tu et al. 2007; Gou et al. 2007; Wu et al. 2007b; Taliercio et al. 2007). Using a PCR-selected cDNA subtractive analysis and differential screening, 172 differentially expressed genes were identified between Xuzou142 fiber and fuzzless-lintless during fiber elongation (Ji et al. 2003); 292 preferentially expressed genes were identified between 10 and 20 DPA in fiber cells and nonfiber tissues (Liu et al. 2006), and 645 were identified between different stages of Pima 3-79 ovules or fibers (Tu et al. 2007).

With the implementation of large-scale EST sequencing, fiber elongation was studied using high-throughput DNA microarray platforms (Table 14.3). Shi et al. (2006) reported a 12K cDNA microarray platform (GEO accession: GPL2610) containing 11962 uniESTs from 5 to 10 DPA Xuzhou142 fibers. They were the first to demonstrate that ethylene plays an

TABLE 14.3

Cotton Microarray Platforms Related to Cotton Fiber Development

Platforms	Array Name	Probes	Experimental Materials	References
GPL2610	12K cDNA Array	11,719 UniESTs	Xuzhou 142 and the fl mutant	Shi et al. 2006
GPL2610	12K cDNA Array	11,719 UniESTs	TM-1 and Hai7124 (5, 10, 15, 20, 25 DPA fiber)	Chen et al. 2012
GPL3035	Cotton Ovule cDNA MicroArray ghA	10,410 cDNA clones	DP16, Xuzhou 142, and six lintless lines mutant	Wu et al. 2007b
GPL3641	5K cDNA array	5,122 unique EST clones	Xuzhou-142 (3–18 DPA fibers)	Gou et al. 2007
GPL4043	CSIRO cotton whole plant cDNA array	24,286 cDNA clones	Siokra V-15 and Pima S7 (7, 11, and 21 DPA fibers)	Al-Ghazi et al. 2009
—	Cotton 70-mer oligos array (Chen Lab)	1,334 oligos cotton EST	TM-1 and N1N1 mutant (multiple tissues)	Lee et al. 2007
GPL4305	Cotton_ oligonucleotide_ array_v1.0	13,158 oligonucleotide probes	TM1 (2, 7, 10, 20, and 25 DPA ovules or fibers)	Hovav et al. 2008d
—	SNP-specific platform	7,574 homoeolog-specific probe sets (35-mer)	A and D wild accessions, F_1 hybrid and cultivars (petals)	Flagel et al. 2008
—	High-throughput SNP-specific platform	11,350 genome-specific probes (35-mer + 60-mer)	TM1, TX2094, Pima S-7, K101 (5, 10, and 20 DPA fibers)	Hovav et al. 2008a
—	High-throughput SNP-specific platform	11,350 genome-specific probes (35-mer + 60-mer)	A2, D5 accessions, TM1 and TX2094 (5, 10, and 20 DPA fiber)	Hovav et al. 2008b
GPL4739	*G. hirsutum* agilent array version 1.0	~11,000 oligonucleotide probes (40–60 nt)	DES119 and ST4793R (1 or 10 DPA fiber)	Taliercio and Boykin 2007
—	Agilent gene chips (Wilkins lab)	22,000 oligonucleotide probes (60-mers)	TM-1 and Li-1 mutant (15 or 24 DPA fiber)	Bolton et al. 2009
GPL4808	Cotton_ oligonucleotide_ array_v2.0	22,000 oligonucleotide probes	Wild accessions A, F, and cultivars (5, 10, 20, and 25 DPA fibers)	Hovav et al. 2008c
GPL5476	32K Cotton cDNA array	31,401 UniESTs from 375,441 cotton ESTs	Xuzhou 142 and fl mutant (10-DPA ovules or fibers)	Pang et al. 2010a
GPL6937	Chen Lab Cotton 25k oligo array	23,171 oligonucleotide probes	Unpublished	Unpublished
GPL6989	Udall global cotton16 EST assembly array	~290,000 expression probes	K101 and Pima S-7 (2, 10, 20 DPA fibers)	Chaudhary et al. 2008
GPL6989	Udall global cotton16 EST assembly array	~290,000 expression probes	Seven *Gossypium* wild accessions and cultivars (2 DPA ovules)	Chaudhary et al. 2009

TABLE 14.3 (Continued)

Cotton Microarray Platforms Related to Cotton Fiber Development

Platforms	Array Name	Probes	Experimental Materials	References
GPL6917	12,227 Cotton oligonucleotide microarrays	12,227 oligonucleotide probes (70-mers)	TM-1 (10 and 24 DPA fibers)	Arpat et al. 2004
GPL6917	Cotton Fiber Oligonucleotide Microarray.V1	12,227 oligonucleotide probes (70-mers)	Pima and TM-1 (5, 8, 10, 14, 17, 21, 24 DPA fibers)	Alabady et al. 2008
GPL8062	Cotton Array Version 2	~25,344 oligonucleotide probes	MD 52ne and MD 90ne (8, 12, 16, 20, 24 DPA fibers)	Hinchliffe et al. 2010
GPL8569	28K cotton cDNA array	28,178 UniESTs from Y.X. Zhu and X.Y. Chen' lab	TM-1 and Hai7124 (15 and 25 DPA fiber)	Chen et al. 2012

essential role in promoting fiber cell elongation by activating fiber-specific genes, such as *SUS*, *EXP1*, *EXP2*, and *TUB1*, that are important for cell wall biosynthesis, wall loosening, and cytoskeleton rearrangement. Qin et al. (2007b) further demonstrated that very long-chain fatty acids (*VLCFAs*) promote cotton fiber and *Arabidopsis* cell elongation by activating the ethylene biosynthesis gene *ACOs*. The *GhAPX1* gene has been shown to be involved in hydrogen peroxide (H_2O_2) homeostasis during cotton fiber development; H_2O_2 production is promoted by ethylene, and H_2O_2 induces ethylene production by a feedback regulatory mechanism, which together modulate cotton fiber development (Li et al. 2007a; Qin et al. 2008b). Additionally, Pang et al. (2010b) reported a new cotton 32K cDNA microarray (GPL5476) containing 31401 UniESTs, indicating that the biosynthesis of pectic precursors is important for cotton fiber and *Arabidopsis* root hair elongation by comparative proteomics. Based on biosynthesis and signaling pathways research, Qin and Zhu (2011) suggested a novel signaling pathway for the linear cell-growth mode of interactions among H_2O_2, Ca^{2+}-dependent protein kinase (*CPK*), *VLCFAs*, ethylene, and pectin during fiber cell elongation.

Gou et al. (2007) constructed a 5K cDNA array (GPL3641) covering 5122 unique ESTs from a cDNA library of *G. hirsutum* L. cv. Xuzhou142 using −3 to 5 DPA ovules and 6 to 24 DPA fibers. They identified 633 differentially regulated genes during cell elongation and SCW synthesis, which indicated that auxin signaling, wall loosening, and lipid metabolism are highly active during fiber elongation, whereas cellulose biosynthesis is predominant in the SCW synthesis stage. Liu et al. (2012a) employed microarray technology and quantitative real-time PCR to compare transcriptomes of Ligon lintless-1 mutant and the normal wild-type TM-1 at 0, 3, and 6 DPA; the results showed that 6 DPA is probably a key phase determining fiber elongation. Transcriptome analysis reveals critical genes and key pathways for early cotton fiber elongation in Ligon lintless-1 mutant.

Large-scale cotton EST sequencing also provides a powerful platform for predicting microRNAs, which will increase our understanding of mechanistic roles in regulating fiber development. Qiu et al. (2007) used bioinformatics approaches to identify microRNAs and their targets from the *G. hirsutum* ESTs database in NCBI, and Kwak et al. (2009) further enriched a set of microRNAs for fiber development. Wang et al. (2012b) revealed 7 fiber initiation-related and 36 novel miRNAs in developing cotton ovules by comparative miRNAome analysis. However, the role of small RNAs, especially microRNAs, in cotton fiber cell development is underexplored.

14.6 Proteome Analyses of Fiber Development

Gene expression at the mRNA level does not reveal the exact functions of genes in cells, and therefore direct research on protein expression patterns and functional models has become an inevitable trend in life sciences. The term proteome was coined to describe the set of proteins encoded by a given genome (Wilkins et al. 1996). Protein profiling is one of the most important recent developments in proteomics; it offers multiple advantages and complements other functional genomics approaches such as transcript profiling.

After an extraction protocol for 2-D electrophoresis (2-DE) was optimized (Yao et al. 2006), a proteomic analysis of cotton fibers during cell elongation was conducted (Yang et al. 2008). It identified differentially expressed proteins from mass spectrometry, which match 66 unique protein species involved in different cellular and metabolic processes, with obvious functional tendencies toward energy/carbohydrate metabolism, protein turnover, cytoskeleton dynamics, cellular responses, and redox homeostasis. This provides a global view of the development-dependent protein changes in cotton fibers and offers a framework for further functional research that targets proteins associated with fiber development. Using a comparative proteomics approach, Pang et al. (2010b) identified 104 proteins from 10 DPA cotton ovules, with 93 preferentially accumulating in the wild-type and 11 accumulating in the fuzzless-lintless mutant, and identified nucleotide sugar metabolism as the most significantly upregulated biochemical process during fiber elongation. Seven protein spots potentially involved in pectic cell wall polysaccharide biosynthesis specifically accumulated in wild-type samples at both the protein and transcript levels.

Zhao et al. (2010) identified 81 differentially expressed proteins from Ligon lintless (Li_1) fibers assigned to different functional categories through 2-DE combined with local EST database-assisted MS/MS analysis; 54 of these proteins were downregulated and 27 upregulated. Three novel aspects can be illustrated:

1. Over half of the downregulated proteins are mainly involved in protein folding and stabilization, nucleocytoplasmic transport, signal transduction, and vesicular-mediated transport.

2. A number of cytoskeleton-related proteins showed a remarkable decrease in protein abundance in the Li_1 fibers. Accordingly, the architecture of the actin cytoskeleton was severely deformed and microtubule organization was moderately altered, accompanied by dramatic disruption of vesicle trafficking.

3. The expression of several proteins involved in unfolded protein response was activated in Li_1 fibers, which indicated that the deficiency of fiber cell elongation was related to endoplasmic reticulum (ER) stress. Collectively, these findings significantly advanced our understanding of mechanisms associated with cotton fiber elongation.

Liu et al. (2012b) revealed the mechanisms governing cotton fiber differentiation and initiation by comparative proteomic analysis. Comparative proteomic analysis based on 2-DE and MS/MS technology was conducted between the fuzzless-lintless mutant (fl) and its parental wild-type (WT). Forty-six differentially expressed proteins were identified in ovules at −3 and 0 DPA, at the times of cotton fiber differentiation and initiation, respectively. A strong burst of ROS was detected on the surface of −3 and −2 DPA fl ovules, and the concentrations of several carbohydrates at 0 DPA were lower in the fl mutant than in the WT ovules. These findings suggest that ROS homeostasis may be a central regulatory mechanism for cotton fiber morphogenesis and that posttranscriptional and posttranslational regulation may be pivotal in this process.

14.7 Future Prospects

In reviewing the status of cotton omics, it is clear that scientists have made significant progress in the fields of genetic map construction, fiber QTL mapping, genetical genomics, transcriptome analysis, and proteome analyses of fiber; however, additional efforts are needed to further develop omics resources and approaches in order to fully and effectively use them in cotton genetic improvements and biological research. In particular, the following areas of cotton omics research should be emphasized.

14.7.1 Transcriptomics, Proteomics, and Metabolomics

The attractiveness of cotton as a model of single-cell development has been acknowledged, and efforts are underway worldwide to elucidate the genetic features that are key to generating superior fiber species. In the postgenomic era, various studies have focused on connecting gene function and gene expression with resulting phenotype through complex networks of DNA → RNA → protein → metabolite → phenotype. A large number of studies have demonstrated that cotton fiber development involves complex molecular mechanisms, and cotton fiber cell activities require complex patterns of gene transcription and protein expression, as well as related metabolic pathways. Therefore, cotton proteomics and metabolomics are important directions for postgenomics cotton research aimed at understanding molecular mechanisms of cotton fiber development because they bridge roles between gene expression and phenotype. Using the Next Generation Sequencing (NGS) technology for RNA profiling, we can discover more novel tags that are differentially expressed. Thus, in order to dissect cotton fiber developmental mechanisms more deeply, continued efforts should be made in transcriptomics, proteomics, and metabolomics.

14.7.2 Integrating Omics

Future directions will also include the integration of different omics in cotton fiber development. The trend in biological investigations is shifting from individual omics toward integrated omics and systems biology. The integration of molecular profiling technologies into plant developmental biology has just begun, and many exciting developments can be anticipated in the near future (Hennig 2007). Gou et al. (2007) have developed a preliminary transcriptome integrated with a metabolome in cotton fiber development studies, and have demonstrated that signaling and metabolic pathways are coordinated to promote cell elongation in the early stage and to support cellulose synthesis in later stages. Therefore, with high-throughput data acquisition by genomic projects, it is possible and necessary to better integrate multiomics technologies and systems approaches that will generate many intriguing insights into cotton fiber development.

14.7.3 Quantitative Genetics

Quantitative genetics in the age of omics will expand in cotton. Genetical genomics, which combines genetics with large-scale expression profiling to provide expression QTLs (eQTLs), has been applied in cotton (Chen et al. 2012). Similar approaches can be followed with data derived from other omics technologies such as proteomics (pQTLs) and metabolomics (mQTLs) (Keurentjes et al. 2008; Joosen et al. 2009). Genetic regulatory networks have shown the usefulness of combining quantitative genetics and large-scale omics analyses (Keurentjes et al. 2007, 2008). Using these approaches, we will be able to integrate genetic, transcriptomic, proteomic, and metabolomic data (eQTL, pQTL, and mQTL) in order to understand molecular mechanisms and construct regulatory networks that underlie complex cotton fiber qualities.

Acknowledgments

We thank all faculties and graduate students at the Cotton Research Institute, Nanjing Agricultural University, and the other members of the cotton genomics and breeding communities for their contributions, and apologize for not citing many more enlightening papers owing to space limitations. This study was supported by grants from 973 (2011CB109300), Jiangsu Province Key Project (BE2012329), and the Priority Academic Program Development of Jiangsu Higher Education Institutions.

References

Alabady MS, Youn E, Wilkins TA. (2008) Double feature selection and cluster analyses in mining of microarray data from cotton. *BMC Genomics* 9: 295.

Al-Ghazi Y, Bourot S, Arioli T, et al. (2009) Transcript profiling during fiber development identifies pathways in secondary metabolism and cell wall structure that may contribute to cotton fiber quality. *Plant Cell Physiol* 50: 1364–1381.

Arpat A, Waugh M, Sullivan JP, et al. (2004) Functional genomics of cell elongation in developing cotton fibers. *Plant Mol Bio* 54: 911–929.

Basra AS, Malik CP. (1984) Development of the cotton fiber. *Int Rev Cytol* 89: 65–113.

Blenda A, Fang D, Rami JF, et al. (2012) A high consensus genetic map of tetraploid cotton that integrates multiple component maps through molecular marker redundancy check. *PLoS ONE* 7: e45739.

Bolton JJ, Soliman KM, Wilkins TA, et al. (2009) Aberrant expression of critical genes during secondary cell wall biogenesis in a cotton mutant, Ligon lintless-1 (Li_1). *Comp Funct Genomics* 2009 (659301): 8.

Chaudhary B, Hovav R, Rapp R, et al. (2008) Global analysis of gene expression in cotton fibers from wild and domesticated *Gossypium barbadense*. *Evol Dev* 10 (5): 567–582.

Chaudhary B, Hovav R, Flagel L, et al. (2009) Parallel expression evolution of oxidative stress-related genes in fiber from wild and domesticated diploid and polyploid cotton (*Gossypium*). *BMC Genomics* 10: 378.

Chee P, Rong J, Williams-Coplin D, et al. (2004) EST derived PCR-based markers for functional gene homologues in cotton. *Genome* 47: 449–462.

Chee P, Draye X, Jiang CX, et al. (2005a) Molecular dissection of interspecific variation between *Gossypium hirsutum* and *Gossypium barbadense* (cotton) by a backcross-self approach: I. Fiber elongation. *Theor Appl Genet* 111: 757–763.

Chee PW, Draye X, Jiang CX, et al. (2005b) Molecular dissection of phenotypic variation between *Gossypium hirsutum* and *Gossypium barbadense* (cotton) by a backcross-self approach: III. Fiber length. *Theor Appl Genet* 111: 772–781.

Chen H, Qian N, Guo WZ, et al. (2009) Using three overlapped RILs to dissect genetically clustered QTL for fiber strength on Chro.D8 in upland cotton. *Theor Appl Genet* 119: 605–612.

Chen XD, Guo WZ, Liu BL, et al. (2012) Molecular mechanisms of fiber differential development between *G. barbadense* and *G. hirsutum* revealed by genetical genomics. *PLoS ONE* 7 (1): e30056.

Clark G, Torres J, Finlayson S, et al. (2010) Apyrase (nucleoside triphosphate-diphosphohydrolase) and extracellular nucleotides regulate cotton fiber elongation in cultured ovules. *Plant Physiol* 152: 1073–1083.

Claverie M, Souquet M, Jean J, et al. (2012) cDNA-AFLP-based genetical genomics in cotton fibers. *Theor Appl Genet* 124: 665–683.

Deng F, Tu L, Tan J, et al. (2012) *GbPDF1* is involved in cotton fiber initiation via the core cis-element HDZIP2ATATHB2. *Plant Physiol* 158: 890–904.

Divya K, Jami SK, Kirti PB. (2010) Constitutive expression of mustard annexin, AnnBj1 enhances abiotic stress tolerance and fiber quality in cotton under stress. *Plant Mol Biol* 73: 293–308.

Draye X, Chee P, Jiang CX, et al. (2005) Molecular dissection of interspecific variation between *Gossypium hirsutum* and *G. barbadense* (cotton) by a backcross-self approach: II. Fiber fineness. *Theor Appl Genet* 111: 764–771.

Fan L, Shi WJ, Hu WR, et al. (2009) Molecular and biochemical evidence for phenylpropanoid synthesis and presence of wall-linked phenolics in cotton fibers. *J Integr Plant Biol* 51: 626–637.

Feng JX, Ji SJ, Shi YH, et al. (2004) Analysis of five differentially expressed gene families in fast elongating cotton fiber. *Acta Biochim Biophys Sin* 36 (1): 51–56.

Flagel L, Udall J, Nettleton D, et al. (2008) Duplicate gene expression in allopolyploid *Gossypium* reveals two temporally distinct phases of expression evolution. *BMC Biol* 6: 16.

Frelichowski Jr JE, Palmer MB, Main D, et al. (2006) Cotton genome mapping with new microsatellites from Acala 'Maxxa' BAC-ends. *Mol Genet Genomics* 275: 479–491.

Gao W, Saha S, Ma DP, et al. (2012) A cotton-fiber-associated cyclin-dependent kinase a gene: Characterization and chromosomal location. *Int J Plant Genomics* 2012: 613812.

Gou JY, Wang LJ, Chen SP, et al. (2007) Gene expression and metabolite profiles of cotton fiber during cell elongation and secondary cell wall synthesis. *Cell Res* 17: 422–434.

Guan XY, Li QJ, Shan CM, et al. (2008) The HD-Zip IV gene *GaHOX1* from cotton is a functional homologue of the *Arabidopsis GLABRA2*. *Physiol Plant* 134: 174–182.

Guan XY, Yu N, Shangguan XX, et al. (2007) Arabidopsis trichome research sheds light on cotton fiber development mechanisms. *Chin Sci Bull* 52: 1734–1741.

Guan X, Lee JJ, Pang M, et al. (2011) Activation of Arabidopsis seed hair development by cotton fiber-related genes. *PLoS ONE* 6: e21301.

Guo WZ, Zhang TZ, Shen XL, et al. (2003) Development of SCAR marker linked to a major QTL for high fiber strength and its usage in molecular-marker assisted selection in upland cotton. *Crop Sci* 43 (6): 2252–2256.

Guo WZ, Cai CP, Wang CB, et al. (2007) A microsatellite-based, gene-rich linkage map reveals genome structure, function and evolution in *Gossypium*. *Genetics* 176: 527–541.

Guo WZ, Cai CP, Wang CB, et al. (2008) A preliminary analysis of genome structure and composition in *Gossypium hirsutum*. *BMC Genomics* 9: 314.

Han ZG, Guo WZ, Song XL, et al. (2004) Genetic mapping of EST-derived microsatellites from the diploid *Gossypium arboreum* in allotetraploid cotton. *Mol Genet Genomics* 272: 308–327.

Han ZG, Wang CB, Song XL, et al. (2006) Characteristics, development and mapping of *Gossypium hirsutum* derived EST-SSRs in allotetraploid cotton. *Theor Appl Genet* 112: 430–439.

Han X, Xu X, Fang DD, et al. (2012) Cloning and expression analysis of novel Aux/IAA family genes in *Gossypium hirsutum*. *Gene* 503 (1): 83–91.

He DH, Lin ZX, Zhang XL, et al. (2007) QTL mapping for economic traits based on a dense genetic map of cotton with PCR-based markers using the interspecific cross of *Gossypium hirsutum* × *G. barbadense*. *Euphytica* 153: 181–197.

He DH, Lin ZX, Zhang XL, et al. (2008a) Dissection of genetic variance of fiber quality in advanced generations from an interspecific cross of *Gossypium hirsutum* and *G. barbadense*. *Plant Breed* 127: 286–294.

He XC, Qin YM, Xu Y, et al. (2008b) Molecular cloning, expression profiling, and yeast complementation of 19 beta-tubulin cDNAs from developing cotton ovules. *J Exp Bot* 59: 2687–2695.

He YJ, Guo WZ, Shen XL, et al. (2008c) Molecular cloning and characterization of a cytosolic glutamine synthetase gene, a fiber strength-associated gene in cotton. *Planta* 228: 473–483.

Hennig L. (2007) Patterns of beauty-omics meets plant development. *Trends Plant Sci* 12 (7): 287–293.

Hinchliffe DJ, Meredith WR, Yeater KM, et al. (2010) Near-isogenic cotton germplasm lines that differ in fiber-bundle strength have temporal differences in fiber gene expression patterns as revealed by comparative high-throughput profiling. *Theor Appl Genet* 120 (7): 1347–1366.

Hovav R, Chaudhary B, Udall JA, et al. (2008a) Parallel domestication, convergent evolution and duplicated gene recruitment in allopolyploid cotton. *Genetics* 179 (3): 1725–1733.

Hovav R, Udall JA, Chaudhary B, et al. (2008b) Partitioned expression of duplicated genes during development and evolution of a single cell in a polyploid plant. *Proc Natl Acad Sci USA* 105 (16): 6191–6195.

Hovav R, Udall JA, Chaudhary B, et al. (2008c) The evolution of spinnable cotton fiber entailed prolonged development and a novel metabolism. *PLoS Genet* 4: e25.

Hovav R, Udall JA, Hovav E, et al. (2008d) A majority of cotton genes are expressed in single-celled fiber. *Planta* 227 (2): 319–329.

Hu Y, Guo WZ, Zhang TZ. (2009) Construction of a bacterial artificial chromosome library of TM-1, a standard line for genetics and genomics in upland cotton. *J Integr Plant Biol* 51: 107–112.

Ji SJ, Lu YC, Feng JX, et al. (2003) Isolation and analyses of genes preferentially expressed during early cotton fiber development by subtractive PCR and cDNA array. *Nucl Acids Res* 31: 2534–2543.

Jiang C, Wright RJ, El-zik KM, et al. (1998) Polyploid formation created unique avenues for response to selection in *Gossypium* (cotton). *Proc Natl Acad Sci USA* 95: 4419–4424.

Jiang YJ, Guo WZ, Zhu H, et al. (2012) Overexpression of *GhSusA1* increases plant biomass and improves cotton fiber yield and quality. *Plant Biotechnol J* 10 (3): 301–312.

Joosen RV, Ligterink W, Hilhorst HW, et al. (2009) Advances in genetical genomics of plants. *Curr Genomics* 10: 540–549.

Keurentjes JJ, Fu J, Terpstra IR, et al. (2007) Regulatory network construction in Arabidopsis by using genome-wide gene expression quantitative trait loci. *Proc Natl Acad Sci USA* 104: 1708–1713.

Keurentjes JJ, Koornneef M, Vreugdenhil D. (2008) Quantitative genetics in the age of omics. *Curr Opin Plant Biol* 11: 123–128.

Kim HJ, Triplett BA. (2001) Cotton fiber growth in planta and in vitro. Models for plant cell elongation and cell wall biogenesis. *Plant Physiol* 127 (4): 1361–1366.

Kim HJ, Triplett BA. (2004) Cotton fiber germin-like protein. I. Molecular cloning and gene expression. *Planta* 218 (4): 516–524.

Kohel RJ, Yu J, Park YH, et al. (2001) Molecular mapping and characterization of traits controlling fiber quality in cotton. *Euphytica* 121 (2): 163–172.

Kwak PB, Wang QQ, Chen XS, et al. (2009) Enrichment of a set of microRNAs during the cotton fiber development. *BMC Genomics* 10: 457.

Lacape JM, Nguyen TB, Thibivilliers S, et al. (2003) Combined RFLP-SSR-AFLP map of tetraploid cotton based on a *Gossypium hirsutum* × *Gossypium barbadense* backcross population. *Genome* 46: 612–626.

Lacape JM, Nguyen TB, Courtois B, et al. (2005) QTL analysis of cotton fiber quality using multiple *Gossypium hirsutum* × *Gossypium barbadense* backcross generation. *Crop Sci* 45: 123–140.

Lacape JM, Jacobs J, Arioli T, et al. (2009) A new interspecific, *Gossypium hirsutum* × *G. barbadense*, RIL population: Towards a unified consensus linkage map of tetraploid cotton. *Theor Appl Genet* 119 (2): 281–292.

Lacape JM, Llewellyn D, Jacobs J, et al. (2010) Meta-analysis of cotton fiber quality QTLs across diverse environments in a *Gossypium hirsutum* × *G. barbadense* RIL population. *BMC Plant Biol* 10: 132.

Lee JJ, Woodward AW, Chen ZJ. (2007) Gene expression changes and early events in cotton fiber development. *Ann Bot* 100 (7): 1391–1401.

Lee J, Burns TH, Ligh G, et al. (2010) Xyloglucan endotransglycosylase/hydrolase genes in cotton and their role in fiber elongation. *Planta* 232 (5): 1191–1205.

Li XB, Xiao YH, Luo M, et al. (2005a) Cloning and expression analysis of two Rac genes from cotton (*Gossypium hirsutum* L.). *Acta Genetica Sinica* 32 (1): 72–78.

Li XB, Fan XP, Wang XL, et al. (2005b) The cotton *ACTIN1* gene is functionally expressed in fibers and participates in fiber elongation. *Plant Cell* 17: 859–875.

Li YL, Sun J, Xia GX. (2005c) Cloning and characterization of a gene for an LRR receptor-like protein kinase associated with cotton fiber development. *Mol Genet Genomics* 273: 217–224.

Li HB, Qin YM, Pang Y, et al. (2007a) A cotton ascorbate peroxidase is involved in hydrogen peroxide homeostasis during fibre cell development. *New Phytol* 175: 462–471.

Li L, Wang XL, Huang GQ, et al. (2007b) Molecular characterization of cotton *GhTUA9* gene specifically expressed in fibre and involved in cell elongation. *J Exp Bot* 58 (12): 3227–3238.

Li YJ, Liu DQ, Tu LL, et al. (2010a) Suppression of *GhAGP4* gene expression repressed the initiation and elongation of cotton fiber. *Plant Cell Rep* 29 (2): 193–202.

Li XR, Wang L, Ruan YL. (2010b) Developmental and molecular physiological evidence for the role of phosphoenolpyruvate carboxylase in rapid cotton fibre elongation. *J Exp Bot* 61 (1): 287–295.

Li Y, Ning H, Zhang Z, et al. (2011) A cotton gene encoding novel MADS-box protein is preferentially expressed in fibers and functions in cell elongation. *Acta Biochim Biophys Sin* 43 (8): 607–617.

Lin ZX, He DH, Zhang XL, et al. (2005) Linkage map construction and mapping QTL for cotton fiber quality using SRAP, SSR and RAPD. *Plant Breed* 124: 180–187.

Lin L, Pierce GJ, Bowers JE, et al. (2010) A draft physical map of a D-genome cotton species (*Gossypium raimondii*). *BMC Genomics* 11: 395.

Liu DQ, Zhang XL, Tu LL, et al. (2006) Isolation by suppression-subtractive hybridization of genes preferentially expressed during early and later fiber development stages in cotton. *Mol Biol* 40 (5): 741–749.

Liu DQ, Tu LL, Li YJ, et al. (2008) Genes encoding fasciclin-like arabinogalactan proteins are specifically expressed during cotton fiber development. *Plant Mol Bio Rep* 26 (2): 98–113.

Liu HW, Wang XF, Pan YX, et al. (2009) Mining cotton fiber strength candidate genes based on transcriptome mapping. *Chn Sci Bull* 54 (24): 4651–4657.

Liu K, Sun J, Yao L, et al. (2012a) Transcriptome analysis reveals critical genes and key pathways for early cotton fiber elongation in Ligon lintless-1 mutant. *Genomics* 100 (1): 42–50.

Liu K, Han ML, Zhang CJ, et al. (2012b) Comparative proteomic analysis reveals the mechanisms governing cotton fiber differentiation and initiation. *J Proteomics* 75 (3): 845–856.

Lorenzo A, Jun K, Haggag A, et al. (2008) Functional analysis of cotton orthologs of GA signal transduction factors *GID1* and *SLR1*. *Plant Mol Biol* 68: 1–16.

Luo M, Xiao YH, Li XB, et al. (2007) *GhDET2*, a steroid 5alpha-reductase, plays an important role in cotton fiber cell initiation and elongation. *Plant J* 51: 419–430.

Luo M, Tan K, Xiao Z, et al. (2008) Cloning and expression of two sterol C-24 methyltransferase genes from upland cotton (*Gossypium hirsuturm* L.). *J Genet Genomics* 35 (6): 357–363.

Ma GJ, Zhang TZ, Guo WZ. (2006) Cloning and characterization of cotton *GhBG* gene encoding beta-glucosidase. *DNA Seq* 17: 355–362.

Machado A, Wu Y, Yang Y, et al. (2009) The MYB transcription factor *GhMYB25* regulates early fibre and trichome development. *Plant J* 59 (1): 52–62.

Mei M, Syed NH, Gao W, et al. (2004) Genetic mapping and QTL analysis of fiber-related traits in cotton (*Gossypium*). *Theor Appl Genet* 108: 280–291.

Mei WQ, Qin YM, Song WQ, et al. (2009) Cotton *GhPOX1* encoding plant class III peroxidase may be responsible for the high level of reactive oxygen species production that is related to cotton fiber elongation. *J Genet Genomics* 36: 141–150.

Michailidis G, Argiriou A, Darzentas N, et al. (2009) Analysis of xyloglucan endotransglycosylase/ hydrolase (*XTH*) genes from allotetraploid (*Gossypium hirsutum*) cotton and its diploid progenitors expressed during fiber elongation. *J Plant Physiol* 166 (4): 403–416.

Munis MF, Tu L, Deng F, et al. (2010) A thaumatin-like protein gene involved in cotton fiber secondary cell wall development enhances resistance against *Verticillium dahliae* and other stresses in transgenic tobacco. *Biochem Biophys Res Commun* 393 (1): 38–44.

Nguyen TB, Giband M, Brottier P, et al. (2004) Wide coverage of the tetraploid cotton genome using newly developed microsatellite markers. *Theor Appl Genet* 109: 167–175.

Pang CY, Wang H, Song WQ, et al. (2010a) The cotton ATP synthaseδ1 subunit is required to maintain a higher ATP/ADP ratio that facilitates rapid fibre cell elongation. *Plant Biology* 12 (6): 903–909.

Pang CY, Wang H, Pang Y, et al. (2010b) Comparative proteomics indicate that biosynthesis of pectic precursors is important for cotton fiber and *Arabidopsis* root hair elongation. *Mol Cell Proteomics* 9: 2019–2033.

Paterson TA, Saranga Y, Menz M, et al. (2003) QTL analysis of genotype × environment interactions affecting cotton fiber quality. *Theor Appl Genet* 106: 384–396.

Paterson AH, Wendel JF, Gundlach H, Guo H, Jenkins J, et al. (2012) Repeated polyploidization of Gossypium genomes and the evolution of spinnable cotton fibers. *Nature* 492: 423–427.

Preuss ML, Kovar DR, Lee YR, et al. (2004) A plant-specific kinesin binds to actin microfilaments and interacts with cortical microtubules in cotton fibers. *Plant Physiol* 136: 3945–3955.

Pu L, Li Q, Fan XP, et al. (2008) The R2R3 MYB transcription factor *GhMYB109* is required for cotton fiber development. *Genetics* 180: 811–820.

Qin YM, Zhu YX. (2011) How cotton fibers elongate: A tale of linear cell-growth mode. *Curr Opin Plant Biol* 14 (1): 106–111.

Qin YM, Pujol FM, Shi YH, et al. (2005) Cloning and functional characterization of two cDNAs encoding NADPH-dependent 3-ketoacyl-CoA reductased from developing cotton fibers. *Cell Res* 15: 465–473.

Qin YM, Pujol FM, Hu CY, et al. (2007a) Genetic and biochemical studies in yeast reveal that the cotton fibre-specific *GhCER6* gene functions in fatty acid elongation. *J Exp Bot* 58: 473–481.

Qin YM, Hu CY, Pang Y, et al. (2007b) Saturated very-long-chain fatty acids promote cotton fiber and *Arabidopsis* cell elongation by activating ethylene biosynthesis. *Plant Cell* 19: 3692–3704.

Qin HD, Guo WZ, Zhang YM, et al. (2008a) QTL mapping of yield and fiber traits based on a four-way cross population in *Gossypium hirsutum* L. *Theor Appl Genet* 117: 883–894.

Qin YM, Hu CY, Zhu YX. (2008b) The ascorbate peroxidase regulated by H_2O_2 and ethylene is involved in cotton fiber cell elongation by modulating ROS homeostasis. *Plant Signal Behav* 3: 194–196.

Qiu CX, Xie FL, Zhu YY, et al. (2007) Computational identification of microRNAs and their targets in *Gossypium hirsutum* expressed sequence tags. *Gene* 395: 49–61.

Ramsay NA, Glover BJ. (2005) MYB-bHLH-WD40 protein complex and the evolution of cellular diversity. *Trends Plant Sci* 10: 63–70.

Reinisch AJ, Dong JM, Brubaker CL, et al. (1994) A detailed RFLP map of cotton, *Gossypium hirsutum × Gossypium barbadense*: Chromosome organization and evolution in a disomic polyploid genome. *Genetics* 138 (3): 829–847.

Rong J, Feltus FA, Waghmare VN, et al. (2007) Meta-analysis of polyploid cotton QTL shows unequal contributions of subgenomes to a complex network of genes and gene clusters implicated in lint fiber development. *Genetics* 176: 2577–2588.

Rong JK, Abbey C, Bowers JE, et al. (2004) A 3347-locus genetic recombination map of sequence-tagged sites reveals features of genome organization, transmission and evolution of cotton (*Gossypium*). *Genetics* 166: 389–417.

Ruan YL, Llewellyn DJ, Furbank RT. (2001) The control of single-celled cotton fiber elongation by developmentally reversible gating of plasmodesmata and coordinated expression of sucrose and K+ transporters and expansion. *Plant Cell* 13 (1): 47–60.

Ruan YL, Llewellyn DJ, Furbank RT. (2003) Suppression of sucrose synthase gene expression represses cotton fiber cell initiation, elongation, and seed development. *Plant Cell* 15: 952–964.

Ruan XM, Luo F, Li DD, et al. (2011) Cotton BCP genes encoding putative blue copper-binding proteins are functionally expressed in fiber development and involved in response to high-salinity and heavy metal stresses. *Physiol Plant* 141 (1): 71–83.

Saha S, Raska DA, Stelly DM. (2006) Upland cotton (*Gossypium hirsutum* L.) x Hawaiian cotton (*G. tomentosum* Nutt. ex. Seem) F1 hybrid hypoaneuploid chromosome substitution series. *J Cotton Sci* 10: 146–154.

Shangguan XX, Xu B, Yu ZX, et al. (2008) Promoter of a cotton fibre MYB gene functional in trichomes of Arabidopsis and glandular trichomes of tobacco. *J Exp Bot* 59: 3533–3542.

Shappley ZW, Jenkins JN, Zhu J, et al. (1998) Quantitative trait loci associated with agronomic and fiber traits of upland cotton. *J Cotton Sci* 4: 153–163.

Shen XL, Guo WZ, Zhu XF, et al. (2005) Molecular mapping of QTLs for qualities in three diverse lines in upland cotton using SSR markers. *Mol Breed* 15: 169–181.

Shen XL, Zhang TZ, Guo WZ, et al. (2006) Mapping fiber and yield QTLs with main, epistatic, and QTL × environment interaction effects in recombinant inbred lines of upland cotton. *Crop Sci* 46 (1): 61–66.

Shen XL, Guo WZ, Lu QX, et al. (2007) Genetic mapping of quantitative trait loci for fiber quality and yield trait by RIL approach in upland cotton. *Euphytica* 155: 371–380.

Shi YH, Zhu SW, Mao XZ, et al. (2006) Transcriptome profiling, molecular biological, and physiological studies reveal a major role for ethylene in cotton fiber cell elongation. *Plant Cell* 18: 651–664.

Shi HY, Wang XL, Li DD, et al. (2007) Molecular characterization of cotton *14-3-3L* gene preferentially expressed during fiber elongation. *J Genet Genomics* 34: 151–159.

Smart LB, Vojdani F, Maeshima M, et al. (1998) Genes involved in Osmoregulation during turgor-driven cell expansion of developing cotton fibers are differentially regulated. *Plant Physiol* 116: 1539–1549.

Song XL, Wang K, Guo WZ, et al. (2005) A comparison of genetic maps constructed from haploid and BC_1 mapping populations from the same crossing between *Gossypium hirsutum* L. and *Gossypium barbadense* L. *Genome* 48: 378–390.

Song WQ, Qin YM, Saito M, et al. (2009) Characterization of two cotton cDNAs encoding trans-2-enoyl-CoA reductase reveals a putative novel NADPH-binding motif. *J Exp Bot* 60: 1839–1848.

Sun Y, Fokaeer M, Asaml T, et al. (2004) Characterization of the brassinosteroid insensitive genes of cotton. *Plant Mol Biol* 54: 221–232.

Suo JF, Liang XE, Pu L, et al. (2003) Identification of *GhMYB109* encoding a R2R3 MYB transcription factor that expressed specifically in fiber initials and elongating fibers of cotton (*Gossypium hirsutum* L.). *Biochim Biophys Acta* 1630: 25–34.

Tai FJ, Wang XL, Xu WL, et al. (2008) Characterization and expression analysis of two cotton genes encoding putative UDP-Glycosyltransferases. *Mol Biol (Mosk)* 42 (1): 50–58.

Taliercio EW, Boykin D. (2007) Analysis of gene expression in cotton fiber initials. *BMC Plant Biol* 7: 22.

Tu LL, Zhang XL, Liang SG, et al. (2007) Genes expression analyses of Sea-island cotton (*Gossypium barbadense* L.) during fiber development. *Plant Cell Rep* 26: 1309–1320.

Ulloa M, Meredith WR. (2000) Genetic linkage map and QTL analysis of agronomic and fiber quality traits in an intraspecific population. *Cotton Sci* 4: 161–170.

Ulloa M, Saha S, Jenkins J N, et al. (2005) Chromosomal assignment of RFLP linkage groups harboring important QTLs on an intraspecific cotton (*Gossypium hirsutum* L.) joinmap. *J Hered* 96: 132–144.

Walford SA, Wu Y, Llewellyn DJ, et al. (2012) Epidermal cell differentiation in cotton mediated by the homeodomain leucine zipper gene, *GhHD-1. Plant J* 71: 464–478.

Wan Q, Zhang ZS, Hu MC, et al. (2007) T_1 locus in cotton is the candidate gene affecting lint percentage, fiber quality and spiny bollworm (*Earias* spp.) resistance. *Euphytica* 158: 241–247.

Wang S, Wang JW, Yu N, et al. (2004) Control of plant trichome development by a cotton fiber MYB gene. *Plant Cell* 16: 2323–2334.

Wang HY, Yu Y, Chen ZL, et al. (2005) Functional characterization of *Gossypium hirsutum* profilin 1 gene (*GhPFN1*) in tobacco suspension cells. Characterization of in vivo functions of a cotton profilin gene. *Planta* 222: 594–603.

Wang BH, Wu YT, Huang NT, et al. (2006a) QTL mapping for plant architecture traits in upland cotton using RILs and SSR markers. *Acta Genetica Sinica* 33 (2): 161–170.

Wang BH, Guo WZ, Zhu XF, et al. (2006b) QTL mapping for fiber quality in an elite hybrid derived-RIL population of upland cotton. *Euphytica* 152 (3): 367–378.

Wang K, Song XL, Han ZG, et al. (2006c) Complete assignment of the chromosomes of *Gossypium hirsutum* L. by translocation and fluorescence *in situ* hybridization mapping. *Theor Appl Genet* 113: 73–80.

Wang K, Guo WZ, Zhang TZ. (2007a) Development of one set of chromosome-specific microsatellite-containing BACs and their physical mapping in *Gossypium hirsutum* L. *Theor Appl Genet* 115: 675–682.

Wang BH, Wu YT, Guo WZ, et al. (2007b) QTL analysis and epistasis effects dissection of fiber qualities in an elite cotton hybrid grown in second generation. *Crop Sci* 47: 1384–1392.

Wang HY, Wang J, Gao P, et al. (2009) Down-regulation of *GhADF1* gene expression affects cotton fibre properties. *Plant Biotechnol J* 7 (1): 13–23.

Wang HH, Guo Y, Lv FN, et al. (2010a) The essential role of *GhPEL* gene, encoding a pectate lyase, in cell wall loosening by depolymerization of the de-esterified pectin during fiber elongation in cotton. *Plant Mol Biol* 72: 397–406.

Wang J, Wang HY, Zhao PM, et al. (2010b) Overexpression of a profilin (*GhPFN2*) promotes the progression of developmental phases in cotton fibers. *Plant Cell Physiol* 51 (8): 1276–1290.

Wang L, Li XR, Lian H, et al. (2010c) Evidence that high activity of vacuolar invertase is required for cotton fiber and *Arabidopsis* root elongation through osmotic dependent and independent pathways, respectively. *Plant Physiol* 154 (2): 744–756.

Wang KB, Wang ZW, Li FG, et al. (2012a) The draft genome of a diploid cotton *Gossypium raimondii*. *Nat Genet* 44: 1098–1103.

Wang ZM, Xue W, Dong CJ, et al. (2012b) A Comparative miRNAome analysis reveals seven fiber initiation-related and 36 novel miRNAs in developing cotton ovules. *Mol Plant* 5 (4): 889–900.

Wendel JF, Cronn RC. (2003) Polyploidy and the evolutionary history of cotton. *Adv Agron* 78: 139–186.

Wilkins MR, Pasquali C, Appel RD, et al. (1996) From proteins to proteomes: Large scale protein identification by two-dimensional electrophoresis and amino acid analysis. *Biotechnol* 14: 61–65.

Wu YT, Liu JY. (2005) Molecular cloning and characterization of a cotton glucuronosyltranferase gene. *J Plant Physiol* 162: 573–582.

Wu Y, Machado AC, White RG, et al. (2006) Expression profiling identifies genes expressed early during lint fibre initiation in cotton. *Plant Cell Physiol* 47: 107–127.

Wu AM, Lv SY, Liu JY. (2007a) Functional analysis of a cotton glucuronosyltransferase promoter in transgenic tobaccos. *Cell Res* 17: 174–183.

Wu Y, Llewellyn DJ, White R, et al. (2007b) Laser capture microdissection and cDNA microarrays used to generate gene expression profiles of the rapidly expanding fibre initial cells on the surface of cotton ovules. *Planta* 226: 1475–1490.

Wu Y, Xu W, Huang G, et al. (2009) Expression and localization of *GhH6L*, a putative classical arabinogalactan protein in cotton (*Gossypium hirsutum*). *Acta Biochim Biophys Sin* 41 (6): 495–503.

Xiao YH, Li DM, Yin MH, et al. (2010) Gibberellin 20-oxidase promotes initiation and elongation of cotton fibers by regulating gibberellin synthesis. *J Plant Physiol* 167: 829–837.

Xu WL, Wang XL, Wang H, et al. (2007) Molecular characterization and expression analysis of nine cotton *GhEF1A* genes encoding translation elongation factor 1A. *Gene* 389: 27–35.

Xu Z, Kohel RJ, Song G, et al. (2008) An integrated genetic and physical map of homoeologous chromosomes 12 and 26 in upland cotton (*G. hirsutum* L.). *BMC Genomics* 9: 108.

Yang SS, Cheung F, Lee JJ, et al. (2006) Accumulation of genome-specific transcripts, transcription factors and phytohormonal regulators during early stages of fiber cell development in allotetraploid cotton. *Plant J* 47: 761–775.

Yang YW, Bian SM, Yao Y, et al. (2008) Comparative proteomic analysis provides new insights into the fiber elongating process in cotton. *J Proteome Res* 7: 4623–4637.

Yao Y, Yang YW, Liu JY. (2006) An efficient protein preparation for proteomic analysis of developing cotton fibers by 2-DE. *Electrophoresis* 27: 4559–4569.

Yin JM, Guo WZ, Yang LM, et al. (2006) Physical mapping of the *Rf₁* fertility-restoring gene to a 100 kb region in cotton. *Theor Appl Genet* 112: 1318–1325.

Yu J, Kohel RJ, Smith CW. (2010) The construction of a tetraploid cotton genome wide comprehensive reference map. *Genomics* 95: 230–240.

Yu Y, Yuan D, Liang S, et al. (2011) Genome structure of cotton revealed by a genome-wide SSR genetic map constructed from a BC_1 population between *Gossypium hirsutum* and *G. barbadense*. *BMC Genomics* 12: 15.

Yu JZ, Kohel RJ, Fang DD, et al. (2012) A high-density simple sequence repeat and single nucleotide polymorphism genetic map of the tetraploid cotton genome. *G3 (Bethesda)* 2 (1): 43–58.

Zhang T, Yuan Y, Yu J, et al. (2003) Molecular tagging of a major QTL for fiber strength in upland cotton and its marker-assisted selection. *Theor Appl Genet* 106: 262–268.

Zhang ZS, Xiao YH, Luo M, et al. (2005) Construction of a genetic linkage map and QTL analysis of fiber-related traits in upland cotton (*Gossypium hirsutum* L.). *Euphytica* 144: 91–99.

Zhang HB, Li Y, Wang BH, et al. (2008) Recent advances in cotton genomics. *Int J Plant Genomics* 2008: 742304.

Zhang ZT, Zhou Y, Li Y, et al. (2010a) Interactome analysis of the six cotton *14-3-3s* that are preferentially expressed in fibres and involved in cell elongation. *J Exp Bot* 61 (12): 3331–3344.

Zhang F, Zuo K, Zhang J, et al. (2010b) An L1 box binding protein, *GbML1*, interacts with *GbMYB25* to control cotton fibre development. *J Exp Bot* 61 (13): 3599–3613.

Zhang F, Liu X, Zuo K, et al. (2011) Molecular cloning and expression analysis of a novel *SANT/MYB* gene from *Gossypium barbadense*. *Mol Biol Rep* 38 (4): 2329–2336.

Zhao PM, Wang LL, Han LB, et al. (2010) Proteomic identification of differentially expressed proteins in the Ligon lintless mutant of upland cotton (*Gossypium hirsutum* L.). *J Proteome Res* 9: 1076–1087.

Zhou L, Duan J, Wang XM, et al. (2011) Characterization of a novel annexin gene from cotton (*Gossypium hirsutum* cv CRI 35) and antioxidative role of its recombinant protein. *J Integr Plant Biol* 53 (5): 347–357.

Zhao L, Lv YD, Cai CP, et al. (2013) Toward allotetraploid cotton genome assembly: Integration of a high-density molecular genetic linkage map with DNA sequence information. *BMC Genomics* 14: 170.

Zhu YQ, Xu KX, Luo B, et al. (2003) An ATP-binding cassette transporter GhWBC1 from elongating cotton fibers. *Plant Physiol* 133: 580–588.

Zuo K, Zhao J, Wang J, et al. (2004) Molecular cloning and characterization of *GhlecRK*, a novel kinase gene with lectin-like domain from *Gossypium hirsutum*. *DNASeq* 15 (1): 58–65.

15

Forestry and Engineered Forest Trees

Mohammed Ellatifi, PhD

CONTENTS

15.1 Fundamentals about Forests 492
    15.1.1 Forests in the World...492
    15.1.2 Major Problems Faced by Forests 495
15.2 Natural Forests and Planted Forests 495
    15.2.1 Extension of Planted Forests 495
    15.2.2 Functions of Planted Forests 497
15.3 Biotechnology and Forest Trees Engineering..497
    15.3.1 Genetic Modifications of Forest Trees...498
    15.3.2 Research Activity on Genetic Modifications of Forest Trees.........499
    15.3.3 Species Subject to GM Research 500
15.4 Toward Shifting Timber and Industrial Wood Harvesting from
    Natural Forests to Planted Forests 500
    15.4.1 Wood Removal from the World's Forests 500
    15.4.2 Consequences of Translating Timber and Industrial Wood
            Harvesting from Natural Forests to Planted Forests.......................501
15.5 Beneficial Traits Subject to Genetic Modifications 502
    15.5.1 Herbicide Tolerance 502
    15.5.2 Resistance to Bacterial, Fungal, and Viral Pathogens.....................504
    15.5.3 Insect Resistance 504
    15.5.4 Resistance to Abiotic Stress 504
    15.5.5 Wood and Growth Characteristics...504
    15.5.6 Flowering Modification...506
    15.5.7 Phytoremediation 506
15.6 Field Trials of Transgenic Forest Trees 507
15.7 Commercial Releases of Genetically Modified Trees 508
15.8 Concerns Regarding Transgenic Forest Trees 511
    15.8.1 Potential Environmental Risks of Transgenic Forest Trees 512
    15.8.2 Environment Risk Assessment..512
    15.8.3 Persistence and Invasiveness..513
    15.8.4 Horizontal Gene Transfer (Out-Crossing)...513
    15.8.5 Vertical Gene Transfer (Out-Crossing) 513
    15.8.6 Effects on Human Health 514
            15.8.6.1 Potential Hazards...514
            15.8.6.2 Crops versus Trees...514
    15.8.7 Effects on the Food Chain 514
            15.8.7.1 Potential Hazards...514
            15.8.7.2 Crops versus Trees...514

15.8.8 Effects of Cultivation and Management..515
     15.8.8.1 Potential Hazards..515
     15.8.8.2 Potential Impact of Cultivation and Management of
           Transgenic Forest Plantations 515
15.9 Conclusion 515
References...518

15.1 Fundamentals about Forests

The term *forest* can be defined as an aggregate of ecosystems in which trees, arborescent shrubs, and other woody species grow, with a dominance of trees. The Food and Agriculture Organization of the United Nations (FAO) defines a forest as land with tree crown cover of more than 10% and an area of more than 0.5 ha (FAO, 1998, 2005). To be distinct from a shrub, a tree must have a minimum height of 5 meters, at maturity, *in situ* (Figures 15.1 through 15.7).

15.1.1 Forests in the World

The total area covered by forests on Earth is around 4 billion hectares (or 40 million square kilometers). This corresponds to an average rate of 0.6 ha per capita, at global level (Ellatifi, 2002).

In a biotope as fragile as the Mediterranean, and even in humid and sub-humid areas, forests are essential to perform various protective functions. They protect soil and water resources, harbor over 50% of the terrestrial biodiversity, and mitigate climate change by reducing carbon emissions generated by deforestation and forest degradation, and through increasing carbon uptake by afforestation, reforestation, and sustainable forest management. Forests also provide employment and livelihoods for over 10 million people who are directly dependent on them, produce wood and nonwood products, and represent human beings' cultural and patrimonial heritage (Ellatifi, 1984, 2005).

FIGURE 15.1
Natural forest of Atlas cedar (*Cedrus atlantica* Manetti) in the Ajdir forest, in the Middle Atlas, Morocco.

FIGURE 15.2
Natural forest of holm oak (*Quercus ilex* L.) in Ajdir Forest, in the Middle Atlas, Morocco.

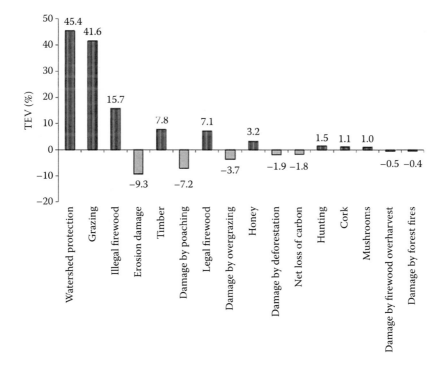

FIGURE 15.3
Major estimated components of the forest total economic value (TEV) in Morocco. (From Ellatifi, M., In Merlo, M. and Croitoru, L. (eds), *Valuing Mediterranean Forests: Towards the Total Economic Value*, p. 85, CABI Publishing, 2005.)

FIGURE 15.4
Afforestation with pine (*Pinus* sp.).

FIGURE 15.5
Afforestation with eucalypt (*Eucalyptus* sp.).

FIGURE 15.6   (See color insert)
Poplar plantation.

FIGURE 15.7   (See color insert)
Pulp pine plantation.

15.1.2 Major Problems Faced by Forests

Throughout the world, and particularly in developing countries, forests are subject to many pressures, such as: deforestation, forest degradation, forest fires, overgrazing, wood over-harvesting, and poor or unsustainable management (Ellatifi, 1984, 1989, 1991a,b, 2005, 2012).

The case of Morocco is significant as both a Mediterranean and a developing country. Table 15.1 highlights the most important forest functions in Morocco and their relationship with watershed protection. This protective function accounts for €278.7 million (+45% of the TEV). This is followed by forage for grazing with around +42%, firewood (illicit) with +16%, erosion with –9%, timber with nearly +8%, and firewood (licit) with +7%.

It is important to point out that not all of the TEV components are independent. For example, overcollection of firewood and overgrazing also trigger forest degradation, soil erosion, and water depletion. This is the case in the Mediterranean basin in general, and in Morocco in particular (Ellatifi, 2005).

15.2 Natural Forests and Planted Forests

15.2.1 Extension of Planted Forests

Out of the 4 billion hectares of forest on Earth, about 264 million hectares are planted forests, of which 75% are grown for wood production, and 25% for environmental protection. Therefore, almost 93% of the forested areas in the world are natural forests and almost 7% are planted forests (FAO, 2005, 2010b). Table 15.2 shows the unequal repartition of planted forests in the world.

Table 15.3 highlights the total afforested and reforested areas, per year, in the world by region (/subregion).

The five countries with the highest annual increase in planted forests are China, the United States of America, Canada, India, and the Russian Federation. The top five countries regarding area of afforestation per year are China, Indonesia, Vietnam, the United States of America, and Turkey. And the five countries with the highest area of reforestation per year are India, the United States of America, Brazil, the Russian Federation, and China (FAO, 2010b).

TABLE 15.1

Overall Ranking of the 15 Major Estimated Components of the Total Economic Value (TEV) of the Forests in Morocco (2001)

Rank	Estimated Component of the TEV	Value (€,000)	Percentage of the TEV
1	Watershed protection	+278,717	+45.4
2	Grazing	+255,000	+41.6
3	Firewood (illegal)	+96,030	+15.7
4	Erosion damage due to poor or no forest management	−57,230	−9.3
5	Timber	+47,724	+7.8
6	Damage caused by poaching	−43,892	−7.2
7	Firewood (legal harvest)	+43,650	+7.1
8	Damage due to overgrazing	−22,479	−3.7
9	Honey	+19,400	+3.2
10	Damage due to deforestation	−11,536	−1.9
11	Net loss of carbon/ emission	−11,316	−1.8
12	Hunting	+8,924	+1.5
13	Cork (all categories)	+6,817	+1.1
14	Mushrooms	+6,111	+1.0
15	Damage due to firewood overcollection	−3,388	−0.5
16	Damage due to forest fires	−2,488	−0.4
Total estimated direct use value		+487,154	+79.4
Total estimated indirect use value		+278,717	+45.4
Total estimated negative externalities		−152,329	−24.8
Total economic value (TEV)		**613,542**	**100.0**

Source: Ellatifi, M., In Merlo, M. and Croitoru, L. (eds), *Valuing Mediterranean Forests: Towards the Total Economic Value*, p. 84, CABI Publishing, 2005.

TABLE 15.2

Breakdown of Planted Forests in the World (by Region/Subregion)

Region/Subregion	Area of Planted Forest	
	1,000 ha	Percentage of the Total Planted Forest
Africa	15,409	5.8
Asia	122,775	46.5
Europe	69,318	26.3
Northern and Central America	38,661	14.7
Southern America	13,821	5.2
Oceania	4,100	1.5
World	**264,084**	**100.0**

Source: Adapted from FAO, Global Forest Resources Assessment 2010: Key Findings, Rome, Italy (Brochure), 2010b.

TABLE 15.3

Total Afforested and Reforested Areas, per Year, in the World (by Region/Subregion)

Region/Subregion	Afforestation[a] (ha/year)	Percentage (%)	Reforestation[b] (ha/year)	Percentage (%)
Africa	160,113	2.9	237,123	4.4
Asia	4,925,668	87.6	2,478,801	46.3
Europe	169,657	3.0	992,540	18.6
Northern and Central America	203,735	3.6	876,205	16.4
Southern America	103,879	1.9	722,527	13.5
Oceania	59,386	1.0	40,819	0.8
World	**5,622,438**	**100.0**	**5,348,017**	**100.0**

Source: Adapted from FAO, Global Forest Resources Assessment 2010: Key Findings, Rome, Italy (Brochure), 2010b.

[a] Afforestation is defined as the establishment of a forest by planting or seeding or both on land that was previously not considered as forest.

[b] Reforestation is defined as a re-creation of a forest by planting or seeding or both on land that was previously considered as forest.

The global area of planted forests increased by 5 million hectares per year between 2000 and 2010. Introduced species account for almost 25% of all planted species, while the remaining 75% consist of native species (FAO, 2010b).

15.2.2 Functions of Planted Forests

If the forest species are well selected and the plantation stands are well managed, planted forest, like natural forests, can perform vital functions in both rural and urban zones. They can provide industrial roundwood, fiber, fire wood, and nonwood forest products (Ellatifi, 2000).

They can also protect soil and water resources, harbor and conserve biodiversity, participate in climate change mitigation, combat land degradation and desertification, rehabilitate degraded land, raise the standard of living of rural communities, ensure food security, and contribute to overseeing alleviation and sustainable development (Ellatifi, 2005; FAO, 2006, 2010a).

Even if planted forests started their expansion during the 1900s, most the 264 million hectares of planted forests which grow in the world today were realized during the last 30 years.

It is estimated that planted forests will be the source of approximately 50% of the world's industrial wood production by the year 2025, and around 75% by the year 2050 (Sohngen et al., 1999). Such a situation will be beneficial for natural forests, which will be considerably relieved of their industrial logging pressures (Friedman and Charbley, 2004).

15.3 Biotechnology and Forest Trees Engineering

The term *biotechnology* is defined as "any technological application that uses biological systems, living organisms, or derivatives thereof, to make or modify products or processes for specific use" (FAO, 2001, Paragraph 1.3 [Definitions], line 1).

TABLE 15.4

Major Selection Technologies Experienced in the Forest Sector since the 1890s

Type of Technology	Period
Development of Mendel's laws on blending inheritance, by Gregor Mendel	Between 1856 and 1863
Development of planted forests with targeted major objectives	1920s to present
Management of planted forests toward targeted objectives	1930s to present
Selection of forest "trees plus" by traditional breeding techniques and their use in afforestation/reforestation	Late 1960s to present
Cloning forest "trees plus" and their use in afforestation/reforestation	Late 1980s to present
Field trials on genetically engineered (GE) forest trees	1990s to present
Commercial forest trees planting, using GE materials	2000 to future

15.3.1 Genetic Modifications of Forest Trees

Genetic modification (GM), or genetic engineering (GE), has been defined as "the alteration of genetic material in an organism in a way that does not occur naturally by mating and/or natural recombination" (European Commission, 2001, p. 2). It is one of the numerous breeding techniques that can be used to improve the productivity and quality of commercially planted forest species.

Genetic engineering involves the alteration of the species genome by insertion, via techniques applied in a laboratory, of genes which may come from the same plant species or from a different species. The modified plant is called a transgenic. In natural genetic modification, some bacteria are able to transfer some of their genes into trees and establish those genes in the tree's genome. Table 15.4 reports the major selection technologies which have been applied to forest trees since the 1890s.

Among the techniques used for GM, *Agrobacterium* is, so far, the most used transformation tool, accounting for 80% of the transgenic plants produced (Broothaerts et al., 2005). *Agrobacteria* are plant pathogenic organisms that infect several dicotyledonous species, causing tumoric diseases on infected host plant species (Li et al., 2008b).

Agrobacterium tumefaciens and *A. rhizogenes* possess an extrachromosomal genetic component called tumor-inducing (Ti) or root-inducing (Ri) plasmid. During infection, *Agrobacterium* inserts a region of this plasmid, known as the transferred-DNA (T-DNA), into plant cells, and this T-DNA piece is then integrated into the plant species' chromosomes.

The T-DNA contains genes which cause aberrant growth of plant cells through the synthesis of growth hormones. These genes can be replaced by any gene(s) of interest without loss of DNA transfer and integration functions, and therefore these novel genes will be transferred to tree cells during *A. tumefaciens* infection.

The use of this method is limited to the host-range of these bacteria because some tree species cannot be infected. Coniferous species are more difficult to modify with *A. tumefaciens* compared to hardwood species, and, in general, mature tissues are more recalcitrant to the infection (Strauss, 2001, 2009).

In 1986, the first regeneration of a genetically modified forest tree was achieved in *Populus*. Since then, this genus has become a model for genetic modification and related biotechnology research studies. The first genetic modification of a conifer tree genus (*Larix*) was achieved in 1991 (Huang et al., 1991; Chen et al., 2001).

Today, forest biotechnology research and application is being carried out on more than 140 genera in the world, with special focus on six genera, that is, *Pinus*, *Eucalyptus*, *Picea*, *Populus*, *Quercus*, and *Acacia* (Franclet, 1970).

Field trials of genetically modified (GM) forest trees are primarily directed toward *Populus*, *Pinus*, *Liquidambar*, and *Eucalyptus* (FAO, 2010a). Forestry research and its field applications are following a very similar path to that followed by agriculture several decades before. The difference between these ways is that, unlike a lot of agricultural crops, forest trees have, since only a few decades ago, been partly domesticated for the production of industrial/commercial wood production (Franclet, 1970).

In developing countries, even where forests are under various pressures, industrial forest plantations are continuously increasing (Ellatifi, 1984, 1991a,b, 2000, 2005, 2012).

For the case of Morocco (9 million hectares of forests, of which approximately 7% are planted forests), forest biomass contributes around 17% to total energy consumption. Around 9.7 million cubic meters of firewood are harvested every year for local consumption (84% in rural and 16% in urban areas) (Ellatifi, 1998, 2001). This creates heavy pressure on the forest, calculated as over 111,000 ha of forest which is degraded every year (Ellatifi, 2005, 2012). In spite of this and various other constraints (overgrazing, deforestation), planted forests continue to extend, with particular planting of the *Pinus* and *Eucalyptus* genera, mainly for the production of paper pulp.

The first hybridization trials on *Eucalyptus* were carried out in Morocco in the 1950s (Franclet, 1956). In 1975, the Moroccan hybridization techniques were transferred, on a large scale, to Brazil (Martin, 2003). The technique was also transferred to China by the FAO (Campinhos, 1987; Water and Menzies, 2010).

15.3.2 Research Activity on Genetic Modifications of Forest Trees

According to the United Nations Food and Agriculture Organization review (FAO, 2004), more than 35 countries regularly carry out genetic modification of forest trees, including all registered laboratory research and field trials. This GM research is mainly found in North America (48%) and Europe (32%). The research share of the other world regions is as follows: Asia (near 14%), Oceania (near 5%), South America (1%), and Africa (near 0.5%) (Figure 15.8).

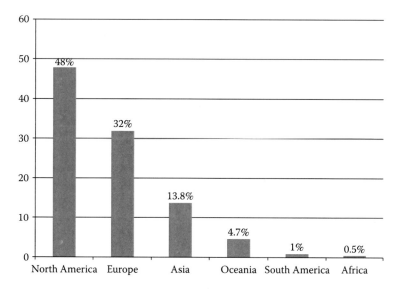

FIGURE 15.8
Share in genetic modification research of the continents (in percentage). (From FAO, Preliminary review of biotechnology in forestry, including genetic modification, Forest Resources Division, Rome, Italy, 2004. http://www.fao.org/docrep/008/ae574e/ae574e00.htm)

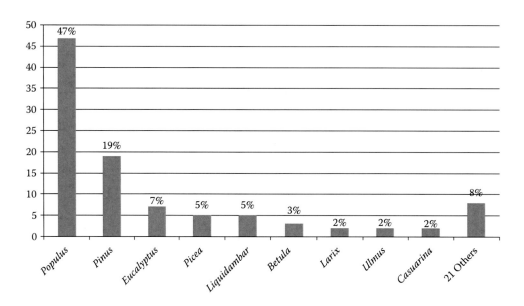

FIGURE 15.9
Share of laboratory research on forest tree species (in percentage). (From FAO, Preliminary review of biotechnology in forestry, including genetic modification, Forest Resources Division, Rome, Italy, 2004. http://www.fao.org/docrep/008/ae574e/ae574e00.htm)

15.3.3 Species Subject to GM Research

The first transgenic forest tree was realized in the United States of America in 1986. It concerned a poplar (*Populus alba* × *P. grandidentata*) (Sederoff, 2007). At the international level, about 80% of the laboratory research and field trials related to forest trees concerned five genera: *Populus* (47%), *Pinus* (19%), *Eucalyptus* (7%), *Picea* (5%), and *Liquidambar* (5%) (Figure 15.9), with a special interest for *Populus*. This became the most studied forest tree genus and the model genus for forest tree genetics (Sedjo, 2004).

The remaining 8% concerned the following genera: *Abies, Acacia, Actinida, Calocedrus, Carica, Castanea, Citrus, Diospyros, Hevea, Juglans, Liriodendron, Malus, Morus, Olea, Poncirus, Prunus, Pseudotsuga, Pyrus, Quercus, Robinia,* and *Tsuga*.

Today, the number of transgenic forest trees continues to increase every year, and the first commercial releases have already taken place.

15.4 Toward Shifting Timber and Industrial Wood Harvesting from Natural Forests to Planted Forests

15.4.1 Wood Removal from the World's Forests

The global volume of wood (all categories) removed from forests was 3.2 billion cubic meters in 2005 (FAO, 2010b). About 55% of this volume was industrial roundwood, and 45% was wood fuel. This total was unequally distributed between the regions or subregions as follows: around 24.3% in Northern and Central America, 23.3% in Asia, 22.5% in

TABLE 15.5

Breakdown of Global Wood Removals (2005)

Region and Subregion	Industrial Round Wood		Wood Fuel		Total	
	Volume (10⁶ c.m.)	Percentage (%)	Volume (10⁶ c.m.)	Percentage (%)	Volume (10⁶ c.m.)	Percentage (%)
Africa	63	3.6	495	34.2	554	**17.4**
Asia	201	11.4	547	37.8	748	**23.3**
Europe	560	31.8	164	11.3	723	**22.5**
Northern and Central America	705	40.0	76	5.2	781	**24.3**
Southern America	178	10.1	167	11.5	344	**10.7**
Oceania	55	3.1	—	—	57	**1.8**
World	**1,762**	**100**	**1,449**	**100**	**3,211**	**100**
Percentage (%)	**54.9**		**45.1**		**100**	

Source: Adapted from FAO, Global Forest Resources Assessment 2010: Key Findings, Rome, Italy (brochure), 2010b.

Note: 10⁶ c.m., million cubic meters.

Europe, 17.4% in Africa, and 1.8% in Oceania. Table 15.5 gives the distribution of the global wood removals per region and subregion.

These global wood removals account only for 0.7% of the global growing stock. Fuelwood constitutes almost 45% of it. This partially reflects the socioeconomic functions that forests provide for local communities, but forests also provide other environmental services which have a higher share in the forest's total economic value (Ellatifi, 2005, 2012).

In Morocco, for example (cf. Table 15.1), timber in the fifth largest contributor to total economic value (TEV) of the Moroccan forests, with a modest share value of 7.8% of the TEV, behind watershed protection (45.4%), grazing (41.6%), firewood (illegal) (15.7%), and erosion damage due to poor or no forest management (9.3%) (Ellatifi, 2005, 2012).

15.4.2 Consequences of Translating Timber and Industrial Wood Harvesting from Natural Forests to Planted Forests

Due to intensive management of planted forests, it is a fact that, today, planted forests are becoming an important supplier of industrial roundwood, although their share in the world's industrial wood production is still around 33%. Based on this trend, it is statistically permissible to forecast for planted forests a share of around 60% of the global industrial wood production by 2025, and around 80% by 2050.

In these conditions, the world will assist in a real transition in the production of roundwood, from natural forests to planted forests, within the next 15 years. Obviously, the role played by genetically engineered (GE) forest trees is the major facilitator of this move, but it is also necessary to enhance the choice of planted forest species, in the different planting sites, and to avoid the causes of degradation (overgrazing, forest fire, deforestation, disease attacks, etc.) in order to facilitate the shift to industrial roundwood from natural forests to planted forests (Carle et al., 2002).

With such a smooth and continuous shift of roundwood production from natural forests to planted forests, it will be possible to expect, in the near future, to:

Produce all necessary industrial roundwood in a relatively smaller and geographically concentrated area, outside the natural and slow-growing forests. Such genetically engineered forest trees are highly desirable to lumber companies because their rapid growth, straightness, and density enable them to produce more pulp and paper since they make less lignin and more cellulose (Chen et al., 2001).

Enable the natural forests to focus on the provision of their noble and vital environmental functions, such as: conservation of soil and water resources; harboring and protection of biodiversity; combating water and aeolian erosion, and desertification; participation in climate change mitigation, through carbon sequestration in forest tree stems, canopies, roots, and forest soil; provision of nonwood products (NWFPs); beautification of landscapes and participation in the development of ecological tourism, and so on.

Participate in the mitigation of deforestation in the world. In order to totally overcome this phenomenon, the aforementioned shift needs to be accompanied by political consensus, involving all countries' decision makers and all forest stakeholders, upon an effective way to eradicate deforestation in all countries, with the necessary involvement of rural and poor communities whose daily existence is totally or partially reliant on forests and their ecosystems (Ellatifi, 2005, 2012; Sedjo, 2005a, 2006).

Optimize planted forest productivity through using trees in fast rotation through intense management, for quick sequestration of carbon, while they are more resilient to climate change.

Research into genetic modification is expected to stimulate the sequencing of all genomes in forest trees genera, such as *Populus*, *Eucalyptus*, and *Pinus* (FAO, 2004).

15.5 Beneficial Traits Subject to Genetic Modifications

In forest trees engineering, genetic modification is focused on herbicide tolerance, insect resistance, stem lignin modification, tree flowering control, drought and cold tolerance, disease tolerance, wood density, tree stem straightness, fiber production, heavy metal accumulation, and increased growth for biomass production (FAO, 2004). In the wood pulp industry, for example, roundwood with a reduced lignin component is easier to process and provides environmental and economic benefits.

The introduction of desired traits into the already "tree plus" provides great promise for increasing the productivity and quality of trees and beneficially opening up many marginal lands and climatic conditions. This therefore makes forest planting profit-earning projects, and would attract more investors to the sector.

On the other hand, in arid zones as well as under cold climates, drought tolerance and cold tolerance are essential for forest plants. The reduction of fiber in the tree is very beneficial for the wood pulp industry. Regarding flower control, it would be necessary to avoid/limit gene escape into nature. Tables 15.6 through 15.9 highlight the various benefits of genetically modified (GM) trees.

15.5.1 Herbicide Tolerance

In forest silviculture, tolerance for herbicides such as *glyphosete* has always been an important target in forest genetic engineering (Straus et al., 2011). For *Eucalyptus grandis* (Jain and Minoche, 2000) and *Larix deciduas* (Shin et al., 1994), hybrid trees with a high percentage of *glyphosate* were created. In the year 2002 the first poplar hybrids were created, with high levels of *glyphosete*, by insertion of the *C4P* gene (Meilan, 2002; Packer and Clay, 2003).

TABLE 15.6

Summary of Genetic Modifications in Forest Trees Targeting Resistance to Herbicides

Gene	Modification	Effects
aroA (EPSP gene)	Overexpressed in *Populus alba* × *P. grandidentata*	First record of insertion and expression of a foreign gene of agronomic importance in woody plants; slight resistance to glyphosate
bar (PAT gene)	Overexpressed in *P. alba* × *tremula* and *P. trichocarpa* × *deltoides*	Transgenics did not accumulate NH_4^+ when treated with Basta®
crs 1-1	Overexpressed in *P. tremula* × *P. alba*	Resistance to chlorsulphuron in greenhouse tests
aroA	Overexpressed in *P. alba* × *P. grandidentata*	Conferred resistance to glyphosate
als, pat	Overexpressed in *P. tremula* × *P. alba*	Conferred herbicide resistance to calli
aroA	Overexpressed in *Larix decidua*	Conferred resistance to glyphosate at moderate treatment levels
bar	Overexpressed in *Eucalyptus camaldulensis*	Conferred resistance to herbicide at over twice the normal field application rate
bar	Overexpressed in *Pinus radiata* and *Picea abies*	Conferred resistance to glufosinate in both species in greenhouse testing
γ-ECS	Overexpressed in *Populus tremula* × *P. alba*	Conferred resistance to acetochlor and metolachlor present in soil. Plants with cytosolic expression were more resistant than those with chloroplastic expression
GOX, CP4	Overexpressed in various poplar hybrids, including *P. trichocarpa* × *P. deltoides*	Genes shown to confer resistance up to 66% at low glyphosate levels, with 25% of lines showing increased growth following herbicide treatment. Lack of damage was attributed to CP4, as GOX was suspected to cause undesirable side effects. Twelve lines expressing only CP4 had similar herbicide tolerance, but grew better and had less damage in response to treatment
γ-ECS	Overexpressed in *P. tremula* × *P. alba*	No change in tolerance to paraquat
GS	Overexpressed in *P. tremula* × *P. alba*	Conferred resistance to PPT
bar	Overexpressed in *P. alba*	Conferred resistance to Basta® at normal field dosage; poplar still tolerant at twice the normal field dosage

Source: Adapted from FAO, Forests and Genetically Modified Trees, Rome, Italy, Available at http://www.fao.org/docrep/013/i1699e/i1699e00.htm, 2010a.

On the other hand, GM research in laboratories succeeded in creating transgenic poplars which were completely tolerant of the herbicide *phosphinothricin* (Confalonieri et al., 2000). Tolerance to herbicides was also achieved by modification with a microbial gene-encoding enzyme for the detoxication of the herbicide. This was the case in *Pinus radiate* and *Picea abies* with the bar gene from *streptomyces hygroscopicus* species providing herbicide tolerance (Bishop et al., 2001; Vengdesan et al., 2006; Botin, 2003).

Table 15.6 gives a summary of genetic modifications to forest trees targeting resistance to herbicides.

15.5.2 Resistance to Bacterial, Fungal, and Viral Pathogens

Likewise, in horticulture genetic engineering has recently developed fungal and bacterial resistance in forest trees through the use of genes of different origins. For example, for *Populus nigra* × *P; maximowiczii* (hybrid poplar) and *Picea mariana* (black spruce), insertion of the *endochitinase* gene (ech 42) from the fungus *Trichoderma harzianum* resulted in a significant increase in resistance to several pathogens (Noël et al., 2005).

15.5.3 Insect Resistance

Defoliation of forest trees due to insects can result in serious damage and significant economic losses. This is the case, for example, in the Maâmora cork oak forest, which is the largest in the world. Various insects are found in this forest, including *Lymantria dispar*, *Coroebus bifas-ciatus*, *Cerambix scutellaris*, *Platypus cylindrus*, and *Tortris viridana*. The first is a dangerous defoliating insect which, during drought years, is able to cause enormous damage to cork oak trees, jeopardizing their longevity.

Insect-infected forest trees, in general, can have their growth and survival limited (Pena and Seguin, 2001). Trees' resistance to insects can be significantly increased through modification of the trees with *cry* genes from the soil bacterium *Bacillus thuringiensis*, coding for delta-endotoxins (Bt toxin). Bt toxins possess an efficient combating action against the insect groups Coleoptera, Lepidoptera, and Diptera. The Bt toxin binds to specific receptors in the intestine, killing the insects through pore formation. This was successfully tested in poplar (Tian et al., 2000; Hu et al., 2001) and in various conifer species (Lachance et al., 2007). However, insects can develop resistance to Bt toxins (MacIntosh, 2009).

15.5.4 Resistance to Abiotic Stress

Many forest tree species cannot survive extreme abiotic stresses, such as high or low temperatures, drought, salinity, or chemical toxicity. Recently, successful adaptation to cold and frost has been achieved in a transgenic hybrid larch (*Larix* × *leptoeuropea*). This forest tree was modified to express a *Vigna aconififolia* gene for pyrroline 5-carboxylate, the rate-limiting step in proline synthesis. The resistance to cold, salt, and frost stresses was significantly increased due to elevated levels of proline in the hybrid tree tissue (Gleeson et al., 2005). Successful modification results are also reported for tolerance in transgenic *Populus nigra* (Hu et al., 2005; Junghans et al., 2006) and cold tolerance in eucalypt hybrids (*Eucalyptus grandis* × *E. urophylla*) (Federal Register, 2007). Table 15.7 gives a summary of genetic modifications in forest trees targeting resistance to abiotic stresses.

15.5.5 Wood and Growth Characteristics

According to Chiang (2006), structure and wood quality are very important traits which are subject to genetic engineering in forest trees. With reduced lignin content, wood needs fewer chemicals during pulp processing, which brings significant economic and environmental benefits (Chiang, 2006). This lignin content reduction can be achieved by introducing antisense genes which suppress enzymes that are highly necessary in lignin biosynthesis, such as 4-coumarate: CoA ligase (*4 CL*) or cinnamyl alcohol dehydrogenase (*CAD*).

TABLE 15.7

Summary of Genetic Modifications in Forest Trees Targeting Resistance to Abiotic Stress

Gene	Modification	Effects
GSS	Overexpressed in *Populus tremula* × *alba*	No changes in response to ozone stress; ozone sensitivity related to leaf development stage
FeSOD	Overexpressed in *Populus tremula* × *alba*	No change in response to high light and photoinhibition of PSII; suggests rate of conversion of superoxide to hydrogen peroxide is not a rate-limiting factor in protection against or repair of photoinhibition
mtlD	Overexpressed in *Populus* sp.	Transgenic plants grew significantly better with a higher survival rate
Bet-A	Overexpressed in *Populus* sp.	Conferred salt resistance
Phospholipase D	Antisense expression in *Populus* sp.	Conferred salt tolerance
PsG6PDH and PsAFP	Overexpression in *Populus* sp.	Freezing resistance tests under way
mt1D and *gutD*	Overexpression in *Pinus taeda*	Increased salt tolerance at both calli and plantlet stage; accumulated mannitol and glucitol
codA	Overexpressed in *Eucalyptus camaldulensis*	Increased salt stress tolerance
mt1D	Overexpressed in *Populus tomentosa*	Increased mannitol; increased salt tolerance *in vitro* and in hydroponic culture; decreased growth in the absence of salt
prxC1a	Overexpressed in *P. sieboldii* × *P. grandidentata*	Increased growth rate; elevated peroxidase activities; calli resistant to oxidative stress imposed by hydrogen peroxide
vhb	Overexpressed in *P. alba*	No change in growth pattern, or chlorophyll and protein contents; no change in stress resistance
dreb1a and citrate synthase	Overexpressed in *Eucalyptus grandis* × *E. urophylla*	Conferred resistance to drought and acid soil tolerance
GS1	Overexpressed in *P. tremula* × *P. alba*	Higher photosynthetic assimilation and stomatal conductance at all levels of water availability; increased photo-assimilation allows increased allocation of resources to photoprotective mechanisms
P5CS	Overexpressed in *Larix leptoeuropaea*	Increased proline; increased resistance to cold, salt, and freezing stress
CaPF1	Overexpressed in *Pinus strobus*	Increase in tolerance to drought, freezing, and salt stress

Source: Adapted from FAO, Forests and Genetically Modified Trees, Rome, Italy, Available at http://www.fao.org/docrep/013/i1699e/i1699e00.htm, 2010a.

In transgenic trees, the lignin content of the wood may be reduced by as much as 45% (Chiang, 2006). Field trials with this type of GM tree have been taking place for many years in the European Union.

Because it leads to increased growth of forest trees, forest engineering participates in the sequestration of higher volumes of carbon in forest trees stems, branches, and roots, and therefore in higher mitigation of climate change. It also participates in the production of more green energy, that is, neutral biofuel.

TABLE 15.8

Summary of Genetic Modifications in Forest Trees Affecting Growth

Gene	Modification	Effects
4CL	Antisense inhibition in *Populus*	Increased plant growth; structural integrity maintained
GS1	Overexpressed in *Populus*	Increased node and leaf number, larger leaves; increased growth; enhanced nitrogen assimilation and increased growth under both high and low nitrogen conditions
Xyloglucanase	Overexpressed in *Populus*	Increased stem length and internode length
cel1	Overexpressed in *Populus tremula*	Increased growth, larger leaves; increased stem diameter
ugt and *acb*	Overexpressed in *Populus*; sense and antisense expression of *acb*	*ugt* plants show increased growth; *ugt* and *acb* lower growth than *ugt* alone; sense *acb* show increased growth; antisense *acb* show decreased growth
vhb	Overexpressed in *Populus*	Increased height and stem diameter
PttEXPA1	Overexpressed in *Populus*	No change in height; increased internode length, fiber diameter and vessel element length; increased leaf expansion

Source: Adapted from FAO, Forests and Genetically Modified Trees, Rome, Italy, Available at http://www.fao. org/docrep/013/i1699e/i1699e00.htm, 2010a.

Increased growth of transgenic forest *Populus tricocarpa* that was modified to overexpress glutamine synthetase (GS), a key enzyme in nitrogen assimilation, has been observed. Such a modification augmented the nitrogen-use efficiency of the tree and resulted in higher growth rates (Li et al., 2005; Fu et al., 2005). Research studies are being carried out on another forest tree (*Jatropha curcas*) to increase its seeds yield for the production of fuelwood (Li et al., 2008b; Ellatifi, 2005). Table 15.8 gives a summary of genetic modifications in forest trees affecting growth.

15.5.6 Flowering Modification

Male and female sterility has been considered to be an important solution for lessening gene escape of transgenes into the nontransgenic tree population. The sterility of male and female poplar trees has been successfully achieved (Li et al., 2000; Hoenika et al., 2006). To reduce the generation of the *BpMADS4* gene in silver birch, another method has been used based on the earliest flowers, which were produced 11 days after rooting, when the seedlings were only 3 cm tall (Elo et al., 2007). This early flowering method was also used on other forest species, such as poplar (Table 15.9).

15.5.7 Phytoremediation

Phytoremediation refers to the use of living green plants for *in situ* risk reduction and removal of contaminants from contaminated soil, water, sediments, and air. It uses plants (particularly forest trees) to remove, sequester, or detoxify polluting agents, including pesticides and heavy metals. For this purpose, rapidly growing transgenic poplars have been engineered with genes that remove heavy metals such as copper, mercury, arsenic, and cadmium (Merkle, 2006), and detoxify pesticides and other pollutants, such as volatile hydrocarbons like chloroform and benzene (Doty et al., 2007).

TABLE 15.9

Summary of Genetic Modifications in Forest Trees Targeting Hormonal Regulation

Gene	Modification	Effects
ipt	Overexpressed in *Populus tremula* × *P. alba*	Increased branching shoots, short internodes, unable to root
iaaH and *iaaM*	Overexpressed in *P. tremula* × *P. tremuloides*	Smaller trees with elevated free and conjugated IAA; decreased axillary bud release following decapitation
rolC	Overexpressed in *P. tremula* × *P. tremuloides*	Reduced apical dominance, increased axillary shooting, fascinated apices
rolC	Overexpressed in *P. tremula* × *P. tremuloides*	Alterations in hormone levels with a decrease of ABA in pre-dormant buds and during resting; earlier flushing
OSH1	Overexpressed in *P. nigra* L. var. *italica*	Alterations in phenotypes with three relatively distinct phenotypes identified: I – slender leaves, II – dwarfed, III – multiple shoot apices with tiny leaves
rol genes	Overexpressed in *P. tremula*	Shorter, but more numerous internodes; axillary shooting; decreased shoot:root ratios; delayed dormancy
GA-20 oxidase	Overexpressed in *P. tremula* × *P. tremuloides*	Increased height and diameter; increased internode length; longer, broader leaves; increased number of cells; significantly longer xylem cells
rolC	Overexpressed in *P. tremula* × *P. tremuloides*	Reduced shoot growth; early bud break; no change in timing of wood formation; more numerous tyloses formed; lacked thick walled late wood
iaaM	Overexpressed in *P. tremula* × *P. tremuloides*	35% increase in IAA, but no change in radial distribution pattern; increased internode length; decreased occurrence of axillary bud break following decapitation

Source: Adapted from FAO, Forests and Genetically Modified Trees, Rome, Italy, Available at http://www.fao.org/docrep/013/i1699e/i1699e00.htm, 2010a.

15.6 Field Trials of Transgenic Forest Trees

Genetically, forest trees field trials are opposed to greenhouse trials, which are considered as willful releases into the environment. They are evaluated under the control of the national biosafety regulations of the countries in which they are developed. By 2004, such field trials had been established in 30 countries. There were over 200 field trials, with the majority in the United States of America, that by May 2009 had approved 187 trials with poplars. At the international level, field trials of transgenic forest trees are reserved for the genera *Populus*, *Eucalyptus*, and *Pinus*. The first release trial of GM poplars in the European Union was established in Belgium (Frankenhuyzen and Beardmore, 2004).

Today, field trials of transgenic forest trees are found in France (10), Germany (4), the UK (2), Sweden, Norway, Belgium, and Spain. Between 1993 and 2008, some 50 field trials of engineered forest trees were approved by national authorities in Europe.

TABLE 15.10

Cumulative Number of Approved Field Trials with Genetically
Modified Trees in the USA, by May 2009

Forest Trees	Species/Genus	Releases
Allegeny serviceberry	*Amelanchier laevis*	1
American elm	*Ulmus americana*	3
American sweetgum	*Liquidambar sp.*	30
Black cottonwood	*Populus tricocarpa*	1
Common aspen	*Populus tremula*	1
Eastern cottonwood	*Populus deltoides*	37
Eucalypt	*Eucalyptus sp*	61
Gray poplar	*Populus tremula* × *P. alba*	15
Hybrid aspen	*Populus tremula* × *P. tremuloides*	6
Hybrid poplar	*Populus euramericana*	3
Loblolly pine	*Pinus taeda*	94
Monterrey pine	*Pinus radiata*	1
Pich pine × loblolly pine	*Pinus rigida* × *P. taeda*	45
Poplar	*Populus sp*	121
White spruce	*Picea glauca*	1

Source: U.S. International Society of Biomechanics (ISB), Available at http://
www.isbweb.org/isb-congresses, 2009.

Table 15.10 shows the cumulative number of approved field trials with genetically modified trees in the USA by May 2009. Table 15.11 reports the forest species NCBI records as of
February 2013.

15.7 Commercial Releases of Genetically Modified Trees

Releases of engineered forest trees on the commercial market remain very limited. Only
some examples of commercial releases are reported. So far, the only GM forest trees which
are deployed on a large scale are engineered poplars in China (Farnum et al., 2007).

The first pilot plantation of 80 ha was established in 1999, with GG insect-resistant
Populus nigra. This pilot plantation was commercialized in 2002, and 14 million trees were
planted on an area of 300–500 ha (FAO, 2004). Also in China, about 400,000 cuttings of the
GM hybrid poplar clone 741 (*Populus alba* × [*P. alba* × *P. simonii*] × *P. tomentosa*), which is
resistant to leaf-eating insects due to insertion of Bt *cry1*, an *AP1*gene (Yang et al., 2003),
were propagated and planted in 2003. Table 15.12 gives a summary of the species and status of R&D on genetic modifications in forest trees in China.

Regarding the current situation of biotechnology research on forest trees, it is expected
that only a limited number of forest tree taxa will be commercialized in the near future.
The taxa concerned are *Populus, Eucalyptus, Pinus radiata, P. taeda*, and *P. pinaster*.

Nevertheless, an increased trend in forestry toward the production of timber plantations
is expected (Carle and Holmgren, 2008; Sepala, 2007), with significant interest in GM forest
trees. In this context, some American private companies plan to develop three GM varieties:
fast-growing loblolly pine plantations, low-lignin *Eucalyptus* species for use in South America,
and cold-hardy *Eucalyptus* species in the southern part of the U.S. (Farnum et al., 2007).

TABLE 15.11

Forest Species NCBI Records as of February 2013

Forest Tree	Common Name	Taxonomy ID	Family	Nucleotide	Nucleotide EST	Protein	Structure	Genome	Pop Set	GEO Data Sets	Uni STS	Probe	Bio Project	Pub-Med Central
Pinus sylvestris	Scots pine	3349	*Pinaceae*	80,630	19,484	5,031	3	1	176	114	73	263	7	444
Pinus radiata	Monterey pine	3347	*Pinaceae*	11,972	8,717	1,997		1	3,410	56	112	26	17	210
Pinus pinaster	Maritime pine	71647	*Pinaceae*	2,970	34,649	1,416		1	34	55	101	327	11	158
Pinus pinea	Parasol pine	3346	*Pinaceae*	83	326	122		1	6		4	21	3	41
Pinus canariensis	Canary Island pine	49510	*Pinaceae*	18		74		1	5					8
Pinus halepensis	Aleppo pine	71633	*Pinaceae*	927		957		1	15			1		38
Cedrus atlantica	Atlantic cedar	123597	*Pinaceae*	37		22		1	11				1	21
Populus trichocarpa	Black cottonwood	3694	*Salicaceae*	93,129	89,943	80,443	3	1	227	332	407	245	21	1,113
Populus deltoides	Cottonwood	3696	*Salicaceae*	371	14,661	223		1	45	183	21	128	7	187
Eucalyptus grandis	Rose gum	7139	*Myrtaceae*	5,407	42,576	452		1	4	131	878	924	8	156
Eucalyptus camaldulensis	Murray red gum	34316	*Myrtaceae*	41,719	58,584	50		1	6	14		2	2	77
Eucalyptus globulus	Blue gum	34317	*Myrtaceae*	426	26,262	144		1	29	5	247	364	4	205
Eucalyptus globulus subsp. Globulus		71271	*Myrtaceae*	102	2	332		1	2			1		
Eucalyptus melliodora		183838	*Myrtaceae*	14		9			7					7
Eucalyptus occidentalis		229548	*Myrtaceae*	6		6								2
Eucalyptus robusta		627158	*Myrtaceae*	2		2								1

(continued)

TABLE 15.11 (Continued)

Forest Species NCBI Records as of February 2013

Forest Tree	Common Name	Taxonomy ID	Family	Nucleotide	Nucleotide EST	Protein	Structure	Genome	Pop Set	GEO Data Sets	Uni STS	Probe	Bio Project	Pub-Med Central
Acacia tortilis		138046	*Mimosoideae*	65		29			23					25
Acacia saligna		420411	*Mimosoideae*	116		13			14					9
Acacia arabica		658885	*Mimosoideae*	15		10								9
Acacia senegal	Gum arabic	138043	*Momosoideae*	63		8			14					28
Acacia mellifera	Hookthorn	381101	*Mimosoideae*	37		1			12					10
Juniperus thurifera		177241	*Cupressaceae*	27		5			10					3
Juniperus phoenicea		61308	*Cupressaceae*	39		30			13					13
Abies pinsapo	Hedgehog fir	56046	*Pinaceae*	14		7			4					3
Quercus suber	Cork oak	58331	*Fagaceae*	361	6,690	126			7		13			76
Cupressus sempervirens		13469	*Cupressaceae*	406		374			26					37
Jatropha curcas		180498	*Euphorbiaceae*	14,359	46,862	680		1	3		104	135	10	92
Tetraclinis articulata		13717	*Cupressaceae*	29		17			27					7

Source: Adapted from Wegrzyn, J.L., et al. *Int. J. Plant Genom.,* 2008, 412875, 2008.

TABLE 15.12

Summary of Species and Status of R&D on Genetic Modifications in Forest Trees in China

Tree Species	Status of Project[a]	Traits Targeted	Gene(s) Inserted
Betula platyphylla	R	Insect resistance	Spider insecticidal peptide gene
Eucalyptus camaldulensis	R	Lowering lignin content	C4H
Eucalyptus urophylla	R	Resistance to disease caused by *Pseudomonas solanaceanum*	cecropin D
Poplar hybrid 741 (*P. alba* × [*P. davidiana* + *P. simonii*] × *P. tomentosa*)	E, C	Resistance to leaf-eating insects	*Bt Cry1* and API
Populus × *xiaozhuanica* W.Y. Hsu et Liang cv. Balizhuangyang-zhongtian	E	Salt tolerance	*mtlD*
Populus deltoides	R	Insect resistance	*Bt*
Populus deltoides × *P. cathayana*	R	Resistance to leaf-eating insects	*mtlD/gutD*
Populus deltoides × *P. simonii* (N-106)	R	Resistance to leaf-eating insects	*AaIT*
Populus nigra	E, C	Resistance to leaf-eating insects	*Bt*
Populus simonii × *P. nigra*	R	Salt tolerance	*Bet-A*
Populus tomentosa	R	Resistance to disease and stress	*NP*-1 (rabbit alexin)

Source: Adapted from FAO, Forests and Genetically Modified Trees, Rome, Italy, Available at http://www.fao.org/docrep/013/i1699e/i1699e00.htm, 2010a.

[a] R, Research; E, Environmental release; C, Commercial planting.

15.8 Concerns Regarding Transgenic Forest Trees

> *In theory there is no difference between theory and practice. In practice there is.*
>
> **Yogi Berra**

Tough engineered trees are supposed to increase forest productivity, but there are concerns regarding probable risks, particularly environmental risks, resulting from this. The objective of regulation is to ensure the safety of transgenic services. So far, no country has clearly approved the commercialization of engineered forest trees. China has released a transgenic poplar, but the corresponding deregulation status already appears to be in place.

Several forest trees are actually being used in field trials in many countries where deregulation is likely to occur soon. Through the Cartagena Protocol on Biosafety and the United Nations Industrial Development Organization (UNIDO), an urgent need to adopt national regulations to control the release of GM organisms into the environment is recognized (Pachico, 2003; Sedjo, 2005b).

Meanwhile, in some countries there is a general agreement regarding the fact that while existing procedures offer the basic process for deregulation, specific protocols and procedures may need to be adopted for commercial industrial wood produced by transgenic trees.

The most important reasons for the regulation of engineered forest trees are concerns about human health, safety, and environmental risks.It has been internationally agreed that the problem areas for forest trees are mainly environmental (Mullin and Bertrand, 1998; Ellatifi, 2005). Regulators are expected to behave as if transgenic introduction may create new environmental risks.

15.8.1 Potential Environmental Risks of Transgenic Forest Trees

As in agriculture, the introduction of GM forest trees brings potential environmental risks which have to be assessed prior to tree planting. However, due to their specific biology, engineered forest trees may imply different environmental risks. Forest trees have, for example, a long lifecycle which might have a long-spanning impact on ecosystem functioning, and they also potentially have long-distance dispersal of seeds and pollen. Little research has been conducted on this issue. Knowledge and significant data on the behavior of engineered forest trees in their environment are still lacking (Farnum et al., 2007). Meanwhile, the available results of existing short-term empirical studies may be extrapolated using predictive models. These models encompass different factors, such as seed and vegetative propagule establishment, hybrid seedlings competitiveness in the wild, mating success, and pollen movement.

Nevertheless, we should be very careful about the statistical value of the results obtained from empirical studies and extrapolated results* (DiFazio et al., 1999; Kuparinen and Schurr, 2007).

15.8.2 Environment Risk Assessment

According to the European Union (EU) directive 2001/18/EC (European Commission, 2001) on the deliberate release of GM organisms into the environment, with regard to the release of GMOs (including forest trees), the environmental risk assessment should address the following items:

Likelihood of the GM tree becoming more persistent than the recipient trees in agricultural habitats or more invasive in natural habitats.

Any selective advantage or disadvantage conferred by the GM tree.

Potential for gene transfer to the same or other sexually compatible tree species under conditions of planting the GM tree, and any selective advantage or disadvantage conferred to those tree species.

Possible immediate and delayed environmental impact resulting from direct and indirect interactions between the GM tree and target organisms, such as predators, parasitoids, and pathogens (if applicable).

Possible immediate and delayed environmental impact resulting from direct and indirect interactions between the GM tree with nontarget organisms (also taking into account organisms which interact with target organisms), parasites, and pathogens.

Possible immediate and delayed effects on human health resulting from potential direct and indirect interactions of the GM tree and persons working with, coming into contact with, or in the vicinity of the GM tree release(s).

Possible immediate and delayed effects on animal health and consequences for the feed/food chain resulting from consumption of the GMO and any products derived from it, if it is intended to be used as animal feed.

Possible immediate and delayed effects on biogeochemical processes resulting from potential direct and indirect interactions of the GMO and target and nontarget organisms in the vicinity of the GMO release(s).

---

* Garbage in, garbage out!

Possible immediate and delayed direct and indirect environmental impacts of the specific cultivation, management, and harvesting techniques used for the GM tree where these are different from those used for non-GM trees.

15.8.3 Persistence and Invasiveness

Invasiveness is a common process with introduced species. Several introduced species have the potential to become invasive, either by seed dispersal or by vegetative spread. A well-known example in Yemen is *Prosopis juliflora*, the seeds of which rapidly invade surrounding areas (Ellatifi, 1996). Another very invasive tree in Morocco and other countries is *Lantana camara*, which also spreads by seeds.

Unlike forest trees, annual crops produce their pollen in the year in which they are planted. Forest trees produce their pollen only after having reached reproductive maturity, which varies from one species to another. For example, for Atlas cedar (*Cedrus atlantica*) flowering begins at around 20 years of age.

For most poplars, flowering begins between 10 and 15 years of age. Consequently, forest trees have a limited risk of becoming invasive. On the other hand, vegetative reproduction could increase the spread of forest trees even before they reach maturity. This constitutes an important mechanism of spread for most species, in several genera. The longer life-cycle of forest trees means that the seed source remains in the same place for many years, whereas for agricultural crops the seed source returns to the same place on the farm only after 3–5 years (crop rotation).

Furthermore, most forest tree species possess an abundant seed production—particularly in mast fruiting years—and long-distance pollen dispersal, ranging up to several kilometers for Mediterranean and other temperate conifers (O'Connell et al., 2007; Ellatifi, 1977). Added to that, most forest trees possess extensive seed longevity (e.g., *Prosopis juliflora*) (Ellatifi, 1996). This means that the transgenic seeds of forest trees may potentially be stored in the seedbank for decades (Kumar and Fladung, 2001).

15.8.4 Horizontal Gene Transfer (Out-Crossing)

The out-crossing of transgenes to nontransgenic relatives potentially enhances the area within which any possible adverse effects related to genetic modification can occur. Successful dispersal of pollen and development of seed is required in order to establish the new trait through sexual reproduction (vertical gene transfer, in forest trees). The transgenes might be transferred to nontransgenic relatives or to wild relatives through out-crossing (Li et al., 2008a).

Transgene escape can be predicted with simulation models such as *AMELIE* (Kuparinen and Schurr, 2007) and *STEVE* (DiFazio et al., 1999).

Horizontal gene transfer (HGT) is another genetic transfer mechanism. This results in the nonsexual transfer of genes between species, for example between tree roots and soil microorganisms.

15.8.5 Vertical Gene Transfer (Out-Crossing)

The majority of tree species used in forest plantations are undomesticated out-crossers that, depending on their specific biology, may easily interbreed with related species (Irwin and Jones, 2006). It is very common for many native eucalypt species to hybridize in the wild (Florence, 1996).

On the other hand, many trees possess abundant production of wind-borne pollen that can travel long distances and remain viable. This enhances the likelihood of vertical gene transfer through pollination of nontransgenic trees. Though there is no trustworthy evidence of gene transfer from transgenic to nontransgenic forest trees, an introgression of genes of a cultivated poplar species (*Populus deltoids*) has been observed in the offspring of natural *Populus nigra* populations (Vanden Broeck et al., 2004).

Nonetheless, out-crossing is not a hazard in its own right. The environmental hazard associated with out-crossing depends on the transgenic trait which is transferred by out-crossing, and the properties of the resulting GM tree.

15.8.6 Effects on Human Health

15.8.6.1 Potential Hazards

Potential effects on human health resulting from genetic modification in forest trees may consist of pollen allergies, such as increased incidence of hay fever, or may occur through consumption of parts of transgenic forest trees. This situation would be of concern if, as a consequence of GM, pollen production or the number of allergic substances is augmented.

To reduce pollen production (sterility), modifications could be employed reduce this effect. Possible risks to human health caused by consumption of transgenic forest tree parts could be either by incidental consumption of parts of the engineered tree itself or by consumption of parts of other forest trees that have hybridized with the transgenic tree and have introgressed the transgenes which lead to the production of toxic or allergy-inducing tree fragments.

15.8.6.2 Crops versus Trees

No justification exists, so far, that human health risks by incidental consumption of GM tree fragments are higher in forest trees than in annual or perennial agricultural crops. Furthermore, allergies to pollen may be higher in forest trees, in comparison with crops, due to the greater production of pollen and long-distance pollen dispersal observed in several tree species, which could cause a larger range of allergy incidence.

Nonetheless, increased pollen production is not one of the traits that are being engineered in GM forest trees. On the contrary, male sterility or the exclusive use of female trees is among the more probable traits for GM forest trees. On the other hand, grasses and agricultural crops are also notable for their rich pollen production and long-distance pollen dispersal.

15.8.7 Effects on the Food Chain

15.8.7.1 Potential Hazards

After consumption of a transgenic forest tree product, the probability of becoming ill depends on whether the consumed piece is antinutritional or has allergy-inducing or toxic properties.

15.8.7.2 Crops versus Trees

In general, except during periods when the ground is completely covered by snow, forest leaves and branches are not consumed by grazing animals. Therefore, the potential effects of transgenic forest trees on grazing cattle are negligible. Nevertheless, some fast-growing trees, such as poplars, willows, plane-trees, and Acacia species, could be capable of dispersing their seedlings or sprouts on pasture lands, making possible the consumption of parts of transgenic forest trees by grazing animals. On the other hand, grazing animals can also ingest waste fruits in orchards or (parts of) fruits in forest lands.

15.8.8 Effects of Cultivation and Management

Today, forest tree engineering is well known. It is commonly used in laboratories and field tests throughout the world. Nevertheless, the commercial use of transgenic forest trees still requires alternative management, compared to that of nontransgenic forest trees. Forest managers and forest harvesters have noticed certain changes in GM forest wood harvesting. For example:

Due to their low lignin content, transgenic tree stems are more easily broken than those of nontransgenic trees.

The application of herbicide-tolerant trees leads to alternative herbicide use, as compared to that of nontolerant trees.

The application of transgenic forest trees modified to kill pest insects leads to a change in insecticide use.

15.8.8.1 Potential Hazards

The ability of transgenic trees to spread may lead to increased monitoring/management of the surrounding vegetation in order to mitigate invasiveness. Additionally, transgenic forest trees which produce chemical substances are likely to lead to more extensive management. However, the release of such substances may harm the surrounding biodiversity.

In transgenic *Bt Pinus nigra* plantations, the incidences of pupa and leaf damage are reduced to the extent that the plantation would be better without chemical protection measures.

The use of pest-resistant GM forest trees could be beneficial for the environment due to a consequent decrease in pesticide use

15.8.8.2 Potential Impact of Cultivation and Management of Transgenic Forest Plantations

Resistance to insects and to harmful microorganisms may increase the commercial attractiveness of certain GM trees.

Herbicide-tolerant forest trees may have negative environmental effects due to the increase in herbicide use, whereas pest-resistant forest trees could have positive impacts on the environment due to the reduced use of pesticides.

A summary of potential environmental risks, as identified by EU directive 2001/18/EC, is given in Table 15.13.

15.9 Conclusion

Plantation forests result from afforestation or reforestation. They were started, in the 1940s, on very limited areas. The plantation effort increased from the mid-1970s onward.

TABLE 15.13

Summary of Potential Environmental Risks, as Identified by the EU Directive 2001/18/EC

Environmental Risk	Crops versus Forest Trees
Persistence and invasiveness	Risks are smaller in trees until reproductive maturity is reached; risks are potentially higher in reproductive trees (e.g., through mast fruiting)
Selective (dis)advantage	No clear difference; possibly smaller risks in trees until reproductive maturity is reached, and potentially higher in reproductive trees
Out-crossing and HGT	No clear difference; varies widely between species and genera. Potential for horizontal gene transfer (HGT)—higher in trees than in crops
Impact on target organisms	Risk potentially higher in trees due to large number of associated organisms
Impact on nontarget organisms	Risks potentially higher in trees due to large number of associated organisms
Effects on human health	No clear difference, but potentially larger risk of pollen allergies in trees (high pollen production and long-distance dispersal since the majority of trees have wind dispersal pollination)
Effects on the feed chain	Risks potentially higher in crops as most animal fodder consists of annual crops such as soy and maize
Effects on biogeochemical cycles	Build-up of litter and soil much lower (negligible) in animal crops
Effects on cultivation and management	No clear difference; potentially higher in trees due to large number of associated organisms (higher incidence of cascading ecological effects), but could also be smaller in trees due to less intensive management

Source: European Commission, Directive 2001/18/EC of European Parliament and of the Council of 12 March 2001 on the deliberate release into the environment of genetically modified organisms and repealing Council Directive 90/220/EEC, *Official Journal of the European Communities*, L106, 101–139, 2001.

Countries that have demonstrated significant success with fast-growing forest plantations have a high probability of gaining additional benefits from genetically engineered (GE) trees.

Today, forest resources and the forest industry are increasingly affected by the large-scale transition in roundwood production from natural forests to forest trees plantations. Field tests using engineered forest trees have been carried out in various countries. Nevertheless, transgenic trees are not yet commercialized, with the exception of insect-resistant poplar in China. The techniques for applying genetic engineering to forest trees are well known and are being used in laboratories and field tests throughout the world.

The major traits that are subject to genetic modification in forest tree breeding encompass resistance to insects, herbicide tolerance, tolerance to fungal, viral, or bacterial pathogens, and to abiotic stresses, flowering modification, wood growth characteristics, and phytoremediation.

At the international level, the main genetic modification activities that have been carried out on forest tree species have been restricted to five genera: *Populus*, *Pinus*, *Eucalyptus*, *Picea*, and *Liquidambar*.

So far, *Populus* remains, by far, the most studied tree genus for modification purposes, but the effective use of GM forest tree species is still increasing. Field trials for transgenic forest trees have mainly been carried out on *Populus*, *Eucalyptus*, and *Pinus*.

The intentional release of engineered forest trees into the environment, either in field trials or commercially, could cause probable risks that need to be assessed as part of the environmental risk assessment.

Forest trees are often seen as the drivers of terrestrial biodiversity because of the high number of organisms associated with them.

TABLE 15.14

Summary of Genetic Modifications in Forest Trees Targeting Lignin Biosynthesis

Gene	Modification	Effects
COMT	Downregulated in *Populus tremula* × *alba*	4-year field trial; no dramatic ecological/biological impacts
	Downregulated in *P. tremula* × *alba*	Decreased lignin content, decreased S:G
	Downregulated in *P. tremula* × *alba*	Increased G, lower pulping efficiency
	Downregulated in *P. tremula* × *alba*	4-year field trial; normal growth
	Downregulated in *P. tremuloides*	No change in lignin content, S:G decreased, more coniferaldehyde
	Downregulated in *P. tremula* × *alba*	No change in lignin content, decreased S, increased G units
	Downregulated in *P. tremula* × *alba*	Lignin contains 5-hydroxyconiferyl alcohol and benzodioxane units
F5H	Overexpressed in *P. tremula* × *alba*	Increased S units
	Overexpressed in *P. tremula* × *alba*	No change in lignin content, increased S:G, decreased kappa
4CL	Downregulated in *P. tremuloides*	Decreased lignin content, no S:G changes
	Downregulated in *P. tremuloides*	Decreased lignin content, S:G increase
	Downregulated in *P. tomentosa*	Decreased lignin content
	Downregulated in *P. tremuloides*	Decreased lignin content, no S:G changes
HCT	Downregulated in *Pinus radiate*	Decreased lignin content, altered monolignol composition
C3H	Downregulated in *P. alba* × *grandidentata*	RNAi down regulation, reduced lignin, increased H units, decreased G units
CCR	Downregulated in *P. tremula* × *alba*	Downregulated in *P. tremula* × *alba*
	Downregulated in *Picea abies*	Downregulated in *Picea abies*
Cald5H	Downregulated in *P. tremuloides*	Decreased lignin content, S:G increased
	Downregulated in *P. tremuloides*	No change in lignin content, increased S:G
CcoAOMT	Downregulated in *P. tremuloides*	Decreased lignin content, slightly increased S:G
	Downregulated in *P. tremula* × *alba*	Decreased lignin content
	Downregulated in *P. tremula* × *alba*	Decreased lignin content
CAD	Downregulated in *P. tremula* × *alba*	Slightly decreased lignin content, increased aldehydes
	Downregulated in *P. tremula* × *alba*	4-year field trial; no dramatic ecological or biological impacts
	Downregulated in *P. tremula* × *alba*	Decreased lignin content, higher free phenolics, easier pulping
	Downregulated in *P. tremula* × *alba*	4-year field trial; normal growth, lower kappa, higher yield, no change in insect interactions
	Downregulated in *Eucalyptus camaldulensis*	No change in lignin content, quality, composition
	Downregulated in *P. taeda*	Reduced lignin, brown wood, lignin contains dihydroconiferyl alcohol, increased aldehydes
Laccase	Downregulated in *P. trichocarpa*	No change to lignin content or composition, deformed xylem, increased phenolics
Peroxidase	Downregulated in *Populus sieboldii* × *grandidentata*	Decreased lignin content, increased S:G ratio

Source: Adapted from FAO, Forests and Genetically Modified Trees, Rome, Italy. Available at http://www.fao.org/docrep/013/i1699e/i1699e00.htm, 2010a.

TABLE 15.15

Species, Genome Properties, and Genomic Resources in Seven Genera of Forest Trees Used in Genomic Research Programs

Family	Genus	Growth	Wood Properties	Biotic Stress	Abiotic Stress
Pinaceae	*Pinus*	Stem growth	Wood density, lignin, and cellulose content	Pathogenic fungi	Cold hardiness, drought tolerance
		Embryogenesis and root formation in loblolly pine	Xylem and secondary cell walls	Pathogenic fungi	Cold acclimation
	Pseudotsuga	Bud phenology	Wood density	N/A*	Cold hardiness
Salicaceae	*Populus*	Stem and leaf growth	Wood density	Pathogenic fungi	Osmotic potential
		Roots, stems, leaves, and during annual growth cycle	Xylem and cambial meristem	Pathogenic fungi	Response to water deficit
Myrtaceae	*Eucalyptus*	Stem growth	Wood density, microfibril angle	Fungal diseases	Frost tolerance
		N/A*	Xylem, wood-forming tissues	Resistance genes	Cold tolerance
Fagaceae	*Quercus*	Bud phenology	N/A*	N/A*	Drought tolerance
				N/A*	Osmotic stress
	Castanea	Bud phenology	N/A*	Pathogenic fungi	Drought tolerance
			N/A*	N/A*	Osmotic stress
			N/A*	Resistance gene	Cold acclimation

Source: Adapted from Neale, D.B. and Kremer, A., *Nature Reviews Genetics*, 12, 111–122, 2011.
Note: N/A* = not currently available.

The main aspects specific to the environmental risk analysis of transgenic trees are:

Potential environmental impacts could increase, or be longer lasting, due to the trees longevity.

Impacts upon large groups of tree-associated organisms which are potentially exposed to engineered trees (insects, birds, litter biota, symbiotic fungi, etc.).

Potential environmental impact as a result of transgenes out-crossing among trees.

Potential impact that engineered forest trees could have on long-term biogeochemical process, such as the decomposition of organic material, due to the long rotation period, and the use of traits that alter the wood composition.

References

Bishop-Hureley, S.L. et al. 2001. Conifer genetic engineering: Transgenic *Pinus radiata* (D. Dom) and *Picea abies* (Karst) plants are resistant to the herbicide Buster. *Plant Cell Reports*, 20(3), 235–243.

Botin, D. 2003. Comments on the environmental effects of transgenic trees. Paper presented to a Conference of Biotechnology in Forestry, July 22–27, 2001. Stevenson, Washington.

Campinhos, E. 1987. Introduction of new techniques for vegetative propagation of Eucalyptus in China. Field document. FAO Technical Cooperation Programme, Fo:TCP/CPR/6653.

Carle, J. and Holmgren, P. 2008. Wood from planted forests: A global outlook 2005–2030. *Forest Products Journal*, 58(12), 6–18.

Carle, J., Vuorinen, P., and Del Lungo, A. 2002. Status and trends in global forest plantation development. *Journal of Forest Products*, 52(7), 12–13.

Chen, C. et al. 2001. Biotechnology in trees: Towards improved paper pulping by lignin engineering. *Euphytica*, 118, 185–195.

Chiang, V.L. 2006. Monolignol biosynthesis and genetic engineering of lignin in trees, a review. *Environmental Chemistry Letters*, 4, 143–146.

DiFazio, S.P. et al. 1999. Assessing potential risks of transgenic escape from fiber plantations. In Lutman, P.W. (ed.), *Geneflow and Agriculture: Relevance for Transgenic Crops*. British Crop Protection Council Symposium Proceedings, 72, 171–176.

Doty, S.L. et al. 2007. Enhanced phytoremediation of volatile environmental pollutants with transgenic trees. *Proceedings of the National Academy of Sciences of the United States of America*, 104, 16816–16821.

Ellatifi, M. 1977. Ecological research in Ajdeer Forest (Middle Atlas, Morocco). PhD thesis in plant biology/forest ecology. Henri Poincaré University, Nancy, France.

Ellatifi, M. 1984. Man and the lack of balance in Moroccan forests. Paper presented at the international symposium on Human Impacts on Forests, 17–22 September, Strasbourg, France.

Ellatifi, M. 1989. The bathroom's consumption of fire wood in Morocco. Invited Oral presentation at the 1st Meeting on Forest Energy, 18–27 January, 1989, ENFI, Rabat, Morocco.

Ellatifi, M. 1991a. Opération reboisement: Réussite ou echec? Communication présentée au XXème Congrès Forestier Mondial, tenu du 17 au 26 septembre 1991, Paris, France, et in: Bulletin de Liaison de des Ingénieurs des Eaux et Forêts, lauréats de l'Ecole Nationale d'Ingénieurs Forestiers (ALENFI), Rabat, Maroc. [Operation reforestation: Success or failure? In: Proceedings of the XXth World Forestry Congress, 17–26 September 1991, Paris, France, and in: The Linkage Bulletin of the ENFI College of Forestry Laureates, Rabat, Morocco.]

Ellatifi, M. 1991b. Reflections on the accounting of the forest plantation success. The case of Morocco. In the *Proceedings of the XXth World Forestry Congress*, 17–26 September, 1991, Paris, France.

Ellatifi, M. 1996. Early growth of forestry species on sand dunes in Yemen. In *Proceedings of the IUFRO Subject Group S4.04, S1.05, S 1.07, P1.10, and S4.11 Conference on Modeling Regeneration Success and Early Growth of Forest Stands*, 10–13 June, Copenhagen, Denmark.

Ellatifi, M. 1998. Fuelwood consumption: A major constraint for the sustainable management of the forest ecosystem in developing countries. The example of Morocco. In *Proceedings of Foresea Miyazaki 1998 International Symposium on Global Concerns for Forest Resource Utilization—Sustainable Use and Management*. Miyazaki, Japan.

Ellatifi, M. 2000. The situation of non-wood forest products in Morocco. In *Proceedings of the Joint FAO/ECE/ILO Committee on Forest Technology, Management and Training Seminar Proceedings on Harvesting of Non-Wood Forest Products*, 2–8 October. Menemen-Izmir, Turkey.

Ellatifi, M. 2001. Bioenergy and sustainable forest management in developing countries. The case of Morocco. In Proceedings of the IEA Bioenergy Task 31 *"Conventional Forestry Systems for Sustainable Production of Bioenergy"*. International Workshop on *"Principles and Practice of Forestry in Densely-Populated Regions,"* held 16–21 September, Veluwe Area, Gelderland, The Netherlands.

Ellatifi, M. 2002. Forests in the biosphere. In *Encyclopedia of Life Support Systems (EOLSS)*. EOLSS Publishers. Available at http://www.eolss.net/.

Ellatifi, M. 2005. Morocco. In Merlo, M. and Croitoru, L. (eds), *Valuing Mediterranean Forests: Towards the Total Economic Value*, pp. 69–87. CABI Publishing.

Ellatifi, M. 2012. Economics of forest and forest products in Morocco: Balance and perspective. PhD Thesis in Economics, Montesquieu University-Bordeaux IV, Bordeaux, France. Available at: http://www.theses.fr/en/2012BOR40007.

Elo, A. et al. 2007. BpMADS5 has a central role in florescence in silver birch (*Betula pendula*). *Physiologia Plantarum*, 131, 149–158.

European Commission. 2001. Directive 2001/18/EC of European Parliament and of the Council of 12 March 2001 on the deliberate release into the environment of genetically modified organisms and repealing Council Directive 90/220/EEC. *Official Journal of the European Communities*, L106, 101–139.

FAO. 2001. *The State of the World's Forests*. Rome, Italy.

FAO. 2004. Preliminary review of biotechnology in forestry, including genetic modification. Genetic Resources Working paper FGR/59E, Forest Resources Division. Rome, Italy. Available at http://www.fao.org/docrep/008/ae574e/ae574e00.htm.

FAO. 2005. Global Forest Assessment. Available at http://www.fao.org/forestry/index.jsp.

FAO. 2006. Global planted forests thematic study, results and analysis. Planted Forests and Trees, Working Paper FP38E. Rome, Italy.

FAO. 2010a. Forests and genetically modified trees. Rome, Italy. Available at http://www.fao.org/docrep/013/i1699e/i1699e00.htm.

FAO. 2010b. Global forest resources assessment 2010: Key findings. Rome, Italy (brochure).

Farnum, P.A. et al. 2007. Ecological and population genetics research imperatives for transgenetic trees. *Tree Genetics and Genomes*, 3, 119–133.

Federal Register. 2007. Available at http://www.fsis.usda.gov/regulations/2007_Notices_Index/index.asp.

Franclet, A. 1956. Premiers travaux d'amélioration génétique des Eucalyptus. *Annales de la Station de Recherche et d'Expérimentation forestière (Rabat, Maroc)*, pp. 65–85.

Franclet, A. 1970. Techniques de bouturage des Eucalyptus camaldulensis—Tunis. Institut national de Recherches forestières, Note technique no 12.

Frankenhuyzen, K. and Beardmore, T. 2004. Current status and environmental impact of 1081 transgenic forest trees. *Canadian Journal of Forest Research*, 34, 1163–1180.

Friedman, S. and Charbley, S. 2004. Environmental and social aspects of the intensive plantation/reserve debate. *Journal of Sustainable Forestry*, 21(4), 59–73.

Fu, D.Q. et al. 2005. Virus-induced gene silencing in tomato fruit. *Plant Journal*, 43, 299–308.

Gleesen, D. et al. 2005. Overproduction of proline in transgenic hybrid larch (*Larix × leptoeuropeae* (Dengler)) cultures renders them tolerant to cold, salt and frost. *Molecular Breeding*, 15, 21–29.

Hu, J.J. et al. 2001. Field evaluation of insect-resistant transgenic *Populous nigra* trees. *Euphytica*, 121, 123–127.

Hu, J.J. et al. 2005. Overexpression of mtlD gene in transgenic *Populus tomentosa* improves salt tolerance through accumulation of mannitol. *Tree Physiology*, 25, 1273–1281.

Irwin, R. and Jones, P.B.C. 2006. Biosafety of transgenic trees in the United States. *Landscapes, Genomics and Transgenics Conifers*, pp. 245–261. Springer, The Netherlands.

ISB. 2009. Information Systems for Biotechnology—A National Resource in Agbiotech Information. Available at http://www.isb.vt.edu/.

Kumar, S. and Fladung, M. 2001. Gene stability in transgenic aspen (*Populus*). Molecular characterization of variable expression of transgene in wild hybrid aspen. *Planta*, 213, 731–740.

Kuparinen, A. et al. 2007. Air-mediated pollen flow from genetically modified to conventional crops. *Ecological Applications*, 17, 431–440.

Kuparinen, A. and Schurr, F.M. 2007. A flexible modeling framework linking the spatio-temporal dynamics of plant genotypes and populations: Application to gene flow from transgenic forests. *Ecological Modelling*, 202, 476–486.

Lachance, D. et al. 2007. Expression of a *Bacillus thurengiensis* cry IAb gene in transgenic budworm (*Choristoneura fumiferana*). *Terre Genetics and Genomes*, 3, 153–167.

Li, J. et al. 2000. A study on the introduction of male sterility of anti-insect transgenic *Populus itigru* by TA29-Bamase gene. *Scientia Sila Sinicae*, 36, 28–32.

Li, J. et al. 2008a. Stability of transgenes in trees: Expression of two reporter genes in poplar over three field seasons. *Tree Physiology*, 29, 299–312.

Li, J. et al. 2008b. Establishment of an *Agrobacterium tumefaciens*-mediated cotyledon disc transformation method for *Jatropha curcas*. *Plant Cell Tissue and Organ Culture*, 92, 173–181.

MacIntosh, S.C. 2009. Managing the risk of insect resistance to transgenic insect control traits: Practical approaches in local environments. *Pest Management Science*, 66, 100–106.

Martin, B. 2003. L'Eucalyptus: Un arbre forestier stratégique. *Revue forestière française*, 2, 141–154.

Meilan, R. 2002. The CP4 transgene provides high levels of tolerance to Roundup (R) herbicide in field-grown hybrid poplars. *Canadian Journal of Forest Research/Revue Canadienne de Recherche Forestière*, 32, 967–976.

Merkle, S.A. 2006. Engineering forest trees with heavy metal resistance genes. *Silvae Genetica*, 55, 2663–2668.

Mullin, T.J. and Bertrand, S. 1998. Environmental release of transgenic trees in Canada. Potential benefits and assessment of biosafety. *The Forestry Chronicle*, 74(2), 203–219.

Neale, D.B. and Kremer, A. 2011. Forest tree genomics: Growing resources and applications. *Nature Reviews Genetics*, 12, 111–122.

Noël, A.C. et al. 2005. Enhanced resistance to fungal pathogens in forest trees by genetic transformation of black spruce and hybrid poplar. *Physiological and Molecular Plant Pathology*, 67, 92–99.

O'Connel, L.M. et al. 2007. Extensive long-distance pollen dispersal in a fragmented landscape maintains genetic diversity in white spruce. *Journal of Heredity*, 98, 640–645.

Pachico, D. 2003. Regulation of transgenic crops: An international comparison. Paper delivered at the International Consortium of Agriculture Biotechnology Research: 7th International Conference on Public Goods and Public Policy for Agricultural Biotechnology. 29 June–3 July, Ravello, Italy. Available at http://www.economia.uniroma2.it/conferenze/icabr2003/.

Packer, A. and Clay, K. 2003. Soil pathogens and *Prunus serotina* and sapling growth near conspecific trees. *Ecology*, 84, 108–119.

Pena, L. and Seguin, A. 2001. Recent advances in the genetic transformation of trees. *Trends in Biotechnology*, 19, 500–506.

Sederoff, R. 2007. Regulatory science in forest biotechnology. *Tree Genetics and Genomes*, 3, 71–74.

Sedjo, R.A. and Botkin, D. 1997. Using forest plantations to spare natural forests. *Environment*, 39(10), 15–20; 30.

Sedjo, R.A. 2004. Genetically engineered trees: Promise and concerns (RFF Report). Resources for the Future, Washington, DC.

Sedjo, R.A. 2005a. Will developing countries be the early adopters of genetically engineered forests? *Journal of Agrobiotechnology Management and Economics*, 8(4), 205–212.

Sedjo, R.A. 2005b. Genetically engineered forests: Financial and economic assessment to the future (draft reports). FAO, Rome.

Sedjo, R.A. 2006. GMO trees: Substantial promise but obstacles to commercialization. *Silvae Genetica*, 55, 241–252.

Sohngen, B. et al. 1999. Forest management, conservation, and global timber markets. *American Journal of Agricultural Economics*, 81(1), 1–13.

Strauss, S.H. et al. 2001. Genetically modified poplars in context. *The Forestry Chronicle*, 77, 271–279.

Strauss, S.H. et al. 2009. Strangled at birth? Forest biotech and the convention on biological diversity. *Nature Biotechnology*, 27, 519–527.

TiaWn, Y.C. et al. 2000. Studies of transgene hybrid poplar 741 carrying two insect-resistant genes. *Acta Botanica Sinica*, 42, 263–268.

U.S. ISB. 2009. Available at http://www.isbweb.org/isb-congresses.

Vanden Broeck, A. et al. 2004. Gene flow between cultivated poplars and native black poplar (*Populus nigra* L.): A case study along the river Meuse on the Dutch–Belgian border. *Forest Ecology and Management*, 197, 307–310.

Vengadesan, G.S. et al. 2006. Transgenic *Acacia sinuata* from *Agrobacterium tumefaciens*: Mediated transformation of hypocotyls. *Plant Cell Reports*, 25, 1174–1180.

Walter, C. and Menzies, M. 2010. Genetic modification as a component of forest biotechnology. In: *The Science of Genetic Modifications in Forest Trees*, pp. 3–17. FAO, Rome, Italy.

Wegrzyn, J.L., et al. 2008. *International Journal of Plant Genomics*, 2008, 412875.

Yang, M.S. et al. 2003. Stability of transgenic hybrid poplad clone 741 carrying two insect resistant genes. *Silvae Genetica*, 52, 5–6.

16

Application of Omics Technologies in Forage Crop Improvement

Suresh Kumar, PhD and Vishnu Bhat, PhD

CONTENTS

16.1 Introduction..523
16.2 Forage Crop Genomics...525
16.3 Transcriptomics in Forage Crops...528
16.4 Proteomics of Forages 529
16.5 Metabolomics Studies in Forages 530
16.6 Applications of Transgenomics in Forages 532
    16.6.1 Deregulation of "Roundup Ready" Alfalfa: A Case Study 534
16.7 Epigenomics of Forage Plants 536
16.8 Scientific Challenges in Forage Crop Omics..539
16.9 Conclusions..540
References..540

16.1 Introduction

Crops and livestock are the two main components of the mixed farming system in agriculture, and they influence agricultural economics and allow sustainability. India has the largest livestock holding in the world, presenting livestock inventories exceeding 529 million heads. Analysis of global trends in animal production indicates that meat and milk consumption will grow at 2.8% and 3.3% per annum, respectively, in developing countries (Anonymous, 2009). With the improving per capita income of the country, the demand for livestock products in India is expected to increase by 3% per annum. This means that we require improved breeds of animals and sufficient high-quality fodder with which to feed them. This also indicates increased pressure on the available land, most of which is currently used for food production. Therefore, future food and forage production will increasingly be affected by competition for natural resources, particularly land and water. Forages are the foundation upon which good dairy nutritional programs are built. The digestibility and intake of forage by livestock directly affect their meat and milk production as well as their rumen function and health. Thus, forages indirectly make a significant contribution to food security by providing the feed requirements of ruminants for meat and milk production (Kumar, 2011). Though information on the availability of forages varies widely, it has been estimated that in India there is a net deficit of 61% in green fodder, 22% in dry fodder, and 64% in feeds for animal production (Anonymous, 2009). Meeting the increasing demands will require increased productivity of the existing fodder crops and the utilization of untapped feed resources and problematic soils for forage production.

Forage crop breeding has conventionally been based on phenotypic selection, followed by the sexual recombination of the natural genetic variations found between and within the ecotypes. Though considerable successes have been achieved in improving crop productivity, the current need is to maximize the productivity of food and fodder crops despite the decreasing availability of land for agricultural production. More importantly, due to increasing demand for food grains, forage crops are being restricted to the leftover land and problematic soils not suitable for growing food crops. The increasing demands of feed and fodder, and the limited area available for forage cultivation have drawn the attention of scientists aiming to develop high-yielding varieties of forage crops with improved fodder quality and tolerance to any stress conditions they are likely to encounter. However, the limited genetic diversity, narrow genetic base, and complex breeding systems constrain the improvement of forage crops. Because of these factors, conventional breeding methods have had limited success in developing improved varieties of forage crops. However, technological developments in the areas of molecular biology and biotechnology have empowered scientists to tailor-make a plant that meets our requirements. These developments have created unprecedented opportunities, not only for the genetic manipulation of biological systems, but also for undertaking studies to understand the fundamental processes of life (Kumar, 2008). It is only in the past decade that major efforts have been made in the field of omics to unravel the underlying causes of the responses of forage species to environmental stimuli and to develop tools for selection in breeding programs (Barth, 2012). The rapidly developing omics technologies are making remarkable headway in forage crop improvement.

German botanist Professor Hans Winkler first coined the word genome in 1920 by blending two words: "gene" and "ome" (the latter from chromosome) to describe the complete set of an organism's genes. The suffix -ome is a neologism referring to a broad field of study in biology, implying completeness of the field. The word genomics was invented in the late 1980s when researchers were planning the Human Genome Project to map the 3.3 billion base pairs of DNA that make up the human genome. Omics informally refers to the fullness and completeness of the field of study. For example, functional genomics aims at identifying the functions of as many genes of an organism as possible. To accomplish this, different omics techniques such as transcriptomics and proteomics with saturated mutant collections are combined (Holtorf et al., 2002). With the advent of sophisticated molecular biology techniques, omics technologies have generated huge amounts of data in different fields of biology from gene sequencing and protein expression to metabolite signatures in different organs.

Omics research in forages has lagged behind that of major food/cash crops. Forages are widely grown, but are probably the least appreciated commodity. When considered in terms of their direct and indirect benefits, it becomes obvious that research and development in this agricultural sector have largely been neglected (Kumar, 2011). The cash value of forages is realized through animals and animal products, thus the public may not realize the direct connection between forage production and the diversity of forage-dependent commodities (meat, milk, wool, etc.). Since livestock productivity largely depends on their forage utilization, the value of forages may be estimated by using the feed costs associated with livestock production. Based on this model, the calculated value of forages far exceeds the cash value of any other crop in the United States (Wang, 2003). Due to the complexity of forage species and the associated difficulties in traditional plant breeding methods, the potential of omics technologies for the development of improved forage cultivars has been recognized.

The omics studies may be broadly categorized into four kinds: genomics, transcriptomics, proteomics, and metabolomics. While genomics is the study of the whole genome of an organism, transcriptomics, proteomics, and metabolomics are the studies of the functional or applied aspects of the genome. Genomics may be further divided into structural genomics, the structure of every gene/protein encoded by the genome; functional genomics, gene and protein functions and interactions; cognitive genomics, the changes in cognitive processes associated with genetic profiles; comparative genomics, the relationship of genome structure and function across different biological species; metagenomics, metagenomes, genetic material recovered directly from environmental samples; and epigenomics, the complete set of epigenetic modifications of the genetic material of a cell (epigenome).

16.2 Forage Crop Genomics

Traditional plant breeding has been based on phenotypic selection and subsequent progeny testing followed by reselection, which used to be a slow process. In order to speed up and increase the precision of the process, genetic markers for the identification and marker-assisted selection (MAS) of quantitative trait loci (QTLs) are being developed (Zhang et al., 2006; Mouradov, 2010). Genomics research is generating new tools, such as functional molecular markers, and knowledge about inheritance patterns that could increase the efficiency and precision of crop improvement (Zhang and Mian, 2003). Over the past two decades, major advances have been made in the genomics of model plants in the areas of structural, comparative, and functional studies. Genomics research achieved greater significance after the success of the Human Genome Project in completely sequencing the genomic DNA of human beings. Soon, plant species followed suit, and completion of the genome sequence of a model plant, *Arabidopsis thaliana*, in 2000 was followed by developments in high-throughput omics technologies which stimulated the sequencing of crop genomes (Mukhopadhyay et al., 2012).

Rice genome sequencing was initiated as a reference species for grasses, and an annual relative of the cultivated alfalfa, *Medicago truncatula* (commonly known as barrel medic), was used as the reference species for legumes. *M. truncatula* has number of features that make it an excellent model system for legumes. Being a member of the Papilionoideae subfamily, *M. truncatula* is closely related to the majority of crops and pasture legumes. Its nearest cousin, alfalfa (*M. sativa*), is the most important forage legume (Choi et al., 2004b). *M. truncatula* has a relatively small, diploid genome (haploid size < 500 Mbp), making it useful for both genetics and genomics studies (Young et al., 2005). Importantly, it is self-fertile and produces a large number of seeds on a plant of relatively small stature, making it amenable to high-density cultivation (Barker et al., 1990). Finally, it is amenable to genetic transformation, which makes it suitable for reverse-genetics experiments (Chabaud et al., 2003; Crane et al., 2006). Since the adoption of *M. truncatula* as a model species, a number of useful tools and resources have been developed, including high-density genetic and physical maps (Thoquet et al., 2002; Choi et al., 2004a), different kinds of mutant populations (Tadege et al., 2008), and protocols for transcriptome, proteome, and metabolome analysis (Gallardo et al., 2003; Watson et al., 2003; Broeckling et al., 2005). The most recent development in the functional genomics toolkit for legumes is the commercialization of Affymetrix GeneChip, which contains probe sets for the majority of genes from

M. truncatula as well as probe sets for all the genes of its symbiotic, nitrogen-fixing partner *S. meliloti.*

The genome sequence data, combined with high-throughput machinery and data analysis, have helped to reveal the organization, evolution, synergic relationships, and gene functions of plant genomes. The genomic data and high performance data analyses using bioinformatics tools have paved the way for a determination of gene expression, their functions, and the subsequent utilization of integrative omics approaches toward metabolic engineering. This may lead to the deployment of new and innovative methods for improving cultivated forages.

Traditional genetic research involved a "one gene at a time" approach to the study of gene functions. Though such research led to breakthroughs in genetics, many plant species have a wide range of genome size (100 to 100,000 Mbp) with large repeated sequences present in the genomes. High-density genetic linkage maps are important tools for QTL fine-mapping, map-based cloning of desirable genes, comparative genome analysis, and integration of genetic and physical maps. Genetic linkage maps for some forage species are now available and have proved valuable for QTL mapping. Different molecular markers have helped in increasing the density of genetic linkage maps. Single nucleotide polymorphism (SNP) markers have attracted much interest, mainly because of their abundance (Rafalski, 2002) and occurrence at regular intervals in the genome (Ponting et al., 2007). SNPs are highly suitable for multiplexed genotyping assays on mass spectrometry, microarray or beadarray-based platforms, and high-throughput data analysis (Gupta et al., 2008). In forage species where a complete genome sequence has not been established, strategies for large-scale SNP discovery are being adopted. Despite the labor-intensive nature of amplicon cloning and sequencing, it has been the method of choice for SNP discovery in ryegrasses (Cogan et al., 2006). The number and quality of publicly available EST data for forage species are often limited. The recent advances in NGS opened up the opportunity for whole genome sequencing as an extremely powerful strategy for *in silico* SNP discovery at appropriate sequence coverage. However, *de novo* assembly of short NGS reads is difficult in many of the outbreeding forage and turf species with highly heterozygous, large, and complex genomes containing a high degree of repetitive elements. Recently, Studer et al. (2012) presented a high-density genetic linkage map of ryegrass, locating candidate genes for agronomically important traits and thus providing a starting point for QTL fine mapping, LD-based association mapping, and map-based cloning (Figure 16.1).

Genes are not situated in isolation, and gene interactions determine all aspects of biology in an organism. Thus, functional genomics helps in understanding how the genes of an organism work together by assigning new functions to unknown genes. This approach requires systematic analyses of gene expression, protein synthesis/degradation, and changes in the metabolism of plants. Comprehensive genomics studies provide us with large amounts of information on genes that regulate traits of agricultural importance, such as yield, stress tolerance, and disease resistance.

Functional genomics of forage and turf lag far behind other major crop species. The establishment of synteny among grass genomes may help in the molecular analysis and manipulation of grass genomes (Wang et al., 2001a). Unfortunately, many of the basic tools and techniques required for the forage and turf genomics revolution do not exist or are incomplete. Filling these critical gaps in forage genomics will allow geneticists and breeders to take advantage of the existing genetic variation, exploit advances in the genomics of other grasses, and make genomics information useful for the improvement of forage species.

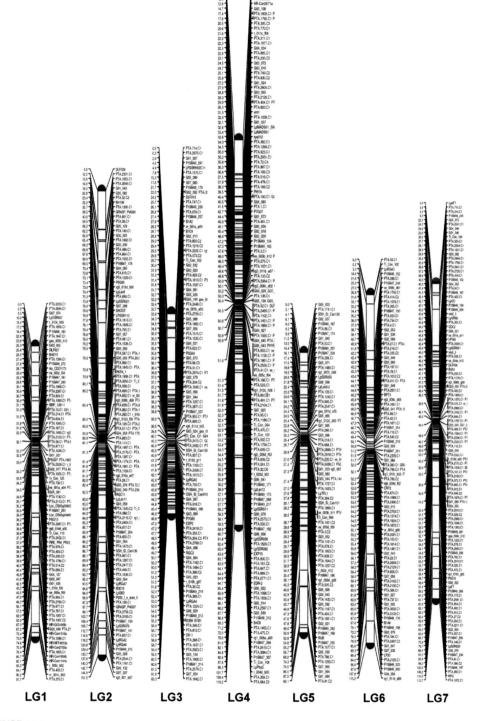

FIGURE 16.1
Transcriptome map of perennial ryegrass (*Lolium perenne* L.) showing the seven linkage groups (LG 1–7). (Redrawn from Studer, B., et al., *BMC Genomics*, 13, 140, 2012.)

16.3 Transcriptomics in Forage Crops

Transcriptomics constitutes a step forward in (functional) genomics. A transcriptome is the set of all RNA molecules, including mRNA, rRNA, tRNA, and other noncoding RNAs, produced in one or a population of cells. The transcribed portions of plant genomes can be studied by randomly selecting and sequencing a large set of cDNA clones and putting together the sequence fragments of the expressed genes. Rice has been used as a model crop plant because of its small genome size (400 Mb). While the first expressed sequence tag (EST) data were submitted to GenBank in 1998 (Frugier et al., 1998), construction of an EST library for a forage species *M. sativa* was initially reported by Hays and Skinner (2001). There are only 19,968 EST sequences available in the public database for alfalfa, as compared to 1,279,343 ESTs for a well-studied species, *Oryza sativa* (http://www.ncbi.nlm. nih.gov; as on 30 March 2013).

In order to obtain insight into the transcribed portions of forage plant genomes, a number of large-scale transcriptome sequencing projects have been successfully launched. For example, a large-scale tall fescue (*Festuca arundinacea* Schreb, an important forage species) EST project was initiated at the Samuel Roberts Noble Foundation in 2000 with the objective of cataloging the majority of the expressed genes of tall fescue, including tissue-specific, developmental stage-specific, and stress-specific genes. This EST database would be useful in gene expression and regulation studies in tall fescue as well as the related *Festuca-Lolium* species. So far, approximately 65,252 ESTs have been submitted to the GenBank dbEST database. EST data from different grass genomes are constantly being compared and refined due to their high conservation. Compared with food crop species, forage species have smaller numbers of ESTs available in public databases. Table 16.1 enlists the number of ESTs in public databases for species commonly used as forage and turf.

In the case of complicated plant species (forages and turf), transcriptome sequencing is a better strategy for complexity reduction where the expressed genes are targeted and highly repetitive nontranscribed genomic regions are excluded (Barbazuk et al., 2007; Trick et al., 2009). This also helps in the high-throughput acquisition of gene-associated SNPs (Barbazuk and Schnable, 2011; Milano et al., 2011). Based on the EST-derived SNPs, Studer et al. (2012) mapped 495 ryegrass unigenes in the VrnA mapping population. The total map length of 750 cM contains 838 DNA markers, with 767 EST-derived SSRs, SNPs, or CAPS markers and an average marker distance of less than 0.9 cM (Figure 16.1). The map provides new anchor points for detailed studies of comparative grass genomics that may prove useful for future ordering and orientation of scaffolds into pseudomolecules during the assembly of a ryegrass reference genome (Studer et al., 2012).

A recent development in the functional genomics of forage legumes has been the commercialization of Affymetrix GeneChip, which contains probe sets for the majority of genes from *M. truncatula* (Tesfaye et al., 2006). Yang et al. (2008) demonstrated the potential of using *M. truncatula* genes for the genetic improvement of alfalfa. Yang et al. (2009) used Medicago Affymetrix GeneChip (Affymetrix, Santa Clara, CA) for a cross-species platform for detecting single-feature polymorphism (SFP) in tetraploid alfalfa. SFP using GeneChip, originally developed for gene expression analysis, is a rapid and cost-effective approach for genome-wide polymorphism discovery, particularly for tetraploid alfalfa. Transcriptome analysis presents a powerful approach for evaluation of the dynamics of

TABLE 16.1

Number of ESTs from Forage and Turf Species in
GenBank dbEST Database (30 March 2013)

Scientific Name	Common Name	ESTs
Medicago truncatula	Barrel medic	286,175
Medicago sativa	Alfalfa	19,968
Sorghum bicolor	Sorghum	210.892
Sorghum halepense	Johnson grass	1,965
Cenchrus ciliaris	Buffelgrass	21,733
Pennisetum glaucum	Bajra	1,381
Cynodon dactylon	Bermuda grass	20,826
Poa pratensis	Kentucky bluegrass	25
Poa secunda	Big bluegrass	36
Festuca arundinacea	Tall fescue	65,252
Tripsacum dactyloides	Eastern gamagrass	7
Panicum virgatum	Switchgrass	20,590
Panicum maximum	Guinea grass	196
Lolium perenne	Ryegrass	21,930
Trifolium repens	White clover	53,732
Trifolium pratense	Red clover	38,353

stress response in forages and the formulation of hypotheses for mechanisms of enhanced
resistance to biotic and abiotic stresses.

16.4 Proteomics of Forages

Proteome means the entire complement of proteins, including the modifications made to
a particular set of proteins, produced by an organism or system. Proteomics is the study
of a set of proteins expressed at a given time or under certain environmental conditions
(Wilkins et al., 1997), and can be further divided into three main areas:

1. Large-scale identification of proteins
2. Identification of proteins' response to biological variations
3. Studies of protein–protein interactions (Holtorf et al., 2002)

For example, a proteome reference map for *M. truncatula* root proteins was established
using two-dimensional gel electrophoresis combined with peptide mass fingerprinting
to aid the dissection of nodulation and root developmental pathways (Mathesius et al.,
2001). Subsequently, a proteomic approach was used to investigate seed development in *M.
truncatula*. Gallardo et al. (2003) investigated seed development in *M. truncatula* at specific
stages of seed filling corresponding to the acquisition of germination capacity and protein
deposition. Li et al. (2009) used Affymetrix *M. truncatula* GeneChip for comparative gene
expression analysis between heterotic and nonheterotic hybrids of tetraploid alfalfa and

found that alfalfa hybridized to approximately 47% of the *M. truncatula* probe sets. Due to a high degree of conservation of gene and synteny between alfalfa and *M. truncatula*, it is expected that the proteomics studies conducted in *M. truncatula* could equally be used for alfalfa improvement.

Heat stress has been found to induce heat-shock proteins (HSPs) in plants (Vierling, 1991). Park et al. (1996) first detected HSPs in heat-tolerant and nontolerant variants of creeping bentgrass, a major cool-season turf species. Heat stress leads to the accumulation of several HSPs. He et al. (2005) found that upregulation of HSPs is a typical response of perennial grasses to heat stress. They also found that heat acclimation improves the heat tolerance of plants due to lower electrolyte leakage in the leaves of heat-acclimated plants, and the induction of two HSP60 proteins during heat acclimation was correlated with enhanced thermotolerance in perennial forage and turf grasses. Now evidence is available on the association of early induction and persistent maintenance of HSPs under elevated temperature/heat stress in grasses. Manipulating the genes controlling HSP production may be beneficial for breeding heat-tolerant grass. Similarly, cold acclimation improves freezing tolerance in plants, including perennial grasses. Espevig et al. (2012) studied protein changes in crowns of velvet bentgrass (*Agrostis canina*) during cold acclimation in association with freezing tolerance. Proteins upregulated after cold acclimation included methionine synthase, serine hydroxymethyltransferase, aconitase, UDP-D-glucuronate decarboxylase, and putative glycine-rich protein. Cold acclimation-responsive proteins involved in amino acid metabolism, energy production, stress defense, and secondary metabolism could contribute to the improved freezing tolerance induced by cold acclimation in velvet bentgrass (Espevig et al., 2012).

16.5 Metabolomics Studies in Forages

Metabolomics studies the dynamic multiparametric metabolic response of a living organism to pathophysiological stimuli or genetic modification. Thus, metabolomics offers opportunities in the phenotyping and characterization of plants (Rasmussen et al., 2012). While other omics technologies such as genomics, transcriptomics, and proteomics analyze only one type of biochemical compound (i.e., DNA, RNA, or proteins) and require only a small number of analytical platforms, metabolomics deals with a multitude of biochemicals and physicochemical properties. In order to provide a holistic view of the metabolites, metabolomics utilizes a combination of several extraction, separation, and detection techniques. The basic instruments required include mass spectrometry (MS), gas chromatography (GC), liquid chromatography (LC), and nuclear magnetic resonance (NMR). All of these techniques produce data that require extensive computational processing and mining in order to interpret the results and deliver meaningful information which can be utilized in plant breeding experiments.

The majority of metabolomics studies carried out so far have been focused on model plant species by academic institutions for basic understanding, or cereal crop and vegetable species with higher commercial values in the breeding industry (Kliebenstein, 2009; Fernie and Klee, 2011). Metabolomics studies in tomato identified QTL for flavored volatile emissions (Schauer et al., 2006; Tieman et al., 2006) and flavonols in poplar

(Morreel et al., 2006). Successful applications of metabolomics studies in potato, tomato, rice, and other fruits and cereals have been recently reviewed by Stewart et al. (2011). Though metabolomics is still very expensive and technologically demanding, it is considered to be one of the important omics technologies of the future in plant improvement research. Certainly, it could play a critical role in the improvement of forage and turf species.

Grasses with higher sugar content in their leaves have been bred to increase energy supply to the rumen microorganisms, decrease nitrogen losses from rumen protein degradation, and increase nitrogen availability, thereby reducing nitrogen losses in urine and nitrous oxide emissions from the fodder (Edwards et al., 2007; Parsons et al., 2011). An example of metabolic profiling has been the identification of QTL related to high sugar content in the leaves of *L. perenne* (Turner et al., 2006). Development of LC–MS-based methods for the rapid analysis of polymeric fructans up to 100 degrees of polymerization (DP) revealed that high-sugar grasses mainly differed in large polymeric fructans (Harrison et al., 2009, 2011, 2012). Generally, the high-DP fructans are difficult to separate and quantify using the analytical methods commonly used for fructan analysis; however, MS analysis may help reveal additional and larger QTL for high sugar in grasses. Another promising area for metabolomics studies in forage breeding is the development of MS-based analysis of grass-fiber composition. Low digestibility of grass cell walls negatively affects feed intake and fiber composition, and has been the target of forage breeding programs (Casler et al., 2008). Current extraction and detection methods for fibers are difficult, time-consuming, and costly, which has hampered the development of forages with improved digestibility. Recently, a mild extraction method with subsequent chromatographic separation of larger complexes of lignin, polysaccharides, and hydroxycinnamic acids and detection by MS has been described (Faville et al., 2010). A recent study on seasonal changes in the metabolic composition of 21 cultivars of grasses and legumes using NMR revealed a high correlation of sugar concentrations, fibers, and *in vitro* organic matter digestibility with the NMR fingerprints, mainly due to the differences in spectral intensities of malic acid, choline, and glucose (Bertram et al., 2010). These studies highlight the fact that a metabolomics-based approach enables high-throughput detection and selection of forage genotypes with improved fodder quality and can help overcome some of the limitations of forage breeding programs. This clearly demonstrates that metabolomics would be powerful in detecting unknown metabolic consequences of forage breeding programs.

Metabolic engineering allows the incorporation and controlled expression of genes to produce a new biochemical phenotype that may not exist in the available genetic diversity of the species (Dixon and Steele, 1999). Isoflavonoids, flavonols, anthocyanins, and condensed tannins are derived from phenylpropanoids and share common precursors (Winkel-Shirley, 2001; Figure 16.2). Franzmayr et al. (2012) demonstrated successful application of metabolomics in forages for the expression of a white clover isoflavone synthase in tobacco. They demonstrated that the gene *IFS2_12* isolated from *T. repens* encodes an isoflavone synthase and paves the way for engineering white clover plants with higher levels of isoflavonoids than are naturally found in this species. White clover plants accumulating high levels of isoflavonoids would be useful in pastoral animal production systems as these compounds can protect clover plants from insect attack. Conventional breeding of white clover with high levels of isoflavonoids has not been successful in the past, and metabolic engineering with known pathway genes might be an alternative strategy.

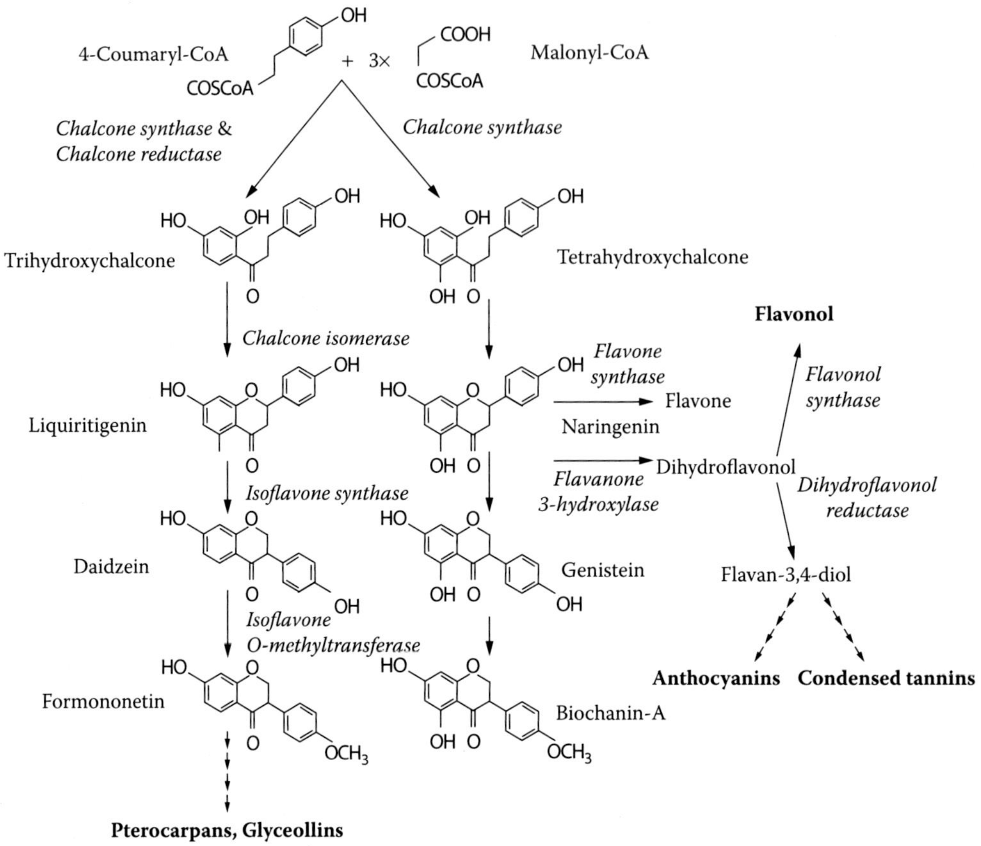

FIGURE 16.2
Isoflavonoid biosynthetic pathway in legumes showing parallel processing of 5-reduced chalcones (*left side*) and 5-hydroxylated (*right side*) chalcones and isoflavones. Naringenin is the key branch point for the production of isoflavonoids or other flavonoids such as flavonols, anthocyanins, and condensed tannins. Multiple arrows indicate multienzyme/multistep reactions. (Redrawn from Franzmayr, B.K., Rasmussen, S., Fraser, K.M., and Jameson, P.E., *Ann. Bot.*, 110, 1291, 2012.)

16.6 Applications of Transgenomics in Forages

The transfer of alien, specific, and useful genes into a plant of choice followed by their inheritance and expression profiling is referred to as transgenomics. Scientists using this approach have already shown success in introducing genes which make many important crops resistant to insects, viruses, and herbicides. It has also been very useful in basic research into understanding physiological and biochemical pathways in plants. Once a gene of interest is identified, it is isolated and sequenced. After determining the protein coded by it and its function, the gene is incorporated into a plant variety that has a proven agronomic performance and standardized protocols for tissue culture and transformation. Performance of independent transgenic lines is rigorously tested and evaluated in the laboratory and greenhouse, as well as under field conditions, for several generations. Further exhaustive tests for yield and overall performance, nutritional value, environmental effects/allergenicity (biosafety), and other specific qualities, if any, are needed before the release of a newly developed transgenic cultivar (Kumar, 2011). The transgenomics technology, as developed in recent decades, has been successfully applied for the improvement of several crop species.

The science of transgenics came of age in 1996 with the first-ever large-scale commercial cultivation of transgenic crops. This milestone was achieved after years of intensive work devoted to the development of reliable systems for plant regeneration from cultured cells and methods for introduction and stable integration of foreign genes (transformation) into cultured cells. Transgenic soybean, corn, cotton, and canola with insect resistance or herbicide tolerance have been adopted by U.S. farmers. The estimated global area of transgenic crops for the year 2012 was about 170 million hectares (James, 2012; Figure 16.3).

Transgenomics allows the introduction of foreign genes from unrelated species and the downregulation or upregulation of endogenous genes, hence it has also been used for the improvement of forages and turf species. Genetic transformation of forage alfalfa was first accomplished in the mid 1980s (Deak et al., 1986; Shahin et al., 1986). Since the production of the first transgenic forage-type tall fescue plant (Wang et al., 1992), tremendous progress has been made in the area of forages and turf transgenomics during the last two decades (Kumar, 2011; Kumar et al., 2012). Input traits to improve agronomic performance and output traits to improve forage quality or to produce novel industrial/pharmaceutical proteins are the focus of current transgenomics research in alfalfa. The first generation GM traits aimed at incorporating a transgene for protein that is directly responsible for a value-added trait. Transgenic alfalfa producing significant quantities of a feed enzyme phytase was developed by Austin-Phillips and Ziegelhoffer (2001). Thus, some of the

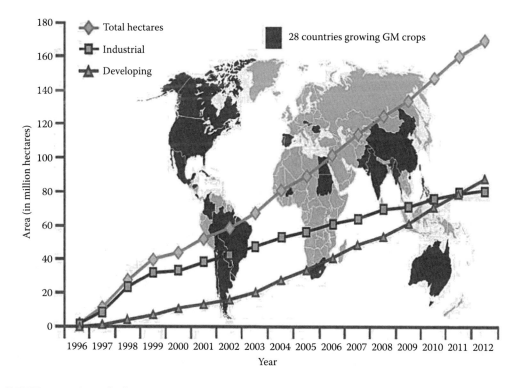

FIGURE 16.3 (See color insert)
Global spread of GM crops, 1996–2012. (From James, C., Global Status of Commercialized Biotech/GM Crops: 2012. ISAAA Brief No. 44. ISAAA, Ithaca, NY.)

high-biomass-producing forages may become a biological factory for the production of high value transgenic proteins (Khoudi et al., 1999).

Now the emphasis is on second generation GM plants in order to explore the possibilities of genetic engineering for metabolic pathways. This has been tested in several model systems, including alfalfa. Mesfin et al. (2001) reported that GM alfalfa for malate dehydrogenase (MDH) showed increased tolerance to acidity and high Al content in soil. The Nobel Foundation Plant Biology Group used alfalfa for transgenic knockouts as a tool to better understand the lignin biosynthesis pathway. Some of the achievements in the genetic engineering of forage grasses and legumes are listed in Table 16.2. The transgenomics approach has been employed to improve *in vitro* dry matter digestibility in alfalfa, tall fescue, and perennial ryegrass (Guo et al., 2001; Chen et al., 2003, 2004; Reddy et al., 2005; Tu et al., 2010). Transgenic alfalfa with downregulated comethyl transferasecaffeic acid 3-O-methyltransferase (COMT) and caffeoyl CoA 3-O-methyltransferase (CCOMT) showed altered lignin content and composition with improved fiber digestibility (Guo et al., 2001). Other success stories include enhanced drought tolerance in alfalfa, white clover, creeping bentgrass, and bahiagrass (Zhang et al., 2005, 2007b; Fu et al., 2007; Jiang et al., 2009, 2010; Xiong et al., 2010); increased phosphorus acquisition in white clover and alfalfa (Ma et al., 2009, 2012); enhanced salt tolerance, cold tolerance, or freezing tolerance in perennial ryegrass, tall fescue, and creeping bentgrass (Hisano et al., 2004; Hu et al., 2005; Wu et al., 2005; Li et al., 2010); delayed flowering in red fescue (Jensen et al., 2004); development of hypoallergenic perennial and Italian ryegrasses (Petrovska et al., 2004); enhanced aluminum tolerance in alfalfa (Tesfaye et al., 2001; Barone et al., 2008); delayed leaf senescence in alfalfa (Calderini et al., 2007; Zhou et al., 2011a); virus resistance in perennial ryegrass and white clover (Xu et al., 2001; Ludlow et al., 2009); increased disease resistance in tall fescue and creeping bentgrass (Fu et al., 2005; Dong et al., 2007, 2008; Zhou et al., 2011b); improved turf quality in bahiagrass (Agharkar et al., 2007; Zhang et al., 2007a); sulfur-rich protein accumulation in subterranean clover and tall fescue (Rafiqul et al., 1996; Wang et al., 2001b); production of polyhydroxybutyrate in switchgrass (Somleva et al., 2008); increased sugar release in alfalfa and switchgrass (Chen and Dixon, 2007; Jackson et al., 2008; Fu et al., 2011a,b; Saathoff et al., 2011); and increased biomass in switchgrass (Fu et al., 2012).

Although 170 million hectares was occupied by GM crops in 2012, about 99% of that area was occupied by GM maize, soybean, cotton, and canola. Besides these four major cash crops, only a few other species have been deregulated, including papaya, squash, rice, potato, and alfalfa. Herbicide resistance has consistently been the dominant trait, followed by insect resistance and virus resistance. In contrast to the large number of reports on transgenomics studies, very limited number of GM crops have been deregulated and only a few traits have been successfully used in commercial GM crops. In fact, the deregulation process has become so complicated and costly that only large multinational companies can afford to do it. The costs of meeting the regulatory requirements and market restrictions guided by the regulatory criteria are substantial impediments to the commercialization of GM crops (Bradford et al., 2005).

16.6.1 Deregulation of "Roundup Ready" Alfalfa: A Case Study

Although considerable efforts have been made toward the genetic engineering of forage and turf species, to date the only deregulated forage crop is "Roundup Ready" alfalfa. In early 2004, the U.S. Department of Agriculture's Animal and Plant Health Inspection Service (USDA-APHIS) received a petition from Monsanto and Forage Genetics International

TABLE 16.2

Application of the Transgenomics Approach in Forage and Turf Species Improvement

Scientific Name	Common Name	Gene	Trait	Reference
Medicago sativa	Alfalfa	Comethyl transferase, caffeic acid 3-O-methyltransferase (COMT) and caffeoyl CoA 3-O-methyltransferase (CCOMT), cytochrome P450	Improved *in vitro* dry matter digestibility	Guo et al., 2001; Reddy et al., 2005
Festuca arundinacea	Tall fescue	Cinnamyl alcohol dehydrogenase, COMT		Chen et al., 2003, 2004
Lolium perenne	Perennial ryegrass	COMT, cinnamoyl-CoA-reductase		Tu et al., 2010
Medicago sativa	Alfalfa	WXP1, WXP2	Enhanced drought tolerance	Zhang et al., 2005, 2007a
Trifolium repens	White clover	WXP1		Jiang et al., 2010
Agrostis stolonifera	Creeping bentgrass	hva1		Fu et al., 2007
Paspalum notatum	Bahiagrass	HvWRKY38		Xiong et al., 2010
Trifolium repens	White clover	Phytase and acid phosphatase genes	Increased phosphorus acquisition	Ma et al., 2009,
Medicago sativa	Alfalfa	Phytase and Acid phosphatase genes		Ma et al., 2012
Lolium perenne	Perennial ryegrass	Wheat fructosyltransferase genes	Freezing tolerance	Hisano et al., 2004; Wu et al., 2005
		Na+/H + antiporter gene	Enhanced salt tolerance	
Festuca arundinacea	Tall fescue	ipt gene	Cold tolerance or freezing tolerance	Hu et al., 2005
Agrostis stolonifera	Creeping bentgrass	*Arabidopsis* H+-pyrophosphatase	Enhanced salt tolerance	Li et al., 2010
Festuca rubra	Red fescue	Flowering repressor gene from Lolium	Inhibition of flowering	Jensen et al., 2004
Medicago sativa	Alfalfa	Malate dehydrogenase, bacterial citrate synthase gene	Aluminum tolerance	Tesfaye et al., 2001; Barone et al., 2008
Medicago sativa	Alfalfa	ipt gene, STAY-GREEN gene	Delayed leaf senescence	Calderini et al., 2007; C. Zhou et al., 2011
Lolium perenne	Perennial ryegrass	Posttranscriptional gene silencing	Virus resistance	Xu et al., 2001
Trifolium repens	White clover			Ludlow et al., 2009
Festuca arundinacea	Tall fescue	Bacteriophage T4 lysozyme gene	Disease resistance	Dong et al., 2007, 2008
Agrostis stolonifera	Creeping bentgrass	Rice TLPD34, penaeidin4–1		Fu et al., 2005; M. Zhou et al., 2011
Trifolium subterraneum	Subterranean clover	Seed albumin gene from sunflower	Sulfur-rich protein accumulation	Rafiqul et al., 1996
Festuca arundinacea	Tall fescue			Wang et al., 2001b
Panicum virgatum	Switchgrass	miR156	Increased biomass	Fu et al., 2012

requesting a determination of nonregulated status for alfalfa lines tolerant to the herbicide glyphosate. This GM alfalfa was obtained by transgenic expression of the 5-enolpyruvyl-shikimate-3-phosphate synthase (*cp4-epsps*) gene from bacterium. The nonselective herbicide glyphosate inhibits an essential step in the synthesis of aromatic amine in plants by blocking the action of the natural EPSPS enzymes in plant. However, the bacterial EPSPS protein is not inhibited by glyphosate, and thus any plant expressing sufficient levels of this protein becomes tolerant to glyphosate. After assessment of the plant pest risks posed by the use of transgenic alfalfa, APHIS prepared an environmental assessment report and, effective from 14 June 2005, APHIS deregulated RR alfalfa.

But, after just nine months, a group of organic alfalfa growers and the Center for Food Safety filed a lawsuit in the Northern District of California challenging APHIS's decision to grant nonregulated status to RR alfalfa. In February 2007, the court found that APHIS's Environmental Assessment failed to adequately consider certain environmental and economic impacts, and the court overruled APHIS's decision to grant nonregulated status to RR alfalfa. The court also ordered the USDA to prepare an Environmental Impact Statement (EIS) on RR alfalfa. Though the existing RR alfalfa crops were allowed to be harvested, used, and sold, new planting of RR alfalfa was banned. When the RR alfalfa case was brought to the US Supreme Court, in June 2010, the Supreme Court issued a ruling on this matter in favor of Monsanto, saying that the district court had overreached itself procedurally in halting the plantings. The ruling allowed the USDA-APHIS to take appropriate action to allow further planting while they completed the EIS. APHIS produced the final EIS on 16 December 2010, stating that RR alfalfa is safe for food and feed purposes and is unlikely to pose plant pest risks. On 27 January 2011, the USDA announced full deregulation of RR alfalfa without any restrictions. Thus, after a four-year court-imposed ban, US farmers can again grow GM alfalfa.

Most of the important forages and turf species are cross-pollinated and perennial in nature. These can readily cross with wild or feral relatives. Some species adapt well to marginal lands and possess invasiveness. These are important issues of biosafety concerns of the regulatory agencies, and hence require additional scrutiny for these perennial forage grasses and legumes (Strauss et al., 2010). Thus, the major challenge now is how to apply GM technology to improve the productivity and quality of food/feed crops in a way that satisfies the regulatory requirements. Of course, the development of EIS for alfalfa and the deregulation of herbicide-tolerant alfalfa paved the way for future transgenic improvement of this important forage legume, as well as for other crops. Wang and Brummer (2012) recently presented an overview of these issues and discussed whether transgenomics is an option for the development of improved forage, turf, and bioenergy plants.

16.7 Epigenomics of Forage Plants

Epigenomics is the study of the complete set of epigenetic modifications on the genetic material of a cell. Epigenetic modifications on DNA and histones that affect gene expression and regulation may be reversible. Nucleic acids (DNA and RNA) are composed of repeating units of nucleotides, and each nucleotide is composed of a phosphate group, a five-carbon pentose sugar, and a cyclic nitrogen-containing base of either adenine (A), cytosine (C), guanine (G), or thymine (T) in DNA, while T is replaced by Uracil (U) in RNA (Figure 16.4). DNA may also contain 5-methylcytosine (5-mC), N^4-methylacytosine,

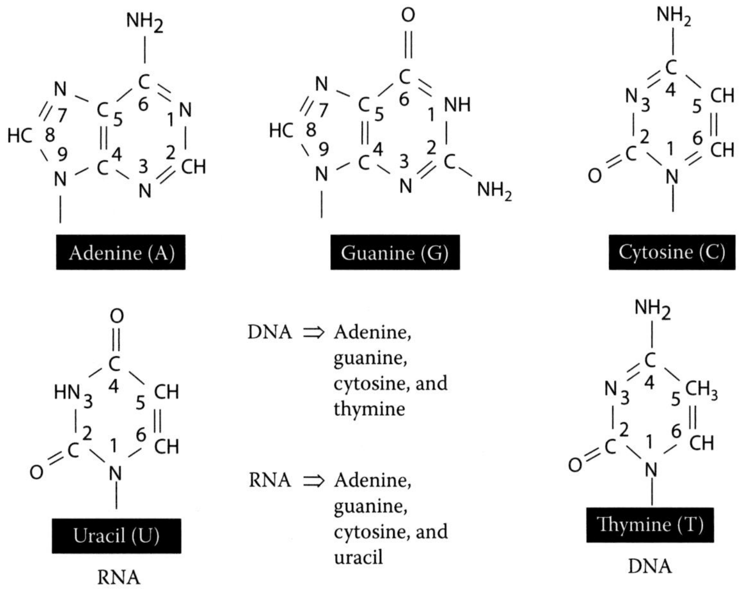

FIGURE 16.4
The four nitrogen bases in nucleic acids. Deoxyribonucleic acid (NA) contains A, G, C, and T, while ribonucleic acid (RNA) contains A, G, C, and U.

and N^6-methylacytosine in small amounts. The 5-mC (also known as the fifth base) is the most common among these, and it was identified long before DNA was recognized as the genetic material (Johnson and Coghill, 1925). The genetic information in a gene is concealed in the nucleotide (A, T, C, G) sequence; however, epigenetic mechanisms (DNA methylation, histone modification, and noncoding RNA activity) are responsible for changes in the gene activity without making any alteration in the DNA sequence. A well-studied epigenetic modification is DNA methylation, which refers to the addition of a methyl group to cytosine, wherein C residue in nuclear DNA gets methylated at 5′ site by the action of an enzyme DNA methyltransferase as a postreplicative event. Deamination of C residue converts it to U, which either gets repaired by the action of Uracil DNA glycosylase enzyme in the course of the DNA repair process, or gets replaced by T during the DNA replication (Laird and Jaenisch, 1994). Hydrolytic deamination of 5-mC converts it to T (Figure 16.5).

The content of 5-mC in genomic DNA varies considerably among eukaryotes. The nuclear genome of higher plants may contain more than 50% 5-mC of the total cytosine residues (Shapiro, 1976; Doerfler, 1983) and it may be found at three different nucleotide sequence contexts: CG, CHG, and CHH (where H = C, T, or A). CHG and CHH methylation is predominantly found in transposable elements (TEs), whereas CG methylation is abundant in both TEs and genes (Law and Jacobsen, 2010; Feng et al., 2010). Methylation was initially considered to be a part of the host's defense system in prokaryotes; later it was also found to be present in eukaryotes performing different roles (Zhang et al., 2011). Now, DNA methylation is considered to be crucial for a wide range of cellular functions in plants, including genome defense from TEs and maintaining transgenerational genome integrity (Bender, 2004). DNA methylation in the promoter region represses gene transcription directly, by interfering with the binding of transcriptional activators, and

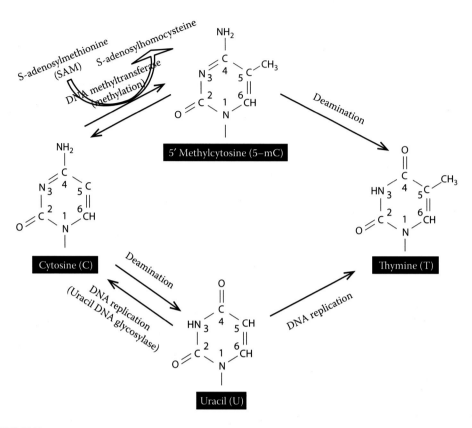

FIGURE 16.5
Modification of pyrimidine bases by methylation and deamination, and replacement during DNA repair and replication processes.

indirectly, by favoring the formation of repressive chromatin by methyl DNA-binding proteins (Bird, 2002). Inheritance of DNA methylation constitutes one of the ways that plants maintain genome stability. Meiotic inheritance of gene silencing is common in plants (Law and Jacobsen, 2010), and plants are not known to undergo genome-wide waves of demethylation in germ cells, as occurs in animals. However, large-scale reprogramming may occur in non-germ-line reproductive cells. One of the ways to actively reprogram the epigenome is by removal of methylated cytosines. The DNA glycosylase enzyme is capable of removing 5-mC (Law and Jacobsen, 2010). One of these enzymes, DEMETER (DME), is expressed in the *Arabidopsis* central cell before fertilization, leading to extensive hypomethylation of the maternal genome (Gehring et al., 2009; Hsieh et al., 2009). This methylation difference between the maternal and paternal genomes in endosperm causes differential expression of a number of genes, depending on parent of origin, which is known as genomic imprinting.

Tissue and developmental stage-specific, as well as stress-induced, variations in DNA methylation have been recorded in plants (Messeguer et al., 1991; Tsaftaris and Polidoros, 2000). Gene imprinting is another well-studied example of the differential expression of genes in tissues due to differences in methylation levels. Zhang et al. (2011) observed tissue-specific differentially methylated regions in Sorghum and reported that DNA methylation plays an important role in regulating tissue-specific or tissue-preferential gene expression.

Xiao et al. (2006) demonstrated that MET1 influences gene expression during embryogenesis in Arabidopsis, and is crucial for normal development of the embryo and seed viability. Loss of DNA methylation derepresses silenced transposons (Kakutani et al., 2004) which could insert into the genes necessary for early embryogenesis. It is now well established that cytosine methylation plays regulatory roles in the expression of imprinted genes in endosperm, and hence normal embryonic development in *Arabidopsis*.

DNA methylation, a generator of epialleles, could have important implications for plant breeders. Epigenetic changes in plants can be inherited over generations in the form of epigenetic alleles (Kakutani, 2002). Such heritable epigenetic alleles are considered as a source of polymorphism and may produce novel phenotypes. Assessing the importance of epialleles in plant breeding requires determination of the extent of variation in methylation among individuals, the degree to which methylation affects phenotype, and the extent to which methylation-linked superior phenotypes are stably inherited. Although these are challenging tasks, the technical potential does exist to assess methylation-pattern differences between individuals and to estimate the levels of methylation-associated epiallelic and phenotypic diversity. Data also indicate that hybrids are generally less methylated than their parental inbreds. The possible role of methylation in the expression of maize genes and performance of hybrids under different growth conditions has been examined in experiments with maize inbreds and hybrids (Tsaftaris and Kafka, 1998; Tani et al., 2005). Exploring the epigenetic mechanisms of gene regulation may reveal the mysteries behind apomixis, that is, an asexual mode of reproduction through seeds where embryos develop without meiosis and double fertilization leading to the production of fertile progenies identical to the mother plant (Koltunow et al., 2003; Yadav et al., 2012). In our studies, it was found that certain regions of the *Cenchrusciliaris* genome are hypermethylated in obligate sexual plants compared to those in obligate apomictic plants. We are working to understand the molecular biology of apomixis in grasses utilizing the molecular markers, gene discovery and epigenetic approaches. If apomixis is successfully deployed to commercial crops, hybrid vigor can be maintained indefinitely, thus overcoming the current limitations of plant breeding and maintaining hybrid vigor in more than one generation.

16.8 Scientific Challenges in Forage Crop Omics

Limited genetic diversity, narrow genetic base, polyploidy, and complex breeding systems are some of the constraints that complicate application of the omics technologies to improvement of forage crops. Use of markers has not been successful in predicting heterosis, but it has been more effective in the identification of heterotic groups required to capture heterosis (Riday et al., 2003). The association mapping approach would also be complicated due to the polyploidy nature of many of the forage species. Successful application of the omics techniques for improving forage species may be challenging, but many of the hurdles can be overcome with concerted scientific efforts.

To date there have been more than 2000 patents in the field of transgenomics, and there are very few promoters characterized for efficient gene expression in forage species. The CaMV35S promoter, widely used as a constitutive promoter in plant genetic engineering research, is only moderately effective in alfalfa (Samac, 2009). Therefore, constitutive, tissue-, and temporal-specific promoters are required for optimized expression of foreign genes in forage species. Due to the polyploidy and perennial nature of forage species, the transgene integration and trait-stacking strategies developed for diploid crop species such as corn, soybean, and cotton have proved to be less appropriate. Better protein quality, improved digestibility, and reduced or no antiquality factors would be desirable features for improved forages. Improved abiotic and biotic stress tolerance, better water-use efficiency, and increased biomass production are the other desirable characteristics to be incorporated when designing a better forage plant. Although each of these traits would be desirable in an ideal forage plant, their genetic characterization and the genetic relationships among them are, so far, not well understood. Many of these traits are quantitative in nature and controlled by many genes; hence, QTL mapping and marker-assisted selection would be required for pyramiding of the compatible traits.

16.9 Conclusions

Forages are a basic requirement for livestock production. To harness the efficiency of highly productive breeds of livestock, the genetic improvement of forages is essential. Quality traits are the major targets of forage crop improvement. Years of research on forages has identified lignin as an impediment to digestibility, many anti-palatability factors reducing their efficiency, and poor nutritional composition in their efficient utilization. Enhanced nutrient utilization would result in improved performance of milk- and meat-producing animals. Meeting the current and future demands of the dairy industry for new forage cultivars with superior agronomic characteristics in order to increase animal production and reduce environmental impacts will be a challenge for plant breeders. Omics technologies such as genomics, transcriptomics, proteomics, and metabolomics can be very powerful tools for large-scale genotyping and phenotyping, and for further improving our understanding of complex traits to manipulate them. In particular, the combination of genomics and metabolomics is very promising for forage breeding. Forage species are mostly outbreeding perennials which have evolved long-term strategies for survival and the propagation of vegetative material. Furthermore, the harvested component of forage plants is the vegetative photosynthetically active tissue, the source of new carbon for plant growth. Since forage yield is a trade-off between removing leaves to feed animals while sustaining enough leaf material for photosynthesis, improvements in forage yield are therefore difficult to achieve and sustain in forage plants. These special features necessitate further research on forage species to assist forage breeders in overcoming the current limitations. The prospects and potential of omics technologies are well understood by forage scientists; hence, rather than working with individual omics technologies in different laboratories, networking between various omics technologies is very much required for the development of desirable forage plants.

References

Agharkar M, Lomba P, Altpeter F, Zhang H, Kenworthy K, and Lange T (2007) Stable expression of AtGA2ox1 in a low-input turf grass (*Paspalumnotatum* Flugge) reduces bioactive gibberellin levels and improves turf quality under field conditions. *Plant Biotechnol J* 5: 791–801.

Anonymous (2009) Forage crops and grasses. *In Handbook of Agriculture.* Indian Council of Agricultural Research, New Delhi. pp. 1353–1417.

Austin-Phillips S and Ziegelhoffer T (2001) The production of value-added proteins in transgenic alfalfa. *In Molecular Breeding of Forage Crops,* (ed G Spangenberg) Kluwer Academic, Dordrecht. pp. 285–301.

Barbazuk WB, Emrich SJ, Chen HD, Li L, and Schnable PS (2007) SNP discovery via 454 transcriptome sequencing. *Plant J* 51: 910–918.

Barbazuk WB and Schnable PS (2011) SNP discovery by transcriptome pyrosequencing. *Methods Mol Biol* 729: 225–246.

Barker DG, Bianch S, Blondon F, Dettee Y, Duc G, Essad S, Flament P, et al. (1990) *Medicago truncatula,* a model plant for studying the molecular genetics of the *Rhizobium*-legume symbiosis. *Plant Mol Biol Rep* 8: 40–49.

Barone P, Rosellini D, LaFayette P, Bouton J, Veronesi F, and Parrott W (2008) Bacterial citrate synthase expression and soil aluminum tolerance in transgenic alfalfa. *Plant Cell Reports* 27: 893–901.

Barth S (2012) Breeding strategies for forage and grass improvement. *Ann Bot* 110: 1261–1262.

Bender J (2004) DNA methylation and epigenetics. *Annu Rev Plant Biol* 55: 41–68.

Bertram HC, Weisbjerg MR, Jensen CS, Pedersen MG, Didion T, Petersen BO, Duus JO, Larsen MK, and Nielsen JH (2010) Seasonal changes in the metabolic fingerprint of 21 grass and legume cultivars studied by nuclear magnetic resonance-based metabolomics. *J Agric Food Chem* 58: 4336–4341.

Bird AP (2002) DNA methylation patterns and epigenetic memory. *Genes Dev* 16: 6–21.

Bradford KJ, Van Deynze A, Gutterson N, Parrott W, and Strauss SH (2005) Regulating transgenic crops sensibly: Lessons from plant breeding, biotechnology and genomics. *Nature Biotechnol* 23: 439–444.

Broeckling CD, Huhman DV, Farag MA, Smith JT, May GD, Mendes P, Dixon RA, and Sumner LW (2005) Metabolic profiling of *Medicago truncatula* cell cultures reveals the effects of biotic and abiotic elicitors on metabolism. *J Exp Bot* 56: 323–336.

Calderini O, Bovone T, Scotti C, Pupilli F, Piano E, and Arcioni S (2007) Delay of leaf senescence in Medicago sativa transformed with the ipt gene controlled by the senescence-specific promoter SAG12. *Plant Cell Rep* 26: 611–615.

Casler MD, Jung HG, and Coblentz WK (2008) Clonal selection for lignin and etherified ferulates in three perennial grasses. *Crop Sci* 48: 424–433.

Chabaud M, de Carvalho-Niebel F, and Barker DG (2003) Efficient transformation of *Medicago truncatula* cv. Jemalong using the hypervirulent *Agrobacterium tumefaciens* strain AGL1. *Plant Cell Rep* 22: 46–51.

Chen F and Dixon RA (2007) Lignin modification improves fermentable sugar yields for biofuel production. *Nature Biotechnol* 25: 759–761.

Chen L, Auh C, Dowling P, Bell J, Lehmann D, and Wang Z-Y (2004) Transgenic down-regulation of caffeic acid O-methyltransferase (COMT) led to improved digestibility in tall fescue (*Festuca arundinacea*). *Functional Plant Biol* 31: 235–245.

Chen L, Auh C, Dowling P, Bell J, Chen F, Hopkins A, Dixon RA, and Wang ZY (2003) Improved forage digestibility of tall fescue (*Festuca arundinacea*) by transgenic down-regulation of cinnamyl alcohol dehydrogenase. *Plant Biotechnology J* 1: 437–449.

Choi HK, Kim DJ, Uhm T, Limpens E, Lim H, Mun JH, Kalo P, et al. (2004a) A sequence-based genetic map of *Medicago truncatula* and comparison of marker colinearity with *M. sativa*. Genetics 166: 1463–1502.

Choi HK, Mun JH, Kim DJ, Zhu H, Baek J.M, Mudge J, Roe B, et al. (2004b) Estimating genome conservation between crop and model legume species. *Proc Natl Acad Sci USA* 101: 15289–15294.

Cogan NOI, Ponting RC, Vecchies AC, Drayton MC, George J, Dracatos PM, Dobrowolski MP, et al. (2006) Gene-associated single nucleotide polymorphism discovery in perennial ryegrass (*Lolium perenne L.*). *Mol Gen Genomics* 276: 101–112.

Crane C, Dixon RA, and Wang ZY (2006) *Medicago truncatula* transformation using root explants. In *Agrobacterium Protocols* (ed K Wang; 2nd edition). Vol 1, Humana Press, Totowa, NJ. pp. 137–142.

Deak M, Kiss GB, Koncz C, and Dudits D (1986) Transformation of *Medicago* by *Agrobacterium* mediated gene transfer. *Plant Cell Rep* 5: 97–100.

Dixon RA and Steele CL (1999) Flavonoids and isoflavonoids – a gold mine for metabolic engineering. *Trends in Plant Sci* 4: 394–400.

Doerfler W (1983) DNA methylation and gene activity. *Annu Rev Biochem* 52: 93–124.

Dong S, Shew HD, Tredway LP, Lu J, Sivamani E, Miller ES, and Qu R (2008) Expression of the bacteriophage T4 lysozyme gene in tall fescue confers resistance to gray leaf spot and brown patch diseases. *Transgenic Res* 17: 47–57.

Dong S, Tredway LP, Shew HD, Wang G, Sivamani E, and Qu R (2007) Resistance of transgenic tall fescue to two major fungal diseases. *Plant Science* 173: 501–509.

Edwards GR, Parsons AJ, Rasmussen S, and Bryant RH (2007) High sugar grasses for livestock systems in New Zealand. *Proc New Zealand Grassland Assoc* 69: 161–172.

Espevig T, Xu C, TS, DaCosta M, and Huang B (2012) Proteomic responses during cold acclimation in association with freezing tolerance of Velvet Bentgrass. *J Am Society Hort Sci* 137: 391–399.

Faville MJ, Richardson K, Gagic M, Mace W, Sun XZ, Harrison S, Knapp K, et al. (2010) Genetic improvement of fibre traits in perennial ryegrass. *Proc New Zealand Grassland Assoc* 72: 71–78.

Feng S, Cokus SJ, Zhang X, Chen PY, Bostick M, Goll MG, Hetzel J, et al. (2010) Conservation and divergence of methylation patterning in plants and animals. *Proc Natl Acad Sci USA* 107: 8689–8694.

Fernie AR and Klee HJ (2011) The use of natural genetic diversity in the understanding of metabolic organization and regulation. *Frontiers Plant Sci* 2: 1–10.

Franzmayr BK, Rasmussen S, Fraser KM, and Jameson PE (2012) Expression and functional characterization of a white clover isoflavone synthase in tobacco. *Ann Bot* 110: 1291–1301.

Frugier F, Kondorosi A, and Crespi M (1998) Identification of novel putative regulatory genes induced during alfalfa nodule development with a cold-plaque screening procedure. *Mol Plant Microbe Interac* 11: 358–366.

Fu C, Mielenz JR, Xiao X, Ge Y, Hamilton CY, Rodriguez M Jr, Chen F, et al. (2011a) Genetic manipulation of lignin reduces recalcitrance and improves ethanol production from switchgrass. *Proc Natl Acad Sci USA* 108: 3803–3808.

Fu C, Sunkar R, Zhou C, Shen H, Zhang J, Wolf J, Tang Y, et al. (2012) Overexpression of miR156 in switchgrass (*Panicum virgatum* L.) results in various morphological alterations and leads to improved biomass production. *Plant Biotechnol J* 10: 443–452.

Fu C, Xiao X, Xi Y, Ge Y, Chen F, Bouton J, Dixon RA, and Wang Z-Y (2011b) Downregulation of cinnamyl alcohol dehydrogenase (CAD) leads to improved saccharification efficiency in switchgrass. *Bioenergy Res* 4: 153–164.

Fu D, Huang B, Xiao Y, Muthukrishnan S, and Liang G (2007) Overexpression of barley hva1 gene in creeping bentgrass for improving drought tolerance. *Plant Cell Rep* 26: 467–477.

Fu D, Tisserat NA, Xiao Y, Settle D, Muthukrishnan S, and Liang GH (2005) Overexpression of rice TLPD34 enhances dollar-spot resistance in transgenic bentgrass. *Plant Sci* 168: 671–680.

Gallardo K, Signor CL, Vandekerckhove J, Thompson RD, and Burstin J (2003) Proteomics of *Medicago truncatula* seed development establishes the time frame of diverse metabolic processes related to reserve accumulation. *Plant Physiol* 133: 664–682.

Gehring M, Bubb KL, and Henikoff S (2009) Extensive demethylation of repetitive elements during seed development underlies gene imprinting. *Science* 324: 1447–1451.

Guo D, Chen J, Wheeler J, Winder J, Selman S, Peterson M, and Dixon RA (2001) Improvement of in rumen digestibility of alfalfa forage by genetic manipulation of lignin O-methyltransferases. *Transgenic Res* 10: 457–464.

Gupta PK, Rustgi S, and Mir RR (2008) Array-based high-throughput DNA markers for crop improvement. *Heredity* 101: 5–18.

Hall RD (2011) Plant metabolomics in a nutshell: Potential and future challenges. *Annu Plant Rev* 43: 1–24.

Harrison S, Fraser K, Lane G, Hughes D, Villas-Boas S, and Rasmussen S (2011) Analysis of high-molecular-weight fructan polymers in crude plant extracts by high-resolution LC-MS. *Anal Bioanal Chem* 401: 2955–2963.

Harrison S, Xue H, Lane G, Villas-Boas S, and Rasmussen S (2012) Linear ion trap MS^n of enzymatically synthesized ^{13}C-labeled fructans reveals differentiating fragmentation patterns of β (1–2) and β (1–6) fructans and provides a tool for oligosaccharide identification in complex mixtures. *Anal Chem* 84: 1540–1548.

Harrison SJ, Fraser K, Lane GA, Villas-Boas S, and Rasmussen S (2009). A reverse-phase liquid chromatography/mass spectrometry method for the analysis of high-molecular-weight fructooligosaccharides. *Anal Biochem* 395: 113–115.

Hays DB and Skinner DZ (2001) Development of an expressed sequence tag (EST) library for *Medicago sativa*. *Plant Sci* 161: 517–526.

He Y, Liu X, and Huang B (2005) Protein changes in response to heat stress in acclimated and nonacclimated creeping bentgrass. *J Am Soc Hort Sci* 130: 521–526.

Hisano H, Kanazawa A, Kawakami A, Yoshida M, Shimamoto Y, and Yamada T (2004) Transgenic perennial ryegrass plants expressing wheat fructosyltransferase genes accumulate increased amounts of fructan and acquire increased tolerance on a cellular level to freezing. *Plant Science* 167: 861–868.

Holtorf H, Guitton M-C, and Reski R (2002) Plant functional genomics. *Naturwissenschaften* 89: 235–249.

Hsieh TF, Ibarra CA, Silva P, Zemach A, Eshed-Williams L, Fischer RL, and Zilberman D (2009) Genome-wide demethylation of *Arabidopsis* endosperm. *Science* 324: 1451–1454.

Hu Y, Jia W, Wang J, Zhang Y, Yang L, and Lin Z (2005) Transgenic tall fescue containing the *Agrobacterium tumefaciens* ipt gene shows enhanced cold tolerance. *Plant Cell Rep* 23: 705–709.

Jackson L, Shadle G, Zhou R, Nakashima J, Chen F, and Dixon R (2008) Improving saccharification efficiency of alfalfa stems through modification of the terminal stages of monolignol biosynthesis. *Bioenergy Res* 1: 180–192.

James C (2012) Global status of commercialized biotech/GM crops. ISAAA Brief No. 44. ISAAA, Ithaca, NY.

Jensen CS, Salchert K, Gao C, Andersen C, Didion T, and Nielsen KK. 2004. Floral inhibition in red fescue (*Festuca rubra* L.) through expression of a heterologous flowering repressor from Lolium. *Mol Breeding* 13: 37–48.

Jiang Q, Zhang J, Guo X, Bedair M, Sumner L, Bouton J and Wang Z (2010) Improvement of drought tolerance in white clover (*Trifolium repens*) by transgenic expression of a transcription factor gene WXP1. *Functional Plant Biol* 37: 157–165.

Jiang Q, Zhang J-Y, Guo X, Monteros M, and Wang Z-Y (2009) Physiological characterization of transgenic alfalfa (*Medicago sativa*) plants for improved drought tolerance. *International J Plant Sci* 170: 969–978.

Johnson TB and Coghill RD (1925) Researches on pyrimidines. C111. The discovery of 5-methylcytosine in tuberculinic acid, the nucleic acid of the tubercle bacillus. *J Am Chem Soc* 47: 2838–2844.

Kakutani T (2002) Epi-alleles in plants: Inheritance of epigenetic information over generations. *Plant Cell Physiol* 43: 1106–1111.

Kakutani T, Kato M, Kinoshita T, and Miura A (2004). Control of development and transposon movement by DNA methylation in *Arabidopsis thaliana*. *Cold Spring Harb Symp Quant Biol* 69: 139–143.

Khoudi H, Laberge S, Ferullo JM, Bazin R, Bazin R, Darveau A, Castonguay Y, Allard G, Lemieux R, and Vézina LP (1999) Production of a diagnostic monoclonal antibody in perennial alfalfa plants. *Biotechnol Bioeng* 64: 135–143.

Kliebenstein DJ (2009) Advancing genetic theory and application by metabolic quantitative trait loci analysis. *The Plant Cell* 21: 1637–1646.

Koltunow AM and Grossniklaus U (2003) Apomixis: A developmental perspective. *Annu Rev Plant Biol* 54: 547–574.

Kumar S (2008) Biotechnology and its utilization for forage crop improvement. In Environment, Agroforestry and Livestock Management (eds SS Kundu, OP Chaturvedi, JC Dagar, SK Sirohi). IBDC, Lucknow. pp. 559–594.

Kumar S (2011) Biotechnological advancements in alfalfa improvement. *J Appl Genet* 52: 111–124.

Kumar S, Tiwari R, Chandra A, Sharma A, and Bhatnagar RK (2012) In vitro direct plant regeneration and *Agrobacterium*-mediated transformation of lucerne (*Medicago sativa* L.). *Grass Forage Sci* doi: 10.1111/gfs.12009.

Laird PW and Jaenisch R (1994) DNA Methylation and Cancer. *Hum Mol Genet* 3: 1487–1495.

Law JA and Jacobsen SE (2010) Establishing, maintaining and modifying DNA methylation patterns in plants and animals. *Nat Rev Genet* 11: 204–220.

Li X, Weil Y, Nettleton D, and Brummer EC (2009) Comparative gene expression profiles between heterotic and non-heterotic hybrids of tetraploid *Medicago sativa*. *BMC Plant Biol* 9: 107.

Li Z, Baldwin CM, Hu Q, Liu H, and Luo H (2010) Heterologous expression of Arabidopsis H+-pyrophosphatase enhances salt tolerance in transgenic creeping bentgrass (*Agrostis stolonifera* L.). *Plant Cell Environ* 33: 272–289.

Ludlow EJ, Mouradov A, and Spangenberg GC (2009) Post-transcriptional gene silencing as an efficient tool for engineering resistance to white clover mosaic virus in white clove mosaic virus in white clover (*Trifolium repens*). *J Plant Physiol* 166: 1557–1567.

Ma X-F, Tudor S, Butler T, Ge Y, Xi Y, Bouton J, Harrison M, and Wang ZY (2012) Transgenic expression of phytase and acid phosphatase genes in alfalfa (Medicago sativa) leads to improved phosphate uptake in natural soils. *Mol Breeding* 30: 377–391.

Ma X-F, Wright E, Ge Y, Bell J, Xi Y, Bouton JH, and Wang Z-Y (2009) Improving phosphorus acquisition of white clover (*Trifolium repens* L.) by transgenic expression of plant-derived phytase and acid phosphatase genes. *Plant Sci* 176: 479–488.

Mathesius U, Keijzers G, Natera SHA, Weinman JJ, Djordjevic MA, and Rolfe B (2001) Establishment of a root proteome reference map for the model legume *Medicago truncatula* using the expressed sequence tag database for peptide mass fingerprinting. *Proteomics* 1: 1424–1440.

Mesfin T, Temple SJ, Allan DL, Vance CP, and Samac DA (2001) Overexpression of malate dehydrogenase in transgenic alfalfa enhances organic acid synthesis and confers tolerance to aluminum. *Plant Physiol* 127: 1836–1844.

Messeguer R, Ganal MW, Steffens JC, and Tanskley SD (1991) Characterization of the level, target sites and inheritance of cytosine methylation in tomato nuclear DNA. *Plant Mol Biol* 16: 753–770.

Milano I, Babbucci M, Panitz F Ogden R, Nielsen RO, Taylor MI, Helyar SJ, et al. (2011): Novel tools for conservation genomics: Comparing two high-throughput approaches for SNP discovery in the transcriptome of the European hake. *PLoS ONE* 6: e28008.

Morreel K, Goeminne G, Storme V, Sterck L, Ralph J, Coppieters W, Breyne P, et al. (2006) Genetical metabolomics of flavonoid biosynthesis in Populus: A case study. *Plant J* 47: 224–237.

Mouradov AZ (2010) Designer pasture plants: From single cells to the field. Proceedings of the National Academy of Sciences of Azerbaijan Republic: *Biological Sciences* 65: 205–211.

Mukhopadhyay B, Blum K, and Ganguly NK (2012) The decade of omics. *IIOAB Journal* 3: 29–31.

Park SY, Shivaji R, Krans JV and Luthe DS (1996) Heat-shock response in heat-tolerant and nontolerant variants of *Agrostis palustris* Huds. *Plant Physiol* 111: 515–524.

Parsons AJ, Edwards GR, Newton PCD, Chapman DF, Caradus JR, Rasmussen S, and Rowarth JS (2011). Past lessons and future prospects: Plant breeding for yield and persistence in cool-temperate pastures. *Grass Forage Sci* 66: 153–172.

Petrovska N, Wu X, Donato R, Wang Z, Ong E-K, Jones E, Forster J, et al. (2004) Transgenic ryegrasses (*Lolium* spp.) with down-regulation of main pollen allergens. *Mol Breeding* 14: 489–501.

Ponting RC, Drayton MC, Cogan NOI, Dobrowolski MP, Spangenberg GC, Smith KF, and Forster JW (2007) SNP discovery, validation, haplotype structure and linkage disequilibrium in full-length herbage nutritive quality genes of perennial ryegrass (*Lolium perenne* L.). *Mol Gen Genom* 278: 585–597.

Rafalski A (2002) Applications of single nucleotide polymorphisms in crop genetics. *Curr Opin Plant Biol* 5: 94–100.

Rafiqul M, Khan I, Ceriotti A, Tabe L, Aryan A, McNabb W, Moore A, Craig S, Spencer D, and Higgins TJV (1996) Accumulation of a sulphur-rich seed albumin from sunflower in the leaves of transgenic subterranean clover (*Trifolium subterraneum* L.). *Transgenic Res* 5: 179–185.

Rasmussen S, Parsons AJ, and Jones CS (2012) Metabolomics of forage plants: A review. *Ann Bot* 110: 1281–1290.

Reddy MSS, Chen F, Shadle G, Jackson L, Aljoe H, and Dixon RA (2005) Targeted down-regulation of cytochrome P450 enzymes for forage quality improvement in alfalfa (*Medicago sativa* L.). *Proc Natl Acad Sci USA* 102: 16573–16578.

Riday H, Brummer EC, Campbell TA, Luth D, and Cazcarro PM (2003) Comparisons of genetic and morphological distance with heterosis between *Medicago sativa* subsp. sativa and subsp. falcata. *Euphytica* 131: 37–45.

Saathoff AJ, Sarath G, Chow EK, Dien BS, and Tobias CM (2011) Downregulation of cinnamyl-alcohol dehydrogenase in switchgrass by RNA silencing results in enhanced glucose release after cellulase treatment. *PLoS ONE* 6: e16416.

Samac DA (2009) Promoters for constitutive and tissue-specific expression of transgenes in alfalfa. Available at http://www. naaic.org/TAG/TAGpapers/samac/samac.html.

Schauer N, Semel Y, Roessner U, Gur A, Balbo I, Carrari F, Pleban T, et al. (2006) Comprehensive metabolic profiling and phenotyping of interspecific introgression lines for tomato improvement. *Nat Biotechnol* 24: 447–454.

Shahin E, Spielmann A, Suhkapinda K, Simpson RB, and Yasher M (1986) Transformation of cultivated alfalfa using disarmed *Agrobacterium tumefaciens*. *Crop Sci* 26: 1235–1239.

Shapiro HS (1976) *Handbook of Biochemistry and Molecular Biology*, CRC, Boca Raton, FL. pp 258–262.

Somleva M, Snell K, Beaulieu J, Peoples O, Garrison B, and Patterson N (2008) Production of polyhydroxybutyrate in switchgrass, a value-added co-product in an important lignocellulosic biomass crop. *Plant Biotechnol J* 6: 663–678.

Stewart D, Shepherd LVT, Hall RD, and Fraser PD (2011) Crops and tasty, nutritious food: How can metabolomics help? *Annu Plant Rev* 43: 181–217.

Strauss SH, Kershen DL, Bouton JH, Redick TP, Tan H, and Sedjo RA (2010) Far-reaching deleterious impacts of regulations on research and environmental studies of recombinant DNA-modified perennial biofuel crops in the United States. *Bioscience* 60: 729–741.

Studer B, Byrne S, Nielsen RO, Panitz F, Bendixen C, Islam MS, Pfeifer M, Lübberstedt T, and Asp T (2012) A transcriptome map of perennial ryegrass (*Lolium perenne* L.). *BMC Genomics* 13:140.

Tadege M, Wen J, He J, Tu H, Kwak Y, Eschstruth A, Cayrel A, et al. (2008) Large scale insertional mutagenesis using the *Tnt1* retrotransposon in the model legume *Medicago truncatula*. *Plant J* 54: 335–347.

Tani E, Polidoros AN, Nianiou-Obeidat I, and Tsaftaris AS (2005) DNA methylation patterns are differently affected by planting density in maize inbreds and their hybrids. *Maydica* 50: 19–23.

Tesfaye M, Silverstein KAT, Bucciarelli B, Samac DA, and Vance CP (2006) The Affymetrix Medicago GeneChip® array is applicable for transcript analysis of alfalfa (*Medicago sativa*). *Funct Plant Biol* 33: 783–788.

Tesfaye M, Temple SJ, Allan DL, Vance CP, and Samac DA (2001) Overexpression of malate dehydrogenase in transgenic alfalfa enhances organic acid synthesis and confers tolerance to aluminum. *Plant Physiol* 127: 1836–1844.

Thoquet P, Ghérardi M, Journet E-P, Kereszt A, Ané JM, Prosperi JM, and Huguet T (2002) The molecular genetic linkage map of the model legume *Medicago truncatula*: An essential tool for comparative legume genomics and the isolation of agronomically important genes. *BMC Plant Biol* 2: 1–13.

Tieman DM, Zeigler M, Schmelz E, Taylor MG, Bliss P, Kirst M, and Klee HJ (2006) Identification of loci affecting flavour volatile emissions in tomato fruits. *J Exp Bot* 57: 887–896.

Trick M, Long Y, Meng J, and Bancroft I (2009) Single nucleotide polymorphism (SNP) discovery in the polyploid *Brassica napus* using Solexa transcriptome sequencing. *Plant Biotechnol J* 7: 334–346.

Tsaftaris AS and Kafka M (1998) Mechanisms of heterosis in crop plants. *J Crop Prod* 1: 95–111.

Tsaftaris AS and Polidoros AN (2000) DNA methylation and plant breeding. *Plant Breed Rev* 18: 87–176.

Tu Y, Rochfort S, Liu Z, Ran Y, Griffith M, Badenhorst P, Louie GV, et al. (2010) Functional analyses of caffeic acid O-methyltransferase and cinnamoyl-CoA-reductase genes from perennial ryegrass (*Lolium perenne*). *Plant Cell* 22: 3357–3373.

Turner LB, Cairns AJ, Armstead IP, Ashton J, Skøt K, Whittaker D, and Humphreys MO (2006) Dissecting the regulation of fructan metabolism in perennial ryegrass (*Lolium perenne*) with quantitative trait locus mapping. *New Phytologist* 169: 45–58.

Vierling E (1991) The roles of heat shock proteins in plants. *Annu Rev Plant Physiol Plant Mol Biol* 42: 579–620.

Wang Z, Hopkins A, and Mian R (2001a). Forage and turf grass biotechnology. *Crit Rev Plant Sci* 20: 573–619.

Wang Z-Y (2003) Biotechnology has potential for forage improvement. AG News and Views: September 2003 Available at http://www.noble.org/ag/research/potentialofbiotech/.

Wang Z-Y and Brummer EC (2012) Is genetic engineering ever going to take off in forage, turf and bioenergy crop breeding? *Ann Bot* 110: 1317–1325.

Wang Z-Y, Ye XD, Nagel J, Potrykus I, and Spangenberg G (2001b) Expression of a sulphur-rich sunflower albumin gene in transgenic tall fescue (*Festuca arundinacea* Schreb.) plants. *Plant Cell Rep* 20: 213–219.

Wang Z-Y, Takamizo T, Iglesias VA, Osusky M, Nagel J, Potrykus I, and Spangenberg G (1992) Transgenic plants of tall fescue (*Festucaarundinacea* Sc hreb.) obtained by direct gene transfer to protoplasts. *Nat Biotechnol* 10: 691–696.

Watson B, Asirvatham V, Wang L, and Sumner LW (2003) Mapping the proteome of *Medicago truncatula*. *Plant Physiol* 131: 1104–1123.

Wilkins MR, Williams KL, Appel RD, and Hochstrasser DF (1997). *Proteome Research: New Frontiers in Functional Genomics*. Springer, Berlin.

Winkel-Shirley B (2001) Flavonoid biosynthesis. A colorful model for genetics, biochemistry, cell biology, and biotechnology. *Plant Physiol* 126: 485–493.

Wu YY, Chen QJ, Chen M, and Chen J (2005) Salt-tolerant transgenic perennial ryegrass (*Lolium perenne* L.) obtained by *Agrobacterium tumefaciens*-mediated transformation of the vacuolar Na+/H + antiporter gene. *Plant Science* 169: 65–73.

Xiao W, Custard KD, Brown RC, Lemmon BE, Harada JJ, Goldberg RB, and Fischer RL, et al. (2006). DNA methylation is critical for Arabidopsis embryogenesis and seed viability. *Plant Cell* 18: 805–814.

Xiong X, James V, Zhang H, and Altpeter F (2010) Constitutive expression of the barley HvWRKY38 transcription factor enhances drought tolerance in turf and forage grass (*Paspalum notatum* Flugge). *Mol Breeding* 25: 419–432.

Xu JP, Schubert J, and Altpeter F (2001) Dissection of RNA-mediated ryegrass mosaic virus resistance in fertile transgenic perennial ryegrass (*Lolium perenne* L.). *Plant J* 26: 265–274.

Yadav CB, Anuj, Kumar S, Gupta MG, and Bhat V (2012) Genetic linkage maps of the chromosomal regions associated with apomictic and sexual modes of reproduction in *Cenchrus ciliaris*. *Mol Breeding* 30: 239–250.

Yang S, Gao M, Xu C, Gao J, Gao J, Deshpande S, Lin S, Roe BA, and Zhu H (2008) Alfalfa benefits from Medicago truncatula: The RCT1 gene from M. truncatula confers broad-spectrum resistance to anthracnose in alfalfa. *Proc Natl Acad Sci* 105: 12164–12169.

Yang SS, Xu WW, Tesfaye M, Lamb FS, Jung HJG, Samac DA, Vance CP, and Gronwald JW (2009) Single-Feature Polymorphism discovery in the transcriptome of tetraploid alfalfa. *Plant Genome* 2: 224–232.

Young ND, Cannon SB, Sato S, Kim D, Cook DR, Town CD, Roe BA, and Tabata S (2005) Sequencing the gene spaces of *Medicago truncatula* and *Lotus japonicus*. *Plant Physiol* 137: 1174–1181.

Zhang H, Lomba P, and Altpeter F (2007b) Improved turf quality of transgenic bahiagrass (*Paspalum notatum* Flugge) constitutively expressing the ATHB16 gene, a repressor of cell expansion. *Mol Breeding* 20: 415–423.

Zhang J-Y, Broeckling C, Sumner LW, and Wang Z-Y (2007a) Heterologous expression of two Medicago truncatula putative ERF transcription factor genes, WXP1 and WXP2, in Arabidopsis led to increased leaf wax accumulation and Improved drought tolerance, but differential response in freezing tolerance. *Plant Mol Biol* 64: 265–278.

Zhang J-Y, Broeckling CD, Blancaflor EB, Sledge M, Sumner LW, and Wang Z-Y (2005) Overexpression of WXP1, a putative *Medicago truncatula* AP2 domain-containing transcription factor gene, increases cuticular wax accumulation and enhances drought tolerance in transgenic alfalfa (*Medicago sativa*). *Plant J* 42: 689–707.

Zhang M, Xu C, von Wettstein D, and Liu B (2011) Tissue-Specific Differences in Cytosine Methylation and their Association with Differential Gene Expression in *Sorghum bicolar*. *Plant Physiol* 156: 1955–1966.

Zhang Y, Mian MAR, and Bouton JH (2006) Recent molecular and genomic studies on stress tolerance of forage and turf grasses. *Crop Sci* 46: 497–511.

Zhang Y and Mian R (2003) Functional genomics in forage and turf: Present status and future prospects. *Afr J Biotechnol* 2: 521–527.

Zhou C, Han L, Pislariu C, Nakashima J, Fu C, Jiang Q, Quan L, et al. (2011a) From model to crop: Functional analysis of a STAY-GREEN gene in the model legume Medicago truncatula and effective use of the gene for alfalfa (*M. sativa*) improvement. *Plant Physiol* 157: 1483–1496.

Zhou M, Hu Q, Li Z, Li D, Chen C-F, and Luo H (2011b) Expression of a novel antimicrobial peptide penaeidin4–1 in creeping bentgrass (*Agrostis stolonifera* L.) enhances plant fungal disease resistance. *PLoS ONE* 6: e24677.

17

Bioenergy Crops Enter the Omics Era

Atul Grover, PhD; Patade Vikas Yadav, PhD; Maya Kumari, PhD;
Sanjay Mohan Gupta, PhD; Mohommad Arif, PhD; and Zakwan Ahmed, PhD

CONTENTS

17.1 Introduction 549
17.2 Molecular and General Physiology of Yield 550
17.3 Genome Structure and Global Picture of Gene Expression 552
17.4 Candidate Genes 554
17.5 Population Genomics 556
References..557

17.1 Introduction

The discovery of alternative fuels that can be blended with traditional fossil fuels is an urgent global priority. The fossil fuel reserve is gradually depleting, leading to a global oil crisis, which is manifested by price rises and dwindling economies. Currently, biofuels obtained from materials derived from biological sources appear to be one of the most feasible options to meet the growing oil demand. However, various hurdles such as land-use requirements and the limitations of present-day technologies exist in a complete switchover from current fossil fuels to biofuels (Rubin, 2008). While a number of these barriers can be removed by changes at the social and administrative levels, rapid advancements in science and technology will play a significant role in this direction. Genomics is an area of research lying at the bottom of science and technologies that can significantly contribute to fuel research from biological sources.

The last 20 years has seen a revolution in the field of genomics, expanding our knowledge of the molecular basis of biological processes, thereby enabling us to answer more questions on biology and offering more solutions to the global problems than ever before. With the recent developments in sequencing and other high-throughput technologies, functional and evolutionary genomics can be well addressed in nonmodel species also. The nonmodel species are often ecologically and economically more important and provide more solutions to the problems of mankind and the planet, compared to the model species. Here, we review the recent upsurge in the genomics of bioenergy plants, with more emphasis on two plants with oil-yielding potential, *Jatropha* and *Camelina*, and two plants with biomass potential, *Miscanthus* and switchgrass, all of which have the capability to grow well in marginal lands.

Leading worldwide funding agencies have invested a lot of money in the development of genomic resources for these plants. As a result, in recent years, genomic and transcriptomic studies, including high-throughput sequencing, have been carried out (Swaminathan et al., 2010, 2012; Liu et al., 2011, 2012; Natarajan and Madasamy, 2011; Saski et al., 2011; Sato et al. 2011; Wang et al., 2011, 2012; Ma et al., 2012; Sharma et al., 2012; Sun et al., 2012a). Economically important traits such as oil and biomass yield have received substantial interest from the scientific community (Hutcheon et al., 2010; Liu et al., 2011; Casler, 2012). Additionally, research lines relating to adaptation and acclimatization to abiotic stresses have also been explored (Karp and Shield, 2008; Oliver et al., 2009; Song et al., 2012; Sun et al., 2012b), especially in light of the complete domestication and adoption of these crops throughout the world in the rapidly changing climate. The conservation of the existing genetic diversity in these bioenergy crops, particularly at those loci that are involved in adaptation to local climates (Kirkpatrick and Barton, 2006), is therefore crucial to maintain the adaptive capacity of these plants in newer locations in the years to come. This goal can be achieved by a thorough examination of their genomic architecture, especially in relation to local adaptation, the number of genes affecting adaptive and agronomic traits, the linkages between genetic and epigenetic factors, epistatic interactions, and so on.

17.2 Molecular and General Physiology of Yield

In agroeconomic terms, the yield of bioenergy plants may refer to the oil yield in seeds or to the total biomass. The oil and biomass yields are also directly or indirectly dependent on contributing factors (e.g., number of branches, number of fruits per branch, number of fruits per cluster, and number of seeds per cluster), which can also be modified for the final yield enhancement of oil and biomass. Many of the genes involved in oil accumulation in seeds have been identified in model crops belonging to the family Brassicaceae (Nambisan, 2007). As most metabolic pathways are conserved across plant species, techniques in comparative genomics may be applied for mining the candidate genes or manipulating the metabolic pathways in the target organisms. Particularly in the case of *Camelina sativa*, its genetic relatedness to both *Arabidopsis thaliana* (Beilstein et al., 2008) and *Brassica napus* may prove to be a boon for advancing the genomics research in this plant species. Experiences from *Brassica* and *Arabidopsis* suggest that overexpression of *glycerol-3-phosphate acyltransferase* (*GPAT*), *lysophosphatidic acid acyltransferase* (*LPAT*), and *diacyl glycerol acyl transferase* (*DGAT*) genes can enhance the seed oil yield (Jain et al., 2000; Sharma et al., 2008; Maisonneuve et al., 2010).

A natural goal in the improvement of *Jatropha* and *Camelina* as bioenergy crops is to modify the seed oil composition for superior biodiesel production. Both plants bear seeds that are rich in polyunsaturated fatty acids, and *Camelina* in particular also has very long-chain fatty acids (Akbar et al., 2009; Perkins, 2010). An ideal biodiesel blend would be richer in oleic acid (18:1) (Durrett et al., 2008). The candidate genes for manipulation thus include *stearoyl-ACP desaturase* (Aghoram et al., 2006) and *fatty acid desaturase 2* (*FAD2*) (Hutcheon et al., 2010) for achieving oil that is rich in saturated fatty acids, and genes such as *3-keto-acyl-ACP synthase II* (Knutzon et al., 1992) and *fatty acid elongase* (*FAE1*) (Hutcheon et al., 2010) for 16-carbon fatty acids enriched oil. Further, knowledge is needed on the reproductive biology of these plants. Currently, a basic understanding of this biology exists, but the finer details on their phenology, mating patterns, pollinators, breeding system, and so on, are not known in a quantum necessary to exploit in their domestication program. *Jatropha* is an outcrosser-monoecious plant: male and female flowers occurring on the same plant

(Chang-wei et al., 2007; Achten et al., 2010). The male to female flower ratio is predominantly biased toward male flowers (25:1) (Arif and Ahmed, 2009).

The biomass yield is a complex trait that is influenced by many plant characteristics, such as plant height, tiller number and density, stem thickness. The propagation method, crop establishment and its invasiveness, as well as disease resistance are also important traits in biomass crops. Currently, the potential of energy grasses, particularly switchgrass and *Miscanthus*, both C4 plants, looks limited, as the selection of specific plant varieties yielding a high biomass has not yet been made on the same scale as for food and forage grasses (Karp and Shield, 2008; Donnison et al., 2009). Nevertheless, C4 plants convert energy more efficiently into biomass than C3 plants and have up to 60% higher water and nutrient use efficiency (Jakob et al., 2009).

Polyploidy is an essential trait when biomass as a whole is being seen as the yield component. In this regard, the ploidy level of switchgrass (*Panicum virgatum*; family Panicoidea) can vary from being tetraploid ($2n = 4x = 36$) to octoploid ($2n = 8x = 72$) (Karp and Shield, 2008), and *Miscanthus* from being diploid ($2n = 2x = 38$) to tetraploid ($2n = 4x = 76$) (Karp and Shield, 2008). The utilization of these energy grasses either depends on the thermal conversion of biomass to an energy-convertible form or through biological (enzymatic) routes. This choice is further determined by the calorific value of the biomass, its moisture content, and its constituent structural carbohydrates. When comparing switchgrass and *Miscanthus*, the calorific values and moisture contents of both these grasses are identical (Karp and Shield, 2008); however, *Miscanthus* has a higher cellulose to lignin ratio as compared to switchgrass (Figure 17.1).

The cell wall composition is an important trait requiring an improvement in energy grasses (Jakob et al., 2009). The bulk of the photosynthetic free energy gets deposited in the form of polymers in the cell walls in plants; therefore, an ideal biomass plant should have an enhanced total cell wall content per plant (Bush and Leach, 2007). Lignin, cellulose, and hemicellulose together constitute the plant cell wall, and their biosynthesis involves complicated biochemical pathways. In recent decades, the major focus has been on understanding the chemistry

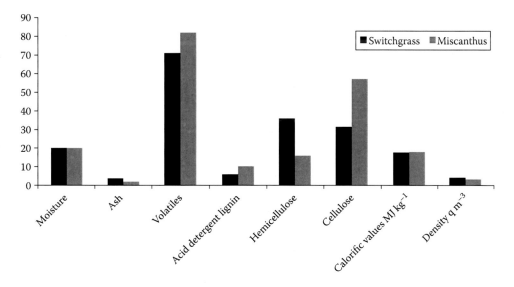

FIGURE 17.1
Attributes in switchgrass and *Miscanthus* influencing the choice of dry biomass for bioenergy purposes. Units not mentioned are % dry matter. (Adapted from Karp, A. and Shield, I., *New Phytologist*, 179, 15–32, 2008.)

and synthesis of lignin (Baucher et al., 1999; Hu et al., 1999; Pilate et al., 2002; Li et al., 2003; Barrière et al., 2004; Ralph et al., 2004; Reddy et al., 2005) and on genetic modification to reduce the lignin content in the cell walls (Jakob et al., 2009). Unfortunately, predicting the accurate measurements of the cell wall components is a challenging task (Kelley et al., 2008; Labbé et al., 2008), and thus there is no consensus on what should be the ideal ratio of lignin, hemicelluloses, and cellulose (Jakob et al., 2009). Nevertheless, it is recognized that a higher carbon content of the cell walls is a desired trait in biomass crops (Bush and Leach, 2007).

Understanding the reproductive biology of biomass crops is also a prime area of research. The flowering time affects the biomass production as the transition from vegetative phase to reproductive phase negatively impacts the biomass accumulation. Thus, for maximum biomass yields, flowering is undesired. Thus, sterile hybrids of *Miscanthus* (*Miscanthus × giganteus*) (Hodkinson et al., 2002) are successful as a biofuel crop in Europe (Clifton-Brown et al., 2001).

The availability of nitrogen and water to switchgrass is a limiting factor in its productivity (Karp and Shield, 2008). Mature switchgrass plants dedicate a large amount of their photosynthates toward the maintenance of their large root system (Karp and Shield, 2008). In general, 50 kg ha^{-1} N without the application of phosphorus (P) and potassium (K) has been found sufficient to obtain the desired yields (McLaughlin and Adams Kszos, 2005). Further, McLaughlin and Adams Kszos (2005) have reported significant differences in photosynthesis, stomatal conductance, transpiration, and water use efficiencies related to yields of switchgrass, but the balance of carbon (C) assimilated per unit of water transpired affected the yield most. Similarly, the water use efficiency of *Miscanthus* has also been reported as high, and drought-stressed *Miscanthus* plants dedicate more biomass to their root and rhizome systems compared to their aerial parts (Clifton-Brown and Lewandowski, 2000).

17.3 Genome Structure and Global Picture of Gene Expression

Only a handful of studies have been carried out to understand the mechanisms of yield and associated traits in bioenergy crops. As discussed earlier, the demand for biodiesel is huge and it cannot be met simply by harvesting the plants with their natural habits in their natural habitats. Since land is one of the limiting natural resources, more land needs to be cultivated, which is possible through developing abiotic stress-tolerant crops with the intervention of genetic engineering for better productivity. In the recent past, gene discoveries and transcriptome characterization efforts have been made in *Jatropha* using expressed sequence tag (EST) sequencing (Costa et al., 2010; Natarajan et al., 2010; Natarajan and Madasamy, 2011) and whole-genome sequencing (Sato et al., 2011). In addition, in 2009, Synthetic Genomics Inc. (SGI), California, USA, and Asiatic Centre for Genome Technology (ACGT), Kuala Lumpur, Malaysia, announced the completion of a whole-genome sequencing effort. However, the generated data are not in the public domain. Together, these studies have helped to identify several genes involved in the metabolic pathways leading to the synthesis and accumulation of fatty acids in the seeds. A summary of the major findings of the EST/RNA Seq/NGS projects is given in Table 17.1. The *Jatropha curcas* genome (286 Mbp) is made up of nearly 41,000 protein-coding genes (Sato et al., 2011), with 4% specific to the Euphorbiaceae family. In a recent study in this laboratory, we have found >40% of *Jatropha* ESTs showing matches with high scores (≥200) with castor genes (unpublished).

TABLE 17.1

Major Transcriptome Sequencing Efforts in Bioenergy Plants

Species	Source of cDNA	ESTs Generated	Unigenes/ Contigs Developed	References
Jatropha curcas	Developing seeds	12,084	7,009	Natarajan et al. (2010)
Jatropha curcas	Seeds	13,249	7,283	Costa et al. (2010)
Jatropha curcas	Leaf and callus	991,050	21,225	Sato et al. (2011)
Jatropha curcas	Seeds	2,200	931	Gomes et al. (2010)
Panicum viragatum	Callus, crown, and seedling	61,585	Not discussed	Tobias et al. (2008)
Panicum viragatum	Germinating seedlings, emerging tillers, flowers, and dormant seeds	910,397	243,601	Wang et al. (2012)

Genomic studies in *C. sativa* are in their infancy, as the species is still considered an underutilized and lesser-known crop. Based on the southern hybridization of well-known single-copy genes in the plants *FAD2*, *FAE1*, and *LFY*, and supported by their experiments involving 454 sequencing analysis and flow cytometry, Hutcheon et al. (2010) concluded that *C. sativa* genome is an allohexaploid.

Miscanthus belongs to a subtribe of grasses (Saccharinae), which include recently derived polyploids arising from a common diploid progenitor. Therefore, understanding and exploiting the genetic mechanisms in *Miscanthus* are relatively difficult. However, Swaminathan et al. (2010) have reported a high level of transcriptomic similarity with other grasses. Recently, Swaminathan et al. (2012) carried out an RNAseq analysis in *Miscanthus* using the Illumina GA II platform for the development of single-nucleotide polymorphisms (SNPs) using GoldenGate genotyping and simple sequence repeat (SSR) markers, in order to prepare a linkage map of the species using the 221 progeny from a full sibling F1 mapping population. The group successfully prepared 19 linkage groups corresponding to haploid chromosomes, employing 658 SNPs and 210 SSR markers. Two C4-pyruvate, phosphate dikinase (C4-PPDK) loci in *Miscanthus* syntenic to a single paralogous locus in *Sorghum* also demonstrated genome-wide duplication in *Miscanthus* compared to *Sorghum*. Interestingly, 12 families of repeats including transposons or centromeric repeats comprise over 95% of the *Miscanthus* × *giganteus* genome (Swaminathan et al., 2010), which are thought to have given rise to the majority of 24-nucleotide small RNAs.

Currently, switchgrass is the most well-studied biomass crop among the upcoming bioenergy crops and it is also subject to whole-genome sequencing (Wang et al., 2012). There have been few reports on generating ESTs for this species or the construction of bacterial artificial chromosome (BAC) libraries in the AP13 clone (Tobias et al., 2008; Sharma et al., 2012; Wang et al., 2012). Based on the analysis of 330,297 BAC end sequences, the switchgrass genome has been suggested to be rich in tandem and interspersed repeats (Sharma et al., 2012; Wang et al., 2012). Interestingly, the structural features of the switchgrass genome, that is, its repeat density, its percentage of guanine–cytosine (%GC) content, its gene density, and so on, are more similar to those of the rice genome. The GC-rich trinucleotide microsatellite repeats are most abundant in the switchgrass genome (Sharma et al., 2012), as in the case of rice (Grover et al., 2007; Roorkiwal et al., 2009), though its gene sequences are more similar to those of *Sorghum* (Tobias et al., 2008; Sharma et al., 2012). The overall character of length versus the abundance of the microsatellite distribution in switchgrass is also reported to be similar to that of rice (Sharma et al., 2012). According to

Wang et al. (2012), a 70%–80% overlap exists between the ESTs of switchgrass and other monocots, such as rice, *Sorghum*, maize, and *Brachypodium*. Interestingly, the distribution and the density of interspersed elements in switchgrass were also similar to those in rice (Sharma et al., 2012). A gypsy group of long terminal repeats (LTRs) was not only reported as the most abundant and repetitive element in switchgrass, but their relative abundance compared to their copia elements was also similar to that in rice (Sharma et al., 2012).

17.4 Candidate Genes

Gigantism in the harvested parts, determinate growth, and an increased harvest index are parts of a "domestication syndrome," as described by Hammer (1984). In order to establish the species with bioenergy potential as successful crops, candidate genes for the improvement of these traits should be targeted. Further, it is desired that a combination of all of these characteristics is present together in a crop. In the recent past, several candidate genes for the enhancement of oil/biomass yield have been identified in different crops, as discussed earlier. The important oil biosynthesis genes identified in *Jatropha* are listed in Table 17.2.

Worldwide attempts are being made to clone these genes for their functional validation. Other genes such as *acyl-CoA oxidase, enoyl-CoA hydratase, β-hydroxyl acyl-CoA dehydrogenase,* and *acyl-CoA acetyl transferase* are involved in β-oxidation, and can thus be exploited for the enhancement of the oil content, oil degradation, and plant growth in both oil- and biomass-yielding plants (Goepfert and Poirier, 2007).

It is important to remember that, globally, it is proposed to grow bioenergy crops in areas that are not suitable for food crops, namely, waste or degraded lands, sides of railway tracks, water-scarce lands, and so forth. Thus, these crops need to withstand a number of abiotic stresses. Plants respond to abiotic stresses by modulating the expression of certain stress-responsive genes, which also include the genes coding for proteins that are responsible for the detoxification of oxidizing ion species and thereafter the recovery of plant growth (Wang et al., 2003; Grafi et al., 2011). Limited studies have been carried out in plants such as *Jatropha* wherein the introduction of a gene has been shown to enhance the plant's tolerance to drought, heat, and salt stresses (Zhang et al., 2008). A number of candidate genes for stress tolerance in *Jatropha* have also been identified, including *phosphoethanolamine N-methyltransferase, Na^+/H^+ antiporter, trehalose-6-phosphate synthase, glutathione peroxidase, glutathione S-transferase, spermidine synthase, ethylene-responsive transcription factors, ascorbate peroxidase, late embryogenesis abundant proteins, aquaporin*. (Natarajan et al., 2010). Further knowledge of candidate genes, especially for their stress-responsive traits, can be derived from plants such as the poplar (Song et al., 2012), a model tree.

For biomass-accumulating crops, lignin modification is of primary interest. The brown-midrib (*bm*) mutants of maize are the primary source of knowledge on lignification and their digestibility as well as their degradability (Barrière et al., 2004). The lignin contents in these mutants are lower than their wild-type plants, and thus they are easier to degrade by microbial, enzymatic, or catalytic methods. Interestingly, several mutant enzymes independently result in the same *bm* phenotype (Hisano et al., 2009). Thus, there should be several candidate genes whose expression modulation can produce the desired effect. Cinnamoyl CoA reductase (CCR; EC 1.2.1.44) catalyzing the conversion of cinnamoyl CoA esters to corresponding cinnamaldehydes and cinnamyl alcohol dehydrogenase (CAD; EC

TABLE 17.2

Candidate Genes for Oil Yield Identified in *Jatropha curcas*

S. No.	Biological Process	Encoded Protein	References
1.	Fatty acid biosynthesis	Acyl-ACP thioesterase A	Costa et al. (2010); Gomes et al. (2010); Natarajan et al. (2010)
2.		Acyl-ACP thioesterase B	
3.		Acyl carrier protein	
4.		Acetyl-CoA carboxylase	
5.		Biotin carboxyl carrier protein of ACCase	
6.		Carboxyl transferase of ACCase β subunit	
7.		Enoyl-ACP reductase	
8.		Geraryl pyrophosphate synthase	
9.		Hydroxyacyl-ACP dehydrase	
10.		Ketoacyl-ACP reductase	
11.		Ketoacyl-ACP synthase I	
12.		Ketoacyl-ACP synthase II	
13.		Ketoacyl-ACP synthase	
14.		3-Ketoacyl-CoA thiolase B	
15.		Malonyl-CoA ACP transacylase	
16.		Oleoyl-ACP desaturase	
17.		Palmitoyl-CoA hydrolase	
18.		Stearoyl-ACP desaturase	
19.	Triacylglycerol biosynthesis	1-Acylglycerol-3-phosphate-O acyltransferase	Costa et al. (2010); Gomes et al. (2010)
20.		Acyl-CoA:diacyl glycerol acyltransferase	
21.		Glycerol-3-phosphate acyltransferase	
22.		Lysophosphatidic acid acyl transferase	
23.		Oleosin	
24.		Phospholipid:diacyl glycerol acyltransferase	
25.		Phosphotidate phosphatase	
26.		Palmoityl-acyl carrier protein thioesterase	
27.	Triacylglycerol degradation	Acyl CoA binding protein	Costa et al. (2010); Natarajan et al. (2010)
28.		Monoacylglycerol lipase	
29.		Peroxysomal long-chain acyl-CoA synthetase	
30.		Peroxisomal fatty acid/acyl-CoA transporter	
31.		Triacylglycerol lipase	
32.	Fatty acid degradation	Acetyl-CoA acyltransferase	Costa et al. (2010)
33.		Acetyl-CoA C-acetyltransferase	
34.		Alcohol dehydrogenase	
35.		Acyl-CoA dehydrogenase	
36.		Acyl-CoA oxidase	
37.		Aldehyde dehydrogenase (NAD+)	
38.		Dienoyl-CoA reductase	
39.		Enoyl-CoA hydratase	
40.		Long-chain acyl-CoA synthetase	
41.	Desaturation of fatty acids	ω-3-Fatty acid desaturase	Natarajan et al. (2010)
42.		ω-6-Fatty acid desaturase	

1.1.195) catalyzing the conversion of these cinnamaldehydes to the corresponding alcohols are key enzymes in lignin biosynthesis (Paiva et al., 2011). The downregulation of these genes not only reduces the lignin content, but their extractability also becomes easier (Paiva et al., 2011). In maize *bm1* genotypes, CAD activity was significantly reduced by 60%–70% in the stem tissue (Halpin et al., 1998). Chen et al. (2004) have already reported the improved digestibility of CAD transgenics. The brown-midrib mutants have also been induced in other monocots such as *Sorghum* and pearl millet with a modified lignin composition (Cherney et al., 1991; Barrière and Argiller, 1993), thus raising hopes for a closely related *Miscanthus* and switchgrass as well. With the rapid developments in the field of plant molecular biology, gene silencing techniques, such as antisense, RNA interference (RNAi), and virus-induced gene silencing (VIGS), can be easily used to knock down or silence the target genes (Chen and Dixon, 2007). The downregulation of lignin genes can significantly lead to the development of genotypes with improved characteristics for biofuel production.

17.5 Population Genomics

Even though functional genomics presents a detailed and wholesome account of the role of genes at a cellular level, often this knowledge is unexplored for crop improvement purposes. Many crops, especially tree species, show recalcitrance to transformation and have longer generation periods. Grasses and other annuals have complex genomes. Further, there are very few bioenergy crops for which whole-genome sequence information is available. Thus, it is difficult to understand the complexity of the yield traits. A major target in the case of bioenergy crops is to maintain their genetic diversity, which can directly or indirectly contribute to the harvest. From a breeder's point of view, understanding the significance of the expression of a gene is of little practical interest, unless the polymorphism in the gene sequence is responsible for the differential expression.

Population genomics facilitates an understanding of the genomic basis for local adaptation, thereby guiding the breeding strategies for traits related to local adaptation. A high level of genetic diversity exists within and across the populations in terms of the yield in most bioenergy plants. Elucidating the ecological and quantitative genetics of these traits remains a long-lasting goal. There are several ways by which these can be exploited, and population genomics in particular can explain the uniform or nonuniform trait variation by quantitative trait loci (QTLs) identification and mapping to individual loci. The diversity in gene expression underlies the phenotypic diversity among individuals for most of the agronomic traits. An analysis of the determinants of candidate gene expression, an integral aim of genomics, can help our understanding of the mechanisms for phenotypic variation, which can be exploited through the molecular mapping of QTL and implementing it in breeding programs. In order to implement marker-assisted selection (MAS) for the genetic improvement of bioenergy crops, the molecular basis of biomass/oil yield and quality needs to be understood, upon identifying the genomic regions where these QTLs are localized. Molecular and genetic markers have made it possible to detect the QTLs that are significantly associated with the traits of interest (Liu et al., 2011). In order to conduct a QTL analysis, an appropriate mapping population is required, showing significant genetic variations at the genotypic and phenotypic levels among the siblings. QTL analyses for total oil content have been made in a number of food and energy crops including *Jatropha* and *Camelina* (Csanadi et al., 2001; Burns et al., 2003; Gehringer et al., 2006; Liu et al., 2011). Similarly, for a number of agronomic characters including biomass, QTLs have been identified and mapped in switchgrass and *Miscanthus* (Atienza et al., 2003; Serba et al., 2011). In addition, *Jatropha* has also been subjected to QTL analysis for higher seed yields and growth rates (Sun et al., 2012a). Eleven linkage groups have been identified in *Jatropha* (Liu et al., 2011, Sun et al., 2012). Recently, Sun et al. (2012) mapped 28 QTLs related to growth and seed traits using a backcross population of *Jatropha*.

Jansen and Nap (2001) proposed an approach wherein mRNA expression levels were recognized as quantitative traits in a segregating population. Mapping of the determinants that control these expression levels (eQTL) has been called *genetical genomics* (Jansen and Nap, 2001), which can identify genomic regions influencing the expression level of almost any gene assessed in a segregating population. Obviously, eQTLs map to the same genetic location as the gene whose transcript is being measured. The term *cis-QTL* has been coined to indicate the *cis*-acting factors that regulate the expression of the gene among siblings. eQTLs that map distant to the location of the gene being assayed identify the location of *trans*-acting regulators (*trans*-eQTL). The genetical genomics approach has been employed for identifying eQTL regulating gene expressions for *oleosin* genes in *Jatropha* (Liu et al., 2011).

Another attractive population genomics approach includes association mapping. It quickly helps us to identify the molecular markers tightly linked to the traits of interest, thereby enabling MAS for accelerating the breeding programs. The approach becomes especially effective in the case of polyploid species, whose genetics are complicated to understand and exploit. Despite its advantages, association mapping has scarcely been used so far, especially in the case of bioenergy crops. Recently, Sharma and Chauhan (2012) have been successful in cross-amplifying candidate gene regions from the castor bean to *Jatropha* and have subsequently carried out an association analysis with high oil content in *Jatropha*.

Overall, studies of the quantitative genetic variation within populations are sparse for all bioenergy crops. The quantitative genetic analyses that have been made are mostly restricted to an analysis of the characters at a particular site, thereby making it difficult to separate the effect of the genotype from the effect of the local environment. For example, the oil content in *Jatropha* has shown a variation in a broad range of 18%–40% (Chikara and Jaworsky, 2007; Kaushik et al., 2007; Rao et al., 2008; Arif and Ahmed, 2009); however, no data are available on the effect of the environment on this variation (Achten et al., 2010).

References

Achten, W.M.J., Nielsen, L.R., Aerts, R., Lengkeek, A.G., Kjaer, E.D., Trabucco, A., Hansen, J.K., et al. 2010. Towards domestication of *Jatropha curcas*. *Biofuels* 1:91–107.

Aghoram, K., Wilson, R.F., Burton, J.W., and Dewey, R.E. 2006. A mutation in a 3-Keto-AcylACP Synthase II Gene is associated with elevated Palmitic acid levels in Soybean seeds. *Crop Science* 46:2453–2459.

Akbar, E., Yakoob, Z., Kamarudin, S.K., Ismail, M., and Salimon, J. 2009. Characteristic and composition of *Jatropha curcas* oil seed from Malaysia and its potential as biodiesel feedstock. *European Journal of Scientific Research* 91:162–163.

Arif, M. and Ahmed, Z. 2009. *Bio-diesel:* Jatropha curcas (*A Promising Source*). Satish Serial Publishing House, Delhi.

Atienza, S.G., Satovic, Z., Petersen, K.K., Dolstra, O., and Martin, A. 2003. Identification of OTLs influencing agronomic traits in *Miscanthus sinensis* Andress I. Total height, flag-leaf height and stem diameter. *Theoretical and Applied Genetics* 107:123–129.

Barrière, Y.O. and Argiller, O. 1993. Brown-midrib genes of maize: A review. *Agronomie* 13:865–876.

Barrière, Y., Ralph, J., Méchin, V., Guillaumie, S., Grabber, J.H., Argillier, O., Chabbert, B., and Lapierre, C. 2004. Genetic and molecular basis of grass cell wall biosynthesis and degradability. II. Lessons from brown-midrib mutants. *Comptes Rendus Biologies* 327:847–860.

Baucher, M., Bernard-Vailhé, M.A., Chabbert, B., Besle, J.M., Opsomer, C., Van Montagu, M., and Botterman, J. 1999. Down-regulation of cinnamyl alcohol dehydrogenase in transgenic alfalfa (*Medicago sativa* L.) and the effect on lignin composition and digestibility. *Plant Molecular Biology* 39:437–447.

Beilstein, M.A., Al-Shehbaz, I.A., Mathews, S., and Kellogg, E.A. 2008. Brassicaceae phylogeny inferred from phytochrome A and *ndh*F sequence data: Tribes and trichomes revisited. *American Journal of Botany* 95:1307–1327.

Burns, M., Barnes, S., Bowman, J., Clarke, M., Werner, C., and Kearsey, M. 2003. QTL analysis of an intervarietal set of substitution lines in *Brassica napus*: (i) Seed oil content and fatty acid composition. *Heredity* 90:39–48.

Bush, D.R. and Leach, J.E. 2007. Translational genomics for bioenergy production: There's room for more than one model. *The Plant Cell* 19:2971–2973.

Casler, M.D. 2012. Switchgrass breeding, genetics and genomics. In: Monti, A. (ed.) *Switchgrass Green Energy and Technology*. Springer Verlag, London, pp. 29–53.

Chang-wei, L., Kun, L., Youc, C., and Yong-yu, S. 2007. Floral display and breeding system of *Jatropha curcas* L. *Forestry Studies in China* 9:114–119.

Chen, F. and Dixon, R.A. 2007. Lignin modification improves fermentable sugar yields for biofuel production. *Nature Biotechnology* 25:759–761.

Chen, L., Auh, C., Dowling, P., Bell, J., Lehmann, D., and Wang, Z.Y. 2004. Transgenic down-regulation of caffeic acid O-methyltransferase (COMT) led to improved digestibility in tall fescue (*Festuca arundinacea*). *Functional Plant Biology* 31:235–245.

Cherney, J.H., Cherney, D.J.R., Akin, D.E., and Axtell, J.D. 1991. Potential of brown-midrib, low-lignin mutants for improving forage quality. *Advances in Agronomy* 46:157–198.

Chikara, J. and Jaworsky, G. 2007. The little shrub that could—maybe. *Nature* 449:652–655.

Clifton-Brown, J.C. and Lewandowski, I. 2000. Water use efficiency and biomass partitioning of three different *Miscanthus* genotypes with limited and unlimited water supply. *Annals of Botany* 86:191–200.

Clifton-Brown, J.C., Lewandowski, I., Andersson, B., Basch, G., Christian, D.G., Bonderup-Kjeldsen, J., Jørgensen, U., et al. 2001. Performance of 15 *Miscanthus* genotypes at five sites in Europe. *Agronomy Journal* 93:1013–1019.

Costa, G.G.L., Cardoso, K.C., Del Bem, L.E.V., Lima, A.C., Cunha, M.A.S., de Campos-Leite, L., Vicentini, R., et al. 2010. Transcriptome analysis of the oil-rich seed of the bioenergy crop *Jatropha curcas* L. *BMC Genomics* 11:462.

Csanadi, G., Vollmann, J., Stift, G., and Lelley, T. 2001. Seed quality QTLs identified in a molecular map of early maturing soybean. *Theoretical and Applied Genetics* 103:912–919.

Donnison, I.S., Farrar, K., Allison, G.G., Hodgson, E., Adams, J., Hatch, R., Gallagher, J.A., Robson, P.R., Clifton-Brown, J.C., and Morris, P. 2009. Functional genomics of forage and and bioenergy quality traits in the grasses. In: Yamada, T. and Spangenberg, G. (eds) *Molecular Breeding of Forage and Turf*. Springer, New York, pp. 111–124.

Durrett, T.P., Benning, C., and Ohlrogge, J. 2008. Plant triacylglycerols: As feedstocks for the production of biofuels. *The Plant Journal* 54:593–607.

Gehringer, A., Freidt, W., Luhs, W., and Snowdon, R.J. 2006. Genetic mapping of agronomic traits in false flax (*Camelina sativa* subsp. *sativa*). *Genome* 49:1555–1563.

Goepfert, S. and Poirier, Y. 2007. ß-Oxidation in fatty acid degradation and beyond. *Current Opinion in Plant Biology* 10:245–251.

Gomes, K.A., Almeida, T.C., Gesteira, A.S., Lôbo, I.P., Guimarães, A.C.R., de Miranda, A.B., Van Sluys, M-.A., da Cruz, R.S., Cascardo, J.C.M., and Carels, N. 2010. ESTs from seeds to assist the selective breeding of *Jatropha curcas* L. for oil and active compounds. *Genomics Insights* 3:29–56.

Grafi, G., Chalifa-Caspi, V., Nagar, T., Plaschkes, I., Simon, B., and Ransbotyn, V. 2011. Plant response to stress meets dedifferentiation. *Planta* 233:433–438.

Grover, A., Aishwarya, V., and Sharma, P.C. 2007. Biased distribution of microsatellite motifs in the rice genome. *Molecular Genetics and Genomics* 277:469–480.

Halpin, C., Holt, K., Chojecki, J., Oliver, D., Chabbert, B., Monties, B., Edwards, K., Barakate, A., and Foxon, G.A. 1998. Brown-midrib maize (*bm1*): A mutation affecting the cinnamyl alcohol dehydrogenase gene. *The Plant Journal* 14:545–553.

Hammer, K. 1984. Das Domestikationssyndrom. *Kulturpflanze* 32:11–34.

Hisano, H., Nandakumar, R., and Wang, Z.Y. 2009. Genetic modification of lignin biosynthesis for improved biofuel production. In Vitro *Cellular and Developmental Biology-Plants* 45:306–311.

Hodkinson, T.R., Chase, M.W., Takahashi, C., Leitch, I.J., Bennett, M.D., and Renvoize, S.A. 2002. The use of DNA sequencing (ITS and *trnL*-F); AFLP; and fluorescent in situ hybridization to study allopolyploid *Miscanthus* (Poaceae). *American Journal of Botany* 89:279–286.

Hu, W.J., Harding, S.A., Lung, J., Popko, J.L., Ralph, J., Stokke, D.D., Tsai, C.J., and Chiang, V.L. 1999. Repression of lignin biosynthesis promotes cellulose accumulation and growth in transgenic trees. *Nature Biotechnology* 17:808–812.

Hutcheon, C., Ditt, R.F., Beilstein, M., Comal, L., Schroeder, J., Goldstein, E., Shewmaker, C.K., Nguyen, T., De Rocher, J., and Kiser, J. 2010. Polyploid genome of *Camelina sativa* revealed by isolation of fatty acid synthesis genes. *BMC Plant Biology* 10:233.

Jain, R.K., Coffey, M., Lai, K., Kumar, A., and MacKenzie, S.L. 2000. Enhancement of seed oil content by expression of glycerol-3-phosphate acyltransferase genes. *Biochemical Society Transactions* 28:958–961.

Jakob, K., Zhou, F., and Paterson, A.H. 2009. Genetic improvement of C4 grasses as cellulosic biofuel feedstocks. *In Vitro Cellular and Developmental Biology-Plant* 45:291–305.

Jansen, R. and Nap, J. 2001. Genetical genomics: The added value from segregation. *Trends in Genetics* 17:388–391.

Karp, A. and Shield, I. 2008. Bioenergy from plants and the sustainable yield challenge. *New Phytologist* 179:15–32.

Kaushik, N., Kumar, K., Kumar, S., Kaushik, N., and Roy, S. 2007. Genetic variability and divergence studies in seed traits and oil content of *Jatropha* (*Jatropha curcas* L.) accessions. *Biomass and Bioenergy* 31:497–502.

Kelley, S.S., Rowell, R.M., Davis, M., Jurich, C.K., and Ibach, R. 2008. Rapid analysis of the chemical composition of agricultural fibers using near infrared spectroscopy and pyrolysis molecular beam mass spectrometry. *Biomass and Bioenergy* 27:77–88.

Kirkpatrick, M. and Barton, N.H. 2006. Chromosome inversions, local adaptation and speciation. *Genetics* 173:419–434.

Knutzon, D.S., Thompson, G.A., Radke, S.E., Johnson, W.B., Knauf, V.C., and Kridlt, J.C. 1992. Modification of Brassica seed oil by antisense expression of a stearoyl-acyl carrier protein desaturase gene. *Proceedings of National Academy of Sciences of the United States of America* 89:2624–2628.

Labbé, N., Ye, P.X., Franklin, J.A., Womac, A.R., Tyler, D.D., and Rials, T.G. 2008. Analysis of switchgrass characteristics using near infrared techniques. *BioResearch* 3:1329–1348.

Li, L., Zhou, Y., Cheng, X., Sun, J., Marita, J.M., Ralph, J., and Chiang, V.L. 2003. Combinatorial modification of multiple lignin traits in trees through multigene cotransformation. *Proceedings of National Academy of Sciences of the United States of America* 100:4939–4944.

Liu, P., Wang, C.M., Li, L., Sun, F., Liu, P., and Yue, G.H. 2011. Mapping QTLs for oil traits and eQTLs for olesin genes in *Jatropha*. *BMC Plant Biology* 11:132.

Liu, L., Wu, Y., Wang, Y., and Samuels, T. 2012. A high-density simple sequence repeat based genetic linkage map of switchgrass. *Genes Genomes Genetics* 2:357–370.

Ma, X.F., Jensen, E., Alexandrov, N., Troukhan, M., Zhang, L., Thomas-Jones, S., Farrar, K., et al. 2012. High resolution genetic mapping by genome sequencing reveals genome duplication and tetraploid genetic structure of the diploid *Miscanthus sinensis. PloS One* 7:e33821.

Maisonneuve, S., Bessoule, J.J., Lessire, R., Delseny, M., and Roscoe, T.J. 2010. Expression of rapeseed microsomal lysophosphatidic acid acyltransferase isozymes enhances seed oil content in Arabidopsis. *Plant Physiology* 152:670–684.

McLaughlin, S.B. and Adams Kszos, L. 2005. Development of switchgrass (*Panicum virgatum*) as a bioenergy feedstock in the United States. *Biomass and Bioenergy* 28:515–535.

Nambisan, P. 2007. Biotechnological intervention in *Jatropha* for biodiesel production. *Current Science* 93:1347–1348.

Natarajan, P. and Madasamy, P. 2011. De novo assembly and transcriptome analysis of five major tissues of *Jatropha curcas* L. using GS FLX titanium platform of 454 pyrosequencing. *BMC Genomics* 12:191.

Natarajan, P., Kanagasabapathy, D., Gunadayalan, G., Panchalingam, J., Shree, N., Sugantham, P.A., Singh, K.K., and Madasamy, P. 2010. Gene discovery from *Jatropha curcas* by sequencing of ESTs from normalized and full-length enriched cDNA library from developing seeds. *BMC Genomics* 11:606.

Oliver, R.J., Finch, J.W., and Taylor, G. 2009. Second generation bioenergy crops and climate change: A review of the effects of elevated atmospheric CO_2 and drought on water use and the implications on for yield. *Global Change Biology Bioenergy* 1:97–114.

Paiva, J.A.P., Prat, E., Vautrin, S., Santos, M.D., San-Clemente, H., Brommonschenkel, S., Fonseca, P.G.S., et al. 2011. Advancing Eucalyptus genomics: Identification and sequencing of lignin biosynthesis genes from deep-coverage BAC libraries. *BMC Genomics* 12:137.

Perkins, J. 2010. REG tests 36 different feedstocks for blending potential. *Biofuel Journal* 10:10–11.

Pilate, G., Guiney, E., Holt, K., Petit-Conil, M., Lapierre, C., Leplé, J.-C., Pollet, B., et al. 2002. Field and pulping performances of transgenic trees with altered lignification. *Nature Biotechnology* 20:607–612.

Ralph, J., Guillaumie, S., Grabber, J.H., Lapierre, C., and Barrière, Y. 2004. Genetic and molecular basis of grass cell-wall biosynthesis and degradability. III. Towards a forage grass ideotype. *Comptes Rendus Biologies* 327:467–479.

Rao, G.R., Korwar, G.R., Shanker, A.K., and Ramakrishna, Y.S. 2008. Genetic associations, variability and diversity in seed characters, growth, reproductive phenology and yield in *Jatropha curcas* (L.) accessions. *Trees-Structure and Function* 22:697–709.

Reddy, M.S., Chen, F., Shadle, G., Jackson, L., Aljoe, H., and Dixon, R.A. 2005. Targeted down-regulation of cytochrome P450 enzymes for forage quality improvement in alfalfa (*Medicago sativa* L.). *Proceedings of National Academy of Sciences of the United States of America* 102:16573–16578.

Roorkiwal, M., Grover, A., and Sharma, P.C. 2009. Genome-wide analysis of conservation and divergence of microsatellites in rice. *Molecular Genetics and Genomics* 282:205–215.

Rubin, E.M. 2008. Genomics of cellulosic biofuels. *Nature* 454:841–845.

Saski, C.A., Li, Z., Feltus, F.A., and Luo, H. 2011. New genomic resources for switchgrass: A BAC library and comparative analysis of homeologous genomic regions harbouring bioenergy traits. *BMC Genomics* 12:369.

Sato, S., Hirakawa, H., Isobe, S., Fukai, E., Watanabe, A., Kato, M., Kawashima, U., et al. 2011. Sequence analysis of the genome of an oil bearing tree, *Jatropha curcas* L. *DNA Research* 18:65–76.

Serba, D.D., Vidya-Saraswathi, D., Saha, M.C., and Bouton, J.H. 2011. Mapping of QTLs for biomass, plant composition, and agronomic traits in switchgrass. Plant & Animal Genomes XIX Conference, January 15–19, San Diego, California.

Sharma, A. and Chauhan, R.S. 2012. Identification and association analysis of castor bean orthologous candidate gene-based markers for high oil content in *Jatropha curcas. Plant Molecular Biology Reporter* 30:1025–1031.

Sharma, N., Anderson, M., Kumar, A., Zhang, Y., Giblin, E.M., Abrams, S.R., Zaharia, L.I., Taylor, D.C., and Fobert, P.R. 2008. Transgenic increases in seed oil content are associated with the differential expression of novel *Brassica*–specific transcripts. *BMC Genomics* 9:619.

Sharma, M.K., Sharma, R., Cao, P., Jenkins, J., Bartley, L.E., Grimwood, J., Schmutz, J., Rokhsar, D., Ronald, P.C., and Qualls, M. 2012. A genome-wide survey of switchgrass genome structure and organization. *PloS One* 7:e33892.

Song, Y., Wang, Z., Bo, W., Ran, Y., Zhang, Z., and Zhang, D. 2012. Transcriptional profiling by cDNA-AFLP analysis showed differential transcript abundance in response to water stress in *Populus hopeiensis*. *BMC Genomics* 13:286.

Sun, F., Liu, P., Ye, J., Lo, Y.C., Cao, S., Li, L., Yue, G.H., and Wang, C.M. 2012a. An approach for *Jatropha* improvement using pleiotropic QTLs regulating plant growth and seed yield. *Biotechnology for Biofuels* 5:42.

Sun, G., Stewart Jr, N., Xiao, P., and Zhang, B. 2012b. MicroRNA expression analysis in the cellulosic biofuel crop switchgrass (*Panicus virgatum*) under abiotic stress. *PloS One* 7:e32017.

Swaminathan, K., Alababy, M.S., Varala, K., De Paoli, E., Ho, I., Rokhsar, D.S., Arumuganthan, A.K., et al. 2010. Genomic and small RNA sequenicng of *Miscanthus* × *giganteus* shows the utility of sorghum as a reference genome sequence for Andropogoneae species. *Genome Biology* 11:R12.

Swaminathan, K., Chae, W.B., Mitros, T., Varula, K., Xie, L., Barling, A., Glowacka, K., et al. 2012. A framework genetic map for *Miscanthus sinensis* from RNAseq-based marker shows recent tetraploids. *BMC Genomics* 13:142.

Tobias, C.M., Sarath, G., Twigg, P., Lindquist, E., Pangilinan, J., Penning, B.W., Barry, K., McCann, M.C., Carpita, N.C., and Lazo, G.R. 2008. Comparative genomics in switchgrass using 61,585 high quality expressed sequence tags. *Plant Genome* 1:111–124.

Wang, W., Vinocur, B., and Altman, A. 2003. Plant responses to drought, salinity and extreme temperatures: Towards genetic engineering for stress tolerance. *Planta* 218:1–14.

Wang, C.M., Liu, P., Yi, C., Gu, K., Sun, F., Li, L., Lo, L.C., et al. 2011. A first generation microsatellite and SNP-based linkage map of *Jatropha*. *PloS One* 6:e23632.

Wang, Y., Zeng, X., Iyer, N.J., Byrant, D.W., Mockler, T.C., and Mahalingam, R. 2012. Exploring the switchgrass transcriptome using second generation sequencing technology. *PloS One* 7:e34225.

Zhang, F.L., Niu, B., Wang, Y.C., Chen, F., Wang, S.H., Xu, Y., Jiang, L.D., et al. 2008. A novel betaine aldehyde dehydrogenase gene from *Jatropha curcas*, encoding an enzyme implicated in adaptation to environmental stress. *Plant Science* 174:510–518.

18

Molecular Farming in the Decades of Omics

Dinesh K. Yadav, PhD; Neelam Yadav, PhD; and S.M. Paul Khurana, PhD

CONTENTS

18.1 Introduction 564
18.2 Molecular Farming 565
18.3 Plants as Recombinant Protein Production Systems: Benefits and Drawbacks 566
    18.3.1 Production Cost...566
    18.3.2 Development of Timescale 566
    18.3.3 Product Authenticity 567
18.4 Important Considerations for Pharmaceutical Expression in Plants 567
    18.4.1 Expression Cassette Design 569
    18.4.2 Construction of Synthetic Gene with Host Plant Preferred Codon 569
    18.4.3 Use of UTRs 569
    18.4.4 Role of Promoters 570
    18.4.5 Posttranslational Modification...571
    18.4.6 Transgene Copy Number and Site of Integration 573
    18.4.7 Subcellular Protein Targeting and Stability..574
    18.4.8 Protein Authenticity and Safety 574
    18.4.9 Protein Storage in Tissue, Seeds, and Tubers..575
    18.4.10 Mucosal Adjuvants 575
18.5 Gene Transfer to Plants...576
    18.5.1 Transient Gene Expression in Plants Using Viral Vectors 576
    18.5.2 Agroinfiltration: Expression with a Bacterial Vector.....................................576
    18.5.3 Recombinant Viral Vectors...577
    18.5.4 Stable Nuclear Plant Transformation 577
    18.5.5 Plastid Transformation..577
    18.5.6 Magnifection...578
18.6 Choice of Production/Host Systems 579
    18.6.1 Crops with Major Foliar Biomass 579
    18.6.2 Vegetable and Fruit Crops 580
    18.6.3 Seed-Based Production Systems..581
    18.6.4 Nonvascular Lower Plants 582
        18.6.4.1 Mosses...582
        18.6.4.2 Microalgae..583
18.7 Alternative Plant-Based Production Systems 583
18.8 Downstream Processing and Recovery of Recombinant Proteins..............................583
    18.8.1 Oleosin Partitioning Technology...584
    18.8.2 Affinity-Based Tags 584

18.9 Pharmaceutically Valuable Proteins Produced in Plants.................................584
    18.9.1 Therapeutic Antibody Production in Plants.....................................584
    18.9.2 Vaccines from Plants 587
    18.9.3 Edible Vaccines...587
18.10 Commercial Aspect of Molecular Farming and Clinical Trials of
    Plant-Produced Pharmaceuticals 589
18.11 Regulatory Issues 589
    18.11.1 Product Safety...592
    18.11.2 Environmental Impact..593
18.12 Concluding Remarks 593
Acknowledgments...594
References...594

18.1 Introduction

Since the dawn of human civilization, plants have been used as medicine and the modern pharmaceutical industry relies on the derivatives and compounds of natural products. The use of plants in medicine stretches back to the earliest stages of civilization. As early as 1600 BC, the Egyptians compiled a list of more than 700 medicinal plants. The active ingredients in many of these plants have now been identified, and close to one-quarter of prescription drugs are still of plant origin. Modern biotechnology is extending the use of plants in medicine well beyond their original boundaries.

The use of biotechnology in the manufacturing of pharmaceutical products from crops has been one of the promised benefits of plant genetic engineering. In the 1980s, the development of techniques to introduce foreign DNA into plants initiated a new era in biotechnology. Such techniques allowed the use of plants for the production of specific heterologous proteins. The advancements in molecular biology, immunology, and plant biotechnology have changed the paradigm of plants as a food source to so-called plant bioreactors to produce valuable recombinant proteins. Thus, plants are now emerging as a promising, inexpensive, and convenient alternative farming system to conventional systems for the large-scale production of valuable therapeutic recombinant molecules (proteins), including therapeutic or diagnostic monoclonal antibodies, interleukins, interferons, and vaccines.

Molecular farming is the production of pharmaceutically important and commercially valuable proteins outside their natural source, namely, plants (Franken et al. 1997). It harnesses plants as heterologous protein expression systems, for the large-scale production of therapeutically valuable recombinant proteins. Molecular farming was developed based on the pretext that many pharmaceutically active proteins have been identified; however, without the means to produce such proteins in sufficient quantities, an evaluation of their therapeutic potential is unavailable. The purpose of molecular farming is to produce large amounts of an active and safe pharmaceutical protein at an affordable price. In principal, molecular farming is the production of pharmaceutically valuable proteins in a plant expression system and not the genetic modification of the expression system to express the pathogen resistance protein to protect the plant itself. For example, the creation of a pathogen-resistant plant by the expression of an antipathogen antibody is only an application of the molecular farming technology, it is not molecular farming. Molecular farming also

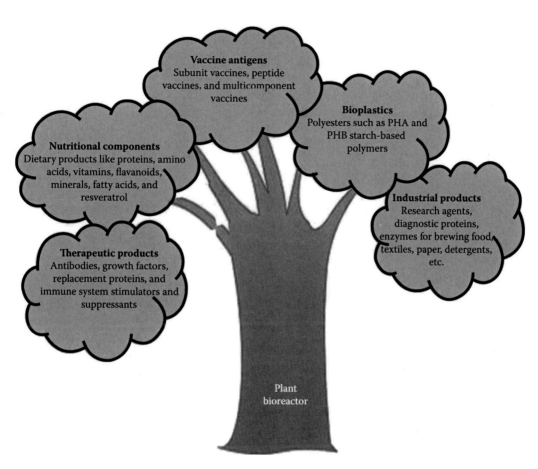

FIGURE 18.1 **(See color insert)**
Plant bioreactor tree showing biotechnological advances.

implies that the large-scale production of a recombinant protein is possible and economically feasible as well. Thus, the purpose of molecular farming is to produce large amounts of a biologically active and safe pharmaceutical protein at an affordable cost (Figure 18.1).

18.2 Molecular Farming

Molecular farming represents an unprecedented approach for the production of pharmaceutically important and commercially valuable proteins in plants. The prospect of producing modern medicines in plants is extremely attractive. A clear advantage is that the concept and its benefits are immediately obvious to the public and nonscientists. However, the introduction of any technology, particularly into a traditionally conservative area such as the pharmaceutical industry, is inevitably cautious. It was entirely predictable that the first plant-derived pharmaceuticals (Newcastle disease vaccine and glucocerebrosidase) would be produced using technologies that closely mimic the existing approved production platforms. Such incremental progress allows the introduction of new approaches within the context of an existing regulatory framework without posing major challenges to the regulatory authorities.

Plant production platforms are best suited for medical targets that are required in very large quantities for passive immunization or topical microbicides. The future perspective of molecular farming has four important areas where plant production systems could contribute to a change in pharmaceutical production, resulting in significant benefits for global health care.

The first important aspect is the production of specifically designed custom molecules in plants. These include vaccines or antibodies with a specific functional activity and displaying specific glycoforms, which provide enhanced particular effector functions (Strasser et al. 2009). The second aspect is the administration of a plant-derived oral vaccine in the context of a heterologous prime-boost strategy. Oral boosting in systemically vaccinated individuals bypasses the issue of inducing oral tolerance. It is also a simple and convenient way to provide second, third, and fourth booster immunizations, which will help to improve the success of any vaccine program, particularly in developing countries. Downstream processing is the third aspect of plant production systems that contributes significantly to production costs and the simplification of this step would greatly enhance the economic feasibility of many products. The regulatory issues relate to the consistency of the product, and this applies not only to the active pharmaceutical ingredient, but also to the coadministered plant ingredients. The final important impact area for plants could be the widening of the participation in pharmaceutical production, particularly in less-developed countries, with an emphasis on addressing local health issues. Molecular farming offers a simplified and inexpensive approach to pharmaceutical production and it would be of great interest to developing and underdeveloped countries struggling with health issues.

18.3 Plants as Recombinant Protein Production Systems: Benefits and Drawbacks

18.3.1 Production Cost

One of the most important factors determining the commercial viability of molecular farming is the achievement of adequate recombinant protein yields. Virtually unlimited amounts of recombinant protein could be grown either in open fields or in contained glasshouses. Although production scale-up can be readily achieved by increasing acreage, it is still necessary to achieve purification levels that are compatible with recombinant protein expression technologies. The production of recombinant proteins in transgenic plants is mainly based on stable nuclear transformation that leads to the expression of the inheritable transgene after its integration into the nuclear genome.

18.3.2 Development of Timescale

Proteins produced in plants accumulate to high levels (Verwoerd et al. 1995; Ziegler et al. 2000) and plant-derived pharmaceuticals are functionally equivalent to those produced by conventional technologies (Hiatt et al. 1989; Voss et al. 1995). However, the expression level of foreign proteins is generally less than 1% of the total soluble protein (TSP) observed in nuclear transgenic plants (Kusnadi et al. 1998). Immunoglobulin G (IgG) antibodies have been reported at levels of around 1% of the TSP, and the more complex secretory Ig has been reported to express at up to 5%–8% of the TSP (Hiatt et al. 1989; De Neve et al. 1993; Ma et al. 1995). Various monomeric polypeptide antigens have been shown to express at 0.0005%–0.38% of the TSP (Haq et al. 1995; Yu and Langridge 2001; Ashraf et al. 2005; Mishra et al. 2006; Yadav et al. 2012). Various strategies have been utilized to improve the expression levels of different antigens (Richter et al. 2000; Gil et al. 2001), including codon optimization, use of alternative promoters, and plastidial expression of proteins (Staub et al. 2000), but further optimization strategies are still required. It is estimated that recombinant proteins can be produced in plants at 2%–10% of the cost of microbial fermentation systems and at 0.1% of the cost of mammalian cell cultures. Varying expression levels of recombinant proteins have been reported.

18.3.3 Product Authenticity

Product authenticity is another important aspect of the success of molecular farming. According to the Swiss-Prot database, more than 50% of the proteins in eukaryotes (Apweiler et al. 1999), as well as one-third of approved biopharmaceuticals are glycoproteins (Walsh and Jefferis 2006). The biological activity of many therapeutic glycoproteins, namely, antibodies, blood factors, and interferons, is dependent on their glycosylation status. So, the heterologous expression systems aiming for molecular farming should have the capabilities to produce biopharmaceuticals with proper glycosylation. The key limitations of bacterial expression systems include their inability to produce correctly folded and assembled protein complexes, their lack of posttranslational modifications, and their requirement for culturing facilities. On the other hand, mammalian expression systems can correctly synthesize and process heterologous proteins, but they are expensive to culture and maintain. Also, their tissue culture conditions must be strictly regulated to avoid disease and cross-contamination (Skerra 1993; Taticek et al. 1994). The absolute requirement of several proteins for N-glycosylation supports the use of transgenic plants over classical production systems, such as *Escherichia coli* or yeast, owing to their ability to perform complex N-glycosylation processes that are required to ensure the effective bioactivity and pharmacokinetics of recombinant proteins in a therapeutic context. Plants offer a promising alternative as a cheap, safe, and efficient expression system for the production of functionally active biopharmaceutical proteins because they offer significant advantages over the classical expression (Table 18.1).

An effective plant production system for recombinant antibodies requires the appropriate plant expression machinery with an optimal combination of transgene expression regulatory elements, control of posttranslational protein processing, and efficient purification methods to recover the product. Hence, appropriate approaches and perspectives are required in plant systems to produce pharmaceutically important proteins.

18.4 Important Considerations for Pharmaceutical Expression in Plants

The successful development of vaccine antigens against human and animal pathogens in plants requires the selection and design of genes and promoters of important biopharmaceuticals that would express them at a high level in the target plant tissue.

TABLE 18.1

Features of Different Expression Systems for the Production of Recombinant Biopharmaceuticals

Cost	Bacteria	Yeast	Mammalian Cell Culture	Transgenic Plants	Transgenic Animals	Plant Cell Culture	Plant Viruses
Cost and Timescale							
Costs	Low	Medium	High	Low	High	Medium	Low
Storage cost	Moderate	Moderate	Expensive	Inexpensive	Expensive	Moderate	Not specified
Productivity	Medium	Medium	Medium	High	High	Medium	Not specified
Timescale	Short	Medium	Medium	Controversial	Very long	Medium	Low
Risks and Ethical Concerns							
Risks	Toxins	Low	Viruses, prions	Low	Virus, prions	Low	Not specified
Public perception of risks	Low	Medium	Medium	High	High	Not specified	High
Safety	Low	Unknown	Medium	High	High	Not specified	High
Contamination risks	Endotoxins	Low	Viruses, prions, oncogenic DNA	Low	Viruses, prions, oncogenic DNA	Low	Not specified
Therapeutic risks	Yes	Unknown	Yes	Unknown	Yes	Not specified	Unknown
Ethical concerns	Low	Low	Medium	Medium	High	Low	Not specified
Scale-Up							
Scale-up capacity	High	High	Very low	Very high	Low	Medium	Not specified
Scale-up cost	High	High	High	Low	Medium	High	Low
Technical Factors							
Gene size	Unknown	Unknown	Limited	Not limited	Limited	Not specified	Limited
Protein yield	Medium	High	Medium-high	High	High	Not specified	Very high
Propagation	Easy	Easy	Limited	Easy	Medium	Easy	Not specified
Multimeric protein assembly	No	No	No	Yes	Yes	Not specified	No
Protein folding accuracy	Low	Medium	High	High	High	Not specified	High
Glycosylation	None	Incorrect	Correct	Minor differences	Correct	Minor differences	Minor differences
Product quality	Low	Medium	High	High?	High	High	Not specified
Protein homogeneity	Low	Medium	Medium	High	High	Not specified	Medium
Distribution	Feasible	Feasible	Difficult	Easy	Difficult	Not specified	Easy
Storage temperature	−20°C	−20°C	Liquid nitrogen	Room temperature	Liquid nitrogen	−20°C	Not specified
GMP conformity	Possible	Possible	Possible	Difficult	Possible	Possible	Not specified

18.4.1 Expression Cassette Design

The expression of genes is regulated by a number of complex systems consisting of inter-acting elements/factors, such as promoter strength, *cis*- and *trans*-acting regulatory elements (transcription factors), cell growth stage, the expression level of RNA polymerase associated factors, and other gene-level regulation. Although the extent of the interdependence between the different factors is not completely understood, several strategies and mechanisms of particular interest have proven to increase recombinant protein yield.

18.4.2 Construction of Synthetic Gene with Host Plant Preferred Codon

The rate of translation can be optimized by ensuring that any mRNA instability sequences present in the target gene sequence are eliminated from the transgene construct, and that the translational start site matches with the Kozak consensus for plants (Gutiérrez et al. 1999; Kawaguchi and Bailey-Serres 2002). In the majority of eukaryotes, the translation initiation of mRNAs depends on the 5′-cap and involves ribosomal scanning of the 5′ untranslated region (UTR) for the initiation codon. An increasing number of upstream open reading frames (ORFs) have been found to play a key role in translational regulation. The first ATG codon has been shown to be in optimal context and is considered as the exclusive translation initiation site even in the presence of a second initiation codon just a few bases downstream. The sequences surrounding the initiation codon also play an important role in the translation initiation, and AACAAUGGC, UAAACAAUGGCU, and GCCAUGGCG have been identified as optimal contexts for plant genes (Lutcke et al. 1987).

The high expression of a transgene can be maximized by preventing transcriptional or posttranscriptional transgene silencing, which can inhibit the expression of even structurally intact transgenes. The potent sources for silencing include prokaryotic DNA sequences (such as those found on the vector backbone) and sequences that have the potential to form hairpin secondary structures at the DNA level, or double-stranded RNA when expressed.

Codon optimization is an effective and necessary step for the *de novo* designing of genes for recombinant protein production. It is well established that the genomes of different organisms, and the different genomes of single organisms, employ codon biases as mechanisms for optimizing and regulating protein expression (Gustafsson et al. 2004). In most cases of heterologous gene expression, it has been observed that the use of the target host crop-preferred/destined codon optimization of transgenes increases the expression efficiency by increasing the efficient translation rates and also eliminating cryptic introns and instability sequences (Koziel et al. 1996). It is the single most important determinant of protein expression (Lithwick and Margalit 2003; Surzycki et al. 2009), and it is necessary for a commercially viable level of expression of transgenes (Franklin et al. 2002; Mayfield et al. 2003; Mayfield and Schultz 2004). Codon optimization is also helpful in reducing the susceptibility of transgenes to posttranscriptional gene silencing (Heitzer et al. 2007). The Codon Adaptation Index (CAI) is used as a quantitative tool to predict heterologous gene expression levels based on their codon usage.

18.4.3 Use of UTRs

The 5′ and 3′ UTRs or leader sequences are very important for translation initiation and termination, and play a critical role in enhancing the expression levels of heterologous proteins. The use of the 5′ UTR of the rice polyubiquitin gene *RUBI3* along with its promoter (Lu et al. 2008; Samadder et al. 2008), the untranslated leader sequences of the alfalfa mosaic virus mRNA 4, or the tobacco etch potyvirus (Datla et al. 1993; Gallie et al. 1995) was reported to enhance the expression at the mRNA level as well as the translational level, suggesting that the 5′ UTR plays an important role in gene expression. The 3′ UTR also plays a critical role in gene expression as it contains a polyadenylation signal sequence for mRNA polyadenylation. Polyadenylation is directly related to mRNA stability (Chan and Yu 1998). Some A/U-rich sequence elements in the 3′ UTRs have been identified as destabilizing and causing rapid degradation of mRNA (De Rocher et al. 1998). Therefore, such destabilizing elements containing AT-rich nucleotides should be avoided for the expression of a heterologous transgene. Useful strategies to maximize the expression in designing the DNA sequence of the gene include the use of appropriate nonrelated UTRs on the 5′ end and 3′ end, which enhances the expression levels. The TA ending codons should be avoided as these are energetically less stable and are used less often in plants. The putative transcription termination signals (AAUAAA and its variants), the mRNA instability element (ATTTA), and the potential splice sites should be eliminated and long hairpin loops should be avoided (Ashraf et al. 2005).

18.4.4 Role of Promoters

The advancements in promoter technology provide a framework for designing an expression cassette that may not only provide precise control of transgene activity but also modulate the expression of a transgene in various contexts. A number of strong constitutive promoters have been derived from plants. For example, plant promoters such as CaMV 35S and 19S (Odell et al. 1985), the figwort mosaic virus (FMV) for full-length transcription (Gowda et al. 1989), the nopaline synthase promoter (An et al. 1990), the octopine synthase promoter (Ellis et al. 1987), and the polyubiquitin promoter (Hernandez-Garcia et al. 2009) are well characterized and extensively used for the expression of plant genes in a variety of plants (Odell et al. 1985). The transgenes can also be expressed in specific tissues of a plant, leaving all other tissues unaffected. This strategy is very helpful to concentrate the transgenic product in certain organs, such as seeds or fruits, to limit any possible negative effect on plant growth and improve the harvesting efficiency. Several such promoters have been used to target the expression of foreign biomolecules to specific organs of plants. The potato tuber-specific patatin promoter (Jefferson et al. 1990) has been extensively used to target several molecules, including cholera toxin B (CTB), heat-labile enterotoxin B (LTB), HBsAg, dextran, mutan, and alternansucrase, to potato tubers. A fruit-specific E8 promoter was identified from the tomato (Deikman et al. 1992) and used for the expression of various antigens in fruits (Jiang et al. 2007; Ramírez et al. 2007; Sandhu et al. 2000). Several promoters have been characterized that restrict the foreign biomolecule expression in the seeds only (Lau and Sun 2009). These include the arcelin promoter (Osborn et al. 1988), the maize globulin-1 promoter (Belanger and Kriz 1991), the maize zein promoter (Marks et al. 1985; Russell and Fromm 1997), the 7S globulin promoter (Fogher 2000), the rice glutelin promoter (Wu et al. 1998), and the soybean P-conglycinin α-subunit promoter (Chen et al. 1986).

The constitutive expression of a foreign protein may interfere with the normal growth and development of a plant. This problem can be overcome by using tissue-specific, chemical-inducible promoters. They offer high inducibility and specificity for regulated gene expression. These expression systems are quiescent in the absence of inducers. The use of chemical-inducible promoters in combination with the chemical responsive transcription factor can further restrict the target transgene expression to specific organs, tissues, or even cell types (Zuo and Chua 2000). To maintain the high efficiency of the inducible gene control system, the inducer should have high specificity to the promoter, a fast response upon induction, and the ability to switch off rapidly upon withdrawal; it should be nontoxic to the plant and easily applicable, and it should not be intrinsic to the plant. Previously, artificial promoters have also been constructed by the combinatorial engineering of *cis*-elements, which include enhancers, activators, and repressors, upstream of the core promoter. It has been found that the promoter strength depends on the motif copy numbers and the spacing between them (Gurr and Rushton 2005; Rushton et al. 2002). Chaturvedi et al. (2006) constructed an artificial bidirectional promoter that is composed of multiple *cis*-regulatory DNA sequence elements, arranged to give a transcription initiation module. It activates transcription simultaneously in both directions at comparable levels.

18.4.5 Posttranslational Modification

The majority of pharmaceutical proteins undergo some form of cotranslational or post-translational covalent modifications before they reach their final destination to perform their natural function. These may include enzymatic modifications such as glycosylation, phosphorylation, methylation, adenosine diphosphate (ADP)-ribosylation, oxidation, acylation, and proteolytic cleavage of the polypeptide backbone, and nonenzymatic modifications such as deamidation, glycation, racemization, and spontaneous changes in protein confirmation (Gomord and Faye 2004). Among these, N-glycosylation is often considered important when dealing with pharmaceutically important recombinant protein production in plants. N-glycosylation is often essential for the stability, solubility, folding, and biological activity of any protein. The biological activity of most therapeutic peptides—notably, antibodies, blood factors, and interferons—is dependent on their glycosylation status. The absolute requirement of several proteins for N-glycosylation also supports the use of transgenic plants over classical production systems such as *E. coli* or yeast, owing to their ability to perform the complex N-glycosylation processes that are required to ensure the effective bioactivity and pharmacokinetics of recombinant proteins in a therapeutic context. Plants have the property to posttranslationally modify a protein and have N-glycan biosynthesis similar to mammals (Figure 18.2).

However, plants are unable to exactly reproduce human-type glycosylation patterns in biopharmaceuticals (Gomord and Faye 2004; Faye et al. 2005), and, at this stage, plant-specific glycosylation is considered a major limitation for the use of plant-made pharmaceuticals (PMP) in human therapy.

In recent years, several approaches have been developed and rapid progress has been made in redesigning and humanizing plant glycosylation patterns. The first approach is based on the fact that the endoplasmic reticulum (ER) of plants and humans has a similar type of glycan structure (Figure 18.3) (Faye et al. 2005; Pagny et al. 2000). Retaining the recombinant glycoprotein in the ER avoids further modification of the glycoprotein in the Golgi apparatus, where plant-specific oligosaccharides are added (Yadav et al. 2012; Lerouge et al. 1998). The fusion of hexapeptide, SEKDEL, or KDEL/HDEL, the ER retrieval signal on the C-terminus of peptides, is often used to retrieve the protein back to the ER (Fujiyama et al. 2009; Gomord et al. 1997, 1995; Ko et al. 2003).

The second approach is based on gene inactivation or the silencing of host plant-specific glycosyltransferases to reduce or eliminate the activity. Using gene knockout strategies, Koprivova et al. (2004) successfully produced human IgG lacking β-1,2 xylose or α-1,3 fucose on mature glycan chains instead of glycoproteins with a core heptasaccharide identical to that found in human IgG. Recently, RNA interference (RNAi) strategies have been devised to knock out α-1,3 fucosyl transferase and β-1,2 xylosyl transferase activities in different species (Cox et al. 2006; Sourrouille et al. 2008; Strasser et al. 2008).

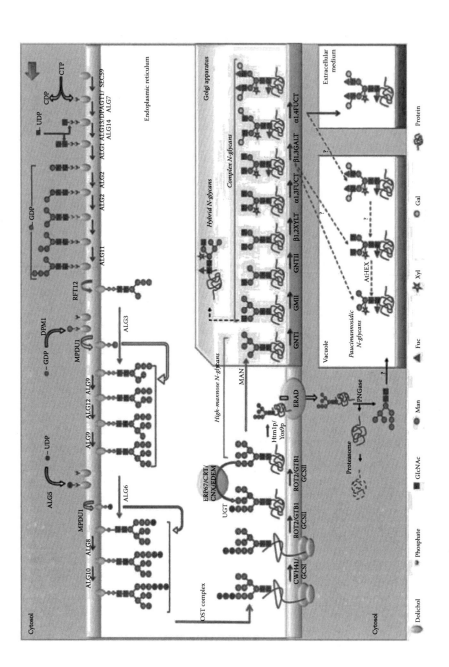

FIGURE 18.2

Reference pathway for protein N-glycosylation in plants. The assembly of the lipid-linked oligosaccharide precursor begins on the cytoplasmic side of the endoplasmic reticulum (ER) membrane. The resulting Man5GlcNAc2-PP-Dol then flips into the ER lumen, where four more Man residues and three glucose residues are added to produce the complete Glc3Man9GlcNAc2-PP-Dol. After the transfer of the oligosaccharide precursor from the lipid carrier onto the nascent protein by the action of the multisubunit oligosaccharyltransferase complex, the newly formed N-glycoprotein enters the calnexin–calreticulin (CNX–CRT) cycle. The alternate action of glucosidase II and UDP-glucose:glycoprotein glucosyl transferase drives the glycoprotein through this cycle until it is correctly folded and exported from the ER to the Golgi apparatus (GA). Terminally, misfolded proteins are retrotranslocated from the ER to the cytosol by the ER-associated degradation machinery for proteasomal degradation. Most host-specific maturations occur on the N-glycan when the glycoprotein is transported through the GA. The major glycan-processing steps leading to the biosynthesis of high mannose-, complex-, hybrid-, and paucimannose-type N-glycans in plant cells are illustrated. (From Gomord, V., Fitchette, A., Menu-Bouaouiche, L., Saint-JoreDupas, C., Plasson, C., Michaud, D., and Faye, L., *Curr Opin Plant Biol, 7,* 171–181, 2010).

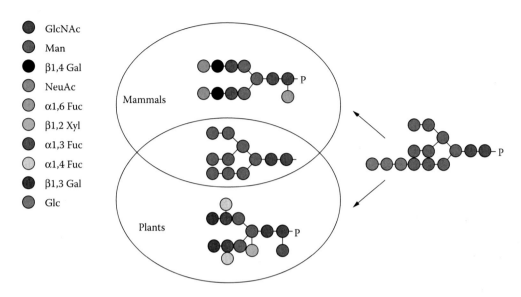

FIGURE 18.3 **(See color insert)**
Mammalian and plant glycan structures.

The third approach is based on the fact that most circulating human glycoproteins have terminal neuraminic acid (NeuAc) oligosaccharide residues on penultimate β-1,4 galactose residues (Figure 18.2). The absence of terminal sialic acid residues on glycosylated proteins reduces their clinical efficiency because of their rapid clearance via the galactose-specific hepatic asialoglycoprotein receptor (Stockert 1995). The half-life of recombinant desialylated erythropoietin (EPO) was estimated to be 2 min, unlike the 5–6 h for a recombinant sialylated variant of EPO in the blood plasma (Erbayraktar et al. 2003). To overcome this problem, plants were engineered for the *de novo* production of β-1,4 galactosyl transferase and sialyl transferase that add β-1,4 galactose and sialic acid residues to the terminus of N-glycosylated proteins to humanize the N-glycans. When expressed in transgenic plants, the galactose-processing enzyme hβ-1,4 GalT was shown to add β-1,4 galactose residues to plant N-glycans (Palacpac et al. 1999; Bakker et al. 2001; Fujiyama et al. 2001; Misaki et al. 2003). This most promising approach for targeting recombinant proteins and glycosyltransferases to specific subcellular locations could greatly improve their expression level and authenticity.

18.4.6 Transgene Copy Number and Site of Integration

Generally, an increase in the transgene copy number in transgenic plants should result in an increase in the recombinant protein expression level; however, in practice, multiple copies do not always result in higher expression levels. The integration of multiple copies of transgenes often results in rearrangements and transgene silencing (Hobbs et al. 1993; Linn et al. 1990; Svitashev et al. 2002). Therefore, single-copy insertions are preferred for a more predictable transgene expression and are less preferred when targeted for gene silencing (Jones et al. 2005; Kohli et al. 2003). Alternatively, independent single-copy transgenic lines with high transgene expression levels can be crossed and transgenic plants with an increased copy number as well as an increased expression level can be generated, reducing the silencing because of multiple copies (Streatfield 2007). The site of the transgene integration directly influences its expression level. A transgene is arbitrarily integrated into a highly transcribed region or a nontranscribed/least transcribed region. If the transgene integrates into a highly transcribed region, it will have a high rate of transgene expression. Further, the relative tolerance of a particular region to an invasion by foreign DNA directly affects the transgene expression. Heterochromatin regions, intercalary heterochromatin, and repetitive DNA stretches have been shown to be associated with transgene silencing in plant genomes (Matzke and Matzke 1998). In higher eukaryotes, the integration of DNA occurs mainly by illegitimate recombination at nonspecific sites in the genome (De Buck et al. 1999; Hohn and Puchta 2003). Homologous recombination can be used for the site-directed insertion of a transgene at a predetermined locus. A zinc finger nuclease containing a DNA recognition domain has been used to create site-specific chromosomal breaks, thereby enhancing the frequency of localized homologous recombination in plants. The result of the localized homologous recombination is very promising, showing a 10^4-fold enhanced site-directed insertion of transgenes (Wright et al. 2005).

18.4.7 Subcellular Protein Targeting and Stability

The expression levels and the quality of the recombinant pharmaceutical proteins produced in plants can be enhanced by the exploitation of the secretory pathway consisting of innate protein sorting and targeting mechanisms. Plant cells use these mechanisms to target host proteins to subcellular organelles. The subcellular targeting of recombinant proteins for secretion into the secretory pathway instead of the cytosol leads to a significant increase in their expression level. The primary site of the secretory pathway is the ER, which provides an oxidizing environment and an abundance of molecular chaperones, while there are a few proteases. These are likely to be the most important factors affecting protein folding and assembly.

In the absence of further targeting signals, the recombinant proteins are secreted into the apoplast. The secretion of recombinant proteins into the apoplast has advantages for downstream processing and also leads to significant levels of expression, but ER retention can give tenfold to a hundredfold higher yields (Conrad and Fiedler 1998). The stability of the recombinant proteins in the apoplast is lower than in the lumen of the ER. Therefore, heterologous protein expression levels can be increased even further if the proteins are retrieved to the ER lumen using an H/KDEL C-terminal tetrapeptide ER retrieval signal. In addition to an increase in their expression levels, the recombinant proteins that are retrieved to the ER are not modified in the Golgi apparatus, that is, they do not have plant-specific xylose and fucose residues; however, they have high-mannose glycans analogous to the mammalian system. Recombinant proteins have been targeted to various compartments of plant cells, namely, the intercellular space, chloroplasts, and the ER (Düring et al. 1990; Firek et al. 1993; Ma et al. 1994; Artsaenko et al. 1995; Voss et al. 1995; Baum et al. 1996; De Wilde et al. 1996; Schouten et al. 1996). The overall expression levels of different antibodies in stably transformed plants vary from 0.35% (van Engelen et al. 1994) to 6.8% of the TSP (Fiedler et al. 1997). Active human somatotropin was expressed in the transgenic plastids of tobacco and reached 7% of the TSP (Staub et al. 2000). This indicates that a chloroplast-based expression is genetically stable in subsequent generations and may be useful for the expression of pharmaceutical proteins.

18.4.8 Protein Authenticity and Safety

Plants are considered much safer than microbes and animals for the production of recombinant pharmaceutical proteins because they lack human pathogens, oncogenic DNA sequences, and endotoxins (Commandeur et al. 2003). However, it is important to consider the structural and functional properties of the recombinant protein, since the improved safety would be of no use if the transgenic product were unable to fulfill its intended biological activity. The function of a protein depends not only on its primary structure, but also on its tertiary and quaternary structures, as well as any posttranslational modifications. Plants can fold and assemble proteins at least as well as mammalian cells, as demonstrated by their ability to synthesize full-size antibodies (four polypeptide chains, each composed of multiple domains; Hiatt et al. 1989).

18.4.9 Protein Storage in Tissue, Seeds, and Tubers

Recombinant pharmaceutical proteins can also be targeted to and stored in harvested tissues, tubers, and seeds. This property emphasizes the versatility of plants as an expression system. Seeds are protein-rich storage organs that can be stored almost indefinitely at room temperature without loss of their biological activity (Fiedler and Conrad 1995; Conrad and Fiedler 1998; Conrad et al. 1998). Potato tubers have also been used as storage containers with their expression levels reaching 2% of the TSP in cold storage for 18 months (Artsaenko et al. 1998). The most promising approach for protein expression is to target the protein to the ER, which provides increased expression levels, reduced protease activity, and authenticity due to posttranslational modifications; for long-term storage, they should be targeted to and expressed in seeds.

18.4.10 Mucosal Adjuvants

Initially, molecular farming in plants was proposed for the production of edible vaccines by engineering edible plants for consumption (Arntzen et al. 2005; Prakash 1996). The idea of a plant vaccine is coming closer to reality. In recent years, a large number of antigens have been produced in plants and have been shown to activate the immune response against the antigen in animal models. Dow AgroSciences has received the first-ever regulatory approval for a plant-made veterinary vaccine. Several antigenic determinants belonging to various pathogens causing a variety of bacterial and viral diseases have been produced in plants (Sharma et al. 2008). Several antigens can be targeted to the mucosal system as chimeric products and the use of adjuvants, such as the CTB subunit or the LTB subunit as one of the components, makes the antigen delivery more efficient. The oral administration of CTB linked to antigens has been shown to trigger acceptable peripheral and mucosal immune responses that were otherwise unattainable when the antigen was given alone (Dertzbaugh and Elson 1993). Several antigens that have been produced in plants as a fusion partner of the CTB subunit or the LTB subunit have been shown to assemble in functionally active oligomers.

Adjuvants are important for enhancing the immune response to antigens. The addition of adjuvants to vaccines sustains and directs the immunogenicity and modulates the appropriate immune responses. This reduces the amount of antigen required and improves the efficacy of vaccines. One of the targeting strategies involves linking antigens to molecules that bind well to the immune system components, such as the M cells in the intestinal lining. The CTB subunit and the *E. coli* LTB subunit are potent mucosal immunogens and adjuvants. They both bind directly to the GM1-ganglioside receptor molecules on the M cells by fusing antigens from other pathogens to these subunits.

18.5 Gene Transfer to Plants

Following the appropriate designing of a recombinant gene, a suitable method is required to transfer the gene expression construct for its expression in whole plants, plant tissues, or cell suspension culture in a suitable expression host. However, the development of an efficient transformation procedure/technique for all major crop species remains a big challenge.

18.5.1 Transient Gene Expression in Plants Using Viral Vectors

The transient gene expression in plants has several advantages. It can be used to test the prototype design and function of a complete expression construct before progressing to large-scale production, but the transient expression itself can be used for protein production. Transient gene expression systems are rapid and flexible, making transient gene expression suitable for the verification of the prototype gene design and the functionality of the resulting gene product before moving on to large-scale production in transgenic plants (Kapila et al. 1996). Transient expression methods often use either agrobacterial or viral vectors. There are three basic transient expression procedures to transfer a gene to plant cells: (a) particle bombardment system—in this method, gold particles are used as the projectiles, which are coated with "naked DNA"; (b) vacuum infiltration of the intact host tissue with agrobacteria containing a recombinant plasmid (agroinfiltration); and (c) infection with modified viral vectors. The overall level of transformation efficiency varies between these three systems. Usually, the particle bombardment method only delivers the recombinant naked DNA to a few cells and, for transcription, the delivered DNA has to reach the cell nucleus (Christou 1996). Agroinfiltration transforms many more cells than the particle delivery system and the tumor–inducing DNA (T-DNA) containing the gene of interest is actively transferred into the nucleus with the aid of several agrobacterial proteins (Kapila et al. 1996).

18.5.2 Agroinfiltration: Expression with a Bacterial Vector

In agroinfiltration, the agrobacteria carrying the expression vector are delivered into the leaf tissue by vacuum infiltration. Agrobacterial proteins then catalyze the transfer of the gene of interest into the host cells and the recombinant protein expression can be observed 3 days postinfiltration. The genes of interest are cloned into binary vectors and are transformed into suitable *Agrobacterium* strains. These agrobacterial suspensions are then used for the vacuum infiltration of leaves. Agrofiltration does not require any antibiotic selection marker to identify transformed cells, since the leaf tissue is only used for transient protein production.

In agroinfiltration, the transferred T-DNA does not integrate into the host chromosome but it is present in the nucleus, where it is transcribed, leading to the transient expression of the gene of interest (Kapila et al. 1996). The major advantages of this technique are that the multiple genes present in the different clones of the agrobacteria can be coexpressed and the assembly of multimeric protein complexes, namely, transiently expressed single-chain antibody fragments (scFvs), diabodies, and individual heavy and light chains, can be evaluated *in planta* (Vaquero et al. 1999). It can also be scaled up to produce tens of milligrams of recombinant protein and significantly accelerate preclinical trials.

18.5.3 Recombinant Viral Vectors

A modified viral vector, retaining its naturally wide host range of infection properties but not its virulence, can systemically infect most cells of a natural/related host plant and can transiently express foreign proteins (Scholthof et al. 1996). Here, the complete expression containing the gene of interest is cloned into the genome of a viral plant pathogen. This recombinant infectious viral genome is used to infect plants. The transcription of the introduced gene into RNA viruses takes place by natural viral replication in the cytoplasm, which transiently generates many transcripts of the gene of interest because of the high level of multiplication during virus replication (Porta and Lomonossoff 1996). Recombinant viral vectors have been used to express single-chain antibodies (Franconi et al. 1999; Hendy et al. 1999; McCormick et al. 1999) and the heavy and light chain of a full-size antibody from two different viral vectors using a tobacco mosaic virus (TMV)-based vector (Verch et al. 1998).

18.5.4 Stable Nuclear Plant Transformation

Stable nuclear transformation is the most attractive strategy for the long-term production of recombinant therapeutic proteins producing stable transgenic plants. Stable nuclear transformation can be defined as the integration of a target gene into a host plant genome. The generation of transgenic plants uses two principle technologies: *Agrobacterium*-mediated gene transfer to dicots, such as tobacco and pea (Horsch et al. 1985), and biolistic delivery of genes to monocots, such as wheat and corn (Christou 1993).

The *Agrobacterium*-mediated transformation method is the most widely used technique for dicot transformation, but it has a restricted host range and does not efficiently infect monocots. However, rice and some other monocots can be transformed by *Agrobacterium* (Chan et al. 1993; Hiei et al. 1994, 1997). For the stable transformation of plants, the gene of interest is cloned into a binary vector. The *Agrobacterium* that is transformed with a binary vector delivers the target gene into the host cell genome. The selection of the transformed cells with stably integrated copies of the target gene is performed by a selectable antibiotic resistance gene that is introduced in the expression vector. The production of stable transgenic plants depends on the plant variety and it may take 3–7 months for plants to be available for the characterization of the target protein.

High-intensity agriculture can produce surprisingly large amounts of biomass. Table 18.2 summarizes the biomass production (yield/hectare) of classical/staple crop species.

18.5.5 Plastid Transformation

In addition to its advantages, the impinging limitations associated with position effects due to random gene integration, low expression levels, and safety due to the environmental dissemination of genes by pollen have hampered the expediency of a stable nuclear transformation. For commercial exploitation, high expression levels of the recombinant therapeutic proteins are required, and plastid transformation provides a valuable alternative to nuclear transformation.

The introduction of foreign DNA into the plastid genome of a plant provides an alternative and effective strategy with the unique advantage of high yields of the recombinant product. The biolistic transformation technique is used to engineer the plastids. The fundamental difference between plastid transformation and nuclear transformation is the insertion of the foreign gene at a defined position in the plastid genome due to

TABLE 18.2

Yields of Several Plant Species in Tons per Hectare

Crop	Crop Yield
Banana	16.6
Lettuce	33.1
Peas	9.1
Rice	6.6
Cabbage	24.3
Maize	8.4
Potato	124.5
Tobacco	2.2
Tomato	59.4
Eggplant	26.9
Peanut	2.9
Rapeseed	1.5
Wheat	2.7
Tobacco	170–200

Source: Fischer, R. and Emans, N., *Transgenic Res*, 9, 279–299, 2000.

homologous recombination. This phenomenon often results in very high protein yields. Each plant cell contains ~100 chloroplasts with about 100 identical genomes, thus a single gene is represented 10,000 times within one cell. The copy number of the gene encoded in two inverted repeat regions of higher plants can reach up to 20,000 copies. The absence of positional effects and gene silencing, the ease of multigene engineering in a single trans-formation event, the transgene containment via maternal inheritance in most crops, and the lack of pleiotropic effects are the advantages of plastid engineering. The attainment of high expression levels of foreign proteins in plastids is a major breakthrough, with levels as high as 46% of the TSP (De Cosa et al. 2001). Some chloroplast transgenes have even been reported to accumulate to levels of 72% of the TSP (Ruhlman et al. 2010), making this system ideal for the large-scale production of recombinant protein. Over 40 transgenes have been stably integrated and expressed via the tobacco chloroplast genome to confer important agronomic traits as well as to produce industrially valuable biomaterials and therapeutic proteins (Grevich and Daniell 2005). Thus, if plants are intended for use as bioreactors for the large-scale production of recombinant proteins, then chloroplasts should be targeted for genetic modification. However, there is a single limitation of the plastid expression due to its prokaryotic nature, as chloroplasts lack the complex posttranslational modifications machinery necessary to perform glycosylation of proteins. So, proteins that do not require posttranslational modifications can only be expressed in chloroplast expression systems.

18.5.6 Magnifection

Magnifection is a newly developed technique that uses *Agrobacterium* cell machinery to deliver viral replicons (Gleba et al. 2005). The unique advantage of this technique is that different 5′ modules carrying different organelle-targeting signals can fuse in-frame with the transgene after recombination. Nuclear processing allows easier testing of the recom-binant protein accumulation in different compartments of a cell. This technique has the advantages of vector efficiency, efficient systemic DNA delivery of the *Agrobacterium*, the speed and expression level of a plant RNA virus, and the posttranslational capabilities and low production costs of plants. This process has been validated through the expression of a number of antigens of microbial and viral origin and antibodies (Gleba et al. 2005). The hepatitis B core antigen (HBcAg) was expressed at levels of 7% of the TSP after 7 days of postinfection in tobacco leaves (Huang et al. 2006). Similarly, plague antigens were expressed in tobacco leaves at levels of 2 mg/g fresh weight of leaves and exhibited a protective efficacy to an aerosol challenge when purified antigens were used to immunize guinea pigs subcutaneously (Santi et al. 2006). However, one of the limitations of this process could be posttranslational modifications, especially the glycosylation pattern. Similar problems are associated with other alternative expression systems. The choice of method of gene transfer depends on the host systems and the type of recombinant protein that needs to be produced.

18.6 Choice of Production/Host Systems

A number of factors must be considered when choosing a production/crop system for molecular farming. An optimal system would offer flexibility, ease of genetic manipulation, high protein yields, rapid scale-up, low production and handling costs, free from antinutritional factors, amenable breeding procedures, and *in situ* stability of the transgene pharmaceutical product. No single model crop species has all these characteristics. However, a large number of species are amenable to molecular farming, including model plants with major foliar biomass (tobacco, *Arabidopsis*, and alfalfa), seed-based cereal and legume crops (rice, wheat, maize, pea, and soybean), and vegetables and fruit crops (tomato, potato, cabbage, and banana). The system of choice has to be evaluated on a case-by case basis according to the global agronomic feasibility, the production area, and the value of the recombinant protein. Representative crop species, which have been used as host systems for the expression of recombinant proteins, are discussed in the following sections.

18.6.1 Crops with Major Foliar Biomass

The major leafy crops that are being investigated for molecular farming include tobacco, alfalfa, soybean, and lettuce. Tobacco is the first and obviously successful choice of crop system for molecular farming. A well-established technology for gene transfer and expression, a high leafy biomass, and a prolific seed production (3000 seeds/capsule) make tobacco an ideal candidate for molecular farming. The major advantages of tobacco include the existence of a large-scale processing infrastructure. Because tobacco is neither a food nor a feed crop, there is little risk that tobacco material will contaminate either the food or the feed chain. Further, the availability of a wide range of germplasm makes tobacco widely customizable to remove metabolites such as nicotine. The important biopharmaceuticals produced in tobacco are listed in Table 18.3.

Alfalfa is a perennial forage crop that can remain productive for up to 5 years with the advantage of using atmospheric nitrogen through nitrogen fixation, thereby reducing the need for chemical fertilizers. It can be propagated by stem cutting with a strong regenerative capacity, thus enabling the production of large clonal populations within a limited time period and it can be harvested up to nine times a year. Alfalfa is particularly useful

TABLE 18.3

Production of Biopharmaceuticals in Transgenic Tobacco

Biopharmaceutical	Use/Therapy	Expression Levels
Human serum albumin	Blood substitute	0.025% TSP
Human protein C	Anticoagulant	0.002% TSP
Human granulocyte-macrophage colony-stimulating factor	Neutropenia	0.5% TSP in seeds
Human somatotropin	Hypopituitary dwarfism in children	7% TSP (chloroplasts)
Human erythropoietin	Regulation of erythrocytes mass/anemia	0.01% TSP
Human epidermal growth factor	Wound repair and cell proliferation	—
Human interferon alpha	Hepatitis B and C, hairy cell leukemia	—
Human interferon beta	Multiple sclerosis	—
Human IL-2	Renal carcinoma	25–350 ng/g callus
Human IL-4	Immune and inflammatory responses modulation	1.1 µg/g callus
Human hemoglobin alpha, beta	Blood substitute	0.05% seed protein
Human homotrimeric collagen	Natural biomaterial	0.1% µg/g FW
Angiotensin-converting enzyme 1	Hypertension	100 µg/g FW
Alpha trichosanthin	AIDS	2% TSP
Glucocerebrosidase	Gaucher's disease	1%–10% TSP
Human alpha lactalbumin	Nutraceutical	0.2% TSP
Human lactoferrin	Antimicrobial	0.3% TSP

Source: Sunil Kumar, G.B., Ganapathi, T.R., Srinivas, L., and Bapat, V.A., *Applications of Plant Metabolic Engineering*, Springer, Germany, 2007.

Note: —, not reported; TSP, total soluble protein; FW, fresh weight.

because it has a large dry biomass yield per hectare. The low content of the secondary metabolites in its leaves makes it an ideal crop over tobacco for molecular farming. Lettuce is also being investigated as a production host for edible recombinant vaccines and has been used in one series of clinical trials for a vaccine against the hepatitis B virus (Kapusta et al. 1999). The disadvantages of leafy crops are that the recombinant proteins are often unstable in dilute and aqueous environments, so the leaves must be dried or frozen or processed soon after they are harvested or they must be harvested during nighttime hours to ensure the lowest possible temperatures when extracting the product. The presence of phenolic compounds that are released during protein extraction can interfere with downstream processing.

18.6.2 Vegetable and Fruit Crops

A major advantage of recombinant protein expression in fruit and vegetable crops is that the edible part of the transgenic plant can be consumed uncooked, unprocessed, or as a partially processed material. This feature makes fruit and vegetable crops ideal for the production of recombinant subunit vaccines, food additives, nutraceuticals, and antibodies designed for topical immunotherapy (Ma et al. 2003).

The potato (*Solanum tuberosum*) is one of the important staple food crops with high nutritional value and yield potential. It is the largest noncereal food and provides approximately one-half of the world's annual production of all roots and tubers. Potatoes have been widely used for the production of plant-derived vaccines and have been administered to humans in most of the clinical trials. The potato tubers have enormous potential as a bulk-production system for the production of antibodies. They have been used for the production of glucanases (Dai et al. 2000), diagnostic antibody-fusion proteins (Schunmann et al. 2002), and human milk proteins (Ma et al. 2003). As potato tubers are the storage organs, they can be stored for long periods of time without any processing before being consumed, thus foreign proteins may be stable for an extended period of time.

The tomato is the second most popular crop in the world, next to the potato. It is grown worldwide and is more palatable than potatoes. It has additional advantages, including a high biomass yield (~68,000 kg/ha), and is a short-duration crop. Tomatoes were first used for the production of the candidate rabies vaccine (McGarvey et al. 1995; Ma et al. 2003). The banana is another crop that is largely produced throughout the year in the tropics and subtropics. It is an ideal vehicle for edible vaccines as it offers advantages such as a rich nutrient source, digestibility and palatability for infants, and it is easily available where economical vaccines are needed. Bananas are vegetatively propagated and do not set seeds. It is an ideal candidate for gene containment and there is no segregation of the transgene. The first report on hepatitis B vaccine production in banana fruit was recently published (Sunil Kumar et al. 2005). Some suitable host systems for biopharmaceutical production are shown in Table 18.4.

18.6.3 Seed-Based Production Systems

Contrary to crops with major foliar biomass, the expression of proteins in seeds enables their long-term storage, even at room temperature, because seeds have the appropriate biochemical environment to promote stable protein accumulation. Seeds have the ability to create biochemically specialized storage compartments derived from the secretory pathway for the accumulation of proteins in protein bodies and storage vacuoles. Seeds are also desiccated, which reduces the nonenzymatic hydrolysis and protease degradation of stored proteins. Stoger et al. (2000) demonstrated that the antibodies expressed in seeds remain stable for at least 3 years at ambient temperatures with no detectable loss of activity. Cereal seeds also lack the polyphenolic substances that are abundantly present in tobacco leaves, thereby increasing the efficiency of downstream processing (Ma et al. 2003). Several different crops, including cereals (rice, wheat, and maize) and legumes (pea and soybean), have been investigated for seed-based production (Stoger et al. 2002; Hood 2002; Vierling and Wilcox 1996).

Rice, wheat, pea, and tobacco have been used to compare the expression merit of each production system using the same scFv antibody (Stoger et al. 2002). Rice plants showed the highest yield per unit biomass, with the use of optimal promoters (the enhanced CaMV 35S promoter for tobacco and the ubi-1 promoter for rice). The levels were lower in wheat and pea seeds. The rice-based human lactoferrin (Anzai et al. 2000) and an oral vaccine expressing the CTB subunit (Nochi et al. 2007) were successfully expressed. The recombinant antibody of a scFv against the carcinoembryonic antigen was produced in rice and wheat. Maize was chosen by Prodigene (http://www.prodigene.com/) as the first

TABLE 18.4

Host Systems for Biopharmaceutical Production

Crop Species	Recombinant Protein(s) Expressed	Company
Alfalfa	Recombinant antibodies	Medicago, Canada
Corn	Industrial enzymes, vaccines, recombinant antibodies	Large Scale Biology Corporation, USA; Meristem Therapeutics, France
Rice	Lactoferrin, lysozyme	Ventria Biosystems Inc., USA
Barley	Diagnostic antibody, cellulase, human serum albumin, lactoferrin, lysozyme, thaumatin, human growth factor	Ventria Biosystems Inc., USA; ORF Genetics, Iceland; Maltagen, Germany
Safflower/rapeseed	Hirudin, insulin	SemBioSys Genetics Inc., Canada
Tobacco	Large numbers of recombinant proteins	Large Scale Biology Corporation, USA
Moss	Antibodies, vascular endothelial growth factor (VEGF)	Greenovation Biotech Gmbh, Germany
Chlamydomonos	Monoclonal antibodies, vaccines and bioremediant delivery systems	Rincopharma, USA; Phycotransgenics, USA

Source: Sunil Kumar, G.B., Ganapathi, T.R., Srinivas, L., and Bapat, V.A., *Applications of Plant Metabolic Engineering*, Springer, Germany, 2007.

plant species for commercial molecular farming. Maize has been used for the commercial production of recombinant avidin, β-glucuronidase, technical enzymes such as laccase, trypsin, and aprotinin, and the production of recombinant antibodies is being explored. Barley is another potentially useful crop as the producer price is low and its seed protein content is high. Wheat has rarely been used for the production of pharmaceutical proteins, but it has a low producer price compared with rice, making it potentially attractive. Soybean has the advantage of a high protein content in its seeds. Oil seeds offer a unique advantage as they facilitate the targeting of the recombinant proteins to the oil bodies by the fusion of oleosin. The fusion protein can be recovered from the oil bodies using a simple extraction procedure and the recombinant protein can be recovered from its fusion partner by endoprotease digestion (Stoger et al. 2005). The limitation with soybean is that transformation procedures are more time consuming and its high oil content may limit its use.

18.6.4 Nonvascular Lower Plants

18.6.4.1 Mosses

Mosses are phylogenetically ancient organisms and are a potential host system for the production of recombinant proteins. *Physcomitrella patens* is an excellent expression host system for biopharmaceutical production. Mosses are unique among all the multicellular plants analyzed so far. They exhibit a very effective homologous recombination in their nuclear DNA and allow a targeted knockout of genes. They also facilitate the manipulation of the glycosylation pathway unlike higher plants. Engineering of the main haploid phase of their life cycle results in a desired phenotype without the requirement of time-consuming crossing steps. Strains of *P. patens* with excellent genetic stability, transgenic for recombinant human vascular endothelial growth factor (VEGF), were subcultured for several years without any selection pressure.

18.6.4.2 Microalgae

Microalgae represent the best combination of a high growth rate and ease of cultivation of microorganisms with the ability to perform posttranscriptional and posttranslational modifications of plants. Recent success in the expression of recombinant proteins in *Chlamydomonas reinhardtii* has shown the possibility of exploring novel expression systems. Lower eukaryotes such as microalgae can be employed to produce therapeutic proteins as they are capable of carrying out posttranslational modifications. This alga has beneficial attributes for the production of recombinant proteins. Nuclear, chloroplast, and mitochondrial DNA can be transformed easily and in a relatively short period of time between the initial transformants. It can synthesize, properly fold, and secrete glycosylated proteins and it can be cultured, ranging from 100 mL to 500,000 L, in a cost-effective manner. These characteristic features make *Chlamydomonas* an attractive system for the expression of recombinant therapeutic proteins.

18.7 Alternative Plant-Based Production Systems

Plant cell suspension cultures can also be used as an alternative to transgenic plants or transient expression systems to produce recombinant proteins. Cell suspension cultures are particularly advantageous when defined, and sterile production conditions are required, since these conditions are particularly applicable to the production of therapeutic proteins (Fischer et al. 1999b). Recombinant proteins expressed in plant cell suspension cultures can be secreted into the culture supernatant or retained within the cells. A wide range of recombinant antibodies, including full-size immunoglobulins, Fab fragments, single chains, bispecific antibody fragments, and fusion proteins, have been expressed in BY-2 tobacco and rice-cell suspension cultures (Doran 2000; Torres et al. 1999).

18.8 Downstream Processing and Recovery of Recombinant Proteins

In all heterologous production systems, the recombinant protein will have to be extracted and purified from the endogenous proteins of the producing organism. Protein purification is a key step in the preparation of a biopharmaceutical and it could be more than 80% of its production cost. The cost breakdown for a highly purified therapeutic protein from transgenic corn shows that the transgenic raw material constitutes only 5%–10% of the total manufacturing cost, whereas extraction and purification make up as much as 90% of the total cost. By contrast, the cost of purification from transgenic corn for the production of industrial protein amounts to only 35%–40% of the total manufacturing cost (Nikolov and Hammes 2002). Thus, the bioprocessing of the pharmaceuticals produced in transgenic plants contributes most to the cost involved in its production.

18.8.1 Oleosin Partitioning Technology

Recombinant antibodies are generally purified from plant protein extracts using their property to specifically interact with *Staphylococcus aureus* protein A or *Streptococcus* sp. protein G antigen. The targeted expression of recombinant proteins into specific subcellular compartments also offers plant-specific strategies to simplify the purification. For example, in oleaginous plants, the recombinant protein can be fused to a protein (oleosin) anchored in the membrane of the lipid bodies (oleosomes), which accumulate in the seed during maturation. By combining this fusion with an expression targeted to the seed, the recombinant protein is recovered with the lipids after a centrifugation step. The recombinant protein, cleaved by targeted proteolysis, is then extracted from the lipid fraction by a second-phase partitioning. This strategy has been used in rapeseed for the production of hirudin (Parmenter et al. 1995; Boothe et al. 1997). Recently, it has been adapted for the production of recombinant antibodies in the seeds of another oilseed plant, safflower (Seon et al. 2002).

18.8.2 Affinity-Based Tags

Many more expression tags, such as His_6 tags, MYC-tags, GST-tags, and TRX-tags, are utilized in the recovery of the recombinant protein from the soup. However, the disadvantage is that the tag has to be removed from the protein before parenteral use.

18.9 Pharmaceutically Valuable Proteins Produced in Plants

Medical and biological research has increased our understanding of many diseases through the sequencing of the human genome. It has shown the involvement of many novel proteins that could be used as cures. Recombinant antibodies, plasma proteins, vaccines, and diagnostic reagents are the target molecules for expression in plants. The use of plants for the production of pharmaceutically important proteins, vaccine antigens, and antibodies has been extensively explored over the last two decades. The very first pharmaceutically important human protein produced in a plant expression system, tobacco and sunflower, was human serum albumin (Barta et al. 1986). Since then, plants have been successfully used for the production of recombinant antibodies (Ma and Hein 1995a, 1995b), enzymes (Hogue et al. 1990; Verwoerd et al. 1995), hormones, interleukins (Magnuson et al. 1998), plasma proteins (Sijmons et al. 1990), and vaccines (Mason and Arntzen 1995; Walmsley and Arntzen 2000). Also, a large number of pharmaceutical proteins have been expressed in plant systems (Table 18.5) with proven efficacy and competitiveness, including antibodies, plasma proteins, human enzymes, and recombinant vaccines. For example, the annual global demand for purified human serum albumin is approaching 550 tons. This demand can be supplied by the production of human serum albumin in plant expression systems.

18.9.1 Therapeutic Antibody Production in Plants

Antibodies are essential therapeutic and diagnostic tools in medicine, human and animal health care, the life sciences, and biotechnology (Winter and Milstein 1991). Antibodies are

TABLE 18.5

Selected Pharmaceutical Proteins Expressed in Transgenic Plants

Year	Transgenically Expressed Pharmaceutical Proteins	Plant Expression System
1986	Human growth hormone	*Nicotiana tabacum* and *Helianthus annus*
1990	Human serum albumin	*N. tabacum* and *Solanum tuberosum*
1993	Human epidermal growth factor	*N. tabacum*
1994	Trout growth factor	*N. tabacum*
1995	Malaria parasite antigen	Virus particle
1994	Human a-interferon	*Oryza sativa*
1995	Hirudin	*N. tabacum* (Suspension cells)
1995	Rabies virus glycoprotein	*Lycopersicon esculentum*
1995	Erythropoietin	*N. tabacum* (suspension cells)
1996	Glucocerebrosidase, human protein C serum protease	*N. tabacum*
1997	Human a and b hemoglobin	*N. tabacum*
1997	Diabetes-associated autoantigen	*N. tabacum* and *S. tuberosum*
1997	Human muscarinic cholinergic receptors	*N. tabacum*
1997	Murine granulocyte-macrophage colony-stimulating factor	*N. tabacum*
1997	Mink enteritis virus epitope	Virus particle
1998	Interleukin-2 and interleukin-4	*N. tabacum* (suspension cells)
1998	Foot and mouth disease virus VP1 structural protein	*Arabidopsis thaliana*
1999	Human placental alkaline phosphatase	
1999	Human a1-antitrypsin	*O. sativa* (suspension cells)
1999	Human cytomegalovirus glycoprotein B	*N. tabacum*
2000	Human growth hormone (somatotrophin)	*N. tabacum* (seeds)
2000	Human growth hormone (somatotrophin)	*N. tabacum* (chloroplasts)
2000	Collagen	*N. tabacum*
2001	Human acetylcholinesterase	*L. esculentum*
2002	Bovine aprotinin	*Zea mays*
2003	Lactoferrin	*N. tabacum*
2003	Interleukin 18	*N. tabacum*
2003	Human granulocyte-macrophage colony-stimulating factor	*O. sativa*

Source: Fischer, R. and Emans, N., *Transgenic Res*, 9, 279–299, 2000.

classical of the expression of pharmaceutically valuable proteins in plants (Hiatt 1990). The expression of a functional antibody in tobacco leaves was a key breakthrough in making molecular farming in plants a reality (Hiatt et al. 1989; Düring et al. 1990). Modern recombinant DNA technologies and antibody engineering have broadened the applications for recombinant antibodies.

After the successful demonstration of the expression of functional full-size rAbs in transgenic plants, a wide range of different recombinant antibody formats have been successfully expressed in many plant species (Table 18.6).

These formats include full-size antibodies (Ma et al. 1994; Voss et al. 1995; Baum et al. 1996; De Wilde et al. 1996), Fab fragments (De Neve et al. 1993), scFvs (Owen et al. 1992),

TABLE 18.6

Recombinant Antibodies Expressed in Transgenic Plants

Year	Antibody Class	Antigen	Plant Organ	Cellular Location	Transformed Plant
1989	IgG1	Phosphonate ester	Leaf	ER	*Nicotiana tabacum*
1990	IgM	NP hapten	Leaf	ER chloroplast	*N. tabacum*
1991	VH domain	Substance P (neuropeptide)	Leaf	Intra- and extracellular	*N. benthamiana*
1992	scFv	Phytochrome	Leaf	Cytosol	*N. tabacum*
1993	IgG1 Fab	Human creatine kinase	Leaf	Nucleolus	*N. tabacum*
1993	scFv	AMCV	Leaf	Cytosol	*N. benthamiana*
1994	IgG	Fungal cutinase	Root	Apoplast	*N. tabacum*
1994	IgG1	*Streptococcus mutans* adhesion	Leaf	Apoplast	*N. tabacum*
1995	IgA/G	*S. mutans* adhesion	Leaf	Apoplast	*N. tabacum*
1995	IgG	TMV	Leaf	Apoplast	*N. tabacum*
1996	scFv	Cutinase	Leaf	ER	*N. tabacum*
1996	IgM	RKN secretion	Leaf root	Apoplast	*N. tabacum*
1996	scFv	BNYVV	Leaf	Apoplast	*N. benthamiana*
1996	scFv	Human creatine kinase	Leaf	Cytoplasm	*N. tabacum*
1997	scFv	b-1,4-endoglucanase	Root	Cytosol	*Solanum tuberosum*
1997	scFv	Oxazolone	Leaf	ER	*N. tabacum*
1998	scFv	Abscisic acid	Leaf	ER	*N. tabacum*
1997	scFv-IT	CD-40	Plant	Apoplast	*N. tabacum*
1998	scFv	Oxazolone	Tuber	ER	*S. tuberosum*
1998	Humanized IgG1	HSV-2	Plant	Secretory pathway	*Glycine max*
1998	scFv	Dihydroflavonol 4-reductase	Leaf	Cytosol	P hybrid
1999	IgG	Human IgG	Plant	Apoplast	*Medicago sativa*
1999	scFv	CEA	Leaf	Transient expression	*N. tabacum*
1999	scFv	Tospoviruses	Plant	ER, apoplast	*N. benthamiana*
1999	bi-scFv	TMV	Leaf	ER, apoplast	*N. tabacum* (suspension cells)
1999	scFv	TMV	Plant	Cytosol	*N. tabacum*
1999	scFv	CEA	Cell	ER, apoplast	*O. sativa* (suspension cells)
1999	scFv	38C13 mouse B cell lymphoma	Leaf	Apoplast	*N. benthamiana*
2000	scFv	CEA	Plant	ER, apoplast	*O. sativa* and *Triticum aestivum*
2000	scFv	TMV	Leaf	Apoplast, membrane	*N. tabacum*

Source: Fischer, R. and Emans, N., *Transgenic Res*, 9, 279–299, 2000.

Note: CEA, carcinoembryonic antigen; ER, endoplasmic reticulum; AMCV, artichoke mottle crinkle virus; TMV, tobacco mosaic virus; RKN, root knot nematode; BNYVV, beet necrotic yellow vein virus; HSV-2, herpes simplex virus-2; scFv-IT, scFv-bryodin-immunotoxin.

bispecific scFv fragments (Fischer et al. 1999a), membrane-anchored scFvs (Schillberg et al. 2000), and chimeric antibodies (Vaquero et al. 1999).

A recombinant scFv antibody (scFvT84.66) and a full-size mouse/human chimeric antibody (cT84.66), derived from the parental murine mAbT84.66 specific for the human carcinoembryonic antigen, were engineered into a plant expression vector. The *in vivo* assembly of a full-size cT84.66 was achieved by the simultaneous expression of the light and heavy chains after vacuum infiltration of tobacco leaves with two populations of recombinant *Agrobacterium*. Upscaling the transient system permitted the purification of significant amounts of functional recombinant antibodies from tobacco leaf extracts within a week (Vaquero et al. 1999).

Secretory immunoglobulin A (sIgA) is a complex multisubunit antibody with great potential for use in topical immunotherapy. IgA is the major antibody found in mucosal secretions. It was first expressed in tobacco by sequentially crossing plants expressing its individual polypeptide components and proved that plants could be used as a production system for sIgA. Ma et al. (1998) showed that recombinant sIgA specific for adhesion proteins from the oral pathogen *S. mutans* could prevent oral streptococcal colonization in human volunteers for up to 4 months.

18.9.2 Vaccines from Plants

Vaccination is the most effective form of health care to prevent, control, and eradicate infectious viral and bacterial diseases. Mass vaccination programs in developing and underdeveloped countries have been severely hampered due to their high cost and the availability of effective vaccine dosages. The worldwide need for the production of contamination-free, safe, and affordable vaccines with minimum manufacturing and processing costs has warranted the mass production of life-saving vaccines using the plant as a bioreactor to improve human life expectancy and health (World Health Organization: http://www.who.int/gpv/). A plant expression system for the production of affordable, contamination-free, and safe vaccines represents an alternative, combining the merits of whole-plant systems with microbial and mammalian cell culture expression systems. Curtis and Cardineau (1990) expressed the *S. mutans* surface protein A antigen in tobacco plants. Since then, a substantial number of immunogenic subunit antigens have been produced in plants to fulfill the demand of developing countries (Table 18.7).

18.9.3 Edible Vaccines

The greatest advantage of the production of vaccines in plants is that they can be used as edible vaccines. Use of edible vaccines in mass immunization would reduce the prerequisite for a cold chain, the need for equipment such as needles, and the requirement of trained personnel to administer the vaccine. Their use would also significantly reduce the cost of mass immunization. Several studies have been performed to establish the effectiveness of oral plant-produced vaccines.

Nochi et al. (2007) showed that transgenic rice could be used to provide oral immunoprotection against the cholera toxin even after being stored at room temperature for 1.5 years. In another oral vaccination study, the progeny of female mice, fed on a transgenic alfalfa-expressing antigen for a class of rotavirus, was detected to have specific cytotoxic T lymphocytes that provided long-term immunoprotection against rotavirus infection. An

TABLE 18.7

Recombinant Vaccines Expressed in Plants

Year	Vaccine Antigen	Transformed Species
1986	Hepatitis virus B surface	*Nicotiana tabacum*
1990	Malaria parasite antigen	Virus particle
1995	Rabies virus glycoprotein	*Lycopersicon esculentum*
1995	*Escherichia coli* heat-labile enterotoxin	*N. tabacum* and *Solanum tuberosum*
1996	Human rhinovirus 14 (HRV-14) and human immunodeficiency virus type (HIV-1) epitopes	Virus particle
1996	Norwalk virus capsid protein	*N. tabacum* and *S. tuberosum*
1997	Diabetes-associated autoantigen	*N. tabacum* and *S. tuberosum*
1997	Hepatitis B surface proteins	*S. tuberosum*
1997	Mink enteritis virus epitope	Virus particle
1997	Rabies and HIV epitopes	Virus particle
1998	Foot and mouth disease virus VP1 structural protein	*Arabidopsis thaliana*
1998	Rabies virus	Virus particle
1998	Cholera toxin B subunit	*S. tuberosum*
1998	Human insulin-cholera toxin B subunit fusion protein	*S. tuberosum*
1999	Foot and mouth disease virus VP1 structural protein	*Medicago sativa*
1999	Hepatitis B virus surface antigen	*Lupinus luteus* and *Lactuca sativa*
1999	Human cytomegalovirus glycoprotein B	*N. tabacum*
1999	Diabetes-associated autoantigen	*N. tabacum* and *Daucus carota*
2001	Rotavirus	*N. tabacum*
2005	Parvovirus	*N. tabacum*
2005	Rabies virus glycoprotein subunit	*N. tabacum*

Source: Fischer, R. and Emans, N., *Transgenic Res*, 9, 279–299, 2000.

effective edible hepatitis B vaccine has been generated using transgenic lupin and lettuce plants expressing the hepatitis B surface antigen. Mice and humans fed transgenic plant material produced hepatitis B-specific antibodies (Kapusta et al. 1999). In another recent study, the transactivating regulatory protein (Tat), necessary for the production of the HIV virus, was produced in transgenic tomato plants and elicited the production of anti-Tat antibodies in mice that ingested plant tissue containing the antigen (Ramírez et al. 2007). The human papillomavirus type 11 (HPV11) L1 major capsid protein expressed in potato plants resulted in an anti-virus-like particle (VLP) immune response in mice fed on transgenic potatoes (Warzecha et al. 2003). This shows the immense potential for orally administered plant-made vaccines to combat even the most virulent and complex pathogens. A major concern related to the gastric proteolysis of oral antigens is reduced by the fact that expressed antigens are encapsulated by the plant tissue, hence they resist immediate breakdown by gastric peptidases.

18.10 Commercial Aspect of Molecular Farming and Clinical Trials of Plant-Produced Pharmaceuticals

The commercial interest in molecular farming is due to its ability to produce recombinant proteins at a significantly lower cost than alternative and classical expression systems. Attention is now shifting toward the commercial exploitation of molecular farming. The U.S. Patent and Trade Organization (USPTO) and the European Patent Office (EPO) have issued a large number of patents covering plant molecular farming (Table 18.8). Plant molecular farming patents are categorized into five major areas: pharmaceuticals and nutraceuticals, antibodies, industrial molecules, vaccines, and post-translationally glycosylated proteins.

The progression of molecular farming products into human clinical trials is an important priority for the field. After the first report of recombinant plant-derived pharmaceutical, human serum albumin, produced in transgenic tobacco and potato protein (Sijmons et al. 1990), the first technical proteins produced in transgenic plants were on the market. This established that the molecular farming technology is moving forward from basic scientific proof of concept studies toward clinical trials of transgenic plant products, such as antibodies, blood products, cytokines, growth factors, hormones, recombinant enzymes, human and veterinary vaccines, and other pharmaceutical proteins. Several plant-produced pharmaceutical products for the treatment of human diseases are under clinical trial (Ma et al. 2003) (Table 18.9).

The first clinical trial of the plant-based immunotherapeutic agent, CaroRx, was reported by Planet Biotechnology Inc. (Mountain View, CA). It is a plant-produced sIgA antibody that can effectively eliminate the oral infection of *S. mutans*, which causes dental caries. Planet Biotechnology Inc. is also developing and clinically evaluating novel antibody-based therapeutics to treat infectious diseases and toxic conditions affecting the oral, respiratory, gastrointestinal, genital, and urinary mucosal surfaces and the skin. The Monsanto Company created a corn line producing human antibodies. A pharmaceutical partner of Monsanto plans to begin injecting cancer patients with doses of up to 250 mg of an antibody-based cancer drug purified from corn seeds. The company is also cultivating transgenic soybeans that produce humanized antibodies against herpes simplex virus 2 (HSV-2). ProdiGene (College Station, TX) and EPIcyte Pharmaceuticals (San Diego, CA) are in the process of producing human mucosal antibodies for passive immunization. The Large Scale Biology Corporation (Vacaville, CA) and Stanford University have developed a technology to produce a tumor-specific vaccine for the treatment of malignancies. Several veterinary vaccines are also in the pipeline; Dow AgroSciences (USA) recently announced the production of plant-based vaccines for the veterinary health industry. A good number of industrial proteins from molecular farming are commercially available for research purposes (Table 18.10).

18.11 Regulatory Issues

Transgene expressions are successful technological demonstrations, but it is necessary to consider the wider impact of pharmaceutical plants and plant products on human and environmental health.

TABLE 18.8

Patents on Molecular Farming

Patent	Issue Date	Original Assignee	Title
US4956282	September 11, 1990	Calgene Inc.	Mammalian peptide expression in plant cells
US5629175	May 13, 1997	Calgene Inc.	Molecular farming
US5929304	July 27, 1999	Croptech Development Corporation Virginia Tech Intellectual Properties Inc.	Production of lysosomal enzymes in plant-based expression systems
US6040498	March 21, 2000	North Caroline State University	Genetically engineered duckweed
US6096547	August 1, 2000	Calgene, LLC	Method and transgenic plant for producing mammalian peptides
US6740526	May 25, 2004		Quantitative transient protein expression in plant tissue culture
US6753459	June 22, 2004	National Research Council of Canada Dow Agrosciences LLC	Transgenic plants and methods for production thereof
US6774283	August 10, 2004	Calgene LLC	Molecular farming
US7161064	January 9, 2007	North Carolina State University	Method for producing stably transformed duckweed using microprojectile bombardment
US7238854	July 3, 2007	E. I. du Pont de Nemours and Company	Method of controlling site-specific recombination
US7265267	September 4, 2007	CropDesign N.V.	Cyclin-dependent kinase inhibitors and uses thereof
US7282625	October 16, 2007	The Scripps Research Institute	Method of producing single-chain protein in plant cells
US7285650	October 23, 2007	Seita Groupe Altadis Stallergenes	Cloning and sequencing of the allergen Dac g5 of _Dactylis glomerata_ pollen, its preparation and its use
US7361331	April 22, 2008	Her Majesty The Queen in Right of Canada, as represented by the Minister of Agriculture and Agri-Food London Health Sciences Centre Research Inc. Terry Delovitch	Plant bioreactors
US7371935	May 13, 2008	National Research Council of Canada Dow Agrosciences LLC	Transgenic plants and methods for production thereof
US7547821	June 16, 2009	SemBioSys Genetics Inc.	Methods for the production of insulin in plants
US7786352	August 31, 2010	Sembiosys Genetics Inc. UTI Ltd. Partnership	Methods for the production of apolipoproteins in transgenic plants
US7803991	September 28, 2010	Auburn University	Universal chloroplast integration and expression vectors, transformed plants and products thereof
US7901691B2	March 8, 2011	Council of Scientific and Industrial Research, New Delhi (India); Unichem Laboratories Ltd., Mumbai (India); Indian Veterinary Research Institute, Izatnagar, India	Chimeric G protein-based rabies vaccine
US7915483	March 29, 2011	Biolex Therapeutics Inc.	C-terminally truncated interferon
US7959910	June 14, 2011	Biolex Therapeutics Inc.	C-terminally truncated interferon alpha variants
US8058512	November 15, 2011	David N. Radin, Carole L. Cramer	Gene expression and production of TGF-β proteins including bioactive mullerian-inhibiting substance from plants
US8182803	May 22, 2012	Biolex Therapeutics Inc.	C-terminally truncated interferon alpha variants

TABLE 18.9

Plant-Derived Pharmaceutical Proteins under Clinical Trials

Product	Class	Indication	Company/ Organization	Crop	Status
AB	Vaccine	AB cancer vaccine	Large Scale Biology Corporation, USA	Tobacco	Phase II
Various scFv antibody fragments	Antibody	Non-Hodgkin's lymphoma	Large Scale Biology Corporation, USA	Viral vectors in tobacco	Phase I
CaroRx	Antibody	Dental caries	Planet Biotechnology Inc.	Transgenic tobacco	Phase II
DoxoRx		Side effect of cancer therapy	Planet Biotechnology Inc.	Not specified	Phase I
Escherichia coli heat-labile toxin	Vaccine	Diarrhea	Prodigene Inc. Arntzen Group (Tacket et al. 1998)	Transgenic maize; transgenic potato	Phase I; Phase I
Gastric lipase	Therapeutic enzyme	Cystic fibrosis, pancreatitis	Meristem Therapeutics	Transgenic maize	Phase II
Hepatitis B virus surface antigen	Vaccine	Hepatitis B	Arntzen Group (Richter et al. 2000) Thomas Jefferson University/Polish Academy of Sciences	Transgenic potato; Transgenic lettuce	Phase I; Phase I
Human glucocerobrosidase (prGCD)	Therapeutic enzyme	Gaucher's disease	Protalix Biotherapeutics, Israel	Plant tissue culture	Phase III
Human intrinsic factor	Dietary	Vitamin B12	Cobento Biotech AS	Transgenic arabidopsis	Phase II
Insulin	Therapy	Diabetes	SemBiosys, Canada	Safflower	Phase I
Lactoferrin	Dietary	Gastrointestinal infections	Meristem Therapeutics	Transgenic maize	Phase I
RhinoRx	Therapy	Cold viruses	Planet Biotechnology Inc.	Tobacco	Phase I
Lactoferon (α-interferon)	Therapy	Hepatitis C	Biolex, USA	Lemna	Phase II
Norwalk virus capsid protein	Vaccine	Norwalk virus infection	Arntzen Group (Tacket et al. 2000)	Transgenic	Phase I
Rabies glycoprotein	Vaccine	Rabies	Yusibov et al. (2002)	Viral vectors in spinach	Phase I
Antigen	Vaccine	Feline parvovirus	Large Scale Biology Corporation, USA	Tobacco	Phase I
Antigen	Vaccine	Papillioma virus	Large Scale Biology Corporation, USA	Tobacco	Phase I
HN protein	Vaccine	Poultry vaccine	Dow Agro Sciences, USA	Plant cell culture	Approved by USDA

A farm-grown recombinant product has multiple demands, such as the efficacy and authenticity of the product and the containment of the plants. In addition, there are several safety and environmental issues that need to be addressed before harnessing the displayed potential by molecular farming. Two major safety issues are often raised in this context.

TABLE 18.10

Commercially Available Industrial Proteins from Molecular Farming

Product	Crop	Source of Genes	Industrial Purpose	Company/ Organization	Marketing Company
Aprotinin	Maize	Cow	R&D	ProdiGene	NA
Aprotinin	Tobacco	Cow	R&D	Kentucky BioProcessing, LLC	Sigma Chemicals (A6103)
Avidin	Maize	Chicken	R&D	ProdiGene	Sigma Chemicals (A8706)
β-Glucuronidase	Maize	Bacteria	R&D	ProdiGene	NA
Trypsin	Maize	Cow	R&D	ProdiGene	Sigma Chemicals (T3568 and T3449)
Lactoferrin	Rice	Human	Research	Ventria Biosciences	Sigma Chemicals (T3568 and T3449)
Lysozyme	Rice	Human	Research	Ventria Biosciences	Sigma Chemicals (L1667)

Source: Spok, A. and Kramer, S., Plant molecular farming: Opportunities and challenges, JRC scientific and technical report, 2007.

18.11.1 Product Safety

Genetically modified foods and health products from plants have associated technical difficulties, which cannot be dealt with under the guidelines drafted for plants or for animals. The major regulatory concerns of pharmaceutical transgenic crops are essentially categorized as the safety of the product and the safety of the environmental impact of the production system. The stringency of the regulatory oversight governing PMPs depends on the type of protein and its intended use. An injectable vaccine antigen for human use is likely to require stricter regulation than a monoclonal antibody used in the oral cavity. Proteins intended to be used strictly for veterinarian applications or industrial processes are likely to be less-intensely regulated than those administered directly to humans (Sparrow et al. 2007). An entirely new protein transgenically produced for the first time in plants will require a completely new regulatory profile.

Apart from oral vaccine candidates, pharmaceutical crops are not intended for consumption, hence, the risk of inadvertent mixing in the food and feed chains must be considered. Allergenicity, adverse reactions to the transgenic plant or its product, and quantitative dietary exposure are the potential risks associated with transgenic crops. Cereals are an exception as they have been consumed for millennia. A vast amount of information related to the allergenicity of cereals and dietary exposure is available for risk assessment.

In cases where direct oral administration of the pharmaceutical is desirable, edible transgenic plants are clearly preferred. There is an essential need to ensure the consumption of the specific quantity that delivers the required dose of the desired antigen to impart the appropriate immune response. Edible vaccines can undoubtedly simplify and speed up vaccine programs; however, it is important to realize that edible vaccines will not be delivered as fresh produce, as is often suggested. A regulated product requires controlled delivery of standardized doses, so some level of processing of the edible plant material would be required.

18.11.2 Environmental Impact

The two main environmental safety issues pertaining to the use of cereals crops are the potential for a horizontal gene flow from transgenic crops to nontransgenic crops or wild species by pollen and the dissemination of seeds through physical mixing. The cross-pollination of nontransgenic maize by neighboring transgenic varieties is possible in commercial agriculture. Many strategies can be adopted to reduce the gene lateral flow in cereals, including biological mitigation measures and barrier methods. Other generally applicable strategies include the use of male sterile varieties. Additional safety measures include the use of contained greenhouses and the development of phenotypic markers, such as green or purple fruit colors in tomatoes, as a label to identify transgenic lines expressing pharmaceuticals. The use of an inducible expression system that requires the application of a chemical inducer to switch on the transgene expression can also be an attractive strategy. To avoid the inadvertent entry of recombinant pharmaceutical into the food chain, the use of nonfood crops such as tobacco is the obvious option.

The proven safety, efficacy, and functional equivalence of recombinant pharmaceuticals will steer the future development and research in plant-based preventive and therapeutic technologies. The equally important rationale is the formative regulatory framework, which will streamline effective guidelines to maximize the potential of plants as a production system while maintaining the expected level of safety.

18.12 Concluding Remarks

This chapter discussed the rationale of plants as an alternative, valuable, and efficient expression system for the production of pharmaceutical and industrial proteins. At the current rate of progress in biotechnology related to the knowledge of gene regulation and protein synthesis and its structural and functional characterization of genes, plants are emerging as the leading heterologous expression system. This is mainly because plants can synthesize and posttranslationally modify the recombinant proteins. Parallel modifications to introduce desired glucosyl transferases and knock out undesired glucosyl transferases enable the plants to produce authentic pharmaceutical products. The identification and characterization of constitutive/inducible/regulated strong promoters, codon optimization, and organelle-specific expression of recombinant proteins significantly increase the expression of recombinant proteins at feasible levels. The availability of a wide range of production/host systems allows the production of pharmaceuticals in several food crops as well. It can also aid in the diagnosis of many diseases using recombinant antibodies as a diagnostic tool leading to the discovery of new therapies. The production of recombinant pharmaceuticals, especially edible vaccines, greatly reduces the costs involved in organizing mass immunization programs and the successful eradication of several dreadful diseases. The reduction in the cost of plant-produced pharmaceutical proteins makes them an attractive industrial-scale expression system with a minimal investment in their cultivation. These properties anticipate plants as a premier expression system for the production of diagnostic, therapeutic, and industrial proteins. Plant expression systems have the potential to be as abundant tomorrow as prescription drugs are today. We foresee that molecular farming will provide a basket full of medicines for every disease.

Acknowledgments

We have liberally borrowed from the articles of the following learned scientists: Fischer and Emans (2000), Sunil Kumar et al. (2007), Spok and Kramer (2007), and Gomord et al. (2010). Our thanks and appreciation are due to them.

References

An, G., Costa, M.A., Ha, S.B. 1990. Nopaline synthase promoter is wound inducible and auxin inducible. *Plant Cell* 2:25–33.

Anzai, H., Takaiwa, F., Katsumata, K. 2000. Production of human lactoferrin in transgenic plants. In: Shimazaki, K., Tsuda, H., Tomita, M. (eds) *Lactoferrin: Structure, Function and Applications*, pp. 265–271. New York: Elsevier Science.

Apweiler, R., Hermjakob, H., Sharon, N. 1999. On the frequency of protein glycosylation, as deduced from analysis of the SWISS-PROT database. *Biochim Biophys Acta* 1473:4–8.

Arntzen, C., Plotkin, S., Dodet, B. 2005. Plant-derived vaccines and antibodies: Potential and limitations. *Vaccine* 23:1753–1756.

Artsaenko, O., Kettig, B., Fiedler, U., Conrad, U., Düring, K. 1998. Potato tubers as a biofactory for recombinant antibodies. *Mol Breeding* 4:313–319.

Artsaenko, O., Peisker, M., zur Nieden, U., Fiedler, U., Weiler, E.W., Müntz, K., Conrad, U. 1995. Expression of a single-chain Fv antibody against abscisic acid creates a wilty phenotype in transgenic tobacco. *Plant J* 8:745–750.

Ashraf, S., Singh, P.K., Yadav, D.K., Shahnawaz, M., Mishra, S., Sawant, S.V., Tuli, R. 2005. High level expression of surface glycoprotein of rabies virus in tobacco leaves and its immunoprotective activity in mice. *J Biotechnol* 119:1–14.

Bakker, H., Bardor, M., Molthoff, J., Gomord, V., Elbers, I., Stevens, L., Jordi, W., Lommen, A., Faye, L., Lerouge, P., Bosch, D. 2001. Galactose-extended glycans of antibodies produced by transgenic plants. *Proc Natl Acad Sci USA* 98:2899–2904.

Barta, A., Sommergruber, K., Thomson, D., Hartmuth, K., Matzke, M.A., Matzke, A.J.M. 1986. The expression of a nopaline synthase—human growth hormone chimeric gene in transformed tobacco and sunflower callus tissue. *Plant Mol Biol* 6:347–357.

Baum, T.J., Hiatt, A., Parrott, W.A., Pratt, L.H., Hussey, R.S. 1996. Expression in tobacco of a functional monoclonal antibody specific to stylet secretions of the root-knot nematode. *Mol Plant Microbe Interact* 9:382–387.

Belanger, F.C., Kriz, A.L. 1991. Molecular basis for allelic polymorphism of the maize globulin-1 gene. *Genetics* 129:863–872.

Boothe, J.G., Saponja, J.A., Parmenter, D.L. 1997. Molecular farming in plants: Oilseed as vehicles for the production of pharmaceutical proteins. *Drug Dev Res* 42:172–178.

Chan, M.T., Chang, H.H., Ho, S.L., Tong, W.F., Yu, S.M. 1993. *Agrobacterium*-mediated production of transgenic rice plants expressing a chimeric alpha-amylase promoter/beta-glucuronidase gene. *Plant Mol Biol* 22:491–506.

Chan, M., Yu, S. 1998. The 3′ untranslated region of a rice α-amylase gene functions as a sugar-dependent mRNA stability determinant. *Proc Natl Acad Sci USA* 95:6543–6547.

Chaturvedi, C.P., Sawant, S.V., Kiran, K., Mehrotra, R., Lodhi, N., Ansari, S.A., Tuli, R. 2006. Analysis of polarity in the expression from a multifactorial bidirectional promoter designed for high-level expression of transgenes in plants. *J Biotechnol* 123:1–12.

Chen, Z.L., Schuler, M.A., Beachy, R.N. 1986. Functional analysis of regulatory elements in a plant embryo-specific gene. *Proc Natl Acad Sci USA* 83:8560–8564.

Christou, P. 1993. Particle gun-mediated transformation. *Curr Opin Biotech* 4:135–141.

Christou, P. 1996. Transformation technology. *Trends Plant Sci* 1:423–431.

Commandeur, U., Twyman, R.M., Fischer, R. 2003. The biosafety of molecular pharming in plants. *AgBiotechNet* 5:110.

Conrad, U., Fiedler, U. 1998. Compartment-specific accumulation of recombinant immunoglobulins in plant cells: An essential tool for antibody production and immunomodulation of physiological functions and pathogen activity. *Plant Mol Biol* 38:101–109.

Conrad, U., Fiedler, U., Artsaenko, O., Phillips, J. 1998. High level and stable accumulation of single chain Fv antibodies in plant storage organs. *J Plant Physiol* 152:708–711.

Cox, K.M., Sterling, J.D., Regan, J.T., Gasdaska, J.R., Frantz, K.K., Peele, C.G., Black, A., et al. 2006. Glycan optimization of a human monoclonal antibody in the aquatic plant *Lemna minor*. *Nat Biotechnol* 24:1591–1597.

Curtis, R., Cardineau, G. 1990. Oral immunization by transgenic plants. In: *World Patent Application*. USA: Washington University.

Dai, Z.Y., Hooker, B.S., Anderson, D.B., Thomas, S.R. 2000. Improved plant-based production of E1 endoglucanase using potato: Expression optimization and tissue targeting. *Mol Breed* 6:277–285.

Datla, R.S., Bekkaoui, F., Hammerlindl, J.K., Pilate, G., Dunstan, D.I., Crosby, W.L. 1993. Improved high-level constitutive foreign gene expression in plants using an AMV RNA4 untranslated leader sequence. *Plant Sci* 94:139–149.

De Buck, S., Jacobs, A., Van Montagu, M., Depicker, A. 1999. The DNA sequences of T-DNA junctions suggest that complex T-DNA loci are formed by a recombination process resembling T-DNA integration. *Plant J* 20:295–304.

De Cosa, B., Moar, W., Lee, S.B., Miller, M., Daniell, H. 2001. Overexpression of the Bt Cry-2 Aa2 operon in chloroplast leads to formation of insecticidal crystals. *Nat Biotechnol* 19:71–74.

De Neve, M., De Loose, M., Jacobs, A., Van Houdt, H., Kaluza, B., Weidle, U., Van Montagu, M., Depicker, A. 1993. Assembly of an antibody and its derived antibody fragment in Nicotiana and Arabidopsis. *Transgenic Res* 2:227–237.

De Rocher, E.J., Vargo-Gogola, T.C., Diehn, S.H., Green, P.J. 1998. Direct evidence for rapid degradation of Bacillus thuringiensis toxin mRNA as a cause of poor expression in plants. *Plant Physiol* 117:1445–1461.

Dertzbaugh, M.T., Elson, C.O. 1993. Comparative effectiveness of the cholera toxin B subunit and alkaline phosphatase as carriers for oral vaccines. *Infect Immun* 61:48–55.

De Wilde, C., De Neve, M., De Rycke, R., Bruyns, A.M., De Jaeger, G., Van Montagu, M., Depicker, A., Engler, G. 1996. Intact antigen-binding MAK33 antibody and Fab fragment accumulate in intercellular spaces of *Arabidopsis thaliana*. *Plant Sci* 114:233–241.

Deikman, J., Kline, R., Fischer, R.L. 1992. Organization of ripening and ethylene regulatory regions in a fruit-specific promoter from tomato (Lycopersicon esculentum). *Plant Physiol* 100:2013–2017.

Doran, P.M. 2000. Foreign protein production in plant tissue cultures. *Curr Opin Biotechnol* 11:199–204.

Düring, K., Hippe, S., Kreuzaler, F., Schell, J. 1990. Synthesis and self assembly of a functional monoclonal antibody in transgenic *Nicotiana tabacum*. *Plant Mol Biol* 15:281–293.

Ellis, J.G., Llewellyn, D.J., Walker, J.C., Dennis, E.S., Peacock, W.J. 1987. The ocs element: A 16 base pair palindrome essential for activity of the octopine synthase enhancer. *EMBO J* 6:3203–3208.

Erbayraktar, S., Grasso, G., Sfacteria, A., Xie, Q., Coleman, T., Kreilgaard, M., Torup, L., et al. 2003. Asialoerythropoietin is a nonerythropoietic cytokine with broad neuroprotective activity in vivo. *Proc Natl Acad Sci USA* 100:6741–6746.

Faye, L., Boulaflous, A., Benchabane, M., Gomord, V., Michaud, D. 2005. Protein modifications in the plant secretory pathway: Current status and practical implications in molecular pharming. *Vaccine* 23:1770–1778.

Fiedler, U., Conrad, U. 1995. High-level production and long-term storage of engineered antibodies in transgenic tobacco seeds. *Biotechnology* 13:1090–1093.

Fiedler, U., Philips, J., Artsaenko, O., Conrad, U. 1997. Optimisation of scFv antibody production in transgenic plants. *Immunotechnology* 3:205–216.

Firek, S., Draper, J., Owen, M.R.L., Gandecha, A., Cockburn, B., Whitelam, G.C. 1993. Secretion of a functional single-chain Fv protein in transgenic tobacco plants and cell suspension cultures. *Plant Mol Biol* 23:861–870.

Fischer, R., Emans, N. 2000. Molecular farming of pharmaceutical proteins. *Transgenic Res* 9:279–299.

Fischer, R., Schumann, D., Zimmermann, S., Drossard, J., Sack, M., Schillberg, S. 1999a. Expression and characterization of bispecific single chain Fv fragments produced in transgenic plants. *Eur J Biochem* 262:810–816.

Fischer, R., Schuster, F., Hellwiq, S., Drossard, J. 1999b. Towards molecular farming in the future: Using plant-cell-suspension cultures as bioreactors. *Biotechnol Appl Biochem* 30:109–112.

Fogher, C. 2000. A synthetic polynucleotide coding for human lactoferrin, vectors, cells and transgenic plants containing it. Gene bank acc no AX006477, Patent: WO0004146.

Franconi, R., Roggero, P., Pirazzi, P., Arias, F.J., Desiderio, A., Bitti, O., Pashkoulov, D., Mattei, B., Milne, R.G., Benvenuto, E. 1999. Functional expression in bacteria and plants of an scFv antibody fragment against tospoviruses. *Immunotechnology* 4:189–201.

Franken, E., Teuschel, U., Hain, R. 1997. Recombinant proteins from transgenic plants. *Curr Opin Biotech* 8:411–416.

Franklin, S., Ngo, B., Efuet, E., Mayfield, S.P. 2002. Development of a GFP reporter gene for *Chlamydomonas reinhardtii* chloroplast. *Plant J* 30:733–744.

Fujiyama, K., Misaki, R., Sakai, Y., Omasa, T., Seki, T. 2009. Change in glycosylation pattern with extension of endoplasmic reticulum retention signal sequence of mouse antibody produced by suspension-cultured tobacco BY2 cells. *J Biosci Bioeng* 107(2):165–172.

Fujiyama, K., Palacpac, N.Q., Sakai, H., Kimura, Y., Shinmyo, A., Yoshida, T., Seki, T. 2001. In vivo conversion of a glycan to human compatible type by transformed tobacco cells. *Biochem Biophys Res Commun* 289(2):553–557.

Gallie, D.R., Tanguay, R.L., Leathers, V. 1995. The tobacco etch viral 5′ leader and poly(A) tail are functionally synergistic regulators of translation. *Gene* 165:233–238.

Gil, F., Brun, A., Wigdorovitz, A., Catala, R., Martinez-Torrecuadrada, J.L., Casal, I., Salinas, J., Borca, M.V., Escribano, J.M. 2001. High yield expression of a viral peptide vaccine in transgenic plants. *FEBS Lett* 488:13–17.

Gleba, Y., Klimyuk, V., Marillonnet, S. 2005. Magnifection—A new platform for expressing recombinant vaccines in plants. *Vaccine* 23:2042–2048.

Gomord, V., Faye, L. 2004. Plant-specific glycosylation patterns in the context of therapeutic protein production. *Plant Biotechnol J* 8:564–587.

Gomord, V., Denmat, L.A., Fitchette-Lainé, A.C., Satiat-Jeunemaitre, B., Hawes, C., Faye, L. 1997. The C-terminal HDEL sequence is sufficient for retention of secretory proteins in the endoplasmic reticulum (ER) but promotes vacuolar targeting of proteins that escape the ER. *Plant J* 11:313–325.

Gomord, V., Fitchette, A., Menu-Bouaouiche, L., Saint-JoreDupas, C., Plasson, C., Michaud, D., Faye, L. 2010. Posttranslational modification of therapeutic proteins in plants. *Curr Opin Plant Biol* 7:171–181.

Gowda, S., Wu, F.C., Herman, H.B., Shepherd, R.J. 1989. Gene VI of figwort mosaic virus (caulimovirus group) functions in post-transcriptional expression of genes on the full-length RNA transcript. *Proc Natl Acad Sci USA* 86:9203–9207.

Grevich, J.J., Daniell, H. 2005. Chloroplast genetic engineering: Recent advances and future perspectives. *Critical Rev Plant Sci* 24:83–107.

Gurr, S.J., Rushton, P.J. 2005. Engineering plants with increased disease resistance: How are we going to express it? *Trends Biotechnol* 23:283–290.

Gustafsson, C., Govindarajan, S., Minshull, J. 2004. Codon bias and heterologous protein expression. *Trends Biotechnol* 22:346–353.

Gutiérrez, R.A., MacIntosh, G.C., Green, P.J. 1999. Current perspectives on mRNA stability in plants: Multiple levels and mechanisms of control. *Trends Plant Sci* 4:429–438.

Haq, T.A., Mason, H.S., Clements, J.D., Arntzen, C.J. 1995. Oral immunisation with a recombinant bacterial antigen produced in transgenic plants. *Science* 268:714–716.

Heitzer, M., Eckert, A., Fuhrmann, M., Griesbeck, C. 2007. Influence of codon bias on the expression of foreign genes in microalgae. In: León, R., Gaván, A., Fernández, E. (eds), *Advances in Experimental Medicine and Biology*. Transgenic Microalgae as Green Cell Factories, Vol. 616. pp. 46–53. New York: Springer.

Hendy, S., Chen, Z.C., Barker, H., Santa Cruz, S., Chapman, S., Torrance, L., Cockburn, W., Whitelam, G.C. 1999. Rapid production of single-chain Fv fragments in plants using a potato virus X episomal vector. *J Immunol Meth* 231:137–146.

Hernandez-Garcia, C.M., Martinelli, A.P., Bouchard, R.A., Finer, J.J. 2009. A soybean (Glycine max) polyubiquitin promoter gives strong constitutive expression in transgenic soybean. *Plant Cell Rep* 28:837–849.

Hiatt, A. 1990. Antibodies produced in plants. *Nature* 344:469–470.

Hiatt, A., Cafferkey, R., Bowdish, K. 1989. Production of antibodies in transgenic plants. *Nature* 342:76–78.

Hiei, Y., Komari, T., Kubo, T. 1997. Transformation of rice mediated by *Agrobacterium tumefaciens*. *Plant Mol Biol* 35:205–218.

Hiei, Y., Ohta, S., Komari, T., Kumashiro, T. 1994. Efficient transformation of rice (*Oryza sativa L.*) mediated by *Agrobacterium* and sequence analysis of the boundaries of the T-DNA. *Plant J* 6:271–282.

Hobbs, S.L., Warkentin, T.D., DeLong, C.M. 1993. Transgene copy number can be positively or negatively associated with transgene expression. *Plant Mol Biol* 21:17–26.

Hogue, R.S., Lee, J.M., An, G. 1990. Production of a foreign protein product with genetically modified plant cells. *Enzyme Microb Technol* 12:533–538.

Hohn, B., Puchta, H. 2003. Some like it sticky: Targeting of the rice gene Waxy. *Trends Plant Sci* 8:51–53.

Hood, E.E. 2002. From green plants to industrial enzymes. *Enzyme Microb Technol* 30:279–283.

Horsch, R.B., Fry, J.E., Hoffmann, N.L., Eicholtz, D., Rogers, S.G., Fraley, R.T. 1985. A simple and general method for transferring genes into plants. *Science* 227:1229–1231.

Huang, Z., Santi, L., LePore, K., Kilbourne, J., Arntzen, C.J., Mson, H.S. 2006. Rapid, high-level production of hepatitis B core antigen in plant leaf and its immunogenicity in mice. *Vaccine* 24:2506–2513.

Jefferson, R., Goldsbrough, A., Bevan, M. 1990. Transcriptional regulation of a patatin-1 gene in potato. *Plant Mol Biol* 14:995–1006.

Jiang, X.L., He, Z.M., Peng, Z.Q., Qi, Y., Chen, Q., Yu, S.Y. 2007. Cholera toxin B protein in transgenic tomato fruit induces systemic immune response in mice. *Transgenic Res* 16:169–175.

Jones, H.D., Doherty, A., Wu, H. 2005. Review of methodologies and a protocol for the *Agrobacterium*-mediated transformation of wheat. *Plant Methods* 1:5.

Kapila, J., De Rycke, R., van Montagu, M., Angenon, G. 1996. An *Agrobacterium* mediated transient gene expression system for intact leaves. *Plant Sci* 122:101–108.

Kapusta, J., Modelska, A., Figlerowicz, M., Pniewski, T., Letellier, M., Lisowa, O., Yusibov, V., Koprowski, H., Plucienniczak, A., Legocki, A.B. 1999. A plant-derived edible vaccine against hepatitis B virus. *FASEB J* 13:1796–1799.

Kawaguchi, R., Bailey-Serres, J. 2002. Regulation of translational initiation in plants. *Curr Opin Plant Biol* 5:460–465.

Ko, K., Tekoah, Y., Rudd, P.M., Harvey, D.J., Dwek, R.A., Spitsin, S., Hanlon, C.A., Rupprecht, C., Dietzschold, B., Golovkin, M., Koprowski, H. 2003. Function and glycosylation of plant derived antiviral monoclonal antibody. *Proc Natl Acad Sci USA* 100:8013–8018.

Kohli, A., Twyman, R.M., Abranches, R., Wegel, E., Stoger, E., Christou, P. 2003. Transgene integration, organization and interaction in plants. *Plant Mol Biol* 52:247–258.

Koprivova, A., Stemmer, C., Altmann, F., Hoffmann, A., Kopriva, S., Gorr, G., Reski, R., Decker, E.L. 2004. Targeted knockouts of *Physcomitrella* lacking plant-specific immunogenic N-glycans. *Plant Biotech J* 2:517–523.

Koziel, M.G., Carozzi, N.B., Desai, N. 1996. Optimizing expression of transgenes with an emphasis on post-transcriptional events. *Plant Mol Biol* 32:393–405.

Kusnadi, A.R., Evangelista, R.L., Hood, E.E., Howard, J.A., Nikolov, Z.L. 1998. Processing of transgenic corn seed and its effect on the recovery of recombinant betaglucuronidase. *Biotechnol Bioeng* 60:44–52.

Lau, O.S., Sun, S.S.M. 2009. Plant seeds as bioreactors for recombinant protein production. *Biotechnol Adv* 27:1015–1022.

Lerouge, P., Cabanes-Macheteau, M., Rayon, C., Fischette-Lainé, A.C., Gomord, V., Faye, L. 1998. N-glycoprotein biosynthesis in plants: Recent developments and future trends. *Plant Mol Biol* 38:31–48.

Li, J.T., Fei, L., Mou, Z.R., Wei, J., Tang, Y., He, H.Y., Wang, L., Wu, Y.Z. 2006. Immunogenicity of a plant-derived edible rotavirus subunit vaccine transformed over fifty generations. *Virology* 356:171–178.

Linn, F., Heidmann, I., Saedler, H., Meyer, P. 1990. Epigenetic changes in the expression of the maize A1 gene in Petunia hybrida: Role of numbers of integrated gene copies and state of methylation. *Mol Gen Genet* 222:329–336.

Lithwick, G., Margalit, H. 2003. Hierarchy of sequence-dependent features associated with prokaryotic translation. *Genome Res* 13:2665–2673.

Lu, J., Sivamani, E., Azhakanandam, K., Samadder, P., Li, X., Qu, R. 2008. Gene expression enhancement mediated by the 5′ UTR intron of the rice rubi3 gene varied remarkably among tissues in transgenic rice plants. *Mol Genet Genomics* 279:563–572.

Lutcke, H.A., Chow, K.C., Mickel, F.S., Moss, K.A., Kern, H.F., Scheele, G.A. 1987. Selection of AUG initiation codons differs in plants and animals. *EMBO J* 6:43–48.

Ma, J., Hein, M. 1995a. Immunotherapeutic potential of antibodies produced in plants. *Trends Biotechnol* 13:522–527.

Ma, J., Hein, M. 1995b. Plant antibodies for immunotherapy. *Plant Physiol* 109:341–346.

Ma, J.K., Hikmat, B.Y., Wycoff, K., Vine, N.D., Chargelegue, D., Yu, L., Hein, M.B., Lehner, T. 1998. Characterization of a recombinant plant monoclonal secretory antibody and preventive immunotherapy in humans. *Nat Med* 4:601–606.

Ma, J.K., Drake, P.M., Christou, P. 2003. The production of recombinant pharmaceutical proteins in plants. *Nat Rev Genet* 4:794–805.

Ma, J.K.C., Lehner, T., Stabila, P., Fux, C.I., Hiatt, A. 1994. Assembly of monoclonal antibodies with IgG1 and IgA heavy chain domains in transgenic tobacco plants. *Eur J Immunol* 24:131–138.

Ma, J.K.C., Hiatt, A., Hein, M., Vine, N.D., Wang, F., Stabila, P., Van Dolleweerd, C., Mostov, K., Lehner, T. 1995. Generation and assembly of secretory antibodies in plants. *Science* 268:716–719.

Magnuson, N.S., Linzmaier, P.M., Reeves, R., An, G., Hay Glass, K., Lee, J.M. 1998. Secretion of biologically active human interleukin-2 and interleukin-4 from genetically modified tobacco cells in suspension culture. *Protein Expr Purif* 13:45–52.

Marks, M.D., Lindell, J.S., Larkins, B.A. 1985. Quantitative analysis of the accumulation of Zein mRNA during maize endosperm development. *J Biol Chem* 260:16445–16450.

Mason, H.S., Arntzen, C.J. 1995. Transgenic plants as vaccine production systems. *Trends Biotechnol* 13:388–392.

Matzke, A.J., Matzke, M.A. 1998. Position effects and epigenetic silencing of plant transgenes. *Curr Opin Plant Biol* 1:142–148.

Mayfield, S.P., Schultz, J. 2004. Development of a luciferase reporter gene, luxCt, for *Chlamydomonas reinhardtii* chloroplast. *Plant J* 37:449–458.

Mayfield, S.P., Franklin, S.E., Lerner, R.A. 2003. Expression and assembly of a fully active antibody in algae. *Proc Natl Acad Sci USA* 100:438–442.

McCormick, A.A., Kumagai, M.H., Hanley, K., Turpen, T.H., Hakim, I., Grill, L.K., Tuse, D., Levy, S., Levy, R., 1999. Rapid production of specific vaccines for lymphoma by expression of the tumor-derived single-chain Fv epitopes in tobacco plants. *Proc Natl Acad Sci USA* 96:703–708.

Misaki, R., Kimura, Y., Palacpac, N.Q., Yoshida, S., Fujiyama, K., Seki, T. 2003. Plant cultured cells expressing human b1,4-galactosyltransferase secrete glycoproteins with galactose-extended N-linked glycans. *Glycobiology* 13:199–205.

Mishra, S., Yadav, D.K., Tuli, R. 2006. Ubiquitin fusion enhances cholera toxin B subunit expression in transgenic plants and the plant-expressed protein binds GM1 receptors more efficiently. *J Biotechnol* 127:95–108.

Nikolov, Z., Hammes, D. 2002. Production of recombinant proteins from transgenic crops. In: Hood, E.E., Howard, J. (eds), *Plants as Factories for Protein Production*, pp. 159–174. Dordrecht: Kluwer Academic Publishers.

Nochi, T., Takagi, H., Yuki, Y., Yang, L., Masumura, T., Mejima, M., Nakanishi, U., et al. 2007. Rice-based mucosal vaccine as a global strategy for cold-chain- and needle-free vaccination. *Proc Natl Acad Sci USA* 104:10986–10991.

Odell, J.T., Nagy, F., Chua, N.H. 1985. Identification of DNA sequences required for activity of the cauliflower mosaic virus 35S promoter. *Nature* 313:810–812.

Osborn, T.C., Burow, M., Bliss, F.A. 1988. Purification and characterization of arcelin seed protein from common bean. *Plant Physiol* 86:399–405.

Owen, M., Gandecha, A., Cockburn, B., Whitelam, G. 1992. Synthesis of a functional anti-phytochrome single-chain Fv protein in transgenic tobacco. *Biotechnology* 10:790–794.

Pagny, S., Cabanes-Macheteau, M., Gillikin, J.W., Leborgne-Castel, N., Lerouge, P., Boston, R.S., Faye, L., Gomord, V. 2000. Protein recycling from the Golgi apparatus to the endoplasmic reticulum in plants and its minor contribution to calreticulin retention. *Plant Cell* 12:739–756.

Palacpac, N.Q., Yoshida, S., Sakai, H., Kimura, Y., Fujiyama, K., Yoshida, T., Seki, T. 1999. Stable expression of human b1,4-galactosyltransferase in plant cells modifies N-linked Glycosylation patterns. *Proc Natl Acad Sci USA* 96:4692–4697.

Parmenter, D.L., Boothe, J.G., van Rooijen, G.J., Yeung, E.C., Moloney, M.M. 1995. Production of biologically active hirudin in plant seeds using oleosin partitioning. *Plant Mol Biol* 29:1167–1180.

Porta, C., Lomonossoff, G.P. 1996. Use of viral replicons for the expression of genes in plants. *Mol Biotech* 5:209–221.

Prakash, C.S. 1996. Edible vaccines and antibody producing plants. *Biotechnol Dev Monit* 27:10–13.

Ramírez, Y.J.P., Tasciotti, E., Gutierrez-Ortega, A., Torres, A.J.D., Flores, M.T.O., Giacca, M., Lim, M.A.G. 2007. Fruit-specific expression of the human immunodeficiency virus type 1 Tat gene in tomato plants and its immunogenic potential in mice. *Clin Vaccine Immunol* 14:685–692.

Richter, L.J., Thanavala, Y., Arntzen, C.J., Mason, H.S. 2000. Production of hepatitis B surface antigen in transgenic plants for oral immunisation. *Nat Biotechnol* 18:1167–1171.

Ruhlman, T., Verma, D., Samson, N., Daniell, H. 2010. The role of heterologous chloroplast sequence elements in transgene integration and expression. *Plant Physiol* 152(4):2088–2104.

Rushton, P.J., Reinstadler, A., Lipka, V., Lippok, B., Somssich, I.E. 2002. Synthetic plant promoters containing defined regulatory elements provide novel insights into pathogen- and wound-induced signaling. *Plant Cell* 14:749–762.

Russell, D.A., Fromm, M.E. 1997. Tissue-specific expression in transgenic maize of four endosperm promoters from maize and rice. *Transgenic Res* 6:157–168.

Samadder, P., Sivamani, E., Lu, J., Li, X., Qu, R. 2008. Transcriptional and post-transcriptional enhancement of gene expression by the 5′ UTR intron of rice *rubi3* gene in transgenic rice cells. *Mol Genet Genomics* 279:429–439.

Sandhu, J.S., Krasnyanski, S.F., Domier, L.L., Korban, S.S., Osadjan, M.D., Buetow, D.E. 2000. Oral immunization of mice with transgenic tomato fruit expressing respiratory syncytial virus-F protein induces a systemic immune response. *Transgenic Res* 9:127–135.

Santi, L., Giritch, A., Roy, C.J., Marillonet, S., Klimyuk, V., Gleba, Y., Webb, R., Arntzen, C.J., Mason, H.S. 2006. Protection conferred by recombinant *Yersinia pestis* antigen produced by a rapid and highly scalable plant expression system. *Proc Natl Acad Sci USA* 103:861–866.

Schillberg, S., Zimmermann, S., Findlay, K., Fischer, R. 2000. Plasma membrane display of anti-viral single chain Fv fragments confers resistance to tobacco mosaic virus. *Mol Breeding* 6:317–326.

Scholthof, H., Scholthof, K., Jackson, A. 1996. Plant virus gene vectors for transient expression of foreign proteins in plants. *Annu Rev Phytopathol* 34:299–323.

Schouten, A., Roosien, J., van Engelen, F.A., de Jong, G.A.M., Borst-Vrenssen, A.W.M., Zilverentant, J.F., Bosch, D., Stiekema, W.J., Gommers, F.J., Bakker, J. 1996. The C-terminal KDEL sequence increases the expression level of a single-chain antibody designed to be targeted to both cytosol and the secretory pathway in transgenic tobacco. *Plant Mol Biol* 30:781–793.

Schunmann, P.H.D., Coia, G., Waterhouse, P.M. 2002. Biopharming the SimpliRED™ HIV diagnostic reagent in barley, potato and tobacco. *Mol Breed* 9:113–121.

Seon, J.H., Szarka, S., Moloney, M.M. 2002. A unique strategy for recovering recombinant proteins from molecular farming: Affinity capture on engineered oil bodies. *J Plant Biotechnol* 4:95–101.

Sharma, M.K., Jani, D., Thungapathra, M., Gautam, J.K., Meena, L.S., Singh, Y., Ghosh, A., Tyagi, A.K., Sharma, A.K. 2008. Expression of accessory colonization factor subunit A (ACFA) of Vibrio cholerae and ACFA fused to cholera toxin B subunit in transgenic tomato (Solanum lycopersicum). *J Biotechnol* 135:22–27.

Sijmons, P.C., Dekker, B.M., Schrammeijer, B., Verwoerd, T.C., van den Elzen, P.J., Hoekema, A. 1990. Production of correctly processed human serum albumin in transgenic plants. *Biotechnology* 8:217–221.

Skerra, A., 1993. Bacterial expression of immunoglobulin fragments. *Curr Opin Biotech* 5:256–262.

Sourrouille, C., Marquet-Blouin, E., D'Aoust, M.A., Kiefer-Meyer, M.C., Seveno, M., Pagny-Salehabadi, S., Bardor, M., Durambur, G., Leorouge, P., Vezina, L., Gomord, V. 2008. Down-regulated expression of plant-specific glycoepitopes in alfalfa. *Plant Biotechnol J* 6(7):702–721.

Sparrow, P.A.C., Irwin, J.A., Dale, P.J., Twyman, R.M., Ma, J.K.C. 2007. Pharma-Planta: Road testing the developing regulatory guidelines for plant-made pharmaceuticals. *Transgenic Res* 16:147–161.

Spok, A., Kramer, S. 2007. Plant molecular farming: Opportunities and challenges. JRC scientific and technical report.

Staub, J., Garcia, B., Graves, J., Hajdukiewicz, P., Hunter, P., Nehra, N., Paradkar, V., Schlittler, M., Carroll, J., Spatola, L., Ward, D., Ye, G., Russell, D. 2000. High-yield production of a human therapeutic protein in tobacco chloroplasts. *Nat Biotechnol* 18:333–338.

Stockert, R.J. 1995. The asialoglycoprotein receptor: Relations between structure, function and expression. *Physiol Rev* 75:591–609.

Stoger, E., Ma, J.K.C., Fischer, R., Christou, P. 2005. Sowing the seeds of success: Pharmaceutical proteins from plants. *Curr Opin Biotechnol* 16:167–173.

Stoger, E., Vequero, C., Torres, E., Sack, M., Nicholson, L., Drossard, J., Williams, S., Keen, D., Perrin, Y., Christou, P., Fischer, R. 2000. Cereal crops as viable production and storage systems for pharmaceutical scFv antibodies. *Plant Mol Biol* 42:583–590.

Stoger, E., Sack, M., Perrin, Y., Vaquero, C., Torres, E., Twyman, R.M., Christou, P., Fischer, R. 2002. Practical considerations for pharmaceutical antibody production in different crop systems. *Mol Breed* 9:149–158.

Strasser, R., Castilho, A., Stadlmann, J., Kunert, R., Quendler, H., Gattinger, P., Jez, J., Rademacher, T., Altmann, F., Mach, L., Steinkellner, H. 2009. Improved virus neutralization by plant-produced anti-HIV antibodies with a homogeneous beta1,4-galactosylated N-glycan profile. *J Biol Chem* 284:20479–20485.

Strasser, R., Stadlmann, J., Schahs, M., Stiegler, G., Quendler, H., Mach, L., Glossl, J., Weterings, K., Pabst, M., Steinkellner, H. 2008. Generation of glyco-engineered Nicotiana benthamiana for the production of monoclonal antibodies with a homogeneous human-like N-glycan structure. *Plant Biotechnol J* 6:392–402.

Streatfield, S.J. 2007. Approaches to achieve high-level heterologous protein production in plants. *Plant Biotechnol J* 5:2–15.

Sunil Kumar, G.B., Ganapathi, T.R., Revathi, C.J., Srinivas, L., Bapat, V.A. 2005. Expression of hepatitis B surface antigen in transgenic banana plants. *Planta* 222:484–493.

Sunil Kumar, G.B., Ganapathi, T.R., Srinivas, L., Bapat, V.A. 2007. Plant molecular farming: Host systems, technology and products. In: Verpoorte, R., Alfermann, A.W., Johnson, T.S. (eds), *Applications of Plant Metabolic Engineering*, pp. 45–77. Dordrecht: Springer.

Surzycki, R., Greenham, K., Kitayama, K., Dibal, F., Wagner, R., Rochaix, J.D., Ajam, T., Surzycki, S. 2009. Factors effecting expression of vaccines in microalgae. *Biologicals* 37:133–138.

Svitashev, S.K., Pawlowski, W.P., Makarevitch, I., Plank, D.W., Somers, D.A. 2002. Complex transgene locus structures implicate multiple mechanisms for plant transgene rearrangement. *Plant J* 32:433–445.

Tacket, C.O., Mason, H.S., Losonsky, G., Clements, J.D., Levine, M.M., Arntzen, C.J. 1998. Immunogenicity in humans of a recombinant bacterial antigen delivered in a transgenic potato. *Nat Med* 4:607–609.

Taticek, R.A., Lee, C.W.T., Shukler, M.L. 1994. Large scale insect and plant cell culture. *Curr Opin Biotech* 5:165–174.

Torres, E., Vaquero, C., Nicholson, L., Sack, M., Stoger, E., Drossard, J., Christou, P., Fischer, R., Perrin, Y. 1999. Rice cell culture as an alternative production system for functional diagnostic and therapeutic antibodies. *Transgenic Res* 8:441–449.

van Engelen, F.A., Schouten, A., Molthoff, J.W., Roosien, J., Salinas, J., Dirkse, W.G., Schots, A., Bakker, J., Gommers, F.J., Jongsma, M.A., Bosch, D., Stiekema, W.J. 1994. Coordinate expression of antibody subunit genes yields high levels of functional antibodies in roots of transgenic tobacco. *Plant Mol Biol* 26:1701–1710.

Vaquero, C., Sack, M., Chandler, J., Drossard, J., Schuster, F., Monecke, M., Schillberg, S., Fischer, R. 1999. Transient expression of a tumor-specific single chain fragment and a chimeric antibody in tobacco leaves. *Proc Natl Acad Sci* 96:11128–11133.

Verch, T., Yusibov, V., Koprowski, H. 1998. Expression and assembly of a full-length monoclonal antibody in plants using a plant virus vector. *J Immunol Meth* 220:69–75.

Verwoerd, T.C., van Paridon, P.A., van Ooyen, A.J.J., van Lent, J.W.M., Hoekema, A., Pen, J. 1995. Stable accumulation of *Aspergillus niger* phytase in transgenic tobacco leaves. *Plant Physiol* 109:1199–1205.

Vierling, R.A., Wilcox, J.R. 1996. Microplate assay for soybean seed coat peroxidase activity. *Seed Sci Technol* 24:485–494.

Voss, A., Niersbach, M., Hain, R., Hirsch, H., Liao, Y., Kreuzaler, F., Fischer, R. 1995. Reduced virus infectivity in *N. tabacum* secreting a TMV specific full size antibody. *Mol Breeding* 1:39–50.

Walmsley, A., Arntzen, C. 2000. Plants for delivery of edible vaccines. *Curr Opin Biotech* 11:126–129.

Walsh, G., Jefferis, R. 2006. Post-translational modifications in the context of therapeutic proteins. *Nat Biotechnol* 24:1241–1252.

Warzecha, H., Mason, H.S., Lane, C., Tryggvesson, A., Rybicki, E., Williamson, A.L., Clements, D.J., Rose, R.C. 2003. Oral immunogenicity of human papillomavirus-like particles expressed in potato. *J Virol* 77:8702–8711.

Winter, G., Milstein, C. 1991. Man-made antibodies. *Nature* 349:293–299.

Wright, D.A., Townsend, J.A., Winfrey Jr, R.J., Irwin, P.A., Rajagopal, J., Lonosky, P.M., Hall, B.D., Jondle, M.D., Voytas, D.F. 2005. High frequency homologous recombination in plants mediated by zinc-finger nucleases. *Plant J* 44:693–705.

Wu, C.Y., Suzuki, A., Washida, H., Takaiwa, F. 1998. The GCN4 motif in a rice glutelin gene is essential for endosperm-specific gene expression and is activated by Opaque-2 in transgenic rice plants. *Plant J* 14:673–683.

Yadav, D.K., Ashraf, S., Singh, P.K., Tuli, R. 2012. Localization of rabies virus glycoprotein into the endoplasmic reticulum produces immunoprotective antigen. *Protein J* 31:447–456.

Yu, J., Langridge, W.H. 2001. A plant-based multicomponent vaccine protects mice from enteric diseases. *Nat Biotechnol* 19:548–552.

Ziegler, M., Thomas, S., Danna, K. 2000. Accumulation of a thermostable endo-1,4-β-D-glucanase in the apoplast of Arabidopsis thaliana leaves. *Mol Breeding* 6:37–46.

Zuo, J., Chua, N.H. 2000. Chemical-inducible systems for regulated expression of plant genes. *Curr Opin Biotechnol* 11:146–151.

19

Natural Pesticidome Replacing Conventional Pesticides

Daiane Hansen, PhD

CONTENTS

19.1 Chemical Pesticide: Utilization and Consequences...603
19.2 Pest Control: Natural Alternatives 605
    19.2.1 Microbial 606
        19.2.1.1 Bacterial...606
        19.2.1.2 Entomopathogenic Fungi...608
        19.2.1.3 Baculoviruses..608
    19.2.2 Plant Extract or Derivate..609
        19.2.2.1 Plant Essential Oils 610
        19.2.2.2 Plant Extracts and Their Bioactive Compounds............................. 612
19.3 Final Remark 614
References.. 614

19.1 Chemical Pesticide: Utilization and Consequences

Since the dawn of humanity, humans have learned to plant in order to ensure their own livelihood and, later, to gain income. With the passage of time and the gradual increase of the population, crops were grown in order to meet the need to feed more people.

Along with this increase, the constant migration of man from the countryside to the city, and the requirement to adapt to modern life came pests and vegetable predators of various types and shapes, which, as a means of survival, began to devastate entire crops and compromise both small and large producers.

To combat agricultural pests, people have used pesticides for thousands of years to try to control plant diseases. As examples, the Sumerians used sulfur to control insects and mites 5000 years ago. The Greeks and Romans used oil, ash, sulfur, and other materials to protect themselves, their livestock, and their crops from various pests. The Chinese used mercury and arsenic compounds to control body lice and other pests. Further, people in various cultures have used smoke, salt, spices, and insect-repelling plants to preserve food and keep pests away (Cunningham and Saigo, 2001).

Plant diseases can be caused by a variety of organisms. Viruses, bacteria, fungi, and nematodes are responsible for a large array of plant diseases, with the vast majority of them being successfully controlled by chemicals nowadays. The evolution of chemical sciences has brought incalculable benefits to various sectors of production, including agriculture, but since the 1960s the use of chemical pesticides has been recognized as having negative impacts upon ecosystems and human health (Carson, 1962).

The development of agriculture is closely related to the application of pesticides, and the excessive use of pesticides has indeed prevented harm by pests and has largely improved the production of crops, but it has also resulted in high groundwater pollution risks, as well as the contamination of soil, food, and feedstuffs. An application of pesticide, depending on crop stage, formulation, intended target, application technique, and weather conditions, is distributed between soil, plant foliage, and crop residues, and losses occur due to drift: when pesticides are applied from an aircraft, up to 50% may drift out of the target area. When a spray boom is used, losses due to drift are smaller but still significant, with the percentages differing according to the literature, being between 1%–10% and 10%–30% (Van der Werf, 1996).

Studies have focused on understanding the toxic mechanisms of pesticides, and concern is increasing worldwide about the damage to human health caused by their excessive use. Table 19.1 shows a summary of some of the main neurological diseases and cancers registered in farm workers and related to exposure to pesticides.

Inadequate education, training, and regulations on the use of pesticides lead to accidents, haphazard application, and overuse. Access to medical treatment is limited and most farmers rely on unsuitable homemade remedies, thus increasing the severity and duration of illnesses. Poor health and diet are other factors that are believed to increase the incidence of illnesses from exposure to pesticides in developing countries. Inadequate or nonexistent storage facilities, poor living conditions, and water supplies contaminated with pesticides also affect the health of families (Ahmed et al., 2011).

The data in Table 19.1 from studies involving pesticides, their related diseases, and consequences of their use are negligible, but briefly expose the dangers involved. Fortunately, nations are increasingly restricting pesticide purchases and are increasing the control and supervision of pesticide use. Laws and standards have been established for the acquisition and use of chemical pesticides, with the purpose of controlling the excess, contributing to the clarification of the dangers involved, and encouraging the use of sustainable and natural solutions to pest control. Despite these efforts, it is known that many chemicals are sold illegally in clandestine markets.

The Food and Agriculture Organization of the United Nations (FAO) has published provisional guidelines on tender procedures for the procurement of pesticides, mainly for agriculture (FAO, 1994). These guidelines also provide useful information for the procurement of public health pesticides. The World Health Organization (WHO) published guidelines for the purchase of public health pesticides in 2000, and 12 years later an updated and expanded version was published: *Guidelines for Procuring Public Health Pesticides*. Large quantities of public health pesticides are procured annually through national or international tender procedures. It is estimated that an average of 4429 tons of active ingredient of organochlorines, 1375 tons of organophosphates, 30 tons of carbamates, and 414 tons of pyrethroids were used annually for global vector control alone during the period 2000–2009 in WHO's six regions (WHO, 2000).

Pesticide procurement is a highly specialized and complex subject. This and other reasons have led producers and consumers to currently have different visions of how to control or combat plant pests. Entities related to sustainability, nature conservation, organic farming, natural products, and their impacts on human health arise daily; herewith, discussions that elucidate these subjects are encouraged.

The aim of this chapter is to present some of the alternatives that researchers and companies, often in interaction with farmers, are seeking in order to combat agricultural pests while using fewer pesticides of chemical origin. New pesticides, including those based

TABLE 19.1

Reports of Neurological and Cancer Diseases Registered among Farm Workers and Related to Pesticide Exposure

Associated Disease	Type	Geographic Localization	Reference
Cancer	Gastric	California, U.S.	Mills and Yang, 2007
	Prostate, soft tissue, larynx, leukemia, lip, esophagus, brain	Brazil (11 states)	Meyer et al., 2003; Chrisman et al., 2009; Miranda-Filho et al., 2012
	Gastric	Province of Forlì, Italy	Bucchi et al., 2004
	esophagus, stomach, liver, lung, pancreas	Korea	Lee et al., 2010
	NI	Australia	MacFarlane et al., 2010
	lymphatic tissues, leukemia, brain	Québec	Godon et al., 1989
Neurologic	neuro-development anomalies among children (attention deficit/hyperactivity disorder)	U.S.	Bouchard et al., 2010; Landrigan, 2001
	Parkinson's disease, depression, organophosphate-induced delayed polyneuropathy	Korea	Lee et al., 2010
	Parkinson's disease	12 different countries of North America (3), Europe (7) and others	Van Maele-Fabry et al., 2012
	Alzheimer's disease, Parkinson's disease, amyotrophic lateral sclerosis, cognitive and psychiatric disorders	NI	Blanc-Lapierre et al., 2012; Malek et al., 2012
	Depression	Colorado, U.S.	Stallones and Beseler, 2002, 2004; Beseler and Stallones, 2008
	Disturbing emotional symptoms	U.S.	Reidy et al., 1992
	Neurobehavioral impairments	Rio de Janeiro, Brazil	Eckerman et al., 2007

Note: NI, not informed.

on natural products, are being discovered and developed to replace chemical pesticides, either due to their toxicity or to combat the evolution of resistance.

19.2 Pest Control: Natural Alternatives

Approaching the subject of natural alternatives in agriculture is not an easy task. Dayan and collaborators (2009), in their review related to this theme, clarified that not every natural product or compound may be legally used in every country for organic agriculture. They explained that the rules regarding what is accepted for organic agriculture vary among countries, and even among states within a country, and do not always have a scientific rationale for inclusion or exclusion. In general, organic agriculture does not accept synthetic versions of natural compounds, including herbicides. Organic farmers need to consult with their certification agency or program to be sure that any material they use is "certified" or acceptable as organic (Dayan et al., 2009).

Plants were the most important source of natural pesticides for centuries, until very recently when the immense potential of bacteria and other microorganisms for the production of biologically active substances emerged. In fact, most of the new pest control agents commercialized since the middle of the twentieth century have come from microbial sources. Exceptions are those botanical preparations for which standardization of the active compounds by modern analytical methods only became feasible in recent decades, thus making the manufacture of reliable products possible (Ujváry, 2001).

A number of biopesticides (bacteria, fungi, virus, pheromones, plant extracts) are already being used to control various types of insects responsible for the destruction of forests and agricultural crops. Botanical pesticides are shown to have little impact upon natural enemies and therefore they have the potential to be used in combination with biological control in the development of an integrated pest management system. The use of plant extracts in agroecosystems is emerging as one of the prime means of protecting crop production and the environment from synthetic pesticide pollution (Charleston et al., 2005). Natural pest and disease control, either directly or indirectly, using natural plant products/botanicals (including essential oils) holds much promise.

The diversity found in natural products is huge and needs to be exploited to generate new classes of compounds in traditional synthetic programs; there is little overlap between the known molecular sites affected by synthetic and natural phytotoxins. Plant-derived secondary compounds may provide a source of environmentally friendly herbicides with novel molecular sites of action (Duke et al., 2000; Dayan et al., 2009). Apparently, the secondary metabolites present in plants function as defense (toxic), which inhibits reproduction and other processes. Phytochemical biomolecules could be used for maximizing the effectiveness and specificity in future insecticide design with specific or multiple target sites, while ensuring economic and ecological sustainability (Rattan, 2010).

The topics below briefly describe studies involving microbe-based pesticides and plant-based pesticides.

19.2.1 Microbial

Various microorganisms among bacteria, fungi, and yeasts have successfully been used against plant diseases caused by pathogens. A biological pesticide is effective only if it has a potentially major impact on the target pest, market size, variability of field performance, cost effectiveness, end-user feedback, and a number of technological challenges, namely fermentation, formulation, and delivery systems (Brar et al., 2006). A lot of innovations are being carried out on that basis. Table 19.2 relates some examples of recent patents involving microbial pesticides.

19.2.1.1 Bacterial

As a major bacterial example, *Bacillus thuringiensis* (Bt)-based biopesticides are of utmost importance and occupy almost 97% of the world biopesticide market. Bt has been extensively used for four decades in biopesticidal formulations due to its safe environmental and human health records. Bt is a gram-positive spore-forming soil bacterium that has insecticidal properties (also called *entomotoxicity*) which affect a selective range of insect orders, namely Lepidoptera, Diptera, and Coleoptera (Brar et al., 2006). The insecticidal activity of Bt is due to the action of d-endotoxin. This toxin, also named *crystal protein*, is a complex formed by three different subunits. The Bt toxins kill the larvae of certain species of insects after being ingested by the larvae.

TABLE 19.2

Recent Patents Involving Microbial Pesticides and Their Agricultural Applications

Patent No. (Year)	Summary	Microbial Species	Reference
US8,237,021 (2012)	Relates to the field of plant pest control, provides new nucleic acid sequences derived from *Bacillus thuringiensis* (Bt) strains, encoding insecticidal proteins	*Bacillus thuringiensis*	Arnaut et al., 2012
US8,226,938 (2012)	Relates to formulations for treating arachnids, especially *Varroa destructor*, using isolates of the fungus *Beauveria bassiana*	*Beauveria bassiana*	Meikle and Nansen, 2012
US8,101,568 (2012)	Relates to the use of peptide fragments of cadherins (including cadherin-like proteins)	*Bacillus thuringiensis*	Adang et al., 2012
US8,183,025 (2012)	Relates to biological methods and products useful for the control of *Solenopsis invicta*; directed to a novel *Solenopsis invicta* virus	*Solenopsis invicta* virus	Valles et al., 2012
US8,168,172 (2012)	Relates to a process for the production of organic formulation of a biopesticide containing *Pseudomonas fluorescens*	*Pseudomonas fluorescens*	Rao, 2012
WO/2009/093257 (2009)	Relates to a biopesticide based on one or more entomopathogenic fungi, which is formulated as a novel dispersible tablet resulting in longer shelf life, preventing contamination with saprophytic fungi	*Beauveria bassiana*	Divi et al., 2009
US7,449,428 (2008)	Relates to the fungal bioherbicide for weed control.	*Pyricularia setariae*	Peng and Byer, 2008
US7,429,477 (2008)	Relates to fungal/bacterial antagonist combinations for controlling plant pathogens	*Trichoderma virens*, *Bacillus amyloliquefaciens*	Johnson, 2008
WO/2005/055724 (2005)	Relates to biopesticidal compositions based on the bacterium *P. luminescens* for controlling and eradicating various agricultural, horticultural, and forestry pests	*Photorhabdus luminescens*	Bhatnagar et al., 2005
WO/2004/049808 (2004)	Relates to the biopesticide composition comprising a bacterium of the genus *Serratia*	*Serratia* spp.	Kogel and Van Overbeek, 2004

In many cases, genetic engineering may play a complementary role in the development of more efficacious formulations by facilitating greater toxin production, broadening the host range, and enhancing germination, sporulation, and expanding the Bt spectrum (Brar et al., 2006). The first generation of insect-resistant crops was introduced onto the market in 1996, based on the expression in plants of proteins isolated from Bt (Arnaut et al., 2012). The insecticidal Bt Cry proteins are produced during the sporulation stage of Bt strains and the proteins accumulate in large cytoplasmic crystals within the bacterium. When taken up by insects, a typical Lepidopteran-toxic Bt Cry protein is solubilized and processed in the insect midgut into an active form of about 60–65 kDa. The active protein exerts its toxic effect by binding to the midgut epithelial cells, causing pore formation in the cell membrane, which leads to osmotic lysis of the cells (Gill et al., 1992). The use of individual Bt proteins is often limited, as most Bt proteins are active against only a relatively small number of the numerous insect pests.

Using PCR amplification, Mohammedi et al. (2006) showed the presence of the Cry IA gene in 12 Bt strains, of which ten strains were isolated from sludges collected from wastewater treatment plants and two strains were isolated from dead Tortricidae larvae samples collected from a forest in Quebec, Canada. Cry IA and Cry IC genes code toxins active against Lepidoptera species; thus this study confirmed that these Bt strains encode an entomotoxic protein targeting the Lepidoptera family and therefore all strains could be used as potential biopesticidal agents. Moreover, five of them also possessed the Cry IC gene that is found on the reference strain Bt subsp. *aizawai*.

19.2.1.2 Entomopathogenic Fungi

Entomopathogenic fungi can also be used for the development of biopesticides because, like many conventional chemical insecticide active ingredients, they act through contact. Studies on insect fungal pathogens such as *Metarhizium anisopliae* and *Beauveria bassiana* have a long history in the context of agricultural pests. These fungal species are capable of infecting a broad range of insect hosts, and several biopesticide products have been developed for use in horticulture and agriculture. Modern molecular techniques have enabled the characterization, detection, and tracking of fungal isolates in the environment. Bell et al. (2009) reported the real-time quantitative PCR (qPCR) for the analysis of candidate fungal biopesticides against malaria. Three qPCR assays were successfully developed for counting fungal genomes: "specific" assays capable of distinguishing two well-characterized fungal entomopathogens, *M. anisopliae* var. acridum and *B. bassiana*, both of which have previously proven to be virulent to *Anopheles* mosquitoes. This provided a sensitive, target-specific, and robust quantification technique for fungal genomes (Bell et al., 2009).

Regarding the *M. anisopliae* virulence mechanisms, Kershaw et al. (1999) proposed that two different mechanisms are active when isolates of the fungi invade and kill insects: a toxin strategy and a growth strategy. With the toxin strategy insects are killed by toxins such as the destruxins produced by some isolates of *M. anisopliae*, and death is caused by the growth of hyphae throughout the insect body cavity. With the growth strategy insects are killed when fungal hyphae proliferate through the hemocoel, using up nutrients. Leemon and Jonsson (2008) proposed a third virulence mechanism occurring when *M. anisopliae* invades ticks: an integument breakdown strategy. Ticks belong to a different class of arthropods to insects, and their integument has a different structure and composition from that of insects. It is possible that different mechanisms of virulence result from the actions of the extracellular enzymes produced by *M. anisopliae* on the integuments of ticks and blowflies (Leemon and Jonsson, 2012). Leemon and Jonsson (2012) showed that *M. anisopliae* is a good candidate for the development of a fungal biopesticide for cattle tick and sheep blowflies due to its virulence toward these pests.

In a recent study, *B. bassiana* was highly virulent to bed bugs, causing rapid mortality (3–5 days) following short-term exposure to spray residues (Barbarin et al., 2012). The human bed bug *Cimex lectularius* is a hematophagous insect that requires blood meals for growth and development throughout its life cycle. Insecticide resistance, together with concerns over extensive use of chemicals in the domestic environment, creates a need for safe alternative methods of bed bug control.

19.2.1.3 Baculoviruses

Baculoviruses are arthropod-specific viruses which have long been considered as potentially useful microbial agents for insect pest control. Since the first baculovirus genome sequence from *Autographa californica* multiple nucleopolyhedrovirus (AcMNPV) was published (Ayres et al., 1994), the number of complete genomes available has risen considerably

(De Castro-Oliveira et al., 2006). However, industrial interest in pursuing the development of baculoviruses as commercially available pesticide products is low, due in part to the slow rate at which baculoviruses kill their hosts compared to chemical pesticides (Prikhod'ko et al., 1998). For many years, research focused on genomics has been undertaken to understand and overcome this deficiency.

Black et al. (1997) showed that the incorporation of an insect-selective neurotoxin gene into the genome of the AcMNPV can dramatically reduce the time required for the virus to kill or otherwise debilitate its insect host, thereby making the virus more attractive for the development of a biopesticide.

Prikhod'ko et al. (1996) described the effects of expressing three PsynXIV-promoted toxin genes, *mag4*, *sat2*, and *ssh1*, that encode secretable and potent neurotoxins (μ-Aga-IV from the spider *Agelenopsis aperta* Gertsch, As II from the sea anemone *Anemonia sulcata* Pennant, and Sh I from the sea anemone *Stichadactyla helianthus* Ellis, respectively) on the ability of AcMNPV to effectively kill its host. Later, it was shown that m-Aga-IV and As II act at distinct sites on the voltage-sensitive sodium channels of insects and synergistically promote channel opening. Moreover, these toxins had synergistic insecticidal activity against the blowfly *Lucilia sericata* and the fall armyworm *Spodoptera frugiperda* (Prikhod'ko et al., 1998).

Anticarsia gemmatalis MNPV (AgMNPV) is another important baculovirus system under study today and its use in the control of the velvet bean caterpillar, *A. gemmatalis*, is the most successful example of a virus used as a biological pesticide. Since the initial isolation of AgMNPV from infected larvae in soybean fields in Brazil, its production, commercialization, and field application have increased constantly. De Castro-Oliveira et al. (2006) presented the complete genome sequence of AgMNPV-2D and its genetic organization, which is similar to that of the *Choristoneura fumiferana* defective MNPV (CfDefNPV). After the DNA was sequenced, several studies were triggered related to its gene content and function, unravelling its genetic regulatory network and monitoring genetic changes and the evolution of the genome (Bilen et al., 2007; De Castro-Oliveira et al., 2008).

19.2.2 Plant Extract or Derivate

For thousands of years, people have recognized that plants are efficient producers of chemical compounds that can be used in defense against herbivore attacks. Botanical compounds were used in several parts of the world long before the arrival of synthetic pesticides. Usually, the large-scale production and commercialization of natural plant products are hampered by difficulties in either isolation from natural sources, which is usually limited by low abundance, or total chemical synthesis, which is sometimes commercially unfeasible considering the complex structures.

With advances in technologies such as DNA sequencing, recombinant DNA, and heterologous expression, in addition to sophisticated tools, equipment, and software and abundant commercial kits, many of the biosynthetic pathways responsible for the production of these valuable compounds are being elucidated. In their review work, Marienhagen and Bott (2012) presented recent advancements in the metabolic engineering of microorganisms for the production of plant natural products, showing the main metabolites biosynthesis pathways and their respective precursors through colored and detailed schematic representations.

Fortunately, companies and research institutes are looking for new compounds from plant sources for use in agriculture. Table 19.3 shows a list of examples of recent patents related to control of pests using natural alternatives involving at least one plant.

TABLE 19.3

Recent Patents Involving Plant-Based Biopesticides and Their Agricultural Applications

Patent No. (Year)	Summary	Plant Species	Reference
US8,202,552 (2012)	Plant extract exhibits biopesticidal activity against *Leptinotarsa decemlineata*	*Bifora radianos* Bieb., *Arctium lappa* L., *Verbascum songaricum*, *Xanthium strumarium* L., *Humulus lupulus*	Gokce et al., 2012
US8,273,389 (2012)	Relates to synergistic combinations of a biopesticide obtained from cinnamon oil (and its component cinnamaldehyde), and diallyl disulfide	*Cinnamomum* spp.	Belkind et al., 2012
US20120128648 (2012)	Relates to the isolation and characterization of the novel biopesticide obtained from *Eucalyptus* species capable of serving as effective biocontrol agents and pest control management agents	*Eucalyptus* spp.	Kaushik, 2012
US7,892,581 (2011)	Relates to the application of essential oils, or components thereof, to protect harvested fruits against pathogenic fungi	*Cinnamomum cassia* Pres., *Brassica nigra, Thymus vulgaris, Myristica fragrans, Eucalyptus citriodora* Hook	Kvitnitsky et al., 2011
CN102246827 (2011)	Relates to a compound pesticide, in particular to a pesticide prepared by compounding *Azadirachta indica* extracts	*Azadirachta indica*	Gao et al., 2011
WO/2008/060136 (2008)	Relates to a pesticide made from natural sources. The pesticide is a blend of neem oil extracted from the seeds of neem plant, jetropha oil extracted from the seeds of *Jetropha curcus*, and turpentine oil harvested from pine trees	*Jetropha curcus, Azadirachta indica*	Sivakumaran, 2008
CN101283690 (2008)	Relates to a method for extracting total alkaloids of plant-based pesticides	*Sophora flavescens*	Li et al., 2008
US7,195,788 (2007)	Relates to a natural pesticide produced from the biomass of plants	*Prunus* spp.	Roberts, 2007
US6,582,712 (2003)	Relates to a nontoxic aqueous pesticide comprising at least one surfactant and at least one high terpene-containing natural oil	*Citrus sinensis*	Pullen, 2003
US20030194455 (2003)	Relates to a biopesticide comprised of whole ground oriental mustard for controlling soil pathogens such as fungi and damaging nematodes	*Brassica juncea*	Taylor, 2003

The topics below describe some examples of promising studies involving plant-based pesticides.

19.2.2.1 Plant Essential Oils

It is well known that plants have built-in natural defenses against insects and other pests, which have evolved over time. Many plant essential oils exhibit antimicrobial, insecticidal, fungicidal, and herbicidal activities. Plant essential oils which do not present any known risk to humans or to the environment are qualified for exemption as minimum-risk pesticides (Belkind et al., 2012).

Plant essential oils are obtained from nonwoody parts of the plant, particularly foliage, through steam or hydrodistillation. They are complex mixture of mainly terpenoids, particularly monoterpenes (C10) and sesquiterpenes (C15), and a variety of aromatic phenols, oxides, ethers, alcohols, esters, aldehydes, and ketones that determine the characteristic aroma and odor of the donor plant (Langenheim, 1994). The presence of volatile monoterpenes or essential oils in the plants provides an important defense strategy, particularly against herbivorous insect pests and pathogenic fungi. Monoterpenoids, for example, are 10-carbon compounds composed of two isoprene units connected in a head-to-end manner, and are found in the essential oils of many plants including mints, pine, cedar, citrus, eucalyptus, and spices. Several monoterpenoids are commercially available for use as flea control on pets and carpets, control of insects on house plants, fumigation of parasitic mites in honey bee colonies, and insect repellency, such as citronella candles (Coats et al., 2001; Taylor, 2003).

Among essential oils, *Eucalyptus* spp. oil is particularly useful as it is commercially extractable (industrial value) and possesses a wide range of desirable properties worth exploiting for pest management. Eucalyptus oil is a complex mixture of a variety of monoterpenes and sesquiterpenes, and aromatic phenols, oxides, ethers, alcohols, esters, aldehydes, and ketones; however, the exact composition and proportion of these varies with species. The pesticidal activity of eucalyptus oils is due to components such as 1,8-cineole, citronellal, citronellol, citronellyl acetate, p-cymene, eucamalol, limonene, linalool, a-pinene, g-terpinene, a-terpineol, alloocimene, and aromadendrene (Batish et al., 2008).

The essential oil of species of the Apiaceae (Umbeliferae) family has been revealed to exhibit potent larvicidal activities against mosquitoes. In a recent study, Evergetis et al. (2013) reported the potent larvicidal properties of *Anethum graveolens* essential oil, suggesting its potential use as bioremediation cultivar in mosquito-thriving areas and the development of novel biocides for the efficient control of mosquitoes. The monoterpene α-phellandrene was the major constituent (59%) revealed in *A. graveolens* essential oil.

The toxicity of 98 plant essential oils against third instars of the cecidomyiid gall midge *Camptomyia corticalis* (Loew) (Diptera: Cecidomyiidae) was examined by Kim et al. (2012) and the results were compared against the conventional insecticide dichlorvos or 2,2-dichlorovinyl dimethyl phosphate. Results showed that all essential oils were less toxic than their control counterpart.

Chu et al. (2012) studied the essential oils derived from flowering aerial parts of *Artemisia giraldii* Pamp. and *A. subdigitata* Mattf. (Family: Asteraceae) against the maize weevil (*Sitophilus zeamais* Motsch.) and their results indicated that these two essential oils show potential in terms of fumigant and contact toxicity against grain storage insects. Members of the Asteraceae family belonging to the *Artemisia* species, which have shown efficiency in use against stored product insects, phytopathogenic and toxigenic fungi, arthropods, and as phytotoxic, deserve special attention. Due to their nontoxic nature and wide spectrum of biological activities, the essential oil of the *Artemisia* species and its pure components have been used to control fungi, insects, and bacteria. In their review, Felicio et al. (2012b) reported the constituents and biological activities of the *Artemisia* species and their possible use in animal and plant health.

The biome of the northeast of Brazil is extremely rich. The repellent and toxicity activities and the composition of the essential oils from both unripe and ripe fruits of *Schinus terebinthifolius*, a northeastern Brazilian plant, were analyzed. The major compound identified in both oils was limonene, and both oils exhibited significant acaricidal activity, suggesting that these oils, in association with their toxicity, could be a great advantage for the integrated management of *Tetranychus urticae* (Do Nascimento et al., 2012).The essential oil of leaves of *Piper aduncum* L., which grows wild in a fragment of the Atlantic Rainforest biome in the northeast of Brazil, also presented acaricidal activity against *Tetranychus urticae* (Araújo et al., 2012).

19.2.2.2 Plant Extracts and Their Bioactive Compounds

Plants in general produce a great variety of secondary metabolites that do not have any apparent function in physiological or biochemical processes; these compounds (or allelochemicals) are important in mediating interactions between plants and their biotic environment (Céspedes et al., 2006).

Botanical pesticides from plants within the Meliaceae family, in particular the Indian neem tree *Azadirachta indica* (A. Juss.) and the syringa tree *Melia azedarach* L. (native to eastern Asia), have proved to be particularly promising and, for some time, great investments in terms of research have been devoted to the study of these two species.

The seeds from the *A. indica* are the source of two types of neem-derived botanical insecticides: neem oil and medium-polarity extracts. Neem seeds contain numerous azadirachtin analogs, but the major form is the tetranortriterpenoid azadirachtin, or azadirachtin A, and the remaining minor analogs likely contribute little to the overall efficacy of the extracts (Isman, 2006). Azadirachtin has been effectively used against >400 species of insects, including many key crop pests, and has proved to be one of the most promising plant ingredients for integrated pest management at the present time. Neem-based insecticides have been tested and used for the management of *Plutella xylostella* and other pest insects on cabbage (Liu et al., 2006).

The fruit extracts from *M. azedarach* have long been known to show insecticidal activities, in some cases comparable to those of neem. A wide range of bioactive limonoids have been isolated from the seeds of this tree, including salannin, meliacarpins, and an analog of volkensin. The most notable constituent, however, is toosendanin, which occurs in the bark at concentrations as high as 0.5% (Isman, 1999).

In the study by Charleston et al. (2005), both syringa extracts and neem products did not have a detrimental effect on the longevity or behavior of the *Cotesia plutellae* (Kurdjumov) (Hymenoptera: Braconidae) or the *Diadromus collaris* (Gravenhorst) (Hymenoptera: Ichneumonidae), two of the most abundant parasitoid species found in the fields of South Africa.

For decades, farmers in India and China have used botanical extracts to battle major agricultural pests (Chiu, 1993, 1995; Singh and Singh, 1996). Those botanical insecticides include pyrethrum obtained from the flower of *Chrysanthemum cinerariifolium* (Trevir.) Vis., nicotine from *Nicotiana tabacum* L., and rotenone from *Lonchocarpus nicou* (Aublet), *Derris indica* (Lam.) Bennet, *Amorpha fruticosa* L., and *Tephrosia purpurea* (Liu et al., 2006). Species of the genus *Tephrosia* (Fabaceae), such as *T. candida* DC., *T. purpurea*, *T. vogelii*, and *T. noctiflora* Bojer ex. Baker, have been used as insecticide. The bark of *Tephoria purpurea* (Dil.) Pers has insecticidal activity against the third instar larvae of *Plutella xylostella* and *Corcyra cephalonica*, and the roots and seeds have insecticidal, piscicidal, and vermifugal properties (You-Zhi et al., 2011; Jasmine et al., 2012). In their studies on biosafety evaluation, Jasmine et al. (2012) verified that a stem-based formulation of *T. purpurea* can be incorporated along with reduviid predators in Bt cotton pest management.

Within the family Cruciferae ("crucifer"), glucosinolates are another class of natural pesticides. Glucosinolates are a group of over 90 secondary metabolites that occur in only

11 families of dicotyledonous plants, mostly in the family Cruciferae. It has been shown that higher concentrations of glucosinolates correlate with less severe insect attacks. Of interest is the fact that although glucosinolates are reportedly toxic to livestock, they are clearly safe for humans. Specifically, within the crucifer family, glucosinolates are found in plants such as cabbages, radishes, turnips, mustard, collard greens, rape, broccoli, kale, and crambe (*Crambe abyssinica* L.). It is the breakdown products of glucosinolates which are responsible for the pungent odor and biting taste of these plants (Coats et al., 2001; Taylor, 2003).

A lot of Brazilian research groups are dedicated to studying plant extract antifungal properties and effects against agricultural pests, with the intention of characterizing a new generation of natural pesticides. The large biodiversity within the Brazilian territory puts the country in a strategic position to develop a rational and sustained exploration of new metabolites with agronomic value. Brazil's landmass covers a wide range of climates, soil types, and altitudes, providing a unique set of selective pressures for the adaptation of plant life in these habitats. Plant chemical diversity is also driven by these forces, in an attempt to best fit the plant to the particular abiotic stresses, fauna, and microbes that coexist in the environment. Certain areas of vegetation, such as the Amazonian Forest, Atlantic Forest, Cerrado (Brazilian Savanna), and Caatinga, are rich in biodiversity, and are therefore of great interest for the discovery of natural compounds with biological activity (Basso et al., 2005; Hansen et al., 2010). Several examples of this wealthy flora are presented below.

The genus *Pterodon* spp. belonging to the Fabaceae (=Leguminosae) family is popularly known as *sucupira branca*, and these trees frequently grow in the Brazilian Cerrado. Extracts of the diterpene 14,15-epoxy geranylgeraniol from *P. emarginatus* fruits presented activity against *Cladosporium* spp., a phytopathogenic fungi genus (Hansen et al., 2010). Extracts and terpenic fractions from *P. polygalaeflorus* fruits showed significant larvicidal activity against the *Aedesaegypti* mosquito (De Omena et al., 2006; Pimenta et al., 2006).

Peraglabrata (Schott) Baill. is a species of the family Euphorbiaceae, which in Brazil is represented by 72 genera and about 1100 species present in all types of vegetation. Cardoso-Lopes et al. (2009) demonstrated that the leaves of *P. glabrata* presented antifungal activity against *Cladosporium* spp. and contain a great amount of caffeine, which may act as a chemical defense against fungal pathogens.

The fungi *Anternaria solani* and *Colletotrichum acutatum* cause serious damage to tomato and strawberry production, respectively. *Sclerotium rolfsii* is a soil pathogen, capable of provoking severe damage to several agricultural species. Domingues et al. (2009) evaluated the action of plant extracts, including the species *Ruta graveolens*, *Allamanda cathartica*, *Impatiens walleriana*, and *Lavandula augustifolia*, in the inhibition of these phytopathogens. The extracts of *I. walleriana* and *L. augustifolia* were the only ones that led to the total absence of conidial germination of *C. acutatum*. The plants *R. graveolens* and *I. walleriana* completely inhibited the sclerotial germination of *S. rolfsii*, and none of these extracts inhibited the conidial germination of *A. solani*.

The fungi of the genus *Aspergillus*, producer of mycotoxins, deserves special attention as the consumption of mycotoxin-contaminated foods has been associated with several cases of human poisoning, or mycotoxicosis, sometimes resulting in death. *Ageratum conyzoides* L. (Asteraceae), commonly known in Brazil as *mentrasto*, is an annual branching herb that is very common in tropical America and has now spread to various tropical and subtropical parts of the world. The essential oil of *A. conyzoides* shows an inhibitory effect on *A. flavus* growth, inhibits aflatoxin biosynthesis, and causes irreversible cellular changes (Nogueira et al., 2010).

Gonçalez et al. (2001) studied the action of the biflavonoids 6,6''-bigenkwanin, amenthoflavone, 7,7''-dimethoxyagastisflavone, and tetradimethoxybigenkwanin isolated from the *Ouratea* (Ochnaceae) species on inhibitory activity of *A. flavus* cultures. The plant species of the *Ouratea* genus are typical of Brazilian Cerrado vegetation. The four biflavonoids showed inhibitory activity on aflatoxin B1 and B2 production, but did not inhibit fungal growth.

When it comes to natural alternatives for the inhibition of *Aspergillus*, Pinto et al. (2001) and Gonçalez et al. (2003) reported the inhibition of aflatoxin production by aqueous and ethanolic extracts from *Polymnia sonchifolia* when added to *A. flavus* culture. *P. sonchifolia* (Asteraceae), commonly known as *yacon* and traditionally cultivated in the northern and central Andes, has recently attracted worldwide attention because of its wide range of uses. Compounds isolated from *P. sonchifolia*, the flavonoid 3',5,7-trihydroxy-3,4'-dimethoxyfla-vone, and a mixture of enhydrin and uvedalin, both sesquiterpenes lactones, inhibited fungal growth and aflatoxin B1 production (Pak et al., 2006).

As previously mentioned, the genus *Artemisia* spp., another member of the Asteraceae family, has shown potential in the use against stored product insects and phytopathogenic and toxigenic fungi. In a general context, the use of natural compounds from essential oils or plant extracts with lower toxicity than synthetic products can be a good alternative for mycotoxins control. Felicio et al. (2012a) presented a list of extracts and essential oils from plants species with antifungal activity and inhibition of mycotoxins production.

19.3 Final Remark

The literature has long shown the effects of using chemical pesticide in crops, and biopesticides are a recognized pest management tool in plant protection.

Pesticides, whether of synthetic origin, biopesticides, or organic, must be used, handled, and stored with care. The collaboration of all stakeholders (researchers, companies, farmers, governments) is extremely important in order to define pest management practices.

The entire world has turned its attention to plants and their active compounds, and this is partially due to the extremely wide and rich biodiversity of certain regions, which plays an important role in the popular culture and in the daily life of the population. Unfortunately, many of these natural compounds have been discovered and patented for use as pesticides, yet are not commercially available for numerous reasons.

In this chapter, we wanted to show new alternatives of natural character, based on plant-compounds, to replace or minimize the use of chemical pesticides. The use of nontoxic plant-origin essential oils, made of highly volatile substances, in the control of pests can greatly assist in reducing chemical residues in the environment and in agricultural products, as well as in protecting the agricultural environment.

Greater commitment, interest, and investment in research areas involving biomonitoring studies and plant-compound isolation and characterization are required. In spite of the immense biodiversity of our planet, the number of studies found in the literature is still insignificant.

References

Adang, M.J., Hua, G., Chen, J., and Abdullah, M.A.F. 2012. Peptides for inhibiting insects. US Patent 8,101,568.

Ahmed, A., Randhawa, M.A., Yusuf, M.J., and Khalid, N. 2011. Effect of processing on pesticide residues in food crops: A review. *Journal of Agricultural Research* 49(3):379–390.

Araújo, M.J., Câmara, C.A., Born, F.S., Moraes, M.M., and Badji, C.A. 2012. Acaricidal activity and repellency of essential oil from *Piper aduncum* and its components against *Tetranychusurticae*. *Experimental and Applied Acarology* 57(2):139–155.

Arnaut, G., Boets, A., De Rudder, K., Vanneste, S., and Van Rie, J. 2012. Insecticidal proteins derived from *Bacillus thuringiensis*. US Patent 8,237,021.

Ayres, M.D., Howard, S.C., Kuzio, J., Lopez-Ferber, M., and Possee, R.D. 1994. The complete DNA sequence of *Autographa californica* nuclear polyhedrosis virus. *Virology* 202:586–605.

Barbarin, A.M., Jenkins, N.E., Rajotte, E.G., and Thomas, M.B. 2012. A preliminary evaluation of the potential of *Beauveria bassiana* for bed bug control. *Journal of Invertebrate Pathology* 111:82–85.

Basso, L.A., Pereira da Silva, L.H., Fett-Neto, A.G., Azevedo Jr, W.F., Moreira, I.S., Palma, M.S., Calixto, J.B., et al. 2005. The use of biodiversity as source of new chemical entities against defined molecular targets for treatment of malaria, tuberculosis, and T-cell mediated diseases: A review. *Memrias do Instituto Oswaldo Cruz* 100(6):575–606.

Batish, D.R., Singh, H.P., Kohli, R.K., and Kaur, S. 2008. *Eucalyptus* essential oil as a natural pesticide. *Forest Ecology and Management* 256: 2166–2174.

Belkind, B.A., Shammo, B., Dickenson, R., Rehberger, L., and Heiman, D.F. 2012. Compositions comprising cinnamon oil (and/or its component cinnamaldehyde) and diallyl disulfide, their formulations, and methods of use. US Patent 8,273,389.

Bell, A.S., Blanford, S., Jenkins, N., Thomas, M.B., and Read, A.F. 2009. Real-time quantitative PCR for analysis of candidate fungal biopesticides against malaria: Technique validation and first applications. *Journal of Invertebrate Pathology* 100:160–168.

Beseler, C.L. and Stallones, L. 2008. A cohort study of pesticide poisoning and depression in Colorado farm residents. *Annals of Epidemiology* 18(10):768–774.

Bhatnagar, R.K., Rajagopal, R., and Rao, N.G.V. 2005. Biopesticide composition. Patent WO/2005/055724.

Bilen, M.F., Pilloff, M.G., Belaich, M.N., Da Ros, V.G., Rodrigues, J.C., Ribeiro, B.M., Romanowski, V., Lozano, M.E., and Ghiringhelli, P.D. 2007. Functional and structural characterization of AgMNPV ie1. *Virus Genes* 35(3):549–62.

Black, B.C., Brennan, L.A., Dierks, P.M., and Gard, I.E. 1997. Commercialization of baculovirus insecticides. In: *The Baculoviruses*, ed. L.K. Miller, pp. 341–387. Plenum Press, New York.

Blanc-Lapierre, A., Bouvier, G., Garrigou, A., Canal-Raffin, M., Raheriso, C., Brochard, P., and Baldi, I. 2012. Effets chroniques des pesticides sur le système nerveux central: État des connaissances épidémiologiques. *Revue d'Épidémiologie et de Santé' Publique* 60(5):389–400.

Bouchard, M.F., Bellinger, D.C., Wright, R.O., and Weisskopf, M.G. 2010. Attention deficit/hyperactivity disorder and urinary metabolites of organophosphate pesticides. *Pediatrics* 125(10):2009–3058.

Brar, S.K., Verma, M., Tyagi, R.D., and Valéro, J.R. 2006. Recent advances in downstream processing and formulations of *Bacillus thuringiensis* based biopesticides. *Process Biochemistry* 41:323–342.

Bucchi, L., Nanni, O., Ravaioli, A., Falcini, F., Ricci, R., Buiatti, E., and Amadori, D. 2004. Cancer mortality in a cohort of male agricultural workers from northern Italy. *Journal of Occupational and Environmental Medicine* 46(3):249–56.

Cardoso-Lopes, E.M., de Paula, D.M.B., Barbo, F.E., de Souza, A., Blatt, C.T.T., Torres, L.M.B., and Young, M.C.M. 2009. Chemical composition, acetylcholinesterase inhibitory and antifungal activities of *Peraglabrata* (Schott) Baill. (Euphorbiaceae). *Revista Brasileira de Botânica* 32(4):819–825.

Carson, R.L. 1962. *Silent Spring*. Riverside Press, Cambridge, MA.

Céspedes, C.L., Avila, J.G., Marin, J.C., Dominguez, M., Torres, P., and Aranda, E. 2006. Natural compounds as antioxidant and molting inhibitors can play a role as a model for search of new botanical pesticides. In: *Naturally Occurring Bioactive Compounds*, eds. M. Rai and M.C. Carpinella, pp. 1–27. Elsevier, Amsterdam.

Charleston, D.S., Kfir, R., Dicke, M., and Vet, L.E.M. 2005. Impact of botanical pesticides derived from *Melia azedarach* and *Azadirachta indica* on the biology of two parasitoid species of the diamondback moth. *Biological Control* 33:131–142.

Chiu, S.F. 1993. Investigations on botanical insecticides in south China: An update. Botanical pesticides in integrated pest management. *Indian Society of Tobacco Science* 134–137.

Chiu, S.F. 1995. *Melia toosendan* Sieb, and Zucc. In: *The Neem Tree*, ed. H. Schmutterer, pp. 642–646.
VCH Press, Weinheim.

Chrisman, J.R., Koifman, S., Sarcinelli, P.N., Moreira, J.C., Koifman, R.J., and Meyer, A. 2009. Pesticide
sales and adult male cancer mortality in Brazil. *International Journal of Hygiene and Environmental
Health* 212:310–321.

Chu, S.S., Liu, Z.L., Du, S.S., and Deng, Z.W. 2012. Chemical composition and insecticidal activity
against *Sitophilus zeamais* of the essential oils derived from *Artemisia giraldii* and *Artemisia subdigitata*. *Molecules* 17(6):7255–7265.

Coats, J.R., Peterson, C.J., Tsao, R., Eggler, A.L., and Tylka, G.L. 2001. Biopesticides related to natural
sources. US Patent 6,207,705.

Cunningham, W. and Saigo, B. 2001. Pest Control. In: *Environmental Science*. 6th ed. McGraw-Hill,
New York.

Dayan, F.E., Cantrell, C.L., and Duke, S.O. 2009. Natural products in crop protection. *Bioorganic and
Medicinal Chemistry* 17:4022–4034.

De Castro-Oliveira, J.V., de Melo, F.L., Romano, C.M., Iamarino, A., Rizzi, T.S., Yeda, F.P., Hársi, C.M.,
Wolff, J.L., and de Andrade-Zanotto, P.M. 2008. Structural and phylogenetic relationship of
ORF 31 from the *Anticarsia gemmatalis* MNPV to poly (ADP-ribose) polymerases (PARP). *Virus
Genes* 37(2):177–184.

De Castro-Oliveira, J.V., Wolff, J.L.C., Garcia-Maruniak, A., Ribeiro, B.M., de Castro, M.E.B., de Souza,
M.L., Moscardi, F., Maruniak, J.E., and de Andrade-Zanotto, P.M. 2006. Genome of the most
widely used viral biopesticide: *Anticarsiagemmatalis* multiple nucleopolyhedrovirus. *Journal of
General Virology* 87:3233–3250.

De Omena, M.C., Bento, E.S., De Paula, J.E., and Santana, A.E. 2006. Larvicidal diterpenes from
Pterodonpoly galaeflorus. *Vector-Borne and Zoonotic Diseases* 6(2):216–222.

Divi, S., Reddy, P.N., Khan, P.A.A., and Koduru, U.D. 2009. Formulation of entomopathogenic fungus for use as a biopesticide. Patent WO/2009/093257.

Do Nascimento, A.F., da Camara, C.A., de Moraes, M.M., and Ramos, C.S. 2012. Essential oil composition and acaricidal activity of *Schinus terebinthifolius* from Atlantic Forest of Pernambuco,
Brazil against *Tetranychus urticae*. *Natural Product Communications* 7(1):129–132.

Domingues, R.J., de Souza, J.D.F, Töfoli, J.G., and Matheus, D.R. 2009. Ação "in vitro" de extratos
vegetais sobre *Colletotrichum acutatum*, *Alternaria solani* e *Sclerotiumrolfsii*. *Arquivos do Instituto
de Biologia Vegetal* 76(4):643–649.

Duke, S.O., Romagni, J.G., and Dayan, F.E. 2000. Natural products as sources for new mechanisms of
herbicidal action. *Crop Protection* 19:583–589.

Eckerman, D.A., Gimenes, L.S., de Souza, R.C., Galvão, P.R.L., Sarcinelli, P.N., and Chrisman, J.R.
2007. Age related effects of pesticide exposure on neurobehavioral performance of adolescent
farm workers in Brazil. *Neurotoxicology and Teratology* 29(1):164–175.

Evergetis, E., Michaelakis, A., and Haroutounian, S.A. 2013. Exploitation of *Apiaceae* family essential oils as potent biopesticides and rich source of phellandrenes. *Industrial Crops and Products*
41:365–370.

FAO. 1994. Provisional guidelines on tender procedures for the procurement of pesticides. Rome,
Food and Agriculture Organization of the United Nations, 1994. Available at: http://www.fao.
org/ag/AGP/AGPP/Pesticid/Code/

Felicio, J.D., Aquino, S., and Gonçalez, E. 2012a. Methods for preventing and controlling mycotoxins
and mycotoxigenic fungi. In: *Mycotoxicoses in Animals Economically Important*, eds. E. Gonçalez,
J.D. Felicio, and S. Aquino, pp. 143–183. Nova Science Publishers, New York.

Felicio, J.D., Soares, L.B., Felicio, R.C., and Gonçalez, E. 2012b. *Artemisia* species as potential weapon
against agents and agricultural pests. *Current Biotechnology* 1:249–257.

Gao, B., Li, A., Tian, H., Shen, D., and Liu, D. 2011. Compound pesticide. CN102246827.

Gill, S.S., Cowles, E.A., and Pietrantonio, P.V. 1992. The mode of action of *Bacillus thuringiensis* endotoxins. *Annual Review of Entomology* 37:615–636.

Godon, D., Lajoie, P., Thouez, J.P., and Nadeau, D. 1989. Pesticides et cancers en milieu rural agricole au Quebec: Interpretation geographique. *Social Science and Medicine* 29(7):819–833.

Gokce, A., Whalon, M.E., Demirtas, I., and Goren, N. 2012. Insecticidal compositions and uses thereof. US Patent 8,202,552.

Gonçalez, E., Felicio, J.D., and Pinto, M.M. 2001. Biflavonoids inhibit the production of aflatoxin by *Aspergillus flavus*. *Brazilian Journal of Medical and Biological Research* 34:1453–1456.

Gonçalez, E., Felicio, J.D., Pinto, M.M., Rossi, M.H., Medina, C., Fernandes, M.J.B., and Simoni, I.C. 2003. Inhibition of aflatoxin production by *Polymnia sonchifolia* and its "in vitro" cytotoxicity. *Arquivos Do Instituto Biologico* 70(2):159–163.

Hansen, D., Haraguchi, M., and Alonso, A. 2010. Pharmaceutical properties of "sucupira" (*Pterodon* spp.). *Brazilian Journal of Pharmaceutical Sciences* 46(4):607–616.

Isman, M.B. 1999. Neem and related natural products. In: *Methods in Biotechnology. Biopesticides: Use and Delivery*, eds. F.R. Hall and J.J. Menn, pp. 139–153. Humana Press, Totowa, NJ.

Isman, M.B. 2006. Botanical insecticides, deterrents and repellents in modern agriculture and an increasingly regulated world. *Annual Review of Entomology* 51:45–66.

Jasmine, C.A., Sundari, S., Kombiah, P., Kalidas, S., and Sahayaraj, K. 2012. Biosafety evaluation of *Tephrosia purpurea* stem-based formulation (Telp 3% EC) against three *Rhynocoris* species. *Asian Journal of Biological Sciences* 5(4):216–220.

Johnson, T.D. 2008. Controlling plant pathogens with bacterial/fungal antagonist combinations. US7,429,477.

Kaushik, N. 2012. Novel biopesticide compositions and method for isolation and characterization of same. US Patent 20120128648.

Kershaw, M.J., Moorhouse, E.R., Bateman, R., Reynolds, S.E., and Charnley, A.K. 1999. The role of destruxins in the pathogenicity of *Metarhizium anisopliae* for three species of insect. *Journal of Invertebrate Pathology* 74(3):213–223.

Kim, J.R., Haribalan, P., Son, B.K., and Ahn, Y.J. 2012. Fumigant toxicity of plant essential oils against *Camptomyia corticalis* (Diptera: Cecidomyiidae). *Journal of Economic Entomology* 105(4):1329–1334.

Kogel, W.J. and Van Overbeek, L.S. 2004. Biopesticide Composition. WO/2004/049808.

Kvitnitsky, E., Ben-Arie, R., Paluy, I., and Semenenko, O. 2011. Compositions and methods for protection of harvested fruits from decay. US Patent 7,892,58.

Landrigan, P.J. 2001. Pesticides and polychlorinated biphenyls (PCBs): An analysis of the evidence that they impair children's neurobehavioral development. *Molecular Genetics and Metabolism* 73:11–17.

Langenheim, J.H. 1994. Higher plant terpenoids: A phytocentric overview of their ecological roles. *Journal of Chemical Ecology* 20:1223–1280.

Lee, W.J., Cha, E.S., and Moon, E.K. 2010. Disease prevalence and mortality among agricultural workers in Korea. *The Korean Academy of Medical Sciences* 25:112–118.

Leemon, D.M. and Jonsson, N.N. 2012. Comparison of bioassay responses to the potential fungal biopesticide *Metarhizium anisopliae* in *Rhipicephalus* (*Boophilus*) *microplus* and *Luciliacuprina*. *Veterinary Parasitology* 185:236–247.

Leemon, D.M. and Jonsson, N.N., 2008. Laboratory studies on Australian isolates of *Metarhizium anisopliae* as a biopesticide for the cattle tick *Boophilus microplus*. *Journal of Invertebrate Pathology* 97:40–49.

Li, D., Li, G., Li, H., Li, S., Tian, C., Wang, Y., Zhang, S., et al. 2008. Method for extracting plant source pesticides total alkaloids of *Sophora flavescens*. Patent CN101283690.

Liu, T.X., Xu, H.H., and Luo, W.C. 2006. Opportunities and potentials of botanical extracts and products for management of insect pests in cruciferous vegetables. In: *Naturally Occurring Bioactive Compounds*, eds. M. Rai and M.C. Carpinella, pp. 171–197. Elsevier, Amsterdam.

MacFarlane, E., Benke, G., Monaco, A.D., and Sim, M.R. 2010. Causes of death and incidence of cancer in a cohort of Australian pesticide-exposed workers. *Annals of Epidemiology* 20(4):273–280.

Maele-Fabry, G.V., Hoet, P., Vilain, F., and Lison, D. 2012. Occupational exposure to pesticides and Parkinson's disease: A systematic review and meta-analysis of cohort studies. *Environment International* 46(1):30–43.

Malek, A.M., Barchowsky, A., Bowser, R., Youk, A., and Talbott, E.O. 2012. Pesticide exposure as a risk factor for amyotrophic lateral sclerosis: A meta-analysis of epidemiological studies: Pesticide exposure as a risk factor for ALS. *Environmental Research* 117:112–119.

Marienhagen, J. and Bott, M. 2012. Metabolic engineering of microorganisms for the synthesis of plant natural products. *Journal of Biotechnology* 163:166–178.

Meikle, W.G. and Nansen, C. 2012. Biocontrol of Varroa mites with *Beauveria bassiana*. US Patent 8,226,938.

Meyer, A., Chrisman, J., Moreira, J.C., and Koifman, S. 2003. Cancer mortality among agricultural workers from Serrana Region, state of Rio de Janeiro, Brazil. *Environmental Research* 93:264–271.

Mills, P.K. and Yang, R.C. 2007. Agricultural exposures and gastric cancer risk in Hispanic farm workers in California. *Environmental Research* 104:282–289.

Miranda-Filho, A.L., Monteiro, G.T.R., and Meyer, A. 2012. Brain cancer mortality among farm workers of the State of Rio de Janeiro, Brazil: A population-based case-control study, 1996–2005. *International Journal of Hygiene and Environmental Health* 215:496–501.

Mohammedi, S., Subramanian, S.B., Yan, S., Tyagi, R.D., and Valéro, J.R. 2006. Molecular screening of *Bacillus thuringiensis* strains from wastewater sludge for biopesticide production. *Process Biochemistry* 41:829–835.

Nogueira, J.H.C., Gonçalez, E., Galleti, S.R., Facanali, R., Marques, M.O.M., and Felício, J.D. 2010. *Ageratum conyzoides* essential oil as aflatoxin suppressor of *Aspergillusflavus*. *International Journal of Food Microbiology* 137:55–60.

Pak, A., Gonçalez, E., Felicio, J.D., Pinto, M.M., Rossi, M.H., Simoni, I.C., and Lopes, M.N. 2006. Inhibitory activity of compounds isolated from *Polymnia sonchifolia* on aflatoxin production by *Aspergillus flavus*. *Brazilian Journal of Microbiology* 37:199–203.

Peng, G. and Byer, K.N. 2008. Control of weed with a fungal pathogen. US Patent 7,449,428.

Pimenta, A.T.A., Santiago, G.M.P., Arriaga, A.M.C., Menezes, G.H.A., and Bezerra, S.B. 2006. Estudo fitoquímico e avaliação da atividade larvicida de *Pterodon polygalaeflorus* Benth (Leguminosae) sobre *Aedes aegypti*. *Revista Brasileira Farmácia* 16(4):501–505.

Pinto, M.M., Gonçalez, E., Rossi, M.H., Felicio, J.D., Medina, C.S., Fernandes, M.J.B., and Simoni, I.C. 2001. Activity of the aqueous extract from *Polymnia sonchifolia* leaves on growth and production of aflatoxin B1 by *Aspergillus flavus*. *Brazilian Journal of Microbiology* 32:127–129.

Prikhod'ko, G.G., Popham, H.J.R., Felcetto, T.J., Ostlind, D.A., Warren, V.A., Smith, M.M., Garsky, V.M., Warmke, J.W., Cohen, C.J., and Miller, L.K. 1998. Effects of simultaneous expression of two sodium channel toxin genes on the properties of baculoviruses as biopesticides. *Biological Control* 12:66–78.

Prikhod'ko, G.G., Robson, M., Warmke, J.W., Cohen, C.J., Smith, M.M., Wang, P., Warren, V., Kaczorowski, G., Van der Ploeg, L.H.T., and Miller, L.K. 1996. Properties of three baculovirus-expressing genes that encode insect-selective toxins: μ-Aga-IV, As II, and Sh I. *Biological Control* 7:236–244.

Pullen, E.M. 2003. Controlling insects and parasites on plants using surfactants and high terpene oil. US Patent 6,582,712.

Rao, M.S. 2012. Process for the production of organic formulation of bio-pesticide *Pseudomonas fluorescens*. US Patent 8,168,172.

Rattan, R.S. 2010. Mechanism of action of insecticidal secondary metabolites of plant origin. *Crop Protection* 29:913–920.

Reidy, T.J., Bowler, R.M., Rauch, S.S., and Pedroza, G.I. 1992. Pesticide exposure and neuropsychological impairment in migrant farm workers. *Archives of Clinical Neuropsychology* 7(1):85–95.

Roberts, D.D. 2007. Pesticidal compositions from *prunus*. US Patent 7,195,788.

Singh, R.P. and Singh, S. 1996. Neem for management of insect pests: Advantages and disadvantages. In: *Recent Advances in Indian Entomology*, ed. O.P. Lal, pp. 67–82. APC Publications, Trivandrum, India.

Sivakumaran, S. 2008. Biopesticide. Patent WO/2008/060136.

Stallones, L. and Beseler, C.L. 2002. Pesticide poisoning and depressive symptoms among farm residents. *Annals of Epidemiology* 12(6):389–394.

Stallones, L. and Beseler, C.L. 2004. Safety practices and depression among farm residents. *Annals of Epidemiology* 14(8):571–578.

Taylor, T.D. 2003. Whole ground oriental mustard biopesticide. US Patent 20030194455.

Ujváry, I. 2001. Pest control agents from natural products. In: *Handbook of Pesticide Toxicology*, eds. R.I. Krieger and W.C. Krieger, pp. 109–179. 2nd edn., Vol 1. Principles. Academic Press, San Diego.

Valles, S.M., Strong, C.A., and Hashimoto, Y. 2012. *Solenopsis invicta* virus. US Patent 8,183,025.

Van der Werf, H.M.G. 1996. Assessing the impact of pesticides on the environment. *Agriculture, Ecosystems and Environment* 60:81–96.

WHO. 2000. Guidelines for the purchase of public health pesticides for use in public health. Geneva, World Health Organization (WHO/CDS/WHOPES/2000.1). Available at: http://whqlibdoc. who.int/hq/1998/CTD_WHOPES_98.5.pdf.

You-Zhi, L., Guan-Hua, L., Xiao-Yi, W., Zhong-Hua, L., and Han-Hong, X. 2011. Isolation and identification of insecticidal compounds from *Tephrosia purpurea* (Fabaceae) bark and their insecticidal activity. *Acta Entomologica Sinica* 54:1368–1376.

20

Intellectual Property Rights in Plant Biotechnology: Relevance, Present Status, and Future Prospects

Dinesh Yadav, PhD; Gautam Anand, PhD; Sangeeta Yadav, PhD;
Amit K. Dubey, PhD; Naveen C. Bisht, PhD; and Bijaya K. Sarangi, PhD

CONTENTS

20.1 Intellectual Property Rights: General Concept and Relevance 621
20.2 IPR and Its Different Forms 622
    20.2.1 Patents 622
    20.2.2 Copyright 623
    20.2.3 Trademarks 624
    20.2.4 Trade Secrets 624
20.3 Technological Innovations in Agriculture and Emergence of Plant Biotechnology 624
20.4 Transgenic Crops: Present Status and Future Prospects 625
20.5 International Treaties and Forms of IPR Applicable to Agriculture and Plant
    Biotechnology 627
    20.5.1 International Treaties for Promoting Agricultural Innovations 627
        20.5.1.1 Trade-Related Aspects of Intellectual Property Rights (TRIPS) 627
        20.5.1.2 Convention on Biological Diversity (CBD) and Its Impact on
               Plant IPRs 627
        20.5.1.3 International Treaty on Plant Genetic Resources for Food
               and Agriculture (IT-PGRFA) 628
        20.5.1.4 Public Intellectual Property Resources for Agriculture (PIPRA) 628
    20.5.2 Different Forms of IPR Relevant to Agriculture 629
        20.5.2.1 Plant Patents 629
        20.5.2.2 PVP 629
        20.5.2.3 Farmers' Rights 630
        20.5.2.4 Utility Patents 631
20.6 Innovations in Plant Biotechnology and IPR Issues 631
20.7 General Considerations for Patent Filing in Plant Biotechnology 664
20.8 Conclusion 665
References 666

20.1 Intellectual Property Rights: General Concept and Relevance

Intellectual Property Right (IPR) is a legal term that covers innovations, novel ideas, thoughts, and information having commercial values. An IPR gives the innovator a right to protect his or her creation from being used by others. Such creations include a wide range of intangible properties ranging from information to inventions. With globalization, the concept of IPR has become more relevant and this generic tag is often used for a group of legal regimes, each of which, to different degrees, confers rights of ownership in a particular subject matter. IPRs generally include patents, copyrights, and trademarks, providing legal rights to protect ideas, the expression of ideas, and the names, logos, and marks used to identify the business of a specific product (Marshall 1997). An IPR bears a similarity with a physical property in the sense that it can be bought and sold or rented. If the IPR is granted, the owner has the exclusive right and is thus protected from infringement. In order to claim for an IPR, the novelty of an idea is very important and the idea should be unknown to anyone else. The novelty, however, does not have to be absolute and it is very important that at the time of claiming for the IPR, the idea is thought to be generally unknown. Intellectual property (IP) law serves a variety of societal goals, including fostering innovation and promoting economic and cultural development. In the modern world, knowledge capital is considered to be more important than physical capital and this ultimately drives the economy of a country. IPRs promote innovations in diverse fields and also protect innovations having commercial values by giving exclusive right to the innovator to prevent others using his or her innovation without permission. These legal rights may vary from country to country, but, in general, this creates competitiveness and also leads to innovations in different sectors with the potential for commercialization. These exclusive rights prevent others from free riding on the innovator's investment and enable the rightful owner to exploit his or her knowledge or creativity in the market, thus creating incentives to innovate. Further, there is a need to disclose an innovation to the public to promote further developments (Wendt and Izquierdo 2001).

The IPR has relevance to each and every field of development, such as health, agriculture, education, trade, biodiversity management, biotechnology, information technology, the entertainment and media industries, and so on. Further, with the advent of the IPR, the gap between developed, developing, and least-developed countries is widening and there is a need for its stringent regulation worldwide. There has been substantial advancement in scientific research owing to collaborations but the issue of IPR needs to be addressed (Hane 2013).

20.2 IPR and Its Different Forms

The common forms of IPRs are patents, copyrights and related rights, industrial designs, trademarks, trade secrets, plant breeders' rights (PBRs), geographical indications, and rights to the layout designs of integrated circuits. Of these, patents, copyrights, and trademarks are arguably the most economically significant. Inventions, literary works, artistic works, designs, and trademarks formed the subject matter of early IP law. IPRs also include specific marks on products to indicate their difference from similar products sold by their various competitors. The concept of IP law, its diverse applications, and its controversies from a scientific point of view have been reviewed (Brown 2003).

20.2.1 Patents

It is a legal right to protect a new innovation, invention, idea, discovery, or concept having a commercial value. The sole purpose of a patent is to encourage and stimulate inventors to bring new and useful products to the marketplace for the benefit of mankind. It is a time-limited legal right assigned to an inventor by the government of a country where the patent is filed to protect the invention from being misused. The patent gives an exclusive right to the patentee (patent owner) to exclude others from making, using, selling, or importing the invention for a limited time and, in doing so, the patentee needs to describe the invention in detail for the benefit of the public or others working in that specific field. Once the patent expires, it can be used by anyone or, more precisely, the invention is in the public domain. Since a patent is granted by a government, if the inventor seeks a patent in a different country, he or she needs to apply for the patent in that country, that is, a filed patent is not universal or we can assume that there is no global patent. It is a misperception that a patent promotes a monopoly as it merely permits the inventor to stop someone else doing or using his or her invention. The patentee may assign the patent rights to another or the patentee may retain the ownership rights and license such rights to others, so that they can use the invention without owning it. In biotechnology, the Cohen–Boyer and PCR patent is a well-known example, having a very high commercial value (Dickson 1993; Lehrman 1993a).

The basic criteria for the patentability of an innovation are that it should be novel, nonobvious, and useful (Barton 2000). Three types of patents, namely, utility patents, design patents, and plant patents, exist under U.S. law. A utility patent has broad coverage for any invention or discovery that is a new and useful composition of matter, process, product, or machine. Design patents give legal protection to the appearance of an article, while plant patents include inventions in plants accomplished by crossbreeding or the discovery of new plants, which may include cultivated mutants, hybrids, and seedlings. There are two ways of patent filing to consider: either first to file or first to invent. In most countries, the first-to-file system is prevalent. The Leahy–Smith America Invents Act, which was passed in late 2011, has changed the U.S. patent regime to a first-to-file system for all patents, including plant patents and all utility patents applicable to plants (Pardey et al. 2013).

The patenting of genetically modified (GM) microorganisms came into existence in 1980 after the U.S. Supreme Court's ruling in *Chakrabarty* for *Pseudomonas* bacteria capable of cleaning oil (Wade 1980). The judgment in this particular matter was highlighted as "anything under the sun that is made by man" may be patentable under U.S. law. Similarly, in 1988, the U.S. Patent Office was first to grant a patent on transgenic animals, that is, a mouse having an activated mouse oncogene (*myc*), popularly referred to as the Harvard Oncomouse (Booth 1988). This was the beginning for the patenting of transgenic animals and several U.S. patents have been granted on transgenic animals thereafter (Lehrman 1993b), in contrast to the European Patent Office (EPO) (Abbott 1993; Spillmann-Furst 1990). Two highly significant developments in biotechnology need to be mentioned here with reference to patents. One was the Cohen–Boyer patent (U.S. Patent No. 4,237,224) on methods of gene cloning and expression, which expired in 1997, resulting in Stanford University and the University of California, San Francisco, where Stanley Cohen and Herbert Boyer were employed, making millions of dollar simply by selling this invention to different biotechnology industries. In contrast, the British Medical Research Council (BMRC) could not benefit from the invention of monoclonal antibodies by their employees Kohler and Milstein, while several patents were granted thereafter on specific monoclonal antibodies and diverse applications (Uhr 1984; Greene and Duft 1990).

20.2.2 Copyright

Copyright is the legal right of an artist, author, or any person who attempts to make any creative work in any field, such as paintings, designs, patterns, fiction, and photographs, which is fixed in a tangible medium such as paper, film, canvas, floppy disk, and the like. The purpose of copyright is that the work remains that of its creator and it should not be copied without his or her consent. This is not applicable to protecting ideas but only the form used to express the ideas. This copyright law is important for scientists and academicians who are associated with writing books. As compared to patents, a copyright need not be registered and it is comparatively weaker than patent law.

20.2.3 Trademarks

The sole purpose of a trademark is to highlight the source or origin of a good or service, which ultimately provides the consumer with a warranty of the quality of the good or service associated with the mark. This is applicable to all fields including the scientific field. The legal protection for trademarks is provided by a well-developed system of registration as in the case of patents. A trademark is generally denoted by a symbol, a word or a series of words, logos, or any other marking that distinguishes one product from another, or, in other words, it is a symbol for that specific product that cannot be copied by others in the same field. Trademarks signify the source of the product and hence are viewed as a marketing strategy, while patents and copyrights are based on creative products or ideas. Trademarks have an indefinite life span and they need not be registered.

20.2.4 Trade Secrets

A very simple way of protecting one's invention is by simply maintaining its secrecy, not disclosing it to anyone, and solely using it for commercialization. One excellent example of a trade secret is the formula for Coca Cola, which still exists with its owner. The main advantage of the trade secret is that it does not have an expiry time; however, it does need to be maintained by the owner and if it is disclosed, there is no means of protecting it under existing laws. It is quite applicable in case of hybrids, where the identification of the hybrid's parental lines is considered to be a trade secret belonging to the innovator (Moschini 2001; Tandon and Yadav 2012). As a result, the parental lines are kept a trade secret and are not accessible to anyone unless they are disclosed by the innovator, whereas in the case of patents, they have a limited time period after which they are made public and can be accessed by anyone.

20.3 Technological Innovations in Agriculture and Emergence of Plant Biotechnology

Conventional plant breeding has significantly contributed to the development of agriculture worldwide by releasing several varieties of different crops with desirable traits that emphasize more on the yield and biotic stress tolerance. The well-known concept of the *green revolution*, a term coined in 1968 by William Gaud, the director of the U.S. Agency of International Development, and manifested by Nobel laureate Dr. Norman E. Borlaug and Indian agricultural scientists Dr. M.S. Swaminathan and Dr. G.S. Khush, was based on conventional plant breeding for high-yielding dwarf varieties of wheat and rice with efficient irrigation and effective fertilizers and pesticides applications. Another revolution in agriculture is needed to feed the growing human population, which is expected to reach more than 9 billion by 2025. Several constraints, such as urbanization, industrialization, climate change, various biotic (pests, diseases, and weeds) and abiotic stresses (salinity, drought, heat, etc.), and the shrinkage of arable lands, demand newer technology for the development of crops with enhanced yield and tolerance to the various biotic and abiotic stresses. The major limitation with conventional plant breeding is the nonavailability of diverse genetic resources with the desired traits for the selection of parents to undergo relevant crossing for the development of suitable varieties of plants. Another limitation is that even though the genetic resources are available, they cannot be used beyond species, that is, the problem of sexual incompatibility and also the time required to develop a variety ranges from 8 to 10 years. Globally, the need for biotechnological interventions to attain global food security is being addressed (Mittler and Blumwald 2010; Yadav et al. 2010a).

The advent of genetic engineering technology and the recent advances in plant genomics with the deciphering of the genome sequences of important crops such as rice, maize, sorghum, soybean, pigeonpea, barley, potato, chickpea, watermelon, and melon (IRGSP 2005; Schnable et al. 2009; Paterson et al. 2009; Schmutz et al. 2010; Potato Genome Sequencing Consortium 2011; Singh et al. 2012; Varshney et al. 2012a, 2013; International Barley Genome Sequencing Consortium 2012; Garcia-Mas et al. 2012; Guo et al. 2013) have opened up new avenues for crop improvement. The availability of genomic sequences has contributed to sequence-based marker development for traits leading to the emergence of molecular breeding and relevant concepts such as marker-assisted selection (MAS) and genomic-assisted selection (GAS) with immense potential for crop improvement programs worldwide (Mittler and Blumwald 2010). This development has also resulted in allele mining for crop improvement (Kumar et al. 2010). In recent years, several review articles reflecting the recent development in agricultural biotechnology and its implications in attaining global food and nutritional security have been published (White and Broadear 2009; Fleury et al. 2010; Kumar et al. 2010; Cook and Varshney 2010; Varshney et al. 2012b; Morrell et al. 2012; Farre et al. 2011; Mir et al. 2012; Ahmad et al. 2012).

Several other state-of-the-art technologies in plant breeding, namely, zinc-finger nuclease (ZFN) technology (Townsend et al. 2009; Shukla et al. 2009), oligonucleotide-directed mutagenesis (ODM) (Zhu et al. 2000), *cis*-genesis and intragenesis (Schouten and Jacobsen 2008), RNA-dependent DNA methylation (RdDM) (Aufsatz et al. 2002), grafting (on GM rootstock) (Stegemann and Bock 2009), reverse breeding (Dirks et al. 2009), and agroinfiltration (Vezina et al. 2009), have shown great potential for crop improvements. The role of epigenetics in crop improvement cannot be ignored. Epigenetics significantly influences the natural variations in crops and its regulation affects the transgene integration, expression, and stability (Springer 2012). The various constraints on the agricultural sector and the potential of several technological innovations for attaining food and nutritional security are shown in Figure 20.1.

20.4 Transgenic Crops: Present Status and Future Prospects

Transgenic technology takes advantage of the vast gene pool available without any hindrance or sexual barrier, using the tools of genetic engineering to develop crops with desirable traits in a time span that is comparatively less than that for conventional plant breeding. The creation of transgenic crops involves the following steps: (i) the isolation, characterization, and manipulation of genes conferring the desired traits and making a gene construct having a gene of interest with suitable promoters,

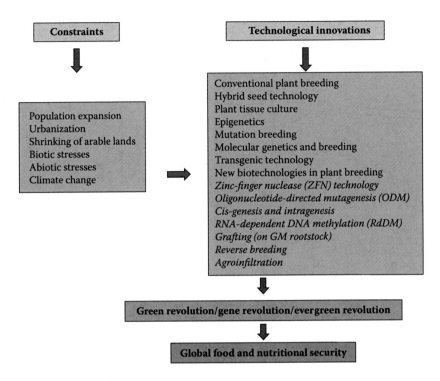

FIGURE 20.1
Technological innovations in agricultural sciences for overcoming the constraints for attaining global food and nutritional security.

terminators, and a selectable marker or reporter gene; (ii) the transfer of the gene construct by different methods, such as biolistic or *Agrobacterium* mediated in specific cells or tissues amenable to *in vitro* regeneration; (iii) the development of an *in vitro* regeneration protocol using suitable explants of targeted crops for the development of transgenic lines; and (iv) the screening and validation of transgenes for a few generations for their integrity, stability, and expression level. Each step in creating transgenic crops has the potential for patentability and, to date, several patents have been granted for the same.

Transgenic crops are also known by several names such as genetically modified crops, biotech crops, or designer crops. Over the years there has been a substantial increase in the cultivation of biotech crops, from 1.7 million hectares in 1996 to 160 million hectares in 2011, and currently, 29 countries are commercializing these crops worldwide (Clive 2011). The latest developments and statistics on biotech crops can be retrieved from a website (http://www.isaaa.org) that is maintained by the International Service for Acquisition of Agribiotech Applications (ISAAA), a not-for-profit international organization that is associated with sharing the benefits of crop biotechnology in different sectors. Among the different crops, the biotech soybean continued to be the principal biotech crop in 2011, occupying 75.4 million hectares or 47% of the global biotech area, followed by biotech maize (51 million hectares or 32%), biotech cotton (24.7 million hectares or 15%), and biotech canola (8.2 million hectares or 5%) of the global biotech crop area. With reference to the traits in biotech crops subjected to commercialization from 1996 to 2011, herbicide tolerance has consistently been the dominant trait. In 2011, the herbicide tolerance deployed in soybean, maize, canola, cotton, sugarbeet, and alfalfa occupied 59% or 93.9 million hectares of the global biotech area of 160 million hectares.

A total of 1045 approvals have been granted for 196 events for 25 crops (Clive 2011). The first- and second-generation transgenic crops with traits such as herbicides resistance, insect resistance, and nutritional quality have been successfully produced and now third-generation transgenic crops associated with molecular farming are also being attempted (Azadi and Ho 2010; Peters and Stoger 2011; Farre et al. 2011; Ahmad et al. 2012; Bhullar and Gruissem 2013). The molecular strategies for gene containment in transgenic crops, the concern for biosafety, and their biodiversity have also been addressed in recent reviews (Daniell 2002; Lee and Natesan 2006; Hill et al. 2007; Penna and Ganapathi 2010; Raven 2010; Manimaran et al. 2011; Mehrotra and Goyal 2013).

20.5 International Treaties and Forms of IPR Applicable to Agriculture and Plant Biotechnology

The innovation in biotechnology has led to several issues related to IPRs (Gold et al. 2002). The innovation in agriculture and plant biotechnology is also a subject matter of IPRs and, in general, plant-based research, plant variety protection (PVP), and utility patents or patents (in the United States) are a few of the forms of IPR (Sechley and Schroeder 2002).

20.5.1 International Treaties for Promoting Agricultural Innovations

20.5.1.1 Trade-Related Aspects of Intellectual Property Rights (TRIPS)

The genesis of IPRs in agriculture was the global intellectual property treaty agreement of TRIPS (http://www.wto.org/english/docs_e/legal_e/27-trips-pdf). This was the outcome of a multilateral agreement between the 134 World Trade Organization (WTO) member countries, which was negotiated during the 1986–1994 Uruguay round of the General Agreement on Tariffs and Trade (GATT). The minimum criteria for patentability were novelty, nonobviousness, usefulness, and enablement, and any inventions, whether products or processes, in all fields of technology can be patented. According to TRIPS, microorganisms and microbiological processes are patentable subject matter. Article 27(2) emphasizes the exclusion of inventions whose commercial exploitation would threaten public order or morality. These include diagnostic, therapeutic, and surgical methods for the treatment of humans or animals, life-forms other than microorganisms, and processes for the production of plants or animals. This provision was made optional and not universal and it varies from country to country. Article 27(3) of TRIPS states that all plants may be excluded from patentability, provided that the member country adopts alternative IP legislation, such as PBRs, or any other effective *sui generis* system to include plant varieties.

20.5.1.2 Convention on Biological Diversity (CBD) and Its Impact on Plant IPRs

This international treaty came into existence on December 29, 1993, as an outcome of the Earth Summit in Rio de Janeiro on June 5, 1992, with the goal of the conservation and sustainable use of biological diversity. More than 186 countries, including all of the developed countries except the United States, are party to the convention. The objectives of this convention are the conservation of biological diversity, the sustainable use of its components, and the fair and equitable sharing of the benefits arising out of the utilization of genetic resources, including by appropriate access to genetic resources and by appropriate transfer of relevant technologies taking into account all rights over those resources and technologies, and by appropriate funding. It increasingly focused on biotechnology, technology transfer, and IPRs (Dutfield 2002), and the Cartagena Protocol on Biosafety (CBD) was one major outcome. The important features of this treaty are national sovereignty over biological resources, the establishment of property rights for both indigenous knowledge and genetic resources, the promotion of the rights of indigenous populations over access to biodiversity, and the provision of access and benefit-sharing legislation (Diaz 2000; Boisvert and Caron 2000). The CBD has initiated the IPR debate between developing countries with rich natural resources and environmentalists mainly from developed countries with reference to the conservation of biodiversity, which has led to the emergence of issues such as biopiracy (Odek 1994; Shiva 1997; Hamilton 2006). The CBD guarantees the protection of IPRs under existing international laws and does not intend to restrict the availability of genetic resources. However, the convention rejects the free flow of resources and emphasizes signing the equivalent of a Material Transfer Agreement so that it is assured that the donor nation receives a share of any profits that may be realized from the material.

20.5.1.3 International Treaty on Plant Genetic Resources for Food and Agriculture (IT-PGRFA)

The IT-PGRFA came into existence on November 2, 2001, and it was adopted by 116 countries excluding the United States and Japan. Before the treaty comes into effect, it must be ratified by 40 countries. This treaty recommends that the raw materials used to develop new crop varieties should remain in the public domain. It promotes the conservation of plant genetic resources for food and agriculture. It exists in a legal and political space between the CBD and the TRIPS agreements and promotes the free exchange of germplasm as fundamental to global food security. It has made provisions for farmers' rights, but legally, these are not well defined. The main aim of this treaty is to develop a multilateral system comprising an aggregate of genetic material from the member countries, so that, after paying a fee, members can have access to the genetic material. The provision of IPRs is not adequately defined in this treaty.

20.5.1.4 Public Intellectual Property Resources for Agriculture (PIPRA)

PIPRA was created in 2004 by the Rockefeller Foundation to address the IP issues in agriculture that influence public investment, especially in developing countries. The delay in utilizing newer agricultural biotechnology, high transaction costs, and limited access to advanced technologies for the benefit of developing countries were the major concerns (Atkinson et al. 2003; Delmer et al. 2003; Grimes et al. 2011). Currently, PIPRA has a network that extends to 60 universities and research institutions in 17 countries worldwide. Since the public sector is the sole player in agricultural research and development in developing countries, PIPRA has a vital role to play in encouraging public sector involvement in agricultural research in the era of IPRs. There is an immediate need for sharing of intellectual property in agricultural biotechnology with references to the transgenic technology for easy accessibility to developing countries for attaining food and nutritional security (Chi-Ham et al. 2012). In addition to PIPRA, there are several other public sector initiatives for promoting agricultural research worldwide taking into consideration the growing awareness of IPRs. Some of these are:

- Centre for Application of Molecular Biology to International Agriculture (CAMBIA)
- Public Interest Property Advisors (PIPA)
- African Agricultural Technology Foundation (AATF)
- International Service for the Acquisition of Agri-biotech Applications (ISAAA)

20.5.2 Different Forms of IPR Relevant to Agriculture

20.5.2.1 Plant Patents

In the United States, a unique form of protection in the form of a plant patent exists, which includes plants that reproduce through asexual reproduction with tuber-propagated plants as an exception. This is framed legally under the Plant Patent Act (PPA) of 1930. It provides a 20-year patent protection for inventions derived from asexually reproduced varieties (Fuglie et al. 1996). This largely includes ornamental plants and fruits. The plant patent in the United States requires no yearly maintenance fee to remain in effect. In the United States, it is possible to get dual protection for plant varieties through a utility patent and a plant patent, or a PVP certificate and a utility patent, but not a PVP certificate and a plant patent (Pardey et al. 2013). In Europe, the patenting of plant varieties is excluded and the European Patent Convention (EPC) excludes the patenting of "plant or animal varieties or essentially biological processes for the production of plants or animals," though this is not applicable to microbiological processes or the products thereof (Blakeney 2012).

20.5.2.2 PVP

The preexisting patent laws were not considered suitable for covering innovations related to the development of new plant varieties using conventional plant breeding techniques. The International Union for the Protection of New Varieties of Plants (UPOV) global agreement came into existence in 1961 and it was later revised in 1972, 1978, and 1991 to frame a minimum standard for the protection of plant varieties similar to TRIPS. Up to April 4, 2011, a total of 69 countries have signed the UPOV Convention. These rights are distinct from patent protection, and are focused on the specific needs and interests of the plant breeding and propagation processes, originally based on traditional plant breeding methods. The members signing this agreement have the right to frame legislation in their respective countries with the inclusion of the minimum standards highlighted in the UPOV. This offers protection for plant varieties that are new, distinct, uniform, and stable. A plant variety is considered distinct if it is clearly distinguishable from any other variety whose existence is a matter of common knowledge. It is considered to be uniform if it shows its relevant characteristics on propagation, and stable if its relevant characteristics remain unchanged after repeated propagation.

Further, the agreement protects not only the plant variety and its propagating materials, but also the harvested product, which may include the entire plant or parts of the plant. Not only is there provision for the granting of exclusive rights to the owner/innovator, but the agreement also gives access to other innovators to use the protected materials for research purposes and allows farmers to save the seeds for planting in subsequent generations. The protection only applies to new plant varieties as such, which are generally defined as the lowest level of taxonomy (or classification) within the plant kingdom—that is, plant varieties that are distinct variations within a given species. These rights are granted by the state to plant breeders to exclude others from producing or commercializing the material of a specific plant variety and are generally for 25 years in the case of trees and vines and 20 years for any other variety. A country can develop its own system of protection, referred to as a *sui generis* system, that is, a system of rights designed to fit a particular context and need that is a unique alternative to standard patent protection (Mauria 2000). The developments and implications of plant varietal rights (PVRs) in the United States since their inception in 1930–2008 have recently been reviewed (Pardey et al. 2013).

PVRs are also manifested in the form of PBRs, which allow innovators/agricultural scientists to use the protected materials for research purposes, and farmers' privilege, which gives farmers the right to save seeds for planting in subsequent seasons. Great flexibility exists for farmers' privilege and it is solely a national option; a nation may allow farmers' privilege as an exception under certain circumstances. The latest version of the UPOV (1991) has expanded the scope of breeders' rights, it has introduced the concept of "essentially derived" varieties, and it has considered farmers' privilege to be a national option (Kolady and Lesser 2009).

India is rather inclined to have a *sui generis* system of legislation, which is nonpatent based. The various legislative systems concerned with providing IPRs in agriculture biotechnology in India are:

- The Protection of Plant Varieties and Farmers' Rights Act (PPVFR), 2001
- Biological Diversity Act, 2002
- Seeds Act, 1996
- Plant Quarantine (Regulation of Import into India) Order, 2003
- Rules for Manufacture, Use, Import, Export and Storage of Hazardous Microorganisms and Genetically Engineered Organisms or Cells, 1989

The PPVFR was passed by the Indian Government in 2001. The act covers all categories of plants, except microorganisms. India's law is unique in that it simultaneously aims to protect both breeders and farmers. The act grants PVP on new varieties, extant varieties, and essentially derived varieties. Extant varieties include farmers' varieties, varieties in the public domain, and varieties about which there is common knowledge. Different rights have been given to farmers under the act including the rights to save, exchange, and sell seeds and propagating material; the right to register varieties; the right to recognition and reward for the conservation of varieties; the right to benefit from sharing; the right to information about the expected performance of a variety; the right to compensation for the failure of a variety to perform; the right to the availability of seeds of a registered variety; the right to free services for registration; the right to conduct tests on varieties; the right to legal claims under the act; and the right to protection from infringement. The main objectives of the act are as follows:

- To provide for the establishment of an effective system for the protection of plant varieties
- To provide for the rights of farmers and plant breeders
- To stimulate investment for research and development and to facilitate the growth of the seed industry
- To ensure the availability of high-quality seeds and planting materials of improved varieties to farmers

20.5.2.3 Farmers' Rights

India's ability to be one of the first countries in the world to forge a national legislation on farmers' rights is a significant landmark. The act recognizes the farmer not just as a cultivator but also as a conserver of the agricultural gene pool and a breeder who has bred several successful varieties. The act makes provisions for such farmers' varieties to be registered, with the help of nongovernmental organizations (NGOs) so that they are protected against being scavenged by formal sector breeders. The rights of rural communities are also duly acknowledged. The farmers' rights of the act define the privilege of farmers and their right to protect the varieties that they have developed or conserved. Farmers can save, use, sow, resow, exchange, share, and sell farm produce of a protected variety except that for sale under a commercial marketing arrangement (branded seeds) [Section 39 (1), (i)–(iv)]. Farmers' rights were first formulated in Resolution 5/89 of a 1989 conference of the Food and Agriculture Organization (FAO) of the United Nations. Farmers' rights means "rights arising from the past, present and future contributions of farmers in conserving, improving, and making available plant genetic resources, particularly those in the centers of origin/diversity." The pupose of farmers' rights was to recognize the role of farmers and members of indigenous rural or traditional communities in creating, domesticating source of agricultural varieties and diversity for food and agriculture. Ironically, to date, farmers' rights have not been given any practical legal expression. Although Article 27(3) of the TRIPS agreement of April 1994 envisaged special types of legal systems for protecting plant and animal varieties, these have not yet been defined.

Further, farmers have also been provided with the protection of innocent infringement when, at the time of the infringement, a farmer is not aware of the existence of breeder rights. This formulation allows the farmer to sell seed in the way he or she has always done, with the restriction that this seed cannot be branded with the breeder's registered name. In this way, both the farmer's and the breeder's rights are protected. The breeder is rewarded for his or her innovation by having control of the commercial marketplace but without being able to threaten the farmer's ability to independently engage in his or her livelihood, and supporting the livelihood of other farmers.

20.5.2.4 Utility Patents

The *Diamond v. Chakrabarty* (447US303, 1980) U.S. Supreme Court decision in 1980 was a landmark decision that gave way to the patenting of life-forms (Chakrabarty 1988, 2010). The utility patents can be used for the patenting of plants or higher life-forms (HLFs) in many countries. The *Pioneer Hi-Bred International, Inc., v. J.E.M. AG Supply, Inc. et al.* (2001) was an important decision for the consideration of utility patents for plants. The scope of the protection offered by a utility patent is broader as compared to PVP or plant patents and it covers many kinds of different innovations in plant biotechnology, such as transformation processes, vectors and their various components, namely, promoters, selectable markers, genes of interest, as well as organisms and their parts. The basic requirements in utility patents are novel, useful, nonobvious, and prior art, while in PVP they are distinct, uniform, and stable.

20.6 Innovations in Plant Biotechnology and IPR Issues

In recent years, the success of transgenic technology and its potential for commercialization have led to the filing of several patents by both public and private sector-based research organizations globally. Genetic engineering is perceived as an important tool that can cater to the future food, feed, and energy requirements globally, but there are several issues and concerns for the commercialization of genetically engineered crops (Rommens 2010). The IPR has a significant role to play in the context of plant biotechnology, biodiversity, and biopiracy in the era of globalization (Blakeney et al. 1999; Sechley and Schroeder 2002; Kowalski et al. 2002; Gold et al. 2002; Boettiger et al. 2004; Yadav et al. 2010b). There has been a substantial increase in the patents granted to plant biotechnology-based innovations. More than a thousand patents have been granted for transgenic plants in the United States (Koo et al. 2004; Lacroix et al. 2008). Some of the patents granted to plant biotechnology-based innovations are listed in Table 20.1. Recently, transgenic drought-resistant maize and soybean with stacked traits of improved fatty acid profiles in the United States by the Monsanto Company and a transgenic bean resistant to bean yellow mosaic virus in Brazil by the Brazilian Agricultural Research Corporation have been approved for commercialization (Marshall 2012). The plant transformation technology comprising different components ranging from vectors, promoters, genes of interest, methods of gene transfer, selectable marker genes, *in vitro* regeneration protocols, and transformation stability/heritability are all subject matters of patents (Dunwell 2005). Tables 20.2 and 20.3, list some of the patents granted for selectable markers and transformation technology, respectively. The highly acclaimed Golden rice, a transgenic rice with enhanced levels of β-carotene for alleviating severe vitamin A deficiency, has 72 patents claimed by 40 different organizations (Ye et al. 2000; Kryder et al. 2000). This includes patents for the phytoene trait genes, promoters, terminators, selectable markers, and transit peptides that are used in its construct. The patents granted on several plant biotechnological innovations for some important crops, namely, rice, maize, potato, *Brassica*, cotton, sugarcane, tea, and rose, are shown in Table 20.4. The patenting of DNA, genes, sequences, and gene technologies influencing agricultural, biomedical, and several industrial sectors is a recent development with the deciphering of several plants, microbial and fungal, and animal genome sequences (Yadav et al. 2012). Whole-genome sequencing does not infringe the gene patent; the "gene patent" is actually a misnomer, as gene patent holders do not actually own the genes that are the subject of their patents. They have the right to exclude others from making, using, selling, or importing a nonnaturally occurring, man-made product or process, as defined by the patent's claims (Holman 2012).

Agricultural innovation can also reorient plant breeding to generate smart crops with an enhanced yield using fewer inputs and the ability to tolerate the changing climatic conditions (Mba et al. 2012). The MAS in breeding has been very successful in reducing the number of generations for evaluating breeding materials with enhanced efficiency for both monogenic and polygenic traits (Eathington et al. 2007; Choudhary et al. 2008). The Consultative Group on International Agricultural Research (CGIAR) has developed the integrated breeding platform (IBP) of the Generation Challenge Program (Clive 2011) to make available the MAS techniques for developing elite varieties of food security crops in developing countries. Multinational companies have already implemented MAS in various crops but it is still not routinely used in public crop improvement programs due to its high set-up costs and IPR restrictions. The following is a list of the emerging biotechnology techniques influencing plant breeding, which have recently been reviewed (Lusser et al. 2012).

1. ZFN technology—designed to create site-specific mutations or gene inactivation for a desired phenotype; ZFNs are basically synthetic restriction endonucleases consisting of a zinc-finger domain that recognizes specific DNA sequences and nucleases domains that assist in cutting.

2. ODM—associated with creating targeted mutations of one or a few nucleotides and also known as targeted gene repair, genoplasty, and chimeraplasty.

TABLE 20.1

List of Some Patents in Plant Biotechnology

S. No.	Patent No.	Inventor	Assignee	Title	Description	Filing Date	Issue Date
1	EP0513884 A1	Josephus Nicholaas Maria Mol, Ingrid Maria Van Der Meer, Andrianus Jhohannes Van Tunen	Mogen International N.V.	Male-sterile plants, methods for obtaining male-sterile plants and recombinant DNA for use therein	It provides male-sterile plants, characterized in that the said plants have a recombinant polynucleotide integrated into their genome, essentially comprising an inhibitory gene, which, upon proper expression in the anthers of the plant, is capable of inhibiting expression of one or more genes encoding an enzyme involved in the synthesis of chalcone, or one of its precursors	4/15/1992	11/19/1992
2	US5917117	Burt D. Ensley, Michael J. Blaylock, Slavik Dushenkov, Nanda P.B.A. Kumar, Yoram Kapulnik/ Phytotech Inc.	Phytotech Inc.	Inducing hyperaccumulation of metals in plant shoots	It provides methods by which the hyperaccumulation of metals in plant shoots is induced by exposure to inducing agents such as low pH, chelators, herbicides, and high levels of heavy metals	3/21/1996	6/29/1999
3	US5947041	Louis A. Licht	Ecolotree Inc. USA	Methods for using tree crops as pollutant control	A method of naturally removing or inhibiting pollutants is described. Tree stems from trees having preformed root initials and a perennial root system are planted adjacent to the area where water is to be decontaminated. The stems are densely planted to achieve maximum pollution control and environmental effects. Row planting is used and the crop is harvested on a rotating basis	2/4/1991	9/7/1999

(continued)

TABLE 20.1 (Continued)

List of Some Patents in Plant Biotechnology

S. No.	Patent No.	Inventor	Assignee	Title	Description	Filing Date	Issue Date
4	US610092	Mykola Borysyuk, Lyudmyla Borysyuk, Ilya Raskin	Rutgers, The State University of New Jersey	Materials and methods for amplifying polynucleotides in plants	The products include nucleic acids containing a plant active amplification promoting sequence (APS) and the methods exploit these products in amplifying target nucleic acids	6/15/1998	8/8/2000
5	US6159270	Ilya Raskin, Nanda P.B.A. Kumar, Slavik Douchenkov	Edenspace Systems Corporation	Phytoremediation of metals	The process is based on manipulating the growth of crop and crop-related members of the plant family Brassicaceae in metal-containing soils so that the metal in the soils is made more available to the plants	3/18/1998	12/12/2000
6	EP0889691 B1	Jr G. Bradford Crandall, W. Ralph Emerson	Proguard Inc.	Use of aromatic aldehydes as pesticides	Methods and compositions based on natural compounds, including balsam, cinnamic aldehyde, α-hexyl cinnamic aldehyde, and coniferyl aldehyde are provided, which find use as pesticides	3/25/1997	1/29/2003
7	EP1346622 A1	Andy Dwayne Beck, Stephen Micheal Faivre, Geord Larcheid, et al.	Deere & Company	Methods and system for automated tracing of an agricultural product	The invention refers to a system for and a method of tracing a crop by electronic regarding of various information, forming a data profile, segregating the crops from intermixing, electronically recording a storage identifier, and associating the formed data profile with the storage identifier	3/19/2003	9/24/2003
8	EP1463811A2	Haviva Eilenberg, Silvia Schuster, Aviah Zilverstein	Ramot at Tel Aviv University Ltd.	Chitinases, derived from carnivorous plants polynucleotide sequences encoding thereof, and methods of isolating and using same	It provides an enzymatic composition comprising at least one protein isolated from a tissue or soup of a carnivorous plant, with at least one protein being characterized with an endochitinase activity	6/17/2002	10/6/2004

	Patent number	Inventors	Title	Assignee	Description		
9	EP1551985 A2	Newell Bascomb, Mark Bossie, Melissa Campo Andrei Golovko, Gerald Hall, Lynne Hirayama, Thomas Petty, Marina	Commercial use of *Arabidopsis* for the production of human and animal therapeutic and diagnostic proteins	Icon Genetics Inc.	It provides methods that make it possible to take advantage of the various growth parameters of *Arabidopsis* in order to grow dense populations of the plant in controlled indoor environments for the purpose of harvesting the biomass and isolating proteins, particularly recombinant proteins suitable for pharmaceutical applications	7/26/2002	7/13/2005
10	EP1671534A1	Claudio Cerboncini, Heide Schnabl, Ralf Theisen	Methods for altering levels of phenolic compounds in plant cells	Stiftung Caesar Centre of Advanced European Studies and Research	It provides a plant-derived extract comprising inhibitory activity against HIV integrase. In addition, methods for the isolation of enzymes and their encoding genes, which are involved in the biosynthesis of selected phenolic compounds such as depsides are provided as well as transgenic organisms transformed therewith	12/16/2004	6/21/2006
11	EP1613796A4	Hyoung-Joon Jin, L. David Kaplan, Ung-Jin Kim, Jaeyung Park	Concentrate aqueous silk fibroin solution and use thereof	Tufts University	It provides for concentrated aqueous silk fibroin solutions and an all-aqueous mode for the preparation of concentrated aqueous fibroin solutions that avoids the use of organic solvents, direct additives, or harsh chemicals. The invention further provides for the use of these solutions in the production of materials, for example, fibers, films, foams, meshes, scaffolds, and hydrogels	4/12/2004	8/15/2007
12	EP1848265A2 US2009011973 US20090300802	Alisa Huffaker, L. Gregory Pearce, A. Clarence Ryan, Yube Yamaguchi	Plant defense signal peptides	Washington State University Research Foundation	AtPtpl plays an important role as a signaling component of the innate immune system of *Arabidopsis*. AtPcpl and its seven paralogs and orthologs play important roles as endogenous signals to amplify innate immunity	1/24/2006	10/31/2007

(continued)

TABLE 20.1 (Continued)

List of Some Patents in Plant Biotechnology

S. No.	Patent No.	Inventor	Assignee	Title	Description	Filing Date	Issue Date
13	EPI885861A2	Jason bull, David Butruille, Sam Eathington, Marlin Edwards, Anju Gupta, Richard Johnson, Wayne Kennard, Jennifer Rinehart, Kunsheng Wu	Monsanto Technology LLC	Methods and compositions to enhance plant breeding	It relates to the field of plant breeding and plant biotechnology, in particular to a transgene inserted into the genetic linkage with a genomic region of a plant, and to the use of the transgene/genomic region to enhance the germplasm and to accumulate other favorable genomic regions in breeding populations	5/26/2006	2/13/2008
14	EP1750497A4	L. Stephen Goldman, V. Sairam Rudrabhatla	The University of Toledo	A method for producing direct *in vitro* flowering and viable seed from cotyledons, radicle, and leaf explants and plants produced there from	It relates to a method of reprogramming plant development that allows flower buds and seeds to arise *de novo*, directly from a cotyledon or radicle explants or from shoots produced on a cotyledon or radicle. The present invention also provides for an improved culturing media that provide for *in vitro* flowering	5/27/2005	8/19/2009
15	EP2187730A1	Filippa Brugliera	International Flower Developments Pty Ltd.	Genetically modified chrysanthemums	It relates to genetically modified chrysanthemum plants expressing altered inflorescence. The chrysanthemum flavonoid pathway is manipulated to produce plants with blue or violet inflorescence. Blue delphinidin pigments are produced by the expression of flavonoid 3'5' hydroxylase and (optionally) the suppression of flavonoid 3' hydroxylase activity	11/14/2008	5/26/2010

No.	Patent number	Inventors	Assignee	Title	Abstract		
16	EP2199304A1 EP1974049A2	Astrid Blau, Beate Kamlage, Ralf Looser, Gunnar Plesch, Piotr Puzio, Oliver Schmitz, Bright Wendel	Metanomics GmbH	Process for the control of production of fine chemicals	It relates further to a process for the control of the production of fine chemical in a microorganism, a plant cell, a plant, a plant tissue, or in one or more parts thereof. The invention furthermore relates to nucleic acid molecules, polypeptides, nucleic acid constructs, vectors, antisense molecules, antibodies, host cells, plant tissue, propagation material, harvested material, plants, microorganisms, as well as agricultural compositions and to their use	12/19/2005	6/23/2010
17	EP2344644A1	Markus Frank, Gunnar Plesh, Piotr Puzio	BASF Plant Science GmbH	Methods for producing transgenic plant cell, a plant or a part thereof with increased resistance to biotic stress	It relates to the control of pathogens. Disclosed herein are methods of producing transgenic plants with increased pathogen resistance, expression vectors comprising polynucleotides encoding for functional proteins, and transgenic plants and seeds generated thereof	9/28/2009	7/20/2011
18	EP2390256A1	Steven Fabijanski, Carl Parez, Edward Pekins	Agrisoma Inc., Calyx Bio-Ventures Inc.	Plant artificial chromosomes, use thereof and methods of preparing plant artificial chromosomes	Methods for preparing cell lines that contain plant artificial chromosomes, methods for the preparation of plant artificial chromosomes, methods for the targeted insertion of heterologous DNA into plant artificial chromosomes, and methods for the delivery of plant chromosomes to selected cells and tissues are provided	5/30/2002	11/30/2011

(continued)

TABLE 20.1 (Continued)

List of Some Patents in Plant Biotechnology

S. No.	Patent No.	Inventor	Assignee	Title	Description	Filing Date	Issue Date
19	WO 2010091248 US2010204921 EP2339214	Byrum J. Clarke JDV, Guo Z, Gutierrez RLA, Kishore VK, Li M, Wang D	Syngenta Participations (Basel)	Method for selecting statistically validated candidate genes	A method of selecting one or more markers associated with a trait of interest in a species of interest, comprising identifying markers associated with the trait of interest in a population of the species using a suitably programmed computer to perform genome-wide association mapping	2/6/2009, 2/4/2010	8/12/2010, 12/28/2011
20	US20120035354	Bloksberg LN, Havukkala I	ArborGen (Summerville, SC, USA), Rubicon Forests Holdings (Auckland, NZ)	Materials and methods for the modification of plant lignin content	A polynucleotide isolated from *Eucalyptus grandis* and *Pinus radiata*, useful for modifying the content, structure, and composition of lignin in target organisms such as plants; useful for wood processing for producing paper, genome mapping, physical mapping, positional cloning of genes, and designing oligonucleotide probes and primers	3/9/2011	2/9/2012
21	EP242372A2	Niranjan Ramanlal Gandhi, Palmer Victoria Skebba, A. Gary Strobbel/	Jeneil Bio surfactant Company LLC	Antimicrobial compositions and related methods of use	Antimicrobial compositions comprising one or more compound components generally recognized as safe for human consumption, and related methods of use, such compositions and methods as can be employed in a wide range of agricultural, industrial, building, pharmaceutical, and personal care products and applications	4/27/2010	3/7/2012

22	EP1874938B1	Christian Dammann, Christina E. Roche, Hee-Sook Song	BASF Plant Science GmbH	Starchy-endosperm and germinating embryo-specific expression in monocotyledonous plants	The present invention relates to the field of agricultural biotechnology. Disclosed herein are expression constructs with the expression specificity for the starchy endosperm and the germinating embryo, transgenic plants comprising such expression constructs, and methods of making and using such DNA constructs and transgenic plants	4/13/2006 — 4/4/2012
23	US20120119080 WO2012068217	Hazabroek J, Janni J, Lightner J	Pioneer Hi-Bred (Johnston, IA, USA)	Prediction of phenotypes and traits based on the metabolome	A method for establishing an unbiased model using the metabolic profile, phenotype profile, and trait profile of two groups of plants, involving separating and extracting metabolites from two groups of plants by chromatography to generate a data set and comparing this to another data set generated using mass spectrometry	11/16/2011 — 5/17/2012, 5/24/2012
24	CN102478563	Chang Y, Lu X, Xu G, Zhao C, Zhao Y, Zhou J	Dalian Institute of Chemical Physics, Chinese Academy of Sciences (Dalian, China)	A method for researching metabolic differences between transgenic rice and nontransgenic rice, involving analyzing rice seed extracted by a liquid-phase chromatography-mass spectrometry technology for obtaining a rice metabolic profile spectrum	It claims an analysis method is simple and fast and the repeatability is good and suitable for a real sample of mass analysis	11/25/2010 — 5/30/2012

(continued)

TABLE 20.1 (Continued)

List of Some Patents in Plant Biotechnology

S. No.	Patent No.	Inventor	Assignee	Title	Description	Filing Date	Issue Date
25	EP2477477A1	Benjamin Edgar Cahoon, G. Jan Jaworski, Nii Patterson, P. Oliver Peoples, D. Kristi Snell, Jihong Tang, Wenyu Yang	Donald Danforth Plant Science Centre Metabolix Inc.	Generation of high polyhydroxybutyrate producing oilseeds	Transgenic oilseed plants synthesize (poly)3-hydroxybutyrate (PHB) in the seed. Genes utilized include phaA, phaB, and phaC, all of which are known in the art. The genes can be introduced in the plant, plant tissue, or plant cell using conventional plant molecular biology techniques	9/15/2010	7/25/2012
26	US8247541B2	Ken W. Richards. Wesley G. Taylor	Her Majesty the Queen in Right of Canada, as represented by The Minister of Agriculture and Agri-food	Plant compositions enriched in dehydrosoyasaponin I (D-I) and methods of producing such compositions	It provided a method of producing a plant composition comprising dehydrosoyasaponin I (D-I), the method comprising the steps of extracting a plant flour with a solvent capable of extracting soyasaponins to produce an extract, and treating the extract with light. Also provided are compositions produced according to the method of the present invention	12/14/2007	8/21/2012
27	US20120220464A1 EP2482647A1	Sabine Giessler-Blank, Martin Schillin, Ewald Sieverding	Evonik Goldschmidt GmbH	Use of sophorolipids and derivatives thereof in combination with pesticides as adjuvant/additive for plant protection and the industrial noncrop field	Use of sophorolipids as adjuvants in combination with pesticides as a tank mix additive and as a formulation additive for crop protection and for the industrial noncrop sector	8/30/2010	8/30/2012
28	US2012225923	Thomas Himmler, Thomas Seitz, Ulrike Wachendorff-Neumann	Bayer CropScience AG	Dithiinetetra(thio) carboximides	Novel dithiinetetra(thio) carboximides for the control of harmful microorganisms in crop protection and in the protection of materials and as plant growth regulators	9/3/2011	9/6/2012

29	US20120227132A1	Mei Guo, Howard P. Hershey, Carl R. Simmons	Pioneer Hi-Bred International Inc.	Cell number polynucleotides and polypeptides and methods of use thereof	It provides polynucleotides and related polypeptides of the protein CNR. The invention provides genomic sequence for the CNR gene. CNR is responsible for controlling cell numbers	5/15/2012	9/6/2012
30	EP2501219A1	Benjamin Greame Cocks, Matthew Knight, Aidyn Mouradov, German Spangenberg, Jianghui Wang	Agriculture Victoria Services Pty Ltd.	Angiogenin expression in plants	It relates to plant-produced angiogenins, to related plant cells, plant calli, plants, seeds, and other plant parts and products derived therefrom and to uses of plant-produced angiogenins. The present invention also relates to the expression of angiogenin genes in plants and to related nucleic acids, constructs, and methods	11/18/2010	9/26/2012
31	US828351982	Robert A. Creelman, Neal I. Gutterson, Jacqueline E. Heard, et al.	Mendel Biotechnology Inc.	Plant transcriptional regulators of abiotic stress	It relates to plant transcription factor polypeptides, polynucleotides that encode them, homologs from a variety of plant species, variants of naturally occurring sequences, and methods of using the polynucleotides and polypeptides to produce transgenic plants	5/4/2004	10/9/2012
32	US20120272408A1	Rudy Maor	Maor Rudy	Compositions and methods for enhancing plants resistance to abiotic stress	A method of increasing the tolerance of a plant to an abiotic stress or increasing the biomass, vigor, or yield of a plant by upregulating within the plant an exogenous polynucleotide of a micro-RNA	12/6/2010	10/25/2012
33	US20120272353A1	Mei Guo, Dale F. Loussaert, Yonjhong Wu	Pioneer Hi-Bred International Inc.	Methods to increase crop grain yield utilizing complementary paired growth and yield genes	The specific genes increase female reproductive organs and are paired with genes responsible for modifying the growth of nonyield-specific plant tissues	6/25/2012	10/25/2012

(continued)

TABLE 20.1 (Continued)
List of Some Patents in Plant Biotechnology

S. No.	Patent No.	Inventor	Assignee	Title	Description	Filing Date	Issue Date
34	US20120284876A1	Howard P. Hershey, Dale Loussaert, Carl R. Simmons	Pioneer Hi-Bred International Inc.	Genes for enhancing nitrogen utilization efficiency in crop plants	It provides isolated nitrogen utilization efficiency (NUE) nucleic acids and their encoded proteins. The present invention provides methods and compositions relating to altering nitrogen utilization and uptake in plants. The invention further provides recombinant expression cassettes, host cells and, transgenic plants	7/17/2012	11/8/2012
35	US20120291154A1	Scott Anderson, James Crowley, Brandon J Fabbri, Bo-Xing Qui, Steven E. Screen	Scott Anderson, James Crowley, Brandon J Fabbri, Bo-Xing Qui, Steven E. Screen	Corn plants and seeds enhanced for asparagine and protein	It relates to DNA constructs that provide expression in the transgenic corn cells of an asparagine synthetase enzyme. The DNA constructs are used in a method to produce transgenic corn plants and seeds and to select plants and seeds with enhanced levels of protein and amino acids	12/27/2011	11/15/2012
36	US8318436B2	Stephen Mayfield, Michael Mendez, Bryan O' Neill, Yan Poon	Sapphire Energy Inc., The Scripps Research Institute	Use of genetically modified organism to generate biomass degrading enzymes	It provides methods of producing one or more proteins, including biomass-degrading enzymes in a plant	10/7/2011	11/27/2012
37	US20120324597A1	Yaakov Tadmor, Yosef Burger, Nurit Katzir, et al.	The State of Israel, Ministry of Agriculture and Rural Development Research Organization	Melon plants comprising tetra-*cis*-lycopene	A *Cucumis melo* plant is disclosed, wherein the flesh of the fruit of the plant comprises tetra-*cis*-lycopene (prolycopene). Methods of generating same are also disclosed	2/22/2011	12/20/2012

TABLE 20.2

List of Patents for Selectable Markers Used in Transgenic Technology

S. No.	Patent No.	Inventor	Assignee	Title	Description	Filing Date	Issue Date
1	EP0289478A2	Maria Burmaz Hayford, Harry John Klee, Stephen Gary Rogers	Monsanto	Gentamicin marker genes for plant transformation	It involves the use of a selectable plant marker gene encoding a gentamicin-3-N-acetyltransferase enzyme. The invention also provides transformed plant cells that contain the gentamicin marker genes as well as differentiated plants containing transformed plant cells	4/26/1988	11/2/1988
2	EP0800583A1 US5633153	M. Virginia Ursin	Calgene Inc.	Aldehyde dehydrogenase markers for plant transformation	It relates to a method of plant transformation in which plant cells are transformed with an aldehyde dehydrogenase gene capable of detoxifying a phytotoxic aldehyde selective agent. The gene construct is linked to another gene construct of interest for expression in plant cells, wherein the aldehyde dehydrogenase gene acts as a selectable marker for transgenic plant cells	10/12/1995	11/15/1997
3	EP1090134A1 US6284956	Ning Huang, L. Raymond Rodriguez	Applied Phytologics Inc.	Plant selectable marker and plant transformation method	It relates to plant transformation expression cassettes with a selectable marker gene. The cassette contains a DNA promoter sequence from the rice beta-glucanase 9 (*gns9*) gene, a selectable marker gene and a 3' untranslated terminator region in 5'-3' direction	6/25/1999	4/11/2001
4	EP0698106B1 EP0698106A1 US5962768	Marcus Cornelissen, Veronique Gossele, Arlette Reynaerts, Roel Van Aarssen	Aventis CropScience N.V.	Marker gene	The application discloses a method to select and identify transformed plant cells by expressing a chimeric gene encoding an aminoglycoside-6'-N-acetyltransferase in the plant cells in the presence of an aminoglycoside antibiotic	5/11/1994	8/1/2001

(continued)

TABLE 20.2 (Continued)

List of Patents for Selectable Markers Used in Transgenic Technology

S. No.	Patent No.	Inventor	Assignee	Title	Description	Filing Date	Issue Date
5	EP1171620A1 US7148398	V. Oleg Bougri, T.M Caius Rommens, Neelam Srivastava, M. Kathleen Swords	Monsanto Technology LLC	Acquired resistance genes in plants	It describes new acquired resistance genes in plants. A method of using the genes to make transgenic plants that are resistant to disease is also provided	5/12/2000	1/16/2002
6	EP1368484A2	Jeffrey L. Dangle, Thomas Eulgen, Jane Glazebrook, Xun Wang, Tong Zhu		Plant genes, the expression of which are altered by pathogen infection	Methods to identify genes, the expression of which are altered in response to pathogen infection, are provided, as well as the genes identified thereby	9/14/2001	12/10/2003
7	EP1370650A2 US2005081267	Alfred Puhler	Bayer CropSciences N.V.	Novel genes for conditional cell ablation	It relates to novel DNA molecules encoding a protein having the biological activity of a deacetylase. These genes are particularly useful for the production of transgenic plants with plant parts that can be destroyed by treatment with N-acetyl-PPT	3/12/2002	12/17/2003
8	EP1781821A2 US7507874	Feng Han, Bradley Hedges, Hong Lu, Scott Sebastian, Debra Steiger	Pioneer Hi-Bred International Inc.	Genetic loci associated with *Phytophthora* tolerance in soybean	It relates to methods and compositions for identifying soybean plants that are tolerant, have improved tolerance, or are susceptible to *Phytophthora* root rot infection. The methods use molecular genetic markers to identify, select, and construct disease-tolerant plants or identify and counterselect disease-susceptible plants	8/8/2005	5/9/2007
9	EP1871879A1	Ronald Koes, Francesca Quattrocchio, Kees Spelt, Walter Verweij	International Flower Development Proprietary Ltd.	Plant genetic sequences with vacuolar pH and uses thereof	It provides genetic and proteinaceous agents capable of modulating or altering the level of acidity or alkalinity in a cell, a group of cells, an organelle, a part, or a reproductive portion of a plant	4/4/2006	1/2/2008

10	EP0960209B1	Helaine Carrer, Sumita Chaudhary, Pal Maliga	Rutgers, The State University of New Jersey	Editing-based selectable plastid marker genes	It relates to novel DNA constructs for selecting plastid transformants in higher plants. Also disclosed are editing-based selectable marker genes that require editing at the transcriptional level for the expression of the selectable marker gene	6/13/1997	5/21/2008
11	EP2102364A2 US20080178325 US20110041214 US7872170	Sadik El Sayed, Eric Hoeft, Zenglu Li, Lomas Tulseiram	Pioneer Hi-Bred International Inc.	Genetic markers for *Orobanche* resistance in sunflower	Methods for identifying sunflower plants or germplasm that display resistance, improved resistance, or susceptibility to *Orobanche cumana* are provided. Sunflower plants or germplasm that are resistant or have improved resistance to *Orobanche cumana* are created	12/27/2007	9/23/2009
12	EP2121982A2	Pascal Delage, Denis Lespinasse, Jean-Paul Muller, Michel Ragot	Syngenta Participation AG	Maize plants characterized by quantitative trait loci (qtl)	It relates to maize plants with a genome comprising a unique allele profile associated with the corresponding QTLs contributing to the expression of a variety of phenotypic traits of economic interest selected from the group of grain yield, grain moisture at harvest, early and late root lodging, stalk lodging, common smut incidence, fusarium ear rot incidence, solcotrione resistance, and tassel architecture	1/18/2008	11/25/2009
13	EP2152063A1 EP1949785A1 EP1949785A9 US20100146662	Henricus Maria Clemens Nicolaas De Vetten, Evert Jacobsen, Gerard Andries Edwin Van der Vossen, Franciscus Gerardus Richard Visser, Agnes Maria Anna Wolters		Use of R-genes as a selection marker in plant transformation and use of *cis*-genes in plant transformation	It discloses plant transformation of Solanaceae, potato in particular. Potato plant with functional R-genes to provide resistance against *Phytophthora infestans*, wherein the said R-gene can be used as a selectable marker	1/28/2008	2/17/2010

(*continued*)

TABLE 20.2 (Continued)

List of Patents for Selectable Markers Used in Transgenic Technology

S. No.	Patent No.	Inventor	Assignee	Title	Description	Filing Date	Issue Date
14	EP1521835B1 EP1521835A1 EP2202311A2 EP2292768A1 US2006174366	Mariette Andersson, Per Hofvander, Adelina Trifonova	BASF Plant Science GmbH	Use of *ahas* (acetohydroxy acid synthase) mutant genes as a selection marker in potato transformation	It is about the mutated AHAS genes conferring resistance to herbicides and providing an efficient system for the selection of transgenic potato lines	7/3/2003	3/10/2010
15	EP2210951A2	Carl F. Beher, Gregoery R. Heck, Catherine Hironaka, Jinsong You	Monsanto Technology LLC	Corn event PV-ZMGT32 (NK603) and compositions and methods for detection thereof	It is about a DNA construct that provides tolerance to transgenic corn plant. It also provide assays for detecting the presence of the PV-ZMGT32(nk603) corn event based on the DNA sequence of the recombinant construct inserted into the corn genome and of genomic sequences flanking the insertion site	6/15/2001	7/28/2010
16	EP2356243A1 US20100122370 US20110283415	M. Stephen Allen, Singh Kanwarpal Dhugga Susanne Groh, Victor Llaca, Stanley Luck, Bernhard Rietmann	E.I. du Pont de Nemours and Company, Pioneer Hi-Bred International Inc.	Genetic loci associated with cell wall digestibility in maize	It is about methods and compositions for identifying maize plants with increased cell wall digestibility. It uses molecular markers to identify and select plants with increased cell wall digestibility or to identify and counterselect plants with decreased cell wall digestibility	11/6/2009	8/17/2011
17	EP1545190B1	Fang-Ming Lai, S. Benoit Landry	BASF Plant Science GmbH DNA landmarks Inc.	Male sterility restoration as a selectable marker in plant transformation	It provides for the fertility restorer genes to restore fertility in *Brassica napus*. It also includes vectors that have male sterility genes flanked by recombinase sites, a fertility restorer gene, and a nucleotide sequence of interest flanked by recombinase sites, and methods of using such vectors to produce transgenic plants, using the restoration of male fertility as the selection for transformation events	8/25/2003	6/20/2012

18	US8278505B2 EP2145007A2 US20100251432	Justin M. Lira, Donald J. Merlo, Andrew E. Robinson, Erika Megan Snodderley, Terry R. Wright	Dow Agrosciences	Herbicide resistance genes for resistance to aryloxyalkanoate herbicides	It provides novel plants that are resistant to 2,4-D and pyridyloxyacetate herbicide. The invention also provides novel methods of preventing the development of, and controlling, strains of weeds that are resistant to one or more herbicides such as glyphosate. The main enzyme and gene for use are referred to as AAD-13 (aryloxyalkanoate dioxygenase)	5/9/2008	10/2/2012
19	EP2515630A2	Paul Altendorf, John Arbuckle, William Briggs, Christine Chaulk-Grace, Dallas Joseph Clarke, Gayle Dace, Molly Dunn, David Foster, Sonali Gandhi, Andres Libardo Gutierrez Rojas, Krishna Venkata Kishore, Vance Cary Kramer, Denise Kari Kust, Meijuan Li, Lynn Robert Miller, Martin Nicolas, Joseph Thomas Prest Tucker Jon Aaron Reinders, Allen Sessions, Wayne Dale Skalla, Daolong Wang, Todd Warner, Chris Zinselmeier	Syngenta Participations AG	Genetic markers associated with drought tolerance in maize	It provides methods and compositions for identifying, selecting, and producing drought-tolerant maize plants or germplasm. The subject matter relates to maize lines, such as *Zea mays* lines, with one or more improved water optimization genotypes, and methods for breeding that involve genetic marker analysis and nucleic acid sequence analysis	12/23/2010	10/31/2012

(continued)

TABLE 20.2 (Continued)

List of Patents for Selectable Markers Used in Transgenic Technology

S. No.	Patent No.	Inventor	Assignee	Title	Description	Filing Date	Issue Date
20	EP2531601A1	Flavie Coulombier, Helene Eckert, Yannick Favre, Bernard Pellisier	Bayer CropScience AG	Soybean transformation using hydroxyphenylpyruvate dioxygenase (HPPD) inhibitors as selecting agents	It relates to the *Agrobacterium*-mediated transformation of soybean organogenic tissue using a gene or genes for tolerance to HPPD inhibitors as selection marker. The methods for regenerating transgenic soybean plants from the said transformed soybean cells or tissue are also covered in the application	2/1/2011	12/12/2012

TABLE 20.3

List of Patents for Plant Genetic Transformation Technology

S. No.	Patent No.	Inventor	Assignee	Title	Description	Filing Date	Issue Date
1	EP0275069 A2	Charles J. Arnizen, Lorin R. DeBonte, Jr., David A. Evans, Willie H. Loh, Joan T. O'Dell	E.I. DuPont de Nemours, DNA Plant Technology Inc.	Pollen-mediated gene transformation in plants	It relates to a novel method for efficiently carrying out pollen-mediated gene transformation of flowering plants by utilizing novel DNA constructs incorporating exogenous DNA fragments coding for specific enzymes	1/11/1988	7/20/1988
2	US4945050	John C. Sanford, Edward D. Wolf, Nelson K Allen	Cornell Research Foundation Inc.	Method for transporting substances into living cells and tissues and apparatus there for	It relates to inert or biologically active particles that are propelled at cells at a speed whereby the particles penetrate the surface of the cells and become incorporated into the interior of the cells. The apparatus for propelling the particles toward target cells or tissues are also disclosed	11/13/1984	7/31/1990
3	US5015580	Paul Christou, Dennis McCabe, William F. Swain, Kenneth A. Barton	Agracetus	Particle-mediated transformation of soybean plants and lines	It relates to the method and apparatus for the genetic transformation of soybean plants by particle-mediated transformation. Foreign genes are introduced into regenerable soybean tissues by coating on carrier particles that are physically accelerated into plant tissues	5/12/1988	5/14/1991
4	US512466	Anne-Marie Stomp, Arthur K. Weissinger, Ronald R. Sederoff	North Carolina State University	Ballistic transformation of conifers	A method of transforming conifers with a DNA construct comprising an expression cassette is disclosed herein. It involves propelling the DNA construct to the plant tissue target at a velocity sufficient to pierce the cell walls and deposit the DNA construct within a cell of the target tissue	6/13/1989	6/16/1992

(continued)

TABLE 20.3 (Continued)

List of Patents for Plant Genetic Transformation Technology

S. No.	Patent No.	Inventor	Assignee	Title	Description	Filing Date	Issue Date
5	US5169770	Paula P. Chee, Stephen L. Goldman, Anne C.F. Graves, Jerry L. Slightom	The University of Toledo	*Agrobacterium*-mediated transformation of germinating plant seeds	The present invention relates to a nontissue culture process using *Agrobacterium*-mediated vectors to produce transgenic plants from seeds of such plants as the common bean and soybean	6/21/1990	12/8/1992
6	US5231019	Jerzy Paszkowski, Ingo Potrykus, Barbara Hohn, Raymond D. Shillito, Thomas Hohn, Michael W. Saul, Vaclav Mandak	Ciba-Geigy Corporation	Transformation of hereditary material of plants	It comprises a novel method for direct foreign gene transfer to a plant cell. The method provides for placing a gene under the control of plant expression signals and transferring it, by contact with protoplasts without the aid of natural systems for infecting plants, directly to the plant cells from which genetically transformed plants can subsequently be derived	2/23/1990	7/27/1993
7	US5371003	Lynne E. Murry, Ralph M. Sinibaldi, Paul S. Dietrich, Sharon C.H. Alfinito	Sandoz Ltd.	Electrotransformation process	It relates to a novel processes for introducing DNA into plant material utilizing nonpulsed electric current, and plant cell lines, differentiated plant tissues, and plants produced by said processes	9/23/1993	12/6/1994
8	US5584253	Richard A. Krzyzek, Cheryl R. M. Laursen, Paul C. Anderson	DeKalb Genetics Corporation	Genetic transformation of maize cells by electroporation of cells pretreated with pectin-degrading enzymes	It provides a method to increase the susceptibility of cultured *Zea mays* cells to stable transformation with recombinant DNA via electroporation, by pretreating the *Zea mays* cells with certain pectin-degrading enzymes	12/28/1990	1/24/1995
10	US5538877	Ronald C. Lundquist, David A. Walters	DeKalb Genetics Corporation	Method for preparing fertile transgenic corn plants	Fertile transgenic *Zea mays* plants that stably express heterologous DNA and a process for producing said plants is disclosed. The process comprises the microjectile bombardment of friable embryogenic callus from the plant to be transformed	11/10/1992	7/23/1996

11	US5563055	Jeffrey A. Townsend, Laurie A. Thomas	Pioneer Hi-Bred International Inc.	Method of *Agrobacterium*-mediated transformation of cultured soybean cells	A method for producing transgenic soybean plants is disclosed. The method employs conditions necessary for genotype-independent, *Agrobacterium*-mediated transformation of soybean explants and the utilization of a specialized medium to cause root induction	3/28/1994	10/8/1996
12	US5591616	Yokoh Hiei, Toshihiko Komari	Japan Tobacco Inc.	Method for transforming monocotyledons	It provides a method for transforming a monocotyledon by transforming a cultured tissue during a dedifferentiation process or a dedifferentiated cultured tissue of said monocotyledon with a bacterium belonging to the genus *Agrobacterium* containing a desired gene	5/3/1994	1/7/1997
13	US5681730	David E. Ellis	Wisconsin Alumni Research Foundation	Particle-mediated transformation of gymnosperms	It involves accelerated particle transformation of gymnosperms. Somatic embryos are produced by the callus and then subjected to an accelerated particle transformation process. The treated embryos are then induced to form embryogenic callus cultures and selected for the presence of gene products coded by the introduced genes	12/21/1993	10/28/1997
14	US5693512	John J. Finer, Harold N. Trick	The Ohio State Research Foundation	Method for transforming plant tissue by sonication	The plant sample to be transformed is sonicated in the presence of a vector, preferably *Agrobacterium*, containing the gene of interest. Then the sample is cultured to induce morphogenesis to form the transformed plant	3/1/1996	12/2/1997
15	US5859327	S.B. Dev. Yasuhiko Hayakawa	Genetronics Inc.	Electroporation-mediated molecular transfer in intact plants	It provides a method for producing a genetically modified plant by introducing a polynucleotide to an intact plant or plant cell(s) by electroporation, in the absence of cell wall-degrading enzymes	8/22/1995	1/12/1999

(continued)

TABLE 20.3 (Continued)

List of Patents for Plant Genetic Transformation Technology

S. No.	Patent No.	Inventor	Assignee	Title	Description	Filing Date	Issue Date
16	US5932782	Dennis Bidney	Pioneer Hi-Bred International Inc.	Plant transformation method using *Agrobacterium* species adhered to microprojectiles	It provides for high rates of stable transformation when bacteria of the species *Agrobacterium* are applied to particles that are used in a typical particle gun in a manner that retains their viability after the dry-down process involved in microparticle bombardment	11/14/1990	8/3/1999
17	US636929	Tishu Cai, Dorothy A. Pierce, Laura A. Tagliani, Zuo-Yu Zhao	Pioneer Hi-Bred International Inc.	*Agrobacterium*-mediated transformation of sorghum	The method involves infection with *Agrobacterium*, particularly those comprising a super-binary vector. In this manner, any gene of interest can be introduced into the sorghum plant	4/7/1998	4/9/2002
18	US6384301	Brian J. Martinell, Lori S. Julson, Carol A. Emler, Yong Huang, Dennis E. McCabe, Edward J. Williams	Monsanto Technology LLC	Soybean *Agrobacterium* transformation method	The method is based on an *Agrobacterium*-mediated gene delivery to individual cells in a freshly germinated soybean meristem, wherein cells are induced directly to form shoots and give rise to transgenic plants. This method does not involve a callus-phase tissue culture and is rapid and efficient	1/14/2000	5/7/2002
19	US6455761	Viktor Kuvshinov, Kimm o Koivu, Anne Kanerva, Eija Pehu	Helsinki University Licensing Ltd.	*Agrobacterium*-mediated transformation of turnip rape	It relates to a novel transformation protocol for obtaining transgenic turnip rape plants with *Agrobacterium*-mediated transformation. In the protocol, an internode section of the inflorescence-carrying stem of mature turnip rape is used as explant	3/10/2000	9/24/2002

20	US6822144	Zuo-Yu Zhao, Weining Gu, Tishu Cai, Dorothy A. Pierce	Pioneer Hi-Bred International Inc.	Methods for *Agrobacterium*-mediated transformation	It relates to methods for improving the transformation frequency of *Agrobacterium*-mediated transformation of maize embryos by contacting at least one immature embryo from a maize plant with *Agrobacterium* capable of transferring at least one gene to said embryo; cocultivating the embryos with *Agrobacterium*; culturing the embryos in medium comprising N6 salts, an antibiotic capable of inhibiting the growth of *Agrobacterium*, and a selective agent to select for embryos expressing the gene; and regenerating plants expressing the gene	11/3/1997 11/23/2004
21	US7064248	Ronald C. Lundquist, David A. Walters, Julie A. Kirihara	DeKalb Genetics Corporation	Method of preparing fertile transgenic corn plants by microprojectile bombardment	A method of preparing fertile transgenic *Zea mays* (corn) plants that stably express DNA encoding a *Bacillus thuringiensis* endotoxin, so as to impart insect resistance to the fertile transgenic plants is disclosed. It involves bombarding regenerable corn cells with DNA-coated microprojectiles and regenerating fertile transgenic plants from the transformed cells	3/27/2001 6/20/2006
22	US7161064	Anne-Marie Stomp, Nirmala Rajbhandari	North Carolina State University	Method for producing stably transformed duckweed using microprojectile bombardment	Efficient transformation of duckweed by ballistic bombardment is disclosed. Transformed duckweed plant tissue culture and methods of producing recombinant proteins and peptides from transformed duckweed plants are also included	10/18/2002 1/9/2007

(continued)

TABLE 20.3 (Continued)

List of Patents for Plant Genetic Transformation Technology

S. No.	Patent No.	Inventor	Assignee	Title	Description	Filing Date	Issue Date
23	US7611898	Zengyu Wang, Yaxin Ge	The Samuel Roberts Noble Foundation	*Agrobacterium* transformation of stolons	It provides methods for transforming monocotyledonous plants with *Agrobacterium* using stolons as a target tissue. The invention allows the creation of transgenic plants without the need for a callus as a target tissue for transformation, thus providing a rapid method for the production of transgenic plants	5/9/2006	11/3/2009
24	US8030076	Brian J. Martinell, Lori S. Julson, Carol A. Emler, Yong Huang, Dennis E McCabe, Edward J. Williams	Monsanto Technology LLC	Soybean transformation method	It is about *Agrobacterium*-mediated germ line genetic transformation of soybean. The method is based on *Agrobacterium*-mediated gene delivery to individual cells in a freshly germinated soybean meristem, from which cells can be induced directly to form shoots that give rise to transgenic plants	11/12/2008	10/4/2011

TABLE 20.4

List of Patents Granted for Biotechnological Innovations in Some Important Crops

S. No.	Patent No.	Inventor	Assignee	Title	Description	Date of Filing	Date of Issue
Rice							
1	EP1539949 A4 EP1539949A2 US6956115	B. Manuel Sainz, John Salmeron	Syngenta Participations AG	Nucleic acid molecules from rice encoding rar1 disease resistance proteins and uses	It relates to methods of enhancing the expression of resistance genes, disease resistance signal transduction genes, genes involved in mediating disease resistance or those involved in the synthesis of molecules mediating disease resistance	11/27/2002	11/30/2005
2	EP1960528 A2 US2009032824B	Jan Theodoor Michiel De Both et al.	Keygene N.V	Constitutive plant promoters	Strong, constitutive plant promoters referred as AA6 promoters under biotic and abiotic stress conditions. It also provides the methods for expressing nucleic acid sequences using AA6 promoters	12/12/2006	8/27/2008
3	US20120240292 A1 EP2471808A1	Xinjie Xia	Xia Xinjie	Proteins relating to grain shape and leaf shape of rice, coding genes and uses thereof	Transgenic rice overexpressing *OsXCL* gene present phenotypes as increases of grain length, grain weight, and number of grains per panicle, and leaf rolling and also includes a method for obtaining transgenic plants	7/8/2010	9/20/2012
4	US8283536 B1	Billy Gene Jordan, Jose Vicente Re	Ricetec AG	Rice hybrid XP753	The invention relates to the seeds of rice hybrid XP753, to the plants of rice hybrid XP753 and to methods for producing a rice plant produced by crossing the hybrid XP753 with itself or another rice plant	10/10/2011	10/9/2012
5	US8318636 B2	Emily Alff, Harsh Bais, Darla Janine Sherrier	University of Delaware	Compositions and methods for improving rice growth and restricting arsenic uptake	It relates to administering one or more rice rhizosphere isolates to a plant, particularly a rice plant, to the seed of the plant, or to the soil surrounding the plant in an amount effective to inhibit infection by a plant pathogen, particularly rice blast, to increase the biomass of the plant, and to decrease arsenic uptake by the plant	03/01/2011	11/27/2012

(continued)

TABLE 20.4 (Continued)

List of Patents Granted for Biotechnological Innovations in Some Important Crops

S. No.	Patent No.	Inventor	Assignee	Title	Description	Date of Filing	Date of Issue
Maize							
1	EP0721509A1	C. Paul Anderson et al.	DeKalb Genetics Corporation	Fertile, transgenic maize plants and methods for their production	It relates to a reproducible system for the production of stable, genetically transformed maize cells, and to methods of selecting cells that have been transformed	8/24/1994	7/17/1996
2	EP1685242A2	Catherine Anderson et al.	Agrigenetics Inc.	Generation of plants with improved drought tolerance	It is directed to plants that display a drought tolerance phenotype due to altered expression of a DRO5 nucleic acid. The invention is further directed to methods of generating plants with a drought tolerance phenotype	6/23/2004	4/20/2011
3	EP2503872A1	C.C. Paul Feng et al.	Monsanto Technology LLC	Transgenic maize event mon 87427 and the relative development scale	It provides transgenic maize event MON 87427, its nucleotide, roundup hybridization system (RHS), and relative development stages useful for monitoring and determining the reproductive development in maize	11/16/2010	10/3/2012
4	US8330006 B2	Alain Murigneux et al.	Biogemma	Maize with good digestibility and disease resistant	It relates to the field of the improvement of the digestibility and the tolerance of maize to fungal pathogens and especially to fusariosis by modification of the *C4H* gene	10/23/2007	12/11/2012
5	US8338677B1	Gustavo Marcelo Garcia	Pioneer Hi-Bred International Inc.	Inbred maize variety PH13C9	A novel maize variety designated PH13C9, its method of production and its genetic material, one or more traits introgressed into PH13C9 through backcross conversion and transformation, and to the maize seed, plant, and plant part produced thereby	5/25/2010	12/25/2012

Potato

#	Patent no.	Inventor	Assignee	Title	Description		
1	EP0375092B1	Dr. Wolf-Bernd Frommer et al.	Institut Fuer Genbiologische Forschung Berlin GmbH	Potato tuber specific transcriptional regulation	Use of the regulatory region (promoter) of the 1527 kb long DraI/DraI fragment which is located on the KpnI/HindIII-fragment of the patatin gene *B33*	12/18/1989	1/24/1996
2	EP1425958A1	The designation of the inventor has not yet been filed	Coöperatieve Verkoop- en Productievereniging van Aardappelmeel en Derivaten	Potatoes with increased protein content	It provides a potato plant or part derived thereof having at least one *amf*-allele, the said potato plant or part is further provided with an increased capacity to store a protein as characterized by the increased protein content of its tubers	12/4/2002	6/9/2004
3	EP1734123A1	Per Hofvander et al.	BASF Plant Science GmbH	Genetically engineered modification of potato to form amylopectin-type starch	The function of the *GBSS* gene and thus the amylose production in potato are inhibited by using completely new antisense constructs. The genomic *GBSS* gene is used as a basis in order to achieve an inhibition of *GBSS* and consequently of the amylose production	12/20/1991	12/20/2006
4	US20120311734	Ian S. Curtis et al.	The Texas A&M University System	Potato transformation compositions, systems, methods, microorganisms, and plants	A method of transforming and transfecting an Atlantic potato plant, its growth, removal of leaf sections, its cultivation and transformation through *Agrobacterium*	6/4/2012	12/6/2012
5	US8330005B2	Robert W. Hoopes	Frito-Lay North America Inc.	Potato cultivar FL 2137	A potato cultivar designated FL 2137 produced by crossing potato cultivar FL 2137 with itself or with another potato variety	3/6/2008	12/11/2012

(continued)

TABLE 20.4 (Continued)

List of Patents Granted for Biotechnological Innovations in Some Important Crops

S. No.	Patent No.	Inventor	Assignee	Title	Description	Date of Filing	Date of Issue
Brassica							
1	EP2144490A1	R. Wayne Leitch et al.	Pioneer Hi-Bred International Inc.	High oil hybrid *Brassica* line 46p50	The *Brassica* hybrid produces seeds having an average weight of oil per gram of mature dried seed that is between about 2.7% and 3.3% points higher than that produced by current commercial hybrids when grown under the same environmental condition	4/24/2008	1/20/2010
2	US7741541	Naveen Chandra Bisht, Arun Jagannath, Vibha Gupta et al.	Viterra Inc. Dhara Vegetable Oil and Food Corporation Ltd.	A novel method for obtaining improved fertility restorer lines for transgenic male-sterile crop plants and a DNA construct for use in said method	The invention relates to the simultaneous use of two different gene sequences encoding the same protein product, one being the naturally occurring wild-type sequence and the other sequence being generated by modification of the wild-type sequence for expression in crop plants by using codon degeneracy. Each of the said sequences being placed under independent transcriptional control of different overlapping plant tissue-specific regulatory elements in the same DNA construct	7/7/2003	6/22/2010
3	EP2480065A1	William Briggs et al.	Syngenta Participations AG	*Brassica oleracea* plants resistant to *Albugo candida*	It relates to the methods of making such plants and for producing seeds. It also includes molecular markers and their use in marker-assisted breeding and for identifying the *Albugo candida* resistance trait in *Brassica oleracea* plants	9/19/2010	8/1/2012
4	US8304610	Daryl Males et al.	Viterra Inc.	*Brassica juncea* lines with high oleic acid profile in seed oil	It provides *Brassica juncea* plants, seeds, cells, nucleic acid sequences, and oils. Edible oil derived from plants of the invention may have significantly higher oleic acid content than other *B. juncea* plants	2/25/2008	11/6/2012

5	US20120304338	Igor Falak et al.	Pioneer Hi-Bred International Inc.	*Sclerotinia*-resistant *Brassica*	It provides *Brassica* plants and lines having an improved *Sclerotinia sclerotiorum* disease incidence (SSDI%) score and is represented by, or descended from, ATCC accession number PTA-6779 or PTA-6778	8/6/2012	11/29/2012
6	EP2543731A1	Kening Yao et al.	Viterra Inc.	Herbicide-resistant *Brassica* plants and methods of use	It provides transgenic or nontransgenic plants with improved levels of tolerance to AHAS-inhibiting herbicides	4/3/2008	1/9/2013

Cotton

1	EP1917358 A2	Jason Barnett et al.	Syngenta Participations AG	Transgenic cotton insecticide CE44-69D expressing cry1ab	It relates to a specific event, designated CE44-69D. The application also relates to polynucleotides that are characteristic of the CE44-69D event, plants comprising of the said polynucleotides, and methods of detecting the CE44-69D event	5/15/2006	5/7/2008
2	EP2035459 B1	Bernard Bizzini et al.	Franzoni Filati S.P.A.	Covalent conjugates of cotton and substitutes (viscose, modal cotton) with bioactive substances having antiseptic, sanitizing, acaricidal, and insect repellent activity, and a method for obtaining them	The conjugates obtained by the new process are characterized by high stability, while maintaining in the long-term the antiseptic, sanitizing, acaricidal, and insect repellent activity imparted by the procedure described	5/22/2007	12/16/2009
3	EP2333082	John W. Pellow et al.	Dow AgroSciences LLC	Cry1F and Cry1AC transgenic cotton lines and event-specific identification thereof	It relates to plant breeding and the protection of plants from insects. The present invention provides DNA and related assays for detecting the presence of certain insect resistance events in cotton. The assays are based on the DNA sequences of recombinant constructs inserted into the cotton genome and of the genomic sequences flanking the insertion sites	10/13/2004	6/15/2011

(continued)

TABLE 20.4 (Continued)

List of Patents Granted for Biotechnological Innovations in Some Important Crops

S. No.	Patent No.	Inventor	Assignee	Title	Description	Date of Filing	Date of Issue
4	EP1417312B1	Marc De Beuckeleer et al.	Bayer Bioscience N.V.	Herbicide-tolerant cotton plants and methods for producing and identifying same	It pertains to transgenic cotton plants, plant material, and seeds, characterized by harboring a specific transformation event, particularly by the presence of a gene encoding a protein that confers herbicide tolerance, at a specific location in the cotton genome	7/19/2002	9/14/2011
5	US20120255050 A1	Ronald J. Brinker et al.	Ronald J. Brinker et al	Cotton transgenic event mon 88701 and methods of use thereof	It provides cotton event MON 88701, and plants, plant cells, seeds, plant parts, and commodity products comprising event MON 88701. The invention also provides polynucleotides specific for event MON 88701 and plants, plant cells, seeds, plant parts, and commodity products comprising polynucleotides specific for event MON 88701	4/13/2012	10/4/2012
Sugarcane							
1	USPP10839	David G. Holder	U.S. Sugar Corporation	Sugarcane variety CL77-797	A cross between a female variety known as CL61-620 with a mixture of male varieties has produced an improved variety of sugarcane	2/8/1996	3/23/1999
2	USPP18826	Kenneth A et al.	Board of Supervisors of Louisiana State University and Agricultural and Mechanical College	Sugarcane variety named L99-233	A new variety of sugarcane, identified as L99-233, is disclosed as having superior sugarcane rust-disease resistance, excellent ratooning ability, and high sugar/sucrose content and cane yield characteristics	11/15/2006	5/20/2008
3	EP2456893 A1	Pierluigi Barone et al.	Syngenta Participations AG	Sugarcane centromere sequences and mini-chromosomes	It is generally related to sugarcane mini-chromosomes and recombinant chromosomes containing sugarcane centromere sequences; it also includes sugarcane mini-chromosomes with novel compositions and structures are used to transform sugarcane cells, which are, in turn, used to generate sugarcane plants	7/23/2010	5/30/2012

4	US8252976 B2	Elke Hellwege, Karola Knuth	Bayer CropScience AG	Sugarcane plants with an increased storage carbohydrate content	It relates to a method for increasing the storage carbohydrate content of sugarcane plants	8/24/2010	8/28/2012

Tea

1	US6930227	Misako Mizuno et al.	Mitsui Chemicals Inc.	*Camellia sinensis* gene encoding a caffeine synthesis associated with N-methyl transferase	It makes it possible to efficiently produce an N-methyl transferase with 7-methylxanthine N3 methyl transferase, theobromine N1 methyl transferase, and paraxanthine N3 methyl transferase activities that can be utilized as an industrial, food, or medical enzyme	5/25/2000	8/16/2005
2	EP1349560 B1	Yu Kuang Chen et al.	Rutgers, The State University of New Jersey	Black tea extract for prevention of disease	Compositions and methods for preventing and treating disease are provided. The compositions are extracts of black tea that include a mixture of theaflavin-3-gallate and theaflavin-3'-gallate	11/14/2001	7/11/2012
3	EP2491939 A1	Michael Koganov	Michael Koganov	Bioactive compositions from *Theacea* plants and processes for their production and use	It relates to methods for isolating bioactive fractions derived from cell juice or a cell wall component of a *Theacea* plant and its isolation, the formulation of bioactive compositions and uses	1/12/2005	8/29/2012
4	EP2510014 A1	Ruud Albers et al.	Hindustan Unilever Ltd.	Polysaccharide suitable to modulate the immune response	It provides such polysaccharides obtained from the species *Camellia sinensis*, which comprise a rhamnogalacturonan-I core, and wherein the molar ratio of galacturonyl acid residues to rhamnosyl residues in the backbone of the polysaccharide is close to 1:1	11/16/2010	10/17/2012

(continued)

TABLE 20.4 (Continued)

List of Patents Granted for Biotechnological Innovations in Some Important Crops

S. No.	Patent No.	Inventor	Assignee	Title	Description	Date of Filing	Date of Issue
Catharanthus roseus							
1	EP0425597 A1	Normand Brisson et al.	National Research Council of Canada	A tryptamine producing tryptophan decarboxylase gene of plant origin	Isolation and cloning of the cDNA sequence of the tryptophan decarboxylase gene from *Catharanthus roseus* and the development of the cDNA sequence in a plasmid vector capable of transforming cell lines that will produce the tryptophan decarboxylase enzyme	2/21/1990	5/8/1991
2	EP0710240 B1	Christian Berrier et al.	Pierre Fabre Medicament	Novel antimitotic binary alkaloid derivatives extracted from *Catharanthus roseus*	Novel fluorinated derivatives of the vinblastine and vinorelbine family of general formula and the therapeutically acceptable salts of these molecules. The invention also concerns the application of said compounds in therapy and their methods of preparation	7/19/1994	12/18/1996

3	EP0740504 B1	Robert M. Bowman	Goldsmith Seeds Inc.	*Phytophthora* resistance gene of catharanthus and its use	It relates to a *Catharanthus* plant having resistance to the fungal disease *Phytophthora*. The invention also relates to an increased level of resistance to aphids and other pests and an increased level of total alkaloid content	1/18/1995	1/2/2003
4	USPP18315	Sushil Kumar et al.	Council of Scientific and Industrial Research	Plant variety of *Catharanthus roseus* named lli	It relates to the development of a unique inflorescence-bearing mutant plant-type lli/lli (LEAF-LESS INFLORESCENCE). Further, the present invention relates to the development of a unique inflorescence bearing mutant plant type lli/lli (LEAF-LESS INFLORESCENCE) through chemical mutagenesis	5/4/2004	12/18/2007
5	EP1934355 A2	Alain Goossens et al.	Universiteit Gent	Means and methods to enhance the production of vinblastine and vincristine in *Catharanthus roseus*	It relates to the production of the anticancer metabolites vincristine and vinblastine. The invention provides novel polynucleotide sequences derived from *Catharanthus roseus* and the use of said polynucleotide sequences to stimulate the production of vinblastine and vincristine in plants of *Catharanthus roseus* and plant cell lines derived thereof	9/14/2006	6/25/2008

3. Cisgenesis and intragenesis—similar to transgenic technology but the DNA/ genes transferred belong to the same species of the transformed plant, or to a cross-compatible species.

4. RdDM—associated with gene silencing by methylation of the promoter sequences and it provides breeders with an opportunity to modify the gene expression epigenetically.

5. Grafting on GM rootstock—achieved by simply grafting a non-GM scion onto a GM rootstock for the desired improvements.

6. Agroinfiltration—*Agrobacterium* sp. containing the gene of interest is used to infiltrate plant tissues, mostly leaves, so that the gene is locally expressed at a high level, without being integrated into the plant genome. This can be used for screening for plants with valuable phenotypes for breeding programs.

7. Reverse breeding—used to rapidly generate suitable transgene-free homozygous parental lines for elite heterozygous genotypes by silencing the genes involved in the meiotic recombination process.

8. Synthetic genomics—not directly related with breeding techniques but they are generally used for constructing viable minimal genomes that can serve as platforms for the biochemical production of chemicals such as biofuels and pharmaceuticals.

These techniques have been used to introduce variable traits such as herbicide tolerance, fungal, bacterial, and viral resistances, male sterility, modified starch content, reduced flower pigmentation, and the production of hepatitis B vaccine in crops such as maize, tobacco, rice, oilseed rape, potato, apple, melon, petunia, grapevine, watermelon, cucumber, plum, walnut, and the like. Since 2000, more than 84 patents have been filed on these newer techniques of plant breeding (Lusser et al. 2012).

The U.S. Food and Drug Administration has approved the first plant-made drug, Elelyso (taliglucerase alfa), an enzyme produced by genetically engineered carrot cells for treating type 1 Gaucher's disease (Fox 2012). In the case of the plant-made pharmaceuticals (PMP) sector, public-funded research has led to more than 50% patented technologies as compared to the industrial contribution, reflecting the academic contribution to patented research in the biopharmaceutical sector (Thangaraj et al. 2009). The inclination for patenting by academic institutions has been facilitated by the U.S. Bayh–Dole Act and similar legislation elsewhere that permits universities to gain a return by patenting their innovations.

In India, there has been a substantial increase in patent applications related to herbal drugs for diabetes, cancer, cardiovascular diseases, asthma, and arthritis (Sahoo et al. 2011). In environmental biotechnology, several patents have been granted on phytoremediation, which is an eco-friendly approach to the remediation of contaminated soil and water using plants (Suresh and Ravishankar 2004).

20.7 General Considerations for Patent Filing in Plant Biotechnology

In general, patent filing is a long process and involves several steps before the patent is granted. A filed patent is not universal, that is, it will be applicable to the country in which it is filed and if one desires to have patents in different countries, then such patents have to be applied for separately according to the norms of that country's patent rules. Patent filing involves the following steps:

- Identification of the innovation/invention by the innovator
- Prior art search (patents, periodicals, books, products, etc.)
- Preparation of patent specification (claim should be defined properly and extensively)
- Clarifying the ownership of the patent filed
- Patent filing (self or with the help of an advocate)
- Prosecuting through the patent office (depends on the country where the patent is filed)
- Patent registration
- Patent issued (if found suitable based on the claims)

Patent searching is crucial and one needs to have a sound search strategy so that one can claim for a patent on a new innovation. The search will be more effective if one defines it properly based on the innovation. The relevant literature in the form of research papers, reviews, products and processes in commercial use, and various internet search engines should be properly consulted prior to patent filing. There is a provision for online patent searching by various national patent offices, such as the U.S. Patent and Trademark Office (USPTO), the EPO, and the Japanese Patent Office (JPO). The USPTO has two separate databases (http://patft.uspto.gov/) for granted patents (PatFT) and patent applications (Ap-pFT). Similarly, the JPO has a provision for both granted patents and patent applications (http://www.jpo.go.jp). The EPO (http://www.espacenet.com) has a publicly available database that includes patents administered by the World Intellectual Property Organization (WIPO) representing over 80 countries. Similarly, another database, PatentScope (http://www.wipo.int/patentscope/search/en/search.jsf), is widely used for patent searches and one major advantage of this database is its ability to search the claims field independently of the full patent specifications/text. The patent searching strategy with reference to plant biotechnology innovations has recently been reviewed (Parisi et al. 2013).

20.8 Conclusion

IPRs in plant biotechnology are a matter of debate especially with reference to plant varieties and plant breeding methods. The emerging new technologies in the agricultural sciences have immense potential for crop improvement to attain global food and nutritional security. The increasing world population, the limitation of arable lands, and the influence of biotic and abiotic stresses are major constraints that require innovative techniques to supplement the tools of conventional plant breeding for developing suitable varieties. Biotech crops are quite promising as evident by the recent increase in their production and also their acceptance by several countries, including developing countries. IPRs have greatly influenced agricultural research in recent years with the development of several private sector companies with the sole aim to commercialize their products. The research in the public sector has also been influenced by the IPR regimes and has led to several patents. An integrated strategy needs to be developed between science, technology, and the market in the context of growing IPR issues in the agricultural sector globally. There is a need for developing countries to establish an effective system to protect IPRs that are relevant to agricultural innovations and to utilize the global germplasm available. There is some concern that IPRs do not recognize or reward the contribution of communities of farmers who have developed, over long periods of time, the landraces that form the basis of the pedigrees of modern crop varieties. The strengthening of IPRs has undoubtedly stimulated the research and development that have made genetically engineered varieties of major crops available to farmers. The legal control of key enabling technologies for transgenic crop production in the form of several patents granted to giant agricultural companies is hampering the development of other genetically engineered crops, in particular small-acreage crops that are relevant to developing countries. International collaboration in the form of treaties is needed to safeguard the genetic resources of each country and to develop an amicable solution for using them globally in the development of superior crop varieties for the benefit of mankind.

References

Abbott, A. 1993. Protestors target European animal patents. *Nature* 361:103.

Ahmad, P., Ashraf, M., Younis, M., Hu, X., et al. 2012. Role of transgenic plants in agriculture and biopharming. *Biotechnology Advances* 30:524–540.

Atkinson, R.C., Beachy, R.N., Conway, G., Cordova, F.A., et al. 2003. Public sector collaborations for agricultural IP management. *Science* 301:174–175.

Aufsatz, W., Mette, M.F., van der Winden, J., Matzke, A.J., and Matzke, M. 2002. RNA-directed DNA methylation in Arabidopsis. *Proceedings of the National Academy Science USA* 99(4):16499–16506.

Azadi, H. and Ho, P. 2010. Genetically modified and organic crops in developing countries: A review of options for food security. *Biotechnology Advances* 28:160–168.

Barton, J.H. 2000. Reforming the patent system. *Science* 287:1933–1934.

Bhullar, N.K. and Gruissem, W. 2013. Nutritional enhancement of rice for human health: The contribution of biotechnology. *Biotechnology Advances* 31:50–57.

Blakeney, M. 2012. Patenting of plant varieties and plant breeding methods. *Journal of Experimental Botany* 63(3):1069–1074.

Blakeney, M., Cohen, J.I., and Crespi, S. 1999. Intellectual property rights and agricultural biotechnology. In Cohen, J.I. (ed.), *Managing Agricultural Biotechnology: Addressing Research Program Needs and Policy Implications*, pp. 209–227. CAB International: Wallingford, UK.

Boettiger, S., Graff, G.D., Pardey, P.G., Dusen, E.V., and Wright, B.D. 2004. Intellectual property rights for plant biotechnology: International aspects. In Christou, P. and Klee, H. (eds), *Handbook of Plant Biotechnology*, pp. 1089–1113. Wiley: Chichester.

Boisvert, V. and Caron, A. 2000. The convention on biological diversity: An ambivalent attempt to reconcile communal rights and private property. Presented at Constituting the Commons: Crafting Sustainable Commons in the New Millennium, the Eighth Conference of the International Association for the Study of Common Property, Bloomington, Indiana, USA, May 31–June 4.

Booth, W. 1988. Animals of invention. *Science* 240:546.

Brown, W.M. 2003. Intellectual property law: A primer for scientists. *Molecular Biotechnology* 23:213–224.

Chakrabarty, A.M. 1988. *Diamond v Chakrabarty*: A historical perspective. In Chisum, D.S. Nard, C.A. Schwartz, H.F. Newman, P. and Kieff, F.S. (eds), *Principles of Patents Law*, pp. 783–788. Foundation Press: New York, NY.

Chakrabarty, A.M. 2010. Bioengineered bugs, drugs and contentious issues in patenting. *Bioengineered Bugs* 1(1):2–8.

Chi-Ham, C.L., Boettiger, S., Figueroa-Balderas, R., Bird, S., Geoola, J.N., et al. 2012. An intellectual property sharing initiative in agricultural biotechnology: Development of broadly accessible technologies for plant transformation. *Plant Biotechnology Journal* 10:501–510.

Choudhary, K., Choudhary, O.P., and Shekhawat, N.S. 2008. Marker assisted selection: A novel approach for crop improvement. *American-Eurasian Journal of Agronomy* 1:26–30.

Clive, J. 2011. Global status of commercialized biotech/GM crops: 2011.ISAAA Brief No. 43.ISAAA: Ithaca, NY. (ISBN: 978-1-892456-52-4).

Cook, D.R. and Varshney, R.K. 2010. From genome studies to agricultural biotechnology: Closing the gap between basic plant science and applied agriculture. *Current Opinion in Plant Biology* 13(2):115–118.

Daniell, H. 2002. Molecular strategies for gene containment in transgenic crops. *Nature Biotechnology* 20(6):581–586.

Delmer, D.P., Nottenburg, C., Graff, G.D., and Bennett, A.B. 2003. Intellectual property resources for international development in agriculture. *Plant Physiology* 133:1666–1670.

Diaz, C.L. 2000. Regional approaches to implementing the convention on biological diversity: The case of access to genetic resources. In *Proceedings of Concerted Action Programme on the Effectiveness of International Environmental Agreements and EU Legislation*, Barcelona.

Dickson, D. 1993. Licenses sought from PCR users in Britain. *Nature* 361:291.

Dirks, R., van Dun, K., de Snoo, C.B., van den Berg, M., et al. 2009. Reverse breeding: A novel breeding approach based on engineered meiosis. *Plant Biotechnology Journal* 7:837–845.

Dunwell, J.M. 2005. Review: Intellectual property aspects of plant transformation. *Plant Biotechnology Journal* 3:371–384.

Dutfield, G. 2002. *Trade, Intellectual Property and Biogenetic resources: A Guide to the International Regulatory Framework*. ICTSD: Geneva.

Eathington, S.R., Crosbie, T.M., Reiter, R.S., and Bull, J.K. 2007. Molecular markers in a commercial breeding program. *Crop Science* 47(suppl 3):S154–S163.

Farre, G., Twyman, R.M., Zhu, C., Capell, T., and Christou, P. 2011. Nutritionally enhanced crops and food security: Scientific achievements versus political expediency. *Current Opinion in Biotechnology* 22(2):245–251.

Fleury, D., Jefferies, S., Kuchel, H., and Langridge, P. 2010. Genetic and genomic tools to improve drought tolerance in wheat. *Journal of Experimental Botany* 61(12):3211–3222.

Fox, J.L. 2012. First plant-made biologic approved. *Nature Biotechnology* 30(6):472.

Fuglie, K., Ballenger, N., Day, K., and Klotz, C. 1996. Agricultural research and development: Public and private investments under alternative markets and institutions, Agricultural Economics Report 735. U.S. Department of Agriculture Economic Research Service: Washington, D.C.

Garcia-Mas, J., Benjak, A., Sanseverino, W., Bourgeois, M., et al. 2012. The genome of melon (*Cucumis melo* L.). *Proceedings of the Natural Academy Science USA* 109(29):11872–11877.

Gold, E.R., Castle, D., Cloutier, L.M., Daar, A.S., and Smith, P.J. 2002. Needed: Models of biotechnology intellectual property. *Trends in Biotechnology* 20(8):327–329.

Greene, H.E. Jr. and Duft, B.J. 1990. Disputes over monoclonal antibodies. *Nature* 347:117–118.

Grimes, H.D., Payumo, J., and Jones, K. 2011. Opinion: Food security needs sound IP. *The Scientist*. 20 July. Available at: http://the-scientist.com/2011/07/20/opinion-food-security-needs-sound-ip/.

Guo, S., Zhang, J., Sun, H., Salse, J., et al. 2013. The draft genome of watermelon (*Citrullus lanatus*) and resequencing of 20 diverse accessions. *Nature Genetics* 45(1):51–58.

Hamilton, C. 2006. Biodiversity, biopiracy and benefits: What allegations of biopiracy tell us about intellectual property. *Developing World Bioethics* 6(3):158–173.

Hane, F. 2013. Intellectual property rights and user facility agreements. *Nature Biotechnology* 31(2):116–117.

Hill, M.J., Hall, L., Arnison, P.G., and Good, A.G. 2007. Genetic use restriction technologies (GURTs): Strategies to impede transgene movement. *Trends in Plant Science* 12(4):177–183.

Holman, C.M. 2012. Debunking the myth that whole-genome sequencing infringes thousands of gene patents. *Nature Biotechnology* 30(3):240–244.

International Barley Genome Sequencing Consortium. 2012. A physical, genetic and functional sequence assembly of the barley genome. *Nature* 491(7426):711–716.

International Rice Genome Sequencing project. 2005. The map-based sequence of rice genome. *Nature* 436(7052):793–800.

James, C. 2011. Executive summary global status of commercialized biotech/GM crops: 2011. In *ISAAA Brief 2011*, vol. 2011, p. 30. ISAAA: Ithaca, NY.

Jong, S.-C. and Cypess, R.H. 1998. Managing genetic material to protect intellectual property rights. *Journal of Industrial Microbiology and Biotechnology* 20:95–100.

Kolady, D. and Lesser, W. 2009. But are they meritorious? Genetic productivity gains under plant intellectual property rights. *Journal of Agricultural Economics* 60(1):62–79.

Koo, B., Nottenburg, C., and Pardey, P.G. 2004. Intellectual property. Plants and intellectual property: An international appraisal. *Science* 306:1295–1297.

Kowalski, S.P., Ebora, R.V., Kryder, R.D., and Potter, R.H. 2002. Transgenic crops, biotechnology and ownership rights: What scientists need to know. *The Plant Journal* 31(4):407–421.

Kryder, R.D., Kowalski, S.P., and Krattiger, A.F. 2000. The intellectual and technical property components of pro-vitamin A rice (GoldenRiceTM): A preliminary freedom-to-operate review. *ISAAA Briefs* 20:1–56.

Kumar, G.R., Sakthivel, K., Sundaram, R.M., Neeraja, C.N., et al. 2010. Allele mining in crops: Prospects and potentials. *Biotechnology Advances* 28(4):451–461.

Lacroix, B., Kozlovsky, S.V., and Citovsky, V. 2008. Recent patents on *Agrobacterium*-mediated gene and protein transfer, for research and biotechnology. *Recent Patents on DNA Gene Sequences* 2:69–81.

Lee, D. and Natesan, E. 2006. Evaluating genetic containment strategies for transgenic plants. *Trends Biotechnology* 24(3):109–114.

Lehrman, S. 1993a. Stanford seeks life after Cohen-Boyer patent expires. *Nature* 363:574.

Lehrman, S. 1993b. Ruling narrows US view of animal patents. *Nature* 361:103.

Lusser, M., Parisi, C., Plan, D., and Rodriquez-Cerezo, E. 2012. Deployment of new biotechnologies in plant breeding. *Nature Biotechnology* 30(3):231–239.

Manimaran, P., Ramkumarm, G., Sakthivel, K., Sundaram, R.M., et al. 2011. Suitability of non-lethal marker and marker-free systems for development of transgenic crop plants: Present status and future prospects. *Biotechnology Advances* 29(6):703–714.

Marshall, P. 1997. Guarding the wealth of nations. Intellectual property, copyright, and international trade and relations. *Wilson Quarterly* 21:64–100.

Marshall, A. 2012. Existing agbiotech traits continue global march. *Nature Biotechnology* 30(3):207.

Mauria, S. 2000. DUS testing of crop varieties, a synthesis on the subject for new PVP-opting countries. *Plant Varieties and Seeds* 13:69–90.

Mba, C., Guimaraes, E.P., and Ghosh, K. 2012. Re-orienting crop improvement for the changing climatic conditions of the 21st century. *Agricultural and Food Security* 1:7.

Mehrotra, S. and Goyal, V. 2013. Evaluation of designer crops for biosafety-A scientist's perspective. *Gene* 515(2):241–248.

Mir, R.R., Zaman-Allah, M., Sreenivasulu, N., Trethowan, R., and Varshney, R.K. 2012. Integrated genomics, physiology and breeding approaches for improving drought tolerance in crops. *Theoretical and Applied Genetics* 125(4):624–645.

Mittler, R. and Blumwald, E. 2010. Genetic engineering for modern agriculture: Challenges and perspectives. *Annual Review of Plant Biology* 61:443–462.

Morrell, P.L., Buckler, E.S., and Ross-Ibarra, J. 2012. Crop genomics: Advances and applications. *Nature Reviews Genetics* 13:85–96.

Moschini, G. 2001. Patents and Other Intellectual Property Rights. Staff General Research Paper. Iowa State University: Ames, Iowa, USA.

Odek, J. 1994. Bio-piracy: Creating proprietary rights in plant genetic resources. *Intellectual Property Law* 2(1):141–146.

Pardey, P., Koo, B., Drew, J., Horwich, J., and Nottenburg, C. 2013. The evolving landscape of plant varietal rights in the United States, 1930–2008. *Nature Biotechnology* 31(1):25–29.

Parisi, C., Rodriguez-Cerezo, E., and Thangaraj, H. 2013. Analysing patent landscapes in plant biotechnology and new plant breeding techniques. *Transgenic Research* 22:15–29.

Paterson, A.H., Bowers, J.E., Bruggmann, R., Dubchak, I., et al. 2009. The *Sorghum bicolor* genome and the diversification of grasses. *Nature* 457(7229):551–556.

Penna, S. and Ganapathi, T.R. 2010. Engineering the plant genome: Prospects of selection systems using non-antibiotic marker genes. *Genetically Modified Crops* 1(3):128–136.

Peters, J. and Stoger, E. 2011. Transgenic crops for the production of recombinant vaccines and antimicrobial antibodies. *Human Vaccines* 7(3):367–374.

Pioneer Hi-Bred International, Inc., v. J.E.M. AG Supply, Inc., Farm Advantage, Inc., et al., December 10, 2001 No.99–1996 (US Supreme Court) 18.

Pioneer Hi-Bred. (http://www.pioneer.com/home/site/about/research/plant-breeding/).

Potato Genome Sequencing Consortium. 2011. Genome sequence and analysis of tuber crop potato. *Nature* 475:189–195.

Raven, P.H. 2010. Does the use of transgenic plants diminish or promote biodiversity? *New Biotechnology* 27(5):528–533.

Rommens, C.M. 2010. Barriers and paths to market for genetically engineered crops. *Plant Biotechnology Journal* 8:101–111.

Sahoo, N., Manchikanti, P., and Dey, S.H. 2011. Herbal drug patenting in India: IP potential. *Journal of Ethnopharmacology* 137:289–297.

Schmutz, J., Cannon, S.B., Schlueter, J., Ma, J., et al. 2010. Genome sequence of the palaeopolyploid soybean. *Nature* 463(7278):178–183.

Schnable, P.S., Ware, D., Fulton, R.S., Stein, J.C., et al. 2009. The B73 maize genome: Complexity, diversity and dynamics. *Science* 326(5956):1112–1115.

Schouten, H.J. and Jacobsen, E. 2008. Cisgenesis and intragenesis, sisters in innovative plant breeding. *Trends in Plant Science* 13:260–261.

Sechley, K.A. and Schroeder, H. 2002. Intellectual property protections of plant biotechnology inventions. *Trends in Biotechnology* 20(11):456–461.

Shiva, V. 1997. *Bio-piracy: The Plunder of Nature and Knowledge.* South End Press: Boston, MA.

Shukla, V.K., Doyon, Y., Miller, J.C., Dekelver, R.C., et al. 2009. Precise genome modification in the crop species Zea mays using zinc finger nucleases. *Nature* 459:437–441.

Singh, N.K., Gupta, D.K., Jayaswal, P.K., Mahato, A.K., et al. 2012. The first draft of the pigeonpea genome sequence. *Journal of Plant Biochemistry and Biotechnology* 21(1):98–112.

Spillmann-Furst, I. 1990. Europe edges closer to GMO patent harmonization. *Nature Biotechnology* 17:842–843.

Springer, N.M. 2012. Epigenetics and crop improvement. *Trends in Genetics* 29(4):241–247.

Stegemann, S. and Bock, R. 2009. Exchange of genetic material between cells in plant tissue grafts. *Science* 324:649–651.

Suresh, B. and Ravishankar, G.A. 2004. Phytoremediation-A novel and promising approach for environmental cleanup. *Critical Reviews in Biotechnology* 24(2–3):97–124.

Tandon, N. and Yadav, S. 2012. Integrating intellectual property rights and development policy in agriculture, biotechnology and genetic resources. *IOSR Journal of Pharmacy and Biological Sciences* 2(1):18–24.

Thangaraj, H., van Dolleweerd, C.J., McGowan, E.G., and Ma, J.K.C. 2009. Dynamics of global disclosure through patent and journal publications for biopharmaceutical products. *Nature Biotechnology* 27:614–618.

Townsend, J.A., Wright, D.A., Winfrey, R.J., Fu, F., et al. 2009. High-frequency modification of plant genes using engineered zinc-finger nucleases. *Nature* 459:442–445.

Uhr, J.W. 1984. The 1984 Nobel prize in medicine. *Science* 226:1025–1028.

US Patent No. 4,237,224. Process for producing biologically functional molecular chimeras.

Varshney, R.K., Bansal, K.C., Aggarwal, P.K., Datta, S.K., and Craufurd, P.Q. 2011. Agricultural biotechnology for crop improvement in a variable climate: Hope or hype? *Trends in Plant Science* 16(7):362–371.

Varshney, R.K., Chen, W., Li, Y., Bharti, A.K., et al. 2012a. Draft genome sequence of pigeon pea (*Cajanus cajan*), an orphan legume crop of resource-poor farmers. *Nature Biotechnology* 30(1):83–92.

Varshney, R.K., Ribaut, J.M., Buckler, E.S., Tuberosa, R., Rafalski, J.A., and Langridge, P. 2012b. Can genomics boost productivity of orphan crops? *Nature Biotechnology* 30(12):1172–1176.

Varshney, R.K., Song, C., Saxena, R.K., Azam, S., et al. 2013. Draft genome sequence of chickpea (*Cicer arietinum*) provides a resource for trait improvement. *Nature Biotechnology* 31, 240–246.

Vezina, L.P., Loic, F., Lerouge, P., Aoust, M.D., et al. 2009. Transient co-expression for fast and high-yield production of antibodies with human-like N-glycans in plants. *Plant Biotechnology Journal* 7:442–455.

Wade, N. 1980. Supreme Court hears argument on patenting life forms. *Science* 208:31–32.

Wendt, J. and Izquierdo, J. 2001. Biotechnology and developments: A balance between IPR protection and benefit-sharing. *Electronic Journal of Biotechnology* 4(3):15–16.

White, P.J. and Broadley, M.R. 2009. Biofortification of crops with several mineral elements often lacking in human diets-iron, zinc, copper, calcium, magnesium, selenium and iodine. *New Phytologist* 182(1):49–84.

Yadav, D., Yadav, S., and Yadav, M.K. 2010a. Biotechnological intervention for attaining global food security. In Kumar, A. and Das, G. (eds), *Biodiversity to Biotechnology: Intellectual Property Rights*, pp. 109–116. Narosa Publishing House: New Delhi, India.

Yadav, D., Anand, G., Gupta, S., Dubey, A.K., Yadav, S., and Sarangi, B.K. 2010b. Intellectual Property Right (IPR) in context with biotechnology, biodiversity and biopiracy: An overview. In Kumar, A. and Das, G. (eds), *Biodiversity, Biotechnology and Traditional Knowledge: Understanding Intellectual Property Rights*, pp. 196–204. Narosa Publishing House: New Delhi, India.

Yadav, D., Anand, G., Dubey, A.K., Gupta, S., and Yadav, S. 2012. Patents in the era of genomics: An overview. *Recent Patents on DNA and Gene Sequences* 6:127–144.

Ye, X., Al-Babili, S., Kloti, A., Zhang, J., Lucca, P., Beyer, P., and Potrykus, I. 2000. Engineering the provitamin A (β-carotene) biosynthetic pathway into (carotenoid-free) rice endosperm. *Science* 287:289–296.

Zhu, T., Mettenburg, K., Peterson, D.J., Tagliani, L., and Baszcznski, C.L. 2000. Engineering herbicide-resistant maize using chimeric RNA/DNA oligonucleotides. *Nature Biotechnology* 18:555–558.

Index

A

Abiotic stress tolerance
  forest trees, 504–505
  ornamental crops, 422–423
  potato transgenesis, 154
  rice genomics, 15–16
    cold-sensitive, 17
    phosphorus-uptake, 17–18
    salinity, 16
    submergence, 16–17
  rice proteomics
    cold and heat stresses, 24
    drought and salinity, 23
    heavy metal toxicity, 25
  sugarcane, 210–212
AB-QTL, *see* Advanced backcross QTL analysis
    (AB-QTL)
Abscisic acid (ABA), 54, 78, 178, 262, 293,
    359, 473
  dependent TFs, 50, 85
  independent TFs, 50, 85
Advanced backcross QTL analysis (AB-QTL),
    7–8
  pigeonpea, 122
  problem associated with, 8
Affinity-based tags, 584
Afforestation, 495
  eucalypt (*Eucalyptus* sp.), 494
  pine (*Pinus* sp.), 494
Affymetrix GeneChip, 206, 525, 528
Affymetrix soybean genome arrays, 112
African Agricultural Technology Foundation
    (AATF), 629
Agrobacterium-mediated gene transformation
  dicots, 577
  ornamental crops, 411
  potato, 140, 142
  pulses, 122, 126
  sugarcane, 222
Agroinfiltration, 576, 664
Agromorphological traits genomics
  Brassica crops, 172
  rice
    grain yield, 14–15
    plant architecture, 14
    plant height, 13–14
    tillers and flower development, 14

    maize, 75–76
    sugarcane, 219, 226
Ajmalicine, 327, 338, 350–352, 356, 358, 363
Alkaloidomics, *Catharanthus roseus*, 363–364
Alkaloids, 326; *see also* Terpenoid indole
    alkaloids (TIAs)
Allele mining, 8, 625
All India Coordinated Research Project
    (AICRP), 80
Amplified fragment length polymorphism
    (AFLP) markers, 111–113, 300
  advantages, 432
  disadvantages, 432
  mango, 300
  ornamental crops, 431–432
  pulses, 111–113
  sugarcane, 194
  tea, 449
Anthocyanin biosynthetic pathway, 413
Antibodies, 584–587
Anticarsia gemmatalis MNPV (AgMNPV), 609
APETALA 2/ethylene response factor (AP2/
    ERF), 51
Aphids, 332
Apple
  ESTs, 259
  genome sequencing, 254, 257
  metabolomics, 272
  microarray transcriptomics, 261–262
  molecular markers, 257–258
  proteomics, 267–268
  RNA sequencing, 265
Arabidopsis Reactome, 213
Arabidopsis thaliana
  Brassica species, 167
  forage crops, 525
  fruits genome sequencing, 252
  ornamental crops, 417, 423
  pomegranate, 307
  potato, 144
  rice metabolomics, 27
Artificial miRNAs (amiRNAs), 12
Ashwagandha, *see Withania somnifera*
Asiatic Centre for Genome Technology (ACGT),
    552
Association mapping
  bioenergy crops, 557
  Brassica crops, 175–177

maize, 78
rice, 8–9
Autographa californica multiple
nucleopolyhedrovirus (AcMNPV),
608–609
Automated Sanger method of sequencing, 5,
118, 205, 292, 304
AutoSNPdb, 171
Avidin, 153, 582

B

BAC-anchored SSR markers, 168
Bacillus thuringiensis (Bt)-based biopesticides,
606
Bacterial artificial chromosome (BAC)
libraries
Brassica species, 167, 170–171, 174
chickpea, 110
cotton fiber, 468–469
cowpea and lentil, 111
maize, 82
pigeonpea, 111
sugarcane, 198
switchgrass, 553
tea, 441
Bacterial blight resistance (*R*) genes, 19
Bacterial pesticides, 606–608
Baculoviruses, 608–609
Banana
genomics, 287–289
metabolomics, 289–290
molecular farming, 581
proteomics, 289–290
BASF-Plant Science, 152
Basic leucine zipper (bZIP), 50
Biochemical markers
ornamental crops, 427–428
sugarcane, 216
tea, 442
Bioenergy plants
biomass yield, 551
candidate genes, 554–556
cell wall composition, 551–552
genome structure, 552–554
oil yield, 550–551
polyploidy, 551
population genomics
association mapping, 557
QTL mapping, 556–557
quantitative genetic analyses, 557
transcriptome sequencing efforts, 553
Biofuels, barriers, 549

Biological Diversity Act, 2002, 630
Bioluminescent orchids, 423
Biotech crops, *see* Transgenic crops
Biotic stress tolerance
maize proteomics, 85–86
potato transgenesis
Colorado potato beetle, 153
late blight, 151–152
nematode pests, 153–154
snakin-1 (SN1) gene overexpression, 154
viruses, 152–153
rice genomics
qualitative resistance, 18–19
quantitative resistance, 19–20
rice proteomics, 25
sugarcane, 215–216
Blast resistance genes, 18
Blue roses, 423–424
B. oleracea C-Genome Sequencing Project, 168
Bph14 gene, 20
BrassEnsembl browser, 171
Brassica crops
genomics, 171–177
molecular markers
comparative mapping, 167
genetic linkage maps, 164–167
patents, 658–659
proteomics, 178–180
transcriptomics, 168–171, 177–180
Brassica database (BRAD), 171
Brassica EST Database (BrED), 169
Brassica Genome Gateway, 171
Brazilian Agricultural Research Corporation,
632
Brazilian sugarcane EST project (SUCEST), 193,
203–205, 208, 220, 229
Bt-cry3A, 153
Bulked segregant analysis (BSA), 55–56, 115,
117, 220

C

Caffeoyl CoA 3-O-methyltransferase (CCOMT),
534
CaMV35S promoter, *see* Cauliflower mosaic
virus (CaMV) 35S promoter
Candidate genes
Brassica species, 167, 170, 174, 179
Catharanthus roseus, 342, 365
cotton fiber, 468, 470, 473
fruit crops 295, 304
Jatropha curcas, 555

maize, 77, 78
pulses, 112, 120
rice, 8, 17, 27, 57, 61
sugarcane, 200, 204, 207, 220
Withania somnifera, 394
Carnation, 410, 414–415
vase life, 420
CaroRx, 589
Cartagena Protocol on Biosafety, 511, 628
Catharanthine, 326, 337, 352, 354, 358, 360, 366
Catharanthus roseus
agriculture
farmyard manure (FYM), 330
growth regulators, 331
harvesting and yields, 331–332
pests and diseases, 332–333
soil and climatic conditions, 330
vegetative propagation, 330–331
weeding and irrigation, 330
alkaloidomics, 363–364
botany, 329
distribution and habitat, 327
future research prospects, 366–367
genomics, 341
metabolomics, 362–363
natural and induced variability, 333
patents, 662–663
pests and diseases
bacteria, 332
fungus, 332
virus, 332–333
phytochemistry
essential oils, 339
folkloric usage, 333–334, 338–339
nonalkaloidal constituents, 340
TIAs, 334–338
proteomics, 344–362
taxonomic hierarchy, 328–329
taxonomic serial number (TSN), 329
TIAs
biological activity, 337–338
detection methods, 337
extraction procedure, 334–337
intermediates, sequestration of, 367
occurrence and distribution, 334
transcriptomics, 341–344
vernacular names, 327–328
vincristine and vinblastine, 338–339
Cauliflower mosaic virus (CaMV) 35S promoter
forages, 539
ornamental crops, 417, 425
rice, 59
sugarcane, 224

cDNA-amplified fragment-length
polymorphism (cDNA-AFLP)
approach
Catharanthus roseus, 341–342
cotton fiber, 473
fruit crops, 305
sugarcane, 199, 204
cDNA libraries, 259
Brassica crops, 169
fruit crops, 259
tea, 452
centiMcClintocks (cMC), 74
Centre for Application of Molecular Biology
to International Agriculture
(CAMBIA), 629
Chemical pesticides, 609, 614
neurological and cancer diseases, 605
procurement guidelines, 604
Chemometric methods, 88
Chickpea (*Cicer arietinum* L.)
BAC library, 110
genetic linkage maps, 113
MABC, 121
mapping populations, 107
MARS, 122
molecular markers, 111–112
production constraints, 103
taxonomy and genome organization, 103
trait mapping, 115
transcriptomics, 118
whole-genome sequence, 120
Chickpea transcriptome database (CTDB), 118
ChillPeach cDNA microarray platform, 263
Chinese Academy of Agricultural Sciences
(CAAS), 470
Chrysanthemum, 410, 412, 415–416, 422,
424–426
Cinnamyl alcohol dehydrogenase (CAD), 83, 504
Cisgenesis, 664
potato, 154–155
Citrus
genomics, 291–293
metabolomics, 293
proteomics, 293
Cleaved amplified polymorphic sequence
(CAPS), 112
Cm-ETR1/H69A gene, 425
Codon adaptation index (CAI), 569
Cohen–Boyer patent, 623
Colorado potato beetle (CPB), 153
Comethyl transferasecaffeic acid
3-O-methyltransferase (COMT), 534
Common-parent-specific (CPS) markers, 78

Comparative genomics
    bioenergy plants, 550
    drought-tolerant rice, 58
    fruits, 258, 304
Composite interval mapping (CIM) analysis,
    118
Conserved intron-spanning primer (CISP)
    makers, 112, 217
Conserved orthologous sequence (COS)
    markers, 5, 118
Conserved primers (CPs), 119
Consultative Group on International
    Agricultural Research (CGIAR), 632
Consultative Group on International Agricultural
    Research (CGIAR)–Generation
    Challenge Program (GCP), 111
Controlled deterioration treatment (CDT), 86
Conventional plant breeding
    constraints, 624–625
    ornamental plants, 421
    rice, 4, 61
    tea, 441
    transgenic crops, 625
Convention on Biological Diversity (CBD),
    627–628
Copyrights, 623–624
CottonDB, 468
Cotton genome-wide comprehensive reference
    map (CRM), 468
Cotton (*Gossypium* spp.)
    expressed genes, 474–477
    genetical genomics, 472–473
    metabolomics, 481
    microarray platforms, 478–479
    molecular linkage maps, 468
    patents, 659–660
    physical maps, 468–470
    proteomics, 480–481
    quantitative genetics, 482
    transcriptomics, 474–479
Cowpea (*Vigna unguiculata* L. Walpers)
    BAC library, 111
    genetic linkage maps, 114
    mapping populations, 108
    MAS, 122
    molecular markers, 112
    production constraints, 105
    taxonomy and genome organization, 105
    trait mapping, 117
    transcriptomics, 119
    whole-genome sequence, 120
Cowpea Genespace Sequences Knowledge base
    (CGKB), 119

Cowpea Genomics Initiative (CGI), 119
Cowpea genomics knowledge base (CGKB), 112
C-repeat/dehydration-responsive element
    binding factor 1 (*CBF1*) genes, 154
Cut flowers, 410
Cytological markers, 443
Cytoplasmic male sterility (CMS), 179
Cytoplasmic genetic male sterility (CGMS)-
    derived hybrid system, 107

D

Deep transcriptome sequencing, 342, 366
DE-ETIOLATED1 (*DET1*), 270
Defoliation, 504
Deletion mutants, 11
Dendrobium, 411
Derived CAPS (dCAPS), 112
Designer crops, *see* Transgenic crops
Design patents, 623
Direct selection for high grain yield, 50
Distortionless enhancement by polarization
    transfer (DEPT) technique, 392
Diversity array technology (DArT) markers,
    170, 217
DNA markers, *see* Molecular markers
DNA methylation, 537–539
Dow AgroSciences, 589
Drought-inducible proteins
    chickpea, 115
    maize
        metabolomics, 88–89
        proteomics, maize, 85
    sugarcane, 215
Drought-tolerant landraces, 48–49
DWARF3 (*D3*) gene, 14
DWARF10 (*D10*) gene, 14
DWARF27 gene, 14
DWARF88 gene, 14

E

EcoTILLING, 8, 198
Edible vaccines, 150, 587–588
EMB564 protein, 86–87
Enhancer trap, 12
Entomopathogenic fungi, 608
EPIcyte Pharmaceuticals, 589
Epigenetics/epigenomics, 625
    drought-tolerant rice, 54–55
    forage crops, 536–539
eQTL, *see* Expression QTL (eQTL)
EST profile, 11

EST–SSRs, 112, 118–120
Ethylene-sensitive flower senescence, 419
European Patent Convention (EPC), 629
European Patent Office (EPO), 589, 623, 665
European Union (EU) directive 2001/18/EC, 512, 516
Expressed sequence tags (ESTs), 11, 75, 82, 112, 195
   Brassica crops, 169
   Catharanthus roseus, 341
   chickpea, 118
   forage crops, 528–529
   fruits, 258–260
   Jatropha curcas, 552
   ornamental crops, 433
   pigeonpea, 118–119
   sugarcane, 193
   Withania somnifera, 394, 396, 400
Expression QTL (eQTL), 6–7

F

Farmers' rights, 630–631
Farmyard manure (FYM), 330
F2 mapping populations, 108
Fingerprinted contigs (FPC) software, 111
First International Congress on Plant Metabolomics, 87
Flavonoids, 412–414
FL-cDNA overexpression (FOX) genehunting system, 11
Fluorescence *in situ* hybridization (FISH), 468
Food and Agriculture Organization of the United Nations (FAO), 492, 604
Forage crops
   cash value, 524
   conventional breeding, 524
   epigenomics, 536–539
   food security, 523
   genomics, 525–527
   isoflavonoid biosynthetic pathway, 532
   metabolomics, 530–532
   proteomics, 529–530
   transcriptomics, 528–529
   transgenomics, 532–536
Forests
   definition, 492
   genetic modifications, *see* Genetically modified (GM) forest trees
   global wood removals, 500–501
   problems, 495
   total economic value (TEV), 493, 496

Fortuna potato, 152
French–Italian Public Consortium for Grapevine Genome Characterization, 253, 296
Fruit ripening, 251
Fruits
   genomics, 252–258
   metabolomics, 270–273
   proteomics, 266–269
   ripening, 251
   transcriptomics, 258–265
Full-size mouse/human chimeric antibody (cT84.66), 587
Functional genomics
   applications, 11–13
   forage crops, 526
   limitations, 9–10
   maize
      full-length cDNA, 83
      gene expression, 82
      mutator transposons, 83
   rice
      abiotic stresses, 15–18
      agromorphological traits, 13–15
      biotic stresses, 18–20
Functional markers, 5, 112

G

Gene-indexed mutants, 11
Gene pyramiding, 6
General Agreement on Tariffs and Trade (GATT), 627
Gene silencing, 12
Gene-space sequence reads (GSRs), 119
Gene-targeted markers (GTMs), 341
Genetically engineered ornamental crops
   abiotic stress tolerance, 422–423
   disease and insect tolerance, 421–422
   floral fragrance, 417–418
   flower color
      flavonoids, 412–414
      modification, 414–416
   flower morphology, 417
   plant architecture, 416–417
   vase life, 418–420
Genetically modified crops, *see* Transgenic crops
Genetically modified (GM) forest trees
   abiotic stress resistance, 504–505
   commercial releases, 508
   early flowering method, 506–507
   environmental risks
      cultivation and management, 515
      food chain, 514–515

horizontal gene transfer, 513
human health, 514
out-crossing, 513–514
persistence and invasiveness, 513
vertical gene transfer, 513–514
field trials, 507–508
fungal and bacterial resistance, 504
herbicide tolerance, 502–503
insect resistance, 504
NCBI records, 509–510
phytoremediation, 507
R&D Status, China, 511
structure and wood quality, 504–506
Genetically modified (GM) potatoes, 147
public acceptance, 155
Genetic linkage mapping
banana, 288
Brassica crops, 164–167
chickpea, 113
citrus, 291–292
cowpea, 114
fruits, 252–253
grape, 295
lentil, 114–115
maize, 75–76
mango, 301
papaya, 304
pigeonpea, 113–114
pomegranate, 307–308
sugarcane, 217–219
Genetical genomics, 472–473
Gene traps, 11–12
Genome organization
chickpea, 103
cowpea, 105
lentil, 106
pigeonpea, 104
Genome sequencing
apple, 254
maize, 81–82
pear, 257
sugarcane, 197–198
tomato, 253
Genome Survey Sequence (GSS) database,
398
Genome tiling arrays, 12
Genome-wide association analysis (GWAS), 9,
65, 175
Genome-wide expression profile, 11
Genomics
banana
genetic linkage mapping, 288
molecular markers, 287–288

transcriptomes, 289
whole genome sequencing, 288–289
Brassica crops
association mapping, 175–177
QTL mapping, 172–174
whole genome resequencing, 174–175
Camelina sativa, 553
Catharanthus roseus, 341
citrus
genetic linkage mapping, 291–292
molecular markers, 291
transcriptomes, 292–293
whole genome sequencing, 292
drought-tolerant rice, 55–58
forage crops, 525–527
fruits
comparative genomics, 258
genetic linkage maps, 252–253
molecular markers, 257–258
whole-genome sequencing, 253–257
grape
genetic linkage mapping, 295
molecular markers, 294–295
transcriptomes, 296
whole genome sequencing, 295–296
maize
agronomic traits, 74–75
association mapping, 78
functional genomics, 82–83
genetic and physical maps, 75–76
genome sequencing, 81–82
marker-assisted selection, 78–81
quality traits, 76–78
transgenics, 83–84
mango
genetic linkage mapping, 301
molecular markers, 300–301
transcriptomes, 301
papaya
genetic linkage mapping, 304
molecular markers, 303–304
transcriptomes, 305
whole genome sequencing, 304–305
pomegranate
genetic linkage mapping, 307–308
molecular markers, 307
rice
advanced backcross QTL analysis, 7–8
allele mining, 8
association mapping, 8
EcoTILLING, 8
expression genetics, 6–7
workflow, 7

sugarcane
    DNA markers, 216–217
    genetic linkage maps, 217–219
    molecular diagnostic tests, 221
    QTLs, 219–221
tea
    functional genomics, 450–453
    marker-based approaches, 444–450
    transgenomics, 450
    Withania somnifera, 394, 396
Geraniol, 348–349
gfp gene, 140–143
Gid1 gene, 13
Gid2 gene, 13
Ginseng, *see Withania somnifera*
Global wood removals, 500–501
Golden potato, 148
G–P map, 28
GRAIN INCOMPLETE FILLING1 (GIF1)
    gene, 15
Grain number1 (Gn1) gene, 14
Grain-weight-related gene (*GW2*), 15
Gramene Maps Database, 3, 5
Grape
    genomics, 294–296
    metabolomics, 272–273
    microarray transcriptomics, 263–264
    proteomics, 296–297
Green revolution, 1, 13, 58, 624
Green rosette disease, 332
GS-FLX 454 technology-based deep
    transcriptome sequencing, 118
Guava (*Psidium guajava* L.)
    conventional breeding, 297
    genetic linkage mapping, 298–299
    molecular markers, 298

H

Haplotype map (HapMap), 165, 181
HarvEST: Cowpea version 1.27, 119
Hepatitis B core antigen (HBcAg), 579
Histone deacetylases (HDACs), 54–55
Horizontal gene transfer (HGT), 513
Human papillomavirus type 11 (HPV11), 588
Hybridization-based markers, 430

I

Illumina 1536 GolenGate assay, 112, 114
Immunoglobulin G (IgG) antibodies, 567
Indian Council of Agricultural Research
    (ICAR), 111

Industrial roundwood, 497, 500, 501
Inflated calyx syndrome (ICS), 401
Insensitive nuclei enhanced by polarization
    transfer (INEPT) technique, 392
Integrated breeding platform (IBP) of the
    Generation Challenge Program, 632
Integrated Taxonomic Information System
    (ITIS), 329
Intellectual property rights (IPR)
    concept, 621–622
    copyrights, 623–624
    goals, 622
    patents, 622–623
    plant biotechnology
        farmers' rights, 630–631
        international treaties, 627–629
        patent filing, 664–665
        plant patents, 629
        PVP, 629–630
        utility patents, 631
    plant genetic transformation technology,
        649–654
    relevance, 622
    strengthening of, 666
    trademarks, 624
    trade secret, 624
International Grape Genome Program (IGGP),
    295–296
International Initiative for Pigeonpea Genomics
    (IIPG) website, 120
International Maize and Wheat Improvement
    Center, 77
International Plant Phenomics Network
    (IPPN), 28
International Rice Functional Genomics
    Consortium (IRFGC), 10
International Rice Research Institute (IRRI),
    50
International RosBREED SNP Consortium
    (IRSC) Malus array, 258
International Service for Acquisition of
    Agribiotech Applications (ISAAA),
    626, 629
International Solanaceae Genome Project, 253
International Treaty on Plant Genetic Resources
    for Food and Agriculture (IT-PGRFA),
    628
International Union for the Protection of New
    Varieties of Plants (UPOV), 629
Inter-simple sequence repeat (ISSR) markers
    pulses, 113
    sugarcane, 194, 216
    tea, 448

Intragenesis, 664
Intron-spanning region (ISR) markers, 118, 119
Intron-targeted amplified polymorphic (ITAP) markers, 119
IPR, *see* Intellectual property rights (IPR)
Irrigated rice, 47–48
Isozyme markers, *see* Biochemical markers

J

Japanese Patent Office (JPO), 665
Jasmonic acid (JA), 26, 209, 361
Jatropha curcas
    candidate genes, 555
    EST sequencing, 552

K

KaPPA-view, 213
KBGP-24K microarray, 170
KBGP-50K microarray, 170
KEGG PLANT, 213
KEGG, 213
Kernel screening assay (KSA), 86
Korea *Brassica* Genome Project (KBrGP), 169

L

Laloo-14, 48
Large-scale cotton EST sequencing, 479
Late embryogenesis abundant (LEA) proteins, 85
Lateral-shoot inducing factor (LIF), 417
Leaf mosaic disease, 332–333
Leahy–Smith America Invents Act, 623
Legume information system (LIS) website, 119
Lentil (*Lens culinaris* Medik.)
    BAC library, 111
    genetic linkage maps, 114–115
    mapping populations, 108, 110
    molecular markers, 112–113
    production constraints, 106
    taxonomy and genome organization, 106
    trait mapping, 117–118
    transcriptomics, 106
    whole-genome sequence, 120
Linkage disequilibrium (LD), *see* Association mapping
Loop-mediated isothermal amplification (LAMP)-based method, 341
luc gene, 140

M

Madagascar periwinkle, *see Catharanthus roseus*
Magnifection, 578–579
Maize (*Zea mays* L.)
    genomics, 74–84
    metabolomics, 87–90
    molecular farming, 581–582
    patents, 656
    proteomics, 84–87
Maize Genetics and Genomics Database (MaizeGDB), 75, 82
Maize Genome Sequencing Project, 81
Maize rough dwarf disease (MRDD), 85–86
Maize rough dwarf virus (MRDV), 86
Malate dehydrogenase (MDH), 534
Mango (*Mangifera indica* L.)
    genomics, 300–301
    metabolomics, 301–302
    proteomics, 301–302
Map-based cloning, 11
MapMan, 213
Mapping populations
    chickpea, 107
    cowpea, 108
    lentil, 108, 110
    pigeonpea, 107–108
Marker-assisted backcrossing (MABC), 62
    chickpea, 121
    drought-tolerant rice, near-isogenic lines (NILs), 58
Marker-assisted breeding (MAB)
    cowpea, 117
    drought-tolerant rice, 62–63
    pulses, 121–122
Marker-assisted QTL pyramiding (MAQP), 62–63
Marker-assisted recurrent selection (MARS), 122
Marker-assisted selection (MAS), 63, 632
    bioenergy crops, 556
    cowpea, 122
    maize
        genetic gain and cost efficiency, 79
        QPM hybrids, 79–81
    rice, 4–6
Massive parallel signature sequencing (MPSS), 12
Matrix-assisted laser desorption/ionization/ time-of-flight mass spectrometry (MALDI-TOF MS)
    peach fruit, 268
    sugarcane, 214–215
    Withania somnifera, 396–397

Max Planck Institute, Germany, 88
Medicago Affymetrix GeneChip, 528
Messenger RNA sequencing, 170–171
Metabolome Tomato Database (MoTo DB), 270
Metabolomics
  banana, 289–290
  Catharanthus roseus, 362–363
  citrus, 293
  cotton fiber, 481
  drought-tolerant rice, 53–54
  forage crops, 530–532
  fruits
    apple, 272
    grape, 272–273
    peach, 272
    strawberry, 273
    tomato, 270–271
  maize
    abiotic stress, 88–89
    mass spectrometry, 87–88
    NMR and PCA, 88
    N use efficiency (NUE), 89–90
  mango, 301–302
  papaya, 305–306
  pomegranate, 308
  sugarcane
    analytical techniques, 212–213
    carbohydrate profiling, 214
    compressive plant chemical profiling, 212
  tea, 454
  Withania somnifera, 387–388
Methylation filtration (MF)-based sequencing, 119
Microarrays, 12, 195
  cotton fiber, 478–479
  fruit transcriptomics, 259, 261–264
Microarray-based eQTL analyses, 6–7
Microbial pesticides
  bacteria, 606–608
  baculoviruses, 608–609
  entomopathogenic fungi, 608
  patents, 607
Microprojectile-mediated transformation, 224
Microsatellites, *see* Simple sequences repeats
    (SSRs)
Minimal tiling path (MTP), 111
Mismatch repair (MMR) system, 143–145
Molecular farming
  alfalfa, 579
  antibodies, 584–587
  areas, 566
  barley, 582
  cell suspension cultures, 583
  cereal seeds, 581

clinical trials, 589
commercially available industrial proteins,
    589, 592
downstream processing, 583–584
  affinity-based tags, 584
  cost breakdown, 583
  oleosin partitioning, 584
edible vaccines, 587–588
gene expression systems
  cassette design, 568
  codon optimization, 569
  features, 568
  mucosal adjuvants, 575
  posttranslational modification, 571–573
  promoters, 570–571
  protein authenticity and safety, 574–575
  protein storage, 575
  site of integration, 573–574
  subcellular targeting, 574
  transgene copy number, 573
  UTRs, 569–570
maize, 581–582
male sterility, 593
microalgae, 583
mosses, 582–583
patents, 590
plant expression systems, transformation
    procedures, 576–579
potato, 581
product authenticity, 567
production cost, 566
production/crop system
  fruit and vegetable crops, 580–581
  nonvascular lower plants, 582–583
  seeds, 581–582
purpose, 564–565
regulatory issues
  environmental impact, 593
  product safety, 592
rice, 581
soybean, 582
sugarcane, 227
timescale development, 566–567
tobacco, 579–580
tomato and banana, 581
transformation procedures, gene expression
  agroinfiltration, 576
  magnifection, 578–579
  plastids, 577–578
  recombinant infectious viral genome, 577
  stable nuclear transformation, 577
  transient gene expression, 576
vaccination, 587

Molecular marker-facilitated QTL mapping, 75
Molecular markers
    banana, 287–288
    Brassica crops
        comparative mapping, *Arabidopsis
            thaliana*, 167
        genetic linkage maps, 164–167
    chickpea, 111–112
    citrus, 291
    cowpea, 112
    desirable properties, 429–430
    fruits, 257–258
    grape, 294–295
    lentil, 112–113
    mango, 300–301
    ornamental crops, 428–433
    papaya, 303–304
    pigeonpea, 112
    pomegranate, 307
    tea, 444–447
MONOCULM1 (*MOC1*) gene, 14
Morphological markers
    Catharanthus roseus, 341
    lentil, 114, 117
    ornamental crops, 426–427
    papaya, 304
    tea, 442
Morphological trait genomics, *see*
            Agromorphological traits genomics
MPSS, *see* Massive parallel signature
            sequencing (MPSS)
Mucosal adjuvants, 575
Multidimensional protein identification
            technology (MudPIT), 21
Multinational *Brassica* Genome Project (MBGP),
            167–168
Multiple drug resistance (MDR), 339

N

Nagina-22, 48, 57
National Biotechnology Board, India, 410
National Botanical Research Institute (NBRI),
            398
National Center for Biotechnology Information
            (NCBI) EST database, 118, 297
    forest trees, 509–510
    sugarcane, 201
    Withania somnifera, 398
National Science Foundation (NSF), 10, 81, 111
Natural forests
    Atlas cedar (*Cedrus atlantica* Manetti), 492
    holm oak (*Quercus ilex* L.), 493

Natural pesticides
    microbe-based
        bacteria, 606–608
        baculoviruses, 608–609
        entomopathogenic fungi, 608
        patents, 607
    plant-based
        botanical extracts, 612–614
        essential oils, 610–612
        patents, 610
NCBI, *see* National Center for Biotechnology
            Information (NCBI) EST database
Nested association mapping (NAM), 78
New Millennium Indian Technology
            Leadership Initiative (NMITLI)
            program, 393
Next-generation sequencing (NGS) technology, 5
    Brassica crops, 165, 168
    cotton fiber, 481
    pigeonpea, 118, 120
    tea, 453
N-glycosylation, 567, 571
NGS technology, *see* Next-generation
            sequencing (NGS) technology
Nongenetically modified organism (GMO), 155
nptII gene, 142–143
NSF Rice Multi-Platform Microarray Search
            tool, 10
Nuclear magnetic resonance (NMR)
            spectroscopy, 54, 88–89, 213, 272, 306,
            363, 391–393, 530–531

O

Octadecanoid-responsive *Catharanthus* AP2-
            domain transcription factors (ORCAs),
            366
Oleosin partitioning technology, 584
Oligonucleotide-directed mutagenesis (ODM),
            625, 632
Oligonucleotide tiling arrays, 11
"One gene at a time" approach, 526
Online patent searching, 665
Opaque-16 gene, 77
Opaque2 gene, 77
Oriental Beauty, 453
Ornamental Bioscience GmbH, 422
Ornamental crops
    anthocyanin biosynthetic pathway, 413
    bioluminescent orchids, 423
    blue roses, 423–424
    breeding objectives, 410
    carnation varieties, 410

chrysanthemum plants
  herbivore repellence, 424–425
  male sterility, 425–426
flavonoids, 412–414
genetic engineering approaches
  abiotic stress tolerance, 422–423
  disease and insect tolerance, 421–422
  floral fragrance, 417–418
  flower color, 412–414, 414–416
  flower morphology, 417
  plant architecture, 416–417
  vase life, 418–420
genetic transformation, 411
horticulture, 410
markers
  biochemical, 427–428
  molecular, 428–433
  morphological, 426–427
Oncidium and *Odontoglossum* orchid species,
    425
somaclonal variation, 423
traditional breeding, 411–412
in vitro flowering, 424
winter flowers, 411
Orphan crops, 102
The *Oryza* Map Alignment Project (OMAP), 4
OsBRI1 gene, 13–14
OsCRY1 gene, 14
OsIAA1 gene, 14
OsMADS34 gene, 14
OsMADS6 gene, 14
OsTB1/FINE CULM1 (*FC1*) gene, 14

P

Papaya (*Carica papaya* L.)
  genomics, 303–305
  metabolomics, 305–306
  proteomics, 305–306
Patent filing, 664–665
Patents, 622–623
  biotechnological innovations, 655–663
  Brassica, 658–659
  Catharanthus roseus, 662–663
  cotton, 659–660
  genetically modified (GM) microorganisms,
    623
  maize, 656
  molecular farming, 590
  phytoremediation, 664
  plant biotechnology, 633–642
  plant genetic transformation technology,
    649–654

potato, 657
rice, 655
sugarcane, 660–661
tea, 661
transgenic markers, 643–648
PatentScope, 665
Patent searching, 665
Peach
  ESTs, 259
  genetic linkage map, 258
  genome sequencing, 257
  metabolomics, 272
  microarray transcriptomics, 262–263
  proteomics, 268–269
  RNA sequencing, 265
Pesticide procurement, 604
Pest management
  Catharanthus roseus, 332–333
  organic agriculture, 605–614
  ornamental crops, 422–423
Phenomics
  Brassica species 181
  rice
    definition, 28
    phenotyping tools, 28–29
Phosphorus stress transcriptomics, *Brassica*
    crops, 178–180
Phosphorus uptake 1 (*Pup1*), 18
Phylogenomics, 13
Phytoene synthase (*y1*) gene, 77
Phytophthora infestans (*Pi*), 151
Phytoremediation, 507, 516, 664
Pigeonpea (*Cajanus cajan* L. Millsp.)
  AB-QTL, 122
  BAC library, 111
  genetic linkage maps, 113–114
  mapping populations, 107–108
  molecular markers, 112
  production constraints, 104
  taxonomy and genome organization, 104
  trait mapping, 115–117
  transcriptomics, 118–119
  whole-genome sequence, 120
Pigeonpea genomics initiative (PGI), 111
Pima cotton (*Gossypium barbadense* L.), 467
Plant-based pesticides
  botanical extracts, 612–614
  essential oils, 610–612
  patents, 610
Plant bioreactor, 565
Plant breeders' rights (PBRs), 622
Plant cell suspension cultures, 583
Plant Cyc, 213

Planted forests
  functions, 497
  total afforested and reforested areas, 497
  unequal repartition, 496
Plant-made pharmaceuticals (PMP), 571
Plant Patent Act (PPA), 629
Plant patents, 623, 629
Plant production platforms, *see* Molecular
    farming
Plant Quarantine (Regulation of Import into
    India) Order, 2003, 630
Plant varietal rights (PVRs), 630
Plant variety protection (PVP), 629–630
Plastid transformation, 577–578
Polymerase chain reaction (PCR) patent, 623
Polymerase chain reaction (PCR)-based markers
  ornamental crops, 430
  rice, 4
Polyploidy, 551
Pomegranate (*Punica granatum* L.)
  genomics, 307–308
  metabolomics, 308
Poplar plantation, 494
Population genomics
  association mapping, 557
  QTL mapping, 556–557
  quantitative genetic analyses, 557
Positional cloning, *see* Map-based cloning
Postgenomics markers, 443, 448–450
Posttranscriptional gene silencing (PTGS),
    152–153
Potato
  cisgenesis, 154–155
  molecular farming, 581
  patents, 657
  transgenesis
    abiotic stress tolerance, 154
    biotic stress tolerance, 151–154
    edible plant vaccines, 150
    Gfp reporter gene, 140–143
    mismatch repair (MMR) system, 143–145
    tuber quality traits, 145–150
Potato Genome Sequencing Consortium
    (PGSC), 139
Potato leaf roll virus (PRLV), 152
Potato potyvirus A (PVA), 152
Potato potyvirus Y (PVY), 152
Pregenomics markers, 442–443
Principal component analysis (PCA), 88–89, 306,
    363
ProdiGene Pharmaceuticals, 589
Production constraints
  chickpea, 103

cowpea, 105
lentil, 106
pigeonpea, 104
PROG1 gene, 14
Promoter trap, 12
Protection of Plant Varieties and Farmers'
    Rights Act (PPVFR), 630
Proteins associated with resistance (RAPs), 86
Proteomics
  banana, 289–290
  Brassica crops
    male sterility, 179–180
    plant growth and development, 178
    seed oil content, 179
    stress tolerance, 180
  Catharanthus roseus, 344–362
  citrus, 293
  cotton fiber, 480–481
  definition, 84
  drought-tolerant rice, 51–53
  forage crops
    cold acclimation, 530
    heat stress, 530
    reference map, *Medicago truncatula*, 529
  fruits
    apple, 267–268
    peach, 268–269
    strawberry, 269
    tomato, 266–267
  grape, 296–297
  maize
    abiotic stress, 74–84
    applications, 84
    biotic stress, 85–86
    endosperm development, 87
    seed viability, 86–87
  mango, 301–302
  papaya, 305–306
  rice
    2D-PAGE, 21
    abiotic stresses, 23–25
    biotic stresses, 25
    brassinosteroids, 26
    functional analysis, 21–22
    genome annotation process, 20–21
    mass spectrometry, 21
    metabolic and developmental processes,
        22–23
    plant hormones, 25–26
    steps, 22
  sugarcane
    abiotic stresses, 215
    biotic stresses, 215–216

tea, 453–454
Withania somnifera, 396–398
Protocorm-like bodies (PLBs), 425
Provitamin A biofortified maize, 77
Public Intellectual Property Resources for
    Agriculture (PIPRA), 628–629
Public Interest Property Advisors (PIPA), 629
Pulp pine plantation, 495
Pulses
    genomic resources, 106–120
    MAB, 121–122
    transgenic approach, 122–126
    whole-genome sequencing, 120
PVP, *see* Plant variety protection (PVP)

Q

QTL mapping, *see* Quantitative trait loci (QTL)
    mapping
Quality protein maize (QPM), 77
Quality traits
    apple, 257, 272
    drought-tolerant rice, 49
    fiber, 471
    forage, 540
    guava, 299
    maize, 76–78
Quantitative genetics, 482
Quantitative trait loci (QTL) mapping
    bioenergy crops, 556–557
    Brassica crops
        flowering time, 172–173
        glucosinolates, 173–174
        morphological diversity, 172
        whole genome resequencing, 174–175
    chickpea, 115
    cotton fiber, 471–472
    cowpea, 117
    drought-tolerant rice
        genomics, 55–58
        markers, 61–62
    forage crops, 526
    lentil, 117–118
    papaya, 304
    pigeonpea, 115–117
    sugarcane, 219–221

R

Rainfed rice, 48
Randomly amplified polymorphic DNA
    (RAPD)
    advantages, 431

disadvantages, 431
    fruit crops, 300
    ornamental crops, 431
    potato, 144
    pulses, 111, 113, 115
    rice, 4
    sugarcane, 194, 216
    tea, 448
Real-time quantitative RT-PCR (qRT-PCR)
    analysis
    cotton, 473
    sugarcane, 198–199, 204
Recombinant inbred lines (RILs)
    cotton fiber, 473
    maize, 78
    pulses, 107–108, 112, 114
    rice, 6, 8
Recombinant pharmaceutical proteins
    antibodies, 584–587
    clinical trials, 589
    commercially available industrial proteins,
        589, 592
    downstream processing
        affinity-based tags, 584
        cost breakdown, 583
        oleosin partitioning, 584
    edible vaccines, 587–588
    gene expression systems
        bacteria and mammals, 567
        cassette design, 568
        codon optimization, 569
        features, 568
        mucosal adjuvants, 575
        posttranslational modification, 571–573
        promoters, 570–571
        protein authenticity and safety, 574–575
        protein storage, 575
        site of integration, 573–574
        subcellular targeting, 574
        transgene copy number, 573
        UTRs, 569–570
    plant-derived proteins, 591
    product authenticity, 567
    production cost, 566
    regulatory issues
        environmental impact, 593
        product safety, 592
    timescale development, 566–567
    transformation procedures, gene
        expression
        agroinfiltration, 576
        magnification, 578–579
        plastids, 577–578

recombinant infectious viral genome, 577
    stable nuclear transformation, 577
    transient gene expression, 576
    vaccination, 587
Recombinant scFv antibody (scFvT84.66), 587
Recombinant viral vectors, 577
Reporter genes, 12
    potato transgenesis
        gfp gene, 140–143
        uidA and *luc* genes, 140
Restriction fragment length polymorphism
    (RFLP) markers
    advantages, 430
    Brassica crops, 165
    disadvantages, 431
    ornamental crops, 430–431
    rice, 4
    pulses, 112, 113
    sugarcane, 194
    tea, 448
Reverse breeding, 625, 664
Reverse northern-dot blots technology, *see*
    Microarray
Reverse phenomics, 27
Rice (*Oryza sativa* L.)
    cereal biology, model for, 3
    drought tolerance
        comparative genomics, 58
        conventional approach, 49–50
        direct selection for high grain yield, 50
        epigenomics, 54–55
        genomics, 55–58
        landraces, 48–49
        metabolomics, 53–54
        molecular breeding strategies, 61–65
        proteomics, 51–53
        rainfed areas, 48
        transcriptomics, 50–51
        transgenics, 58–61
    genomics
        functional genomics, 9–20
        sequencing, 3–9
    global statistics, 1–2
    metabolomics, 26–27
    molecular farming, 581
    patents, 655
    phenomics, 27–28
    proteomics, 20–26
    whole-genome sequencing, 4
Rice black-streaked dwarf virus (RBSDV), 86
Rice Genome Annotation Project database
    (RGAP 7), 4
Rice Kinase Database (RKD), 13

Rice Proteome Database, 51
Rice protoplast transient assay system, 12–13
RIL populations, *see* Recombinant inbred lines
    (RILs)
RNA-dependent DNA methylation (RdDM),
    625, 664
RNAi, *see* RNA interference (RNAi)
RNA interference (RNAi) technology, 12, 83,
    150, 199, 207, 261, 359, 452, 556, 571
RNA sequencing
    fruits, 264–265
    Miscanthus, 553
Roche 454 GS-FLX titanium technology, 120
Rose, 415, 425
"Roundup Ready" alfalfa, 534, 536
Rpi genes, 151
Rules for Manufacture, Use, Import, Export and
    Storage of Hazardous Microorganisms
    and Genetically Engineered
    Organisms or Cells, 1989, 630

S

SAGE, *see* Serial analysis of gene expression
    (SAGE)
Secologanin, 349–350
Secretory immunoglobulin A (sIgA), 587
Seeds Act, 1996, 630
Semi dwarf1 (*sd1*) gene, 13
Sequence-characterized amplified region
    (SCAR), 113
Sequence-related amplified polymorphisms
    (SRAPs), 195
Sequence-tagged microsatellite site (STMS)
    markers,
    Catharanthus roseus, 341
    pulses, 113
    tea, 449
Sequence tagged site (STS) markers, 4, 115, 420
Serial analysis of gene expression (SAGE), 12, 82
Serpentine, 327, 337–338, 351, 356, 359, 362–363
Shotgun proteogenomics, 21
Simple sequence repeat (SSR) markers, 4, 5, 113,
    114, 120, 217, 258, 300
    advantages, 432
    Brassica crops, 165
    disadvantages, 432
    ornamental crops, 432
    sugarcane, 194
    tea, 449
Single-feature polymorphism (SFP) markers, 112
Single nucleotide polymorphism (SNP) markers,
    3, 9, 11, 63, 74, 112, 118, 120, 217

Brassica crops, 165–166
forage crops, 526
sugarcane, 194
tea, 449
Single-strand conformation polymorphisms
(SSCPs), sugarcane, 194
S-linalool synthase (LIS), 418
Slr1 gene, 13
SMD resistance, 115
SNAC1 gene, 50
SNAC2 gene, 16
Snowdrop lectin gene, 226
Solexa transcriptome sequencing technology,
166
Soybean Bowman-Birk inhibitor (SBBI), 226
Soybean Kunitz trypsin inhibitor (SKTI), 226
Stable nuclear transformation, 577
Strawberry
ESTs, 259
genome sequencing, 257
metabolomics, 273
microarray transcriptomics, 264
proteomics, 269
Strictosidine, 338, 350–352, 366
Strictosidine β-D-glucosidase, 351–352
StrucEluc computer-aided structure elucidation
(CASE) program, 392
Sugarcane
biofactory, 227
conventional breeding, 221
genomics, 214–216
metabolomics, 212–214
patents, 660–661
physical and chemical mutagens, 229
proteomics, 216–221
somaclonal variation, 228
taxonomy, 192
transcriptomes, 200–212
transgenics, 221–226
Sugarcane-assembled sequences (SASs),
sugarcane, 193, 204–205
Sugarcane Genome Sequencing Initiative
(SUGESI), 197–198
Suppression subtractive hybridization (SSH), 452
Swiss-Prot database, 567
Synthetic genomics, 664
Synthetic Genomics Inc. (SGI), 552

T

Tabersonine, 352, 358, 363, 366
Targeting-induced local lesions in genomes
(TILLING), 11

Target region amplification polymorphisms
(TRAPs), 195, 217, 219
Taxonomy
chickpea, 103
cowpea, 105
lentil, 106
pigeonpea, 104
sugarcane, 192
Tea (*Camellia sinensis*)
biotechnological studies, time line of, 440
genomic resources, 440–442
genomics
functional genomics, 450–453
marker-based approaches, 444–450
transgenomics, 450
metabolomics, 454
patents, 661
proteomics, 453–454
Terpenoid indole alkaloids (TIAs)
alkaloidomics, 363–364
biological activity, 337–338
biosynthetic pathway, *Catharanthus roseus*
cellular and subcellular
compartmentation, 355–357
enzymology, 345–355
regulation, 357–362
detection methods, 337
extraction procedure, 334–337
occurrence and distribution, 334
Thrips, 332
TIAs, *see* Terpenoid indole alkaloids (TIAs)
Tiling arrays, 12
Tomato
genome sequencing, 253
metabolomics, 270–271
microarray transcriptomics, 261
molecular farming, 581
proteomics, 266–267
RNA sequencing, 265
Total soluble protein (TSP), 566–567
chloroplast transgenes, 578
Trademarks, 624
Trade-Related Aspects of Intellectual Property
Rights (TRIPS), 627
Trade secret, 624
Transcript assembly contigs (TACs), 118
Transcription factors (TFs), 16, 50
Transcriptomics, 13
banana, 289
Brassica crops, 168–171
cDNA libraries, 169
DArT, 170
ESTs, 169

expression microarrays, 169–170
messenger RNA sequencing, 170–171
plant growth and development, 177–178
stress tolerance, 180
Catharanthus roseus
cDNA-AFLP, 341–342
deep transcriptome sequencing, 342
ESTs, 341
genes encoding, 342, 344
chickpea, 118
citrus, 292–293
cotton fiber
cell elongation, 477–479
expressed genes, 474–477
cowpea, 119
drought-tolerant rice, 50–51
forage crops, 528–529
fruits
ESTs, 258–260
microarray, 259, 261–264
RNA sequencing, 264–265
grape, 296
lentil, 106
mango, 301
papaya, 305
perennial ryegrass, 527
pigeonpea, 118–119
sugarcane
abiotic stresses, 210–212
biotic stresses, 208–210
DNA microarray systems, 203–204
microRNAs, 206
polymorphisms, 207–208
SASs, 204–205
sources, 200–203
sucrose accumulation, 207
Withania somnifera, 398
bioinformatics studies, 398, 401
docking and molecular dynamics
simulation studies, 401–402
TPX2-Aurora A complex, cancer cells, 402
Transgene copy number, 573
Transgenic alfalfa, 533–534
Transgenic crops
creation steps, 625–626
soybean, 626–627
traits, 627
Transgenic forest trees, *see* Genetically modified
(GM) forest trees
Agrobacterium, 498
field trials, 498–499
involved species, 500

Moroccan hybridization techniques, 499
research share, 499
Transgenics/transgenomics, 625
forage crops, 532–536
maize
RNAi, 83
virus-induced gene silencing (VIGS),
83–84
ornamental crops, 411
blue roses, 423–424
commercialization, 433–434
herbivore repellence, 424–425
male sterility, 425–426
in vitro flowering, 424
potato
abiotic stress tolerance, 154
biotic stress tolerance, 151–154
edible plant vaccines, 150
Gfp reporter gene, 140–143
mismatch repair (MMR) system,
143–145
public acceptance, 155
tuber quality traits, 145–150
pulses, 122–126
sugarcane, 228
Agrobacterium-mediated transformation,
222
field trials, 226
microprojectile-mediated transformation,
224
proteinase inhibitors (PIs), 225–226
sucrose content, 226
viral disease resistance, 225
tea, 450
Transient assay systems, 12–13
Transient gene expression systems, 576
Transplastomic plants, 126
Tryptamine, 344, 349–351, 356, 360, 363, 367
Tryptophan, 27, 80, 89, 345, 348, 364
Tuber quality traits, potato transgenesis
biofuels, 149–150
carotenoid content, 148
glycoalkaloids, 145–148
inulin, 149
protein and amino acids, 148
vitamin E, 149
Turf species, *see* Forage crops
Two-dimensional polyacrylamide gel
electrophoresis (2D-PAGE)
rice, 21
Withania somnifera, 396–397
Type 1 Gaucher's disease, 664

U

uidA gene, 140
Undeveloped tapetum1 (*Udt1*) gene, 14
United Nations Industrial Development
 Organization (UNIDO), 511
Upland cotton (*G. hirsutum* L.), 467
U.S. Bayh–Dole Act, 664
U.S. Department of Agriculture's Animal
 and Plant Health Inspection Service
 (USDA-APHIS), 534, 536
U.S. Patent and Trademark Office (USPTO), 589,
 665
U.S. Patent Office, 623
Utility patents, 623, 631

V

Vaccination, 587
Vandana, 57
Vertical gene transfer, 513–514
VIGS, *see* Virus-induced gene silencing (VIGS)
Vinblastine, 338–339, 354–355
Vincristine, 338–339, 354–355
Vindoline, 352–354
Virus-induced gene silencing (VIGS), 83–84
Vivek QPM 21, 80–81
Vivek QPM 9, 79–80

W

Whole-genome eQTL analysis, 6
Whole-genome sequencing
 banana, 288–289
 Brassica crops, 174–175
 chickpea, 120
 citrus, 292
 cotton fiber, 470
 cowpea and lentil, 120
 fruits, 253–257

grape, 295–296
papaya, 304–305
patent, 632
pigeonpea, 120
pulses, 120
switchgrass, 553–554
Whole-genome shotgun (WGS) sequencing,
 120, 253
Whole transcriptome shotgun sequencing, *see*
 Messenger RNA sequencing
Wild potato species (*Solanum chacoense*),
 143–144
Winter cherry, *see Withania somnifera*
Winter flowers, 411
Withaferin A, 386, 398, 401–403
Withania somnifera
 genomics, 394, 396
 metabolomics, 387–388
 pharmacological value, 386
 secondary metabolites, 386–387
 transcriptomics, 398, 401–402
Withanolides
 biosynthesis pathway, 395
 deposited entries, 392
 isolation and analysis, 389–390
 occurrence, 390–391
 phytochemical studies, 390
 screening, 389
 structural characterization, 391–394
 structures, 388
World Health Organization (WHO), 604
World Intellectual Property Organization
 (WIPO), 665
WsFGT, 398, 399

Z

Zinc-finger nuclease (ZFN) technology, 625, 632
Zinc fingers, 24, 25, 50, 205, 305, 361, 394, 417, 574

For Product Safety Concerns and Information please contact our EU
representative GPSR@taylorandfrancis.com Taylor & Francis Verlag GmbH,
Kaufingerstraße 24, 80331 München, Germany

Printed and bound by CPI Group (UK) Ltd, Croydon, CR0 4YY

01/05/2025

01858611-0001